T0304612

Modern Impact and Penetration Mechanics

Do you want to understand how projectiles are stopped by, or go through, armors and other materials? Master the fundamentals of impact mechanics through the use of analytical modeling, large-scale numerical simulations, and experiments with this practical text. This book spans topics including continuum mechanics, waves and shocks, and the high-strain-rate and large deformation constitutive and failure modeling of solids, and addresses the mechanics of materials in extreme dynamic environments. It also covers the stress and strain tensors and provides understanding of how they are used in modeling large, high-rate deformations. Offering both a qualitative and a quantitative understanding, with an emphasis on solid mechanics, this is an essential text for graduates. In addition, it is perfect as a reference for academic researchers and professionals interested in wave motion, impact, and penetration.

Sales points

- Follows a step-by-step methodology
- Features end-of-chapter exercises
- Appropriate as a supplement to or as a required text for graduate level courses in solid and continuum mechanics and impact mechanics

Dr. James D. Walker is Director of the Engineering Dynamics Department in the Mechanical Engineering Division at Southwest Research Institute in San Antonio, Texas, where he is also an Institute Scientist. He is a Fellow of the American Institute of Aeronautics and Astronautics and the American Society of Mechanical Engineers and a Ballistics Science Fellow of the International Ballistics Society. His awards and honors include the Holley Medal from ASME for his work in support of the space shuttle *Columbia* accident investigation, NASA Group Achievement Awards for work on the space shuttle return to flight and the New Horizons mission to Pluto, and *Popular Science*'s 2004 list of "Brilliant 10" scientists to watch. He is also an adjunct professor of mechanical engineering at the University of Texas at San Antonio, where he teaches graduate-level mechanical engineering courses.

Modern Impact and Penetration Mechanics

JAMES D. WALKER
Southwest Research Institute

CAMBRIDGE
UNIVERSITY PRESS

University Printing House, Cambridge CB2 8BS, United Kingdom

One Liberty Plaza, 20th Floor, New York, NY 10006, USA

477 Williamstown Road, Port Melbourne, VIC 3207, Australia

314–321, 3rd Floor, Plot 3, Splendor Forum, Jasola District Centre, New Delhi – 110025, India

79 Anson Road, #06–04/06, Singapore 079906

Cambridge University Press is part of the University of Cambridge.

It furthers the University's mission by disseminating knowledge in the pursuit of education, learning, and research at the highest international levels of excellence.

www.cambridge.org
Information on this title: www.cambridge.org/9781108497107
DOI: 10.1017/9781108684026

First published 2021

A catalogue record for this publication is available from the British Library.

Library of Congress Cataloging-in-Publication Data
Names: Walker, James D., author.
Title: Modern impact and penetration mechanics / James Walker.
Description: Cambridge ; New York, NY : Cambridge University Press, 2021. |
    Includes bibliographical references and index.
Identifiers: LCCN 2020007789 (print) | LCCN 2020007790 (ebook) |
    ISBN 9781108497107 (hardback) | ISBN 9781108684026 (epub)
Subjects: LCSH: Impact. | Penetration mechanics.
Classification: LCC TA354 .W35 2020 (print) | LCC TA354 (ebook) |
    DDC 620.1/125–dc23
LC record available at https://lccn.loc.gov/2020007789
LC ebook record available at https://lccn.loc.gov/2020007790

ISBN 978-1-108-49710-7 Hardback

To Debbie, Brynn, and Tess

Contents

Preface

This is an applied mechanics book about impact and penetration and includes the requisites in continuum mechanics, waves and shocks, and the high strain rate and large deformation constitutive and failure modeling of solids. It addresses the mechanics of materials in extreme dynamic environments. It is easy to envision impact and penetration applications, and these applications have the nice feature that the effects of nonlinear material response, inertia, large deformations, large rotations, and shock behavior can be quantified and appreciated at the macroscopic scale.

The first course I taught in shock waves was an internal course at Southwest Research Institute (SwRI) in 1992 using Dennis Hayes' notes. The first course I taught in plasticity theory was at the University of Texas at San Antonio (UTSA) in 1994. Thanks to Harry Millwater for encouraging me to return to teaching at UTSA in 2005. Chapters in this book have been used in a variety of graduate courses including Advanced Solid Mechanics, Continuum Mechanics, Theoretical and Computational Inelasticity, and Combustion. Penetration mechanics topics have been included in the SwRI Penetration Mechanics Short Course, which has been taught continuously since 1985, with my involvement beginning in the 1990s. The original concept was a joint book with Charlie Anderson, but as I used chapters of the book in my teaching at UTSA and as it grew in scope, it fell upon me to pursue the effort.

This book can be used as a stand-alone text for Continuum Mechanics, Solid Mechanics, and Plasticity courses. Since this book presents many practical computational results, it can be used as a text for Theoretical and Computational Inelasticity when supplemented with a numerical text, papers, or software. The book is written for graduate students in Engineering, Physics, and Mathematics. The big conceptual step is the stress and strain tensors and understanding how they are used in modeling large, high rate deformations. The text and appendices contain the mathematical background required to understand and pursue this material. The mathematical level is a bit uneven, as I chose to present things in the best logical order, which is not a mathematical sophistication order. If the mathematics seems too tedious, the reader is encouraged to skim to the conclusion of the argument or to move on to the next section. In teaching the material, Chapters 13 and 14 and Appendix A can be covered at any point after Chapter 2; I have used various orders, including Chapters 13 and 14 immediately following Chapter 2, then moving on to Chapter 3, etc., depending on when I wanted to present certain material and what background a course project required.

Three individuals have had a significant influence on my work: Charlie Anderson, Sidney Chocron, and Jack Riegel. Charlie hired me out of graduate school and we have worked together for many years. Sidney and I have worked together and co-taught courses at the University of Texas at San Antonio. Jack Riegel asked me important questions about impact modeling as I began my career, and the process of answering them had a large influence on my research directions. Thank you Charlie, Sidney, and Jack, for our work together and your support and encouragement. Thank you also to my Ph.D. adviser at the University of Utah, Tim Folias, who steered my early work in wave propagation and fracture.

This book has taken two decades to write. I asked Werner Goldsmith how he finished his book *Impact* [**77**]. His reply was that the publisher asked him to do a second edition, which he vehemently refused to do, informing them the first edition had cost three years of his life and one marriage. Fortunately, for me, the cost has not been as high.

I have the privilege of working at Southwest Research Institute, a unique nonprofit research center in San Antonio, Texas. It was founded in 1947 and currently has a staff of 2,700 on 1,500 acres. We are supported entirely by government and commercial research contracts. SwRI colleagues I've worked with include Charlie Anderson, Nathan Andrews, Janet Banda, Andrew Barnes, Hakan Başağaoğlu, Steve Beissel, Rory Bigger, Joe Bradley, Tim Brockwell, Ray Burghamy, Alex Carpenter, Sidney Chocron, Mary Ann Clark, Derrik Coffin, Hervé Couque, Kathryn Dannemann, Dan Durda, Joe Elizondo, Chris Freitas, Charles Gerlach, Walt Gray, Matt Grimm, Don Grosch, Tim Holmquist, Walter Huebner, Gordon Johnson, Ryan Keedy, Trent Kirchdoerfer, Kris Kozak, Jim Lankford, David Littlefield, John MacFarland, James Mathis, Katie McLoud, Michael Moore, Bruce Morris, Nick Mueschke, Scott Mullin, Art Nicholls, Dan Pomerening, Carl Popelar, Jack Riegel, David Riha, Scott Runnels, Erick Sagebiel, Nikki Scott, Dick Sharron, Diane Steiner, Mark Tapley, Ben Thacker, Suzanne Timmons, Randy Tullos, Hunter Waite, Wendy Walding, Greg Wattis, Carl Weiss, Greg Willden, and Bob Young. SwRI staff I reference but did not work with include John Gehring and William Ko. I thank two vice presidents at SwRI who provided a good working environment early in my career, Ulric Lindholm (Engineering and Materials Sciences Division) and Mel Kanninen (Structural Systems and Technology Division). My time at SwRI, particularly the early years, has felt like a golden age in mechanics.

People who helped with the book in particular include my wife, Charlie Anderson, Sidney Chocron, Mike Shearn, Dick Sharron, and Janet Banda.

Reviews of the text in its later stages with many helpful comments were by Sidney Chocron, Deb Deffenbaugh, Rebecca Brannon, an anonymous Cambridge University Press reviewer, and Alicia McAuley (Cambridge). Thank you to the many students who read the chapter drafts and provided feedback. Other colleagues at UTSA and elsewhere who read chapter drafts include John Foster, Hai-Chao Han, Jerry Liu, Dennis Orphal, James Wilbeck, and Justin Wilkerson.

In 2006–2007 my wife Debbie converted the draft from Word to LaTeX. She helped with other aspects of the text, including material property tables and proofreading. She was also, over the many years, very supportive of the effort. The book is dedicated to her and our two daughters in small consolation for the time the book received instead of them.

Finally, thanks to God for this beautiful planet and the breath of life.

James D. Walker

San Antonio, Texas, 2000—2020

1

Introduction

Material is in motion all around us. Sometimes the relative motion leads to collisions, either accidental or intentional. The purpose of this book is to describe the mechanics of these collisions and the impact or penetration that follows, and to provide tools for determining the forces and deformation involved. Representative speeds of interest are shown in Table 1.1.

This book is an applied mechanics text, meaning it develops the mathematical tools in physics and engineering that are required to solve impact and penetration problems. Our primary interest is in solid materials. Since impacts can lead to large forces, there will be large deformations, and so our mathematical tools and our material models will address large deformation. A big step is understanding the stress tensor – the relationship of the stress tensor to the strain tensor contains information about the stiffness and strength (resistance to shear) of solids. We will explore how metals deform, flow, and break. We will explore how yarns and fabrics undergo large deflections. Modern armors are made from metals, ceramics, fabrics, explosives, and space. Armors are interesting in that they are designed to be as light weight as possible, and during an impact event the armor material is utilized through large deflection and deformation all the way to material failure (material separation).

The general framework we use is *continuum mechanics*. Continuum mechanics is the study of materials that can be viewed as a continuous material. This means that there is a smallest scale that it can reasonably address – on the order of tens of nanometers; otherwise atoms must be modeled. Our interest is typically in much larger scales, in macroscopic objects that are usually on the order of millimeters to meters. The basic equations of continuum mechanics will be developed – equations of conservation of mass, momentum, and energy. Then they will be applied. We will study waves in detail. All information in dynamic mechanical systems is conveyed through mechanical waves. In metals, the low pressure acoustical waves have a typical speed of 5 to 6 km/s, which is 15 to 18 times the speed of sound in air. High pressure shocks can travel faster than these acoustical waves.

In modern mechanics we have a threefold approach to understanding, namely experiments, analytical modeling, and large-scale numerical simulations. As a preliminary step, basic material tests are performed and the response of materials is either fit to analytic forms or stored in tables. The material response is typically referred to as equation of state and constitutive models. These material models are then used in analytical modeling and large-scale numerical simulations.

When it comes to applications to mechanics problems in impact and penetration, the analytical modeling approach makes assumptions about the geometry of the system response that reduce the problem to a handful of ordinary differential equations that are solved either explicitly or numerically. Large-scale numerical

1. INTRODUCTION

TABLE 1.1. Impact speeds in nature and technology.

Origin	Speed	Comments
	(m/s)	
One-meter springboard	4.4	Drop $v = \sqrt{2gh}$, $g = 9.8 \text{ m/s}^2$
Human marathon	5.7	Dennis Kimetto, 42.195 km
		(26.2 mile) in 2:02:57 (2014)
Three-meter springboard	7.7	Drop
Human sprint	10.4	Usain Bolt, 100 m in 9.58 s (2009)
Ten-meter platform	14	Drop
Upward hull motion	5–15	Armored vehicle/buried explosive
Car speeds	27	Car collisions (60 mph)
Baseball fastball pitch	47.0	Aroldis Chapman, 105.1 mph (2010)
Paintball rounds	91	Serious eye injury (300 ft/s)
Bird strike (commercial)	154	Low-flying jets (300 knots)
Foam strike on *Columbia*	235	Loss of space shuttle (2003)
Jetliner cruise	246	550 mph (Mach 0.85)
Bird strike (military)	283	Low-flying jets (550 knots)
Sound speed in air	332	0°C
Handgun bullets	350–450	Handguns have short barrels
N_2 gas molecule speed	510	Average in air at 20°C
Waterjet speed	750	In metal-cutting machines
Rifle bullets	750–950	Rifles have long barrels
	(km/s)	
Anti-tank rounds	1.2–1.6	KE projectiles fired by tanks
Sound speed in water	1.48	
Warhead fragment speeds	1–2	
Explosively formed pen.	2.5–3.0	EFPs invert, not collapse
Shear wave speed in metals	2.5–3.2	Transverse wave c_s
Sound speed in metals	5–6	Longitudinal wave c_L
Missile defense closing rate	7	Midcourse intercept
Low Earth orbit speed	7.5	Leads to orbital debris impacts
Shaped-charge jet tip	8–12	Stagnation point collapse
Earth escape velocity	11.2	Minimal meteor strike
Cassini flybys of Enceladus	5–18	Ice grain impacts
Chelyabinsk meteor	19	Blast damage in Russia (2013)
Earth orbit around sun	29.8	Meteor speeds
Solar escape velocity	42.1	From 1 AU; comet speeds
Jupiter escape velocity	60	Shoemaker–Levy 9 comet strike,
		fastest recorded large impact (1994)

simulations are computations where the materials and/or space are discretized into a large number of computational cells or elements and the partial differential equations of mass, momentum, and energy conservation are numerically solved. The distinction between the two modeling techniques will become clear as we develop analytical models and compare their results to those of both experiments and large-scale numerical computations. The three approaches – experiments, analytical

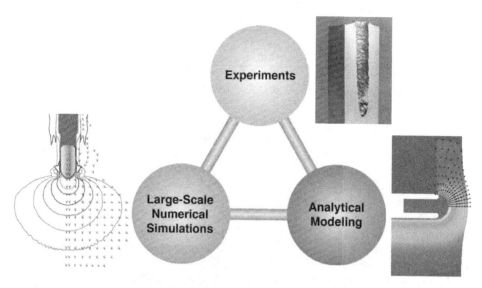

FIGURE 1.1. Three-fold approach to understanding: experiments, large-scale numerical simulations, and analytical modeling.

modeling, and large-scale numerical simulations – work towards an understanding of impact and penetration events (Fig. 1.1). A particular comment about analytical modeling: it is an important part of understanding physical phenomena since development of analytical models requires the assumption of certain physical processes. Therefore, if an analytical model matches the experimental results over a wide range of, say, impact velocities, then the relevant physics has been included in the model and the physical mechanisms governing the event are understood. Large-scale numerical simulations are able to address complicated geometries and complicated material models. As computational speeds have increased, storage costs have come down, and as parallelization has been embraced, powerful computational tools have become widely available. Graphics technology now produces beautiful images and movies of simulated impact and penetration events.

In this book experimental data, which are the final arbiter of truth, are presented both for their own sake and for direct comparison to models.

1.1. Launchers

There are two primary techniques used to launch impactors to high speeds in a research environment. The first technique is through the expansion of a high-pressure gas and the second technique is through electromagnetic force. The expansion of a gas technique relies on the fact that, as a gas expands, its pressure does not drop very much. Hence, as expansions get larger in the launch tube, significant pressures are still applied to the back of the launch package so that it continues to accelerate. Such behavior can be contrasted with liquids, where a high fluid pressure quickly drops with expansion. The reason why the electromagnetic approach works is because the electromotive force can be applied over a significant distance.

Low speed launchers in a research environment are simply a pressure vessel, a fast acting valve, and a launch tube or gun barrel. The projectile is placed near

TABLE 1.2. Laboratory launcher speeds in study of impact and penetration.

Launcher	Speeds (km/s)	Comments
Leading edge of SHPB	0–0.12	Split-Hopkinson pressure bar
Compressed gas gun	0–0.3	Air or nitrogen as driving gas
Compressed gas gun	0–1.0	Hydrogen as driving gas
Powder gun	0.2–2.1	Propellant gases drive projectile
Rocket powered sled	0.5–3	Extended rail track to guide sled
Electromagnetic rail gun	1–4.5	Current through contact armature
Laser launched flyer	0.7–6.0	Spalled by laser pulse
Explosively launched flyer	2–6.5	Explosive lens launched plate
Large two-stage light gas gun	2–7	50 gram, 2 to 4 cm diameter
Small two-stage light gas gun	2–10	50 milligram, 2 to 4 mm diameter
Inhibited shaped charge	8–12	1-gram fragments
Three-stage gun	9–14	Two-stage gun with spall pillow
Magnetically launched flyer	15–44	Huge electric currents
Nuclear explosive launched flyer	30–60	No longer available

the valve at the rear of the barrel. The pressure vessel is taken to a high pressure. The valve is opened and the projectile is accelerated down the barrel. The speed at which the gas drives the projectile is limited by the sound speed of the gas, because if the projectile moves faster than the sound speed, information about the fact that the projectile has moved on cannot be communicated to the gas, so it cannot move to apply pressure to the back of the projectile. In the laboratory, typically air or nitrogen is used as the gas to propel projectiles up to around 300 m/s. If higher velocities are required, helium or hydrogen is used. Since the sound speed of a gas is given by $\sqrt{\gamma p/\rho}$ (this formula will be derived later in the text), going from N_2 of molecular weight 28 to H_2 of molecular weight 2 increases the sound speed by a factor of $\sqrt{14} = 3.74$, allowing speeds of around 1,100 m/s to be reached. A large gas gun that has barrel diameters of up to 30 cm is shown in Fig. 1.2. This gas gun has been used to launch bird carcasses and simulants for aircraft certification and turbine blades and pieces of rotors for containment studies (i.e., containment of parts after an aircraft, train, or power turbine fails, to ensure the safety of nearby people). It was also used to launch insulating foam into a wing leading edge test fixture during the space shuttle *Columbia* accident investigation.

The next step up in speed is accomplished by using a solid or liquid propellant or gun powder to produce the gas. In these guns, there is a breech attached to a barrel. The powder is placed in the breech and ignited. As the powder or propellant burns, it transitions to a gas, which is hot. It is confined and large pressures result. The expanding gas then pushes the launch package down the barrel. This is the way most firearms work, where the cartridge holds the gunpowder, a primer on the bottom (which produces a spark or flame when struck to ignite the powder), and a bullet at the top, held to the cartridge with a crimp, to ensure the powder burn reaches an appropriate level before releasing the projectile (the "fast-acting valve"). A powder gun from the laboratories at Southwest Research Institute (SwRI) is shown in Fig. 1.3. The 50-mm powder gun has launched projectiles up to 2.1 km/s. The propellant is placed in the breech and a "spit tube" is used to extend the

FIGURE 1.2. The large gas gun at SwRI. A cylindrical pressure vessel is to the left, followed by a fast-acting valve, followed by a long barrel. This photograph was taken before an impact test against the space shuttle leading edge test article, July 2003. High-speed cameras and lights surround the target.

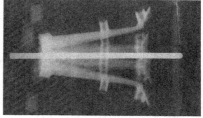

FIGURE 1.3. Left: 50-mm powder gun at SwRI. The breech is to the left; orthogonal flash X-ray heads are near the target. Right: A flash X-ray of a launch package shot from the gun at 1.5 km/s, comprising a threaded tungsten projectile held by a four-petal aluminum sabot, which is opening up and separating owing to aerodynamic drag (the launch package is moving left to right).

initiation sparks the full length of the propellant. High-velocity shots launching 400-gram launch packages to 2 km/s take about 1,300 grams of a modern propellant or gun powder. The pressure in the breech reaches 350 MPa.

The next step up in speed is accomplished by using two stages. The first stage is a powder gun, which launches a polyethylene piston to compress a gas for the second

FIGURE 1.4. Large two-stage light gas gun at SwRI during assembly. The high pressure powder breech is to the left; the pump tube is in two parts; the high-pressure section connector is missing. Then come the launch tube and finally the evacuated target chambers. The gun can launch 60 grams to 7 km/s.

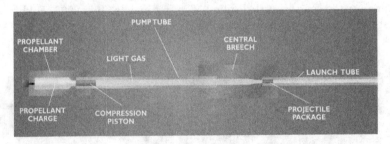

FIGURE 1.5. Two-stage light gas gun schematic.

stage. The piston speed is not particularly high (up to 1 km/s), but it is massive and can compress the gas in the second stage. The second stage gas is typically hydrogen because that is the gas with the highest sound speed. It is compressed to pressures of 800 MPa and temperatures of 1,750 K in the high-pressure coupling section. High temperatures are important, because temperature controls the sound speed in the gas (Exercise 5.17). Also, the light gas is in motion, so that the sound speed is augmented by the speed of the gas. The fast-acting valve, in this case, is a disk of steel that has been scored with an X to different depths so that it fails and peels back, opening up to allow the compressed hydrogen gas to drive the launch package. Speeds in these two-stage light gas guns, as they are called, can reach 7 to 10 km/s, depending on the diameter of the barrel or launch tube. The large two-stage light gas gun at SwRI is shown in Fig. 1.4, with a schematic in Fig. 1.5. It has a 38-mm diameter, 10-meter long launch barrel and a 11.4-cm diameter, 9-meter long pump tube. Before shots, the barrel, flight chamber, and target chambers are evacuated, typically to a pressure of 5 to 20 Torr. The burst

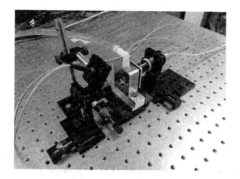

FIGURE 1.6. Laser-launched flyer arrangement at SwRI. Laser pulse comes in from the left through the lens into the evacuated chamber. Fibers to right are the particle displacement velocimetry (PDV) laser measurement system for measuring particle velocity on back surface of target.

disk is part of the pressure seal. The high-pressure steel coupling, where the larger diameter pump tube is gradually reduced to the small diameter launch tube, has an outer diameter of 56 cm and a length of 97 cm, and weighs 1,800 kg.

The next step up in speed is accomplished by placing a graded-density plate or pillow at the muzzle end of the two-stage light gas gun. A graded-density impactor then strikes the plate or pillow, and a high-pressure stress wave transits the pillow and reflects off the free surface. When it returns in the direction of the gun and meets the reflected wave off the back of the projectile, it launches some of the plate or pillow (see Section 4.10 for a description of this spall process). Speeds achieved by this process have been 9 to 14 km/s. In this case, the projectile is typically a flat flyer plate. This is called a three-stage light gas gun. It marks the end and upper limit of what has been accomplished so far with pressure-gun technology.

There are other ways to produce high-pressure gases beyond burning and compressing. A fast laser pulse can vaporize material and, when placed in contact with a thin flyer, like aluminum foil, it can launch the flyer. Figure 1.6 shows the set-up from SwRI's laser lab. Launch speeds of 6 km/s have been achieved with this technique, but the mechanism is the same – a high-pressure gas is produced that then drives a flyer. In this case the flyer is thin and the confinement and short "barrel-like" behavior is produced by nearby material through the localization of the laser pulse.

In some ways the National Ignition Facility at Lawrence Livermore National Laboratory uses this technique to compress deuterium and tritium fuel pellets for fusion studies. Huge banks of high-powered lasers provide a pulse that produces high-energy X-rays that penetrate a thin layer on the outside of the fuel pellet. The layer vaporizes, which then drives the spherically shaped surface inward, producing very high pressures in a spherical compression. Collapse speeds of 350 km/s have been reported.

Finally there is one other approach, and that is to use electromagnetism. In this case, a current flows through a closed circuit. The launch package has an armature that touches two rails. Current flows around this loop, producing a magnetic field that wants to expand. The expansion accelerates the projectile touching the rails. These rail guns, as they are called, require a high-power electrical storage source that can unload of its energy quickly. This is done mechanically – for example, by rapidly stopping a flywheel (homopolar generator) or through capacitor banks. Speeds for rail guns have been reported at 6 km/s. A large struggle with rail guns is that the projectile needs to maintain good contact with the rails to maintain the electrical current. This leads to gouging of the rails at high speeds, greatly limiting their life span. To date, however, the focus on rail guns has been their development; they have not been used where the focus was impact studies.

The next step in electromagnetic driving has been accomplished with the Z-machine at Sandia National Laboratories. Here a huge machine provides a charge of 6 million amps over a short time span, which produces motion in flyers due to magnetic forces. These forces have launched flyer plates up to 44 km/s [20]. There is a smaller laboratory device based on the same principle.

Electrostatic launchers are used to launch dust. Here, in a vacuum, micron-sized dust particles are ionized and a voltage potential is applied across separated, perforated plates to accelerate the dust, similar to a cathode ray tube. The dust particles can reach tens of kilometers per second, but they are very small.

1.2. Launch Packages

In research settings, most launches occur with sub-caliber impactors or projectiles, meaning that the projectile diameter is less than the diameter of the barrel. The projectile is held in a *sabot*, which is the French word for clog or outer shoe with a thick wooden sole. The sabot is the diameter of the barrel and provides the pressure seal with the barrel. In particular, the sabot produces a full-barrel cross sectional area footprint for the driving force on the launch package, thus allowing sub-caliber high-density projectiles to be launched. The sabot can also provide structural support to the projectile or impactor, to prevent buckling of long projectiles, for example. When the launch package exits the gun barrel (the launch end of the gun is called the muzzle), the sabot is separated from the projectile. This is accomplished in several fashions. First, aerodynamic drag on the sabot petals causes them to separate from the projectile and fly on a different trajectory. Typically the projectile will fly through a hole in what is called a stripper plate; the sabot petals strike this plate, as they don't fly through the hole, and are stopped. Second, a rifled barrel (meaning there are spiral scores in the interior sides, to impart a spin during launch) so that the launch package emerges with a rotational velocity that separates the sabot from the projectile. Third, for lower speed launches it is possible to have a constricting cage attached to the muzzle, so that the sabot is decelerated and stopped by its interaction with the cage and the projectile continues on. An X-ray of an aluminum sabot separating from a cylindrical tungsten alloy projectile is shown in Fig. 1.3.

Sometimes a sabot cannot be efficiently separated from the projectile. For example, when we launch foam in our labs we do not use a sabot, but produce a specialty barrel for each foam cross section of interest.

1.3. Diagnostics

Historically the main diagnostics in impact and penetration mechanics have been flash X-rays, shorting pins, strain and pressure gages, and laser interferometry techniques (VISAR and PDV) to measure speed or displacement of a point on a surface. However, the world of diagnostics is being overtaken by fast electronic optical cameras and digital image correlation. These cameras and software tools allow a determination of the displacement of a surface at various points in time. Very fast cameras can record at a rate of millions of frames per second, though typically such cameras have a limited number of frames, such as 16, as the speed is achieved by each electronic image plane only recording one or two images. Slower cameras with "unlimited" numbers of frames, where the electronic image plane is stored and reset for each recorded frame, can run at speeds up to 100,000 frames per second (and faster for reduced image size).

1.4. Nonlinearities and Confinement

A theme of this book is explicitly addressing and quantitatively computing nonlinear response. We will see a variety of origins and results of nonlinearity:

(1) When solid materials undergo large compressions, the pressure vs. volumetric strain is nonlinear. At small strains, it is linear, but at large strains it exhibits higher order terms that greatly increase the pressure and limit further compression.

(2) The nonlinear compressive response gives rise to impulsive loads forming shocks in the material, where the shock speed is supersonic with respect to the ambient sound speed of the material. A shock is a discontinuous change in material properties across a moving propagation front.

(3) When a material is compressed by a shock wave, not only does lattice compression occur, but energy is deposited in the material. This leads to thermal motions of the atoms, which in turn leads to additional stress. Thermal motion leading to additional stress is due to nonlinearity in atomic interactions.

(4) Metals begin to plastically deform when their yield stress is exceeded. At small stress and before yield, the stress-strain response of metals is elastic and linear, but when the stress is large enough yield occurs and the subsequent response is nonlinear. Modeling with this nonlinear strength response is central to predicting penetration events.

(5) In one-dimensional penetration events, nonlinearity of the target material can produce the surprising result that deforming impactors penetrate more deeply than nondeforming impactors.

(6) The nonlinear stress–strain response of materials gives rise to a two-wave structure in shock waves at lower pressures. There is a fast running elastic precursor followed by the wave front where plastic deformation occurs.

(7) Large motions lead to acceleration having a significant nonlinear term due to advection. This nonlinearity is the origin of inertial stress terms in penetration resistance.

(8) For metals, higher strength typically implies smaller failure strain, and this reciprocal relationship between strength and failure strain produces interesting results in finite target penetration.

(9) Liberation of material during cratering has nonlinear size scaling and exhibits a transition in size scale behavior indicative of the saturation of a damage process.

(10) Yarns and fabrics undergo small strains but large rotations in impact events and their effectiveness in stopping impactors is through large deflections. When materials undergo large rotation, it is necessary to use large strain formulations that are nonlinear to model the behavior.

(11) Rubber-like materials have, for low pressures, relative incompressibility and nonlinear response at large strains, which leads to experimentally observed nonuniqueness and bifurcation of solutions.

All these behaviors are interesting. We will pursue them to obtain explicit expressions and fully understand the origin and effects of the various terms.

Another factor that has a large influence on impact and penetration events is confinement. Confined motions lead to higher stresses. One-dimensional motion is highly confined and leads to strong shocks. Sometimes the confinement is provided by boundary conditions, sometimes by impact geometry, and sometimes by inertia. Stresses are lower in penetration events than in one-dimensional impact events since the confinement is less and material is able to move laterally, out of the way of the projectile. Of course, there is some lateral confinement resisting the motion, and the details of how the material moves and the material's constitutive response will allow us to explicitly compute the resistance to penetration.

1.5. Sources

Some other books that address impact and terminal ballistics – that is, what happens when projectiles impact targets (as opposed to interior ballistics and exterior ballistics, which discuss projectile motion inside a gun and in flight, respectively) – are the monograph *Impact* by Goldsmith [**77**], and the collections of survey chapters in *High Velocity Impact Phenomena* [**119**], *Impact Dynamics* [**219**], *High Velocity Impact Dynamics* [**218**], the Springer series *High-pressure Shock Compression of Solids I – VIII* [**21, 37, 53, 72**], *Dynamic Behavior of Materials* by Meyers [**147**], Rosenberg and Dekel's *Terminal Ballistics* [**165**], Grady's *Physics of Shock and Impact* [**81**], and the SwRI Penetration Mechanics Short Course notes [**18**].

There are three major conferences held in impact and penetration, with corresponding professional societies and published proceedings. The International Symposium on Ballistics is held every 18 months. It began in 1974 and falls under the auspices of the International Ballistics Society. The Hypervelocity Impact Symposium has occurred in two spurts. The first was from 1955 to 1969 where there were 8 symposia. With the success of the Apollo missions, interest waned. A new series of Hypervelocity Impact Symposia began in 1986, spurred by the Strategic Defense Initiative for ballistic missile defense. That series is still running, held every two or three years, and many of its proceedings have appeared in the *International Journal of Impact Engineering*. The American Physical Society has a topical group in the Shock Compression of Condensed Matter that meets every other year. Participation in one of these or related symposia is highly recommended for anyone interested in the field of impact and penetration. These three offer competitive incentives for students that cover conference registration fees and some travel expenses.

Exercise

(1.1) Look up when the three symposia mentioned in the text next occur. What are the details of their student programs? When are applications due?

Conservation Laws and the Hugoniot Jump Conditions

Quantitatively understanding impact and penetration requires the use of the physical laws of mass, momentum, and energy conservation. In this chapter, the conversation laws are written in three forms: an integral form, a differential form, which implies certain differentiability conditions, and an algebraic form, relating properties on either side of a steadily moving front. This latter form is called the Hugoniot jump conditions. These equations allow the analysis of dynamic wave and impact events. An ideal gas is used as a simple example of an equation of state and to review some thermodynamic assumptions and notation. Required boundary conditions are discussed.

2.1. Conservation Laws in the Eulerian Reference Frame

In physics and mechanics, three equations play a pivotal role in mathematically describing the behavior of a system: conservation of mass, conservation of momentum, and conservation of energy. This chapter presents these balance equations in an Eulerian or laboratory frame. Appendix A contains background on vector and tensor notation. The Einstein summation notation, reviewed below, will be used extensively.

Before we get to the specific equations, it is important to clearly define terms and assumptions. First, for the majority of this book, the equations of motion are written in terms of the current configuration and coordinates (in the Eulerian or laboratory frame of reference), typically denoted in lower case – for example, x, y, z, or x_1, x_2, x_3. This assumption is the one typically made. Another possibility is to write the equations in terms of the initial configuration and coordinates (in the Lagrangian frame of reference), typically denoted in upper case – for example, X, Y, Z, or X_1, X_2, X_3, which we will do when we examine fabrics and composites (Chapter 13). For small strain, small displacement, small rotation behavior the distinction is unimportant. However, for large displacement, large rotation behavior, the distinction is important. This chapter, then, is written in the Eulerian or laboratory frame of reference, which is the typical frame used in physics analysis.

An irony of our approach is that we begin with integral statements of conservation laws, from them develop differential versions of the conservation laws which require differentiability of the relevant values, and then use the differential versions to develop, through integration, the shock jump conditions where some of these values are not only not differentiable but are not even continuous. It is possible to develop shock jump conditions directly from integral conservation statements of conservation laws, and we do so for the simplest case; however, we extensively use the differential statements of conservation laws, so we need them, too.

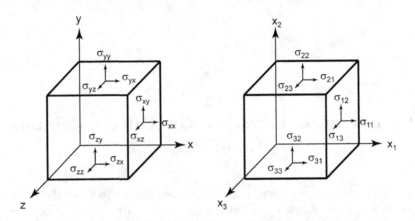

FIGURE 2.1. Tractions at surfaces due to stress.

2.2. Tractions and the Cauchy Stress Tensor

The difference between solids and fluids vs. point particles is that solids and fluids have internal forces. These forces arise from the deformation of the material and they act to resist deformation. Consider a small cube of material, depicted in Fig. 2.1. On each face of this cube there will be a force vector due to the internal forces within the solid or fluid material, . When we divide that force vector by the area of each face, we obtain a *traction vector* on each face of the cube. The traction is force per unit area. The (remarkable) Cauchy stress tensor is obtained by writing each of these tractions as a row vector,

$$\sigma = \begin{pmatrix} \sigma_{11} & \sigma_{12} & \sigma_{13} \\ \sigma_{21} & \sigma_{22} & \sigma_{23} \\ \sigma_{31} & \sigma_{32} & \sigma_{33} \end{pmatrix} = \begin{pmatrix} \vec{t}_1^T \\ \vec{t}_2^T \\ \vec{t}_3^T \end{pmatrix} \quad \text{where} \quad \vec{t}_i = \begin{pmatrix} \sigma_{i1} \\ \sigma_{i2} \\ \sigma_{i3} \end{pmatrix}. \tag{2.1}$$

Here, $\vec{t}_i$ is the traction vector on the face of the cube whose surface normal is given by $\hat{e}_i$. The superscript T means transpose. The component j of the traction vector for each face i is given by σ_{ij}; i.e., the leading index refers to the face defined by the normal $\hat{e}_i$ and the trailing index refers to $\hat{e}_j$-th component.

An assumption of continuum mechanics is that this stress tensor applies at every point – that is, we can take the limit as the cube in Fig. 2.1 is reduced in size to a point, so that the traction is $\vec{t} = \lim_{A \to 0} \vec{f}/A$, the force divided by the area in the limit as the face area A goes to zero. Our modern understanding of the physics of solids and fluids is that they are made of discrete particles (electrons and atomic nuclei) that move in mostly "empty space" and interact through the electromagnetic force propagator, the photon. The atomic size (nuclei plus bound electrons) is around 1.5 angstroms (an angstrom is 10^{-10} m) and so we expect the continuum approximation to break down on the order of tens of nanometers (1 nm is 10^{-9} m). However, for the problems we work in this book, we are in size scales where the continuum approximation applies to our materials.

The traction on an arbitrary surface element with normal vector $\vec{n}$ pointing outwards is

$$t_i = n_j \sigma_{ji} \quad \text{or} \quad \vec{t}^T = \vec{n}^T \sigma, \tag{2.2}$$

where the Einstein summation convention is used in the above equation and used throughout the book: repeated Latin indices are summed from 1 to 3,

$$t_i = n_j \sigma_{ji} = \sum_{j=1}^{3} n_j \sigma_{ji}. \tag{2.3}$$

The left multiplication by the transpose of $\vec{n}$ is nothing "deep," it is simply due to the standard stress tensor indices convention of Fig. 2.1. (Note $\vec{t}^{\,T} = \vec{n}^T \sigma$ says $\vec{t} = \sigma^T \vec{n}$.) The stress tensor is essentially true by definition, though it is necessary to show that Eq. (2.2) follows from Eq. (2.1). The standard argument uses a tetrahedron and $\vec{f} = m\vec{a}$. The volume of the tetrahedron goes as length cubed and the surface area goes as length squared, so as the size goes to zero one concludes that there is a force balance on the tetrahedron's faces; hence Eq. (2.2) holds (details are worked in Exercise 2.1). With the traction vector and the surface normal vector transforming correctly under a coordinate change, it is straightforward to show that stress transforms correctly and hence that stress is a tensor (Exercise 2.2).

If V is a volume of material enclosed in a surface S (the notation for such a surface $S = \partial V$ will sometimes be used), the force $\vec{f}$ is given by

$$f_i = \int_S n_j \sigma_{ji} dS + \int_V \rho b_i dV \quad \text{or} \quad \vec{f}^{\,T} = \int_S \vec{n}^T \sigma dS + \int_V \rho \vec{b}^T dV. \tag{2.4}$$

Here, ρ is the material density and $\vec{n}$ is the unit-length surface normal. Application of the divergence theorem yields

$$f_i = \int_V \left(\frac{\partial \sigma_{ji}}{\partial x_j} + \rho b_i \right) dV \quad \text{or} \quad \vec{f} = \int_V \left(\vec{\nabla} \cdot \sigma + \rho \vec{b} \right) dV. \tag{2.5}$$

The divergence of the stress holds the component i fixed while summing the derivative over each traction vector (or coordinate direction). For example, for the x direction the x component of the traction from faces of the cube are involved in forming the force, namely σ_{11}, σ_{21}, and σ_{31}.

In this book, the stress σ_{ij} is positive in tension. This is the sign convention that is typically used in solid mechanics. However, impact and penetration usually involve compression, and it seems odd to have a negative stress for compression – some of the equations just don't look right with minus signs, especially to those who have worked in the field for a long time. Because we occasionally want the stress to be positive in compression, we introduce a second convention: stresses denoted $\tilde{\sigma}_{ij}$ (with the tilde on top) are positive in compression. Thus, to emphasize:

DEFINITION 2.1. Stress written σ_{ij} is positive in tension.

DEFINITION 2.2. Stress written $\tilde{\sigma}_{ij}$ is positive in compression.

Hence $\tilde{\sigma}_{ij} = -\sigma_{ij}$. The same convention will be used for strain: ε_{ij} is positive in tension and $\tilde{\varepsilon}_{ij}$ is positive in compression, with $\tilde{\varepsilon}_{ij} = -\varepsilon_{ij}$. Pressure is defined as

$$p = -\frac{1}{3}(\sigma_{11} + \sigma_{22} + \sigma_{33}) = -\frac{\sigma_{ii}}{3}. \tag{2.6}$$

Pressure is always positive in compression (notice it is opposite in sign to σ_{ij}). The right-hand side uses the Einstein summation convention, where repeated Latin indices are summed from 1 to 3.

The Cauchy stress is symmetric,[1]

$$\sigma_{ij} = \sigma_{ji}. \tag{2.7}$$

This result is due to conservation of angular momentum (Appendix A, page 604). However, when we discuss fabrics and composites we will work in a different frame of reference (the Lagrangian frame) and there the stress tensor we use (the first Piola–Kirchhoff stress) is not symmetric. At that time we will rederive the appropriate conservation equations (Chapter 13).

2.3. Conservation of Mass

The total mass in a system is given by the volume integral of the density. It is assumed that mass is neither created nor destroyed. For a fixed stationary (Eulerian) volume V in space, the change in mass is given by how much mass flows out through the surface S of the volume V. With density ρ and material velocity $\vec{v}$, we have

$$\frac{d}{dt}\text{mass}_V = \frac{d}{dt}\int_V \rho dV = \int_V \frac{\partial \rho}{\partial t}dV = -\int_S \rho\vec{v}\cdot\vec{n}dS, \tag{2.8}$$

where $\vec{n}$ is the unit-length surface normal and the last integral on the right is the amount of mass flowing out of the surface, since $\vec{v}\cdot\vec{n}$ gives the normal material speed across the surface boundary. The divergence theorem says

$$\int_S \rho\vec{v}\cdot\vec{n}dS = \int_V \text{div}(\rho\vec{v})dV. \tag{2.9}$$

Combining the previous two equations gives

$$\int_V \left(\frac{\partial \rho}{\partial t} + \text{div}(\rho\vec{v})\right) dV = 0, \tag{2.10}$$

and since this holds for any arbitrary volume V we might choose, we conclude

$$\frac{\partial \rho}{\partial t} + \frac{\partial}{\partial x_i}(\rho v_i) = 0 \quad \text{or} \quad \frac{\partial \rho}{\partial t} + \text{div}(\rho\vec{v}) = 0. \tag{2.11}$$

The result has been written in both Cartesian coordinates (left) and in a general form (right). The terms "conservation of mass" and "mass balance" refer to the same thing and will be used interchangeably. This form of the conservation equation, where everything is within derivatives, is called *conservation form*. Its usefulness will become clear later in this chapter.

Use of the divergence theorem and taking arbitrarily small volumes require the differentiability of the density ρ and the velocity $\vec{v}$. This implies that when materials change or a material is sitting next to a void, these changes must be addressed through boundary conditions, and that such boundary conditions may be internal

---

[1]There are constitutive models that produce stress tensors that are referred to as Cauchy that are not symmetric, but they are not part of classical continuum mechanics. These models fall under names such as couple stress or micropolar stress. What is occurring in these theories is that it is felt there is some mechanics (usually referred to as micromechanics) that is not being resolved at the physical scale of analysis in the stress tensor, so additional behaviors are added. Typically it is assumed there are material motions that $\vec{x} = \vec{x}(\vec{X}, t)$ does not completely capture, such as a rotation of the material about $\vec{x}$ that is independent of the average rotation implicit in a differentiable $\vec{x}(\vec{X}, t)$. This is directly akin to the way beams are modeled in strength of materials, where the concept of bending moment is introduced, which is an additional detail of stress beyond the stress tensor's longitudinal, transverse, and shear stress components. We will not use any of these models, but see the discussion in Section 2.12 and Exercises 2.17 and 14.41.

to the body. This is because the density is not continuous, let alone differentiable, across such a boundary. Also, two solid materials that are not connected can touch and be sliding along each other, and hence velocities need not be continuous nor differentiable across these internal boundaries, and such boundaries must be handled with boundary conditions and cannot be handled with the differential conservation equation.

We now clarify what the temporal derivative of density means. As written, we have $\partial\rho/\partial t$. Since the density is a function of spatial location and time, $\rho = \rho(x, y, z, t)$, the partial derivative with respect to time shows how the density changes as we stay at that point in space and consider ρ as a function of time at that stationary point. Suppose, however, we want to know how the density is changing if we are a point moving with the deformation and flow of the material. In that case, we want to determine what is labeled D/Dt, referred to as the total time derivative, and not the partial with respect to time. D/Dt is also called the material derivative or the substantial derivative. Let f be a function (in this case density, but we are interested in the general case) of space and time, $f = f(x, y, z, t)$. Since we are moving with the flow, at the next instant of time the function f is time $t + \Delta t$ and has moved to a new spatial position, $(x + \Delta x, y + \Delta y, z + \Delta z)$; hence f has value $f(x + \Delta x, y + \Delta y, z + \Delta z, t + \Delta t)$. For small Δt, the motion is given by $\Delta x \approx v_x \Delta t$, and so the new value of f is

$$f(x + v_x\Delta t, y + v_y\Delta t, z + v_z\Delta t, t + \Delta t). \tag{2.12}$$

Subtracting the new value of f from the original value of f, the Taylor expansion gives

$$f(x+\Delta x, y+\Delta y, z+\Delta z, t+\Delta t)-f(x, y, z, t) \approx \frac{\partial f}{\partial x}v_x\Delta t+\frac{\partial f}{\partial y}v_y\Delta t+\frac{\partial f}{\partial z}v_z\Delta t+\frac{\partial f}{\partial t}\Delta t. \tag{2.13}$$

Dividing by Δt and taking the limit as Δt goes to zero gives the material derivative:

DEFINITION 2.3. (material derivative) The *material derivative of f* is

$$\frac{Df}{Dt} = \frac{\partial f}{\partial t} + v_x\frac{\partial f}{\partial x} + v_y\frac{\partial f}{\partial y} + v_z\frac{\partial f}{\partial z} = \frac{\partial f}{\partial t} + v_i\frac{\partial f}{\partial x_i} \quad \text{or} \quad \frac{Df}{Dt} = \frac{\partial f}{\partial t} + \vec{v}\cdot\text{grad}(f).$$

$$\tag{2.14}$$

where the left is in Cartesian coordinates and the right is general ($\vec{\nabla} = \text{grad}$).

Equation (2.14) gives rise to many of the nonlinearities of fluid mechanics and is central to understanding dynamic behavior. D/Dt measures the total change in time that occurs to a variable associated with a piece of material that is in motion. The velocity terms arise because of material motion in the Eulerian frame. Thus we have computed how the density changes in time for a point moving with the flow, $D\rho/Dt = \partial\rho/\partial t + v_i\partial\rho/\partial x_i$.

Coming back to the conservation of mass, we now have

$$\frac{\partial\rho}{\partial t} + \frac{\partial}{\partial x_i}(\rho v_i) = \frac{\partial\rho}{\partial t} + \rho\frac{\partial v_i}{\partial x_i} + v_i\frac{\partial\rho}{\partial x_i} = \frac{D\rho}{Dt} + \rho\frac{\partial v_i}{\partial x_i} = 0. \tag{2.15}$$

Hence we have another form of the conservation of mass, this time written in terms of the material derivative,[2]

$$\frac{D\rho}{Dt} + \rho\frac{\partial v_i}{\partial x_i} = 0 \quad \text{or} \quad \frac{D\rho}{Dt} + \rho\,\text{div}(\vec{v}) = 0. \tag{2.16}$$

To clarify the notation, in Einstein summation convention and in Cartesian coordinates,

$$\text{div}(\vec{v}) = \frac{\partial v_i}{\partial x_i} = \frac{\partial v_1}{\partial x_1} + \frac{\partial v_2}{\partial x_2} + \frac{\partial v_3}{\partial x_3} = \frac{\partial v_x}{\partial x} + \frac{\partial v_y}{\partial y} + \frac{\partial v_z}{\partial z}, \tag{2.17}$$

where the subscript i takes on the values of 1, 2, 3 (where $x_1 = x$, $x_2 = y$ and $x_3 = z$), and

$$\frac{D}{Dt} = \frac{\partial}{\partial t} + \vec{v}\cdot\vec{\text{grad}}() = \frac{\partial}{\partial t} + v_i\frac{\partial}{\partial x_i} = \frac{\partial}{\partial t} + v_x\frac{\partial}{\partial x} + v_y\frac{\partial}{\partial y} + v_z\frac{\partial}{\partial z} \tag{2.18}$$

is the material derivative.

2.4. Conservation of Momentum

Momentum is mass times velocity. Conservation of momentum is Newton's second law. Since momentum is a vector, three equations exist, one for each coordinate direction; these equations are commonly referred to as the equations of motion. Newton's second law states that a change in motion is proportional to the forces exerted. There can be external forces, such as gravity, represented by $\vec{b}$, as well as the forces arising from within the material itself because of to its desire to resist deformation, represented by the stress tensor σ_{ij}. In impact problems these internal forces are far greater than the force that gravity exerts. Figure 2.1 depicts the components of the stress tensor on the surface of the cubic volume. The gradient of the stress is the force per unit volume. For a fixed stationary (Eulerian) volume V, the change in momentum of the material in the volume V is given by the amount of momentum that flows out (advects) through the surface of the volume and the momentum change due to the force. In Cartesian coordinates,

$$\frac{d}{dt}\int_V \rho v_i dV = \int_V \frac{\partial}{\partial t}(\rho v_i)dV = -\int_S \rho v_i \vec{v}\cdot\vec{n}dS + f_i = -\int_V \text{div}(\rho v_i\vec{v})dV + f_i, \tag{2.19}$$

where the divergence theorem was used. Thus,

$$\int_V \left(\frac{\partial}{\partial t}(\rho v_i) + \text{div}(\rho v_i \vec{v})\right) dV = \int_V \left(\frac{\partial}{\partial t}(\rho v_i) + \frac{\partial}{\partial x_j}(\rho v_i v_j)\right) dV = f_i, \tag{2.20}$$

and using the force from Eq. (2.5) and the fact that we have an arbitrary volume,

$$\frac{\partial}{\partial t}(\rho v_i) + \frac{\partial}{\partial x_j}(\rho v_i v_j) = \frac{\partial \sigma_{ji}}{\partial x_j} + \rho b_i. \tag{2.21}$$

Momentum conservation in conservation form is thus

$$\frac{\partial(\rho v_i)}{\partial t} + \frac{\partial}{\partial x_j}(\rho v_j v_i - \sigma_{ji}) = 0 \quad \text{or} \quad \frac{\partial(\rho\vec{v})}{\partial t} + \vec{\nabla}\cdot(\rho\vec{v}\vec{v}^T - \sigma) = 0 \tag{2.22}$$

---

[2]Appendix A presents a derivation of the conservation equations where a moving surface encloses the same material and that volume is followed in the Eulerian frame, rather than being a fixed stationary volume as in this chapter. The resulting conservation equations are, of course, the same. However, the fixed stationary control volume leads immediately to the conservation form of the equation while the control volume following the material leads first to the material derivative form.

where we have dropped the body force. There are three equations here, as the index i ranges from 1 to 3.

Expanding the left-hand side, using the definition of the material derivative and mass conservation,

$$\frac{\partial}{\partial t}(\rho v_i) + \frac{\partial}{\partial x_j}(\rho v_i v_j) = \rho \frac{\partial v_i}{\partial t} + v_i \frac{\partial \rho}{\partial t} + \rho v_j \frac{\partial v_i}{\partial x_j} + v_i \frac{\partial}{\partial x_j}(\rho v_j) = \rho \frac{Dv_i}{Dt}. \quad (2.23)$$

Thus, another form of momentum conservation is the "graduate school version" of $\vec{f} = m\vec{a}$,

$$\frac{\partial \sigma_{ji}}{\partial x_j} + \rho b_i = \rho \frac{Dv_i}{Dt} \quad \text{or} \quad \vec{\nabla} \cdot \sigma + \rho \vec{b} = \rho \frac{D\vec{v}}{Dt}. \quad (2.24)$$

The right-hand side has the material derivative of the velocity, which is the acceleration for a particle moving with the flow. For completeness we have included the body force (such as gravity) b_i; however, for the problems we will consider, the body force will be negligible. The form of the equation being used here is where x, y, z are the fixed coordinates in space. In that frame of reference, the stress is the Cauchy stress.

Use of the divergence theorem and taking arbitrarily small volumes require the differentiability of the stress σ. The integral form of force term in the momentum conservation shows that the boundary condition between two materials is an assumed continuity of the tractions $\vec{t}$ for the materials at the boundary. Continuity of the traction does not imply continuity of the stress tensor, as the traction vector does not completely define the stress tensor. In fact, it is typically the case that the stress tensor is discontinuous at material boundaries and exterior boundaries in solids, even for static problems.

As another example of the material derivative of a function f, consider the material derivative of the velocity $\vec{f} = \vec{v}$, which yields the acceleration $D\vec{v}/Dt$. If the material motion is "steady state," meaning that the velocity field does not change with respect to time ($\partial \vec{v}/\partial t = 0$), there can still be acceleration of particles as they traverse a path in the velocity field with spatially changing velocities. The material derivative captures these accelerations with $Dv_i/Dt = \partial v_i/\partial t + v_j \partial v_i/\partial x_j = v_j \partial v_i/\partial x_j \neq 0$ (in Cartesian coordinates). This is the acceleration in the Eulerian or laboratory (fixed in space) reference frame. Alternatively, if the coordinate system used is the original configuration (Lagrangian), so $f(X, Y, Z, t)$, where (X, Y, Z) is the original spatial location of the material which does not change, then $f(X, Y, Z, t)$ goes to $f(X, Y, Z, t + \Delta t)$; hence the derivative is $\partial f/\partial t$ and so the acceleration is $\partial \vec{v}/\partial t$. In the Eulerian frame, if the motion is small (as it often is for elastic deformation of a solid body), the nonlinear velocity terms in the total derivative can be dropped, yielding $D\vec{v}/Dt \approx \partial \vec{v}/\partial t$.

2.5. Conservation of Energy

It will be taken as true that energy is not lost in the system. Energy in a material is comprised of internal energy, which may be recoverable because of deformation (elastic stored energy) and may include thermal energy (heat), which may not be recoverable, and there is a bulk kinetic component due to motion. This bulk kinetic component is not the local thermal motion of the molecules (which is reflected in the internal energy) but is based on the overall translational speed.

The energy is

$$\tilde{e} = \rho e = \frac{1}{2}\rho v_i v_i + \rho E. \tag{2.25}$$

When working with internal variables such as energy, it is often useful to work with the specific internal variable:

DEFINITION 2.4. (specific) The *specific* form of an internal quantity is the quantity per unit mass of material.

The specific form (per unit mass) is to be distinguished from the other common form, which is per unit volume. E is the internal energy per unit mass, or specific internal energy, of the material; ρE is the internal energy per unit volume. Defining e to be the sum of specific kinetic and specific internal energies, hence the total specific energy, yields

$$e = \frac{1}{2}v_i v_i + E = \frac{\tilde{e}}{\rho}. \tag{2.26}$$

Here, e is a specific energy – energy per unit mass – which is why there is no density term multiplying the velocity squared.

For a fixed stationary (Eulerian) volume V of material, the change of total energy in the volume is given by the amount that advects out of the volume, plus inputs related to the work performed by the tractions on the surface of the volume $\vec{t} \cdot \vec{v}$, mechanical work done internal to the volume by body forces $\vec{b} \cdot \vec{v}$, energy deposited in the volume r (say because of a chemical reaction or radiation absorption), and heat flux q through the boundary,

$$\frac{d}{dt}\int_V \tilde{e}\,dV = \int_V \frac{\partial \tilde{e}}{\partial t}dV = -\int_S \tilde{e}\vec{v}\cdot\vec{n}dS + \int_S \vec{t}\cdot\vec{v}dS + \int_V \rho\vec{b}\cdot\vec{v}dV + \int_V \rho r\,dV - \int_S \vec{q}\cdot\vec{n}dS. \tag{2.27}$$

Use of the divergence theorem, the stress tensor ($\vec{t} \cdot \vec{v} = n_i\sigma_{ij}v_j$), and the fact that the volume is arbitrary yields

$$\frac{\partial \tilde{e}}{\partial t} + \frac{\partial}{\partial x_i}(\tilde{e}v_i) - \frac{\partial}{\partial x_i}(\sigma_{ij}v_j) - \rho b_i v_i - \rho r + \frac{\partial q_i}{\partial x_i} = 0. \tag{2.28}$$

Dropping the body force work, the internal source and heat flux, the energy conservation in conservation form is

$$\frac{\partial \tilde{e}}{\partial t} + \frac{\partial}{\partial x_i}(\tilde{e}v_i - \sigma_{ij}v_j) = 0 \qquad \text{or} \qquad \frac{\partial \tilde{e}}{\partial t} + \text{div}\,(\tilde{e}\vec{v} - \sigma\vec{v}) = 0. \tag{2.29}$$

Though the conservation form of the energy balance is what we will use for the jump conditions, it is not entirely satisfactory as written from an understanding perspective, since $\tilde{e}$ is not invariant under a Galilean transformation (i.e., a steady reference frame at a different velocity). Going back to energy conservation in Eq. (2.28), when $\tilde{e}$ and all the partial derivatives are expanded out, two applications of mass conservation and one application of momentum conservation remove many of the terms in the sum (including the $v_j\partial\sigma_{ij}/\partial x_i$ term and the body force), and we are left with the conservation of energy as

$$\rho\frac{DE}{Dt} = \sigma_{ij}\frac{\partial v_j}{\partial x_i} + \rho r - \frac{\partial q_i}{\partial x_i}. \tag{2.30}$$

The derivative of the velocity with respect to the spatial components is called the *velocity gradient* L, where $L = \partial v_i/\partial x_j$. It is a tensor that contains the gradient of each component of the velocity. This tensor can be partitioned into symmetric and skew-symmetric parts,

$$L_{ij} = \frac{1}{2}(L+L^T)_{ij} + \frac{1}{2}(L-L^T)_{ij} = \frac{1}{2}\left(\frac{\partial v_i}{\partial x_j} + \frac{\partial v_j}{\partial x_i}\right) + \frac{1}{2}\left(\frac{\partial v_i}{\partial x_j} - \frac{\partial v_j}{\partial x_i}\right) = D_{ij} + W_{ij}.$$

(2.31)

The curl of a vector field quantifies the rotation in the vector field (see derivation of the curl in Appendix A). The rotation rate in radians per second and rotation axis are given by $\vec{w} = \frac{1}{2}\mathrm{curl}(\vec{v})$. An examination of W shows

$$W = \frac{1}{2}\left(\frac{\partial v_i}{\partial x_j} - \frac{\partial v_j}{\partial x_i}\right) = \frac{1}{2}\begin{pmatrix} 0 & \left(\frac{\partial v_1}{\partial x_2} - \frac{\partial v_2}{\partial x_1}\right) & \left(\frac{\partial v_1}{\partial x_3} - \frac{\partial v_3}{\partial x_1}\right) \\ \left(\frac{\partial v_2}{\partial x_1} - \frac{\partial v_1}{\partial x_2}\right) & 0 & \left(\frac{\partial v_2}{\partial x_3} - \frac{\partial v_3}{\partial x_2}\right) \\ \left(\frac{\partial v_3}{\partial x_1} - \frac{\partial v_1}{\partial x_3}\right) & \left(\frac{\partial v_3}{\partial x_2} - \frac{\partial v_2}{\partial x_3}\right) & 0 \end{pmatrix}$$

(2.32)

$$= \frac{1}{2}\begin{pmatrix} 0 & -\mathrm{curl}(\vec{v})_3 & \mathrm{curl}(\vec{v})_2 \\ \mathrm{curl}(\vec{v})_3 & 0 & -\mathrm{curl}(\vec{v})_1 \\ -\mathrm{curl}(\vec{v})_2 & \mathrm{curl}(\vec{v})_1 & 0 \end{pmatrix} = -\varepsilon_{ijk}w_k = -\frac{1}{2}\varepsilon_{ijk}\mathrm{curl}(\vec{v})_k.$$

Here, ε_{ijk} is the permutation symbol (see page 50). W stores the curl of the velocity, so $W = \frac{1}{2}(L - L^T)$ is called the *spin tensor*. W is skew symmetric, meaning that $W^T = -W$. With the rotation of the material removed, the remaining symmetric term in the partition of L is called the *rate of deformation tensor* $D = \frac{1}{2}(L + L^T)$,

$$D_{ij} = \frac{1}{2}\left(\frac{\partial v_i}{\partial x_j} + \frac{\partial v_j}{\partial x_i}\right),$$

(2.33)

since it contains the information of how the material is deforming. Since the Cauchy stress tensor is symmetric, $\sigma_{ij}W_{ij} = 0$, and the energy balance can be written

$$\rho\frac{DE}{Dt} = \sigma_{ij}D_{ij} + \rho r - \frac{\partial q_i}{\partial x_i} \quad \text{or} \quad \rho\frac{DE}{Dt} = \sigma_{ij}D_{ij} + \rho r - \mathrm{div}(\vec{q}).$$

(2.34)

Here, r is an energy release rate within the material (such as the chemical reaction that occurs when explosives detonate) and q_i is the heat flux. Typically, the thermal conduction term is on a much longer time scale than the penetration event and we will ignore it in this book (see page 65). Equation (2.30) says that the amount of energy stored and/or dissipated within the material is due to its deformation while under stress (i.e., the power is force times velocity), heat flux or energy sources. This equation does not say how the internal energy is stored within the material.

Use of the divergence theorem and taking arbitrarily small volumes require the differentiability of the internal energy E and heat flux $\vec{q}$.

2.6. Conserved Quantities

Examination of the above equations shows that the conservation form of each of the mass, momentum in each direction, and energy equations can be written

$$\frac{\partial A}{\partial t} + \mathrm{div}(\vec{B}) = 0$$

(2.35)

in vector notation, where the A and $\vec{B}$ are various functions of density, velocity, and stress, and the divergence of a vector field in Cartesian coordinates is

$$\mathrm{div}(\vec{B}) = \frac{\partial B_i}{\partial x_i}.$$

(2.36)

We now consider the volume integral of A over a volume V fixed in space that contains all the material such that no material touches the boundary. For example, V could be all of space,

$$\int_V A dV = \int_{-\infty}^{\infty} \int_{-\infty}^{\infty} \int_{-\infty}^{\infty} A dx dy dz. \tag{2.37}$$

Differentiating with respect to time, we have

$$\frac{\partial}{\partial t} \int_V A dV = \int_V \frac{\partial A}{\partial t} dV = -\int_V \text{div}(\vec{B}) dV = -\int_S \vec{B} \cdot \vec{n} dS \tag{2.38}$$

where S is the bounding surface of the region containing all the material and $\vec{n}$ is the unit normal to the surface. Thus, since there is no material touching the boundary, the right-hand side of this equation equals zero and we obtain the result that the spatial volume integral of the quantity A is independent of time and is thus conserved. In particular, given material of finite extent, we obtain the result that mass is conserved, momentum in each of the three coordinate directions is conserved, and the sum of kinetic and internal energy is conserved,

$$\int_V \rho \, dV = \text{mass} = \text{constant},$$

$$\int_V \rho v_i \, dV = \text{momentum in the } i \text{th direction} = \text{constant}, \tag{2.39}$$

$$\int_V \tilde{e} \, dV = \int_V \left(\frac{1}{2} \rho v_i v_i + \rho E \right) dV = \text{total energy} = \text{constant}.$$

2.7. The Rankine–Hugoniot Jump Conditions

Of particular interest to us are dynamic solutions to the equations of motion and, in particular, waves traveling through the material. Traveling wave fronts that have a finite jump in values across the wave front are referred to as shock waves.

2.7.1. Simplest Case.
In this section we derive the equations that relate the properties within the material in front of the wave front and behind the wave front for a general wave traveling through a material. The simplest wave to imagine is one where the wave moves from left to right and the material on the right has initial conditions of being at rest, being stress free, and with initial density. The wave front marks a change in all three of these initial properties: density, stress, and particle velocity (Fig. 2.2). The wave is assumed to move with velocity U. First, for this simple case, we derive the expressions of mass, momentum, and energy conservation. In each case the area of the cross section will not be included, since it is the same on both sides of the wave front.

For mass conservation, for a given time Δt, the amount of material encompassed by the wave front is $\rho_0 U \Delta t$. Behind the wave front, this same material is now compressed into a smaller region because of the material velocity u (also called the particle velocity) of the material. The mass behind the wave front is $\rho_1 (U - u)\Delta t$. These two values of mass are equal, yielding

$$\frac{\rho_0}{\rho_1} = 1 - \frac{u}{U}. \tag{2.40}$$

This is the mass conservation jump condition.

For momentum conservation, the force applied to the wave front is $\tilde{\sigma}$ over a time Δt. The momentum in front of the wave front is zero, since the material

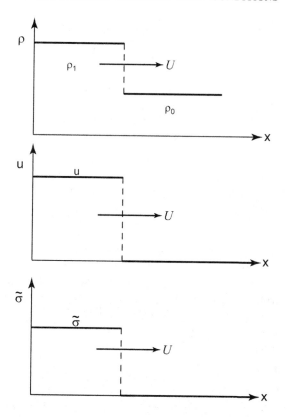

FIGURE 2.2. Density, velocity, and stress in front of and behind the shock front.

is at rest, and behind the wave front the momentum is $(\rho_1(U - u)\Delta t)u$. By the mass conservation derivation above, this equals $(\rho_0 U \Delta t)u$. Thus, the change in momentum with respect to time is this quantity divided by Δt. Setting the time rate of change of momentum equal to the force gives

$$\tilde{\sigma} = \rho_0 U u. \tag{2.41}$$

This is the momentum conservation jump condition.

Finally, we consider energy. The work performed by the passage of the wave front is the force times the distance: $\tilde{\sigma} u \Delta t$. The energy within the material is given by a change in internal energy ΔE and a kinetic energy term $\frac{1}{2}\rho_1 u^2$. The mass of material behind the wave front is $(\rho_1(U - u)\Delta t)u$, and so the total change in energy (using the mass jump condition) is $\rho_0 U \Delta t(\Delta E + \frac{1}{2}u^2)$. Setting this equal to the work done gives

$$\Delta E + \frac{1}{2}u^2 = \frac{\tilde{\sigma} u}{\rho_0 U}. \tag{2.42}$$

If we directly use the momentum balance jump condition, we get the surprising (but correct) result that $\Delta E = \frac{1}{2}u^2$, a result that is only true for this simple case. It says that energy change is equally partitioned between a change in kinetic energy and a change in internal energy. More typically, the momentum and mass conservation jump conditions are used at this point to write the change in internal energy as

$$\Delta E = \frac{1}{2} \tilde{\sigma} \left(\frac{1}{\rho_0} - \frac{1}{\rho_1} \right). \tag{2.43}$$

Here $1/\rho$ is the specific volume or volume per unit mass occupied by the material.[3] (The mass of the material is given by the volume times the density, so the volume divided by the mass leaves $1/\rho$.) This is the energy conservation jump condition. These, then, are the three jump conditions for the simplest wave.

2.7.2. General Case. We now address the general case of a discontinuous wave. To do so, the conservation form of the balance equations will be used. The first step will be the calculation of the jump condition for the arbitrary equation in conservation form

$$\frac{\partial A}{\partial t} + \frac{\partial B}{\partial x} = 0. \tag{2.44}$$

Figure 2.3 shows the region in space under consideration, defined by a wave front traveling at speed U from left to right. The conditions in front of the wave front (to the right of the wave) will be denoted with a subscript r and the conditions behind the wave front (to the left of the wave) will be denoted with a subscript ℓ. We are going to perform an area integral over the box with corners (x, t) and $(x + \Delta x, t + \Delta t)$. The relation between Δx and Δt is given by $\Delta x = U \Delta t$. First we perform the area integral for the A term,

$$\int_{x}^{x+\Delta x} \int_{t}^{t+\Delta t} \frac{\partial A}{\partial t} dt dx = \int_{x}^{x+\Delta x} (A(t+\Delta t) - A(t)) \, dx = (A_\ell - A_r) \Delta x, \tag{2.45}$$

where the evaluation of the x integral was based on the values to the left and right of the wave front. Similarly, evaluation of the area integral of the B term gives

$$\int_{t}^{t+\Delta t} \int_{x}^{x+\Delta x} \frac{\partial B}{\partial x} dx dt = \int_{t}^{t+\Delta t} (B(x+\Delta x) - B(x)) \, dt = (B_r - B_\ell) \Delta t. \tag{2.46}$$

The integral of the right-hand side of Eq. (2.44) (i.e., zero) is zero and so combining these two expressions with the wave speed yields the general form of the jump conditions for a conservation law,

$$(B_\ell - B_r) = U (A_\ell - A_r). \tag{2.47}$$

In books on the mathematical theory of shocks, this equation is often written

$$[B] = U[A] \tag{2.48}$$

where the square bracket says to take the difference of the quantities across the shock front.

Now that we have developed the general form of the shock jump conditions, we will apply it to each of our conservation laws. The first is mass conservation. Here we have (from Eq. (2.11), where owing to convention in the field we write velocity as u rather than v)

$$A = \rho, \quad B = \rho u, \tag{2.49}$$

and therefore

$$\rho_\ell u_\ell - \rho_r u_r = U(\rho_\ell - \rho_r). \tag{2.50}$$

---

[3]The use in the shock-physics community of $v = 1/\rho$ for specific volume is one of the reasons that u is used for velocity. However, we will exclusively refer to the specific volume as $1/\rho$, since v will be a velocity in our penetration modeling.

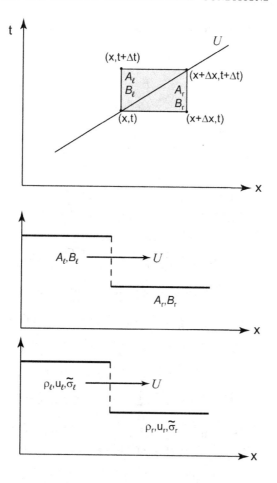

FIGURE 2.3. General shock front.

This can be manipulated into two more useful forms:

$$\frac{\rho_r}{\rho_\ell} = \frac{U - u_\ell}{U - u_r} \tag{2.51}$$

and

$$\frac{\rho_r}{\rho_\ell} = 1 - \frac{u_\ell - u_r}{U - u_r}. \tag{2.52}$$

This latter equation is the typical form of the mass jump condition. When $u_r = 0$ it simplifies to the simple case derived above (Eq. 2.40).

Next we tackle the momentum balance. From Eq. (2.22) we have

$$A = \rho u, \quad B = \rho u^2 + \tilde{\sigma}. \tag{2.53}$$

The jump condition is thus

$$\rho_\ell u_\ell^2 + \tilde{\sigma}_\ell - \rho_r u_r^2 - \tilde{\sigma}_r = U(\rho_\ell u_\ell - \rho_r u_r). \tag{2.54}$$

Some manipulation and use of the mass balance equation gives the momentum jump in its traditional form

$$\tilde{\sigma}_\ell - \tilde{\sigma}_r = \rho_r(U - u_r)(u_\ell - u_r). \tag{2.55}$$

Setting the particle velocity and stress in front of the wave equal to zero gives the above simple-case result.

Finally, for the energy balance we have

$$A = \tilde{e}, \quad B = \tilde{e}u + \tilde{\sigma}u, \tag{2.56}$$

which yields

$$\tilde{e}_\ell u_\ell + \tilde{\sigma}_\ell u_\ell - \tilde{e}_r u_r - \tilde{\sigma}_r u_r = U(\tilde{e}_\ell - \tilde{e}_r). \tag{2.57}$$

If we go back to the original definition of $\tilde{e} = \rho E + (1/2)\rho u^2$ and write out all the terms, we obtain

$$\left(\rho_\ell E_\ell + \frac{1}{2}\rho_\ell u_\ell^2\right)(u_\ell - U) + \tilde{\sigma}_\ell u_\ell - \left(\rho_r E_r + \frac{1}{2}\rho_r u_r^2\right)(u_r - U) - \tilde{\sigma}_r u_r = 0. \tag{2.58}$$

Application of the mass jump condition (Eq. 2.52) gives

$$E_\ell + \frac{1}{2}u_\ell^2 - E_r - \frac{1}{2}u_r^2 + \frac{\tilde{\sigma}_\ell u_\ell - \tilde{\sigma}_r u_r}{\rho_r(u_r - U)} = 0. \tag{2.59}$$

Writing both stress terms using the momentum jump condition as

$$\tilde{\sigma}_\ell = \frac{1}{2}\tilde{\sigma}_\ell + \frac{1}{2}\{\tilde{\sigma}_r + \rho_r(U - u_r)(u_\ell - u_r)\} \tag{2.60}$$

$$\tilde{\sigma}_r = \frac{1}{2}\tilde{\sigma}_r + \frac{1}{2}\{\tilde{\sigma}_\ell - \rho_r(U - u_r)(u_\ell - u_r)\} \tag{2.61}$$

leads to

$$E_\ell - E_r + \frac{1}{2}\frac{(\tilde{\sigma}_\ell + \tilde{\sigma}_r)(u_\ell - u_r)}{\rho_r(u_r - U)} = 0. \tag{2.62}$$

A second use of the mass jump condition to remove U from the equation produces the traditional form of the energy jump condition,

$$E_\ell - E_r = \frac{1}{2}(\tilde{\sigma}_\ell + \tilde{\sigma}_r)\left(\frac{1}{\rho_r} - \frac{1}{\rho_\ell}\right). \tag{2.63}$$

An interesting observation of this form of the energy jump condition is that it does not include either u or U, and is thus dependent only on internal variables of the material. Thus, the thermodynamic state of a jump can be determined with an equation of state of a material and the energy jump condition.

In the presence of an initial stress, the change in internal energy is more complicated than just the addition of $\frac{1}{2}u^2$, as occurred in the simple case. Two applications of the momentum balance yield

$$E_\ell - E_r = \frac{1}{2}\frac{\tilde{\sigma}_\ell + \tilde{\sigma}_r}{\tilde{\sigma}_\ell - \tilde{\sigma}_r}(u_\ell - u_r)^2. \tag{2.64}$$

The relative change in internal energy is invariant under a Galilean transformation, but the change in specific kinetic energy is not,

$$\Delta E_k = \frac{1}{2}u_\ell^2 - \frac{1}{2}u_r^2 = \frac{1}{2}(u_\ell - u_r)(u_\ell + u_r). \tag{2.65}$$

This is one of the reasons why we always work with the internal energy change (which is frame independent) rather than the total energy change.

Here, at the end of this section, we list the three Rankine–Hugoniot jump conditions again. They are central to studying shock waves.

THEOREM 2.5. *(Rankine–Hugoniot jump conditions, Eulerian frame) The relationship between density, stress, energy, and particle velocity u in front of and behind a discontinuous shock moving at velocity U is given by*

$$\text{Mass:} \qquad \frac{\rho_r}{\rho_\ell} = 1 - \frac{u_\ell - u_r}{U - u_r}, \qquad (2.66)$$

$$\text{Momentum:} \quad \tilde{\sigma}_\ell - \tilde{\sigma}_r = \rho_r(U - u_r)(u_\ell - u_r), \qquad (2.67)$$

$$\text{Energy:} \qquad E_\ell - E_r = \frac{1}{2}(\tilde{\sigma}_\ell + \tilde{\sigma}_r)\left(\frac{1}{\rho_r} - \frac{1}{\rho_\ell}\right). \qquad (2.68)$$

These are algebraic equations for the locus of end states. A differential equation for the energy along the Hugoniot can be found in Section 5.10.

We step aside in this paragraph for a technical mathematical argument. Our derivation of the shock jump conditions above assumed that the jump was discontinuous and instantaneous. However, a weaker assumption can be made. Assume the density, stress, and energy are uniformly bounded. If the wave front – that is, the region where the material states transition from values in front of the wave to values behind the wave – is of finite extent and does not grow in time or in space (i.e., the wave front is always of thickness less than a given thickness and it takes less than a fixed amount of time to occur; for example, these assumptions hold for a *steady wave*, a wave that does not change shape as it travels) and the wave front travels at a constant speed U, then the shock jump conditions still apply in defining the relationship between the material states in front of and behind the wave front. This result can be seen by assuming that the extent of the wave front is less than a given thickness in space (δ_x) and in time (δ_t). Since outside of this region the values are constant, we can let Δx and Δt in Eqs. (2.45) and (2.46) become very large, so that $\delta_x \ll \Delta x$ and $\delta_t \ll \Delta t$ and in the limit the δ_x and δ_t terms are negligible and the same jump conditions result. Thus,

THEOREM 2.6. *(finite transition region) If a wave front moving at a constant velocity U has a finite, uniformly bounded transition region in space and in time, then the Rankine–Hugoniot jump conditions hold connecting the conditions in front of the wave and behind the wave.*

The main application of this theorem is that in the physical world we do not expect shocks to be infinitely steep; rather, we expect there to be a finite rise time, and hence the shock front will be of finite extent. However, as long as the front moves at one speed, then the Hugoniot jump conditions apply relating initial and final states.

If we have a two-wave structure (as will be encountered in Chapter 5) where an initial shock passes at one velocity followed by a second shock at a different velocity, then the state in front of the first shock and behind the second shock are *not* related by the Hugoniot jump conditions, as the assumptions in this theorem do not hold: the two-shock front separates in time and thus is not uniformly bounded in either time or space. For two shocks traveling at different velocities it is necessary to use two separate applications of the jump conditions to determine the relations between the conditions in front of the first shock and behind the second shock.

The question might arise: what if the wave moves to the left, rather than to the right? If Fig. 2.3 is redrawn to account for the passage of the wave from the right to the left (i.e., the wave velocity $U < 0$), then redoing the integrals leads to the conclusion that

$$[B] = U[A].\tag{2.69}$$

Thus, the above equations work for waves traveling both to the left and to the right. Of course, these expressions are rather awkward for the right-to-left-moving wave since they were derived with the idea that to the right of the wave front was the initial condition. Since Eq. (2.69) also says that

$$(B_r - B_\ell) = U (A_r - A_\ell),\tag{2.70}$$

we can just swap indices in the above equations, giving

$$\text{Mass:}\qquad \frac{\rho_\ell}{\rho_r} = 1 - \frac{u_r - u_\ell}{U - u_\ell},$$

$$\text{Momentum:}\quad \tilde\sigma_r - \tilde\sigma_\ell = \rho_\ell(U - u_\ell)(u_r - u_\ell),\tag{2.71}$$

$$\text{Energy:}\qquad E_r - E_\ell = \frac{1}{2}(\tilde\sigma_r + \tilde\sigma_\ell)\left(\frac{1}{\rho_\ell} - \frac{1}{\rho_r}\right).$$

Both sets of equations work for waves moving to either the left or the right; the most useful set depends on which conditions are known best, to the left or right.

One other item bears discussion. The jump conditions have been derived in Eulerian coordinates, meaning that the spatial coordinates are fixed and the material moves through the fixed coordinates. This fact is so important that we highlight it:[4]

REMARK 2.7. The Hugoniot jump conditions are in Eulerian coordinates, meaning a fixed laboratory frame. Thus the wave speed U may be due, in part, to the material motion with particle velocity u.

It is often an algebraic simplification to consider the jump conditions in coordinates where the material in front of the wave front is at rest. This frame is achieved by performing a Galilean transformation with velocity u_r (if the wave front is moving to the right). Transforming to the coordinate system where the material is at rest in front of the wave front allows one to speak of an absolute wave speed through the material, with no qualifiers about the material motion. We will use a hat to indicate the speeds after the Galilean transformation has been applied that brings the material to rest in front of the wave front. Thus, for example, for a wave front moving to the right,

$$\hat U = U - u_r,\tag{2.72}$$

$$\hat u = u_\ell - u_r,\tag{2.73}$$

and the two Hugoniot jump conditions can be written (the energy jump condition is unchanged)

$$\text{Mass:}\qquad \frac{\rho_r}{\rho_\ell} = 1 - \frac{\hat u}{\hat U},\tag{2.74}$$

$$\text{Momentum:}\quad \tilde\sigma_\ell - \tilde\sigma_r = \rho_r \hat U \hat u.\tag{2.75}$$

Thus, a simplification in the form of the equations occurs. We will sometimes use this transformation.

---

[4]The Hugoniot jump condition in Lagrangian coordinates, or in terms of the original configuration, will be presented in Chapter 13.

Since we are summarizing the jump conditions, we also write down the original jump conditions before we worked on them for a wave traveling in the x_i direction at speed U:

Mass: $[\rho v_i] = U[\rho],$

Momentum: $[\rho v_i v_j - \sigma_{ij}] = U[\rho v_j] \quad j = 1,\ 2$ and $3,$

Energy: $\left[\left(\frac{1}{2}\rho v_j v_j + \rho E\right) v_i - \sigma_{ij} v_j\right] = U\left[\frac{1}{2}\rho v_j v_j + \rho E\right].$

$$(2.76)$$

There is another jump condition that deals with entropy. It is a requirement that the entropy s is nondecreasing across a shock front,

$$[s] = s_\ell - s_r \geq 0. \tag{2.77}$$

The entropy jump condition is used in situations where there are two or more solutions to the mass, momentum, and energy jump conditions and it is necessary to determine which one is physical. Entropy will be further discussed in Chapter 4. We will show there that the physical assumption that the rarefaction wave does not fall behind the shock front implies the entropy jump condition. The mass, momentum, and energy jump conditions are symmetric in time, meaning that the shock can run both forward and backward in time (t can be replaced by $-t$). A nonlinear physical response combined with the entropy jump condition allows only one of these solutions, thus breaking the time reversal symmetry.

2.8. Differential Jump Conditions

Though the purpose in deriving the Hugoniot jump conditions is to deal with finite changes in material properties across a wave front, it is of interest to consider very small amplitude waves, commonly called sound waves. To determine their properties, we derive a differential form of the jump conditions. If we consider a wave moving from the left to the right, and the properties on the left-hand side to be given by $\rho_\ell = \rho_r + \Delta\rho$, $\tilde\sigma_\ell = \tilde\sigma_r + \Delta\tilde\sigma$, $E_\ell = E_r + \Delta E$, and $u_\ell = u_r + \Delta u$, for the mass balance we have

$$\frac{\rho_r}{\rho_r + \Delta\rho} = 1 - \frac{\Delta u}{U - u_r - \Delta u_r}. \tag{2.78}$$

For the momentum balance,

$$\Delta\tilde\sigma = \rho_r(U - u_r)\Delta u. \tag{2.79}$$

For the energy balance,

$$\Delta E = \frac{1}{2}(2\tilde\sigma_r + \Delta\tilde\sigma)\left(\frac{1}{\rho_r} - \frac{1}{\rho_r + \Delta\rho}\right). \tag{2.80}$$

As the Δ terms get small, assuming $U - u \neq 0$, these equations give

$$\frac{d\rho}{\rho} = \frac{du}{U - u}, \tag{2.81}$$

$$d\tilde\sigma = \rho(U - u)du, \tag{2.82}$$

$$dE = \tilde\sigma\frac{d\rho}{\rho^2} = -\tilde\sigma d(1/\rho). \tag{2.83}$$

These are the differential jump conditions. When the material is at rest in front of the wave ($u = 0$), the wave speed is referred to as the local sound speed $c = |\hat U|$. The general solution is $U = \pm c + u$. These equations will be discussed in Chapter 4.

We mention here, and it will be explored there, that in our notation the definition of entropy is $dE = Tds - \tilde{\sigma}d(1/\rho)$, and hence the differential jump condition for energy implies

$$ds = 0 \tag{2.84}$$

and the solution to the differential jump conditions is isentropic. Note that the differential jump conditions are not the derivative of the jump conditions (Exercise 2.8).

2.9. Cylindrical and Spherical Shock Fronts

In addition to one-dimensional waves as described above, we will be examining waves propagating outward with cylindrical and spherical symmetry when we examine the cavity expansion solution in Chapter 6. In preparation for that analysis, we show that the Hugoniot jump conditions, as written without any modification, apply in cylindrical coordinates for a wave front with cylindrical symmetry (the spherical case will be left as an exercise – the same ideas apply). Consider the mass and momentum balances with cylindrical symmetry:

$$\rho\left(\frac{\partial v}{\partial r} + \frac{v}{r}\right) = -\frac{D\rho}{Dt}, \tag{2.85}$$

$$\frac{\partial \sigma_{rr}}{\partial r} + \frac{\sigma_{rr} - \sigma_{\theta\theta}}{r} = \rho\frac{Dv}{Dt}. \tag{2.86}$$

By manipulation the mass balance can be rewritten

$$\frac{\partial \rho}{\partial t} + \frac{1}{r}\frac{\partial(r\rho v)}{\partial r} = 0. \tag{2.87}$$

This equation can be shown to satisfy the Hugoniot jump conditions just as was done in general earlier, only now the integral is performed with a volume element $rdrdt$ rather than $dxdt$. In other words, defining

$$A = \rho, \qquad B = \rho v, \tag{2.88}$$

our equation can be written

$$\frac{\partial A}{\partial t} + \frac{1}{r}\frac{\partial(rB)}{\partial r} = 0. \tag{2.89}$$

We take a rectangular region in the $r-t$ plane and integrate over the region. The first term is

$$\int_r^{r+\Delta r}\int_t^{t+\Delta t}\frac{\partial A}{\partial t}rdtdr \approx \int_r^{r+\Delta r}(A(t+\Delta t) - A(t))\,rdr \approx r\left(A_\ell - A_r\right)\Delta r, \tag{2.90}$$

where we are assuming that r is strictly greater than zero and Δr is small compared to r, thus allowing us to pull the r out of the integral. Also, we are assuming that A and B are continuous on either side of the discontinuity, but not necessarily constant values, as was assumed in the one-dimensional case. That is why the first equality is only approximately equal. In cylindrical and spherically expanding waves, owing to the geometry, the stress, density, and velocity are not constant on either side of the discontinuity. Thus, our integration region is very small (and, of course, in the limit it has zero area). The second integral is

$$\int_t^{t+\Delta t} \int_r^{r+\Delta r} \frac{1}{r}\frac{\partial(rB)}{\partial r} r\,dr\,dt$$

$$\approx \int_t^{t+\Delta t} ((r+\Delta r)B(r+\Delta r) - B(r))\,dt \approx r\,(B_r - B_\ell)\,\Delta t, \quad (2.91)$$

where we have discarded the $\Delta r\Delta t$ term. Thus, upon placement back into the original Eq. (2.89), the r terms cancel and the result is

$$[B] = U[A], \quad (2.92)$$

where $U = \Delta r/\Delta t$ is again the shock velocity. Thus, exactly the same Hugoniot jump condition results for mass conservation,

$$\rho_\ell v_\ell - \rho_r v_r = U(\rho_\ell - \rho_r). \quad (2.93)$$

Similarly, through manipulation and use of the mass balance, the momentum balance can be written as

$$\frac{\partial(\rho v)}{\partial t} + \frac{1}{r}\frac{\partial}{\partial r}\left\{r(\rho v^2 - \sigma_{rr})\right\} = -\frac{\sigma_{\theta\theta}}{r}. \quad (2.94)$$

The previous result, with

$$A = \rho v, \qquad B = \rho v^2 + \tilde{\sigma}_{rr}, \quad (2.95)$$

can now be applied, save for the term on the right-hand side of the equation. Notice, however, that when the term is integrated over the region,

$$\int_r^{r+\Delta r} \int_t^{t+\Delta t} \frac{\sigma_{\theta\theta}}{r} r\,dt\,dr \approx \sigma_{\theta\theta}\Delta t\Delta r. \quad (2.96)$$

As the region becomes very small, after division by Δt this term goes to zero. Notice that $\sigma_{\theta\theta}$ can be discontinuous – for the integral to go to zero, it need only be bounded. The same equations apply and exactly the same Hugoniot jump condition results (Eq. 2.55). Thus, we have shown that the same jump conditions apply for a cylindrically symmetric expanding wave. Exercise 2.9 asks the reader to perform the same analysis for a spherical wave.

2.10. Equations of State

To solve impact and wave propagation problems, it is necessary to define a relationship between the stress, energy, and density. In thermodynamics, for a material at equilibrium the relationship between pressure, energy, density, entropy, temperature, and other variables is referred to as an *equation of state*. The simplest equation of state is that describing the behavior of an ideal gas. Though this book mainly deals with solids, some ideas will be presented in this section using the ideal gas equation of state for simplicity and clarity. It should be pointed out that the ideal gas equation of state models the behavior of gases well except at low temperatures or high pressures.

The equation of state that is memorized in high schools is

$$pV = nRT, \quad (2.97)$$

where p is the pressure, V the volume of gas, n the number of moles of gas, R the universal gas constant, and T the absolute temperature. For our purposes, in this form the equation of state is not useful for two reasons: it deals with an arbitrary volume of material; and the Hugoniot jump condition is in terms of energy, not temperature. To address the first point, with m the weight of one mole of gas (i.e., the mass of Avogadro's number of molecules of the gas), then the equation can be rewritten

$$p = \frac{R}{m}\frac{m}{V}T = \frac{R}{m}\rho T = R'\rho T. \qquad (2.98)$$

This notation explicitly identifies the universal gas constant in terms of mass instead of moles with $R' = R/m$, though typically in the literature that is not done and it is assumed that the individual using a formula is checking units to make sure they have the right form of R.

To address the second point it is necessary to connect the temperature to the internal energy of the material. Through arguments that can be found in thermodynamics texts (such as [161]), Eq. (2.97) implies that the energy is a function of the temperature alone. The connection between them is the heat capacity,[5]

$$C_v = \left(\frac{\partial E}{\partial T}\right)_\rho. \qquad (2.99)$$

The constant volume or constant density heat capacity C_v applies where the volume of material is not allowed to change as heat (energy) is added to the material. The heat capacity is a function of temperature, and for ideal gases is a constant. Experimentally, or based on arguments from the kinetic theory of gases, example heat capacities are

$$\text{ideal monatomic gas (noble gases like He): } C_v = \frac{3}{2}R',$$

$$\text{ideal diatomic gas (such as } H_2 \text{ or } N_2\text{): } C_v = \frac{5}{2}R'. \qquad (2.100)$$

(It can be shown using the kinetic theory of gases that in a low-density gas each degree of freedom provides $\frac{1}{2}R$ to the heat capacity. For an ideal monatomic gas there are three translational degrees of freedom and for an ideal diatomic gas there are three translational degrees of freedom plus two rotational degrees of freedom.) With a constant heat capacity, we can integrate to get (where the integration constant is chosen to be zero)

$$E = C_v T. \qquad (2.101)$$

In addition to the constant volume heat capacity, there are other heat capacities such as the constant pressure heat capacity. To compute it, we note from the first law of thermodynamics

$$dQ = dE + pd(1/\rho), \qquad (2.102)$$

where Q is heat or energy coming into the system, that by differentiating the ideal gas law

$$d(p/\rho) = d(R'T) \quad \Rightarrow \quad pd(1/\rho) = -dp/\rho + R'dT \qquad (2.103)$$

we obtain

$$C_p = \left(\frac{dQ}{dT}\right)_p = \left(\frac{dE}{dT} + R' - \frac{1}{\rho}\frac{dp}{dT}\right)_p = C_v + R'. \qquad (2.104)$$

---

[5]Placing the ρ as a subscript outside the parenthesis is the standard thermodynamics notation necessitated by the assumption that only two state variables are independent, in this case T in the partial derivative and ρ being held constant. Thus, if E were explicitly written in terms of another state variable that was not ρ or T, it is understood that that state variable is also a function of ρ and T and so it will need to be differentiated accordingly and not be held constant during the partial differentiation with respect to T.

The heat capacity results are combined to give

$$p = R'\rho T = (C_p - C_v)\rho\frac{E}{C_v} \tag{2.105}$$

or finally

$$p(\rho, E) = (\gamma - 1)\rho E \tag{2.106}$$

where γ is the ratio of the heat capacities,

$$\gamma = \frac{C_p}{C_v}. \tag{2.107}$$

This form of the ideal gas law is more useful for us. The pressure is given in terms of two independent variables, ρ and E. This equation can be rearranged to give $\rho(p, E)$ and $E(\rho, p)$. These expressions are all forms of the same equation of state. An experimental observation of thermodynamics is that there are only two independent parameters describing the equilibrium state of the material. Using the equation of state it is possible to explore the behavior of the material under defined loading conditions.

Suppose the material is compressed in an isothermal environment, where the temperature and hence specific energy are held constant during compression. Only the density changes, and the pressure is

$$p_T = (\gamma - 1)\rho E_0 = \frac{\rho}{\rho_0}p_0. \tag{2.108}$$

The subscript T identifies isothermal compression (T is held constant). The pressure depends linearly on the density. Next, suppose the material is compressed isentropically, so that the entropy does not change. Given the thermodynamic expression $dE = T\,ds - p\,d(1/\rho)$, the change in energy is given by the mechanical work (nothing is lost to a change in entropy) and

$$d(\rho E)_s = E\,d\rho + \rho\,dE = E\,d\rho - \rho p\,d(1/\rho) = (E + p/\rho)d\rho. \tag{2.109}$$

For the ideal gas $E = p/\rho(\gamma - 1)$ and so, undergoing isentropic compression,

$$\left(\frac{dp}{d\rho}\right)_s = (\gamma - 1)\left(\frac{d(\rho E)}{d\rho}\right)_s = (\gamma - 1)(E + \frac{p}{\rho}) = \gamma\frac{p}{\rho}. \tag{2.110}$$

Integration produces the famous expression for the isentropic compression or expansion of an ideal gas,

$$p_s = p_0\left(\frac{\rho}{\rho_0}\right)^\gamma. \tag{2.111}$$

The pressure increases nonlinearly as a function of density. The temperature and energy increase as

$$\frac{T_s}{T_0} = \frac{E_s}{E_0} = \left(\frac{\rho}{\rho_0}\right)^{\gamma-1}. \tag{2.112}$$

Finally, consider compression by a shock. The stress for the ideal gas is given by

$$\tilde{\sigma}_{ij} = p\delta_{ij} \quad \text{or} \quad \tilde{\sigma} = pI \tag{2.113}$$

for the stress tensor, or $\tilde{\sigma} = p$ in one dimension. $I = (\delta_{ij})$ is the identity matrix. The equation-of-state expression for the energy is placed in the energy jump condition. Straightforward algebraic manipulation produces

$$p_H = p_0\frac{(1+\gamma)\rho/\rho_0 + 1 - \gamma}{(1-\gamma)\rho/\rho_0 + 1 + \gamma}. \tag{2.114}$$

The subscript H refers to Hugoniot, which means the locus of end states for the jump from the initial conditions. This topic is explored in great detail in Chapters 4 and 5. The temperature and energy along the Hugoniot are

$$\frac{T_H}{T_0} = \frac{E_H}{E_0} = \frac{(1-\gamma)\rho_0/\rho + 1 + \gamma}{(1-\gamma)\rho/\rho_0 + 1 + \gamma}. \tag{2.115}$$

One can show that for compression

$$p_T(\rho) < p_s(\rho) < p_H(\rho), \quad \rho > \rho_0. \tag{2.116}$$

The Hugoniot information can be represented by the relationship between the shock velocity U and the particle velocity u. Using the mass and momentum jump conditions, and with a bit of algebra, we obtain

$$U = \frac{\gamma+1}{4}u + \sqrt{c_0^2 + \left(\frac{\gamma+1}{4}\right)^2 u^2}, \quad c_0 = \sqrt{\frac{\gamma p_0}{\rho_0}}. \tag{2.117}$$

Here, c_0 is the speed of sound in the gas at its initial conditions. We will learn how to compute the sound speed in Chapter 5. For really high compression, $U \to (\gamma+1)u/2$.

It is interesting to ask how difficult it is to compress the gas. The measure of difficulty is referred to as the bulk modulus – the larger it is the stiffer the material, hence the more difficult it is to compress. The modulus is defined as the change in pressure in terms of the change in volumetric strain. There are various definitions of strain, but the correct one here is a logarithmic strain, $\tilde{\varepsilon}_v = \ln(\rho/\rho_0)$ (Exercise 5.21). When this strain is used, the modulus definition is

$$\text{modulus} = \frac{\partial p}{\partial \tilde{\varepsilon}_v} = \frac{\partial p}{\partial \ln(\rho/\rho_0)} = \rho \frac{\partial p}{\partial \rho}. \tag{2.118}$$

There are two bulk moduli, one corresponding to compression where the temperature is held constant and one where the entropy is held constant,

$$K_T = \rho \left(\frac{\partial p}{\partial \rho}\right)_T = p, \quad K_s = \rho \left(\frac{\partial p}{\partial \rho}\right)_s = \gamma p. \tag{2.119}$$

The general definitions of the isothermal and the isentropic bulk modulus are given, and then the specific results for the ideal gas bulk moduli are given. There is no modulus corresponding to the Hugoniot because the Hugoniot represents a set of end states and not a loading curve. However, it is very difficult to produce really large compressions of nonlinear materials with a shock; examination of Eq. (2.114) reveals a maximum densification that a shock can produce in an ideal gas (Exercise 5.17).

A final piece of the equation of state is the entropy. With our expression for the entropy,

$$ds = \frac{dE}{T} + \frac{p}{T}d(1/\rho) = C_v \frac{dT}{T} + R'\rho \, d(1/\rho) = C_v \frac{dT}{T} - R'\frac{d\rho}{\rho}. \tag{2.120}$$

Integrating from a reference point gives

$$s - s_0 = C_v \ln\left(\frac{T}{T_0}\right) + R' \ln\left(\frac{\rho_0}{\rho}\right)$$

$$= C_v \ln\left\{\frac{T}{T_0}\left(\frac{\rho_0}{\rho}\right)^{\gamma-1}\right\} = C_v \ln\left\{\frac{p}{p_0}\left(\frac{\rho_0}{\rho}\right)^{\gamma}\right\} = C_v \ln\left\{\left(\frac{T}{T_0}\right)^{\gamma}\left(\frac{p_0}{p}\right)^{\gamma-1}\right\}. \tag{2.121}$$

With the other independent variable held constant, entropy increases with increasing temperature and decreases with increasing density.

Solids are much more complicated that ideal gases. In particular, there is not one value that quantifies "strain" or deformation, but rather a tensor does. For many solids in their nonlinear response regime, it is necessary to know the deformation history, not just the current deformation. (Having said that, some researchers are adamant that there are "state variables" that contain all the deformation history information – i.e., it is not necessary to know the deformation history, just to know the current values of the state variables. One therefore needs to develop evolution equations that describe how the state variables change during deformation.) For more complicated materials, the relation between stress, strain, strain rate, strain history, and temperature is referred to as a constitutive model. The next chapter begins the study of solid material response.

2.11. Initial and Boundary Conditions

To specify a problem, in addition to material properties it is necessary to specify initial conditions and boundary conditions. Initial conditions for mechanics problems include the initial location and velocity distribution of all materials as well as the initial state of the material, such as stress, strain, and temperature. The initial conditions need to be consistent with the constitutive model and other initial conditions such as body forces.

Boundary conditions are defined at all material boundaries. In three dimensions, at each boundary point the boundary condition is specified by three items of information. The boundary condition can be a defined material location, traction, or a mixture of the two. Boundary velocity is another way of designating boundary location, given an initial location. Prescribing the displacement or velocity is referred to as a kinematic boundary condition as it focuses on the motion regardless of the mass or force (from the Greek *kinema*, meaning motion).

As as example that illustrates a variety of possible boundary conditions, consider a piston problem in cylindrical coordinates. This geometry will be a common geometry in our study of wave propagation. For initial conditions, there is a cylinder of material of radius R with a central axis coinciding with the z-axis and running from $z = 0$ to its length $z = L$. At time $t = 0$ the material is at rest. At the $z = 0$ disk boundary, a set velocity V is applied in the $\hat{z}$ direction. Three pieces of information for the boundary are specified as each component of the velocity $\vec{v}(r, \theta, z) = V\hat{z}$, $0 \le r \le R$, $0 \le \theta < 2\pi$, $z = Vt$ is specified. The boundary on the outer surface of the cylinder at $r = R$ will be mixed. It is assumed the material cylinder is not allowed to expand; hence $v_r(r, \theta, z) = 0$, $r = R$, $0 \le \theta < 2\pi$, $Vt \le z < L$. This condition sets one component of the boundary condition. The next two components are in the $\hat{\theta}$ and $\hat{z}$ directions, where the tractions in these directions will be zero. With $\hat{r}^T\sigma = \vec{t}^T = \sigma_{rr}\hat{r} + \sigma_{r\theta}\hat{\theta} + \sigma_{rz}\hat{z}$, specifying that the $\hat{\theta}$ and $\hat{z}$ directions are traction free says that $\sigma_{r\theta}(r, \theta, z) = \sigma_{rz}(r, \theta, z) = 0$, $r = R$, $0 \le \theta < 2\pi$, $Vt \le z < L$. For the far end of the cylinder we assume the surface is traction free, hence $\hat{n}^T\sigma = 0$. Notice that the boundary condition is traction free, not "stress free." The stress tensor can be nonzero; it is only the tractions on the free surface that are zero. At the initial time the surface normal is $\hat{n} = \hat{z}$, and so this condition corresponds to $\hat{z}^T\sigma(r, \theta, z) = \sigma_{zr}\hat{r} + \sigma_{z\theta}\hat{\theta} + \sigma_{zz}\hat{z} = \vec{0}$ for $0 \le r \le R$, $0 \le \theta < 2\pi$, $z = L$. The stress components $\sigma_{r\theta}$ and σ_{rr} need not be zero. However, the surface can deform owing to the wave interactions at the free surface (it will certainly move) and so the surface normal $\vec{n}$ may not always align with $\hat{z}$.

Traction free is not the only possibility for the outer surface of the cylinder. There could be a frictional force. If the frictional force per area is given by $f(v)$, where v is the speed of the material, then the traction part of the boundary is $\sigma_{r\theta}(r,\theta,z) = (v_\theta/v)f(v)$, $\sigma_{rz}(r,\theta,z) = (v_z/v)f(v)$, $r = R$, $0 \leq \theta < 2\pi$, $Vt \leq z$, where $v = \sqrt{v_\theta^2 + v_z^2}$. This condition mixes velocity and traction in the same boundary condition, not explicitly specifying either. The third condition on the cylinder is still that no fluid exits, $v_r(r,\theta,z) = 0$.

Finally, if the problem is solved in two-dimensional axisymmetric geometry, there is an implied boundary condition on the centerline. In this case $v_r(0,\theta,z) = 0$ and the two traction components t_θ and t_z are zero along the axis. The traction condition corresponds to $\sigma_{r\theta}(0,\theta,z) = \sigma_{rz}(0,\theta,z) = 0$.

In the example here, the directions of each of the specified components are orthogonal, but they need not be; the requirement is that the directions of each of the specified components need to span the three-dimensional space. If they do not span three dimensions, then in one or more directions the boundary condition is overspecified and in another direction the boundary condition has not been specified.

In solid mechanics, simplified boundary conditions are often used that in some sense match the actual boundary conditions (integrated tractions over a portion of the surface match and moments of those integrals called resultants also match). It is a general result that when one is far enough away from the boundary the resulting deformation due to the simplified boundary conditions match those of the actual boundary conditions. This is done particularly in static elasticity, and the general principle of the validity of these approximations is referred to as *Saint-Venant's principle*. For example, when we consider a tensile test, the test specimen is usually threaded, but we do not model the threaded boundary with the testing machine but use a simplified traction condition at the top and bottom of the specimen.

In the examples here the boundary conditions were stated in the Eulerian reference frame, consistent with our treatment of the jump conditions. For solid materials they can be specified in the original reference frame (Lagrange frame) instead, as is sometimes done when it is simpler.

2.12. Comments on Waves and Classical Continuum Mechanics

When we use the word *wave* we always mean a moving front as described by the Hugoniot jump conditions or a rarefaction fan to be described in a later chapter.[6] The moving front is stable and moves at a speed that is sonic or supersonic relative to the material in front of it ($U \geq c_0$). It does not disperse. It only decreases in magnitude when it interacts with rarefaction waves.

In traditional analysis of linear waves, a periodic oscillation is assumed that is propagating in a specified direction. To clarify what we are *not* talking about in this book, the wave is typically written [48]

$$u(x,t) = A\cos(kx - \omega t), \tag{2.122}$$

---

[6]This section uses some concepts and terminology not yet introduced, but this is the appropriate place for this discussion; the reader may wish to return to it later after reading Chapters 3–5.

where ω is the angular frequency in radians per second and k is the wave number, which is related to the peak-to-peak wavelength λ as $k = 2\pi/\lambda$. The speed at which a given frequency moves is given by the phase speed $v_\varphi = \omega/k$ and the speed a wave packet or information moves is given by the group velocity $v_g = d\omega/dk$. The relationship $\omega = \omega(k)$ is called a dispersion relation. We will not use these terms, because they are not relevant to our moving fronts. Our fronts move at the shock speed U, and that is the speed at which information and energy is transferred.

However, to more clearly understand some assumptions in continuum mechanics, let us consider linear springs with force $F(x) = ax$ and length L connecting masses m arrayed in a line. The equation of motion of mass number i is

$$m\ddot{u}_i = F(u_{i+1} - u_i) - F(u_i - u_{i-1}) = a(u_{i+1} - 2u_i + u_{i-1}), \qquad (2.123)$$

where the dot means derivative with respect to time. If the displacement assumption in Eq. (2.122) is placed into this expression, with some algebra and trigonometric angle sum formulas (e.g., thinking of $\cos(k(x+L) - \omega t) = \cos((kx - \omega t) + kL)$), we obtain

$$-m\omega^2 = 2\cos(kL) - 2 \quad \Rightarrow \quad \omega = \sqrt{4a/m}\sin(kL/2). \qquad (2.124)$$

This gives

$$v_\varphi = \frac{\omega}{k} = \sqrt{\frac{4a}{m}}\frac{\sin(kL/2)}{k}, \qquad v_g = \frac{d\omega}{dk} = \sqrt{\frac{4a}{m}}\frac{L}{2}\cos(kL/2). \qquad (2.125)$$

If the wavelength is large compared to L, so that kL is small, then $v_\varphi \approx v_g \approx \sqrt{4a/m}L/2$, which is a constant. This wave speed is what we will obtain in our classical continuum mechanics model, where we homogenize the mass-spring response and say that $p = K\tilde{\varepsilon}$, which will produce a constant wave speed of $\sqrt{K/\rho}$ for all frequencies. This continuum model will not produce any of the features that the mass-spring system produces for wavelengths near L, such as frequency-dependent wave speeds and neighboring masses moving in opposite directions for $\lambda = 2L$, so no information propagates.

In classical continuum mechanics (which is the approach taken in this book), it is assumed that the constitutive relationship between stress and strain is *local*. In particular, local means that the stress is a function of the first derivative of the displacement $\varepsilon = \partial u/\partial x$ and the strain rate or, equivalently, the rate of deformation, $\dot{\varepsilon} = \partial\varepsilon/\partial t = \partial v/\partial x = D_{11}$ (we use small strains here, but even large strain only uses the first spatial derivative of the displacement – see Eq. (3.13)). The first derivative of the displacement is the first quantification of deformation, since a uniform rigid displacement contains no deformation. When the constitutive model is such that the stress σ is a function of ε and $\dot{\varepsilon}$, then the mass spring model described above cannot be represented in continuum mechanics as a material that is the same at each point. However, if we are willing to use higher derivatives of either strain or stress, then we can represent the mass-spring dispersion behavior in continuum mechanics. For example, with the momentum balance $\partial\sigma/\partial x = \rho\,\partial^2 u/\partial t^2$, $\varepsilon = \partial u/\partial x$, and the assumed wave solution Eq. (2.122), a constitutive model with a higher order strain term and one with a higher order stress term yield

$$\sigma = a_1\varepsilon + a_3\partial^2\varepsilon/\partial x^2 \quad \Rightarrow \quad v_\varphi = \sqrt{(a_1/\rho)}\sqrt{1 - (a_3/a_1)k^2}, \qquad (2.126)$$

$$\sigma - b_2\,\partial^2\sigma/\partial x^2 = a_1\varepsilon \quad \Rightarrow \quad v_\varphi = \sqrt{(a_1/\rho)}/\sqrt{1 + b_2k^2}. \qquad (2.127)$$

Including the higher order derivative in the constitutive model gives rise to *strain gradient elasticity* and *stress gradient elasticity* theories. Matching Taylor series with Eq. (2.125) gives $a_3/a_1 = b_2 = L^2/12$, and we see that the nonclassical continuum model includes the length scale of the spaced masses. This matches better the dispersion behavior, and using additional higher order derivatives allows an even better match. (It is interesting to note that the static tensile test produces a linear result, since uniform strain gives $\sigma = a_1\varepsilon$.) This constitutive model is *nonlocal* because including higher order derivatives contains more information about deformation in the neighborhood than just at the point of interest. Owing to the higher order derivatives it is referred to as a nonclassical elasticity model. Other approaches to nonlocality include using for the strain or stress a weighted integral of the strain or stress over a specified volume rather than just at the point in question. Both approaches (using higher order derivatives, using integrals) can be developed in an essentially identical fashion (Exercise 2.15), though the numerical implementation of the integral version is typically simpler than the higher derivative version. The constitutive models we use in this book only depend on the first spatial derivative of the displacement and the corresponding strain rate (and their history). Higher order derivatives require additional boundary conditions (Exercise 2.16).

If these wave dispersion (and other scale relevant) behaviors are of interest, and classical constitutive models are used, it is necessary to model at the scale of the microstructure of interest. In today's parlance such modeling is referred to as *mesoscale modeling*. If methods for studying microstructure are based on the above nonlocal approach or approaches suggested in the footnote on page 14, the corresponding nonclassical continuum mechanics fall under the broad heading *generalized continua* [60, 140, 169].

The spreading of our rarefaction fans is not due to a frequency dependence of the wave speed; rather, it is due to the nonlinear compressibility of the material, which leads to a pressure-dependent sound speed; our sound speeds are not frequency dependent.

If the springs in the mass-spring model are nonlinear, oscillatory motions of the masses that are below the classical continuum model scale enter into our equations of state as energy, as they represent thermal motions (see page 210).

Any damage model that incorporates the notion of damage propagation, where a coarse scale damage variable represents phenomena like cracks or shear bands, is inherently nonlocal, as propagation requires being aware of the surrounding neighborhood (e.g., in a computational setting, the computational cell or element needs to know if neighboring cells or elements already have cracks or shear bands in them that can propagate into that cell).

To see another approach to microstructure leading to generalized continua, outlined in the footnote on page 14, consider the conservation of angular momentum for a rigid cube of edge length $2a$, with the center of the cube at the origin and edges aligned with axes directions. We assume there is a linear traction distribution on the cube surface, for example

$$\vec{t}_1(x,y,z) = \vec{t}_1|_+ + \left.\frac{\partial \vec{t}_1}{\partial y}\right|_+ y + \left.\frac{\partial \vec{t}_1}{\partial z}\right|_+ z \qquad (2.128)$$

for the front face, where subscript $+$ means evaluated at $(x,y,z) = (a,0,0)$. The traction distribution leads to force moments on the face of the cube. The first moment is defined as $\vec{M} = \int_{\text{face}} \vec{x} \times \vec{t}\, dA$, where $\times$ is the cross product. Similar to our notation for the stress tensor, we define a moment matrix M_{ij} where the index i refers to the face of the cube and j refers to the axis of rotation for the moment. Computing the moments for the 1 face of the cube gives (where the 2 component of $\vec{t}_1$ is σ_{12}, etc.)

$$M_{11} = \int_{-a}^{a}\int_{-a}^{a} \{y\,\sigma_{13}(a,y,z) - z\,\sigma_{12}(a,y,z)\}dydz = \frac{4a^4}{3}\left(\left.\frac{\partial\sigma_{13}}{\partial y}\right|_{+} - \left.\frac{\partial\sigma_{12}}{\partial z}\right|_{+}\right),$$

$$M_{12} = \int_{-a}^{a}\int_{-a}^{a} \{z\,\sigma_{11}(a,y,z) - a\,\sigma_{13}(a,y,z)\}dydz = -4a^3\sigma_{13}|_{+} + \frac{4a^4}{3}\left.\frac{\partial\sigma_{11}}{\partial z}\right|_{+},$$

$$M_{13} = \int_{-a}^{a}\int_{-a}^{a} \{a\,\sigma_{12}(a,y,z) - y\,\sigma_{11}(a,y,z)\}dydz = 4a^3\sigma_{12}|_{+} - \frac{4a^4}{3}\left.\frac{\partial\sigma_{11}}{\partial y}\right|_{+}.$$
$$(2.129)$$

The direct stress terms are due to the standoff of the face from the origin; those terms are not present when the moments are computed with respect to the center of the face rather than the center of the cube. We separate these and performing similar computations for all the faces leads to

$$M^{\text{fc}} = \frac{4a^4}{3}\begin{pmatrix} \partial\sigma_{13}/\partial y - \partial\sigma_{12}/\partial z & \partial\sigma_{11}/\partial z & -\partial\sigma_{11}/\partial y \\ -\partial\sigma_{22}/\partial z & \partial\sigma_{21}/\partial z - \partial\sigma_{23}/\partial x & \partial\sigma_{22}/\partial x \\ \partial\sigma_{33}/\partial y & -\partial\sigma_{33}/\partial x & \partial\sigma_{32}/\partial x - \partial\sigma_{31}/\partial y \end{pmatrix}.$$
$$(2.130)$$

We can use our moment computation to compute the torque on the body, where torque is defined as $\vec{\tau} = \int_{\text{surface}} \vec{x} \times \vec{t}\, dA$. Going through the arithmetic yields

$$\vec{\tau} = \sum_{\text{faces } i\pm} M_{ij} = 8a^3\varepsilon_{ijk}\hat{e}_i\sigma_{jk} + \sum_{\text{faces } i\pm} M^{\text{fc}}_{ij}\hat{e}_j \approx 8a^3\varepsilon_{ijk}\hat{e}_i\sigma_{jk} + 2a\frac{\partial M^{\text{fc}}_{ij}}{\partial x_i}\hat{e}_j. \quad (2.131)$$

The last step approximates differences from opposite faces with $\partial f/\partial x \approx (f(a,y,z) - f(-a,y,z))/2a$, for example. From this we define the *couple stress tensor* $m_{ij} = 2aM^{\text{fc}}_{ij}/8a^3 = M^{\text{fc}}_{ij}/4a^2$, which is the moment per unit area on the face (the couple stress on a surface with normal $\vec{n}$ is $m_i = n_j m_{ji}$). Alternatively, it is the torque at the face center per unit area (recall the stress σ_{ij} is the force at the face per unit area). The equation of motion for a rigid body leads to

$$\frac{d}{dt}(I\vec{\omega}) = \vec{\tau} \quad \Rightarrow \quad \rho\frac{2a^2}{3}\frac{d\vec{\omega}}{dt} = \frac{\vec{\tau}}{8a^3} = \varepsilon_{ijk}\hat{e}_i\sigma_{jk} + \frac{\partial m_{ij}}{\partial x_i}\hat{e}_j, \quad (2.132)$$

where $\vec{\omega}$ is the angular velocity. The moment of inertia tensor I for a cube, with origin the center of the cube, is $I_{ij} = \frac{16}{3}\rho a^5\delta_{ij}$. This fact was used on the right.

We now go to the continuum. While doing so is a limit as $a \to 0$, we assume that the couple stress does not go to zero and that the inertia tensor per unit volume, either a tensor J_{ij} or assumed diagonal with magnitude J, does not go to zero. We are left with a continuum equation reflecting some sort of microstructure,

$$J_{ij}\frac{D\omega_j}{Dt} = \varepsilon_{ijk}\sigma_{jk} + \frac{\partial m_{ji}}{\partial x_j} + \rho m_i^c \quad \text{or} \quad J\frac{D\vec{\omega}}{Dt} = \varepsilon_{ijk}\hat{e}_i\sigma_{jk} + \vec{\nabla}\cdot m + \rho\vec{m}^c,$$
$$(2.133)$$

where a body couple stress $\vec{m}^c$ (such as could be introduced through interactions with a magnetic field) has been included. The inclusion of local rotation with angular velocity $\vec{\omega}$, not related to the global rotation arising from the deformation gradient F, allows for a nonsymmetric Cauchy stress tensor. Required boundary conditions now include specifying the couple stress $\vec{m}$ over the surface of the body. This angular momentum equation is an additional equation: the momentum balance Eq. (2.24) still applies, though now the Cauchy stress tensor need not be symmetric. This approach again falls under the heading of generalized continua, as it is not classical continuum mechanics, and we will not pursue it [60, 140, 169].

2.13. Sources

The Hugoniot jump conditions are part of classical mechanics; original papers are collected in Johnson and Chéret [110]. The presentation here has been influenced by Courant and Friedrichs [47], Hayes [91], Lax [129] and LeVeque [131]. Good introductions to continuum and solid mechanics that present the conservation equations include Malvern [138] and Lin and Segal [132].

Exercises

(2.1) Showing the traction on an arbitrary surface is $\vec{t}^{\,T} = \vec{n}^T\sigma$, thus showing that the stress tensor has the properties presented in Section 2.2, is typically argued using Cauchy's tetrahedron. In this exercise you will work through the details. For simplicity, assume the point of interest is the origin (you can translate your coordinate system to that point). Suppose you are given a unit normal vector $\vec{n}$ to an arbitrary plane. Cauchy's tetrahedron has vertices at the origin and on each of the coordinate axes at h/n_1, h/n_2, and h/n_3, where h is an arbitrary positive length. Show that the plane passing through the three nonorigin vertices has normal $\vec{n}$. Show that the area of the tetrahedron face that touches the origin and is normal to the x-axis is $\frac{1}{2}h^2/(n_2n_3)$, and the other two are similar with appropriate indices. Show that the area of the face that is normal to $\vec{n}$ is $\frac{1}{2}h^2/(n_1n_2n_3)$. Show that the volume of the tetrahdron is $\frac{1}{6}h^3/(n_1n_2n_3)$ (recall the volume of any cone is $\frac{1}{3}$(base area)×height). Now with $\vec{f} = m\vec{a}$ for the tetrahedron, divide by h^2 and show the acceleration term and any body force terms go to zero as $h \to 0$, leaving the surface force term equalling zero. Show for the surface force (area×traction) terms to sum to zero gives $\vec{t} = n_1\vec{t}_1 + n_2\vec{t}_2 + n_3\vec{t}_3$, where $\vec{t}$ is the traction on the large face and $\vec{t}_i$ are the tractions on the origin-touching faces. Show this is equivalent to $t_i = n_j\sigma_{ji}$.

(2.2) If Q is the transformation between two coordinate systems, and the traction and surface normal vectors transform correctly, $\vec{t}^* = Q\vec{t}$ and $\vec{n}^* = Q\vec{n}$, show using Eq. (2.2) that $\sigma^* = Q\sigma Q^{-1}$ and hence that stress is a tensor.

(2.3) Show that if the stress tensor is symmetric then $\sigma_{ji}\partial v_i/\partial x_j = \sigma_{ij}D_{ij}$.

(2.4) Derive the jump conditions for a shear wave propagating in the 1 direction where the material motion is transverse in the 2 direction. In particular, assume that $u_1 = 0$.

(2.5) Repeat Exercise 2.4, assuming that u_1 is constant but nonzero.

(2.6) Now compute the Hugoniot jump conditions, assuming that there is both a longitudinal (compression) and transverse (shear) change associated with the wave front, again assuming the wave propagates in the 1 direction and that the transverse velocity occurs in the 2 direction. (Hint: in this case there are two nontrivial jump conditions associated with the momentum balance.)

(2.7) Demonstrate the following theorem:

THEOREM 2.8. *When a compressive shock travels into a compressed (meaning $\tilde{\sigma}_r > 0$) material that is at rest ($u_r = 0$), the change in internal energy across the shock front is always greater than the change in kinetic energy.*

(2.8) Differentiate the Hugoniot jump conditions and show that the results are not the same as the differential jump conditions.

(2.9) Show that for a spherically symmetric wave the Hugoniot jump conditions apply without modification.

(2.10) In studying wave fronts where chemical reactions occur it is often useful to consider the enthalpy instead of the internal energy. The enthalpy is defined as $H = E + p/\rho$. Replacing p with $\tilde{\sigma}$ and using the energy jump condition, show that the jump condition for the enthalpy is

$$H_\ell - H_r = \frac{1}{2}(\tilde{\sigma}_\ell - \tilde{\sigma}_r)\left(\frac{1}{\rho_r} + \frac{1}{\rho_\ell}\right).$$

(2.11) Suppose a shock wave is moving at a speed U in an arbitrary direction $\vec{n}$. There are two perpendiculars to the direction of the wave propagation; let them be given by $\vec{p}$ and $\vec{q}$ ($\vec{n}$, $\vec{p}$, and $\vec{q}$ are unit vectors). Using the notation that the components of the material velocity related to the direction of propagation are $v_n = v_i n_i$, $v_p = v_i p_i$, and $v_q = v_i q_i$, show that the general jump conditions relating to Eq. (2.76) are

Mass: $\qquad [\rho v_n] = U[\rho],$

Momentum: $[\rho(v_n)^2 - n_i\sigma_{ij}n_j] = U[\rho v_n],$

Momentum: $[\rho v_n v_p - n_i\sigma_{ij}p_j] = U[\rho v_p],$

Momentum: $[\rho v_n v_q - n_i\sigma_{ij}q_j] = U[\rho v_q],$ $\qquad$ (2.134)

Energy: $\left[\left(\frac{1}{2}\rho|v|^2 + \rho E\right)v_n - n_i\sigma_{ij}v_j\right] = U\left[\frac{1}{2}\rho|v|^2 + \rho E\right].$

Notice that $n_i\sigma_{ij}n_j$, $n_i\sigma_{ij}p_j$, and $n_i\sigma_{ij}q_j$ are the components of the traction vector ($t_j = n_i\sigma_{ij}$) corresponding to the n, p, and q directions; $n_i\sigma_{ij}v_j$ is the work rate given by the inner product of traction (force/area) with the velocity. The square of the velocity $|v|^2 = v_j v_j$ or alternatively $|v|^2 = (v_n)^2 + (v_p)^2 + (v_q)^2$.

(2.12) Suppose a fluid is incompressible (ρ = constant), has constant stress (σ = constant), and all components of the velocity $\vec{u}$ only depend on x. With these assumptions, show the conservation of momentum in the x direction is given by

$$\frac{Du}{Dt} = \frac{\partial u}{\partial t} + \frac{\partial}{\partial x}\left(\frac{1}{2}u^2\right) = 0,$$ $\qquad$ (2.135)

where $u = u_x(= v_x)$. This equation is the *inviscid Burgers' equation*.

(a) What are A and B? Write down the corresponding jump condition. Explicitly solve for U for a given u_r and u_ℓ where $u_\ell > u_r > 0$. What is the conserved quantity?

(b) Show that the following equation is equivalent to the inviscid Burgers' equation if u is differentiable and $u \neq 0$,

$$\frac{\partial}{\partial t}\left(\frac{1}{2}u^2\right) + \frac{\partial}{\partial x}\left(\frac{1}{3}u^3\right) = 0.$$ $\qquad$ (2.136)

However, show that it is not the same conversation law, by writing A and B and writing down the corresponding jump condition. Explicitly solve for U for a given u_r and u_ℓ where $u_\ell > u_r > 0$. Show that this shock speed differs from that in part (a) and they are not equal even if $u_r = 0$. What is the conserved quantity?

(c) Write down yet another conservation law that, assuming u is differentiable and nonzero, is equivalent to the inviscid Burgers' equation. What is the conserved quantity? Thus, determining conservation laws cannot be done strictly by examining differential equations – there also needs to be some physical reasoning behind what is conserved.

(d) Since ρ is constant, the mass conservation jump condition does not apply. To understand how mass is conserved, suppose we had a smooth velocity $u(x,t) = f(x - Ut)$. Incompressibility says (δ is the spatial dimension)

$$\frac{\partial u_x}{\partial x} + (\delta - 1)\frac{\partial u_y}{\partial y} = 0 \Rightarrow u_y = \frac{dy}{dt} = -\frac{yf'(x - Ut)}{(\delta - 1)} \Rightarrow \frac{y_\ell}{y_r} = \exp\left\{\frac{u_\ell - u_r}{(\delta - 1)U}\right\}.$$
(2.137)

Thus, there is a finite lateral displacement at the shock front. Show this.

(2.13) Assume a fluid is incompressible and that the stress is given by $\sigma = -\hat{p}I + \eta D$, where D is the rate of deformation tensor, η the viscosity, and $\hat{p}$ is a constant. In one dimension, show the *viscous Burgers' equation* results,

$$\rho_0 \frac{Du}{Dt} = \frac{\partial}{\partial x}(\eta D_{11}) \quad \Rightarrow \quad \frac{\partial u}{\partial t} + \frac{\partial}{\partial x}\left(\frac{1}{2}u^2\right) = \nu \frac{\partial^2 u}{\partial x^2},$$
(2.138)

where $\nu = \eta/\rho_0 > 0$ is the *kinematic viscosity*. For the Riemann initial value problem of $u_r = 0$ for $x > 0$ and $u_\ell = u_0 > 0$ for $x < 0$, an exact analytic solution has been found [35],

$$u(x,t) = u_0 \cdot \left\{1 + \exp\left\{\frac{u_0}{2\nu}\left(x - \frac{1}{2}u_0 t\right)\right\} \frac{\text{erfc}\{-x/(2\sqrt{\nu t})\}}{\text{erfc}\{(x - u_0 t)/(2\sqrt{\nu t})\}}\right\}^{-1},$$
(2.139)

where the complementary error function is

$$\text{erfc}(x) = \frac{2}{\sqrt{\pi}}\int_x^{+\infty} e^{-y^2}\,dy, \qquad \frac{d}{dx}\text{erfc}(x) = -\frac{2}{\sqrt{\pi}}e^{-x^2},$$
(2.140)

with $\text{erfc}(-\infty) = 2$, $\text{erfc}(0) = 1$, and $\text{erfc}(+\infty) = 0$. For all $t > 0$, this solution is smooth (in fact, infinitely differentiable) because of the fact that if the center term of Eq. (2.138) is ignored what remains is the heat equation, which smoothes solutions. Also notice for all $t > 0$ that $u(x,t)$ does not equal 0 or u_0; that is, there has been an infinite speed of information transfer, which is also due to the same (unphysical) behavior of the heat equation.

(a) Show that Eq. (2.139) satisfies the stated initial conditions and that the limit on any straight line in (x,t) to the point $(0,0)$ gives $u(0,0) = \frac{1}{2}u_0$. (Hint: for $x \neq 0$ showing that it satisfies the initial conditions is done by fixing either $x < 0$ or $x > 0$ and taking $t \to 0$.)

(b) Owing to the smoothing process of the diffusive viscosity term, there is not a jump in u for $t > 0$. There appear to be three wave speeds in Eq. (2.139), $U = 0$, $U = \frac{1}{2}u_0$, and $U = u_0$. However, show the nice result along the curve $x - \frac{1}{2}u_0 t = 0$ that $u = \frac{1}{2}u_0$, so the "mid-point of the rise" moves at a speed $U = \frac{1}{2}u_0$.

(c) Show that the *vanishing viscosity* solution found by taking the limit as $\nu \to 0$ of Eq. (2.139) produces the solution of part (a) of the previous exercise on the inviscid Burgers' equation. Thus, the viscous Burgers' equation selects the inviscid form, Eq. (2.135), of the conservation law. Presumably this is due to the fact that the viscous Burgers' equation can be written in conservation form with $B = \frac{1}{2}u^2 - \nu\frac{\partial u}{\partial x}$, and there does not exist an $f(u, \frac{\partial u}{\partial x})$ such that $\frac{\partial f}{\partial x} = u^{\alpha}\frac{\partial^2 u}{\partial x^2}$ for $\alpha \neq 0$.

(d) Show that for large time the viscous solution approaches

$$u(x,t) \approx u_0 \left\{1 + \exp\left\{\frac{u_0}{2\nu}\left(x - \frac{1}{2}u_0 t\right)\right\}\right\}^{-1} = \frac{u_0}{2}\left\{1 - \tanh\left\{\frac{u_0}{4\nu}\left(x - \frac{1}{2}u_0 t\right)\right\}\right\}.$$
(2.141)

Thus, at large time the width of the rise or front and the rise time approach a constant value. Compute these by determining $\partial u/\partial x$ along $x = \frac{1}{2}u_0 t$ and showing

$$\text{front width} \approx \frac{u_0}{|\partial u/\partial x|} = \frac{8\nu}{u_0}, \quad \text{rise time} = \frac{\text{front width}}{U} = \frac{16\nu}{u_0^2}. \quad (2.142)$$

The front width and rise time increase with viscosity, but decrease with increased jump height.

(e) (Extra Credit) Show that Eq. (2.139) solves Eq. (2.138) for $t > 0$ (showing this is straightforward but tedious).

The front speed in (b) agrees with the shock speed for the inviscid Burgers' equation of the previous exercise because of Theorem 2.6. There does not exist a shock solution to the viscous Burgers' equation (i.e., constant u on either side of the shock front): the derivation breaks down because the $B = \frac{1}{2}u^2 - \nu\frac{\partial u}{\partial x}$ term is not well behaved – the $\partial u/\partial x$ term in B leads to a singular B for any finite jump in u, and our derivation of $[B] = U[A]$ does not apply. In Section 4.11, we solve the wave propagation problem for a constitutive model with viscosity. These strain rate and viscosity terms smooth out shocks; the above solution applies at early time solution when strains are small. However, we ignore these effects as the shock width is typically small (see Section 5.8).

(2.14) Figure 2.4 shows a sophisticated equation of state for deuterium D_2 built by Kerley compared to shock compression data [117]. Initially the deuterium is a liquid and is shocked to a gas. Dissociation means that the diatomic D_2 dissociates into monatomic D. Compare these results to the ideal gas Hugoniot result in this chapter (Eq. 2.114). Based on the data in the figure compared to the ideal gas equation of state, does a high pressure shock dissociate deuterium? Are the ideal gas Hugoniots with dissociation and without dissociation qualitatively the same as the equations of state Hugoniots shown in the figure? Assume $\rho_0 = 0.165$ g/cm^3 at 20 K as the initial state (Table 5.7). (Hint: you will need to infer an initial state "gas" pressure by matching the curve; you also need to know γ for diatomic and monatomic ideal gases.)

(2.15) To see a relation between a nonlocal integral approach and higher order derivatives, suppose the weighting $\phi(x, L)$ is a hat function composed of straight lines, $\phi = 0$ for $x < -L$, ϕ is the straight line connecting $(-L, 0)$ and $(0, 1/L)$, then the straight line connecting $(0, 1/L)$ and $(L, 0)$, and then $\phi = 0$ for $x > L$. Suppose that the strain can be expanded in a Taylor series as $\varepsilon(x) = \varepsilon(0) + \varepsilon^{(1)}(0)x + \frac{1}{2}\varepsilon^{(2)}(0)x^2 + \cdots$. Show that using as the strain an

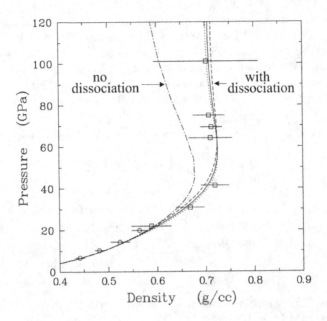

FIGURE 2.4. Sophisticated equation of state Hugoniots for deuterium compared to shock data. (From Kerley [117].)

integral average $\bar{\varepsilon}(x) = \int \varepsilon(y)\phi(x - y, L)dy$, rather than the strain at just a point, yields

$$\bar{\varepsilon}(0) = \int_{-\infty}^{\infty} \varepsilon(x)\phi(x, L)dx = \sum_{n \text{ even}} \frac{2L^n\varepsilon^{(n)}(0)}{(n+2)!}, \tag{2.143}$$

which has the interesting result that the coefficient multiplying $\varepsilon^{(2)}(0)$ is $L^2/12$, to compare with the discussion following Eq. (2.126).

(2.16) For the nonclassical elasticity model in Eq. (2.126), write down the static one-dimensional equilibrium equation and show the displacement solution is

$$u(x) = c_1 \cos(k_1 x) + c_2 \sin(k_1 x) + c_3 x + c_4, \qquad k_1^2 = a_1/a_3. \tag{2.144}$$

Also determine the general solution when $a_3 = 0$ (classical elasticity). For classical elasticity, specifying both endpoint displacements determines a unique solution, while for nonclassical elasticity ($a_3 \neq 0$), specifying both endpoints does not determine a unique solution. Thus, for higher order nonlocal nonclassical theories, additional boundary conditions, beyond what we typically expect, need to be identified and applied. What boundary conditions might you suggest? Why?

(2.17) Show that the following angular momentum balance leads to Eq. (2.133) assuming the momentum balance holds and using Reynolds Transport Theorem,

$$\frac{d}{dt}\int_V (\vec{x} \times \rho\vec{v} + J\vec{\omega})dV = \int_V (\vec{x} \times \rho\vec{b} + \rho\vec{m}^c)dV + \int_{\partial V} (\vec{x} \times \vec{t} + \vec{m})dS. \tag{2.145}$$

Don't worry about, or gloss over, a ρ term in $J\vec{\omega}$, as this is an approximate way to address microstructure. This angular momentum balance holds for an arbitrary fixed point as $\vec{x}$ can be replaced by $\vec{x} - \vec{x}_0$ and the result still holds.

Elastic-Plastic Solids

Quantitatively understanding impact and penetration requires understanding how solids behave. This chapter introduces the basics of elasticity and elastic-plastic response of solids (metals, for example) and thermal effects.

3.1. Strain

Constitutive models relate stress to strain, strain rate, temperature, and deformation history. In this section the basics of strain are presented. A more detailed presentation of strain and deformation occurs in Chapter 13, where the Lagrangian, or large, strain and deformation gradient are extensively used.

The original configuration of the material is denoted $\vec{X}$ and the current configuration (where the material currently is, in the laboratory frame) is denoted $\vec{x}$. The current configuration is a function of the original configuration and time; hence

$$x_i = x_i(X_j, t) \qquad \text{or} \qquad \vec{x} = \vec{x}(\vec{X}, t). \tag{3.1}$$

The expression on the left is indicial notation, and the expression on the right is a vector notation. They mean the same thing – that the point of material originally at point $\vec{X}$ is now at point $\vec{x}$. Writing this out in detail in Cartesian coordinates,

$$\begin{aligned} x &= x(X, Y, Z, t) \\ y &= y(X, Y, Z, t) \\ z &= z(X, Y, Z, t). \end{aligned} \tag{3.2}$$

Strain relates to length. We are interested in understanding how a small segment in the initial configuration, denoted $d\vec{X}$, is rotated and stretched or shortened to a small segment in the current configuration, $d\vec{x}$, when deformation occurs. Differentiating the current configuration, which is a function of the initial configuration, gives

$$dx = \frac{\partial x}{\partial X}dX + \frac{\partial x}{\partial Y}dY + \frac{\partial x}{\partial Z}dZ, \tag{3.3}$$

and similarly for dy and dz,

$$d\vec{x} = \begin{pmatrix} dx \\ dy \\ dz \end{pmatrix} = \begin{pmatrix} \frac{\partial x}{\partial X} & \frac{\partial x}{\partial Y} & \frac{\partial x}{\partial Z} \\ \frac{\partial y}{\partial X} & \frac{\partial y}{\partial Y} & \frac{\partial y}{\partial Z} \\ \frac{\partial z}{\partial X} & \frac{\partial z}{\partial Y} & \frac{\partial z}{\partial Z} \end{pmatrix} \begin{pmatrix} dX \\ dY \\ dZ \end{pmatrix} = F d\vec{X}, \tag{3.4}$$

where the equations define the *deformation gradient F*. In indicial notation and matrix notation in Cartesian coordinates, the deformation gradient is

$$F_{ij} = \frac{\partial x_i}{\partial X_j} \qquad \text{or} \qquad F = (F_{ij}) = \left(\frac{\partial x_i}{\partial X_j} \right). \tag{3.5}$$

Here, (F_{ij}) indicates the entire matrix, a notation that will sometimes be used.

Strain is a measure of a change in length over a gage section compared to a characteristic length of that gage section. In one dimension, the simplest measure of strain over an initial length L_0 is to consider the change in length $\Delta L = L - L_0$ and look at the ratio $\varepsilon = \Delta L / L_0$. This measure is the strain. The challenge we now face is translating this simple one-dimensional notion to a three-dimensional setting. As a first step, the length of our increments in the previous paragraph can be determined by taking the inner product of the two small vectors, in particular

$$ds^2 = d\vec{x} \cdot d\vec{x} = d\vec{x}^T d\vec{x} = dx_i dx_i, \qquad dS^2 = d\vec{X} \cdot d\vec{X} = d\vec{X}^T d\vec{X} = dX_i dX_i, \quad (3.6)$$

where the superscript T means matrix or vector transpose. Now consider

$$\varepsilon = \frac{L - L_0}{L_0} = \frac{ds - dS}{dS} = \frac{ds - dS}{dS} \frac{ds + dS}{ds + dS} \approx \frac{ds^2 - dS^2}{2dS^2}, \qquad (3.7)$$

where the last approximation assumes that the change in length is not large so that $ds \approx dS$. Thus, for small amounts of deformation we can recover the classical definition of strain without taking a square root. This is fortuitous, and the last term yields our formal definition of the strain,

$$ds^2 - dS^2 = d\vec{x}^T d\vec{x} - d\vec{X}^T d\vec{X} = d\vec{X}^T F^T F d\vec{X} - d\vec{X}^T d\vec{X} = d\vec{X}^T (F^T F - I) d\vec{X}. \quad (3.8)$$

Comparing Eqs. (3.7) and (3.8) leads to the definition of the Green strain tensor or the Lagrangian strain tensor as

$$(E_{ij}) = \frac{1}{2}(F^T F - I) \qquad \text{or} \qquad E_{ij} = \frac{1}{2}(F_{ik}^T F_{kj} - \delta_{ij}). \qquad (3.9)$$

The displacement $\vec{u}$ is the difference between the current location of a point of the material and its original location,

$$x_i = X_i + u_i \qquad \text{or} \qquad \vec{x} = \vec{X} + \vec{u}. \qquad (3.10)$$

In Cartesian coordinates, in terms of displacement the deformation gradient is

$$F_{ij} = \frac{\partial(X_i + u_i)}{\partial X_j} = \delta_{ij} + \frac{\partial u_i}{\partial X_j}, \qquad (3.11)$$

where δ_{ij} is the Kronecker delta

$$\delta_{ij} = \begin{cases} 1 & i = j, \\ 0 & i \neq j, \end{cases} \qquad I = (\delta_{ij}). \qquad (3.12)$$

The Lagrangian strain tensor is then

$$E_{ij} = \frac{1}{2} \left\{ \left(\delta_{ki} + \frac{\partial u_k}{\partial X_i} \right) \left(\delta_{kj} + \frac{\partial u_k}{\partial X_j} \right) - \delta_{ij} \right\} = \frac{1}{2} \left(\frac{\partial u_i}{\partial X_j} + \frac{\partial u_j}{\partial X_i} + \frac{\partial u_k}{\partial X_i} \frac{\partial u_k}{\partial X_j} \right).$$
$$(3.13)$$

In metals and ceramics, the elastic strain is typically less than 1%. Thus, a common assumption is of small strains and small rotations, which is met when all the $\partial u_i / \partial X_j$ are small in magnitude compared to 1, so the nonlinear terms on the right of (E_{ij}) can be neglected. When they are dropped, the small strain is obtained as

$$\varepsilon_{ij} = \frac{1}{2} \left(\frac{\partial u_i}{\partial x_j} + \frac{\partial u_j}{\partial x_i} \right), \qquad (3.14)$$

where the lower case x in the partial derivatives means the derivative is taken with respect to the current Eulerian configuration, permissible since $\partial x_i / \partial X_j \approx \delta_{ij}$.

Small displacement does not imply small strain, as $u = A\cos(kx)$ with small A and large k demonstrates.

3.2. Small Strain Linear Elasticity

Now that we have the equations of motion, we need to determine how material deformation gives rise to stress. Initially our focus is on metals (and, to a lesser extent, on ceramics – that is, our focus is on hard solids), which we assume are isotropic (the same in all directions). Though metals and ceramics are made of small crystals which are not isotropic themselves, the individual crystals, called grains, are randomly oriented, thus forming a homogenized material that is isotropic at the scale we use in our analysis. Grain sizes typically range from 10 to 100 μm (10 to 100 microns), much smaller than our millimeter to meter scale of interest.

There are two types of behavior in metals: elastic and plastic. An elastic deformation is such that after the deformation, the body will return to its original shape if all the forces are removed. During plastic deformation, a permanent set or deformation occurs, and when the forces are released the body does not return to its original shape and typically residual stresses remain in the body.

For small strains, solids behave in a linear elastic fashion, and can be modeled with Hooke's law, which says, in indicial notation and tensor notation,

$$\sigma_{ij} = \lambda\varepsilon_{kk}\delta_{ij} + 2\mu\varepsilon_{ij} \qquad \text{or} \qquad \sigma = \lambda\text{trace}(\varepsilon)I + 2\mu\varepsilon, \qquad (3.15)$$

where λ and μ are the Lamé constants and δ_{ij} is the Kronecker delta. This equation only assumes isotropy and linear response: all solids obey it for small enough strains (it is easier to show with more theory – Exercise 14.4). The reason there is a 2 in front of the μ is that historically the off-diagonal shear strains were written $\gamma_{ij} = 2\varepsilon_{ij}$, and $\sigma_{ij} = G\gamma_{ij}$, $i \neq j$; hence, $\mu = G$ where G is called the shear modulus. In Cartesian coordinates the small strain tensor ε_{ij} is given by

$$\varepsilon_{ij} = \frac{1}{2}\left(\frac{\partial u_i}{\partial x_j} + \frac{\partial u_j}{\partial x_i}\right). \qquad (3.16)$$

In this equation u_i is the displacement in the i direction. (In this chapter u is displacement and in the previous and next chapter u is velocity – these are the conventions in the field.) This definition of strain follows the same sign convention as stress: it is positive in tension and negative in compression.

Another way to write Hooke's law is

$$\begin{pmatrix} \sigma_{11} \\ \sigma_{22} \\ \sigma_{33} \\ \sigma_{23} \\ \sigma_{31} \\ \sigma_{12} \end{pmatrix} = \begin{pmatrix} \lambda+2\mu & \lambda & \lambda & 0 & 0 & 0 \\ \lambda & \lambda+2\mu & \lambda & 0 & 0 & 0 \\ \lambda & \lambda & \lambda+2\mu & 0 & 0 & 0 \\ 0 & 0 & 0 & \mu & 0 & 0 \\ 0 & 0 & 0 & 0 & \mu & 0 \\ 0 & 0 & 0 & 0 & 0 & \mu \end{pmatrix} \begin{pmatrix} \varepsilon_{11} \\ \varepsilon_{22} \\ \varepsilon_{33} \\ 2\varepsilon_{23} \\ 2\varepsilon_{31} \\ 2\varepsilon_{12} \end{pmatrix} \qquad (3.17)$$

The matrix of elastic constants is sometimes referred to as the *stiffness*. The larger the stiffness, the smaller the strain for a given stress. With some arithmetic it can be inverted to give the *compliance,* which is the coefficient matrix that gives the strain in terms of the stress (where we initially focus on the upper-left quarter of the matrix)

$$\begin{pmatrix} \varepsilon_{11} \\ \varepsilon_{22} \\ \varepsilon_{33} \end{pmatrix} = \frac{1}{2\mu(3\lambda + 2\mu)} \begin{pmatrix} 2(\lambda + \mu) & -\lambda & -\lambda \\ -\lambda & 2(\lambda + \mu) & -\lambda \\ -\lambda & -\lambda & 2(\lambda + \mu) \end{pmatrix} \begin{pmatrix} \sigma_{11} \\ \sigma_{22} \\ \sigma_{33} \end{pmatrix}. \tag{3.18}$$

The larger the compliance, the larger the strain for a given stress. Elastic constants are typically determined by *uniaxial stress* tests performed in a tension or compression machine. A long, thin specimen is pulled or a stubby cylinder is pushed. The lateral stress is zero ($\sigma_{22} = \sigma_{33} = 0$). The strain state of the material can be determined through the compliance as

$$\varepsilon_{11} = \frac{\lambda + \mu}{(3\lambda + 2\mu)\mu}\sigma_{11}, \quad \varepsilon_{22} = \varepsilon_{33} = -\frac{\lambda}{2\mu(3\lambda + 2\mu)}\sigma_{11}. \tag{3.19}$$

The Young's modulus E is defined as the stiffness in a uniaxial stress test, $\sigma_{11} = E\varepsilon_{11}$, and the Poisson's ratio ν is defined as the contraction in the lateral direction in a uniaxial stress test in terms of strain in the axial direction, $\varepsilon_{22} = \varepsilon_{33} = -\nu\varepsilon_{11}$. Thus,

$$E = \frac{(3\lambda + 2\mu)\mu}{\lambda + \mu}, \quad \nu = \frac{\lambda}{2\mu(3\lambda + 2\mu)}E = \frac{\lambda}{2(\lambda + \mu)}. \tag{3.20}$$

These definitions allow the compliance equation to be written

$$\begin{pmatrix} \varepsilon_{11} \\ \varepsilon_{22} \\ \varepsilon_{33} \\ \varepsilon_{23} \\ \varepsilon_{31} \\ \varepsilon_{12} \end{pmatrix} = \begin{pmatrix} 1/E & -\nu/E & -\nu/E & 0 & 0 & 0 \\ -\nu/E & 1/E & -\nu/E & 0 & 0 & 0 \\ -\nu/E & -\nu/E & 1/E & 0 & 0 & 0 \\ 0 & 0 & 0 & 1/2\mu & 0 & 0 \\ 0 & 0 & 0 & 0 & 1/2\mu & 0 \\ 0 & 0 & 0 & 0 & 0 & 1/2\mu \end{pmatrix} \begin{pmatrix} \sigma_{11} \\ \sigma_{22} \\ \sigma_{33} \\ \sigma_{23} \\ \sigma_{31} \\ \sigma_{12} \end{pmatrix}. \tag{3.21}$$

Notice that

$$1 + \nu = 1 + \frac{\lambda}{2(\lambda + \mu)} = \frac{3\lambda + 2\mu}{2(\lambda + \mu)} = \frac{E}{2\mu}, \tag{3.22}$$

and the compliance relationship can be written

$$\varepsilon_{ij} = -\frac{\nu}{E}\sigma_{kk}\delta_{ij} + \frac{1+\nu}{E}\sigma_{ij} \quad \text{or} \quad \varepsilon = -\frac{\nu}{E}\text{trace}(\sigma)I + \frac{1+\nu}{E}\sigma. \tag{3.23}$$

For a linear elastic material, the pressure is

$$p = -\frac{\sigma_{ii}}{3} = -\frac{(3\lambda + 2\mu)(\varepsilon_{11} + \varepsilon_{22} + \varepsilon_{33})}{3} = -K(\varepsilon_{11} + \varepsilon_{22} + \varepsilon_{33}) = -K\varepsilon_{ii}, \tag{3.24}$$

where K is the bulk modulus.

We have presented five elastic coefficients based on relations between stress and strain, and have determined E, ν, and K in terms of λ and μ. Only two are independent. Based on these relations, the rest of the relations are

$$\lambda = K - \frac{2}{3}G = \frac{(E-2G)G}{3G-E} = \frac{E\nu}{(1+\nu)(1-2\nu)} = \frac{2G\nu}{1-2\nu} = \frac{3K\nu}{1+\nu},$$

$$\mu = G = \frac{E}{2(1+\nu)} = \frac{3K(1-2\nu)}{2(1+\nu)} = \frac{\lambda(1-2\nu)}{2\nu},$$

$$K = \lambda + \frac{2}{3}G = \frac{EG}{3(3G-E)} = \frac{2G(1+\nu)}{3(1-2\nu)} = \frac{E}{3(1-2\nu)} = \frac{\lambda(1+\nu)}{3\nu},$$

$$E = \frac{9KG}{3K+G} = 2G(1+\nu) = 3K(1-2\nu) = \frac{G(3\lambda+2G)}{\lambda+G} = \frac{\lambda(1+\nu)(1-2\nu)}{\nu},$$

$$\nu = \frac{3K-2G}{2(3K+G)} = \frac{\lambda}{2(\lambda+G)} = \frac{E}{2G} - 1.$$

$$(3.25)$$

The shear modulus G and the bulk modulus K can be any nonnegative value – independent of each other – otherwise the material would be unstable (the stress would decrease as the strain increased). The other elastic constants have permissible values based on the values of bulk and shear moduli. For example, examining denominators in Eq. (3.25) for G and K shows that $-1 \le \nu \le \frac{1}{2}$ (to date, all naturally found materials also have $\nu > 0$), and the expressions for E show that $0 \le E \le \min(3G, 9K)$. The Lamé constant λ can be any real number; however, to date all naturally found materials have $\lambda > 0$ ($\lambda < 0$ implies $\nu < 0$). Experimentally measured K, G, and ν are provided in Appendix C for most elements. E and λ can then be computed.

As an example, consider the case of uniaxial strain. Here the material only exhibits strain in one direction, ε_{11}; hence $\varepsilon_{22} = \varepsilon_{33} = 0$. In this case the stress is given by

$$\sigma_{11} = (\lambda + 2\mu)\varepsilon_{11},$$
$$\sigma_{22} = \lambda\varepsilon_{11} = \sigma_{33}.$$

$$(3.26)$$

In this case stress must be applied to the sides of the material in order to maintain uniaxial strain – the computed value for σ_{22} shows how much. In this case, the slope of the σ_{11} stress vs. strain is always larger than the case of uniaxial stress, save for when the slopes are equal when Poisson's ratio equals zero ($\lambda + 2\mu \ge E$),

$$\lambda + 2\mu = K + \frac{4}{3}G = 2\frac{1-\nu}{1-2\nu}G = 3\frac{1-\nu}{1+\nu}K = \frac{E(1-\nu)}{(1+\nu)(1-2\nu)} = \lambda\frac{1-\nu}{\nu}. \quad (3.27)$$

Since there is no permanent deformation in elasticity, when the stress goes to zero, so does the strain. Unloading occurs along the same path as loading (Fig. 3.1).

3.3. Metal Plasticity

Initial deformation of metals is elastic. When metals deform beyond their elastic limit, they plastically deform, meaning part of the deformation is permanent. We will examine what is referred to as *incremental plasticity*, since that is what the large-scale numerical codes (hydrocodes) use, though in our analytical modeling not all parts of the theory will be exercised. Plasticity theory is highly mathematical: the majority of this monograph can be understood without following the entire discussion of plasticity in this chapter. One should, however, understand the

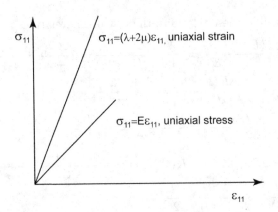

FIGURE 3.1. Stress-strain curves for elastic uniaxial stress and uniaxial strain.

examples of uniaxial stress (Section 3.4), uniaxial strain (Section 3.5), and rigid plasticity (Section 3.7).

It is assumed that an increment of deformation (strain) can be partitioned into elastic and plastic parts,

$$d\varepsilon_{ij} = d\varepsilon_{ij}^e + d\varepsilon_{ij}^p. \tag{3.28}$$

The elastic part gives rise to the stress in the body, through Hooke's law, and the plastic part of the deformation does not give rise to stress. The form written here is with increments, but it also can be written in rate form as

$$\dot{\varepsilon}_{ij} = \dot{\varepsilon}_{ij}^e + \dot{\varepsilon}_{ij}^p \tag{3.29}$$

where the dot over the ε terms means d/dt or differentiation with respect to time. There are various definitions of strain that can be used in these expressions. In rigid plasticity and in most transient computational software (often referred to as *hydrocodes* in the impact community), rather than the small-strain strain rate $\dot{\varepsilon}$ that is used in small strain elasticity, the rate of deformation tensor D is used as the strain rate, and the partition is written as

$$D_{ij} = D_{ij}^e + D_{ij}^p, \tag{3.30}$$

again with v_i the velocity in the i direction; in Cartesian coordinates the rate of deformation tensor D is

$$D_{ij} = \frac{1}{2}\left(\frac{\partial v_i}{\partial x_j} + \frac{\partial v_j}{\partial x_i}\right). \tag{3.31}$$

D and the small-strain strain rate $\dot{\varepsilon}$ are not the same, but for small strain they are nearly equal (the exact relationship is in Exercise 14.23). We will use both the small-strain strain rate $\dot{\varepsilon}$ and D.

Mathematically, plasticity is modeled using two assumptions:

(1) The stress remains within a *yield surface* $f(\sigma_{ij}) \leq 0$ and is on the yield surface $f(\sigma_{ij}) = 0$ during plastic flow.
(2) A *flow rule* is specified, giving the direction of plastic deformation. Often the flow rule is associated with the yield surface, and the flow is then

called associated flow

$$d\varepsilon_{ij}^p = d\Lambda \frac{\partial f}{\partial \sigma_{ij}}, \tag{3.32}$$

where the value $d\Lambda$ depends on the specifics of the loading situation. Since $d\Lambda$ is a real number, Eq. (3.32) says the plastic strain increment $d\varepsilon_{ij}^p$ is in the direction of the stress gradient of the yield surface $\partial f/\partial \sigma_{ij}$.

It has been experimentally observed that metal plasticity is independent of pressure – that is, when experiments have been performed under large pressure, the plastic portion of the stress-strain behavior of the material does not change. This experimental result implies that there is no volumetric component to plasticity since if, for example, plastic flow involved a reduction in volume, then high pressures would help drive plastic flow. Thus, for metals,

$$d\varepsilon_{ii}^p = 0. \tag{3.33}$$

This result implies for the pressure dependence

$$p = -K\varepsilon_{ii}^e = -K(\varepsilon_{ii}^e + \varepsilon_{ii}^p) = -K\varepsilon_{ii}. \tag{3.34}$$

Later, when high pressure response of metals is considered, the pressure will be written as a function of material density and energy. It will not be necessary to include plasticity behavior in the pressure response since there is no change in volume, and hence no change in density, due to plastic flow.

In addition, no pressure dependence implies that the yield function $f(\sigma_{ij})$ is independent of pressure. Thus, the stress is separated into a pressure and a deviatoric component,

$$\sigma_{ij} = s_{ij} - p\delta_{ij} \quad \text{or} \quad \sigma = s - pI. \tag{3.35}$$

Pressure was defined earlier, $p = -\sigma_{ii}/3$ (Eq. 2.6). It will be noted that the trace of the stress deviator tensor is zero: $s_{ii} = 0$. As an aside, in a similar fashion it is possible to introduce the strain deviator tensor, defined as

$$e_{ij} = \varepsilon_{ij} - \frac{1}{3}\varepsilon_{kk}\delta_{ij}. \tag{3.36}$$

For the elastic stress-strain relation, placement of the deviator into Hooke's law yields, for example,

$$s_{11} = \sigma_{11} + p = (\lambda + 2\mu)\varepsilon_{11}^e + \lambda(\varepsilon_{22}^e + \varepsilon_{33}^e) - (3\lambda + 2\mu)(\varepsilon_{11}^e + \varepsilon_{22}^e + \varepsilon_{33}^e)/3 = 2\mu\varepsilon_{11}^e. \tag{3.37}$$

This form holds for all components along the diagonal; it directly holds for the off-diagonals. Hence, the relationship holds for all components of the stress deviator,

$$s_{ij} = 2\mu e_{ij}^e \quad \text{or} \quad s = 2\mu e^e. \tag{3.38}$$

The Cauchy stress tensor is symmetric and has three real eigenvalues. These are referred to as the principal stresses. Under the action of a change in coordinate system, either a translation or a rotation (e.g., $Q\sigma Q^{-1}$), the eigenvalues do not change[1] and hence are referred to as an invariant of the stress state – in other words, regardless of the orientation of the coordinate system, the principal stresses are the same. Since solving for the eigenvalues involves solving a cubic polynomial (the characteristic equation) whose leading term has coefficient 1, the other coefficients

---

[1]When $\sigma\vec{x} = \lambda\vec{x}$, $\vec{x} \neq 0$ (this is the definition of *eigenvalue* λ), then $Q\sigma Q^{-1}\vec{y} = \lambda\vec{y}$, where $\vec{y} = Q\vec{x}$. Hence, the same numerical value λ is an eigenvalue of $Q\sigma Q^{-1}$. Thus λ is called an *invariant* of σ under the action of rotation.

of the eigenvalue equation are also invariant under that action. The coefficients of the polynomial equation are more typically worked with and are referred to as the three *invariants* of the stress tensor.[2] We have

$$\det(\lambda I - \sigma) = \det(\lambda \delta_{ij} - \sigma_{ij}) = \lambda^3 - I_1 \lambda^2 - I_2 \lambda - I_3 = 0,$$

$$I_1 = \sigma_{11} + \sigma_{22} + \sigma_{33} = \sigma_{ii} = \text{trace}(\sigma) = -3p,$$

$$I_2 = -\sigma_{11}\sigma_{22} - \sigma_{22}\sigma_{33} - \sigma_{33}\sigma_{11} + \sigma_{12}\sigma_{21} + \sigma_{23}\sigma_{32} + \sigma_{13}\sigma_{31}$$

$$= \frac{1}{2}\sigma_{ij}\sigma_{ji} - \frac{1}{2}(I_1)^2 = \frac{1}{2}\left\{\text{trace}(\sigma^2) - (\text{trace}(\sigma))^2\right\}, \tag{3.39}$$

$$I_3 = \det(\sigma) = \varepsilon_{ijk}\sigma_{1i}\sigma_{2j}\sigma_{3k} = \varepsilon_{ijk}\sigma_{i1}\sigma_{j2}\sigma_{k3}$$

$$= \frac{1}{3}\text{trace}(\sigma^3) - \frac{1}{2}\text{trace}(\sigma) \cdot \text{trace}(\sigma^2) + \frac{1}{6}(\text{trace}(\sigma))^3,$$

where λ is the eigenvalue (using traditional notation) and I_i are the three invariants. ε_{ijk} is the permutation symbol, where $\varepsilon_{ijk} = 1$ if the ordering is an even permutation of the indices ($ijk = 123, 231$ or 312), $\varepsilon_{ijk} = -1$ if the ordering is odd ($132, 213$ or 321), and $\varepsilon_{ijk} = 0$ if any index repeats. The three invariants J_i for the stress deviator tensor are

$$\det(\lambda I - s) = \lambda^3 - J_1\lambda^2 - J_2\lambda - J_3 = 0,$$

$$J_1 = s_{ii} = \text{trace}(s) = 0,$$

$$J_2 = \frac{1}{2}s_{ij}s_{ji} = \frac{1}{2}\text{trace}(s^2), \tag{3.40}$$

$$J_3 = \det(s) = \frac{1}{3}\text{trace}(s^3).$$

Another way the determinant is written is

$$J_3 = \det(s_{ij}) = \begin{vmatrix} s_{11} & s_{12} & s_{13} \\ s_{21} & s_{22} & s_{23} \\ s_{31} & s_{32} & s_{33} \end{vmatrix}. \tag{3.41}$$

The way we have written the invariants here ($s_{ij}s_{ji}$) does not assume the stress tensor is symmetric; however, below we will make that assumption. J_2 and J_3 are invariants of the stress tensor, not just the stress deviator tensor (Exercise 3.50).

There are lots of invariants, since every function $f(I_1, I_2, I_3)$ is again invariant. Looking at Eq. (3.39) we immediately see, by linear combinations, that $\text{trace}(\sigma)$, $\text{trace}(\sigma^2)$, and $\text{trace}(\sigma^3)$ are invariants. However, for 3×3 symmetric matrices there are only three independent invariants, as that is the number of eigenvalues, each of which is independent. (To see that I_1, I_2, and I_3 are independent, see Exercise 3.19.) We have written the scalar invariants in terms of two functions, the trace and the determinant. Each has a nice multiplicative property: $\text{trace}(AB) = A_{ij}B_{ji} = \text{trace}(BA)$ and $\det(AB) = \det(A)\det(B)$ (this last property is shown in the footnote on page 511). It is easy to see that all these scalars are invariants under a rotation since $\text{trace}(Q\sigma Q^T \cdots Q\sigma Q^T) = \text{trace}(Q\sigma^n Q^T) = \text{trace}(Q^T Q\sigma^n) = \text{trace}(\sigma^n)$.

Since the material is isotropic (the same in all directions), the yield function can be written in terms of the three tensor invariants. As was mentioned for metal plasticity, experimentally the strength is independent of pressure. If we remove the pressure from the stress tensor, we are left with the stress deviator tensor s. Recall

---

[2]There are only three invariants when the tensor is symmetric, which will always be the case in our discussion of tensor invariants. However, see Exercise 3.20 and invariants of F on page 560.

the definition of the stress tensor, which is the tractions on each face of the small cube written as a row of the tensor. Removing the pressure means that the normal component of the traction from each of the three faces sum to zero – hence, J_1 is always zero – so there are really only two invariants of interest, J_2 and J_3. A measure of magnitude of the remaining traction on each face is the sum of the square of the components – for example, $s_{1j}s_{1j}$ for face 1. If we sum up the magnitude of all six traction deviator vectors, from each face of the cube (the opposite faces differ by a sign and have the same magnitude), we obtain $2s_{ij}s_{ij}$. This value measures the forces acting to deform or distort the material in the cube (not compress or expand it). It is nice that this value is a multiple of an invariant we have already derived, namely it is $4J_2$. Hence, J_2 is a scalar measure of the magnitude of the stress deviator acting at a point. In von Mises plasticity, the yield function is taken as a function of J_2. A useful and convenient algebraic expression is

$$J_2 = \frac{1}{2}s_{ij}s_{ij} = \frac{1}{6}\left\{(\sigma_{11} - \sigma_{22})^2 + (\sigma_{22} - \sigma_{33})^2 + (\sigma_{33} - \sigma_{11})^2\right\} + \sigma_{23}^2 + \sigma_{31}^2 + \sigma_{12}^2.$$
(3.42)

To use J_2 in a yield expression, it is necessary to determine what it equals at yield.

To determine the physical meaning of J_3, recall $J_3 = \det(s)$. First consider the case when $J_3 = 0$. If we change to coordinates that diagonalize the stress tensor, then, since $J_1 = s_{11} + s_{22} + s_{33} = 0$ and at least one eigenvalue is zero,

$$s = \begin{pmatrix} s_{11} & 0 & 0 \\ 0 & -s_{11} & 0 \\ 0 & 0 & 0 \end{pmatrix}.$$
(3.43)

Further rotating $\theta = 45°$ about the z axis,

$$Q = \begin{pmatrix} \cos(\theta) & -\sin(\theta) & 0 \\ \sin(\theta) & \cos(\theta) & 0 \\ 0 & 0 & 1 \end{pmatrix} \quad \Rightarrow \quad QsQ^T = \begin{pmatrix} 0 & s_{11} & 0 \\ s_{11} & 0 & 0 \\ 0 & 0 & 0 \end{pmatrix}.$$
(3.44)

Thus, the stress state is one of pure shear, as the only nonzero entries in this coordinate system are the two shear components.

If J_3 does not equal zero, then when we rotate to coordinates where s is diagonal, the diagonal entries are all nonzero. Since trace$(s) = 0$, either two eigenvalues are positive and one is negative, or one is positive and two are negative. As they sum to zero, the "other"-signed eigenvalue is larger in magnitude than each of the eigenvalues that share a sign. If two eigenvalues are negative, then the largest eigenvalue in magnitude is positive and hence the stress state is qualitatively one of tension. Since it is a positive times two negatives, $J_3 > 0$. If two eigenvalues are positive and one negative, the largest eigenvalue in magnitude is negative and the stress state is qualitatively one of compression. With two eigenvalues positive and one negative, $J_3 < 0$.

To summarize, since the determinant of s equals the volume of the parallelepiped with edges defined by the three row vectors in the stress deviator tensor s, when the volume is positive ($J_3 > 0$), it is qualitatively a state of *tension*, when the volume is negative ($J_3 < 0$), it is qualitatively a state of *compression*, and when the volume is zero ($J_3 = 0$), it is a stress state of *pure shear*.

We can determine the largest that J_3 can be by fixing s_{11}; the diagonal is then $\{s_{11}, s_{22}, -s_{11} - s_{22}\}$ and $J_3 = \det(s) = -s_{11}s_{22}(s_{11} + s_{22})$. With s_{11} fixed, we differentiate J_3, set it equal to zero, and obtain $s_{11}(s_{11} + 2s_{22}) = 0$. Since s_{11} is

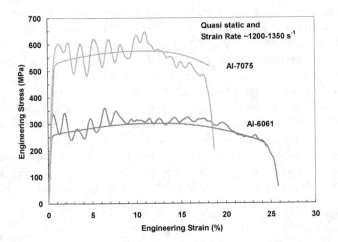

FIGURE 3.2. Two aluminum alloys' quasi-static and high-rate uniaxial stress tensile stress-strain behavior. (Chapter 5 describes the
origin of the oscillations in the high-rate test.)

arbitrary, the extremum is when $s_{22} = -\frac{1}{2}s_{11}$, yielding maximum in magnitude
$J_3 = \frac{1}{4}s_{11}^3 = \pm\frac{2}{27}(3J_2)^{3/2}$, where s_{11} is the maximum eigenvalue in magnitude.
The sign of this J_3 matches the sign of the maximum eigenvalue.

Figure 3.2 shows the behavior of two actual metals during a tensile uniaxial
stress test. A cylindrical tensile specimen of the metal is pulled. Initially the
response is elastic and is modeled by Hooke's law. Then the metal hits yield, the
stress at which permanent deformation first occurs. This theoretical definition of
yield is hard to apply. In engineering, yield is defined to be the stress where, if
unloading were to occur with the same elastic slope E, a 0.2% permanent offset
occurs in the strain *after unloading* (i.e., a strain of 0.002). Hence, the definition
is tied to a small but permanent strain. As the specimen continues to be pulled,
the stress typically increases, exhibiting what is referred to as work hardening. It
is also possible for metals to show a strain-rate dependence – that is, the yield
value is higher for higher rate deformation. In addition, as metals approach their
melting temperature they exhibit thermal softening, meaning that their strength
decreases owing to the high temperatures. An elastic-perfectly plastic material is
an idealization where the flow stress is constant, as shown in Fig. 3.3. We will be
using elastic-perfectly plastic material models in our analytic models – thus, a flow
stress (the stress beyond yield at which the material flows) should be chosen to be
representative of the material's flow behavior in the strain and strain-rate regime
of interest, including the effects of work hardening, strain-rate strengthening, and
thermal softening. When large-scale numerical simulations are run, typically the
material models include these effects, and we write

$$Y = Y(\varepsilon_p, \dot{\varepsilon}_p, T). \tag{3.45}$$

In this expression, ε_p is the equivalent plastic strain, to be defined below. When
transient rate dependence is included in the yield, the response is *viscoplastic*.

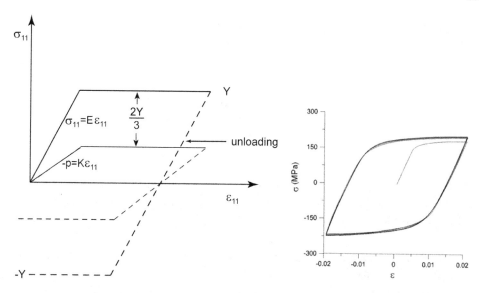

FIGURE 3.3. Uniaxial stress loading for an elastic-perfectly plastic material, showing the stress and negative of the pressure (left); uniaxial stress cyclic loading curve, traversed clockwise, for an aluminum alloy (right). (From Chung, Lee, and Pan [45].)

The type of work hardening discussed in this text is *isotropic hardening*, where the yield surface expands (Y gets larger) but remains centered at the origin of stress space. Movement of the center of the yield surface in stress space (referred to as the back stress β_{ij}, with yield surface $f(\sigma_{ij} - \beta_{ij}) = 0$) is referred to as *kinematic hardening*. Kinematic hardening requires a rule for the movement or evolution (kinematics) of the back stress. Figure 3.3 (right) shows closed uniaxial stress cycles for an aluminum. The rounded unloading (rather than sharp corners) in the upper left and lower right are the *Bauschinger effect*, which can be modeled with kinematic hardening. Kinematic hardening and isotropic hardening can be combined for even more complicated plastic response. When an impact or penetration event is being examined the deformation is usually loading, not unloading; however, in dynamic structural events, unloading can be an important part of the deformation.

3.4. Uniaxial Stress

We now work through some examples. The first is uniaxial stress of an elastic-perfectly plastic material, meaning that the flow stress Y is a constant (Fig. 3.3). During loading in uniaxial stress, initially the deformation is elastic and the stress tensor is given by

$$(\sigma_{ij}) = \begin{pmatrix} E\varepsilon_{11} & 0 & 0 \\ 0 & 0 & 0 \\ 0 & 0 & 0 \end{pmatrix}. \tag{3.46}$$

Elastic deformation continues until yield is reached when $\sigma_{11} = E\varepsilon_{11} = Y$; then

$$(\sigma_{ij}) = \begin{pmatrix} Y & 0 & 0 \\ 0 & 0 & 0 \\ 0 & 0 & 0 \end{pmatrix}. \tag{3.47}$$

Thus, at yield, the pressure in the specimen is $p = -Y/3$ and the stress deviator is

$$(s_{ij}) = \begin{pmatrix} 2Y/3 & 0 & 0 \\ 0 & -Y/3 & 0 \\ 0 & 0 & -Y/3 \end{pmatrix}. \tag{3.48}$$

Placing this stress state into the expression for J_2 we obtain

$$J_2 = \frac{Y^2}{3}. \tag{3.49}$$

Because of this result, the effective stress is defined as follows:

DEFINITION 3.1. (effective stress or equivalent stress) The *effective stress* or *equivalent stress* is a measure of the deviatoric, or shear, stress state of a material and is defined as

$$\sigma_{\text{eff}} = \sqrt{3J_2}. \tag{3.50}$$

It is based on the second invariant of the stress deviator tensor and is scaled so as to equal the axial stress in a uniaxial stress condition.

For J_2 or von Mises plasticity, the J_2 or von Mises yield surface is defined through a function of J_2 only, that is $f(J_2) = 0$. Two examples are

$$f(\sigma_{ij}) = J_2 - \frac{Y^2}{3} \quad \text{and} \quad f(\sigma_{ij}) = \sqrt{3J_2} - Y. \tag{3.51}$$

When $f = 0$, the stress is on the yield surface and plastic flow can occur. If $f < 0$, any increment of strain is elastic. The situation of $f > 0$ is not allowed – the stress state is never allowed to be outside the yield surface. The effective stress is a measure of how close the stress state is to yielding. When $\sigma_{\text{eff}} = Y$, yielding occurs. For other yield criteria, such as Tresca, σ_{eff} does not necessarily equal Y during plastic flow. The associated flow direction is defined by Eq. (3.32). To calculate it, three intermediate results are computed first,

$$\frac{\partial p}{\partial \sigma_{ij}} = \frac{\partial}{\partial \sigma_{ij}} \left(-\frac{\sigma_{kk}}{3} \right) = -\frac{1}{3} \delta_{ik} \delta_{jk} = -\frac{1}{3} \delta_{ij},$$

$$\frac{\partial s_{kl}}{\partial \sigma_{ij}} = \frac{\partial}{\partial \sigma_{ij}} (\sigma_{kl} + p\delta_{kl}) = \frac{\partial \sigma_{kl}}{\partial \sigma_{ij}} + \frac{\partial p}{\partial \sigma_{ij}} \delta_{kl} = \delta_{ik} \delta_{jl} - \frac{1}{3} \delta_{ij} \delta_{kl},$$

$$\frac{\partial J_2}{\partial \sigma_{ij}} = \frac{\partial}{\partial \sigma_{ij}} \left(\frac{1}{2} s_{kl} s_{lk} \right) = \frac{1}{2} \left\{ s_{kl} \frac{\partial s_{lk}}{\partial \sigma_{ij}} + s_{lk} \frac{\partial s_{kl}}{\partial \sigma_{ij}} \right\} \tag{3.52}$$

$$= \frac{1}{2} \left\{ s_{kl} \left(\delta_{il} \delta_{jk} - \frac{1}{3} \delta_{ij} \delta_{lk} \right) + s_{lk} \left(\delta_{ik} \delta_{jl} - \frac{1}{3} \delta_{ij} \delta_{kl} \right) \right\}$$

$$= \frac{1}{2} \left\{ s_{ji} - \frac{1}{3} s_{kk} \delta_{ij} + s_{ji} - \frac{1}{3} s_{kk} \delta_{ij} \right\} = s_{ji}.$$

For symmetric stress, the associated flow direction is then

$$\frac{1}{d\Lambda} d\varepsilon_{ij}^p = \frac{\partial f}{\partial \sigma_{ij}} = \frac{\partial}{\partial \sigma_{ij}} \left(J_2 - \frac{Y^2}{3} \right) = s_{ij}. \tag{3.53}$$

The result is simple and states that the plastic flow direction for associated von Mises plasticity is in the same direction as the stress deviator. An application of the chain rules shows that the associated flow direction is the same for any function of J_2, since $\partial f(J_2)/\partial \sigma_{ij} = (df/dJ_2)\partial J_2/\partial \sigma_{ij} = (df/dJ_2)s_{ij}$.

Let's apply this result to the case of elastic-perfect plasticity for uniaxial stress. At yield, the stress deviator is given in Eq. (3.48), and so the plastic flow is given by

$$(d\varepsilon_{ij}^p) = d\Lambda \begin{pmatrix} 2Y/3 & 0 & 0 \\ 0 & -Y/3 & 0 \\ 0 & 0 & -Y/3 \end{pmatrix}. \tag{3.54}$$

Thus $d\varepsilon_{22}^p = d\varepsilon_{33}^p = -(1/2)d\varepsilon_{11}^p$. At and after yield, since σ_{11} does not change, all the axial strain is plastic deformation,

$$(d\varepsilon_{ij}^p) = \begin{pmatrix} d\varepsilon_{11} & 0 & 0 \\ 0 & -d\varepsilon_{11}/2 & 0 \\ 0 & 0 & -d\varepsilon_{11}/2 \end{pmatrix}. \tag{3.55}$$

(Given the assumed isotropy and symmetry before and during plastic flow and the assumed incompressibility of plastic flow, there is little else this plastic strain increment tensor could be. Hence, though we have chosen to work out the details using the flow rule, the cases of uniaxial stress and next section's uniaxial strain can be viewed as scenarios that confirm the flow rule for the specific loading conditions.) For uniaxial stress, once yield occurs the stress state remains the same, as given in Eq. (3.47). Since stress arises from elastic deformation, at and after yield the elastic portion of the strain is simply given by

$$(\varepsilon_{ij}^e) = \begin{pmatrix} Y/E & 0 & 0 \\ 0 & -\nu Y/E & 0 \\ 0 & 0 & -\nu Y/E \end{pmatrix}. \tag{3.56}$$

For a uniaxial stress state, $d\varepsilon_{22}^e = d\varepsilon_{33}^e = -\nu d\varepsilon_{11}^e$, but since σ_{11} does not change, $d\varepsilon_{11}^e = 0$. For the total axial strain ε_{11} where yielding has occurred (i.e., $\varepsilon_{11} > Y/E$), the plastic strain is given by[3]

$$(\varepsilon_{ij}^p) = \begin{pmatrix} \varepsilon_{11} - Y/E & 0 & 0 \\ 0 & -(1/2)(\varepsilon_{11} - Y/E) & 0 \\ 0 & 0 & -(1/2)(\varepsilon_{11} - Y/E) \end{pmatrix}. \tag{3.57}$$

The plastic strain tensor has zero trace; hence there is no volumetric plastic flow, as assumed. Addition of the elastic and plastic strain tensor gives the total strain. The plastic strain is a tensor, but it is typically not explicitly computed in codes. It is often desirable to have a scalar measure of plastic strain as a magnitude, and this scalar measure is computed incrementally. Since the plastic strain is deviatoric, the natural choice for a scalar magnitude is the second invariant of the plastic strain increment tensor, $(1/2)d\varepsilon_{ij}^p d\varepsilon_{ij}^p$. For uniaxial stress, we expect the plastic strain

---

[3]In this example, as in the rest in this chapter, the deformation has no rotation (either it aligns with the loading axes or only the rate of deformation tensor is specified, which lacks the spin information contained in the antisymmetric part of the velocity gradient) and so it is possible to calculate the plastic strain tensor by direct componentwise time integration of the rate of deformation tensor. The additive decomposition $\varepsilon = \varepsilon^e + \varepsilon^p$ is a logarithmic strain, behaving like $\ln(L_{\text{total}}/L_0) = \ln(L_e/L_0) + \ln(L_{\text{total}}/L_e) = \varepsilon^e + \varepsilon^p$. If rotations are not small, complications arise. In the incremental approach, it is necessary to use a corotational stress rate for computing the stress (see Section 14.8) as well as appropriately rotating the accumulating plastic strain tensor during integration. If, instead of the incremental theory, finite strain theory is used, one multiplicatively factors the deformation gradient: $F = F_e F_p$. When strains and rotations are small, this factoring agrees with the additive decomposition: letting $F = I + \hat{F}$, then the Lagrangian (Green) strain is $(E_{ij}) = \frac{1}{2}(F^T F - I) = \frac{1}{2}(F_p^T F_e^T F_e F_p - I) \approx \frac{1}{2}(\hat{F}_e + \hat{F}_e^T + \hat{F}_p + \hat{F}_p^T) = \varepsilon^e + \varepsilon^p$. (Finite strain theory, the deformation gradient, and Lagrangian strain are defined in Chapter 13.)

increment to equal the total strain increment – that is, $d\varepsilon_p = d\varepsilon_{11}$. Evaluation of the second invariant gives $(1/2)d\varepsilon_{ij}^p d\varepsilon_{ij}^p = (3/4)(d\varepsilon_{11})^2$ and based on this result the equivalent plastic strain is defined to be

$$\varepsilon_p = \int d\varepsilon_p = \int \sqrt{\frac{2}{3} d\varepsilon_{ij}^p d\varepsilon_{ij}^p}. \tag{3.58}$$

If these increments are divided by dt to produce rates, this expression defines the equivalent plastic strain rate, where the dot means differentiation with respect to time:

DEFINITION 3.2. (equivalent plastic strain, equivalent plastic strain rate) The *equivalent plastic strain rate* is a scalar measure of plastic strain rate tensor,

$$\dot{\varepsilon}_p = \sqrt{\frac{2}{3} \dot{\varepsilon}_{ij}^p \dot{\varepsilon}_{ij}^p}. \tag{3.59}$$

Integrated over time it provides the *equivalent plastic strain* $\varepsilon_p = \int \dot{\varepsilon}_p dt$ (Eq. 3.58).

This definition is based on the second invariant of the plastic strain rate tensor and is scaled so as to give the correct strain value for the uniaxial stress response of an elastic-perfectly plastic material. The definition implies that the plastic strain increment is deviatoric (see Eq. (3.40) and Exercise 3.46).

Upon unloading, initially the unloading is elastic with modulus E. When zero stress is reached, a permanent offset is observed (Fig. 3.3). Once a stress of $-Y$ is reached, a constant stress is maintained as now the strain is again plastic.

3.5. Uniaxial Strain

Next we examine the case of uniaxial strain for an elastic-perfectly plastic solid. Initially the deformation will be elastic and the stress is

$$(\sigma_{ij}) = \begin{pmatrix} (\lambda + 2\mu)\varepsilon_{11} & 0 & 0 \\ 0 & \lambda\varepsilon_{11} & 0 \\ 0 & 0 & \lambda\varepsilon_{11} \end{pmatrix}. \tag{3.60}$$

The pressure is

$$p = -\frac{3\lambda + 2\mu}{3}\varepsilon_{11} = -K\varepsilon_{11} \tag{3.61}$$

and the stress deviator tensor is

$$(s_{ij}) = \begin{pmatrix} 4\mu\varepsilon_{11}/3 & 0 & 0 \\ 0 & -2\mu\varepsilon_{11}/3 & 0 \\ 0 & 0 & -2\mu\varepsilon_{11}/3 \end{pmatrix}. \tag{3.62}$$

The effective stress is

$$\sigma_{\text{eff}} = \left\{ \frac{3}{2}(s_{11}^2 + s_{22}^2 + s_{33}^2) \right\}^{1/2} = 2\mu |\varepsilon_{11}|. \tag{3.63}$$

As the uniaxial strain loading progresses, initial yield occurs when $\sigma_{\text{eff}} = Y$ or when

$$\varepsilon_{11} = \varepsilon_{\text{HEL}} = \frac{Y}{2\mu}. \tag{3.64}$$

The axial stress at yield is given by

$$\sigma_{\text{HEL}} = (\lambda + 2\mu)\frac{Y}{2\mu} = \left(\frac{K}{2G} + \frac{2}{3}\right)Y = \frac{4G - E}{2(3G - E)}Y = \frac{1 - \nu}{1 - 2\nu}Y. \tag{3.65}$$

The uniaxial strain yield point is referred to as the Hugoniot elastic limit (HEL) for reasons that will be explained in Chapter 5. For Poisson's ratios greater than zero, the axial stress at HEL is greater than the uniaxial stress yield (i.e., $\sigma_{HEL} > Y$ when $\nu > 0$).

Once yield is reached, subsequent deformation is comprised of both elastic and plastic deformation. As to the plastic flow direction, the stress deviator direction for uniaxial strain is the same as the stress deviator direction for uniaxial stress (Eq. 3.48). In particular, when the stress deviator tensor is evaluated at the yield point for the uniaxial strain condition, it equals the stress deviator tensor at the yield point for the uniaxial stress condition. Thus, the plastic flow direction for uniaxial strain will be the same as it was for uniaxial stress, namely $d\varepsilon_{22}^p = d\varepsilon_{33}^p = -\frac{1}{2}d\varepsilon_{11}^p$. This is a requirement on the strain components. There are additional requirements, and the fact that the 22 and 33 directions are equivalent will be used in the following expressions. The requirement that the stress stays on the yield surface while flow occurs says $f(\sigma_{ij}) = 0$ or

$$3J_2 = (\sigma_{11} - \sigma_{22})^2 = Y^2. \tag{3.66}$$

Since Y is constant, this implies that $d\sigma_{11} = d\sigma_{22}$ as loading progresses, which says

$$(\lambda + 2\mu)d\varepsilon_{11}^e + 2\lambda d\varepsilon_{22}^e = \lambda d\varepsilon_{11}^e + 2(\lambda + \mu)d\varepsilon_{22}^e, \tag{3.67}$$

which gives $d\varepsilon_{11}^e = d\varepsilon_{22}^e$. We thus now have four equations for the incremental strain terms: the two relationship between the strain increments for uniaxial strain (Eq. 3.28), the relationship between elastic strains from the yield surface, and the relationship between the plastic strains from the flow rule. Written down, they are

$$\begin{aligned} d\varepsilon_{11} &= d\varepsilon_{11}^e + d\varepsilon_{11}^p = \text{given}, \\ d\varepsilon_{22} &= d\varepsilon_{22}^e + d\varepsilon_{22}^p = 0, \\ d\varepsilon_{11}^e &= d\varepsilon_{22}^e, \\ d\varepsilon_{11}^p &= -2d\varepsilon_{22}^p. \end{aligned} \tag{3.68}$$

The solution of this linear system is

$$d\varepsilon_{11}^e = \frac{1}{3}d\varepsilon_{11}, \quad d\varepsilon_{11}^p = \frac{2}{3}d\varepsilon_{11}, \quad d\varepsilon_{22}^e = \frac{1}{3}d\varepsilon_{11}, \quad d\varepsilon_{22}^p = -\frac{1}{3}d\varepsilon_{11}. \tag{3.69}$$

The equivalent plastic strain increment is computed to be

$$d\varepsilon_p = |d\varepsilon_{11}^p| = \frac{2}{3}|d\varepsilon_{11}|. \tag{3.70}$$

Of particular interest is the slope of the stress-strain curve during uniaxial loading during plastic flow,

$$\frac{d\sigma_{11}}{d\varepsilon_{11}} = \frac{(\lambda + 2\mu)d\varepsilon_{11}^e + 2\lambda d\varepsilon_{22}^e}{d\varepsilon_{11}} = \frac{3\lambda + 2\mu}{3} = K. \tag{3.71}$$

Thus, the loading slope is equal to the bulk modulus K.

Let us examine the implications of this solution. Since the elastic increments are all the same, during loading

$$\begin{aligned} ds_{11} &= d\sigma_{11} + dp = (\lambda + 2\mu)d\varepsilon_{11}^e + 2\lambda d\varepsilon_{22}^e - K d\varepsilon_{11} = 0, \\ ds_{22} &= d\sigma_{22} + dp = \lambda d\varepsilon_{11}^e + 2(\lambda + \mu)d\varepsilon_{22}^e - K d\varepsilon_{11} = 0. \end{aligned} \tag{3.72}$$

Thus, the stress deviators do not change during plastic flow and so the plastic flow direction does not change. Though the above derivations were for initial yield, they

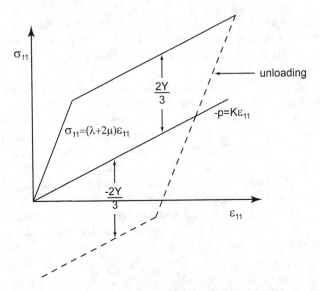

FIGURE 3.4. Uniaxial strain loading for an elastic-perfectly plastic material, showing stress and negative of the pressure.

now extend to the complete uniaxial strain loading regime, thus solving the entire case of uniaxial strain loading. The stress deviators remain as in Eq. (3.48) during elastic-plastic flow and the following important result is obtained (Fig. 3.4),

$$\sigma_{11} = -p \pm \frac{2Y}{3}. \tag{3.73}$$

The above derivation is for either tension $(+)$ or compression $(-)$. In impact situations the typical loading is in compression. For compression the signs of the strains and stresses are all negative, so that $\varepsilon_{11} = -\tilde{\varepsilon}_{HEL}$ at yield. For compression the previous result during plastic flow is

$$\tilde{\sigma}_{11} = p + \frac{2Y}{3}. \tag{3.74}$$

Upon unloading, the initial response is elastic with a slope of $\lambda + 2\mu$. During this unloading the axial stress deviator term will transition (for the tensile case) from $2Y/3$ to $-2Y/3$. Once the axial stress deviator reaches $-2Y/3$, unloading follows the slope of the hydrostat (modulus K), as shown in Fig. 3.4.

As above, the pressure–density relation is referred to as the hydrostat. The linear hydrostat is $p(\varepsilon) = -K\varepsilon$. In general, the hydrostat may be nonlinear $p(\varepsilon) = f(\varepsilon)$ and the stress deviators then add to the hydrostat to give the stress. An examination of Fig. 3.4 shows that with the combined effects of loading and unloading the whole region between $\pm 2Y/3$ of the hydrostat is achievable stress states. Thus, for a given strain ε, any stress σ between $p - 2Y/3$ and $p + 2Y/3$ can be realized by an appropriate choice of the loading path. This observation demonstrates that for a material exhibiting plasticity the stress state is path dependent (or history dependent) and cannot be defined solely in terms of current strain. For uniaxial strain loading, the stress state can be uniquely identified in terms of the current strain ε_{11} and the (signed) plastic strain ε_{11}^p. Notice that the stress state is not uniquely identified by the current strain ε_{11} and the equivalent plastic strain ε_p;

the equivalent plastic strain is the accumulation of the absolute value of the plastic strain increment and thus does not distinguish loading and unloading information.

Other solved examples in elastic-plastic response will be presented as they are needed in this book, but we summarize with the results for uniaxial strain:

THEOREM 3.3. *For an elastic-perfectly plastic material, in uniaxial stress the initial loading occurs with a slope of the Young's modulus E. Plastic yield occurs at stress $\sigma_{11} = Y$. After yield, the stress tensor remains constant and the entire increment of strain is plastic. In uniaxial strain, the initial loading occurs with slope $\lambda + 2\mu$, where $\lambda + 2\mu \geq E$. Yield is reached at the Hugoniot elastic limit $\sigma_{HEL} = \{(1-\nu)/(1-2\nu)\}Y$ with strain $\varepsilon_{HEL} = Y/(2\mu)$. Post yield loading occurs with a slope of the bulk modulus $K \leq \lambda + 2\mu$. One-third of the deformation increment is elastic and two-thirds is plastic.*

3.6. Various Yield Surfaces

Though we called our yield criterion in previous sections a yield surface, we only really used some yield points. To address combined states of stress that arise in complicated deformation, it is necessary to have a yield surface to describe how the various components of the stress tensor are "weighted" with each other to allow a single value for a yield criterion. The yield surface discussed above was the von Mises yield surface, which was shown to be represented by the equation (Eq. 3.51)

$$f(\sigma_{ij}) = J_2 - \frac{Y^2}{3} \quad \text{or} \quad f = \sqrt{3J_2} - Y, \quad \text{where again} \quad J_2 = \frac{1}{2}s_{ij}s_{ij}. \quad (3.75)$$

The von Mises yield surface depends only on J_2; $f(\sigma_{ij}) = f(J_2(\sigma_{ij}))$.

Historically, the first yield surface to be discussed was the Tresca yield surface, which states that plastic flow occurs when the maximum shear stress reaches a predefined value. This approach was taken because it was known that shear, for example with the displacement given by $\vec{u} = u_x(z)\hat{e}_z$ driven by a shear stress σ_{xz}, transitioned from elastic deformation to plastic flow at a certain stress and the subsequent flow was in shear, meaning it maintained the same displacement form. Thus the stress tensor and the strain tensor look something like

$$\sigma = \begin{pmatrix} 0 & 0 & Y_{\text{shear}} \\ 0 & 0 & 0 \\ Y_{\text{shear}} & 0 & 0 \end{pmatrix}, \quad \varepsilon = \begin{pmatrix} 0 & 0 & \varepsilon_{13} \\ 0 & 0 & 0 \\ \varepsilon_{13} & 0 & 0 \end{pmatrix}, \quad (3.76)$$

where we have chosen x and z as variables and directions to be consistent with the next paragraph and $\varepsilon_{13} = \frac{1}{2}\partial u_x/\partial z$. As the deflection continues, the strain is plastic, but it stays in the shearing form.

Our intent is to apply this observation to a general stress state. This paragraph is a rather tedious computation to arrive at Eq. (3.83) and can be skipped as only the final result is used subsequently. We begin with a reference frame where the stress is principal,

$$\sigma = \begin{pmatrix} \sigma_{11} & 0 & 0 \\ 0 & \sigma_{22} & 0 \\ 0 & 0 & \sigma_{33} \end{pmatrix}. \quad (3.77)$$

The three principal values are distinct; otherwise it becomes a two-dimensional calculation that we perform at the end of the paragraph. In what coordinate frame does this stress tensor appear with a maximum shear, which we will interpret as the frame with the largest off-diagonal stress component? The traction for a normal vector $\vec{n}$ is given by $\vec{t}^T = \vec{n}^T \sigma$, and we can separate this into a normal component and a shear component, $\vec{t} = t_n \vec{n} + t_s \vec{s}$, where $\vec{n} \cdot \vec{s} = 0$. These components are given by

$$t_n = \vec{n} \cdot \vec{t} = \vec{n}^T \sigma \vec{n} = \sum_{i=1}^{3} n_i^2 \sigma_{ii}, \quad t_s^2 = |\vec{t}|^2 - t_n^2 = \sum_{i=1}^{3} n_i^2 \sigma_{ii}^2 - \left(\sum_{i=1}^{3} n_i^2 \sigma_{ii} \right)^2. \quad (3.78)$$

We wish to maximize t_s with respect to $\vec{n}$ subject to the constraint that $\vec{n}$ is a unit vector. This is most easily accomplished through Lagrange multipliers, where we find the location where the gradient of the function to be maximized points in the same direction as the normal to the constraint surface. Hence, $\partial t_s^2 / \partial n_j = \lambda \partial |\vec{n}|^2 / \partial n_j$, where λ is an arbitrary constant. This provides three equations (one for each i),

$$n_i \{ \sigma_{ii}^2 - 2 (\sum_{j=1}^{3} n_j^2 \sigma_{jj}) \sigma_{ii} - \lambda \} = 0 \quad \text{(no sum on } i). \quad (3.79)$$

If any $n_i = 1$, then since $|\vec{n}| = 1$, we conclude $t_s = 0$, which is not of interest (no shear). So at least two components of $\vec{n}$ are nonzero. We want to show, in fact, that *only* two components are nonzero. Assume all three n_i are nonzero. Subtracting the i and j equations after multiplying each by the opposite n_i yields

$$n_i n_j (\sigma_{ii} - \sigma_{jj}) \{ \sigma_{ii} + \sigma_{jj} - 2 n_i^2 \sigma_{ii} - 2 n_j^2 \sigma_{jj} \} = 0 \quad \text{(no sum on } i, j). \quad (3.80)$$

The expression in parentheses is nonzero, since we assumed the principal stresses are distinct, so the expression in brackets is zero. There are three such equations,

$$\begin{pmatrix} 1 - 2n_1^2 & 0 & 1 - 2n_3^2 \\ 1 - 2n_1^2 & 1 - 2n_2^2 & 0 \\ 0 & 1 - 2n_2^2 & 1 - 2n_3^2 \end{pmatrix} \begin{pmatrix} \sigma_{11} \\ \sigma_{22} \\ \sigma_{33} \end{pmatrix} = 0. \quad (3.81)$$

The only way this can occur is if the determinant of the coefficient matrix is zero,

$$\det(\text{above matrix}) = 2(1 - 2n_1^2)(1 - 2n_2^2)(1 - 2n_3^2). \quad (3.82)$$

Thus, at least one $n_i = \pm \frac{1}{\sqrt{2}}$. If one n_i equals that value, then another n_j must equal that to satisfy Eq. (3.80), and from that we conclude that since $|\vec{n}| = 1$ the third $n_k = 0$, which contradicts our assumption that all n_i are nonzero. Thus, one n_k equals zero and the problem of maximization decouples the coordinate directions into a rotation in the 1–2, 1–3, or 2–3 plane. With $Q = \begin{pmatrix} \cos(\theta) & -\sin(\theta) \\ \sin(\theta) & \cos(\theta) \end{pmatrix}$ the off-diagonal shear term is $(\sigma_{ii} - \sigma_{jj}) \sin(\theta) \cos(\theta)$ and the maximum is at $\pm 45°$ rotation with maximum value $|\sigma_{ii} - \sigma_{jj}|/2$. Thus,

$$\text{maximum shear stress} = \frac{1}{2} \max \{ |\sigma_{11} - \sigma_{22}|, |\sigma_{22} - \sigma_{33}|, |\sigma_{33} - \sigma_{11}| \}. \quad (3.83)$$

This argument can also be made and visualized with Mohr's circles.

For uniaxial stress at yield, $\sigma_{11} = Y$, $\sigma_{22} = \sigma_{33} = 0$, and the maximum shear stress is $Y/2$. This defines the Tresca yield criterion,

$$\text{maximum shear stress} = \frac{Y}{2}, \quad (3.84)$$

which produces the Tresca yield function for the principal stres tensor,

$$f(\sigma_{ij}) = \max \{ |\sigma_{11} - \sigma_{22}|, |\sigma_{22} - \sigma_{33}|, |\sigma_{33} - \sigma_{11}| \} - Y, \quad (3.85)$$

where $f = 0$ on the yield surface.

Let us examine the associated flow rule. In three dimensions at yield, if we rotate to a principal stress orientation with $\sigma_{11} \geq \sigma_{22} \geq \sigma_{33} = \sigma_{11} - Y$, the stress there and the stress in the maximum shear frame is

$$
\sigma = \begin{pmatrix} \sigma_{11} & 0 & 0 \\ 0 & \sigma_{22} & 0 \\ 0 & 0 & \sigma_{11} - Y \end{pmatrix}, \qquad Q = \begin{pmatrix} \frac{1}{\sqrt{2}} & 0 & -\frac{1}{\sqrt{2}} \\ 0 & 1 & 0 \\ \frac{1}{\sqrt{2}} & 0 & \frac{1}{\sqrt{2}} \end{pmatrix},
$$

$$
Q\sigma Q^T = \begin{pmatrix} \frac{1}{2}(\sigma_{11} + \sigma_{33}) & 0 & \frac{1}{2}(\sigma_{11} - \sigma_{33}) \\ 0 & \sigma_{22} & 0 \\ \frac{1}{2}(\sigma_{11} - \sigma_{33}) & 0 & \frac{1}{2}(\sigma_{11} + \sigma_{33}) \end{pmatrix} = \begin{pmatrix} \sigma_{11} - Y/2 & 0 & Y/2 \\ 0 & \sigma_{22} & 0 \\ Y/2 & 0 & \sigma_{11} - Y/2 \end{pmatrix}.
$$

$$(3.86)$$

The Tresca yield surface is

$$
f(\sigma_{ij}) = \sigma_{11} - \sigma_{33} - Y \tag{3.87}
$$

and the associated flow rule says

$$
\frac{\partial f}{\partial \sigma_{ij}} = \begin{pmatrix} 1 & 0 & 0 \\ 0 & 0 & 0 \\ 0 & 0 & -1 \end{pmatrix}, \qquad Q\frac{\partial f}{\partial \sigma_{ij}}Q^T = \begin{pmatrix} 0 & 0 & 1 \\ 0 & 0 & 0 \\ 1 & 0 & 0 \end{pmatrix}. \tag{3.88}
$$

Thus we see that the plastic flow direction given by the associated flow is in the direction expected based on the simple shear argument. In the 45° rotated frame, the deformation is shear; when 45° shearing or failure planes are observed, it is indicative of a Tresca condition or some other J_3 dependence. While the von Mises yield condition only depends on J_2, the Tresca yield surface depends on J_2 and J_3. It can be written (Lubliner [136])

$$
4J_2^3 - 27J_3^2 - 9J_2^2Y^2 + 6J_2Y^4 - Y^6 = 0 \tag{3.89}
$$

(see Exercise 3.30 for restrictions). $J_3 = \det(s)$ is the third invariant of the stress deviator tensor. As described earlier, J_3 is positive for tensile stress states, negative for compressive stress states, and zero for pure shear stress states. An expression for the maximum shear stress in terms of J_2 and J_3 can be found in Exercise 3.22. The Tresca criterion is easy to work with when the symmetry of the problem allows the stress tensor to always be principal; in a general stress state it is more tedious as it is necessary to solve the eigenvalue problem and determine the rotation to the principal coordinate system. We use a Tresca state when the stresses are principal and it is algebraically simpler to consider the difference $\sigma_{11} - \sigma_{33}$ rather than $\frac{1}{2}s_{ij}s_{ij}$ and it is not necessary to determine the third principal stress σ_{22}.

The π plane projection is taken by looking down the principal stress space pressure axis (Exercise 3.21). In this view, which has removed the pressure, the von Mises yield surface is a circle and the Tresca yield surface is a hexagon, as in Fig. 3.5. The principal axes each appear at 120° apart. The von Mises circle touches the Tresca hexagon at the six stress values corresponding to uniaxial stress in compression and tension in each principal direction, as both the von Mises and Tresca yield criteria have been calibrated so that they produce the correct yield stress for uniaxial stress loading. Associated flow is not well defined for the Tresca yield surface at the corner points. However, owing to the symmetry of the uniaxial state, the flow rule at those points is radial; hence the von Mises and Tresca

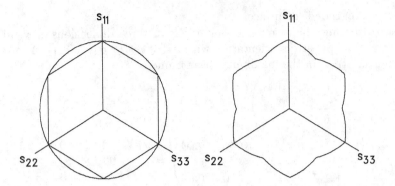

FIGURE 3.5. Von Mises (cirle) and Tresca (hexagon) yield surfaces in the π plane (left) and a possible yield surface for an aluminum alloy (right).

yield criteria both produce the same answer for the specific examples worked in Sections 3.4 and 3.5 above.

However, they are different and they produce different predictions of yield for a pure shear stress state and for combined stress states. If all the stress tensor components are zero save s_{12}, then $|s_{12}|$ is the maximum shear stress and the yield shear stresses are

$$\text{von Mises:} \quad |s_{12}| = \frac{Y}{\sqrt{3}}, \tag{3.90}$$

$$\text{Tresca:} \quad |s_{12}| = \frac{Y}{2}. \tag{3.91}$$

By using hollow specimens that were simultaneously deformed by extension and twist, Taylor and Quinney showed for three polycrystalline alloys (copper, aluminum, and mild steel) that yield occurred with the von Mises criterion for combined stress states [185]. But there is no theoretical reason why there would not be a J_3 dependence, and also shown in Fig. 3.5 is a postulated yield surface for Al 6061-T651 that has a $|J_3|$ dependence [212] (see Eq. 3.273). For this particular yield surface, tension and compression are the same and the entire yield surface is defined by an arc covering 30° in the π plane. If yield in tension and compression are not the same, then the yield surface must be J_3 or pressure dependent. If it depends on J_2 and J_3 but not pressure, then the entire yield surface is defined by a 60° arc in the π plane.

Some materials exhibit a pressure dependence in strength and so there will be occasion to use yield criteria that include pressure. The one that will be used in this text is referred to as a Drucker–Prager criterion, where the second invariant of the stress is assumed to be pressure dependent. In particular, we write it as

$$\sqrt{3J_2} = Y_0 + bp. \tag{3.92}$$

Another commonly referenced pressure-dependent yield criterion is the Mohr–Coulomb criteria, which comes from the geometry of Mohr's circles and says that yield occurs when the Mohr's circle touches and is tangent to the line passing through $(\sigma_n, \tau) = (0, c)$ and making angle $-\phi$ with the σ_n-axis. The equation representing this line is $\tau = c - \sigma_n \tan(\phi)$. Following Lubliner, if the principal stresses have been computed $\sigma_{11} \geq \sigma_{22} \geq \sigma_{33}$, the criterion can be written (where $c = Y/2$)

$$\sigma_{11} - \sigma_{33} + (\sigma_{11} + \sigma_{33})\sin(\phi) = Y\cos(\phi), \tag{3.93}$$

where ϕ is the "angle of internal friction." This can be written

$$\sigma_{11} - \sigma_{33} + \left\{\frac{1}{3}(\sigma_{11} - \sigma_{22} + \sigma_{33} - \sigma_{22}) + \frac{2}{3}(\sigma_{11} + \sigma_{22} + \sigma_{33})\right\}\sin(\phi) = Y\cos(\phi). \tag{3.94}$$

The sum of the principal stresses is simply $-3p$, so this expression can be written

$$\sigma_{11} - \sigma_{33} + \frac{1}{3}(\sigma_{11} - \sigma_{22} + \sigma_{33} - \sigma_{22})\sin(\phi) = Y\cos(\phi) + 2\sin(\phi)p. \tag{3.95}$$

The left-hand side is a function of stress deviators only and so can be written as a function of J_2 and J_3. The right-hand side shows the dependence on pressure. The Mohr–Coulomb criterion is different than the Drucker–Prager criterion even though both have a linear dependence on pressure. This can be seen most easily by noting that if there is no pressure dependence ($b = 0$ or $\phi = 0$), the Drucker–Prager criterion collapses to a von Mises criterion and the Mohr-Coulomb criterion collapses to a Tresca criterion.

Given different yield surfaces, the question also arises whether there will be different flow rules. Often for simplicity the von Mises flow rule is used for all these yield surfaces. Otherwise, the Tresca yield surface has undefined flow direction for associated flow, since the yield surface is not differentiable with respect to stress at the corners of the hexagon. If associated flow is assumed, for the Drucker–Prager yield surface the computed flow direction is (see Eqs. 3.52)

$$\frac{1}{d\Lambda}d\varepsilon_{ij}^p = \frac{\partial}{\partial\sigma_{ij}}\left(\sqrt{3J_2} - bp - Y_0\right) = \sqrt{\frac{3}{4J_2}}\,s_{ij} + \frac{b}{3}\delta_{ij}. \tag{3.96}$$

This associated flow rule is seen to have a volumetric component to plastic strain when b is nonzero. For associated flow rules that include J_3, the derivative of J_3 with respect to the stress component σ_{ij} is the corresponding subdeterminant of the stress deviator tensor plus $J_2\delta_{ij}/3$,

$$\frac{\partial J_3}{\partial\sigma_{11}} = s_{22}s_{33} - s_{23}s_{32} + \frac{J_2}{3}, \qquad \frac{\partial J_3}{\partial\sigma_{12}} = -(s_{21}s_{33} - s_{23}s_{31}), \tag{3.97}$$

etc. There is no volumetric component in this flow direction as $\partial J_3/\partial\sigma_{ii} = 0$.

3.7. Rigid Plasticity

Elastic-plastic problems are typically difficult to solve as the current stress depends on the complete deformation history. For a full elastic-plastic problem, it is necessary to solve for the rate of deformation tensor and then separate it into elastic and plastic parts. The separation depends on the deformation history and is very tedious to perform. That is why today we use large-scale numerical simulations for problems of this type. In this text, in addition to the above, we also explicitly solve a few other elastic-plastic problems, most notably the cavity expansion problem. However, on a handful of occasions, in order to overcome the difficulty of solving the full elastic-plastic problem, the rigid plastic approximation is employed.

As a first step, however, assuming that the plastic deformation rate $\dot{\varepsilon}_{ij}^p = D_{ij}^p$ is known, where again

$$D_{ij} = \frac{1}{2}\left(\frac{\partial v_i}{\partial x_j} + \frac{\partial v_j}{\partial x_i}\right) = D_{ij}^e + D_{ij}^p, \qquad (3.98)$$

for a von Mises yield surface it is possible to determine the deviatoric stress tensor. In particular, for associated flow, Eq. (3.53) says

$$D_{ij}^p = \Lambda s_{ij}, \qquad (3.99)$$

where Λ is a constant (using $\dot{\Lambda}$ as our constant would be more consistent with the $d\Lambda$ notation above, but we will just use Λ, as it is, after all, a constant). The von Mises yield surface is given by

$$J_2 = \frac{1}{2}s_{ij}s_{ij} = \frac{Y^2}{3}. \qquad (3.100)$$

These two equations combine to give

$$s_{ij}s_{ij} = \frac{1}{\Lambda^2}D_{ij}^p D_{ij}^p = \frac{2Y^2}{3} \quad \Rightarrow \quad \frac{1}{\Lambda^2} = \frac{2Y^2}{3D_{ij}^p D_{ij}^p}. \qquad (3.101)$$

Now we can explicitly write the stress deviators in terms of plastic strain rate,

$$s_{ij} = \frac{1}{\Lambda}D_{ij}^p = \frac{\sqrt{\frac{2}{3}}Y D_{ij}^p}{\sqrt{D_{ij}^p D_{ij}^p}} = \frac{2Y}{3}\frac{\dot{\varepsilon}_{ij}^p}{\dot{\varepsilon}_p}. \qquad (3.102)$$

Since the plastic strain rate is divergence free for von Mises flow ($D_{ii}^p = 0$), the stress deviator is deviatoric, as expected.

Now let us consider rigid plasticity. In rigid plasticity the assumption is that *all deformation is plastic* – that is,

$$D_{ij}^p = D_{ij}. \qquad (3.103)$$

An equivalent assumption is that the elastic moduli are infinite – thus, as soon as there is any deformation, the stresses are on the yield surface. If one has a velocity field v_i where the deformation is incompressible,

$$\operatorname{div}(\vec{v}) = \frac{\partial v_x}{\partial x} + \frac{\partial v_y}{\partial y} + \frac{\partial v_z}{\partial z} = 0, \qquad (3.104)$$

then by assumption in rigid plasticity $D_{ij}^p = D_{ij}$ and thus one can explicitly compute the deviatoric stress tensor from the velocity field. We explore the accuracy of this assumption in Section 3.16. Note that regarding the total stress tensor σ_{ij}, the pressure $p(\vec{x})$ can be a function of the spatial location and is not determined by the rigid plasticity assumption; it is typically determined by boundary conditions.

3.8. Energy Dissipation through Plastic Flow

In Chapter 2 the energy balance was presented as (Eq. 2.34)

$$\rho\frac{DE}{Dt} = \sigma_{ij}D_{ij} + \rho r - \frac{\partial q_i}{\partial x_i}. \qquad (3.105)$$

The amount of energy change that is due to deformation is given by the first term on the right-hand side. The change in energy is a work rate, or

$$\frac{dW}{dt} = \sigma_{ij}D_{ij}. \qquad (3.106)$$

Since the rate of deformation tensor can be separated into elastic and plastic parts,

$$D_{ij} = D_{ij}^e + D_{ij}^p, \tag{3.107}$$

the energy dissipated due to plastic deformation can be written

$$\frac{dW_p}{dt} = \sigma_{ij} D_{ij}^p. \tag{3.108}$$

In general it is not easy to determine this value, since it requires a complete solution to the problem. Plastic flow is assumed dissipative ($\sigma_{ij} D_{ij}^p > 0$ if any $D_{ij}^p \neq 0$): if the yield surface is convex and associated flow is used, it is always dissipative; otherwise it is necessary to check.

For a material satisfying a von Mises yield surface with associated flow, placing the result for the stress deviators from Eq. (3.102) into the expression for plastic work rate gives

$$\frac{dW_p}{dt} = (-p\delta_{ij} + s_{ij}) D_{ij}^p = s_{ij} D_{ij}^p = Y \sqrt{\frac{2}{3} D_{ij}^p D_{ij}^p} = Y \dot{\varepsilon}_p. \tag{3.109}$$

To use this result it is necessary to compute the equivalent plastic strain for the specific problem. The work rate is energy dissipated by the plastic flow. Most, if not all, of this energy appears as heat in the deforming body.[4] Assuming complete conversion to heat, it is possible to compute the temperature rise using $\Delta E = C_v \Delta T$, where E is the specific energy. Assuming a constant flow stress, the result is

$$\frac{Y \varepsilon_p}{\rho_0} = C_v \Delta T \quad \Rightarrow \quad \Delta T = \frac{Y \varepsilon_p}{\rho_0 C_v}. \tag{3.110}$$

As examples, for steel with a flow stress of 1 GPa, $\rho_0 = 7.85 \text{ g/cm}^3$, and $C_v = 450 \text{ J/kg K}$ the temperature rise per unit plastic strain is 283 K or 2.83 K per 1% plastic strain; for aluminum with a flow stress of 380 MPa, $\rho_0 = 2.70 \text{ g/cm}^3$, and $C_v = 890 \text{ J/kg K}$ the temperature rise per unit plastic strain is 158 K or 1.58 K per 1% plastic strain.

Heat is conducted by metals. Our studies are of high-speed, short-time events, and we should wonder about the role of heat conduction. Conduction is given by $\vec{q} = -k \operatorname{grad}(T)$, where k is heat conductivity. Putting this expression into the energy balance, and just focusing on heat conduction, says

$$\rho \frac{\partial E}{\partial t} = -\operatorname{div}(-k \operatorname{grad}(T)) \quad \Rightarrow \quad \rho C_v \frac{\partial T}{\partial t} = k \nabla^2 T \quad \Rightarrow \quad \rho C_v \frac{\partial T}{\partial t} = k \frac{\partial^2 T}{\partial x^2}, \tag{3.111}$$

where the last equation is for one dimension, which will have the slowest conduction. The heat kernel in one dimension is [35]

$$\phi(x,t) = \frac{1}{2a\sqrt{\pi t}} \exp\left\{ -\frac{x^2}{4a^2 t} \right\}, \qquad a^2 = \frac{k}{\rho C_v}. \tag{3.112}$$

---

[4]The amount of energy from plastic deformation converted to heat is a research topic. The fraction of plastic deformation energy that goes into heat is sometimes denoted β and is typically thought to be between 0.85 and 1. Recent research has implied that β may be path and rate dependent. If $\beta < 1$ the implication is that there are plastic deformation mechanisms where the stored energy is in some sense recoverable, though they don't give rise to stress. See [184, 213].

With an initial hot spot of T_0 over the extent of $(-L/2, L/2)$ and zero elsewhere, the temperature vs. time can be obtained by convolution

$$T(x,t) = \int_{-L/2}^{L/2} T_0 \phi(x - \xi, t)d\xi = \frac{T_0}{2}\left\{ \mathrm{erf}\left(\frac{L/2 - x}{2a\sqrt{t}}\right) + \mathrm{erf}\left(\frac{L/2 + x}{2a\sqrt{t}}\right)\right\}$$ (3.113)

$$= T_0\psi(L, x, t), \qquad\qquad \mathrm{erf}(x) = \frac{2}{\sqrt{\pi}}\int_0^x e^{-t^2} dt.$$

Let's consider the center point, $x = 0$. The temperature there is

$$\frac{T(0,t)}{T_0} = \psi(L, 0, t) = \mathrm{erf}\left(\frac{L}{4a\sqrt{t}}\right).$$ (3.114)

In two dimensions, the heat kernel is $\phi(x, y, t) = \phi(x, t)\phi(y, t)$ and in three dimensions the heat kernel is $\phi(x, y, z, t) = \phi(x, t)\phi(y, t)\phi(z, t)$. If we choose to have, in two dimensions, an initial square of side length L at T_0 and everywhere else zero, and in three dimensions an initial cube with edge length L at T_0 and everywhere else zero, then because the heat kernel is multiplicative, the solution is too, giving

$$T(x, y, t) = T_0\psi(L, x, t)\psi(L, y, t), \quad T(x, y, z, t) = T_0\psi(L, x, t)\psi(L, y, t)\psi(L, z, t).$$ (3.115)

Let's focus on how long it takes the temperature at the center point to drop to half its initial value. The temperature vs. time is given, where $d = 1$, 2, and 3 is the dimension of space,

$$T = T_0\psi(L, 0, t)^d = T_0\left\{\mathrm{erf}\left(\frac{L}{4a\sqrt{t}}\right)\right\}^d.$$ (3.116)

If we ask how long it takes for the temperature to drop to half its initial value, we obtain

$$t = \left(\frac{L}{4a\,\mathrm{erf}^{-1}((T/T_0)^{1/d})}\right)^2 = \frac{\rho C_v L^2}{16k}\left(\frac{1}{\mathrm{erf}^{-1}((T/T_0)^{1/d})}\right)^2.$$ (3.117)

For $T/T_0 = \frac{1}{2}$, the parenthesis squared term on the right is 4.34 for one dimension, 1.83 for two dimensions, and 1.26 for three dimensions.[5] Thus, for steel ($k = 50$ W/m K, where 1 watt $= 1$ joule per second), the times it takes for a localized square temperature spike of width $L = 1$ mm to cool to half its original temperature for the respective dimensions are

$$t = 19\,\mathrm{ms} \quad (1\mathrm{D}), \quad t = 8\,\mathrm{ms} \quad (2\mathrm{D}), \quad t = 5.5\,\mathrm{ms} \quad (3\mathrm{D}).$$ (3.118)

For aluminum ($k = 230$ W/m K) the respective times are 2.8, 1.2, and 0.85 ms. If $L = 1$ cm, the times are 100 times greater than these. Since most events we discuss in this book are over in less than 1 millisecond, we view impact events as being fast enough that heat does not conduct and thus stays locally where it is produced or deposited. Typically tensile tests for strength characterization are performed either slowly or at high rates. The only temperature concern about low-rate tests, which are usually viewed as isothermal (meaning the heat is conducted away), is what happens right at material failure, since there may be rapid localized strain at that time.

---

[5]For the error function, $\mathrm{erf}(0.48) = 0.50$, $1/\sqrt{2} \approx 0.71$ and $\mathrm{erf}(0.74) = 0.71$, $1/2^{1/3} \approx 0.79$ and $\mathrm{erf}(0.89) = 0.79$.

3.9. Energy Stored in Elastic Deformation

The energy stored in elastic deformation can be found by computing the work done in deforming the material. This energy is recoverable, and as such is often referred to as potential energy. If it is assumed that we have a box of material with rigid walls that can only deform in one direction, then mass conservation says

$$\rho LA = \rho_0 L_0 A, \tag{3.119}$$

where L is the length of the box and A is the constant cross section. As material is deformed, the work is given by integrating the force times the displacement over the displacement from 0 to x,

$$W = \int_0^x (\text{force})d(\text{displacement}). \tag{3.120}$$

For our box of material, the displacement is given by $L - L_0$ and the force is given by σA,

$$W = \int_0^x \sigma A d(L - L_0) = \int_{L_0}^L \sigma A dL = \rho_0 A L_0 \int_{\rho_0}^\rho \sigma d(1/\rho). \tag{3.121}$$

The term in front of the integral is the material mass in the box, and thus the specific work done (work per unit mass), which is the specific stored (or potential) elastic energy E_e, is given by

$$E_e = \frac{W}{\rho_0 A L_0} = \int_{\rho_0}^\rho \sigma d(1/\rho). \tag{3.122}$$

It should be noted that the integral is in terms of $1/\rho$, the specific volume.

For small strains the energy is often written in terms of the linear strain with $\varepsilon = \rho_0/\rho - 1$, giving $d\varepsilon = \rho_0 d(1/\rho)$, so that

$$E_e = \frac{1}{\rho_0} \int_0^\varepsilon \sigma d\varepsilon. \tag{3.123}$$

Thus, for a small volumetric change where the stress is given by $p = -K\varepsilon$, the stored elastic energy is

$$\rho_0 E_e = \int_0^\varepsilon K\varepsilon d\varepsilon = \frac{1}{2}K\varepsilon^2 = \frac{1}{2}p\varepsilon = \frac{1}{2}\frac{p^2}{K}, \tag{3.124}$$

where we have written $\rho_0 E_e$ since in small-strain elasticity theory the energy is usually written as energy per initial unit volume rather than specific energy (energy per unit mass). In a similar fashion, the amount of energy that is stored for the elastic strain tensor can be computed. With the stress given by $\sigma_{ij} = \lambda \varepsilon_{kk}^e \delta_{ij} + 2\mu\varepsilon_{ij}^e$, the elastic energy is

$$\rho_0 E_e = \int_0^{\varepsilon_{ij}} \sigma_{ij} d\varepsilon_{ij}^e = \frac{1}{2}\lambda(\varepsilon_{kk}^e)^2 + \mu\varepsilon_{ij}^e\varepsilon_{ij}^e = \frac{1}{2}\frac{p^2}{K} + \frac{s_{ij}s_{ij}}{4\mu}. \tag{3.125}$$

If an elastic-plastic material obeys a von Mises yield condition, then while yielding the stress deviators are on the yield surface given by $3J_2 = (3/2)s_{ij}s_{ij} = Y^2$. Thus, the stored elastic energy in the deviatoric part of the deformation is

$$\rho_0 E_e = \frac{s_{ij}s_{ij}}{4\mu} = \frac{J_2^2}{2\mu} = \frac{Y^2}{6\mu}. \tag{3.126}$$

This deviatoric energy is then added to the spherical (pressure) energy term to provide the total stored elastic energy.

3.10. Thermal Terms

In the elastic and elastic-plastic examples so far, there has been a relation between stress and strain of the material. Energy was not part of the relation. However, the effect of thermal expansion is usually considered as part of elasticity theory. Changes in temperature produce expansion of a material and, if the material is confined, lead to stress.

The classical experimental result of thermal expansion of metals is that, given an initial length L_0 for a metal bar, with a temperature change ΔT the new length of the metal bar will be

$$L = (1 + \alpha \Delta T)L_0, \tag{3.127}$$

where α is the linear coefficient of thermal expansion. We ask the question: for a material of length L, which was originally at length L_0, how does the thermal expansion affect the strain that gives rise to stress? Such expansion affects the strain as follows:

$$\varepsilon = \frac{L - L_0}{L_0}$$

$$\rightarrow \frac{L}{(1 + \alpha \Delta T)L_0} - 1 \approx \frac{L}{L_0}(1 - \alpha \Delta T) - 1 = (\varepsilon + 1)(1 - \alpha \Delta T) - 1 \approx \varepsilon - \alpha \Delta T. \tag{3.128}$$

Thus, as the temperature goes up, the effective strain is reduced because of thermal expansion. For linear elasticity, the adjustment due to temperature is

$$\sigma_{11} = (\lambda + 2\mu)\varepsilon_{11} + \lambda\varepsilon_{22} + \lambda\varepsilon_{33}$$
$$\rightarrow (\lambda + 2\mu)(\varepsilon_{11} - \alpha\Delta T) + \lambda(\varepsilon_{22} - \alpha\Delta T) + \lambda(\varepsilon_{33} - \alpha\Delta T) \tag{3.129}$$
$$= (\lambda + 2\mu)\varepsilon_{11} + \lambda\varepsilon_{22} + \lambda\varepsilon_{33} - (3\lambda + 2\mu)\alpha\Delta T$$

since the thermal expansion occurs in all directions. With $3\lambda + 2\mu = 3K$, for each σ_{ii} (no sum) term there is a $3K\alpha\Delta T$ term and thus the adjustment to the pressure is

$$p = -\frac{1}{3}\sigma_{ii} \rightarrow -\frac{1}{3}\sigma_{ii} + 3K\alpha\Delta T. \tag{3.130}$$

The general effect of increasing the temperature is to increase the compressive stress by the value shown. Including thermal terms, the elastic stress can be written

$$\sigma_{ij} = \lambda\varepsilon_{kk}\delta_{ij} + 2\mu\varepsilon_{ij} - 3K\alpha\Delta T\delta_{ij}. \tag{3.131}$$

3.11. A Discussion of Strain

Strain is a measure of deformed length relative to original length. In the laboratory, when a specimen is deformed, strain is typically measured over a gage length L_0, comparing the final length L to the gage length. There are various definitions of strain, all of which are the same for infinitesimal strains. The "engineering strain" or "Lagrangian strain" is given by the change in length divided by the original gage length,

$$\varepsilon_{11} = \frac{\Delta L}{L_0} = \frac{L}{L_0} - 1. \tag{3.132}$$

Another definition of strain is the "Eulerian strain," with the change of length compared to the final length of the gage,

$$\varepsilon_{11}^* = \frac{\Delta L}{L} = 1 - \frac{1}{L/L_0}. \tag{3.133}$$

For small strain, ΔL should be small compared to L_0 and so these values are roughly the same. Between these two definitions of strain lies a third. If $L(t)$ is the current gage length, then the "natural strain" or "logarithmic strain" is given by

$$\varepsilon_{11} = \int_{L_0}^{L_0+\Delta L} \frac{dL}{L} = \ln\left(1 + \frac{\Delta L}{L_0}\right). \tag{3.134}$$

It is straightforward to show that

$$\frac{\Delta L}{L_0 + \Delta L} \leq \ln\left(1 + \frac{\Delta L}{L_0}\right) \leq \frac{\Delta L}{L_0} \tag{3.135}$$

and that the inequality is strict if $|\Delta L| > 0$. For small strain, all these expressions are the same. Each of these strains can be written as a function of the other two, since all depend on L/L_0. Fundamentally, our measure of deformation is always the ratio of current length to initial length. Choice of a given strain expression is based on convenience and on the one that provides the simplest relation between the measured stress and strain.

If the strains are computed from the rate of deformation tensor D_{ij}, the question arises as to which strain is being computed. The rate of deformation tensor is evaluated in the Eulerian frame. If we take a length L_0 of material, fix one end at $x = 0$ and pull on the other end at a constant velocity V, the length of the bar is given by $L(t) = L_0 + Vt$. The velocity distribution within the bar is $v_x = Vx/L(t)$. Thus, the rate of deformation is

$$D_{11} = \frac{\partial v_x}{\partial x} = \frac{V}{L(t)}. \tag{3.136}$$

The strain is determined by integrating this with respect to time,

$$\varepsilon_{11} = \int_0^t D_{11} dt = \int_0^t \frac{V}{L(t)} dt = \ln\left(\frac{L(t)}{L_0}\right). \tag{3.137}$$

Thus, strains computed in this fashion are logarithmic strains or natural strains. In particular, the equivalent plastic strain is a logarithmic strain.

For high pressures, in material response modeling usually the pressure terms are separated from the stress deviator terms, the former being referred to as the "equation of state" and the latter the "constitutive model." Such a separation necessitates an expression for the volumetric strain ε_v,

$$\varepsilon_v = \frac{\Delta V}{V} \approx \varepsilon_{ii} = \varepsilon_{11} + \varepsilon_{22} + \varepsilon_{33}, \tag{3.138}$$

where V is a representative small volume of interest and ΔV is its change. Strains can be defined in an analogous manner to the above. In particular, mass conservation says

$$V\rho = (V_0 + \Delta V)\rho = V_0\rho_0 \tag{3.139}$$

and the three definitions of strain give

$$\varepsilon_v = \frac{\Delta V}{V_0} = \frac{\rho_0}{\rho} - 1, \tag{3.140}$$

$$\varepsilon_v = \int_{V_0}^{V_0 + \Delta V} \frac{dV}{V} = \ln\left(\frac{\rho_0}{\rho}\right), \tag{3.141}$$

$$\varepsilon_v = \frac{\Delta V}{V_0 + \Delta V} = 1 - \frac{\rho}{\rho_0}. \tag{3.142}$$

Therefore, the volumetric strain can be written in terms of density, and it is possible to write the stress as

$$\sigma_{ij} = -p(\rho)\delta_{ij} + s_{ij} + \text{thermal term}. \tag{3.143}$$

For example, for linear elasticity

$$\begin{aligned}
\sigma_{ij} &= (\lambda + 2\mu/3)\varepsilon_{kk}\delta_{ij} + 2\mu e_{ij}^e - 3K\alpha\Delta T\delta_{ij} \\
&= K\varepsilon_{kk}\delta_{ij} + 2\mu e_{ij}^e - 3K\alpha\Delta T\delta_{ij},
\end{aligned} \tag{3.144}$$

where again e_{ij}^e is the elastic deviator strain tensor, using the first strain above gives

$$\sigma_{ij} = K\left(\frac{\rho_0}{\rho} - 1\right)\delta_{ij} + 2\mu e_{ij}^e - 3K\alpha\Delta T\delta_{ij}. \tag{3.145}$$

This expression now writes the pressure p as a function of density,

$$p(\rho) = K\left(1 - \frac{\rho_0}{\rho}\right). \tag{3.146}$$

In the next two chapters there will be further discussion of strain and of equations of state.

3.12. Characterization of Real Materials

So far in this chapter, we have discussed the classical elasticity and plasticity *models* of material response. Real materials are not so simple. As an example of real materials, their complexity, and how they fit into the above models, we consider the constitutive characterization of four materials: 6061-T6 and 7075-T6 aluminum, which are used in armors, and a hard steel and a copper alloy used in the 7.62-mm (0.30-cal) APM2 armor piercing bullet. There will also be a discussion of damage modeling and a specific damage model will be presented. However, damage modeling is still in its infancy and, unlike the plasticity models which are fairly robust and widely used, the use of damage models is not well established.

The characterization of two aluminums, Al 6061-T6 and Al 7075-T6, highlights the property differences between two alloys. Tests were conducted on extruded material in the extrusion direction (i.e., longitudinal orientation). (In theory metals are isotropic, but due to their work-hardening behavior the extrusion process can introduce a slight anisotropy. We only mention this fact and do not pursue it.) Both quasi-static and dynamic material test results are presented. Dynamic tension and compression tests utilized the split-Hopkinson pressure bar (see Chapter 5 for discussion of this experiment). Dynamic tests were conducted at two different strain rates: approximately 200–300 s^{-1} and 1,000–1,300 s^{-1}. Quasi-static tension tests (approximate strain rate = 10^{-3} s^{-1}) were performed. Torsion tests were conducted at low strain rates (10^{-3} s^{-1}). Figure 3.6 shows specimen geometries

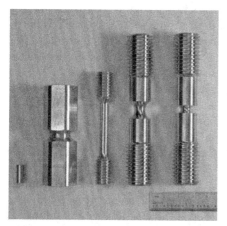

FIGURE 3.6. Material test specimens, left to right: split-Hopkinson pressure bar compression specimen, high rate torsion specimen, three round tensile test specimens with different notches to produce different stress triaxiality during test, tensile specimen and failed tensile specimen, showing necking.

used in the presented curves (left three specimens). An example of the increasing stress triaxiality caused by the notches is given in Table 3.5.

The presentation of the data for these tests requires some precursor discussion about what is measured and reported. In a tension test, the force at the ends of the specimen is measured and the elongation over a certain gage length is measured, though sometimes a strain gage is attached directly to the specimen (strain gages typically can only measure strains to 2 or 3% strain). The "engineering strain" is found by measuring the change in length divided by the initial length, $\varepsilon_{\text{eng}} = \Delta L / L_0$. The "engineering stress" is found by dividing the force by the original cross section of the specimen in the gage (narrow) region, $\sigma_{\text{eng}} = F / A_0$, which for small strain is an accurate representation of the stress at the gage section. However, it is not accurate for large strain, because as the test specimen stretches in the longitudinal direction it contracts laterally. When the engineering stress-engineering strain is plotted, the maximum value of the stress is referred to as the *ultimate tensile strength* (UTS, i.e., the maximum load divided by original cross-sectional area). The engineering stress begins to decrease when the specimen necks. "True stress" is defined as the force divided by the narrowest cross section, $\sigma = F / A$. If the gage section is plastically flowing, if necking or strain localization has not occurred, and if it is assumed that plastic flow is incompressible (volume conserving), then with A being the cross section and L the gage length, we have $AL = A_0 L_0$. The cross section at the gage location is $A = A_0 L_0 / L = A_0 / (1 + \varepsilon_{\text{eng}})$. Thus, the often reported "true stress" is given by multiplying the engineering stress by $A_0 / A = 1 + \varepsilon_{\text{eng}}$. If the engineering strain is 20%, then the engineering stress is multiplied by 1.20 to give the actual stress in the gage section. This derivation assumed there was no localization of the strain, or "necking;" if localization occurs, determination of the stress requires measurement of the cross section in the localized region. Similarly, to measure the true strain after necking it is necessary to measure the local deformation. Sometimes in the engineering literature, rather than engineering stress-engineering strain, data are presented as true stress-true strain. Unless localization is measured, true stress typically means $\sigma = \sigma_{\mathrm{eng}}(1 + \varepsilon_{\mathrm{eng}})$ and true strain is the logarithmic strain $\varepsilon = \ln(1 + \varepsilon_{\mathrm{eng}})$. If the cross-sectional area A at the neck or localized region is measured, then true stress is F/A and true strain is $\varepsilon = \ln(L/L_0) = \ln(A_0/A)$. These latter are what is reported, but these are approximations to the actual stress state, since after localization neither the stress nor strain is uniform across the gage section, as will be discussed later in this section. For torsion data, the engineering shear strain is twice the tensorial shear strain: $\gamma_{12} = 2\varepsilon_{12}$. We report the data this way since that is the way they are usually handled.

If we write the flow stress as $Y(\varepsilon_p)$, then the load or force during the tensile test is $F = AY = A_0 \exp(-\varepsilon_p)Y$. At some point a maximum load is reached – hence $\partial F/\partial \varepsilon_p = 0 = A(dY/d\varepsilon_p - Y)$; hence the elastic-plastic modulus equals the strength. Before maximum load, the loading is unique and uniform, because if at any point along the gage length the deformation localized, then it could support more load than its neighboring points and would stop deforming till the rest of the gage length caught up to it. After ultimate load, however, since the total load supported decreases, when any point localizes (necks) it now reduces the total load that can be supported by the gage section, so the rest of the gage length ceases to deform plastically and all deformation is localized at the initially necked point. Before the ultimate or maximum load, the deformation in the gage section is uniform and unique; after ultimate load, the solution bifurcates or is no longer unique and localization can occur at any point along the uniform gage length.

Figure 3.2 shows the results of tension tests at two strain rates for the two aluminums. The quasi-static and high-strain-rate curves are quite similar, showing that aluminum does not have a great deal of strain-rate dependence in its plastic flow response. The Al 7075-T6 alloy shows higher strength and somewhat lower ductility than the Al 6061-T6. Where the loading curves end is where the material in the tension test had localized necking and then failure. These materials work harden, meaning the strength increases with strain. Over these strains, the primary reason for the downturn of the curve is the geometric gage section area decrease described above (compare with Fig. 3.13 to see just the effect of work hardening and thermal softening with no geometric effect). Representative dynamic compressive stress-strain curves are shown in Fig. 3.7 for each alloy. As with the tensile data, the strain rates listed are approximate. In the tensile tests the specimen was tested to failure, but in the compressive test the length of the striker bar determined the amount of loading. In this case the resulting maximum strain was 9%. In a compression test the specimen bulges outward rather than narrows during the test. However, the same comments about the actual stress apply, except that now the strain is negative; thus, the multiplier on the engineering stress is 0.91 to give true stress. The torsion test results are given in Fig. 3.8. They are for samples oriented in the longitudinal direction (i.e. parallel to the extrusion axis). The samples were tested to failure. Measured tensile properties for the alloys are summarized in Table 3.1. The values listed are average values based on multiple tests per alloy. Performing both uniaxial stress and shear stress tests allows selection of a von Mises or a Tresca yield surface for modeling, though it is expected that neither will exactly match the data (Exercise 3.31).

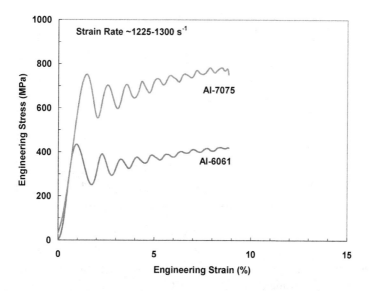

FIGURE 3.7. Comparison of the dynamic compressive stress-strain behavior for the aluminum alloys.

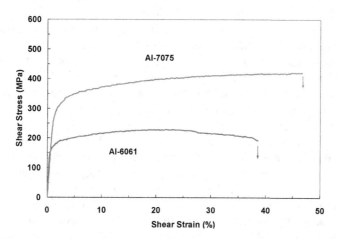

FIGURE 3.8. Comparison of the torsion behavior for the aluminum alloys. The tests are at low strain rates (i.e., 10^{-3} s^{-1}).

TABLE 3.1. Quasi-static and dynamic tension, compression, and torsion test results for the aluminum alloys.

Material	Tension Quasi-static $(10^{-3}\,\text{s}^{-1})$			Tension Dynamic $(1{,}200\,\text{s}^{-1})$		Compression Dynamic $(1{,}300\,\text{s}^{-1})$		Torsion Quasi-static $(10^{-3}\,\text{s}^{-1})$	
	Y_0 (MPa)	UTS (MPa)	ε_f (%)	σ_{max} (MPa)	ε_f (%)	σ_{max} (MPa)	ε_{max} (%)	τ_{max} (MPa)	γ_f (%)
Al 6061-T6	262	300	25	338	26	417	9	231	36
Al 7075-T6	538	559	19	565	18	793	9	407	46

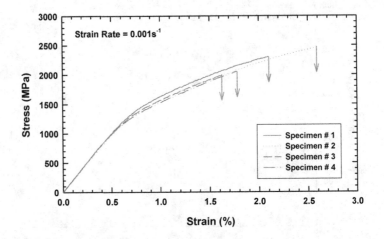

FIGURE 3.9. Tensile engineering stress-strain data for the APM2 steel core. (From Chocron et al. [**38**].)

For the two materials used in the APM2, the material properties were found at the relevant scale by performing quasi-static tension tests using test specimens machined directly from the rounds. The 1070 tool steel core is hardened steel (Rockwell-C hardness Rc 62–65). Tensile stress-strain data for the steel core are shown in Fig. 3.9. The downward arrows indicate when the core material fractured in tension. The material has very high strength and very low ductility (i.e., the fracture strain is small). Because the ductility is low there is no necking or localization in determining the stresses and strains. (It is not possible to tell from a stress-strain curve whether localized necking occurred; such localization is determined by examining the test specimen.) Given that there was little ductility and the small strains, it is not necessary to distinguish between true stress and strain and engineering stress and strain. For the gilding metal, nonstandard specimens were fabricated directly from the jacket of the round. Tensile stress-strain data for the gilding metal are shown in Fig. 3.10. This material showed necking/localization in the test sections; the resulting engineering stress-strain relationship shows small strains, but the actual strains once the localization begins are quite large. A microscopic examination of the fracture/failure surfaces gave an average through-section fracture area for test specimens of $A_f = 1.55$ mm^2. Measurements of the initial cross section gave an area of $A_0 = 5.29$ mm^2. Assuming that the strain before failure was plastic and thus the volume was conserved, we roughly have $A_f L_f = A_0 L_0$ (roughly because the deformation field is in fact quite complex). This assumption can be combined with the logarithmic strain described above to estimate the equivalent plastic strain at the time of fracture,

$$\varepsilon_p^f = \ln\left(\frac{L_f}{L_0}\right) = \ln\left(\frac{A_0}{A_f}\right) = 1.23. \qquad (3.147)$$

Thus, the actual strain at the time of failure is much larger than the engineering strain $(0.04 - 0.05)$ from the gage attached to the specimen.

We now return to the tensile test for very large strains. Bridgman performed extensive experiments looking at the deformation of metals under high pressures.

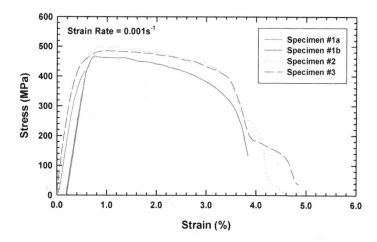

FIGURE 3.10. Tensile engineering stress-strain data for the APM2 gilding metal (jacket material).

He noted that when subjected to hydrostatic pressures on the order of 2 to 3 GPa, steel became much more ductile, with the area of necking down in a cylindrical tensile specimen sometimes exceeding a logarithmic strain of 4 (i.e., $\ln(A_0/A) > 4$, or $A < A_0/e^4 = A_0/55$, or a 98% reduction in cross-sectional area). In his book collecting much of this work, the first chapter is a detailed discussion of the tensile specimen and the stress distribution within the specimen [31].

For small strains and no necking, the stress state and strain state within the specimen can be approximated by our solution to the uniaxial stress problem. However, as strains become larger (typically around 0.20 strain for steels) the cylindrical tensile specimen necks down, and the stress and strain behavior becomes more complex. The maximum strain occurs at the location of maximum necking, a symmetry plane at the center of the neck that divides the specimen in half. It will be assumed the necked region maintains its cylindrical symmetry (Bridgman states that beyond a logarithmic strain of 4 this symmetry is lost) with the radius of the neck given by a. Bridgman approximates the shape of the neck by fitting a circle of radius R into the necked region, and suggests one way of measuring this radius is to take a gently sloping cone of material and slide it along the neck till it extinguishes back-lighting right at the neck, and then R is the radius of that cross section of the cone. Based on this shape of the neck and using von Mises plasticity, Bridgman presents an approximate stress distribution on the symmetry plane within the necked region as

$$\sigma_{zz} = Y + Y \ln\left(\frac{a^2 + 2aR - r^2}{2aR}\right) - p_a, \quad \sigma_{rr} = Y \ln\left(\frac{a^2 + 2aR - r^2}{2aR}\right) - p_a = \sigma_{\theta\theta},$$

$$(3.148)$$

where p_a is the applied external pressure and Y is the flow stress. When $p_a = 0$ (no applied external pressure), this solution has an additional tension component to the stress in the region of the neck. (Computations have shown that the $\sigma_{rr} = \sigma_{\theta\theta}$ assumption in developing the approximate solution is satisfied in the interior of the specimen (it is always true on the axis), but as one approaches the free surface σ_{rr} goes to zero (its boundary condition) and $\sigma_{\theta\theta}$ deviates from σ_{rr} and goes into compression at the free surface. See [209, 212] and Fig. 3.11.) When integrated over the necked region for $p_a = 0$, the load divided by the actual area is

$$\frac{\text{load}}{\pi a^2} = \frac{1}{\pi a^2} \int_0^a \sigma_{zz} 2\pi r \, dr = Y \left(1 + \frac{2R}{a}\right) \ln \left(1 + \frac{a}{2R}\right). \qquad (3.149)$$

This load over area is greater than Y, and so the measured "true stress" needs to be adjusted downward to provide the actual flow stress in the necked region. Bridgman noted, with scatter, an empirical relationship between the logarithmic strain and a/R for metals. A curve fit to his data is

$$a/R = 0.94 \left(\ln(A_0/A) - 0.1\right) - 0.1034 \left(\ln(A_0/A) - 0.1\right)^2. \qquad (3.150)$$

Using this relation, it is possible to obtain the flow stress given the load and the neck radius a. Table 3.2 is Bridgman's data for 1045 steel hardened to Rc 40.3. The applied external pressure is p_a, and the strain and load were measured. The average stress computed from the load divided by the cross section is shown, and then the flow stress as adjusted according to Eq. (3.149) is given. For many of the interrupted tests, a/R was measured. Where it was not measured, Eq. (3.150) supplies a/R for the computation of flow stress from the load. The first set of data in the table consists of tests that were stopped without failure. To show the work hardening and the relative independence of the pressure on the work hardening, these tests are listed in increasing order of strain. The work hardening is roughly linear, with $Y = 1.24 + 0.69\varepsilon_p$ GPa, independent of pressure.

The second set consists of specimens taken to failure under different loading pressures. These are ordered in terms of applied external pressure to show the effect of applied external pressure on failure strain. As can be seen, large strains are achieved before failure when large external pressures are applied. Bridgman demonstrated for steels that there is a linear dependence between logarithmic failure strain and the applied pressure: this steel shows such behavior, with roughly $p_a = -0.76 + 0.90\varepsilon_p^f$ GPa. However, p_a is an external applied pressure, and we are more interested in the local interior stress state at the time of failure, since the specimen interior only knows about its own stress and strain history and not how that stress state was externally achieved. Since the material is flowing and the von Mises criterion is satisfied, $\sigma_{zz} - \sigma_{rr} = Y$ always. Thus, we look to the local hydrostatic pressure to understand the local pressure effect on failure strain. Bridgman states, "Fracture without doubt is initiated on the axis." The hydrostatic pressure within the material in the necking region on the axis is

$$p = p_a - Y \left\{\frac{1}{3} + \ln \left(1 + \frac{a}{2R}\right)\right\}. \qquad (3.151)$$

Examination of the data shows little correlation with this pressure. However, the material has strengthened through work hardening, and so the last column in the table presents stress triaxiality $\eta = -p/Y$. This expression shows more of a correlation with failure strain; η will be discussed more in upcoming sections. (Since the stress state is uniaxial tension, the other stress state variable to be discussed later has value $\zeta = 1$.)

As another example, and to elucidate the stress state in a tensile specimen after necking, we present results of a large-strain numerical simulation by Wilkins, Streit, and Reaugh [212]. A tensile test for Al 6061-T6 was performed and the beginning of failure (void formation at the center) occurred when the local (necked) reduction in

TABLE 3.2. Tensile tests performed under large static pressures for 1045 steel quenched in water from 855°C and drawn to 425°C, Rc 40.3 (Bridgman's "9-2" [31]). Values for a/R in parentheses are from Eq. (3.150); values not in parentheses are experimental measurements.

p_a (GPa)	$\ln(A_0/A)$	$\frac{\text{Load}_{max}}{\pi a^2}$ (GPa)	a/R	$\frac{\text{Load}_{final}}{\pi a^2}$ (GPa)	Y (GPa)	p (GPa)	η
In the tests below, the sample did not fracture and the stress and strain values represent the values when the test was stopped.							
1.50	0.19		0.10	1.50	1.47	0.95	−0.64
0.655	0.22		0.19	1.45	1.38	0.12	−0.09
2.46	0.22		0.13	1.42	1.38	1.93	−1.40
2.52	0.57		0.53	2.21	1.97	1.49	−0.75
2.41	0.75	1.45	(0.57)	2.23	1.97	1.26	−0.64
0.793	0.88		0.76	2.03	1.74	−0.29	0.17
1.52	1.02		0.84	2.31	1.95	0.23	−0.12
2.46	1.09		0.92	2.45	2.04	1.07	−0.53
0.800	1.35		1.18	2.40	1.93	−0.64	0.33
2.38	1.63	1.41	(1.20)	3.14	2.49	0.38	−0.15
1.50	1.92		1.45	3.07	2.37	−0.52	0.22
2.28	2.35		1.66	4.20	3.15	−0.62	0.20
2.39	2.43		1.67	3.57	2.68	−0.10	0.04
2.72	3.34	1.45	(1.96)	4.52	3.28	−0.61	0.19
The tests below were taken to failure, and the stress values correspond to values at failure.							
Atmos	0.85		(0.65)	2.32	2.03	−1.25	0.61
Atmos	0.88	1.26	(0.67)	2.10	1.82	−1.13	0.62
Atmos	0.89		(0.68)	2.04	1.78	−1.11	0.63
0.414	1.28		(0.97)	2.65	2.19	−1.18	0.54
0.772	1.63		(1.20)	2.63	2.10	−0.91	0.43
0.896	1.54		(1.14)	2.87	2.31	−0.92	0.40
1.19	2.11		(1.47)	3.93	3.03	−1.49	0.50
1.43	2.57		(1.69)	3.96	2.97	−1.38	0.46
1.85	2.98		(1.85)	5.49	4.05	−2.15	0.53
1.88	2.76		(1.77)	4.69	3.49	−1.49	0.43
2.64	3.73		(2.05)	5.33	3.84	−1.35	0.35
2.72	3.95		(2.09)				

tensile specimen radius was $R_f/R_0 = 0.772$, ($2R_0 = 1.585$ cm) ($R_f = a$ in Bridgman's notation). This corresponds to a logarithmic strain of $\varepsilon = \ln(A_0/A_f) = -2\ln(R_f/R_0) = 0.52$. Assuming von Mises plasticity, the load (force) vs. displacement curve from the tensile test could be matched by using $Y = 284(1 + 125\varepsilon_p)^{0.1}$ MPa. (The maximum load of 65 kN was at a strain of $\ln(A_0/A) = 0.10$; see Exercise 3.16.) The result of the computation on the central plane perpendicular to the loading axis, as a function of radial distance from the central loading axis, is shown in Fig. 3.11. By symmetry, $\sigma_{r\theta} = \sigma_{z\theta} = 0$ everywhere and $\sigma_{rz} = 0$

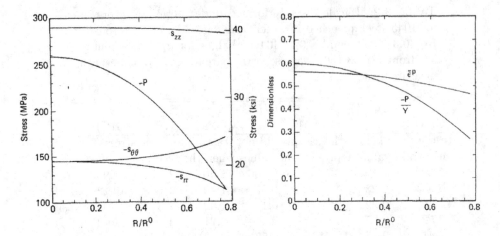

FIGURE 3.11. Computational determination of the stress tensor and equivalent plastic strain on center symmetry plane in a tensile test of Al 6061-T6 at the strain just before failure. (From Wilkins, Streit, and Reaugh [212], redrawn in [210].)

on the central plane. The left figure shows the stress deviators $(s_{zz}, -s_{rr}, -s_{\theta\theta})$ and the negative of the pressure. Since $\sigma_{zz} = s_{zz} - p$, the maximum axial stress on the center is 550 MPa, much larger than the flow stress of $Y(0.56) = 435$ MPa. The outer boundary is traction free since $\sigma_{rr} = s_{rr} - p = 0$. The hoop stress on the outer boundary is -55 MPa, not zero as in uniaxial stress. The right figure shows that the equivalent plastic strain is fairly uniform across the center plane of symmetry; hence Y is nearly constant. The stress triaxiality $\eta = -p/Y = 0.60$ at the center of the specimen due to the large tensile pressure (the same value seen in Table 3.2 for steel specimens at failure with no external pressure applied), while $\eta = 0.28$ at the outer free surface, closer to the $1/3$ value in tensile uniaxial stress.

Finally, a few comments about the material failure. Figure 3.12 shows the failure surface for three uniaxial stress tests. The first shows the typical "cup and cone" failure surface of a tensile test with a steel. As strain increases, nucleation of voids leads to failure first on the central axis and material separation first occurs there. As the interior failure surface expands laterally, at some point it transitions to a 45° failure surface, leading to complete separation of the tensile specimen into two parts. At the scale of this photo there is a clear difference between the rough, flat failure surface at the center and the smooth, 45° failure surface around the rim. However, scanning electron microscope images show that typically both surfaces have dimples indicative of ductile failure; hence the 45° failure surface is a ductile fracture running at an angle, and not a shear band failure (e.g., see Fig. 9.6 and see discussion and images in [187]). The next failure shown is a 45° failure surface for an aluminum alloy in the same tensile test. The third shows a slow compression test of an aluminum alloy showing multiple 45° failure planes, whose origin could be either ductile fracture, as in the tensile specimens, or shearing failure (localized shear bands).[6] All three tests began as uniaxial stress states and at some

---

[6]Shear bands that form rapidly are referred to as *adiabatic shear bands*. Shear failures can form because of strain leading to temperature rise, leading to thermal softening, leading to strain

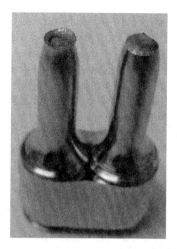

FIGURE 3.12. Failure in three uniaxial stress tests: the "cup and cone" from a tensile test of a steel (left), a 45°shearing plane with a tensile test of an aluminum (center), and 45° shearing planes in a compression test of an aluminum (right). (Figures courtesy of Sidney Chocron.)

point lost their symmetry. With associated flow the Tresca yield surface has 45° shearing planes for roughly uniaxial stress (Eq. 3.86 with $\sigma_{11} = Y$ and $\sigma_{22} = 0$), where a small perturbation moves it from the symmetry radial flow direction to the 45° flow direction near the yield surface corner (Eq. 3.88, Fig. 3.5). When the symmetry is broken, the flow is along the 45° direction, which can localize to planes. In our damage modeling discussion below, we will use a von Mises yield and flow criterion for plasticity but a Tresca criterion for failure. As a final comment, Bridgman showed that as the applied exterior pressure increases the total tensile specimen cross section decreases and the size of the initial failure surface decreases, but there is still a (smaller) cup and cone failure surface. When the applied pressure gets high enough, the material necks down to a point and there is no obvious failure surface.

3.13. Constitutive Models for Yield and Flow Stress

For metals, constitutive models address the question of flow stress or strength while damage models address the question of ductility or the strain to material failure. The material fails when it separates. Separation typically occurs through coalescence of growing ductile voids (ductile failure) or crack propagation (brittle failure). The two aspects of material response, flow stress and ductility, are distinct. However, they are often related for a given metal in that higher strength is usually accompanied by reduced ductility.

In some sense, these models are our level of empiricism in metal plasticity, in that the constitutive models are typically functional forms where the constants are

---

localization, as in Tresca's observation of luminous Xs on the sides of forged platinum alloy blocks in the 1870s [112, 213]. There are other mechanisms for adiabatic shear band formation with minimal thermal role, such as dynamic recrystallization [163]. A conclusion is that thermal properties need not be the origin of the length scale required for this type of material failure.

determined through testing of specimens at the 0.1 to 10 cm scale. Efforts to go to a lower scale than we use are referred to as crystal plasticity or dislocation modeling. In our subsequent modeling and computations of impact events we use these empirical fits to mechanical tests to provide the material response. Our discussion first focuses on the most widely used constitutive model in impact modeling, the Johnson–Cook model. It is popular because of the relative ease of fitting a small number of constants to data and because of the fact that many materials have been characterized with this model.

In the Johnson–Cook constitutive model, the von Mises plasticity model is used with isotropic hardening and a flow stress Y that is a function of plastic strain, strain rate, and temperature [105],

$$Y(\varepsilon_p, \dot{\varepsilon}_p, T) = \{A + B(\varepsilon_p)^n\}\{1 + C\ln(\dot{\varepsilon}^*)\}\{1 - (T^*)^m\}. \tag{3.152}$$

There are five material constants: $A, B, n, C,$ and m. The expression in the first set of brackets gives the work hardening with coefficient n and constant B, where ε_p is the equivalent plastic strain. The expression in the second set of brackets gives the strain rate hardening, which is assumed transient (i.e., it is assumed that the strain under a higher strain rate does not lead to increased permanent hardening beyond the strain-based work hardening); $\dot{\varepsilon}^* = \max(\dot{\varepsilon}, \dot{\varepsilon}_p^c)/(1 \text{ s}^{-1})$ is the dimensionless strain rate, where typically the equivalent plastic strain rate $\dot{\varepsilon}_p$ is used for $\dot{\varepsilon}$ (Exercise 3.28). There needs to be a low rate cutoff for the strain rate $(\dot{\varepsilon}_p^c)$ since otherwise a material at rest has a strain rate of zero, and if $C \neq 0$ the at-rest strength is undefined. A value of $\dot{\varepsilon}_p^c = 1 \text{ s}^{-1}$ leads to A equaling the low-rate yield stress. If $\dot{\varepsilon}_p^c < 1 \text{ s}^{-1}$ and $C > 0$, then A is greater than the yield stress. Because of this rate behavior, when fitting constants even the low-rate material tests need to have a defined rate (or else the low strain rate cutoff value $\dot{\varepsilon}_p^c$ is implicitly assumed). However, it is expected that for impact and penetration problems the strain rates are such that the specific value of $\dot{\varepsilon}_p^c$ will not influence the results. (For example, current computer codes have $\dot{\varepsilon}_p^c = 0.002 \text{ s}^{-1}$ in CTH and $\dot{\varepsilon}_p^c = 0.0001 \text{ s}^{-1}$ in EPIC.) The final set of brackets gives the thermal softening. T^* is referred to as the homologous temperature, which is normalized to room temperature,

$$T^* = \frac{T - T_{\text{room}}}{T_{\text{melt}} - T_{\text{room}}}, \quad 0 \leq T^* \leq 1. \tag{3.153}$$

Room temperature in the model is typically $70°F = 21.1°C$. The basic form of the model is readily adaptable to most computer codes since it uses variables that are computed in the codes. It is also robust in that it is clear how it extrapolates beyond available data. Often, models developed to address a certain observed behavior do not extrapolate well to regimes beyond that behavior.

Given this constitutive model, the thermal softening exponents, m, for Al 6061-T6 and Al 7075-T6 were determined from handbook data for elevated temperatures. A more detailed description of how the constants are obtained is provided in [106]. The resulting constants for the two aluminum alloys are listed in Table 3.3. Adiabatic stress-strain relationships are shown in Fig. 3.13, computed by assuming that all the work goes into heating of the material and hence

$$T - T_{\text{room}} = \int_0^{E_p} \frac{1}{\rho_0 C_v} dE_p = \int_0^{\varepsilon_p} \frac{Y(\varepsilon_p, \dot{\varepsilon}_p, T)}{\rho_0 C_v} d\varepsilon_p. \tag{3.154}$$

TABLE 3.3. Johnson–Cook strength and failure constants for the aluminum alloys and for the APM2 steel core and gilding metal (jacket material).

Material	A (MPa)	B (MPa)	n	C	m	D_2
Al 6061-T6	324	114	0.42	0.002	1.34	1.11
Al 7075-T6	496	310	0.30	0.000	1.20	0.65
Steel core	1,030	18,130	0.64	0.005	1.00	0.025
Gilding metal	500	0	1.00	0.025	—	2.00

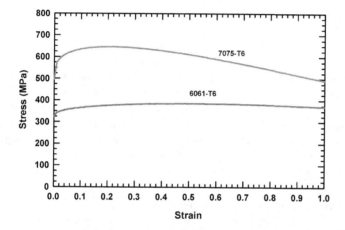

FIGURE 3.13. Stress-strain plots for Al 6061-T6 and Al 7075-T6 obtained using the Johnson–Cook strength model with the constants derived from the characterization tests.

The plots show Y vs. equivalent plastic strain from the yield function, and as such correspond to true stress-true strain. The decrease in the strength as a function of strain is strictly due to thermal softening. The 7075-T6 alloy is about twice as strong as the 6061-T6 alloy. Al 7075-T6 softens more than Al 6061-T6 as a function of equivalent plastic strain since the flow stress (and hence work) performed is greater, while the specific heat and melting temperature are the same.

For the APM2 steel core, the yield and strain hardening constants – A, B, n – are determined from the test data in Fig. 3.9 and are also given in Table 3.3. The dimensionless strain rate constant $C = 0.005$ is estimated from those of other high-strength steels as is the thermal softening constant $m = 1.00$. The strength model for the gilding metal, from Fig. 3.10, is a constant $A = 500$ MPa. The assumed strain rate effect is $C = 0.025$ and no thermal softening, with these values based on other materials' behavior and not these specific tests. Constants for the two models are summarized in Table 3.3.

There are some features of the Johnson–Cook model that do not agree with all materials. For example, the Johnson–Cook model multiplicatively combines work hardening and strain rate effects (i.e., the increase in strength due to the strain rate multiplies the current flow stress due to work hardening, $Y(\varepsilon_p, \dot{\varepsilon}_p) = \{A + B(\varepsilon_p)^n\}\{1 + C\ln(\dot{\varepsilon}^*)\}$). It is observed in some metals, such as some tungsten alloys, a better form is additively combining the hardening effects $Y(\varepsilon_p, \dot{\varepsilon}_p) =$

$A + B(\varepsilon_p)^n + C\ln(\dot{\varepsilon}^*)$. Room temperature in the model has been viewed by some researchers as another parameter: for example, absolute zero has been used instead [86]. The question becomes what values of T_{room} and m, as well as A, B, and C, best fit the available thermal softening data. Other modifications to the Johnson–Cook model have been made, including adjustments to include the J_3 invariant in the yield surface [41] and to include the observation that some materials above strain rates of 5×10^3 s^{-1} have a stronger strain rate dependence (Johnson et al. change the strain rate term to $1 + C\ln(\dot{\varepsilon}^*) + C_2(\ln(\dot{\varepsilon}^*))^{C_3}$ [107]).

There has been extensive research aimed at understanding plastic flow at the atomic and crystal grain/grain boundary level. This work has motivated other constitutive models. For example, in some materials plastic shear follows an Arrhenius form (e.g., [133]),

$$\dot{\varepsilon} = \dot{\varepsilon}_0 \exp\left(-\frac{v^* \cdot (Y_0 - \sigma_{\text{eff}})}{k_B T}\right), \tag{3.155}$$

where $\dot{\varepsilon}_0$, v^*, and Y_0 are material constants and k_B is Boltzmann's constant. This thermal activation behavior suggests a relationship between the yield, temperature, and strain rate, of the form $Y = f(T\ln(\dot{\varepsilon}_0/\dot{\varepsilon}))$. Follansbee [63] provides multiple examples of this behavior, with some specific examples being iron, 1018 steel, and niobium whose behavior is roughly bilinear with $Y = 800$ MPa for $T\ln(\dot{\varepsilon}_0/\dot{\varepsilon}) = 2,000$ K, $Y = 200$ MPa for $T\ln(\dot{\varepsilon}_0/\dot{\varepsilon}) = 6,000$ K, and $Y = 100$ MPa for $T\ln(\dot{\varepsilon}_0/\dot{\varepsilon}) = 10,000$ K, where $\dot{\varepsilon}_0 = 10^8$ s^{-1} (data for the steel ranged over 148 K $< T < 823$ K, 10^{-2} s^{-1} $< \dot{\varepsilon} < 3.8 \times 10^3$ s^{-1}). These metals have body-centered cubic (bcc) crystal structures. Lower temperature has the same effect on yield as higher strain rate. This activation-energy-like thermal behavior has been included in the Zerilli–Armstrong constitutive model [217], which has the following forms depending on the bcc or face-centered cubic (fcc) crystal structure,

$$\begin{aligned} Y &= Y_G + c_1 \exp(-c_4 T\ln(\dot{\varepsilon}_0/\dot{\varepsilon})) + c_5(\varepsilon_p)^n + k/\sqrt{\ell}, & \text{(bcc)}, \\ Y &= Y_G + c_2\sqrt{\varepsilon_p}\exp(-c_4 T\ln(\dot{\varepsilon}_0/\dot{\varepsilon})) + k/\sqrt{\ell}, & \text{(fcc)}, \end{aligned} \tag{3.156}$$

where $\dot{\varepsilon}_0 = \exp(c_3/c_4)$ in their original notation and ℓ is the average grain diameter. Notice that the functional form of f is $\exp(-c_4 T\ln(\dot{\varepsilon}_0/\dot{\varepsilon}))$ in this model. The increase in strength as the reciprocal square root of the grain diameter is known as the *Hall–Petch* relation, and is an experimentally observed behavior for many metals. The constants Y_G, c_1–c_5, and k are determined experimentally. Zerilli and Armstrong's paper lists constants for two materials: Armco iron (bcc), $Y_G = 0$, $c_1 = 1.033$ GPa, $c_4 = 4.15 \times 10^{-4}$ K^{-1}, $\dot{\varepsilon}_0 = 2.02 \times 10^7$ s^{-1}, $c_5 = 266$ MPa, $n = 0.289$, $k = 22$ MPa$\sqrt{\text{mm}}$, and $\ell = 111$ μm, and OFHC copper (fcc), $Y_G = 46.5$ MPa, $c_2 = 890$ MPa, $c_4 = 1.15 \times 10^{-4}$ K^{-1}, $\dot{\varepsilon}_0 = 3.75 \times 10^{10}$ s^{-1}, $k = 5$ MPa$\sqrt{\text{mm}}$, and $\ell = 73$ μm. The grain diameters are average for one of their particular examples.

In the Mechanical Threshold Stress (MTS) model [63], the mechanical threshold stress, defined as $\hat{\sigma} = \lim_{T \to 0} f(T\ln(\dot{\varepsilon}_0/\dot{\varepsilon}))$, is the 0 K *yield stress* of the material in its *current worked state*.[7] This stress $\hat{\sigma}$ is viewed as an internal state variable, which contains information about the work hardening of the material. As the simplest example of the MTS model we consider pure copper, where Y is the flow stress

---

[7]In Chapter 5 we will consider the 0 K cold compression curve of a material as a characterization of the elastic high-pressure response of the lattice; it is not related to the mechanical threshold stress, which is a measure of hardening, hence defects, and hence plastic response in the lattice.

and μ is the shear modulus, which is temperature dependent,

$$\frac{Y}{\mu} = \frac{Y_a}{\mu} + \frac{\hat{\sigma}}{\mu_0}\left\{1 - \left[\frac{kT}{g_0\mu b^3}\ln\left(\frac{\dot{\varepsilon}_0}{\dot{\varepsilon}}\right)\right]^{1/q}\right\}^{1/p}, \quad \mu = \mu_0 - D_0(\exp(T_0/T) - 1)^{-1},$$

$$\frac{d\hat{\sigma}}{d\varepsilon} = \theta_{II}\cdot\left\{1 - \frac{\hat{\sigma}}{\hat{\sigma}_s}\right\}^{\kappa}, \quad \theta_{II} = A_0 + A_1\ln(\dot{\varepsilon}) + A_2\sqrt{\dot{\varepsilon}}, \quad \hat{\sigma}_s = \hat{\sigma}_{s0}\left(\frac{\dot{\varepsilon}_s}{\dot{\varepsilon}_{s0}}\right)^{kT/g_{s0}\mu b^3}.$$

$$(3.157)$$

The main purpose of listing this is to show the form of the equation, though care must be exercised as the model does not apply for high temperatures or very low strain rates. The upper row of Eq. (3.157) shows the influence on the flow stress of the strain rate and temperature at a given instant in time. The lower row shows the evolution equation for the hardening variable $\hat{\sigma}$, the mechanical threshold stress. It can accumulate at a rate that is not just a constant times the strain rate, since $d\hat{\sigma} \sim \theta_{II}d\varepsilon_p$; depending on the values of A_1 and A_2, more permanent hardening occurs during higher rate plastic strain, as is seen in some materials. This behavior cannot be captured by the Johnson–Cook and Zerilli–Armstrong model, since their strain rate dependence is transitory, so the amount of permanent hardening only depends on the final amount of plastic strain, not the rate at which it was accumulated. There is a saturation limit (subscript s) at which no additional hardening can occur. Constants for copper are $G = 45.78$ GPA, $D_0 = 3.0$ GPA, $T_0 = 180$ K, $Y_a = 40$ MPa, $\hat{\sigma}_{s0} = 710$ MPa, $p = 2/3$, $q = 1$, $\kappa = 2$, $A_0 = 2.39$ GPa, $A_1 = 12$, $A_2 = 1.696$ s$^{-1/2}$, $\dot{\varepsilon}_0 = 10^7$ s^{-1}, $\dot{\varepsilon}_{s0} = 10^8$ s^{-1}, $g_0 = 1.6$, $g_{s0} = 0.301$, the Burgers vector is $b = 0.256$ nm, and k is the Boltzmann constant. For a fully annealed pure copper, $\hat{\sigma} = 0$ at $t = 0$. For other materials there are additional terms based on postulated hardening mechanisms.

Another constitutive model used in impact calculations is the Steinberg–Guinan–Lund constitutive model [174, 175]. In this model, the shear modulus is assumed to have a pressure and temperature dependence of the form

$$G(\rho, p, T) = G_0\{1 + Ap(\rho_0/\rho)^{1/3} - B(T - T_{\text{room}})\}. \quad (3.158)$$

We mention, for example, that the implementation of the Johnson–Cook model in CTH also has a pressure-dependent shear modulus since the Poisson's ratio is assumed to be constant and the bulk modulus depends on the current density and energy state of the material (see Section 5.13 and Exercise 5.42). The flow stress is

$$Y = \{\min(Y_T(\dot{\varepsilon}_p, T), Y_p) + \min(Y_A(1 + \beta(\varepsilon_p + \varepsilon_i))^n, Y^*_{\text{max}})\}\frac{G(\rho, p, T)}{G_0}, \quad (3.159)$$

where the strain rate hardening term and the work hardening term are additive, and both terms have maximum values. The strain rate term is found in the equation

$$\dot{\varepsilon}_p = \left\{\frac{1}{C_1}\exp\left[\frac{2U_K}{kT}\left(1 - \frac{Y_T}{Y_p}\right)^2\right] + \frac{C_2}{Y_T}\right\}^{-1}, \quad (3.160)$$

which needs to be implicitly solved for Y_T in conjunction with the algorithm that is performing the partition of the strain rate into elastic and plastic parts ($\dot{\varepsilon}_p$ depends on the yield surface value Y). The melt temperature is determined independently,

$$T_m = T_{m0}\exp\left\{2a\left(1 - \frac{\rho_0}{\rho}\right)\right\}\left(\frac{\rho}{\rho_0}\right)^{2(\Gamma_0 - a - 1/3)}, \quad (3.161)$$

and when the melt temperature is exceeded, G and Y are set to zero. To match release slopes on flyer plate impact tests, a Bauschinger effect is included (different unloading vs. loading) by reducing the shear modulus during unloading. Representative numbers for tantalum are $Y_p = 0.88$ GPa, $C_1 = 0.71 \times 10^6$ s^{-1}, $C_2 = 12$ kPa s, $U_K = 3,600$ K, $Y_A = 0.375$ GPa, $Y^*_{\max} = 0.45$ GPa, $\beta = 22$, $n = 0.283$, and $\varepsilon_i = 0$. The rate-independent form of the Steinberg–Guinan model has $Y_p = 0$ and the flow stress is given by $Y = Y_0(1 + \beta\varepsilon_p)^n G/G_0, \leq Y_{\max}$.

3.14. Damage, Failure, and Stress State

The constitutive models presented above model flow stress. We now discuss damage and failure. In armor mechanics, to achieve the lightest weights, we utilize materials all the way through failure. We begin with some definitions.

DEFINITION 3.4. (damage) *Damage* occurs when material has failed at a scale below which we are modeling it. The material failure manifests itself as a reduction of elastic moduli and/or material strength.

DEFINITION 3.5. (failure) *Failure* is when 1) the material loses all strength and can no longer support shear stress (i.e., $\sigma_{\text{eff}} = 0$), or 2) the material separates and hence can no longer be rigorously modeled as a continuum. Because of the material separation, the words *fracture* and *failure* are often used synonymously.

Damage modeling is a whole field of study unto itself. For ductile material failure, it is expected that after a certain amount of plastic strain the material will fail. The simplest model is simply to state that there is an equivalent plastic failure strain, ε^f_p, and that when the material reaches that strain, failure occurs. The next step in complexity is to assume that when a certain amount of energy is dissipated in a material, it will fail. Thus,

$$W_p = \int \sigma_{ij} d\varepsilon^p_{ij} = \int Y d\varepsilon_p \leq W^f_p, \qquad (3.162)$$

and when W_p reaches W^f_p, the material fails. If the flow stress is constant, $\varepsilon^f_p = W^f_p/Y$. This can be used to represent the reduced ductility of a high-strength version of a material, since as the material strength increases, the failure strain decreases. At failure the material can no longer support a shear stress (hence Y is set equal to zero) nor a tensile stress ($p \leq 0$ is not allowed and material separation occurs rather than a decrease in material density). The material can, when confined in some fashion, support compressive loads like a fluid. Some damage models, when the material is partially damaged but before failure, reduce the elastic moduli and the strength.

These simple strain models do not take into account the state of stress of the material. In some materials the stress state affects the failure strain. If a proportional stress state could be maintained during loading, meaning the ratios of the various components of the stress tensor remain the same as the load is increased, we expect the amount of plastic strain to depend upon the stress state – that is, we would expect $\varepsilon^f_p = \varepsilon^f_p(\sigma_{ij})$. For an isotropic material, the dependence would only be on the three stress invariants, so this dependence simplifies to $\varepsilon^f_p = \varepsilon^f_p(p, J_2, J_3)$. For metals we assume ductile failure, meaning that the material is undergoing large plastic flow prior to failure, so the stress is also on the yield surface $f(p, J_2, J_3) = 0$, and so really there are only two independent invariants. If the yield surface is

TABLE 3.4. Various stress states.

Stress state	Triaxiality η	Lode parameter ζ	Lode angle θ
Pure pressure tension	$+\infty$	undefined	undefined
Biaxial stress tension	$2/3$	-1	$\pi/3 = 60°$
Uniaxial stress tension	$1/3$	1	0
Pure shear	0	0	$\pi/6 = 30°$
Uniaxial stress compression	$-1/3$	-1	$\pi/3 = 60°$
Biaxial stress compression	$-2/3$	1	0
Pure pressure compression	$-\infty$	undefined	undefined

dominated by J_2 (and for von Mises flow it is strictly a function of J_2), we conclude that the equivalent plastic strain to failure for a ductile material should depend only on p and J_3. Given the fact that there may be work hardening (i.e., J_2 increases during plastic flow), we will write this dependence as dimensionless ratios of p and J_2 and J_3 and J_2. We define (the mean stress is $\sigma_{\mathrm{mean}} = \frac{1}{3}I_1 = -p$)

$$\eta = -\frac{p}{\sigma_{\mathrm{eff}}} = \frac{\sigma_{\mathrm{mean}}}{\sigma_{\mathrm{eff}}} = \frac{\sigma_{ii}/3}{\sqrt{3J_2}},$$

$$\zeta = \frac{27}{2}\frac{J_3}{\sigma_{\mathrm{eff}}^3} = \frac{27}{2}\frac{J_3}{(3J_2)^{3/2}}. \tag{3.163}$$

Using principal stresses it is straightforward to show that $-1 \leq \zeta \leq 1$ (the origin of the $27/2$ factor). Both η and ζ are invariants since they are composed of invariants. η is referred to as the stress triaxiality. ζ is a Lode parameter which is related to the Lode angle $\theta = \cos^{-1}(\zeta)/3$ (named after W. Lode). Example values are given in Table 3.4. Thus, we expect the equivalent plastic strain to failure to be of the form

$$\varepsilon_p^f = \varepsilon_p^f(\eta, \zeta). \tag{3.164}$$

As a first approach to understanding the effect of stress state on failure, we consider a spherical void inside a plastic region. If the material is pulled in tension in all directions, it will expand. If the inner void radius is a and the radial speed of the expanding void wall is $da/dt = V$, then the spherically symmetric incompressible flow field and its corresponding strain rate are

$$v_r = V\left(\frac{a}{r}\right)^2 \quad \Rightarrow \quad D_{rr} = -2V\frac{a^2}{r^3}, \quad D_{\theta\theta} = D_{\varphi\varphi} = V\frac{a^2}{r^3},$$

$$\dot{\varepsilon}_p(r) = \sqrt{\frac{2}{3}D_{ii}D_{ii}} = 2V\frac{a^2}{r^3} \quad \Rightarrow \quad \frac{da}{a} = \frac{1}{2}\left(\frac{r}{a}\right)^3 d\varepsilon_p(r). \tag{3.165}$$

If we assume that this relative expansion of the void is somehow correlated with damage, then for a damage variable D we have $dD \sim da/a$. We assume that damage accumulates as follows,

$$D = \int \frac{1}{\varepsilon_p^f(\sigma_{ij}, \dot{\varepsilon}_p, T)} d\varepsilon_p. \tag{3.166}$$

Thus, if proportional loading were to occur, the denominator would not change and a value of $D = 1$ would occur when $\varepsilon_p = \varepsilon_p^f$ and that is when the material would fail. In actual loading conditions, ε_p^f changes along the loading path, and the integral accumulates how much strain occurred at each material state relative to its failure strain at that state. Back to our spherical void, we have

$$dD = \frac{d\varepsilon_p}{\varepsilon_p^f} \sim \frac{da}{a} = \frac{d\varepsilon_p(r)}{2(a/r)^3} \quad \Rightarrow \quad \varepsilon_p^f \sim 2(a/r)^3. \tag{3.167}$$

In the plastic region around the void, it is possible to explicitly compute the radial stress. With our spherically symmetric expanding flow, $V > 0$, $D_{rr} < 0$, and $D_{\theta\theta} = D_{\varphi\varphi} = -\frac{1}{2}D_{rr} > 0$. Thus $s_{rr} = -2Y/3$ and $s_{\theta\theta} = s_{\varphi\varphi} = Y/3$. Assuming the acceleration is small, the momentum balance becomes

$$\frac{\partial \sigma_{rr}}{\partial r} = -\frac{1}{r}(2\sigma_{rr} - \sigma_{\theta\theta} - \sigma_{\varphi\varphi}) = -\frac{2(s_{rr} - s_{\theta\theta})}{r} = \frac{2Y}{r}. \tag{3.168}$$

This equation can be explicitly integrated to give

$$\sigma_{rr}(r) = 2Y \ln\left(\frac{r}{a}\right) \quad \Rightarrow \quad \frac{r}{a} = \exp\left(\frac{\sigma_{rr}(r)}{2Y}\right), \tag{3.169}$$

where at the lower limit $r = a$ the interior void surface is traction free. Thus,

$$\varepsilon_p^f \sim 2(a/r)^3 = 2\exp\left(-3\frac{\sigma_{rr}(r)}{2Y}\right) = 2\exp\left(-3\frac{-2Y/3 - p(r)}{2Y}\right) = 2e\exp\left(-\frac{3}{2}\eta(r)\right). \tag{3.170}$$

Following this form, and based on their data, Hancock and Mackenzie [90] postulated the failure strain for metals to be

$$\varepsilon_p^f = \varepsilon_p^n + \alpha\exp(-\frac{3}{2}\eta), \tag{3.171}$$

where ε_p^n is the plastic strain at which voids nucleate, and showed that this equation could roughly fit data gathered for three steels. Thus, we have now incorporated triaxiality of the stress state into the strain to failure. It is interesting, in the context of scaling to be discussed in the next section, that they note "it is not a sufficient condition for failure initiation that a strain which is a function of stress-state is exceeded at a point. Failure initiation must involve a minimum amount of material which is characteristic of the scale of physical events involved."

The form of Eq. (3.171) was used as a framework for the failure model of Johnson and Cook [106]. They refer to their model as a fracture model. We refer to it as a failure model, rather than a damage model, since the internal damage parameter D does not affect the constitutive response until $D = 1$, at which point the strength of the material is set equal to zero, so complete failure occurs. The general expression for the strain at failure in the Johnson–Cook failure model is defined in regions depending on the stress. Then

$$\varepsilon_p^f(\sigma_{ij}, \dot{\varepsilon}_p, T) = \{D_1 + D_2\exp(D_3\eta)\}\{1 + D_4\ln(\dot{\varepsilon}^*)\}\{1 + D_5 T^*\}, \quad \text{for } \eta \le 1.5, \tag{3.172}$$

where the terms $\dot{\varepsilon}^*$ and T^* are as above. To try to match spall where high triaxiality is expected, they define $\eta_{\text{spall}} = \sigma_{\text{spall}}/\sigma_{\text{eff}}$, a value that changes depending on the value of σ_{eff}, and use linear interpolation (assuming $\eta_{\text{spall}} > 1.5$)

$$\varepsilon_p^f = (\varepsilon_p^f|_{\eta=1.5})\frac{\eta_{\text{spall}} - \eta}{\eta_{\text{spall}} - 1.5} + \varepsilon_{\text{min}}^f\frac{\eta - 1.5}{\eta_{\text{spall}} - 1.5}, 1.5 < \eta \le \eta_{\text{spall}}; \varepsilon_p^f = \varepsilon_{\text{min}}^f, \eta > \eta_{\text{spall}}, \tag{3.173}$$

where $\varepsilon_p^f|_{\eta=1.5}$ is the evaluation of Eq. (3.172) with $\eta = 1.5$ but all other values as exist in the current state.

The model requires constants D_1 through D_5, σ_{spall}, and $\varepsilon^f_{\min}$. The strain rate term controlled by D_4 and the thermal softening term controlled by D_5 are similar to those in the Johnson–Cook strength model. The smaller ε^f_p is, the more quickly damage accumulates (Eq. 3.166). The first part of the expression gives a failure strain dependence on the triaxiality of the current stress state, the ratio of the mean stress to the effective shear stress. For any damage to accumulate, Eq. (3.166) requires that there be plastic deformation: thus by definition $\sqrt{3J_2} = Y > 0$ and so η is well defined. Based on Eq. (3.171), typically $D_3 = -\frac{3}{2}$ and $D_2 > 0$. The more tensile the pressure state (the more negative p), the more positive η, then, as $D_3 < 0$, the more negative the exponential term; therefore, the failure strain is smaller for larger tensile stress states, as one might expect from ductile failure for metals. In a similar fashion, for compressive stress states ($p > 0$), the exponential term is larger and it takes a large amount of plastic strain to produce failure. A purely hydrostatic or spherical tensile stress state ($\sigma_{ij} = -p\delta_{ij}$) does not produce any damage since the stress is not on the yield surface and so no plastic flow is occurring.

We now consider the Johnson–Cook failure model for the four materials for which we presented Johnson–Cook constitutive models in the previous section. A more detailed description of how the constants are obtained is provided with the original description of the model [106]. The plastic strain to failure is strongly dependent on the hydrostatic tension that exists during the deformation. The torsion tests provide failure data with essentially no hydrostatic tension. The tension tests provide failure data at moderate hydrostatic tensions. Various levels of hydrostatic tension are typically achieved by tensile tests with notched specimens, and such was done for the aluminum alloys. Notched tensile specimens produce different values of triaxiality at failure by producing higher hydrostatic tensions. These tests require large-scale finite-element-type calculations to determine the stress state in the specimen. As such, the tests are not presented here. If it is assumed that $D_1 = D_4 = D_5 = 0$ and $D_3 = -\frac{3}{2}$ in Eq. (3.172), these assumptions yield a simplified form of the failure model that does not include temperature or strain-rate effects,

$$\varepsilon^f_p = D_2 e^{-3\eta/2}. \tag{3.174}$$

Based on all the tests, for the 6061-T6 alloy, $D_2 = 1.11$, and for the 7075-T6 alloy $D_2 = 0.65$ (see Exercise 3.32). The ductility of Al 7075-T6 is less than Al 6061-T6.

For the APM2 steel core failure model it was again assumed that $D_1 = D_4 = D_5 = 0$ and $D_3 = -\frac{3}{2}$. It is then possible to determine D_2 from the strains in Fig. 3.9. The elastic strain is roughly 0.6%. The average engineering strain at failure is 2%. Thus, the plastic strain at failure is roughly 1.4% (it is in fact somewhat less than this, since some of the subsequent strain is elastic, as is reflected in the work hardening). If the failure is assumed to occur in a tensile uniaxial stress state, then $\eta = \frac{1}{3}$. This yields $\varepsilon^f_p = D_2 e^{-1/2} = 0.014$, which gives $D_2 = 0.023 \approx 0.025$. For the gilding metal, with the above-computed failure strain (1.23), assuming a uniaxial stress state at the time of failure, the failure constant D_2 is determined in a similar manner to that used for the core material (see Exercise 3.32). Values of the constant D_2 are listed in Table 3.3.

We now take our first step in considering more general stress states. Bao and Wierzbicki [25] argue, based on their data and a careful evaluation of Bridgman's extensive data on metal flow and fracture under high pressures [31], that ductile failure or fracture does not occur if $\eta \leq -\frac{1}{3}$. However, the Bridgman data typically consist of compressive tests where ζ is close to -1. Our perspective is that it takes a tensile (positive) principal stress for ductile void growth to occur. This approach is in line with Cockcroft and Latham's study on workability of metals, where they assume that damage accumulates like $dD \sim (\sigma_{\max})_+ \cdot d\varepsilon_p$, where $\sigma_{\max}$ is the most positive principal stress and damage accumulation only occurs when $\sigma_{\max} > 0$ [46]. Hence, our interest is to determine when tensile principal stresses occur and when they do not. In particular, we look for the demarcation surface between stress states where damage can occur and stress states where it cannot. For large pressures, compared to the flow stress, η is a negative number and all principal stresses are negative. When all principal stresses are negative, $I_3 = \det(\sigma) < 0$. As pressure decreases, η increases, and we end up in a region where there is at most one tensile principle stress. When there is one positive principle tensile stress, and the other two are negative, $I_3 = \det(\sigma) > 0$. Equation (3.284) from Exercise 3.50 says

$$I_3 = \det(\sigma) = J_3 + pJ_2 - p^3 = J_3 + Y^3\eta\left(\eta^2 - \frac{1}{3}\right), \qquad (3.175)$$

where the last equality assumes plastic flow is occurring. Thus setting $I_3 = 0$ in Eq. (3.175) is the demarcation equation between a region where no failure occurs (or no damage accumulates) and where failure can occur (damage accumulates). Having found the equation, we need to understand to what η it applies. Differentiating,

$$\frac{dI_3}{d\eta} = 3Y^2\left(\eta^2 - \frac{1}{9}\right), \qquad (3.176)$$

we see that I_3 is an increasing function of η for $\eta < -\frac{1}{3}$. Equation (3.163) implies that $-\frac{2}{27}Y^3 \leq J_3 \leq \frac{2}{27}Y^3$. If we plug in the lower bound, we find the first solution from below to $I_3 = 0$ is $\eta = -\frac{1}{3}$. If we plug in the upper bound, we find the first solution from below is $\eta = -\frac{2}{3}$. Thus, there are no tensile principal stresses for $\eta < -\frac{2}{3}$ and there is always a tensile principal stress state for $\eta > -\frac{1}{3}$. Therefore it is only for $-\frac{2}{3} \leq \eta \leq -\frac{1}{3}$ that we need concern ourselves with J_3 when discussing the failure/no failure demarcation. There is a tensile principal stress when

$$I_3 = \det(\sigma) = J_3 + pJ_2 - p^3 = J_3 + Y^3\eta\left(\eta^2 - \frac{1}{3}\right) > 0 \quad \text{and} \quad \eta > -\frac{2}{3}. \quad (3.177)$$

There is no positive principal stress when $I_3 < 0$ and $\eta < -\frac{1}{3}$.

Bao, Wierzbicki, Lee, and Bai have looked at ductile fracture of 2024-T351 aluminum to examine the failure strain dependence on the stress state $\varepsilon_p^f(\eta, \zeta)$ [23, 24, 208]. They used different test specimens to lead to different stress states at the fracture/failure point. The aluminum was characterized to have an initial yield in the vicinity of 325 MPa with a work hardening to nearly 700 MPa; a fit they provide to their true stress-true strain based on matching the round tensile specimen loading data is

$$Y = Y_1\varepsilon^n = 744\varepsilon^{0.153} \text{ MPa}. \qquad (3.178)$$

They performed computations with the finite element code ABAQUS using this work hardening form and von Mises (J_2) isotropic plasticity to model the deformations seen in the tests. They report good agreement. When the material failure or fracture was first observed (through a drop in load or visually observing material

TABLE 3.5. Analysis of various tests to failure with 2024-T351 aluminum [23, 208].

Specimen description	η_{av}	ζ_{av}	ε_p^f
Compression			
Cylinder ($L/D = 2$)	−0.278	−0.91	0.45
Cylinder ($L/D = 1.25$)	−0.234	−0.81	0.38
Cylinder ($L/D = 1$)	−0.233	−0.82	0.356
Cylinder ($L/D = 2/3$)	−0.224	−0.80	0.341
Round notched	−0.248	−0.84	0.62
Shear			
Flat dog-bone tensile	0.0124	0.055	0.21
Flat	0.117	0.50	0.26
Tension			
Plate with large circular hole	0.343	1.0	0.31
Dog bone	0.357	0.979	0.48
Hollow cylinder	0.356	0.984	0.33
Rectangular bar	0.369	1.0	0.36
Round (standard tensile test specimen)	0.40	1.0	0.46
Round, large radius notch	0.63	1.0	0.28
Round, small radius notch	0.93	1.0	0.17
Flat, grooved	0.61	0.097	0.21

separation) the deflection was noted. At the same deflection in the computations, the equivalent plastic strain at the location where failure began was recorded and denoted as the failure strain ε_p^f. (Notice that these computations do not require a failure model – only a description of plastic flow is required.) The history of the Lagrangian material location where failure occurred was examined from the beginning of the computation through the time of failure, and the average values

$$\eta_{av} = \frac{1}{\varepsilon_p^f} \int_0^{\varepsilon_p^f} \eta \, d\varepsilon_p, \qquad \zeta_{av} = \frac{1}{\varepsilon_p^f} \int_0^{\varepsilon_p^f} \zeta \, d\varepsilon_p, \qquad (3.179)$$

were computed. Values for η and ζ can vary considerably over the loading path, especially for actual test specimens undergoing plastic deformation, but we will consider them as constants for our discussion. Table 3.5 and Fig. 3.14 summarize their results. The upper left and right figures show the analyzed data. The lower left figure shows where the results lie in the η–ζ plane.[8] Figure 3.6 contains images of round tensile specimens with and without notches; note the stress triaxiality η and failure strain ε_p^f progression.

---

[8]Wilkins, Streit, and Reaugh performed a numerical simulation looking at a 1.27 cm thick Al 6061-T6 compact tension specimen used for determining fracture toughness [212, 210]. In the center of the specimen, near the tip of the fatigue crack, the pressure was −800 MPa and the flow stress around 350 MPa. This stress triaxiality, $\eta = -p/Y = 2.3$, is much larger than any of the tests in Table 3.5, showing that the compact tension test at plane strain is at a much higher relative hydrostatic tension. The crack is, in some sense, the small radius limit of a notch.

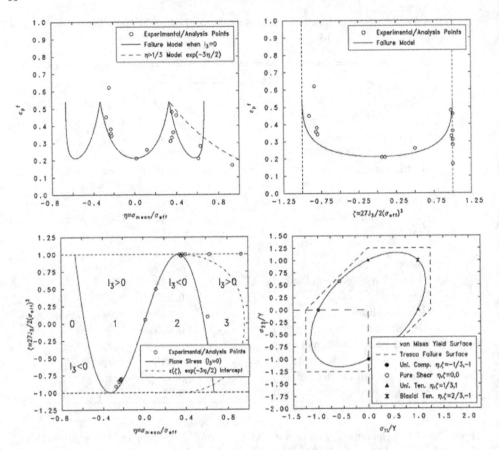

FIGURE 3.14. Analyzed experimental data points compared to the plane stress model employing maximum shear stress as the failure criteria for the work hardening material Al 2024-T351 (analyzed data from [**23, 208**]).

The data show considerable structure, especially the peak at $\eta = \frac{1}{3}$. Obviously the form $\varepsilon_p^f = D_1 + D_2 \exp(D_3 \eta)$ cannot produce the local maximum exhibited by the data when viewed as a function of η only. As a possible explanation for these data and to develop a damage model, following Wierzbicki et al., we examine plane stress deformation with von Mises plasticity and with a stress-based Tresca failure surface. We will assume isotropic hardening of the yield surface; during proportional loading the stress will increase from initial yield until it arrives at the failure surface, and we will compute the equivalent plastic strain that occurs during that deformation, hence computing the strain to failure. We will then compare the failure strains to the experimental results. We now proceed with the details.

We assume plane stress deformation where $\sigma_{33} = \sigma_{13} = \sigma_{23} = 0$, and we also set $\sigma_{12} = 0$. This means there are only two stress components that are nonzero (σ_{11} and σ_{22}), so when the yield surface is imposed there is only a curve of stress states in stress space that can be achieved. Arithmetic shows that the von Mises yield criterion for plane stress and no shear stresses is

$$\sigma_{11}^2 - \sigma_{11}\sigma_{22} + \sigma_{22}^2 = Y^2 \quad \Rightarrow \quad \frac{3}{4}(\sigma_{11} - \sigma_{22})^2 + \frac{1}{4}(\sigma_{11} + \sigma_{22})^2 = Y^2. \quad (3.180)$$

Thus, the yield surface in the σ_{11}–σ_{22} plane is an ellipse, rotated $45°$ from the horizontal (Fig. 3.14, lower right). We need to parameterize the yield surface, and a natural parameter is

$$\gamma = \frac{\sigma_{11} + \sigma_{22}}{\sigma_{11} - \sigma_{22}}. \quad (3.181)$$

When this expression is inserted into the yield surface,

$$\sigma_{11} - \sigma_{22} = \pm \frac{2Y}{\sqrt{\gamma^2 + 3}}, \quad \sigma_{11} + \sigma_{22} = \pm \frac{2\gamma Y}{\sqrt{\gamma^2 + 3}}, \quad (3.182)$$

and hence

$$\sigma_{11} = \pm \frac{\gamma + 1}{\sqrt{\gamma^2 + 3}}Y, \quad \sigma_{22} = \pm \frac{\gamma - 1}{\sqrt{\gamma^2 + 3}}Y. \quad (3.183)$$

The $+$ corresponds to the lower-right side of the ellipse and the $-$ to the upper-left side of the ellipse (Fig. 3.14). This parameterization of the yield surface has some nice properties: when $\gamma = \pm 1$ the stress state is uniaxial stress; when $\gamma = 0$ the stress state is pure shear; when $\gamma = \pm\infty$, the stress state is biaxial stress ($\sigma_{11} = \sigma_{22} = \pm Y$). During proportional loading, when σ_{11}/σ_{22} is constant, γ and s_{11}/s_{22} are also constant.

We now parameterize the Tresca failure surface. The failure surface is given by the maximum shear stress condition as

$$\text{maximum shear stress} = \frac{Y_f}{2} = \frac{1}{2}\max\{|\sigma_{11} - \sigma_{22}|, |\sigma_{22} - \sigma_{33}|, |\sigma_{33} - \sigma_{11}|\}, \quad (3.184)$$

where Tresca failure stress Y_f is a constant. When isotropic work hardening occurs for the von Mises yield surface, where $Y = Y(\varepsilon_p)$, it expands to intersect the failure surface. We can parameterize the failure surface in terms of γ and the strain to failure ε_p^f for that γ. Thus, inserting the stresses from Eq. (3.183),

$$Y_f = \frac{Y(\varepsilon_p^f)}{\sqrt{\gamma^2 + 3}}\max\{2, |\gamma - 1|, |\gamma + 1|\}. \quad (3.185)$$

Writing the γ terms on the right-hand side as $F(\gamma)$ and solving for the max expression gives

$$F(\gamma) = \frac{1}{\sqrt{\gamma^2 + 3}}\begin{cases} 2 & -1 \le \gamma \le 1, \\ |\gamma| + 1 & |\gamma| > 1. \end{cases} \quad (3.186)$$

We can explicitly solve for the equivalent plastic strain to failure as

$$\varepsilon_p^f = Y^{-1}\left(\frac{Y_f}{F(\gamma)}\right) = \left(\frac{Y_f}{Y_1}\frac{1}{F(\gamma)}\right)^{1/n}, \quad (3.187)$$

where the far right-hand side is for the specific work hardening mentioned above for this aluminum, with $Y_1 = 744$ MPa and $n = 0.153$. The right-hand side can be viewed as an equation with three unknowns that are available to fit the data points: Y_f, Y_1, and n, though for our specific example the yield and work hardening data has already set Y_1 and n so there is only one parameter left to match the observed ductile failure, namely Y_f. To be explicit, we have derived a strain to failure for a given γ of

$$\varepsilon_p^f = \left(\frac{Y_f}{Y_1}\right)^{1/n} (\gamma^2 + 3)^{1/2n} \begin{cases} 1/2^{1/n} & -1 \leq \gamma < 1, \\ 1/(1+|\gamma|)^{1/n} & |\gamma| > 1. \end{cases} \qquad (3.188)$$

We choose to match the pure shear failure strain: this produces $Y_f = 677$ MPa to match the pure shear test result (Table 3.5, $\varepsilon_p^f = 0.21$ at $\eta \approx 0$, $\zeta \approx 0$, which corresponds to $\gamma = 0$). Though the stress-strain curve in this case is in terms of total strain, for ductile failure the failure strains are large and the elastic portion is small. We mention that one can instead work in the π plane and show the equivalent (Exercise 3.25)

$$\varepsilon_p^f = \left\{ \frac{\sqrt{3}}{2} \frac{Y_f}{Y_1} \frac{1}{\cos(\theta - \pi/6 - n\pi/3)} \right\}^{1/n}, \qquad n\frac{\pi}{3} \leq \theta \leq (n+1)\frac{\pi}{3}, \quad n = 0, ..., 5, \qquad (3.189)$$

where the Lode angle $\theta = \frac{1}{3}\cos^{-1}(\zeta)$.

To compare with the data, it is necessary to determine the corresponding invariants η and ζ. Dividing through by Y^2, the equation for the plane stress von Mises yield surface is

$$\left(\frac{\sigma_{11}}{Y}\right)^2 - \frac{\sigma_{11}\sigma_{22}}{Y^2} + \left(\frac{\sigma_{22}}{Y}\right)^2 = 1. \qquad (3.190)$$

Noting for plane stress $p = -(\sigma_{11} + \sigma_{22})/3$, we find η from its definition

$$\eta = -\frac{p}{Y} = \frac{\sigma_{11} + \sigma_{22}}{3Y}. \qquad (3.191)$$

Squaring yields

$$9\eta^2 = \left(\frac{\sigma_{11}}{Y}\right)^2 + \frac{2\sigma_{11}\sigma_{22}}{Y^2} + \left(\frac{\sigma_{22}}{Y}\right)^2. \qquad (3.192)$$

Subtracting Eqs. (3.190) and (3.192) gives

$$\frac{\sigma_{11}\sigma_{22}}{Y^2} = 3\eta^2 - \frac{1}{3}. \qquad (3.193)$$

This intermediate result is now used to determine J_3 in terms of η,

$$\frac{J_3}{Y^3} = \frac{s_{11}s_{22}s_{33}}{Y^3} = \left(\frac{\sigma_{11}}{Y} - \eta\right)\left(\frac{\sigma_{22}}{Y} - \eta\right)(0 - \eta) = -\eta\left(\frac{\sigma_{11}\sigma_{22}}{Y^2}\right)$$
$$-\eta\frac{\sigma_{11} + \sigma_{22}}{Y} + \eta^2 = -\eta\left(3\eta^2 - \frac{1}{3} - 3\eta^2 + \eta^2\right) = -\eta\left(\eta^2 - \frac{1}{3}\right). \qquad (3.194)$$

Thus, for plane stress, with ζ as defined in Eq. (3.163),

$$\zeta = \frac{27}{2}\frac{J_3}{(3J_2)^{3/2}} = -\frac{27}{2}\eta\left(\eta^2 - \frac{1}{3}\right), \qquad (3.195)$$

which is just Eq. (3.175) with $\frac{27}{2}I_3 = \zeta + \frac{27}{2}\eta(\eta^2 - \frac{1}{3}) = 0$, since $I_3 = 0$ means at least one principal stress is zero, hence plane stress. This curve is shown on the lower left-hand side of Fig. 3.14 and the number in each region corresponds to the number of principal stress components that are positive. The plane stress assumption produces a cubic polynomial curve in the η–ζ plane that passes through the uniaxial stress states of compression $(\eta, \zeta) = (-\frac{1}{3}, -1)$ and tension $(\eta, \zeta) = (\frac{1}{3}, 1)$. On the plane stress yield surface, values of $\eta > \frac{2}{3}$ or $\eta < -\frac{2}{3}$ cannot occur – the magnitude of the pressure is large enough that a component of the stress cannot equal zero.

Also, based on its definition, $-1 \leq \zeta \leq 1$. Thus, the cubic relation has end points at $(\eta, \zeta) = (-\frac{2}{3}, 1)$ (biaxial stress compression) and $(\eta, \zeta) = (\frac{2}{3}, -1)$ (biaxial stress tension). The fact that the data points fall on this cubic (lower left-hand figure) implies that the average stress state in the test specimens is in accord with the plane stress assumption. The two invariants, in terms of the parameterization, are now calculated to be

$$\eta = \pm\frac{2}{3}\frac{\gamma}{\sqrt{\gamma^2 + 3}}, \qquad \zeta = \mp\frac{\gamma(\gamma^2 - 9)}{(\gamma^2 + 3)^{3/2}}. \tag{3.196}$$

This information allows us to compare the model expression for plastic failure strain to the data. (The related plasticity problem is completely solved in Exercise 3.41, though it is not necessary to do so to determine the failure strains.)

We can now determine the equivalent plastic failure strain ε_p^f (Eq. 3.188) in terms of the invariants η and ζ (Eq. 3.196). With $Y_f = 677$ MPa to match the pure shear test result $\varepsilon_p^f = 0.21$ at $\eta \approx 0$, $\zeta \approx 0$, $\gamma = 0$ (Table 3.5), the predicted equivalent plastic failure strain is shown in Fig. 3.14 (upper figures). When plotted as a function of η and ζ, the failure strain surface looks like the snowboarder's halfpipe, and the plane stress path is like a snowboarder's path back and forth from rim to rim. The minimum failure strain is for shear stress states, where the yield surface is nearest to the maximum stress failure surface (lower-right figure), and the maximum failure strain is when the stress state is either tensile or compressive. The agreement between the model and the data (upper left figure) for $-\frac{1}{3} < \eta < \frac{1}{3}$ is remarkable, especially considering that the only data used are the work hardening curve and one failure point (pure shear). In the ζ plot the $\varepsilon_p^f(\zeta)$ curve traces over itself: if $\varepsilon_p^f(\zeta)$ is plotted for $-\infty \leq \gamma \leq +\infty$, the same curve in the ζ-ε_p^f plane is traced three times, back and forth between $\zeta = -1$ and $\zeta = 1$.

As yet, our derivation is only for the plane stress curve. However, based on the curve tracing over itself in the ζ-ε_p^f plane and the fact that the derivation used J_2 flow and a Tresca failure criterion, neither of which depend on the pressure, we conjecture that the computed failure strain is explicitly a function of ζ and not of η for certain range of η (i.e., we will write $\varepsilon_p^f(\eta, \zeta) = \varepsilon_p^f(\zeta)$). Assuming this conjecture is correct, and because the best agreement in Fig. 3.14 is in the region $-\frac{1}{3} < \eta < -\frac{1}{3}$, we focus on that region to describe a general failure criterion for arbitrary stress states. Writing down from before, the failure expression is

$$\varepsilon_p^f(\gamma) = Y^{-1}\left(\frac{Y_f}{2}\sqrt{\gamma^2 + 3}\right) = \left(\frac{Y_f}{2Y_1}\sqrt{\gamma^2 + 3}\right)^{1/n}, \tag{3.197}$$

$$\zeta = \mp\frac{\gamma(9 - \gamma^2)}{(\gamma^2 + 3)^{3/2}}, \qquad -1 \leq \gamma \leq 1. \tag{3.198}$$

For a given ζ, by solving Eq. (3.198) for γ (say with Newton's method) or by using the very good approximation to $\gamma(\zeta)$ given by

$$\gamma(\zeta) = \mp\mathrm{sign}(\zeta)\left\{1 - \sqrt{1 - |\zeta|}\right\}, \tag{3.199}$$

we can then evaluate the failure strain so we have $\varepsilon_p^f(\zeta) = \varepsilon_p^f(\gamma(\zeta))$. This failure strain will be assumed to hold for $-\frac{1}{3} \leq \eta \leq \frac{1}{3}$. For the $\eta > \frac{1}{3}$ region, we link with the spherical void tensile pressure result of Eq. (3.170). A question arises as to whether the curved ζ dependence should be promulgated into the $\eta > \frac{1}{3}$ region in conjunction with the exponential η decay; however, the data point for the flat-grooved tension specimen (Table 3.5) does not show a decrease over the flag-dog-bone tensile shear specimen. Thus, we will take the minimum of the $\exp(-D_3\eta)$ dependence and the above model rather than some additive or multiplicative blending. The uniaxial stress ($\zeta = 1$) strain to failure is given by $\varepsilon_p^f = (Y_f/Y_1)^{1/n}$, and we assume $\varepsilon_p^f(\eta) = (Y_f/Y_1)^{1/n}\exp(-\frac{3}{2}(\eta - \frac{1}{3}))$ for $\eta > \frac{1}{3}$. This expression is shown in the upper left of Fig. 3.14 as a dashed line, demonstrating that it reasonably reflects the notched tension tests depicted on the plot (and if D_3 were suitably chosen, rather than $\frac{3}{2}$, we could get it to pass right through the data points). For large pressures we will continue with the von Mises/Tresca model until there is no positive principal stress, and hence no damage. These three damage regions are linked together to give

$$\varepsilon_p^f(\eta,\zeta) = \begin{cases} +\infty, & \zeta + \frac{27}{2}\eta\left(\eta^2 - \frac{1}{3}\right) < 0 \text{ and } \eta < -\frac{1}{3}, \\ \varepsilon_p^f(\gamma(\zeta)), & \zeta + \frac{27}{2}\eta\left(\eta^2 - \frac{1}{3}\right) > 0 \text{ and } \eta < -\frac{1}{3} \text{ or } -\frac{1}{3} \le \eta \le \frac{1}{3}, \\ \min\{\varepsilon_p^f(\gamma(\zeta)), \varepsilon_p^f(\frac{1}{3},1)\sqrt{e}\exp(-\frac{3}{2}\eta)\}, & \eta > \frac{1}{3}. \end{cases}$$

$$(3.200)$$

For the specific work hardening form, this expression is

$$\varepsilon_p^f(\eta,\zeta) = \left(\frac{Y_f}{Y_1}\right)^{1/n} \begin{cases} +\infty, & \zeta + \frac{27}{2}\eta\left(\eta^2 - \frac{1}{3}\right) < 0 \text{ and } \eta < -\frac{1}{3}, \\ \left(\frac{1}{2}\sqrt{\gamma(\zeta)^2 + 3}\right)^{1/n}, & \zeta + \frac{27}{2}\eta\left(\eta^2 - \frac{1}{3}\right) > 0 \text{ and } \eta < -\frac{1}{3}, \\ \left(\frac{1}{2}\sqrt{\gamma(\zeta)^2 + 3}\right)^{1/n}, & -\frac{1}{3} \le \eta \le \frac{1}{3}, \\ \min\left\{\left(\frac{1}{2}\sqrt{\gamma(\zeta)^2 + 3}\right)^{1/n}, \sqrt{e}\exp(-\frac{3}{2}\eta)\right\}, & \eta > \frac{1}{3}. \end{cases}$$

$$(3.201)$$

The intersection of the two surfaces in the $\eta > \frac{1}{3}$ region is given by

$$\eta = \frac{1}{3} - \frac{2}{3}\ln\left(\frac{\varepsilon_p^f(\gamma(\zeta))}{\varepsilon_p^f(\frac{1}{3},1)}\right) = \frac{1}{3} - \frac{1}{3n}\ln\left(\frac{\gamma(\zeta)^2 + 3}{4}\right) \qquad (3.202)$$

and is shown in the lower left of Fig. 3.14. Damage is accumulated as $dD = (1/\varepsilon_p^f)d\varepsilon_p$, an equation that can also be used for nonproportional loadings. Again, what is remarkable upon comparing with the data in Fig. 3.14 is that the whole data comparison is achieved by only matching one failure point, the equivalent plastic strain to failure in shear where $\eta = \zeta = 0$. In Chapter 9 we will present an additional rate-dependent term.

Xue and Wierzbicki [208] developed a different form for the equivalent plastic strain to failure based on examining their experiments; they suggest

$$\varepsilon_p^f(\eta,\zeta) = C_1 e^{-C_2\eta} - (C_1 e^{-C_2\eta} - C_3 e^{-C_4\eta})(1 - \zeta^{1/n})^n, \qquad \eta \ge -\frac{1}{3}. \qquad (3.203)$$

The model has four constants that are determined by comparing with the experimental data. The ζ term at the end gives rise to the halfpipe behavior seen in their data. However, in this case, there is pressure dependence in the half-pipe region due to the η term. Two other damage models are presented in the exercises. Two papers that discuss a variety of failure models in the context of impact are Teng and Wierzbicki [186] and Dey et al. [55].

This discussion has taken us through an interesting application of plasticity theory to ductile failure of metals. In analytical modeling, with geometrical assumptions, typically there is one damage variable and it needs to be adjusted or calibrated based on experimental ballistic data. As can be seen with this discussion, such a calibration can be viewed as a realization that the strain to failure is stress state and stress history dependent, and that it is thus necessary to find the failure strain for a similar stress history to the problem of interest.

GENERAL PRINCIPLE 3.6. *The equivalent plastic strain to failure for a material undergoing ductile failure depends upon the stress state, both the stress triaxiality and J_3. Thus, to apply a failure model requiring strain to failure requires knowledge of the stress state and the material behavior as a function of stress, or the performance of a similar-stress-state test to calibrate the model.*

We point out that there is considerable numerical evidence that the von Mises yield surface does not produce the localization seen in many failures, including tests described in this section, even when a failure model that includes J_3 is used. The implication is that the yield surface itself needs a J_3 dependence. However, it then becomes necessary to redo the computations that gave rise to the equivalent plastic strains in Table 3.5, and thus it is not possible with the data presented here to determine the appropriateness of a different yield surface.

From a "deeper" perspective, failure involves material separation, which means that the material being modeled is no longer a continuum – that is, continuum mechanics no longer applies unless each new surface is modeled with boundary conditions, and typically failure begins at a smaller scale than the numerical mesh resolution and then progresses to scales on par with the mesh resolution. Typical techniques assume a continuum material after failure with homogenized material properties. It may be that methods of addressing separated materials that do not satisfy the assumptions of classical continuum mechanics (e.g., generalized continua) will speed progress in modeling failure. Otherwise one can appeal to brute force – finer and finer zoning of computations as computational capabilities increase.

However, often the experience is that sophisticated damage models do not produce better results in ballistic impact calculations: simple strain-to-failure or work-to-failure models are robust and are often used. Typically the failure strain used in this approach is larger than what has been measured in tests, though questions of scale come into play.

3.15. Effects of Scale

Nearly all the data in this book, both ballistic and material property, are from test samples with a scale on the order of 0.1 to 10 cm. All physical expressions can be dimensionally expressed in terms of length, mass, and time. If L is a characteristic length scale (such as the length or diameter of the projectile, or the thickness of the target), t is a characteristic time, and m a characteristic mass, then traditional linear scaling is

$$L^* = \lambda L, \quad t^* = \lambda t, \quad m^* = \lambda^3 m, \qquad (3.204)$$

where λ is the scaling factor and the asterisk marks the scaled size. Straightforward dimensional computations show that for velocity ($V = L/t$), density ($\rho = m/L^3$), and stress ($\sigma = m/Lt^2$),

$$V^* = V, \quad \rho^* = \rho, \quad \sigma^* = \sigma. \qquad (3.205)$$

Based on the scaling, when length and time are scaled, the velocity, density, and strength values are the same, and so we expect that regardless of the scale size at which a penetration test is performed, for the same materials and impact velocity, the normalized penetration $P^*/L^* = \lambda P/\lambda L = P/L$ to be the same. However, experimentally that is not what is observed. Typically the larger the target scale, the larger the P/L when everything else is purportedly "the same." This scaling behavior is important in applications:

GENERAL PRINCIPLE 3.7. *(material size scale effect) As size increases, materials become softer and weaker.*

As will be shown in Chapters 8 and 10, projectile penetration is more sensitive to target strength than to projectile strength, so

GENERAL PRINCIPLE 3.8. *(armor size scale effect) As armor size increases, its relative performance decreases.*

Three explanations are typically given to explain the scale effect, all of which argue that larger targets are weaker. The first reason argues that the scale effect is an apparent scale effect because the material used in tests at different scales are not the same. The second and third reasons say that larger materials are softer owing to material properties. The reasons are

(1) Most targets are strong metals, and it is difficult to carry out hardening techniques for thicker plates. Hence, large targets are weaker. As a specific example, Table 3.6 shows the hardness requirements in Mil-Spec-A-12560 for rolled homogeneous armor (RHA, "Armor plate, steel, wrought, homogeneous, for use in combat vehicles and for ammunition testing"). As the armor plate gets thicker, less hardness is required. (As a comparison, high hard armor, which typically only comes in thicknesses less than 2.54 cm, has a hardness range of HB 477–534 (nominally Rc 49.6–53.5) required by Mil-Spec-A-46100.)

(2) The flow stress for plastic deformation is typically rate dependent, with a larger flow stress for higher rates of deformation. Since the strain rate has units s^{-1}, the strain rate scales as $1/\lambda$. Hence, a target that is $\lambda = 10$ times larger than a baseline target will, during impact, have strain rates one-tenth those of the baseline target. Lower strain rates lead to lower flow stress and thus larger targets appear weaker.

(3) Some theories of damage in a material invoke either an inherent length scale (such as a grain boundary or a flaw size) or an inherent time for damage to occur. This length or time is independent of material sample size. Thus, since the large scale event takes longer, damage occurs relatively faster, compared to a small target, and thus the larger target appears weaker. Similarly, if damage depends on an initial flaw distribution, then a larger target will have a more extensive and diverse distribution of flaws and will appear weaker.

The theory of fracture mechanics, which deals with pre-existing sharp cracks of a given length being loaded by distant boundary conditions, has a scale size dependence, as the stress intensity behaves as the square root of the length of the crack. As to item (2), we note that computations performed with a strain-rate effect in the material strength models typically do not show a large enough effect to reproduce observed penetration differences due to scale. As to item (3), in Chapter 9 we present a model where failure is due to slippage of a shear band

TABLE 3.6. Hardness specification for Class 1 and Class 3 RHA in terms of thickness (see Exercises 6.16 and 6.17 for descriptions of the Brinell and Rockwell C hardness tests).

Thickness range (cm)	Brinell hardness 3,000 kg load BHN = HB	Brinell indentation diameter (mm)	Rockwell C hardness HRC = Rc
0.635–1.27	341–388	3.30–3.10	36.6–41.8
1.27–1.91	331–375	3.35–3.15	35.4–40.5
1.91–3.18	321–375	3.40–3.15	34.3–40.5
3.18–5.08	293–331	3.55–3.35	30.9–35.4
5.08–10.2	269–311	3.70–3.45	27.7–33.0
10.2–17.8*	241–277	3.90–3.65	22.7–28.7
17.8–22.9*	223–262	4.05–3.75	19.2–26.7
22.9–30.5*	212–248	4.15–3.85	16.4–24.2

*Beyond 15.2 cm RHA is only for ammunition testing (Class 3). Class 1 is wrought armor plate that is heat treated to develop maximum resistance to penetration and is used as armor on vehicles. Class 2 is wrought armor plate that is heat treated to develop maximum resistance to shock (0.635 to 3.175 cm has BHN 277–321). Class 3 is hardened to specific values for ammunition testing.

over a certain finite distance, that with certain assumptions translates into a strain to failure that depends on strain rate to the 2/3rds power. It is able to replicate nonlinear size scale behavior in cratering. The difficulties related to modeling scale effects are that size scale effects are most often related to material failure, which is the most challenging aspect of materials to model.

3.16. Exact Solution for an Arbitrary Strain Increment

Our plasticity solutions so far have taken advantage of geometric symmetries of the problem. However, even in the absence of such symmetries there is an exact solution to the problem of a finite increment of strain from an arbitrary initial stress state for von Mises J_2 plasticity with no work hardening [124]. To simplify some of the notation, the finite strain step will be written as a strain rate tensor multiplied by an increment of time Δt. Since the von Mises yield surface depends only on the stress deviator, only the strain deviator e_{ij} will be required. (If there is a change in volumetric strain, the pressure change is elastic and is not affected by the plastic step.) The finite deviatoric strain increment is

$$\Delta e_{ij} = \dot{e}_{ij}\Delta t = (\dot{e}^e_{ij} + \dot{e}^p_{ij})\Delta t. \tag{3.206}$$

Solving the problem means determining how the strain increment is partitioned into elastic and plastic parts. It will be assumed that the initial stress deviator state s^0_{ij} is on the yield surface (otherwise, subtract out the elastic portion of the deformation required to place the stress on the yield surface). Thinking of the stress deviator tensor and the strain rate deviator tensor as vectors, the angle between the stress and the strain rate will be defined by

$$|s||\dot{e}|\cos(\psi) = s_{ij}\dot{e}_{ij}, \tag{3.207}$$

where $||$ is the norm of the vector – for example, $|s| = \sqrt{s_{ij}s_{ij}}$. In order that the deformation be plastic, it is necessary that $s_{ij}\dot{e}_{ij} \geq 0$, otherwise the initial deformation would be elastic (it would point into the interior of the region defined by the yield surface). The initial angle will be denoted ψ_0, where ψ_0 is chosen such that $0 \leq \psi_0 \leq \frac{\pi}{2}$. The final stress state will be s_{ij}^1 with angle ψ_1. The elastic strain rate deviator is given by the difference of the total strain rate deviator and the plastic strain rate deviator, which is a multiplier times the current stress deviator according to the von Mises flow rule. Thus, the elastic strain rate deviator can be written in terms of the total strain rate deviator and the stress deviator. Hooke's law says $\dot{s}_{ij} = 2G\dot{e}_{ij}^e$, which then implies that the stress will always lie in the plane defined by s_{ij}^0 and $\dot{e}_{ij}$, viewed as vectors. To be specific, throughout the deformation, the stress state can be written as a linear combination of the initial stress state and strain rate,

$$s_{ij} = a_0 s_{ij}^0 + a_1 \frac{R}{|\dot{e}|}\dot{e}_{ij}, \qquad (3.208)$$

where $R = \sqrt{2/3}Y$ is the radius of the yield surface and a_0 and a_1 are functions of time, though below they specifically represent their values at the end of the time step Δt. Since the stress always lies on the yield surface, $s_{ij}s_{ij} = R^2$. Taking the inner product of Eq. (3.208) at the end of the step (when $s_{ij} = s_{ij}^1$) with s_{ij}^0, s_{ij}^1, and $R\dot{e}_{ij}/|\dot{e}|$ yields three equations,

$$s_{ij}^1 s_{ij}^0 = a_0 R^2 + a_1 R^2 \cos(\psi_0),$$
$$R^2 = a_0 s_{ij}^1 s_{ij}^0 + a_1 R^2 \cos(\psi_1), \qquad (3.209)$$
$$R^2 \cos(\psi_1) = a_0 R^2 \cos(\psi_0) + a_1 R^2.$$

Setting the two $s_{ij}^0 s_{ij}^1$ terms equal to each other and combining with the third equations, plus the third equation, leads to two equations for a_0 and a_1,

$$a_0^2 + 2a_0 a_1 \cos(\psi_0) + a_1^2 = 1, \qquad (3.210)$$
$$a_0 \cos(\psi_0) + a_1 = \cos(\psi_1).$$

Solving both these equations for a_1 and setting them equal gives

$$-a_0 \cos(\psi_0) \pm \sqrt{1 - a_0^2 \sin^2(\psi_0)} = \cos(\psi_1) - a_0 \cos(\psi_0). \qquad (3.211)$$

This gives

$$a_0 = \frac{\sin(\psi_1)}{\sin(\psi_0)}. \qquad (3.212)$$

Equation (3.210) then gives

$$a_1 = \frac{\sin(\psi_0 - \psi_1)}{\sin(\psi_0)}. \qquad (3.213)$$

Though we said the derivation was for the last time step, that fact was never explicitly used, and so in fact we can write the stress over time as

$$s_{ij} = \frac{\sin(\psi)}{\sin(\psi_0)}s_{ij}^0 + \frac{\sin(\psi_0 - \psi)}{\sin(\psi_0)}\frac{R}{|\dot{e}|}\dot{e}_{ij} \qquad (3.214)$$

and, in particular, this holds for the state of stress at the end of the strain step with $\psi = \psi_1$.

The previous result is purely geometric, viewing the s_{ij} and $\dot{e}_{ij}$ as vectors. To determine ψ it is necessary to include some of the mechanics of the deformation. Since $\dot{e}_{ij}$ is constant, differentiation of Eq. (3.207) gives

$$-|s||\dot{e}|\sin(\psi)\dot{\psi} = \dot{s}_{ij}\dot{e}_{ij}. \tag{3.215}$$

The deformation rate is comprised of both plastic and elastic parts,

$$\dot{e}_{ij} = \dot{e}^p_{ij} + \dot{e}^e_{ij}, \tag{3.216}$$

and the stresses are due to the elastic strains, so

$$\dot{s}_{ij} = 2G\dot{e}^e_{ij} = 2G(\dot{e}_{ij} - \dot{e}^p_{ij}). \tag{3.217}$$

The inner product of this with $\dot{e}$ is

$$\dot{s}_{ij}\dot{e}_{ij} = 2G\left(|\dot{e}|^2 - \dot{e}^p_{ij}\dot{e}_{ij}\right). \tag{3.218}$$

From the von Mises flow rule,

$$\dot{e}^p_{ij} = \Lambda s_{ij}, \tag{3.219}$$

which gives on the yield surface,

$$\dot{e}^p_{ij}s_{ij} = \Lambda R^2. \tag{3.220}$$

During plastic deformation, the stress stays on the yield surface $f(\sigma_{ij}) = 0$, and differentiation gives

$$\frac{df}{dt} = \frac{\partial f}{\partial \sigma_{ij}}\frac{d\sigma_{ij}}{dt} = s_{ij}(-\dot{p}\delta_{ij} + \dot{s}_{ij}) = s_{ij}\dot{s}_{ij} = 2Gs_{ij}\dot{e}^e_{ij} = 0, \tag{3.221}$$

saying that as vectors s_{ij} is perpendicular to $\dot{e}^e_{ij}$, and so Eq. (3.220) is equivalent to

$$\dot{e}_{ij}s_{ij} = \Lambda R^2. \tag{3.222}$$

Thus, we have

$$\dot{e}^p_{ij} = \Lambda s_{ij} = \frac{\dot{e}_{kl}s_{kl}}{R^2}s_{ij}, \tag{3.223}$$

and placing this expression in Eq. (3.218) gives

$$\dot{s}_{ij}\dot{e}_{ij} = 2G\left(|\dot{e}|^2 - \frac{(\dot{e}_{ij}s_{ij})^2}{R^2}\right). \tag{3.224}$$

Equations (3.207), (3.215) and (3.224) combine to yield

$$-|s||\dot{e}|\sin(\psi)\dot{\psi} = 2G\left(|\dot{e}|^2 - \frac{|s|^2|\dot{e}|^2\cos^2(\psi)}{R^2}\right) \tag{3.225}$$

or

$$\frac{\dot{\psi}}{\sin(\psi)} = \frac{-2G|\dot{e}|}{R}. \tag{3.226}$$

This can be integrated to give

$$\ln\left(\frac{\sin(\psi)}{1 + \cos(\psi)}\right)\bigg|_{\psi_0}^{\psi_1} = -\frac{2G|\dot{e}|}{R}\Delta t. \tag{3.227}$$

The half angle formula and the summation formula for the sine give

$$\frac{\sin(\psi)}{1 + \cos(\psi)} = \frac{2\sin(\psi/2)\cos(\psi/2)}{2\cos^2(\psi/2)} = \tan\left(\frac{\psi}{2}\right). \tag{3.228}$$

Exponentiation and using various definitions gives

$$\tan\left(\frac{\psi_1}{2}\right) = \exp\left(\frac{-\sqrt{6}G\,|\Delta e|}{Y}\right)\tan\left(\frac{\psi_0}{2}\right). \tag{3.229}$$

This final expression gives us the angle at the end of the elastic-plastic step. As the step proceeds, ψ decreases, and the stress deviator becomes more aligned with the strain step direction. As $0 \le \psi_0 \le \frac{\pi}{2}$, the largest the initial half-angle tangent term will be is 1, and since $G/Y \approx 70$ for strong metals, a strain of 3% reduces $\psi_0 = 90°$ to $\psi_1 = 0.67°$, thus showing that with relatively small strains the stresses will align with the direction of the deformation. This result shows that the rigid plastic approach we use in later chapters to determine the stresses produces very accurate stresses.

THEOREM 3.9. *For an elastic-perfectly plastic von Mises material, the stress deviators approach the rigid plastic stresses exponentially fast as a function of strain.*

It is possible to write the new stress by taking the ψ_1 calculated here and by inserting the angle into Eq. (3.214). In particular, a little algebra yields (Exercise 3.37)

$$a_0 = \frac{\sin(\psi_1)}{\sin(\psi_0)} = \frac{2r}{1+(1-r^2)\cos(\psi_0)+r^2}, \quad \text{where } r = \exp\left(\frac{-\sqrt{6}G\,|\Delta e|}{Y}\right),$$

$$a_1 = \frac{\sin(\psi_0-\psi_1)}{\sin(\psi_0)} = \frac{1+(1-r)^2\cos(\psi_0)-r^2}{1+(1-r^2)\cos(\psi_0)+r^2}. \tag{3.230}$$

The elastic step can now be computed as $\Delta e^e_{ij} = (s^1_{ij} - s^0_{ij})/2G$ and the plastic step as $\Delta e^p_{ij} = \Delta e_{ij} - \Delta e^e_{ij}$.

As another use of this result, consider the plastic dissipation rate,

$$\dot{W}_p = s_{ij}\dot{\varepsilon}^p_{ij} = s_{ij}\dot{\varepsilon}_{ij} = |s||\dot{\varepsilon}|\cos(\psi). \tag{3.231}$$

As time increases, $\dot{W}_p$ tends towards its maximal value of $|s||\dot{\varepsilon}| = \sqrt{2/3}\,Y|\dot{\varepsilon}|$. Notice that if the stress is inside the yield surface, then a formal calculation of $s_{ij}\dot{\varepsilon}^p_{ij}$ will also be less than $\sqrt{2/3}\,Y|\dot{\varepsilon}|$. Thus, the stress tends towards a state that maximizes the plastic dissipation. In some presentations of plasticity a *postulate of maximum dissipation* is presented, but we are not doing that here as the author sees no reason why nature would maximize dissipation. However, the observation that the stress tends towards a state that maximizes dissipation allows us to use the idea of maximizing dissipation as an approach to finding approximate solutions to plasticity problems.

The *dynamic plasticity* approach is as follows. We assume that the plasticity flowing region is parameterized by parameters α_i that are functions of time; in other words, the velocity field in the plastically flowing region can be written as $\vec{v}(\vec{x}, \alpha_1(t), ..., \alpha_n(t))$. These parameters define the extent, shape, and magnitude of the velocity field and hence the extent of the plastically flowing region. The intent is to determine how the velocity field evolves over time, hence to determine $\alpha_i(t)$ as a function of time. Assuming the α_i are known at a given time (i.e., the velocity is known at time t), that translates into determining $\dot{\alpha}_i$ at time t. We assume rigid plasticity so that we can explicitly compute the plastic rate of deformation tensor

D_{ij}^p from this assumed velocity field. We next compute the acceleration of this velocity field, $\vec{a} = D\vec{v}/Dt$. For example, the acceleration in Cartesian coordinates for the parameterized velocity field is

$$a_i = \frac{Dv_i}{Dt} = \frac{\partial v_i}{\partial t} + v_j \frac{\partial v_i}{\partial x_j} = \frac{\partial v_i}{\partial \alpha_1}\dot{\alpha}_1 + \cdots + \frac{\partial v_i}{\partial \alpha_n}\dot{\alpha}_n + v_j \frac{\partial v_i}{\partial x_j}. \tag{3.232}$$

The $\dot{\alpha}_i$ appear in the acceleration linearly. This linear dependence is also true in other coordinate systems. Next, we formally solve the momentum balance

$$\frac{\partial \sigma_{ji}^a}{\partial x_j} = \rho a_i = \rho \frac{Dv_i}{Dt}. \tag{3.233}$$

The formal stress σ^a need not be constitutively related to D^p, it just needs to formally satisfy the momentum balance. The stress σ^a depends linearly on the $\dot{\alpha}_i$. Further, to be an admissible stress (hence the superscript a), σ^a needs to lie on or within the yield surface and satisfy the appropriate boundary conditions. Now we maximize the plastic dissipation $\sigma_{ij}^a D_{ij}^p$, where again the stress being admissible means the maximization is subject to the constraints that the formal stress is within or on the yield surface and subject to the appropriate boundary conditions (b.c.),

$$\max_{\text{admissible } \sigma_{ij}^a(\dot{\alpha}_k)} \sigma_{ij}^a(\dot{\alpha}_k)D_{ij}. \tag{3.234}$$

D^p is known because the velocity at the current time is known. Since σ^a is linear in the unknowns (the $\dot{\alpha}_i$), the specific maximal value of the dissipation will occur for $\dot{\alpha}_i$ determined by the constraints, either the yield surface or boundary conditions. Symbolically,

$$\vec{v}(x_i, \alpha_i(t)) \xrightarrow{\rho D/dt} \rho\vec{a} \xrightarrow{(\partial/\partial x_j)^{-1}, \sqrt{3J_2(\sigma^a)} \le Y, \text{ b.c.}} \sigma^a \xrightarrow{\max \sigma_{ij}^a D_{ij}^p} \dot{\alpha}_i.$$

We use this maximization approach twice in this text to find the extent of the plastically flowing field (determined by $\alpha_i(t)$). The dynamic plasticity approach is typically quite tedious to use because of the role of the constraints (for other examples see [190]).

We finish this section by describing how computations are performed in most plasticity algorithms in finite element and finite volume codes. The breakthrough in numerically modeling plasticity was Mark Wilkins' "radial return" method. In it, one is given a deformation increment Δe_{ij}, the stress deviator update is computed assuming elastic response, $s_{ij}^0 + 2\mu\Delta e_{ij}$, and then, if the computed stress is outside the yield surface, the stresses are reduced on a radial line connecting to the point in stress deviator space to the origin until the stress is on the yield surface. Essentially all solution algorithms in numerical plasticity are variations on this theme. A feature of this approach is that the algorithm predicts the same result if the beginning point is on or within the yield surface; thus, it is not necessary to determine when the stress first reaches the yield surface. However, to compare the radial return method to the exact solution, we assume the starting point is on the yield surface for both. The equation for the radial return method is then, assuming $\cos(\psi_0) \ge 0$,

$$(s_{ij}^1)_{rr} = Y \frac{s_{ij}^0 + 2\mu\Delta e_{ij}}{\sqrt{3J_2(s_{ij}^0 + 2\mu\Delta e_{ij})}} = \frac{s_{ij}^0 + 2\mu\Delta e_{ij}}{\sqrt{1 + 2x\cos(\psi_0) + x^2}}, \quad x = \frac{2\mu|\Delta e|}{R}. \tag{3.235}$$

This solution is a linear combination of s^0 and Δe, just as the exact solution is. Thus, a reasonable way to compare them is to compare the arc length on the sphere in stress deviator space connecting s^0 and s^1, since for both solutions the beginning and end stress are on the yield surface. The arc length is given by $R\cos^{-1}(s^0_{ij}s^1_{ij}/R^2)$, and some arithmetic yields

$$
\begin{aligned}
\text{dist}_{\text{exact}} &= R\cos^{-1}(a_0 + a_1\cos(\psi_0)) \\
&= R\cos^{-1}\left\{\frac{2r + (1-r)\{1+r+(1-r)\cos(\psi_0)\}\cos(\psi_0)}{1+r^2+(1-r^2)\cos(\psi_0)}\right\}, \quad r = e^{-x}, \\
&= R\cos^{-1}(1 - \frac{1}{2}\sin^2(\psi_0)x^2 + \frac{1}{2}\cos(\psi_0)\sin^2(\psi_0)x^3 + \cdots) \\
&= R\sin(\psi_0)(x - \frac{1}{2}\cos(\psi_0)x^2 + \cdots), \\
\text{dist}_{rr} &= R\cos^{-1}\left\{\frac{1+x\cos(\psi_0)}{\sqrt{1+2x\cos(\psi_0)+x^2}}\right\} \\
&= R\cos^{-1}(1 - \frac{1}{2}\sin^2(\psi_0)x^2 + \cos(\psi_0)\sin^2(\psi_0)x^3 + \cdots) \\
&= R\sin(\psi_0)(x - \cos(\psi_0)x^2 + \cdots),
\end{aligned}
$$

$$(3.236)$$

where

$$
\cos^{-1}(z) = \sqrt{2(1-z)}\left(1 + \frac{1}{6}(1-z) + \frac{3}{160}(1-z)^2 + \cdots\right) \tag{3.237}
$$

has been used. The radial return agrees with the exact solution to first order in the size of the strain increment, and it is seen that the distance along the yield surface is larger for the exact solution than for the radial return. Once this computation is performed, the elastic and plastic partition of the strain step is immediately available since $\Delta e^e = (s^1 - s^0)/2\mu$.

3.17. Sources

Linear elasticity as presented above has been studied since the early 1800s and is an important part of mechanics. Plasticity theory was developed in the 1900s, with work still continuing today. Some references are Malvern [138], Fung [74], Sadd [169], Love [135], Hill [93], and Lubliner [136].

Exercises

(3.1) When a material is incompressible, $K \to \infty$. What are the other four elastic constants in this case? (No actual materials are incompressible.) It is possible for $\nu = 1/2$ and the material *not* be incompressible. What are the other four elastic constants in this case?

(3.2) Show that $\lambda + 2\mu \geq E$ and that equality only holds for $\nu = 0$.

(3.3) Show that the two expressions for J_2 in Eq. (3.42) are equivalent.

(3.4) Show that if $J_2 = 0$ then $J_3 = 0$.

(3.5) The name elastic-*perfectly plastic* arises from the response of the material model in uniaxial stress. For constant flow stress in uniaxial stress, what is "perfect" about the plasticity? What is "imperfect" about a work hardening material undergoing uniaxial stress loading?

(3.6) Assume that an elastic-perfectly plastic steel is deformed in uniaxial stress to a strain of 5% and then unloaded back to zero stress. Assume $Y = 1$ GPa and $E = 205$ GPa. What is the final strain ε_{11}? What are the final plastic strains ε_{11}^p, ε_{22}^p and ε_p? If all the plastic dissipation was converted into heat, what is the final temperature of the steel if the deformation was performed adiabatically at room temperature ($25°C$)? Assume $C_v = 450$ J/kg K. With the thermal expansion coefficient $\alpha = 11.7 \times 10^{-6}$/K, how much strain results from heating if the strain was performed adiabatically (show that it is negligible)?

(3.7) Repeat the previous problem (same material properties, again to 5% axial strain), but now for uniaxial strain. Assume $\nu = 0.29$. As an intermediate step, show that the axial strain required during loading so that upon unloading zero axial stress is reached when the stress state reaches the bottom side of the yield surface (i.e., the point at which any further unloading would be plastic) is

$$\varepsilon_{11} = \varepsilon_{\text{HEL}} + \frac{\tilde{\sigma}_{\text{HEL}}}{K} = \frac{4 - 2\nu}{1 + \nu}\varepsilon_{\text{HEL}} \tag{3.238}$$

(hint: in uniaxial strain, to go from a zero stress state to the yield surface is done in an axial stress change of σ_{HEL}, so to go from one side of the yield surface to the other is done in an axial stress change of $2\sigma_{\text{HEL}}$) and the final residual strain upon unloading is

$$\varepsilon_{11} = \frac{2 - 4\nu}{1 + \nu}\varepsilon_{\text{HEL}}. \tag{3.239}$$

Also, for this case, determine the final elastic strain components and the transverse stress (σ_{22}).

(3.8) Describe how the equivalent plastic strain can differ from the second invariant of the plastic strain tensor.

(3.9) Under uniaxial stress deformation of an elastic-plastic material with yield Y, for a given strain ε_{11} and a given stress σ_{11} such that $-Y \leq \sigma_{11} \leq Y$, explicitly describe a loading path to achieve the given stress for the given strain, thus demonstrating the path dependent nature of the stress-strain state. Write the final stress state in terms of ε_{11} and ε_{11}^p. Sketch ("color in") the graph of achievable stress-strain states for uniaxial stress loading.

(3.10) In uniaxial stress, show by considering a closed loading path that includes both plastic deformation in loading and plastic deformation in unloading that a large equivalent plastic strain ε_p can occur while ε_{11}^p is still small. In particular, you are demonstrating that, for complicated loading histories, there may be little relation between the current plastic strain state and the accumulated equivalent plastic strain.

(3.11) A common form of work hardening is the linear work hardening assumption, where for uniaxial stress the elastic part has modulus E and the work hardening region has modulus H, where $0 < H < E$. Thus, if Y_0 and $\varepsilon_0 = Y_0/E$ are the yield stress and strain at yield, the stress for a uniaxial stress is

$$\sigma_{11} = Y_0 + H(\varepsilon_{11} - \varepsilon_0) = Y_0 + B\varepsilon_p. \tag{3.240}$$

Solve the elastic-plastic incremental deformation problem after yield for the uniaxial stress loading condition assuming linear work hardening. Show that $B = H/(1 - H/E)$. In a sentence, why is B larger than H? (Hint: show $d\sigma_{11} = Hd\varepsilon_{11} = Bd\varepsilon_p$ and then solve for the rest of the incremental terms.)

(3.12) (kinematic hardening) As in the previous exercise, assume a linear uniaxial stress-strain response in the elastic-plastic region with modulus H. Now assume kinematic hardening rather than isotropic hardening, so that $f(\sigma) = \sqrt{3J_2(\sigma - \beta)} - Y_0$. Show that for uniaxial stress the yield function is $f = |\sigma_{11} - \beta_{11}| - Y_0$. Write equations for β_{11} in terms of ε_{11} and in terms of ε_{11}^p (as an aside, from the previous exercise we know $\varepsilon_{11}^p = \varepsilon_p$ for uniaxial stress). Write the corresponding incremental forms. (Hint: use Eq. 3.240 to show $\beta_{11} = H(\varepsilon_{11} - \varepsilon_0)$ during plastic flow and proceed from there.) Show that a general expression for the evolution of the back stress deviator that agrees with your incremental β_{11} is $d\beta'_{ij} = \frac{2}{3}Bd\varepsilon_{ij}^p$, where B is given in the previous exercise. (Hint: since f is a function of J_2, only the deviator of the back stress actually enters in, so compute it given that the stress state is uniaxial stress.) Now assume $H = \frac{1}{3}E$. Plot the uniaxial stress stress-strain curve from an initial stress free condition to $\sigma_{11} = 2Y_0$. Then unload to zero strain. Do the same for the isotropic hardening model of the previous exercise and compare to the kinematic hardening model. What is the stress after unloading to zero strain for each hardening model?

(3.13) Solve the elastic-plastic incremental deformation problem after yield for the uniaxial strain loading condition assuming linear work hardening (as described in the previous two exercises). Show that the slope of the loading curve in the plastic regime is given by

$$K + \frac{4\mu}{3}\left\{\frac{B}{B + 3\mu}\right\}$$

and show that this modulus lies between K and $\lambda + 2\mu$.

(3.14) Another form for representing work hardening in metals is $Y = Y_0(1 + \beta\varepsilon_p)^n$, where the 1 inside the parenthesis avoids an infinite tangent modulus for initial yield as is seen in the Johnson–Cook form for $n < 1$. For different heat treats of 4340 steel, fits to this form given by Reaugh et al. [160] are

Intended Rc	Measured Rc	Y_0 (GPa)	β	n	Final strain
35	34	1.03	125	0.0700	0.86
40	40	1.29	282	0.0389	0.67
45	43	1.32	580	0.0350	0.66

where true strain was measured by optical means of the diameter D in the gage section, $\varepsilon_{\text{true}} = -2\ln(D/D_0)$. The final strain is the final true strain they reported. For this exercise, plot these three curves as well as the Johnson–Cook curve in Appendix C. Discuss these flow stress curves in the context of the RHA and high hard armor descriptions in Table 3.6 and show that Eq. (7.2) refers to vicinity of yield and not work hardened flow stress.

(3.15) This exercise examines the expansion or compression of a cylinder of rigid plastic material. Assume the z-axis is the axis of symmetry, and that the velocity distribution along the z-axis is given by $v_z(r, z) = Vz/L$, where $L(t) = L_0 + Vt$ is the length of the cylinder vs. time with V the constant velocity at the $z = L(t)$ end. Assume the deformation is axially symmetric, meaning there is no θ dependence to the problem and the θ component of the velocity is zero. Show from volume conservation $D_{rr} + D_{\theta\theta} + D_{zz} = 0$ that $v_r = -(1/2)Vr/L$. From the velocity field show the equivalent plastic strain rate is uniform throughout the cylinder and equal to $\dot{\varepsilon}_p = |V|/L$. Integrated over time, show that the equivalent plastic strain is

$$\varepsilon_p = \int_0^t \dot{\varepsilon}_p dt = \int_0^t \frac{|V|}{L} dt = \int_0^t \frac{|V|}{L_0 + Vt} dt = \left| \ln \left(\frac{L}{L_0} \right) \right| = |\ln(1 + \varepsilon_{\text{eng}})|, \quad (3.241)$$

confirming that the equivalent plastic strain is the logarithmic strain.

(3.16) The plasticity theory presented in this chapter is referred to as incremental plasticity. There is another plasticity theory, referred to as *deformation plasticity*, which is really a nonlinear elasticity theory. A nonlinear stress-strain relation is given, and it is assumed that the loading is proportional (i.e., the ratios of the components of the stress tensor remain the same through loading) and that only loading occurs, as unloading would require a different stress-strain relation (notice that incremental plasticity is applicable when both these assumptions do not hold). Show the following nice result of deformation plasticity: for a uniaxial stress test of a cylindrical specimen that conserves volume with no necking during loading and that has an axial stress-strain relation $\sigma = E\varepsilon^n$, the hardening exponent n is equal to the logarithmic strain at the maximum load (the load is cross-sectional area times stress) – that is, $n = \varepsilon_{\text{max load}} = \ln(L_{\text{max load}}/L_0)$, where L is the gage length.

(3.17) In our approach to plasticity, the stress always lies within or on the yield surface and we include rate effects by causing the yield surface to depend on the strain rate. Another approach in plasticity theory, referred to as viscoplasticity, is to allow the stress to exceed the yield surface and then relax back to the yield surface (the terminology is that there is an overstress relaxing back to the yield surface). Let's explore this approach in uniaxial stress under a constant strain rate deformation. An example equation for the stress is, assuming the relaxation is a linear function of the overstress (the amount by which the stress exceeds the yield surface),

$$\frac{d\sigma_{11}}{dt} = E\dot{\varepsilon}_{11} - \frac{(\sigma_{11} - Y)_+}{\eta}, \quad (3.242)$$

where stress grows as $E\dot{\varepsilon}_{11} > 0$ and where η is a viscosity of some sort that controls the stress relaxation rate back to the yield value of Y. Assume $\sigma_{11}(0) = Y$ and solve this equation as a function of time. Assuming the strain carries on to late time, what is the limiting value of the stress? (Ans: the limiting increase in stress due to the strain rate is $\eta E\dot{\varepsilon}_{11}$.) Notice that this limiting value can be determined without solving the differential equation by assuming a steady state situation. Now suppose we had a general case where instead of $(\sigma_{11} - Y)_+/\eta$ for the relaxation function we have $(g(\sigma_{11} - Y))_+/\eta$ where $g(0) = 0$ at yield, $dg/d\sigma > 0$, and $g(\sigma) > 0$ means yield is occurring. A common experimental fit to rate dependence is a logarithm in strain rate, such as in Johnson–Cook $Y(\dot{\varepsilon}) = f(\dot{\varepsilon}) = Y_0(1 + C \ln(\dot{\varepsilon}/\dot{\varepsilon}_0))$. Show the steady state limiting value leads to solving the equation $f(\dot{\varepsilon}) = Y + g^{-1}(\eta E\dot{\varepsilon})$. Solve this equation for the logarithmic/Johnson–Cook form. There is a problem. What is it? (Notice that $g(0) \neq 0$.) Thus the overstress/relaxation approach cannot exactly produce a logarithmic form of strain-rate dependence. Assume $g(\sigma - Y) = (\sigma - Y)^b$. Show the interesting behavior (observed for some materials) that for $0 < b < 1$ there is a small increase in stress at low strain rates but there is a large increase in the strain-rate effect for high strain rates. See also Eq. (5.113).

(3.18) For the spherical stress state $\sigma_{ij} = \sigma\delta_{ij} = -p\delta_{ij}$, show that the conservation of energy presented in Chapter 2 of $\rho(DE/Dt) = \sigma_{ij}D_{ij}$ is equivalent to $E = \int \sigma d(1/\rho) = -\int p\, d(1/\rho)$.

(3.19) Matrix invariants can be combined in many functional ways to produce other invariants, but there can at most be three independent invariants for a 3×3 symmetric matrix. But consider the following, the functions $\{1, x, x^2, x^3\}$ would be considered independent in the context of linear independence, since they cannot be multiplied by nonzero constant coefficients and summed to zero. However, they are not independent in our context of independent functions, since they are all functions of x, namely x^n; for a given value of x one never gets a different value of x^2, hence x and x^2 are not independent. How do we know that $\{I_1, I_2, I_3\}$ are independent – for example, how do we know that $I_3 \neq f(I_1, I_2)$ for some function f? The most direct approach to show independence is to determine stress states where I_1 and I_2 are the same, but I_3 differs, thus showing that no function f exists. Do this for the three matrix invariants by using the diagonal matrix pairs of $\pm(2, -1, -1)$; $\pm(1, 0, 0)$; and $(0, 0, 0)$ and $(1, -1, 0)$. These three pairs show $\{I_1, I_2, I_3\}$ are independent. Which two pairs show that J_2 and J_3 are independent? Why do you need both?

(3.20) If a 2×2 matrix is not symmetric then there are more than two invariants for the action of rotation. For the matrix $\begin{pmatrix} a & b \\ c & d \end{pmatrix}$ show through direct calculation that $b - c$ is invariant. Show by judicious choice of 2×2 matrices that the three invariants trace, determinant, and $b - c$ are independent, as described in the previous exercise.

(3.21) When one looks down the axis in principal stress space defined by $\sigma_{11} = \sigma_{22} = \sigma_{33}$ (principal means $\sigma_{12} = \sigma_{23} = \sigma_{31} = 0$), the view (projection) of the stress is referred to as the π plane. This plane has the three axes of principal deviator stress located $120°$ apart. If θ measures the angle of the stress from the s_{11}-axis, then the principal stress deviators can be written

$$s_{11} = r\cos(\theta),$$
$$s_{22} = r\cos(\theta - 2\pi/3), \qquad\qquad (3.243)$$
$$s_{33} = r\cos(\theta - 4\pi/3),$$

where r is related to the magnitude of the stress. Confirm that $s_{ii} = 0$. Using this form show that $r = (2/3)\sqrt{3J_2}$ and

$$J_3 = s_{11}s_{22}s_{33} = \frac{1}{4}r^3\cos(3\theta) = \frac{2}{27}(3J_2)^{3/2}\cos(3\theta). \qquad (3.244)$$

(Hints: the algebra is simpler if you use $s_{33} = -s_{11} - s_{22}$ to get rid of s_{33} before inserting the trigonometric functions; or multiply it out using complex exponentials recalling $1 + \exp(2\pi i/3) + \exp(-2\pi i/3) = 0$.) As an aside, notice $\zeta = \cos(3\theta)$. Thus, if $-\pi/6 < \theta < \pi/6$ then J_3 is positive and the material is in a tensile stress state, if $\theta = \pm\pi/6$ then $J_3 = 0$ and the stress state is pure shear, and if $\pi/6 < \theta < \pi/2$ then J_3 is negative and the stress state is compressive. This pattern repeats as one continues around the π plane. θ, the Lode angle, is an invariant of the stress tensor, and can be used instead of J_3 as one of the three invariants of the stress tensor. Going around the circle in the π plane there are three values of θ where the stress state is uniaxial tension, three values where the stress state is uniaxial compression, and three values where the stress state is pure shear. The stress principal deviators can be viewed as a projection of the principal stress into the deviatoric plane by

$$\begin{pmatrix} s_{11} \\ s_{22} \\ s_{33} \end{pmatrix} = \frac{1}{3} \begin{pmatrix} 2 & -1 & -1 \\ -1 & 2 & -1 \\ -1 & -1 & 2 \end{pmatrix} \begin{pmatrix} \sigma_{11} \\ \sigma_{22} \\ \sigma_{33} \end{pmatrix}. \tag{3.245}$$

The orientation of the (x, y) plane in Fig. 3.5 corresponds to the π plane coordinates

$$x = \frac{s_{33} - s_{22}}{\sqrt{3}}, \quad y = s_{11} \quad \Longleftrightarrow \quad s_{11} = y, \quad s_{22} = -\frac{\sqrt{3}}{2}x - \frac{y}{2}, \quad s_{33} = \frac{\sqrt{3}}{2}x - \frac{y}{2}. \tag{3.246}$$

In the π plane, the von Mises yield surface has radius $r = \sqrt{x^2 + y^2} = 2Y/3$.

(3.22) Using the results of Exercise 3.21, it is possible to write down an explicit expression for the maximum shear stress. Using the standard range for the arccosine, Eq. (3.244) yields for the Lode angle (ζ is from Eq. 3.163)

$$\theta = \frac{1}{3} \cos^{-1}\left(\frac{27 J_3}{2(3J_2)^{3/2}}\right) = \frac{1}{3} \cos^{-1}(\zeta) \quad \text{where} \quad 0 \le \theta \le \frac{\pi}{3}. \tag{3.247}$$

Using Eqs (3.83) and (3.243), show that

$$\text{maximum shear stress} = \frac{1}{2}\sqrt{3J_2}\left(\cos(\theta) + \frac{\sqrt{3}}{3}\sin(\theta)\right). \tag{3.248}$$

It is not possible to avoid computational expense, but it can be reduced from trigonometric evaluations to solving a simple cubic. In particular, show $\cos(3\phi) = 4\cos^3(\phi) - 3\cos(\phi)$ and hence, with $x = \cos(\theta)$, if one solves

$$4x^3 - 3x - \zeta = 0 \quad \text{where} \quad \frac{1}{2} \le x \le 1, \tag{3.249}$$

then Eq. (3.248) can be used with $\cos(\theta) = x$ and $\sin(\theta) = \sqrt{1 - x^2}$.

(3.23) Using the results of Exercise 3.21 and the previous exercise, write down explicit expressions for the principal stress deviators and the principal stresses in terms of the three invariants p, J_2, and J_3. Order them. Also show that $3^3|J_3|^2 \le 2^2|J_2|^3$. (Hint: with $x = \cos(\theta)$ from the previous exercise, the principal stress deviators are $(1/3)\sqrt{3J_2} \cdot \{2x, -x+\sqrt{3}\sqrt{1-x^2}, -x-\sqrt{3}\sqrt{1-x^2}\}$.)

(3.24) Sometimes the effects of J_3 are written in terms of the Lode parameter

$$L = \frac{2\sigma_{22} - \sigma_{11} - \sigma_{33}}{\sigma_{11} - \sigma_{33}}, \tag{3.250}$$

where $\sigma_{11} \ge \sigma_{22} \ge \sigma_{33}$ are the principal stresses. Show that

$$L = 3\frac{\sqrt{3}\tan(\theta) - 1}{\sqrt{3}\tan(\theta) + 3} = \sqrt{3}\tan\left(\theta - \frac{\pi}{6}\right), \tag{3.251}$$

where θ is given in Eq. (3.247). Determine L for uniaxial stress tension, pure shear, and uniaxial stress compression.

(3.25) Show that, in the π plane, the Tresca yield criterion can be written

$$r\cos(\theta - \pi/6 - n\pi/3) = Y/\sqrt{3}, \quad n\pi/3 \le \theta \le (n+1)\pi/3, \quad n = 0, ..., 5. \tag{3.252}$$

Using this result, show Eq. (3.189).

(3.26) If a material has a yield Y_t in uniaxial stress tension and Y_c in uniaxial stress compression, there are two possible origins of the difference.

(a) There could be a J_3 dependence in the yield surface. As an example, find A and B so that

$$f(J_2, J_3) = A\sqrt{3J_2} + B\left(\frac{27J_3}{2}\right)^{\frac{1}{3}} - Y_t \qquad (3.253)$$

produces a valid yield surface. Sketch the yield surface in the π plane.

(b) There could be a pressure dependence in the yield surface. As an example, find Y_0 and b for the Drucker–Prager (Eq. 3.92) yield criterion that matches the tension and compression data.

(c) Show that both these yield surfaces produce the same yield stress in pure shear. Thus, to determine the origin of the difference it is necessary to perform tests under pressure.

(3.27) Using Eq. (3.125), show for a linear elastic material (i.e., $\varepsilon_{ij}^e = \varepsilon_{ij}$) that the stresses are derivable from the stored elastic energy as a potential with

$$\sigma_{ij} = \frac{\partial(\rho_0 E_e)}{\partial \varepsilon_{ij}}. \qquad (3.254)$$

In this context, σ and ε are conjugate variables.

(3.28) In Eqs. (3.152) and (3.172), the strain rate term was written as the typical definition in the community, namely as the equivalent plastic strain rate. This definition is used because of its availability in computational codes. However, this definition is not what is used by Johnson and Cook. They use the second invariant of the total deviatoric strain rate, defined as

$$\dot{e} = \sqrt{\frac{2}{3}\dot{e}_{ij}\dot{e}_{ij}}. \qquad (3.255)$$

Since the plastic strain is deviatoric, this definition of strain rate using the total strain is very close to the equivalent plastic strain. Given that the original authors use a different definition than almost everyone else in the community when they implement the Johnson–Cook model in their numerical codes, the question is: how much difference does it make? To simplify things and to be specific, consider the flow stresses $Y_1(\varepsilon_p, \dot{e}) = Y_0 + H\varepsilon_p + M\dot{e}$ and $Y_2(\varepsilon_p, \dot{\varepsilon}_p) = Y_0 + H\varepsilon_p + M\dot{\varepsilon}_p$, where Y_0, H, and M are constants. If the material is deformed at a constant strain rate $\dot{\varepsilon}$, what is the difference in response for the two different flow stresses? (Hint: separate the strain into elastic and plastic parts, $\varepsilon = \varepsilon^e + \varepsilon_p$, and write the stress as a function of total strain and total strain rate for each case.)

(3.29) This exercise is a follow-on from the previous one. Show for a work hardening von Mises material with a flow stress that is a function of the equivalent plastic strain only (i.e., $Y = Y(\varepsilon_p)$ and $\partial Y/\partial \varepsilon_p \geq 0$) that the total deviatoric strain rate is always greater than or equal to the equivalent plastic strain rate ($\dot{e} \geq \dot{\varepsilon}_p$). Is it ever greater?

(3.30) In this exercise we explore Eq. (3.89) as a descriptor of the Tresca yield surface. In the π plane, the straight line segments of the Tresca hexagon yield surface can be represented by $s_i - s_j = \pm Y$, where $i \neq j$. Defining

$$\bar{f} = \left((s_1 - s_2)^2 - Y^2\right)\left((s_2 - s_3)^2 - Y^2\right)\left((s_3 - s_1)^2 - Y^2\right), \qquad (3.256)$$

show that $\bar{f} = 0$ includes the Tresca yield surface hexagon. Now show that $\bar{f}$ equals the left-hand side of Eq. (3.89). Since the lines in Eq. (3.256) extend to infinity, the solution to $\bar{f} = 0$ includes more than the hexagon. Sketch a picture of $\bar{f} = 0$. Now, shade the regions where $\bar{f} > 0$. Related to this sketch, show for a uniaxial stress loading condition that $\bar{f} \leq 0$ regardless of the magnitude of the loading stress. Thus, to use $\bar{f}$ to define the yield surface requires additional information to ensure the stress stays within the hexagon (such as also examining the von Mises criterion). Finally, show that the gradient of $\bar{f}$ at the hexagon corners is zero, and hence does not define a plastic flow direction for associated flow. (Hint: look at Eq. 3.256, using the product rule and noting that a corner is defined by two intersecting lines.)

(3.31) Given both uniaxial stress (tension/compression) and pure shear (torsion) test data, it is possible to determine whether a von Mises yield criteria or a Tresca yield criteria might be superior. For the two aluminum alloys in Table 3.1, which is better?

(3.32) The damage constants D_2 for the two aluminum alloys are based on the tests presented in the text as well as additional tests. Also, not all the data are being presented – for example, the final geometry of the failure surface to provide necking information, such as is seen in the gilding metal. However, for simplicity for this exercise, assume the strains in Table 3.1 were good (i.e., there is minimal necking before failure) and compute what D_2 would be for the two aluminums based first on the torsion tests and then on the tensile tests. Finally, if D_1 is also assumed nonzero, find D_1 and D_2 that fit both the tension and the torsion data (still assume $D_3 = -3/2$).

(3.33) Compute D_2 for the APM2 gilding metal based on the assumptions in the text.

(3.34) Plot ε_p^f vs. η for Bridgman's data for 1045 steel in Table 3.2. Do these data argue for their being a relationship between these two variables? Discuss the three outliers. They can either be due to bad assumptions about the stress distribution, measurement error, variations in material, or that η isn't the best way to quantify failure strain. Comment on these points.

(3.35) Bridgman notes that for ductile metals, failure (rupture) of an internally pressurized thick-walled cylinder always originates on the outer surface (though there may be slip lines from the inner surface that do not lead to complete failure) [30]. The elastic solution for the thick-walled cylinder is similar to that for the thick-walled sphere, and shows that yielding always occurs first on the inner surface. Show that if the hoop strain is $\varepsilon_{\theta\theta} = r/r_0 - 1$, and plastic flow is volume conserving, then for a thick-walled cylinder of initial inner radius a_0 and outer radius b_0, the hoop strain relation is

$$\varepsilon_{\theta\theta}(r) = -1 + \sqrt{1 + (a_0/r_0)^2(2 + \varepsilon_{\theta\theta}(a))\varepsilon_{\theta\theta}(a)} \approx (a_0/r_0)^2\varepsilon_{\theta\theta}(a), \qquad (3.257)$$

and so $\varepsilon_{\theta\theta}(b) < \varepsilon_{\theta\theta}(a)$. Thus, both the stress and strain state are more severe on the inner surface. Explore the possible role of stress state for why final failure occurs on the outside by examining η. Assume an internal pressure such that the cylinder is entirely plastic and static, that $\sigma_{\theta\theta} = \sigma_{zz}$, and show that

$$\eta = \frac{2}{3} + \ln(r/b). \qquad (3.258)$$

Plot the hoop strain vs. η and compare this to the behavior of Fig. 3.14. Comment. What is the effect of work hardening?

(3.36) When the exact arbitrary step solution was developed in the last section, we did not, as a first step, go to principal stress space. Why not?

(3.37) Derive Eqs. (3.230). (Hint: using Eq. (3.228), let $a = r\sin(\psi_0)/(1+\cos(\psi_0))$ and using the same trigonometric identity show $\sin(\psi_1) = 2a/(1+a^2)$ and $\cos(\psi_1) = (1-a^2)/(1+a^2)$.)

(3.38) Assuming the material is at rest in front of a shock front and using the engineering strain $\varepsilon = 1 - \rho_0/\rho$, write the Hugoniot jump conditions in terms of strain instead of density.

(3.39) For the case of a transverse (shear) shock with the material at rest in front of the shock, write the Hugoniot jump conditions in terms of shear strain. (Hint: the first step is determining the strain caused by the passage of the wave – write the transverse displacement as $\Delta x_2 = u_2(t - x_1/U)$, $t \geq x_1/U$, where u_2 is the transverse velocity.)

(3.40) In this exercise we explore the stress state near the surface of a spherical void. We will demonstrate that the $\eta = -p/Y = \pm 2/3$ behavior is strictly due to being near the inner free surface of a spherical void at yield. Since the surface of the spherical void is traction free, $\sigma_{rr} = \sigma_{r\theta} = \sigma_{r\varphi} = 0$. If the loading is spherically symmetric, then $\sigma_{\theta\theta} = \sigma_{\varphi\varphi}$ and $\sigma_{\theta\varphi} = 0$. Show (a simple computation) that $p = -2\sigma_{\theta\theta}/3$ and $\sqrt{3J_2} = |\sigma_{\theta\theta}|$, thus yielding $p/Y = \pm 2/3$ always. Now, suppose the spherical void is traction free, but the local loading conditions are not assumed spherically symmetric. Then it can no longer be assumed that $\sigma_{\theta\theta} = \sigma_{\varphi\varphi}$ and $\sigma_{\theta\varphi} = 0$. Show that

$$J_2 = \frac{1}{6}\{(\sigma_{\theta\theta} - \sigma_{\varphi\varphi})^2 + \sigma_{\theta\theta}^2 + \sigma_{\varphi\varphi}^2\} + \sigma_{\theta\varphi}^2. \tag{3.259}$$

Show that the yield condition (envelope) can be written

$$\sigma_{\theta\theta}^2 - \sigma_{\theta\theta}\sigma_{\varphi\varphi} + \sigma_{\varphi\varphi}^2 \leq Y^2 - 3\sigma_{\theta\varphi}^2, \tag{3.260}$$

which describes an ellipse in the $\sigma_{\theta\theta}$ - $\sigma_{\varphi\varphi}$ plane. Show the maximum or minimum of $\sigma_{\theta\theta} + \sigma_{\varphi\varphi}$ is given when $\sigma_{\theta\theta} = \sigma_{\varphi\varphi} = \pm Y$ and that the maximum or minimum of $\sigma_{\theta\theta}$ or $\sigma_{\varphi\varphi}$ alone occurs when $\sigma_{\theta\theta} = 2\sigma_{\varphi\varphi} = \pm 2Y/\sqrt{3}$. (Hint: use Lagrange multipliers, where given a constraint function $h(x,y) = c$ and a function $f(x,y)$ to maximize or minimize, the maximum or minimum occurs when $\mathrm{grad}(h) = \lambda\mathrm{grad}(f)$, where λ is an arbitrary constant.) Thus, you have demonstrated that at the surface of the spherical void

$$|\sigma_{\theta\theta}| \leq \frac{2}{\sqrt{3}}\bar{Y} = 1.1547\bar{Y}, \qquad |\sigma_{\varphi\varphi}| \leq \frac{2}{\sqrt{3}}\bar{Y}, \qquad |\sigma_{\theta\varphi}| \leq \sqrt{3}Y,$$

$$-\frac{2}{3}\frac{\bar{Y}}{Y} \leq \eta = -\frac{p}{Y} \leq \frac{2}{3}\frac{\bar{Y}}{Y}, \qquad \text{where } \bar{Y} = \sqrt{Y^2 - 3\sigma_{\theta\varphi}^2}. \tag{3.261}$$

This result shows the pressure state cannot exceed the value that occurs for spherically symmetric loading, but it can be anything less than that. Of course, if less it is unlikely that the yield condition will hold on the entire surface of the sphere. (These results give the maximum values of the ellipse in Fig. 3.14.)

(3.41) In this exercise you will completely solve the plasticity problem of the von Mises flow in plane stress as set up in Section 3.14. We assume the initial

elastic deformation has already occurred to place the stress state on the von
Mises yield surface. The material is deformed in such a way that the stresses
are proportional (i.e., σ_{11}/σ_{22} is a constant) – solving the plasticity problem
means determining the strain increments required to do this and how the
elastic and plastic strain increment is partitioned. Show that J_2 plasticity
means that the plastic strain increments are related as

$$
\begin{aligned}
d\varepsilon_{11}^p &= s_{11}d\Lambda = (2\sigma_{11} - \sigma_{22})d\Lambda/3, \\
d\varepsilon_{22}^p &= s_{22}d\Lambda = (2\sigma_{22} - \sigma_{11})d\Lambda/3, \\
d\varepsilon_{33}^p &= s_{33}d\Lambda = -(\sigma_{11} + \sigma_{22})d\Lambda/3.
\end{aligned}
\tag{3.262}
$$

A computation shows that $d\varepsilon_p = 2Yd\Lambda/3$. (Since plastic flow is dissipative,
$dW_p = s_{ij}d\varepsilon_{ij}^p = (s_{11}^2 + s_{22}^2 + s_{33}^2)d\Lambda > 0$ and hence $d\Lambda > 0$.) In the presence
of work hardening $Y = Y(\varepsilon_p)$, differentiating the yield surface produces

$$
2\sigma_{11}d\sigma_{11} - \sigma_{11}d\sigma_{22} - \sigma_{22}d\sigma_{11} + 2\sigma_{22}d\sigma_{22} = 2YdY = 2Y\frac{dY}{d\varepsilon_p}d\varepsilon_p.
\tag{3.263}
$$

Assuming proportional loading gives

$$
d\left(\frac{\sigma_{11}}{\sigma_{22}}\right) = 0 \quad \Rightarrow \quad \frac{d\sigma_{11}}{\sigma_{11}} = \frac{d\sigma_{22}}{\sigma_{22}}.
\tag{3.264}
$$

Equations (3.263) and (3.264) combine to say that

$$
\frac{d\sigma_{11}}{\sigma_{11}} = \frac{d\sigma_{22}}{\sigma_{22}} = \frac{dY}{Y} = \frac{1}{Y}\frac{dY}{d\varepsilon_p}d\varepsilon_p.
\tag{3.265}
$$

For the elastic deformation show

$$
\begin{pmatrix} d\sigma_{11} \\ d\sigma_{22} \\ d\sigma_{33} \end{pmatrix} = \begin{pmatrix} \lambda + 2\mu & \lambda & \lambda \\ \lambda & \lambda + 2\mu & \lambda \\ \lambda & \lambda & \lambda + 2\mu \end{pmatrix} \begin{pmatrix} d\varepsilon_{11}^e \\ d\varepsilon_{22}^e \\ d\varepsilon_{33}^e \end{pmatrix}.
\tag{3.266}
$$

With the plane stress assumption $d\sigma_{33} = 0$, this yields

$$
d\sigma_{11} = \frac{2\mu(d\varepsilon_{11}^e + \nu d\varepsilon_{22}^e)}{1 - \nu}, \quad d\sigma_{22} = \frac{2\mu(\nu d\varepsilon_{11}^e + d\varepsilon_{22}^e)}{1 - \nu}, \quad d\varepsilon_{33}^e = -\frac{\nu(d\varepsilon_{11}^e + d\varepsilon_{22}^e)}{1 - \nu},
\tag{3.267}
$$

and inverting the expressions yields

$$
d\varepsilon_{11}^e = \frac{1}{E}(d\sigma_{11} - \nu d\sigma_{22}), \quad d\varepsilon_{22}^e = \frac{1}{E}(d\sigma_{22} - \nu d\sigma_{11}), \quad d\varepsilon_{33}^e = -\frac{\nu}{E}(d\sigma_{11} + d\sigma_{22}).
\tag{3.268}
$$

Placing the result of Eq. (3.265) into these expressions yields

$$
d\varepsilon_{11}^e = \frac{1}{E}(\sigma_{11} - \nu\sigma_{22})\frac{dY}{Y}, \quad d\varepsilon_{22}^e = \frac{1}{E}(\sigma_{22} - \nu\sigma_{11})\frac{dY}{Y}, \quad d\varepsilon_{33}^e = -\frac{\nu}{E}(\sigma_{11} + \sigma_{22})\frac{dY}{Y}.
\tag{3.269}
$$

Now all the terms have been computed and you are in a position to explicitly write down all the strains in terms of the equivalent plastic strain. Show

$$d\varepsilon_{11}^e = \frac{\sigma_{11} - \nu\sigma_{22}}{Y}\frac{Y'}{E}d\varepsilon_p = \pm\frac{((1-\nu)\gamma + 1 + \nu)}{\sqrt{\gamma^2 + 3}}\frac{Y'}{E}d\varepsilon_p,$$

$$d\varepsilon_{22}^e = \frac{\sigma_{22} - \nu\sigma_{11}}{Y}\frac{Y'}{E}d\varepsilon_p = \pm\frac{((1-\nu)\gamma - 1 - \nu)}{\sqrt{\gamma^2 + 3}}\frac{Y'}{E}d\varepsilon_p,$$

$$d\varepsilon_{33}^e = -\nu\frac{\sigma_{11} + \sigma_{22}}{Y}\frac{Y'}{E}d\varepsilon_p = \mp\frac{2\gamma\nu}{\sqrt{\gamma^2 + 3}}\frac{Y'}{E}d\varepsilon_p,$$

$$d\varepsilon_{11}^p = \frac{2\sigma_{11} - \sigma_{22}}{2Y}d\varepsilon_p = \pm\frac{\gamma + 3}{2\sqrt{\gamma^2 + 3}}d\varepsilon_p, \qquad\qquad (3.270)$$

$$d\varepsilon_{22}^p = \frac{2\sigma_{22} - \sigma_{11}}{2Y}d\varepsilon_p = \pm\frac{\gamma - 3}{2\sqrt{\gamma^2 + 3}}d\varepsilon_p,$$

$$d\varepsilon_{33}^p = -\frac{\sigma_{11} + \sigma_{22}}{2Y}d\varepsilon_p = \mp\frac{\gamma}{\sqrt{\gamma^2 + 3}}d\varepsilon_p,$$

where $Y' = dY/d\varepsilon_p$. The direction of the elastic deformation is not the same as the plastic deformation (Exercise 3.42). Because of the proportional loading, these equations can be explicitly integrated to give the plastic and elastic strains, where $\varepsilon_p^f(\gamma)$ is the failure strain for the given γ,

$$\varepsilon_{11}^e = \pm\frac{(1-\nu)\gamma + 1 + \nu}{\sqrt{\gamma^2 + 3}}\cdot\frac{Y(\varepsilon_p^f(\gamma))}{E}, \qquad \varepsilon_{22}^e = \pm\frac{(1-\nu)\gamma - 1 - \nu}{\sqrt{\gamma^2 + 3}}\cdot\frac{Y(\varepsilon_p^f(\gamma))}{E},$$

$$\varepsilon_{33}^e = \mp\frac{2\gamma\nu}{\sqrt{\gamma^2 + 3}}\cdot\frac{Y(\varepsilon_p^f(\gamma))}{E},$$

$$\varepsilon_{11}^p = \pm\frac{\gamma + 3}{2\sqrt{\gamma^2 + 3}}\varepsilon_p^f(\gamma), \qquad \varepsilon_{22}^p = \pm\frac{\gamma - 3}{2\sqrt{\gamma^2 + 3}}\varepsilon_p^f(\gamma), \qquad \varepsilon_{33}^p = \mp\frac{\gamma}{\sqrt{\gamma^2 + 3}}\varepsilon_p^f(\gamma).$$

$$(3.271)$$

The elastic strain is small compared to the plastic strain; the elastic strain specifically accumulated during plastic flow is given by the above, replacing $Y(\varepsilon_p^f(\gamma))$ with $Y(\varepsilon_p^f(\gamma)) - Y(0)$.

(3.42) For the solution of the plane stress plasticity problem, plot the angle between the elastic step and the plastic step for $-10 \le \gamma \le 10$. (The angle here is defined to be $\cos(\theta) = d\varepsilon_{ij}^e d\varepsilon_{ij}^p/\sqrt{(d\varepsilon_{kl}^e d\varepsilon_{kl}^e)(d\varepsilon_{mn}^p d\varepsilon_{mn}^p)}$.) Do the elastic and plastic step (assume there is work hardening: $Y' > 0$) ever point in the same direction?

(3.43) Another yield surface including a J_3 dependence was presented by Wilkins, Streit, and Reaugh [212] (the same Wilkins as that of the radial return and the JWL equation of state for explosives). They consider

$$A = \max\left(\frac{s_{22}}{s_{33}}, \frac{s_{22}}{s_{11}}\right) = \frac{2\tan(\frac{1}{3}\cos^{-1}|\zeta|)}{\sqrt{3} - \tan(\frac{1}{3}\cos^{-1}|\zeta|)}, \qquad \zeta = \frac{27J_3}{2(3J_2)^{3/2}}, \qquad (3.272)$$

and where $s_{11} \le s_{22} \le s_{33}$ are the principal stress deviators (eigenvalues) in increasing order. The first equality is their definition; you are to show the second equality using the π plane and Exercise 3.21. Using this expression show that $0 \le A \le 1$. Plot A and show $A \approx 1 - |\zeta|$ (good near $\zeta = 0$, not so good near $\zeta = 1$). They write a work hardening yield surface as

$$Y = Y_T(\varepsilon_p)A^\lambda + Y_S(\varepsilon_p)(1 - A^\lambda), \tag{3.273}$$

where Y_T is the work hardening of the material in a tensile test, Y_S is the work hardening of the material in a shear (torsion) test, and λ is a constant. (For Al 6061-T651 they use stress-strain curves from a tension test to define Y_T, from a hollow cylinder torsion test to define Y_S, and $\lambda = 0.7$.) Of interest to us is their damage function, which is

$$D = \int \frac{(2 - A)^\beta}{(1 + ap)^\alpha} d\varepsilon_p, \tag{3.274}$$

where a, α, and β are constants. Given there is a critical value of damage D_c, from this expression we can define a failure strain for a given loading condition,

$$\varepsilon_p^f = D_c \frac{(1 + ap)^\alpha}{(2 - A)^\beta}. \tag{3.275}$$

Notice there is always a pressure dependence and a J_3 dependence. For $\beta > 0$, the failure strain is less for a shear strain state than for a uniaxial stress loading state. To study this formulation, fit this equation to the data for 2024-T351 aluminum as shown in Table 3.5 and Fig. 3.14 – in other words, find values of the constants a, α, β, and D_c that reasonably fit the data, if they exist. You will need the work hardening $Y(\varepsilon_p)$ to determine the pressure from $p = -Y\eta$; use the one in Eq. (3.178) or the Johnson–Cook values in Appendix C. Then with ε_p^f, η, and ζ from the table you can determine p and A, at which point it becomes a fitting exercise (ignore the fact that it is not quite proportional loading). By taking logarithms, three of the variables can be fit with linear least squares, leaving just a as your nonlinear parameter to deal with in some other way (see [208] for one such fit).[9]

(3.44) Write the failure model of Eq. (3.201) in terms of the plastic strain to fail in shear and the work hardening coefficient. What is the implied relation between failure strains in pure shear and uniaxial tension (assuming $\eta = 1/3$)?

(3.45) Make a three-dimensional plot of the failure surface Eq. (3.201) in $(\eta, \zeta, \varepsilon_p^f)$ space. Plot the plane stress curve on the surface.

(3.46) In 1977 Gurson introduced a yield surface that attempts to include porosity in its formulation. The surface is

$$f(\sigma_{ij}, \phi) = f(I_1, J_2, \phi) = 3J_2 + Y^2[2\phi \cosh(I_1/2Y) - 1 - \phi^2] = 0, \tag{3.276}$$

---

[9]Wilkins, Streit, and Reaugh do not state that failure occurs when D reaches D_c. Rather, in line with our quote from Hancock and MacKenzie on page 86, they say "Fracture begins when the cumulative damage D exceeds a critical damage D_c over a critical distance r_c." This model was developed to understand fracture behavior in metals. The implementation of the length scale aspect of this model is quite complex, as can be seen by reading [210, 212]. A set of five Al 6061-T651 geometrically self-similar compact tension specimens were produced and used for J-curve fracture analysis. The thicknesses were 1.905, 1.27, 0.635, 0.476, and 0.318 mm. When tested, the resulting J_{Ic} values were 10.7, 11.9 (a second evaluation method produced 6.5), 7.0, 5.6, and 4.4 kPa m, respectively, showing there is a size effect, as otherwise the J_{Ic} values would be the same [141]. Computational results were only reported for the 12.7 mm thick specimen, where the model produced a value of $J_{Ic} = 7.0$ kPa m [212]. Corresponding fracture toughnesses are given by $K_{Ic} = \sqrt{J_{Ic}E}$ and range from 27 to 17 MPa$\sqrt{\text{m}}$. The model constants from the report are $D_c = 0.67$, $r_c = 0.08$ mm, $a = 1.33$ GPa^{-1}, $\alpha = 1.8$, $\beta = 0.75$, and $Y = 285(1 + 125\varepsilon_p)^{0.1}$ MPa.

where $I_1 = \sigma_{ii} = -3p$ is the first invariant of the stress and ϕ is the porosity or void fraction in the metal, where $\phi = 0$ means no porosity and $\phi = 1$ means no metal [109]. We will associate porosity ϕ with damage; once a certain porosity is reached, the metal will be assumed to fail. In the following steps, we use associated flow to compute a volumetric plastic strain and then link the void growth with the volumetric plastic strain.

(a) Determine the direction of the associated flow. Show that it has a non-zero volumetric component – that is, there is a volumetric plastic strain.

(b) If the stress deviators are zero and plastic flow is occurring, what is the pressure?

(c) The next step is to link the porosity with the volumetric plastic strain. Do this by assuming that all the plastic volume strain is related to change in porosity and not density (i.e., we still assume that plastic flow is incompressible, so a computed volumetric plastic strain must be due to void growth and not change in material density). Assume that $\rho = (1 - \phi)\rho_0$ and that $\varepsilon_{ii}^p \approx -(1 - \rho_0/\rho)$. Show that

$$d\phi = -\frac{d\rho}{\rho_0} = (1 - \phi)^2 d\varepsilon_{ii}^p. \qquad (3.277)$$

Further show that if $J_2 = 0$ then

$$d\phi = \frac{3Y}{2}(1 - \phi)^3(1 + \phi)d\Lambda, \qquad (3.278)$$

where $d\Lambda$ is the arbitrary constant from the strain increment. Thus, porosity (or damage) can accumulate in a purely hydrostatic tensile stress state.

(d) Now let's ask another question that brings up an interesting point. Essentially in our definition of equivalent plastic strain we assumed that the plastic strain increment was deviatoric. Now it can have a volumetric component. Let's define the deviatoric equivalent plastic strain increment as (similar to Eq. 3.58)

$$de_p = \sqrt{\frac{2}{3}de_{ij}^p de_{ij}^p}, \qquad (3.279)$$

where de_{ij}^p is the deviatoric plastic strain increment. Assuming the material is yielding and $J_2 \neq 0$ (which implies de_{ij}^p is nonzero; why?), show that

$$\frac{d\phi}{3\phi} = \frac{1}{2}(1 - \phi)^2 \frac{Y}{\sqrt{3J_2}}\sinh(I_1/2Y)de_p. \qquad (3.280)$$

Thus, the porosity can change in shearing deformation if the pressure is nonzero, but no damage accumulates in pure shear. (Hint: you can divide the plastic strain direction equation into deviatoric and spherical parts. Solve the deviatoric version for $d\Lambda$ and place this term in the spherical equation.)

(e) Suppose the metal was filled with spherical voids of uniform radius R. If V is the volume the metal occupies, then the density is given by

$$\rho = \rho_0 \frac{V - \Sigma \frac{4}{3}\pi R^3}{V}. \qquad (3.281)$$

Show that this equation gives $\phi = \Sigma \frac{4}{3} \pi R^3 / V$, which implies

$$\frac{dR}{R} = \frac{d\phi}{3\phi}, \tag{3.282}$$

which, through Eq. (3.280), relates the rate of void growth to the deviatoric equivalent plastic strain rate.

(3.47) (For those with some linear algebra.) Show that for an isotropic material and a yield surface with no pressure dependence, the assumption of a normal to the yield surface flow rule is equivalent to assuming that the plastic strain rate or increment is minimized. For a Drucker–Prager flow rule, compute the corresponding direction of plastic flow based on minimizing the plastic strain increment. Based on the statement in the text that the volumetric strain is typically too large, does this idea make things better or worse? (Ans: worse.)

(3.48) The Frobenius norm or vector norm of a matrix is $|A| = \sqrt{A_{ij} A_{ij}}$. For a 3×3 symmetric matrix A, show that the norm can be written as a function of the invariants I_1 and I_2 and hence is an invariant itself.

(3.49) As an exercise in indicial notation and invariant definitions, for the identity matrix I, the scaler α, and any other matrix A show

$$I_1(\alpha I + A) = 3\alpha + I_1(A),$$
$$I_2(\alpha I + A) = -3\alpha^2 - 2\alpha I_1(A) + I_2(A), \tag{3.283}$$
$$I_3(\alpha I + A) = \alpha^3 + \alpha^2 I_1(A) - \alpha I_2(A) + I_3(A).$$

(3.50) This exercise has some index manipulation for the stress and stress deviator tensor and includes writing expressions in a form that may be easier to remember. First, using the previous exercise, show that

$$I_2 = J_2 - \frac{1}{3} I_1^2 = J_2 - 3p^2,$$

$$I_3 = J_3 - \frac{I_1}{3} J_2 + \frac{I_1^3}{27} = J_3 + p J_2 - p^3, \tag{3.284}$$

$$J_3 = I_3 + \frac{I_1 I_2}{3} + \frac{2 I_1^3}{27}.$$

These equations show that J_2 and J_3 are invariants of the stress tensor, not just the stress deviator tensor. Next, if we define the "differentiate by a tensor" notation

$$\left(\frac{\partial f}{\partial \sigma} \right)_{ij} = \frac{\partial f}{\partial \sigma_{ij}}, \tag{3.285}$$

then show that our results for differentiating with respect to the stress can be written

$$\frac{\partial I_1}{\partial \sigma} = I, \qquad \frac{\partial p}{\partial \sigma} = -\frac{1}{3} I, \qquad \frac{\partial J_2}{\partial \sigma} = s^T, \qquad \frac{\partial J_3}{\partial \sigma} = J_3 s^{-T} + \frac{J_2}{3} I. \tag{3.286}$$

Use the above to show that

$$\frac{\partial I_2}{\partial \sigma} = \sigma^T - I_1 I = s^T + 2p I. \tag{3.287}$$

For the final similar expression it is best to use a cofactor expansion (Eq. A.268) to show

$$\frac{\partial I_3}{\partial \sigma} = I_3 \sigma^{-T} = \det(\sigma) \sigma^{-T}. \tag{3.288}$$

(3.51) Given the usefulness of additively decomposing the stress tensor into spherical and deviatoric parts, $\sigma = -pI + s$, where trace$(s) = 0$, one might wonder if s can be further additively decomposed by removing the "J_2 part." However, in three dimensions show that for a symmetric deviatoric tensor B, $J_2(B) = 0$ implies $B = 0$. Thus, if $s = A + B$ and trace$(B) = 0$, then further requiring $J_2(B) = 0$ means $B = 0$. Thus, there is no decomposition.

(3.52) In a book-length exposition on materials failure theory [**43**], Christensen argues that, similarly to elasticity, there should only be two experimentally determined values that control a universal failure surface. He takes those to be the yield or failure in uniaxial stress tension Y_t and the yield or failure in uniaxial stress compression Y_c. He then posits that the failure surface for isotropic materials should be a power series expansion of the stress tensor invariants,

$$f(\sigma) = -1 + AI_1 + BI_1^2 + CJ_2 + DI_1^3 + EI_1J_2 + FJ_3 + \cdots , \qquad (3.289)$$

where we have all the first powers in stress, then second powers in stress, then third powers in stress, etc. Here, $f = 0$ is the yield or failure surface. Notice that for $Y_c \neq Y_t$ it is necessary to have an odd power of I_1 or J_3. Why? He also has the additional requirement that the maximum principal tensile stress is less than or equal to Y_t. The first power in stress only has one constant, A, while the second power in stress has three constants, A, B, and C. To remove one of these on theoretical grounds he states "For the materials of interest that do not fail in hydrostatic compression it is necessary to take $B = 0$" (his page 36, our notation), giving the yield or failure surface

$$f(\sigma) = -1 + AI_1 + CJ_2. \qquad (3.290)$$

Determine the constants A and C. Show that Eq. (3.290) is neither a Drucker–Prager nor a Mohr–Coulomb surface. (Hint: a simple comparison of powers of stress and pressure will suffice.)

However, let's explore the disposition of B. His statement clearly implies that $B > 0$. Starting with Eq. (3.92), square both sides and determine A, B, and C in terms of Y_0 and b. What is the sign of B? Something has happened here, though, in that squaring both sides loses sign information, which leads to an additional allowable region. Argue that if the material starts in a stress free configuration, then the new region cannot be reached without failing or yielding. However, given three constants, it now requires three experimentally determined material values to determine the yield or failure surface, not two. (Drucker–Prager assumes linearity to only have two experimental material values).

Now show that Eq. (3.289) can produce the Mohr–Coulomb yield surface, though unfortunately it also takes more than two experimentally determined material values. If we assume that we can truncate the series at the highest power needed, how many constants are required for a Mohr–Coulomb? (Don't determine the constants, just determine how many. Hint: Eq. 3.89 gives you the highest stress power required, so expand the series to that power in stress and count terms.)

Mechanical Waves, Shocks, and Rarefactions

This chapter describes waves traveling in an elastic material and the impact of elastic solids. In this chapter it will be assumed that the response of the material is purely mechanical, meaning that the stress is a function of the density or strain only and there are no thermal terms. Elastic means that near-static loading and unloading stress-strain response are the same. The elastic response may be non-linear. There is extensive use of the Hugoniot jump conditions. The Hugoniot is defined as the locus of solutions to the Hugoniot jump conditions. The rarefaction fan is presented, as is the rarefaction shock approximation. Specific problems are explicitly solved, including our first calculated depth of penetration, which produces a surprising result.

4.1. Linear Elastic Waves; Pushing on a Half Space

If the material in front of the wave front is at rest, then the mass jump condition is $\rho_0/\rho = 1 - u/U$. In the previous chapter the engineering strain was presented (Eq. 3.140). Written positive in compression (indicated by the tilde), it is

$$\tilde{\varepsilon} = 1 - \rho_0/\rho. \tag{4.1}$$

These combine to give

THEOREM 4.1. *(strain in a wave) For ambient material initially at rest, the engineering strain (positive in compression) behind a wave front is given by*

$$\tilde{\varepsilon} = \frac{u}{U} \tag{4.2}$$

or, in words, the strain is equal to the material particle velocity divided by the shock velocity. The same result also holds for transverse (shear) waves, using the engineering strain γ and the transverse particle velocity.

The last statement is a result of Exercise 3.39. Thus, in some sense, engineering strain is the natural strain for shock physics. The momentum jump condition can be written in terms of strain as

$$\tilde{\sigma} = \rho_0 U^2 \tilde{\varepsilon}. \tag{4.3}$$

The wave speed U is

$$U = \sqrt{\frac{\tilde{\sigma}(\tilde{\varepsilon})}{\rho_0 \tilde{\varepsilon}}}, \tag{4.4}$$

where we are explicitly showing that the stress depends on the strain and we have assumed that $\tilde{\sigma}(0) = 0$. For linear elasticity, in uniaxial strain

$$\sigma_{11} = (\lambda + 2\mu)\varepsilon_{11} \quad \Rightarrow \quad \tilde{\sigma} = (\lambda + 2\mu)\tilde{\varepsilon}, \tag{4.5}$$

giving the uniaxial wave speed

$$U = c_L = \sqrt{\frac{\lambda + 2\mu}{\rho_0}}. \tag{4.6}$$

Similarly, for elastic shear waves, $\sigma_{12} = G\gamma_{12}$ and so

$$U = c_s = \sqrt{\frac{\mu}{\rho_0}} = \sqrt{\frac{G}{\rho_0}}. \tag{4.7}$$

These are the two waves that travel in bulk linear-elastic solids: longitudinal waves, where the particle motion is aligned with the wave propagation direction, and shear waves, where the material motion is perpendicular to the motion of the wave. Example values are computed in Exercise 4.1 and are found in Appendix C. In seismology, longitudinal waves from an earthquake are referred to as p-waves, for primary, since they arrive at a recording station first, and shear waves are referred to as s-waves, for secondary, since they arrive second.

Uniaxial stress conditions are traditionally dealt with in the same way, using $\sigma_{11} = E\varepsilon_{11}$ and thus producing a bar wave speed

$$U = c_E = \sqrt{\frac{E}{\rho_0}}. \tag{4.8}$$

However, in dynamical problems uniaxial stress is an idealization and not a reality unless Poisson's ratio is zero. There is material motion in the transverse directions and thus there are radial inertial effects as waves travel down a bar of elastic material. Because of these radial motions, uniaxial stress waves are dispersive and have much more complicated behaviors than do waves where the particle motion is only in one direction. The next chapter's discussion of the split-Hopkinson pressure bar will show a specific example of what happens with uniaxial stress waves in an elastic bar. For elastic materials, when $\nu = 0$, $\lambda + 2\mu = 0 + 2(E/2) = E$ and uniaxial stress and uniaxial strain specify the same condition. No known metals have $\nu = 0$, but some distended materials such as foam exhibit a zero Poisson's ratio. (To study uniaxial stress waves, it is necessary to work with the jump conditions written in terms of strain and not density, as Eq. (4.1) no longer holds when $\nu \neq 0$.)

We also note that there is another wave speed in this same vein, namely

$$c_K = \sqrt{\frac{K}{\rho_0}}, \tag{4.9}$$

where K is the bulk modulus and c_K is referred to as the bulk sound speed. However, in a solid material with shear strength, no elastic wave travels at this speed. In a fluid, $c_L = c_K$ and $c_s = 0$, and the bulk wave corresponds to the longitudinal wave. (The only fluid to support transverse waves is the superfluid (zero viscosity) helium-3, by way of exotic quantum effects.)

Relationships between these various wave speeds come from Eqs. (3.25) and (3.27):

$$c_L = \sqrt{\frac{\lambda + 2\mu}{\rho_0}} = \sqrt{c_K^2 + \frac{4}{3}c_s^2} = \sqrt{2\frac{1-\nu}{1-2\nu}}\, c_s = \sqrt{\frac{(1-\nu)}{(1+\nu)(1-2\nu)}}\, c_E$$

$$= \sqrt{3\frac{1-\nu}{1+\nu}}\, c_K,$$

$$c_s = \sqrt{\frac{G}{\rho_0}} = \sqrt{\frac{3}{4}(c_L^2 - c_K^2)} = \sqrt{\frac{1 - 2\nu}{2(1 - \nu)}} \, c_L = \frac{c_E}{\sqrt{2(1 + \nu)}} = \sqrt{\frac{3(1 - 2\nu)}{2(1 + \nu)}} \, c_K,$$

$$c_E = \sqrt{\frac{E}{\rho_0}} = c_s \sqrt{\frac{3c_L^2 - 4c_s^2}{c_L^2 - c_s^2}} = 3c_K \sqrt{\frac{c_L^2 - c_K^2}{c_L^2 + 3c_K^2}} = \frac{3c_K c_s}{\sqrt{3c_K^2 + c_s^2}}$$

$$= \sqrt{\frac{(1 + \nu)(1 - 2\nu)}{1 - \nu}} \, c_L = \sqrt{2(1 + \nu)} \, c_s = \sqrt{3(1 - 2\nu)} \, c_K,$$

$$c_K = \sqrt{\frac{K}{\rho_0}} = \sqrt{c_L^2 - \frac{4}{3}c_s^2} = \sqrt{\frac{1 + \nu}{3(1 - \nu)}} \, c_L = \sqrt{\frac{2(1 + \nu)}{3(1 - 2\nu)}} \, c_s = \frac{c_E}{\sqrt{3(1 - 2\nu)}}.$$

$$(4.10)$$

Values for c_L, c_s and $c_0 \approx c_K$ can be found in Appendix C and Exercise 4.1.

We are now in a position to write down the solution for the initial value problem of pushing on a linear elastic half-space of material. This problem will be solved in detail for nonlinear elastic materials in the next section. If we push at a velocity u then the resulting stress is given by

$$\tilde{\sigma}(u) = \rho_0 U u = \rho_0 c u = \rho_0 c^2 \tilde{\varepsilon} = \rho_0 c^2 \left(1 - \frac{\rho_0}{\rho}\right), \qquad (4.11)$$

where the appropriate c is used. The strain in the material behind the wave front is

$$\tilde{\varepsilon} = 1 - \frac{\rho_0}{\rho} = \frac{u}{U} = \frac{u}{c} \qquad (4.12)$$

and the particle velocity is

$$u = c\tilde{\varepsilon} = \frac{\tilde{\sigma}}{\rho_0 c}. \qquad (4.13)$$

Typically the mechanical relationship between stress and strain is referred to as a stress-strain relation. When the material motion is constrained to be uniaxial, Eq. (4.1) gives a relationship between strain and density. Thus, the stress-strain relation also defines the relationship between stress and density, which is what we will use in solving the general Hugoniot jump conditions.

4.2. Compressive Shocks

The first problem geometry we study is that of pushing on an elastic half space with either a prescribed velocity or a prescribed stress. This problem is often called the piston problem, referring to a piston pushing at the end of a radially constrained column of material at a constant velocity (Fig. 4.1). A wave is formed that travels to the right into undisturbed material. The specific conditions of the wave are

$$\begin{aligned} \rho_\ell &=? & \rho_r &= \rho_0 \\ \tilde{\sigma}_\ell &=? & \tilde{\sigma}_r &= \tilde{\sigma}_r \text{ in general; } \tilde{\sigma}_r = 0 \text{ for a specific case} \\ u_\ell &=? & u_r &= u_r \text{ in general; } u_r = 0 \text{ for a specific case.} \end{aligned} \qquad (4.14)$$

The wave or shock speed U is an unknown. As we have stated, either $\tilde{\sigma}_\ell$ or u_ℓ is given as an initial condition. There are three equations: the mass conservation jump condition, the momentum conservation jump condition, and the relationship between stress and strain. The energy jump condition is not an additional constraint since our expression for stress does not explicitly depend on the energy, and thus the energy jump condition is a stand-alone equation. With these three

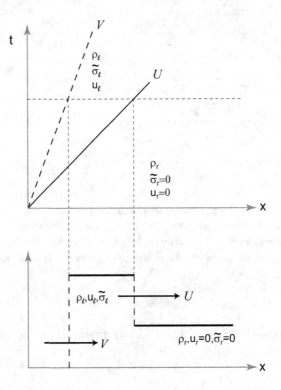

FIGURE 4.1. Pushing an elastic half space at constant velocity.

equations there are three unknowns: whichever one of $\tilde{\sigma}_\ell$ or u_ℓ not specified; the density behind the wave front ρ_ℓ; and the wave speed U.

Solving the mass balance jump condition for u_ℓ gives in general

$$\frac{\rho_r}{\rho_\ell} = 1 - \frac{u_\ell - u_r}{U - u_r} \quad \Rightarrow \quad u_\ell = u_r + \left(1 - \frac{\rho_r}{\rho_\ell}\right)(U - u_r) \qquad (4.15)$$

and for the specific initial conditions $u_r = 0$, $\tilde{\sigma}_r = 0$,

$$u_\ell = (1 - \rho_0/\rho_\ell)U. \qquad (4.16)$$

Inserting this result into the momentum balance $\tilde{\sigma}_\ell - \tilde{\sigma}_r = \rho_r(U - u_r)(u_\ell - u_r)$ gives

$$\tilde{\sigma}_\ell - \tilde{\sigma}_r = \rho_r(U - u_r)^2(1 - \rho_r/\rho_\ell). \qquad (4.17)$$

Using the relationship between stress and strain, the wave speed is thus

$$U = u_r + \sqrt{\frac{\tilde{\sigma}(\rho_\ell) - \tilde{\sigma}(\rho_r)}{\rho_r(1 - \rho_r/\rho_\ell)}}. \qquad (4.18)$$

The particle velocity is given by

$$u_\ell = u_r + \left(1 - \frac{\rho_r}{\rho_\ell}\right)(U - u_r)$$

$$= u_r + \left(1 - \frac{\rho_r}{\rho_\ell}\right)\sqrt{\frac{\tilde{\sigma}(\rho_\ell) - \tilde{\sigma}(\rho_r)}{\rho_r(1 - \rho_r/\rho_\ell)}} = u_r + \sqrt{(\tilde{\sigma}(\rho_\ell) - \tilde{\sigma}(\rho_r))\left(\frac{1}{\rho_r} - \frac{1}{\rho_\ell}\right)}.$$

$$(4.19)$$

Assuming $u_r = 0$, if we apply Eq. (4.18) to the linear stress-strain relation $\tilde{\sigma} = (\lambda + 2\mu)\tilde{\varepsilon}$, we obtain for uniaxial stress $U = c_L = \sqrt{(\lambda + 2\mu)/\rho_0}$ as before. It is clear why the engineering strain is natural to shock physics and simplifies the algebra: the $\tilde{\varepsilon} = 1 - \rho_0/\rho_\ell$ term in the stress-strain relation cancels with the same term in the denominator of the square root in Eqs. (4.18) and (4.19) for the linear material. The shock velocity for the linear material does not depend on how fast or how hard the free surface is being pushed, as long as the response remains elastic and can be modeled with the linear stress-strain relation. The sign of the applied stress or velocity has not been specified; thus, both compression and tensile waves travel at the same speed into the half space. In the rest of this section, for the linear stress-strain response the wave speed U will simply be written as the constant c and the expression relating stress and strain is

$$\tilde{\sigma} = \rho_0 c^2 (1 - \rho_0/\rho),$$

$$(4.20)$$

where $\rho_0 c^2$ is the modulus at zero strain and zero stress.

We are now in a position to introduce an important concept. Consider all the values of stress, density, particle velocity, and shock speed that solve the Hugoniot jump conditions. With four physical values and three relating expressions (the mass jump condition, the momentum jump condition, and the stress-strain relationship for the material), the resulting solution is a curve in the four-dimensional space. This solution curve is the Hugoniot:

DEFINITION 4.2. (Hugoniot) The locus of solutions of the Hugoniot jump conditions, beginning at an initial ambient state, is referred to as the *Hugoniot*, the *Hugoniot curve*, or the *Hugoniot locus*.

Rather than consider the four-dimensional space, however, it is easier to look at projections of the solution into various planes. For example, for our linear stress-strain relation that we just solved, consider six ways the solutions can be characterized:

(1) In the (ρ, u) plane by the curve $u = c(1 - \rho_0/\rho)$. We learn from this expression that the fastest the material can be driven is at the velocity c, since as the density of the compressed material approaches infinity, u approaches c. The infinite density prohibits acceleration beyond that velocity.

(2) In the $(\rho, \tilde{\sigma})$ plane by the curve $\tilde{\sigma} = \rho_0 c^2 (1 - \rho_0/\rho)$. From this representation we learn that there is an upper bound on the stress, namely that the maximum stress is $\tilde{\sigma} = \rho_0 c^2$. The fact that the stress has an upper bound also can be seen directly by observing that the largest the engineering strain $\tilde{\varepsilon} = 1 - \rho_0/\rho$ can be is 1.

(3) In the $(1/\rho, \tilde{\sigma})$ plane (the stress vs. specific volume plane), the representation is a straight line segment of $\tilde{\sigma} = \rho_0 c^2 (1 - \rho_0/\rho)$ connecting the points $(1/\rho_0, 0)$ and $(0, \rho_0 c^2)$.

(4) In the (ρ, U) plane by the straight line given by the equation $U(\rho) = c$.

(5) In the $(u, \tilde{\sigma})$ plane by the straight line segment given by equation $\tilde{\sigma} = \rho_0 c u$ which connects the points $(0, 0)$ and $(c, \rho_0 c^2)$.

(6) In the (u, U) plane, by the straight line segment given by the equation $U(u) = c$ connecting the points $(0, c)$ and (c, c), since, regardless of the driving wall velocity, the shock velocity is c.

It turns out that the most useful representations for studying waves are the (u, U) plane (often called the U_s–u_p plane for the shock velocity–particle velocity plane), the $(u, \tilde{\sigma})$ plane, and the $(1/\rho, \tilde{\sigma})$ plane. Though all these are different representations, they all represent the same object, the Hugoniot.

The Hugoniot is a locus of endpoints of solutions to the Hugoniot jump conditions. It must be pointed out that for the elastic material, *mathematically there is no stress-strain "path" connecting the initial material state in the front of the shock and the final material state behind the shock.* The jump is discontinuous. In reality there is obviously a physical path that the material takes, but this is only to the extent that the shock front does not actually solve the jump conditions on a (likely) molecular scale, and hence not a "path" in the sense of being part of the material's bulk stress-strain response.

For many people it is desirable to think of a "path" that the shock takes. A conceptual path arises from the energy jump condition. The energy jump condition is (Eq. 2.63)

$$E_\ell - E_r = \frac{1}{2}(\tilde{\sigma}_\ell + \tilde{\sigma}_r)\left(\frac{1}{\rho_r} - \frac{1}{\rho_\ell}\right). \tag{4.21}$$

The change in internal energy due to loading along a stress-strain curve is (Eq. 3.122)

$$E = -\int_{1/\rho_r}^{1/\rho_\ell} \tilde{\sigma} d(1/\rho). \tag{4.22}$$

If we connect the beginning and ending point for the shock in the $(1/\rho, \tilde{\sigma})$ plane with a straight line, we see that the change in internal energy obtained by integrating the region beneath that straight line equals the change in internal energy from the Hugoniot jump condition for energy connecting the initial state in front of the shock front with the final state behind the shock front. This straight line has a name.

DEFINITION 4.3. (Rayleigh line) The straight line in the $(1/\rho, \tilde{\sigma})$ plane connecting an initial state with the material state defined by the Hugoniot jump conditions is referred to as the *Rayleigh line* (Fig. 4.2).

Fixing the initial and end states, which fixes U, the equation for the Rayleigh line is

$$\tilde{\sigma} - \tilde{\sigma}_r = \rho_r^2 (U - u_r)^2 (1/\rho_r - 1/\rho), \qquad \tilde{\sigma} - \tilde{\sigma}_r = \rho_r (U - u_r)(u - u_r). \tag{4.23}$$

Similarly, from the mass jump condition, $1/\rho = (1/\rho_r)(1 - (u - u_r)/(U - u_r))$, so u and $1/\rho$ are linearly related; hence the Rayleigh line is also a straight line in the $(u, \tilde{\sigma})$ plane, as indicated in the second equation for the Rayleigh line.

Though the Rayleigh line is not really the shock-loading path, it has meaning in our understanding of shock waves. In particular, we now begin a chain of technical

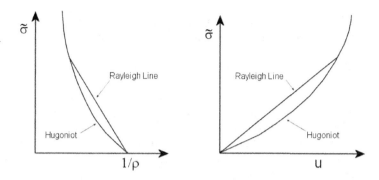

FIGURE 4.2. The Rayleigh line.

arguments that will show the equivalence of many characterizations of a shock. Essentially, if the Hugoniot locus lies below the collection of all the Rayleigh lines, then a single shock will form. We will show that this behavior of the Hugoniot locus and the Rayleigh lines is equivalent to the statement that the rarefaction does not fall behind the shock and also that the entropy is nondecreasing across the shock front.

To begin, we determine the local sound speed of the material. Sound waves are very small amplitude waves. Their speed can be found from our solution for the wave speed (Eq. 4.18) under the assumptions that the material is at rest in front of the wave and that the change in density of the sound wave goes to zero,

$$
\begin{aligned}
c(\rho) &= \lim_{\rho \to \rho_r} U = \lim_{\rho \to \rho_r} \sqrt{\frac{\tilde{\sigma}(\rho) - \tilde{\sigma}(\rho_r)}{\rho_r(1 - \rho_r/\rho)}} = \lim_{\rho \to \rho_r} \sqrt{\left(\frac{\rho}{\rho_r}\right) \frac{\tilde{\sigma}(\rho) - \tilde{\sigma}(\rho_r)}{\rho - \rho_r}} \\
&= \sqrt{\frac{d\tilde{\sigma}(\rho)}{d\rho}}.
\end{aligned}
\tag{4.24}
$$

This result could also have been obtained immediately by combining the differential mass and momentum jump conditions of Chapter 2 (Eqs. 2.81 and 2.82),

$$
c(\rho) = U_{\text{Eqs. 2.81, 2.82}} = \frac{1}{\rho} \frac{d\tilde{\sigma}}{du} = \sqrt{\frac{d\tilde{\sigma}(\rho)}{d\rho}}.
\tag{4.25}
$$

These results give

DEFINITION 4.4. (local sound speed) The speed at which a very small amplitude wave travels in a material at rest for a mechanical stress-strain relation is the *local sound speed*, given by

$$
c(\rho) = \sqrt{\frac{d\tilde{\sigma}(\rho)}{d\rho}}.
\tag{4.26}
$$

The sound speed at pressure will be referred to as $c(\rho)$, showing the explicit density or pressure dependence, and the sound speed at ambient conditions will be referred to as c with no subscripts or functional dependence,

$$
c = c(\rho_0) = U(\rho_0).
\tag{4.27}
$$

The reason we do not use "c_0" for the sound speed at ambient conditions is that c_0 is historically reserved for the approximate low-pressure bulk wave speed.

Typically the sound speed depends on the amount of compression in the material. For the linear stress-strain relation, a simple computation gives

$$c(\rho) = \frac{\rho_0 c}{\rho}.$$
(4.28)

This result says that as the compression of the material increases, the local sound speed goes down. For the linear material, the modulus (i.e., stiffness) of the material decreases (Exercise 5.22) and the density increases, reducing the sound speed; both these effects combine in the same amount to reduce the sound speed. To see this, suppose the modulus of interest is the bulk modulus, then the bulk modulus at pressure is given by

$$K(\rho) = \rho c(\rho)^2 = \frac{\rho_0^2}{\rho} c^2.$$
(4.29)

(There is more discussion of high pressure moduli in the next chapter.) The sound speed at pressure is

$$c(\rho) = \sqrt{\frac{K(\rho)}{\rho}} = \sqrt{\frac{\rho_0^2}{\rho} c^2 \frac{1}{\rho}} = \frac{\rho_0 c}{\rho}.$$
(4.30)

Thus, a multiplicative decrease of $\sqrt{\rho_0/\rho}$ is due to the decrease in modulus and a multiplicative decrease of $\sqrt{\rho_0/\rho}$ is due to the increase in density.

Now we make a physical assumption – that is, an assumption of our expectations regarding the physics of a shock wave. We expect that the material state behind the shock front is what drives the shock front. For the material state behind the shock front to drive the shock front, the shock front needs to have information about the material state behind it. Small amplitude waves transport information, so we expect that *the shock front cannot outrun the sound waves*. (Another way to say this, after study of the next section, is that the shock wave cannot outrun the subsequent rarefaction wave.) We will show below that this physical assumption is equivalent to assuming that the entropy is nondecreasing across a shock front, but the physical assumption about the sound wave seems more intuitive. Thus, the local sound speed always needs to be at least as great as the shock velocity in the frame of reference where the material behind the shock front is at rest, or in an equation

$$c(\rho) \geq U - u.$$
(4.31)

Let us consider what the shock speed is behind a shock front for the linear material in the reference from where the material is at rest behind the shock front (i.e., the material frame in which we just computed the sound speed). For the linear material, the shock speed is a constant $U = c$ in the frame where the material is at rest in front of the shock. Thus, the shock speed in the frame of reference where the material behind the shock is at rest is given by the shock speed in the laboratory (Eulerian) frame minus the particle motion (Eq. 4.19),

$$U - u_\ell = c - \left(1 - \frac{\rho_0}{\rho_\ell}\right) c = \frac{\rho_0 c}{\rho_\ell} = c(\rho_\ell).$$
(4.32)

Thus, for the linear stress-strain relation, with respect to the compressed material behind the shock front, the shock front travels at the local sound velocity. This result is a feature of the linear stress-strain relation *only*. In fact, the linear stress-strain relation is a special case that demarcates certain types of behavior.

Equations (4.18), (4.19), and (4.31) combine to give

$$c(\rho_\ell) = \sqrt{\left.\frac{d\tilde{\sigma}(\rho)}{d\rho}\right|_{\rho_\ell}} \geq U - u_\ell = U - u_r - (u_\ell - u_r)$$

$$= U - u_r - \left(1 - \frac{\rho_0}{\rho_\ell}\right)(U - u_r) = \frac{\rho_0}{\rho_\ell}(U - u_r) \qquad (4.33)$$

$$= \frac{\rho_0}{\rho_\ell}\sqrt{\frac{\tilde{\sigma}(\rho_\ell) - \tilde{\sigma}(\rho_0)}{\rho_0(1 - \rho_0/\rho_\ell)}}.$$

Both sides are positive, so squaring both sides, we have translated the sound speed requirement into a relationship that the stress-strain relationship must satisfy, where we now write the density to the left ρ_ℓ as ρ to simplify the notation, with the understanding that that is the value at which $d\tilde{\sigma}(\rho)/d\rho$ is evaluated,

$$\frac{d\tilde{\sigma}(\rho)}{d\rho} \geq \frac{1}{\rho}\left(\frac{\tilde{\sigma}(\rho) - \tilde{\sigma}(\rho_0)}{\rho/\rho_0 - 1}\right). \qquad (4.34)$$

It is natural to write this inequality in terms of specific volume $1/\rho$, noting that $d(1/\rho) = -(1/\rho^2)\,d\rho$,

$$\frac{d\tilde{\sigma}(\rho)}{d(1/\rho)} \leq \frac{\tilde{\sigma}(\rho) - \tilde{\sigma}(\rho_0)}{1/\rho - 1/\rho_0}. \qquad (4.35)$$

The right-hand side of this equation is the slope of the Rayleigh line (which is negative). The left-hand side is the slope of the Hugoniot locus. Thus, the slope of the Hugoniot locus is more negative (i.e., steeper descent) than the Rayleigh line whenever they intersect. Since the Hugoniot and all the Rayleigh lines start at the point $(1/\rho_0, 0)$, we conclude that *the Hugoniot curve always lies on or below the Rayleigh line*. This nice geometrical interpretation is shown in Fig. 4.2. Hence,

LEMMA 4.5. *The physical requirement that the shock front not outrun the sound speed* $(c(\rho) \geq U - u)$ *is equivalent to the Hugoniot curve lying below or on the Rayleigh line.*

Given this result, one immediately should ask: what happens if the Hugoniot goes above the Rayleigh line? Essentially, if this occurs the physical condition is not satisfied and the shock is unstable. There exists an intermediate-density raising shock that has a higher shock speed than the shock speed that produces the total density increase with one shock (otherwise the Hugoniot curve would not be above the Rayleigh line). The total density jump will be achieved by two or more shocks or a collection of compressive shocks connected by compressive fans (depending on the material stress-strain response). Though we will not discuss compressive fans, after the next section on rarefaction fans we will have a good idea of their properties (Exercise 4.27). Thus another intermediate result:

LEMMA 4.6. *When the Hugoniot curve lies on or below the Rayleigh line, a single compressive shock wave is formed.*

As will become apparent below, this argument is equivalent to choosing the solution to the Hugoniot jump conditions that has the largest increase in entropy. In the event that there are multiple shocks, the shock wave analysis is performed in each region of the stress-strain response which are, by themselves, single shock forming, and the results in this section apply in those regions separately.

We now wish to rejoin the discussion about inequalities to obtain a similar geometric interpretation for the $(u, \tilde{\sigma})$ plane. Combination of the mass and momentum jump conditions gives

$$\rho_0 (u - u_0)^2 = (1 - \rho_0/\rho)(\tilde{\sigma}(\rho) - \tilde{\sigma}(\rho_0)). \tag{4.36}$$

Differentiation with respect to ρ gives

$$2\rho_0(u - u_0)\frac{du}{d\rho} - \frac{\rho_0}{\rho^2}(\tilde{\sigma} - \tilde{\sigma}_0) = (1 - \rho_0/\rho)\frac{d\tilde{\sigma}}{d\rho} \geq \frac{\rho_0}{\rho^2}(\tilde{\sigma} - \tilde{\sigma}_0), \tag{4.37}$$

where we used the inequality in Eq. (4.34), yielding

$$\frac{du}{d\rho} \geq \frac{\tilde{\sigma} - \tilde{\sigma}_0}{\rho^2(u - u_0)} = \frac{u - u_0}{\rho(\rho/\rho_0 - 1)}. \tag{4.38}$$

The right-hand side comes with a second application of Eq. (4.36). This expression can also be written

$$\frac{du}{d(1/\rho)} \leq \frac{u - u_0}{1/\rho - 1/\rho_0} = -\frac{\tilde{\sigma} - \tilde{\sigma}_0}{u - u_0}. \tag{4.39}$$

If now we differentiate Eq. (4.36) with respect to u,

$$(1 - \rho_0/\rho)\frac{d\tilde{\sigma}}{du} - 2\rho_0(u - u_0) = -\frac{\rho_0}{\rho^2}(\tilde{\sigma} - \tilde{\sigma}_0)\frac{d\rho}{du} \geq -\rho_0(u - u_0), \tag{4.40}$$

where the minus sign and reciprocal have combined to keep the inequality from Eq. (4.38) in the same direction. Thus,

$$\frac{d\tilde{\sigma}}{du} \geq \frac{\rho_0(u - u_0)}{(1 - \rho_0/\rho)} = \frac{\tilde{\sigma} - \tilde{\sigma}_0}{u - u_0}, \tag{4.41}$$

where again Eq. (4.36) has been used to obtain the right-hand side. Equation (4.41) gives a geometrical description of the relationship between the Hugoniot locus and the Rayleigh lines in the $(u, \tilde{\sigma})$ plane, namely that the Hugoniot lies on or below the Rayleigh line, just as in the $(1/\rho, \tilde{\sigma})$ plane.

While we are here, we write down one more equivalent expression based on the shock velocity U. Viewing the shock as parameterized by u, notice that

$$\begin{aligned}
\frac{d\tilde{\sigma}(u)}{du} &= \frac{d\{\rho_0(U - u_0)(u - u_0)\}}{du} = \rho_0(u - u_0)\frac{dU}{du} + \rho_0(U - u_0) \\
&= \frac{\tilde{\sigma} - \tilde{\sigma}_0}{U - u_0}\frac{dU}{du} + \frac{\tilde{\sigma} - \tilde{\sigma}_0}{u - u_0} \geq \frac{\tilde{\sigma} - \tilde{\sigma}_0}{u - u_0}.
\end{aligned} \tag{4.42}$$

With $U - u_0 > 0$ this provides

$$\frac{dU}{du} \geq 0. \tag{4.43}$$

We have now developed a visual image and some inequalities describing the behavior in a stress-strain relation that produces the expected sound speed/shock velocity relationship. At this point, we need terminology to describe this behavior. Intuitively, the terminology is *concave up*, but it turns out concave up is a stronger requirement than we have. Though not explicitly stated so far, we are assuming that stress-strain relations are continuous and piecewise differentiable. Mathematically, a convex curve is one where, if a straight line is drawn between any two points on the curve, the curve values lie below or on the straight line [168]. (In an equation, if f is a function that is convex on domain (a, b), then $f((1-\lambda)x+\lambda y) \leq (1-\lambda)f(x)+\lambda f(y)$ for $a \leq x < y \leq b$ and $0 \leq \lambda \leq 1$.) However, convex is not quite the right definition since we are really only looking at straight lines that begin at the lower end point (our Rayleigh lines). Thus, we introduce a definition related to our geometrical picture for the stress-strain relationship and the Rayleigh lines:

DEFINITION 4.7. (ρ_0-convex stress-strain relation) A stress-strain relation $\tilde{\sigma}(\rho)$ is said to be ρ_0-*convex* if, in the $(1/\rho, \tilde{\sigma})$ plane, the stress-strain curve (to the left) lies either below or on the straight line connecting any point on the stress-strain curve with the initial ambient state $(1/\rho_0, \tilde{\sigma}(\rho_0))$. (To the left means for $1/\rho < 1/\rho_0$.)

The same definition holds in the $(u, \tilde{\sigma})$ plane (but to the right) when the stress-strain curve is combined with the Hugoniot jump conditions. (As to other mathematical terminology, this definition differs from *star convex* in that we are specifying the reference point and the direction for the convexity.)

We point out that a stress-strain relation that is convex is ρ_0-convex, since the lower end point is one of the points where it needs to satisfy the strain-line bounding criterion. Thus, if a stress-strain relation was such that for any density $\rho \geq \rho_0$ the stress-strain relation was ρ-convex, then the stress-strain relation would be convex and the sound speed would be greater than the shock velocity for shocks formed from any initial density above ρ_0. Convex functions have a nondecreasing derivative f'. Thus, we use our calculus descriptions of curves for the following definition:

DEFINITION 4.8. (concave up stress-strain relation) A stress-strain relation is said to be *concave up* if

$$\frac{d^2\tilde{\sigma}}{d(1/\rho)^2} \geq 0 \quad \text{or, equivalently,} \quad \frac{d^2\tilde{\sigma}}{d\tilde{\varepsilon}^2} \geq 0 \tag{4.44}$$

where $\tilde{\varepsilon}$ is the engineering strain.

This definition is typically easier to check for a given stress-strain relation. Concave up stress-strain relations are convex. Thus,

LEMMA 4.9. *Concave up stress-strain relations are ρ_0-convex, and thus lie on or beneath the Rayleigh line.*

The converse need not be true (i.e., ρ_0-convex stress-strain relations need not be concave up, but they still lie on or beneath the Rayleigh line).

Let us come back to the comment made above that the linear stress-strain response is in a sense a special case. The local sound speed exactly equaling the shock velocity exemplifies this fact. We would like to compare a stress-strain response to the linear stress-strain response, and a natural definition for whether a stress-strain response is functionally more steep than the linear one is to consider the derivative of the ratio of the stress-strain response to the linear stress-strain response, which is just a constant times the engineering strain,

$$\frac{d}{d\tilde{\varepsilon}}\left(\frac{\tilde{\sigma}(\tilde{\varepsilon})}{\tilde{\varepsilon}}\right) \geq 0. \tag{4.45}$$

When this derivative is positive, then the stress-strain relationship grows more quickly than the linear case. (Keep in mind that we are allowing the curve to be piecewise differentiable, so there may be a handful of points where the derivative is not defined.) If the above inequality is an equality, then the stress-strain relation recovered from solving the differential equation is the linear one, and is unique given the initial condition that $\tilde{\sigma}(\rho_0) = 0$. These facts provide the following:

DEFINITION 4.10. (Superlinear stress-strain response, strictly superlinear stress-strain response.) A mechanical stress-strain response is said to be *superlinear* if it satisfies

$$\frac{d}{d\rho}\left(\frac{\tilde{\sigma}(\rho) - \tilde{\sigma}(\rho_0)}{1 - \rho_0/\rho}\right) \geq 0 \tag{4.46}$$

and the inequality is strict at at least one density greater than ρ_0. If the inequality is strict for all densities greater than ρ_0 or all velocities $u > 0$, then the stress-strain response is said to be *strictly superlinear*.

Equation (4.46) is equivalent to Eq. (4.45). Differentiation of this definition gives

$$\frac{d}{d\rho}\left(\frac{\tilde{\sigma}(\rho) - \tilde{\sigma}(\rho_0)}{1 - \rho_0/\rho}\right) = \frac{d\tilde{\sigma}/d\rho}{1 - \rho_0/\rho} - \frac{\tilde{\sigma}(\rho) - \tilde{\sigma}(\rho_0)}{(1 - \rho_0/\rho)^2}\frac{\rho_0}{\rho^2} \geq 0 \tag{4.47}$$

and that says that this inequality is

$$\frac{d\tilde{\sigma}(\rho)}{d\rho} \geq \frac{1}{\rho}\left(\frac{\tilde{\sigma}(\rho) - \tilde{\sigma}(\rho_0)}{\rho/\rho_0 - 1}\right). \tag{4.48}$$

This is the same criterion as for single shock forming. Thus, our inequalities from the single shock forming definition allow a comparison to the linear case.

LEMMA 4.11. *Superlinear is equivalent to ρ_0-convex except that for superlinear the inequality is strict at at least one point.*

All these facts put together provide the following:

THEOREM 4.12. *(single shock forming, near equivalence of ρ_0-convex and superlinear) For a mechanical stress-strain relation and the Hugoniot jump conditions, the following inequalities are equivalent for $\rho > \rho_0$:*

$$c(\rho) \geq U - u, \tag{4.49}$$

$$\frac{d}{d\rho}\left(\frac{\tilde{\sigma}(\rho) - \tilde{\sigma}(\rho_0)}{1 - \rho_0/\rho}\right) \geq 0, \tag{4.50}$$

$$\frac{d(\tilde{\sigma}(\rho))}{d\rho} \geq \frac{1}{\rho}\left(\frac{\tilde{\sigma}(\rho) - \tilde{\sigma}(\rho_0)}{\rho/\rho_0 - 1}\right), \tag{4.51}$$

$$\frac{d(\tilde{\sigma}(\rho))}{d(1/\rho)} \leq \frac{\tilde{\sigma}(\rho) - \tilde{\sigma}(\rho_0)}{1/\rho - 1/\rho_0}, \tag{4.52}$$

$$\frac{d(\tilde{\sigma}(u))}{du} \geq \frac{\tilde{\sigma}(u) - \tilde{\sigma}_0}{u - u_0}, \tag{4.53}$$

$$\frac{d(U(u))}{du} \geq 0, \tag{4.54}$$

$$\frac{d(u(\rho))}{d\rho} \geq \frac{u(\rho) - u_0}{\rho(\rho/\rho_0 - 1)} = \frac{\tilde{\sigma}(u) - \tilde{\sigma}_0}{\rho^2(u - u_0)}, \tag{4.55}$$

$$\frac{d(u(\rho))}{d(1/\rho)} \leq \frac{u(\rho) - u_0}{1/\rho - 1/\rho_0} = -\frac{\tilde{\sigma}(u) - \tilde{\sigma}_0}{u - u_0}, \tag{4.56}$$

$$\frac{d(s(\rho))}{d\rho} \geq 0, \tag{4.57}$$

$$\frac{d(s(\rho))}{d(1/\rho)} \leq 0, \tag{4.58}$$

$$\frac{d(s(u))}{du} \geq 0. \tag{4.59}$$

Any stress-strain relation satisfying these inequalities forms a single shock wave in compression. Further,

(1) *A stress-strain relation that satisfies these inequalities is ρ_0-convex.*
(2) *A stress-strain relation that is concave up, $\partial^2\tilde{\sigma}/\partial(1/\rho)^2 \geq 0$, is ρ_0-convex and hence single shock wave forming.*
(3) *For the linear stress-strain relation $\tilde{\sigma}(\rho) = \rho_0 c^2(1 - \rho_0/\rho)$, all these inequalities are equalities.*
(4) *A stress-strain relation that satisfies these inequalities and the inequality is strict in at least one point is superlinear.*
(5) *A stress-strain relation that satisfies these inequalities in their strict form except at the ambient state is strictly superlinear.*
(6) *The only difference between ρ_0-convex and superlinear is that the inequality is strict at some point for superlinear.*
(7) *A ρ_0-convex stress-strain relation is either linear or superlinear.*
(8) *Linear, superlinear, and strictly superlinear stress-strain relations are ρ_0-convex.*
(9) *Strictly superlinear stress-strain responses are superlinear.*
(10) *Entropy is nondecreasing across a shock front and can increase for superlinear stress-strain relations and does increase for strictly superlinear stress-strain relations.*
(11) *As a particular important case, the stress-strain response giving rise to*

$$U(u) = c + ku, \text{ where } c > 0 \text{ and } k > 0 \text{ are constants} \tag{4.60}$$

is strictly superlinear.

Three of these inequalities written in terms of engineering strain can be found in Exercise 4.33.

Many metals are characterized by the behavior of the specific example, and the fact that it is superlinear is trivial from the relation $dU/du = d(c+ku)/du = k > 0$.

The last three inequalities in the theorem deal with entropy, which we have yet to discuss. They will now be demonstrated. Entropy in thermodynamics is defined by "$dU = TdS - PdV$," which in our notation, with a change to specific variables, is written

$$dE = Tds - \tilde{\sigma}d(1/\rho). \tag{4.61}$$

Rearranging gives

$$ds = \frac{1}{T}(dE + \tilde{\sigma}d(1/\rho)). \tag{4.62}$$

The entropy is related to the difference between the internal energy that is produced by the passage of the shock and the internal energy that is produced by just straining the material. The last term on the right-hand side is the negative of the specific strain energy (Eq. 4.22). The term dE is the energy change due to the shock, which is found by the energy jump condition,

$$E - E_0 = \frac{1}{2}(\tilde{\sigma}(\rho) + \tilde{\sigma}(\rho_0))(1/\rho_0 - 1/\rho). \tag{4.63}$$

Differentiation of this expression gives

$$dE = \frac{1}{2}(d\tilde{\sigma}(\rho))(1/\rho_0 - 1/\rho) - \frac{1}{2}(\tilde{\sigma}(\rho) + \tilde{\sigma}(\rho_0))d(1/\rho). \tag{4.64}$$

Placement in the expression for entropy gives

$$T ds = dE + \tilde{\sigma}d(1/\rho) = \frac{1}{2}(d\tilde{\sigma}(\rho))(1/\rho_0 - 1/\rho) + \frac{1}{2}(\tilde{\sigma}(\rho) - \tilde{\sigma}(\rho_0))d(1/\rho). \tag{4.65}$$

Rewriting gives

$$T\frac{ds}{d(1/\rho)} - \frac{1}{2}(\tilde{\sigma}(\rho) - \tilde{\sigma}(\rho_0)) = \frac{1}{2}\frac{d\tilde{\sigma}(\rho)}{d(1/\rho)}(1/\rho_0 - 1/\rho). \tag{4.66}$$

The general case of Theorem 4.12 (for when $\tilde{\sigma}(\rho_0) \neq 0$) says

$$\frac{d(\tilde{\sigma}(\rho))}{d(1/\rho)} \leq \frac{\tilde{\sigma}(\rho) - \tilde{\sigma}(\rho_0)}{1/\rho - 1/\rho_0}, \tag{4.67}$$

which gives

$$T\frac{ds}{d(1/\rho)} - \frac{1}{2}(\tilde{\sigma}(\rho) - \tilde{\sigma}(\rho_0)) \leq -\frac{1}{2}(\tilde{\sigma}(\rho) - \tilde{\sigma}(\rho_0)) \tag{4.68}$$

and, since $T > 0$,

$$\frac{ds}{d(1/\rho)} \leq 0, \tag{4.69}$$

which immediately yields $ds/d\rho \geq 0$ and $ds/du \geq 0$, since $du/d\rho \geq 0$. Thus, the entropy is nondecreasing across a shock front and is increasing across a shock front for a strictly superlinear stress-strain relation.

It is important to understand the nature of the Hugoniot locus. Since in this chapter we are assuming that there is a mechanical relation between stress and strain – that is, that the stress is a function of density only and not of temperature or internal energy – it is in fact the case that the stress for the Hugoniot locus for a given density is the same as the stress for an isentrope at the same given density. However, the energy Hugoniot jump condition defines the material energy state, and the energy state on the Hugoniot locus is not the same as the energy state along the isentrope, even though the density and stress are the same. In particular, the energy is higher along the Hugoniot than along the isentrope, and thus there is an entropy increase.

Though it is not necessary to explicitly know the entropy, it is possible to compute the entropy change due to the shock if we make a few simplifying assumptions. Initially, it is not entirely clear how to compute the entropy because of the temperature term. Mathematically speaking, the material changes across a shock front do not follow any path, they change instantaneously from one state to another. Thus, the temperature to use in the evaluation of the entropy is unknown. Since entropy is a state variable, it is possible to compute it piecewise, since thermodynamic theory says that the equation of state is determined by only two independent variables. If we use the density and temperature as our two independent variables, then it is possible to compute the entropy using Eq. (4.62) by taking a path that involves two steps: the first step is from the ambient state along an isentrope to the density that is achieved by the shock; the second step raises the temperature the rest of the way to that achieved by the shock while holding the density constant. Along the isentrope we have $Tds = 0$ and

$$dE = -\tilde{\sigma}d(1/\rho) \quad \Rightarrow \quad \Delta E_{\text{isentrope}} = -\int_{1/\rho_0}^{1/\rho} \tilde{\sigma}d(1/\rho). \tag{4.70}$$

We can compute this integral since we have a mechanical stress-strain relation and the stress is only a function of density. Also, since there are no thermal terms we are saying that all the energy stored in the isentropic compression is all potential energy and there is no corresponding temperature change (see Section 5.11 for the more general case). We need to assume a relationship between energy change and temperature, in particular that we are in a region where the heat capacity is constant and

$$C_v dT = dE. \tag{4.71}$$

For the second step of the path, holding the density fixed (an isochor) and changing the temperature gives

$$Tds = dE. \tag{4.72}$$

Equations (4.71) and (4.72) combine to give

$$ds = C_v \frac{dT}{T} \quad \Rightarrow \quad \Delta s_{\text{isochor}} = C_v \ln\left(\frac{T_H(\rho)}{T_s(\rho)}\right), \tag{4.73}$$

where $T_H(\rho)$ is the temperature on the Hugoniot and $T_s(\rho)$ is the temperature along the isentrope. Since there is no change in entropy along the isentrope, attaching the two parts of the path is simply the observation that $\Delta s_{\text{shock}} = \Delta s_{\text{isochor}}$. Thus, for a mechanical stress-strain response,

$$\begin{aligned} T_H(\rho) &= \frac{1}{C_v}\left\{\Delta E_{\text{shock}} - \Delta E_{\text{isentrope}}\right\} + T_s \\ &= \frac{1}{C_v}\left\{\frac{1}{2}(\tilde{\sigma}(\rho) + \tilde{\sigma}(\rho_0))(1/\rho_0 - 1/\rho) + \int_{1/\rho_0}^{1/\rho} \tilde{\sigma}d(1/\rho)\right\} + T_s \end{aligned} \tag{4.74}$$

and T_s is just the initial ambient temperature T_0 (again, for the purely mechanical stress-strain relation). For the linear stress-strain relation, the two energy terms are equal and there is no temperature change. Of course, this conclusion already followed from Theorem 4.12, since the inequalities are all equalities for the linear stress-strain relation and thus $ds = 0$ along the Hugoniot.

Returning to Eq. (4.66) for the entropy, the derivatives of the entropy on the Hugoniot are

$$\frac{ds}{d(1/\rho)} = \frac{1}{2T}\left(\tilde{\sigma}(\rho) - \tilde{\sigma}(\rho_0) + \frac{d\tilde{\sigma}(\rho)}{d(1/\rho)}(\frac{1}{\rho_0} - \frac{1}{\rho})\right),$$

$$\frac{d^2s}{d(1/\rho)^2} = \frac{d(1/2T)}{d(1/\rho)}\left(\tilde{\sigma}(\rho) - \tilde{\sigma}(\rho_0) + \frac{d\tilde{\sigma}(\rho)}{d(1/\rho)}(\frac{1}{\rho_0} - \frac{1}{\rho})\right) + \frac{1}{2T}\frac{d^2\tilde{\sigma}(\rho)}{d(1/\rho)^2}(\frac{1}{\rho_0} - \frac{1}{\rho}),$$

$$\begin{aligned} \frac{d^3s}{d(1/\rho)^3} &= \frac{d^2(1/2T)}{d(1/\rho)^2}\left(\tilde{\sigma}(\rho) - \tilde{\sigma}(\rho_0) + \frac{d\tilde{\sigma}(\rho)}{d(1/\rho)}(\frac{1}{\rho_0} - \frac{1}{\rho})\right) \\ &+ \frac{d(1/T)}{d(1/\rho)}\frac{d^2\tilde{\sigma}(\rho)}{d(1/\rho)^2}(\frac{1}{\rho_0} - \frac{1}{\rho}) + \frac{1}{2T}\left(\frac{d^3\tilde{\sigma}(\rho)}{d(1/\rho)^3}(\frac{1}{\rho_0} - \frac{1}{\rho}) - \frac{d^2\tilde{\sigma}(\rho)}{d(1/\rho)^2}\right). \end{aligned} \tag{4.75}$$

Evaluation at ρ_0 gives

$$\left.\frac{ds}{d(1/\rho)}\right|_{\rho=\rho_0} = 0, \qquad \left.\frac{d^2s}{d(1/\rho)^2}\right|_{\rho=\rho_0} = 0, \qquad \left.\frac{d^3s}{d(1/\rho)^3}\right|_{\rho=\rho_0} = -\frac{1}{2T}\left.\frac{d^2\tilde{\sigma}(\rho)}{d(1/\rho)^2}\right|_{\rho=\rho_0}. \tag{4.76}$$

This result gives

$$s = s_0 - \frac{1}{12T}\left.\frac{d^2\tilde{\sigma}(\rho)}{d(1/\rho)^2}\right|_{\rho=\rho_0} (1/\rho - 1/\rho_0)^3 + O((1/\rho - 1/\rho_0)^4), \tag{4.77}$$

which says that the isentrope and the Hugoniot are quite close for small strains. In particular,

THEOREM 4.13. *(the Hugoniot and the isentrope are close) The change in entropy along the Hugoniot from the initial ambient $(\rho_0, \tilde{\sigma}_0)$ is third order in the change in specific volume $1/\rho$.*

Another way to say this is that the difference between the Hugoniot and the isentrope (adiabat) only becomes apparent for strong shocks. As noted in the statement of the theorem, the only reference point this result applies to is the initial ambient state (i.e., unshocked material state) that is used in the definition of the Hugoniot. It does not apply to an arbitrary point on the Hugoniot curve. The fact that it only applies to the deviations from the ambient state shows itself when we talk about sound speeds for equations of state with thermal terms in Chapter 5.

As a second example of a stress-strain relation, and to complete our discussion, consider

$$\tilde{\sigma} = \rho_0 c^2(\rho/\rho_0 - 1) = \rho c^2(1 - \rho_0/\rho). \tag{4.78}$$

For very small strain, this stress-strain relation is essentially the same as the linear stress-strain relation. A simple computation shows that

$$\frac{d}{d\rho}\left(\frac{\tilde{\sigma}(\rho)}{(1 - \rho_0/\rho)}\right) = \frac{d}{d\rho}\left(\frac{\rho_0 c^2(\rho/\rho_0 - 1)}{(1 - \rho_0/\rho)}\right) = c^2 > 0; \tag{4.79}$$

hence, it is single shock forming and strictly superlinear. The shock speed is given by

$$U = \sqrt{\frac{\tilde{\sigma}(\rho_\ell)}{\rho_0(1 - \rho_0/\rho_\ell)}} = c\sqrt{\frac{\rho_\ell/\rho_0 - 1}{1 - \rho_0/\rho_\ell}} = c\sqrt{\frac{\rho_\ell}{\rho_0}}. \tag{4.80}$$

Thus, this assumption on the stress-density relation gives rise to a wave where the shock speed increases with increasing wall velocity. The properties behind the wave affect the wave speed: the larger the stress applied to the left boundary, the larger the compression on the left side of the wave and hence the faster the wave travels. Placing the wave speed back into the mass jump condition yields

$$\frac{\rho_\ell}{\rho_0} - \frac{u}{c}\sqrt{\frac{\rho_\ell}{\rho_0}} - 1 = 0. \tag{4.81}$$

This equation is solved to give

$$\sqrt{\frac{\rho_\ell}{\rho_0}} = \sqrt{1 + \left(\frac{u}{2c}\right)^2} + \frac{u}{2c}, \tag{4.82}$$

which gives

$$U = \sqrt{c^2 + u^2/4} + \frac{1}{2}u, \tag{4.83}$$

directly showing that as the pressure and particle velocity prescribed on the end are increased, the wave speed increases. The particle velocity is given by

$$u_\ell = c\left(\sqrt{\frac{\rho_\ell}{\rho_0}} - \sqrt{\frac{\rho_0}{\rho_\ell}}\right). \tag{4.84}$$

Notice that for this stress-strain relation, there are no bounds on ρ, U, u or $\tilde{\sigma}$ – the Hugoniot locus has infinite extent in these variables. The local sound speed is

$$c(\rho) = \sqrt{\frac{\partial \tilde{\sigma}}{\partial \rho}} = c, \tag{4.85}$$

which is constant regardless of material density. Thus, from the reference frame of the material being at rest behind the shock front, we have

$$U - u_\ell = c\sqrt{\frac{\rho_\ell}{\rho_0}} - c\left(\sqrt{\frac{\rho_\ell}{\rho_0}} - \sqrt{\frac{\rho_0}{\rho_\ell}}\right) = c\sqrt{\frac{\rho_0}{\rho_\ell}} \le c = c(\rho). \tag{4.86}$$

The local sound speed is greater than or equal to the shock velocity, as is true for superlinear stress-strain relations.

At this stage, let us collect our results in a general statement:

THEOREM 4.14. *(compressive shock waves) A compressive shock wave is a large amplitude wave that increases the density of the material from an ambient density ρ_0 to a larger density ρ_ℓ. The passage of the shock wave imparts a velocity to the material in the direction of the shock wave propagation. With respect to the ambient material in front of the shock, the shock speed is supersonic, $U \ge c(\rho_0)$, and with respect to a frame of reference where the material behind the shock wave is at rest, the shock speed is subsonic, $U - u \le c(\rho_\ell)$. For the specific case of the linear stress-strain response $\tilde{\sigma} = \rho_0 c^2(1 - \rho_0/\rho)$, the inequalities are equalities and the relationships are sonic. For superlinear stress-strain relations, at some point the inequalities are strict, and for strictly superlinear stress-strain relations, the inequalities are strict for all densities above ρ_0 and $u > 0$.*

4.3. Rarefaction Fans and the Rarefaction Shock Approximation

We now consider the same piston problem as in the previous section, only now we pull the piston so that it has a negative velocity $u < 0$. There is a qualitative difference in this problem in that it is possible to pull the piston quickly enough that there is a separation of the piston wall from the material. If the material is not attached to the piston wall, this separation occurs when the stress is zero. (Such a separation even occurs for an ideal gas – see Exercise 5.17.) In fact this latter question, namely the release or rarefaction of a compressed material by a free surface, is in some ways of more interest, and so we will explicitly solve it.

Consider the following initial condition: a half space $x > 0$ of material has an initial prescribed compressive stress. The left boundary is stress free. Beginning at time $t = 0$, a wave will propagate to the right that relieves the stress. In this section we explicitly solve for the particle velocity and stress distribution of this wave. It turns out that the stress distribution forms a fan in the (x, t) plane. Specifically, the conditions to the right and left of the wave are

$$\begin{aligned} \rho_\ell &= \rho_0 & \rho_r &= \text{given} \\ \tilde{\sigma}_\ell &= 0 & \tilde{\sigma}_r &= \text{given} \\ u_\ell &= ? & u_r &= 0. \end{aligned} \tag{4.87}$$

A wave will move into the material, relieving the stress and thus reducing the density. This wave is called a rarefaction, which by definition means to make less dense.

Since we have the Hugoniot jump conditions, one approach would be to obtain a rarefaction solution to the shock jump conditions. However, in nature, for large amplitude shock waves, rarefaction shocks are rarely seen.[1] Rather, rarefactions result in a fan-like dispersive wave and the change in stress and particle velocity is continuously spread out.

To understand the physics, it is easiest to imagine the material compression caused by the passing of a shock. As was shown in the previous section, the sound speed behind the shock front in the frame of reference where the compressed material is at rest is typically higher than the shock front speed. Thus, small amplitude rarefaction waves catch up with the shock front and reduce its amplitude in a non-uniform matter. Given this thinking, we can imagine that for each density there is a speed at which a very small amplitude rarefaction front is traveling. In the laboratory frame, this speed is given by the sum of the sound speed at that specific material density and the particle velocity caused by the rarefaction to that point:

$$\text{small amplitude rarefaction speed in the laboratory frame} = c(\rho) + u(\rho). \quad (4.88)$$

To compute the particle velocity $u(\rho)$ due to the rarefaction wave we integrate the differential jump conditions. In particular, from Chapter 2 (Eq. 2.81) the differential mass jump condition is

$$c(\rho)d\rho = \rho du \quad (4.89)$$

and the differential momentum jump condition is (Eq. 2.82)

$$d\tilde{\sigma} = \rho c(\rho) du. \quad (4.90)$$

These can be integrated to give

$$u(\rho) = \int_{\rho_r}^{\rho} \frac{c(\rho)}{\rho} d\rho = \int_{\tilde{\sigma}_r}^{\tilde{\sigma}} \frac{d\tilde{\sigma}}{\rho c(\rho)} \leq 0. \quad (4.91)$$

This integral, which is referred to as the Riemann integral, accumulates the particle velocity associated with the sound wave passage through the various densities. The value of the integral is negative since the density and stress are decreasing.

At this point we need to introduce another notion to pursue the solution. We have loosely used the term "fan" since that is what we expect the solution to look like, in a qualitative sense. This argues that we should look for solutions where the various variables such as density, material velocity, and stress are constant along lines of constant x/t – notice that these lines form a fan originating at the origin or center of our problem initial conditions. Thus, we assume that all the variables are functions of $\xi = x/t$. Such an assumption is called a similarity solution. It will be used again in a slightly different form in Chapter 8. Here, with $\xi = x/t$, it is necessary to compute some derivatives in order to write the equations of motion in terms of ξ,

$$\frac{\partial \xi}{\partial t} = -\frac{x}{t^2} = -\frac{\xi}{t}, \qquad \frac{\partial \xi}{\partial x} = \frac{1}{t} = \frac{\xi}{x}. \quad (4.92)$$

---

[1]Data for fused quartz in [**139**], with $\rho_0 = 2.204$ g/cm^3, $c_L = 5.96$ km/s, and $c_s = 3.77$ km/s, has a fit of $U = 5.25 - 0.11u$ (U and u in km/s) for $0 < u < 2.1$ km/s and $0 < \tilde{\sigma} < 23.4$ GPa, implying rarefaction shocks and compressive fans in that regime, since $k = -0.11 < 0$.

For example, if ρ is our variable of interest, then $\rho = \rho(\xi)$ by assumption and

$$\frac{\partial \rho}{\partial t} = \frac{d\rho}{d\xi} \frac{\partial \xi}{\partial t} = -\frac{\xi \rho'}{t}, \tag{4.93}$$

where the prime denotes differentiation with respect to ξ. Similarly,

$$\frac{\partial \rho}{\partial x} = \frac{d\rho}{d\xi} \frac{\partial \xi}{\partial x} = \frac{\xi \rho'}{x}. \tag{4.94}$$

The conservation of mass and momentum equations from Chapter 2, applied to one dimension, become the ordinary differential equations

$$\frac{\partial \rho}{\partial t} + u\frac{\partial \rho}{\partial x} + \rho\frac{\partial u}{\partial x} = 0 \quad \Rightarrow \quad -\frac{\xi \rho'}{t} + u\frac{\xi \rho'}{x} + \rho\frac{\xi u'}{x} = 0 \quad \Rightarrow \quad (-\xi + u)\rho' + \rho u' = 0,$$

$$-\frac{\partial \tilde{\sigma}}{\partial x} = \rho\frac{\partial u}{\partial t} + \rho u\frac{\partial u}{\partial x} \quad \Rightarrow \quad -\frac{\xi \tilde{\sigma}'}{x} = -\rho\frac{\xi u'}{t} + \rho u\frac{\xi u'}{x} \quad \Rightarrow \quad -\tilde{\sigma}' = (-\xi + u)\rho u'. \tag{4.95}$$

Let us solve each of these equations for $-\xi + u$, recalling the differential jump conditions – that is, $u(\rho)$ comes from the Riemann integral, Eq. (4.91) –

$$-\xi + u = -\frac{\rho u'}{\rho'} = -\rho\frac{du}{d\rho} = -c(\rho),$$

$$-\xi + u = -\frac{\tilde{\sigma}'}{\rho u'} = -\frac{1}{\rho}\frac{d\tilde{\sigma}}{du} = -c(\rho). \tag{4.96}$$

Both these equations imply

$$c(\rho) + u(\rho) = \xi = x/t. \tag{4.97}$$

Thus, we have found our explicit solution for the rarefaction fan in terms of the material density as the rarefaction wave travels through the material – the highest compression traveling the fastest and the lowest compression traveling the slowest (Fig. 4.3):

$$\rho(x, t) = \begin{cases} \rho_r & c(\rho_r) \leq x/t, \\ (c + u)^{-1}(x/t) & c(\rho_0) + u(\rho_0) \leq x/t \leq c(\rho_r), \\ \rho_0 & c(\rho_0) + u(\rho_0) \geq x/t. \end{cases} \tag{4.98}$$

In this expression, $(c + u)^{-1}$ means the inverse of $c(\rho) + u(\rho)$ and may be quite tedious to compute.

The above solution was derived on a mechanical basis – that is, it used only the mass and momentum balance. We now ask about the energy balance. In the similarity transformation it is

$$\rho\frac{\partial E}{\partial t} + \rho u\frac{\partial E}{\partial x} = -\tilde{\sigma}\frac{\partial u}{\partial x} \quad \Rightarrow \quad -\rho\frac{\xi E'}{t} + \rho u\frac{\xi E'}{x} = -\tilde{\sigma}\frac{\xi u'}{x} \quad \Rightarrow \quad \rho(-\xi + u)E' = -\tilde{\sigma}u'. \tag{4.99}$$

Since our solution has $-\xi + u(\rho) = -c(\rho)$ and solves the differential jump condition $\rho du = c(\rho)d\rho$ by construction, the energy balance becomes

$$E' = \frac{\tilde{\sigma}}{\rho c(\rho)}u' = \frac{\tilde{\sigma}}{\rho c(\rho)}\frac{c(\rho)}{\rho}\rho' = -\tilde{\sigma}(1/\rho)'. \tag{4.100}$$

This says that $dE = -\tilde{\sigma}d(1/\rho)$ for the rarefaction fan. Thus,

$$T ds = dE + \tilde{\sigma}d(1/\rho) = 0 \tag{4.101}$$

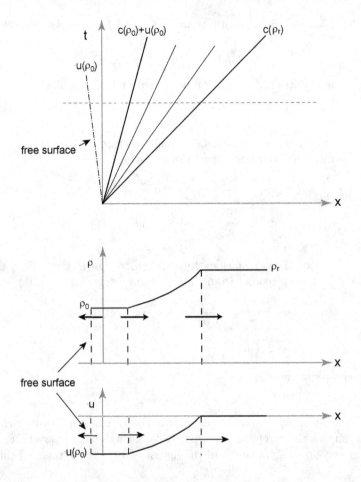

FIGURE 4.3. The rarefaction fan in position–time and the density
and particle velocity in terms of position.

and we conclude that the rarefaction fan is an isentrope.

Now let us examine this solution to understand its implications. To begin with, the solution assumes that for $\rho_2 < \rho_1$ that

$$c(\rho_2) + u(\rho_2) < c(\rho_1) + u(\rho_1) \tag{4.102}$$

since if this does not hold a shock is formed (a lower density wave travels faster than a higher density wave, thus catching up with it, leading to a finite drop in density). This requirement can be written

$$c(\rho_2) - c(\rho_1) < u(\rho_1) - u(\rho_2) = \int_{\rho_r}^{\rho_1} \frac{c(\rho)}{\rho} d\rho - \int_{\rho_r}^{\rho_2} \frac{c(\rho)}{\rho} d\rho = \int_{\rho_2}^{\rho_1} \frac{c(\rho)}{\rho} d\rho. \tag{4.103}$$

If we divide both sides of this equation by $\rho_2 - \rho_1 < 0$, we obtain

$$\frac{c(\rho_2) - c(\rho_1)}{\rho_2 - \rho_1} > -\frac{1}{\rho_2 - \rho_1} \int_{\rho_1}^{\rho_2} \frac{c(\rho)}{\rho} d\rho. \tag{4.104}$$

Taking limits gives

$$\frac{d(c(\rho))}{d\rho} > -\frac{c(\rho)}{\rho}. \tag{4.105}$$

For a rarefaction fan to form this relation must hold for the sound speed. To see what this relation means for a mechanical stress-strain relation, we insert the definition of sound speed $c(\rho) = \sqrt{d\tilde{\sigma}/d\rho}$ to obtain

$$\frac{d^2\tilde{\sigma}}{d\rho^2} > -2\frac{c^2(\rho)}{\rho} = -\frac{2}{\rho}\frac{d\tilde{\sigma}}{d\rho}. \tag{4.106}$$

We now write

$$\begin{aligned}
\frac{d^2\tilde{\sigma}}{d(1/\rho)^2} &= -\rho^2\frac{d}{d\rho}\left(-\rho^2\frac{d\tilde{\sigma}}{d\rho}\right) \\
&= \rho^2\left\{2\rho\frac{d\tilde{\sigma}}{d\rho} + \rho^2\frac{d^2\tilde{\sigma}}{d\rho^2}\right\} \\
&> \rho^2\left\{2\rho\frac{d\tilde{\sigma}}{d\rho} + \rho^2\left(-\frac{2}{\rho}\frac{d\tilde{\sigma}}{d\rho}\right)\right\} \\
&= 0.
\end{aligned} \tag{4.107}$$

Thus, we have shown

LEMMA 4.15. *Rarefaction fans form when the stress-strain relation is strictly concave up.*

The condition for the formation of a rarefaction fan is more restrictive in its allowable stress-strain relations than that for the formation of a compressive shock since the rarefaction essentially slides down all density values and every density, not just the initial density, must be such that it will not lead to a rarefaction shock. Materials that do not have a strictly concave up stress-strain relation over the region of interest will produce rarefaction shocks or a combination of rarefaction shocks connecting rarefaction fans.

As examples, let's compute the various terms for two stress-strain relations. First we look at the linear stress-strain relation, realizing that the rarefaction solution probably should not apply because the linear stress-strain relation is not concave up. For the linear stress-strain relation, we previously computed $c(\rho) = \rho_0 c/\rho$, giving

$$u(\rho) = \int_{\rho_r}^{\rho}\frac{c(\rho)}{\rho}d\rho = \int_{\rho_r}^{\rho}\frac{\rho_0 c}{\rho^2}d\rho = \rho_0 c\left(\frac{1}{\rho_r} - \frac{1}{\rho}\right). \tag{4.108}$$

The particle velocity $u(\rho)$ is less than zero, as we would expect – the wave that reduces the density moves material in an opposite direction to its motion. This result gives

$$c(\rho) + u(\rho) = \frac{\rho_0 c}{\rho} + \rho_0 c\left(\frac{1}{\rho_r} - \frac{1}{\rho}\right) = \frac{\rho_0 c}{\rho_r} = \text{constant.} \tag{4.109}$$

Since $c(\rho) + u(\rho) = x/t$ is a constant, the rarefaction fan collapses to a discontinuous jump in density with the wave front traveling at speed $x/t = \xi = \rho_0 c/\rho_r$. In fact, the linear stress-strain relation supports rarefaction shocks.

As an example which does lead to a rarefaction fan, consider the stress-strain relation $\tilde{\sigma} = \rho_0 c^2(\rho/\rho_0 - 1)$. This relation is concave up. We previously computed

$c(\rho) = c$, and so

$$u(\rho) = \int_{\rho_r}^{\rho} \frac{c(\rho)}{\rho} d\rho = \int_{\rho_r}^{\rho} \frac{c}{\rho} d\rho = c\ln(\rho/\rho_r). \tag{4.110}$$

Notice again that since the rarefaction reduces the density, the particle velocity is negative. In this case

$$c(\rho) + u(\rho) = c + c\ln(\rho/\rho_r). \tag{4.111}$$

It is possible to invert this expression to obtain the density in the fan region of the rarefaction,

$$\rho(x,t) = (c+u)^{-1}(x/t) = \rho_r \exp\left(\frac{x}{ct} - 1\right), \tag{4.112}$$

which gives for the whole solution

$$\rho(x,t) = \begin{cases} \rho_r & c \le x/t, \\ \rho_r \exp(x/ct - 1) & c(1 + \ln(\rho_0/\rho_r)) \le x/t \le c, \\ \rho_0 & c(1 + \ln(\rho_0/\rho_r)) \ge x/t. \end{cases} \tag{4.113}$$

The velocity is given by

$$u(x,t) = \begin{cases} 0 & c \le x/t, \\ x/t - c & c(1 + \ln(\rho_0/\rho_r)) \le x/t \le c, \\ c\ln(\rho_0/\rho_r) & c(1 + \ln(\rho_0/\rho_r)) \ge x/t. \end{cases} \tag{4.114}$$

In the region of the fan, $c + u(x,t) = x/t$, as assumed in the solution. For a large enough density change, namely $\rho_0/\rho_r < 1/e$, it is possible for $c(\rho) + u(\rho) < 0$. Such a situation simply means that in the laboratory reference frame the particle velocity exceeds the sound speed and so material is moving to the left more quickly than it is being unloaded, though in the end it will all be unloaded. In particular, the fan expands in both directions.

Though the rarefaction fan is an explicit, analytic solution, it is difficult to work with when we are looking for simple wave propagation solutions because there is not a single arrival time for the complete rarefaction. In modeling, it is often useful to approximate rarefactions as shock waves as a simplification. Thus, a definition

DEFINITION 4.16. (rarefaction shock approximation) The *rarefaction shock approximation,* which is exact for the linear stress-strain response $\tilde{\sigma} = \rho_0 c^2(1 - \rho_0/\rho)$, is to use the Hugoniot jump conditions to calculate a rarefaction wave. Such an assumption violates the requirement that entropy increase across a shock front.

Proceeding with the rarefaction shock approach, the shock jump conditions are

$$\frac{\rho_r}{\rho_0} = 1 - \frac{u_\ell}{U} \Rightarrow u_\ell = U\left(1 - \frac{\rho_r}{\rho_0}\right), \tag{4.115}$$

$$\tilde{\sigma}_r - \tilde{\sigma}_0 = \rho_r U u_\ell, \tag{4.116}$$

and these combine to give

$$\hat{U}_R = \sqrt{\frac{\tilde{\sigma}(\rho_r) - \tilde{\sigma}(\rho_0)}{\rho_r(\rho_r/\rho_0 - 1)}}, \tag{4.117}$$

where the hat denotes the frame in which the material in front of the rarefaction is
at rest. This wave speed is not the same as the compressive shock wave speed. For
a compressive shock traveling into ambient material (Eq. 4.18),

$$U_0 = \sqrt{\frac{\tilde{\sigma}(\rho_\ell) - \tilde{\sigma}(\rho_0)}{\rho_0(1 - \rho_0/\rho_\ell)}}, \tag{4.118}$$

where we've applied the subscript 0 on U to distinguish it from the rarefaction
shock; the subscript 0 on U refers to the fact that it applies to the wave propagation
case where the material was initially unstressed (i.e., it is at density ρ_0). Notice
that if it is assumed that the stress to the right in our current example is equal to
the applied stress in the previous example, then

$$\hat{U}_R = \sqrt{\frac{\tilde{\sigma}(\rho_r) - \tilde{\sigma}(\rho_0)}{\rho_r(\rho_r/\rho_0 - 1)}} = \sqrt{\frac{\rho_0}{\rho_r^2} \frac{\tilde{\sigma}(\rho_r) - \tilde{\sigma}(\rho_0)}{1 - \rho_0/\rho_r}} = \frac{\rho_0}{\rho_r} U_0. \tag{4.119}$$

This result is quite stunning. It says that the rarefaction shock traveling into the
compressed material travels at a slower speed than the compressive shock traveling
into the uncompressed material. In this derivation, we have not used a specific
stress-strain relation; thus, it always applies. The result says that *the time it takes
to travel through material is preserved*: with mass conservation, a given length of
material will satisfy $L\rho = L_0\rho_0$. The time it takes the compressive shock to go
through the uncompressed material is $t = L_0/U_0$. The time it takes the rarefaction
shock to travel through the compressed material is

$$t = \frac{L}{U} = \frac{L_0\rho_0/\rho_r}{\rho_0 U_0/\rho_r} = \frac{L_0}{U_0}. \tag{4.120}$$

We summarize with a theorem:

THEOREM 4.17. *(equal travel time) In the rarefaction shock approximation
(which is strictly true for a linear stress-strain response), the transit time for a
compressive shock through a given amount of material is equal to the transit time
of a rarefaction shock through that same material taking the shock-compressed material back to the ambient state.*

Another consequence of the rarefaction shock approximation is obtained from
the mass jump condition for the rarefaction shock moving into the compressed
material:

$$\hat{u}_\ell = \hat{U}_R\left(1 - \frac{\rho_r}{\rho_0}\right) = \hat{U}_R - U_0 = -\sqrt{(\tilde{\sigma}(\rho_r) - \tilde{\sigma}(\rho_0))\left(\frac{1}{\rho_0} - \frac{1}{\rho_r}\right)} \le 0. \tag{4.121}$$

Thus, the particle speed of the material is simply the difference of the rarefaction
shock speed into the compressed material and the compressive shock speed into
the uncompressed material. Similarly, the particle velocity behind the compressive
shock is (Eq. 4.19)

$$u_{\ell 0} = \left(1 - \frac{\rho_0}{\rho_\ell}\right)\sqrt{\frac{\tilde{\sigma}(\rho_\ell) - \tilde{\sigma}(\rho_0)}{\rho_0(1 - \rho_0/\rho_\ell)}} = U_0 - \hat{U}_R = -\hat{u}_\ell. \tag{4.122}$$

Thus, in the rarefaction shock approximation, the particle velocity change due to
the passage of the compressing shock equals the negative of the particle velocity
change due to the passage of the unloading rarefaction shock. These results solve the square pulse

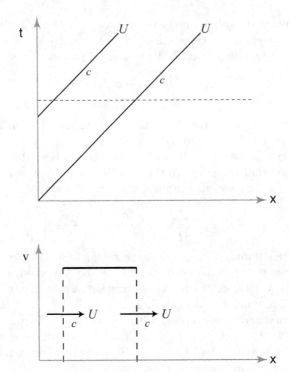

FIGURE 4.4. A square pulse.

problem (Fig. 4.4). The compressive wave has $U = U_0$ with the particle velocity $u_{\ell 0}$ behind the shock front. The rarefaction shock has $U = \hat{U}_R + u_{\ell 0} = U_0$, so it travels at the same speed as the compressive shock in the laboratory frame. Behind the pulse the particle speed is $u = \hat{u}_\ell + u_{\ell 0} = 0$; hence at rest. The length of the pulse remains the same, as required by momentum conservation.

Equation (4.122) can be written

$$\hat{U}_R = \frac{\rho_0}{\rho_r} U_0 = U_0 - u_{\ell 0}. \tag{4.123}$$

We will often use this expression to write down the rarefaction shock speed. It turns out that for the linear stress-strain relation, a much more general result is true. If a shock has passed through the material with an associated particle velocity $u_{\ell 0}$, then in the frame of reference where the material behind the shock is at rest, the next wave taking the material to any density travels at

$$\hat{U} = \sqrt{\frac{\tilde{\sigma}(\rho_\ell) - \tilde{\sigma}(\rho_{\ell 0})}{\rho_{\ell 0}(1 - \rho_{\ell 0}/\rho_\ell)}} = \sqrt{\frac{\rho_0 c^2(\rho_0/\rho_{\ell 0} - \rho_0/\rho_\ell)}{\rho_{\ell 0}(1 - \rho_{\ell 0}/\rho_\ell)}} = \frac{\rho_0}{\rho_\ell} c = U_0 - u_{\ell 0}. \tag{4.124}$$

Thus, for the linear stress-strain relation, the Lagrangian wave speed for the next wave can be immediately written down if it is known what particle velocity led to the current compression of the material. Equation (4.124) does not just apply to rarefactions, and this result gives a useful result that applies to the linear stress-strain relation:

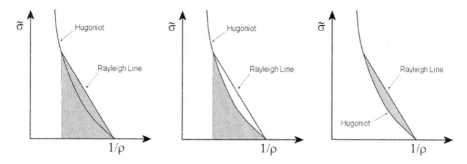

FIGURE 4.5. The internal energy deposited in the material by the passage of the shock (left), the amount of internal energy required to perform the rarefaction (center), and the residual amount of internal energy deposited in the material by the passage of the shock followed by the rarefaction (right).

THEOREM 4.18. *For the linear stress-strain relation $\tilde{\sigma} = \rho_0 c^2(1 - \rho_0/\rho) = \rho_0 c^2 \tilde{\varepsilon}$, the wave speed is determined by the conditions of the material in front of the wave – in particular the particle velocity u with respect to the laboratory frame and the density ρ, stress $\tilde{\sigma}$, or engineering strain $\tilde{\varepsilon}$. The laboratory frame wave speed is*

$$U = \pm \frac{\rho_0}{\rho} c + u = \pm c(1 - \tilde{\varepsilon}) + u = \pm \left(c - \frac{\tilde{\sigma}}{\rho_0 c} \right) + u, \qquad (4.125)$$

where the $(+)$ is for a wave moving through the material to the right and the $(-)$ is for a wave moving through the material to the left.

In general, when a shock passes through a material and is followed by a rarefaction that reduces the material to the initial ambient stress state, residual internal energy remains in the material. Figure 4.5 shows that the residual internal energy deposited by the shock front is equal to the energy deposited by the passage of the shock minus the energy required to expand the shocked material back to the ambient state (the energy stored in elastic compression). The energy stored in the material due elastic compression only equals that deposited by the shock if the stress-strain response is linear – that is, if it is a straight line in the $(1/\rho, \tilde{\sigma})$ plane. For a superlinear material, the shock deposits more internal energy in the material behind the shock front than is stored in elastic strain of the material. In a loose sense, to choose the largest entropy solution for a shock wave is to choose the wave or waves that lead to the largest area under the curve in the right of Fig. 4.5. This gives another view of why, for a material whose Hugoniot crossed the Rayleigh line, there would be a collection of Rayleigh lines with higher residual energy, hence higher entropy. In this figure the material response is assumed to be purely mechanical, since often the density after unloading is less than the initial density due to thermal terms.

Given the explicit solution for the rarefaction, it is possible to assess the quality of the rarefaction shock assumption. One measure of validity of the rarefaction shock approximation is made by examining the difference of the wave speeds of the leading edge and trailing edge of the rarefaction fan, and seeing how large the fan is when the rarefaction interacts with an interface or other region of interest in a problem (Exercise 4.21). Another measure is to compare the rarefaction shock

TABLE 4.1. Particle velocities from rarefaction fan solution vs. rarefaction shock approximation. The angle comparison is discussed in Exercise 4.22.

ρ_r/ρ_0	$u(\rho_0)/u_\ell$	Fan angle α $(°)$
1.1	0.99962	2.86
1.2	0.99862	5.73
1.3	0.99714	8.59
2.0	0.98026	27.9
3.0	0.95143	50.6
4.0	0.92420	66.1
5.0	0.89970	76.4

approximation particle velocity u_ℓ with the actual rarefaction fan particle velocity $u(\rho_0)$. For the $\tilde{\sigma} = \rho_0 c^2(\rho/\rho_0 - 1)$ stress-strain relation, the rarefaction shock particle velocity is

$$u_\ell = \left(1 - \frac{\rho_r}{\rho_0}\right)\sqrt{\frac{\tilde{\sigma}(\rho_r)}{\rho_r(\rho_r/\rho_0 - 1)}} = c\left(1 - \frac{\rho_r}{\rho_0}\right)\sqrt{\frac{\rho_0}{\rho_r}} = c\left(\sqrt{\frac{\rho_0}{\rho_r}} - \sqrt{\frac{\rho_r}{\rho_0}}\right).$$
(4.126)

The above solutions yield the ratio

$$\frac{u_{\text{rarefaction fan}}}{u_{\text{rarefaction shock}}} = \frac{u(\rho_0)}{u_\ell} = \frac{\ln(\rho/\rho_r)}{\sqrt{\rho_0/\rho_r} - \sqrt{\rho_r/\rho_0}}.$$
(4.127)

Table 4.1 gives values for the ratio for various ratios of initial compressed density to final ambient density after the release. The table also shows how wide the rarefaction fan is, as defined in Exercise 4.21.

It is possible to provide an upper bound on the particle velocity produced by a rarefaction for the case of a purely mechanical stress-strain response. This result does not apply to a stress-strain response that includes thermal terms, but, for the purely mechanical case, note that

$$-u(\rho_0) = \int_{\rho_0}^{\rho_r} \frac{c(\rho)}{\rho}d\rho = \int_{\rho_0}^{\rho_r} \frac{1}{\rho}\sqrt{\frac{d\tilde{\sigma}}{d\rho}}d\rho = \int_{1/\rho_r}^{1/\rho_0} \sqrt{-\frac{d\tilde{\sigma}}{d(1/\rho)}}d(1/\rho).$$
(4.128)

The Cauchy-Schwartz inequality for functions f and g says [168]

$$\int_a^b fg\,dx \le \sqrt{\int_a^b f^2\,dx}\sqrt{\int_a^b g^2\,dx}.$$
(4.129)

Setting

$$f = \sqrt{-\frac{d\tilde{\sigma}}{d(1/\rho)}}, \quad g = 1, \quad x = 1/\rho$$
(4.130)

yields

$$-u(\rho_0) = \int_{1/\rho_r}^{1/\rho_0} \sqrt{-\frac{d\tilde{\sigma}}{d(1/\rho)}}d(1/\rho) \le \sqrt{\int_{1/\rho_r}^{1/\rho_0} -\frac{d\tilde{\sigma}}{d(1/\rho)}d(1/\rho)}\sqrt{\int_{1/\rho_r}^{1/\rho_0} d(1/\rho)}$$

$$= \sqrt{\tilde{\sigma}(\rho_r)}\sqrt{1/\rho_0 - 1/\rho_r} = -u_\ell.$$
(4.131)

Since both $u(\rho_0)$ and u_ℓ are negative, we have

$$|u(\rho_0)| \leq |u_\ell|. \tag{4.132}$$

Thus, in magnitude, the particle velocity upon unloading of the actual smooth rarefaction is less than that predicted by the rarefaction shock approximation. This result also follows from a consideration of the energy balance, namely that the amount of kinetic energy available upon unloading must be less than the amount of energy deposited by the passage of a shock that increased the material density and stress to the state from which unloading occurs. This argument also does not transfer to materials with thermal terms or changes in stress-strain response upon unloading, such as phase changes.

This last observation says that the passage of a shock and subsequent rarefaction through a material can leave the material with positive particle velocity. It only occurs when there is a rarefaction fan. Since the rarefaction fan overtakes the shock, the shock propagation into the material will be of finite spatial extent. Thus, there is not an infinite extent of material put in residual motion, which would violate momentum conservation. In contrast, for a linear material the material behind the rarefaction is at rest and, in theory, the square shock pulse continues on forever into the material.

Let us collect our results about rarefactions in a general statement:

THEOREM 4.19. *(rarefactions) A rarefaction wave is a wave that decreases the density of the material from an ambient density ρ_r to a lesser density ρ_0. The passage of the rarefaction wave imparts a velocity to the material in the direction opposite of the rarefaction wave propagation. In a frame of reference where the material compressed by a shock is at rest, the leading edge of the rarefaction is traveling at the speed of sound, which is greater than the propagation speed of the shock and is thus able to overtake the shock to dissipate it. The rarefaction looks like a fan bounded between the two speeds $c(\rho_r) + u_{initial}$ at the leading edge and $c(\rho_0) + u(\rho_0) + u_{initial}$, where $u(\rho_0) < 0$, at the trailing edge. The linear stress-strain response $\tilde{\sigma} = \rho_0 c^2 (1 - \rho_0/\rho)$ supports rarefaction shocks. Strictly concave up materials have rarefaction fans only. The rarefaction release is isentropic.*

4.4. Impacting a Rigid Wall

We now examine a series of wave propagation examples. For simplicity, unless stated otherwise, we will use the linear stress-strain response $\tilde{\sigma} = \rho_0 c^2 (1 - \rho_0/\rho)$. The compressive wave speed into ambient at-rest material for the linear stress-strain relation is the constant c, independent of the magnitude of the strain (which can be large).

Consider the case of a projectile traveling to the right and impacting a rigid wall. Figure 4.6 shows the initial conditions,

$$\begin{aligned}
\rho_\ell &= \rho_0 & \rho_r &= ? \\
\tilde{\sigma}_\ell &= 0 & \tilde{\sigma}_r &= ? \\
u_\ell &= V & u_r &= 0,
\end{aligned} \tag{4.133}$$

where V is the impact velocity. The wave speed U is one of the unknowns. Since we know the initial conditions to the left of the wave front, we use the second form

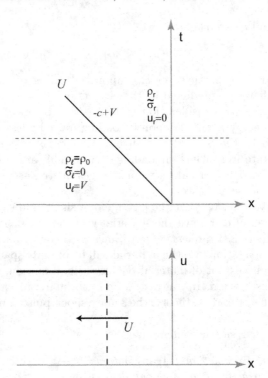

FIGURE 4.6. Impact into a rigid wall.

of the Rankine–Hugoniot jump conditions, typically used when solving for waves moving to the left (Eq. 2.71):

$$\frac{\rho_\ell}{\rho_r} = 1 - \frac{u_r - u_\ell}{U - u_\ell} \quad \Rightarrow \quad \frac{\rho_0}{\rho_r} = 1 + \frac{V}{U - V},$$

$$\tilde{\sigma}_r - \tilde{\sigma}_\ell = \rho_\ell(U - u_\ell)(u_r - u_\ell) \quad \Rightarrow \quad \tilde{\sigma}_r = \rho_0(U - V)(-V). \tag{4.134}$$

Writing the mass jump condition as

$$V = (\rho_0/\rho_r - 1)(U - V) \tag{4.135}$$

and inserting into the momentum balance gives

$$(U - V)^2 = c^2, \tag{4.136}$$

where c is the linear stress-strain response compressive shock speed into ambient at-rest material (in the rest of these examples, this fact will not be explicitly spelled out). Since the wave is moving to the left we take the negative in the square root, yielding

$$U = -c + V. \tag{4.137}$$

Thus, the wave is traveling *within the material* at sound speed c, but in our fixed laboratory (Eulerian) reference frame the wave front is moving at the sound speed plus the material velocity. Putting this result back into the momentum and mass jump conditions gives

$$\tilde{\sigma}_r = \rho_0 c V, \quad \rho_r = \rho_0/(1 - V/c). \tag{4.138}$$

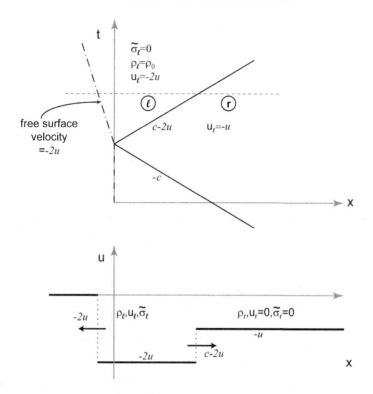

FIGURE 4.7. Reflection off a free surface.

In terms of the engineering strain $\tilde{\varepsilon} = 1 - \rho_0/\rho = u/U = V/c$, we have

$$\tilde{\sigma}_r = \rho_0 c^2 \tilde{\varepsilon}, \qquad \rho_r = \rho_0/(1 - \tilde{\varepsilon}). \tag{4.139}$$

Solving the wave propagation problem where various material properties are specified to the left and right of an interface as in Eq. (4.134) is referred to as solving the *Riemann problem*. We can explicitly write down the solution as our linear elastic response is relatively simple. Each following section of this chapter solves a Riemann problem for linear elastic response. With more complicated constitutive models the solution is often performed numerically, and the algorithm to determine the various wave speeds and material states behind or between the wave fronts is referred to as a *Riemann solver*. In particular, Riemann solvers are used in the Godunov method for numerically solving the conservation laws.

4.5. Reflection off a Free Surface

Consider the case of a compressive stress wave that has been formed in a material, traveling to the left, arriving at a free surface. Upon its arrival, a wave will be formed traveling to the right. The free surface is stress free. The initial conditions are (Fig. 4.7)

$$\begin{aligned}
\rho_\ell &=? & \rho_r &= \rho_0/(1 - u/c) \\
\tilde{\sigma}_\ell &= 0 & \tilde{\sigma}_r &= \rho_0 c u \\
u_\ell &=? & u_r &= -u.
\end{aligned} \tag{4.140}$$

Again, U is an unknown. The first thing we observe is that since the stress is zero to the left of the wave, we also know the density on the left side of the wave: $\rho_\ell = \rho_0$. The Hugoniot jump conditions give

$$\frac{\rho_r}{\rho_\ell} = 1 - \frac{u_\ell - u_r}{U - u_r} \quad\Rightarrow\quad \frac{\rho_0/(1 - u/c)}{\rho_0} = 1 - \frac{u_\ell + u}{U + u} \quad\Rightarrow\quad \frac{u_\ell + u}{U + u} = -\frac{u/c}{1 - u/c},$$

$$\tilde\sigma_\ell - \tilde\sigma_r = \rho_r(U - u_r)(u_\ell - u_r) \quad\Rightarrow\quad -\rho_0 cu = (\rho_0/(1 - u/c))(U + u)(u_\ell + u).$$

$$\tag{4.141}$$

Solving for $U + u$ from the mass jump condition and inserting the result into the momentum jump condition gives

$$\rho_0 cu = \rho_0(u_\ell + u)^2 c/u, \tag{4.142}$$

yielding $u_\ell = -u \pm u$. Since $U > 0$ (the wave travels from left to right), the mass balance shows that the term $u_\ell + u$ must be negative, and therefore we arrive at the famous result

$$u_\ell = -2u. \tag{4.143}$$

The derivation of this result used the rarefaction shock approximation. However, the rarefaction fan solution (Riemann integral) supplies the exact solution for the reflection off a free surface as

$$u_\ell = -u - \int_{\rho_0}^{\rho_r} \frac{c(\rho)}{\rho} d\rho = -u - \int_0^{\tilde\sigma_r} \frac{d\tilde\sigma}{\rho c(\rho)}. \tag{4.144}$$

As was seen in Section 4.3, the particle velocity from the rarefaction fan solution is close to the particle velocity u obtained from the rarefaction shock approximation, and so we have

THEOREM 4.20. *(free surface reflections) When a compressive wave reflects off a free surface, the resulting outward particle velocity is roughly twice the particle velocity caused by the passage of the shock wave internal to the body.*

When measurements are taken of the free surface during plate impact tests, the velocity measured is roughly twice the particle velocity of the wave within the material. Completing this example, the wave speed of the rarefaction is

$$U = c - 2u. \tag{4.145}$$

You will recognize the two parts of this. Since the rarefaction wave is traveling through compressed material caused by the passage of a wave with particle velocity u, the wave speed with respect to material in front of the wave at rest is $c - u$ (Eq. 4.124). The material itself as a whole is in motion to the left, thus contributing the additional $-u$.

4.6. Finite Impactor Striking a Rigid Wall

We are finally in a position to do a full-fledged impact problem. Suppose an elastic impactor has a length L and strikes a rigid wall at velocity V, moving from the left to the right. A compression wave travels from right to left, slowing the projectile. When it reaches the back free surface of the projectile, a rarefaction wave forms, traveling to the right. These waves cause the projectile to bounce away from the rigid wall with its striking velocity. The solution comes from combining the two previous examples. After the initial impact, we have (Fig. 4.8)

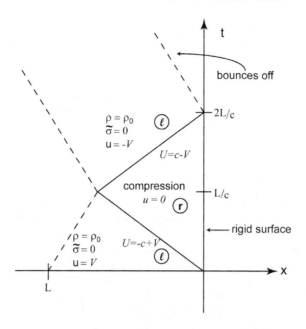

FIGURE 4.8. Finite impactor striking a rigid wall.

$$\begin{array}{ll} \rho_\ell = \rho_0 & \rho_r = \rho_0/(1 - V/c) \\ \tilde{\sigma}_\ell = 0 & \tilde{\sigma}_r = \rho_0 cV \\ u_\ell = V & u_r = 0. \end{array} \tag{4.146}$$

The wave velocity is $U = -c + V$. Next, for the stress release wave coming from the free surface, we have from the previous problem (performing a Galilean transformation, shifting the material velocity and U by $+V$)

$$\begin{array}{ll} \rho_\ell = \rho_0 & \rho_r = \rho_0/(1 - V/c) \\ \tilde{\sigma}_\ell = 0 & \tilde{\sigma}_r = \rho_0 cV \\ u_\ell = -V & u_r = 0. \end{array} \tag{4.147}$$

with $U = c - V$. (Since the impactor material has been brought to rest, U is less than c because of the compression of the material – Eq. 4.123.) Notice that the change in momentum of the projectile is given by $-2L\rho_0 V$, and the impulse on the wall is given by the length of time of the impact multiplied by the stress there. The time of the impact load on the wall includes the time during the initial compression until the compressive wave reaches the back surface and the time for the rarefaction to travel from the back surface to the rigid wall. The time of each is given by the wave speed divided by the distance, recalling that the left side of the projectile moves during the initial compression before the arrival of the compressive wave at the free surface, thus shortening the projectile,

$$\begin{aligned} t_c &= \frac{L - V t_c}{|U_1|} = \frac{L - V t_c}{c - V} \quad \Rightarrow \quad t_c = \frac{L}{c}, \\ t_r &= \frac{L - V t_c}{U_2} = \frac{L - V t_c}{c - V} \quad \Rightarrow \quad t_r = \frac{L}{c}. \end{aligned} \tag{4.148}$$

Alternatively, $t_r = t_c = L/c$ also follows from our "equal travel times" Theorem 4.17. The time of impact is

$$t_{\text{impact}} = t_c + t_r = \frac{2L}{c} = \text{loading time.} \tag{4.149}$$

The load applied to the rigid wall is given by $\tilde{\sigma}_r$ above, and it is seen that multiplying the load by the impact time produces the momentum transferred to the wall, thus agreeing with the velocity change of the projectile,

$$\Delta \text{momentum} = \text{impulse} = 2L\rho_0 V. \tag{4.150}$$

4.7. Finite Impactor with Hysteresis Striking a Rigid Wall

It is our experience that balls and other objects striking a rigid wall bounce back as described above, but some objects do not. In the one-dimensional wave propagation we are examining, the amount of bounce can be explored if the material has hysteresis, meaning that its behavior during unloading differs from its behavior during loading. The behavior might also be termed inelastic, since the impactor does not return to its original shape upon unloading. An initial compressive wave is formed by impacting a rigid wall by a material with with initial density ρ_0 and sound speed c as described above. Suppose, due to the passage of the compressive wave caused by the impact, that the material permanently compresses and thus has a new ambient-state sound speed c_f and a new zero-stress unloaded density ρ_{of}. Since the unloading occurs with less density change, the new modulus is larger and $c_f \geq c$. From Eq. (4.146), the Riemann problem to solve for the release wave moving left to right is

$$\begin{aligned} \rho_\ell &= \rho_{of} & \rho_r &= \rho_0/(1 - V/c) \\ \tilde{\sigma}_\ell &= 0 & \tilde{\sigma}_r &= \rho_0 cV \\ u_\ell &=? & u_r &= 0. \end{aligned} \tag{4.151}$$

The mass and momentum jump conditions are, using $U = (\rho_{of}/\rho_r)c_f \neq c_f - V$ in this case since the material properties have changed,

$$\begin{aligned} \frac{\rho_r}{\rho_{of}} &= 1 - \frac{u_\ell}{(\rho_{of}/\rho_r)c_f}, \\ \tilde{\sigma}_\ell - \tilde{\sigma}_r &= -\tilde{\sigma}_r = \rho_r U u_\ell = \rho_{of} c_f u_\ell. \end{aligned} \tag{4.152}$$

For these to be consistent, ρ_{of} and c_f are related, and some algebra reveals

$$c_f^2 = \frac{cV}{\{1 - (1 - V/c)\rho_{of}/\rho_0\}\rho_{of}/\rho_0}. \tag{4.153}$$

If we write $u_\ell = -qV$, $q \geq 0$, then the solution to the Hugoniot jump conditions is

$$u_\ell = -qV, \quad c_f = \frac{c}{q}\left(1 - \frac{V}{c}(1 - q^2)\right), \quad \frac{\rho_0}{\rho_{of}} = 1 - \frac{V}{c}(1 - q^2). \tag{4.154}$$

An expression for q is

$$q^2 = \frac{c}{V}\left(\frac{\rho_0}{\rho_{of}} - \frac{\rho_0}{\rho_r}\right). \tag{4.155}$$

This equation tells us that $q = 1$ when $\rho_{of} = \rho_0$ – that is, when the impactor returns to its original shape – and that $q = 0$ when $\rho_{of} = \rho_r$. Thus, the less expansion of the material during the rarefaction (notice that $\rho_{of} \leq \rho_r$, since the release wave is a rarefaction), the less the bounce off the wall by the projectile. The amount of elastic rebound is not related to how much compression occurred during the initial impact, but rather how much expansion occurs during release. If there is no expansion, there is no bounce off the wall. The last expression placed in the energy jump condition immediately yields

$$E_\ell - E_r = -\frac{1}{2}V^2 q^2, \tag{4.156}$$

showing the amount of internal energy lost to the projectile by the release. Since the initial compressive wave deposited an internal energy of $\frac{1}{2}V^2$, the amount of internal energy left in the projectile after release is

$$E_f = \frac{1}{2}V^2(1 - q^2). \tag{4.157}$$

To compute the amount of time for the rarefaction to travel the length of the projectile, the compressed length of material is $L/(1 - V/c)$ and so $L/(\rho_{0f}/\rho_r)c_f = qL/c$. Thus, when the release is combined with the initial compression, the impulse and time of loading for this impact are

$$\Delta\text{momentum} = \text{impulse} = (1 + q)L\rho_0 V, \quad \text{loading time} = (1 + q)L/c. \tag{4.158}$$

In the above analysis we assumed linear stress-strain behavior for both loading and unloading. More complicated hysteresis can occur, including load/unload cycles where the final density is the same but loading different paths are traversed in stress-density space. These lead to multiple wave interactions, but, if a final internal energy is known, q can be determined and the velocity coming off the wall can be computed.

4.8. A Finite Projectile Impacting a Material Wall

Consider an impactor of length L, density ρ_1 and sound speed c_1 impacting a semi-infinite wall of density ρ_2 and sound speed c_2. The impact velocity is V. This problem entails at least three waves. Two are produced by the initial impact, one traveling to the left in material 1 and one to the right in material 2. Later, the wave in the impactor will reflect at the impactor free surface and return to the projectile/target interface. Upon arrival at the interface, two things can occur. Either the impactor bounces off or we begin another iteration of three waves.

At impact at the interface between the two materials, the stresses are equal and the particle velocities are equal; between the two wave fronts that are moving away from the impact, the stress is constant and the particle velocity is constant. To keep the notation manageable, there will be two indices, one to indicate material and one to indicate left or right. There are two simultaneous sets of Hugoniot jump conditions to solve. The various knowns and unknowns are (Fig. 4.9)

$$\begin{aligned}
\rho_{1\ell} &= \rho_{10} & \rho_{1r} &=? & \rho_{2\ell} &=? & \rho_{2r} &= \rho_{20} \\
\tilde{\sigma}_{1\ell} &= 0 & \tilde{\sigma}_{1r} &= \tilde{\sigma}_{2\ell} =? & \tilde{\sigma}_{2r} &= 0 \\
u_{1\ell} &= V & u_{1r} &= u_{2\ell} =? & u_{2r} &= 0.
\end{aligned} \tag{4.159}$$

We already know that

$$U_1 = -c_1 + V, \quad U_2 = c_2. \tag{4.160}$$

The momentum jump conditions for each side give

$$\tilde{\sigma}_{1r} = \rho_{10}(-c_1 + V - V)(u_{1r} - V) = \rho_{10}c_1(V - u_{1r}) = \tilde{\sigma}_{2\ell} = \rho_{20}c_2 u_{2\ell}. \tag{4.161}$$

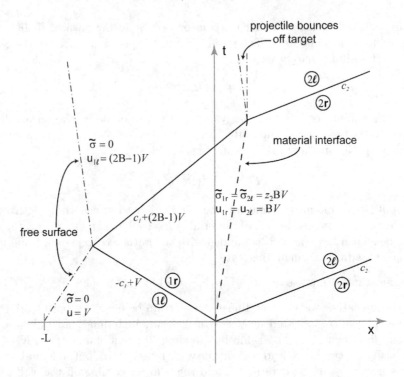

FIGURE 4.9. Projectile 1 of lesser impedance striking target 2 of greater impedance ($\rho_1 c_1 < \rho_2 c_2$) and bouncing off.

We denote the particle velocity at the projectile/target interface by $u = u_{1r} = u_{2\ell}$. The momentum jump conditions can be solved explicitly for the common particle velocity to give

$$u = \frac{\rho_{10} c_1}{\rho_{10} c_1 + \rho_{20} c_2} V. \tag{4.162}$$

Thus, the particle velocity depends only on the initial impact velocity and on the ρc term for each material.

DEFINITION 4.21. (shock impedance, acoustic impedance) The *shock impedance* of a material is given by $\rho_0 U$, where U is on the Hugoniot and is thus final density state dependent. At low pressures, when the response is elastic, the shock impedance is given by the *acoustic impedance* $\lim_{\rho \to \rho_0} \rho_0 U = \rho_0 c$. The acoustic impedance of the material is often denoted $z = \rho_0 c$.

The specific value of c for the acoustic impedance depends on the strain state under study, either c_L, c_S, or c_E. Recall that for a fluid, $c_L = c_K$.

We introduce a piece of notation to simplify some of what comes below,

$$B = \frac{\rho_{10} c_1}{\rho_{10} c_1 + \rho_{20} c_2}, \tag{4.163}$$

where it is seen that $0 < B < 1$ and $u = BV$. If the materials are the same, $B = 1/2$, and we immediately arrive at the result that the particle velocity is half the impact velocity. ($B = 0$ corresponds to material 2, the target, being a rigid wall and $B = 1$ corresponds to no target material.) The acoustic impedance determines stress levels and particle velocities when two materials impact. The stress $\tilde{\sigma} = \tilde{\sigma}_{1r} = \tilde{\sigma}_{2\ell}$ in the region between the two outgoing waves, and hence at the interface between the two materials, is

$$\tilde{\sigma} = \frac{\rho_{10}c_1\rho_{20}c_2}{\rho_{10}c_1 + \rho_{20}c_2}V = \rho_{20}c_2BV. \tag{4.164}$$

Next we have the arrival of the wave traveling to the left in material 1 and reflecting as a release wave. This situation is just our impactor above with a Galilean transformation of $+V$. Thus there will be a reflection at the surface, completely releasing the material from stress and returning it to its initial density. The particle velocity of that reflection is twice $u_{1r} - V$, so the particle velocity after the reflection to the left side of the returning wave is

$$u_{1\ell} = 2(u_{1r} - V) + V = 2u_{1r} - V = (2B - 1)V. \tag{4.165}$$

We are now in a position to repeat solving the two wave problem – we now have the impactor, which is stress free and at its original density but with a reduced velocity about to recollide with the material 2 surface. The initial conditions are

$$
\begin{array}{llll}
\rho_{1\ell} = \rho_{10} & \rho_{1r} =? & \rho_{2\ell} =? & \rho_{2r} = \rho_{20}/(1 - BV/c) \\
\tilde{\sigma}_{1\ell} = 0 & \tilde{\sigma}_{1r} = \tilde{\sigma}_{2\ell} =? & & \tilde{\sigma}_{2r} = \rho_{20}c_2BV \\
u_{1\ell} = (2B - 1)V & u_{1r} = u_{2\ell} =? & & u_{2r} = BV.
\end{array} \tag{4.166}
$$

Since the impactor material is stress free, the laboratory frame wave speed in material 1 is

$$U_1 = -c_1 + (2B - 1)V. \tag{4.167}$$

In material 2, the laboratory frame wave speed is slowed by the compression of the material but increased by the material motion (Eq. 4.124),

$$U_2 = \frac{\rho_{20}}{\rho_{2\ell}}c_2 + u = (c_2 - u) + u = c_2. \tag{4.168}$$

In fact, the waves going to the right in the target material will always propagate at speed c_2 in the laboratory frame. With the wave speeds, we can equate the stresses from the two momentum jump conditions. As before, we introduce the Lagrangian variables

$$\hat{U}_1 = U_1 - (2B - 1)V, \quad \hat{u}_1 = u_{1r} - (2B - 1)V, \tag{4.169}$$

$$\hat{U}_2 = U_2 - BV, \quad \hat{u}_2 = u_{2\ell} - BV. \tag{4.170}$$

The requirement that $u_{1\ell} = u_{2r}$ translates into $\hat{u}_2 = \hat{u}_1 + (B - 1)V$. The jump conditions give

$$\rho_{10}\hat{U}_1\hat{u}_1 = \rho_{20}c_2BV + \frac{\rho_{20}}{1 - BV/c_2}\hat{U}_2(\hat{u}_1 + (B - 1)V). \tag{4.171}$$

Plugging in the various terms and after some algebra, and noting that $\hat{U}_1 = -c_1$, one arrives at

$$\hat{u}_1 = -\frac{\rho_{20}c_2}{\rho_{10}c_1 + \rho_{20}c_2}(2B - 1)V = (B - 1)(2B - 1)V. \tag{4.172}$$

The particle velocity is then found to be

$$u_{1r} = u_{2\ell} = B(2B - 1)V. \tag{4.173}$$

Let us now compute the stress between the two outgoing waves. Choosing the material 1 side, we have

$$\tilde{\sigma}_{1r} = \rho_{10}\hat{U}_1\hat{u}_1 = \rho_{10}c_1(1 - B)(1 - 2B)V. \tag{4.174}$$

This expression is interesting. Recall that B can range from 0 to 1, so the stress in the region between the two wave fronts, where the interface between materials 1 and 2 is located, will be negative (tensile) if $B > 1/2$. Clearly this will never happen, because the materials will just pull apart at the interface. Thus we have arrived at a major conclusion for elastic impact:

(1) If the impedance of the impacting material is greater than the target material ($B > 1/2$), the impactor "sticks" to the target.
(2) If the impedance of the impacting material is less than that of the target material ($B < 1/2$), the impactor bounces off the target and a square wave travels into the target.
(3) If the impedance of the impacting material is the same as that of the target material ($B = 1/2$) the impacting material comes to rest at the surface of the target and a square pulse travels into the target.

Notice that a rigid wall as the target has infinite impedance ($B = 0$) and, as we saw above, the projectile bounces off the wall, in line with our conclusion here. These results are true in a more general setting:

GENERAL PRINCIPLE 4.22. *(relative impedances) For one-dimensional impact, if the projectile impedance $(\rho_0 U)$ exceeds the target impedance, the projectile continues to penetrate the target, while if the projectile impedance equals or is less than the target impedance, the projectile ceases to penetrate after one wave reverberation.*

In the more general setting, the impedance $\rho_0 U$ is a function of impact velocity and must be computed for each interaction. The reason why the conclusion is a general principle and not a theorem is because of complications due to the rarefaction fan (it is strictly true in the rarefaction shock approximation).

If the projectile bounces off, we have already calculated the time of the interaction from our rigid wall case:

$$t_{\text{impact}} = \frac{2L}{c_1}. \tag{4.175}$$

It is possible to compute the load transferred to the wall from the stress at the interface during the loading:

$$\text{impulse} = \frac{\rho_{10}c_1\rho_{20}c_2}{\rho_{10}c_1 + \rho_{20}c_2}\frac{2L}{c_1}V. \tag{4.176}$$

The velocity at which the impactor bounces back does not follow from Eq. (4.173) because that equation assumes the interfaces stayed together (notice that if $B = 0$ the predicted bounce-back velocity is zero, since it essentially has been assumed that the projectile is tied to the rigid surface). Thus, to get the bounce-back velocity we instead use $\tilde{\sigma}_{1r} = 0$, since the interface is now a free surface, and obtain $\hat{u}_{1r} = 0$. This implies

$$u_{1r} = (2B - 1)V. \tag{4.177}$$

The change in momentum of the impactor is

$$\Delta\text{momentum} = \rho_{10}L(V - u_{1r}) = 2\rho_{10}L(1 - B)V, \tag{4.178}$$

which equals Eq. (4.176). Note that when like materials strike, the projectile comes to rest against the surface of the target material and the amount of momentum transferred equals the original momentum in the projectile.

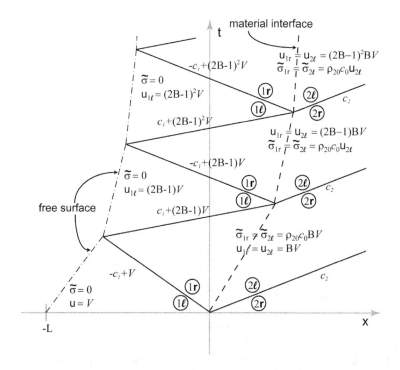

FIGURE 4.10. Projectile 1 of greater impedance striking target 2
of lesser impedance $(\rho_1 c_1 > \rho_2 c_2)$ and sticking.

Having dealt completely with the case of the impactor that rebounds, let us now ask what becomes of the impactor that does not (Fig. 4.10). It will be noted that the new state of the impactor behavior is just as if it were a projectile impacting with the reduced velocity $(2B - 1)V$ (Eq. 4.166). Thus, going through the same algebra will yield the result that, on the next pair of reflections, the velocity is $(2B - 1)^2 V$. The projectile/target interface velocity is given by B times the impact velocity. This pattern continues on forever,

$$
u = \begin{cases}
BV & 0 \le t \le 2L/c_1, \\
(2B - 1)BV & 2L/c_1 \le t \le 4L/c_1, \\
(2B - 1)^2 BV & 4L/c_1 \le t \le 6L/c_1, \\
\cdots & \cdots .
\end{cases}
\tag{4.179}
$$

Since $1/2 < B < 1$, at the end of time the projectile comes to rest, and hence the momentum transferred is equal to that in the projectile originally. It is possible to sum the displacement that arises from each finite time $(2L/c_1)$ spent at the various velocities (it is an infinite series that converges) and thus to compute the final displacement of the interface between the target and the impactor:

$$
\text{final displacement} = BV\frac{2L}{c_1} + (2B - 1)BV\frac{2L}{c_1} + (2B - 1)^2 BV\frac{2L}{c_1} + \cdots . \tag{4.180}
$$

For a geometric series with $-1 < x < 1$,

$$
\frac{1}{1 - x} = 1 + x + x^2 + x^3 + \cdots . \tag{4.181}
$$

Since $0 < 2B - 1 < 1$,

$$1 + 2B - 1 + (2B - 1)^2 + \ldots = \frac{1}{2(1 - B)}. \tag{4.182}$$

Thus we have computed our first depth of penetration,

$$\text{final displacement} = \frac{BV}{(1 - B)\, c_1} L. \tag{4.183}$$

In this case, an infinitely long wave travels into the target, which geometrically decays as one moves along it from right to left. In fact, given that it is our first depth of penetration, it demands some discussion. As a matter of notation, penetration is usually written with a capital P, and the penetration is often normalized by the length of the projectile. Also, the projectile properties are usually indicated with a subscript p and the target properties with a subscript t. Given these conventions and inserting the expression for B,

$$\frac{P}{L} = \frac{BV}{(1 - B)\, c_p} = \frac{\rho_p}{\rho_t} \frac{V}{c_t} = \frac{\rho_p}{\sqrt{\rho_t(\lambda_t + 2\mu_t)}} V. \tag{4.184}$$

Two things are apparent for this specific result. First, the stiffness of the projectile does not affect the depth of penetration – only the projectile mass ($\rho_p L$). This result is surprising in that the penetration velocity u depends on the projectile material properties, but in the end they do not affect the final depth of penetration because the higher interface velocity that occurs when the projectile material is stiffer is exactly offset by the lower transit time, since the sound speed in the projectile material is greater. Second, the stiffness of the target matters – the higher the modulus and sound speed of the target, the less the penetration. This latter conclusion we expect. Stiffer targets are more resistant to penetration.

There is an impact velocity upper limit to Eq. (4.184). The penetration velocity u cannot at any time exceed c_t, which says that $V \le c_t/B$. This translates into a maximum depth of penetration at the limit of applicability of

$$\frac{P}{L} \le \frac{\rho_p}{\rho_t} + \frac{c_t}{c_p}. \tag{4.185}$$

This bound occurs because we have assumed the target material has a linear stress-strain response. In an impact situation with real materials, which are not linear, at high impact velocities a shock will form in the target that will always outrun the penetration velocity.

We are now within reach of a surprising result. It might be thought that a rigid impactor – that is, one that cannot deform, will always penetrate more deeply than an impactor of a material that can deform. We now show that this is not the case.

THEOREM 4.23. *There exist target materials that are penetrated more deeply by a projectile that deforms (is not rigid) than by a nondeforming (rigid) impactor of equal mass at the same impact velocity.*

The proof is by a simple argument. The result is in the context of a one-dimensional impact that corresponds to a blunt impactor.

As a first step, consider what is meant by nondeforming impactor. All real materials deform. By nondeforming we mean a strong elastic material with a large Young's modulus, one that goes to infinity in the limit (infinite stiffness). To compute the penetration of such an impactor, we assume that all that matters is the projectile's mass, which in one dimension is given by the areal density $\rho_p L$. The deceleration is, from Newton's second law,

$$\rho_p L \frac{du}{dt} = -\tilde{\sigma}. \tag{4.186}$$

$\tilde{\sigma}$ will be determined by the Hugoniot jump conditions to give the target's resisting stress vs. penetration speed u. With $du/dt = (du/dx)(dx/dt) = u(du/dx)$,

$$-\rho_p L \frac{u}{\tilde{\sigma}} du = dx. \tag{4.187}$$

This equation can be integrated over the penetration event, from $u = V$ to 0, noting that the integral of dx is $x(0) - x(V) = P - 0 =$ the penetration, to give

$$\frac{P}{L} = \rho_p \int_0^V \frac{u}{\tilde{\sigma}(u)} du = \frac{\rho_p}{\rho_t} \int_0^V \frac{1}{U(u)} du. \tag{4.188}$$

For the target linear stress-strain relation, $\tilde{\sigma}(u) = \rho_t c_t u$ and $U(u) = c_t$, so we immediately see that the rigid impactor penetration matches the linear material impactor penetration result of Eq. (4.184).

The target material is assumed deformable for both impact scenarios of deformable impactor and rigid impactor. Fix an impact velocity V. For the deformable impactor, there is an initial penetration speed (interface velocity) u_{i0}, which is less than V. Now modify the target material by making it superlinear (stiffer) for $u > u_{i0}$. We will refer to the modified target material with a subscript m. Thus, by our modification, $\tilde{\sigma}_m(u) > \tilde{\sigma}(u)$ and $U_m(u) > U(u)$ for $u > u_{i0}$. Thus,

$$\left(\frac{P}{L}\right)_{m,\text{rigid}} < \left(\frac{P}{L}\right)_{\text{rigid}} = \left(\frac{P}{L}\right)_{\text{deforming}} = \left(\frac{P}{L}\right)_{m,\text{deforming}}, \tag{4.189}$$

where the first inequality is due to Eq. (4.188). We saw the central equality in the previous paragraph. The last equality holds because the particle (interface) velocity in the deformable projectile impacting the modified target is always below u_{i0}, and so there is no change to the constitutive model invoked in solving the deformable impactor penetration problem. Thus we have shown what we set out to show, namely that owing to nonlinear deformation of the target material the deformable (soft) impactor penetrates more deeply than the nondeformable (rigid) penetrator.

A little thought about the argument here implies that, for one-dimensional impacts, the result of this theorem is the typical behavior, not the exception.

This result was demonstrated by the author for the space shuttle program when the question was asked whether an ice and ablator penetration model for impacts into thermal tiles [203] would produce bounding results for Felt Reusable Surface Insulation (FRSI) plug impacts. Two 2.54-cm diameter, 4-cm tall cylindrical plugs fell off the Orbiter on flight STS-116 in December 2006. The ablator and ice are relatively stiff. The space shuttle thermal tiles behave like foam. The FRSI plugs are highly compressible with initial density 0.15–0.19 g/cm^3. They penetrate more deeply into tile than do nondeforming or slightly deforming impactors of the same size and mass for the same reason: the rigid impactors are decelerated by much higher stresses and impedances owing to the nonlinearity of the thermal tile stress-strain response, while the highly deformable impactors do not produce such high stresses. Thus, the ablator and ice impact models do not produce bounding penetration results. The specific argument given above does not rigorously apply to the

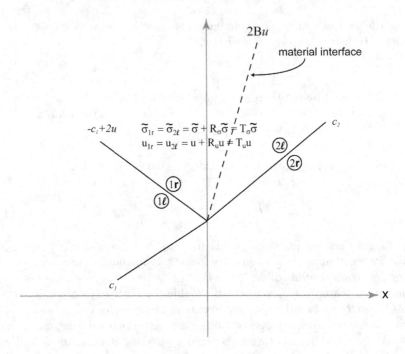

FIGURE 4.11. Wave reflection off an interior material interface.

space shuttle case because the foam and tile materials are not ρ_0-convex, and two-wave structures are produced in the impact, but the underlying reason producing the deeper penetration for the softer material is the same.

4.9. Wave Reflection and Transmission at an Internal Interface

Many armors are composed of layered materials. When an impact occurs on the outer surface, a wave travels into the material and can reflect at internal material boundaries. To examine the behavior of a wave arriving at an internal boundary, producing two waves, a reflected wave to the left and a transmission wave to the right, we consider the initial conditions (Fig. 4.11)

$$
\begin{array}{llll}
\rho_{1\ell} = \rho_{10}/(1 - u/c_1) & \rho_{1r} = ? & \rho_{2\ell} = ? & \rho_{2r} = \rho_{20} \\
\tilde{\sigma}_{1\ell} = \rho_{10}c_1 u & \tilde{\sigma}_{1r} = \tilde{\sigma}_{2\ell} = ? & \tilde{\sigma}_{2r} = 0 & \quad (4.190) \\
u_{1\ell} = u & u_{1r} = u_{2\ell} = ? & u_{2r} = 0.
\end{array}
$$

We have

$$
U_1 = -c_1 + 2u, \quad U_2 = c_2 \tag{4.191}
$$

(for U_1 we have a u due to the particle velocity and a u due to compression). These initial conditions differ from the previous case of impact in that the material to the left is compressed. Putting the various terms into the momentum jump conditions for the two waves gives (with some algebra for the left-going wave)

$$
\tilde{\sigma}_{1r} = \rho_{10}c_1(2u - u_{1r}), \qquad \tilde{\sigma}_{2\ell} = \rho_{20}c_2 u_{2\ell}. \tag{4.192}
$$

The stresses and the particle velocities match in the center section, yielding

$$
u_{1r} = u_{2\ell} = \frac{2\rho_{10}c_1 u}{\rho_{10}c_1 + \rho_{20}c_2} = 2Bu. \tag{4.193}
$$

The stress is

$$\tilde{\sigma}_{1r} = \tilde{\sigma}_{2\ell} = \frac{2\rho_{10}c_1\rho_{20}c_2 u}{\rho_{10}c_1 + \rho_{20}c_2} = 2\rho_{10}c_1(1-B)u = 2\rho_{20}c_2 Bu. \tag{4.194}$$

The respective densities are

$$\rho_{1r} = \rho_{10}/(1-2(1-B)u/c_1), \qquad \rho_{2\ell} = \rho_{20}/(1-2Bu/c_2). \tag{4.195}$$

(Notice these equations say that the incident wave problem for waves $1r$ and 2ℓ is mathematically equivalent to material 1 striking material 2 at an impact speed of $2u$.) When this problem is worked using the traditional wave propagation approach (Exercise 4.31), it is often broken into three elastic waves: the incident, the reflected, and the transmitted. Coefficients are derived describing the amplitude of the reflected and transmitted waves with respect to the incident wave. We will do the same.

There are different coefficients depending on whether the velocity (hence displacement) is being considered or the stress. First we do velocity. The reflected wave is the $r1$ conditions minus the $\ell1$ conditions (the sum of the incident and reflected waves makes the actual wave, so the reflected wave is the actual wave minus the incident), so that the change in velocity is

$$\Delta u = 2Bu - u = (2B-1)u = \frac{\rho_{10}c_1 - \rho_{20}c_2}{\rho_{10}c_1 + \rho_{20}c_2}u = R_u u. \tag{4.196}$$

Here we have defined the reflection coefficient

$$R_u = \frac{\rho_{10}c_1 - \rho_{20}c_2}{\rho_{10}c_1 + \rho_{20}c_2} = 2B - 1. \tag{4.197}$$

Notice that if the materials have the same impedance then the reflection coefficient is zero, as one would expect. Also note that $-1 < R_u < 1$. Next, for the transmitted wave

$$u_{2\ell} = 2Bu = \frac{2\rho_{10}c_1}{\rho_{10}c_1 + \rho_{20}c_2}u = T_u u, \tag{4.198}$$

where the transmission coefficient T_u is

$$T_u = \frac{2\rho_{10}c_1}{\rho_{10}c_1 + \rho_{20}c_2} = 2B. \tag{4.199}$$

Notice that $0 < T_u < 2$ and that $T_u = 1 + R_u$.

The two coefficients for the stress are as follows. For the reflected wave,

$$\Delta\tilde{\sigma} = 2\rho_{10}c_1(1-B)u - \rho_{10}c_1 u = \rho_{10}c_1(1-2B)u = \rho_{10}c_1\frac{\rho_{20}c_2 - \rho_{10}c_1}{\rho_{10}c_1 + \rho_{20}c_2}u = R_\sigma\tilde{\sigma}_{1\ell}, \tag{4.200}$$

where the reflection coefficient R_σ is

$$R_\sigma = \frac{\rho_{20}c_2 - \rho_{10}c_1}{\rho_{10}c_1 + \rho_{20}c_2} = 1 - 2B, \tag{4.201}$$

where again $-1 < R_\sigma < 1$, but notice that $R_\sigma = -R_u$. For the transmitted wave,

$$\tilde{\sigma}_{2l} = 2\rho_{20}c_2 Bu = 2\rho_{20}c_2\frac{\rho_{10}c_1}{\rho_{10}c_1 + \rho_{20}c_2}u = \rho_{10}c_1\frac{2\rho_{20}c_2}{\rho_{10}c_1 + \rho_{20}c_2}u = T_\sigma\tilde{\sigma}_{1\ell} \tag{4.202}$$

where the transmission coefficient T_σ is

$$T_\sigma = \frac{2\rho_{20}c_2}{\rho_{10}c_1 + \rho_{20}c_2} = 2(1-B) \tag{4.203}$$

where again $0 < T_\sigma < 2$ and $T_\sigma = 1 + R_\sigma$.

TABLE 4.2. Wave reflection and transmission coefficients for an internal material boundary. The particle velocity and the stress between the reflection wave front and transmission wave front are given by the transmission wave values, which equal the combined incident and reflected waves. $B = \rho_{10}c_1/(\rho_{10}c_1 + \rho_{20}c_2)$, where the incident and reflected waves are in material 1 and the transmitted wave is in material 2.

	Incident wave	Reflected wave	Transmitted wave
Particle velocity	u	$R_u u = (2B - 1)u$	$T_u u = 2Bu$
Stress	$\tilde{\sigma}$	$R_\sigma \tilde{\sigma} = (1 - 2B)\tilde{\sigma}$	$T_\sigma \tilde{\sigma} = 2(1 - B)\tilde{\sigma}$

The coefficients are summarized in Table 4.2. Notice that the above coefficients have the correct behavior in the limiting cases. If material 2 were a rigid wall, then $B = 0$, and we have $R_u = -1$, meaning no material motion ($u + R_u u = 0$) when we add the two waves, and $R_\sigma = 1$, meaning the stress is doubled, just as in the rigid wall impact solved for above ($\tilde{\sigma} + R_\sigma \tilde{\sigma} = 2\tilde{\sigma}$). If $B = 1/2$, meaning the impedances match (as they do for the same material), then both reflection coefficients are zero and both transmission coefficients are 1. Finally, if $B = 1$, corresponding to a free surface, $R_u = 1$, thus doubling the velocity ($u + R_u u = 2u$), and $R_\sigma = -1$, thus zeroing out the stress ($\tilde{\sigma} + R_\sigma \tilde{\sigma} = 0$), just as with our free surface reflection example above.

4.10. Square Pulse Reflection: Tensile Stress States and Tensile Spall

An interesting situation arises when, starting with the previous section's initial conditions, the incident wave coming from the left has finite length – that is, it is a square pulse that ends with a right-moving rarefaction wave that returns material 1 to its original state. What happens when the left-going reflection from the material interface interacts with the right-going rarefaction at the end of the square pulse? The conditions are (Fig. 4.12)

$$
\begin{aligned}
&\rho_\ell = \rho_{10} \quad \rho_r =? \quad \rho_{1\ell} =? \quad \rho_{1r} = \rho_{10}/(1 - 2(1 - B)u/c_1) \\
&\tilde{\sigma}_\ell = 0 \quad \tilde{\sigma}_r = \tilde{\sigma}_{1\ell} =? \quad \tilde{\sigma}_{1r} = 2\rho_{10}c_1(1 - B)u \\
&u_\ell = 0 \quad u_r = u_{1l} =? \quad u_{1r} = 2Bu.
\end{aligned}
\tag{4.204}
$$

(The properties to the left are denoted ℓ to avoid confusion with the previous section's notation.) When the reflected and rarefaction meet, two waves are produced, one traveling to the left and one traveling to the right. The wave traveling to the left has $U_\ell = -c_1$ since it is traveling into material behind the square pulse that is at rest with zero stress. The wave traveling to the right has $U_r = c_1 + 2Bu - 2(1 - B)u = c_1 + 2(2B - 1)u$, where the first adjustment term ($2Bu$) is the material velocity and the second term is due to compression ($2(1 - B)u$). The jump conditions provide

$$
\begin{aligned}
&\tilde{\sigma}_r = \rho_\ell U_\ell u_r = -\rho_{10}c_1 u_r, \\
&\tilde{\sigma}_{1\ell} = \tilde{\sigma}_{1r} + \rho_{1r}(U_r - u_{1r})(u_{1\ell} - u_{1r}) = \rho_{10}c_1\{2(1 - 2B)u + u_{1\ell}\}.
\end{aligned}
\tag{4.205}
$$

Equating the stresses gives the particle velocities and the stress as

$$
\begin{aligned}
&u_r = u_{1\ell} = (2B - 1)u, \\
&\tilde{\sigma}_r = \tilde{\sigma}_{1\ell} = \rho_{10}c_1(1 - 2B)u.
\end{aligned}
\tag{4.206}
$$

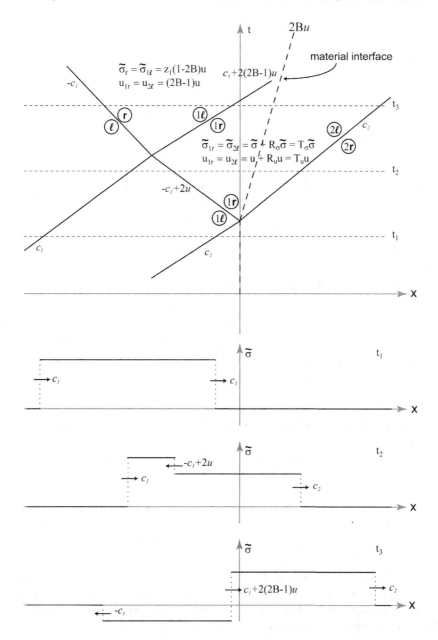

FIGURE 4.12. Square pulse interacting with an interior material interface.

Considering various values of B, if the original interface between materials is rigid ($B = 0$), then the stress drops back down to the stress of the incident wave. If the materials are the same, the stress drops to zero (there was no interaction since there was no reflection). But, if $B > 1/2$, then a tensile stress is developed ($\tilde{\sigma} < 0$); the extreme value is when $B = 1$, corresponding to a free surface: then the tensile stress equals in magnitude the original compressive stress of the stress pulse. In actual material, these tensile stress regions can lead to a tensile failure or separation of the material, a phenomenon referred to as spall. Notice, however, that a tensile stress state requires the release of the initial wave – that is, tensile spall occurs when the rarefaction off the free surface or interior boundary interacts with a rarefaction following a compressive shock:

GENERAL PRINCIPLE 4.24. *(spall requires interacting rarefactions) Tensile stress states occur when two or more rarefaction waves interact.*

Let us explore a square pulse reflecting off a free surface, where $B = 1$. Behind the square pulse moving to the right the material has returned to its ambient condition of zero stress, original density, and at rest. Behind the reflected wave traveling from the free surface to the left, the material has zero stress and original density, but it is moving to the right with a speed $2u$. When these two waves meet is where tension occurs, and if the material does not fail the resulting particle velocity for both is u and the stress is $\tilde{\sigma} = -\rho_{10}c_1u$. However, suppose there was a maximum tensile stress that the material could support, referred to as the spall stress, $\sigma_{\text{spall}} > 0$. We are interested in the particle speed that results in the wave moving to the right. Instead of Eq. (4.204), the conditions for this wave are

$$\begin{aligned} \rho_{1\ell} &=? & \rho_{1r} &= \rho_{10} \\ \tilde{\sigma}_{1\ell} &= -\sigma_{\text{spall}} & \tilde{\sigma}_{1r} &= 0 \\ u_{1\ell} &=? & u_{1r} &= 2u. \end{aligned} \tag{4.207}$$

The resulting wave speed to the right is $U = c_1 + 2u$ owing to the material motion, and so the resulting wave has

$$-\sigma_{\text{spall}} = \rho_{10}c_1(u_{1\ell} - 2u) \quad \Rightarrow \quad u_{1\ell} = 2u - \sigma_{\text{spall}}/(\rho_{10}c_1). \tag{4.208}$$

When this tensile wave reaches the free surface, it reflects as a wave moving to the left with

$$\begin{aligned} \rho_{2\ell} &= \rho_{10}/(1 + \sigma_{\text{spall}}/(\rho_{10}c_1^2)) & \rho_{2r} &= \rho_{10} \\ \tilde{\sigma}_{2\ell} &= -\sigma_{\text{spall}} & \tilde{\sigma}_{2r} &= 0 \\ u_{2\ell} &= u_{1\ell} = 2u - \sigma_{\text{spall}}/(\rho_{10}c_1) & u_{2r} &=? \end{aligned} \tag{4.209}$$

The wave speed traveling to the right is $U = -(c_1 + \sigma_{\text{spall}}/(\rho_{10}c_1)) + u_{1\ell}$ and the momentum balance is

$$\sigma_{\text{spall}} = \rho_{10}(-c_1)(u_{2r} - u_{2\ell}) \quad \Rightarrow \quad u_{2r} = u_{1\ell} - \frac{\sigma_{\text{spall}}}{\rho_{10}c_1} = 2u - 2\frac{\sigma_{\text{spall}}}{\rho_{10}c_1}. \tag{4.210}$$

When experiments are performed, the free surface particle velocity is measured. After the first reflection the free surface speed is $2u$; then the velocity drops to the free surface velocity of u_{2r} defined by the spall stress. The magnitude of the free surface velocity drop (or pullback) is

$$\Delta u_{\text{free surface}} = 2u - u_{2r} = 2\frac{\sigma_{\text{spall}}}{\rho_{10}c_1}, \tag{4.211}$$

thus yielding a way to measure the spall stress,

$$\sigma_{\text{spall}} = \frac{1}{2}\rho_{10}c_1\Delta u_{\text{free surface}}. \tag{4.212}$$

After the spall failure occurs (i.e., the material separates forming internal free surfaces), the observed external free surface velocity increases; for spall that happens very quickly – the velocity approaches the initial value of $2u$; the longer the damage processes require for material separation to occur, the more the late time free

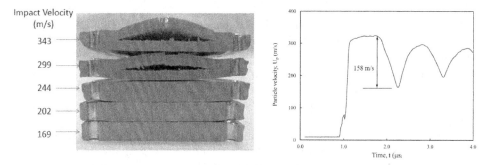

FIGURE 4.13. Left: sectioned 0.638-cm-thick, 5.08-cm-diameter Al 6061-T6 target plates impacted by an aluminum plate at various speeds exhibiting spall (separate tests, stacked for display). Right: back surface velocity for the 343 m/s spalled aluminum target plate.

surface velocity approaches u_{2r}. To be clear, the free surface velocity $u_{1r} = 2u$ represents the unloaded to zero axial stress material; if there were no spall strength, there would be no tensile stress and the free surface velocity would never drop. The drop in free surface speed (pullback) is entirely due to tensile stress in the interior of the target. The $\frac{1}{2}$ in the equation is due to the doubling of speeds at the free surface. Thus, the tensile stress is directly measured by the impedance times the change in particle velocity.

As a specific instance, at SwRI five Al 6061-T6 target plates of thickness 0.638 cm and 5.08 cm in diameter were impacted by 0.319 cm-thick Al flyer plates at various speeds V. The flyer being half the thickness of the target is intentional; the impact produces a pulse that is twice the thickness of the impactor, hence the thickness of the target. When the pulse is entirely in the target, the impactor is left at rest while the compressed target material is traveling with particle velocity $\frac{1}{2}V$. Release waves, with associated particle velocity $\pm\frac{1}{2}V$, now come in from both the front and back free surfaces of the target and meet at the midplane of the target. When they arrive at the midplane, the target has zero axial stress, the right side of the target has particle velocity V, and the left side is at rest. Thus, a compressive impact at speed V has led to a tensile loading at speed V. The waves continue traveling to the other side, and a diamond of tension is produced in the position–time plane. If the material does not fail, the center of this x–t diagram is diamonds of alternating tension and compression, of height $1\,\mu$s, while the right-angle triangles with hypotenuse forming the target boundaries are traction free, of height $1\,\mu$s, and exhibiting a back surface oscillation with period $2\,\mu$s. However, the test is intended to fail the material in tension, to explore the high-triaxiality strength of the material. Failure occurs by material separation at the midplane, and the two pieces of target material separate at a speed less than V. Figure 4.13 (right) shows the target back face motion. The target back face is nearly at the impact speed once the rarefaction wave propagates back into the target. Figure 4.13 shows the sectioned target plates sitting on top of each other. Incipient spall (meaning the onset of ductile void formation in the target plate) can be seen in the 244 m/s impact plate, with the clear development of a separated spall layer for the higher velocities.

In examining spall failure surfaces for metals, they are usually dimpled surfaces, as in Fig. 4.13 and Fig. 9.8. Examination of the surfaces shows the failure (material separation) occurs through voids that grow in the metal while it is in tension and then link up to create extended surfaces of material separation. Spall surfaces for brittle materials appear to be created by coalescence of microcracks.

To perform a quantitative analysis, it is necessary to examine the impact in detail. Due to the elastic-plastic response of aluminum, a two-wave structure is formed by the impact. The details of such impacts are worked out in the next chapter (Section 5.3); we use the formulas to determine quantities here. Appendix C contains properties for Al 6061-T6. Coefficients for the Johnson-Cook model produce a flow stress of $Y = 350$ MPa at a strain of 0.03; we will use this value. Ultrasonically measured sound speeds provide the elastic constants, yielding $c_K = 5,266$ m/s, $K = 75$ GPa, and $G = 26.8$ GPa. When the 343 m/s impact occurs, an elastic precursor is formed with $\tilde{\varepsilon}_{HEL} = 0.0065$, $\tilde{\sigma}_{HEL} = 722$ MPa, and with particle speed $u_{HEL} = 42$ m/s (compare to Fig. 4.13). The bulk wave, which includes elastic and plastic deformation, has the maximum compressive stress of $\tilde{\sigma} = 2.57$ GPa with $\tilde{\varepsilon}_{11}^e = 0.0147$ and $\tilde{\varepsilon}_{11}^p = 0.0165$ for a total strain of $\tilde{\varepsilon}_\ell = 0.0312$. The wave arrives at the target back face, which is a free surface, and unloads, first elastically from the compressive side of the yield surface to the tensile side, then elastic-plastically as plastic flow begins at $\tilde{\varepsilon}_\ell - 2\tilde{\varepsilon}_{HEL} = 0.0181$ and stress 1.12 GPa. After that, the unloading is along the bulk modulus down to zero axial stress at a residual strain of $\tilde{\varepsilon} = \tilde{\varepsilon}_{res} = \frac{4G}{3K}\tilde{\varepsilon}_{HEL} = 0.0031$ with a corresponding residual speed of $u = u_{res} = c_L \tilde{\varepsilon}_{HEL}(c_L/c_K - 1) = 9.0$ m/s (there are residual stresses in the transverse directions). Thus, the speed is not quite doubled at the free surface, but rather goes to $V - u_{res} = 334$ m/s. (Similarly, the impactor is not brought to rest but is moving in the direction of flight at u_{res}.) This plateau can be seen as the maximum back surface velocity in Fig. 4.13 (right); it represents the zero axial stress state. Any reduction of the back surface velocity from this plateau is because of tensile stress within the target. Further unloading beyond the zero axial stress point is tensile, where the elastic-plastic loading continues with modulus K. When the two stress free waves meet each other, the speeds on either side of the midplane of the target are $u_{res} = 9$ m/s to the left and $V - u_{res} = 334$ m/s to the right.[2,3] The appropriate wave speed to use in the stress calculation from the pullback signal is c_K or c_0 (in this case they differ by 1.5%). We use c_0 in the calculation, and the

---

[2]If there is no spall failure, the target remains intact and the resulting wave has particle speed $u_p = \frac{1}{2}V = 171.5$ m/s, with a tensile stress of $\sigma = \rho_0 c_0 \Delta u = 2.35$ GPa, where $\Delta u = V - u_{res} - u_p = u_p - u_{res}$. The tensile wave is released by a compressive wave when it reaches the free surface. The compressive unloading to zero axial stress is initially elastic. The final strain in the target is $\tilde{\varepsilon} = -2\tilde{\varepsilon}_{res}$ with a corresponding residual speed of $u = 2u_{res}$. The back surface velocity is $V - u_{res} - (\frac{1}{2}V - u_{res} + \frac{1}{2}V - 2u_{res}) = 2u_{res}$. The period of the back surface velocity oscillation is $4L_p/c_L = 2$ μs. This pattern of reducing velocity variations is due to plastic dissipation and continues until the oscillations become entirely elastic. The average speed of the target is always $\frac{1}{2}V$.

[3]The noticeable difference in unloading slope (it is shallower) compared to the loading slope of the shock in Fig. 4.13 is primarily because of the Bauschinger effect in the unloading (see Fig. 3.3). In a uniaxial stress test, after a sample has yielded, initially upon unloading the modulus is Young's modulus E. However, in unloading when the stress changes sign, the modulus begins to decrease; hence, more strain occurs for a corresponding change in stress. This effectively leads to a decrease in the modulus to below K for unloading, further spreading out the rarefaction wave (where each strain state has its own wave speed).

spall stress for the free surface pullback speed of 158 m/s is

$$\sigma_{\text{spall}} = \frac{1}{2}\rho_0 c_0 \Delta u_{\text{free surface}} = 1.14\,\text{GPa}. \qquad (4.213)$$

Though not shown here, the 244 m/s impact test, with incipient spall, had a pullback of 167 m/s, corresponding to $\sigma_{\text{spall}} = 1.21\,\text{GPa}$.[4] The two stress state parameters are $\zeta = 1$, because it is a uniaxial tension stress state, and triaxiality $\eta = -p/Y = (\sigma_{\text{spall}} - \frac{2}{3}Y)/Y = 2.6$ and 2.8 for the two measured spall values. This stress state is very tensile and is similar to that near the crack tip in a compact tension specimen (see footnote page 89). The strain in going from zero to the spall stress is $\sigma_{\text{spall}}/K = 0.015$ and the amount of plastic strain is two-thirds of this, $\varepsilon_p = 0.01$.

When the analysis is performed for all the impacts, the compressive stresses from the impacts are 1.33, 1.57, 1.86, 2.26, and 2.57 GPa for the 169, 202, 244, 299, and 343 m/s impacts, respectively, and the maximum tensile stresses in the material if it did not fail are 1.09, 1.33, 1.63, 2.03, and 2.35 GPa, respectively. Since spall did not occur for the two lower-speed impacts, we infer the material supported tensile stresses of at least 1.33 GPa.[5]

These results bring up a that fact needs to be recognized. Suppose the stress went to a tensile level and held there for some amount of time and then the material failed. The wave that carries the tensile load information has speed c_0 because it is tensile loading and the material is undergoing elastic-plastic deformation. Upon failure (material separation), the unloading compression wave is *elastic*, and hence has speed c_L. This wave, which increases the particle velocity up to twice the HEL particle velocity – 84 m/s, in this case – can potentially catch up with the slower wave and decrease the pullback magnitude. The travel time difference from the midplane to the back surface is $\Delta x(1/c_b - 1/c_L) = 0.1\,\mu s$, so if the time the tensile stress holds is less than this, such interactions will occur. The spall stress computed from the pullback signal should be interpreted as a *minimum* tensile stress supported by the material; the actual tensile stress supported by the material may have been, and probably was, larger.

The Johnson-Cook failure model fits itself to spall failure data for high triaxiality (Eq. 3.173). The failure model developed for Al 2024-T351 (Eq. 3.201) is not based on any large η data – in fact, its single point calibration was with $\eta = 0$ shear failure strain. Though for a different alloy, we will apply it, since the shear failure strain for 2024 is $\varepsilon_p^f = 0.21$ and for 6061 is $\varepsilon_p^f = 0.36/\sqrt{3} = 0.208$ (Tables 3.1 and 3.5). We will not consider any damage accumulation in loading (it is small) and consider the damage to begin at zero axial stress. Since the material is in uniaxial strain and in tension, the stress is $\sigma = -p + \frac{2}{3}Y = (\eta + \frac{2}{3})Y$ and $\zeta = 1$. To begin at zero axial stress, we begin with $\eta = -\frac{2}{3}$. The damage model has a constant failure strain of $\varepsilon_p^f = \varepsilon_0^f = 0.54$ for $-\frac{2}{3} < \eta < \frac{1}{3}$. For triaxiality above $\frac{1}{3}$, the failure strain is

$$\varepsilon_p^f(\eta) = \varepsilon_0^f \exp(-\frac{3}{2}\eta + \frac{1}{2}), \qquad \text{where}\quad \eta > \frac{1}{3}, \zeta = 1. \qquad (4.214)$$

---

[4]The spall, though difficult to see in the sectioning, clearly influenced the unloading signal because the back surface velocity oscillation period was $1\,\mu s$, corresponding to half the target thickness.

[5]The back surface velocity was not collected for the 202 and 299 m/s impacts. The two lowest speed impacts showed no spall. The pullback, or the amount the velocity dropped for the 169 m/s impact, was 117 m/s showing a minimum tensile stress of 0.85 GPa.

The loading is uniaxial strain with plastic flow, so the loading modulus is the bulk modulus K. The corresponding relationship between pressure and strain is $dp = -Kd\varepsilon = -\frac{3}{2}Kd\varepsilon_p$, and with $dp = -Yd\eta$, we have $Yd\eta = \frac{3}{2}Kd\varepsilon_p$. We compute the damage as

$$
\begin{aligned}
D &= \int \frac{d\varepsilon_p}{\varepsilon_p^f(\eta)} = \frac{2Y}{3K\varepsilon_0^f}\left(\int_{-2/3}^{1/3} d\eta + \int_{1/3}^{\eta} \exp(\frac{3}{2}\eta - \frac{1}{2})d\eta\right) \\
&= \frac{4Y}{9K\varepsilon_0^f}\left\{\frac{1}{2} + \exp(\frac{3}{2}\eta - \frac{1}{2})\right\}.
\end{aligned}
\tag{4.215}
$$

Failure occurs when $D = 1$, and the spall stress is

$$
\sigma_{\text{spall}} = Y(\eta_{\text{spall}} + \frac{2}{3}), \qquad \text{where} \quad \eta_{\text{spall}} = \frac{2}{3}\ln(\frac{9K\varepsilon_0^f}{4Y} - \frac{1}{2}) + \frac{1}{3}.
\tag{4.216}
$$

The predicted value is $\sigma_{\text{spall}} = 1.65$ GPa with $\eta_{\text{spall}} = 4.04$. The spall stress is relatively insensitive to the failure strain – reducing it to half its value gives $\sigma_{\text{spall}} = 1.48$ GPa.

A fascinating fact about spall is that it is size dependent. For the failure process to have a size dependence, it must have either a fixed length scale or fixed time scale. A model that is used to model spall is the Tuler–Butcher model, where damage accumulates when the stress is more tensile than a minimum value as

$$
D = \frac{1}{D_c}\int_0^t (\sigma_{\text{max}} - \sigma_0)_+^\lambda dt, \qquad (x)_+ = \begin{cases} x, & x > 0, \\ 0, & x < 0, \end{cases}
\tag{4.217}
$$

where σ_{max} is the maximum principal tensile stress, σ_0, λ, and D_c are constants [188]. The integration over time introduces the time scale. Tuler and Butcher give constants of $\lambda = 2.02$, $\sigma_0 = 425$ MPa, and $D_c = 3.80 \times 10^{11}$ Pa$^{2.02}$ s for Al 6061-T6. Using a spall stress of 1.2 GPa and assuming the spall stress is constant over the time to failure, the predicted failure time is $0.42\,\mu$s, which seems reasonable looking at Fig. 4.13. Using an impactor and target that are one half the respective lengths, the model predicts a spall stress of 1.5 GPa, and when the impact and target size are doubled, the prediction is 1 GPa. One approach to adding size dependence for the Al 2024-T351 model in the previous paragraph is to require that $D = 1$ over a specified finite region, not just at a point (see footnote page 113).

The simple wave analysis does not provide a strain rate for spall, as it has an immediate jump to the tensile strain upon the meeting of the rarefactions. However, estimates of the strain rate are given by the strain going from compression to tension and dividing by the pulse time, $2L_p/c_L$. For the test in Fig. 4.13, the pulse width is $1\,\mu$s and the change in strain is roughly 0.045, giving an estimated strain rate of 4.5×10^4 s^{-1}. Assuming the drop occurs over roughly $0.4\,\mu$s (Fig. 4.13), the strain rate is 1.1×10^5 s^{-1}. Another way to include size dependence in the computations is to include strain rate dependence in the failure model, as is developed in Chapter 9 for the Al-2024-T351 failure model (Eq. 9.30 and Exercise 4.45).

More discussion of spall can be found in [19, 53, 101].

4.11. Exact Solution for Viscoelastic Smooth Wave Fronts

In this section we construct solutions where the solid material is viscoelastic, following [29, 81]. For the elastic part, we need a stress-strain relation that is strictly superlinear; we choose the simplest example, $\tilde{\sigma}(\tilde{\varepsilon}) = K_1\tilde{\varepsilon} + K_2\tilde{\varepsilon}^2$. When a wave front moves at a constant speed, without spreading, the Hugoniot jump conditions link the conditions at either end, even if the wave front has a finite rise time (Theorem 2.6). Thus, we use the jump conditions to determine the speed U of the front, solving the problem

$$
\begin{array}{ll}
\rho_\ell = \rho_\ell & \rho_r = \rho_0 \\
\tilde{\varepsilon}_\ell = 1 - \rho_0/\rho_\ell & \tilde{\varepsilon}_\ell = 0 \\
\tilde{\sigma}_\ell = ? & \tilde{\sigma}_r = 0 \\
u_\ell = ? & u_r = 0.
\end{array} \tag{4.218}
$$

The solution using the jump conditions is

$$
U = \sqrt{\frac{K_1 + K_2\tilde{\varepsilon}_\ell}{\rho_0}}, \qquad u_\ell = U\tilde{\varepsilon}_\ell, \qquad \tilde{\sigma}_\ell = \rho_0 U^2 \tilde{\varepsilon}_\ell. \tag{4.219}
$$

We now show something very interesting. First, with some manipulation we can write the mass and momentum conservation differential equations in terms of $\tilde{\varepsilon}$ rather than ρ. The mass and momentum conservation equations, respectively, are

$$
\frac{\partial \tilde{\varepsilon}}{\partial t} + \frac{\partial \tilde{\varepsilon}}{\partial x}u + (1 - \tilde{\varepsilon})\frac{\partial u}{\partial x} = 0, \qquad (1 - \tilde{\varepsilon})\frac{\partial \tilde{\sigma}}{\partial x} + \rho_0\frac{\partial u}{\partial t} + \rho_0 u\frac{\partial u}{\partial x} = 0. \tag{4.220}
$$

Second, for a steady wave that moves without changing shape, $\tilde{\varepsilon}(x,t) = \tilde{\varepsilon}(x - Ut)$, $u(x,t) = u(x - Ut)$, and $\tilde{\sigma}(x,t) = \tilde{\sigma}(x - Ut)$, and mass and momentum conservation become

$$
(u - U)\tilde{\varepsilon}' + (1 - \tilde{\varepsilon})u' = 0, \qquad (1 - \tilde{\varepsilon})\tilde{\sigma}' + \rho_0(u - U)u' = 0, \tag{4.221}
$$

where prime means differentiation. Third, a solution to this equation is

$$
\tilde{\sigma} = \rho_0 U^2 \tilde{\varepsilon}, \qquad u = U\tilde{\varepsilon}, \qquad U = \sqrt{(K_1 + K_2\tilde{\varepsilon}_\ell)/\rho_0}, \tag{4.222}
$$

as is simply shown by placement in the mass and momentum balances. Notice that U is fixed while the first two relations are linear, lying on the Rayleigh line. Since this solution lies on the Rayleigh line, all disturbances propagate at speed U.

However, the compressive stress is given by $\tilde{\sigma}(\tilde{\varepsilon}) = K_1\tilde{\varepsilon} + K_2\tilde{\varepsilon}^2$, not by the Rayleigh line expression $\tilde{\sigma} = \rho_0 U^2 \tilde{\varepsilon}$. We introduce viscosity to make up the difference. In particular, a nonlinear viscoelastic constitutive model that includes compressibility and viscosity is[6]

$$
\tilde{\sigma}(\tilde{\varepsilon}, \dot{\tilde{\varepsilon}}) = K_1\tilde{\varepsilon} + K_2\tilde{\varepsilon}^2 + \eta\,\text{sign}(\dot{\tilde{\varepsilon}})\,|\dot{\tilde{\varepsilon}}|^p, \tag{4.223}
$$

where $p > 0$ is a power and η is a viscosity coefficient. The strain rate $\dot{\tilde{\varepsilon}} = -\frac{\partial u}{\partial x} = -D_{11}$. Assuming $\dot{\tilde{\varepsilon}} \geq 0$, the viscosity is

$$
\eta(\dot{\tilde{\varepsilon}})^p = \eta(-U\tilde{\varepsilon}')^p = \rho_0 U^2 \tilde{\varepsilon} - (K_1\tilde{\varepsilon} + K_2\tilde{\varepsilon}^2) = K_2\tilde{\varepsilon}(\tilde{\varepsilon}_\ell - \tilde{\varepsilon}). \tag{4.224}
$$

The right-hand side is a parabola in $\tilde{\varepsilon}$, with a maximum at $\tilde{\varepsilon} = \frac{1}{2}\tilde{\varepsilon}_\ell$, and is zero at either end, where $\tilde{\varepsilon} = 0$ and $\tilde{\varepsilon} = \tilde{\varepsilon}_\ell$. The parabola is symmetric about $\frac{1}{2}\tilde{\varepsilon}_\ell$. Hence,

---

[6]This specific viscoelastic model has a spring and a dashpot in parallel and is known as the Kelvin–Voigt model, except in our case both spring and dashpot are nonlinear. If the spring and dashpot are in series, it is called a Maxwell model. In fluid mechanics, $\sigma = -pI + 2\eta(D - \frac{1}{3}\text{trace}(D)I) + \kappa\text{trace}(D)I$ is a linear viscosity constitutive model, where D is the rate of deformation tensor. For incompressible flow, $\text{trace}(D) = 0$; the *bulk viscosity* κ term drops out and when $\sigma = -pI + 2\eta D$ is inserted into the momentum balance (again assuming incompressible flow) it leads to the *Navier–Stokes equation*. The *dynamic viscosity* coefficient η is sometimes written μ.

we can solve the spatial distribution of the strain by integrating from the midpoint $(x, \tilde{\varepsilon}) = (0, \frac{1}{2}\tilde{\varepsilon}_\ell)$ at time $t = 0$. The result is

$$\left(\frac{\eta}{K_2}\right)^{1/p} \int_{\tilde{\varepsilon}_\ell/2}^{\tilde{\varepsilon}} \frac{d\tilde{\varepsilon}}{(\tilde{\varepsilon}(\tilde{\varepsilon}_\ell - \tilde{\varepsilon}))^{1/p}} = -\int_0^x \frac{dx}{U} = -\frac{x}{U}. \qquad (4.225)$$

For $p \leq 1$, the rise has infinite spatial extent. As examples, for $p = 1$ we obtain (integration by partial fractions)

$$\int_{\tilde{\varepsilon}_\ell/2}^{\tilde{\varepsilon}} \frac{d\tilde{\varepsilon}}{\tilde{\varepsilon}(\tilde{\varepsilon}_\ell - \tilde{\varepsilon})} = -\frac{1}{\tilde{\varepsilon}_\ell} \ln\left(\frac{\tilde{\varepsilon}_\ell}{\tilde{\varepsilon}} - 1\right) \;\Rightarrow\; \tilde{\varepsilon}(x, t) = \tilde{\varepsilon}_\ell \left\{1 + \exp\left(\frac{K_2 \tilde{\varepsilon}_\ell}{\eta U}(x - Ut)\right)\right\}^{-1}, \qquad (4.226)$$

where x was replaced by $x - Ut$ after the integration. For $p = \frac{1}{2}$,

$$-\int_{\tilde{\varepsilon}_\ell/2}^{\tilde{\varepsilon}} \frac{d\tilde{\varepsilon}}{\tilde{\varepsilon}^2(\tilde{\varepsilon}_\ell - \tilde{\varepsilon})^2} = \frac{2}{\tilde{\varepsilon}_\ell^3} \ln\left(\frac{\tilde{\varepsilon}_\ell}{\tilde{\varepsilon}} - 1\right) - \frac{2\tilde{\varepsilon} - \tilde{\varepsilon}_\ell}{\tilde{\varepsilon}_\ell^2 \tilde{\varepsilon}(\tilde{\varepsilon}_\ell - \tilde{\varepsilon})} = \left(\frac{K_2}{\eta}\right)^2 \frac{x - Ut}{U}, \qquad (4.227)$$

where there is not an explicit form for strain in terms of x. For the aluminum experiments in Fig. 5.12, Grady states that $\eta = 80 \text{ kPa}\sqrt{s}$ gives good agreement with the data [81]. (Exercise 5.33 describes how to determine K_2.) For $p = 2$, the result is

$$\int_{\tilde{\varepsilon}_\ell/2}^{\tilde{\varepsilon}} \frac{d\tilde{\varepsilon}}{\sqrt{\tilde{\varepsilon}(\tilde{\varepsilon}_\ell - \tilde{\varepsilon})}} = -\sin^{-1}\left(1 - 2\frac{\tilde{\varepsilon}}{\tilde{\varepsilon}_\ell}\right) \;\Rightarrow\; \tilde{\varepsilon}(x, t) = \frac{\tilde{\varepsilon}_\ell}{2}\left\{1 - \sin\left(\sqrt{\frac{K_2}{\eta}} \frac{x - Ut}{U}\right)\right\}, \qquad (4.228)$$

where this rise has a finite footprint, with $-\frac{\pi}{2} \leq \frac{x - Ut}{U}\sqrt{\frac{K_2}{\eta}} \leq \frac{\pi}{2}$ (Exercise 4.44).

The maximum strain rate is found by evaluating Eq. (4.224) at $\tilde{\varepsilon} = \frac{1}{2}\tilde{\varepsilon}_\ell$,

$$\dot{\tilde{\varepsilon}}_{\max} = \left(\frac{K_2}{4\eta}\right)^{1/p} \tilde{\varepsilon}_\ell^{2/p} \approx \left(\frac{K_2}{4\eta}\right)^{1/p} \frac{\tilde{\sigma}_\ell^{2/p}}{K_1^{2/p}}. \qquad (4.229)$$

The maximum strain rate is proportional to the maximum stress squared for the linear viscosity and to the maximum stress to the fourth power for the square root viscosity. This latter observation ties into the use of the square root viscosity to model materials' finite rise times as above and discussed in the next chapter. The temporal and spatial extent of the rise can be approximated as

$$\Delta t \approx \frac{\tilde{\varepsilon}_\ell}{\dot{\tilde{\varepsilon}}_{\max}} = \left(\frac{4\eta}{K_2}\right)^{1/p} \tilde{\varepsilon}_\ell^{2/p-1}, \qquad \Delta x \approx U\Delta t \approx U\left(\frac{4\eta}{K_2}\right)^{1/p} \tilde{\varepsilon}_\ell^{2/p-1}, \qquad (4.230)$$

and we see that the rise time and rise extent are linear with the maximum stress for the linear viscosity and the third power of the maximum stress for the square root viscosity. For the quadratic viscosity $p = 2$, the approximate footprint is $\Delta x \approx 2U\sqrt{\eta/K_2}$, while the exact footprint is $\Delta x = \pi U\sqrt{\eta/K_2}$, showing the approximate approach provides a reasonable estimate.

In an impact of a material with viscosity, there are initial transients and then this solution is approached. At early time the viscosity dominates the response. For any time $t > 0$, there are no discontinuities. For the linear viscosity case ($p = 1$), the early time solution is given in Eq. (2.139). Using that solution, the maximum strain rate at early time (which occurs at $x = \frac{1}{2}u_\ell t$) is

$$\dot{\tilde{\varepsilon}}_{\max} \approx \frac{u_\ell}{2}\sqrt{\frac{\rho_0}{\pi\eta t}}. \qquad (4.231)$$

Using $\xi = x/t^q$ and $u = f(\xi)$, analysis of the early time for a general power p viscosity indicates the behavior of the maximum strain rate behaves as $1/t^q$, where $q = \frac{1}{p+1}$; hence the $p = \frac{1}{2}$ case as $t^{-2/3}$ and the $p = 2$ case as $t^{-1/3}$.

In the next chapter, an elastic-plastic with relaxation approach to viscosity will be presented. Typically, we do not include viscosity in our material models. Another boundary condition is to drive the system with a piston, ramping up the speed. When finite (i.e., large) stress values are reached, the wave in the material steepens and forms a shock or, in this case, approaches the steady viscosity solution.

4.12. A Warning about the $(u, \cdot)$ Plane

We have discussed, side by side, the Rayleigh line in the $(1/\rho, \tilde{\sigma})$ plane and the $(u, \tilde{\sigma})$ plane. However, these two representations are not equivalent:

LEMMA 4.25. *The $(u, \cdot)$ plane representations of the Hugoniot only apply to the initial shock from ambient. They are not suitable for determining a second shock or rarefaction.*

This statement is a warning; it is tempting to think that the $(1/\rho, \tilde{\sigma})$ and $(u, \tilde{\sigma})$ or (u, U) representations are the same, but Hugoniot representations such as $U(u) = c_0 + ku$ only apply to the first shock. You will occasionally see in the literature comments that the Hugoniot jump conditions are for small strains. These statements are false. The Hugoniot jump conditions apply to arbitrarily large jumps in strain (the maximum is, of course, an engineering strain of 1). Typically these statements arise because people are trying to solve problems using one of the particle velocity representations of the Hugoniot (such as described in Exercise 5.39).

We will now back up the statements in the previous paragraphs. Let us suppose we were going to solve a compressive shock followed by a second compressive shock in the $(1/\rho, \tilde{\sigma})$ plane. We would connect the Hugoniot $\tilde{\sigma}(\rho) = \tilde{\sigma}_H(\rho)$ states from the initial specific volume $1/\rho_0$ to the next state $1/\rho_1$ with a straight Rayleigh line, and then connect $1/\rho_1$ to the final state $1/\rho_2$, again with a straight Rayleigh line. The behavior of each shock would be computed with the Rayleigh line, yielding both the shock velocities U_1 and $\hat{U}_2$ and particle velocities u_1 and $\hat{u}_2$. (Recall that the hat means the second shock is being computed in the frame of reference where the material in front of the shock is at rest; in the laboratory frame, $U_2 = u_1 + \hat{U}_2$.) Now suppose that, for example, the $(u, \tilde{\sigma})$ plane was a legitimate place to compute the second shock, after the passage of the first one. The notion that it could be solved in the $(u, \tilde{\sigma})$ plane is equivalent to saying that the second shock's particle velocity $\hat{u}_2$ plus the first shock's particle velocity u_1 will give rise to the same stress as would a single shock from ambient with particle velocity $u_0 = u_1 + \hat{u}_2$. Two equations result, the first from the stress and the second from the particle velocities. The first one can be written in terms of the two jumps in stress, and the second one can be written in terms of stress, shock velocity, and density, using the momentum jump condition,

$$\tilde{\sigma}_2 - \tilde{\sigma}_0 = \tilde{\sigma}_2 - \tilde{\sigma}_1 + \tilde{\sigma}_1 - \tilde{\sigma}_0,$$

$$\frac{\tilde{\sigma}_2 - \tilde{\sigma}_0}{\rho_0 U_0} = \frac{\tilde{\sigma}_2 - \tilde{\sigma}_1}{\rho_1 \hat{U}_2} + \frac{\tilde{\sigma}_1 - \tilde{\sigma}_0}{\rho_0 U_1}. \qquad (4.232)$$

But now we have *two* equations for one unknown $(\hat{U}_2)$. (The other two wave speeds U_0 and U_1 are known directly from the $\tilde{\sigma}_H(u)$ representation of the Hugoniot.) For there to be a solution these equations must be the same equation. That occurs if

$$\rho_0 U_0 = \rho_1 \hat{U}_2 = \rho_0 U_1. \tag{4.233}$$

Let's explore what the expression on the right-hand side means. We have

$$\hat{U}_2 = \frac{\rho_0}{\rho_1} U_0. \tag{4.234}$$

Notice that the right-hand side of this equation is independent of ρ_2. Thus, it is a constant for any ρ_2. This observation gives

$$\hat{U}_2 = \sqrt{\frac{\tilde{\sigma}(\rho_2) - \tilde{\sigma}(\rho_1)}{\rho_1(1 - \rho_1/\rho_2)}} = \text{constant.} \tag{4.235}$$

This gives

$$\tilde{\sigma}(\rho_2) = k_1(1 - \rho_1/\rho_2) + \tilde{\sigma}(\rho_1). \tag{4.236}$$

If we write $k_1 = (\rho_0/\rho_1)k_2$, this can be written

$$\tilde{\sigma}(\rho_2) = k_2(1 - \rho_0/\rho_2) + \tilde{\sigma}(\rho_1) - k_2(1 - \rho_0/\rho_1). \tag{4.237}$$

Thus, the equation of state is the linear stress-strain relation, and we recall that $U_0 = U_1 = c$ and $\hat{U}_2 = (\rho_0/\rho_1)c$. The only time it is legitimate to work in the $(u, \tilde{\sigma})$ plane is for the first shock or when $U = c$, a constant; otherwise errors result. In an intuitive sense, the discrepancy occurs because stress is a function of strain, and in terms of density $\tilde{\varepsilon} = 1 - \rho_0/\rho$ while in terms of u, $\tilde{\varepsilon} = u/U$. Thus, it does not just depend on u, but it also depends on U, which can change.

These results bring up an interesting fact about the sound speed. Using the differential jump condition for momentum, a legitimate way to write the sound speed is (Eq. 4.25)

$$c(\rho) = \frac{1}{\rho} \frac{d\tilde{\sigma}}{du}. \tag{4.238}$$

However, for an arbitrary ρ it is *not* legitimate to differentiate $\tilde{\sigma}(u) = \rho_0(c_0 + ku)u$ to obtain the sound speed, because the sound speed at a density $\rho > \rho_0$ is for a "second shock," and this expression for the stress in terms of particle velocity only applies to the first shock.

In the next chapter we will learn that this single-shock-from-ambient limitation applies to all Hugoniot representations where the material response includes thermal terms (Theorem 5.3).

4.13. Sources

Again, sources for the Hugoniot jump conditions are Courant and Friedrichs [47], Hayes [91], Lax [129] and LeVeque [131]. Elastic waves are covered in Kolsky [121] and Rinehart [162]; our approach to elastic waves is unique in that it relies entirely on the Hugoniot jump conditions.

Exercises

(4.1) Fill in the blank entries in the table for 4340 steel, Al 6061-T6, Ti-6Al-4V, 93% W alloy, and OFHC (oxygen free high conductivity) copper.

(4.2) For 10-cm long, 1-cm diameter aluminum, steel, and copper impactors, using the moduli and densities from the preceding exercise, calculate the contact time for an elastic impact with a rigid surface.

Metal	ρ_0 (g/cm^3)	E (GPa)	ν	K (GPa)	G (GPa)	c_L (m/s)	c_E (m/s)	c_K (m/s)	c_s (m/s)
Steel	7.85	205	0.29			5,850			
Al	2.70	72	0.33				5,164		
Ti	4.45	115	0.31					4,761	
W	17.30	410	0.28						3,043
Cu	8.90	110	0.35						

(4.3) Show that the longitudinal wave speed is always greater than or equal to the shear wave speed (i.e., $c_L \geq c_s$).

(4.4) Show that Poisson's ratio is given by

$$\nu = \frac{1}{2} \cdot \frac{c_L^2 - 2c_s^2}{c_L^2 - c_s^2}. \tag{4.239}$$

(4.5) Using the stress-strain relation $\tilde{\sigma} = \rho_0 c^2 (\rho/\rho_0 - 1)$, show that the sometimes heard statement "the shock velocity is the average of the sound velocity in front of and behind the shock" is not strictly correct.

(4.6) Show that the shock velocity equals the sound speed for small amplitudes of deformation – that is, $\lim_{\rho \to \rho_0} U(\rho) = U(\rho_0) = c$.

(4.7) Though not explicitly stated in the solution of the penetration problem for two elastic materials, the particle speed cannot exceed the sound speed in the target. Given this constraint, show that $P/L < (\rho_p/\rho_t)/B$. Based on the derivation, the particle speed $V - u$ cannot exceed the sound speed in the projectile, either. Show that this constraint implies $P/L < B/(1-B)^2$. Show that these constraints are independent (i.e., one does not imply the other; hint: fix $B > 1/2$ and contrive sound speeds and densities that invoke each constraint).

(4.8) Consider a nonlinear elastic equation of state of the form $\tilde{\sigma} = a\tilde{\varepsilon}^b$, where $\tilde{\varepsilon} = 1 - \rho_0/\rho$. Solve the problem of a constant velocity boundary condition (i.e., the piston problem). Check that it agrees with the $b = 1$ case solved for in this chapter.

(4.9) Using the result of the previous problem, how deep will a rigid projectile of density ρ_p and thickness L penetrate into the material (in this one-dimension case, perhaps the word displace is better) with the equation of state of the form $\tilde{\sigma} = a\tilde{\varepsilon}^b$? (Hint: recall that $du/dt = (du/dx)(dx/dt) = u\,du/dx$.) Show that this result agrees for the case of $b = 1$ (Eq. 4.188).

(4.10) Using the $\tilde{\sigma} = a\tilde{\varepsilon}^b$ stress-strain relation and the rarefaction shock approximation, calculate the penetration if the projectile and target materials are the same. (Hints: since the materials are the same, you automatically know the interface velocity. Also, since they are the same, you know there will only be one reverberation.) Is this case deeper or less deep than the rigid penetration?

(4.11) (Convexity) (If you plan to work this problem, do not read the next two exercises until this one is completed.) For a region in the plane, convex means that for any two points in the region, the line connecting the two points is also entirely within the region. For a function it means that given any two points on the curve, the straight line connecting those two points lies above the curve of the function. Our definition of concave up is similar to this in that if one uses the region in the $(u, \tilde{\sigma})$ plane above the Hugoniot curve, then the line connecting the origin $(0, 0)$ with any point in the region is within the region. Thus, instead of two arbitrary points and the line connecting them, we have one defined point; the other arbitrary. Create, using $\sigma = a\varepsilon^b$ and the linear stress-strain response, a concave up stress-strain response that is not convex. If instead of ambient the initial condition were at a nonconvex point on the Hugoniot, what would be the result of an impact?

(4.12) Show that the following two stress-strain relations are superlinear, where $\tilde{\varepsilon} = 1 - \rho_0/\rho$, $b > 1$, $\rho^* > \rho_0$ is fixed, and $\tilde{\varepsilon}^* = 1 - \rho_0/\rho^*$:

$$\tilde{\sigma}(\rho) = \begin{cases} (\lambda + 2\mu)\tilde{\varepsilon} & \rho \leq \rho^* \\ (\lambda + 2\mu)(\tilde{\varepsilon}^*)^{1-b}\tilde{\varepsilon}^b & \rho > \rho^* \end{cases}, \quad \tilde{\sigma}(\rho) = \begin{cases} (\lambda + 2\mu)(\tilde{\varepsilon}^*)^{1-b}\tilde{\varepsilon}^b & \rho \leq \rho^* \\ (\lambda + 2\mu)\tilde{\varepsilon} & \rho > \rho^* \end{cases}.$$
$$(4.240)$$

(4.13) Show in the previous problem (Exercise 4.12) that one of the stress-strain relations is convex and one is not convex. Show that both, however, are ρ_0-convex. (Hint: pick b and draw some pictures.)

(4.14) For stress-strain relation Eq. (4.240), what is the depth of penetration for the impact of this material on itself? What is the depth of penetration for a rigid material of the same mass at the same impact speed into this material? Which is deeper?

(4.15) Describe the Hugoniot in the six planes for the stress-strain relation $\tilde{\sigma} = \rho_0 c^2 (\rho/\rho_0 - 1)$.

(4.16) Show that when the Hugoniot can be represented as $U(u) = c + ku$, if $k < 1$ then the Hugoniot locus has finite extent in the (u, U) and $(u, \tilde{\sigma})$ planes.

(4.17) Using the mass and momentum jump conditions, show that if the Hugoniot can be written $\tilde{\sigma}_H = K_1\tilde{\varepsilon} + K_2\tilde{\varepsilon}^2$, then the U_s–u_p relation is $(c_0 = \sqrt{K_1/\rho_0})$

$$U = \frac{2c_0}{\sqrt{3}} \cos \left\{ \frac{1}{3} \cos^{-1} \left(\frac{3\sqrt{3}K_2 u}{2K_1 c_0} \right) \right\} = c_0 + \frac{K_2}{2K_1} u - \frac{3K_2^2}{8K_1^2 c_0} u^2 + \cdots. \quad (4.241)$$

(Hint: the resulting cubic has no U^2 term; look at Exercise 3.21 to solve it.)

(4.18) Show that the stress-strain relation $\sigma = a\varepsilon^b \Rightarrow \tilde{\sigma} = a(1 - \rho_0/\rho)^b$ is strictly superlinear for constants a and b where $a > 0$ and $b > 1$.

(4.19) Compute the entropy for the stress-strain relation $\tilde{\sigma}(\rho) = \rho_0 c^2 (\rho/\rho_0 - 1)$.

(4.20) Relate the change in entropy for a material with a mechanical stress-strain response to the kinetic energy after the passage of a shock and to the residual kinetic energy after the subsequent passage of a rarefaction.

(4.21) Another way to look at how similar the rarefaction shock approximation as compared with the actual rarefaction fan is to compute the angle that the fan covers. Since the fan occurs in the (x, t) plane, the dimensions are not quite right for the definition of an angle. To better define the problem, suppose the time axis is multiplied by $c(\rho_r)$. Then the units are the same when considering the $(x, c(\rho_r)t)$ plane. Determine the angle that the rarefaction fan covers in this plane. Are there an upper and lower bounds to the angle? If so, what are they?

(4.22) Using the result of the previous problem, what is an equation for the angle of the rarefaction fan for the $\tilde{\sigma} = \rho_0 c^2 (\rho/\rho_0 - 1)$ stress-strain response? Confirm the angles for the ρ_r/ρ_0 density ratios in Table 4.1.

(4.23) Show that in the (x, t) plane the rarefaction shock approximation U wave front lies within the rarefaction fan – that is,

$$c(\rho_o) + u(\rho_o) \leq U \leq c(\rho_r) \qquad (4.242)$$

(4.24) In the text only two solutions of the equation for a rarefaction were produced: the rarefaction fan and the rarefaction shock (which is a solution of the jump conditions except for the entropy jump condition). Given the result of the previous problem, show that there are "a lot" of solutions to the rarefaction problem by showing that the combination of a rarefaction shock to an intermediate density followed by a rarefaction fan is also a solution (except that entropy decreases). (Hint: you need to show that the regions of the solution do not interfere with each other.)

(4.25) The rarefaction fan is isentropic, so that the entropy state in front of the wave equals the entropy state behind the wave. The rarefaction shock decreases the entropy, which is why we conclude that it is not the actual solution. However, given the previous problem, for a strictly superlinear material is it possible to have a rarefaction fan that reduces the density to a density lower than ambient followed by a compressive shock that then increases the shock to ambient? The rarefaction fan has zero entropy change and the compressive shock has positive entropy change, so this solution, if it existed, would have a higher entropy change than the rarefaction fan, and thus should be preferred. Is it a solution? Why?

(4.26) Show that $d(U(\rho))/d\rho \geq 0$ is equivalent to the inequalities in Theorem 4.12 for a compressive shock, and from it conclude that $U(\rho) \geq c(\rho_o)$ for linear and superlinear materials where $\rho \geq \rho_o$.

(4.27) It is also possible to define the notion of "sublinear" stress-strain relation. How would you define it? How can you tell if a stress-strain relation is sublinear? Give a couple of examples. What behavior will a sublinear stress-strain relation have in compression waves and rarefaction waves?

(4.28) What must k be in $U = c_0 + ku$ so that $c(\rho) \geq c(\rho_0)$ for $\rho \geq \rho_o$?

(4.29) Explicitly show, by direct substitution, that the rarefaction fan solves the mass and momentum conservation equations.

(4.30) Write down the equation that must be solved for the general piston boundary condition with piston velocity $V < 0$, assuming the material is initially at ambient density ρ_o and at rest. What variable is being solved for?

(4.31) The traditional approach in studying smooth small amplitude elastic wave propagation is to assume the displacement u, strains, and rotations are small. The strain is then $\varepsilon_{11} = \partial u / \partial x$, where u is the displacement in the x direction. The density is nearly constant. The Lagrangian acceleration is $Dv/Dt = \partial v/\partial t = \partial^2 u/\partial t^2$, where v is the material velocity in the x direction. Assume a linear elastic uniaxial strain stress-strain curve. Place these assumptions in the momentum balance, assuming that $u(x, t)$ is twice differentiable in each variable (i.e., no shocks), to obtain a second order partial differential equation for u in terms of x and t,

$$\frac{\partial^2 u}{\partial x^2} = \frac{1}{c_L^2} \frac{\partial^2 u}{\partial t^2}. \qquad (4.243)$$

This is the one-dimensional linear wave equation. Show that the solution is

$$u(x, t) = f(x - c_L t) + g(x + c_L t), \qquad (4.244)$$

where f and g are twice differentiable, arbitrary functions, with f moving to the right and g moving to the left. (Hint: inserting this expression into the wave equation shows that it is a solution, but to show that it is the complete solution, change the variables to $\xi = x - c_L t$, $\zeta = x + c_L t$, show that the wave equation becomes $\partial^2 u / \partial \xi \partial \zeta = 0$, and integrate.) Determine the wave speed. Show that Theorem 4.1 holds.

(4.32) Show that the square of the sound speed is not the "symmetric finite difference" version of the square of the shock velocity. What is the multiplicative term that is not symmetric? In particular, what density multiplies the square of the shock velocity to give the slope of the Rayleigh line in the $(1/\rho, \tilde{\sigma})$ plane? How does this relate to the sound speed?

(4.33) For Theorem 4.12, find the engineering strain version of Eqs. (4.52),(4.56), and (4.57).

(4.34) Show when equality holds in Eq. (4.107), that the linear stress-strain response results.

(4.35) For a mechanical stress-strain that leads to compressive shocks, show that at the initial density ρ_0 the stress-strain relation must be concave up.

(4.36) Show that over the density region where a mechanical stress-strain curve gives rise to rarefaction fans, it will give rise to a single compressive shock.

(4.37) Write down a stress-strain response that leads to a compressive shock from ambient and then part of the rarefaction is also a shock. (Hint: if you are unwilling to be bold, you can always consider something like Exercise 4.12.)

(4.38) For the linear stress-strain relation, solve the problem of an infinite expanse of material initially at rest, with $\rho = \rho_2 > \rho_0$ for $x < 0$ and $\rho = \rho_0$ for $x \geq 0$. (Hint: a compressive shock moves to the right and a rarefaction shock moves to the left.)

(4.39) For a concave up stress-strain relation, write down the equations that need to be solved for an infinite expanse of material initially at rest, with $\rho = \rho_2 > \rho_0$ for $x < 0$ and $\rho = \rho_0$ for $x \geq 0$. (Hint: a compressive shock moves to the right and a rarefaction fan moves to the left.)

(4.40) (This problem is due to Hayes [91] after Taylor.) Suppose the rarefaction has caught up with the shock. We wish to calculate the attenuation rate, or rate at which the leading edge stress of the shock decreases with distance of run. Assume the shock is moving to the right. In the $(x, \tilde{\sigma})$ plane, there is a jump to the stress at the front of the shock (let's call it $\tilde{\sigma}_{\max}$), followed by the rarefaction that slopes down to zero stress. Fix the time. At this time, the shock front is at location x_0. Call the stress distribution at that time $\tilde{\sigma}(x)$. The stress distribution will look roughly triangular, and notice that $d\tilde{\sigma}/dx > 0$ in the rarefaction region. We expect that the shock strength decreases as the shock advances due to the rarefaction. Write down a differential equation for the decay of $\tilde{\sigma}_{\max}(x)$ in terms of x in terms of slope of the stress in the rarefaction – that is, find $d\tilde{\sigma}_{\max}/dx_0$ in terms of $d\tilde{\sigma}/dx$.

(4.41) During the Apollo missions to the Moon, nine spacecraft were intentionally impacted into the lunar surface to provide seismic pulses to be measured by the seismometers placed at the landing sites. Five Saturn IVB upper stages of the Saturn V rocket (Apollos 13–17), which was the stage that took the astronauts from Earth orbit to the lunar orbit, were impacted into the Moon. The Saturn IVB stage weighed around 14,000 kg and impacted at around

2.55 km/s. Four Lunar Modules (Apollos 12, 14, 15, 17), took the astronauts from the lunar surface back to the command module, were also impacted into the Moon. A Lunar Module weighed around 2,350 kg and impacted at around 1.68 km/s. Figure 4.14 is from the Apollo 14 Preliminary Science Report [**128**], showing primary and secondary wave arrivals in terms of time and distance at various sites. Of course, the Moon is layered, with different layers having different properties. As an approximation, assume there are two layers of relevance: a top layer with a low constant sound speed and a lower layer with a higher constant sound speed. As a function of surface distance between impact site and seismometer, show that the initial arrival time as a function of distance is given by two straight lines – the first corresponding to the wave that travels along the surface directly from the impact site to the seismometer, and the second corresponding to the wave traveling through the upper layer to the lower layer, along the top of the lower layer, and then back up to the surface. Show that the distance and time at which the two lines cross is

$$x = 2T\sqrt{\frac{c_2 + c_1}{c_2 - c_1}}, \quad t = \frac{x}{c_1}, \qquad (4.245)$$

where T is the thickness of the layer, c_1 is the appropriate (longitudinal or shear) upper layer sound speed, and c_2 is the lower layer sound speed. (Hint: the seismic wave expands spherically from the impact site and then interacts with the layer boundary, where there are three-dimensional effects that are examined in seismology texts whereas we only examined one-dimensional reflection and transmission. For the initial arrival time, minimize the travel time over the various paths the seismic wave can take from the impact site to the high speed layer, along the top of the high speed layer, then back up through the low speed layer to the surface seismometer. Note that this expression can be inverted to determine T if the intersection x is known from the data; however, in the figure we have no arrival times from travel strictly through the upper layer.) Why (in words) does the second line not pass through the origin of the plot? From this plot, estimate the longitudinal wave speed, the shear wave speed, and the Poisson's ratio (Exercise 4.4) for this particular layer near the surface of the Moon.

(4.42) As an extension of the previous exercise, use Euler's equation from the calculus of variations, $(\partial f/\partial y) - (d/dx)(\partial f/\partial y') = 0$, assume the sound speed is given as a function of depth $c(x)$ (viewing $+x$ as increasing depth), and then show that the equation for a path $y = y(x)$ where the travel time $t = \int \sqrt{dx^2 + dy^2}/c(x)$ is minimized is

$$\frac{(dy/dx)^2}{1 + (dy/dx)^2} = \text{constant} \cdot c^2(x). \qquad (4.246)$$

Show that for a constant sound speed the solutions are straight lines and that for the case of $c(x) = cx$, where the sound speed increases linearly with depth, the solutions are arcs of circles, $x^2 + (y - y_0)^2 = R^2$. (The latter solution is the half-plane representation of two-dimensional hyperbolic geometry.)

(4.43) For the Tuler–Butcher model, show that if a geometrically self-similar flyer plate spall test is performed with the scale size difference given by a, then it

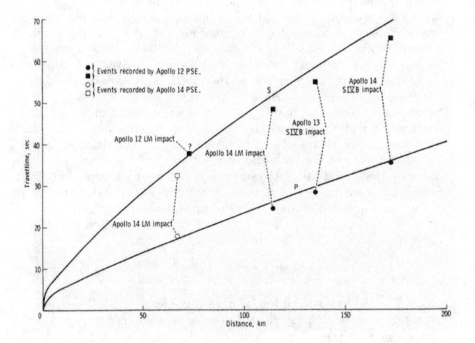

FIGURE 4.14. Arrival times for signals from spacecraft impacts at seismometers during Apollo program. (From Latham et al. [**128**].)

is possible to determine σ_0 without knowing D_c. Show the relationship is

$$\sigma_0 = \frac{b\,\sigma_{\text{spalla}} - \sigma_{\text{spall}}}{b - 1}, \tag{4.247}$$

where $b = a^{1/\lambda}$ if a square pulse is assumed, while $b = a^{1/(\lambda+1)}$ if it is assumed the loading begins at zero and has the unloading slope scaled by $1/a$ (i.e., a triangular shape). Go the other direction and determine, based on the Tuler–Butcher constants given for Al 6061-T6, how large a flyer and target are required to put the spall stress 10% above σ_0, based on the result of the 343 m/s test. Finally, show that if the spall stress at the two scales is the same, then $\sigma_0 = \sigma_{\text{spall}}$ and the model becomes an immediate tensile failure criterion.

(4.44) Work through the details in Eq. (4.228), thus determining the explicit solution for $p = 2$ viscosity.

(4.45) Show that if $\varepsilon_p^f(\eta) = \varepsilon_0^f \exp(-\tfrac{3}{2}\eta + \tfrac{1}{2}) + C_r \dot{\varepsilon}^{2/3}$ for $\eta > \tfrac{1}{3}$ and $\varepsilon_p^f(\eta)$ includes this additive strain rate failure from Eq. (9.30) for all η, then the spall occurs at $(\varepsilon_r^f = C_r \dot{\varepsilon}^{2/3})$

$$\eta_{\text{spall}} = \frac{1}{3} + \frac{2}{3}\ln\left\{\left(1 + \frac{\varepsilon_0^f}{\varepsilon_r^f}\right)\exp\left(\frac{9\varepsilon_r^f}{4}\left(\frac{K}{Y} - \frac{2/3}{\varepsilon_0^f + \varepsilon_r^f}\right)\right) - \frac{\varepsilon_0^f}{\varepsilon_r^f}\right\}. \tag{4.248}$$

Using C_r from Chapter 9 for Al-2024, plot σ_{spall} vs. strain rate for strain rates from $1\ \text{s}^{-1}$ to $10^5\ \text{s}^{-1}$.

Elastic-Plastic Deformation and Shocks

This chapter applies the Hugoniot jump conditions to problems that involve both elastic and plastic deformation. Specifically, the three primary methods for determining dynamic material properties are considered in detail: the Taylor anvil experiment, the split-Hopkinson pressure bar apparatus, and the flyer plate impact experiment. Completing the section is a discussion of the general response of solids as pressures get larger.

5.1. Cylindrical Impactor Striking a Rigid Wall (Taylor Anvil)

Given success with solving wave propagation and impact problems where the materials behave in an elastic fashion, it is now time to analyze problems where permanent deformation (plasticity) is part of the material response. As described in Chapter 3, the difference between elasticity and plasticity is that in plasticity strain occurs which does not give rise to stress.

For the first two examples we work, we solve equations in uniaxial stress with an assumed linear dependence. In Chapter 4, we assumed uniaxial strain, but the results carry over. Now the wave speed is given by the bar wave speed $c_b = c_E = \sqrt{E/\rho_0}$. In uniaxial stress the strain is the engineering strain and does not have the same relation to density as it does in uniaxial strain, due to Poisson's ratio (Exercise 5.10). Thus, for uniaxial stress problems we work with the strain and not the density. For uniaxial stress where elastic-perfectly plastic behavior occurs, the stress-strain relation before unloading is

$$\tilde{\sigma}(\tilde{\varepsilon}) = \begin{cases} -Y & \tilde{\varepsilon} < -Y/E, \\ E\tilde{\varepsilon} & -Y/E \leq \tilde{\varepsilon} \leq Y/E, \\ Y & \tilde{\varepsilon} > Y/E. \end{cases} \tag{5.1}$$

During and after unloading, this relation no longer applies. For more discussion on uniaxial stress, see Section 4.1.

We assume a blunt cylindrical projectile of radius R and length L strikes a rigid wall (the anvil) at velocity V. The stress produced if the impact is elastic is

$$\tilde{\sigma} = \rho_0 c_E V. \tag{5.2}$$

If the stress is less than the yield Y, then the elastic solution in Section 4.4 applies, and the projectile bounces off the wall. If the stress predicted is above Y, then plastic flow occurs, and material at the rigid wall flows radially outward. Because of this radial plastic flow, the problem is not one-dimensional. However, we initially focus on the elastic rear of the projectile, which is amenable to a one-dimensional analysis. The maximum stress the cylinder can support in uniaxial stress is Y, and that is the stress jump associated with the elastic wave moving back through

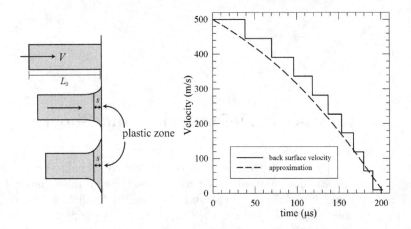

FIGURE 5.1. Aluminum cylinder striking a rigid wall (anvil).

the projectile. The conditions determining the wave moving to the back of the projectile are (Fig 5.1)

$$\begin{aligned} \tilde{\varepsilon}_\ell &= 0 & \tilde{\varepsilon}_r &=? \\ \tilde{\sigma}_\ell &= 0 & \tilde{\sigma}_r &= Y \\ u_\ell &= V & u_r &=? \end{aligned} \tag{5.3}$$

The wave speed in the laboratory frame is $U = -c_E + V$. The jump conditions give

$$Y = \rho_0(U - V)(u_r - V) = -\rho_0 c_E(u_r - V) \tag{5.4}$$

or that

$$u_r = -\frac{Y}{\rho_0 c_E} + V. \tag{5.5}$$

The change in velocity is simply

$$\Delta v = -\frac{Y}{\rho_0 c_E}. \tag{5.6}$$

For example, for a 6061-T6 aluminum with strength $Y = 380$ MPa, $\Delta v = 27$ m/s, and for 4340 steel with strength 850 GPa, $\Delta v = 21$ m/s. Impacts at or above these velocities produce yield in the striking projectile; impacts below these velocities elastically rebound from the rigid wall. In Section 4.5 it was shown that when this wave reaches the free surface at the left end of the projectile it reflects, doubling the change in particle velocity. Thus, the rarefaction that returns the projectile material to its initial conditions has a particle velocity

$$u_\ell = -\frac{2Y}{\rho_0 c_E} + V = -2\Delta v + V. \tag{5.7}$$

This behavior continues until the velocity is such that the stress applied to the rigid wall is less than Y,

$$\rho_0 c_E u_\ell \leq Y. \tag{5.8}$$

The back end of the impactor experiences a stepping down in time. Unfortunately, this solution is rather tedious to work with because the projectile is continually getting shorter and thus the travel time decreases for the elastic wave reflecting back and forth. Also adding to the complexity is the fact that there is a region where the material is plastically deforming of length s (Fig. 5.1). The reflection of the wave occurs at the nonlinear boundary between elastic and elastic-plastic response. After two paragraphs discussing the elastic-plastic interface, we will return to the question of determining s.

The reflection at the elastic-plastic boundary merits discussion. Beginning with a stress-free projectile, the impact produces a wave that travels from the front to the back of the projectile. Reiterating from above, the change in particle velocity in this wave is $\Delta v = -Y/\rho_0 c_E$ and the stress change associated with the wave is $\Delta \tilde{\sigma} = +Y$. This wave is compressive, and decelerates projectile material. The projectile is in a state of uniaxial stress. When the compressive wave reaches the rear of the projectile, which is a free surface, a tensile wave is formed that travels back towards the front of the projectile. This tensile wave reduces the stress state back to its initial state of zero stress and it does so with a change in particle velocity of $\Delta v = -Y/\rho_0 c_E$ and a stress change of $\Delta \tilde{\sigma} = -Y$. In the region near the front of the projectile where plastic flow is occurring, the stress state is no longer uniaxial stress, and along the centerline axis the compressive stress increases above the value of Y as it heads towards the stress at the impact surface where the compressive stress exceeds $+Y$ (otherwise there would be no plastic flow). (Figure 7.11 shows velocity and strain profiles along the centerline of a projectile, showing large gradients in the plastically flowing region near the projectile/target interface where there are large stresses.) As the wave travels towards the rigid interface, there is still a change in stress associated with it on the order of $\Delta \tilde{\sigma} = -Y$ and so the stress behind the wave front along the centerline increases from zero to values greater than zero. Since the rest of the back end of the projectile is stress free, there is now a stress-free region with an applied stress at one end. This boundary condition creates an elastic wave that travels towards the projectile back surface. If the stress at impact surface is high enough, the wave that travels back will have a stress $\Delta \tilde{\sigma} = +Y$. This cycle continues. The reflection at the elastic-plastic boundary is not sharp since the stress rises over a finite distance in the projectile, but it is not a particularly large distance and so fairly steep wave fronts are seen in numerical simulations for the reflection off the elastic-plastic boundary.

Though the last few sentences focused on the formation of the wave that travels back down the length of the impactor, the returning wave continues to travel forward in the impactor. Since the impactor cross section is increasing as it deforms in the plastically flowing region, the change in stress level decreases as $\Delta \tilde{\sigma} = -Y A_0/A(x)$, where A is the area. Though reduced in magnitude as it approaches the rigid anvil surface, this stress change reduces the stress at the rigid surface. Similarly, the wave reduces the particle velocity along the steep particle velocity gradient in the plastically flowing region. If instead of a rigid anvil the impactor was striking a deforming target, the wave would propagate into the target, continuing to reduce the stress, though the amount of the reduction decreases rapidly since in the target the wave expands as a spherical front. In this fashion the target material during a penetration event is informed of the deceleration of the projectile as it occurs (with a delay due to the finite speed of sound in the materials).

Now we determine s. The method is fairly complicated, and the reader not interested in the details can skip ahead to the result in Eq. (5.19). In the penetration model developed in Chapter 10 we determine s using a different approach that does not apply when the target is not deforming.

We use the dynamic plasticity approach as described at the end of Section 3.16, page 100, where we assume a flow field characterized by a geometric parameter and then maximize the resulting plastic dissipation subject to the constraint of an admissible stress that formally satisfies the momentum balance while lying within the yield surface. Our application of the approach will hopefully clarify the method. In thinking about the location of the elastic-plastic boundary s in the impacting cylinder, we expect that $s(0) = 0$, that then s will grow to some maximum value, and then as the velocity of the cylinder decreases the stress and hence s will decrease and end up back at zero at the end of the impact event. Notice that s is not a record of the maximum extent of plastic deformation, but rather it is the current boundary between elastic and elastic-plastic deformation. In cylindrical coordinates with z the axisymmetric axis, the cylinder moves in the positive z direction and strikes a rigid wall at $z = 0$. We have $v \geq 0$ and $s \geq 0$, and we assume the simplest divergence-free flow field (i.e., velocity field) holds in the plastically deforming region $-s \leq z \leq 0$, which is linear in r and z, namely

$$v_z = -\frac{v}{s}z = -\Gamma z, \quad v_r = \frac{1}{2}\frac{v}{s}r = \frac{1}{2}\Gamma r, \quad v_\theta = 0, \quad \text{where} \quad \Gamma = \frac{v}{s}. \tag{5.9}$$

The flow field is parameterized by one geometric parameter, $\Gamma = \Gamma(t)$; hence we have a flow field $\vec{v} = \vec{v}(r, \theta, z, \Gamma)$. At a given time, since v and s are known, this parameter is given. What we are looking for is the time rate of change of this parameter, $\dot{\Gamma} = d\Gamma/dt$. In the plastically flowing region, the rate of deformation tensor is given by (Eq. A.174)

$$D_{rr} = \frac{\partial v_r}{\partial r} = \frac{1}{2}\Gamma, \quad D_{\theta\theta} = \frac{v_r}{r} = \frac{1}{2}\Gamma, \quad D_{zz} = \frac{\partial v_z}{\partial z} = -\Gamma, \quad D_{ij} = 0 \text{ for } i \neq j. \tag{5.10}$$

We see that the assumed flow is incompressible, $D_{ii} = 0$, as constructed, since that is what we expect of plastic flow. The plastic dissipation rate is then

$$\sigma^a_{ij}D_{ij} = \Gamma(\sigma^a_{rr} - \sigma^a_{zz}). \tag{5.11}$$

We will maximize this dissipation.

Now we determine the admissible stress σ^a, which is a stress that satisfies the momentum balance $\partial \sigma^a_{ji}/\partial x_j = a_i$, the yield condition $\sqrt{3J_2(\sigma^a)} \leq Y$, and the boundary conditions. Using Eq. (A.172) we compute the accelerations for the flow field,

$$a_r = \frac{\partial v_r}{\partial t} + v_r\frac{\partial v_r}{\partial r} = \frac{1}{2}(\dot{\Gamma} + \frac{1}{2}\Gamma^2)r, \quad a_\theta = 0, \quad a_z = \frac{\partial v_z}{\partial t} + v_z\frac{\partial v_z}{\partial z} = -(\dot{\Gamma} - \Gamma^2)z. \tag{5.12}$$

It is in the accelerations that $\dot{\Gamma}$ first appears. There are lots of admissible stresses, but we only consider a subspace of the admissible stresses for computational convenience: we will assume that $\sigma^a_{\theta\theta} = \sigma^a_{rr}$ and that $\sigma^a_{ij} = 0$ for $i \neq j$. With this assumption on the stress, the momentum balance says

$$\frac{\partial \sigma^a_{rr}}{\partial r} = \rho a_r \quad \Rightarrow \quad \sigma^a_{rr} = \rho\frac{r^2}{4}(\dot{\Gamma} + \frac{1}{2}\Gamma^2) + f(z),$$

$$\frac{\partial \sigma^a_{zz}}{\partial z} = \rho a_z \quad \Rightarrow \quad \sigma^a_{zz} = -\rho\frac{z^2}{2}(\dot{\Gamma} - \Gamma^2) + g(r), \tag{5.13}$$

where f and g are arbitrary functions of integration. Notice that the admissible stresses are functions of $\dot{\Gamma}$, our unknown. Since admissible stresses satisfy the boundary conditions, and $\sigma_{zz} = -Y$ at the elastic-plastic boundary $z = -s$, we find a g that yields

$$\sigma_{zz}^a = -\rho \frac{z^2 - s^2}{2}(\dot{\Gamma} - \Gamma^2) - Y. \tag{5.14}$$

The admissible stress must also satisfy the yield condition,

$$\sqrt{3J_2} = |\sigma_{rr}^a - \sigma_{zz}^a| = |\rho \frac{r^2}{4}(\dot{\Gamma} + \frac{1}{2}\Gamma^2) + f(z) + \rho \frac{z^2 - s^2}{2}(\dot{\Gamma} - \Gamma^2) + Y| \leq Y. \tag{5.15}$$

We are maximizing the dissipation,

$$\sigma_{ij}^a D_{ij} = \Gamma(\sigma_{rr}^a - \sigma_{zz}^a) = \Gamma \left(\rho \frac{r^2}{4}(\dot{\Gamma} + \frac{1}{2}\Gamma^2) + f(z) + \rho \frac{z^2 - s^2}{2}(\dot{\Gamma} - \Gamma^2) + Y \right). \tag{5.16}$$

By examining Eqs. (5.15) and (5.16), we observe that the maximal $\sigma_{rr}^a - \sigma_{zz}^a = Y$ is achievable by removing the r and z dependence of $\sigma_{rr}^a - \sigma_{zz}^a$ through the appropriate choices $f(z) = -\rho(z^2 - s^2)(\dot{\Gamma} - \Gamma^2)/2$ and

$$\dot{\Gamma} = -\frac{1}{2}\Gamma^2. \tag{5.17}$$

Thus, we have determined $\dot{\Gamma}$. This equation can be integrated to give

$$\Gamma(t) = \frac{2}{t}, \tag{5.18}$$

where the constant of integration is zero, since $\Gamma(0) = V/s(0) = +\infty$. Thus by maximizing the dissipation over admissible stresses we have found $\dot{\Gamma}$, which in this case is directly integrable. Since $\Gamma = v/s$, we have an explicit expression for s,

$$s = \frac{1}{2}vt. \tag{5.19}$$

This has the expected properties at the beginning and end of the impact event, namely no plastically deforming region ($s = 0$). Initially, $\dot{s}(0) = \frac{1}{2}V$. The extent of the plastic zone s achieves a maximum value some time during the event.

To make a final comment about the maximization procedure, the maximizing admissible stress we found is

$$\sigma_{rr}^a = -\frac{3}{4}\rho(s^2 - z^2)\Gamma^2, \quad \sigma_{zz}^a = -\frac{3}{4}\rho(s^2 - z^2)\Gamma^2 - Y. \tag{5.20}$$

Technically the admissible stress should satisfy the traction free boundary condition $\vec{n}^T \sigma^a = \vec{0}$ on the outer radius of the plastically deforming region, but as we are not even trying to track the shape of the deforming cylinder it is doubtful that the boundary condition is rigorously satisfied; hence, our stress is only approximately admissible. Because $\dot{\Gamma}$ does not depend on the initial radius and mushroomed radius of the impacting cylinder, the solution for the final length of the cylinder will also not depend on a radius. The admissible stress of the maximization problem produces a centerline stress at the rigid surface of

$$\tilde{\sigma} = \tilde{\sigma}_{zz}^a|_{z=0} = \frac{3}{4}\rho_0 v^2 + Y, \tag{5.21}$$

where v is the current back end velocity of the projectile.[1]

---

[1]This stress result also follows from the projectile model developed in Chapter 10, when the interface stress in Eq. (10.70) is computed assuming $s = \frac{1}{2}vt$, the elastic wave speed is large compared to v, and the penetration velocity is zero. Notice that if the deceleration were "steady state" with $\dot{\Gamma} = 0$, then the stress would be $\tilde{\sigma} = \frac{1}{2}\rho_0 v^2 + Y$.

Recall the argument here is that we should maximize the energy dissipation rate. We showed in Theorem 3.9 that for large plastic deformation the stress deviators quickly approach the rigid plastic stress. Computing the stress deviators of our admissible stress, we find that

$$\sigma_{rr}^a = \sigma_{\theta\theta}^a = f(z), \quad \sigma_{zz}^a = f(z) - Y \quad \Rightarrow \quad s_{rr}^a = s_{\theta\theta}^a = \frac{Y}{3}, \quad s_{zz}^a = -\frac{2Y}{3}, \quad (5.22)$$

and so the admissible deviators are the rigid plastic stress deviators based on the rate of deformation tensor (Eq 5.10). Hence they have approached (exactly equal in this case) the stresses we expect. These stresses are reached through a maximization since the von Mises yield surface is convex; hence $(\sigma_{ij}^a - \sigma_{ij}^{rp})D_{ij} \leq 0$ because D_{ij} is perpendicular to the von Mises yield surface and the admissible stress is inside or on the yield surface; the larger $\sigma_{ij}^a D_{ij}$ is, the closer the admissible stress deviators are to equaling the rigid plastic stress deviators.

We should discuss what this solution looks like. Let's follow a Lagrangian point (meaning we are going to follow a piece of material) near the front of a cylinder that is impacting a rigid wall at a high enough speed to undergo plastic deformation. Initially the point decelerates with elastic waves traveling back and forth, so its v_z decreases. Then, the point arrives at the elastic-plastic interface at time t_s, when its z coordinate is such that $z(t_s) = -s(t_s)$. At that time the point receives a jolt (a singular, delta-function like acceleration that produces a finite jump in velocity) in the radial direction, giving it a radial speed $v_r = \Gamma(t_s)r(t_s)/2$. Now, with Γ given by Eq. (5.17), the radial acceleration is zero, $a_r = 0$, so this point will maintain this radial speed as long as it is in the plastically deforming region, and its motion is

$$r(t) = r(t_s) + \frac{1}{2}\Gamma(t_s)r(t_s)(t - t_s) = \frac{1}{2}r(t_s)\Gamma(t_s)t, \quad (5.23)$$

where $\Gamma = 2/t$ has been used. This point is decelerating in the z direction as it maintains a uniform motion in the r direction, namely

$$a_z = \frac{3}{2}\Gamma^2 z \quad \Rightarrow \quad a_z = \frac{d^2 z}{dt^2} = \frac{6z}{t^2} \quad \Rightarrow \quad z(t) = z(t_s)\left(\frac{t_s}{t}\right)^2. \quad (5.24)$$

The motion parameterized by time leads to the following curve for the Lagrangian point,

$$z(t)r(t)^2 = z(t_s)r(t_s)^2, \quad (5.25)$$

which can also be arrived at by volume conservation for our particular flow field. As the cylinder continues to decelerate owing to elastic reflections off the back end, at some time the elastic-plastic front will pass back through the point, so $z(t) = -s(t)$ again. At this time the particle again receives a radial jolt, so that now the radial velocity is again zero, and it continues to be decelerated in the axial direction by elastic waves. It is possible to avoid jolts by choosing smoother velocity fields; the mathematics is more tedious and the elastic-plastic boundary may no longer be right at the boundary of the flow field, as discussed in Section 10.16.

We now return to the deceleration of the back end of the cylinder. The analysis is greatly simplified if we take limits to obtain, then solve, a differential equation rather than try to sum the series of discrete wave reflections. The time it takes for the elastic wave to travel from the elastic-plastic interface to the back of the projectile through uncompressed material is $\Delta t_c = (L - s)/c_E$. The back-end location of the projectile at the beginning of the release wave is

$z_p = -L + v\Delta t_c$. The location of the elastic-plastic interface when the rarefaction arrives is $z_s = -s(t + \Delta t_c + \Delta t_r) \approx -s(t) - \dot{s}(t) \cdot (\Delta t_c + \Delta t_r)$, where Δt_r is the time for the return trip. These locations give the distance the rarefaction wave travels in the laboratory frame in traversing from the back free surface to the elastic-plastic interface as $z_s - z_p = L - v\Delta t_c - s - \dot{s} \cdot (\Delta t_c + \Delta t_r)$. In the laboratory frame, the elastic wave speed is $v + c_E - 2\Delta v$. If we assume Δv is small compared to the bar wave speed, the time for the return trip is

$$\Delta t_r = \frac{L - v\Delta t_c - s - \dot{s} \cdot (\Delta t_c + \Delta t_r)}{c_E + v}. \tag{5.26}$$

Solving yields

$$\Delta t_r = \frac{L - s}{c_E} \cdot \frac{c_E - v - \dot{s}}{c_E + v + \dot{s}}. \tag{5.27}$$

Over this time, the projectile velocity changes by $2\Delta v$, and the decrease in velocity can be approximated with a differential equation,

$$\frac{dv}{dt} \approx \frac{2\Delta v}{\Delta t_c + \Delta t_r} = -\frac{Y}{\rho_0(L - s)}\left(1 + \frac{v}{c_E} + \frac{\dot{s}}{c_E}\right). \tag{5.28}$$

This equation represents the decrease in velocity of the striking impactor. Using our expression $s = vt/2$,

$$\frac{dv}{dt} = -\frac{Y}{\rho_0(L - vt/2)}\left(1 + \frac{3v}{2c_E} + \frac{t}{2c_E}\frac{dv}{dt}\right). \tag{5.29}$$

Solving for dv/dt,

$$\frac{dv}{dt} = -\frac{Y}{\rho_0(L - vt/2)}\left(1 + \frac{3v}{2c_E}\right) \Big/ \left(1 + \frac{Yt}{2\rho_0 c_E(L - vt/2)}\right). \tag{5.30}$$

The rate of change of impactor length is simply given by its back-end velocity, since the front of the impactor is at the rigid wall (as per our flow field above, material moves radially to get out of the way),

$$\frac{dL}{dt} = -v. \tag{5.31}$$

These two first-order differential equations describe the behavior of the impacting cylinder, with $L(0) = L_0$ and $v(0) = V$ as initial conditions. These equations are easily solved numerically.

These equations can be combined into a single second order equation for length,

$$\left(1 + \frac{Yt}{2\rho_0 c_E(L + (dL/dt)t/2)}\right)\frac{d^2L}{dt^2} = \frac{Y}{\rho_0(L + (dL/dt)t/2)}\left(1 - \frac{3}{2c_E}\frac{dL}{dt}\right), \tag{5.32}$$

where $L(0) = L_0$ and $dL/dt\,(0) = -V$. We now change to nondimensional variables. Though often physical intuition is lost by such a change, in some cases, this one in particular, the change will allow us to learn more about the solution without explicitly solving the equation. Choosing (recall $V = v(0)$ is a constant)

$$\ell = \frac{L}{L_0}, \quad \tau = \frac{Vt}{L_0}, \tag{5.33}$$

then

$$\frac{dL}{dt} = \frac{d(L_0\ell)}{d(L_0\tau/V)} = V\frac{d\ell}{d\tau}, \quad \frac{d^2L}{dt^2} = \frac{d}{dt}\frac{dL}{dt} = \frac{d}{d(L_0\tau/V)}V\frac{d\ell}{d\tau} = \frac{V^2}{L_0}\frac{d^2\ell}{d\tau^2}, \tag{5.34}$$

and Eq. (5.32) becomes, after some manipulation,

$$\left(2\ell + \frac{d\ell}{d\tau}\tau + \frac{Y}{\rho_0 V^2}\frac{V}{c_E}\tau\right)\frac{d^2\ell}{d\tau^2} = \frac{Y}{\rho_0 V^2}\left(2 - 3\frac{V}{c_E}\frac{d\ell}{d\tau}\right),\tag{5.35}$$

with initial conditions $\ell(0) = 1$ and $d\ell/d\tau(0) = -1$ until the cylinder comes to rest when $d\ell/d\tau = -v/V = 0$. Since Eq. (5.35) has no L_0 in it, we arrive at the remarkable conclusion that *the ratio of the final length to the original length* $\ell_f = L_f/L_0$ *is independent of the original length.* This is really quite amazing, and we will shortly show that this conclusion agrees with experimental data. Further, we have written Eq. (5.35) in such a way as to highlight that ℓ_f only depends on two variables, namely

$$\frac{L_f}{L_0} = f\left(\frac{\rho_0 V^2}{Y}, \frac{V}{c_E}\right).\tag{5.36}$$

When Eq. (5.30) is solved numerically, we can find average values for s, where there are two natural ways to define the average,

$$\langle s \rangle = \frac{1}{L_0 - L_f}\int_{L_f}^{L_0} s\, dL, \qquad \langle s \rangle = \frac{1}{V}\int_0^V s\, dv.\tag{5.37}$$

Numerically we find that both these averages are similar. Using numbers for aluminum to be presented below: at 300 m/s impact speed, $\langle s \rangle/L_0 \approx 0.043$; at 600 m/s, $\langle s \rangle/L_0 \approx 0.11$; and at 1,000 m/s, $\langle s \rangle/L_0 \approx 0.15$. To produce an integrable, closed form solution, we can approximate $s = \frac{1}{2}vt$ with a constant. When we do that, and also assume c_E is much larger than v, then Eq. (5.30) can be approximated as

$$\frac{dv}{dt} = -\frac{Y}{\rho_0(L - \langle s \rangle)}.\tag{5.38}$$

With a standard change of variables in impact problems,

$$\frac{dv}{dt} = \frac{dv}{dL}\frac{dL}{dt} = -v\frac{dv}{dL} \quad \Rightarrow \quad v\frac{dv}{dL} = -\frac{Y}{\rho_0(L - \langle s \rangle)},\tag{5.39}$$

the equation can be explicitly integrated with upper limit $v = V$ and lower limit $v = 0$ to give (technically the lower limit is given by the elastic wave speed $v_e = Y/\rho c_E$, but if v_e is used as the lower limit the resulting term is v_e/c_E, which we have assumed is small)

$$L_f = \langle s \rangle + (L_0 - \langle s \rangle)\exp\left(\frac{-\rho_0 V^2}{2Y}\right)\tag{5.40}$$

or

$$\frac{L_f}{L_0} = \exp\left(\frac{-\rho_0 V^2}{2Y}\right) + \frac{\langle s \rangle}{L_0}\left[1 - \exp\left(\frac{-\rho_0 V^2}{2Y}\right)\right].\tag{5.41}$$

The final length of the projectile is a function of the ratio of initial specific kinetic energy $\frac{1}{2}V^2$ to the specific strength of the projectile material Y/ρ_0. These quantities are important in elastic-plastic impact events. The plastically flowing region s increases the final L_f/L_0. Firing metal cylinders at rigid walls (such as a hard steel or ceramic) was suggested by Taylor [183, 184] as a way to determine the dynamic flow stress Y of the metal. This expression for the Taylor anvil problem was obtained by Wilkins and Guinan [211] assuming a time-independent s. They suggest $\langle s \rangle/L_0 = 0.12$, based on fits to experimental data. Interestingly, Taylor originally assumed $\dot{s} = $ constant; see Exercise 5.6. Figure 5.2 shows experiments of Wilkins and Guinan along with experiments of Anderson et al. [3], showing good

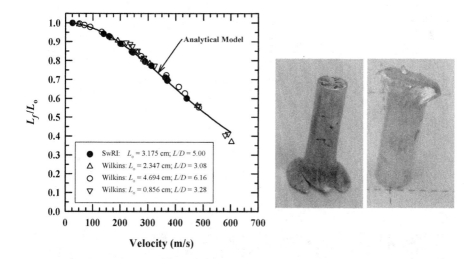

FIGURE 5.2. Comparison of model with data for the Taylor anvil test with 6061-T6 aluminum (left); shear failure mode in two Taylor tests for other aluminum alloys (right).

agreement using Eq. (5.41) with $\langle s \rangle / L_0 = 0.20$ and $Y = 380$ MPa for 6061-Al. Table 5.1 contains the experimental results for four different L/Ds. The photographs in Fig. 5.3 show the Anderson et al. projectiles on a graph of the final length vs. impact velocity. Increasing $\langle s \rangle$ increases L_f/L_0, and increasing $\langle s \rangle$ decreases the Y that best fits the data.

The Al 6061-T6 material shows a ductile failure mode with failure in the direction of largest $\varepsilon_{\theta\theta}$, as can be seen in the photographs in Fig. 5.3. The 45° shear failure mode is seen in some materials, where the near uniaxial stress state is perturbed and leads to a loss of symmetry and subsequent failure along shear planes. Examples are shown in Fig. 5.2. This behavior should be compared to the failure behavior discussed in Chapter 3, page 78.

Numerical methods incorporating constitutive models that include work hardening and strain-rate effects are able to replicate not only the final length of the projectile, but also the outer boundary shape (see, e.g., [211, 106, 3]).

Occasionally one will find modelers trying to work with the full momentum balance of the projectile – in other words, expressions such as $d(\rho L v)/dt$ or variations on this value set equal to a decelerating force divided by some area. However, Eq. (5.28) was derived *based on wave reflections off the back of the projectile* – it has nothing to do with the "total momentum" of the projectile. In the literature, attempts to work with the total momentum of a deforming projectile are often not carried out correctly since great care must be made in dealing with the axial momentum of the projectile material involved in the radial flow.

5.2. The Split-Hopkinson Pressure Bar

The split-Hopkinson pressure bar (SHPB) is an experimental technique that is widely used to determine the high-rate plastic-flow stress of materials. The technique involves a cylindrical material specimen, two cylindrical bars on either

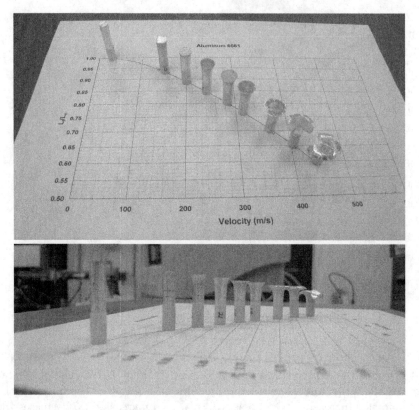

FIGURE 5.3. Photographs of $L/D = 5$ Al 6061-T6 cylinders after striking a rigid anvil.

side of equal length (referred to as the incident bar and the transmitter bar) and a cylindrical striker bar. The striker bar, incident bar, and transmitter bar are all of the same material, typically steel or aluminum. The striker bar of length L_0 is fired at a velocity V into the incident bar and a square wave pulse is formed. The wave speed in the incident bar is c_E, the particle velocity is $V/2$ (since the materials are the same, $B = 1/2$) and a square pulse is formed that is spatially twice the length of the striker ($2L_0$) and in time is of duration $2L_0/c_E$. The pulse travels down the incident bar, where a strain gage, typically located midway on the bar, measures the strain in the pulse. From Theorem 4.1,

$$\tilde{\varepsilon} = \frac{u_\ell}{c_E} = \frac{V}{2c_E}. \tag{5.42}$$

Upon arrival at the specimen, placed between the incident and transmitter bars, the wave partly transmits and partly reflects. The bars are long so that the incident pulse and the reflected pulse do not overlap at the incident bar strain gage. There is a strain gage on the transmitter bar to record the transmitted pulse. For historical reasons it is located the same distance from the specimen as the incident bar strain gage, so that the transmitted and reflected pulses arrive at the same time at their respective gages (this placement allowed the integrations described below to be done electronically in an oscilloscope, something no longer necessary in the digital age). The specimen is of a diameter less than the bars. If the specimen has a

TABLE 5.1. Final length and diameter of four different L/D 6061-T6 aluminum cylinders impacting a rigid anvil (data from [211] with $L/D = 5$ data from [3]). Above 250 m/s, the mushroom begins to develop radial cracks and therefore the end diameter is only approximate. Wilkins and Guinan believe thermal softening effects are important above 500 m/s [211].

L/D	L_0 (cm)	V (m/s)	L_f/L_0	D_f/D_0	L/D	L_0 (cm)	V (m/s)	L_f/L_0	D_f/D_0
5	3.175	26	0.999	1.01	6.16	4.694	275	0.823	
3.08	2.347	50	0.993		3.08	2.347	283	0.806	
6.16	4.694	51	0.995		3.28	0.856	285	0.814	
6.16	4.694	69	0.988		5	3.175	289	0.794	1.85
6.16	4.694	90	0.978		3.08	2.347	306	0.784	
6.16	4.694	132	0.952		3.28	0.856	324	0.773	
5	3.175	139	0.942	1.18	5	3.175	363	0.710	2.32
5	3.175	159	0.929	1.25	6.16	4.694	367	0.722	
5	3.175	159	0.929	1.23	5	3.175	370	0.694	2.54
3.08	2.347	169	0.920		3.08	2.347	373	0.703	
3.08	2.347	191	0.906		6.16	4.694	412	0.660	
5	3.175	201	0.889	1.42	6.16	4.694	434	0.626	
6.16	4.694	207	0.888		5	3.175	440	0.600	3.48
3.28	0.856	222	0.896		3.08	2.347	478	0.562	
3.28	0.856	236	0.888		3.28	0.856	482	0.561	
5	3.175	242	0.848	1.58	6.16	4.694	484	0.552	
3.28	0.856	244	0.876		3.28	0.856	580	0.404	
5	3.175	246	0.843	1.55	3.28	0.856	591	0.412	
3.08	2.347	262	0.844		3.08	2.347	603	0.370	
3.28	0.856	262	0.852						

(rate dependent) flow stress Y_s, then matching forces at the incident bar–specimen boundary provides the following problem to solve (Figs. 5.4 and 5.5)

$$\tilde{\varepsilon}_I = V/2c_E \qquad \tilde{\varepsilon}_r = ?$$
$$\tilde{\sigma}_I = \rho_0 c_E V/2 \qquad \tilde{\sigma}_r = (A_s/A_b)Y_s \qquad (5.43)$$
$$u_I = V/2 \qquad u_r = ?$$

For the left value the subscript I has been used, since it is referred to as the incident pulse; the subscript b means bar and s is for specimen. Notice that setting the forces equal says that the cross-sectional area (denoted A) times the stress must equal, or $A_b\tilde{\sigma}_r = A_s Y_s$. Thus, we have a boundary with a specified stress and we must determine what the reflected wave looks like. From above, we know that the wave speed of the reflected wave traveling to the left is $U = -c_E + V/2 + V/2 = -c_E + V$, adjusted from the bar wave speed by material motion and compression. Placing these values into the momentum jump conditions gives

$$\tilde{\sigma}_r - \tilde{\sigma}_I = \rho_0(U - u_I)(u_r - u_I) \quad \Rightarrow \quad u_r = V - \frac{1}{\rho_0 c_E}\frac{A_s}{A_b}Y_s. \qquad (5.44)$$

Recall that ρ_0 and c_E refer to the bars, not the specimen. To use our results on interior boundary reflection (Section 4.9), we need to define an equivalent B for

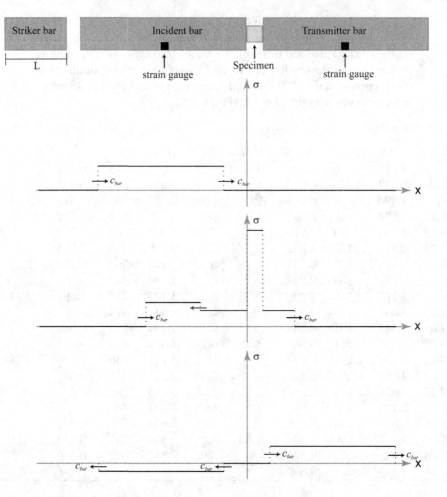

FIGURE 5.4. Schematic of the split-Hopkinson pressure bar experiment ($c_{bar} = c_E$).

the stress boundary condition for the reflected wave. Our specified stress at the right-hand boundary is $\tilde{\sigma}_s = A_s Y_s / A_b$. We have two possible ways to define B. First, the velocity relationship with B implies (Eq. 4.193)

$$u_r = 2Bu, \qquad (5.45)$$

which, when applied to the situation here, says

$$V - \frac{1}{\rho_0 c_E}\tilde{\sigma}_s = 2B\frac{V}{2} \quad \Rightarrow \quad B = 1 - \frac{\tilde{\sigma}_s}{\rho_0 c_E V} = 1 - \frac{\tilde{\sigma}_s}{2\tilde{\sigma}_I}. \qquad (5.46)$$

In particular we have

$$u_r = BV. \qquad (5.47)$$

Another approach would be based on the stress relationship (Eq. 4.194)

$$\tilde{\sigma}_r = 2\rho_0 c_E (1 - B)u, \qquad (5.48)$$

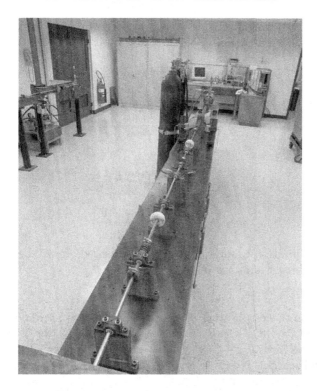

FIGURE 5.5. A split-Hopkinson pressure bar at SwRI. The laboratory is an ASME National Historic Engineering Landmark in recognition of the location where the SHPB concept was reduced to engineering practice in 1962 by Lindholm and Yeakley.

which gives

$$\tilde{\sigma}_s = 2\rho_0 c_E (1 - B)\frac{V}{2} \quad \Rightarrow \quad B = 1 - \frac{\tilde{\sigma}_s}{\rho_0 c_E V} = 1 - \frac{\tilde{\sigma}_s}{2\tilde{\sigma}_I}. \tag{5.49}$$

Both the methods give the same B, and so we have a consistent definition. With B we are able to use Eq. (4.196) to determine what happens after the interaction with the back end of the incident stress pulse, and to determine the stress in the wave that travels back towards the striker after it is no longer superposed on the incident pulse:

$$
\begin{aligned}
u_R &= (2B - 1)V/2, \\
\tilde{\sigma}_R &= \rho_0 c_E (1 - 2B)V/2.
\end{aligned}
\tag{5.50}
$$

The subscript R has been used as the wave is referred to as the reflected pulse.

Next we address the transmitter bar. Here, a stress has been passed through the specimen of magnitude Y_s, and with the area change the subsequent stress is

$$\tilde{\sigma}_T = \tilde{\sigma}_s = \frac{A_s}{A_b}Y_s = \rho_0 c_E (1 - B)V, \tag{5.51}$$

where T stands for transmitter. A wave travels down the transmitter bar with velocity c_E, and the corresponding particle velocity is

$$u_T = \frac{\tilde{\sigma}_T}{\rho_0 c_E} = (1 - B)V. \tag{5.52}$$

Measured in the test are the strains at the gages on the two bars. For each of our three waves, the strain that will be measured at the gage is

$$\tilde{\varepsilon}_I = \frac{\tilde{\sigma}_I}{\rho_0 c_E^2} = \frac{V}{2c_E},$$

$$\tilde{\varepsilon}_R = \frac{\tilde{\sigma}_R}{\rho_0 c_E^2} = \frac{(1 - 2B)V}{2c_E}, \tag{5.53}$$

$$\tilde{\varepsilon}_T = \frac{\tilde{\sigma}_T}{\rho_0 c_E^2} = \frac{2(1 - B)V}{2c_E}.$$

Now we are in a position to interpret what is measured. It will be seen by observation that the three strains are related,

$$\tilde{\varepsilon}_I + \tilde{\varepsilon}_R = \tilde{\varepsilon}_T \tag{5.54}$$

and consistency in this observation is used by some to measure the accuracy of the experiment (see Gray [87] where each side is multiplied by the Young's modulus and different stress measures – "1 wave," "2 wave" or "3 wave" – are discussed, all variations on this check). Nearly everyone computes specimen stress directly from the strain value in the transmitter bar,

$$Y_s = \frac{A_b}{A_s} \rho_0 c_E^2 \tilde{\varepsilon}_T = \frac{A_b}{A_s} E \tilde{\varepsilon}_T. \tag{5.55}$$

Next, to compute strain in the specimen, it is necessary to compute the motion of its two faces. The front face, in contact with the incident bar, has a displacement

$$\Delta_f = \int u_r dt = \int BV dt. \tag{5.56}$$

The displacement at the back face of the sample is given by

$$\Delta_b = \int u_T dt = \int (1 - B)V dt. \tag{5.57}$$

The strain in the specimen is

$$\tilde{\varepsilon}_s = \frac{\Delta_f - \Delta_b}{L_s} = \frac{1}{L_s} \int (2B - 1)V dt, \tag{5.58}$$

where L_s is the original specimen length. From Eq. (5.53), we obtain

$$\tilde{\varepsilon}_s = -\frac{2c_E}{L_s} \int \tilde{\varepsilon}_R dt. \tag{5.59}$$

The strain rate the specimen sees is taken from the strain

$$\frac{d\tilde{\varepsilon}_s}{dt} = -\frac{2c_E}{L_s} \tilde{\varepsilon}_R. \tag{5.60}$$

We have left things in terms of the integrals over time because the equations we have solved can be subdivided into small increments, thus allowing for work hardening. The time integrals are the way the tests are analyzed. If we assume that the flow stress in the specimen is constant, then it is possible to explicitly compute the strain rate and total strain that the specimen sees,

$$\frac{d\tilde{\varepsilon}_s}{dt} = -\frac{(1-2B)V}{L_s} = \frac{1}{L_s}\left(V - \frac{2}{\rho_0 c_E}\frac{A_s}{A_b}Y_s\right), \tag{5.61}$$

$$(\tilde{\varepsilon}_s)_{total} = 2\frac{L_0}{L_s}\frac{(2B-1)V}{c_E} = 2\frac{L_0}{L_s}\left(\frac{V}{c_E} - 2\frac{A_s}{A_b}\frac{Y_s}{E}\right). \tag{5.62}$$

The strain rate is always positive: if Eq. (5.61) predicts a negative value, then the experiment will not yield the specimen. The strain rate the specimen sees depends linearly on the striker impact velocity V. The stronger the specimen is (the larger Y_s), the lower the strain rate. A striker length can be chosen by matching anticipated flow stress with a desired final strain in the specimen – the final strain depends linearly on the length of the striker bar. Intuitively, the strain rate in the test depends on how fast the front face deflects (recall for a free surface, its velocity would be V, twice the particle velocity of the incident wave) divided by the length of the specimen. The higher the strength of the specimen, the more of the incident wave it will reflect, leading to less front-face motion, and the more of the wave it will transmit, leading to more back-face deflection, both of which decrease the shortening of the sample and thus decrease the strain and strain rate.[2]

Figure 5.6 shows the traces from the strain gages on the incident and transmitter bars and the constructed stress-strain curve from an experiment. The oscillations in the stress are because the uniaxial stress is not entirely uniaxial; there are radial motions that give rise to wavelength dependent wave speeds, which give rise to dispersion of the square pulse, which in turn gives rise to the stress oscillations seen in the figure.[3] The oscillations are referred to as Pochhammer–Chree oscillations, due to their examinations of uniaxial stress waves in the late 1800s. As to the practicalities of the test, it is desirable to use as strong a metal as possible for testing strong specimens. We use Vascomax 350 for our steel bars, with $Y = 2.4$ GPa (350 ksi), thus allowing a maximum impact velocity for the striker of 120 m/s. Typically, the radius of the specimen is half the radius of the bars, giving an area ratio of $A_b/A_s = 4$. Thus, the largest flow stress that conceivably could be tested in the bar is $4Y$, or around 9.5 GPa. As an example, an impact of 60 m/s will, for a sample strength of 1 GPa, $A_b/A_s = 4$, and a 2.5 cm sample length, lead to a strain rate of around 1,800 s^{-1}, a typical value achievable in the experiment. Very strong samples can potentially damage the ends of the bars, so either a small piece of hardened steel is placed at the end that can be discarded if indented or a ceramic platen is used to protect the end of the bar. AD-999 alumina nearly matches the acoustic impedance of steel, and is thus a good choice. Typically it will fracture and is replaced after each test.

---

[2]The strain rate described in this paragraph is the plastic strain rate for plastically flowing materials. For brittle materials, the material does not flow, and the strain rate for the experiment is taken to be the elastic strain rate in the rise of the wave front – see the finite rise time in Fig. 5.6. We have not developed analytic tools to compute this rise; it is an average slope measured from the reflected-pulse trace on the incident bar strain gage or preferably on a strain gage mounted directly on the specimen. The strain rate increases with increasing striker bar impact velocity because of both an increase in the maximum strain and a decrease in rise time.

[3]The dispersion in elastic uniaxial stress waves in long, thin bars is a geometrical effect, since the material sound speeds c_L and c_s are constants, independent of wave frequency (see page 34).

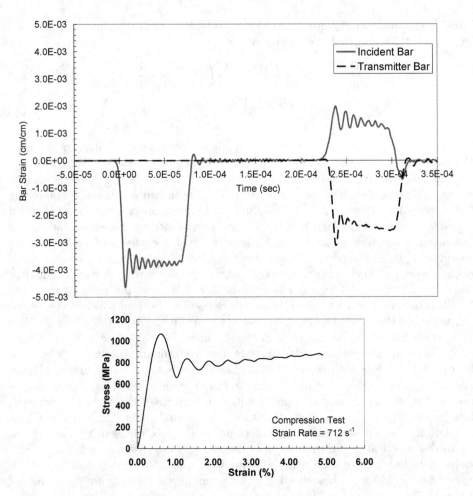

FIGURE 5.6. Strain vs. time at the incident bar and transmitter bar strain gages as well as the reconstructed dynamic stress-strain response for a steel alloy.

5.3. Pushing on an Elastic-Plastic Half Space: The Two-Wave Structure and Flyer Plate Impacts

We return to the uniaxial strain problem, with a boundary condition of pushing with a given velocity or stress at the edge of a half space of material. Here, material is constrained so there is no radial motion. To begin, we write down the explicit stress-strain relation before unloading for uniaxial strain that includes plasticity:

$$\tilde{\sigma}(\rho) = \begin{cases} K(1 - \rho_0/\rho) - 2Y/3 & 1 - \rho_0/\rho < -Y/(2\mu), \\ (K + (4/3)\mu)(1 - \rho_0/\rho) & -Y/(2\mu) \leq 1 - \rho_0/\rho \leq Y/(2\mu) = \tilde{\varepsilon}_{\text{HEL}}, \\ K(1 - \rho_0/\rho) + 2Y/3 & 1 - \rho_0/\rho > Y/(2\mu). \end{cases}$$

$$(5.63)$$

The initial part of the stress-strain relationship is strictly elastic and above the yield the response is elastic-plastic.

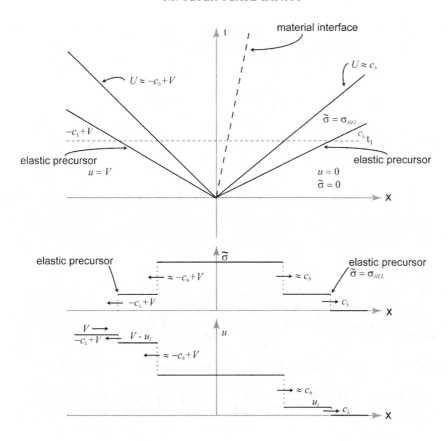

FIGURE 5.7. Impact of two elastic-plastic materials.

Two waves form due to the bilinear behavior of the stress-strain behavior for uniaxial strain (Eq. 5.63). If one considers only the right-hand side of Fig. 5.7, it shows the two regions to solve for. For the rightmost wave, labeled elastic precursor in the figure and with a subscript 2 below, we use the subscript e for elastic. The elastic wave solution has already been developed (Section 4.2) and the results are

$$
\begin{array}{lll}
\rho_{1\ell} = ? & \rho_{1r} = \rho_e = \rho_0/(1 - u_e/c_L) & \rho_{2r} = \rho_0 \\
\tilde{\sigma}_{1\ell} = ? & \tilde{\sigma}_{1r} = \tilde{\sigma}_e = \rho_0 c_L u_e & \tilde{\sigma}_{2r} = 0 \\
u_{1\ell} = ? & u_{1r} = u_e = \tilde{\sigma}_{\text{HEL}}/(\rho_0 c_L) & u_{2r} = 0
\end{array}
\tag{5.64}
$$

The elastic wave has speed $U_e = c_L$. Since it runs in front of the elastic-plastic wave, it is called the *elastic precursor*. This elastic wave has a name, already alluded to in the notation:

DEFINITION 5.1. (Hugoniot elastic limit, HEL) The *Hugoniot elastic limit*, often denoted *HEL*, is the Hugoniot state of the elastic precursor when multiple waves form. The state is characterized by the longitudinal wave speed c_L, a particle velocity u_{HEL}, and the stress $\tilde{\sigma}_{\text{HEL}}$.

Often when people refer to the HEL they mean the stress $\tilde{\sigma}_{\text{HEL}}$. Though in our elastic-perfectly plastic material model the HEL is well defined, in materials and wave propagation tests things are not so well defined (see Exercise 5.26 for an example of results of a flyer plate impact test) and so the HEL is only approximate. Some values are included in Exercise 5.24. For an elastic-perfectly plastic material, expressions for the HEL are in Eqs. (3.64) and (3.65), and from $u = c\tilde{\varepsilon}$ (Eq. 4.13),

$$\tilde{\varepsilon}_{\text{HEL}} = \frac{Y}{2\mu}, \quad u_{\text{HEL}} = c_L \tilde{\varepsilon}_{\text{HEL}} = \frac{c_L Y}{2\mu}, \quad \tilde{\sigma}_{\text{HEL}} = \rho_0 c_L u_{\text{HEL}} = \left(1 + \frac{\lambda}{2\mu}\right) Y.$$
(5.65)

The jump conditions give us, for the elastic-plastic shock (writing a subscript e for elastic precursor, rather than writing out HEL),

$$\tilde{\sigma}_\ell - \tilde{\sigma}_e = \tilde{\sigma}_\ell - \tilde{\sigma}_{\text{HEL}} = \rho_e (U - u_e)(u_\ell - u_e), \qquad \frac{\rho_e}{\rho_\ell} = 1 - \frac{u_\ell - u_e}{U - u_e}.$$
(5.66)

Now we follow the procedure that we have developed, working with the speeds in the frame of reference where the material is at rest behind the precursor,

$$\hat{U} = U - u_e, \qquad \hat{u}_\ell = u_\ell - u_e.$$
(5.67)

Placing these definitions in the jump conditions and solving gives

$$\hat{U}^2 = \frac{\tilde{\sigma}_\ell - \tilde{\sigma}_e}{\rho_e(1 - \rho_e/\rho_\ell)} = \frac{K(1 - \rho_0/\rho_\ell) + 2Y/3 - K(1 - \rho_0/\rho_e) - 2Y/3}{\rho_e(1 - \rho_e/\rho_\ell)} = \frac{K}{\rho_0}\left(\frac{\rho_0}{\rho_e}\right)^2.$$
(5.68)

We obtain the result that the elastic-plastic front moves at a constant velocity,

$$U = u_e + \frac{\rho_0}{\rho_e}\sqrt{\frac{K}{\rho_0}} = u_e + \frac{\rho_0}{\rho_e} c_K = \frac{\tilde{\sigma}_{\text{HEL}}}{\rho_0 c_L} + \left(1 - \frac{\tilde{\sigma}_{\text{HEL}}}{\rho_0 c_L^2}\right) c_K$$

$$= c_K + (c_L - c_K)\tilde{\varepsilon}_{\text{HEL}} = c_K + \left(1 - \frac{c_K}{c_L}\right)\frac{\tilde{\sigma}_{\text{HEL}}}{\rho_0 c_L},$$
(5.69)

where $u_e = \tilde{\sigma}_{\text{HEL}}/\rho_0 c_L$ and $\tilde{\varepsilon}_{\text{HEL}} = u_e/c_L$. With respect to the material in motion owing to the elastic precursor, the elastic-plastic wave front is traveling at a speed just slightly less than $c_K = \sqrt{K/\rho_0}$, while with respect to the initial laboratory frame the elastic-plastic wave front is traveling just slightly faster than c_K. The wave speed is what we would have guessed: it is the elastic precursor material motion plus the wave speed based on the uniaxial strain elastic-plastic strain loading modulus K (see Theorem 3.3), as modified by the material being slightly more dense after the passage of the initial elastic wave. Though the wave speed is referred to as the bulk sound speed, again it must be pointed out that *no elastic wave travels at the bulk sound speed*. Rather, the elastic-plastic front in a uniaxial loading condition travels at nearly (slightly more than) the bulk sound speed c_K. Placing the wave speed result back into the momentum equation allows a calculation of the material velocity,

$$u_\ell = u_e + \frac{\tilde{\sigma}_\ell - \tilde{\sigma}_{\text{HEL}}}{\rho_0 c_K}.$$
(5.70)

This equation can be used to get the velocity of the boundary if a stress is applied, or the stress if the velocity of the boundary is given. The mass jump conditions give the density behind the shock,

$$\frac{1}{\rho_\ell} = \frac{1}{\rho_e} - \frac{1}{\rho_0}\frac{\tilde{\sigma}_\ell - \tilde{\sigma}_{\text{HEL}}}{K}.$$
(5.71)

Continuing the solution, writing $u = u_\ell$ the stress behind the front is

$$\tilde{\sigma}(u) = \tilde{\sigma}_{\text{HEL}} + \rho_e(U - u_e)(u - u_e)$$
$$= \tilde{\sigma}_{\text{HEL}} + \rho_0 c_K(u - u_e)$$
$$= \rho_0 c_K u + \tilde{\sigma}_{\text{HEL}}\left(1 - \frac{c_K}{c_L}\right). \tag{5.72}$$

We have written the stress $\tilde{\sigma}$ as a function of u to be evocative for what we will do next. However, we want to make some observations about this solution. First, the existence of the elastic precursor reduces the strain in the material from what we might expect if there were no elastic strength. In particular, note that

$$1 - \frac{\rho_e}{\rho} = \frac{u - u_e}{U - u_e} = \frac{u - u_e}{(\rho_0/\rho_e)c_K} \quad \Rightarrow \quad \frac{\rho_0}{\rho_e} - \frac{\rho_0}{\rho} = \frac{u - u_e}{c_K}. \tag{5.73}$$

This gives

$$\tilde{\varepsilon} = 1 - \frac{\rho_0}{\rho} = 1 - \frac{\rho_0}{\rho_e} + \frac{\rho_0}{\rho_e} - \frac{\rho_0}{\rho} = \frac{u_e}{c_L} + \frac{u - u_e}{c_K} = \frac{u}{c_K} + \frac{u_e}{c_L} - \frac{u_e}{c_K} < \frac{u}{c_K}. \tag{5.74}$$

Therefore, the increase in stress over the no-strength solution will not be as great as the $2Y/3$ that might be intuitively expected from the uniaxial strain plasticity solution developed in Chapter 3. In particular, the increase in stress is only

$$\Delta\tilde{\sigma} = \tilde{\sigma}(u) - \rho_0 c_K u = \tilde{\sigma}_{\text{HEL}}\left(1 - \frac{c_K}{c_L}\right) = \frac{1 - \nu}{1 - 2\nu}\left(1 - \sqrt{\frac{1 + \nu}{3(1 - \nu)}}\right)Y \approx 0.37Y \tag{5.75}$$

for $0.29 < \nu < 0.34$. The multiplier for $\nu = 0.25$ is 0.38. With our computations of strain, we gather some useful expressions for the total strain and for the partitioning of strain into elastic and plastic parts,

$$\tilde{\varepsilon}_\ell = 1 - \frac{\rho_0}{\rho_\ell} = \tilde{\varepsilon}_{\text{HEL}} + \frac{\tilde{\sigma}_\ell - \tilde{\sigma}_{\text{HEL}}}{K} = \tilde{\varepsilon}_{\text{HEL}} + \frac{u - u_e}{c_K} = \tilde{\varepsilon}_{\text{HEL}}\left(1 - \frac{c_L}{c_K}\right) + \frac{u}{c_K},$$

$$\tilde{\varepsilon}_{11}^e = \tilde{\varepsilon}_{\text{HEL}} + \frac{\tilde{\sigma}_\ell - \tilde{\sigma}_{\text{HEL}}}{3K} = \tilde{\varepsilon}_{\text{HEL}} + \frac{u - u_e}{3c_K} = \tilde{\varepsilon}_{\text{HEL}}\left(1 - \frac{c_L}{3c_K}\right) + \frac{u}{3c_K},$$

$$\tilde{\varepsilon}_{11}^p = \frac{2}{3}\frac{\tilde{\sigma}_\ell - \tilde{\sigma}_{\text{HEL}}}{K} = \frac{2}{3}\frac{u - u_e}{c_K} = -\frac{2c_L}{3c_K}\tilde{\varepsilon}_{\text{HEL}} + \frac{2u}{3c_K} = \varepsilon_p = -2\tilde{\varepsilon}_{22}^p = 2\tilde{\varepsilon}_{22}^e. \tag{5.76}$$

Now we pose the question: what happens if two flyer plates of an elastic-plastic material are impacted into each other at velocity V? Based on our experience with solving problems using the jump conditions, we can easily write down the answer. For all the initial conditions, in impactor material 1, coming from the left, we have an elastic wave traveling to the left and the elastic-plastic front traveling to the left. In target material 2, initially at rest, we have an elastic wave traveling to the right and the elastic-plastic front traveling to the right. The conditions are (Fig. 5.7)

$$\begin{array}{llll}
\rho_{1\ell} = \rho_{1e} & \rho_{1r} =? & \rho_{2\ell} =? & \rho_{2r} = \rho_{2e} \\
\tilde{\sigma}_{1\ell} = \tilde{\sigma}_{1\text{HEL}} & \tilde{\sigma}_{1r} = \tilde{\sigma}_{2\ell} =? & & \tilde{\sigma}_{2r} = \tilde{\sigma}_{2\text{HEL}} \\
u_{1\ell} = V - u_{1e} & u_{1r} = u_{2\ell} =? & & u_{2r} = u_{2e}.
\end{array} \tag{5.77}$$

The wave velocities are

$$U_1 = -\frac{\rho_{10}}{\rho_{1e}}c_{1K} + u_{1e} - V, \qquad U_2 = \frac{\rho_{20}}{\rho_{2e}}c_{2K} + u_{2e}. \tag{5.78}$$

Matching the particle velocities at the material interface gives $u = u_{1r} = u_{2\ell}$. Matching the stresses from the momentum jump conditions for waves 1 and 2 gives $\tilde{\sigma} = \tilde{\sigma}_{1r} = \tilde{\sigma}_{2l}$ or

$$\tilde{\sigma} = \tilde{\sigma}_1(V - u) = \tilde{\sigma}_2(u).$$

This equation is

$$\tilde{\sigma}_{1\text{HEL}} + \rho_{10}c_{1K}(V - u - u_{1e}) = \tilde{\sigma}_{2\text{HEL}} + \rho_{20}c_{2K}(u - u_{2e}) \qquad (5.79)$$

or

$$\rho_{10}c_{1K}(V - u) + \tilde{\sigma}_{1\text{HEL}}\left(1 - \frac{c_{1K}}{c_{1L}}\right) = \rho_{20}c_{2K}u + \tilde{\sigma}_{2\text{HEL}}\left(1 - \frac{c_{2K}}{c_{2L}}\right), \qquad (5.80)$$

which yields a particle velocity and a stress of

$$u = \frac{\rho_{10}c_{1K}(V - u_{1e}) + \rho_{20}c_{2K}u_{2e} + \tilde{\sigma}_{1\text{HEL}} - \tilde{\sigma}_{2\text{HEL}}}{\rho_{10}c_{1K} + \rho_{20}c_{2K}}, \qquad (5.81)$$

$$\tilde{\sigma} = \frac{\rho_{10}c_{1K}\rho_{20}c_{2K}(V - u_{1e} - u_{2e}) + \rho_{20}c_{2K}\tilde{\sigma}_{1\text{HEL}} + \rho_{10}c_{1K}\tilde{\sigma}_{2\text{HEL}}}{\rho_{10}c_{1K} + \rho_{20}c_{2K}}. \qquad (5.82)$$

If, as before, we define the impedances ratio based on the bulk wave speed

$$B = \frac{\rho_{10}c_{1K}}{\rho_{10}c_{1K} + \rho_{20}c_{2K}}, \qquad (5.83)$$

the expressions for particle velocity and stress are somewhat simpler:

$$u = BV - Bu_{1e} + (1 - B)u_{2e} + \frac{\tilde{\sigma}_{1\text{HEL}} - \tilde{\sigma}_{2\text{HEL}}}{\rho_{10}c_{1K} + \rho_{20}c_{2K}}$$

$$= BV + \frac{1}{\rho_{10}c_{1K} + \rho_{20}c_{2K}}\left\{\tilde{\sigma}_{1\text{HEL}}\left(1 - \frac{c_{1K}}{c_{1L}}\right) - \tilde{\sigma}_{2\text{HEL}}\left(1 - \frac{c_{2K}}{c_{2L}}\right)\right\}, \qquad (5.84)$$

$$\tilde{\sigma} = B\rho_{20}c_{2K}(V - u_{1e} - u_{2e}) + (1 - B)\tilde{\sigma}_{1\text{HEL}} + B\tilde{\sigma}_{2\text{HEL}}$$

$$= B\rho_{20}c_{2K}V + (1 - B)\tilde{\sigma}_{1\text{HEL}}\left(1 - \frac{c_{1K}}{c_{1L}}\right) + B\tilde{\sigma}_{2\text{HEL}}\left(1 - \frac{c_{2K}}{c_{2L}}\right).$$

Recall that the HEL stresses are defined in terms of the longitudinal sound speed, not the bulk sound speed, which is why there is not a simpler form of the above equations. These formulas were used in Section 4.9 to analyze spall.

Figure 5.8 shows a typical arrangement for a flyer plate impact test. A flyer plate, held in a sabot, is accelerated in a gun barrel and strikes a target material. The target material is often backed by a well characterized material, which is sometimes transparent. On the back of either the target directly or the backing material the partical velocity is measured. Figure 5.9 shows the results of such a test, in this case the free surface velocity for an iron target where there was no backing. The initial HEL wave can be seen, followed by the elastic-plastic front, and that is followed by a phase transition. The attributes of these data will be described in subsequent sections, and a detailed analysis is left to the student in Exericse 5.26.

5.4. The Question of Path

In the derivation of the two-wave solution, an implicit assumption was made, one that is very important. We assumed that the stress in terms of strain was based on uniaxial strain loading of the elastic-plastic material (Eq. 5.63). However, in the derivation of the Hugoniot jump conditions, there was no discussion of the "path" connecting the initial density, stress, and energy state in front of the shock and the

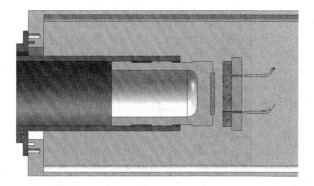

FIGURE 5.8. Schematic of a flyer plate, held in a sabot that is moving left to right within the gun barrel, just about to strike a target plate with backing and timing pins, all within a vacuum chamber.

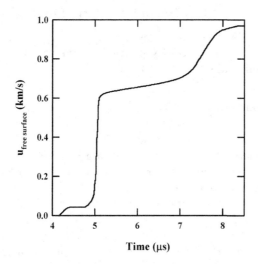

FIGURE 5.9. Free surface velocity from a flyer plate impact of Armco iron into itself at 977 m/s. (Redrawn from [148].)

final density, stress, and energy state behind the shock. In fact, we mentioned in the previous chapter that mathematically there is no stress-strain "path" connecting the initial material state in the front of the shock and the final material state behind the shock for the elastic material. The jump is discontinuous. It would be fine to adopt this attitude if we were only dealing with ideal gases or fluids or elastic solids. Gases and most fluids do not have a "memory" of their deformation, and thus their current pressure state depends on state-dependent variables that do not contain history information. The stress state of elastic solids is uniquely determined by their current strain state, so again no history information is in the stress-strain relation.

However, now we are dealing with elastic-plastic solids. Once solids are deformed beyond their elastic state, they deform plastically and the stress state is dependent on the loading path. The stress is not uniquely defined by state variables such as density and energy, but also depends on history-dependent variables such as plastic strain, as discussed in Chapter 3. Thus, we recognize that for a shock front, in reality there is a physical path that the material takes. The path is likely not an instantaneous jump, and material viscosity (strain-rate terms) may spread out the shock front, though this behavior may be on the molecular scale. The Hugoniot jump conditions still connect the beginning and end states of the density, stress, and energy, even though the wave front may be spread out (Theorem 2.6). For elastic-plastic materials, we will use the stress-strain response that is achieved by uniaxial strain of the material with no unloading to uniquely define the relationship between stress and strain in determining the Hugoniot. We assume (for large enough shocks) that the final stress state is on the yield surface; this assumption implies the stress deviator in compression is $-2Y/3$ after the passage of the shock.

5.5. The Hugoniot Elastic Limit

Though the HEL is often viewed as a material property, the reality is much more complex. For an elastic-perfectly plastic material the state of the material is the initial yield in uniaxial strain (Eq. 3.65),

$$\tilde{\varepsilon}_{\mathrm{HEL}} = \frac{Y}{2\mu}, \quad \tilde{\sigma}_{\mathrm{HEL}} = \left(1 + \frac{\lambda}{2\mu}\right) Y = \frac{1-\nu}{1-2\nu} Y. \tag{5.85}$$

However, most materials are not elastic-perfectly plastic and the yield Y is not constant but is a function of plastic strain, strain rate, and temperature. In flyer plate impact experiments, typically strain rates are on the order of 1×10^6 s^{-1}. A work hardening response, when added to the bulk pressure dependence, produces a curved uniaxial stress-strain curve that can lead to rounded leading edges of the elastic precursor leading to compressive fan behavior, which makes it hard to determine the HEL. The effect of strain rate is hard to determine; strain rate requires knowing something about the loading path across the precursor front and the corresponding strain rates. Strain-rate leads to transient behavior and prevents solving with the Hugoniot jump conditions; the situation is like a viscosity as discussed in the previous chapter and again below, where the viscosity leads to a finite rise time, not a jump. In most numerical techniques, the strain rate is decided by an artificial viscosity, which is an artificial numerical scheme to limit the rise time of fronts and hence has significant influence on the calculated strain rate.

In an almost opposite fashion, many elastic precursors show an initial high HEL stress and then the HEL stress drops off as the wave travels into the material until it reaches a steady value. This reduction in amplitude is referred to as elastic precursor decay. Because of this decay, different sample thicknesses in flyer plate impact tests lead to different measured $\tilde{\sigma}_{\mathrm{HEL}}$: thinner samples produce larger values for the $\tilde{\sigma}_{\mathrm{HEL}}$. These behaviors of the HEL make it difficult to treat the HEL state as a material property for all materials. For some materials, the HEL is well defined and the elastic precursor is a steady wave. In these cases it can be viewed as a material property, providing a dynamic yield stress.

5.6. The Hugoniot and Rarefaction of Real Materials

In Chapter 4 the Hugoniot was presented. At that time the focus was on mechanical stress-strain relations and so only the mass and momentum jump conditions were needed to define the Hugoniot. However, now we are considering real materials, which are more complicated in that they have thermal terms and elastic-plastic effects.

It is straightforward to extend the results of Sections 4.2 and 4.3 to real materials. First, for the computation of the Hugoniot locus we must now include the energy jump condition, but that is the only difference:

DEFINITION 5.2. (Hugoniot) For a given material response $\tilde{\sigma}(\rho, E)$, beginning at an initial ambient state (ρ_0, E_0), the locus of solutions of the Hugoniot jump conditions

$$\frac{\rho_0}{\rho} = 1 - \frac{u}{U},$$

$$\tilde{\sigma} - \tilde{\sigma}_0 = \rho_0 U u, \tag{5.86}$$

$$E - E_0 = \frac{1}{2}(\tilde{\sigma} + \tilde{\sigma}_0)\left(\frac{1}{\rho_0} - \frac{1}{\rho}\right),$$

is referred to as the *Hugoniot*, the *Hugoniot curve*, or the *Hugoniot locus*. The particle velocity u and the shock velocity U are determined as part of the solution.

The solution can be represented $E = E_H(\rho)$, which gives rise to $\tilde{\sigma}_H(\rho) = \tilde{\sigma}(\rho, E_H(\rho))$. All the definitions (such as superlinear) and theorems (such as compressive shocks) in Chapter 4 now apply to this material with $\tilde{\sigma} = \tilde{\sigma}_H$. The added complications due to the two-wave structure mean that the shock results are applied in each region of response. Given an equation of state for the material, this theorem says that only the energy jump condition combined with the equation of state is required to determine $E_H(\rho)$. This remark assumes something we are not going to discuss: that the material behind the shock front is in thermodynamic equilibrium (for a discussion in solids, see Graham [85]; in diatomic mixed gases, there is a finite distance and time behind the shock to reach chemical species equilibrium [32]).

However, now arises a fact about the Hugoniot that was not true for mechanical stress-strain curves considered in the $(1/\rho, \tilde{\sigma})$ plane:

THEOREM 5.3. *(Hugoniots are for one shock only) For a material with thermal terms, the Hugoniot curve applies only to the initial shock from ambient. It cannot be used (except as an approximation) for a second shock or rarefaction from a nonambient point on the Hugoniot curve.*

For the second shock, a whole new Hugoniot curve needs to be constructed. A consideration of the energy jump condition for a strictly superlinear material explains why this is so. If we think of the energy deposited by the shock in terms of the area under the Rayleigh line, then it is clear that one shock has a different energy deposited in the material than two shocks that get the material to the same density and stress. For a mechanical stress-strain relation where the energy does not affect the stress, the Hugoniot curve could then be used to determine a second shock from the intermediate value. But for a material with thermal terms, recalling that the Hugoniot is just a projection of a curve onto the $(1/\rho, \tilde{\sigma})$ plane, it is a different curve that is being projected down onto the plane for a shock starting at the material state formed by the passage of the first shock. It also cannot be used to determine a rarefaction from a nonambient state since a sound wave is a very small amplitude shock, and hence is a second shock from the nonambient point.

Rarefactions are straightforward:

DEFINITION 5.4. (release isentrope) For a given material response $\tilde{\sigma}(\rho, E)$, beginning at an initial state (ρ_0, E_0), the solution of the differential jump conditions

$$\frac{d\rho}{\rho} = \frac{du}{c},$$

$$d\tilde{\sigma} = \rho c\, du, \tag{5.87}$$

$$dE = \tilde{\sigma}\frac{d\rho}{\rho^2} = -\tilde{\sigma}d(1/\rho),$$

gives the *release isentrope*. The material speed $u(\rho)$ and the sound speed $c(\rho, E_s(\rho))$ are determined as part of the solution.

These equations assume that rarefaction occurs quickly enough that thermal conduction can be ignored. Again, the release can be viewed as $E = E_s(\rho)$ (s is an isentrope), which gives rise to a $\tilde{\sigma}(\rho) = \tilde{\sigma}(\rho, E_s(\rho))$ that can be used in the derivations of Section 4.3. (Notice that only the differential energy jump condition combined with the equation of state is required to determine $E_s(\rho)$.) One of the specific results to point out is that the differential jump conditions give rise to the sound speed for an arbitrary state of the material:

DEFINITION 5.5. (adiabatic sound speed) The speed at which very small amplitude waves travel in a material at rest with response $\tilde{\sigma} = \tilde{\sigma}(\rho, E)$ is the *adiabatic sound speed*, found by solving the differential jump conditions. It is

$$c(\rho, E) = \sqrt{\frac{d\tilde{\sigma}}{d\rho}} \quad \text{with} \quad dE = \tilde{\sigma}\frac{d\rho}{\rho^2} = -\tilde{\sigma}d(1/\rho)$$

$$= \sqrt{\left(\frac{\partial\tilde{\sigma}}{\partial\rho}\right)_E + \left(\frac{\partial\tilde{\sigma}}{\partial E}\right)_\rho \left(\frac{dE}{d\rho}\right)_{\text{adiabatic}}}$$

$$= \sqrt{\left(\frac{\partial\tilde{\sigma}}{\partial\rho}\right)_E + \frac{\tilde{\sigma}}{\rho^2}\left(\frac{\partial\tilde{\sigma}}{\partial E}\right)_\rho} \tag{5.88}$$

$$= \sqrt{\left(\frac{\partial\tilde{\sigma}}{\partial\rho}\right)_E + \left(\frac{\partial\tilde{\sigma}}{\partial E}\right)_\rho \left(\frac{dE}{d\rho}\right)_s}$$

$$= \sqrt{\left(\frac{\partial\tilde{\sigma}}{\partial\rho}\right)_s}.$$

The first two lines assume that the differential jump conditions hold to relate energy and density and are the local sound speed introduced in Chapter 4. The last line is often stated as the definition of sound speed. The material velocity comes from the Riemann integral (Eq. 4.91):

DEFINITION 5.6. (Riemann integral) The Riemann integral determines the material velocity, assuming the material in front of the rarefaction is initially at rest and at density ρ_r:

$$u(\rho) = \int_{\rho_r}^{\rho} \frac{c}{\rho}d\rho = \int_{\tilde{\sigma}_r}^{\tilde{\sigma}} \frac{d\tilde{\sigma}}{\rho c} \leq 0, \tag{5.89}$$

where $c = c(\rho, E_s(\rho))$ and $\tilde{\sigma}(\rho, E_s(\rho))$ are along the isentrope.

Immediately upon unloading the internal energy changes and so the release isentrope, for a material with thermal response, is no longer along the Hugoniot. However, recall Theorem 4.13, which says that the Hugoniot and release isentrope are the same to third order in specific volume. Computationally, the difference between the Hugoniot and the release isentrope is that the Hugoniot involves the solution of algebraic equations for the finite jump while the release isentrope involves the solution of differential equations for the infinitesimal jump.

During the 1950s and 1960s many experiments were performed on materials to determine the Hugoniots. A relatively stunning result was obtained for metals, namely that the Hugoniot is nearly always described by a linear relation between the shock velocity and the particle velocity,

$$U = c_0 + ku. \tag{5.90}$$

Figure 5.10 and Tables 5.2 and 5.3 contain experimental results for 6061 aluminum and 304 stainless steel. These results could be obtained from like-on-like material flyer plate impact experiments, namely obtaining the velocity of the arriving shock wave U by measuring the arrival time of the shock at the back surface of the target and its associated particle velocity u by measuring the impact velocity of the flyer plate and using the symmetry to obtain $u = v/2$. However, though some Hugoniot data have been obtained from flyer plate impacts, most have not. Most often the loading comes from explosives. In the experiments, the shock velocity is directly measured from arrival times on the back surface of the test specimen. Determining the particle velocity behind the shock is more complicated. What is typically measured is the free surface velocity, which is the sum of the particle velocity plus the Riemann integral particle velocity for the rarefaction fan. Typically the particle velocity is relatively close to half the free surface velocity, especially for lower pressures (recall that the isentrope is close to the Hugoniot). However, the Riemann integral calculation requires equation-of-state information such as a Mie–Grüneisen parameter. Thus, the equation of state is part of what is determined by the experiments, not just the shock Hugoniot. For a set of standard reference materials, the equations of state were carefully determined (see a list in Appendix C). Given accurate equations of state for reference materials, an impedance matching technique can then be used where the shock arrival times through the reference material alone and through a specimen comprised of the reference backed by the material of interest provide enough information to determine the Hugoniot locus (Exercise 5.54). The data in the tables were obtained through the impedance matching technique, using explosives as the driver.

The figures plot the shock velocity in terms of the particle velocity behind the shock front, and it can be seen that the relationship is nearly linear. The linear result is often referred to as the "linear U_s–u_p" for "shock velocity–particle velocity." There is not a Hugoniot data point for $u = 0$; the value of c_0 is the value obtained by the linear least-squares fit to the data. The term $\rho_0 c_0^2$ very nearly gives the bulk modulus K, but material strength influences the intercept location (because they are nearly the same we label both of them c_0, even though they are different – Exercise 5.12). If the material has no strength, $c_0 = c_K$ exactly. Once again we mention that there is no elastic wave that travels at the speed c_0. At low pressures there are only longitudinal and shear elastic waves. As the impact velocity increases

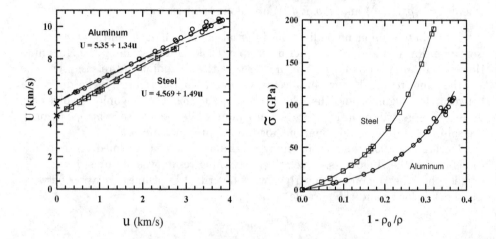

FIGURE 5.10. Hugoniot data in two forms (data from [**139**]): shock velocity–particle velocity (u, U) and pressure–strain $(\tilde{\varepsilon}, \tilde{\sigma})$ for 6061 aluminum and 304 stainless steel. The solid lines on the plots are the values from the $U = c_0 + ku$ fit to the data and the dashed lines on the velocity plot are the sound speeds c_H along the Hugoniot, computed as described in the text.

leading to higher stress, the two-wave structure develops – the longitudinal elastic precursor wave followed by the elastic-plastic shock. The data are for the arrival of the elastic-plastic shock. At higher velocities the elastic-plastic shock overtakes the elastic precursor and again there is only one wave. The particle velocity at which this occurs can be found by setting $U = c_L$ in Eq. (5.90),

$$u = \frac{c_L - c_0}{k}. \tag{5.91}$$

For the stainless steel we get a particle velocity of 865 m/s and for 6061-T6 aluminum we get a particle velocity of 725 m/s. Thus, for like-into-like plate impacts of twice these speeds, we expect that only one wave will be produced. Also shown in Fig. 5.10 is the stress behind the shock front as a function of the strain $1 - \rho_0/\rho$. It can be seen how steep the stress-strain response gets, though recall that the Hugoniot also includes thermal terms.

Tables 5.2 and 5.3 list two columns of experimental data, u and U, and the rest of the columns are computed values. The third column has material density behind the shock front computed from the conservation of mass jump condition,

$$\rho = \frac{\rho_0}{1 - u/U}. \tag{5.92}$$

The fourth column contains the compressive stress or pressure behind the shock front, as computed from the conservation of momentum jump condition,

$$\tilde{\sigma} = \rho_0 U u. \tag{5.93}$$

The fifth column contains engineering strain,

$$\tilde{\varepsilon} = 1 - \frac{\rho_0}{\rho} = \frac{u}{U}. \tag{5.94}$$

TABLE 5.2. Hugoniot data for 6061 aluminum, with $\rho_0 = 2.703 \text{ g/cm}^3$. Least-squares fit to the data is $U = 5.35 \text{ km/s} + 1.34u$ [139]. Sound speed and adiabatic bulk modulus use $\Gamma_0 = 2$. Approximate temperatures use $T_0 = 298 \text{ K}$ and $C_v = 0.89 \text{ J/g K}$. (During unloading, the aluminum can melt: pure aluminum's melt temperature at ambient is 934 K. Upon reaching this temperature, T_u remains at 934 K for the equivalent of $H_F/C_v = 445 \text{ K}$ and then begins to rise again. See Section 5.15.)

u km/s	U km/s	ρ g/cm^3	$\tilde{\sigma}_H$ GPa	$\tilde{\varepsilon}$	$\bar{K}/K_0$	K_s/K_0	c_H km/s	T_H K	T_u K
0.000	5.266[a]	2.703	0.00	0.000	0.969	1.000	5.350	298	298
0.442	5.928	2.921	7.08	0.075	1.228	1.377	6.039	353	304
0.497	5.975	2.948	8.03	0.083	1.247	1.428	6.122	363	307
0.657	6.176	3.025	10.97	0.106	1.333	1.577	6.350	393	318
0.999	6.652	3.181	17.96	0.150	1.546	1.902	6.802	483	358
1.198	6.956	3.265	22.53	0.172	1.690	2.094	7.043	552	391
1.741	7.655	3.499	36.02	0.227	2.047	2.681	7.700	850	539
1.925	7.963	3.565	41.43	0.242	2.215	2.864	7.885	969	597
1.935	7.970	3.570	41.69	0.243	2.219	2.878	7.898	978	602
2.288	8.431	3.710	52.14	0.271	2.483	3.298	8.293	1,298	754
2.473	8.663	3.783	57.91	0.285	2.622	3.533	8.501	1,504	850
2.752	9.146	3.866	68.03	0.301	2.923	3.817	8.740	1,776	973
2.780	9.069	3.898	68.15	0.307	2.874	3.928	8.830	1,890	1,024
2.886	9.289	3.921	72.46	0.311	3.015	4.013	8.898	1,979	1,063
3.120	9.529	4.019	80.36	0.327	3.172	4.380	9.183	2,390	1,242
3.151	9.830	3.978	83.72	0.321	3.376	4.224	9.064	2,210	1,164
3.376	9.649	4.158	88.05	0.350	3.253	4.949	9.596	3,100	1,540
3.435	9.969	4.124	92.56	0.345	3.472	4.806	9.495	2,913	1,462
3.449	9.843	4.161	91.76	0.350	3.385	4.963	9.606	3,119	1,548
3.470	10.269	4.083	96.32	0.338	3.684	4.634	9.371	2,696	1,372
3.510	10.014	4.162	95.01	0.351	3.504	4.966	9.609	3,123	1,549
3.585	10.047	4.203	97.36	0.357	3.527	5.145	9.732	3,364	1,648
3.758	10.301	4.255	104.64	0.365	3.707	5.385	9.894	3,700	1,784
3.758	10.448	4.221	106.13	0.360	3.814	5.229	9.790	3,480	1,695
3.789	10.325	4.270	105.75	0.367	3.725	5.452	9.939	3,796	1,822
3.844	10.377	4.293	107.82	0.370	3.762	5.563	10.012	3,957	1,886

[a]measured from ultrasound, $c_L = 6.40 \text{ km/s}$, $c_s = 3.15 \text{ km/s}$

The sixth and seventh columns give an indication of the increase of the bulk modulus pressure increases – essentially answering the question of how nonlinear is the material response. The sixth column has the average bulk modulus, obtained by dividing the stress on the Hugoniot by the strain,

$$\bar{K} = \frac{\tilde{\sigma}}{\tilde{\varepsilon}}. \tag{5.95}$$

This value is then normalized by the bulk modulus computed by the linear fit to the data, $K_0 = \rho_0 c_0^2$. If there were no nonlinear response, then the value in this column would always be 1. The first entry is not 1 because the $u = 0$ intersection

TABLE 5.3. Hugoniot data for 304 stainless steel, with initial density $\rho_0 = 7.890$ g/cm^3. Least-squares fit to the data is $U = 4.569$ km/s $+ 1.49u$ [139]. Sound speed and adiabatic bulk modulus use $\Gamma_0 = 2.2$. Approximate temperatures use $T_0 = 298$ K and $C_v = 0.45$ J/g K. (For the last line, upon unloading the steel will partially melt: pure iron's melt temperature (hence T_u) at ambient is 1,811 K. See Section 5.15.)

u km/s	U km/s	ρ g/cm^3	$\tilde{\sigma}_H$ GPa	$\tilde{\varepsilon}$	$\bar{K}/K_0$	K_s/K_0	c_H km/s	T_H K	T_u K
0.000	4.507[a]	7.890	0.00	0.000	0.973	1.000	4.569	298	298
0.232	4.925	8.280	9.02	0.047	1.162	1.260	5.006	333	301
0.339	5.056	8.457	13.52	0.067	1.225	1.388	5.200	354	306
0.529	5.355	8.755	22.35	0.099	1.374	1.620	5.521	402	324
0.659	5.577	8.947	29.00	0.118	1.490	1.782	5.727	445	343
0.745	5.651	9.088	33.22	0.132	1.530	1.907	5.878	484	362
0.889	5.891	9.292	41.32	0.151	1.662	2.097	6.097	555	398
0.969	6.011	9.406	45.96	0.161	1.731	2.209	6.219	603	423
1.010	6.080	9.462	48.45	0.166	1.771	2.264	6.279	629	436
1.057	6.152	9.527	51.31	0.172	1.813	2.331	6.349	661	453
1.383	6.639	9.966	72.44	0.208	2.111	2.820	6.827	948	600
1.409	6.732	9.978	74.84	0.209	2.171	2.835	6.840	958	605
1.409	6.734	9.978	74.86	0.209	2.172	2.834	6.840	958	604
1.653	7.007	10.326	91.39	0.236	2.352	3.275	7.228	1,294	770
1.915	7.460	10.615	112.72	0.257	2.666	3.682	7.559	1,663	945
2.334	8.054	11.109	148.32	0.290	3.107	4.477	8.147	2,530	1,337
2.711	8.600	11.522	183.95	0.315	3.543	5.253	8.665	3,539	1,769
2.772	8.667	11.600	189.56	0.320	3.598	5.412	8.766	3,764	1,863

[a]measured from ultrasound, $c_L = 5.77$ km/s, $c_s = 3.12$ km/s

does not equal c_0, but rather is from the measured elastic waves value (see below). This column is obtained directly from the data as

$$\frac{\bar{K}}{K_0} = \frac{1}{K_0}\frac{\tilde{\sigma}}{\tilde{\varepsilon}} = \frac{\rho_0 U^2}{K_0} = \left(\frac{U}{c_0}\right)^2. \tag{5.96}$$

The seventh column contains the "local" bulk modulus, estimating the local adiabatic stiffness (including thermal terms). This value will be calculated below, but the result is (see Section 5.13 and Exercise 5.21)

$$K = \frac{d\tilde{\sigma}}{d\ln(\rho/\rho_0)} = \rho\frac{d\tilde{\sigma}}{d\rho} = -\frac{1}{\rho}\frac{d\tilde{\sigma}}{d(1/\rho)}, \tag{5.97}$$

$$\frac{K_s}{K_0} = (1-\tilde{\varepsilon})\left\{\frac{1 + k\tilde{\varepsilon} - k\Gamma_0\tilde{\varepsilon}^2}{(1-k\tilde{\varepsilon})^3}\right\}. \tag{5.98}$$

It is seen that the materials rapidly stiffens – that is, the response is very nonlinear. The high pressure sound speed for density-energy states along the Hugoniot is shown in column eight (calculated in Section 5.13),

$$c_H = c(\rho, E_H(\rho)) = \frac{\rho_0 c_0}{\rho}\sqrt{\frac{1 + k\tilde{\varepsilon} - k\Gamma_0\tilde{\varepsilon}^2}{(1 - k\tilde{\varepsilon})^3}} = c_0(1 - \tilde{\varepsilon})\sqrt{\frac{1 + k\tilde{\varepsilon} - k\Gamma_0\tilde{\varepsilon}^2}{(1 - k\tilde{\varepsilon})^3}}. \quad (5.99)$$

The high pressure sound speed steadily increases. It can be seen that c_H is close to U, thus easily satisfying the requirement that $c_H > U - u$. A rarefaction wave would quickly catch a shock since in the laboratory frame the shock propagates at speed U and the leading edge of the rarefaction (sound) wave propagates at speed $c_H + u$. The last column contains an estimated temperature of the material behind the shock, computed below in Section 5.11. The result derived there is

$$T_H(\rho) = \frac{1}{C_v}\left[\frac{c_0^2\tilde{\varepsilon}^2}{2(1 - k\tilde{\varepsilon})^2} - \frac{c_0^2}{k^2}\left\{\frac{k\tilde{\varepsilon}}{1 - k\tilde{\varepsilon}} + \ln(1 - k\tilde{\varepsilon})\right\}\right] + T_0\exp(\Gamma_0\tilde{\varepsilon}) \quad (5.100)$$

and the initial temperature is taken as $T = 298\,\mathrm{K}$. The temperature value is computed from $\tilde{\varepsilon} = u/U$. Similarly, the approximate temperature upon unloading (this value is not for the Hugoniot, but is the residual temperature after the shock to density ρ has unloaded back to ambient stress) is given by

$$T_u = \exp(-\Gamma_0\tilde{\varepsilon})T_H(\rho). \quad (5.101)$$

(For very low velocities the temperatures will be slightly higher – on the order of $10\,\mathrm{K}$ for 5% strain in steel – than these computations indicate since this value is based on the U_s–u_p curve fit and not on the two-wave detailed solution, thus missing the details of the elastic-plastic response; see Exercise 3.7.)

The first row of each table has a data point that is determined from ultrasonic tests. The value represents the bulk modulus directly, without any influence of strength. For both materials, the longitudinal and shear wave speeds were measured, and then a bulk wave speed was computed using

$$U = c_K = \sqrt{\frac{K}{\rho_0}} = \sqrt{c_L^2 - \frac{4}{3}c_s^2}. \qu(5.102)$$

The value of u for this data line is arbitrarily set to 0. The other values are computed to show the difference between the elastic measurement of the bulk modulus and the c_0 intercept from the elastic-plastic flyer plate impact tests. The effect of strength produces a c_0 that is greater than c_K (Eq. 5.69 and Exercise 5.12).

Appendix C contains many Hugoniots for many materials. That a linear U_s–u_p relation holds to high impact velocities is a surprise; however, not all materials give a nice linear dependence, and a few listed in Appendix C require the inclusion of a quadratic term for good agreement,

$$U = c_0 + ku + qu^2. \quad (5.103)$$

For really large compressions, Nellis states that experimental data for particle velocities ranging from 4 to 42 km/s for metals Al, Cu, Fe, and Mo all have a linear U_s–u_p curve with values approaching $c_0 = 5.90\,\mathrm{km/s}$ and $k = 1.22$ (pressures above 1 TPa $= 1,000\,\mathrm{GPa}$ are produced by nuclear explosives). (A more recent estimate is $c_0 = 5.8\,\mathrm{km/s}$ and $k = 1.2$, up to $u = 45\,\mathrm{km/s}$.) Theoretical compression computations for these metals in the range of 20 to 100 TPa show a maximum compression on the Hugoniot of 4.6 to 5.7 times the initial density, and with $\rho_0/\rho = 1 - u/U \approx 1 - 1/k$ these compressions correspond to $1.21 < k < 1.28$. Thus, metals under large shock loading produce similar behavior regardless of the element

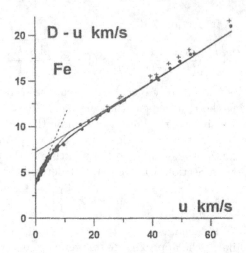

FIGURE 5.11. Shock pressure as a function of particle velocity for iron to very high pressures. The vertical axis is U_s–u_p, the difference between shock velocity and particle velocity (in the Russian literature, D is used for U). (From Kalitkin and Kuzmina [114].)

in question [150, 151]. In particular, Fig. 5.11 shows the U_s–u_p compression curve for iron from [114], which shows the transition from the lower pressure linear fit to the very high pressure linear fit that essentially matches all materials. Notice that $U_s \approx 80$ km/s for $u_p \approx 60$ km/s, so $\rho/\rho_0 \approx 1/(1 - 60/80)) = 4$, which is essentially the maximum compression that can be reasonably achieved by a single shock, with pressure $\tilde{\sigma}_H \approx 38$ TPa.

5.7. Flyer Plate Impact Test

We now have the tools to calculate the particle velocities and pressures from flyer plate impact tests to high velocities. We will first compute the single wave case. This case is the complete solution for velocities high enough that the shock velocity (relative to the material at rest) is greater than the longitudinal wave velocity (Eq. 5.91). We assume material 1 is coming from the left at velocity V and striking material 2, initially at rest. The initial conditions are

$$
\begin{aligned}
\rho_{1\ell} = \rho_{10} \quad & \rho_{1r} = ? \quad \rho_{2\ell} = ? \quad \rho_{2r} = \rho_{20} \\
\tilde{\sigma}_{1\ell} = 0 \quad & \tilde{\sigma}_{1r} = \tilde{\sigma}_{2\ell} = ? \quad \tilde{\sigma}_{2r} = 0 \\
u_{1\ell} = V \quad & u_{1r} = u_{2\ell} = ? \quad u_{2r} = 0
\end{aligned}
\tag{5.104}
$$

The particle velocity matching at the interface of the two materials gives $u = u_{1r} = u_{2\ell}$. The wave velocities come from the linear shock velocity–particle velocity relationship:

$$
U_1 = V - (c_{10} + k_1(V - u)), \qquad U_2 = c_{20} + k_2 u. \tag{5.105}
$$

With the Hugoniot stress in terms of particle velocity u,

$$
\tilde{\sigma}(u) = \rho_0(c_0 + ku)u, \tag{5.106}
$$

matching the stresses $\tilde{\sigma} = \tilde{\sigma}_{1r} = \tilde{\sigma}_{2\ell}$ with the momentum jump conditions $\tilde{\sigma} = \tilde{\sigma}_1(V - u) = \tilde{\sigma}_2(u)$ gives

$$\tilde{\sigma} = \rho_{10}(c_{10} + k_1(V - u))(V - u) = \rho_{20}(c_{20} + k_2 u)u. \qquad (5.107)$$

Notice the signs work out because the momentum jump condition for the left-moving wave is, explicitly,

$$\begin{aligned}
\tilde{\sigma}_{1r} &= \rho_{10}(U - V)(u - V) \\
&= \rho_{10}\{V - (c_{10} + k_1|V - u|) - V\}(u - V) \qquad (5.108) \\
&= \rho_{10}(c_{10} + k_1(V - u))(V - u).
\end{aligned}$$

Solving this requires a solution of the quadratic equation $-Au^2 + Bu - C = 0$,

$$u = \frac{2C}{B + \sqrt{B^2 - 4AC}} \quad \text{or} \quad u = \frac{1}{2A}\left(B - \sqrt{B^2 - 4AC}\right), \qquad (5.109)$$

$$A = k_1 - k_2\rho_{20}/\rho_{10}, \quad B = 2k_1 V + c_{10} + c_{20}\rho_{20}/\rho_{10}, \quad C = c_{10}V + k_1 V^2.$$

The sign leading the square root was chosen because the solution is expected to have material velocity less than V. Both expressions for u solve the quadratic; the first one has the advantage that it works when the materials are the same. Once the particle velocity is found, it can be inserted back into the momentum jump equation to obtain the stress. In a similar fashion, the new density for each material can be computed. The algebra simplifies when a velocity is applied to a half space ($\tilde{\sigma} = \rho_o c_o u + \rho_o k u^2$) or when two like materials impact ($\tilde{\sigma} = \rho_o c_o V/2 + \rho_o k V^2/4$). These equations show that higher stresses result from the linear ku term – notice that $k = 0$ corresponds to the linear stress-strain response.

The complete two-wave solution is simpler than might be expected. It is simpler because the data reduction in arriving at the shock velocity–particle velocity fit does not assume a two-wave solution. Thus, the above analysis always holds for the elastic-plastic shock, and the elastic precursor solution is computed separately, as in Section 5.3.

5.8. Elastic-Plastic Shock Rise Times

The rise time of the second wave has been studied. Swegle and Grady [178] measured rise times from shock profiles and empirically showed that the strain rate of the elastic-plastic shock wave is proportional to the fourth power of the stress rise across the jump. In this section we present some of their data with some analysis to provide a sense of the rise time and rise distance of the shock. In the experiments there were elastic precursors, but they will not be included in the analysis.

If it is assumed there is a rise time for a steady wave, the particle velocity can be written $u = u(x - Ut)$. Differentiation of this expression provides

$$\text{strain rate} = \frac{d\varepsilon}{dt} = \frac{\partial u}{\partial x} = -\frac{1}{U}\frac{\partial u}{\partial t}. \qquad (5.110)$$

Swegle and Grady examined a variety of flyer plate impact VISAR data and examined the elastic-plastic shocks. They took the largest change in velocity in the data over the times at which particle velocity was measured ($\partial u/\partial t$) as the stress rose from the elastic precursor stress level to the elastic-plastic shock, then divided this value by the shock speed to arrive at a strain rate (Fig. 5.12). Results for $d\tilde{\varepsilon}/dt$

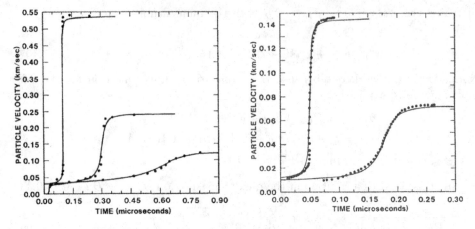

FIGURE 5.12. Shock profiles for aluminum (left) and copper (right) used to estimate rise times; points are data, lines are from model. (From Swegle and Grady [178].)

are shown in Table 5.4 for five metals. Subsequent columns use the Hugoniot jump conditions, starting with

$$\tilde{\sigma}_H = \rho_0(c_0 + ku)u \quad \Rightarrow \quad u = \frac{2\tilde{\sigma}_H/\rho_0}{c_0 + \sqrt{c_0^2 + 4k\tilde{\sigma}_H/\rho_0}}, \tag{5.111}$$

to back out estimates of rise times and rise distances for the shock front, with U_s–u_p relations that are roughly the same as those in Appendix C. Figure 5.12 shows some shock profiles from the original paper. As mentioned, the data shows the interesting property that

$$\dot{\tilde{\varepsilon}}_{\max} = \text{material constant} \times \tilde{\sigma}_{\max}^4 \tag{5.112}$$

(Exercise 5.15). The rise time rapidly decreases as the shock stress rises. There has been considerable subsequent work trying to understand either the materials or mechanics origin of this observation.

In Chapter 4, we used viscosity to produce wave solutions with finite rise time. We showed, with a viscosity $\eta|\dot{\tilde{\varepsilon}}|^p$, that $\dot{\tilde{\varepsilon}}_{\max} = \text{constant} \times \tilde{\sigma}_{\max}^{2/p}$ (Eq. 4.229). Based on that result, we will assume that the viscous stress, which we view to be deviatoric, scales as the square root of the strain rate ($\frac{2}{p} = 4 \Rightarrow p = \frac{1}{2}$), which we take to be the equivalent plastic strain rate: $s_{\text{viscous}} = \text{constant} \times (\dot{\varepsilon}_p)^{1/2}$. Alternatively, the relation between viscous stress and strain rate is $\dot{\varepsilon}_p = \text{constant} \times (s_{\text{viscous}})^2$.

Using the variant of the second invariant of the strain deviator, $\varepsilon_d = (\frac{2}{3}e_{ij}e_{ij})^{1/2}$, for linear elasticity, $\sigma_{\text{eff}} = 3\mu\varepsilon_d^e$, where μ is the elastic shear modulus. The total strain rate equals the elastic strain rate plus the plastic strain rate; hence,

$$\frac{d\sigma_{\text{eff}}}{dt} = 3\mu\dot{\varepsilon}_d^e = 3\mu(\dot{\varepsilon}_d - \dot{\varepsilon}_p) = 3\mu\frac{d\varepsilon_d}{dt} - 2\sqrt{2}\mu A'(\sigma_{\text{eff}} - Y)_+^2, \tag{5.113}$$

where the last term on the right is the empirical form of Swegle and Grady relating the viscous stress to the plastic strain rate, shown at the end of the previous paragraph. Stated values for A' for the materials in Table 5.4 are 100, 50, 300, 30, and 2 for Al, Be, Cu, Fe, and U, respectively, where the units are $(\text{MPa}^2\,\text{s})^{-1}$. If the deformation is elastic (i.e., $\sigma_{\text{eff}} < Y$), then the viscous term is zero and the effective

TABLE 5.4. Shock rise time for five metals (data to left of vertical divider from [178]).

	$\tilde{\sigma}_H$ (GPa)	$d\tilde{\varepsilon}/dt$ (1/s)	u (m/s)	U (km/s)	$\tilde{\varepsilon}$ (%)	Δt (ns)	Δx (μm)
Al	2.10	1.3×10^5	140	5.57	2.5	193	1,075
Al	3.60	1.10×10^6	234	5.69	4.1	37	213
Al	3.70	2.00×10^6	240	5.70	4.2	21	120
Be	10.5	2.90×10^6	650	8.73	7.4	26	224
Be	10.9	3.40×10^6	673	8.75	7.7	23	198
Be	11.4	3.20×10^6	701	8.79	8.0	25	219
Be	16.6	7.40×10^6	985	9.11	10.8	15	133
Be	17.0	1.01×10^7	1,010	9.13	11.0	11	100
Be	17.4	9.00×10^6	1,030	9.15	11.2	12	114
Be	17.5	1.16×10^7	1,030	9.16	11.3	10	89
Be	24.5	7.04×10^7	1,390	9.56	14.5	2	20
Be	25.1	5.32×10^7	1,410	9.59	14.8	3	27
Be	25.2	5.86×10^7	1,420	9.59	14.8	3	24
Be	25.2	4.45×10^7	1,420	9.59	14.8	3	32
Cu	2.60	3.4×10^5	72	4.04	1.8	52	212
Cu	5.60	7.30×10^6	151	4.16	3.6	5	21
Fe	10.3	1.57×10^6	263	4.98	5.3	34	168
Fe	13.2	4.00×10^6	332	5.07	6.5	16	83
U	7.52	2.7×10^5	144	2.77	5.2	193	533
U	9.57	7.3×10^5	180	2.82	6.4	87	246
U	11.6	1.56×10^6	214	2.87	7.4	48	137
U	12.6	1.86×10^6	230	2.90	8.0	43	124
U	14.3	4.14×10^6	258	2.94	8.8	21	62

stress increases as expected for linear elasticity. If $\sigma_{\text{eff}} > Y$, then there is a viscous relaxation back to the yield surface; the stresses are allowed to be beyond the yield surface and relax back to the yield surface as described in Exercise 3.17, but in this case with the rate controlled by Eq. (5.113). This viscoplastic material model is not the same as the nonlinear viscoelastic model of Eq. (4.223); for example, this model has a finite jump in strain and stress at the initial time, where the response is entirely elastic, whereas the solution to Eq. (4.223) initially has zero strain and infinite stress. The same method used to solve Eq. (4.223) can be used to find a steady wave solution for Eq. (5.113); however, the integral is more complicated, being a quartic in $\tilde{\varepsilon}$, thus precluding a nice solution. This behavior is not typically included in our modeling; rather, we use strain-rate dependent yield expressions.

5.9. Thermal Terms

In Chapter 3 there was a discussion of the effect of thermal expansion on the stress. This is an example of thermal, or energy, terms in the equation of state. For a linear thermal expansion, the equation relating stress, strain, and energy derived was (Eq. 3.131)

$$\sigma_{ij} = -K\varepsilon_{kk}\delta_{ij} + 2\mu e_{ij}^e - 3K\alpha\Delta T. \tag{5.114}$$

The first two terms of the right-hand side are the elastic stress vs. strain relation and the third term is the thermal term. Increasing the temperature while confining the material increases the compressive stress. Considering only elastic behavior, the relationship between stress, strain, and temperature in one dimension is

$$\tilde{\sigma}(\rho, T) = \rho_0 c^2 (1 - \rho_0/\rho) + 3K\alpha\Delta T. \tag{5.115}$$

If we assume the relationship between energy and temperature is given by $\Delta E = C_v \Delta T$, where C_v is the constant volume heat capacity (assumed constant), then this equation can also be written in terms of energy:

$$\tilde{\sigma}(\rho, E) = \rho_0 c^2 (1 - \rho_0/\rho) + 3K\alpha\frac{\Delta E}{C_v}. \tag{5.116}$$

For small strains (so that the stored elastic energy can be ignored), with $\Delta E = E - E_R(\rho) = E - E_0$, this equation can be viewed as an equation of state where we deviate from the reference curve $\tilde{\sigma}_R(\rho) = \rho_0 c^2 (1 - \rho_0/\rho)$ and $E_R(\rho) = E_0$.

An equation of state could be defined with respect to any reference curve in the (ρ, E) plane. If we let the subscript R refer to any reference curve, then by fixing density we can expand any smooth equation of state in terms of a Taylor series as

$$p(\rho, E) = p_R(\rho) + \left(\frac{\partial p}{\partial E}\right)_\rho (E - E_R(\rho)) + \frac{1}{2}\left(\frac{\partial^2 p}{\partial E^2}\right)_\rho (E - E_R(\rho))^2 + \cdots, \tag{5.117}$$

where $p_R(\rho) = p(\rho, E_R(\rho))$. We write p for pressure because that is the way it is usually written in equation of state theory. If only the linear term is included, and the higher order ones discarded, this equation of state is referred to as a Mie–Grüneisen equation of state.

DEFINITION 5.7. (Grüneisen parameter) The *Grüneisen parameter* Γ is defined as

$$\Gamma = \frac{1}{\rho}\left(\frac{\partial p}{\partial E}\right)_\rho, \tag{5.118}$$

where Γ is a function of ρ and E and where the subscript on the partial derivative means to keep density constant while taking the partial derivative with respect to internal energy.[4]

The Grüneisen parameter is a dimensionless number that relates the change in internal energy to the change in pressure for a confined material. The larger the value of Γ, the greater the pressure change for a given amount of internal energy change. With this definition, the Mie–Grüneisen equation of state is

$$p(\rho, E) = p_R(\rho) + \rho\Gamma(E - E_R(\rho)). \tag{5.119}$$

---

[4]This is the standard thermodynamics notation necessitated by the assumption that only two state variables are independent. This brings up the important point that historically in shock physics it is assumed that two independent variables suffice – that is, the stress and strain deviators can be neglected in the thermodynamic relations because of the high pressures involved. We will not do so here, but it is possible to develop thermodynamics based not on pressure p and specific volume $1/\rho$, but rather where all appropriately defined stress and strain components are variables and thus there are more than two independent variables (e.g., [34]).

The simple linear thermal expansion has a Grüneisen parameter

$$\Gamma = \frac{3K\alpha}{\rho C_v}. \tag{5.120}$$

For steel at ambient conditions, using the constants in Tables 5.2 and 5.3 and $\alpha = 12.6 \times 10^{-6}/°C$, this equation gives $\Gamma = 1.8$ and for aluminum with $\alpha = 23.4 \times 10^{-6}/°C$ gives $\Gamma = 2.1$. Along a specified reference curve, the Grüneisen parameter depends upon density, and a widely used relationship for the density dependence of Γ within the shocks-in-solids community is

$$\rho\Gamma = \rho_0\Gamma_0. \tag{5.121}$$

With this assumption, the dependence of the equation of state on the energy is strictly linear since the coefficient is a constant. With the Grüneisen parameter, the relation between stress, density, and energy can be written

$$\tilde{\sigma}(\rho, E) = \rho_0 c^2(1 - \rho_0/\rho) + \rho_0\Gamma_0(E - E_0), \tag{5.122}$$

where E_0 is a baseline energy – for example the energy at room temperature. Given this equation, we are now ready to analyze a wave that requires all three jump conditions to solve for it.

5.10. Mie–Grüneisen Equation of State

Can the Hugoniot define an equation of state for a general state of density and energy? The Hugoniot is just a curve through the equation of state surface, but if assumptions about the thermal behavior of the material are made, the Hugoniot can supply the information for an equation of state. We show how a Mie–Grüneisen equation of state can be constructed.

The first step is to write the Hugoniot in terms of the density. Given the linear U_s–u_p relation, compute the relationship between density and pressure along the Hugoniot by inserting the U_s–u_p relation into the mass jump condition,

$$\frac{\rho_0}{\rho} = 1 - \frac{u}{c_0 + ku} \quad \Rightarrow \quad u = \frac{c_0(1 - \rho_0/\rho)}{1 + k(\rho_0/\rho - 1)} = \frac{c_0\tilde{\varepsilon}}{1 - k\tilde{\varepsilon}}, \tag{5.123}$$

where $\tilde{\varepsilon} = 1 - \rho_0/\rho$. Putting this result in the momentum jump condition allows explicit calculation of the pressure along the Hugoniot in terms of the density,

$$\tilde{\sigma}_H(\rho) = \rho_0(c_0 + ku)u = \frac{\rho_0 c_0^2 \tilde{\varepsilon}}{(1 - k\tilde{\varepsilon})^2} = K_0\tilde{\varepsilon} + 2kK_0\tilde{\varepsilon}^2 + \dots . \tag{5.124}$$

We see that k quantifies the nonlinear stress-strain response along the Hugoniot with $2kK_0\tilde{\varepsilon}^2$ being the first nonlinear term; below we will see that $(2k - \frac{1}{2}\Gamma)K_0\tilde{\varepsilon}^2$ is the first nonlinear term for isothermal nonlinear response (Eq. 5.127). Often $p_H(\rho)$ is written instead of $\tilde{\sigma}_H(\rho)$ (since equations of state are pressures), but it should be kept in mind that this is an approximation since the data fit contains strength. Assuming the stress in front of the shock is zero, the conservation of energy jump condition is

$$E_H(\rho) - E_0 = \frac{1}{2}\tilde{\sigma}_H(\rho)\left(\frac{1}{\rho_0} - \frac{1}{\rho}\right), \tag{5.125}$$

which gives

$$E_H(\rho) = E_0 + \frac{\tilde{\varepsilon}}{2\rho_0}\tilde{\sigma}_H(\rho). \tag{5.126}$$

As defined, the Mie–Grüneisen form of the equation of state is

$$\tilde{\sigma}(\rho, E) = \tilde{\sigma}_H(\rho) + \rho\Gamma(E - E_H(\rho))$$

$$= \tilde{\sigma}_H(\rho)\left(1 - \frac{\rho\Gamma\tilde{\varepsilon}}{2\rho_0}\right) + \rho\Gamma(E - E_0)$$

$$= \frac{\rho_0 c_0^2 \tilde{\varepsilon}}{(1 - k\tilde{\varepsilon})^2}\left(1 - \frac{\rho\Gamma\tilde{\varepsilon}}{2\rho_0}\right) + \rho\Gamma(E - E_0) \tag{5.127}$$

$$\approx p(\rho, E).$$

The last line is the assumption made in shock physics work, that the strength effects in the data are negligible. As before,

$$\Gamma = \frac{1}{\rho}\left(\frac{\partial\tilde{\sigma}}{\partial E}\right)_\rho. \tag{5.128}$$

As was mentioned above, a widely used assumption about the density dependence of Γ within the shocks-in-solids community is

$$\rho\Gamma = \rho_0\Gamma_0. \tag{5.129}$$

With this assumption, the dependence of the equation of state on the energy is strictly linear. The equation of state is applied to any pressure and energy of the solid material, not just along the Hugoniot.

Johnson [108] described an elegant mechanical model showing the role of non-linear behavior in the Grüneisen equation of state. Consider a mass m that is held between two identical springs of unstressed length L_0. These springs are nonlinear, with the force $F(x) = ax + bx^2$, where x is the displacement of the spring and a and b are positive. The springs are attached to a wall on the right and left that can move an equal inward displacement Y. The mass can oscillate about the central equilibrium with a deflection $y(t)$. Thus, the equation of motion of the center mass is

$$m\ddot{y} = -F(Y + y) + F(Y - y) = -2ay - 4bYy. \tag{5.130}$$

This equation can be solved for $y(t)$,

$$y(t) = B\sin(\omega t), \quad \omega = \sqrt{2(a + 2bY)/m}. \tag{5.131}$$

The mass speed provides the kinetic energy of the oscillatory motion,

$$\dot{y}(t) = B\omega\cos(\omega t) \quad \Rightarrow \quad e_k(t) = \frac{1}{2}mB^2\omega^2\cos^2(\omega t). \tag{5.132}$$

As the mass oscillates back and forth, the average force applied to the wall is

$$\langle F\rangle = \frac{\omega}{2\pi}\int_0^{2\pi/\omega} F(Y + y)dt = \frac{\omega}{2\pi}\int_0^{2\pi/\omega}\{a(Y + y) + b(Y + y)^2\}dt$$

$$= aY + bY^2 + \frac{1}{2}bB^2 = aY + bY^2 + \frac{2b\langle e_k\rangle}{m\omega^2} = aY + bY^2 + \frac{2b}{\omega^2}\langle E_k\rangle, \tag{5.133}$$

where $\langle E_k\rangle = \langle e_k\rangle/m$ is the specific kinetic energy. We now transform this equation into something that looks similar to our equations of state. Assume the relevant wall cross-sectional area is A, so that $p = \langle F\rangle/A$. A unit cell will be of length $2L_0 - 2Y$ (i.e., both springs and the mass). The density is then $\rho = m/2A(L_0 - Y)$, which gives $Y = L_0(1 - \rho_0/\rho)$, with $\rho_0 = m/2AL_0$. Equation (5.133) becomes

$$p(\rho, E) = \frac{aL_0}{A}\left(1 - \frac{\rho_0}{\rho}\right) + \frac{bL_0^2}{A}\left(1 - \frac{\rho_0}{\rho}\right)^2 + \frac{2b\rho_0 L_0}{a + 2bL_0(1 - \rho_0/\rho)}E. \tag{5.134}$$

The first two terms represent the lattice compression of the material when there are no thermal (or oscillatory) motions; hence they are called the "cold compression curve," and the last term is the portion of the pressure due to thermal energy. The pressure is a linear function of energy. The coefficient leading E could be written $\rho\Gamma(\rho)$, yielding a complicated $\Gamma(\rho)$, but it still is in the form of the Mie–Grüneisen equation of state. If the springs are linear ($b = 0$) then there is no thermal term and $p(\rho)$ is our linear mechanical response. Because of the nonlinear term, when a shock passes through a material represented by this equation of state, energy is deposited (as given by the energy Hugoniot jump condition) and there will be a thermal term in the pressure, even if $E_0 = 0$; in other words, the nonlinear material undergoing a shock cannot stay on the cold compression curve. Determining the final state requires simultaneously solving all three Hugoniot jump conditions and the equation of state. This simple spring model provides a feel for the underlying lattice and thermal atomic motion that gives rise to the Mie–Grüneisen equation of state.[5]

We have yet to describe how Γ is obtained, except for a formula to calculate it at low temperature and pressure (Eq. 5.120). It turns out that for some materials (metals, in particular) it is possible to directly determine Γ by knowing the Hugoniot for the material with differing amounts of initial porosity (i.e., different ρ_0). A metal powder is taken and sintered. As such it will range from around 50% fully dense to fully dense. When it is compressed by a shock, the porosity is compacted out and the rest of the compression is assumed to be with the fully dense material. However, the energy that has been deposited in the material behind the shock is a strong function of the initial density. If the ambient stress is zero, the energy jump condition is (Eq. 2.63)

$$E - E_0 = \frac{1}{2}\tilde{\sigma}(\rho, E)\left(\frac{1}{\rho_0} - \frac{1}{\rho}\right). \tag{5.135}$$

Suppose the initial density of the porous metal were 50% of the fully dense ambient material, and further suppose that the stress is roughly the same behind a shock that, for both, increased the density to 25% higher than ambient fully dense. Then the ratio of the amount of energy deposited by the shock in the porous material to the fully dense metal is $(1/0.5 - 1/1.25)/(1/1 - 1/1.25) = 6$, so roughly six times as much energy is deposited. Here, we assumed the post-shock stress to be roughly the same, but it is not: because of thermal terms, the initially porous material will have a higher stress and so the amount of energy deposited is even larger. Though this large difference in deposited energy may not be intuitive, the reason is because the shock has no "path." It is a discontinuous jump: all that matters are the initial and final states. We assume the same base material equation of state is being accessed by the shock; it is possible to obtain many different post-shock energy states for the same post-shock density of the material. This allows a determination of the thermal term. For example, if the Mie–Grüneisen equation of state is assumed, it can be written as (Eq. 5.119)

$$\tilde{\sigma}(\rho, E) = f(\rho) + \rho_0\Gamma_0(E - E_0). \tag{5.136}$$

---

[5]Actually, there is even more here. In statistical mechanics, the Grüneisen parameter can be shown in certain situations to be (where ω is related to lattice vibrations) $\Gamma = -d\ln(\omega)/d\ln(1/\rho)$, which in this case equals $b(L_0 - Y)/(a + 2bY) = b/A\rho\omega^2$, so $\rho\Gamma = b/A\omega^2$, differing by a factor of 2 from above. Johnson avoids the 2 by using the maximum kinetic energy rather than the average.

Given the same final density ρ after the passage of a shock in both an initially porous sample (1) and an initially fully dense sample (2), with their respective ΔEs from the shock, the Grüneisen parameter is

$$\Gamma_0 = \frac{1}{\rho_0} \frac{\tilde{\sigma}(\rho, E_0 + \Delta E_1) - \tilde{\sigma}(\rho, E_0 + \Delta E_2)}{\Delta E_1 - \Delta E_2}. \tag{5.137}$$

An average Γ_0 is then obtained based on the available data. This procedure has been applied to a number of common metals (see [143] and Exercise 5.43).

Another approach to the Grüneisen parameter is to obtain an isothermal compression curve to compare to the Hugoniot. Historically the pressures achieved were not high enough to provide a high-pressure Grüneisen parameter. Bridgman, for example, collected isothermal compression data on many materials, up to around 10 GPa [30]. Today, much higher pressures can be achieved in the diamond anvil cell (DAC) technique, where the material of interest is surrounded by a soft material in a cell and is compressed between diamond anvils. Strains are measured with X-ray diffraction. It is a challenge to determine the pressure (there is a good pressure standard up to 200 GPa), but if an isotherm can be determined, denoted $p_T(\rho)$, then Eq. (5.137) can be invoked to give

$$\rho\Gamma = \rho_0\Gamma_0 = \frac{\tilde{\sigma}(\rho, E_H(\rho)) - \tilde{\sigma}(\rho, E_0)}{E_H(\rho) - E_0} = \frac{2\rho_0}{\tilde{\varepsilon}}\left(1 - \frac{p_T(\rho)}{\tilde{\sigma}_H(\rho)}\right). \tag{5.138}$$

An average over some large compressions can be used to determine an average Γ_0.

Some of the values listed in Appendix C are found using shock experiments with the initial porosity technique. Most of the values, however, are found by shock experiments providing the density, pressure, and energy behind the shock – theoretical computations providing the pressure that would be achieved by a cold compression of the material to that density, and then the difference in pressure and internal energy allowing the use of Eq. (5.137) to determine Γ.

It is possible to write a differential equation for the Hugoniot, though the equations we typically use are the jump conditions. To write such an equation, we assume the equation of state for the material is known and write $\tilde{\sigma} = \tilde{\sigma}(\rho, E)$. The derivative with respect to density along the Hugoniot, indicated by subscript H, is

$$\left(\frac{\partial\tilde{\sigma}}{\partial\rho}\right)_H = \left(\frac{\partial\tilde{\sigma}}{\partial\rho}\right)_E + \left(\frac{\partial\tilde{\sigma}}{\partial E}\right)_\rho\left(\frac{\partial E}{\partial\rho}\right)_H. \tag{5.139}$$

Differentiating the energy jump condition with respect to density yields

$$\left(\frac{\partial E}{\partial\rho}\right)_H = \frac{1}{2}\left(\frac{1}{\rho_0} - \frac{1}{\rho}\right)\left(\frac{\partial\tilde{\sigma}}{\partial\rho}\right)_H + \frac{\tilde{\sigma} + \tilde{\sigma}_0}{2\rho^2}. \tag{5.140}$$

Combining these expressions produces a differential equation for the energy along the Hugoniot, where all other terms are known when the equation of state is known,

$$\left\{1 - \frac{1}{2}\left(\frac{1}{\rho_0} - \frac{1}{\rho}\right)\left(\frac{\partial\tilde{\sigma}}{\partial E}\right)_\rho\right\}\left(\frac{\partial E}{\partial\rho}\right)_H = \frac{1}{2}\left(\frac{1}{\rho_0} - \frac{1}{\rho}\right)\left(\frac{\partial\tilde{\sigma}}{\partial\rho}\right)_E + \frac{\tilde{\sigma} + \tilde{\sigma}_0}{2\rho^2}. \tag{5.141}$$

Replacing the derivatives of the equation of state using the constant-energy bulk modulus and Grüneisen parameter yields

$$\left\{1 - \left(\frac{1}{\rho_0} - \frac{1}{\rho}\right)\frac{\rho\Gamma}{2}\right\}\left(\frac{\partial E}{\partial\rho}\right)_H = \frac{1}{2\rho^2}\left\{\left(\frac{\rho}{\rho_0} - 1\right)K_E + \tilde{\sigma} + \tilde{\sigma}_0\right\}. \tag{5.142}$$

Solving this differential equation gives the energy along the Hugoniot, $E = E_H(\rho)$, which then allows the stress along the Hugoniot to be explicitly computed. The initial material state is built into the differential equation, further emphasizing that a Hugoniot curve is defined for one initial state, and that a second Hugoniot originating from another point on the first Hugoniot produces a second, different Hugoniot curve.

5.11. Temperature

The internal energy for a solid is comprised of potential energy terms, due to the closeness of the atoms in the lattice, and thermal terms, due to the kinetic motion of the nuclei and electrons. In this sense, solids are much more complicated than an ideal gas where all the internal energy is assumed to be due to thermal (kinetic energy of molecular motion) terms. To understand the separation, consider compressing a steel spring and then placing it in a freezer. Even cold, the spring will still spring back when released, since the energy stored in the spring is due to the steel's atomic lattice deformation and is not due to thermal energy. In this section, we will find the approximate temperature on the Hugoniot and estimate how much of the internal energy on the Hugoniot is due to potential energy and how much is due to thermal energy.

To begin with, we assume that the pressure can be written in Mie–Grüneisen form

$$p(\rho, E) = \tilde{\sigma}_R(\rho) + \rho\Gamma(E - E_R(\rho)), \tag{5.143}$$

where the subscript R refers to an arbitrary reference curve. Further, we will assume that the internal energy can be written as a function of density and temperature in a similar fashion along the reference curve:

$$E(\rho, T) = E_R(\rho) + C_v(T - T_R(\rho)). \tag{5.144}$$

The first term on the right relates to the potential energy and the second term to the thermal (kinetic) energy. These equations imply that the pressure can be written as a function of density and energy as

$$p(\rho, T) = \tilde{\sigma}_R(\rho) + \rho\Gamma C_v(T - T_R(\rho)). \tag{5.145}$$

Directly computing $T_R(\rho)$ along the Hugoniot (i.e., $T_H(\rho)$) is difficult, so we will use an isentrope as our reference curve.

It is possible to explicitly compute the temperature on the isentrope. Assuming that the entropy s can be written as a function of specific volume $1/\rho$ and temperature T, we have

$$ds = \left(\frac{\partial s}{\partial(1/\rho)}\right)_T d(1/\rho) + \left(\frac{\partial s}{\partial T}\right)_{(1/\rho)} dT. \tag{5.146}$$

We now use one of Maxwell's relations (left) as well as one of the entropy relations (right)

$$\left(\frac{\partial s}{\partial(1/\rho)}\right)_T = \left(\frac{\partial p}{\partial T}\right)_{(1/\rho)}, \qquad \left(\frac{\partial s}{\partial T}\right)_{(1/\rho)} = \frac{C_v}{T} \tag{5.147}$$

to give

$$ds = \left(\frac{\partial p}{\partial T}\right)_{(1/\rho)} d(1/\rho) + \frac{C_v}{T} dT. \tag{5.148}$$

On an isentrope, $ds = 0$, and this equation gives

$$\left(\frac{dT}{d(1/\rho)}\right)_s = -\frac{T}{C_v}\left(\frac{\partial p}{\partial T}\right)_{(1/\rho)}. \tag{5.149}$$

Using $\rho\Gamma = \rho_0\Gamma_0$, Eq. (5.145) gives

$$\left(\frac{\partial p}{\partial T}\right)_{(1/\rho)} = \rho_0\Gamma_0 C_v, \tag{5.150}$$

which gives

$$\left(\frac{dT}{d(1/\rho)}\right)_s = -\rho_0\Gamma_0 T. \tag{5.151}$$

Assuming the initial density is ρ_i and the initial temperature is T_i, this integrates to give the temperature along the isentrope,

$$T_s(\rho) = T_i \exp\{\Gamma_0(\rho_0/\rho_i - \rho_0/\rho)\} = T_i \exp\{\Gamma_0(\tilde{\varepsilon} - \tilde{\varepsilon}_i)\}. \tag{5.152}$$

We emphasize that this equation assumes $\rho\Gamma = \rho_0\Gamma_0$. This equation has the expected behavior that if an isentropic compression occurs at $0\,\mathrm{K}$ there is no change in temperature. Such a theoretical compression is referred to as the cold compression curve.

The temperature along the isentrope (Eq. 5.152) raises a question when compression begins at an ambient state of zero pressure. For small strains, the predicted change in temperature along the isentrope is $\Delta T \approx T_0\Gamma_0\tilde{\varepsilon}$, which corresponds to an energy change of $\Delta E = C_v\Delta T \approx C_v T_0\Gamma_0\tilde{\varepsilon}$. For small strains, the work performed in isentropic compression is $\rho_0\Delta E \approx \frac{1}{2}K_s\tilde{\varepsilon}^2$, since $p(\rho, T) \approx K_s\tilde{\varepsilon}$. For small strains, the temperature energy change is much larger (linear in the change in strain vs. quadratic) than the externally applied work. How can this be?

The reason has to do with the way energy is stored in the material. As was described above, for a solid material there is elastic energy in the lattice and there is thermal energy of the motion of the atomic nuclei (for higher energies, the thermal energy in the electrons also becomes important). Thus, the pressure can be viewed as separating into a cold compression curve and a thermal term. In particular, for the Mie–Grüneisen equation above, the equation of state is

$$p(\rho, T) = \tilde{\sigma}_s(\rho) - \rho\Gamma(E_s(\rho) + C_v T_0) + \rho\Gamma C_v T = \tilde{\sigma}_c(\rho) + \rho\Gamma C_v T, \tag{5.153}$$

where $\tilde{\sigma}_c(\rho)$ is the cold compression curve, which is the stress or pressure vs. density in the elastic lattice at $0\,\mathrm{K}$. In particular, note that the cold compression curve is strictly a function of density. When the material is initially at ambient conditions, the temperature is greater than zero and the pressure is zero, $p(\rho_0, T_0) = 0$, so

$$\tilde{\sigma}_c(\rho_0) = -\rho_0\Gamma_0 C_v T_0 < 0. \tag{5.154}$$

Thus, the elastic lattice is in tension: the thermal motion of the nuclei expands the elastic lattice. When the material is compressed, the density increases and there is a potential energy release in the lattice,

$$dE = -\tilde{\sigma}_c(\rho)d(1/\rho) \approx \rho_0\Gamma_0 C_v T_0 d(1/\rho) \quad\Rightarrow\quad \Delta E \approx -\Gamma_0 C_v T_0\tilde{\varepsilon}. \tag{5.155}$$

This energy release in the lattice corresponds to the temperature rise noted above. Thus, though there is only a small amount of external work performed, the resulting release of tensile strain energy by the elastic lattice provides more energy to raise the temperature than is provided by the externally applied work. (For other examples, see Exercises 5.19 and 5.20). The overall change in pressure is due to the change in pressure of the cold compression curve – as it becomes less tensile, the pressure increases, and the slope of the cold compression curve at the ambient point is nearly the adiabatic bulk modulus K_s. Looking at the derivation of the temperature along the isentrope (Eq. 5.152), it will be seen that the cold compression curve – the elastic lattice deformation – completely dropped out of the derivation. The cold compression curve does not contribute to the thermal energy (temperature) of the material; it is elastically stored energy. The temperature along the isentrope is controlled by the Grüneisen parameter. The temperature rise in an elastic material during isentropic tension or compression is sometimes referred to as the elastocaloric or piezocaloric effect [216]. (How large is the tensile lattice strain as a function of temperature at ambient pressure? Using Eq. (5.154), the low pressure expression for the Grüneisen parameter (Eq. 5.120) and the assumption that the cold compression curve is a linear function of strain with the material's bulk modulus yields the estimate $\tilde{\varepsilon} \approx 3\alpha T$. For aluminum, the estimated lattice strain is 2% at room temperature, corresponding to 1.5 GPa.)

Given the temperature along the isentrope, we would also like to know the internal energy along the isentrope. Unfortunately, finding the internal energy is difficult and we will not be able to find a simple expression for it. Along the isentrope

$$\frac{dE}{d(1/\rho)} = -\tilde{\sigma}. \tag{5.156}$$

Thus, directly inserting the Mie–Grüneisen form of the equation of state where the reference curve is the Hugoniot, we have

$$\frac{dE}{d(1/\rho)} = -\tilde{\sigma}_H(\rho) - \rho_0 \Gamma_0 (E - E_H(\rho)). \tag{5.157}$$

With $\tilde{\varepsilon} = 1 - \rho_0/\rho$ and $d\tilde{\varepsilon} = -\rho_0 d(1/\rho)$,

$$dE = \frac{\tilde{\sigma}_H(\tilde{\varepsilon})}{\rho_0} d\tilde{\varepsilon} + \Gamma_0 E d\tilde{\varepsilon} - \Gamma_0 E_H(\tilde{\varepsilon}) d\tilde{\varepsilon}. \tag{5.158}$$

Rearrangement gives

$$dE - \Gamma_0 E d\tilde{\varepsilon} = \frac{\tilde{\sigma}_H(\tilde{\varepsilon})}{\rho_0} d\tilde{\varepsilon} - \Gamma_0 E_H(\tilde{\varepsilon})) d\tilde{\varepsilon}. \tag{5.159}$$

This expression can be explicitly integrated using $\exp(-\Gamma_0 \tilde{\varepsilon})$ as an integrating factor,

$$E_s(\tilde{\varepsilon}) = \exp\{\Gamma_0(\tilde{\varepsilon} - \tilde{\varepsilon}_i)\} E_H(\tilde{\varepsilon}_i) + \exp(\Gamma_0 \tilde{\varepsilon}) \int_{\tilde{\varepsilon}_i}^{\tilde{\varepsilon}} \exp(-\Gamma_0 \tilde{\varepsilon}) \left\{ \frac{\tilde{\sigma}_H(\tilde{\varepsilon})}{\rho_0} - \Gamma_0 E_H(\tilde{\varepsilon}) \right\} d\tilde{\varepsilon}. \tag{5.160}$$

For compression from the initial state ρ_0 to a final density ρ, the limits of this integral are $\tilde{\varepsilon}_i = 0$ and $\tilde{\varepsilon}_f = 1 - \rho_0/\rho$. We mention these limits explicitly because to compute energy on the release from a Hugoniot state, the starting energy is the Hugoniot energy. The initial density will be $\tilde{\varepsilon}_i = 1 - \rho_0/\rho$ and the final density will be $\tilde{\varepsilon}_f < 0$, where the value is determined by the density at which the equation of state gives $\tilde{\sigma}(\tilde{\varepsilon}_f) = 0$. The stress along the isentrope is $\tilde{\sigma}_s = \tilde{\sigma}_H + \rho_0 \Gamma_0 (E_s - E_H)$. By computing E_s it is possible to determine $\tilde{\varepsilon}_f$, which then allows the calculation of temperature along the release isentropic from Eq. (5.152).

An alternate approximate approach for small strains is to assume that the Hugoniot and the isentrope are close (Theorem 4.13). Approximating the energy along the isentrope by using the differential energy equation and assuming the pressure is given by the Hugoniot pressure yields

$$E_s(\rho) \approx - \int_{\rho_0}^{\rho} \tilde{\sigma}_H(\rho) d(1/\rho). \tag{5.161}$$

For example, if the U_s–u_p relation is linear, it is possible to explicitly compute this integral as (Exercise 5.41)

$$E_s(\tilde{\varepsilon}) = \frac{c_0^2}{k^2} \left\{ \frac{k\tilde{\varepsilon}}{1 - k\tilde{\varepsilon}} + \ln(1 - k\tilde{\varepsilon}) \right\}. \tag{5.162}$$

Once the internal energy along the isentrope is known, since we know the internal energy along the Hugoniot from the Hugoniot jump conditions, we obtain

$$T_H(\rho) = \frac{1}{C_v}(E_H(\rho) - E_s(\rho)) + T_s(\rho) \tag{5.163}$$

as the approximate temperature along the Hugoniot. Specifically for the linear U_s–u_p relation and the energy along the isentrope approximation,

$$T_H(\rho) = \frac{c_0^2}{C_v} \left[\frac{\tilde{\varepsilon}^2}{2(1 - k\tilde{\varepsilon})^2} - \frac{1}{k^2} \left\{ \frac{k\tilde{\varepsilon}}{1 - k\tilde{\varepsilon}} + \ln(1 - k\tilde{\varepsilon}) \right\} \right] + T_0 \exp(\Gamma_0 \tilde{\varepsilon}). \tag{5.164}$$

For strong shocks, most of the temperature rise is due to the energy deposited in the material by the shock, which is larger than the energy deposited by the isentrope, represented by the first term; the second term, the temperature of the isentrope, is not large. Example values were given in Tables 5.2 and 5.3 using this approximation to E_s. Table 5.5 shows the difference in T_H between using a numerical method to compute the isentrope energy integral in Eq. (5.160) and then Eq. (5.163), versus the approximation in Eq. (5.162) leading to Eq. (5.164). However, keep in mind that there are approximations involved in both approaches, including an assumed constant heat capacity. Given a density and temperature, it is now possible to write the pressure based on Hugoniot data,

$$p(\rho, T) = \tilde{\sigma}_H(\rho) + \rho \Gamma C_v(T - T_H(\rho)). \tag{5.165}$$

We can now estimate how much of the pressure on the Hugoniot is due to thermal effects. In particular, the fraction that is due to thermal effects is

$$\frac{\rho \Gamma(E_H(\rho) - E_s(\rho))}{\tilde{\sigma}_H(\rho)}. \tag{5.166}$$

For the linear U_s–u_p relationship, inserting the values above this fraction is given by

$$\frac{\rho \Gamma(E_H(\rho) - E_s(\rho))}{\tilde{\sigma}_H(\rho)} = \Gamma_0 \left(\frac{3\tilde{\varepsilon}}{2} - \frac{1}{k} - (1 - k\tilde{\varepsilon})^2 \frac{\ln(1 - k\tilde{\varepsilon})}{k^2 \tilde{\varepsilon}} \right)$$
$$= \frac{\Gamma_0 k \tilde{\varepsilon}^2}{3} \left(1 + \frac{k\tilde{\varepsilon}}{4} + \cdots \right). \tag{5.167}$$

Thus, as the impact speed increases, a higher fraction of the pressure is due to thermal terms. For metals, Γ_0 is typically roughly 2 and k is roughly 1.5 and so, compared to the adiabat,

GENERAL PRINCIPLE 5.8. *For metals, the fraction of the Hugoniot pressure due to thermal effects is roughly given by the square of the engineering strain.*

TABLE 5.5. Temperature for Hugoniot 6061 aluminum, with $\rho_0 = 2.703 \, \text{g/cm}^3$, $U = 5.35 \, \text{km/s} + 1.34u$, $\Gamma_0 = 2$, $T_0 = 298 \, \text{K}$ and $C_v = 0.89 \, \text{J/g K}$. The last column includes phase changes, PM for partial melt, M for complete melt, and V for partial vaporization upon unloading – values based on pure aluminum; see Section 5.15 for the discussion.

u	ρ	$\tilde{\sigma}_H$	T_H 5.160	T_H 5.164	$\tilde{\sigma}_s(\rho_0)$ 5.160	$\tilde{\varepsilon}_f$ 5.160	T_u 5.160	T_u 5.169	$(T_u)_{PC}$ 5.169
km/s	g/cm³	GPa	K	K	GPa		K	K	K
0.00	2.703	0.00	298	298	0.00	0.000	298	298	298
0.50	2.948	8.14	363	363	0.05	−0.001	307	307	307
1.00	3.178	18.08	487	481	0.30	−0.004	358	357	357
1.50	3.395	29.84	719	696	0.87	−0.011	468	463	463
2.00	3.600	43.41	1,100	1,039	1.78	−0.023	638	631	631
2.50	3.793	58.79	1,661	1,534	3.06	−0.041	862	864	864
3.00	3.976	75.98	2,424	2,201	4.71	−0.063	1,125	1,160	934PM
3.50	4.150	94.98	3,408	3,054	6.73	−0.092	1,412	1,521	1,076M
4.00	4.314	115.80	4,627	4,105	9.11	−0.127	1,701	1,945	1,500M
4.50	4.471	138.42	6,092	5,365	11.86	−0.168	1,974	2,433	1,988M
5.00	4.620	162.86	7,812	6,841	14.96	−0.216	2,211	2,983	2,538M
5.50	4.762	189.10	9,793	8,541	18.41	−0.271	2,397	3,597	2,740V

Another inference is that for small pressures, the isotherm is close to the Hugoniot. The Taylor expansion of Mie–Grüneisen gives $p_T \approx K_0 \tilde{\varepsilon} + (2k - \Gamma/2) K_0 \tilde{\varepsilon}^2$ for the isotherm and $p_s \approx p_H \approx K_0 \tilde{\varepsilon} + 2k K_0 \tilde{\varepsilon}^2$ for the adiabat and Hugoniot, showing that the slope of the U_s–u_p curve and the Grüneisen parameter contain the nonlinearity in the compression. Notice that at the maximum compression for the linear U_s–u_p relation, where $\tilde{\varepsilon} = 1/k$, the fraction of the energy that is thermal is given by $\Gamma_0/(2k)$, which implies k should be greater than $\Gamma_0/2$ if the equation of state is valid at that compression.

The approximate temperature upon isentropic unloading from a shock is given by Eq. (5.152). The next column in Table 5.5 shows the isentropic release pressure upon release to the initial density – that is, $\tilde{\sigma}_s(\rho_0) = \rho_0 \Gamma_0 C_v (T_u - T_0)$, where T_u is based on the shock E_s from Eq. (5.160). If instead the unloading is carried out until ambient pressure is achieved, the next column shows $\tilde{\varepsilon}_f$ upon release – that is, the engineering strain at which zero stress is achieved during the isentropic expansion $\tilde{\sigma}(\tilde{\varepsilon}_f) = \tilde{\sigma}_H(\tilde{\varepsilon}_f) + \rho_0 \Gamma_0 (E_s(\tilde{\varepsilon}_f) - E_H(\tilde{\varepsilon}_f)) = 0$, using Eq. (5.160) to compute E_s upon release. (As we approach the bottom of this column, these strains become too large to be reasonable, and it is not surprising to see in the last column of the table that the residual temperature is high enough to melt the aluminum.) The next column shows the unloaded temperature at this computed ambient pressure release state. These temperature estimates assume a constant heat capacity and no phase changes.

If we ignore the lower density attained on the expansion and just release back to ρ_0, we can approximately arrive at the temperature

$$T_u = \exp(-\Gamma_0 \tilde{\varepsilon}) \cdot T_H(\rho). \tag{5.168}$$

Thus, approximating the isentropic energy with the linear U_s–u_p response, we have

$$T_u = \frac{c_0^2 \exp(-\Gamma_0 \tilde{\varepsilon})}{C_v} \left[\frac{\tilde{\varepsilon}^2}{2(1-k\tilde{\varepsilon})^2} - \frac{1}{k^2} \left\{ \frac{k\tilde{\varepsilon}}{1-k\tilde{\varepsilon}} + \ln(1-k\tilde{\varepsilon}) \right\} \right] + T_0, \qquad (5.169)$$

where $\tilde{\varepsilon} = 1 - \rho_0/\rho$ is the compressive strain in the shock. Table 5.5 compares this approximate expression to computing the full isentrope from the previous paragraph, and for lower values of particle velocity the agreement is seen to be good. For small strains, $T_u = (kc_0^2 \tilde{\varepsilon}^3 / 3C_v)(1 + (9k/4 - \Gamma_0)\tilde{\varepsilon}) + T_0$ is a good approximation.

5.12. Reflection and Transmission

In Section 4.9 we computed the reflection and transmission behavior for the linear elastic wave (see Fig. 4.11). For the nonlinear shock, the situation is more complicated. Given two materials in contact, initially at rest and zero pressure, when a shock traveling through the first arrives at the second the impedance determines the behaviors. If the particle velocity in the arriving wave is $u_{1\ell}$ and the stress is $\tilde{\sigma}_{1\ell}$, the immediate question is: what is the Hugoniot stress in material 2 for the particle velocity $u_{1\ell}$? If the stress is the same as in the first material, then the shock impedances match, since $\rho_0 U = \tilde{\sigma}_{1\ell}/u_{1\ell}$, and there will be no reflection of the shock – the wave will be transmitted. If the Hugoniot stress for the second material is less than the first material – that is, $\tilde{\sigma}_{H2}(u_{1\ell}) < \tilde{\sigma}_{1\ell}$ – then the impedance of the second material is less and a rarefaction wave will be reflected back into material 1 and a shock will be transmitted into material 2. Thus, the equations to solve at the interface are the Hugoniot jump conditions for material 2 and

$$\tilde{\sigma}_{S1}(\rho_{1\ell}, \tilde{\sigma}_{1\ell}; \rho_{1r}) = \tilde{\sigma}_{H2}(u_{2\ell}), \quad u_{1r} = u_{1\ell} - \int_{\rho_{1\ell}}^{\rho_{1r}} \frac{c(\rho, E_S(\rho_{1\ell}, \tilde{\sigma}_{1\ell}; \rho))}{\rho} d\rho = u_{2\ell}.$$

$$(5.170)$$

The stress and the particle velocity at the interface are being matched. The notation in this equation is that the material 1 isentrope $\tilde{\sigma}_{S1}(\rho_{1\ell}, \tilde{\sigma}_{1\ell}; \rho)$ is a function of density ρ and passes through state $(\rho_{1\ell}, \tilde{\sigma}_{1\ell})$ (i.e., it passes through the shocked state, and represents a release from that state), and similarly the sound speed c along the isentrope is a function of density beginning at the state $(\rho_{1\ell}, \tilde{\sigma}_{1\ell})$. The simultaneous solution of these equations produces the shock moving to the right in material 1 and the rarefaction moving to the left in material 2.

If, on the other hand, the Hugoniot stress for the second material is greater than the first material - that is, $\tilde{\sigma}_{H2}(u_{1\ell}) > \tilde{\sigma}_{1\ell}$, then the impedance of the second material is greater than the first and a shock wave increasing the stress will be reflected back into material 1. In this case we have a second shock in the first material, and so a Hugoniot from the initial shocked state must be developed to solve the problem. In particular, the equations to solve are the Hugoniot jump conditions for both material 1 (using the new Hugoniot) and 2 (from its initial state) and

$$\tilde{\sigma}_{H1}(\rho_{1\ell}, \tilde{\sigma}_{1\ell}; \rho_{1r}) = \tilde{\sigma}_{H2}(u_{2\ell}), \quad u_{1r} = u_{2\ell}. \qquad (5.171)$$

Again, we are matching the stress and particle velocity at the interface. The notation in this equation is that the material 1 Hugoniot $\tilde{\sigma}_{H1}(\rho_{1\ell}, \tilde{\sigma}_{1\ell}; \rho)$ is a function of density ρ and passes through state $(\rho_{1\ell}, \tilde{\sigma}_{1\ell})$ (i.e., it passes through the shocked state, and represents a second shock from that state). Since these equations are nonlinear, it is not possible to write down a solution as we did in Section 4.9, but once it is determined it is possible to determine the shock impedances and represent the solution in that fashion.

Though a simple exact solution is out of reach, it is interesting to ask the question about the effect of nonlinearity if a shock impinges on a rigid surface. We assume that a shock moves to the right and reflects to the left off the rigid surface. The initial shock has particle velocity $u > 0$. Upon reflection, the particle velocity goes to zero and it is necessary to compute U and $\tilde{\sigma}_{1r}$. The Hugoniot jump conditions are

$$\rho_\ell/\rho_r = 1 - u/|U|, \quad u_{1r} = 0, \quad \tilde{\sigma}_r - \tilde{\sigma}_\ell = \tilde{\sigma}_H(\rho_\ell, \tilde{\sigma}_\ell; \rho_r) - \tilde{\sigma}_\ell = \rho_\ell|U|u, \quad (5.172)$$

where the absolute values are on U since it is negative (the wave moves to the left) and the new Hugoniot must be computed from the initial shocked state of the material. Combining these equations gives

$$(1/\rho_\ell - 1/\rho_r)\{\tilde{\sigma}_H(\rho_\ell, \tilde{\sigma}_\ell; \rho_r) - \tilde{\sigma}_\ell\} = u^2 \quad \Rightarrow \quad (\tilde{\varepsilon}_r - \tilde{\varepsilon}_\ell)\{\tilde{\sigma}_H(\rho_\ell, \tilde{\sigma}_\ell; \varepsilon_r) - \tilde{\sigma}_\ell\} = \rho_0 u^2, \tag{5.173}$$

where for what follows it is useful to write the equation in terms of strain $\tilde{\varepsilon} = 1 - \rho_0/\rho$.

To obtain any more information requires an equation of state. We will assume that the solid material can be represented by $U = c_0 + ku$ and the Mie–Grüneisen form (Eq. 5.127, $\tilde{\sigma}(\tilde{\varepsilon}, E) = f(\tilde{\varepsilon}) + \rho_0\Gamma_0(E - E_0)$). To determine the Hugoniot from the state of the incoming shock (i.e., a second shock), the energy jump condition is used,

$$E_{H2nd}(\tilde{\varepsilon}) - E_\ell = \frac{1}{2\rho_0}(\tilde{\sigma}(\tilde{\varepsilon}_\ell, \tilde{\sigma}_\ell; \tilde{\varepsilon}) + \tilde{\sigma}_\ell)(\tilde{\varepsilon} - \tilde{\varepsilon}_\ell). \tag{5.174}$$

In this equation the subscript H2nd means second shock Hugoniot. The energy for the initial shocked state of the material is $E_l = E_0 + \tilde{\varepsilon}_\ell\tilde{\sigma}_\ell/2\rho_0$, and some algebra yields

$$E_{H2nd}(\tilde{\varepsilon}) = E_0 + \frac{\tilde{\varepsilon}\tilde{\sigma}_\ell + (\tilde{\varepsilon} - \tilde{\varepsilon}_\ell)f(\tilde{\varepsilon})}{\rho_0\{2 - \Gamma_0(\tilde{\varepsilon} - \tilde{\varepsilon}_\ell)\}}. \tag{5.175}$$

Thus, we have explicitly computed the second shock Hugoniot that passes through the state $(\rho_\ell, \tilde{\sigma}_\ell)$ as $\tilde{\sigma}(\tilde{\varepsilon}_\ell, \tilde{\sigma}_\ell; \tilde{\varepsilon}) = f(\tilde{\varepsilon}) + \rho_0\Gamma_0(E_{H2nd}(\tilde{\varepsilon}) - E_0)$. This expression is then placed in Eq. (5.174) and it is possible to solve for the density behind the reflected shock wave, $\tilde{\varepsilon}_r$. Some algebra and the observation that equipartition between kinetic and internal energy of the initial shock (i.e., $\frac{1}{2}\rho_0 u^2 = \frac{1}{2}\tilde{\varepsilon}_\ell\tilde{\sigma}_\ell$) yields

$$f(\tilde{\varepsilon}_r) - \tilde{\sigma}_\ell(1 - \Gamma_0\tilde{\varepsilon}_r) = \rho_0 u^2/(\tilde{\varepsilon}_r - \tilde{\varepsilon}_\ell), \tag{5.176}$$

or, writing out f,

$$\frac{\rho_0 c_0^2 \tilde{\varepsilon}_r}{(1 - k\tilde{\varepsilon}_r)^2}\left(1 - \frac{\Gamma_0}{2}\tilde{\varepsilon}_r\right) - \tilde{\sigma}_\ell(1 - \Gamma_0\tilde{\varepsilon}_r) = \frac{\rho_0 u^2}{\tilde{\varepsilon}_r - \tilde{\varepsilon}_\ell}. \tag{5.177}$$

Initial conditions are $\tilde{\varepsilon}_\ell = u/(c_0 + ku)$ and $\tilde{\sigma}_\ell = \rho_0(c_0 + ku)u$. As a specific example, for Al 6061-T6 with the linear U_s–u_p expression and Grüneisen parameter, Table 5.6 shows incident shocks with particle velocities of 1 through 6 km/s incident on a rigid surface. For a linear elastic reflection, the ratio of the reflected pressure to the incident pressure divided by two ($\tilde{\sigma}_r/2\tilde{\sigma}_\ell$) is 1, so this value provides a measure of the nonlinearity of the reflection. Also, due to the nonlinear nature of the equation of state, the strain ratio $\tilde{\varepsilon}_r/2\tilde{\varepsilon}_\ell$ is less than 1 (for the linear elastic case the ratio is again 1).

TABLE 5.6. Rigid surface reflection for Al 6061-T6 (c_0 = 5.35 km/s, $k = 1.34$, $\Gamma_0 = 2$).

u (km/s)	$\tilde{\varepsilon}_\ell$	$\tilde{\sigma}_\ell$ (GPa)	$\tilde{\varepsilon}_r$	$\tilde{\sigma}_r$ (GPa)	$\tilde{\varepsilon}_r/2\tilde{\varepsilon}_\ell$	$\tilde{\sigma}_r/2\tilde{\sigma}_\ell$
1	0.1495	18.0	0.2561	43.4	0.857	1.20
2	0.2491	43.4	0.3948	117.6	0.793	1.35
3	0.3202	76.0	0.4816	226.8	0.752	1.49
4	0.3735	115.8	0.5397	376.1	0.723	1.62
5	0.4149	162.9	0.5805	570.1	0.700	1.75
6	0.4481	217.2	0.6101	817.7	0.681	1.88

TABLE 5.7. Primary and reflected shock pressures for liquid H_2 and D_2 [139].

ρ_0 g/cm^3	U_ℓ km/s	u_ℓ km/s	ρ_ℓ g/cm^3	p_ℓ GPa	u_r km/s	ρ_r g/cm^3	p_r GPa	$\frac{p_r}{2p_\ell}$
Deuterium, $\rho_0 = 0.165$ g/cm^3, at 20 K								
0.165	7.13	4.27	0.41	5.02	1.31	0.654	14.7[a]	1.47
					0.90	0.680	16.8[b]	1.67
0.165	6.78	4.28	0.474	4.79	1.18	1.075	12.9[a]	1.35
					0.83	1.031	15.2[b]	1.58
0.165	7.97	4.48	0.376	5.88	1.42	0.571	16.3[a]	1.38
					1.07	0.538	20.5[b]	1.74
0.167	8.31	4.99	0.418	6.92	1.80	0.581	22.1[a]	1.60
					1.20	0.654	23.6[b]	1.70
0.165	9.45	5.97	0.45	9.36	2.15	0.694	28.1[a]	1.50
					1.60	0.699	33.8[b]	1.81
0.163	10.97	7.25	0.481	12.96				
Hydrogen, $\rho_0 = 0.072$ g/cm^3, at 20 K								
0.072	6.60	3.49	0.152	1.65	0.65	0.205	6.4[a]	1.94
					0.25	0.474	4.0[b]	1.21
0.072	9.09	5.25	0.170	3.42	1.22	0.234	13.4[a]	1.96
					0.76	0.290	13.8[b]	2.02
0.072	10.57	6.34	0.179	4.80	1.58	0.253	18.7[a]	1.95
					1.06	0.266	20.1[b]	2.09
0.072	11.80	7.67	0.204	6.49	1.61	0.500	19.2[a]	1.48
					1.13	0.465	22.1[b]	1.7

[a]AZ31B magnesium or [b]2024 aluminum was the standard used for the reflected shock measurements.

The last two velocities are extrapolations beyond the experimental range, but an examination of the table shows that using the first shock equation is a reasonable approximation for the second shock (i.e., the 1 km/s reflection stress is well approximated by the 2 km/s initial stress, etc.). Thus, a reasonable approximation of an incident shock in a first material impinging on a second material, both of which can be represented by linear U_s–u_p relations, is

$$\tilde{\sigma}_{1\ell} \approx \rho_{10}(c_{01} + k_1(2u_{1\ell} - u))(2u_{1\ell} - u) = \rho_{20}(c_{02} + k_2 u)u = \tilde{\sigma}_{2r}, \tag{5.178}$$

where $u = u_{1r} = u_{2\ell}$. This expression says that the approximation is the same as considering an impact of the material 1 with twice its initial particle velocity on material 2, $V = 2u_{1\ell}$ or $\tilde{\sigma}_1(2u_{1\ell} - u) = \tilde{\sigma}_2(u)$. This approximate equation can also be used in the event of a rarefaction when the rarefaction shock approximation is used.

Showing data to demonstrate Eq. (5.171) for a reflection is a bit circular, as nearly all data uses the impedance matching (i.e., Eq. 5.171) to determine shock properties. Since large pressures are required to see nonlinear effects in a solid reflection, and since the effect is well known in gases, we show the reflected shock values for two fluids, hydrogen and deuterium at 20 K in Table 5.7 [139]. Most metals are quite complex at the atomistic level in how they give rise to observed elastic and plastic properties, but from a force-between-molecules point of view, hydrogen and deuterium should be the same, except that deuterium has roughly twice the mass of hydrogen. A combination of the mass and momentum balances gives $\tilde{\sigma} = (\sqrt{\rho_0}u)^2/(1 - \tilde{\varepsilon})$, so given that the force between molecules is the same we expect $\tilde{\sigma}_{H_2}(u) = \tilde{\sigma}_{D_2}(\sqrt{\rho_{0H_2}/\rho_{0D_2}}u)$. Figure 5.13 shows both that this holds and the relative accuracy (or inaccuracy) of the $\tilde{\sigma}(2u_1 - u_2)$ approximation for the second shock. All the second shocks represent a *different* Hugoniot, since the first shocks are to different pressures and densities.

One of the approaches to fusion is inertial confinement fusion, where deuterium and tritium in a small sphere are compressed by the rapid vaporization of an outer layer as it is uniformly illuminated by X-rays inside a cavity (called a hohlraum) that is lit through both ends by a laser pulse. The pressures and temperatures required are well above what can be produced by conventional explosive loading. A spherical implosion shape is used instead of a planar shock to achieve extremely large compressions. Though the initial compression wave is a shock, the effort is to form an isentropic compression from that point on since shocks have limited compression (why?). The original estimates for what would be required involve collapse velocities of 350 km/s, compressed densities of 1,000 g/cm^3, pressures on the order of 10^7 GPa (the initial shock is around 150 GPa), and temperatures on the order of 10 keV (10^8 K) [152].

5.13. Additional Comments on Sound Speed and Precursors

Let's compute the sound speed for the linear U_s–u_p relation along the Hugoniot, assuming the material has no thermal terms. We can directly compute it since we know the Hugoniot in terms of density (assuming no thermal terms),

$$c(\rho) = \sqrt{\frac{d\tilde{\sigma}_H(\rho)}{d\rho}} = \sqrt{\frac{d}{d\rho}\left(\frac{\rho_0 c_0^2 \tilde{\varepsilon}}{(1 - k\tilde{\varepsilon})^2}\right)} = \sqrt{\frac{c_0 \rho_0}{\rho}\sqrt{\frac{1 + k\tilde{\varepsilon}}{(1 - k\tilde{\varepsilon})^3}}} = c_0(1 - \tilde{\varepsilon})\sqrt{\frac{1 + k\tilde{\varepsilon}}{(1 - k\tilde{\varepsilon})^3}}, \tag{5.179}$$

where we used $d\tilde{\varepsilon}/d\rho = \rho_0/\rho^2$.

Next, let's compute the sound speed for the Mie–Grüneisen equation for an arbitrary density, energy state assuming the linear U_s–u_p relation and assuming a

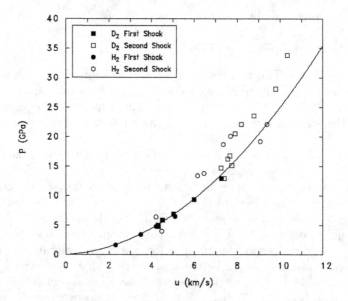

FIGURE 5.13. Shock pressure as a function of particle velocity for liquid (20 K) D_2 and H_2 (scaled by $u_{D_2} = \sqrt{\rho_{0H_2}/\rho_{0D_2}}\, u_{H_2}$). The D_2 $U_s\text{-}u_p$ is $U = 1.34$ km/s $+ 1.38u$ based on a least-squares fit to first shock data of D_2 and scaled H_2. The plot includes both first (solid symbol) and second (open symbol) shocks; the particle velocity for second shocks is $2u_1 - u_2$.

Grüneisen parameter,

$$
\begin{aligned}
c(\rho, E) &= \sqrt{\left(\frac{\partial \tilde{\sigma}}{\partial \rho}\right)_E + \frac{\tilde{\sigma}}{\rho^2}\left(\frac{\partial \tilde{\sigma}}{\partial E}\right)_\rho} \\
&= \frac{1}{\rho}\sqrt{\rho_0\left(\frac{\partial \tilde{\sigma}}{\partial \varepsilon}\right)_E + \tilde{\sigma}\rho_0\Gamma_0} \\
&= \frac{\rho_0 c_0}{\rho}\sqrt{\frac{1 + (k - \Gamma_0)\tilde{\varepsilon}}{(1 - k\tilde{\varepsilon})^3} + \Gamma_0\frac{\tilde{\sigma}}{\rho_0 c_0^2}} \\
&= c_0(1 - \tilde{\varepsilon})\sqrt{\frac{1 + k\tilde{\varepsilon} - k\Gamma_0\tilde{\varepsilon}^2}{(1 - k\tilde{\varepsilon})^3} + \left(\frac{\Gamma_0}{c_0}\right)^2 (E - E_H(\rho))} \\
&= c_0(1 - \tilde{\varepsilon})\sqrt{\frac{1 + k\tilde{\varepsilon} - k\Gamma_0\tilde{\varepsilon}^2 - (1 - k\tilde{\varepsilon})\Gamma_0^2\tilde{\varepsilon}^2/2}{(1 - k\tilde{\varepsilon})^3} + \left(\frac{\Gamma_0}{c_0}\right)^2 (E - E_0)}
\end{aligned}
\tag{5.180}
$$

after some algebra and use of $\rho\Gamma = \rho_0\Gamma_0$. Notice that if we had a given density and a temperature difference from the Hugoniot, we would replace $E - E_H(\rho)$ with $C_v\Delta T = C_v(T - T_H(\rho))$.

Finally, consider the sound speed along the Hugoniot from ambient for the above equation of state, which includes thermal terms, immediately from above with $E = E_H$,

$$c_H = c(\rho, E_H(\rho)) = c_0(1 - \tilde{\varepsilon})\sqrt{\frac{1 + k\tilde{\varepsilon} - k\Gamma_0\tilde{\varepsilon}^2}{(1 - k\tilde{\varepsilon})^3}}. \tag{5.181}$$

The difference from assuming purely mechanical behavior is a reduction in the sound speed by the $-k\Gamma_0\tilde{\varepsilon}^2$ term, which is small compared to 1 for small strains. This implies that it is a reasonably good approximation to use the ambient-state Hugoniot for computing rarefactions and second shocks when the strains are small. The reason Eqs. (5.179) and (5.181) differ is because the sound speed is essentially a "second shock" from a different material state along the Hugoniot, and so the Hugoniot no longer applies (Theorem 5.3 and Section 4.12). To be more explicit, suppose the initial state is $(\rho_0, \tilde{\sigma}_0)$ and the equation of state is Mie–Grüneisen (Eqs. 5.119 and 5.121). Then the adiabatic sound speed is given by

$$c^2(\rho, E) = \left(\frac{\partial\tilde{\sigma}}{\partial\rho}\right)_s = \left(\frac{\partial\tilde{\sigma}}{\partial\rho}\right)_E + \frac{\tilde{\sigma}}{\rho^2}\left(\frac{\partial\tilde{\sigma}}{\partial E}\right)_\rho$$

$$= \left(1 - \frac{\Gamma_0\tilde{\varepsilon}}{2}\right)\frac{d\tilde{\sigma}_H}{d\rho} + \frac{\rho_0\Gamma_0}{2\rho^2}(\tilde{\sigma}_H - \tilde{\sigma}_0) + \rho_0^2\Gamma_0^2(E - E_H). \tag{5.182}$$

It is clear the only time on the Hugoniot ($E = E_H$) that the square of the sound speed equals the derivative of the Hugoniot curve is at the initial state, where $\tilde{\sigma}_H = \tilde{\sigma}_0$ and $\tilde{\varepsilon} = 0$.

We can use this result to compute the adiabatic bulk modulus of the material (see Exercise 5.21),

$$K_s(\rho, E) = \rho\left(\frac{\partial\tilde{\sigma}}{\partial\rho}\right)_s = \rho c^2(\rho, E)$$

$$= \frac{\rho_0^2 c_0^2}{\rho}\left\{\frac{1 + k\tilde{\varepsilon} - k\Gamma_0\tilde{\varepsilon}^2}{(1 - k\tilde{\varepsilon})^3} + \left(\frac{\Gamma_0}{c_0}\right)^2(E - E_H(\rho))\right\}$$

$$= K_0(1 - \tilde{\varepsilon})\left\{\frac{1 + k\tilde{\varepsilon} - k\Gamma_0\tilde{\varepsilon}^2 - (1 - k\tilde{\varepsilon})\Gamma_0^2\tilde{\varepsilon}^2/2}{(1 - k\tilde{\varepsilon})^3} + \left(\frac{\Gamma_0}{c_0}\right)^2(E - E_0)\right\}, \tag{5.183}$$

and along the Hugoniot from ambient,

$$K_s(\rho, E_H(\rho)) = K_0(1 - \tilde{\varepsilon})\left\{\frac{1 + k\tilde{\varepsilon} - k\Gamma_0\tilde{\varepsilon}^2}{(1 - k\tilde{\varepsilon})^3}\right\}. \tag{5.184}$$

There are two regimes of response for an elastic-plastic solid. In the purely elastic regime, the shear modulus is involved, making the material response stiffer and increasing the sound speed. This region of response gives rise to the compressive elastic precursor. For large rarefactions, it also leads to a two-wave structure, the leading wave being an elastic precursor. In the compressive shock, when plastic deformation occurs, $\tilde{\sigma} = p + 2Y/3$, which has the principal stress deviator $s_{11} = -2Y/3$ (compression). Initially upon unloading, the response is purely elastic and the shear terms are involved, giving rise to a change in the principal stress deviator from $s_{11} = -2Y/3$ (compression) to $s_{11} = +2Y/3$ (tension). The change in stress and change in particle velocity associated with the elastic precursor is twice that seen for the compressive elastic precursor, namely

$$\Delta u \approx -2u_{\text{HEL}},$$
$$\Delta \tilde{\sigma} \approx -2\tilde{\sigma}_{\text{HEL}}.$$

(5.185)

We say approximately because the actual values depend on the local bulk and shear moduli and on the yield strength, all of which may be pressure and temperature dependent, and thus differ from the initial ambient conditions of the HEL. In particular, in Chapter 3 we derived (Eq. 3.65)

$$\tilde{\sigma}_{\text{HEL}} = (\lambda + 2\mu)\frac{Y}{2\mu} = \left(\frac{K}{2G} + \frac{2}{3}\right) Y = \frac{1-\nu}{1-2\nu}Y,$$

(5.186)

showing the explicit dependence on K, G, and Y. Subsequent stress release occurs with plastic flow and the modulus related to this is the bulk modulus and exhibits the rarefaction fan.

The initial elastic rarefaction will be purely elastic with a longitudinal wave that is based on the modulus $\lambda + 2\mu$. To determine the sound speed of the initial release, one typically works with the bulk sound speed as a starting point. There are various ways to obtain the longitudinal wave speed from this expression. For example, if it is assumed that the shear modulus is independent of pressure, then since $\lambda + 2\mu = K + (4/3)\mu$, the longitudinal sound speed is

$$c_L(\rho, E) = \sqrt{c^2(\rho, E) + \frac{4\mu}{3\rho}}.$$

(5.187)

If it is assumed that the shear modulus scales with the bulk modulus (i.e., increases with pressure), hence Poisson's ratio is constant, then from the $\lambda + 2\mu = 3((1 - \nu)/(1 + \nu))K$ we have

$$c_L(\rho, E) = c(\rho, E)\sqrt{3\frac{1-\nu}{1+\nu}}.$$

(5.188)

This latter expression is the one typically used, since it is known for some materials that shear modulus increases with pressure.

It is not possible to it a priori determine or bound the final unloading state by looking at the material state before the passage of the loading shock. Upon unloading, because of the thermal terms, the material can expand to densities much lower than the original, unshocked state. The end state of the unloading is a return to the ambient pressure. Thus, it is possible for the unloading to reduce the internal energy of the material to values well below that of the material before the initial shock passage. Hence, it is not possible to bound the energy, density, or final velocity $u(\rho)$ of the material upon unloading. For example, if the thermal behavior is such that the material changes in state from a solid to a gas upon unloading, there will be large expansions of the material as it returns to the ambient pressure state, which can result in large particle velocities.

5.14. Initial Porosity

Continuing Section 5.10's discussion about determining the Mie–Grüneisen coefficient through the use of porous samples, in this section we further discuss the effect of initial porosity. Notationally in this section, the density of the initially porous material is ρ, and note that this density is the density including the pores – that is, the overall density of the foam. We will also be considering the density of the material excluding the pores, and this density will be called ρ_s where the

TABLE 5.8. Data for porous iron [33]. The equilibrium stress behind the shock front $\tilde{\sigma}$ and particle velocity u are from analysis in the original paper; the shock speed U and final density ρ are calculated from the jump conditions.

ρ_0 g/cm^3	$\tilde{\sigma}$ MPa	u m/s	U m/s	ρ_0 g/cm^3	$\tilde{\sigma}$ MPa	u m/s	U m/s	ρ g/cm^3
1.26	24	119	160	4.58	155	27.7	1222	4.686
1.37	117	266	321	4.74	434	120	763	5.625
1.26	212	392	429	4.83	893	240	770	7.016
1.33	237	391	456	4.78	1,722	357	1,009	7.397
1.24	496	620	645	4.53	3,700	568	1,438	7.488
1.35	630	669	698					
				5.82	374	22.4	2,869	5.866
2.17	70	134	241	5.74	662	83	1,390	6.105
2.19	169	230	336	5.74	1,170	174	1,171	6.741
2.11	317	370	406	5.76	1,610	230	1,215	7.105
2.25	822	537	680	5.76	4,270	435	1,704	7.734
2.16	907	525	800	5.86	4,490	430	1,782	7.724
3.3	137	124	335	6.99	610	21.2	4,116	7.026
3.3	461	266	525	7.00	960	62.7	2,187	7.207
3.3	756	358	640	6.99	1,680	130	1,849	7.519
3.3	1,925	565	1,032	6.98	2,900	203	2,047	7.749
				7.00	4,400	273	2,302	7.942

subscript s is for solid; if ϕ is the volume fraction of the voids, then $\rho_s = \rho/(1-\phi)$. As a specific example of porous response, we present data of Butcher and Karnes on porous iron [33]. They explored the dynamic compaction of iron with initial densities $\rho_0 = 1.3, 2.2, 3.3, 4.7, 5.8,$ and 7.0 g/cm^3. The two lowest density materials were iron foams with irregularly shaped voids of size on the order of 0.1 mm, and the higher density materials were sintered iron products with irregularly shaped voids of size on the order of 0.02 mm. A chemical analysis of the 5.8 g/cm^3 sample showed 0.41% by weight carbon and 0.59% by weight oxygen. Before tests, the samples were placed in a vacuum to remove air from the voids. Impact tests were performed that provided the final state of the material. The data are shown in Table 5.8.

To estimate the yield point of the foam (i.e., the stress at which it first plastically deforms), as a first step an estimate of the elastic precursor velocity was obtained from the impact tests and from ultrasound. This information is shown in Table 5.9. The data in Table 5.8 were then plotted in the $(u, \tilde{\sigma})$ plane and a straight line for the elastic precursor was drawn using $\tilde{\sigma} = \rho_0 c_e u$. The intercept, where this line intersected a curve fit of the respective data, was then identified as the $\tilde{\sigma}_e$ yield stress. The results are shown in Table 5.9. The sound speed and yield values are amenable to curve fits as

$$c_e = 5.9 - 5.5 \ln(\rho_{s0}/\rho_0) \text{ km/s} = 5.9 - 5.5 \ln(\alpha_0) \text{ km/s},$$
$$\tilde{\sigma}_e = 850(\rho_{s0}/\rho_0)^{-3.9} \text{ MPa} = (850/\alpha_0^{3.9}) \text{ MPa},$$

(5.189)

TABLE 5.9. Elastic precursors for porous iron [**33**].

| ρ_0 | Yield point $\tilde{\sigma}_e$ | | Wave velocity c_e | |
(g/cm^3)	$(u, \tilde{\sigma})$ plane (MPa)	Transmitted profiles (MPa)	Transmitted profiles (km/s)	Ultrasonic (km/s)
7.0	600	560 ± 30		5.3
5.78	300	280 ± 15	4.7 ± 0.3	4.3
4.75	155	120 ± 10	3.0 ± 0.15	2.95
3.3		5	1.15 ± 0.1	

where ρ_{s0} is the ambient density of solid iron without any porosity. The α notation will be described later.

As with the argument to determine the Grüneisen coefficient in a previous section, we wish to use the nonporous iron Hugoniot to compute an expected stress in the material upon the arrival of the second shock. The above elastic data allow a determination of the particle velocity and foam density for the denser starting materials after the passage of the elastic precursor,

$$u_e = \frac{\tilde{\sigma}_e}{\rho_0 c_e}, \quad \rho_e = \frac{\rho_0}{(1 - \tilde{\sigma}_e / \rho_0 c_e^2)}, \quad E_e = \frac{1}{2}\tilde{\sigma}_e \left(\frac{1}{\rho_0} - \frac{1}{\rho_e} \right). \tag{5.190}$$

Our first use of the nonporous iron Hugoniot is to assume that the second shock completely presses out the voids – that is, that the solid iron equation of state provides the pressure after the passage of the shock. This says that $\tilde{\sigma}_f(\rho, E) = \tilde{\sigma}_s(\rho_s, E)$, where $\rho = \rho_s$ after the passage of the shock. Butcher and Karnes use a Murnaghan form for the reference iron Hugoniot, $\tilde{\sigma}_{sH}(\rho_s) = (K_0/b)\{(\rho_s/\rho_{s0})^b - 1\}$, with $K_0 = 166.7$ GPa, $b = 5.95$, and $\rho_{s0} = 7.83$ g/cm^3, and they use a Grüneisen parameter of 1.67 (note that if a linear U_s–u_p were used, for small strains the Murnaghan form corresponds to $k = (b + 1)/4$). With this information, the solid equation of state is

$$\tilde{\sigma}_s(\rho, E) = \tilde{\sigma}_{sH}(\rho) \left(1 - \frac{1}{2}\rho_{s0}\Gamma_0 \left(\frac{1}{\rho_{s0}} - \frac{1}{\rho} \right) \right) + \rho_{s0}\Gamma_0 E = f_s(\rho) + \rho_{s0}\Gamma_0 E, \tag{5.191}$$

where this equation defines the function f_s. Assuming that all the porosity compresses out, the jump conditions give

$$\frac{\rho_e}{\rho} = 1 - \frac{u - u_e}{U - u_e}, \quad \tilde{\sigma}_s(\rho, E) - \tilde{\sigma}_e = \rho_e(U - u_e)(u - u_e),$$

$$E - E_e = \frac{1}{2}\{\tilde{\sigma}_s(\rho, E) + \tilde{\sigma}_e\} \left(\frac{1}{\rho_e} - \frac{1}{\rho} \right). \tag{5.192}$$

The energy jump condition gives

$$E = \frac{\frac{1}{2}f_s(\rho)(1/\rho_e - 1/\rho) + \frac{1}{2}\tilde{\sigma}_e(1/\rho_0 - 1/\rho)}{1 - \frac{1}{2}\rho_{s0}\Gamma_0(1/\rho_e - 1/\rho)}. \tag{5.193}$$

Given the density and the energy, the stress is then known from the equation of state, and the corresponding particle velocity from the jump conditions is

$$(u - u_e)^2 = (\tilde{\sigma} - \tilde{\sigma}_e)(1/\rho_e - 1/\rho). \tag{5.194}$$

Figure 5.14 shows the prediction of these equations compared with the data for the porous foam. For the model computations, the three lowest densities assume

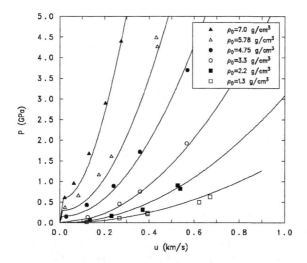

FIGURE 5.14. Hugoniot of porous iron samples, data [**33**] and computations assuming pores completely close by the passage of the shock. For the computations, the initial densities are given in the figure. The three lower initial densities do not have an elastic precursor; the three higher initial densities do, with a sound speed given by Eq. (5.189) and $\tilde{\sigma}_e = 155$, 300, and 600 MPa.

no elastic precursor and the three higher densities use the elastic precursor values given in Table 5.9.

Owing to the denominator in the energy expression Eq. (5.193), there is a qualitative difference in behavior demarcated when the denominator is zero. The qualitative difference occurs when (Exercise 5.44)

$$\rho_e = \frac{\rho_{s0}}{1 + 2/\Gamma_0}. \tag{5.195}$$

In particular, the demarcating porous density for this iron is $\rho_e = 3.56$ g/cm^3. This difference is observed in the development of the curves shown in Fig. 5.14. For the three curves where the initial porous density is below 3.56 g/cm^3 (namely 1.3, 2.2, and 3.3 g/cm^3), the final solid density after the passage of the shock decreases from ρ_{s0} as the pressure increases – that is, the stress is primarily arising from the thermal terms. For the three initial densities greater than 3.56 g/cm^3 (namely 4.75, 5.78, and 7.0 g/cm^3), the density increases from ρ_{s0} as the pressure increases.

Though these curves show reasonable agreement with respect to the stress behind the shock front, it will be noted that for the cases with higher initial density the predicted values are low. Posttest observations of these cases show that the final specimens still have porosity. Thus, the computations where complete compression occurs, and so $\rho > \rho_{s0}$, do not reflect the physical results. In these cases, it is possible to compute from the Hugoniot jump conditions the density behind the shock front. (For the large porosity initial conditions, u and U are close, so the density computation from the data does not yield good values.) However, to use the material density after the shock in the jump conditions leaves the question of what material density should be used in the equation of state, since there is still porosity.

An approach to this question is known as the p–α model. The model is often used in large-scale numerical simulations to model porous materials, particularly metals. The main idea is to assume a relationship between the actual density in the porous material ρ and a corresponding solid (no porosity) density ρ_s to be used in the solid material equation of state,

$$\alpha\rho = \rho_s, \quad \alpha \geq 1. \tag{5.196}$$

The corresponding stress in the porous material is (subscript f for foam)

$$\tilde{\sigma}_f(\rho, E) = \frac{1}{\alpha}\tilde{\sigma}_s(\rho_s, E) = \frac{1}{\alpha}\tilde{\sigma}_s(\alpha\rho, E) = \frac{1}{\alpha}\{f(\alpha\rho) + \rho_{s0}\Gamma_0 E\}, \tag{5.197}$$

where $\tilde{\sigma}_s(\rho_s, E)$ is the equation of state of the material without pores. It is assumed that the internal (thermal) energy is the same in both. The terminology p–α refers to pressure since, if strength effects are ignored, $p = \tilde{\sigma}$. The original formulation of the p–α model, due to Herrmann [92], did not include the $1/\alpha$ multiplier in the pressure: this term was added by Carroll and Holt [36]. The model only addresses the bulk material response, and does not address shear behavior explicitly (i.e., how material strength depends on porosity). As a porous material is compressed, α decreases and ρ_s increases; notice that if only void was being compressed out (i.e., α alone changed and ρ_s remained fixed at its initial value) then the pressure would remain zero, save for a small pressure change due to the thermal term. Thus, even for small-strain elastic behavior, *when ρ changes, both α and ρ_s change.* It is necessary to define the behavior of α. Typically the behavior of α is determined as a function of pressure, though it could be a function of other items such as strain and/or energy. The usual view is to think of $\alpha = \alpha(p)$ having an initial elastic response (meaning α can return to its initial value when unloaded) up to some pressure p_e, and then a region of permanent crush (meaning that α will return to a lesser value than the initial value when unloaded) up to a final upper pressure p_{crush}, where the material has no void and is fully solid (α then permanently remains 1).

Coming back to the porous iron data in question, the energy behind the shock is now given by

$$E = \frac{\frac{1}{2}f(\alpha\rho)\left(1/\rho_e - 1/\rho\right) + \frac{1}{2}\tilde{\sigma}_e(1/\rho_0 - 1/\rho)}{1 - \frac{1}{2}\rho_{s0}\Gamma_0(1/\rho_e - 1/\rho)} \tag{5.198}$$

and since we know the stress after the passage of the shock and we can compute the density, it is possible to explicitly solve for α. To be explicit, the nonlinear equation that needs to be solved for α is

$$\left\{1 - \frac{1}{2}\rho_{s0}\Gamma_0\left(\frac{1}{\rho_e} - \frac{1}{\rho}\right)\right\}\tilde{\sigma}_f \cdot \alpha - f(\alpha\rho) = \frac{1}{2}\tilde{\sigma}_e\left(\frac{1}{\rho_0} - \frac{1}{\rho}\right). \tag{5.199}$$

When solved, these values of α put the Hugoniot curves exactly through the data points. (A value for α_e can be computed by setting $\rho = \rho_e$ in this equation; when this is done, it can occur that $\alpha_0 < \alpha_e$ because strength is not explicitly addressed. In a modeling scenario, the slope $\alpha(0)$ at α_0 is determined as described below, and α_e is found by the intersection of the $\alpha(p)$ curve with the straight line elastic behavior.) Figure 5.15 and Table 5.10 show the values of α for the two largest initial densities (the lower initial densities do not produce consistent values). The curves are due to two sources. The solid curves follow Herrmann, who suggested

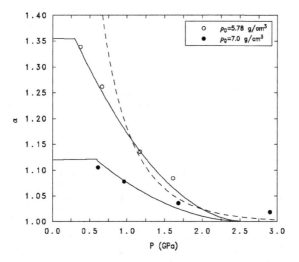

FIGURE 5.15. Values for α for the porous iron material, where the solid curves are the quadratic fits and the dashed curve is the spherical void-closing expression.

polynomial expressions for α as a function of pressure. The figure uses a particularly simple one:

$$\alpha(\tilde{\sigma}) = 1 + (\alpha_e - 1)\left(\frac{\tilde{\sigma}_{crush} - \tilde{\sigma}}{\tilde{\sigma}_{crush} - \tilde{\sigma}_e}\right)^2, \quad \tilde{\sigma}_e \leq \tilde{\sigma} \leq \tilde{\sigma}_{crush}. \tag{5.200}$$

This quadratic expression completely closes out the voids at a finite pressure; in the figure, following Butcher and Karnes, $\tilde{\sigma}_{crush} = 2.6\,\text{GPa}$. The dashed curve in the figure is based on the spherical void computation in Chapter 3, following Carroll and Holt. The spherical shell void collapse problem, with inner spherical void radius a and outer spherical shell radius b, has a void fraction

$$\phi = \left(\frac{a}{b}\right)^3, \quad \alpha = \frac{\text{mass}/(b^3 - a^3)}{\text{mass}/b^3} = \frac{1}{1 - \phi} \quad \Rightarrow \quad \phi = \frac{\alpha - 1}{\alpha}, \tag{5.201}$$

and our previous result of $\sigma_{rr}(b) = -p_\infty = -2Y\ln(b/a)$ (Eq. 3.169, where we now have a minus sign because of the reversal of signs of the deviators due to the compression) gives for compression ($p_\infty > 0$),

$$p_\infty = \frac{2Y}{3}\ln\left(\frac{b}{a}\right)^3 = -\frac{2Y}{3}\ln(\phi) = -\frac{2Y}{3}\ln\left(\frac{\alpha - 1}{\alpha}\right). \tag{5.202}$$

Inverting this expression defines a loading curve, where

$$\alpha(p) = \frac{1}{1 - \exp(-3p/2Y)}, \quad p > p_e. \tag{5.203}$$

Though this relationship between α and pressure is simple, has a nice derivation, and does not depend on Hugoniot data, it has the unfortunate behavior that the porosity is never completely crushed out (i.e, $p_{crush} = \infty$). Comparison to the porous data is shown in Fig. 5.15, where the yield of the nonporous iron was estimated to be $Y = 800\,\text{MPa}$. The last column shows the fraction of the pressure due to the thermal term.

TABLE 5.10. Values for α for porous iron.

ρ_0 g/cm^3	ρ_e g/cm^3	ρ g/cm^3	ρ_s g/cm^3	α_0	α_e	α	$\tilde{\sigma}$ MPa	$\rho_{s0}\Gamma_0 E/\tilde{\sigma}$
4.580	4.595	6.312	7.833	1.710	1.707	1.241	155	0.652
4.740	4.755	5.938	7.848	1.652	1.649	1.322	434	0.346
4.830	4.845	7.366	7.851	1.621	1.618	1.066	893	0.539
4.780	4.795	7.542	7.869	1.638	1.635	1.043	1,722	0.539
4.530	4.545	7.507	7.904	1.728	1.725	1.053	3,700	0.587
5.820	5.837	5.868	7.853	1.345	1.345	1.338	374	0.016
5.740	5.757	6.233	7.864	1.364	1.363	1.262	662	0.116
5.740	5.757	6.936	7.877	1.364	1.363	1.136	1,170	0.236
5.760	5.777	7.276	7.889	1.359	1.359	1.084	1,610	0.273
5.760	5.777	7.776	7.968	1.359	1.359	1.025	4,270	0.311
5.860	5.877	7.754	7.980	1.336	1.336	1.029	4,490	0.288
6.990	7.012	7.112	7.861	1.120	1.121	1.105	610	0.026
7.000	7.022	7.307	7.875	1.119	1.120	1.078	960	0.059
6.990	7.012	7.629	7.902	1.120	1.121	1.036	1,680	0.102
6.980	7.002	7.806	7.950	1.122	1.123	1.018	2,900	0.116
7.000	7.022	7.968	8.005	1.119	1.120	1.005	4,400	0.126

There are two additional topics related to the p–α model to address: sound speeds and how computations are carried out. First, consider the sound speed of the porous material. Computing from our expression (and writing things in terms of $p = \tilde{\sigma}$), we have

$$\left(\frac{\partial p}{\partial \rho}\right)_E = -\frac{p_s}{\alpha^2}\frac{d\alpha}{dp}\left(\frac{\partial p}{\partial \rho}\right)_E + \frac{1}{\alpha}\left(\frac{\partial p_s}{\partial \rho_s}\right)_E \left\{\rho\frac{d\alpha}{dp}\left(\frac{\partial p}{\partial \rho}\right)_E + \alpha\right\},$$

$$\left(\frac{\partial p}{\partial E}\right)_\rho = \frac{1}{\alpha}\left(\frac{\partial p_s}{\partial E}\right)_\rho = \frac{1}{\alpha}\left(\frac{\partial p_s}{\partial E}\right)_{\rho_s}.$$

(5.204)

The last equality holds because of our assumption that α is not a function of energy. Writing the solid material isothermal bulk modulus as $K_{sT} = \rho_s(\partial p_s/\partial \rho)_E$ and the pressure derivative of α as $\alpha' = d\alpha/dp < 0$, with some arithmetic these expressions allow a computation of the sound speed from Eq. (5.88),

$$c^2 = c_s^2 + \frac{K_{sT}}{\rho_s}\frac{(K_{sT} - p_s)\alpha'}{\{\alpha^2 - (K_{sT} - p_s)\alpha'\}},$$

(5.205)

where c_s is the solid material sound speed. As examples, during loading that crushes out pores the slopes for the spherical-void α and the polynomial α are given by, respectively,

$$\alpha'(p) = -\frac{3\alpha^2}{2Y}, \quad \alpha'(p) = -2\frac{\sqrt{(\alpha_e - 1)(\alpha - 1)}}{p_{crush} - p_e}, \quad p > p_e.$$

(5.206)

If we consider the initial ambient condition with $p = 0$, the elastic wave speed in the porous material is

$$c_e^2 = \frac{\alpha_0^2 c_{s0}^2}{\alpha_0^2 - K_{sT0}\alpha'(0)}.$$

(5.207)

Thus, we are able to compute the initial slope $\alpha' = d\alpha/dp$ from the ambient wave speed in the porous material compared to the ambient wave speed in the nonporous material,

$$\alpha'(0) = -\frac{\alpha_0^2}{K_{sT0}}\left\{\left(\frac{c_{s0}}{c_e}\right)^2 - 1\right\}. \tag{5.208}$$

It is expected that $c_e < c_{s0}$ and hence $\alpha'(0) < 0$, but if the sound speeds are the same in the solid material and in the porous material then $\alpha' = 0$. In the elastic region (where the change in α is reversible), we have

$$\alpha(p) = \alpha_0 + \alpha'(0)p, \quad p < p_e. \tag{5.209}$$

Using this expression, the minimum α in the elastic region is $\alpha_e = \alpha(p_e) = \alpha_0 + \alpha'(0)p_e$. Due to the bulk modulus of the solid material in the denominator of Eq. (5.208), α_0 and α_e differ by a very small amount.

The α' expressions in the previous paragraph apply to loading. While unloading the material with porosity, Herrmann suggests basing it on a smooth interpolation of the sound speed in the initial porous material and the bulk sound speed of the solid material with no porosity,

$$c = h(\alpha)c_{s0}, \quad \text{e.g.} \quad h(\alpha) = \frac{\alpha - 1}{\alpha_e - 1}\frac{c_e}{c_{s0}} + \frac{\alpha_e - \alpha}{\alpha_e - 1}. \tag{5.210}$$

Denoting the historical maximum of the pressure by $p_{\max}$, unloading or reloading is assumed to occur on a line connected to the maximum crush point $\alpha = \alpha_{\min} = \alpha(p_{\max})$, $p = p_{\max}$ with the slope from Eq. (5.208)

$$\alpha'(p) = \alpha'(p_{\max}) = -\frac{\alpha_{\min}^2}{K_{sT0}}\left\{\frac{1}{h^2(\alpha_{\min})} - 1\right\}, \quad p \le p_{\max}. \tag{5.211}$$

Thus, $\alpha(p) = \alpha_{\min} + \alpha'(p_{\max})(p - p_{\max})$, and the behavior of α in both loading and unloading has been defined. To implement the p–α model, typically a hydrocode provides a strain or density change $d\rho$ and an energy change dE. Knowing α' as given above (noting that the values differ for loading and unloading) allows a determination of the pressure change dp from the equation of state,

$$dp = -\frac{p\alpha'}{\alpha}dp + \left(\frac{\partial p_s}{\partial \rho_s}\right)_E\left\{\frac{\rho\alpha'}{\alpha}dp + d\rho\right\} + \frac{1}{\alpha}\left(\frac{\partial p_s}{\partial E}\right)_{\rho_s}dE, \tag{5.212}$$

which provides $d\alpha = \alpha'dp$. The change in density is

$$d\rho = d\left(\frac{\rho_s}{\alpha}\right) = \frac{1}{\alpha}d\rho_s - \frac{\rho_s}{\alpha^2}d\alpha = \frac{1}{\alpha}d\rho_s - \frac{\rho_s}{\alpha^2}\alpha'dp, \tag{5.213}$$

which allows the determination of $d\rho_s$.

As a final example, we return our attention to the interesting behavior of highly porous materials. A case where the data clearly displays decreasing density with increasing shock stress for a porous material is the Los Alamos data for uranium dioxide [139]. Table 5.11 contains the experimental results for the material. Figure 5.16 shows these data along with two sets of computations. The solid lines in the figure are given by plotting the least-squares U_s–u_p data through each of the data sets. The lowest density data set shows decreasing density with increasing pressure because the $u = 0$ intercept is negative. The dashed curves are based on the assumption that the solid material Hugoniot is given by the U_s–u_p form and that all the void is compressed out, similar to the first analysis of the porous iron

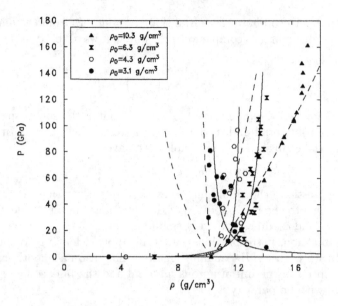

FIGURE 5.16. Hugoniot data for uranium dioxide with differing initial densities (data from [**139**]).

above. In particular, the stress state as a function of final density is (assuming there is no precursor)

$$\tilde{\sigma}_f(\rho) = \tilde{\sigma}_{sH}(\rho) \frac{1 - \frac{1}{2}\rho_{s0}\Gamma_0 \left(1/\rho_{s0} - 1/\rho\right)}{1 - \frac{1}{2}\rho_{s0}\Gamma_0(1/\rho_0 - 1/\rho)}. \tag{5.214}$$

With $\Gamma_0 = 1.6$, the demarcating initial density is $\rho_0 = 4.6\ \mathrm{g/cm^3}$. For the initial density near this value, the final state is almost independent of density (a vertical line on the graph, similar to the snowplow model, Exercise 5.46). The initial density of $3.1\ \mathrm{g/cm^3}$ clearly decreases in density with increasing pressure. It will be seen that these theoretical curves are qualitatively similar to the data results, but the center is ρ_{s0}, whereas the data appear to be centered at a higher density. The dashed curve furthest to the right is the consolidated material response.

5.15. Phase Changes

In addition to plasticity and thermal effects, there are physical processes that affect the propagation of waves in solids. Phase changes can occur. Aluminum melts on the Hugoniot at 125 GPa at an estimated temperature of roughly 4,700 K. As pressures increase, so does the melt temperature; hence the high temperature of melt during shock impact. Based on the properties in Table 5.2, an aluminum plate on plate impact would require an impact velocity of (Eq. 5.107)

$$V = 2u = \frac{\sqrt{c_0^2 + 4k\tilde{\sigma}/\rho_0} - c_0}{k} \tag{5.215}$$

or 8.4 km/s to cause the aluminum behind the shock front to melt. Similarly, iron melts on the Hugoniot at 243 GPa at a temperature of roughly 6,700 K. Again, melting occurs because of to the passage of the compression wave, and a plate on plate impact of 6.4 km/s is required to melt iron. It is possible to tell if the material

TABLE 5.11. Data for uranium dioxide (UO_2) with various initial densities [139] with U_s–u_p fits in km/s.

ρ_0 g/cm^3	U km/s	u km/s	ρ g/cm^3	$\tilde{\sigma}$ GPa	ρ_0 g/cm^3	U km/s	u km/s	ρ g/cm^3	$\tilde{\sigma}$ GPa
$\rho_0 = 10.3$ g/cm^3, $U = 3.72 + 0.83u$					$\rho_0 = 6.3$ g/cm^3, $U = 0.43 + 1.70u$				
10.30	3.985	0.000	10.300	0	6.425	2.152	1.025	12.269	14
10.38	4.141	0.568	12.030	24	6.359	2.081	1.032	12.615	14
10.31	4.059	0.571	11.998	24	6.445	2.585	1.246	12.442	21
10.31	4.276	0.753	12.514	33	6.354	2.571	1.262	12.480	21
10.39	4.642	1.063	13.476	51	6.443	3.234	1.645	13.113	34
10.30	4.573	1.088	13.516	51	6.303	3.193	1.684	13.337	34
10.32	4.627	1.223	14.028	58	6.309	3.444	1.828	13.446	40
10.31	4.783	1.355	14.385	67	6.247	3.891	1.948	12.510	47
10.31	5.086	1.658	15.297	87	6.466	4.142	2.081	12.995	56
10.32	5.168	1.712	15.432	91	6.342	4.397	2.285	13.203	64
10.42	5.308	1.874	16.106	104	6.269	4.532	2.356	13.057	67
10.34	5.312	1.904	16.117	105	6.451	4.809	2.512	13.506	78
10.31	5.390	1.983	16.311	110	6.216	4.758	2.595	13.673	77
10.32	5.637	2.147	16.669	125	6.346	4.865	2.659	13.995	82
10.43	5.754	2.169	16.740	130	6.317	5.254	2.840	13.749	94
10.38	5.962	2.268	16.753	140	6.465	5.373	2.855	13.795	99
10.32	6.123	2.334	16.677	147	6.219	5.567	3.028	13.636	105
10.30	6.277	2.493	17.086	161	6.277	5.883	3.286	14.219	121
$\rho_0 = 4.3$ g/cm^3, $U = 0.12 + 1.51u$					$\rho_0 = 3.1$ g/cm^3, $U = -0.22 + 1.47u$				
4.772	1.563	0.880	10.920	6.6	3.144	1.741	1.355	14.181	7.4
4.281	1.848	1.230	12.801	9.7	3.236	2.298	1.651	11.494	12
4.428	2.468	1.476	11.016	16	3.108	2.968	2.225	12.415	21
4.234	2.337	1.517	12.067	15	2.910	2.971	2.248	11.958	19
4.249	3.042	1.999	12.393	26	3.183	3.273	2.393	11.839	25
4.270	3.287	2.171	12.577	30	3.145	3.728	2.567	10.099	30
4.297	3.750	2.302	11.128	37	3.160	4.259	3.023	10.889	41
4.128	4.277	2.753	11.585	49	3.155	4.426	3.098	10.515	43
4.317	4.404	2.780	11.707	53	3.122	4.699	3.394	11.242	50
4.430	4.827	2.918	11.201	62	2.927	4.746	3.418	10.460	47
4.125	4.651	3.092	12.306	59	3.258	4.800	3.462	11.688	54
4.233	4.752	3.169	12.707	64	3.118	5.269	3.733	10.696	61
4.283	5.206	3.348	12.001	75	2.945	5.295	3.890	11.099	61
4.304	5.535	3.536	11.917	84	3.136	5.682	3.926	10.147	70
4.237	5.918	3.855	12.154	97	3.113	6.116	4.256	10.236	81

has melted by looking at the initial rarefaction after the passage of the shock to see if the rarefaction has an elastic precursor (Eq. 5.188). These data is shown for iron in Fig. 5.17 from [75], where the velocity of the release wave is compared to the computed value for that pressure based on the equation of state and Hugoniot data. Before melt, the release wave velocity c_L is clearly higher than the bulk sound velocity $c_K (= c_B$ in plot) for the higher pressure, indicating there is shear strength

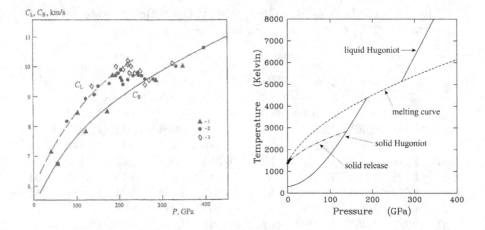

FIGURE 5.17. Left: Rarefaction sound speed measurements in iron after the passage of a shock to the Hugoniot pressure based on data from three research groups, showing melt in the region 200 to 250 GPa. (From Funtikov [75].) Right: Pressure–temperature diagram for copper showing pressure dependent melt and the Hugoniot path from ambient initial conditions. (From Kerley [116].)

and hence the material is solid and not melted. Melting is not at an isolated velocity: incipient melt begins at a certain velocity and, as the impact velocity increases and more thermal energy is deposited in the material, complete melt is reached at a higher velocity, where it is seen that the release wave speed equals the expected bulk sound speed for that pressure. Also shown in Fig. 5.17 is a pressure–temperature diagram for copper. Based on a detailed equation of state developed by Kerley, the pressure-dependent melt curve is shown. The Hugoniot from ambient conditions is shown. Once the Hugoniot reaches incipient melt, the slope changes to follow the melt curve until complete melt is achieved. Once complete melt is achieved, the remainder of the displayed Hugoniot is melted states. Also shown on the curve is one release isentrope, from $\tilde{\sigma}_H = 138$ GPa, which, Kerley reports, Kanel has shown to be the onset of melt upon release. Given a formula for the melt temperature of copper, we can reproduce this curve using our analytic formula, and we can also compute the release from the solid state (Exercise 5.25).

Upon unloading from a shock, the material is at a higher temperature and the final state of the material can be melted due to the passage of the shock. The impact speed that leads to a final state melt is less than that required to melt the material on the Hugoniot because the low-pressure melt temperature is lower than the high-pressure melt temperature. The melt temperature at ambient for pure aluminum is 934 K and for pure iron is 1,811 K. Using our approximate expressions for temperature, symmetric impact of 6061-T6 aluminum will produce incipient melt upon unloading at an impact speed of 5.3 km/s and a symmetric impact of 304 stainless steel will produce to incipient melt upon unloading at an impact speed of 5.5 km/s (Exercise 5.29). Ignoring the difference between melted and solid material on the expansion (i.e., using the same equation of state), the material will be completely melted upon unloading for aluminum at an impact speed of 6.6 km/s and for steel at an impact speed of 6.4 km/s, assuming that the heat of fusion of aluminum is 396 J/g (notice $H_F/C_v = 445$ K) and iron 267 J/g ($H_F/C_v = 593$ K). Continuing to assume the heat capacity remains constant after melt, similar extrapolations can be obtained for incipient vaporization upon unloading. A symmetric impact of aluminum, with a vaporization temperature of 2,740 K and a 12.2 kJ/g heat of vaporization, begins incipient vaporization at 10.3 km/s. A symmetric impact of SS304, with a vaporization temperature of 3,160 K and a 6.29 kJ/g heat of vaporization, begins incipient vaporization at 8.0 km/s. Though we stated the heats of vaporization for informational purposes, we will not estimate complete vaporization, since as the material turns to vapor, an equation of state based on the solid Hugoniot cannot conceivably give good pressure–density values on unloading. All these velocities are approximate; sophisticated equations of state should be used to obtain better estimates. See Table 5.5 for aluminum melt and vaporization upon unloading.

As an example of melting and vaporization, consider the following application [206]. The Cassini spacecraft arrived at Saturn in 2004. While orbiting the planet, Cassini performed over 20 flybys of the moon Enceladus. Plumes have been observed near the surface of the moon that throw primarily water ice into nearby space. During the flybys, the composition of the moon's plume has been measured by the Ion and Neutral Mass Spectrometer (INMS). This instrument measures the molecular weight of molecules in rarefied gases. Neutral particles in the plume first enter the instrument in a spherical titanium antechamber, which has an inner diameter of about 2.5 cm. During high-velocity encounters of over 17 km/s the INMS recorded different compositions than low velocity flybys of 7 to 8 km/s. The question arises whether there is a qualitative difference in ice grain impacts into the walls of the titanium antechamber at low and high impact speeds.

To pursue the question, the impact speeds of ice that first melt and then first vaporize titanium will be computed. The initial density, U_s–u_p relation for higher velocities, and Grüneisen coefficient for pure titanium in Appendix C were used. An initial temperature of 70 K is assumed, representative of the instrument temperature on the spacecraft. Additionally, a constant heat capacity of 524 J/kg K, a melting temperature of 1,941 K, a heat of fusion of 2.96×10^5 J/kg, and a vaporization temperature of 3,560 K were assumed. Equation (5.169) then produces a temperature upon unloading. This procedure produces a particle velocity in the titanium at the onset of melt of 3.15 km/s and for the onset of vaporization of 4.35 km/s. These values are of course approximate.

We now combine the unloading response of the titanium with the impact with ice. Hugoniot behavior for ice is taken from Stewart and Ahrens [177], giving the Hugoniot locus $U = 1.70$ km/s$+1.44u$ with an initial density of 0.932 g/cm^3 (the stated density at 100 K). Using the Hugoniot jump conditions, impacting this ice into the titanium material produces a temperature upon unloading as shown in Fig. 5.18. In particular, the onset of titanium melt occurs at 11.9 km/s impact speed and the onset of titanium vaporization is at 15.6 km/s impact speed. This result gives us an initial indication that there may be a qualitative difference in the Enceladus flybys, as to the formation of titanium vapor in the titanium antechamber. The shocked titanium state for these two impact speeds has a temperature on the Hugoniot of 3,215 K and 7,480 K (ignoring possible melting), respectively;

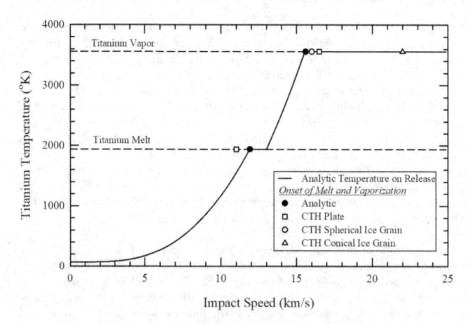

FIGURE 5.18. The temperature upon unloading from ice at 100 K striking 70 K titanium. (From Walker et al. [206].)

however, these are confined temperatures, and the temperatures drop considerably (to the melt and vaporization temperature) during release. The compressed strains $\tilde{\varepsilon}$ at the Hugoniot states are 0.385 and 0.461, respectively. In the figure, the melt and vaporization regions are seen as constant temperature regions in the plot, and the speeds at which partial melt and partial vaporization are first observed are marked. The liquid-vapor plateau lasts till approximately 29 km/s, or 9 km/s particle velocity in the titanium. Also shown on the plot are the results of both one-dimensional and axisymmetric CTH computations with more sophisticated equations of state showing similar behavior. The upshot of the argument is that this high-impact-speed-produced titanium vapor may undergo chemical reactions with other molecules in the antechamber, which then affect the resulting mass spectrum, though at the time of this writing the quantitative aspect of the chemistry is still under analysis.

There is a crystal structure change in iron at 13 GPa at room temperature, from the α-Fe phase, which is body centered cubic, to the ε-Fe phase which is hexagonal close packed. Shock techniques were instrumental in finding and confirming this phase transition (Exercise 5.26). The iron crystal phase and melt transition pressures and temperatures presented in this section, as well as further details on melting and phase changes, can be found in Young [214].

5.16. Detonation of Explosives

Owing to the compression and energy deposited by a shock wave, chemical reactions can be induced in the material. The most dramatic example is the detonation of explosives. In the situation of an exothermic reaction, there are two Hugoniot curves of interest. The first is the Hugoniot of the unreacted material.

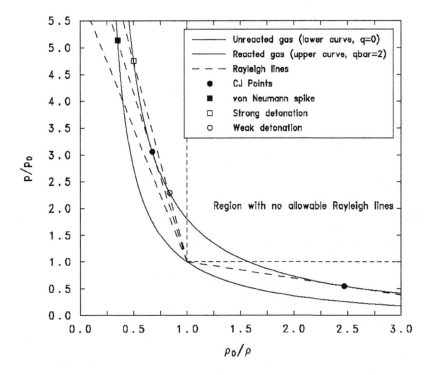

FIGURE 5.19. The unreacted gas and explosive products (reacted) gas Hugoniots, showing Rayleigh lines, the Chapman–Jouguet points, and the von Neumann spike for an ideal gas where $\gamma = 7/5$ before and after the reaction and the energy release is $\bar{q} = 2$.

The second is the Hugoniot of the material after the energy has been released – that is, the reaction products Hugoniot. In this section we will work exclusively in the $1/\rho$–$\tilde{\sigma}$ plane. In this plane, the reacted Hugoniot typically lies above the unreacted Hugoniot, as indicated in Fig. 5.19. We will explore some interesting behavior that can occur in such a situation. Since explosive products are gases, in this section pressure $p = \tilde{\sigma}$ will be used in the expressions.

Since in addition to solid and liquid explosives, gaseous explosive mixtures are of interest, Fig. 5.19 is for gaseous explosives and the initial pressure is greater than zero. If the explosive is a liquid or a solid, then the initial pressure is nearly zero. The figure shows three Rayleigh lines with the final state to the left of the initial state (i.e., compressive). The leftmost one lies completely below the reacted Hugoniot. In this situation, there is no detonation, since there is not a Hugoniot jump that takes the initial unreacted state to a reacted state (the products Hugoniot). The fact that the Rayleigh line lies beneath the reacted Hugoniot shows that there is a minimum detonation velocity. The second curve touches the reacted product tangentially. The point of tangency is the Chapman–Jouguet (CJ) point. The CJ point has the lowest speed at which a detonation can occur. The Rayleigh line is shown to extend to the unreacted Hugoniot. The ZND (Zeldovich, von Neumann, Döring) theory of detonation is that a shock propagates into the unreacted material, and thus the Hugoniot pressure on the unreacted Hugoniot is realized at the leading edge of the shock wave. Behind the shock wave, due to the available energy from the shock compression, the chemical reaction occurs over a fixed, finite distance. The peak pressure is referred to as the von Neumann spike. Over this fixed-length reaction zone the material state moves back along the Rayleigh line from the von Neumann spike to the CJ point. The spike is a very narrow region, and what is typically measured for an explosive is the pressure at the CJ point. The self-sustaining reaction runs at the detonation velocity as given by the Rayleigh line that intersects the CJ point, and this CJ point behavior is what is usually observed in explosive detonation. The third and steepest Rayleigh line shows two intersections with the reaction products Hugoniot. The higher pressure intersection is referred to as a strong detonation and is seen experimentally when the detonation is overdriven; and the lower intersection is referred to as weak detonation and is not observed.

The Rayleigh line being tangent to the Hugoniot has interesting implications. Tangency says (the Chapman–Jouguet subscript CJ is used to identify the point of tangency)

$$\left(\frac{dp}{d(1/\rho)} \right)_{H,CJ} = \frac{p_{CJ} - p_0}{1/\rho_{CJ} - 1/\rho_0}, \tag{5.216}$$

where subscript H means along the Hugoniot. The momentum and mass jump conditions give

$$\left(\frac{dp}{d(1/\rho)} \right)_{H,CJ} = \frac{p_{CJ} - p_0}{1/\rho_{CJ} - 1/\rho_0} = \frac{\rho_0 U_{CJ} u_{CJ}}{1/\rho_{CJ} - 1/\rho_0} = -\rho_0^2 U_{CJ}^2 = -\rho_0^2 D^2, \tag{5.217}$$

where for detonation the detonation velocity D equals shock velocity: $D = U_{CJ}$. The energy jump condition is

$$E - E_0 = \frac{1}{2} (p + p_0) \left(\frac{1}{\rho_0} - \frac{1}{\rho} \right), \tag{5.218}$$

which when differentiated gives

$$\left(\frac{\partial E}{\partial (1/\rho)} \right)_H = \frac{1}{2} \left(\frac{\partial p}{\partial (1/\rho)} \right)_H \left(\frac{1}{\rho_0} - \frac{1}{\rho} \right) - \frac{1}{2} (p + p_0). \tag{5.219}$$

At the point of tangency the energy derivative is

$$\left(\frac{\partial E}{\partial (1/\rho)} \right)_{H,CJ} = \frac{1}{2} \left(\frac{\partial p}{\partial (1/\rho)} \right)_{H,CJ} \left(\frac{1}{\rho_0} - \frac{1}{\rho_{CJ}} \right) - \frac{1}{2} (p_{CJ} + p_0) = -p_{CJ}. \tag{5.220}$$

However, the general thermodynamic expression $dE = Tds - pd(1/\rho)$ tells us that the isentrope behaves as

$$\left(\frac{\partial E}{\partial (1/\rho)} \right)_s = -p \quad \Rightarrow \quad \left(\frac{\partial E}{\partial (1/\rho)} \right)_{s,CJ} = -p_{CJ}. \tag{5.221}$$

At the point of tangency (the CJ point) the energy derivatives match and thus the Hugoniot and the isentrope match. For example, with $p = p(\rho, E)$,

$$\left(\frac{\partial p}{\partial \rho} \right)_{H,CJ} = \left(\frac{\partial p}{\partial \rho} \right)_E + \left(\frac{\partial p}{\partial E} \right)_\rho \left(\frac{\partial E}{\partial \rho} \right)_H \Big|_{CJ}$$

$$= \left(\frac{\partial p}{\partial \rho} \right)_E + \left(\frac{\partial p}{\partial E} \right)_\rho \left(\frac{\partial E}{\partial \rho} \right)_s \Big|_{CJ} = \left(\frac{\partial p}{\partial \rho} \right)_{s,CJ}. \tag{5.222}$$

The sound speed is then

$$c_{CJ}^2 = \left(\frac{\partial p}{\partial \rho}\right)_{s,CJ} = \left(\frac{\partial p}{\partial \rho}\right)_{H,CJ} = -\frac{1}{\rho_{CJ}^2}\left(\frac{\partial p}{\partial(1/\rho)}\right)_{H,CJ} = \frac{\rho_0^2}{\rho_{CJ}^2}U_{CJ}^2, \quad (5.223)$$

or

$$c_{CJ} = \frac{\rho_0}{\rho_{CJ}}U_{CJ} = \frac{\rho_0}{\rho_{CJ}}D. \quad (5.224)$$

This result combines with the mass jump condition to give

$$u_{CJ} + c_{CJ} = U_{CJ}\left(1 - \frac{\rho_0}{\rho_{CJ}}\right) + \frac{\rho}{\rho_{CJ}}U_{CJ} = U_{CJ} = D. \quad (5.225)$$

THEOREM 5.9. *If the Rayleigh line is tangent to the Hugoniot curve, then* $u + c = U$ *and* $c = \rho_0 U/\rho$ *at the point of tangency.*

This result says that a detonation wave running at the Chapman–Jouguet point can be self sustaining, because rarefaction waves, which travel at the local sound speed, do not catch it from behind. (Notice that theorem also trivially holds for a small amplitude wave at the initial density.)

To facilitate explicitly writing a solution, we introduce the adiabatic expansion exponent

$$\gamma = -\left(\frac{\partial \ln(p)}{\partial \ln(1/\rho)}\right)_s = -\frac{1}{\rho p}\left(\frac{\partial p}{\partial(1/\rho)}\right)_s. \quad (5.226)$$

For simplicity of expression we write γ when in particular we mean this expression evaluated at the Chapman–Jouguet point – that is, $\gamma = \gamma_{CJ}$. If the explosive products behave like an ideal gas, then (as is common) γ is the ratio of the specific heats. For gaseous explosive products from solid explosives, γ is typically around 3. Using the slope of Rayleigh line at tangency,

$$\gamma = \frac{1}{\rho_{CJ}p_{CJ}}\left(\frac{p_{CJ} - p_0}{1/\rho_0 - 1/\rho_{CJ}}\right) \quad \Rightarrow \quad \frac{1}{\rho_{CJ}} = \frac{\gamma/\rho_0}{\gamma + 1 - p_0/p_{CJ}}. \quad (5.227)$$

Since the slope of the Rayleigh line also equals $-\rho_0^2 D^2$, some algebra yields the CJ pressure

$$p_{CJ} = \frac{\rho_0 D^2}{\gamma + 1 - p_0/p_{CJ}} = \frac{p_0 + \rho_0 D^2}{\gamma + 1}. \quad (5.228)$$

The CJ density is

$$\rho_{CJ} = \frac{\gamma + 1 - p_0/p_{CJ}}{\gamma}\rho_0 = \frac{\gamma + 1}{\gamma}\frac{\rho_0}{1 + p_0/\rho_0 D^2} \quad (5.229)$$

and the particle velocity at the CJ point is

$$u_{CJ} = D\left(1 - \frac{\rho_0}{\rho_{CJ}}\right) = \frac{1 - p_0/p_{CJ}}{\gamma + 1 - p_0/p_{CJ}}D = \frac{D}{\gamma + 1}\left(1 - \frac{\gamma p_0}{\rho_0 D^2}\right). \quad (5.230)$$

If p_0 is negligible compared to the CJ pressure, as occurs for solid and liquid explosives, these expressions simplify to

$$p_{CJ} = \frac{\rho_0 D^2}{\gamma + 1}, \quad \frac{\rho_{CJ}}{\rho_0} = \frac{\gamma + 1}{\gamma}, \quad u_{CJ} = \frac{D}{\gamma + 1}, \quad c_{CJ} = \frac{\gamma}{\gamma + 1}D. \quad (5.231)$$

Again, for condensed explosives typically $\gamma \approx 3$. If a condensed explosive has explosive product gases behaving as an ideal gas with $p = (\gamma - 1)\rho(E + q)$, where q

is the specific energy released by the chemical reaction, then some algebra shows (where E is the energy deposited by the shock at the CJ point) that

$$p_{CJ} = 2(\gamma - 1)\rho_0 q, \quad D = \sqrt{2(\gamma^2 - 1)q}. \tag{5.232}$$

Thus the energy released is directly proportional to the CJ pressure and behaves as the square of the detonation velocity. These expressions are approximate as, for example, experimentally the detonation velocity D depends on the initial density of the explosive, unlike the above expression (recall q is a specific energy and thus depends on explosive mass, not density).

Other than the previous few sentences, until now we have avoided any discussion of equation of state of the explosive product gases, though such an equation is implicit in γ. One of the most used post-detonation equations of state is the Jones–Wilkins–Lee (JWL) equation of state. The release isentrope through the CJ point for the JWL equation of state is

$$p_s(\rho) = A \exp\left(-R_1 \frac{\rho_0}{\rho}\right) + B \exp\left(-R_2 \frac{\rho_0}{\rho}\right) + C \left(\frac{\rho_0}{\rho}\right)^{-(\omega+1)}, \tag{5.233}$$

where A, B, C, R_1, R_2, and ω are constants. The energy along the isentrope can be found from $dE = Tds - pd(1/\rho) = -pd(1/\rho)$,

$$E_s(\rho) = \frac{1}{\rho_0}\left\{\frac{A}{R_1}\exp\left(-R_1\frac{\rho_0}{\rho}\right) + \frac{B}{R_2}\exp\left(-R_2\frac{\rho_0}{\rho}\right) + \frac{C}{\omega}\left(\frac{\rho_0}{\rho}\right)^{-\omega}\right\} + \text{constant}. \tag{5.234}$$

Using the isentrope as the reference curve, and setting the integration constant to zero to produce ideal gas behavior at low densities, the Mie–Grüneisen equation of state yields

$$p(\rho, E) = A\left(1 - \frac{\Gamma\rho}{R_1\rho_0}\right)\exp\left(-R_1\frac{\rho_0}{\rho}\right) + B\left(1 - \frac{\Gamma\rho}{R_2\rho_0}\right)\exp\left(-R_2\frac{\rho_0}{\rho}\right)$$
$$+ \left(1 - \frac{\Gamma}{\omega}\right)C\left(\frac{\rho_0}{\rho}\right)^{-(\omega+1)} + \rho\Gamma E, \tag{5.235}$$

which is the JWL equation of state. The $C(\rho_0/\rho)^{-(\omega+1)}$ term in the isentrope is the same as the ideal gas adiabatic expansion if $\gamma = \omega + 1$ (here γ is the ideal gas ratio of the specific heats, not the CJ point γ). For large expansion ($\rho < \rho_0/10$) this ideal-gas-like term entirely accounts for the pressure, while at the CJ point it accounts for less than 10% of the pressure. Since the ideal gas equation of state is a Mie–Grüneisen equation of state with $\Gamma = \gamma - 1$, it is natural to assume $\Gamma = \omega$. When this is done, the $C(\rho_0/\rho)^{-(\omega+1)}$ term drops out of the equation of state, yielding the form typically given,

$$p(\rho, E) = A\left(1 - \frac{\omega\rho}{R_1\rho_0}\right)\exp\left(-R_1\frac{\rho_0}{\rho}\right) + B\left(1 - \frac{\omega\rho}{R_2\rho_0}\right)\exp\left(-R_2\frac{\rho_0}{\rho}\right) + \rho\omega E. \tag{5.236}$$

As an example, TNT has $\rho_0 = 1.63$ g/cm^3, $\rho_{CJ} = 2.23$ g/cm^3, $p_{CJ} = 21$ GPa, $D = 6.93$ km/s, and JWL values are $A = 371.2$ GPa, $B = 3.231$ GPa, $C = 1.045$ GPa, $R_1 = 4.15$, $R_2 = 0.95$, $\omega = 0.30$, and $E_{CJ} = 4.32 \times 10^{10}$ cm^2/s^2.[6] These values are

---

[6] Note that $p(\rho_{CJ}, E_{CJ}) = p_{CJ}$ defines E_{CJ}; E_{CJ} is not the chemical energy released by the explosive, nor does it come from the energy jump condition, which says $E_{CJ} - E_0 \approx \frac{1}{2}u_{CJ}^2 =$

typical for high explosives. As to the adiabatic expansion exponent at the CJ point, a computation yields

$$\gamma = -\left(\frac{\partial \ln(p)}{\partial \ln(1/\rho)}\right)_{S,CJ} = \frac{\rho_0}{\rho_{CJ}p_{CJ}}\left\{AR_1\exp\left(-R_1\frac{\rho_0}{\rho_{CJ}}\right) + BR_2\exp\left(-R_2\frac{\rho_0}{\rho_{CJ}}\right)\right.$$

$$\left. +C(1+\omega)\left(\frac{\rho_0}{\rho_{CJ}}\right)^{-(\omega+2)}\right\} = 2.727. \quad (5.237)$$

The values for A, B, R_1, R_2, C, and ω are usually obtained by matching copper cylinder expansion data where a 2.54 cm diameter, 30.5 cm long cylinder of explosive is encased in a 0.26 cm thick copper tube, detonated at one end, and the radial expansion history of the copper tube at a point roughly 20 cm from the detonation point is recorded. Since a typical use of the equation of state is driving metal plates, the JWL equation of state works well as it has been calibrated in a similar fashion. The size of the reaction zone (i.e., the physical distance separating the von Neumann spike from the CJ point at the front of the detonation as the shock propagates) is estimated to be on the order of 0.01 cm.

Let us return to gaseous explosives, where the initial (unreacted) state is an ideal gas given by $p = (\gamma_0 - 1)\rho E$ and the product (reacted) equation of state is given by $p = (\gamma - 1)\rho(E + q)$. Given the well known thermodynamics of gaseous systems, it is possible to compute the energy released by the reaction, as well as the various heat capacities and effects of changes in the number of moles of gas (the different ratio of specific heats indicates the possibility of a different number of moles of gas). The term q includes the heat of reaction as well as any adjustments due to different molalities of the gas (if the reactants and the reaction products have the same number of gas molecules, then q is strictly the energy released in the reaction). The energy jump condition and some algebra yields

$$\frac{p}{2}\left(\frac{\gamma+1}{\gamma-1}\frac{1}{\rho} - \frac{1}{\rho_0}\right) = q + \frac{p_0}{2}\left(\frac{\gamma_0+1}{\gamma_0-1}\frac{1}{\rho_0} - \frac{1}{\rho}\right). \quad (5.238)$$

If we assume that the ratio of the specific heats before and after detonation is the same, then this equation simplifies to

$$\left(\frac{p}{p_0} + \chi\right)\left(\frac{\rho_0}{\rho} - \chi\right) = 1 - \chi^2 + 2\chi\bar{q} = \kappa, \quad \text{where} \quad \chi = \frac{\gamma-1}{\gamma+1} \quad \text{and} \quad \bar{q} = \frac{\rho_0 q}{p_0}. \quad (5.239)$$

Since $\gamma > 1$, $0 < \chi < 1$, and $\kappa > 0$. If $\bar{q} = 0$, this curve is the initial gas equation of state and it is seen to pass through the value $(1,1)$. Figure 5.19 shows both the unreacted equation of state as well as the reacted equation of state for $\gamma = 7/5$ and $\bar{q} = 2$ (this amount of energy release is small, as is chosen so that the figure elements are easier to see). Two CJ points are given by

$$\frac{\rho_0}{\rho_{CJ}} = \chi + \kappa\frac{1 \pm \sqrt{1 - (1-\chi^2)/\kappa}}{1 + \chi}. \quad (5.240)$$

1.73 km^2/s^2. The energy jump condition provides the change in internal energy due to the compression at the shock detonation front and does not include the chemical energy released due to the detonation. Also, we do not know the initial internal energy consistent with the JWL equation of state of the predetonation solid. In most descriptions of the JWL model the energy is energy per unit initial volume rather than specific energy as presented here and the value "E_0" in tables equals our $\rho_0 E_{CJ} = 7$ GPa. (Specific numerical values are for TNT.)

The CJ point to the left of ambient corresponds to detonation and the CJ point to the right of ambient will be discussed below. For the specific γ and $\bar{q}$, the CJ densities ρ_{CJ}/ρ_0 are 1.48 and 0.41, and corresponding CJ pressures p_{CJ}/p_0 are 3.05 and 0.55, and with the ambient sound speed in the unreacted gas $c_0 = \sqrt{\gamma p_0/\rho_0}$ the corresponding CJ detonation velocities D/c_0 are 2.13 and 0.47, where

$$\frac{D}{c_0} = \sqrt{\left(\frac{\gamma+1}{\gamma}\right)\frac{p_{CJ}}{p_0} - \frac{1}{\gamma}}. \tag{5.241}$$

The first velocity is the detonation, and material is compressed at the detonation front.

The above discussion of Fig. 5.19 was strictly about detonation waves that compressed the materials, thus looking to the left of the initial value on the figure in terms of specific volume. What about looking to the right? In a nonreacting Hugoniot, the figure to the right of the initial specific volume corresponds to rarefaction shocks, which as we described are typically not seen in nature because of the curvature of the Hugoniot. If we consider Rayleigh lines looking to the right for a reactive situation, if the slope of the Rayleigh line in the figure is positive, since the slope of the Rayleigh line is $-\rho_0^2 D^2$, the Rayleigh line is not physically realizable, as it would correspond to an imaginary detonation velocity. Thus, there are no acceptable Rayleigh lines until the slope again becomes negative, which occurs to the right when the reacted Hugoniot pressure drops below that of the initial pressure. One CJ state exists to the right. Solutions to the right correspond to deflagration (i.e., burning) of the gaseous mixture, rather than detonation, and in particular allow an estimate of the flame speed for laminar pre-mixed flames. The CJ point to the right corresponds to a drop in pressure and density, clearly different physical phenomena than the detonation event.

5.17. Sources

Good references for detailed shock wave analysis are Courant and Friedrichs [47], Duvall [56, 57], Fickett and Davis [61], Graham [85], Hayes [91], LeVeque [131], McQueen, Marsh, Taylor, Fritz, and Carter [143], and Zel'dovich and Raizer [215].

Exercises

(5.1) Assuming uniaxial strain, solve the three-wave problem of a linearly elastic material impacting an elastic-plastic material. In particular, obtain an expression relating impact velocity to stress at the material interface. Now assume steel is the elastic material and 6061-T6 aluminum is the elastic-plastic material. Compute the velocity at which the HEL is reached for a hard steel (say a yield of 2 GPa). Is the velocity close to the transition velocity where hard steel projectiles begin to fracture/erode while penetrating an aluminum alloy target in Figs. 7.1 and 7.2? Though we have not discussed penetration yet, why do you think they may agree or disagree?

(5.2) A tungsten plate impacts a steel plate at 200 m/s and 500 m/s. Compute the material interface velocity using the Hugoniot jump conditions assuming 1) no strength for either material, 2) a tungsten strength of 1.5 GPa and a soft steel ($Y = 0.5$ GPa), and 3) the same 1.5 GPa tungsten and a hard steel ($Y = 2$ GPa). Compare with results of linear U_s–u_p for these impact speeds.

(5.3) An elastic-perfectly plastic material cannot support an elastic-plastic front in uniaxial stress. However, a work hardening material can. Using the linear work hardening material described in Exercise 3.11 ($Y = Y_0 + H(\varepsilon - \varepsilon_0)$), solve the piston problem of pushing on the end of a thin cylinder of material at a constant velocity. Compute the speed of the elastic-plastic front.

(5.4) Suppose a metal bar in uniaxial stress is axially pre-stressed into the plastic response regime. Now, impact the end of the bar. The previous exercise states that either there is no wave propagation in the case of elastic-perfect plasticity or the wave speed is much less than the elastic bar wave speed in the presence of work hardening (we expect $c = \sqrt{H/\rho_0} \ll c_E$). However, experimentally it is often observed that a small amplitude wave travels down the bar at the bar wave speed c_E (see [176] for experiments with copper). This wave is an elastic precursor that exists due to strain-rate hardening of the material. Given a Johnson–Cook $Y = A(1 + C \ln(\dot{\varepsilon}))$-type strain-rate dependence and assuming $\dot{\varepsilon} > 1\,\mathrm{s}^{-1}$ in the "jump," if the bar is initially at a stress of A compute the change in stress due to this wave and the corresponding change in particle speed. Explain why this wave travels at the speed c_E. Why is the word jump above written in quotations? How might this observation relate to the observation that the HEL is sometimes higher when measured for thin specimens than for thick specimens?

(5.5) Below are data from Wilkins and Guinan for pure copper cylinders striking a rigid anvil (Table 5.12). Based on these data, determine the dynamic flow stress Y. Solve with both the with dynamic plasticity solution for s and assuming constant s with $s/L_0 = 0.12$ and $s/L_0 = 0.20$.

TABLE 5.12. Experimental data for pure copper cylinders, $L_0 = 2.347$ cm and $\rho_0 = 8.9$ g/cm^3, strking a rigid anvil [211].

V (m/s)	L_f/L_0	V (m/s)	L_f/L_0
89	0.895	183	0.716
123	0.835	204	0.669
153	0.780	210	0.643

(5.6) When Taylor first discussed using the Taylor anvil impact experiment to get dynamic stresses, his analysis assumed that the speed of the elastic-plastic front in the impacting cylinder was constant. Subsequent authors have suggested this speed be calculated using the slope of the work hardening – namely $\dot{s} = \sqrt{H/\rho}$, where H is the slope of the stress strain curve after yield, assuming it is bilinear (i.e., composed of two straight lines). Show that if $\dot{s}$ is constant, that $s = 0$ at the time of impact, and defining $\bar{L} = L - s$, then

$$\frac{d\bar{L}}{dv} = -\frac{\rho_0 \bar{L}}{Y} \frac{v + \dot{s}}{1 + v/c_b + \dot{s}/c_b}. \tag{5.242}$$

Consistent with our analysis in the text, show that letting $c_b \to \infty$ gives

$$\frac{d\bar{L}}{dv} = -\frac{\rho_0 \bar{L}}{Y}(v + \dot{s}). \tag{5.243}$$

Then show this equation can be solved to give

$$\frac{L_f}{L_0} = \frac{s_f}{L_0} + \exp\left\{-\frac{\rho_0}{Y}\left(\frac{V^2}{2} + V\dot{s}\right)\right\}. \tag{5.244}$$

The awkwardness in this result is that the final time t_f is not known or, equivalently, $\dot{s}$ is not known.

(5.7) Another approach to solving for the plastic extent in the Taylor anvil problem is to assume the rate of growth of the plastic zone is proportional to the cylinder back-end velocity, $\dot{s} = \alpha v$, where α is a constant. Using this assumption and the assumption that the elastic wave speed is very large, derive an equation for the final length of the elastic-plastic cylinder in terms of impact velocity against the rigid wall. Show that $\alpha/(1+\alpha)$ appears in a similar form to s/L_0 in Eq. (5.41). Show that for the 6061-T6 aluminum, this assumption does not give a much better match to data than the $s = 0$ assumption in Eq. (5.41). Why not? (Mathematically, where is the $1 + \alpha$ term in the exponential?)

(5.8) It seems surprising in the Taylor anvil solution Eq. (5.41) that the length of the plastically deforming region is given in terms of projectile length – that is, $s/L_0 = $ constant. It seems more intuitive that the plastically flowing region would be given in terms of projectile diameter $s/D = $ constant. Given the data for four different L/D projectiles, do we have enough information to conclude, if the mechanics are as presented leading up to Eq. (5.41), that the plastically deforming region scales with initial projectile length and not with initial projectile diameter? For example, choose reasonable values for 6061-T6 al and see what L_f/L_0 would be if $s/D = $ constant rather than $s/L_0 = $ constant.

(5.9) Consider the loading by a fluid projectile of length L_0 and radius R_0 impacting a rigid surface much like the Taylor anvil problem. In this case, since the fluid has no strength, the projectile does not appreciably decelerate during the impact event and the fluid sprays radially. After the impact stresses have died away, the stress (pressure) at the centerline of the impact is given by $(1/2)\rho v^2$. The total load is given by the change in momentum of the projectile vs. time of the event, which is $\pi R_0^2 \rho v L_0/\Delta t = \pi R_0^2 \rho v^2$. Obviously the fluid loads the rigid surface beyond the initial fluid projectile radius, since this load is greater than the cross-sectional area times the centerline pressure. Assuming the pressure can be represented as a cubic polynomial $p(r) = a_0 + a_1 r + a_2 r^2 + a_3 r^3$, find the four constants a_i and the loading radius R. (Hint: you know the pressure on the centerline and on the outer edge of the loading surface (i.e., where $p(R) = 0$), and the total load. You need two additional assumptions to find all the terms; reasonable ones are $dp/dr = 0$ at the centerline and the loading radius.) If you have access to a hydrocode, compare the pressure load it computes vs. this approximation.

(5.10) What is the relationship between density and longitudinal strain for uniaxial stress elastic deformation?

(5.11) For the experiment recorded in Fig. 5.6, given that the bars in the SHPB apparatus are steel, what length striker bar was used? What was the striker impact speed? What is the distance between the strain gage and the specimen? Based on the flow stress reported in the figure, what was the area ratio of the specimen to the transmitter bar? The specimen $L/D = 2$. Using an approximate (constant) flow stress, what does Eq. (5.61) give as the strain and strain rate prediction? Are they close to the values in Fig. 5.6?

(5.12) Material strength leads to a larger c_0 in the U_s–u_p linear relationship than c_K. To estimate the effect of strength on the U_s–u_p relation, consider the explicit solution developed in Section 5.3. Show that the shock velocity can be written

$$U = c_K + \frac{\tilde{\sigma}_{\text{HEL}}}{\rho_0 c_L}\left(1 - \frac{c_K}{c_L}\right) = c_K + \frac{\tilde{\sigma}_{\text{HEL}}}{\rho_0 c_L}\left(1 - \sqrt{\frac{1+\nu}{3(1-\nu)}}\right). \tag{5.245}$$

The amount of upward shift in the shock velocity due to strength is given by the second term of the right-hand side. Next, there is a leftward shift in the particle velocity due to strength. Show that the equation for particle velocity in the same section can be written

$$u_\ell = \frac{p}{\rho_0 c_K} + \frac{\tilde{\sigma}_{\text{HEL}}}{\rho_0 c_L}\left(1 - \sqrt{\frac{3(1-\nu)}{1+\nu}}\right). \tag{5.246}$$

(Hint: you need to use $\tilde{\sigma} = p + 2Y/3$ and the relationship between Y and the HEL.) This shift can correspond to a movement on the $u = 0$ intercept of an amount equal to the second term times $-k$. These two adjustments lead to

$$c_0 - c_K \approx \frac{\tilde{\sigma}_{\text{HEL}}}{\rho_0 c_L}\left\{1 - \sqrt{\frac{1+\nu}{3(1-\nu)}} - k\left(1 - \sqrt{\frac{3(1-\nu)}{1+\nu}}\right)\right\}. \tag{5.247}$$

Given the ultrasonic data given in Tables 5.2 and 5.3 for 6061 aluminum and 304 stainless steel, what is of $c_0 - c_K$? Given the above analysis, what is the implied $\tilde{\sigma}_{\text{HEL}}$ and Y of the aluminum and steel? Are these values reasonable? Given a really strong steel like Vascomax 350 (where the yield is 350 ksi), what percentage difference would you expect between c_0 from the flyer plate experiments and the value of c_K from elastic tests?

(5.13) Show for a material whose Hugoniot can be written $U = c_0 + ku$ that $U = c_0/(1 - k\tilde{\varepsilon})$ along the Hugoniot.

(5.14) From Appendix C, pick any five materials (at least two of them metals) and plot (on the same respective plot) the Hugoniots in the (u, U) plane and in the $(u, \tilde{\sigma})$ plane. Compare and comment.

(5.15) Swegle and Grady also analyzed two plate impact tests for bismuth, finding for $\tilde{\sigma}_H = 1.20$ and 2.45 GPa, that $d\tilde{\varepsilon}/dt = 1.4 \times 10^5$ and $1.89 \times 10^6\,\text{s}^{-1}$, respectively. Compute the wave-front rise times and distances, completing all the entries in Table 5.4, for bismuth. Next, for all the materials in both tables, plot $\log \sigma_H$ vs. $\log$ strain rate on the same plot and show that $\dot{\tilde{\varepsilon}} = a\tilde{\sigma}_H^4$, where a depends on the specific material of interest, confirming their result.

(5.16) (Ideal gas equation of state) The ideal gas equation of state is given by $p = (\gamma - 1)\rho E$, where $\gamma = C_p/C_v$ is the ratio of the specific heats.
    (a) Is this a Mie–Grüneisen form of equation of state?
    (b) What is the Grüneisen parameter Γ?
    (c) How does the ideal gas equation of state differ from our usual assumptions for solids?

(5.17) (Ideal gas Hugoniot and isentrope) The ideal gas equation of state is given by $p = (\gamma - 1)\rho E$, where $\gamma = C_p/C_v$ is the ratio of the specific heats. (Parts of this problem were worked in Chapter 2, so don't repeat work already done; just cite the answers.)

(a) Determine the Hugoniot for the ideal gas. Write it in the (ρ, p) and (ρ, E) planes. Once you know what the sound speed is in the ambient gas, part (c) below, come back and write the Hugoniot in the (ρ, U) and (u, U) planes.

(b) What is the maximum densification ρ/ρ_0 achievable by a shock wave passing through an ideal gas? What is it for a monatomic ideal gas (like helium)? What is it for a diatomic ideal gas (like hydrogen)? (Of course, more physics enters in as compression and hence energy increase; but for the problem assume the equation of state continues to hold.)

(c) Determine the sound speed for the ideal gas (recall that Newton's derivation using Boyle's law, which he knew was incorrect from experimental observation, was $c^2 = \partial p/\partial \rho = p/\rho$, so don't make the same mistake).

(d) Show the surprising result that the sound speed only depends on the specific energy – that is, the temperature.

(e) The Mach number is defined as the speed of the shock divided by the initial sound speed of the gas: $M = U/c_0$. With this definition, show the representations of the Hugoniot ρ_0/ρ, p/p_0 and $E/E_0 = T/T_0$ in terms of γ and M. Surprisingly, no other variables or properties of the gas appear.

(f) Compute the release isentrope, showing the famous result that $p/p_0 = (\rho/\rho_0)^\gamma$, $E/E_0 = T/T_0 = (\rho/\rho_0)^{\gamma-1}$, and $c = c_0(\rho/\rho_0)^{(\gamma-1)/2}$ along the isentrope.

(g) What is the adiabatic bulk modulus of an ideal gas?

(h) Assuming the density of air at standard temperature and pressure is $\rho_0 \approx 0.00129$ g/cm^3, what is the sound speed?

(i) Since you have computed $c = c(p, \rho)$ and you know the pressure as a function of density along the isentrope, you are now able to use the Riemann integral to solve the following problem: what is the free expansion velocity of an ideal gas? In other words, if we have a half space of ideal gas at an initial pressure and density on one side of a barrier and free space on the other, when we remove the barrier instantaneously, what is the leading edge velocity of the gas as it expands into the vacuum? Specifically, what is it for a monatomic gas? A diatomic gas?

(5.18) Assuming the ideal gas equation of state represents air and assuming a constant temperature independent of height above the surface of the Earth, show that the density and pressure distribution for the atmosphere are

$$\rho(h) = \rho_0 \exp\left(-\frac{\rho_0 gh}{p_0}\right), \qquad p(h) = p_0 \exp\left(-\frac{\rho_0 gh}{p_0}\right),$$

where h is the height above the surface of the Earth and subscript 0 refers to the density and pressure at sea level (see Exercise 7.5). What altitude gives half the pressure?

(5.19) (Mechanical solid analog) To better understand the behavior of elastic solids, consider the following mechanical analog. Take a rigid-walled cylinder of cross-sectional area A running along the x-axis with a rigid far wall at $x = L_0$ and place a movable piston at $x = 0$. Connect the piston and far wall with a spring with a spring constant k, so the force on the spring is $F = -kx$. Now fill the cylinder with an ideal gas, $p = (\gamma - 1)\rho E_g$, to a predefined pressure p_0

with density ρ_0. During the gas fill, the piston moves to the left to $x = -u_0$ for static equilibrium, where $ku_0 = Ap_0$. Now compute the force vs. displacement u from the initial equilibrium, where $x = u - u_0$. Show that the force applied to the piston to achieve a displacement u is

$$F(u) = k(u - u_0) + (\gamma - 1)\rho E_g A = k(u - u_0) + (\gamma - 1)\frac{L_0 + u_0}{L_0 + u_0 - u}\rho_0 E_g A, \quad (5.248)$$

where E_g is the energy in the gas. Show that for isothermal loading (constant E_g) the force vs. displacement can be written, with its corresponding zero pressure isothermal modulus,

$$F_T(u) = ku\left(1 + \frac{u_0}{L_0 + u_0 - u}\right), \quad \left.\frac{\partial F_T}{\partial u}\right|_{u=0} = k\left(1 + \frac{u_0}{L_0 + u_0}\right). \quad (5.249)$$

Next, compute adiabatic loading as follows. Show that the system total energy is

$$E = \frac{1}{2}k(u - u_0)^2 + \rho_0 A(L_0 + u_0)E_g - \int_0^u F(u)du = \text{constant}, \quad (5.250)$$

where the integral term is the work due to loading. Solve for the energy of the gas and show, just as in the derivation of Eq. (5.152), that the spring does not directly affect the gas energy (i.e., temperature): $E_g = E_g(0)\left(\frac{L_0+u_0}{L_0+u_0-u}\right)^{\gamma-1}$. Now show that the adiabatic loading force vs. displacement, with its corresponding adiabatic modulus, can be written

$$F_s(u) = ku + ku_0\left\{\left(\frac{L_0 + u_0}{L_0 + u_0 - u}\right)^{\gamma} - 1\right\}, \quad \left.\frac{\partial F_s}{\partial u}\right|_{u=0} = k\left(1 + \frac{\gamma u_0}{L_0 + u_0}\right). \quad (5.251)$$

Comment and compare to the behavior of elastic solids. In particular show that the initial energy rise in the gas is due to potential energy release by the spring and not directly to the external work.

(5.20) Similarly, consider Johnson's spring model and Eq. (5.133) for the average force $\langle F\rangle = aY + bY^2 + \frac{2b}{\omega^2}\langle E_k\rangle$. For $\langle F\rangle = 0$, with positive kinetic energy, $Y < 0$ – that is, the spring is in tension. Show the potential energy in the springs is $\langle e_e\rangle = m\langle E_e\rangle = aY^2 + \frac{2}{3}bY^3 + m\langle E_k\rangle$. Conservation of energy says $\langle E_e\rangle + \langle E_k\rangle = \frac{2}{m}\int\langle F\rangle\,dY$, and show differentiation gives $d\langle E_k\rangle/dY = b\langle E_k\rangle/(a + 2bY) > 0$ (note the spring "cold compression curve" has dropped out), hence the change in temperature (kinetic energy) is positive due to the release of potential energy in the spring, and the change is proportional to ΔY, not $(\Delta Y)^2$. Explicitly determine $\langle E_k\rangle$ as a function of Y. The result differs from Eq. (5.152) since "$\rho\Gamma$" is not constant.

(5.21) The bulk modulus at zero pressure is given by $K_0 = \partial p/\partial\tilde{\varepsilon}$. If we want $c = \sqrt{K(\rho, E)/\rho}$ for all material states, where ρ is the current density and where the bulk modulus is given by $K = dp/d\tilde{\varepsilon}$, then what is the appropriate definition of strain $\tilde{\varepsilon}$?

(5.22) Determine the modulus of the linear stress-strain relation $\tilde{\sigma} = \rho_0 c^2(1 - \rho_0/\rho)$ and the $\tilde{\sigma} = \rho_0 c^2(\rho/\rho_0 - 1)$ stress-strain relation. What stress-strain relation has a constant modulus, and what is it? What is its sound speed?

(5.23) Suppose a 1 cm flyer plate of stainless steel 304 struck a thick block of like material at 3 km/s. How wide in time and space is the initial pulse that travels into the steel block? When (in time and location into the thick steel plate) is the initial shock first caught by the rarefaction fan? Give a bound on when (in time and location in the thick steel block) the wave is decayed to elastic levels. (In working this problem, don't worry about elastic precursors for the two waves, just focus on the elastic-plastic behavior.)

(5.24) Table 5.13 lists some HELs from McQueen et al. [**143**]. Using these values and Poisson's ratios from Appendix C, what is the implied flow stress? How does the Al 6061 value agree with the data in Chapter 3?

TABLE 5.13. Hugoniot elastic limits [**143**].

Material	Sample thickness (cm)	$\tilde{\sigma}_{HEL}$ (GPa)
Aluminum alloys 6061, 2024	1.2	0.54
Magnesium alloy AZ31B	1.2	0.11
Molybdenum	0.3	1.6
Nickel	1.2	1.0
Lead (annealed)	1.2	0
Tantalum	1.0	1.87
Titanium	1.2	1.85
Tungsten	0.5	3.2
Stainless steel 304	1.2	0.23

(5.25) (Hugoniot of copper) In this exercise you will produce three representations of the Hugoniot of copper using data in Appendix C. 1) Plot $U_s - u_p$ for $0 \leq u \leq 4$ km/s. 2) Plot $\tilde{\sigma}_H$ vs. $1 - \rho_0/\rho_H$ for the same range. 3) Using our analytic formula for Hugoniot temperature Eq. (5.100) and your curve fit to the temperature vs. pressure melt curve shown in Fig. 5.17, reproduce the copper temperature-pressure Hugoniot plot in Fig. 5.17 and include the isentropic release from the 138 GPa Hugoniot state using Eq. (5.101). (Hint: recall the Hugoniot is the end state, so first compute the Hugoniot pressure, then the melt temperature at that pressure, then if the analytically predicted Hugoniot temperature is above the melt temperature, subtract the heat of fusion from the temperature (assuming a constant heat capacity) down to the melt temperature.)

(5.26) Figure 5.9 shows the free surface velocity from a flyer plate impact experiment performed by J. W. Taylor at Los Alamos National Laboratory a few days before the author was born (shot #56-64-328). The flyer plate and target were Armco iron, with measured $\rho_0 = 7.87$ g/cm^3, $c_L = 5.94$ km/s and $c_s = 3.26$ km/s. The target thickness was 2.54 cm and the impact velocity was 977 m/s. The plot shows a three-wave structure.

   (a) What is σ_{HEL}? (Hint: the velocity in the plot is the *free surface* velocity measured at the free surface when the wave arrives, not the particle velocity of the original wave formed by the impact. Assume $u_{fs} = 2u$.) What is the corresponding flow stress?

   (b) If the zero time is the time of impact, are the elastic precursor and elastic-plastic shock wave arriving at the correct time? (For the elastic-plastic shock, assume iron below the phase change behaves like stainless steel 304.)

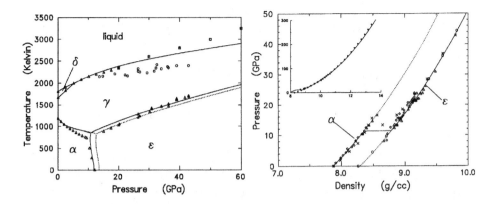

FIGURE 5.20. The phase diagram of iron, showing different solid states and melt (left) and the room temperature isotherm (solid line) with comparisons to diamond anvil cell static tests, showing the α–ε phase transition (right). (From Kerley [115].)

(c) The third wave is due to a phase transition. Estimate the stress at which the phase transition occurs. (Hint: it is the stress behind the second wave front; you don't need to know anything about the third wave, just the details of the elastic precursor and the second wave.)

(d) Is the final particle velocity in the plot correct?

(e) Estimate the final stress state of this experiment.

(f) Estimate an upper bound for the density change (or specific volume change) of the phase transition. (For this, you will need to estimate the third-wave speed. The mass jump condition provides an upper bound because some of the volume change is due to further compression.)

(g) It is known that there is a transition to γ-Fe, a face centered cubic lattice, at atmospheric pressure in the vicinity of 900–1,000 K. Is this the transition? Figure 5.20 left, due to Kerley [115], shows the iron phase diagram. Copy the phase diagram and on it place a rough plot of the Hugoniot by using that of stainless steel ignoring the multiple waves (Table 5.3). Identify the final state point of this experiment on that diagram. Compare your result in (f) with Fig. 5.20 right, also from Kerley's report.

(5.27) Based on the previous problem, explain why the Appendix C Hugoniot values for iron for the U_s–u_p relationship do not give good values for low pressures.

(5.28) For aluminum and iron, from the Appendix C values, compute the melt temperature on the Hugoniot based on the pressures stated in the text. (Hint: since the iron U_s–u_p relationship is quadratic, it will require a numerical integration.)

(5.29) For 6061 aluminum and 304 stainless steel, using the melt temperatures in Appendix C for the pure materials compute an approximate symmetric flyer plate impact speed to produce melt.

(5.30) The inner core of the Earth is solid and the outer core is liquid. The Preliminary Reference Earth Model states that the Inner Core Boundary, located 1,221 km from the center of the Earth, is at pressure 328.9 GPa with a density of 12.16 g/cm³ on the liquid side and 12.75 g/cm³ on the solid side. The core is thought to be primarily iron. Show, however, based on the Hugoniot for iron in Appendix C (and there are two data points at roughly 330 GPa: $(u, U) = (4.05 \text{ km/s}, 10.2 \text{ km/s})$ and $(4.07 \text{ km/s}, 10.35 \text{ km/s})$, [139]) that the core is not *pure* iron.

(5.31) Show that the Hugoniot stress for the quadratic U_s–u_p relation $U = c_0 + ku + qu^2$ is

$$\tilde{\sigma}_H = \frac{\rho_0}{2q^2\tilde{\varepsilon}^3}\left\{(1 - k\tilde{\varepsilon})^2 - 2c_0q\tilde{\varepsilon}^2 - (1 - k\tilde{\varepsilon})\sqrt{(1 - k\tilde{\varepsilon})^2 - 4c_0q\tilde{\varepsilon}^2}\right\}. \quad (5.252)$$

(5.32) Show that for the quadratic Hugoniot U_s–u_p relation $U = c_0 + ku + qu^2$. the maximum densification of the material by a shock wave is

$$\left(\frac{\rho}{\rho_0}\right)_{\text{max}} = \frac{k + 2\sqrt{c_0q}}{k - 1 + 2\sqrt{c_0q}}. \quad (5.253)$$

(5.33) Sometimes the Hugoniot stress is written as a polynomial in terms of strain. Use a Taylor series expansion about 0 to show that for the quadratic U_s–u_p of the previous problem that if

$$\tilde{\sigma}_H = K_1\tilde{\varepsilon} + K_2\tilde{\varepsilon}^2 + K_3\tilde{\varepsilon}^3 \quad (5.254)$$

then

$$K_1 = \rho_0 c_0^2, \quad K_2 = 2k\rho_0 c_0^2, \quad K_3 = (3k^2 + 2qc_0)\rho_0 c_0^2, \quad (5.255)$$

and so given the polynomial coefficients,

$$c_0 = \sqrt{\frac{K_1}{\rho_0}}, \quad k = \frac{K_2}{2K_1}, \quad q = \frac{1}{2c_0}\left(\frac{K_3}{K_1} - 3k^2\right). \quad (5.256)$$

Show that if the Hugoniot stress is written as

$$\tilde{\sigma}_H = B\mu + C\mu^2 + D\mu^3, \quad (5.257)$$

where $\mu = \rho/\rho_0 - 1$, then

$$B = \rho_0 c_0^2, \quad C = (2k - 1)\rho_0 c_0^2, \quad D = (2qc_0 + 3k^2 - 4k + 1)\rho_0 c_0^2, \quad (5.258)$$

and

$$c_0 = \sqrt{\frac{B}{\rho_0}}, \quad k = \frac{1}{2}\left(1 + \frac{C}{B}\right), \quad q = \frac{1}{2c_0}\left(\frac{D}{B} - 3k^2 + 4k - 1\right). \quad (5.259)$$

Finally, show (you may want to do this first to show Eq. 5.258)

$$B = K_1, \quad C = K_2 - K_1, \quad D = K_3 - 2K_2 + K_1,$$
$$K_1 = B, \quad K_2 = B + C, \quad K_3 = B + 2C + D. \quad (5.260)$$

(Hints: do the Taylor expansion of Eq. (5.252) by computing the derivatives. Also, $\tilde{\varepsilon} = \mu/(1 + \mu) = \mu - \mu^2 + \mu^3 - \cdots$.)

As a comment, when working with polynomial forms of equations of state, it is important to be clear as to which curve the polynomial describes. Above, the polynomial describes the Hugoniot. However, it is also possible to describe the isothermal compression with a polynomial. When the polynomial describes the Hugoniot and the Mie–Grüneisen form of the equation of state (Eq. 5.127) is used,

$$\tilde{\sigma}(\rho, E) = \tilde{\sigma}_H(\rho)\left(1 - \frac{\rho\Gamma\tilde{\varepsilon}}{2\rho_0}\right) + \rho\Gamma(E - E_0)$$

$$= (K_1\tilde{\varepsilon} + K_2\tilde{\varepsilon}^2 + K_3\tilde{\varepsilon}^3)\left(1 - \frac{\Gamma_0\tilde{\varepsilon}}{2}\right) + \rho_0\Gamma_0(E - E_0)$$

$$= K_1\tilde{\varepsilon} + \left(K_2 - K_1\frac{\Gamma_0}{2}\right)\tilde{\varepsilon}^2 + \left(K_3 - K_2\frac{\Gamma_0}{2}\right)\tilde{\varepsilon}^3 - K_3\frac{\Gamma_0}{2}\tilde{\varepsilon}^4 + \rho_0\Gamma_0(E - E_0)$$

$$(5.261)$$

and

$$\tilde{\sigma}(\rho, E) = (B\mu + C\mu^2 + D\mu^3)\left(1 - \frac{\Gamma\mu}{2}\right) + \rho\Gamma(E - E_0)$$

$$(5.262)$$

$$= B\mu + \left(C - B\frac{\Gamma}{2}\right)\mu^2 + \left(D - C\frac{\Gamma}{2}\right)\mu^3 - D\frac{\Gamma}{2}\mu^4 + \rho\Gamma(E - E_0).$$

These equations define the relations between the polynomial coefficients along the Hugoniot and the polynomial coefficients of isothermal compression. A reminder is that these comparisons are based on matching at 0 and not on best fits over a range of data.

(5.34) The paragraph following Eq. (5.190) describes the Murnaghan form for equation of state. This equation of state form is based on the assumption that along a specified curve (the Hugoniot for this application) the bulk modulus varies linearly in terms of pressure, $K(p) = K_0 + bp$. Show that this assumption combined with the definition of bulk modulus produces the stated pressure–density relationship and show the small strain relationship between the U_s–u_p slope k and Murnaghan slope b, also given in the paragraph.

(5.35) Show that the stress-strain relation arising from $U = c_0 + ku$ is strictly concave up for $k > 1$ (hence it gives rise to compressive shocks and rarefaction fans) and is concave up for some strains for $0 < k < 1$. (Hint: use Eq. 5.124.)

(5.36) Compute the increase in pressure for a shock in steel reflected from a rigid boundary for impinging shocks of 1, 2, and 3 km/s particle velocity.

(5.37) Using the "twice the initial velocity" approximation of Eq. (5.178), plot the approximate increase in pressure on reflection of an ideal gas from a rigid surface and show that $1 \leq p_r/2p_\ell \leq 2$ – that is, the reflected pressure is at most four times the incident pressure. (Hint: use the Hugoniot expression for U from Exercise 5.17.)

(5.38) The statement that the reflection off a free surface leads to twice the particle velocity need not hold if there has been a phase change. Given what you know about ideal gasses from Exercise 5.17, write down the free surface velocity if on release the residual energy changes the material from a solid to a gas.

(5.39) There is an interesting graphical solution to the flyer plate impact problem. If the Hugoniots vs. the particle velocity are plotted for each material (which can be done, for example for the linear U_s–u_p relation, using the jump condition $\tilde{\sigma}(u) = \rho_0 U u = \rho_0(c_0 + ku + qu^2)u$), then if we make a copy of the chart on a transparency, the Hugoniot can be solved for by flipping the transparent copy over so that the stress on the vertical axis is now to the right of the page instead of the left, placing the $u = 0$ axis of the transparency copy at the impact velocity V of the regular copy, so that the lower particle velocity axes overlay (save that the one on paper is in increasing particle direction and now the one on the transparency is in decreasing particle direction), and then finding the intersection of the curves for the target material (on the paper copy) and the impactor material (on the transparency). The intersection point identifies the impact stress and the velocity on the paper figure is the particle velocity. Assignment: do this for steel impacting aluminum at 5 km/s. Since both of these materials are exhibiting a first shock, the warning regarding the errors introduced beyond the first shock in the $(u, \tilde{\sigma})$ plane does not apply.

(5.40) Derive the various sound speed expressions for the linear U_s–u_p relation, Eqs. (5.179), (5.180), and (5.181).

(5.41) Given that the U_s–u_p relation is linear, assume the pressure vs. density for the isentrope and the Hugoniot are the same to derive Eq. (5.162), which is an approximate energy in isentropic compression.

(5.42) Assuming a constant Poisson's ratio, what is the shear modulus dependence on density, pressure, and temperature, assuming a Mie–Grünesien equation of state based on a linear U_s–u_p relationship?

(5.43) Table 5.14 shows three Hugoniot (from impedance matching) values of initially porous 2024 aluminum from Los Alamos ([**139**]). For initially fully dense 2024 aluminum, $\rho_0 = 2.785$ g/cm^3 and $U = 5.328$ km/s $+ 1.338u$. Use this information to compute Γ_0 for a Mie–Grüneisen equation of state, assuming $\rho\Gamma = \rho_0\Gamma_0$ for each of the three data points.

TABLE 5.14. Initially porous 2024 aluminum Hugoniot data [**139**].

ρ_0 (g/cm^3)	u (km/s)	U (km/s)
1.648	1.747	4.050
1.671	3.302	7.092
1.694	4.567	9.118

(5.44) As discussed in the section on porosity, shocks from porous materials can show interesting behavior when there are thermal terms. Suppose that the fully dense (solid) material can be written as a Mie–Grüneisen equation of state in the form (Eq. 5.136)

$$\tilde{\sigma}_s(\rho, E) = f(\rho) + \rho_{s0}\Gamma_0(E - E_0). \tag{5.263}$$

Assume that $f(\rho_{s0}) = 0$ and that $df/d\rho > 0$ and that when a shock wave passes through the material its subsequent behavior is modeled with this equation of state. The initial density of the porous material will be given by $\rho_0 = \rho_{s0}/\alpha_0$, where $\alpha_0 > 1$ (using p-α model notation). Before the shock the material is not represented by this equation of state as we assume $\tilde{\sigma}_0 = 0$ and $\alpha_0 > 1$.

(a) Show that the Hugoniot is given by

$$\tilde{\sigma}_H(\rho) = \frac{f(\rho)}{1 - (\Gamma_0/2)(\alpha_0 - \rho_{s0}/\rho)}. \tag{5.264}$$

(b) Since the shock stress is compressive, show that there are two possibilities:
(1) the numerator and denominator of Eq. (5.264) are both positive, and
thus

$$\alpha_0 - \frac{2}{\Gamma_0} < \frac{\rho_{so}}{\rho} < 1 \qquad (5.265)$$

for the final density ρ, or (2) the numerator and denominator of Eq. (5.264)
are both negative, and thus

$$1 < \frac{\rho_{so}}{\rho} < \alpha_0 - \frac{2}{\Gamma_0}. \qquad (5.266)$$

(c) These inequalities demarcate two qualitatively different types of behavior.
Show that the value of α_0 that satisfies

$$\alpha_0 = 1 + \frac{2}{\Gamma_0} \qquad (5.267)$$

is the demarcation line. For this value of α_0, the inequalities in (b) imply
that the only allowable post-shock density state is $\rho = \rho_{so}$. What does
this demarcation represent? How do we determine the Hugoniot state?
What is the energy in the Hugoniot state? This is an example of what is
referred to as the *snowplow model* (Exercise 5.46).

(d) The inequality Eq. (5.266) is intriguing. It says that in a shock the
material is never fully compacted. For this case, *show that increasing
values of stress lead to decreasing values of final density* – that is, as you
push the material harder and harder, because of the thermal terms, the
final density is less. The easiest way to show this is to show $d\tilde{\sigma}_H/d\rho < 0$.
Obviously, there is concern about the accuracy of the equation of state in
these regimes, since the energy term is an assumed linear extrapolation.

(5.45) The Hugoniot data for balsa wood in Marsh [**139**] shows an initial density
of 0.123 g/cm^3 and then a post-shock density of (very roughly) 0.5 g/cm^3,
regardless of the strength of the shock. Given these results, what is the
$U = c_0 + ku$ form of the Hugoniot? In general, given a constant post shock
density behavior, what is $U = c_0 + ku$? What kind of equation of state does
this represent?

(5.46) (Snowplow model, from Hayes [**91**]) One way to represent a porous material's
shock behavior is to assume that it behaves like what we visually observe while
shoveling snow: wet fluffy snow seems to crush to wet dense snow, and the
densification front moves out in front of the shovel more quickly that we push
the shovel. The simplest way to represent this in a Hugoniot is to begin with
a porous material has an initial density given by $\rho = \gamma\rho_0$, where $\gamma < 1$, and
then *assume* that the only allowable post-shock density state for the porous
material is $\rho = \rho_0$. This assumption is referred to as the *snowplow model*. (See
Exercise 5.44 to see this behavior arise in an equation of state setting.) This
exercise explores the ramifications of the snowplow model and why foams are
viewed as being good shock attenuators.

(a) Solve the piston problem for the snowplow model, assuming either an
applied stress or an applied velocity.

(b) Suppose the piston (either applied stress or velocity) is applied for only a
finite amount of time, say t_f. What happens then? (Hint: the densified
material is rigid since it cannot be compressed further.) For what distance
will the shock front continue to travel? Write an equation relating the

shock front location x (assume the piston stops driving the material when the shock is at $x = 0$) and time t (assume the piston stops driving the material at time $t = 0$).

(c) What is the particle velocity of the compacted material in terms of the shock front location? What is the stress at the shock front in terms of distance traveled? (Hint: momentum is always conserved.)

(d) What is the stress distribution in the densified material? How does this stress distribution change over time? The drop in stress is the attenuation.

(e) Write the shock front location vs. time and the particle velocity and shock stress in terms of both shock location and time in the impulse limit, where the impulse I_0 is held constant but $L_0 \rightarrow 0$.

(f) What is the behavior of the kinetic energy? Where is the energy going?

(5.47) (More on the snowplow model) We have already solved for a rigid impactor striking the snowplow foam material (it is equivalent in Exercise 5.46 to beginning with an appropriately lengthened amount of post-shock foam in motion at the impact velocity). Now, however, let's impact the foam with an impactor of length L with a linear U_s–u_p material response.

(a) What is the shock impedance of the snowplow foam?

(b) There are two regions of qualitatively different response. What are they? Find the impact velocity that demarcates the two regimes. What specifically is it for a steel foam with $\gamma = 1/2$?

(c) For the regime where the projectile bounces off the target, what is the deepest penetration of the impactor as a function of impact velocity? What is the total motion of the foam?

(d) Suppose you had a steel impactor that you wanted to completely stop with a porous metal. Given the result in Exercise 5.44 implying roughly that $\gamma = 1/2$, what would it take to stop the steel impactor? What is a reasonable starting density for the foam? Assume the impact speeds of interest are 500 m/s, 1 km/s, and 2 km/s. (For these cases, use the behavior for stainless steel given in Table 5.3 to provide the appropriate stress.)

(5.48) (For students of thermodynamics) The volume thermal expansivity is

$$3\alpha = \beta = \rho \left(\frac{\partial (1/\rho)}{\partial T} \right)_p = -\frac{1}{\rho} \left(\frac{\partial \rho}{\partial T} \right)_p. \qquad (5.268)$$

Using this, the thermodynamic definitions of moduli (K_T is the isothermal bulk modulus), Γ, the heat capacities, and standard relationships from thermodynamics, show that

$$\rho \Gamma = \frac{3\alpha K_T}{C_v} = \frac{3\alpha K_s}{C_p} = \frac{3\alpha \rho c_0^2}{C_p}. \qquad (5.269)$$

(Hint: only the last step is trivial; you'll need to know the heat capacity expressions in terms of entropy and temperature and the general thermodynamic derivative relations based on there only being two independent variables.) In particular, show that

$$c_0 = \sqrt{\frac{\Gamma C_p}{3\alpha}}. \qquad (5.270)$$

In the process, you have either shown or could show $K_s/K_T = C_p/C_v = 1 + \beta\Gamma T = 1 + 3\alpha\Gamma T$: thus, a solid has to be quite hot to show much difference between the isothermal and adiabatic bulk moduli. Plug in values from Appendix C for a few materials into this last expression (Eq. 5.270) and compare with bulk sound speeds inferred through ultrasound. Does Eq. (5.270) imply that the sound speed is a thermal property? (Ans: no – the thermal terms Γ and α essentially cancel each other out.)

(5.49) Derive an expression to estimate the tensile strain as a function of temperature in the cold compression curve for an elastic material at ambient pressure. (Hint: use Eq. (5.154) and the low pressure expression for the Grüneisen parameter (Eq. 5.120) and assume that the cold compression curve is a linear function of strain with the material's bulk modulus.) What is the estimate for the lattice strain in aluminum at room temperature? What is the corresponding stress?

(5.50) McQueen et al. [143] make the intriguing statement that the extrapolation of the linear U_s–u_p curve down to $U = 0$ gives a reasonable approximation for the binding energy of the material. Notice that the mass balance equation implies that if U goes to zero then the final state density of the material goes to zero. What does extrapolation to $U = 0$ mean in terms of energy? Write an equation for the binding energy of the material in terms of c_0 and k. From Young [214], the stated values of the cohesive energy ("the heat of vaporization at $0\,K$") is 12.1 kJ/g and 7.40 kJ/g for aluminum and iron, respectively. How do these values compare with the values based on the curve fits to Tables 5.2 and 5.3?

(5.51) Suppose a solid or liquid explosive material, with given p_{CJ}, ρ_{CJ}, and D, has an unreacted linear U_s–u_p response $U = c_0 + ku$. Assuming the ambient pressure is negligible, show that the von Neumann spike density and pressure are given by

$$\tilde{e}_{vN} = \frac{1}{k}\left\{1 - \sqrt{\frac{\rho_0 c_0^2 \tilde{e}_{CJ}}{p_{CJ}}}\right\} = \frac{1}{k}\left\{1 - \frac{c_0}{D}\right\},$$

$$p_{vN} = \frac{\rho_0 D^2}{k}\left\{1 - \frac{c_0}{D}\right\} = \frac{\tilde{e}_{vN}}{\tilde{e}_{CJ}}p_{CJ}, \quad \text{where} \quad \tilde{e}_{CJ} = \frac{p_{CJ}}{\rho_0 D^2}.$$

(5.271)

Explicitly calculate these values for the solid explosive Comp-B ($\rho_0 = 1.717$ g/cm^3, $p_{CJ} = 29.5$ GPa, $D = 7.82$ km/s, $c_0 = 3.08$ km/s and $k = 2.01$, roughly) and the liquid explosive nitromethane ($\rho_0 = 1.128$ g/cm^3, $p_{CJ} = 12.5$ GPa, $D = 6.28$ km/s, $c_0 = 1.65$ km/s and $k = 1.64$). What are the predicted shock speeds in the unreacted materials for this predicted von Neumann spike pressure? Are they close to the detonation velocities?

(5.52) Kerley has questioned the linear U_s–u_p relation, arguing that the data often look linear because a large component of U is actually u. He argues that one should look at $U - u$ vs. u [118]. Whether the material response is linear or not can only be determined by examining the data, but the question to you is a little more basic: which is more fundamental, U or $U - u$?

(5.53) In the report mentioned in the previous exercise, Kerley states that when the energy is high enough for all atoms to be ionized, every material obeys the ideal gas equation of state with $\gamma = 5/3$ – that is, $p = (2/3)\rho E$. Using the energy jump condition and assuming a *really* strong shock so that the energy is high enough that the material is fully ionized, show that $\rho/\rho_0 = 4$. This result is not bounding, since the compression can be higher at lower shock pressures before the material fully ionizes, when it is represented by a different equation of state. Show, however, that for an ideal gas it is bounding, as $(d\rho/dp)_H > 0$.

(5.54) (Impedance matching.)   The impedance matching technique is the source of the majority of Hugoniot data on materials. This exercise works through the basics of the idea, identifying why standard materials need to be completely characterized. Basically, the technique is as follows. Suppose we have a material whose equation of state is completely characterized. We will refer to this material as the standard. We place explosive in contact with the standard, which is then in contact with the material specimen of interest. We detonate the explosive using a wave shaper to get a plane wave. We use multiple sample blocks sitting on the surface of the explosive; some only have the standard material, with no specimen, and from these we measure the arrival time of the shock. Different thicknesses allow different arrival times and thus a determination of the shock speed. Since the standard is completely characterized, knowing the shock speed means we also know the stress and particle velocity in the shock (this assumes the material is strictly superlinear; Why?). Thus, the incident shock on the standard/specimen interface is completely characterized. Next, we measure the arrival times at the back of the specimen samples. Since we know the thickness, we now know the shock velocity through the specimen. This completely determines the shock impedance, since we also know the starting density of the specimen material. Now we need to solve the shock impedance matching problem. After the shock passes through the standard and arrives at the specimen, it is the first shock for the specimen, but the reflection problem is a second shock for the standard. Thus, Theorem 5.3 says that we need to compute either a new Hugoniot from the initial standard shock state or a rarefaction from that shock state. We either shock up or unload down to a point at which the stresses and particle velocities match. The density at which this occurs, in conjunction with the Hugoniot jump conditions, then allows a determination of the particle velocity u in the specimen, which completely defines the Hugoniot state. Thus, thermal information is not needed for the specimen, but thermal information is needed for the standard to compute the second shock or unloading curve. (Notice that if the free surface velocity of the specimen were measured, specimen thermal information would be required to determine u.)

Your problem: show that the technique works by writing down the equations that need to be solved and showing that you can determine the Hugoniot point for the specimen by just knowing its shock speed and the shock speed of the standard. (Hint: essentially we are examining the impact of pre-stressed standard material into ambient specimen material.)

6

The Cavity Expansion

The analog of the piston problem in cylindrical and spherical symmetry is the cavity expansion. There are two cavity expansions of interest: the dynamic motion of a space filled with material driven by an infinitely long expanding cylinder, and the dynamic motion of a space filled with material driven by an expanding sphere. These two problems are referred to as the cylindrical cavity expansion and the spherical cavity expansion, respectively. The cavity expansion solution is, with reasonable assumptions, explicitly solvable through the use of a similarity transformation (the assumption of self-similar motion) that reduces the partial differential equations to ordinary differential equations. The solution includes elastic and strength effects and inertia. This chapter will explicitly derive solutions in one dimension, two dimensions (cylindrical), and three dimensions (spherical). There really is no one-dimensional cavity expansion, but the exercise helps clarify the assumptions and particularly highlights the role of the boundary condition at the elastic-plastic interface. After the solution of the problem assuming linear pressure–volume response, there will be a discussion of how the solution is modified to take into account the nonlinearity of the volumetric response at higher pressures. In the next chapter, the cavity expansion will be used to estimate strength effects in a penetration model. The spherical cavity expansion will be used in Chapter 9 to determine the crater radius. In Chapter 10, where a full analytic penetration model is derived, the cylindrical cavity expansion will be used to determine the extent of plastic flow within the target during penetration.

6.1. The One-Dimensional Cavity Expansion

There really is no one-dimensional cavity expansion, but it is a good discussion point for the approach we plan to take to give us an idea of what we expect the solution to look like. In one dimension, the cavity expansion is to begin at the origin and push on an infinite half space of elastic-plastic material at a constant velocity (or perhaps a more consistent way to visualize the problem is to separate an infinite space of material into two halves, and then push each half in the opposing direction at equal speeds). This geometry is simply the piston problem presented in Section 5.3. The solution has a two-wave structure, an elastic precursor moving at the longitudinal sound speed followed by the elastic-plastic interface moving at roughly the bulk sound speed ($c_0 \approx c_K$; Eq. 5.69; in this chapter c_K and c_0 will be used synonymously). Thus, there are three regions of the solution (Fig. 6.1). To the far right (furthest away from the surface being pushed) is the undisturbed region, where the elastic precursor has not yet arrived. Next is an elastic region, behind the elastic precursor. Finally, the last region is the elastic-plastic region, behind the shock, where the deformation of the material includes some plasticity. The boundary between each region is defined by the Hugoniot jump conditions.

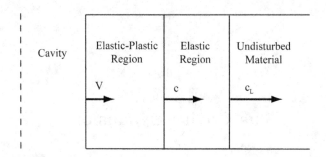

FIGURE 6.1. Geometry for one-dimensional cavity expansion.

The mathematical method that will be used to look for solutions to the cylindrical and spherical cavity expansions will be to look for a self-similar solution, or what is sometimes referred to as a similarity transformation. We successfully applied this technique to find the rarefaction fan in Chapter 4. The similarity transformation assumes that the solution maintains the same form or shape as time progresses. It is assumed that there is a "center" at $r = 0$, and that, as time progresses, the solution expands out from the center, maintaining its shape. Mathematically, given the density $\rho(r,t)$, stress $\sigma(r,t)$ or velocity $v(r,t)$, which will be generically referred to here as a function $f(r,t)$, it is assumed that it is possible to write f as

$$f(r,t) = f(\xi), \quad \text{where} \quad \xi = \frac{r}{ct}. \tag{6.1}$$

Thus, given the time t and the spatial coordinate r, the solution depends only upon the ratio of r and t. In the definition of ξ, c is an arbitrary *constant* wave speed and ξ is dimensionless. We will later choose c to be the speed at which the elastic-plastic interface moves. Thus, the solution is self similar and expands over time – for a larger time, it is spread over a greater extent in space, but the solution values themselves are the same.

In order to proceed with this assumption, it is necessary to compute various derivatives. To begin with,

$$\frac{\partial \xi}{\partial t} = -\frac{r}{ct^2} = -\frac{\xi}{t}; \qquad \frac{\partial \xi}{\partial r} = \frac{1}{ct} = \frac{\xi}{r}. \tag{6.2}$$

Thus,

$$\frac{\partial f}{\partial t} = \frac{df}{d\xi}\frac{\partial \xi}{\partial t} = -\frac{r}{ct^2}f' = -\frac{\xi f'}{t}, \tag{6.3}$$

where the prime denotes differentiation with respect to ξ. The second derivative with respect to time (to be used later) is

$$\frac{\partial^2 f}{\partial t^2} = -\xi f'\frac{\partial}{\partial t}\left(\frac{1}{t}\right) - \frac{1}{t}\frac{d}{d\xi}(\xi f')\frac{\partial \xi}{\partial t} = \frac{\xi f'}{t^2} + \frac{\xi}{t^2}(f' + \xi f'') = \frac{2\xi f' + \xi^2 f''}{t^2}. \tag{6.4}$$

Similarly,

$$\frac{\partial f}{\partial r} = \frac{df}{d\xi}\frac{\partial \xi}{\partial r} = \frac{1}{ct}f' = \frac{\xi f'}{r} \tag{6.5}$$

and

$$\frac{\partial^2 f}{\partial r^2} = \xi f'\frac{\partial}{\partial r}\left(\frac{1}{r}\right) + \frac{1}{r}\frac{d}{d\xi}(\xi f')\frac{\partial \xi}{\partial r} = -\frac{\xi f'}{r^2} + \frac{\xi}{r^2}(f' + \xi f'') = \frac{\xi^2 f''}{r^2}. \tag{6.6}$$

The similarity method reduces the mass and momentum equations from Chapter 2, applied to one dimension, to ordinary differential equations. The partial differential equations for mass and momentum conservation are

$$\frac{D\rho}{Dt} + \rho\frac{\partial v}{\partial r} = 0, \qquad \frac{\partial \sigma}{\partial r} = \rho\frac{Dv}{Dt}. \tag{6.7}$$

The spatial coordinate has been written r instead of x so that the notation throughout this chapter is consistent. Under the similarity assumption, mass conservation becomes

$$\frac{\partial \rho}{\partial t} + v\frac{\partial \rho}{\partial r} + \rho\frac{\partial v}{\partial r} = 0 \quad \Rightarrow \quad -\frac{\xi\rho'}{t} + v\frac{\xi\rho'}{r} + \rho\frac{\xi v'}{r} = 0 \quad \Rightarrow \quad (-c\xi + v)\rho' + \rho v' = 0. \tag{6.8}$$

Momentum conservation becomes

$$\frac{\partial \sigma}{\partial r} = \rho\frac{\partial v}{\partial t} + \rho v\frac{\partial v}{\partial r} \quad \Rightarrow \quad \frac{\xi\sigma'}{r} = -\rho\frac{\xi v'}{t} + \rho v\frac{\xi v'}{r} \quad \Rightarrow \quad \sigma' = (-c\xi + v)\rho v'. \tag{6.9}$$

In each of these expressions there is a $(-c\xi + v)$ term. This term arises from the material derivative. Later, to find analytic solutions, the v will be ignored. This assumption is that $D(\cdot)/Dt \approx \partial(\cdot)/\partial t$. Though this may seem an unusual choice since large deflections occur, our analytic solutions will be compared with numerical solutions that include the nonlinear term of the material derivative and will be shown to produce good answers.

We will only consider the elastic-plastic region here, as an example. In the elastic-plastic region the material response in compression can be written

$$\tilde{\sigma} = K(1 - \rho_0/\rho) + 2Y/3. \tag{6.10}$$

Taking the derivative with respect to ξ, assuming the density change is small compared to the density, yields

$$\sigma' = -\frac{K\rho_0}{\rho^2}\rho' \approx -\frac{K}{\rho}\rho' \approx -\frac{K}{\rho_0}\rho' = -c_0^2\rho'. \tag{6.11}$$

Inserting this expression into the momentum balance yields

$$c_0^2\rho' = -(-c\xi + v)v' \tag{6.12}$$

and when this result is placed back into the mass balance the result is

$$\left(1 - \frac{1}{c_0^2}(-c\xi + v)^2\right)\rho v' = 0. \tag{6.13}$$

Since the density is strictly positive and ξ is variable, it is concluded that $v' = 0$ and therefore the velocity is a constant. A constant velocity implies constant density and therefore constant stress. We already knew that was the solution – namely that all the values are constant in between the elastic-plastic interface and the applied boundary condition – and it is reassuring that the similarity approach provides the correct answer.

In order to compare with solutions to be derived below, it is helpful to obtain an explicit value for the stress at the applied boundary velocity. It is possible to explicitly calculate the stress, but we wish to approximate it as follows. One of the differences between our one-dimensional "cavity expansion" and the cylindrical and spherical cavity expansions is that in one dimension at low velocity only an elastic wave is formed. That is not the case for the cylindrical and spherical cavity expansions – owing to the geometry there is always an elastic-plastic region, and with it an associated minimum stress. The minimum stress to have an elastic-plastic region in one dimension is the HEL. At the upper end of velocities, the velocity of the interface between the elastic and elastic-plastic region is c_0, independent of how fast the boundary is driven. This clearly implies that the maximum velocity for driving the boundary is c_0, and the Hugoniot jump condition gives for the stress at the prescribed velocity boundary

$$\tilde{\sigma} \approx \rho_0 U u = \rho_0 c_0 c_0 = K. \tag{6.14}$$

(This value can also be argued by letting $\rho \to \infty$ in Eq. 6.10, assuming $Y \ll K$.) Thus, the maximum stress is simply the bulk modulus, and the stress depends linearly on velocity with slope $\rho_0 c_0$. Since the stress is linear in terms of the velocity V at which we are driving the boundary, we approximate the stress there as

$$\tilde{\sigma} \approx \tilde{\sigma}_{\text{HEL}} + \rho_0 c_0 (V - u_{\text{HEL}}) = \left(1 - \frac{c_0}{c_L}\right) \tilde{\sigma}_{\text{HEL}} + \rho_0 c_0 V. \tag{6.15}$$

This approximate stress equals the HEL stress when V equals the HEL velocity and is roughly equal to K when $V = c_0$. We define

$$
\begin{aligned}
B_1 &= \left(1 - \frac{c_0}{c_L}\right) \tilde{\sigma}_{\text{HEL}} \\
&= \left(1 - \sqrt{\frac{K}{\lambda + 2\mu}}\right) \left(\frac{1-\nu}{1-2\nu}\right) Y = \left(1 - \sqrt{\frac{1+\nu}{3(1-\nu)}}\right) \frac{1-\nu}{1-2\nu} Y
\end{aligned} \tag{6.16}
$$

and thus write

$$\tilde{\sigma} \approx B_1 + \rho_0 c_0 V \tag{6.17}$$

as the final expression for the stress at the cavity wall.

6.2. The Boundary Condition at the Elastic-Plastic Interface

There is a preliminary result we need before performing the cavity expansions. The question we need to address is: what occurs at a front that is not traveling at the velocity predicted by the Hugoniot jump conditions? Such a situation can occur in the cylindrical and spherical cavity expansions for low expansion velocity. Due to the geometry, as the cavity opens up the material must yield. Thus there is an elastic-plastic region. However, the front of this region can travel at a velocity less than the velocity c_0. What occurs across such a front?

The Hugoniot jump conditions apply. Interest here is between an elastic region and an elastic-plastic region. Thus, near the front the stress will be given by

$$\tilde{\sigma} = \tilde{\sigma}_e + K(1 - \rho_0/\rho), \tag{6.18}$$

where $\tilde{\sigma}_e$ is a constant. The mass and momentum jump conditions are

$$\frac{\rho_r}{\rho_\ell} = 1 - \frac{u_\ell - u_r}{U - u_r}, \tag{6.19}$$

$$\tilde{\sigma}_\ell - \tilde{\sigma}_r = \rho_r (U - u_r)(u_\ell - u_r). \tag{6.20}$$

The assumed form of the stress and the mass jump condition lead to

$$\tilde{\sigma}_\ell - \tilde{\sigma}_r = K \left(\frac{\rho_0}{\rho_r} - \frac{\rho_0}{\rho_\ell}\right) = K \frac{\rho_0}{\rho_r} \left(1 - \frac{\rho_r}{\rho_\ell}\right) = K \frac{\rho_0}{\rho_r} \frac{u_\ell - u_r}{U - u_r}. \tag{6.21}$$

Putting this expression in the momentum jump condition gives

$$K \frac{\rho_0}{\rho_r} \frac{u_\ell - u_r}{U - u_r} = \rho_r (U - u_r)(u_\ell - u_r). \tag{6.22}$$

Using the definition of the bulk sound speed gives

$$\left\{ \left(\frac{\rho_0}{\rho_r} \right)^2 c_0^2 - (U - u_r)^2 \right\} (u_\ell - u_r) = 0. \tag{6.23}$$

Thus, we conclude that if the wave front speed U is not the bulk sound speed c_0 (adjusted for the density and material motion), then the velocity must be continuous – that is, there can be no jump (i.e., $u_\ell = u_r$). This result is quite profound:

THEOREM 6.1. *If a front is not moving at the speed specified by the shock jump conditions combined with the constitutive model and equation of state, then the particle velocity is continuous across the front and, hence, density, stress, and energy are continuous also.*

The physical quantities can have a change in slope – that is, the spatial derivatives of the quantities can be discontinuous – but the quantities themselves must be continuous. This result will be used as a boundary condition below.

6.3. The Compressible Cylindrical Cavity Expansion

The geometry of the problem is that of opening up a cylindrical cavity at a constant radial velocity V located in an infinite expanse of material (Fig. 6.2). Just as with the two-wave solution above, it is assumed that the space divides into three regions: a region where the material is plastically deforming, near the cavity; a region of elastic response; and a region of material that is as yet undisturbed. The elastic-plastic region is assumed to obey a Tresca yield condition. A number of assumptions are made to make the problem tractable.

In cylindrical coordinates, the equations of mass and momentum conservation are

$$\rho \left(\frac{\partial v}{\partial r} + \frac{v}{r} \right) = -\frac{D\rho}{Dt}, \tag{6.24}$$

$$\frac{\partial \sigma_{rr}}{\partial r} + \frac{\sigma_{rr} - \sigma_{\theta\theta}}{r} = \rho \frac{Dv}{Dt}. \tag{6.25}$$

Here σ_{ij} are the stresses, which are positive in tension, ρ is the material density, and v is the radial velocity. These equations will be solved in each region. The material velocities at each of the region boundaries will then be matched. The material velocity at the cavity wall is V, the cavity expansion velocity. In the presentation here it is assumed that the velocity of the elastic-plastic interface c is constant, but this assumption need not be made and a constant velocity falls out as the solution to the problem. The material displacement at the furthest edge of the elastic region is zero. The major assumption is to look for a similarity solution as above. Based on this assumption, with $\xi = r/ct$, the problem regions are defined with their boundary conditions:

(1) *Elastic-plastic region* $(V/c \leq \xi \leq 1)$: the region extends from the cavity boundary to the elastic-plastic interface. At the cavity boundary, the material has a constant material velocity V and $\xi = V/c$. The material is assumed to be compressible with bulk modulus K and flowing in this plastic region, obeying a Tresca yield condition with flow stress Y.

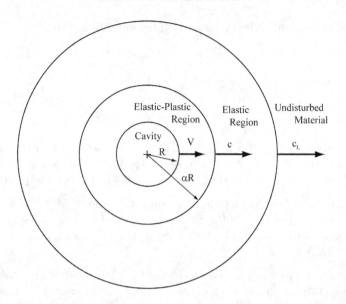

FIGURE 6.2. Geometry for cylindrical cavity expansion. (From Walker and Anderson [199].)

(2) *Elastic region* ($1 \leq \xi \leq c_L/c$), where c_L is the longitudinal elastic wave speed: the material is assumed to have small strains and to obey Hooke's law with Lamé constants λ and μ. The material velocities at the elastic-plastic interface ($\xi = 1$) are assumed to be equal. As was discussed above, this boundary condition is correct if the velocity of the elastic-plastic interface is less than the bulk sound speed c_0. Below, it will be shown where the boundary condition holds and where it does not. However, it turns out our solution works well for either case. When the velocity of the elastic-plastic interface equals the bulk sound speed, the boundary condition is the Hugoniot jump conditions.

(3) *Undisturbed region* ($\xi > c_L/c$): the elastic wave has not yet arrived in this material. The boundary between the elastic region and the undisturbed material $\xi = c_L/c$ propagates at the longitudinal elastic wave speed and has matching material displacements; that is, the material displacement is zero at this interface.

The following paragraphs outline the solution to the problem, with each of the boundary conditions being dealt with in roughly reverse order.

6.3.1. The Elastic Region. In the elastic region, the strains are small, and the stresses are given by Hooke's law,

$$\varepsilon_{rr} = \frac{\partial u}{\partial r}, \qquad \varepsilon_{\theta\theta} = \frac{u}{r}, \qquad \varepsilon_{zz} = 0, \tag{6.26}$$

$$\sigma_{rr} = (\lambda + 2\mu)\,\varepsilon_{rr} + \lambda\varepsilon_{\theta\theta}, \qquad \sigma_{\theta\theta} = \lambda\varepsilon_{rr} + (\lambda + 2\mu)\,\varepsilon_{\theta\theta}, \qquad \sigma_{zz} = \lambda\,(\varepsilon_{rr} + \varepsilon_{\theta\theta}), \tag{6.27}$$

where u is the radial displacement. Putting these expressions into the momentum equation and assuming the convective term is small (so that $Dv/Dt \approx \partial^2 u/\partial t^2$)

gives

$$\frac{\partial}{\partial r} \left(\frac{1}{r} \frac{\partial}{\partial r} (ru) \right) = \frac{1}{c_L^2} \frac{\partial^2 u}{\partial t^2}, \tag{6.28}$$

where $c_L^2 = (\lambda + 2\mu)/\rho$.

In addition to using the similarity transformation, it helps to write the displacement in a nondimensional form. This could be done by dividing it either by r or by ct. Looking at the expression for the second derivative with respect to time (Eq. 6.4), the choice is to divide by ct in the hopes that it will simplify the expression. With $\bar{u} = u/(ct)$,

$$\frac{\partial^2 u}{\partial t^2} = \frac{\partial^2 (ct\, \bar{u})}{\partial t^2} = 2c \frac{\partial \bar{u}}{\partial t} + ct \frac{\partial^2 \bar{u}}{\partial t^2} = 2c \left(-\frac{\xi\, \bar{u}'}{t} \right) + ct \frac{2\xi\, \bar{u}' + \xi^2 \bar{u}''}{t^2} = \frac{c\xi^2 \bar{u}''}{t}. \tag{6.29}$$

Thus, we do indeed get a simplification. For the left-hand side of Eq. (6.28) we have

$$\frac{\partial}{\partial r} \left(\frac{1}{r} \frac{\partial}{\partial r} (r\, u) \right) = ct \left(-\frac{\bar{u}}{r^2} + \frac{1}{r} \frac{\partial \bar{u}}{\partial r} + \frac{\partial^2 \bar{u}}{\partial r^2} \right) = \frac{1}{\xi} \left(-\frac{\bar{u}}{r} + \frac{\xi\, \bar{u}'}{r} + \frac{\xi^2 \bar{u}''}{r} \right). \tag{6.30}$$

With the similarity transformation, Eq. (6.28) thus becomes

$$\xi^2 \left(1 - \beta^2 \xi^2 \right) \bar{u}'' + \xi\, \bar{u}' - \bar{u} = 0, \tag{6.31}$$

where $\beta = c/c_L$. A solution is seen by observation to be $\bar{u} = \xi$. Using this solution, Eq. (6.31) can be reduced in order by the substitution $\bar{u} = \xi f(\xi)$, where f is an arbitrary function, giving

$$2 + \xi \left\{ \ln (f') \right\}' = \frac{-1}{1 - \beta^2 \xi^2}. \tag{6.32}$$

This equation can be solved for f,

$$f = -\ln \left(\frac{1 + \sqrt{1 - \beta^2 \xi^2}}{\beta \xi} \right) + \frac{\sqrt{1 - \beta^2 \xi^2}}{\beta^2 \xi^2} \tag{6.33}$$

up to arbitrary multiplicative and additive constants. This solution yields a second solution to the second order ordinary differential equation, and thus the total solution can be written as a linear combination of ξ and ξf,

$$\bar{u} = c_1 \xi - c_2 \left\{ \beta \xi \ln \left(\frac{1 + \sqrt{1 - \beta^2 \xi^2}}{\beta \xi} \right) - \frac{\sqrt{1 - \beta^2 \xi^2}}{\beta \xi} \right\}, \tag{6.34}$$

where c_1 and c_2 are arbitrary constants. Matching the boundary condition of zero material displacement ($\bar{u} = 0$) at the outer edge of the elastic region ($\xi = c_L/c = 1/\beta$) gives $c_1 = 0$.

We are now in a position to write the cavity expansion solution. The strains in terms of $\bar{u}$ are

$$\varepsilon_{rr} = \frac{\partial u}{\partial r} = ct \frac{\partial \bar{u}}{\partial r} = \bar{u}', \qquad \varepsilon_{\theta\theta} = \frac{u}{r} = \frac{\bar{u}}{\xi}, \tag{6.35}$$

which allows calculation of the stresses. Recalling that $u = ct\bar{u}$, which gives $\partial u / \partial t = c(\bar{u} - \xi \bar{u}')$ for the material velocity, the solution in the elastic region is

$$u = ctc_2 \left\{ \beta \xi \ln \left(\frac{1 + \sqrt{1 - \beta^2 \xi^2}}{\beta \xi} \right) - \frac{\sqrt{1 - \beta^2 \xi^2}}{\beta \xi} \right\},$$

$$\frac{\partial u}{\partial t} = -2cc_2 \frac{\sqrt{1 - \beta^2 \xi^2}}{\beta \xi},$$

$$\sigma_{rr} = 2c_2 \beta \left\{ (\lambda + \mu) \ln \left(\frac{1 + \sqrt{1 - \beta^2 \xi^2}}{\beta \xi} \right) + \mu \frac{\sqrt{1 - \beta^2 \xi^2}}{\beta^2 \xi^2} \right\}, \qquad (6.36)$$

$$\sigma_{\theta\theta} = 2c_2 \beta \left\{ (\lambda + \mu) \ln \left(\frac{1 + \sqrt{1 - \beta^2 \xi^2}}{\beta \xi} \right) - \mu \frac{\sqrt{1 - \beta^2 \xi^2}}{\beta^2 \xi^2} \right\},$$

$$\sigma_{zz} = 2c_2 \beta \lambda \ln \left(\frac{1 + \sqrt{1 - \beta^2 \xi^2}}{\beta \xi} \right).$$

On the elastic-plastic region interface ($\xi = 1$), the material is assumed to be at yield. The Tresca yield criteria is

$$\sigma_{\max} - \sigma_{\min} = \sigma_{\theta\theta} - \sigma_{rr} = Y, \qquad (6.37)$$

where Y is the flow stress, and the fact that the hoop stress is in tension (hence positive) and the radial stress is in compression (hence negative) has been used. The yield condition allows c_2 to be found:

$$c_2 = -\frac{Y}{4\mu} \frac{\beta}{\sqrt{1 - \beta^2}}. \qquad (6.38)$$

To match the boundary condition at the elastic-plastic region interface we need the material velocity at the interface. The velocity is

$$\left. \frac{\partial u}{\partial t} \right|_{\xi=1} = \frac{cY}{2\mu}. \qquad (6.39)$$

This completes the solution in the elastic region.

6.3.2. The Elastic-Plastic Region.

Within the elastic-plastic region, it is assumed that the Tresca yield condition always holds. Thus, the equations are

$$\rho \left(\frac{\partial v}{\partial r} + \frac{v}{r} \right) = - \left(\frac{\partial \rho}{\partial t} + v \frac{\partial \rho}{\partial r} \right), \qquad (6.40)$$

$$\frac{\partial \sigma_{rr}}{\partial r} - \frac{Y}{r} = \rho \left(\frac{\partial v}{\partial t} + v \frac{\partial v}{\partial r} \right). \qquad (6.41)$$

In the plastic region, we assume the stresses are large compared with the deviators and that the relative change in density is small, and so use the approximation $-\sigma'_{rr}/K \approx \rho'/\rho$ (Eq. 6.11), where K is the bulk modulus, to link Eqs. (6.40) and (6.41). Under the similarity change of variables and using this approximation, the mass and momentum conservation become

$$v' + \frac{v}{\xi} = \frac{1}{K} (-c\xi + v) \sigma'_{rr}, \qquad (6.42)$$

$$\sigma'_{rr} - \frac{Y}{\xi} = \rho (-c\xi + v) v'. \qquad (6.43)$$

Again, prime refers to differentiation with respect to ξ. Examination shows that constants for the various terms do not solve the equation – thus, the cylindrical cavity expansion solution is more complicated than the one-dimensional case. These equations combine to give

$$\left(1 - \frac{\rho}{K}(-c\xi + v)^2\right)v' + \left(1 - \frac{Y}{K}\right)\frac{v}{\xi} = -c\frac{Y}{K}. \tag{6.44}$$

If it is assumed that $1 - Y/K \approx 1$ (a very good approximation for metals) and that v can be dropped from the $-c\xi + v$ expression (this is the assumption that $D(\cdot)/Dt \approx \partial(\cdot)/\partial t$; notice that $c\xi$ and v are equal at the cavity boundary, but as one moves into the plastic region $c\xi$ increases while v decreases), then Eq. (6.44) can be written as

$$\left(1 - \frac{\rho c^2}{K}\xi^2\right)v' + \frac{v}{\xi} = -c\frac{Y}{K}. \tag{6.45}$$

If we choose to treat ρ as approximately constant, Eq. (6.45) yields the simple solution

$$v = -\frac{Y}{\rho c\xi} + c_3\frac{\sqrt{K/\rho c^2 - \xi^2}}{\xi}. \tag{6.46}$$

The arbitrary constant c_3 may be determined by applying the boundary condition at the inner wall of the cavity ($\xi = V/c$), where the material velocity is a constant V,

$$c_3 = \frac{Y/\rho + V^2}{\sqrt{K/\rho - V^2}}. \tag{6.47}$$

The velocity at the interface between the elastic and elastic-plastic region is then given by

$$v|_{\xi=1} = -\frac{Y}{\rho c} + \frac{1}{c}\left(\frac{Y}{\rho} + V^2\right)\sqrt{\frac{K/\rho - c^2}{K/\rho - V^2}}. \tag{6.48}$$

6.3.3. The Elastic/Elastic-Plastic Interface.

Now we connect the elastic-plastic and elastic solutions at the elastic-plastic boundary. We use the boundary condition that the material velocities match at the interface, which is true for lower cavity driving velocities. From Eqs. (6.39) and (6.48) and after some algebra, we obtain

$$\left(1 + \frac{\rho V^2}{Y}\right)\sqrt{K - \rho c^2} = \left(1 + \frac{\rho c^2}{2\mu}\right)\sqrt{K - \rho V^2}. \tag{6.49}$$

In Chapter 10, this equation will be used to determine the extent of the plastically flowing region within the target. The argument is as follows. The cavity is assumed to be opening at the velocity at which the projectile is penetrating the target. The cavity (crater) radius is therefore achieved at time

$$t_c = R_c/V, \tag{6.50}$$

where R_c is the crater radius. When the crater radius is achieved, the extent of the plastically flowing region is given by ct_c. Thus, in terms of crater radii, the extent of the plastically flowing region within the target is

$$\alpha = \frac{ct_c}{R_c} = \frac{c}{V}, \tag{6.51}$$

TABLE 6.1. Properties used in cavity expansion computations.

Material	ρ (g/cm^3)	K (GPa)	G (GPa)	Y (GPa)
Steel	7.85	167	77	1.2
Aluminum	2.70	72.8	27.3	0.38

where c is obtained from Eq. (6.49). It is possible to rewrite Eq. (6.49) in terms of α,

$$\left(1 + \frac{\rho V^2}{Y}\right)\sqrt{K - \rho\alpha^2 V^2} = \left(1 + \frac{\rho\alpha^2 V^2}{2\mu}\right)\sqrt{K - \rho V^2}. \tag{6.52}$$

In this equation, α is the unknown. Squaring this equation results in a quadratic equation in α^2. The equation is $A\alpha^4 + B\alpha^2 - C = 0$, and can be solved by the quadratic formula where the coefficients and solution are (Exercise 6.2)

$$A = \frac{\rho V^2(K - \rho V^2)}{4\mu^2}, \quad B = 1 + \frac{K}{\mu} + \frac{\rho V^2}{Y}\left(2 - \frac{Y}{\mu} + \frac{\rho V^2}{Y}\right), \quad C = 1 + \frac{2K}{Y} + \frac{K\rho V^2}{Y^2},$$

$$\alpha = \sqrt{\frac{2C}{B + \sqrt{B^2 + 4AC}}}. \tag{6.53}$$

The coefficients A, B, and C are positive for $0 < \rho V^2 < K$. The α at the lower and upper limits of the velocity domain can be obtained by direct evaluation, namely

$$\alpha_0 = \alpha|_{V=0} = \sqrt{\frac{2/Y + 1/K}{1/\mu + 1/K}} \tag{6.54}$$

at the lower end and $\alpha|_{V=c_0} = 1$ at the upper end. It is straightforward to show

$$\lim_{Y \to 0} \alpha = \frac{c_0}{V}, \tag{6.55}$$

which is what we would expect – namely, that in the absence of strength the wave front expands at the bulk sound speed, $c = c_0$.

Figure 6.3 shows the solution as a function of cavity velocity for steel (properties in Table 6.1). Displayed there is the approximate analytic solution Eq. (6.53) as well as the full numerical solution.[1] By full numerical solution, we mean that the assumption that $D(\cdot)/Dt \approx \partial(\cdot)/\partial t$ is not made, but rather Eqs. (6.42) and (6.43) are solved. Below a critical velocity the interior boundary condition at the elastic-plastic interface is the one we have discussed – namely, the material velocities match at the interface between the elastic and elastic-plastic regions. Above that critical velocity, the numerical solution uses the Hugoniot jump conditions as its interior boundary condition at the elastic-plastic interface. The vertical line in the plot notes where this transition occurs – at about 1,400 m/s or $\alpha = c_0/V = 3.3$. Thus, our approximate analytic solution provides good values of α, not only in the region where the interior boundary condition assumption holds, but also at higher velocities where it does not.

---

[1]The numerical solution was done using a slightly different pressure expression in the elastic-plastic region, namely $p = K(\rho/\rho_0 - 1)$ rather than $p = K(1 - \rho_0/\rho)$ (Eq. 6.10).

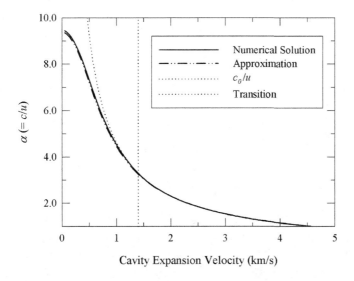

FIGURE 6.3. Elastic-plastic interface velocity for cylindrical cavity expansion in steel.

6.3.4. Stresses. Equation (6.42) can now be integrated using Eq. (6.46) to determine the radial stress. In particular, notice that Eq. (6.42) gives, being consistent with our solution and assuming $D(\cdot)/Dt \approx \partial(\cdot)/\partial t$,

$$\frac{(\xi v)'}{\xi} = -\frac{c\xi}{K}\sigma'_{rr} \tag{6.56}$$

or

$$\sigma'_{rr} = -\frac{K}{c\xi^2}(\xi v)'. \tag{6.57}$$

The velocity within the elastic-plastic region can be written (Eq. 6.46)

$$v\xi = -\frac{Y}{\rho c} + c_3\sqrt{K/\rho c^2 - \xi^2}. \tag{6.58}$$

This can be differentiated to give

$$(v\xi)' = -c_3\frac{\xi}{\sqrt{K/\rho c^2 - \xi^2}}. \tag{6.59}$$

Therefore, the radial stress is given by

$$\sigma'_{rr} = -\frac{c_3 K}{c}\frac{1}{\xi\sqrt{K/\rho c^2 - \xi^2}}. \tag{6.60}$$

This equation can be integrated to give

$$\sigma_{rr} = -c_3\sqrt{\rho K}\left\{\ln\left(2\sqrt{\frac{K}{\rho c^2}}\sqrt{K/\rho c^2 - \xi^2} + \frac{2K}{\rho c^2}\right) - \ln(\xi)\right\} \tag{6.61}$$

up to an additive constant. Evaluating over the region between the elastic-plastic boundary ($\xi = 1$) and the cavity interface ($\xi = V/c$) gives

$$\sigma_{rr}|_{\xi=1} - \sigma_{rr}|_{\xi=V/c} = -c_3\sqrt{\rho K}\left\{\ln\left(\frac{\sqrt{K-\rho c^2} + \sqrt{K}}{\sqrt{K-\rho V^2} + \sqrt{K}}\right) + \ln\left(\frac{V}{c}\right)\right\}. \tag{6.62}$$

From the elastic solution (Eq. 6.36), the stress at the elastic-plastic interface can be computed,

$$\sigma_{rr}|_{\xi=1} = 2c_2\beta \left\{ (\lambda + \mu) \ln \left(\frac{1 + \sqrt{1 - \beta^2}}{\beta} \right) + \mu \frac{\sqrt{1 - \beta^2}}{\beta^2} \right\}. \tag{6.63}$$

Thus, the stress at the cavity wall is given by

$$\sigma_{rr}|_{\xi=V/c} = 2c_2\beta \left\{ (\lambda + \mu) \ln \left(\frac{1 + \sqrt{1 - \beta^2}}{\beta} \right) + \mu \frac{\sqrt{1 - \beta^2}}{\beta^2} \right\}$$

$$+ c_3 \sqrt{\rho K} \left\{ \ln \left(\frac{\sqrt{K - \rho c^2} + \sqrt{K}}{\sqrt{K - \rho V^2} + \sqrt{K}} \right) + \ln \left(\frac{V}{c} \right) \right\}, \tag{6.64}$$

where c_2 and c_3 are defined above. This stress is what is required to open the cavity at the constant velocity V. This expression is somewhat complicated. We wish to make an approximation similar to what was obtained in the one-dimensional case. The first question that arises is: what is the $V \to 0$ limit of the stress on the cavity wall? Recalling that $\beta = c/c_L$ and therefore $\beta \to 0$ as $V \to 0$, taking the limit yields

$$\lim_{V \to 0} \sigma_{rr}|_{\xi=V/c}$$

$$= \lim_{V \to 0} 2c_2\beta \left\{ (\lambda + \mu) \ln \left(\frac{1 + \sqrt{1 - \beta^2}}{\beta} \right) + \mu \frac{\sqrt{1 - \beta^2}}{\beta^2} \right\} - c_3 \sqrt{\rho K} \ln(\alpha_0), \tag{6.65}$$

where α_0, which we derived an expression for above, is the zero velocity limit of the c/V ratio. Using the expression for c_2 and c_3 (Eqs. 6.38 and 6.47) gives

$$\lim_{V \to 0} \sigma_{rr}|_{\xi=V/c} = -\frac{Y}{2\mu} \left\{ (\lambda + \mu) \lim_{\beta \to 0} \beta^2 \ln \left(\frac{1}{\beta} \right) + \mu \right\} - Y \ln(\alpha_0) = -Y \left(\frac{1}{2} + \ln(\alpha_0) \right). \tag{6.66}$$

This stress is the low velocity limit – the stress required to open the cavity at zero speed. The upper limit on the stress is the same as that derived for the one-dimensional case – namely K – since the same Hugoniot jump conditions apply. It turns out that a good approximation to the stress on the cavity wall is given by the following fit between the two extreme values:

$$\tilde{\sigma}_{rr} \approx Y \left(\frac{1}{2} + \ln(\alpha_0) \right) + K \left(\frac{V}{c_0} \right)^{1.65} = Y \left(\frac{1}{2} + \ln(\alpha_0) \right) + \rho_0 c_0^{0.35} V^{1.65}. \tag{6.67}$$

From a theoretical viewpoint it would be nice if the power were $3/2$ rather than 1.65, but the 1.65-power empirical fit to the full numerical solution is much closer than the fit with the power $3/2$. In any event, we do not intend to use this stress expression in actual penetration calculations. We define

$$B_2 = \lim_{V \to 0} \tilde{\sigma}_{rr}|_{\xi=V/c} = Y \left(\frac{1}{2} + \ln(\alpha_0) \right) = \frac{Y}{2} \left(1 + \ln \left\{ \frac{2/Y + 1/K}{1/\mu + 1/K} \right\} \right) \tag{6.68}$$

and thus write

$$\tilde{\sigma}_{rr} \approx B_2 + \rho_0 c_0^{0.35} V^{1.65}. \tag{6.69}$$

Figure 6.4 compares the stress as in Eqs. (6.64) and (6.69) and the full numerical solution.

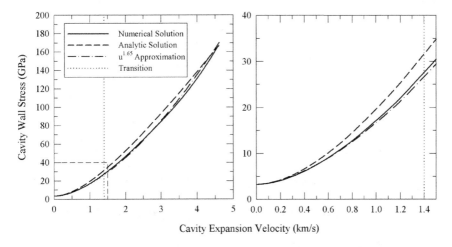

FIGURE 6.4. Cavity wall stress for the cylindrical cavity expansion in steel (full velocity scale left and detail right).

As a side comment, it might seem that a constant stress on the cavity wall throughout the expansion counters our intuition that it should be getting more difficult to expand the cavity as the radius increases. In fact, in terms of force, it is getting harder to expand the cavity as the cavity radius increases: the area of the cavity increases linearly with the cavity radius and, thus, with a constant stress the force required to open the cavity for the cylindrical cavity expansion increases linearly with cavity radius.

Finally, we have three comparisons of the full numerical solution with our approximate analytic one, again for steel. Figure 6.5 shows the particle velocity as a function of ξ for two cavity expansion velocities, $V = 1,000\,\text{m/s}$ and $V = 2,000\,\text{m/s}$. The first velocity is below the critical velocity, and so the velocities are continuous at the elastic-plastic interface (denoted on the graph). The second velocity is above the critical velocity, and a jump in the velocities due to the Hugoniot jump conditions is seen. Similarly, Fig. 6.6 shows the radial stress as a function of ξ for these two velocities. Once again continuity in the stress occurs for the lower velocity and a jump occurs for the higher velocity. Finally, Fig. 6.7 shows density as a function of ξ.

6.4. The Compressible Spherical Cavity Expansion Solution

The geometry of the spherical cavity expansion is that of opening up a spherical cavity at a constant radial velocity V located in an infinite expanse of material (Fig. 6.2), just as with the cylindrical case, and again there are three regions: a region where the material is plastically deforming, near the cavity; a region of elastic response; and a region of material that is as yet undisturbed. The elastic-plastic region is assumed to obey a von Mises yield condition, which in this case is identical with the Tresca yield condition. Again, a number of assumptions are made to make the problem tractable.

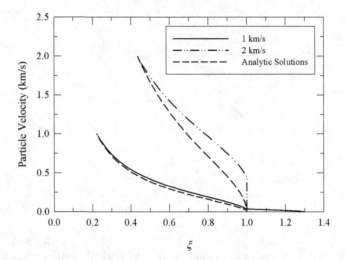

FIGURE 6.5. Material velocity v as a function of ξ for cylindrical cavity expansion in steel for two cavity expansion velocities (1 and 2 km/s).

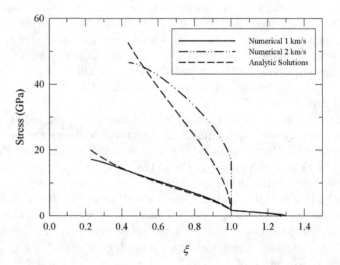

FIGURE 6.6. Radial stress $\tilde{\sigma}_{rr}$ as a function of ξ for cylindrical cavity expansion in steel for two cavity expansion velocities (1 and 2 km/s).

In spherical coordinates, the equations of mass and momentum conservation are

$$\rho \left(\frac{\partial v}{\partial r} + \frac{2v}{r} \right) = -\frac{D\rho}{Dt}, \qquad \frac{\partial \sigma_{rr}}{\partial r} + \frac{2\sigma_{rr} - \sigma_{\theta\theta} - \sigma_{\varphi\varphi}}{r} = \rho \frac{Dv}{Dt}. \qquad (6.70)$$

Here σ_{ij} are the stresses, which are positive in tension, ρ is the material density, and v is the radial velocity. Notice the only essential difference in the equations is the factor 2 over the r for spherical geometry, where the factor was 1 for cylindrical

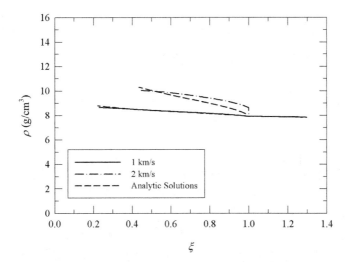

FIGURE 6.7. Density as a function of ξ for cylindrical cavity expansion in steel for two cavity expansion velocities (1 and 2 km/s).

geometry. The regions with the boundary conditions are the same as with the cylindrical cavity expansion. The following paragraphs outline the solution to the problem, with each of the boundary conditions being dealt with in roughly reverse order.

6.4.1. The Elastic Region. In the elastic region, the strains are small, and the stresses are given by Hooke's law:

$$\varepsilon_{rr} = \frac{\partial u}{\partial r}, \quad \varepsilon_{\theta\theta} = \frac{u}{r}, \quad \varepsilon_{\varphi\varphi} = \frac{u}{r}, \quad \sigma_{rr} = (\lambda + 2\mu)\,\varepsilon_{rr} + \lambda(\varepsilon_{\theta\theta} + \varepsilon_{\varphi\varphi}),$$
$$\sigma_{\theta\theta} = \lambda\varepsilon_{rr} + (\lambda + 2\mu)\,\varepsilon_{\theta\theta} + \lambda\varepsilon_{\varphi\varphi}, \quad \sigma_{\varphi\varphi} = \lambda\,(\varepsilon_{rr} + \varepsilon_{\theta\theta}) + (\lambda + 2\mu)\varepsilon_{\varphi\varphi}, \quad (6.71)$$

where u is the radial displacement. Putting these expressions into the momentum equation and assuming the convective term is small (so that $Dv/Dt \approx \partial^2 u/\partial t^2$) gives

$$\frac{\partial}{\partial r}\left(\frac{1}{r^2}\frac{\partial}{\partial r}\left(r^2 u\right)\right) = \frac{1}{c_L^2}\frac{\partial^2 u}{\partial t^2}, \qquad (6.72)$$

where $c_L^2 = (\lambda + 2\mu)/\rho$. Again, the only difference from the cylindrical case are the two powers of 2 on the left-hand side. With $\bar{u} = u/ct$, we get the same expression for the time derivative (see Eq. 6.29)

$$\frac{\partial^2 u}{\partial t^2} = \frac{c\xi^2 \bar{u}''}{t}. \qquad (6.73)$$

For the left-hand side of Eq. (6.72) we now have

$$\frac{\partial}{\partial r}\left(\frac{1}{r^2}\frac{\partial}{\partial r}\left(r^2 u\right)\right) = ct\left(-\frac{2\bar{u}}{r^2} + \frac{2}{r}\frac{\partial\bar{u}}{\partial r} + \frac{\partial^2\bar{u}}{\partial r^2}\right) = \frac{1}{\xi}\left(-\frac{2\bar{u}}{r} + \frac{2\xi\,\bar{u}'}{r} + \frac{\xi^2\bar{u}''}{r}\right)$$
$$(6.74)$$

and with the similarity transformation Eq. (6.72) becomes

$$\xi^2\left(1 - \beta^2\xi^2\right)\bar{u}'' + 2\xi\,\bar{u}' - 2\bar{u} = 0, \qquad (6.75)$$

where $\beta = c/c_L$. Again, a solution is seen by observation to be $\bar{u} = \xi$. Using this solution, Eq. (6.75) can be reduced in order by the substitution $\bar{u} = \xi f(\xi)$, giving

$$2 + \xi \left\{ \ln \left(f' \right) \right\}' = \frac{-2}{1 - \beta^2 \xi^2}. \tag{6.76}$$

(Notice the only difference is the factor of 2 on the right-hand side.) This equation can be solved for f,

$$f = -\frac{1}{3\xi^3} + \frac{\beta^2}{\xi} \tag{6.77}$$

up to arbitrary multiplicative and additive constants. This solution yields a second solution to the second order ordinary differential equation, and thus the total solution can be written as a linear combination of ξ and ξf,

$$\bar{u} = c_1 \xi + c_2 \left\{ -\frac{1}{3\xi^2} + \beta^2 \right\}, \tag{6.78}$$

where c_1 and c_2 are arbitrary constants. Matching the boundary condition of zero material displacement ($\bar{u} = 0$) at the outer edge of the elastic region ($\xi = c_L/c = 1/\beta$) gives

$$\bar{u} = c_2 \left\{ -\frac{1}{3\xi^2} - \frac{2\beta^3 \xi}{3} + \beta^2 \right\}. \tag{6.79}$$

We are now in a position to write the solution to the elastic region. The strains in terms of $\bar{u}$ are

$$\varepsilon_{rr} = \frac{\partial u}{\partial r} = ct \frac{\partial \bar{u}}{\partial r} = \bar{u}', \qquad \varepsilon_{\theta\theta} = \varepsilon_{\varphi\varphi} = \frac{u}{r} = \frac{\bar{u}}{\xi}, \tag{6.80}$$

which allows calculation of the stresses. Recalling that $u = ct\bar{u}$, which gives $\partial u/\partial t = c(\bar{u} - \xi\bar{u}')$ for the material velocity, the solution in the elastic region is

$$u = ctc_2 \left\{ \beta^2 - \frac{2\beta^3 \xi}{3} - \frac{1}{3\xi^2} \right\},$$

$$\frac{\partial u}{\partial t} = cc_2 \left(\beta^2 - 1/\xi^2 \right),$$

$$\sigma_{rr} = c_2 \left\{ \frac{4}{3} \frac{\mu}{\xi^3} - \frac{2}{3}(3\lambda + 2\mu)\beta^3 + 2\lambda \frac{\beta^2}{\xi} \right\}, \tag{6.81}$$

$$\sigma_{\theta\theta} = \sigma_{\varphi\varphi} = c_2 \left\{ -\frac{2}{3} \frac{\mu}{\xi^3} - \frac{2}{3}(3\lambda + 2\mu)\beta^3 + 2(\lambda + \mu) \frac{\beta^2}{\xi} \right\}.$$

On the elastic-plastic region interface ($\xi = 1$), the material is assumed to be at yield. For the spherical geometry, both the von Mises and Tresca yield criteria give

$$\sigma_{\theta\theta} - \sigma_{rr} = \sigma_{\varphi\varphi} - \sigma_{rr} = Y. \tag{6.82}$$

Thus,

$$c_2 = \frac{Y}{2\mu(\beta^2 - 1)} \tag{6.83}$$

which again gives the material velocity at the interface of

$$\left. \frac{\partial u}{\partial t} \right|_{\xi=1} = \frac{cY}{2\mu} \tag{6.84}$$

the same as for the one-dimensional wave propagation (Eq. 5.65) and cylindrical cavity expansion (Eq. 6.39).

6.4.2. The Elastic-Plastic Region. Within the elastic-plastic region, it is assumed that the yield condition always holds, and thus the equations are

$$\rho\left(\frac{\partial v}{\partial r} + \frac{2v}{r}\right) = -\left(\frac{\partial \rho}{\partial t} + v\frac{\partial \rho}{\partial r}\right), \tag{6.85}$$

$$\frac{\partial \sigma_{rr}}{\partial r} - \frac{2Y}{r} = \rho\left(\frac{\partial v}{\partial t} + v\frac{\partial v}{\partial r}\right). \tag{6.86}$$

In the plastic region, we assume the stresses are large compared with the deviators and that the relative change in density is small, and so use the approximation $-\sigma'_{rr}/K \approx \rho'/\rho$ (Eq. 6.11), where K is the bulk modulus, to link Eqs. (6.85) and (6.86). Under the similarity change of variables and using this approximation, the mass and momentum conservation become

$$v' + \frac{2v}{\xi} = \frac{1}{K}\left(-c\xi + v\right)\sigma'_{rr}, \tag{6.87}$$

$$\sigma'_{rr} - \frac{2Y}{\xi} = \rho\left(-c\xi + v\right)v'. \tag{6.88}$$

The 2's are the only difference from the cylindrical expansion. These equations combine to give

$$\left(1 - \frac{\rho}{K}\left(-c\xi + v\right)^2\right)v' + 2\left(1 - \frac{Y}{K}\right)\frac{v}{\xi} = -2c\frac{Y}{K}. \tag{6.89}$$

Assuming that $1 - Y/K \approx 1$ and that v can be dropped from the $-c\xi + v$ expression (this is the assumption that $D(\cdot)/Dt \approx \partial(\cdot)/\partial t$), then Eq. (6.89) can be written as

$$\left(1 - \frac{\rho c^2}{K}\xi^2\right)v' + \frac{2v}{\xi} = -c\frac{2Y}{K}. \tag{6.90}$$

If we choose to treat ρ as approximately constant, Eq. (6.90) can be explicitly solved, though it is fairly tedious (unlike the result in cylindrical coordinates). The solution is

$$v = -\frac{Y}{\rho c\xi} + \frac{Y}{2\rho c^2\xi^2}\sqrt{\frac{K}{\rho}}\left(1 - \frac{\rho c^2\xi^2}{K}\right)\ln\left(\frac{\sqrt{K} + c\xi\sqrt{\rho}}{\sqrt{K} - c\xi\sqrt{\rho}}\right) + c_3\frac{1 - \rho c^2\xi^2/K}{\xi^2}, \tag{6.91}$$

where c_3 is an arbitrary constant. This expression appears somewhat simpler if the bulk wave speed definition $c_0 = \sqrt{K/\rho}$ is used,

$$v = -\frac{Y}{\rho c\xi} + \frac{Y c_0}{2\rho c^2\xi^2}\left(1 - \frac{c^2}{c_0^2}\xi^2\right)\ln\left(\frac{c_0 + c\xi}{c_0 - c\xi}\right) + c_3\frac{1 - c^2\xi^2/c_0^2}{\xi^2}. \tag{6.92}$$

The constant c_3 may be determined by applying the boundary condition at the inner wall of the cavity ($\xi = V/c$), where the material velocity is a constant V,

$$c_3 = \frac{V^2}{c^2}\frac{1}{1 - V^2/c_0^2}\left\{V + \frac{Y}{\rho V} - \frac{Y c_0}{2\rho V^2}\left(1 - \frac{V^2}{c_0^2}\right)\ln\left(\frac{c_0 + V}{c_0 - V}\right)\right\}. \tag{6.93}$$

To simplify some upcoming expressions, we define a function g,

$$g(V) = \frac{Y}{\rho V^2} - \frac{Y c_0}{2\rho V^3}\left(1 - \frac{V^2}{c_0^2}\right)\ln\left(\frac{c_0 + V}{c_0 - V}\right). \tag{6.94}$$

Then c_3 can be written

$$c_3 = \frac{V^2}{c^2}\frac{1}{1 - V^2/c_0^2}\left(V + Vg(V)\right). \tag{6.95}$$

The velocity at the interface between the elastic and elastic-plastic regions, where $\xi = 1$, is then given by

$$v|_{\xi=1} = -cg(c) + c_3 \left(1 - \frac{c^2}{c_0^2}\right) = -cg(c) + \frac{V^2}{c^2}\frac{1 - c^2/c_0^2}{1 - V^2/c_0^2}V\left(1 + g(V)\right). \quad (6.96)$$

6.4.3. The Elastic/Elastic-Plastic Interface.
We now join the elastic-plastic and elastic solutions at the elastic-plastic boundary. Again the material velocities are assumed to be the same on both sides of the interface. Then, from Eqs. (6.84) and (6.96),

$$\frac{cY}{2\mu} = -cg(c) + \frac{V^2}{c^2}\frac{1 - c^2/c_0^2}{1 - V^2/c_0^2}V\left(1 + g(V)\right). \quad (6.97)$$

This result leads to

$$c^3\left(1 - \frac{V^2}{c_0^2}\right)\left(\frac{Y}{2\mu} + g(c)\right) = V^3\left(1 - \frac{c^2}{c_0^2}\right)(1 + g(V)) \quad (6.98)$$

or, written out explicitly,

$$c^3\left(1 - \frac{V^2}{c_0^2}\right)\left(\frac{Y}{2\mu} + \frac{Y}{\rho c^2} - \frac{Yc_0}{2\rho c^3}\left(1 - \frac{c^2}{c_0^2}\right)\ln\left(\frac{c_0 + c}{c_0 - c}\right)\right)$$
$$= V^3\left(1 - \frac{c^2}{c_0^2}\right)\left(1 + \frac{Y}{\rho V^2} - \frac{Yc_0}{2\rho V^3}\left(1 - \frac{V^2}{c_0^2}\right)\ln\left(\frac{c_0 + V}{c_0 - V}\right)\right). \quad (6.99)$$

This equation can be solved numerically using Newton's method, and Fig. 6.8 shows the solution as a function of cavity velocity for steel (properties in Table 6.1). Displayed there is the approximate analytic solution as well as the full numerical solution. By full numerical solution, we mean that the assumption that $D(\cdot)/Dt \approx \partial(\cdot)/\partial t$ is not made, but rather Eqs. (6.87) and (6.88) are solved. Below a critical velocity the boundary condition is the one we have discussed – namely, the material velocities match at the interface between the elastic and plastically flowing regions. Above that critical velocity, the numerical solution used the Hugoniot jump conditions as its boundary condition. The vertical line in the plot notes where this transition occurs – at about 2,550 m/s, or a value of $\alpha = 1.8$. In the case of the spherical cavity expansion, most of the region is at velocities less than the transition. Again, the approximate solution gives good results both below and above the transition region where the Hugoniot jump conditions must be used.

At low velocity the ratio c/V approaches a limit. To calculate the limit requires determining the low velocity limit of the function g. In particular, the Taylor series

$$\ln(1 + x) = x - \frac{x^2}{2} + \frac{x^3}{3} - \frac{x^4}{4} + \frac{x^5}{5} - \dots \quad (6.100)$$

gives

$$\ln\left(\frac{1 + x}{1 - x}\right) = \ln(1 + x) - \ln(1 - x) = 2\left(x + \frac{x^3}{3} + \frac{x^5}{5}\dots\right). \quad (6.101)$$

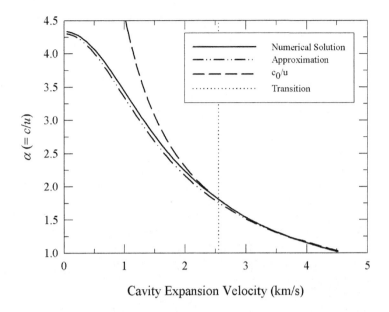

FIGURE 6.8. Elastic-plastic interface velocity for spherical cavity expansion in steel.

Thus, using $x = c/c_0$,

$$
\begin{aligned}
g(c) &= \frac{Y}{\rho c^2} - \frac{Y c_0}{2\rho c^3}\left(1 - \frac{c^2}{c_0^2}\right)\ln\left(\frac{c_0 + c}{c_0 - c}\right) \\
&= \frac{Y}{\rho c^2} - \frac{Y c_0}{2\rho c^3}\left(1 - \frac{c^2}{c_0^2}\right)2\left\{\frac{c}{c_0} + \frac{c^3}{3c_0^3} + \frac{c^5}{5c_0^5} + \dots\right\} \\
&= \frac{Y}{\rho c^2} - \frac{Y c_0}{\rho c^3}\left(\frac{c}{c_0} - \frac{2c^3}{3c_0^3} - \frac{2c^5}{15c_0^5} - \dots\right) \\
&= \frac{2Y}{3\rho c_0^2} + \frac{2Y}{15\rho c_0^4}c^2 + \dots,
\end{aligned}
\tag{6.102}
$$

where the additional terms contain higher powers of c. Thus, in the limit we have

$$
\lim_{c\to 0} g(c) = \frac{2Y}{3\rho c_0^2} = \frac{2Y}{3K}.
\tag{6.103}
$$

From Eq. (6.98), it is now possible to compute the low velocity limit of the α parameter for the spherical cavity expansion,

$$
\begin{aligned}
\alpha_0 &= \lim_{V\to 0}\frac{c}{V} = \lim_{V\to 0}\left\{\frac{1 - c^2/c_0^2}{1 - V^2/c_0^2}\frac{1 + g(V)}{Y/2\mu + g(c)}\right\}^{1/3} \\
&= \left\{\frac{1 + g(0)}{Y/2\mu + g(0)}\right\}^{1/3} = \left\{\frac{2/Y + 4/(3K)}{1/\mu + 4/(3K)}\right\}^{1/3}.
\end{aligned}
\tag{6.104}
$$

Note the similarity to the result found for the cylindrical cavity expansion, namely that the power was $1/2$ for the cylindrical, but now is $1/3$ for spherical, and the bulk modulus K has been replaced by $3K/4$; but otherwise the result is the same.

6.4.4. Stresses. Equation (6.87) can now be integrated using Eq. (6.92) to determine the radial stress. In particular, notice that Eq. (6.87), being consistent with our solution and assuming $D(\cdot)/Dt \approx \partial(\cdot)/\partial t$, gives

$$\frac{(\xi^2 v)'}{\xi^2} = -\frac{c\xi}{K}\sigma'_{rr} \quad \Rightarrow \quad \sigma'_{rr} = -\frac{K}{c\xi^3}(\xi^2 v)'. \tag{6.105}$$

The velocity within the elastic-plastic region can be written (Eq. 6.92)

$$v\xi^2 = -\frac{Y\xi}{\rho c} + \frac{Y c_0}{2\rho c^2}\left(1 - \frac{c^2}{c_0^2}\xi^2\right)\ln\left(\frac{c_0 + c\xi}{c_0 - c\xi}\right) + c_3(1 - c^2\xi^2/c_0^2). \tag{6.106}$$

This can be differentiated to give

$$(v\xi^2)' = -\frac{Y c_0 \xi}{K}\ln\left(\frac{c_0 + c\xi}{c_0 - c\xi}\right) - \frac{2c_3 c^2 \xi}{c_0^2}. \tag{6.107}$$

Therefore, the radial stress is given by

$$\sigma'_{rr} = \frac{Y c_0}{c\xi^2}\ln\left(\frac{c_0 + c\xi}{c_0 - c\xi}\right) + \frac{2c_3 \rho c}{\xi^2}. \tag{6.108}$$

This equation can be integrated to give

$$\sigma_{rr} = -Y\left\{\frac{c_0}{c\xi}\ln\left(\frac{c_0 + c\xi}{c_0 - c\xi}\right) + \ln\left(\frac{c_0^2}{c^2\xi^2} - 1\right)\right\} - \frac{2c_3 \rho c}{\xi} \tag{6.109}$$

up to an additive constant. Evaluating over the region between the elastic-plastic boundary ($\xi = 1$) and the cavity interface ($\xi = V/c$) gives

$$\sigma_{rr}|_{\xi=V/c} = \sigma_{rr}|_{\xi=1} - Y\left\{\frac{c_0}{V}\ln\left(\frac{c_0 + V}{c_0 - V}\right) - \frac{c_0}{c}\ln\left(\frac{c_0 + c}{c_0 - c}\right)\right.$$
$$\left. + \ln\left(\frac{c_0^2}{V^2} - 1\right) - \ln\left(\frac{c_0^2}{c^2} - 1\right)\right\} - 2c_3 \rho c\left(\frac{c}{V} - 1\right). \tag{6.110}$$

From the elastic solution (Eq. 6.81), the stress at the elastic-plastic interface can be computed,

$$\sigma_{rr}|_{\xi=1} = -\frac{Y}{2\mu(1+\beta)}\left\{\frac{4}{3}\mu\left(1 + \beta + \beta^2\right) + 2\lambda\beta^2\right\}. \tag{6.111}$$

Thus, we have determined the stress at the cavity wall. This stress is what is required to open the cavity at the constant velocity V.

Again, to make an approximation similar to what was obtained in the one-dimensional and cylindrical cavity expansion cases, we find the $V \to 0$ limit of the stress on the cavity wall. For the stress at the elastic-plastic boundary, recall that $\beta = c/c_L$ and therefore $\beta \to 0$ as $V \to 0$, yielding

$$\lim_{V\to 0}\sigma_{rr}|_{\xi=1} = -\frac{2Y}{3}. \tag{6.112}$$

Next, there are two terms in Eq. (6.110) of the form

$$\lim_{c\to 0}\frac{c_0}{c}\ln\frac{c_0 + c}{c_0 - c} = \lim_{c\to 0}\frac{c_0}{c}2\left(\frac{c}{c_0} + \frac{c^3}{3c_0^3} + \dots\right) = 2. \tag{6.113}$$

Since both terms yield the same constant in the limit, they cancel, as they are of opposite sign. The final term is of the form

$$\lim_{V\to 0}\ln\left(\frac{c_0^2/V^2 - 1}{c_0^2/c^2 - 1}\right) = \lim_{V\to 0}\ln\left(\frac{c^2}{V^2}\right) = 2\ln(\alpha_0). \tag{6.114}$$

Examination of c_3 shows that for small velocity it behaves like a constant times V, and thus in the $V \to 0$ limit $c_3 = 0$. Thus, the stress to open the cavity at zero velocity is

$$\lim_{V \to 0} \sigma_{rr}|_{\xi=V/c} = -\frac{2Y}{3} - 2Y \ln(\alpha_0). \tag{6.115}$$

This stress is the low velocity limit – the velocity required to open the cavity at zero speed. The upper limit on the stress is the same as that derived for the one-dimensional case – namely K – since the same Hugoniot jump conditions apply. It turns out that a good approximation to the stress on the cavity wall is given by the following fit between the two extreme values,

$$\tilde{\sigma}_{rr} \approx Y \left(\frac{2}{3} + 2\ln(\alpha_0)\right) + K \left(\frac{V}{c_0}\right)^2 = Y \left(\frac{2}{3} + 2\ln(\alpha_0)\right) + \rho V^2. \tag{6.116}$$

The α_0 value is for the spherical expansion. We define

$$B_3 = \lim_{V \to 0} \tilde{\sigma}_{rr}|_{\xi=V/c} = Y \left(\frac{2}{3} + 2\ln(\alpha_0)\right) = \frac{2Y}{3} \left(1 + \ln\left\{\frac{2/Y + 4/(3K)}{1/\mu + 4/(3K)}\right\}\right) \tag{6.117}$$

and thus

$$\tilde{\sigma}_{rr} \approx B_3 + \rho V^2 \tag{6.118}$$

as the final result. Figure 6.9 compares the expressions for the stress at the cavity wall. The value of B_3 is the low velocity limit of a spherical expansion, and so approaches the "static punch pressure" from above. It is directly comparable to the elastic-plastic thick-walled sphere problem solved by Bishop, Hill, and Mott in 1945 in work they performed during World War II [93]. Their approach was to find the largest applied pressure on the interior surface of a thick-walled sphere that admitted a static solution, and hence to approach the static punch stress from the bottom. Their result was

$$\alpha \approx \left(\frac{2/Y}{4/(3K) + 1/\mu}\right)^{1/3} = \left(\frac{E}{3(1-\nu)Y}\right)^{1/3} \tag{6.119}$$

and

$$p = \tilde{\sigma}_{rr} = 2Y \left\{\frac{1}{3} + \ln\left(\frac{2/Y}{4/(3K) + 1/\mu}\right)^{1/3}\right\} = \frac{2Y}{3} \left\{1 + \ln\left(\frac{E}{3(1-\nu)Y}\right)\right\}. \tag{6.120}$$

The only difference between Eq. (6.120) and Eq. (6.117) is the $4/3K$ in the numerator of the logarithm term, and this is negligible compared to $2/Y$.

Finally, we have three comparisons of the full numerical solution with our approximate analytic one, again for steel. Figure 6.10 shows the velocities as a function of ξ for two cavity velocities, $V = 1,500$ m/s and $V = 3,000$ m/s. The first velocity is below the critical velocity, and so the velocities are continuous at the elastic-plastic interface (denoted on the graph). The second velocity is above the critical velocity, and a jump in the velocities owing to the Hugoniot jump conditions can be seen. Similarly, Fig. 6.11 shows the radial stress as a function of ξ for these two velocities. Once again, continuity in the stress occurs for the lower velocity and a jump occurs for the higher velocity. Finally, Fig. 6.12 displays the density as a function of ξ for the two velocities.

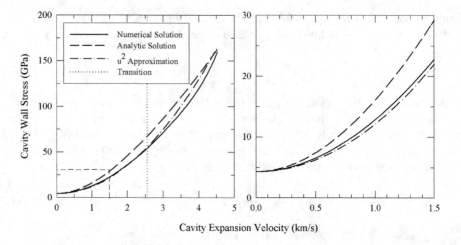

FIGURE 6.9. Cavity wall stress $\tilde{\sigma}_{rr}$ for the spherical cavity expansion in steel (left: full velocity scale; right: detail).

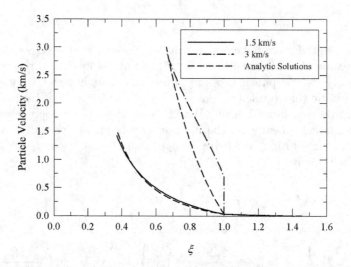

FIGURE 6.10. Material velocity v as a function of ξ for spherical cavity expansion in steel for two cavity expansion velocities (1.5 and 3 km/s).

6.5. Comparing the One-Dimensional, Cylindrical, and Spherical Cavity Expansions

With the development of the cavity expansion solutions, it is interesting to compare the results of the one-dimensional, cylindrical, and spherical cases. To begin, note that the α_0 values for the spherical and cylindrical cavity essentially only differ by a power,

$$(\alpha_0)_{\text{sph}} \approx \{(\alpha_0)_{\text{cyl}}\}^{2/3}, \tag{6.121}$$

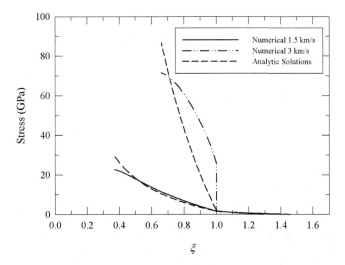

FIGURE 6.11. Radial stress $\tilde{\sigma}_{rr}$ as a function of ξ for spherical cavity expansion in steel for two cavity expansion velocities (1.5 and 3 km/s).

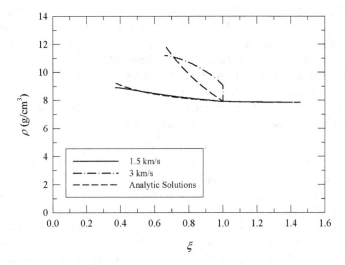

FIGURE 6.12. Density as a function of ξ for spherical cavity expansion in steel for two cavity expansion velocities (1.5 and 3 km/s).

so that

$$(\tilde{\sigma}_{rr}|_{V=0})_{\text{cyl}} = Y\left(\frac{1}{2} + \ln(\alpha_0)_{\text{cyl}}\right), \tag{6.122}$$

$$(\tilde{\sigma}_{rr}|_{V=0})_{\text{sph}} = Y\left(\frac{2}{3} + 2\ln(\alpha_0)_{\text{sph}}\right) \approx \frac{4}{3}Y\left(\frac{1}{2} + \ln(\alpha_0)_{\text{cyl}}\right). \tag{6.123}$$

Recalling the definitions of the zero velocity stress intercepts,

$$B_1 = \left(1 - \sqrt{\frac{1+\nu}{3(1-\nu)}}\right)\frac{1-\nu}{1-2\nu}Y, \tag{6.124}$$

$$B_2 = \frac{Y}{2}\left(1 + \ln\left\{\frac{2/Y + 1/K}{1/\mu + 1/K}\right\}\right), \tag{6.125}$$

$$B_3 = \frac{2Y}{3}\left(1 + \ln\left\{\frac{2/Y + 4/(3K)}{1/\mu + 4/(3K)}\right\}\right), \tag{6.126}$$

another way to write Eq. (6.123) is

$$B_3 \approx \frac{4}{3}B_2. \tag{6.127}$$

The equations for the stress to open the cavity at a constant velocity V are

One-dimensional: $\qquad \tilde{\sigma}_{1D} \approx B_1 + \rho_0 c_0 V,$ $\qquad\qquad$ (6.128)

Cylindrical: $\qquad\qquad \tilde{\sigma}_{rr} \approx B_2 + \rho_0 c_0^{0.35} V^{1.65},$ $\qquad$ (6.129)

Spherical: $\qquad\qquad \tilde{\sigma}_{rr} \approx B_3 + \rho_0 V^2.$ $\qquad\qquad\qquad$ (6.130)

The stress required to open the cavity is shown in Fig. 6.13. The greater stress is required for one dimension, where the density increase is larger, since there is no hoop direction to relieve the compression and the stresses are the highest. Figure 6.14 shows the elastic-plastic interface velocity, again showing that the added confinement leads to a higher elastic-plastic interface velocity, and hence that the spherical case has the lowest velocity. Figure 6.15 displays the stress jump from the numerical solution at the elastic-plastic interface. At lower cavity expansion velocities there is not a discontinuity in the velocity or stress for the cylindrical and cavity expansion, and again the values are least for the spherical case owing to less material confinement. Finally, Fig. 6.16 shows the same trends for the velocity jump at the elastic-plastic interface. We are in a position to summarize:

THEOREM 6.2. *The cavity expansion problem, namely that of opening a cavity at a constant velocity from an initial zero radius, can be solved by a similarity solution, producing elastic and elastic-plastic deformation regions with an elastic-plastic interface between them. The cylindrical and spherical cavity expansion solutions produce an elastic-plastic interface velocity that is less than the shock velocity for low cavity expansion velocities, and the particle velocity, stress, density, and energy are continuous across the elastic-plastic interface. At higher cavity expansion velocities, the elastic-plastic interface forms a shock traveling at the shock velocity, with finite jumps in the particle velocity, stress, density, and energy, where the quantitative values are obtained from the Hugoniot jump conditions. The cavity wall stress that produces the constant velocity cavity expansion is a function of cavity wall velocity only – that is, it does not depend on the spatial extent of the cavity. The cavity wall stress is well approximated by a fairly simple function of the cavity expansion velocity.*

6.6. Effect of Incompressibility and Velocity Bounds for the Cylindrical Cavity Expansion

We now wish to derive some results regarding the cylindrical cavity expansion, because of the way we will use it. The first is the limit in which the plastic region

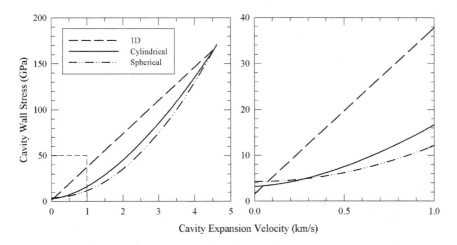

FIGURE 6.13. Cavity wall stress for one-dimensional, cylindrical, and spherical cavity expansions in steel (full velocity range to left, detail to right).

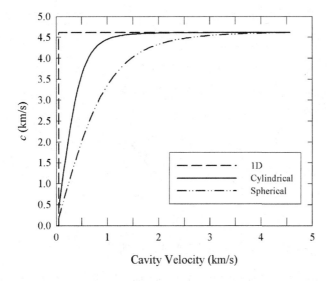

FIGURE 6.14. Elastic-plastic interface velocity for one-dimensional, cylindrical and spherical cavity expansions in steel.

is incompressible. To obtain this, we let $K \to +\infty$ in Eq. (6.53) to obtain

$$\lim_{K \to +\infty} \alpha = \lim_{K \to +\infty} \frac{c}{V} = \sqrt{\frac{2\mu}{Y}}. \tag{6.131}$$

(In a similar fashion, for the spherical cavity expansion the limit is $(2\mu/Y)^{1/3}$; see also Exercise 6.9). This limit is velocity independent. Thus, the extent of the plastic region for a material where the plastic region is incompressible has no velocity dependence. It will turn out that the velocity dependence of α is a way

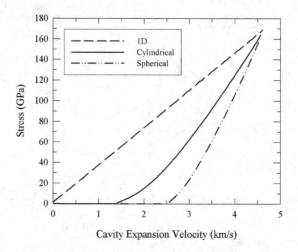

FIGURE 6.15. Stress jump at elastic-plastic interface for one-dimensional, cylindrical and spherical cavity expansions in steel.

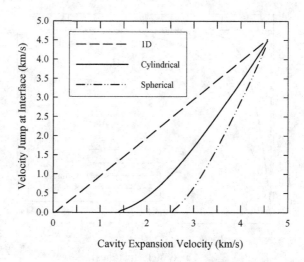

FIGURE 6.16. Velocity jump at elastic-plastic interface for one-dimensional, cylindrical and spherical cavity expansions in steel.

in which compressibility of the target enters into our general penetration model, to be presented in Chapter 10. Notice that this result agrees with taking the $K \to \infty$ limit in Eq. (6.54) of the low velocity limit compressible case.

Now we bound c and show that it is well behaved. Equation (6.49) can be rewritten as

$$\frac{1 + \dfrac{\rho V^2}{Y}}{\sqrt{K - \rho V^2}} = \frac{1 + \dfrac{\rho c^2}{2\mu}}{\sqrt{K - \rho c^2}}. \tag{6.132}$$

If $Y < 2\mu$, which is true for all materials known to date, then

$$\frac{1 + \dfrac{\rho V^2}{2\mu}}{\sqrt{K - \rho V^2}} < \frac{1 + \dfrac{\rho V^2}{Y}}{\sqrt{K - \rho V^2}} = \frac{1 + \dfrac{\rho c^2}{2\mu}}{\sqrt{K - \rho c^2}}. \tag{6.133}$$

We define

$$f(x) = \frac{1 + \dfrac{x}{2\mu}}{\sqrt{K - x}}. \tag{6.134}$$

Eq. (6.133) gives $f(\rho V^2) < f(\rho c^2)$. Differentiating f provides

$$\frac{df}{dx} = \frac{\dfrac{1}{2\mu}}{\sqrt{K - x}} + \frac{1 + \dfrac{x}{2\mu}}{2(K - x)^{3/2}} > 0. \tag{6.135}$$

Since f is a strictly increasing function of x and $f(\rho V^2) < f(\rho c^2)$, we conclude that $V < c$. This result, combined with the requirement that the term under the square root be positive, yields the following bounds,

$$V < c < \sqrt{K/\rho} = c_0. \tag{6.136}$$

Thus, the velocity of the plastic front is greater than the cavity expansion velocity (which we expect) and less than the bulk sound speed of the solid.

6.7. Extending the Cavity Expansion to Address Nonlinear Pressure Response

Because of our material response assumption in this chapter of a linear pressure dependence $p = K(1 - \rho_0/\rho)$ for constant K, we have a maximum shock speed of c_K. However, we know at large compressions that this form does not represent how metals respond.[2] At higher cavity expansion velocities the two-wave structure will be overtaken by a single strong shock. From Chapter 5, for the one-dimensional shock wave, the shock speed can often be written $U = c_0 + ku_p$. Thus, a way to heuristically reflect the nonlinear pressure dependence at higher pressures in our cavity expansion expression is to replace c_0 with some variant of $c_0 + kV$, where V is the cavity expansion velocity. Specifically, in the elastic-plastic region we replace K by $\bar{K}$, where

$$\sqrt{\frac{\bar{K}}{\rho_0}} = \bar{c}_0 = \sqrt{c_0}\sqrt{c_0 + kV} \quad \Rightarrow \quad \bar{K} = K(1 + kV/c_0). \tag{6.137}$$

This bulk modulus arises from a sound speed that is a multiplicative average of the bulk sound speed and the one-dimensional shock velocity. Thus, for the cylindrical or spherical cavity expansion at higher velocities, it provides a stiffer volumetric response (higher pressures for a given volumetric strain). (Clearly any power composition $c_0^{1-\beta}(c_0 + kV)^\beta$ could be used; $\beta = 1/2$ is used for simplicity, though full up numerical computations could be performed to determine the best power for a given velocity regime.) The bulk modulus used in the cavity expansion to compute

---

[2]With $p = K(1 - \rho_0/\rho)$, the pressure equals the bulk modulus when $\rho \to \infty$, or $\varepsilon = 1 - \rho_0/\rho = 1$. For the $U = c_0 + ku$ Mie–Grüneisen equation of state, the strain at which the pressure $p = K$ is reached for cold compression is $\varepsilon = (1 + 2k - \sqrt{1 + 4k - 2\Gamma})/(2k^2 + \Gamma)$, and on the Hugoniot the pressure K is reached when $\varepsilon = (1 + 2k - \sqrt{1 + 4k})/2k^2$. For 304 stainless steel, with the values given in Table 5.3, the strains are 0.36 and 0.30, respectively, very different from 1.

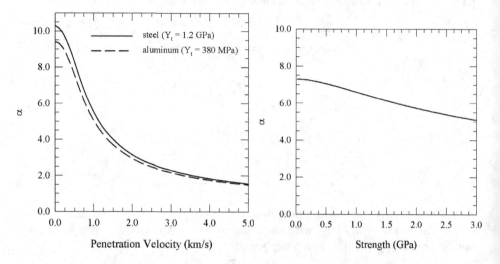

FIGURE 6.17. Left: plastic extent α vs. cavity expansion velocity including stiffened bulk modulus. (From Walker and Anderson [**199**].) Right: plastic extent α vs. strength for steel including stiffened bulk modulus for expansion velocity $V = 700$ m/s.

α from Eq. (6.52) is the stiffened $\bar{K}$, which is a function of the cavity expansion velocity. Given the result of the previous section, it follows that

$$V < c < \sqrt{c_0^2 + kc_0 V}. \tag{6.138}$$

This stiffening provides values of c up to penetration velocities of $V = c_0(k/2 + \sqrt{1 + k^2/4})$ (roughly $2c_0$ for most materials). For higher penetration velocities, one can directly stiffen the material by the one-dimensional result, saying

$$\sqrt{\frac{\bar{K}}{\rho_0}} = \bar{c}_0 = c_0 + kV. \tag{6.139}$$

This equation would correspond to the one-dimensional confinement case, as it returns the experimental result. As such, for the cylindrical, and spherical cavity expansion it would only arise at very high expansion velocities. For this one-dimensional stiffening,

$$V < c < c_0 + kV. \tag{6.140}$$

This provides values of c for all penetration velocities when $k > 1$, which is true for most materials.

Equations (6.52) and (6.137) allow calculation of α. Figure 6.17 (left) shows α vs. cavity expansion velocity for aluminum and steel using the values in Table 6.1. Plastic extent α is seen to have a strong dependence on velocity, especially in the 0 to 2 km/s regime. The right side of the figure shows the strength dependence of α.

The focus in this section has been on the cavity expansion velocity, since that is what is used in the subsequent modeling. For a discussion of the effect of the nonlinear response on the cavity wall stress, see Exercises 6.10–6.12.

6.8. Numerical Implementation

To numerically compute the various cavity expansions, the two situations of continuous solutions and jumps at the elastic-plastic interface are considered separately. For a continuous solution (no jumps), the unknown is the elastic-plastic interface velocity c. In this case, an estimate on the interface velocity is provided, and the Newton method is applied to the function

$$f(c) = \xi_{\text{numerical}} - V/c. \tag{6.141}$$

In this expression, $\xi_{\text{numerical}}$ is the value of ξ where the numerical solution's velocity equals the prescribed cavity expansion velocity. In particular, the elastic solution is used to provide a starting stress, velocity, and density at $\xi = 1$ (notice that the elastic solution depends on c), and then the system

$$\left(1 - \frac{\rho}{K}(-c\xi + v)^2\right)v' = -(\delta - 1)\left\{\left(1 - \frac{Y}{K}\right)\frac{v}{\xi} + c\frac{Y}{K}\right\}, \tag{6.142}$$

$$\sigma'_{rr} = (\delta - 1)\frac{Y}{\xi} + \rho(-c\xi + v)v' \tag{6.143}$$

is integrated numerically for decreasing ξ (again we used the forward Euler method) until the material velocity v equals the cavity expansion velocity, and the value of ξ to achieve that velocity is returned and used in the Newton iteration. In Eq. (6.142), δ is the dimension of the system: $\delta = 2$ for cylindrical and $\delta = 3$ for spherical. Also, $\rho = \rho_0(1 - \sigma_{rr}/K)$ (the other "linear" response, which is not identical to Eq. 6.10) was used to compute ρ, and the ρ calculation lagged one step ($\Delta\xi$) behind. The solution c is found when $f(c) = 0$.

In a similar fashion, for higher velocities where $c = c_0$ and there is a jump, Newton's method is used just as above, only now the unknown is the magnitude of the particle velocity jump. The starting point for the numerical solution is the stress, velocity, and density provided by the elastic solutions, followed by the application of the Hugoniot jump condition for stress with the prescribed jump in velocity. The same approximate relation between density and radial stress is used throughout. The equation for the Newton method is thus

$$f(v_{\text{jump}}) = \xi_{\text{numerical}} - V/c_0, \tag{6.144}$$

where again, $\xi_{\text{numerical}}$ is the value of ξ where the numerical solution's velocity equals the prescribed cavity expansion velocity, and it now depends upon the value of the velocity jump. The magnitude of the velocity jump is found when $f(v_{\text{jump}}) = 0$.

The cavity expansion velocity where the transition from a continuous solution to one with the discontinuity is determined by finding where unphysical behavior begins for the continuous solution (the slope of the cavity stress as a function of cavity velocity begins to drop for increasing velocity) and where a solution begins to exist for the jump case (below the transition velocity, a solution to $f(v_{\text{jump}}) = 0$ does not exist).

6.9. Sources

The cavity expansion approach dates from the 1940s and 1950s. Our derivation follows Hopkins [100], Crozier and Hunter [49] and Forrestal and Luk [69] for the spherical cavity expansion. The cylindrical cavity expansion follows Walker and Anderson [197, 199].

Exercises

(6.1) Calculate α_0 for the cylindrical and spherical cavity expansions for the aluminum and steel used in this chapter.

(6.2) Beginning with Eq. (6.52), derive the solution to the quadratic in α^2 given in Eq. (6.53). (Hint: the standard quadratic formula we memorize in grade school that solves $ax^2 + bx + c = 0$ is derived by completing the square of the a and b terms. In this case it is best to complete the square of the c and b terms. Doing so produces a solution that works when $a = 0$.)

(6.3) Later we will need $d\alpha/dt = (d\alpha/dV)(dV/dt)$. Assuming that the nonlinearity of the bulk response is represented by $K = K(V)$, use the previous exercise to show that (where prime means differentiation with respect to V)

$$A' = \frac{\rho V(2K + K'V - 4\rho V^2)}{4\mu^2}, \quad B' = \frac{K'}{\mu} + \frac{2\rho V}{Y}\left(2 - \frac{Y}{\mu} + \frac{2\rho V^2}{Y}\right),$$

$$C' = \frac{2K'}{Y} + \rho V\frac{2K + K'V}{Y^2}, \quad \frac{d\alpha}{dV} = \alpha' = \frac{C' - \alpha^4 A' - \alpha^2 B'}{2\alpha(2A\alpha^2 + B)}. \tag{6.145}$$

(6.4) Find α for a fluid – that is, in the limit as $Y \to 0$, assuming $u > 0$.

(6.5) In Exercise 4.31 it was stated that in one dimension a general elastic wave going in the positive x direction could be written $f(x - ct)$. Show, for one-dimensional elastic waves with displacement $u_x = f$, that $\varepsilon_{xx} = \frac{\partial u}{\partial x} = f'$ and therefore the particle velocity $\frac{\partial u}{\partial t} = c\varepsilon_{xx}$. Things are not so simple for radial elastic waves – hence the complexity of the derivations in this chapter. However – suppose you could write it that way – that is, $u_r = f(r - ct)$. Show that the principal-axis strain for all three dimensions can be written $\varepsilon_{rr} = f'$, etc. (i.e., look at the displacement expressions in Cartesian, cylindrical, and spherical coordinates). Show that the radial velocity is $c\varepsilon_{rr}$. Looking at the results for all three dimensions for the elastic wave, show that at yield we always have a radial particle velocity of $cY/2\mu$. (For one dimension you need to use the velocity at the wave front – that is, the longitudinal wave speed.) Why? Is the radial stress always the same at yield? Why not?

(6.6) Show that in the cavity expansion the hoop strain is tensile (positive) and so the radial stress at yield on the elastic-plastic interface is less (in magnitude) than the HEL. In particular show that for the spherical cavity expansion,

$$-|\sigma_{\text{HEL}}| < \sigma_{rr}|_{\xi=1} \le -2Y/3. \tag{6.146}$$

(6.7) Compute α using Eq. (6.53) for the cylindrical cavity expansion velocities up to 15 km/s for steel using a constant low-pressure bulk modulus and the two heuristic stiffened bulk moduli, Eqs. (6.137) and (6.139). What is the behavior at higher velocities? Why is it necessary to take into account the nonlinear behavior of the stress-strain curve?

(6.8) A question in damage mechanics is the rate at which spherical voids open in a ductile material. This has been explored by various authors as was mentioned in Chapter 3. We can now address the relationship between applied hydrostatic pressure, strength and void opening speed for a constant speed spherical void opened from zero radius. Suppose we apply a tensile hydrostatic pressure ($p < 0$) to an infinite space of material. Since our plasticity theory is pressure independent, this initial stress state can be added to our cavity expansion solution. By setting an initial applied tensile pressure equal in magnitude to the stress required to open the boundary in our cavity expansion, we produce a stress-free inner cavity boundary ($\tilde{\sigma} + p = 0$ on the cavity boundary). Thus, if the infinite space is initialized with a tensile pressure and a point is allowed to open up into a void, then from our cavity expansion we obtain

$$-p = B_3 + \rho V^2. \tag{6.147}$$

Discuss the implications of this solution. What does the pressure need to be to open up such a void so that it continues to open? What does this say about the ratio p/Y that was used in the Johnson–Cook damage model presented in Chapter 3? Show that the elastic solution to a stress free spherical void of radius R in an infinite space with an applied tensile pressure at infinity is

$$\sigma_{rr} = -p\left\{1 - \left(\frac{R}{r}\right)^3\right\}, \qquad \sigma_{\theta\theta} = \sigma_{\varphi\varphi} = -p\left\{1 + \frac{1}{2}\left(\frac{R}{r}\right)^3\right\}, \tag{6.148}$$

which gives

$$\sqrt{3J_2} = |\sigma_{rr} - \sigma_{\theta\theta}| = \frac{3|p|}{2}\left(\frac{R}{r}\right)^3. \tag{6.149}$$

Show that the initial yield of the elastic solution occurs when $p = -2Y/3$ (note that it is independent of initial void radius R). This stress is in magnitude much lower than that predicted by the cavity expansion solution. What happens when the pressure $p > -2Y/3$? For $p \leq -B_3$ an infinitely expanding, constant wall speed spherical void growth occurs. What do these results imply happens for pressures $-B_3 < p < -2Y/3$?

(6.9) Show that if the elastic-plastic region is assumed incompressible, then α is constant (as stated in Section 6.6), as follows: show that the mass conservation equations in the incompressible region are $v' + v/\xi = 0$ and $v' + 2v/\xi = 0$ for cylindrical and spherical expansions, respectively; solve these equations; use the solution to link the cavity expansion velocity at the cavity surface to the elastic solution velocity at yield; explicitly compute the constant $\alpha = (2\mu/Y)^{1/2}$ and $\alpha = (2\mu/Y)^{1/3}$ for the cylindrical and spherical expansions, respectively.

(6.10) Another approach to the cavity expansion problem is to assume that the material is incompressible. Such an assumption allows an exact integration of the mass and momentum equations instead of the approximation we introduced in the text preceding Eq. (6.45). Thus, one could view the incompressible solution as one way to check on the approximation (at least at one extreme). For the three-dimensional spherical cavity expansion, assume the material is incompressible and solve the problem two ways: first, by just using the solution derived in the chapter and assuming the bulk modulus goes to infinity; and second, assume the density is constant in the elastic-plastic region and from this assumption explicitly integrate the mass conservation equation and then the momentum conservation equation (use the elastic solution from the first part).

To outline what happens for the first approach in the elastic region, since

$$\beta = \frac{c}{c_L} = \frac{c}{\sqrt{(K + (4/3)\mu)/\rho}}, \tag{6.150}$$

as incompressibility is given by $K \to \infty$, it implies $c_L \to \infty$ and $\beta \to 0$. Thus, $c_2 = -Y/2\mu$,

$$\frac{\partial u}{\partial t}(\xi = 1) = -cc_2 = -\frac{Yc}{2\mu},$$

$$\sigma_{rr}(\xi = 1) = c_2 \left\{ \frac{4}{3}\mu + 2\rho c^2 \right\} = -\frac{2Y}{3} - 2\rho c^2 \frac{Y}{2\mu}. \tag{6.151}$$

In the elastic-plastic region, the derived compressible solution yields

$$\lim_{K \to 0} g(V) = \lim_{c_0 \to 0} g(V) = 0. \tag{6.152}$$

This result quickly gives

$$c_3 = \frac{V^3}{c^2}, \quad v|_{\xi=1} = \frac{V^3}{c^2}, \quad \alpha = \frac{c}{V} = \left(\frac{2\mu}{Y}\right)^{1/3}, \tag{6.153}$$

and

$$\sigma_{rr}|_{\xi=V/c} = \sigma_{rr}|_{\xi=1} - 2Y\ln\left(\frac{c}{V}\right) - 2c_3\rho c\left(\frac{c}{V} - 1\right). \tag{6.154}$$

Using the stress from the incompressible elastic solution above, the stress at the elastic-plastic interface is

$$\sigma_{rr}|_{\xi=V/c} = -\frac{2Y}{3}\left\{1 + \ln\left(\frac{2\mu}{Y}\right)\right\} - 2\rho V^2. \tag{6.155}$$

For the second approach, in the elastic-plastic region, assuming incompressibility leads to an explicit integration of the conservation of mass,

$$v' + \frac{2v}{\xi} = 0 \Rightarrow v = \frac{V^3}{c^2\xi^2}. \tag{6.156}$$

The momentum balance becomes

$$\sigma'_{rr} - \frac{2Y}{\xi} = \rho\left(-c\xi + v\right)v' = -2\rho V^3\left(-\frac{1}{c\xi^2} + \frac{V^3}{c^4\xi^5}\right) \tag{6.157}$$

and can also be explicitly integrated,

$$\sigma_{rr}(\xi) = \sigma_{rr}(1) + 2Y\ln(\xi) - 2\rho V^3\left(\frac{1}{c\xi} - \frac{1}{c} - \frac{V^3}{4c^4\xi^4} + \frac{V^3}{4c^4}\right). \tag{6.158}$$

Using the incompressible elastic solution, the cavity wall stress is

$$\sigma_{rr}(V/c) = -\frac{2Y}{3} - 2\rho c^2 \frac{Y}{2\mu} + 2Y\ln(V/c) - 2\rho V^3\left(\frac{3}{4V} - \frac{1}{c} + \frac{V^3}{4c^4}\right)$$

$$= -\frac{2Y}{3}\left\{1 + \ln\left(\frac{2\mu}{Y}\right)\right\} - \left\{\frac{3}{2} + \frac{1}{2}\left(\frac{Y}{2\mu}\right)^{4/3}\right\}\rho V^2. \tag{6.159}$$

The coefficient on ρV^2 in brackets is nearly $3/2$, since $Y \ll 2\mu$. This coefficient can be compared to the value of 2 arrived at from the compressible solution. The difference implies that the approximation leads to a stiffer response. Such a conclusion is in agreement with Fig. 6.9, where the analytical solution is above the numerical solution (and why in the final approximation the coefficient is set equal to 1).[3]

---

[3]In recent years, Forrestal and colleagues have used the $3/2$ coefficient multiplying the ρV^2 term in their modeling, which results from setting a constant of integration equal to $B = 2/3$ rather than the exact $B = 2/3 + (1/2Y)\rho c^2(3Y/2E)^2$ in their development [69]. In [65], they did

(6.11) In some sense, incompressible is the stiffest that the material response can be. Use the result of the previous problem to bound the stress on the cavity wall. Our solution without the Section 6.7 adjustments breaks down when $c = c_0$, since this case leads to the maximum stress the linear stress-strain response can achieve, namely K. From the previous problem, what is the corresponding stress for the incompressible response when $V = c_0$? What is the stress for the one-dimensional plate impact problem for a material that obeys $U = c_0 + ku$? Compare these three values and comment.

(6.12) Repeat Exercise 6.10 for the cylindrical cavity expansion problem. To outline what needs to be done, begin with the solution in the text and let K go to infinity,

$$c_3 \sim \frac{Y/\rho + V^2}{\sqrt{K/\rho}}. \tag{6.160}$$

The velocity at the interface between the elastic and elastic-plastic region is

$$v|_{\xi=1} = \frac{V^2}{c}, \quad \alpha = \frac{c}{V} = \left(\frac{2\mu}{Y}\right)^{1/2} \tag{6.161}$$

and the stress is

$$\sigma_{rr}|_{\xi=1} - \sigma_{rr}|_{\xi=V/c} = -(Y + \rho V^2)\ln(V/c). \tag{6.162}$$

Thus, the stress at the cavity wall is given by

$$\sigma_{rr}|_{\xi=V/c} = \sigma_{rr}(1) - \frac{Y}{2}\ln\left(\frac{2\mu}{Y}\right) - \frac{1}{2}\rho V^2 \ln\left(\frac{2\mu}{Y}\right). \tag{6.163}$$

For the second approach, when the deformation is assumed incompressible, the conservation of mass equation can be integrated directly to give

$$v' + \frac{v}{\xi} = 0 \Rightarrow v = \frac{V^2}{c\xi}. \tag{6.164}$$

The conservation of momentum is then

$$\sigma'_{rr} - \frac{Y}{\xi} = \rho\left(-c\xi + v\right)v' = -\rho V^2\left(-\frac{1}{\xi} + \frac{V^2}{c^2\xi^3}\right). \tag{6.165}$$

This equation can be integrated directly:

$$\sigma_{rr}(\xi) = \sigma_{rr}(1) + Y\ln(\xi) + \rho V^2\left(\ln(\xi) + \frac{V^2}{2c^2\xi^2} - \frac{V^2}{2c^2}\right). \tag{6.166}$$

When this is evaluated at $\xi = V/c$, we obtain

$$\sigma_{rr}(V/c) = \sigma_{rr}(1) - \frac{Y}{2}\ln\left(\frac{2\mu}{Y}\right) - \rho V^2\left(\frac{1}{2}\ln\left(\frac{2\mu}{Y}\right) - \frac{1}{2} + \frac{Y}{4\mu}\right). \tag{6.167}$$

With the values for steel used in this chapter, $(1/2)\ln(2\mu/Y) = 2.43$, and so the two expressions differ in their coefficient on ρV^2 by about 20%, again showing that the approximate solution leads to a somewhat stiffer response.

---

a curve fit to match a numerical compressible cavity expansion solution. The curve fit led to a coefficient of 1.133 and adjusted their leading constant term by 4%.

(6.13) As an opening comment, Bishop, Hill, and Mott also solved for the case where the material is elastically incompressible ($\nu = \frac{1}{2}$) and there is a linear work hardening with slope H [93, 100, 136]. In that case, the result is

$$\tilde{\sigma}_{rr} = \frac{2Y}{3}\left\{1 + \ln\left(\frac{2E}{3Y}\right)\right\} + \frac{2\pi^2 H}{27}. \tag{6.168}$$

Similarly, Forrestal and Luk [70] and Luk, Forrestal and Amos [137] have solved the cavity expansion assuming logarithmic strains and strain hardening and an incompressible elastic-plastic region. The expression they use for the flow stress is $\sigma = Y\left(E\varepsilon/Y\right)^n$, where ε is the strain. For 6061-T651 aluminum, their values are $Y = 276\,\text{MPa}$, $E = 68.9\,\text{GPa}$, $\nu = 1/3$, and $n = 0.051$ [137]. Given this, their derived coefficients are (in our notation)

$$B_2 = \frac{Y}{\sqrt{3}}\left\{1 + \left(\frac{E}{\sqrt{3}Y}\right)^n I(\alpha^2)\right\}, \quad \alpha = \left(\frac{\sqrt{3}E}{2(1+\nu)Y}\right)^{1/2}, \tag{6.169}$$

$$B_3 = \frac{2Y}{3}\left\{1 + \left(\frac{2E}{3Y}\right)^n I(\alpha^3)\right\}, \quad \alpha = \left(\frac{2E}{3Y}\right)^{1/3}, \tag{6.170}$$

$$I(y) = \int_0^{1-1/y} \frac{(-\ln(x))^n}{1-x}dx. \tag{6.171}$$

The integral I is computed numerically (notice in the integrand $x = 0$ and $x = 1$ are singular). For the spherical case with the coefficients above, $I(\alpha^3) = 4.539$ and $B_3 = 1.268\,\text{GPa}$. How does this value compare to that derived above for B_3, given the values in Table 6.1 for Al 6061-T6? If the material is assumed elastic-perfectly plastic (i.e., $n = 0$), what are the two coefficients?

(6.14) If the material into which expansion occurs has finite extent, a similarity solution no longer exists. One approach to estimating the stress required to open the cavity is to assume the material is plastically flowing and then compute the stress required to cause that flow, ignoring inertia. That provides the low velocity B_i term. Do this for both the two- and three-dimensional flows. Estimate the amount of material that provides the same confinement as the semi-infinite case (i.e., the cavity expansion solution).

(6.15) What is the difference between the similarity approach used in Section 4.3 and that used in this chapter? Is there any essential difference? How could you transform the Section 4.3 approach to this chapter's approach?

(6.16) Brinell hardness of steel (BHN or HB) is determined by taking a tungsten carbide sphere of diameter $2R = 10$ mm and pressing it against the surface of the steel with a weight of $m = 3{,}000$ kg (so the force is given by this mass times the acceleration of gravity, or 29,400 kN). After removing the weight, the diameter of the indentation circle on the surface ($2R_i$ across) is measured. Then BNH is computed by dividing the weight mass in kg by the area A in mm^2 of metal that was in contact with the indenter sphere. Show that the surface area of the metal in contact with the sphere is $A = 2\pi R(R - \sqrt{R^2 - R_i^2})$. Then, BHN $= m/A$. For example, a BHN of 255 has units of kg/mm^2 and corresponds to a stress of 2.50 GPa: work through the units to show this. Pick some indentation diameters from Table 3.6 and show that you obtain the correct BHN with the above computation. Assuming the motion in the metal plate during the indent is similar to the zero velocity limit cavity expansion, compare B_3 to the empirical relation between BHN and Y contained in Eq. (7.2). The assumption that the resisting stress comes from the cavity expansion says that BHN should essentially equal B_3 once the units are the same. Are they similar? Notice that the material motion in the test is not as confined as the spherical cavity expansion and we are not considering the elastic rebound once the load is removed. How do these assumptions affect the "multiplier?" What happens if we want to determine the BHN for another material, like tungsten or aluminum? Do we expect the same empirical relationship to hold? How different might they be?

(6.17) Rockwell C hardness (Rc or HRC) is typically determined by taking a diamond cone with angle $2\theta = 120°$, loading it with a 10 kg weight and measuring the penetration P_1 of the cone into the material, loading it with a 150 kg weight, then removing the 150 kg weight and loading the cone with a 10 kg weight and measuring the total penetration P_2. The hardness value is given by Rc $= 100 - (P_2 - P_1)/0.002$ mm. The displacements P_i are measured when the diamond cone is loaded with the 10 kg weight, before and after the indent caused by the heavier weight. P_2 will be greater than P_1 owing to the permanent set caused by the larger load. Unfortunately, we have not developed an expression to compute the elastic rebound that occurs when a weight is removed. However, let us charge ahead and apply our spherical cavity expansion to this conical indention to see if it compares with expected values for steel. Assuming the stress required to push the material is given by B_3, show that the resisting force is $F = \pi R_i^2 B_3 = \pi B_3 (P \tan(\theta))^2$, where P is the depth of penetration and R_i is the radius of the top of the cone. Derive an expression relating Rc to Y. Does it give reasonable values? How would the stated relations between Rc and flow stress for steel change if aluminum or tungsten were indented instead? (Notice, $P = \sqrt{mg/(3\pi B_3)}$, so use the two different masses to produce the different Ps. Not including the elastic rebound leads to Rc values that are less than what is measured, since the rebound decreases P for the second measurement and, of course, there are other approximations, too.)

(6.18) The cavity expansion assumes a constant driving stress (an increasing driving force), producing a self-similar solution. What is the solution of propagation for an elastic wave that is not driven? In three dimensions, if the displacement can be written as the gradient of a potential ϕ, then for a wave that only has radial motion, $u = \partial\phi/\partial r$. Show that the elastic wave equation (Eq. 6.28) becomes

$$\frac{1}{r^2}\frac{\partial}{\partial r}\left(r^2\frac{\partial\phi}{\partial r}\right) = \frac{1}{c_L^2}\frac{\partial^2\phi}{\partial t^2} + F(r), \tag{6.172}$$

where $F(r)$ is an arbitrary function that we will set equal to zero. While $u = f(x - c_L t) + g(x + c_L t)$ solves the general one-dimensional equation (Exercise 4.31), show that

$$\phi(x, t) = \frac{f(r - c_L t) + g(r + c_L t)}{r} \tag{6.173}$$

solves the three-dimensional spherically symmetric wave equation, where f and g are arbitrary functions. If we focus on a wave moving outward from the origin, show that the displacement, velocity, strain, and acceleration are given by

$$u(r,t) = \frac{\partial \phi}{\partial r} = \frac{f'}{r} - \frac{f}{r^2}, \qquad v(r,t) = \frac{\partial u}{\partial t} = c_L \left(\frac{f'}{r^2} - \frac{f''}{r} \right), \qquad (6.174)$$

$$\varepsilon_{rr}(r,t) = \frac{f''}{r} - \frac{2f'}{r^2} + \frac{2f}{r^3}, \qquad a(r,t) = \frac{\partial^2 u}{\partial t^2} = c_L^2 \left(\frac{f'''}{r} - \frac{f''}{r^2} \right). \qquad (6.175)$$

To be specific, consider the quartic polynomial that has a double root at $\pm \Delta r$,

$$f(r) = \begin{cases} a(1 - r/\Delta r)^2 (1 + r/\Delta r)^2 & -\Delta r \leq r \leq \Delta r, \\ 0 & |r| > \Delta r. \end{cases} \qquad (6.176)$$

The function is zero outside its compact support of length $2\Delta r$, hence argue that the wave packet does not spread out as it expands (in this behavior it is quite different from our cavity expansion solution). Show that $|f| \leq a$, $|f'| \leq 8a/(3\sqrt{3}\Delta r)$, $|f''| \leq 8a/(\Delta r)^2$, and that those bounds (maximum values) are achieved. Hence for large propagation distances the above terms decay as $1/r$ (again different from the cavity expansion). This may seem a bit surprising, as we expect in some sense that for an expanding sphere things should decay as $1/r^2$. What decays as $1/r^2$ is the energy. The local energy density is $e = \frac{1}{2}\lambda(\varepsilon_{kk})^2 + \mu\varepsilon_{ij}\varepsilon_{ij} + \frac{1}{2}\rho v^2$. Towards evaluating the total energy, show

$$r^2 (\varepsilon_{kk})^2 = (f'')^2,$$

$$r^2 \varepsilon_{ij}\varepsilon_{ij} = (f'')^2 - 2 \left(\frac{f^2}{r^3} \right)' + 4 \left(\frac{ff'}{r^2} \right)' - 2 \left(\frac{(f')^2}{r} \right)', \qquad (6.177)$$

$$r^2 v^2 = c_L^2 \left\{ (f'')^2 - \left(\frac{(f')^2}{r} \right)' \right\}.$$

With these results, show that energy in the spherically expanding elastic wave packet (i.e., $f(r) = 0$ if $|r| >$constant) is conserved and show, just as for the simple one-dimensional energy jump condition, the energy is evenly partitioned into elastic strain energy and kinetic energy. Show that locally the energy is decaying as $1/r^2$.

(Notice that this solution approach does not work in two dimensions. The assumption that the displacement is given by a gradient in cylindrical symmetry yields the same $u = \partial\phi/\partial r$, which leads to the similar

$$\frac{1}{r} \frac{\partial}{\partial r} \left(r \frac{\partial \phi}{\partial r} \right) = \frac{1}{c_L^2} \frac{\partial^2 \phi}{\partial t^2} + F(r). \qquad (6.178)$$

In this case, however, $\phi(r,t) = h(r)(f(r - c_L t) + g(r + c_L t))$ leads to two equations for $h(r)$: $h + 2rh' = 0$ and $h' + rh'' = 0$. Differentiation of the first equation shows that it is incompatible with the second.)

Penetration

Intuitively, the difference between the penetration event and the impact events studied so far is that in the penetration event the projectile is "smaller" than the target and hence there is material motion in the target that is transverse to the projectile flight direction, which allows target material to get out of the way of the projectile, yielding penetration. In this chapter experiments and numerical simulations are used to develop an understanding of the penetration process over a wide range of velocities. We primarily consider large L/D projectiles, where L is projectile length and D is projectile diameter, because this separates out various phases of penetration which for, say, a spherical projectile, overlap. The understanding we develop will be used in subsequent chapters to understand the Tate–Alekseevskii model and to develop the Walker-Anderson model.

7.1. Steel Projectiles Penetrating Aluminum Targets

As a first example, consider two steel projectiles with an ogive nose: one made of VAR 4340 steel with a hardness of Rc 38, a tensile yield strength of 1,140 MPa, and a fracture toughness of 130 MPa$\sqrt{\text{m}}$; and the other made of AerMet 100 steel, with a hardness of Rc 53, a tensile yield strength of 1,720 MPa, and a fracture toughness of 126 MPa$\sqrt{\text{m}}$. The second steel has roughly 1.5 times the strength of the first, while their fracture toughnesses are similar. The target is 6061-T6511 aluminum, with a yield of 276 MPa. The projectiles are 59.3 mm from the back of the nose to the tail, a nose length of 11.8 mm, radius 7.11 mm, and a caliber radius head of 3 (CRH = 3 means that the projectile nose front-to-back cross section is a circle with radius $3D$ that is tangent to the exterior length surface of the projectile; there is more discussion of this nose geometry in Chapter 12). Hence $L/D = 10$.

The penetration over the impact velocity range from 0.5 km/s to 3.0 km/s is shown in Fig. 7.1 (material properties and penetration data from [156]). At the lower speeds the penetration increases rapidly with increasing impact speed. At these velocities, the projectile maintains its shape – in particular, its pointed nose. As the impact speed reaches 1.5 km/s for the softer steel (the VAR 4340) and 1.8 km/s for the harder steel (the AerMet 100), a large reduction in penetration occurs. The projectile nose begins to lose it shape, fracture, and erode. The transition between nondeforming projectile penetration and eroding projectile penetration is complex. X-ray images from two targets taken after the experiments are shown in Fig. 7.2. At 1,237 m/s the projectile maintained its shape; at 1,977 m/s it did not and its penetration is much less. Andy Piekutowski, who ran the test, feels that the front of the projectile turned shortly after impact and broke, forming the first off-axis jag in the cavity. Then, an additional piece broke off the front of the projectile, leading to the second jag. The remaining projectile continued to penetrate into the target, with debris trailing it, and then came to rest at the bottom of the

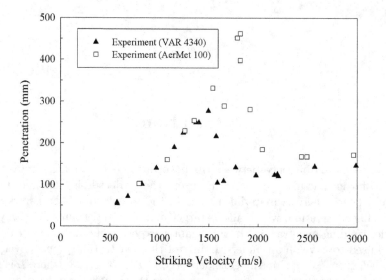

FIGURE 7.1. Penetration vs. impact velocity for steel projectiles into 6061-T6511 aluminum (data from [156]).

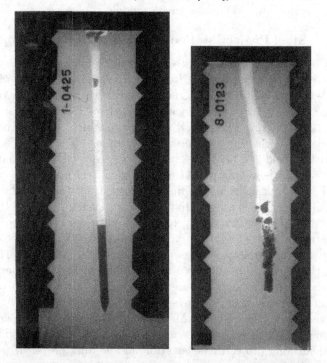

FIGURE 7.2. Post test X-ray radiographs of AerMet 100 steel rods striking 6061-T6511 aluminum: impact velocity 1,237 m/s (left) and 1,977 m/s (right). (Original figures courtesy of Andrew Pieku-towski, from Piekutowski et al. [156].)

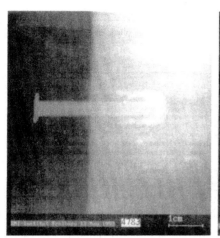

FIGURE 7.3. Flash X-ray images during penetration of an $L/D =$ 12.5 tungsten alloy penetrating a hard steel target at 1.70 km/s; the image on the left is at $20.7\,\mu s$ after impact and the image on the right is at $25.3\,\mu s$ after impact (different tests, identical setup). (From Anderson et al. [**6**].)

crater with a slight rebound. When projectiles penetrate a strong metal target and remain rigid, the radius of the penetration channel, called the crater radius, is equal to the projectile radius. When a projectile erodes while penetrating a target, the crater radius is larger than the projectile radius. When the projectile fractures and penetrates a specific target less at a higher velocity than it does at a lower velocity, the phenomenon is generically referred to as *shatter gap*, since there is a gap in performance. For this specific example, as the impact velocity increases to 3 km/s, the penetrations for the two steel projectiles approach each other. As we will see, at higher velocities the inertia-originating stress at the nose of the projectile increases as the velocity squared and hence the strength of the projectile becomes less important.

7.2. Tungsten Projectiles Penetrating Steel Targets

Our next example is blunt tungsten alloy projectiles striking steel alloy targets from work of Hohler and Stilp and Anderson, Morris, and Littlefield [**10, 95**]. Even though the tungsten alloy is strong and dense, tungsten projectiles erode while penetrating steel at a high velocity, as seen in Fig. 7.3, which shows two flash X-ray images of $L/D = 12.5$ tungsten alloy projectiles penetrating a strong steel target at 1.7 km/s. The erosion of the tungsten projectile material creates a mushroom head as it penetrates the steel.

Next, projectiles in experiments were right-circular cylinders, with a length-to-diameter ratio $L/D = 10$, 20, and 30. The tungsten alloys had hardnesses of BHN 294–303 ($L/D = 10, 20$) and 404 ($L/D = 10, 20, 30$), with respective densities 17.0 and 17.6 g/cm^3. For the $L/D = 10$ targets, the steel targets had four different Brinell hardnesses: BHN 180, 255, 295, and 388; otherwise they were rolled-homogeneous-armor-(RHA)-like steel targets. The depth of penetration P

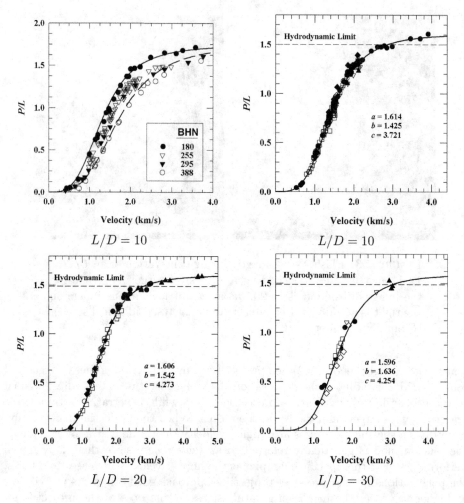

FIGURE 7.4. Penetration data for tungsten alloy projectiles impacting armor steels. BHN refers to target steel. Values for b are in units of km/s. (From Anderson and Walker [13].)

normalized by the initial projectile length L is shown as a function of impact speed in Fig. 7.4. The impact velocities range from 0.5 km/s to 4.0 km/s with the flight in the direction of the surface normal. These figures show a characteristic S-shaped curve as higher velocities are reached. Once penetration begins it increases rapidly as a function of impact velocity. The rapid increase is almost linear. However, as the impact velocity exceeds 2 km/s, the increase in penetration as a function of impact speed decreases. It almost appears that a limit is being approached.

An equation that fits observed data is

$$\frac{P}{L} = \frac{a}{1 + (V/b)^{-c}} = \frac{a\,V^c}{b^c + V^c}. \tag{7.1}$$

Curves from this equation with the specific coefficients for the cases of $L/D = 10$, 20, and 30, are also shown in Fig. 7.4. Values for b increase with increasing L/D, giving that, for the same impact velocity, P/L decreases with increasing L/D. Also plotted on the figures is the hydrodynamic limit, which is an approximate theoretical penetration result if both the projectile and target had no strength. It will be derived later in this chapter and for tungsten into steel is $(P/L)_{\text{hyd}} = \sqrt{\rho_p/\rho_t} \approx 1.5$, which is close to a. It can be shown that b is the impact speed near the center of the rise and the larger c is, the more rapid the rise.

The data in the upper left figure from Hohler and Stilp for $L/D = 10$ projectiles show that P/L depends upon target strength. The target yield strength Y_t is correlated to BHN through the empirical expression for steel[1]

$$Y_t(\text{GPa}) = 3.483 \times 10^{-3} \cdot (\text{BHN} - 11.24). \qquad (7.2)$$

It would be nice if a convenient nondimensional form of the penetration fit could be written – for example, if it were possible to lump the impact velocity and target strength as $\rho_p V^2 / Y_t$. However, good correlation with the data is not obtained using this nondimensional term. Therefore, similar to the approach by Rapacki et al. [157], we write the influence of the strength as $b = b_1 Y_t^{b_2}$, where b_1 and b_2 are found from a least-squares regression to the data. The regression gives

$$a = 1.75, \quad b = 1.58\{Y_t/(1\,\text{GPa})\}^{0.34}\,\text{km/s}, \quad c = 3.44. \qquad (7.3)$$

The solid and dashed curves in the upper left figure of Fig. 7.4 were computed from Eq. (7.3) using the above coefficients for BHN hardnesses of 180 and 388, respectively.

7.3. Projectile Erosion

When projectiles erode as they travel through targets, the projectile material fails (separates) and falls and flakes off. Figure 7.5 shows sectioned 4340 steel targets (Rc 30±2) after tungsten alloy X27X projectiles impact [126] and Fig. 7.6 is an enlargement of one of the pictures. The steel has primarily been moved out of the way through plastic deformation, forming a penetration channel. The tungsten, however, has failed, meaning there has been material separation. The $L/D = 15$ projectiles were $L = 14.86$ cm, $D_p = 0.990$ cm, with mass 206 g. The $L/D = 30$ projectiles were $L = 21.42$ cm, $D_p = 0.714$ cm, with mass 149 g.

The rearward section of the debris in the penetration channel is composed of projectile (tungsten) material. The fact that the tungsten is piling up at the bottom of the crater tells us that the erosion debris speed v_d is greater than zero (see Exercise 7.8). The cavity or channel walls appear to be coated with a smear layer of debris; the shiny appearance of the channel walls in Fig. 7.5 is the result of a microns-thin, integrally bonded tungsten coating of the steel cavity. The tungsten debris, more clearly seen in Fig. 7.6, is composed of an aggregate of macro-size particles integrally bonded together into a void-filled mass. The residual projectile is at or near the front of the debris track. In Fig. 7.6 the residual projectile is rotated 180° (it is loose), and the undeformed rear of the projectile (the elastic region) is now pointing down. By inspecting the polished debris in oblique lighting (Fig. 7.7), it is easy to see the residual projectile; the remainder of the debris is the hybrid solid mass composed of erosion debris in a reconstituted aggregate. The eroded projectile gives rise to a mushroom-shaped profile: the original tail end of

---

[1]This expression and subsequent fit in Eq. (7.3) are for initial yield in steel; see Exercise 3.14 to compare to work hardened flow stress and Exercise 6.16 for a discussion of other metals.

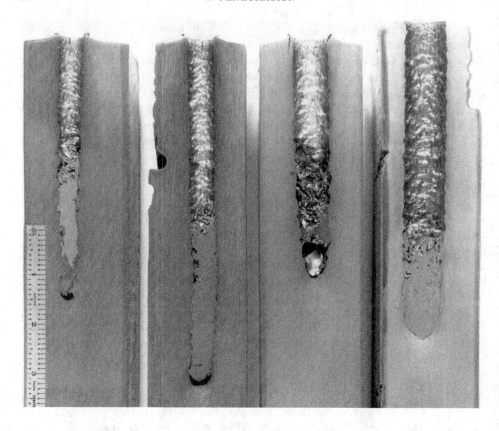

FIGURE 7.5. Sectioned penetration channels, left to right:
$L/D = 30$, $V = 1,449$ m/s, $P/L = 0.652$, $D_c/D_p = 1.84$;
$L/D = 30$, $V = 1,638$ m/s, $P/L = 0.815$, $D_c/D_p = 1.91$;
$L/D = 15$, $V = 1,469$ m/s, $P/L = 0.860$, $D_c/D_p = 1.99$;
$L/D = 15$, $V = 1,630$ m/s, $P/L = 1.013$, $D_c/D_p = 1.95$.
Eroded (flaked) tungsten remains in the channels and the residual projectiles are at the bottom of the crater. The ruler is in
inches; the sectioning is not exactly in the center of the channel.
(From Lankford et al. [126].)

the projectile (with its original diameter) is visible at the bottom of a crater in
Fig. 7.5 and in a sectioned and polished channel in Fig. 7.7, and it can be seen
that the mushroom head and the crater (i.e., channel) diameter are larger than the
projectile diameter. The eroded material consists of dark, lens shaped flake-like
structures immersed in a light matrix.

Tungsten alloys are made of spheres or nodules of tungsten held in a metal
matrix. The tungsten particles range from 10 to 50 microns (10^{-6} m) in typical
alloys. When the post-impact microstructure is examined, there are deformed particles within the microstructure, so the temperature attained did not reach the level
required to melt tungsten. However, it appears that the matrix phase may melt, or,

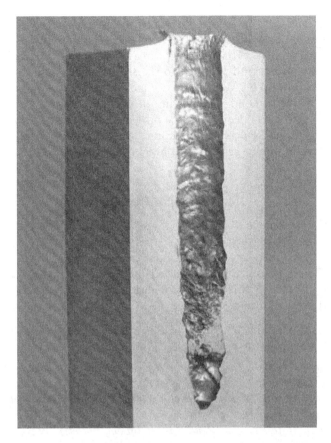

FIGURE 7.6. Another view of $L/D = 15$, $V = 1,469\,\text{m/s}$, eroded (flaked) tungsten in the channel and the residual projectile, which has rotated $180°$ after an elastic rebound.

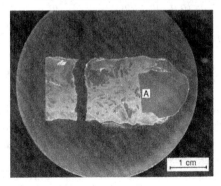
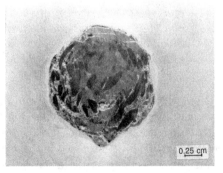

FIGURE 7.7. Debris from crater under oblique lighting, clearly showing residual projectile (left) and cross-sectional cut through penetration channel clogged with erosion products (right). (From Lankford et al. [126].)

if not, it is hot and ductile, as is borne out by the absence of any evidence of cracks or physical separation at the boundaries between the more normal matrix phase and the affected matrix. High temperatures allow a thin layer of the projectile material to become integrally bonded to the steel channel wall, thus giving rise to the shiny appearance [76]. Within the channel itself, high temperatures cause the erosion particulates to bond together into a solid aggregate. These photographs and discussion give some idea of how the projectile material fails and erodes as it is penetrating the target. Tungsten impacting steel will be extensively examined as it is representative of dense, heavy-metal alloy projectiles striking a common armor material.

7.4. Phases of Penetration

For all the penetration events, there are a number of features evident. First, there are initial impact transients. These early transients are followed by a steady state phase. Then, there is a final terminal transient phase when the penetration event ends. These phases were recognized by Allen and Rogers [2], Eichelberger and Gehring [59], and Christman and Gehring [44], and we adopt the following terminology to describe these features:

GENERAL PRINCIPLE 7.1. *(four phases of penetration) Penetration into thick targets where the projectile erodes occurs in four phases:*

    (1) *Initial Transient (Shock) Primary Penetration: this involves initial transients and possibly shocks due to the impact, and ends when the projectile nose has eroded away and been replaced by a hemispherically shaped eroding projectile flow;*

    (2) *Steady State Primary Penetration: though not strictly steady state, in this regime the projectile penetrates the target and erodes, with the penetration velocity and the projectile erosion rate being fairly constant;*

    (3) *Final (Terminal) Primary Penetration: here the projectile either completely erodes away or the stress drops to a low enough value that projectile erosion ceases; penetration continues by the momentum left in the target and by the remaining nondeforming (rigid) projectile material;*

    (4) *Secondary Penetration: following the primary penetration event, it is possible for the eroded projectile debris to penetrate more deeply into the target.*

Often the word "primary" will be dropped when referring to phases of primary penetration, since it will be assumed we are talking about phases of primary penetration unless it is explicitly called out as secondary penetration. Secondary penetration will not be discussed in this chapter, but will be discussed in Chapter 10. The initial transient phase is fairly well defined in duration, but the beginning of the terminal phase is not so well defined and must be estimated, particularly for the lower impact velocities.

7.5. Centerline Momentum Balance and Hydrodynamic Approximation

To assist in analyzing the penetration event, we introduce the centerline momentum balance. The momentum balance $\partial \sigma_{jz}/\partial x_j = \rho D v_z/Dt$ is integrated along the full extent of the axis of cylindrical symmetry (the z-axis), assuming the density of each material changes little. The z velocity along the centerline for the entire penetration event is $v_z(x,t)$. We will see with the numerical simulations that the velocity profile through the target and projectile primarily changes by translating at the penetration velocity, and so we rewrite the velocity as $v_z(x,t) = \hat{v}_z(z - z_i(t), \tau)|_{\tau=t}$, where $z_i(t)$ is the location of the target–projectile interface (i.e., the nose of the projectile, so $dz_i/dt = u = v_z(z_i, t)$). The variable τ is to include small temporal changes in the generally unchanging, translating shape. Now,

$$\frac{Dv_z}{Dt} = \frac{\partial v_z}{\partial t} + v_z \frac{\partial v_z}{\partial z} = -\frac{\partial \hat{v}_z}{\partial z}\frac{dz_i}{dt} + \left.\frac{\partial \hat{v}_z}{\partial \tau}\right|_{\tau=t} + \frac{1}{2}\frac{\partial \hat{v}_z^2}{\partial z} = \frac{\partial}{\partial z}\left\{\hat{v}_z\left(\frac{\hat{v}_z}{2} - u\right)\right\} + \left.\frac{\partial \hat{v}_z}{\partial \tau}\right|_{\tau=t},$$
(7.4)

and with this form of the velocity integrating the acceleration yields

$$\int_{z_p}^{z_i} \rho \frac{Dv_z}{dt} dz = \rho_p\, v_z(z)\left.\left(\frac{v_z(z)}{2} - u\right)\right|_{z_p}^{z_i} + \rho_p \int_{z_p}^{z_i} \left.\frac{\partial \hat{v}_z}{\partial \tau}\right|_{\tau=t} dz$$

$$= \rho_p\left\{-\frac{1}{2}u^2 - v\left(\frac{1}{2}v - u\right)\right\} + \rho_p \int_{z_p}^{z_i} \left.\frac{\partial \hat{v}_z}{\partial \tau}\right|_{\tau=t} dz = -\frac{1}{2}\rho_p(v-u)^2 + \rho_p \int_{z_p}^{z_i} \left.\frac{\partial \hat{v}_z}{\partial \tau}\right|_{\tau=t} dz,$$
(7.5)

where the back end of the projectile $z_p(t)$ has speed $v = dz_p/dt = v_z(z_p, t)$. Similarly, deep in the target $v_z(+\infty) = 0$, and so

$$\int_{z_i}^{+\infty} \rho \frac{Dv_z}{dt} dz = \frac{1}{2}\rho_t u^2 + \rho_t \int_{z_i}^{+\infty} \left.\frac{\partial \hat{v}_z}{\partial t}\right|_{\tau=t} dz.$$
(7.6)

The integral of the stress decomposes into two terms, one for the projectile and one for the target, of the form (this specific one is the projectile side)

$$\int_{z_p}^{z_i} \frac{\partial \sigma_{jz}}{\partial x_j} dz = 2\int_{z_p}^{z_i} \frac{\partial \sigma_{xz}}{\partial x} dz + \int_{z_p}^{z_i} \frac{\partial \sigma_{zz}}{\partial z} dz = 2\int_{z_p}^{z_i} \frac{\partial \sigma_{xz}}{\partial x} dz + \sigma_{zz}|_{z_p}^{z_i}.$$
(7.7)

The back of the projectile is stress free and deep in the target is stress free, hence $\sigma_{zz}(z_p) = \sigma_{zz}(+\infty) = 0$. Thus, the momentum balance on the centerline becomes (adding these terms together)

$$\frac{1}{2}\rho_p(u-v)^2 - \rho_p \int_{z_p}^{z_i} \left.\frac{\partial \hat{v}_z}{\partial \tau}\right|_{\tau=t} dz + 2\int_{z_p}^{z_i} \frac{\partial \sigma_{xz}}{\partial x} dz$$

$$= \frac{1}{2}\rho_t u^2 + \rho_t \int_{z_i}^{+\infty} \left.\frac{\partial \hat{v}_z}{\partial t}\right|_{\tau=t} dz - 2\int_{z_i}^{+\infty} \frac{\partial \sigma_{xz}}{\partial x} dz = \tilde{\sigma}_{zz}(z_i),$$
(7.8)

where we moved everything relating to the projectile to the left-hand side and everything relating to the target to the right-hand side. Both sides equal the stress at the interface. These equalities are approximate in that there has been an assumption that the densities of the respective materials could be viewed as constant. The density times velocity squared terms have arisen from advection and the nonlinear terms in the material derivative of the velocity,

GENERAL PRINCIPLE 7.2. *The terms* $-\frac{1}{2}\rho_p(v-u)^2$ *and* $\frac{1}{2}\rho_t u^2$ *are the centerline rate of change in momentum (inertial terms) arising from advection for the projectile and target, respectively.*

In the next chapter we explore the Tate–Alekseevskii model. The central equation in that model can be realized by assuming the integral terms on either side of the equation are constant,

$$\frac{1}{2}\rho_p(u-v)^2 + Y_p = \frac{1}{2}\rho_t u^2 + R_t = \tilde{\sigma}_{zz}(z_i). \tag{7.9}$$

There are other aspects to the model to be described there.

If the materials had no strength, then the shear stress integrals would be zero, and if the penetration event were "steady state," then the transient terms are zero, and we obtain from Eq. (7.8) the *hydrodynamic approximation*

$$\frac{1}{2}\rho_p(u-v)^2 = \frac{1}{2}\rho_t u^2 = \tilde{\sigma}_{zz}(z_i). \tag{7.10}$$

The terminology "hydrodynamic" originates from the assumption of zero strength. If we assume the integral terms on either side of Eq. (7.8) are small or bounded, then we expect that this hydrodynamic approximation will be reached for high velocities. Recall, however, that in the derivation it was assumed the density change in each material is negligible; hence at some speed we expect this approximation to be less accurate, as that behavior no longer holds. Explicit solutions for the penetration velocity yield (now marked with subscript hyd for hydrodynamic)

$$u_{\text{hyd}} = \frac{v}{1+\sqrt{\rho_t/\rho_p}}, \qquad v - u_{\text{hyd}} = \frac{v}{1+\sqrt{\rho_p/\rho_t}},$$

$$\tilde{\sigma}_{zz}(z_i)_{\text{hyd}} = \frac{1}{2}\rho_t u_{\text{hyd}}^2 = \frac{1}{2}\rho_p(v-u_{\text{hyd}})^2 = \frac{1}{2}\left(\frac{1}{\frac{1}{\sqrt{\rho_t}}+\frac{1}{\sqrt{\rho_p}}}\right)^2 v^2. \tag{7.11}$$

The rate of change in length of the projectile is $dL/dt = -(v-u)$ and taking the square root of both sides and integrating with respect to time, assuming the entire projectile erodes, yields

$$-\sqrt{\rho_p}\,dL = \sqrt{\rho_t}\,dP \qquad \Rightarrow \qquad \left(\frac{P}{L}\right)_{\text{hyd}} = \sqrt{\frac{\rho_p}{\rho_t}} = \frac{u_{\text{hyd}}}{v-u_{\text{hyd}}}. \tag{7.12}$$

This depth is referred to as the hydrodynamic limit. Penetration results where erosion occurs will often be compared to this hydrodynamic result.

7.6. Numerical Simulations

We continue our discussion of eroding penetration by presenting the results of numerical simulations of impacts over a wide range of impact conditions. As the discussion progresses there will be comparison to experimental data.

The numerical methods used in examining impact and penetration events is a topic unto itself. The large-scale numerical computations presented in this book were performed with the Eulerian hydrocode CTH. The terminology "hydrocode" originated as early wave propagation and impact software did not include material strength (hence "hydrodynamic") and is still used even though strength is now included. The numerical computations are carried out in an explicit fashion, which means that the values of the variables at the current time step are written explicitly in terms of the values of the variables at previous time steps. The method solves the equations of conservation of mass, momentum, and energy on a discretized spatial array referred to as a finite volume technique. This step is performed in a Lagrangian fashion, where $\frac{D}{Dt} = \frac{\partial}{\partial t}$ and the mesh deforms with the material motion. This step is typically second order in cell length Δx and time step Δt. In Eulerian codes, the deformed mesh is then mapped back to the original mesh and in arbitrary Lagrangian-Eulerian (ALE) methods the deformed mesh is mapped to another mesh, typically somewhere between the mesh at the beginning of the step and the deformed mesh. The time step is limited by the Courant–Fredrichs–Lewy (CFL) stability criterion, which says that Δt must be less than the amount of time it takes information to cross a computational cell or element in its narrowest direction Δx,

$$\Delta t < \frac{\Delta x}{c + |v|}, \tag{7.13}$$

where c is the material sound speed at the appropriate pressure and v is the material velocity in the appropriate direction. CTH uses a second order advection scheme for the remap step and a sophisticated interface tracker to construct subcell geometry in mixed cells (computational cells or elements that contain more than one material). Artificial viscosity is used to spread compressive shock fronts over a finite number of cells (rather than just one), since otherwise the large jump from one computational cell to the next leads to oscillations when finite differences are performed to approximate derivatives. The stress σ in the conservation equations is replaced with $\sigma - qI$, where the artificial viscosity q resists further compression during compression – hence resisting steep gradients – and is usually taken as positive, hence the minus sign in the stress adjustment. A successful form is [210]

$$q = 4 \cdot \rho(\Delta x)^2 D_{11}^2 + 1 \cdot \sqrt{\rho p} |D_{11}| \Delta x, \quad \text{when } D_{11} < 0, \quad \text{otherwise } q = 0, \tag{7.14}$$

where Δx is the cell size, the rate of deformation tensor component is $D_{11} = \partial v_x / \partial x$, and the direction x is the direction of material acceleration. Using artificial viscosity allows the computations to be based on the differential continuum equations since now the rise at a shock front is not too steep and the finite differences are well behaved. This approach is referred to as shock capturing and, owing to the conservation form of the equations, the correct shock speeds result even though these artificial stresses are introduced. It is "artificial" because the stress term does not come from the material constitutive model. The 4 and 1 are constants that can be adjusted to change the number of cells for the rise of a shock. (An alternative approach, which might seem more in the spirit of this book, is to explicitly track shock fronts and apply the Hugoniot jump conditions across those fronts. This method is referred to as shock tracking or front tracking. It is more complicated in software to track fronts, especially in higher dimensions.) The stress update was performed with the Jaumann corotational rate and plastic flow was handled with the radial return algorithm. CTH was written in the 1980s by a team led by Michael McGlaun and Sam Thompson at Sandia National Laboratories and is extensively used for impact computations in the United States [142].

7.7. Numerical Simulations of $L/D = 10$ Tungsten Impacting Steel

In the following sections, numerical simulations with CTH were performed that assumed axial symmetry to compare to experiments reported in [9]. The projectile was of a tungsten alloy with density 17.4 g/cm^3 and mass 73.3 g. It was 0.817 cm in diameter and 8.17-cm long, including the hemispherical nose; hence $L/D = 10$. The computations were performed with a relatively coarse mesh with an aspect ratio of 2 to 1 with five zones across the radius of the projectile in the vicinity of the target–projectile interface path and outside that region the zoning was geometric, with an average growth rate of 6%. A typical computational run required 1,700

TABLE 7.1. Equation of state parameters (x changes for different alloys).

Material	ρ_0 (g/cm^3)	c_0 (m/s)	k	Γ_0	C_v (J/g K)
Tungsten	17.x	3.85	1.44	1.58	0.13
Steel	7.85	4.50	1.49	2.17	0.44

TABLE 7.2. Constitutive constants in numerical simulations.

Run numbers	Material	A (MPa)	B (MPa)	n	C	m
1	4340 Steel	792	0	0.0	0.0	—
2	4340 Steel	792	510	0.26	0.0	—
3	4340 Steel	792	510	0.26	0.014	—
4,5,7	4340 Steel[a] (Rc 30)	792	510	0.26	0.014	1.03
6,8,9	4340 Steel	735	473	0.26	0.014	1.03
1–4	W alloy	2,000	0	0.0	0.0	—
5–9	W alloy	1,350	0	0.0	0.06	—

[a] Original Johnson–Cook parameters (see Appendix C)

TABLE 7.3. Numerical simulation P/L (experimental $P/L \pm 0.049$).

Run	Velocity (km/s)	Exp. P/L	Comp. P/L	Diff. from Exp. (%)	Diff. from 6,8,9[a] (%)	Change in physics
1	1.5		1.040	+17.8	+21	Target just yield A
2	1.5		0.840	−5.19	−2.1	plus work hardening
3	1.5		0.785	−11.4	−8.5	plus strain rate
4	1.5		0.850	−4.06	−0.9	plus thermal softening
5	1.5		0.825	−6.88	−3.8	Better projectile Y
6[a]	1.5	0.886	0.858	−3.16		Target Rc 27 (not 30)
7	1.2		0.484	−14.3	−3.4	
8[a]	1.2	0.565	0.501	−11.3		Target Rc 27 (not 30)
9[a]	1.7	1.099	1.061	−3.45		

[a] Best estimate to constitutive properties

integration time steps to compute 120 μs. A similar problem was more finely zoned in the axial direction (three times as many zones); this computation gave an answer within 4.4% of the baseline computation. An additional computation was run to 30 μs with the zoning doubled in each direction; the velocity profiles were found to agree well with the baseline computation.

The steel used for the targets was 4340 steel hardened to a Rockwell hardness Rc 27±2. Johnson and Cook have characterized the response of 4340 steel hardened to Rc 30, and the constitutive response for the target material was represented by the Johnson–Cook model. Material constants for the tungsten alloy and 4340 steel alloy are given in Tables 7.1 and 7.2. The stress-strain behavior of the tungsten alloy used in the experiments (90.1% W; 4.47% Ni; 1.95% Fe; 2.6% Co) is also shown in Table 7.2 [**127**]. For this alloy there is no discernible work hardening, and the Johnson–Cook model does a good job reflecting the strain-rate strengthening (Exercise 7.18).

There are many topics we will explore with numerical simulations. The first topic is whether the depth of penetration from the numerical simulation is reasonable. It is the easiest measurement to make in the experiment and so it is our first indication of validity of a simulation. We expect that the constitutive constants given in the preceding paragraph influence the results, and one of the questions we ask in the process of comparing depths is how much influence various parts of the constitutive model have on the final depth of penetration. This parameter exploration is a feature of numerical simulations – it allows us to explore the influence of material properties in a way that is difficult or impossible in experiments.

Initial numerical simulations were performed at 1.5 km/s. The values closest to what we might expect the material properties to be were used in Run 6. However, we start by assuming in Run 1 that the target is elastic, perfectly plastic (i.e., no work hardening or strain rate effects). Each term of the Johnson–Cook constitutive model was "turned on" in sequence; Run 2 included work hardening, Run 3 included work hardening and strain-rate effects, and Run 4 was the complete model using the 4340 steel constants, which includes thermal softening in addition to the strain rate and work hardening terms. The projectile was modeled as an elastic, perfectly plastic material for Runs 1 through 4 using a value of 2.0 GPa as an effective flow stress; this value was chosen to account for dynamic effects in an average manner. Run 5 modeled the projectile with a more realistic yield strength and strain-rate dependence. Finally, since the actual target material had a hardness of Rc 27 instead of Rc 30, the parameters A and B in the Johnson–Cook model were decreased for Run 6. Additional simulations were performed at impact velocities of 1.2 and 1.7 km/s. Table 7.2 lists the values of the parameters used for each simulation.

The depths of penetration normalized by the initial projectile length from the numerical simulations are given in Table 7.3. The table also lists experimental values of P/L computed from a linear least-squares curve fit to the experimental data for the $L/D = 10$ tungsten alloy projectiles striking Rc 27 steel, which was $P/L = 1.068V - 0.716 \pm 0.049$ for $0.85 \leq V \leq 1.77$ km/s.

The difference between the experimental and numerical results is presented in the last column of Table 7.3. It is seen that the elastic-perfectly plastic target with the flow stress equal to the yield stress (Run 1) is too soft. Strain hardening effects (Run 2) decrease the depth of penetration a lot (23% of best result depth), and strain-rate hardening (Run 3) decreases total penetration even further (another 6.4%). However, the heating caused by plastic work softens the target considerably and the final depth of penetration computed increases (by 7.6%, Run 4). Run 5 utilized a better match to the tungsten constitutive properties, which effectively softened the projectile, leading to the depth of penetration decreasing (2.9%). Reducing the initial yield stress of the steel while keeping $B/A = 0.644$ constant (i.e., to reduce the Rc by 3, Run 6) increases the calculated depth of penetration (by 3.8%) and puts it within 3% of the mean of the experimental data, which is within data scatter. The use of constitutive models specifically calibrated for the materials used in the experiments results in agreement between the computations and experiment.

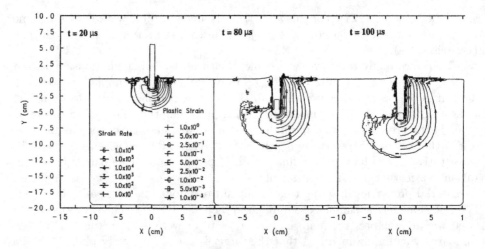

FIGURE 7.8. Plastic strain rate and equivalent plastic strain contours in the target material. (From Anderson and Walker [12].)

We now examine the penetration event in some additional detail, and we use Run 6 for the examination. The geometry of the impact is shown in Fig. 7.8 at two instances of time during the quasi-steady-state penetration phase (20 and $80\,\mu s$) and near the end of penetration ($100\,\mu s$). The projectile is progressively shorter at each time; hence this impact is an eroding impact. Equivalent plastic strain contours and plastic strain rates of the target material are shown. The strain rates are shown on the left-hand side of each target block, and the plastic strain contours are on the right-hand side. An indication of the work the penetrator is doing on the target is represented by the extent of the plastic zone. Using 1% plastic strain as the measure of the extent of the plastic zone, it is seen that the plastic zone increases both axially and radially during the steady-state portion of penetration. At $20\,\mu s$, the axial extent of penetration in terms of projectile radii is approximately $5.6R_p$, and $6.2R_p$ for radial extent. At $80\,\mu s$, the axial extent is $8.6R_p$ and the radial extent is $9.6R_p$. The ratio of the axial extent to radial extent remains approximately 0.90 for the specific impact example examined here until near the end of penetration, when the ratio becomes approximately 1.0. Although the extent of the plastic flow has increased, the very large strains are concentrated near the projectile–target interface. Strains of 10% and larger are confined to within approximately $3.0R_p$ of the projectile–target interface. In addition, the high-strain-rate region, $10^4\,\mathrm{s}^{-1}$ and higher, is within approximately $3.0R_p$ of the interface. The $10^3\,\mathrm{s}^{-1}$ strain-rate contour expands with penetration, and has an extent similar to the 1% plastic strain contour.

The nose or penetration velocity u and the tail velocity v as a function of time are shown in Fig. 7.9 as solid lines. The dashed lines are from the Tate model, which will be described in the next chapter. The fact that the projectile is eroding during the penetration event is evident since $u < v$; in particular, the change in length of the projectile is given by $dL/dt = -(v - u)$. There is an initial transient phase of penetration that persists on the order of 10–$15\,\mu s$, corresponding to 1.0–$1.5\,cm$ of penetration distance, which is equivalent to approximately 1.2–1.8 projectile diameters of penetration. Then there is a relatively steady phase of penetration,

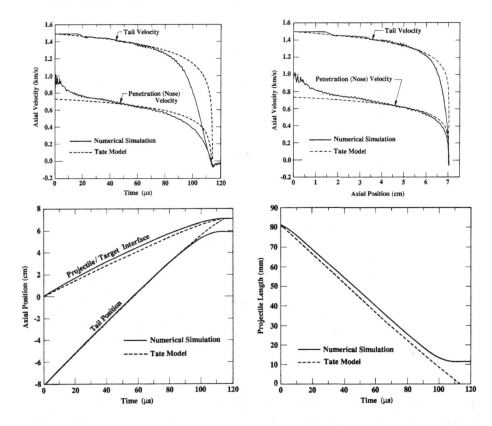

FIGURE 7.9. Tail and nose speed vs. time (upper left) and vs. penetration depth (upper right). Tail and nose position vs. time (lower left) and length of projectile vs. time (lower right). The solid lines are for the numerical simulation (Run 6) and the dashed lines are for the Tate model with $R_t = 4.94$ GPa, $Y_p = 2$ GPa chosen to match the numerical simulations final depth of penetration. (From Anderson and Walker [12].)

where u and v are slowly changing. Following the steady phase of penetration, the final transient phase ends with a slightly negative speed for the crater floor and penetrator tail: both velocities coincide at the end of penetration, since the residual penetrator is now a "rigid" body. This reversal in velocity is the elastic rebound of the crater. It is sometimes observed experimentally, particularly for shallow craters, that the residual projectile is ejected from the crater back towards the muzzle of the gun. The velocity associated with the elastic rebound is approximately 75 m/s for the impact case examined here. For deep craters, the residual projectile strikes eroded projectile material and the crater walls and remains trapped. X-rays of sectioned deep craters often show that the residual projectile rebounds away from the crater floor, as seen in Fig. 7.5.

An elastic wave travels the length of the projectile after impact. The arrival of the elastic wave at the rear of the projectile is clearly evident in Fig. 7.9. The bulk sound speed of the tungsten alloy is 3.85 km/s, and Poisson's ratio is 0.3; using

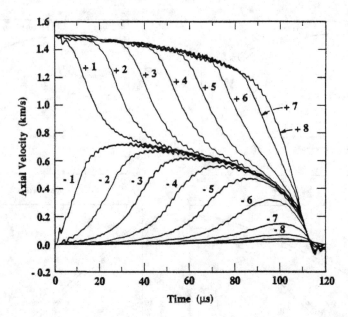

FIGURE 7.10. Velocities of tracer particles along the projectile–target centerline. (From Anderson and Walker [12].)

these values, the bar wave speed $c_E = 4.22$ km/s. The arrival time of the elastic wave is L/c_E; inserting the appropriate values gives an arrival time of $19.4\,\mu$s; the computation is in excellent agreement with this value. The decrease in velocity of the rear of the projectile is (Eq 5.7) $\Delta u = 2Y_p/(\rho_0 c_E)$. Using the yield stress for the back end of the projectile 1.35 GPa (Table 7.2) gives $\Delta u = 37$ m/s, which is in reasonable agreement with the numerical simulation value of approximately 42 m/s. A number of elastic wave transits occur, decreasing both the tail and interface velocities.

We can understand the deceleration of the projectile by examining the velocity profile within the projectile itself. The velocity history of tracers initially located 1 cm apart along the centerline of the projectile and target is shown in Fig. 7.10. This figure is similar to Fig. 7.9, except that tracers within the target and projectile are shown, rather than just at the nose and tail of the projectile. The initial location of each tracer particle is shown in the figure, referenced to the initial interface position of 0. Positive values are in the projectile and negative values are in the target. Because of the axial symmetry of the computation, the tracer stays on the central axis. Tracers located within the projectile move along at a nominally constant velocity until a time when they are quickly decelerated to the penetration (interface) velocity. In a similar fashion, tracers in the target remain stationary until they are accelerated to the interface velocity, although a more gradual acceleration occurs than for projectile material.

The axial velocity profile along the centerline through the projectile and target at $40\,\mu$s is shown in Fig. 7.11 (other times are similar). The position of the tail and the projectile–target interface are given by the vertical dot-dash lines; the projectile is to the left and the target to the right. The penetration velocity $u = 0.74$ km/s at this time. There exists a finite region in the projectile over which deceleration

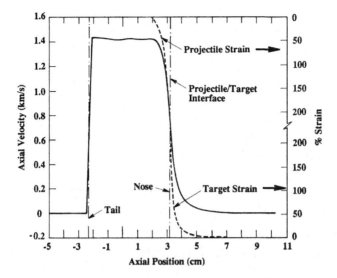

FIGURE 7.11. Centerline velocity and equivalent plastic strain vs. position. (From Anderson and Walker [12].)

occurs. This region is reflected in the equivalent plastic strain in the projectile which is superimposed on the figure. The equivalent plastic strain shows that material in a finite region of the projectile is plastically deforming, which is coincident with the large velocity gradient. The velocity gradient in the target also corresponds to where plastic strain occurs, as shown. Equivalent plastic strain rates and equivalent plastic strain contours within the projectile are shown at 20 and 80 μs in Fig. 7.12. Using 1% plastic strain as an indicator of the extent of the plastic zone, the plastic zone extends 1.4–1.6 projectile diameters from the interface. This region is not large and remains about the same size throughout the penetration event. The plastic region is not large because a condition of uniaxial stress holds for most of the length of the projectile: the lack of confinement (hence uniaxial stress) limits the stress to the yield stress. In the terminal phase of penetration, the projectile deceleration is more complicated and most of the projectile "tail" material identified with this plastic zone region survives (as opposed to erodes).

The velocity along the projectile–target centerline for three velocities using constitutive constants for Run 5 are plotted vs. scaled distance in Fig. 7.13 at a time when the projectile is approximately 50% consumed. The dashed vertical line is the location of the projectile–target interface; the velocity profile to the left is within the projectile and to the right is within the target. The results were shifted so that the interfaces have a common coordinate of $x = 0$ to assist in comparing results. The tail of the projectile is located at approximately $x/L = -0.5$. Moving right from the projectile tail, the velocity in the projectile is relatively constant until approximately 1.5 projectile diameters from the projectile–target interface. The large velocity gradient near the projectile–target interface is coincident with the extent of plastic flow in both the projectile and the target. Rapid deceleration of projectile material occurs when the material enters the plastic zone – that is, when the material begins to see the large stresses at the projectile–target interface. The extent of the plastic zone in the projectile is essentially independent of impact

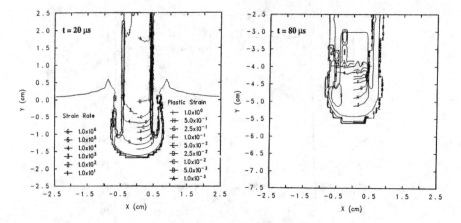

FIGURE 7.12. Plastic strain rates and equivalent plastic strains in the projectile. (From Anderson and Walker [12].)

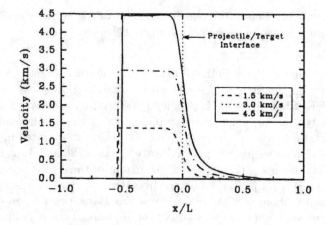

FIGURE 7.13. Velocity through the projectile and target along the centerline for three impact velocities. (From Anderson, Littlefield, and Walker [8].)

velocity. On the other hand, the amount of target material flowing plastically increases with penetration velocity, as inferred from the larger region of target material with a nonzero velocity. However, on this plot it is not possible to see that the crater radius increases as impact speed increases. It turns out that when the extent of plastic flow is measured in terms of crater radius, then the extent decreases as impact speed increases, as will be quantified in a later section.

The penetration u and tail velocities v along the centerline are shown for impact speeds of 1.5, 3.0, and 4.5 km/s (Fig. 7.14). The plots differ from the previous plots in that instead of velocity being plotted in terms of penetration depth it is plotted in terms of time, though the time is scaled by the impact velocity divided by the length. If the penetration velocity were half the impact velocity ($u = V/2$), then the duration of the impact event, in scaled time, would be $tV/L = 2$. In these plots, phases of penetration are evident in that there exists an initial transient

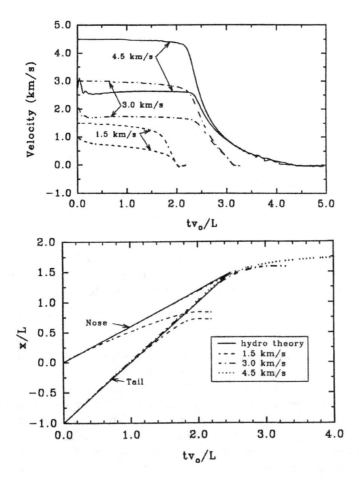

FIGURE 7.14. Penetration and tail velocities and position are plotted vs. scaled time for $L/D = 10$ tungsten projectiles striking steel.
(From Anderson, Littlefield, and Walker [8].)

(shock) phase, a quasi-steady-state phase, and a residual penetration phase. Also,
a small elastic recovery can be observed where the crater rebounds elastically at the
end of penetration. The final transient grows considerably with increasing impact
velocity; as more kinetic energy is deposited in the target, the final transient phase
of penetration takes longer and produces a higher percentage of the penetration.

During the shock phase of penetration, the transient phase in which wave effects dominate the physics and mechanics of penetration, there is an initial rapid
increase in the penetration velocity as rarefaction waves from the free lateral surface
of the projectile and the top surface of the target release the geometric confinement
of impact. The stress release waves allow radial motion of target material, allowing easier penetration, hence the rapid increase in penetration velocity. However,
the lateral motions fall to lower velocities as the high pressures from the shock are
attenuated, and the penetration velocity subsequently decreases. At this point,
the penetration transitions from the transient wave phase to a quasi-steady material flow phase. For the projectile dimensions used in these calculations, it takes

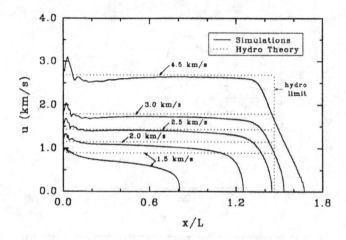

FIGURE 7.15. Penetration velocities vs. normalized depth of penetration $x/L = P(t)/L$ for $L/D = 10$ tungsten projectiles striking steel. (From Anderson, Littlefield, and Walker [8].)

7 to $10\,\mu s$ to achieve quasi-steady penetration. Although the shock pressures are dissipated in a time frame approximately given by D/c_0 (=$1.80\,\mu s$ for the specific dimensions $D = 8.17\,\text{mm}$ and materials used for these computations), the high pressures from the shock sets the target material in motion, and it takes some time for this transient state to decay into a quasi-steady-state penetration. For $L/D = 10$ projectiles, the penetration velocities vs. normalized depth of penetration are plotted in Fig. 7.15 for five impact velocities. The duration of the transient phase is approximately independent of impact velocity and quasi-steady penetration is achieved after 1.2 to $1.5D$ of penetration. (This conclusion is for a tungsten projectile into a steel target. It takes approximately $2.5D$ of penetration before quasi-steady penetration is achieved for tungsten impacting a titanium target of density 4.41 g/cm^3. The density affects the length of time of this phase.) To first order, the duration of the transient phase depends on geometric and material parameters and not the impact velocity.

The hydrodynamic penetration velocities from Eq. (7.10) are also plotted on Fig. 7.15, and the steady-state hydrodynamic penetration depth $(P/L)_{\text{hyd}} = \sqrt{\rho_p/\rho_t}$ = 1.47 is shown. Even at an impact velocity of 4.5 km/s it is seen that the penetration velocity falls below hydrodynamic theory, and the steady-state phase of penetration is completed before the depth of penetration reaches the hydrodynamic limit. This observation implies that even though the interface stresses are substantially higher than the flow stress – such that the hydrodynamic approximation should be valid – strength effects result in a lower penetration velocity and an increased erosion rate compared to hydrodynamic theory. The terminal transient residual penetration phase, resulting from the inertial effect of target material set in motion during earlier phases of penetration, allows penetration to continue towards the hydrodynamic limit and, if the velocity is sufficiently high, beyond. Figure 7.15 shows that the final phase of penetration contributes substantially to the total penetration at the higher impact velocities.

For the 1.5 km/s impact velocity, penetration proceeds sufficiently slowly that a number of elastic wave transits occur, thereby decreasing both the tail and interface velocities. (At higher impact velocities, fewer reflections occur before the projectile erodes, since the step deceleration occurs every $2L/c_E$ time increment.) The value of the elastic step should be independent of impact velocity, and so the relative percentage of the elastic wave deceleration step as compared to the projectile tail velocity goes from approximately 3% for the 1.5-km/s case to only 1% for a 4.5-km/s impact case. Thus, elastic deceleration of the projectile is more important at lower impact speeds and it preserves a residual piece of the projectile, while at higher velocities the projectile is not decelerated quickly enough to prevent nearly complete projectile erosion. The computations predict approximately $1.2D$ of penetrator remaining at the end of penetration for an impact velocity of 1.5 km/s; only $0.3D$ of the projectile remains at 2.0 km/s; and the projectile is completely eroded at impact velocities of 2.5 km/s and greater.

The penetration vs. impact speed saturates, or seems to approach a limiting value, at high velocities. This limit is owing to the fact that the steady-state penetration velocity approaches the hydrodynamic limit and the deceleration of the back of the projectile is less over the penetration event due to the reduced time in the event. The corresponding penetration due to the steady state phase approaches the hydrodynamic limit since the projectile completely erodes away. There is still the terminal phase of primary penetration that leads to a slow increase in penetration depth as impact velocity increases, but the overall appearance is that of an S shape with the top of the S representing a saturation. Thus, the reason penetration saturates is that the projectile erodes away. This observation is important enough for a formal statement,

GENERAL PRINCIPLE 7.3. *For large L/D, final penetration depth saturates at high impact speeds because the projectile completely erodes away.*

7.8. The Stress at the Projectile–Target Interface

From the numerical simulation, the hydrostatic pressure p (the negative of the mean stress) at the projectile–target interface is plotted in Fig. 7.16 as a function of time. Also plotted from the Tate model using the R_t and Y_p to match the depth of penetration are the total "Tate" stress of Eq. (7.9) and the two centerline advected change in momentum terms $\frac{1}{2}\rho_p(v-u)^2$ and $\frac{1}{2}\rho_t u^2$ (sometimes referred to as "fluid" or "dynamic" pressure components). The shock phase is evident from the spike in the pressure from the numerical computations; rarefaction waves originating from the free surface of the projectile reduce the shock pressure very quickly. The duration of this shock spike is on the order of $2R_p/c = 1.80\,\mu$s in the present example, where c is the rarefaction wave speed.

The pressures plotted in Fig. 7.16 are approximately constant for most of the penetration time, since the velocities are relatively constant. The value of $\tilde{\sigma}$ is higher than the hydrostatic pressure obtained from the numerical simulation and the way it partitions into various components is shown. Thus, $\tilde{\sigma} = \tilde{\sigma}_{zz}$ is the axial stress component at the interface centerline. This axial stress consists of a hydrostatic pressure component and a deviator stress due to the material strength. The centerline interface pressure for five different impact velocities is shown in Fig. 7.17, plotted in terms of scale time tv_0/L, with the same projectile geometries and the material properties corresponding to Run 5. The peak values from the

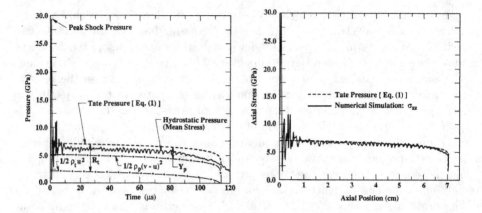

FIGURE 7.16. Centerline pressure (left) at projectile–target interface vs. time from numerical simulation (solid) and Tate model
(dot-dashed curves) and axial stress from numerical simulation and
Tate model (right). (From Anderson and Walker [**12**].)

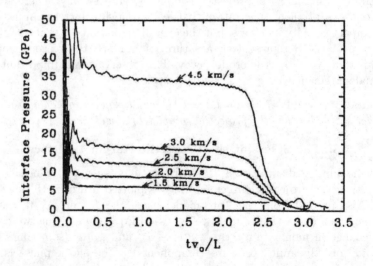

FIGURE 7.17. Projectile–target interface pressure along the centerline vs. scale time. (From Anderson, Littlefield, and Walker
[**8**].)

initial shock and the steady-state penetration pressures are listed in Table 7.4. The
duration of the shock pressure is approximately a constant, since it is a function of
the geometric and material parameters, so that when plotted as a function of scaled
time the "duration" for the 4.5 km/s case is approximately three times longer than
for the 1.5 km/s case. The radial confinement of the initial impact leads to the
clear difference between the impact stress and the penetration stress.

TABLE 7.4. Impact shock pressures and steady state penetration pressures.

Impact velocity (km/s)	Peak pressure (GPa)	Steady state penetration pressure (GPa)
1.5	25.0	6.5
2.0	36.4	8.5
2.5	49.8	12.5
3.0	65.0	16
4.5	115	34.5

TABLE 7.5. Crater radii.

Impact velocity (km/s)	Radius at $x = 0$ R_c/R_p	Minimum radius R_c/R_p	Maximum radius R_c/R_p	Experimental radii R_c/R_p
1.5	1.77	1.66	1.92	$1.7 - 2.4$
2.0	2.10	2.00	2.26	$2.2 - 2.7$
2.5	2.50	2.42	2.80	$2.7 - 3.2$
3.0	2.96	2.92	3.38	$2.9 - 3.5$
4.5	4.04	3.82	5.64	$4.7 - 4.8$

7.9. Crater Radii, Plastic and Elastic Strains, and the Energy Partition

To explore the target response at higher impact velocities, the material properties of Run 5 were used (Table 7.2). Again the projectile geometry was $L/D = 10$,
with a length of 8.17 cm. Square zoning with five zones across the radius of the
projectile was used in the interaction region. The impact velocities considered were
1.5, 2.0, 2.5, 3.0, and 4.5 km/s.

Contours of equivalent plastic strain rate and equivalent plastic strain are shown
in Fig. 7.18 for three impact velocities. The figures are when the projectile is roughly
half of its original length. The impact craters are approximately cylindrical, with
a characteristic radius that is a function of the impact velocity. The minimum and
maximum crater radii along with the crater radius at the original target free surface
are shown in Table 7.5. Experimentally determined crater radii from the data in
Chapter 8 are also tabulated.

The plastic strain contours are similar at the three impact velocities. The
extents of the plastic zone increase with impact velocity; however, the growth of
the plastic zone in the target is a weaker function of velocity than the growth of
the crater radius. Estimates for the extent of the plastic zone in the target along
the centerline, normalized by an estimated crater radius (measured at the target
impact face) were made from the computational results. These values correspond
to the times in Fig. 7.18. Two values were obtained – the 0.2% equivalent plastic
strain contour and the $1\,\mathrm{s}^{-1}$ equivalent plastic strain rate contour – and are shown
in Table 7.6. The extent of the plastic zone is scaled by the crater radius because the
target motion knows about the crater radius and not about the projectile radius,
per se. Thus, the crater radius controls target response in analytical penetration
theories and not the projectile radius. Although the actual physical extent of

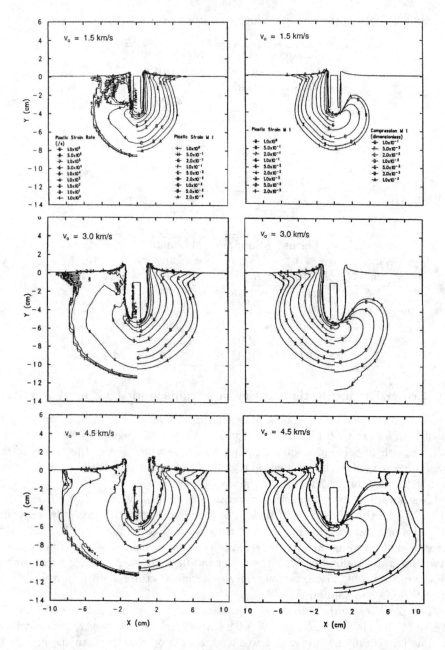

FIGURE 7.18. Equivalent plastic strain rate and equivalent plastic strain contours (left) and equivalent plastic strain and compression contours (right) in the target for 1.5, 3.0, and 4.5 km/s impacts. Contours for the equivalent plastic strain rate are 1, 10, 10^2, 10^3, 10^4, 5×10^4, 10^5, 5×10^5, and $10^6 \, \mathrm{s}^{-1}$, contours for the equivalent plastic strain (towards the center of each plot) are 0.002, 0.005, 0.01, 0.02, 0.05, 0.1, 0.2, 0.5, and 1, and contours for the volumetric strain (compression) $\rho/\rho_0 - 1$ are 0.001, 0.002, 0.005, 0.01, 0.02, 0.05, and 0.1. (From Anderson, Littlefield, and Walker [8].)

TABLE 7.6. Estimated normalized extent of plastic zone in the target (plastically flowing region radius in terms of crater radii).

Impact velocity (km/s)	0.2% equivalent plastic strain contour R_{plastic}/R_c	1 s^{-1} equivalent plastic strain rate contour R_{plastic}/R_c
1.5	5.4	6.4
2.0	5.2	6.2
2.5	5.0	6.0
3.0	4.4	5.2
4.5	3.6	4.0

the flow field increases in the target with penetration velocity, the extent of the flow field decreases when expressed in terms of the crater radius. This behavior appears to be a consequence of compressibility of the target, as discussed when we derived the cavity expansion in Chapter 6. In the Walker-Anderson model developed in Chapter 10, the target resistance depends on the normalized extent of the plastic zone and the target resistance due to target material strength decreases as penetration velocity increases because the model predicts that the normalized plastic zone extent decreases with increasing penetration velocity, as is observed here. The strain rate increases with impact velocity, and has similar qualitative features to the plastic strain contours.

Plastic flow is a dissipative mechanism, converting energy into heat in the material. Energy can also be stored elastically in the target, though this storage mechanism is temporary, as the compression is released after the passage of the projectile. The volumetric strain $\rho/\rho_0 - 1$ is a measure of the elastic compression of the target material and is plotted on the right-hand side in Fig. 7.18. The density of the target material does not change appreciably and, in this sense, incompressibility is a reasonable approximation. However, considerable elastic energy can be stored in a small amount of compression of the target (this is true also of the projectile, but the volume involved is small relative to that of the target). The compression of the target is a function of impact velocity. At low impact velocities – for example, 1.5 km/s – most of the plastic straining occurs within the vicinity (in front and to the side) of the projectile–target interface. After passage of the projectile, there is very little additional plastic strain accumulated in the target. At higher impact velocities, a significant amount of plastic straining occurs when the target unloads from a compressed state. This straining occurs in the target volume "behind" the projectile, and hence does not directly affect penetration. This can be seen in Fig. 7.18; for the 3.0 and 4.5 km/s impact cases, plastic strain rates of 10^4 and 10^5 s^{-1} are evident in regions of the target well behind (above) the projectile–target interface plane. This additional plastic strain is due to the release of compression.

The target volume in the vicinity of the projectile–target interface controls penetration. Because some of the energy transferred from the projectile to the target is temporarily stored in elastic compression, the rate at which energy is dissipated by plastic work near the projectile–target interface is less than if there were no compression. As already discussed in the preceding paragraph, additional plastic work does occur during unloading, but it occurs away from the projectile–target interface and does not directly influence penetration. Therefore, compressibility tends

TABLE 7.7. Comparing volumetric elastic strains and equivalent plastic strains that produce the same energy density, and corresponding temperature changes

Isentropic elastic ΔT (K)	Volumetric elastic strain $1 - \rho_0/\rho$	Energy density (J/cm^3)	Equivalent plastic Strain ε_p	Dissipated plastic ΔT (K)
7	0.010	7.6	0.0076	2
17	0.025	47.5	0.048	14
34	0.050	190	0.19	55
53	0.075	428	0.43	124
72	0.100	760	0.76	220

to decrease the extent of plastic dissipation in the vicinity of the projectile–target interface as compared to an incompressible case.

An estimate of the amount of energy per unit volume stored in elastic compression can be made from Eq. (3.122), using a linear relationship between stress and strain:

$$\rho_0 E_e = \rho_0 \int_{\rho_0}^{\rho} \sigma d(1/\rho) \approx - \int_{\rho_0}^{\rho} K \left(1 - \frac{\rho_0}{\rho} \right) d(\rho_0/\rho) = \frac{K}{2} \left(1 - \frac{\rho_0}{\rho} \right)^2, \quad (7.15)$$

where K is the bulk modulus. This expression is an estimate because only the linear term of the equation of state has been retained. Since the plastic work is approximately the area under the stress-strain curve, given by $Y_t \varepsilon_p$, it is possible to determine the corresponding strains that give rise to the same energy densities, as is done in Table 7.7 using the bulk modulus for steel of 152 GPa and an effective flow stress of 1 GPa. Also shown on the table is the temperature rise due to isentropic compression (Eq. 5.152) assuming $\Gamma_0 = 2.17$ and an initial temperature of 298 K, and the temperature rise due to plastic dissipation assuming $\rho_0 = 7.85$ g/cm^3 and $C_v = 0.44$ J/g K, as in Table 7.1. The temperature rise due to elastic compression would reverse itself when unloading to ambient pressure occurs, while the plastic dissipation produces irreversible heating.

Figure 7.18 provides an estimate of the compression that occurs in the target. At 1.5 km/s, the amount of volumetric strain in the target directly in front of the projectile is 1 to 2%. At 3.0 km/s, the amount of volumetric strain in the target directly in front of the projectile is 2 to 5%, with a very small region exceeding 5% volumetric strain. At 4.5 km/s, the amount of volumetric strain in the target directly in front of the projectile is 5 to 10%, with a small very small region exceeding 10%. Thus, the density does not change a great deal, but the compressions for the high velocities do represent (locally) a large amount of energy. Moving away from the high-pressure region at the projectile–target interface, the pressures quickly decay and so does the compression of the target.

As the projectile penetrates, it does work on the target. Within the target, the energy is partitioned into three modes: kinetic energy in the target, energy dissipated through plastic flow, and energy stored in elastic deformation. To explicitly look at the energy partitioning within the target, a different set of calculations was performed with CTH for an $L/D = 10$ tungsten alloy projectile of length 10 cm impacting a steel target. The tungsten alloy was modeled with a Mie–Grüneisen equation of state as given in Table 7.1, except that the initial density was 17.6 g/cm^3. An elastic-perfectly plastic constitutive model with a flow stress of 1.5 GPa and Poisson's ratio 0.30 was used. The steel was modeled with a Mie–Grüneisen equation of state as given in Table 7.1 and an elastic-perfectly plastic constitutive model with a flow stress of 1.2 GPa and Poisson's ratio 0.29. These computations have a slightly longer projectile and slightly different constitutive models that the previous computations. The various components to the energy are shown in Fig. 7.19 for impact velocities of 1.5, 3.0, and 4.5 km/s. The projectile kinetic energy decreases almost linearly during the impact event, mostly through erosion, which decreases the length of the projectile at a nearly constant rate. The total target energy increases in almost like fashion. The target energy is further broken out into components of elastic, kinetic, and energy dissipated through plastic flow. The total target energy is noticeably less for the 1.5 km/s impact, where the amount of energy lost to plastic deformation of the projectile is about 11%. The total energy dissipated internally within the projectile is nearly independent of impact velocities – about 20 KJ for this projectile – so as the impact velocity goes up the relative percentage goes down. Equating this energy rise to plastic work gives an average strain in the eroded projectile material of 170%. Large amounts of kinetic energy occur in the target at the higher impact velocities. For both the 1.5 and 3.0 km/s impact cases, the amount of kinetic energy in the target reaches steady state, meaning that the amount of energy initially deposited in kinetic energy is balanced by kinetic energy that is being dissipated through plastic flow. For the 4.5 km/s case, the kinetic energy is still increasing when the projectile completely erodes at about 55 μs, showing that more energy is being deposited in kinetic form than is being dissipated. If the projectile had been a larger L/D, at a later time steady state would have been achieved and the curve would have flattened out. The amount of energy in elastic deformation is actually quite small, since the energy is localized to the high-pressure region near the projectile–target interface. For the elastic compressive energy, the amount of energy deposited in this form has reached a balance with the amount transferred to other means (kinetic and plastic dissipation) for all three impact velocities. After energy is initially deposited in the kinetic and compressive energy states, it is then dissipated through plastic flow. In the end, nearly all the energy ends up in internal energy in the form of heat due to plastic deformation of the target and projectile and in kinetic energy due to any late-time projectile debris and full target motion. Little energy is stored in permanent elastic deformation.

7.10. The L/D Effect

As an application of our phases of penetration, we examine the role of the projectile diameter D in the final normalized depth of penetration P/L. We begin with a specific set of computations. A 4340 steel intended to be 1.5 times as strong as Rc 30 was used for the target and a tungsten alloy was used for the projectile to compare to armor penetration experiments. A Mie–Grüneisen equation of state and the Johnson–Cook plasticity model were used with parameters given in Tables 7.1 and 7.8. The 4340 steel constitutive properties for the material intended to be 1.5 times as strong as Rc 30 were obtained by multiplying the original Johnson–Cook

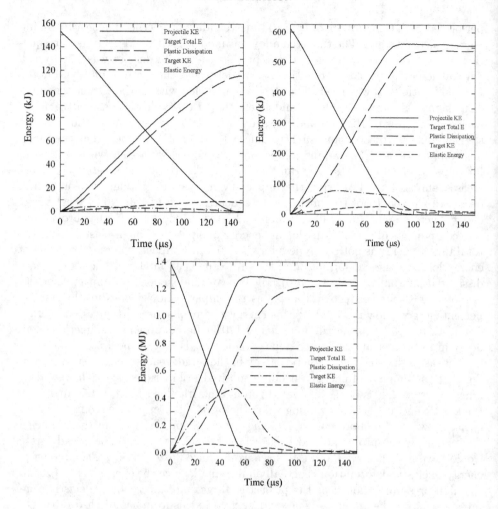

FIGURE 7.19. Energy partition an $L/D = 10$ tungsten alloy projectile of length 10 cm striking a steel target at 1.5 km/s (upper left), 3.0 km/s (upper right), and 4.5 km/s (lower). The projectile kinetic energy is seen to almost linearly decrease as the projectile erodes, and the total target energy increases accordingly. Eventually, most of the energy is internally dissipated through plastic flow, though kinetic and elastic compression energy components are important transient states. (From Walker [**194**].)

A and B (obtained for a 4340 steel with Rc 30) by 1.5; the original values are shown for comparison [**105**]. CTH's tensile stress limiter (fracture) model was used with a maximum tensile stress (fracture stress) of 2 GPa.[2] Geometrically, seven zones were used to resolve the projectile radius and the zoning was square in the

---

[2]This numerical failure model does not allow the stress to become more positive than a stated value through the mechanism of inserting a sufficient volume fracture of void into the computational cell to increase the density of the material so that the material's pressure increases and thus decreases the positive-in-tension total stress to allowable levels.

TABLE 7.8. Constitutive parameters.

Material	A (GPa)	B (GPa)	n	C	m	T_{melt} (K)	ν
Tungsten Alloy[a]	1.51	0.177	0.12	0.016	1.00	1,752	0.30
4340 Steel (1.5×Rc 30)	1.189	0.765	0.26	0.014	1.03	1,793	0.29
4340 Steel (Rc 30)[a]	0.792	0.510	0.26	0.014	1.03	1,793	0.29

[a]These are original Johnson–Cook parameters; see Appendix C

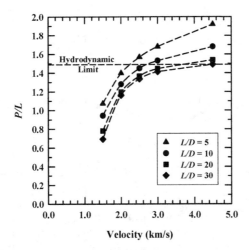

FIGURE 7.20. Final depth of penetration results of numerical simulations for impact velocities 1.5, 2.0, 3.0, and 4.5 km/s and L/Ds 5, 10, 20, and 30. (From Anderson et al. [15].)

TABLE 7.9. P/L from numerical simulations.

L/D	Impact velocity (km/s)				
	1.5	2.0	2.5	3.0	4.5
5	1.073	1.398	1.568	1.680	1.921
10	0.945	1.284	1.448	1.533	1.681
20	0.781	1.197	1.367	1.445	1.539
30	0.680	1.163	1.336	1.412	1.490

interaction region. The projectile L/D ranged from 5 to 30 and the impact velocity varied from 1.5 to 4.5 km/s.

The final depths of penetration for these computations are shown in Fig. 7.20 and Table 7.9. Examining the depths of penetration, it is clear that for all impact speeds the final depth of penetration depends upon the L/D of the projectile. This result is quite interesting, especially in the context of the Tate model to be described in the next chapter, because the Tate model has no projectile diameter dependence. That P/L depends on projectile diameter is referred to as the L/D effect.

To explore where the length and diameter of the projectile enter into the penetration event, we consider the penetration speed as the projectile penetrates into

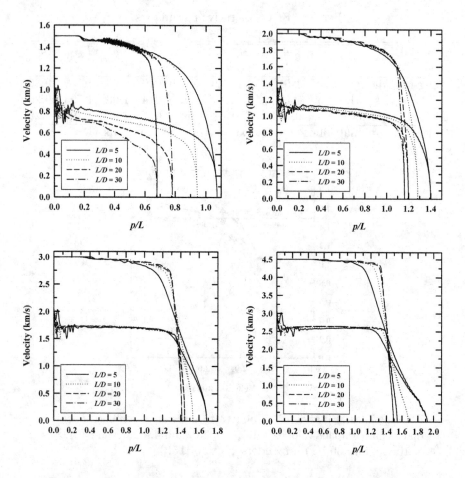

FIGURE 7.21. Penetration and projectile tail velocities vs. depth of penetration normalized by original length $(P(t)/L)$ for impact velocities 1.5, 2.0, 3.0 and 4.5 km/s and L/Ds 5, 10, 20, and 30. Note that the horizontal axis extent differs in each plot. (From Anderson et al. [15].)

the target. Figure 7.21 displays the penetration (nose) and tail velocities along the projectile–target centerline as a function of normalized penetration depth $P(t)/L$ for the computations. Each frame is for a given impact speed; since the penetration is normalized by projectile length, essentially there are four different projectile diameters shown in each frame and the projectiles all have the same length. For each impact speed we see that the elastic wave reflection on the back of the projectile is arriving at the same time (displayed penetration location) for each diameter. Thus, the back end of the projectile deceleration scales off projectile length L. However, it can be seen that at the beginning of the impact event, the penetration speeds do not overlay for the different projectile diameters D. Thus, the early time penetration is not scaling off the projectile length. The terminal phase of penetration is characterized by a rapid decay in both penetration and projectile tail velocities.

TABLE 7.10. P/L from numerical simulations for 1.5 km/s impacts.

L/D	$D = 0.3$ cm P/L	$L = 3$ cm P/L
3	1.139	1.157
6	1.010	1.009
12	0.893	0.887
10	0.923	0.923
20	0.790	0.781
30	0.690	0.680

The projectile tail velocities in Fig. 7.21 do not have the same slopes; thus the terminal phase also is not scaling with the projectile length.

Before the arrival of the first reflected elastic wave from the projectile tail at the projectile–target interface, the only finite geometric extent in the impact event is the projectile diameter D. Thus, at early time the penetration velocities should agree if the depths of penetration are normalized by the projectile diameter. That this behavior indeed occurs can be seen in Fig. 7.22 for impacts at 1.5 km/s: in this figure, the penetration velocities are shown for constant diameter projectiles. At early times before the arrival of the reflected elastic wave from the back end of the projectile, the penetration velocity is the same for all projectiles. Thus, at early time penetration behavior scales with projectile diameter. The terminal phase deceleration of the projectile tail velocities in Fig. 7.22 shows the velocities in the terminal phases are almost parallel to each other, an indication that the terminal phase is scaling with the projectile diameter D. This behavior is what we would expect, since the terminal phase is essentially the penetration of the final $L/D = 1$ residual projectile or the continued motion of the crater after the projectile has completely eroded, and that crater bottom is characterized by its diameter. In a contrasting fashion, Fig. 7.22 shows that at early time the tail velocities do not agree for constant diameter projectiles of differing length. Thus, we see how the two different length scales are appearing in the penetration process, both the length and the diameter of the projectile, and so it is not surprising that both length scales affect the final depth of penetration.[3]

For projectiles of the same diameter, since the penetration velocities are the same before the axial stress relief wave arrives from the rear of the projectile, we also have that the depth of penetration vs. eroded length will be the same (before v changes, both depend only on $\int_0^t u\,dt$). Pursuing this observation, Fig. 7.23 is a plot of $P(t)/D$ vs. the normalized eroded length of the projectile $L_e/D = (L_0 - L)/D$. For a given impact velocity, for most of the penetration these curves overlay each other. The primary component of these curves occurs in the quasi-steady-state phase of penetration. A more detailed plot for 1.5 km/s with more L/Ds is also given in Fig. 7.23. (The bending up of the $P(t)/D$ vs. L_e/D curve at the end of an specific L/D trace occurs during the terminal phase of penetration, and represents essentially rigid-body penetration.) Thus, the $P(t)/D$ curve for quasi-steady-state

---

[3]Final depths for these computations are shown in Table 7.10. They were performed with constant diameter and then constant length projectiles, with seven cells across the radius of the projectile. The Johnson–Cook model includes a strain rate term (Table 7.8), the only source in the numerical computation of a size dependence scaling with D. The variation is at most 1.5%.

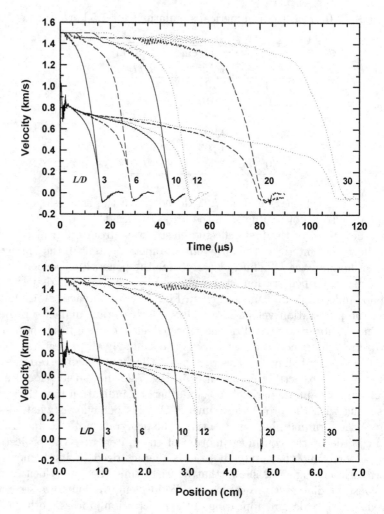

FIGURE 7.22. Penetration and projectile tail velocities vs. time and penetration depth for constant diameter tungsten projectiles striking steel at 1.5 km/s ($D = 0.3$ cm). (From Anderson et al. [14].)

penetration can be written as a function of L_e/D,

$$\left(\frac{P(t)}{D}\right)_{\text{steady state}} = f\left(\frac{L_e(t)}{D}\right). \tag{7.16}$$

This form yields

$$\frac{dP}{dL_e} = \frac{u}{v - u} = f'\left(\frac{L_e}{D}\right), \tag{7.17}$$

since $dP/dt = u$ and $dL_e/dt = v - u$, where the prime means differentiation with respect to the argument. Thus, we expect the dP/dL_e penetration efficiency curves to overlay. That this conjecture is indeed the case is shown in Fig. 7.24, which plots the instantaneous measure of penetration efficiency for eroding penetration dP/dL_e for the four L/Ds at four impact speeds. A qualitative transition is evident as the

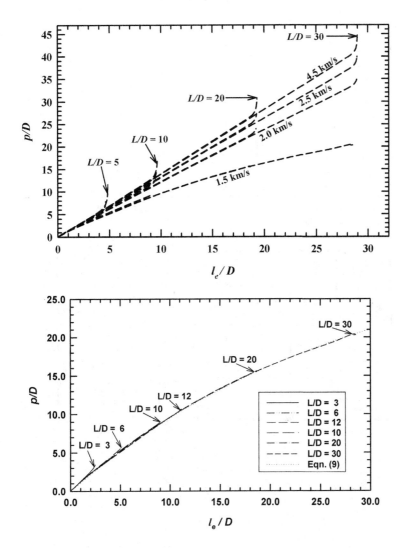

FIGURE 7.23. Depth of penetration vs. projectile eroded length, both in terms of projectile diameters, with the lower plot providing more detail for 1.5 km/s (see also Exercise 7.14). (From Anderson et al. [**14, 15**].)

impact velocity increases. At 1.5 km/s, the penetration efficiency decreases nearly linearly with $P(t)/D$ for all four L/Ds, while for the higher impact velocities the penetration efficiency is nearly a constant. The vertical lines in the figures occur at the terminal phase of penetration, when the projectile ceases eroding – that is, $v = u$. When $f'' \neq 0$, there is an L/D effect in the steady state portion of penetration.

Since the terminal phase deviates from the quasi-steady-state portion of penetration, it suggests that the final penetration depth be written

$$P(v) = P_{\text{steady state}}(v) + P_{\text{terminal}}(v). \tag{7.18}$$

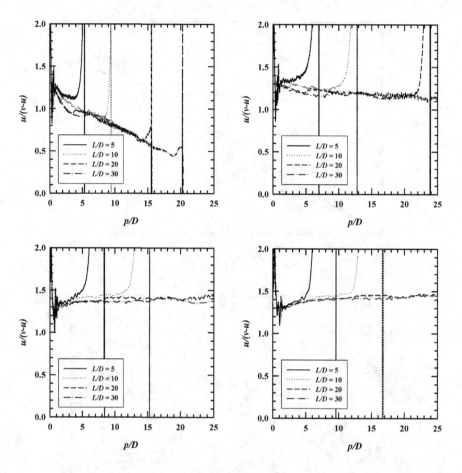

FIGURE 7.24. Penetration efficiency dP/dL_e vs. normalized depth of penetration for L/Ds 5, 10, 20, and 30 at 1.5, 2.0, 3.0, and 4.5 km/s. (From Anderson et al. [15].)

The fact that the $P(t)/D$ vs. L_e/D curves overlay at early times allows us to include the initial transient in the quasi-steady-state penetration phase expression. In Figs. 7.23 and 7.24, the nominal steady-state phase of penetration has two qualitatively different behaviors, depending on the impact velocity. For the low velocities (1.5 km/s and mildly for 2.0 km/s), the $P(t)/D$ vs. eroded length curve is concave down, and dP/dL_e decreasing with penetration depth, indicating a diminishing return on penetration as the projectile is consumed. The low impact velocity cases manifest an L/D effect during the quasi-steady-state portion of penetration. For the higher velocities (2.5, 3.0, and 4.5 km/s) the steady-state portion of the $P(t)/D$ vs. eroded length curve is a straight line (actually, it is slightly concave upwards at 4.5 km/s) and dP/dL_e is constant. Thus, there is no L/D effect in the steady-state portion of the calculation for these higher velocities. Inspection of Fig. 7.21 implies that for high-speed impacts the L/D effect is due to the final transient phase of primary penetration, while for the lower velocities it also depends on the steady-state phase of penetration.

TABLE 7.11. Penetration constants from the numerical simulations.

Velocity (km/s)	a_1	a_2	g
1.5	1.091	−0.0135	0.82
2.0	1.236	−0.00249	1.54
2.5	1.318	—	1.74
3.0	1.37	—	2.05
4.5	1.41	—	3.02

TABLE 7.12. Estimate of penetration velocity at beginning of terminal phase.

	Impact velocity (km/s)				
L/D	1.5	2.0	2.5	3.0	4.5
5	0.741	1.10	1.42	1.73	2.63
10	0.688	1.09	1.42	1.73	2.63
20	0.554	1.07	1.42	1.73	2.63
30	0.367	1.04	1.42	1.73	2.63

Although Eq. (7.16) introduced a generic functional form for the $P(t)/D$ curve, to be more specific for the calculations here, the $(P(t)/D)_{\text{steady state}}$ vs. eroded length curve can be written as

$$\left(\frac{P(t)}{D}\right)_{\text{steady state}} = a_1(v)\frac{L_e}{D} + a_2(v)\left(\frac{L_e}{D}\right)^2, \tag{7.19}$$

where a_1 and a_2 depend on the impact velocity but not on the length or diameter of the rod. For low velocities, $a_2 < 0$, corresponding to the curve being concave down. At higher velocities, $a_2 \approx 0$. Least-squares fits to the quasi-steady-state portions of penetration provide the values of a_1 and a_2 given in Table 7.11. These values tend towards the hydrodynamic limit of $a_1 = \sqrt{\rho_p/\rho_t} = 1.47, a_2 = 0$.

Another observation from Fig. 7.23 is that at the higher velocities, the final penetrations are nearly given by a constant offset from the P/D vs. eroded length curve. Since the terminal phase of penetration scales with D, it is possible to approximate this observation as

$$P_{\text{terminal}} = g(v)D. \tag{7.20}$$

The value for g depends on impact velocity, but not the length or diameter of the rod. More properly, g depends on the velocity at the end of the steady-state penetration as the terminal phase begins, as indicated in Table 7.12, but we are not going to pursue that finer detail. Values for g are given in Table 7.11. Equations (7.16), (7.19), and (7.20) allow the final penetration to be written as

$$P = P_{\text{steady state}} + P_{\text{terminal}} = a_1(v)(L - L_r) + a_2(v)\frac{(L - L_r)^2}{D} + g(v)D, \tag{7.21}$$

where L_r is the residual projectile length at the end of penetration. Though approximate, this equation shows the role of both projectile length and diameter in the penetration.

We have shown that there is a change in mechanism for the L/D effect as velocities transition from lower velocities to higher velocities, with the different mechanism invoked in different phases of penetration.

GENERAL PRINCIPLE 7.4. *(the L/D effect) For eroding penetration, at lower velocities the penetration velocity u deceleration scales with projectile diameter during the "steady state" portion of penetration, and this deceleration is the major component of the L/D effect. As velocities increase, the penetration velocity u becomes constant during the steady state portion of the penetration and the L/D effect is due to the terminal phase of primary penetration.*

To be explicit, looking at the terms in the Walker–Anderson model derived in Chapter 10, the centerline momentum balance roughly states

$$\int \rho \frac{Dv_z}{Dt} dz = \text{constant} \cdot \rho R\dot{u} - \frac{1}{2}\rho_p(v-u)^2 + \frac{1}{2}\rho_t u^2 - Y_p = \int \frac{\partial \sigma_{iz}}{\partial x_i} dz = -\frac{7}{3}\ln(\alpha)Y_t,$$

$$(7.22)$$

and since the crater radius R directly depends on the projectile diameter D, we see that the deceleration u scales with $1/R$ and is the origin of the L/D effect for lower velocity impacts. Further, the penetration speed does not tend to an equilibrium speed because the $\frac{7}{3}\ln(\alpha)Y_t$ term increases with decreasing penetration speed, thus encouraging a more negative $\dot{u}$.

Though we have described the L/D effect by considering numerical simulations, in light of the fact that for many years it was generally believed that the L/D effect saturated for $L/D > 10$ it is helpful to look at penetration data. When looking at "second order effects" like the L/D effect in experiments, care must be exercised regarding the actual materials used, since subtle effects in the penetration process can easily be masked by small variations in target and projectile strength and/or density. Data carefully collected from various tungsten alloy projectiles striking RHA and RHA-like targets show that the effect on penetration due to differences in density (tungsten content) for these alloys are within the experimental scatter, but differences in target hardness and/or projectile flow characteristics (projectile strength) are large enough to compete with the L/D effect. In Fig. 7.25 (left), normalized penetration as a function of impact velocity is compared for L/Ds of 10, 15, and 30 (Woolsey and Magness and Farrand) for the specific tungsten alloy X9C, which is 97% by weight tungsten contained in a Ni-Fe-Co matrix, with $\rho = 18.61$ g/cm^3 and strength 985 MPa. The dashed lines are linear least-squares fits of P/L as a function of impact velocity for $V \leq 1.8$ km/s. Focusing on 1.5 km/s, the experimental results roughly give that when the L/Ds goes from 10 to 15 there is a 4% decrease in P/L, when L/D goes from 15 to 30 there is a 22% decrease in P/L, and when L/D goes from 10 to 30 there is a 26% decrease in P/L. Thus the L/D effect is quite large.

These data and other tungsten alloy striking steel data were used to plot P/L as a function of L/D for three impact velocities: 1.2, 1.5, and 1.8 km/s. The results along with computational results for 1.5 km/s are plotted in Fig. 7.25 (right). The figure shows that P/L is more sensitive to L/D for low L/D. The curves appear to translate upward in P/L with increasing impact velocity without much change in shape. Shown in the figure is a curve fit to the data:

$$\frac{P}{L} = -0.209 + 1.044V - 0.194\ln\left(\frac{L}{D}\right),$$

$$(7.23)$$

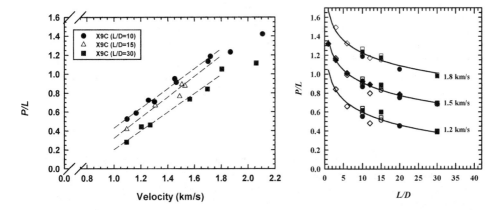

FIGURE 7.25. Left: experimental observation of the L/D effect for L/Ds 10, 15, and 30 projectiles made of X9C tungsten alloy (97% by weight tungsten in a Ni-Fe-Co matrix). Right: P/L vs. L/D for tungsten alloy striking RHA steel; closed diamonds at 1.5 km/s are computations, while all other symbols are data points. The lines are from the curve fit Eq.(7.23). (From Anderson et al. [14].)

where V is in km/s. This fit addresses the main effect of steady-state phase, and should not be used at higher velocities.

7.11. Hypervelocity Impact

As to other features in these numerical simulations, it is evident that for the 1.5 km/s impact and to a lesser extent the 2 km/s impact, the penetration speed decreases noticeably with penetration depth, whereas for the higher velocity impacts, the steady-state phase of penetration has a nearly constant penetration speed. The decrease in u is related to target strength. We use this change in behavior to define the hypervelocity impact regime, because in some sense it demarcates when the target and projectile strength play a primary role in the depth of penetration,

DEFINITION 7.5. (hypervelocity impact) An impact event is referred to as a *hypervelocity impact* when impact velocity is such that the steady-state phase of penetration has nearly constant penetration speed u.

Strength effects still matter, especially with respect to crater radius. In the event of a thin target or a low L/D projectile, replace the target with a thick target of the same material and replace the projectile with a large L/D projectile of the same material, thus making the definition strictly dependent on the impact velocity and the target and projectile material properties and independent of the specific target and projectile geometry. Based on results shown in the previous sections, in the impact community the transition to hypervelocity impact is usually taken as 2.0 km/s. Another reason is that it is at this speed that it is necessary to start using specialty launchers. Hence, when the word "hypervelocity impact" is used, people usually mean that the impact speed is at or above 2 km/s.

7.12. Sources

The analytical curve fits are from Anderson and Walker [13]. The computations in this chapter are from two papers that analyzed eroding penetration, Anderson and Walker [12], Anderson, Littlefield, and Walker [8], two papers discussing the L/D effect, Anderson, Walker, Bless, and Partom [14] and Anderson, Walker, Bless, and Sharron [15], and the energy partition from Walker [194].

Exercises

(7.1) Why does it sting when you slap your hand on the water or do a bellyflop off a diving board, but it does not continue to sting as your hand or body moves through the water? Estimate the stresses involved.

(7.2) If the eroding front of the projectile penetrates the target at a constant velocity u and the rest of the projectile is traveling at a constant velocity v, what is P/L? (Assume penetration ceases when there is no projectile remaining.) Use this result to determine $(P/L)_{\text{hyd}}$.

(7.3) How much must the density of the target material increase to reduce the hydrodynamic penetration by a factor of 2? What does this say about the total weight of the armor to stop the same projectile?

(7.4) The areal density of a target is given by its thickness T times its density. What does the hydrodynamic penetration predict is the optimal areal density of a target to stop a given round (ignore bulge and breakout at the back of the target, so that the required thickness of material is given by P/L)? What is the corresponding density and thickness? Is this result reasonable? Why not?

(7.5) The previous exercise implies that an optimum armor has low density and large thickness. Nature provides this type armor for our protection against meteoroids, namely the atmosphere. What is the areal density of the atmosphere protection layer at sea level? (Hint: with h the height above sea level, pressure at sea level is related to the mass of air above by $p = \int \rho g dh \approx g \int \rho dh$. The areal density is $\int \rho dh = p/g$.) How does this compare to one meter of steel?

(7.6) After the penetrator fails in the steel into aluminum penetration of Fig. 7.1, is the penetration modeled well by the hydrodynamic approximation? (In other words, does the strength of the projectile still matter?)

(7.7) For eroding penetration, as described in Eq. (7.1), penetration continues to increase as the impact velocity increases. Of course, launching projectiles becomes more difficult as the velocity increases. In this exercise, take two approaches to the projectile launch constraint. First assume that projectiles can be launched at a given momentum – that is, $mv = $ constant – so for a higher velocity launch, the mass is less. Second, assume that projectiles can be launched at a given energy, $\frac{1}{2}mv^2 = $ constant. Now assume that the projectile is geometrically self similar – that is, it maintains its shape but is larger or smaller depending on the mass (i.e., L/D is constant but $L \sim m^{1/3}$). Show that, for $P/L = f(v)$, these assumptions say that the maximum depth is achieved when $v^{-1/3}f(v)$ (constant momentum launch) and $v^{-2/3}f(v)$ (constant energy launch) are maximum. For the three L/Ds of tungsten impacting steel in Fig. 7.4, determine the maximum penetration for constant momentum launch and constant energy launch (hint: they are between 2.5 and 3.0 km/s and between 2.0 and 2.5 km/s, respectively).

(7.8) In Chapter 10 it is argued that the eroded debris speed in the Eulerian (laboratory) frame is given by $v_d = 2u - v$. Given that tungsten piles up at the bottom of the crater in Fig. 7.5, we know that $v_d > 0$. However, by looking at, for example, Fig. 7.22, we see there is a problem. What is the problem? The argument in Chapter 10 says to go to a frame moving with the front of the projectile. There, the projectile is coming in at a speed $v - u$ and the target is coming in at a speed $-u$. It seems reasonable that the projectile material will exit with a speed $-(v-u)+v_{loss}$, where $v_{loss} < v-u$ represents energy and friction losses that occur while the projectile material erodes and changes direction. Hence $v_d = 2u - v + v_{loss}$. Estimate v_{loss} for tungsten into steel for the four penetration cases of Fig. 7.5 using computational results in this chapter to estimate u and v. Next, based on your reasoning and on Fig. 7.24, determine whether hypervelocity impacts will see tungsten pile up at the bottom of the crater or not.

(7.9) Assuming RHA strength as a function of thickness is as given in Table 3.6, and P/L as a function of impact speed and target strength is given in Eqs. (7.1) and (7.3), determine the apparent scale effect (i.e., the P/L dependence on projectile length) for an $L/D = 10$ tungsten projectile striking RHA at 1.5 km/s for $L = 1$ cm, $L = 3$ cm, $L = 10$ cm, $L = 20$ cm, and $L = 1$ m.

(7.10) Show for the hydrodynamic approximation the following symmetric forms of the interface stress,

$$\tilde{\sigma}_{zz}(z_i)_{hyd} = \frac{1}{2} \frac{\rho_p \rho_t v^2}{\left(\sqrt{\rho_t} + \sqrt{\rho_p}\right)^2} = \frac{\frac{1}{2}\sqrt{\rho_p \rho_t}\, v^2}{2 + \sqrt{\frac{\rho_t}{\rho_p}} + \sqrt{\frac{\rho_p}{\rho_t}}} = \frac{1}{2}\sqrt{\rho_p \rho_t} \left(\frac{v}{\sqrt[4]{\frac{\rho_t}{\rho_p}} + \sqrt[4]{\frac{\rho_p}{\rho_t}}}\right)^2.$$

$$(7.24)$$

(7.11) In looking at Fig. 7.21, estimate the terminal phase of penetration in terms of projectile diameter. Is it constant (i.e., independent of projectile length)?

(7.12) To explicitly compute the integrals in the centerline momentum balance, it was assumed that the target and project densities were constant so they could be pulled out of the integral – that is, it was assumed the material was incompressible. One way to see if this is a good approximation is to compute the volumetric strain at its maximum – the projectile–target interface – and see how much the density might change, and thus how much the density terms might change. According to Fig. 7.16, what is the pressure at the steel–tungsten interface for a 1.5 km/s impact? Given the bulk modulus of steel, what is the volumetric strain? What is the corresponding compressed density of steel? What is the percentage change? Do you think this amount is a concern?

(7.13) The previous problem considered compressibility at typical tank ordnance velocities, 1.5 km/s. What about higher velocities? For 3 and 5 km/s impact velocities, compute the approximate pressure at the projectile–target interface for a tungsten into steel impact. Then, assuming a linear elastic relationship between the pressure and the volumetric strain, compute the compressed density and the percentage change, as in the previous exercise. Does the fact that the material response is actually nonlinear affect this result very much?

(See Eq. 5.124 and the its following isothermal comment or Table 5.3 to see a minimum compression.)

(7.14) Looking at Fig. 7.24, the penetration efficiency in the steady-state phase of penetration appears to be a linear function of penetration depth, or

$$\frac{u}{v - u} = a - b\frac{P(t)}{D}, \tag{7.25}$$

where a and b are appropriate constants that depend on impact velocity. Assuming $b \neq 0$, integrate this equation to find penetration as a function of eroded projectile length. What is the maximum penetration depth? This curve is plotted for 1.5 km/s in Fig. 7.23 using $a = 1.127$ and $b = 0.03455$ (it is labeled "Eqn. (9)"). Does it give a good final depth of penetration?

(7.15) For both ordnance and hypervelocity impact regimes, what is the expected L/D behavior of the residual length of the projectile?

(7.16) Show that if $\dot{v}$ is a function of projectile residual length and not of projectile diameter, then $\dot{u} = 0$ implies that there is no L/D effect.

(7.17) Suppose the steady-state portion of the penetration at lower velocities could be written as a function of velocity times the eroded length of the projectile with the assumption that the residual projectile had an $L/D = 1$. The total penetration is then

$$P = f(V) \cdot (L - D) + g(V)D. \tag{7.26}$$

Though this equation may seem similar to Eq. (7.21), it is in fact quite different. Obviously f and g depend on the projectile and target strength and density, but for now we assume those properties are constant. We expect that f and g are on the order of 1. Normalizing by the original length of the projectile gives

$$\frac{P}{L} = f(V)\left(1 - \frac{1}{L/D}\right) + \frac{g(V)}{L/D}. \tag{7.27}$$

Show that this equation cannot give rise to the observed L/D effect at lower velocities, and argue that hence the L/D effect at lower velocities cannot be explained by the terminal phase of penetration.

(7.18) The tungsten alloy described in Table 7.2 exhibits little work hardening and has a measured flow stress of 1.35 GPa at static strain rates, 1.90 GPa at 960 s^{-1}, and 2.15 GPa at 5,220 s^{-1}. What are good values for A, B, and C in the Johnson–Cook model? Were good ones used, based on the values in the table?

(7.19) Assume that one has projectiles that are right-circular cylinders of tungsten of constant mass m_p but with various L/D. Show that Eq. (7.23) implies that a maximum absolute depth of penetration P (absolute means the depth not normalized by L) can be achieved by these projectiles. Derive the equation for the L/D at the maximum depth and the maximum depth. For a tungsten projectile of mass 65 g striking at 1.5 km/s, what is the depth of penetration into steel?

The Tate–Alekseevskii Model

The first model with widespread success was for projectiles with a high length-to-diameter (L/D) ratio, developed independently by Tate [179, 180] and Alekseevskii [1]. To begin the discussion, and to see what insights were used in the development of the theory, the Bernoulli equation from fluid mechanics is derived, and then the development of the Tate model (as it is typically called) is presented. The model is used to examine penetration into semi-infinite (that is, very thick with respect to the length of the projectile) targets. The behavior of the Tate model is explored in detail and is compared to penetration data. Many of the data are from the work of Hohler and Stilp at the Ernst-Mach-Institut, in Freiburg, Germany. Their data confirmed the S-curve penetration behavior for strong metals.

8.1. Bernoulli's Equation for Steady Flow

To begin, we develop Bernoulli's equation from fluid mechanics. As a first step, we write the acceleration term in a slightly different way. Since

$$\frac{1}{2}\frac{\partial(v_j v_j)}{\partial x_i} = v_j \frac{\partial v_j}{\partial x_i}, \tag{8.1}$$

the acceleration can be written

$$\frac{Dv_i}{Dt} = \frac{\partial v_i}{\partial t} + v_j \frac{\partial v_i}{\partial x_j} = \frac{\partial v_i}{\partial t} + \frac{1}{2}\frac{\partial(v_j v_j)}{\partial x_i} + v_j\left(\frac{\partial v_i}{\partial x_j} - \frac{\partial v_j}{\partial x_i}\right). \tag{8.2}$$

The parenthesis term on the far right-hand side contains the components of the curl of the velocity. Using the spin tensor W (Eq. 2.32), it equals $v_j 2W_{ij} = 2W_{ij}v_j$. An interesting general feature of a skew symmetric matrix is that

$$2W\vec{v} = \begin{pmatrix} 0 & -w_3 & w_2 \\ w_3 & 0 & -w_1 \\ -w_2 & w_1 & 0 \end{pmatrix}\begin{pmatrix} v_1 \\ v_2 \\ v_3 \end{pmatrix} = \begin{vmatrix} \hat{e}_1 & \hat{e}_2 & \hat{e}_3 \\ w_1 & w_2 & w_3 \\ v_1 & v_2 & v_3 \end{vmatrix} = \vec{w}\times\vec{v} = -\vec{v}\times\vec{w}. \tag{8.3}$$

For the spin tensor $\vec{w} = \frac{1}{2}\text{curl}(\vec{v})$ and hence $(2W_{ij}v_j) = -\vec{v}\times\text{curl}(\vec{v})$. Thus, the acceleration can be written in vector form as

$$\frac{D\vec{v}}{Dt} = \frac{\partial\vec{v}}{\partial t} + \frac{1}{2}\text{grad}(|\vec{v}|^2) - \vec{v}\times\text{curl}(\vec{v}). \tag{8.4}$$

The Bernoulli equation is a statement of the conservation of momentum subject to certain restrictions on the flow field. In particular, the assumptions are

(1) The fluid is inviscid, which means that the stress tensor can be written

$$\sigma_{ij} = -p\delta_{ij}. \tag{8.5}$$

Inviscid means that there are no shear viscosity terms in the fluid's response.

(2) The body force is conservative, which means that it can be expressed in terms of a potential function. In particular, with the potential Ψ,

$$b_i = -\frac{\partial \Psi}{\partial x_i}. \tag{8.6}$$

For example, the gravitational force can be written as

$$\Psi_{\text{grav}} = gz. \tag{8.7}$$

The force per unit mass is then

$$b_x = b_y = 0, \quad b_z = -\frac{\partial}{\partial z}(gz) = -g. \tag{8.8}$$

The force per unit volume is $\rho b_z = -\rho g$.

(3) The flow is incompressible, meaning that $\text{div}(\vec{v}) = 0$ or, equivalently, that $\rho = \rho_0$, a constant.

(4) The flow is steady, meaning that $\partial \vec{v}/\partial t = 0$.

When divided by the density, the momentum balance is

$$-\frac{1}{\rho}\frac{\partial p}{\partial x_i} + b_i = \frac{Dv_i}{Dt}. \tag{8.9}$$

Inserting the acceleration expression and the assumptions above gives

$$-\text{grad}(\frac{p}{\rho_0}) - \text{grad}(\Psi) = \frac{1}{2}\text{grad}(|\vec{v}|^2) - \vec{v} \times \text{curl}(\vec{v}). \tag{8.10}$$

Collecting terms yields

$$\text{grad}(\frac{1}{2}|\vec{v}|^2 + \frac{p}{\rho_0} + \Psi) = \vec{v} \times \text{curl}(\vec{v}). \tag{8.11}$$

Now, a particle of the fluid follows a streamline. Since locally the particle of fluid is always traveling at velocity $\vec{v}$, the streamline has the property that it is tangent to the velocity $\vec{v}$. Because of this, along the streamline, the inner product of the velocity with the term on the right is zero owing to the velocity being in the cross product: $\vec{v} \cdot \vec{v} \times \text{curl}(\vec{v}) = 0$. Thus, along a streamline,

$$\vec{v} \cdot \text{grad}(\frac{1}{2}|\vec{v}|^2 + \frac{p}{\rho_0} + \Psi) = 0. \tag{8.12}$$

Moving along a streamline, the components of the gradient perpendicular to the velocity do not enter into this dot product, which says that the component of the gradient term in the tangential direction must be zero. Thus, the term in parentheses on the left-hand side is a constant along the streamline. Multiplying through by the constant density,

$$\frac{1}{2}\rho_0 v^2 + p + \rho_0 \Psi = \text{constant} \tag{8.13}$$

yields the final form of the Bernoulli equation.

8.2. The Tate Model

We first consider the case of a fluid jet impinging on a rigid wall. We assume the jet is coming in from the left at velocity v. Along the centerline of the jet, far away from the wall, the velocity of the jet is v. Following a fluid particle directly on the centerline, as the fluid particle approaches the wall, it decelerates. Since it started on the centerline, it stays on the centerline. Fluid radially sprays away from the impact point, but for our fluid particle it continues to slow down until it comes to rest right at the rigid wall. Our fluid particle followed a streamline and so the Bernoulli equation holds. Denoting the far condition to the left with ℓ and the condition near the wall (to the right) with r, the Bernoulli equation says

$$\frac{1}{2}\rho v_\ell^2 + p_\ell + \Psi_\ell = \frac{1}{2}\rho v_r^2 + p_r + \Psi_r. \tag{8.14}$$

Since the fluid along the center streamline comes to rest at the rigid wall, the velocity there is zero ($v_r = 0$), and the pressure at the stagnation point (the rigid wall) is

$$p_r = \frac{1}{2}\rho v_\ell^2 + p_\ell + \Psi_\ell - \Psi_r. \tag{8.15}$$

If there are no body forces, $\Psi = 0$, and typically it is assumed that far to the left the pressure is zero, and thus the pressure on the rigid wall is simply

$$p_r = \frac{1}{2}\rho v^2. \tag{8.16}$$

This term is sometimes referred to as the "dynamic pressure."

During a penetration event, one material penetrates another. If the penetration event is steady, then the penetration occurs at a constant velocity u. In the reference frame where the centerline interface between the target and projectile (the stagnation point) is at rest, the pressures on either side of the stagnation point are equal. Applying Bernoulli's equation to the fluid jet, at velocity $v - u$ in the center of the penetration reference frame, and also to the target material, moving towards the center of penetration at velocity $-u$, equating pressures gives

$$\frac{1}{2}\rho_p(v-u)^2 = \frac{1}{2}\rho_t(-u)^2 = \frac{1}{2}\rho_t u^2. \tag{8.17}$$

This equation was the first step in computing penetration velocities, and was applied for high velocity shaped charge jet penetration during World War II [28].

Some thought, however, reveals that there is something lacking in Eq. (8.17). In particular, the target and projectile enter into the equation symmetrically – there is no indication that the projectile is penetrating into the target. The first attempt to address this problem (actually, it was addressed not based on the lack of symmetry but on the lack of accuracy when compared with data) was to consider the form of the Bernoulli equation, and to add a Ψ or a pressure at a distance to the $\frac{1}{2}\rho v^2$, like we see in Eq. (8.15), to represent forces (strength) within the target. Tate [179] states that these ideas were first explored by Hill, Mott, and Pack during the war in a Ministry of Supply report and similarly later suggested by Eichelberger [58], again to better predict the penetration by shaped charge jets,

$$\frac{1}{2}\rho_p(v-u)^2 = \frac{1}{2}\rho_t u^2 + R_t. \tag{8.18}$$

We are using Tate's notation, but essentially the idea is that strength corresponds to a term that corresponds to pressure at a distance or some force that helps hold the target together. The idea is very heuristic, but the results for penetration velocity were much more accurate.

Then Tate and Alekseevksii added a similar term for the projectile,

$$\frac{1}{2}\rho_p(v-u)^2 + Y_p = \frac{1}{2}\rho_t u^2 + R_t. \tag{8.19}$$

Alekseevksii had multiplicative coefficients that might differ from $\frac{1}{2}$ to account for different drag or cross sections, or in his notation

$$k_p\rho_p(v-u)^2 + \sigma_{SD} = k_t\rho_t u^2 + H_D. \tag{8.20}$$

Equation (8.19) is the equation in the Tate–Alekseevskii model that determines the penetration velocity u. The penetration velocity can be explicitly determined, given the densities, projectile velocity, and the terms corresponding to projectile strength and target resistance, Y_p and R_t. In the last chapter we showed that the axial stress at the stagnation point is

$$\tilde{\sigma} = \frac{1}{2}\rho_p(v-u)^2 + Y_p = \frac{1}{2}\rho_t u^2 + R_t. \tag{8.21}$$

Once u is known it is possible to determine the stress at the interface between the projectile and the target.

In addition, Tate and Alekseevskii addressed the behavior of the projectile. The projectile has finite length. Once a finite length has been introduced into the problem, strictly speaking, the Bernoulli equation no longer holds, since the problem cannot be steady state. However, Eq. (8.19) is viewed as approximately correct and is used to determine the penetration velocity u. The behavior of the projectile will determine v. In Chapter 5 we solved for the deceleration of a finite length uniaxial stress cylinder impacting a rigid wall. The deceleration turned out to only depend on the length of the projectile and was based on elastic waves reflecting off the back end of the projectile, reducing its velocity. Thus, the equation for projectile deceleration is (Eq. 5.28, where it is assumed the sound speed is infinite and there is no plastic region)

$$\frac{dv}{dt} = -\frac{Y_p}{\rho_p L}, \tag{8.22}$$

where $L(t)$ is the time dependent length of the projectile. The Tate–Alekseevskii model includes one more equation, namely the rate of change of the length of the projectile to reflect the fact that it is eroding and getting shorter during the penetration event,

$$\frac{dL}{dt} = -(v-u). \tag{8.23}$$

The three equations, Eqs. (8.19), (8.22), and (8.23) are what is referred to as the Tate–Alekseevskii model or the Tate model. The initial conditions are initial velocity $v = V$ and initial length $L = L_0$ of the projectile.

In the Tate model, the penetration velocity only depends on the current back end velocity of the projectile. Explicitly solving Eq. (8.19) for u, the penetration velocity is

$$u = \frac{\rho_p v - \sqrt{\rho_p \rho_t v^2 + 2(\rho_p - \rho_t)(R_t - Y_p)}}{\rho_p - \rho_t}, \qquad \rho_p \neq \rho_t, \tag{8.24}$$

$$u = \frac{v}{2} - \frac{R_t - Y_p}{\rho_t v}, \qquad \rho_p = \rho_t, \tag{8.25}$$

$$u = \frac{\rho_p v^2 - 2(R_t - Y_p)}{\rho_p v + \sqrt{\rho_t \rho_p v^2 + 2(\rho_p - \rho_t)(R_t - Y_p)}}. \tag{8.26}$$

Here, $u < v$ and the projectile erodes (becomes shorter due to mass loss at the front end) as it penetrates. The last line is a more robust way of solving the quadratic that works both when the densities are the same and when they differ. Increasing the target resistance always decreases the penetration velocity, as can be shown by taking the derivative of the penetration velocity with respect to R_t,

$$\frac{\partial u}{\partial R_t} = -\frac{1}{\sqrt{\rho_p \rho_t v^2 + 2(\rho_p - \rho_t)(R_t - Y_p)}} < 0. \tag{8.27}$$

Further examination of Eq. (8.19) shows that there are two critical projectile velocities, depending on the sign of $R_t - Y_p$. The first arises for $R_t > Y_p$ when the penetration velocity u is zero and the axial stress is equal to R_t,

$$v_c = \sqrt{\frac{2(R_t - Y_p)}{\rho_p}}, \qquad R_t > Y_p. \tag{8.28}$$

If the impact velocity is below v_c, then the axial stress is given by $\tilde{\sigma} = (1/2)\rho_p v^2 + Y_p < R_t$. This stress is less than the strength of the target and no penetration occurs. However, the axial stress is above the strength of the projectile and so the projectile will erode, behaving like the Taylor anvil experiments in the previous chapter. For an impact velocity above v_c, penetration occurs with the projectile eroding until the back end of the projectile decelerates to v_c, at which point the projectile ceases to penetrate the target ($u = 0$) but continues to decelerate and erode until the projectile velocity is zero ($v = 0$ – the Taylor anvil problem, with the final length of the projectile $L_f = L_c \exp(-\rho_p v_c^2/2Y_p) = L_c \exp(1 - R_t/Y_p)$, where L_c is the length of the projectile when $v = v_c$).

The second critical velocity occurs for $R_t < Y_p$ when $v = u$ and the axial stress is equal to Y_p,

$$v_r = \sqrt{\frac{2(Y_p - R_t)}{\rho_t}}, \qquad R_t < Y_p. \tag{8.29}$$

If the projectile velocity is below v_r, then the axial stress is given by $\tilde{\sigma} = (1/2)\rho_t v^2 + R_t < Y_p$. The stress is less than the strength of the projectile and so the projectile does not deform. The axial stress is greater than the strength of the target and so penetration occurs. For an impact velocity above v_r, the projectile penetrates in an eroding fashion until the projectile decelerates to a velocity of v_r, and then penetration continues with what is referred to as rigid penetration, meaning that the projectile no longer erodes or deforms. The resisting pressure during the rigid penetration phase is given by

$$\tilde{\sigma} = \frac{1}{2}\rho_t v^2 + R_t \tag{8.30}$$

and the deceleration of the projectile is given by

$$\frac{dv}{dt} = -\frac{\tilde{\sigma}}{\rho_p L} = -\frac{1}{\rho_p L}\left(\frac{1}{2}\rho_t v^2 + R_t\right).$$

(8.31)

The penetration event is over when the projectile comes to rest, $v = 0$. By making use of the change of variable $dv/dt = (dv/dx)(dx/dt) = vdv/dx$ it is straightforward to integrate this expression to obtain the penetration during the rigid phase as

$$P_r = \frac{\rho_p}{\rho_t}L_r \ln\left(1 + \frac{\rho_t v_r^2}{2R_t}\right) = \frac{\rho_p}{\rho_t}L_r \ln\left(\frac{Y_p}{R_t}\right),$$

(8.32)

where L_r is the length of the projectile when $v = v_r$. Thus, when $R_t < Y_p$ the penetration event has two phases: an eroding phase when the velocity v is above v_r and then a rigid penetration phase for when the velocity drops below v_r. The eroding phase of the model needs to be numerically integrated, but the rigid phase can be explicitly integrated.

8.3. Behavior of the Tate Model

We first consider some special cases. If the strength of the projectile is zero, then the rear of the projectile never decelerates and the penetration velocity is a constant. In this case it is possible to explicitly solve for the time of the event, since it is given by the time it takes the projectile to completely erode t_f,

$$L_0 = -\int_0^T \frac{dL}{dt}dt = \int_0^T (v - u)dt = (v - u)t_f.$$

(8.33)

Total penetration is then given by

$$P = \int_0^{t_f} udt = ut_f = \frac{u}{v - u}L_0,$$

(8.34)

where u can be inserted from above. Two special cases can now be obtained. First, if the projectile and target material have the same density, the penetration is

$$\frac{P}{L} = \frac{\frac{1}{2}\rho v^2 - R_t}{\frac{1}{2}\rho v^2 + R_t} \quad \text{if} \quad \rho_p = \rho_t, \quad Y_p = 0, \quad v \geq v_c,$$

(8.35)

which shows the dependence of penetration on target resistance. A second case of considerable interest is when both the target resistance and projectile strength terms are zero, $R_t = Y_p = 0$. This case is referred to as hydrodynamic penetration, since it represents a fluid jet penetrating a fluid target. In this event, the penetration velocity is

$$u_{\text{hyd}} = \frac{1}{1 + \sqrt{\rho_t/\rho_p}}v$$

(8.36)

and the final depth of penetration is

$$\left(\frac{P}{L}\right)_{\text{hyd}} = \sqrt{\frac{\rho_p}{\rho_t}}.$$

(8.37)

This depth is referred to as the hydrodynamic limit. What is extraordinary about this result is that the penetration does not depend on velocity. As the impact velocity increases the penetration stress increases and the target and projectile strength terms become less relevant. The Tate model predicts that penetration approaches the hydrodynamic limit as impact velocity increases.

If the target resistance term is greater than or equal to the projectile strength term $(R_t \geq Y_p)$, then the predicted depth of penetration from the Tate model is always less than the hydrodynamic limit. To demonstrate this fact, consider the instantaneous penetration per amount of projectile eroded to achieve that penetration:

DEFINITION 8.1. (instantaneous penetration efficiency) The *instantaneous penetration efficiency* of an eroding projectile is given by the ratio of the depth of penetration per length of projectile eroded to achieve that penetration. It is given by

$$\frac{dP}{dL_e} \equiv \frac{dP(t)/dt}{dL_e(t)/dt} = \frac{u}{v-u}. \tag{8.38}$$

In Exercise 8.7 a general definition of penetration efficiency is presented that also covers noneroding penetration; however, this definition for eroding penetration is convenient and simple. First we demonstrate that as v decreases so does this measure of penetration efficiency. To that end,

$$\frac{\partial}{\partial v}\left(\frac{dP}{dL_e}\right) = \frac{v(\partial u/\partial v) - u}{(v-u)^2}. \tag{8.39}$$

Differentiating the original Tate model equation yields

$$\rho_p(v-u)\left(1 - \frac{\partial u}{\partial v}\right) = \rho_t u \frac{\partial u}{\partial v} \tag{8.40}$$

or

$$\frac{\partial u}{\partial v} = \frac{\rho_p(v-u)}{\rho_p(v-u) + \rho_t u}. \tag{8.41}$$

Placing this term back in Eq. (8.39) gives

$$\frac{\partial}{\partial v}\left(\frac{dP}{dL_e}\right) = \frac{\rho_p(v-u)^2 - \rho_t u^2}{(\rho_p(v-u) + \rho_t u)(v-u)^2} = \frac{2(R_t - Y_p)}{(\rho_p(v-u) + \rho_t u)(v-u)^2} \geq 0. \tag{8.42}$$

Thus, if the projectile velocity decreases then the penetration per unit length of projectile eroded is less efficient. Since the projectile decelerates when $Y_p > 0$, the maximum instantaneous penetration efficiency occurs at impact velocity, and the penetration efficiency decreases from then on during the penetration event. Since $\partial(dP/dL_e)/\partial v$ is positive, the maximum penetration efficiency is achieved when the impact velocity goes to infinity,

$$\lim_{v \to \infty} \frac{dP}{dL_e} = \lim_{v \to \infty} \frac{u}{v-u} = \lim_{v \to \infty} \frac{\rho_p v - \sqrt{\rho_p \rho_t v^2 + 2(\rho_p - \rho_t)(R_t - Y_p)}}{\sqrt{\rho_p \rho_t v^2 + 2(\rho_p - \rho_t)(R_t - Y_p)} - \rho_t v}$$

$$= \frac{\rho_p - \sqrt{\rho_p \rho_t}}{\sqrt{\rho_p \rho_t} - \rho_t} = \sqrt{\frac{\rho_p}{\rho_t}}. \tag{8.43}$$

Since the projectile length is given and when $R_t \geq Y_p$ the projectile completely erodes in the penetration event, we conclude with the strong result that

THEOREM 8.2. *For the Tate model, when $R_t \geq Y_p$, the final penetration in terms of P/L is less than the instantaneous penetration efficiency at the beginning of the penetration process, which is in turn less than the hydrodynamic penetration:*

$$\left(\frac{P}{L}\right)_{Tate\ model} \leq \frac{dP}{dL_e}\bigg|_{v=Impact\ velocity} < \left(\frac{P}{L}\right)_{hyd} = \sqrt{\frac{\rho_p}{\rho_t}}. \tag{8.44}$$

Thus, the hydrodynamic limit is an upper bound for the Tate model when the target resistance is greater than the strength of the projectile. Further, the initial instantaneous penetration efficiency is an upper bound for the final length-normalized penetration from the model.

8.4. An Example

Now that we have developed the Tate model, in this section we look at a specific example to see how it behaves. In the next section we will look at a wide range of materials and impact velocities. We specifically consider a set of small-scale penetration tests with steel targets and tungsten alloy penetrators conducted by reverse ballistics in a light-gas gun. A pair of flash X-rays provided orthogonal views through the target during penetration yielding the depth of penetration and the penetrator length at two times in each test. In each test, a tungsten alloy penetrator was supported by fine nylon thread and aligned in front of the muzzle. The penetrator diameter was 1.58 mm and the length-to-diameter (L/D) ratio was 20. Penetrators were made of Kennametal W-10, which is composed of 90W-7Ni-3Fe in an 18% cold-worked (swaged) condition with a density of 17.2 g/cm^3. A target assembly was launched from the gun to strike the stationary penetrator at a velocity of 1.5 km/s. This target assembly consisted of a steel core, a titanium alloy confinement sleeve, and a polypropylene sabot. The steel core was 13 mm in diameter and 32 mm in length, and was made of S-7 steel (density of 7.84 g/cm^3), heat treated to a nominal hardness of Rc 30. This core was pressed into a confinement sleeve made of the Ti-6Al-4V alloy, heat treated to a nominal hardness of Rc 36. The outside of the confinement was 25 mm by 25 mm, with a length of 62 mm. This target assembly provided high-strength support for the core, minimized the mass to be launched, and permitted radiography.

Three tests with targets of S-7 steel located the depth of penetration z_i and the projectile back-end location z_p at different times after impact. The final depth of penetration was measured from the recovered targets. Actual impact velocities were 1,485, 1,487, and 1,509 m/s, and the data were adjusted to a common impact velocity of 1.5 km/s. Adjusted data are plotted and listed in Fig. 8.1. The position of the projectile–target interface and the rear of the projectile at other times are shown by simply connecting the data points. The dotted portion of the curve represents a "best guess" extrapolation of the data to the final depth of penetration.

Numerical simulations were performed to compare with these data and the Tate model: two-dimensional axisymmetric computations were performed with the Eulerian wave propagation code CTH. The zoning was fairly coarse, with five computational cells across the radius with an aspect ratio of 2 to 1 maintained throughout the penetration region. The constitutive response for the target material was represented by the Johnson–Cook model (Eq. 3.152), with parameters given in Table 8.1. The value for A for the S-7 steel alloy has been changed from the value reported by Johnson and Cook (see Appendix C) since the alloy they tested had a hardness of Rc 50; the value A shown in Table 8.1 has been reduced proportionally to represent the hardness of the alloy used in the experiment. It has been assumed that the other constitutive parameters do not change with the initial yield stress. Although this may not true, particularly for B, it was the approach taken here.

The depth of penetration (projectile nose position) and the position of the tail along the projectile–target centerline as a function of time are presented in Fig. 8.2.

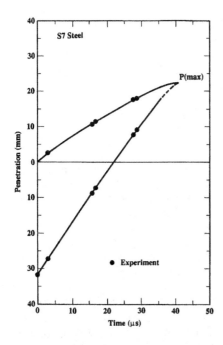

t	z_i	z_p
(μs)	(mm)	(mm)
0.0	0.0	−31.8
3.0	2.6	−27.3
15.8	10.8	−8.7
16.8	11.5	−7.2
27.8	17.7	7.9
28.8	18.0	9.1
40.5[a]	22.3	final depth

[a]Estimated

FIGURE 8.1. Experimental results of an $L/D = 20$ tungsten alloy projectile striking S-7 steel target. (From Anderson, Walker, and Hauver [16].)

TABLE 8.1. Constitutive model parameters.

Material	ρ (g/cm^3)	A (MPa)	B (MPa)	n	C	m
S-7 steel (Rc 30)	7.84	792	470	0.18	0.012	1.00
Tungsten alloy	17.2	1,506	177	0.12	0.016	1.00

The solid lines are from the numerical simulation, the dashed lines are from the Tate model, and the solid circles are the experimental data. For the Tate model result, a value of $R_t = 5.43$ GPa was chosen to *match the final depth of penetration of the numerical simulation*. The numerical simulation, the Tate model, and the data agree fairly well, though the final depth of penetration for the numerical simulation is 25.1 mm vs. the 22.3 mm from the data, making the disagreement with experimental results 12.6%, which is a large disagreement for modeling purposes. In this case, however, the only attempt to provide material parameters for this particular alloy was to adjust the initial yield stress due to the large difference in hardness between the target used in these experiments and the material originally used to calibrate the Johnson–Cook model. Although there might be some inclination to think that the agreement between the experimental data and the Tate model is better than between the numerical simulation and the experimental data, it must be pointed out that the Tate model was calibrated to yield the same depths of penetration as the numerical simulations. The penetration velocity (nose velocity) and tail velocity along the centerline are shown vs. time in Fig. 8.3.

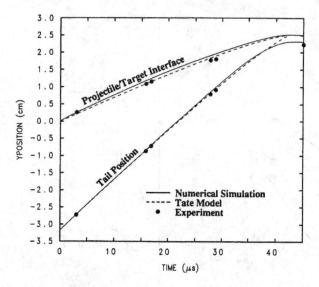

FIGURE 8.2. Projectile nose and tail positions vs. time for penetration of an $L/D = 20$ tungsten projectile into S-7 steel for a numerical simulation, the Tate model, and experimental data. (From Anderson, Walker, and Hauver [16].)

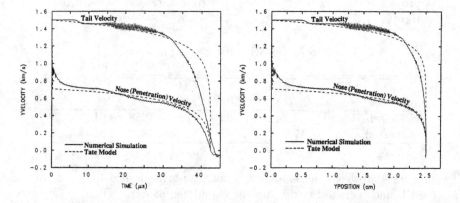

FIGURE 8.3. Penetration and tail velocities vs. time (left) and vs. depth of penetration (right) of an $L/D = 20$ tungsten projectile into S-7 steel for a numerical simulation and the Tate model. (From Anderson, Walker, and Hauver [16].)

In the numerical simulation, there is an initial transient stage and final transient stage that are not included in the Tate model, but in average terms the Tate model captures much of the penetration velocity behavior. The CTH numerical results account for the initial shock pressures, and their subsequent reduction due to rarefaction waves. Although the high pressures from the initial shock exist only for a few microseconds, this high pressure gives the crater floor a high initial impulse resulting in a higher rate of penetration than predicted by the steady-state model. This transient effect of penetration velocity persists on the order of $3 \, \mu$s. In terms

TABLE 8.2. Average and integrated R_t for tungsten striking S-7 steel target from data assuming $Y_p = 2$ GPa.

Penetration range z_i (mm)	$\bar{z}_i$ (mm)	$\bar{z}_i/L$	R_t (GPa)	Integrated R_t (GPa)
0.0 – 2.6	1.3	0.041	2.51	2.51
2.6 – 11.2	6.9	0.216	6.05	5.39
11.2 – 17.8	14.5	0.456	6.46	5.85
17.8 – 22.3	20.1	0.631	6.53	6.11
0.0 – 22.3	22.3	0.701		6.18

of a normalized penetration distance, the tungsten projectile achieves quasi-steady state penetration after approximately 1.5 projectile diameters of penetration into the steel target.

The penetration data and the numerical simulations allow us to examine R_t over the course of the penetration event, by examining

$$R_t - Y_p = \frac{1}{2}\rho_p(v - u)^2 - \frac{1}{2}\rho_t u^2. \tag{8.45}$$

The Tate model assumes a constant R_t and Y_p, and so variations in the difference of these two values identify the departure from idealized behavior. The target resistance can then be computed if a value for the projectile strength is assumed. From the experimental data, it is possible to use finite differencing to approximate v and u to compute R_t. Values are shown in Table 8.2, where the data points close in time have been averaged together. Both an "instantaneous" R_t and an integrated R_t were calculated; the instantaneous R_t, denoted $\bar{R}_t$, represents an average value for the target resistance between two successive data points. The depth of penetration for $\bar{R}_t$ is $\bar{z}_i$, the midpoint of two measured depths of penetration, and the velocities used in Eq. (8.45) are $u = \Delta z_i/\Delta t$ and $v = \Delta z_p/\Delta t$ from the bracketing flash X-rays. The integrated R_t represents the average target resistance over the penetration path (i.e., the velocities are computed from the current positions referenced to the initial positions). Both $\bar{R}_t$ and the integrated R_t are plotted vs. penetration depth normalized by the lengths of the projectile in Fig. 8.4. The instantaneous R_ts are plotted at $\bar{P}/L$ and the integrated R_ts are plotted at the ends of each interval. The value of R_t computed from the Tate model to match the *experimental* final depth of penetration is 6.18 GPa, shown as a solid dot in the figure and included as the last column in the table. The curves connecting the data points in the figures are spline fits to the data. The same computation for R_t can be performed for the numerical simulation and is also shown in Fig. 8.4. The experimentally determined instantaneous R_t values are shown in the figure.

It is seen that R_t varies during penetration. The numerical results lie somewhat below the experimental results, since in the computation the projectile penetrates a little deeper (i.e., the target looks a little softer) than what is observed experimentally. The major observation is not that there is some disagreement, but rather, the numerical and experimental results indicate the same trend in R_t: R_t generally increases during penetration and is not constant over the penetration event. The numerical results were integrated to provide average R_t values over the total

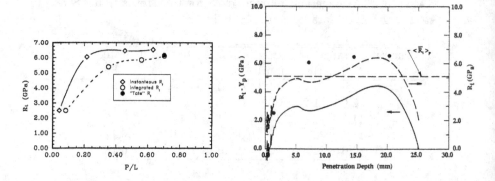

FIGURE 8.4. R_t vs. time from experimental depth of penetration data (left) and from numerical simulations (right) of an $L/D = 20$ tungsten alloy projectile striking S-7 steel at 1.5 km/s, assuming $Y_p = 2$ GPa. (From Anderson, Walker, and Hauver [16].)

penetration event:

$$\langle \bar{R}_t \rangle_T = \frac{1}{T} \int_0^T R_t(t)\, dt = 4.91 \text{ GPa}, \tag{8.46}$$

$$\langle \bar{R}_t \rangle_P = \frac{1}{P} \int_0^P R_t(x)\, dx = 5.05 \text{ GPa}, \tag{8.47}$$

where T and P are the total time interval of penetration and total depth of penetration, respectively. Again, matching the depth of penetration from the numerical simulation using the Tate model led to an R_t of 5.43 GPa. The time-averaged R_t is slightly different than the space-averaged R_t primarily because of the initial shock phase and final deceleration phase. The R_t computed from the Tate model matching the computational final depth of penetration exceeds the two integrated values, but the three different methods for computing a final R_t from the simulation are within 10% of each other.

We also provide a brief comparison of the Tate model to the numerical simulations presented in the previous chapter, where comparisons to the Tate model were shown in Fig. 7.9. To compare the numerical simulations with the Tate model, we need values for Y_p and R_t. For the tungsten alloy projectile, $Y_p = 2.0$ GPa was chosen to be a reasonable value for the dynamic strength of the material. We choose R_t to match the numerical simulation's penetration depth for Run 6. At the impact velocity of 1.5 km/s, a value of $R_t = 4.94$ gives an identical penetration depth of 7.01 cm. Essentially, the Tate model is the steady-state phase of penetration, and it does not include the projectile diameter dependent transients. The position–time histories of the interface and the tail as well as the residual length of the projectile are shown in Fig. 7.9. The tail positions agree extremely well. Since the position of the tail is largely governed by Y_p in the Tate model, we conclude that a value of 2.0 GPa works. The position of the interface is in slight disagreement. Examination of the nose and tail speeds reveals that the Tate model also disagrees with the numerical simulation at early times and near the end of penetration. The early time model discrepancy is due to the shock phase, before hydrodynamic behavior dominates. However, the time involved is quite short. The larger discrepancy between the Tate model and the numerical simulations is in the final deceleration of the projectile. Since $R_t > Y_p$, the Tate model decelerates the tail of the projectile elastically until $v = v_c$ (Eq. 8.28), at which point the target acts as a rigid Taylor anvil and the projectile continues to erode without any further penetration until $v = 0$. However, the numerical simulation indicates that the tail, and to some extent the nose, decelerates sooner and more gradually near the end of penetration than predicted by the Tate model. In the numerical simulation there is a residual penetrator that near the end of the penetration event penetrates as a rigid body. That such a rigid body penetrator occurs at these velocities is evident in Figs. 7.7 and 7.8. The eroded material is clearly seen, and the very rear of the residual projectile shows the original tail of the projectile and plastic deformation is not apparent. Thus, the rear of the projectile is still elastic. However, for the impact condition examined here, the Tate model predicts that essentially the entire projectile erodes. The length of the penetrator is shown in Fig. 7.9. It is seen that the overall length of the penetrator is reasonably predicted by the Tate model until the very end of penetration. The Tate model predicts total erosion of the penetrator, whereas the computations predict a residual penetrator of approximately 0.78 cm ($0.95D$) in length remaining at the bottom of the crater floor.

In summary, the Tate model provides a fairly accurate description of the time-dependent penetration history of the projectile. The model uses a single value for target resistance, but these results show that the target resistance value actually varies during penetration. The target resistance R_t used in the Tate model is usually calculated by matching final depth of penetration. This approach compares well with integrating the time-dependent R_t computed directly from numerical simulations. Thus, the target resistance R_t used in the Tate model can be viewed as an integrated or average value over the course of the penetration event.

8.5. Further Examples with the Tate Model

Examples of the Tate model require values for the target resistance R_t and the projectile strength Y_p. The values have historically been obtained by finding the values of R_t and Y_p that compare best with penetration depth vs. impact velocity data sets. That is what we will present here; later we will discuss ways to compute R_t and Y_p a priori.

The specific examples are due to Tate [**180, 182**] and Hohler and Stilp [**94, 95, 96, 97, 98**]. Table 8.3 lists the value of the parameter $R_t - Y_p$ that best fits, through the Tate model, final P/L data. Penetration depth is the most sensitive to this parameter – it is only weakly dependent on Y_p. Table 8.4 then tries to obtain a consistent set of R_t and Y_p values based on the fits and includes estimates of the initial yield values for the materials based on the experimentally measured hardness properties for comparison.

The penetration data that Tate compared his model to were excellent, mostly obtained by Hohler and Stilp. The data are shown in Tables 8.6–8.13. In all the tests, a blunt-nosed projectile was used and the impact yaw was less than 1°. The specific cases have been presented here for comparison with the Tate model and to the Walker–Anderson model in Chapter 10. In particular, the data set has many features for model validation: it has a range of impactor and target densities (tungsten into tungsten, two steels, aluminum; steel into steels, and aluminum into

TABLE 8.3. Experimental values for $R_t - Y_p$ [97, 180, 182].

Projectile	Target	$R_t - Y_p$ (experimental) (GPa)
D17	Steel 52	3.01
D17	Steel HzB20	3.84
Steel C110W2	Steel 37	2.35
Steel C110W2	Steel 52	3.13
Steel C110W2	Steel HzB20	3.86
D17	D17	4.17
D17	Al alloy 1	−0.36 (0.28)[a]
Al alloy 2	Lead	−1.05

[a]Neither value gives good agreement with the experiments; the first value, given by Tate in his paper, agrees with the data in the 250 to 500 m/s regime; the second value has reasonable agreement from 1 to 1.5 km/s and is used in the plots. It is based on $R_t = 1.83\,\text{GPa}$, which fits well the crater radius data using Tate's equation for crater radius and is consistent with $Y_p = 1.55\,\text{GPa}$ for D17.

TABLE 8.4. Material properties for the projectiles and targets used in the experiments; Brinell hardness, estimated initial yield stress based on the Brinell hardness, and elongation to failure values, save for Al alloy 2, are from [97]. Tate values for target and projectile R_t and Y_p are from fitting the experiments [180, 182]; these values are used to generate the plots.

Material	ρ (g/cm^3)	BHN (kp/mm^2)	Y_0 estimate (GPa)	ε_f (%)	Y_p (GPa)	R_t (GPa)
Steel C110W2	7.85	230 ± 30	0.77		1.1	
Steel 37	7.85	135 ± 20	0.45			3.45
Steel 52	7.85	180 ± 20	0.60	22		4.40
Steel HzB20	7.85	295 ± 35	0.98			5.18
W alloy D17	17.0	294 ± 10	0.76	8	1.55	5.72
Al alloy 1	2.85	75–85	0.31			1.83[a]
Al alloy 2	2.7		0.165[b]		1.38	
Lead	11.2					0.33

[a]See footnote to Table 8.3

[b]This value is an ultimate tensile strength from Tate [180]

lead). (The tungsten into aluminum data set is very interesting and is discussed in Chapter 10.) One type of impactor is striking three different types of steels, thus allowing a direct evaluation of the role of target material strength. Finally, the aluminum into lead data show that for a strong projectile and a weak target, one can have $Y_p > R_t$. When that occurs, our result about the P/L curve being strictly increasing as a function of velocity no longer holds, and the penetration exceeds the hydrodynamic limit in the region of 2 km/s and then approaches the hydrodynamic limit from above as velocity increases.

TABLE 8.5. Hydrodynamic limits.

Projectile	Target	$(P/L)_{\mathrm{hyd}} = \sqrt{\rho_p/\rho_t}$
D17	D17	1
D17	Steel	1.47
D17	Al alloy 1	2.44
Steel	Steel	1
Steel	Al alloy 1	1.66
Al alloy 2	Lead	0.49

TABLE 8.6. Penetration depth and crater radius of $L/D = 10$ tungsten alloy (D17) projectiles striking tungsten alloy (D17) targets, $L = 2.8$ cm [98]. (D17 is 90.5W-6.5Ni-3Fe [94].)

V (m/s)	P/L	R_c/R_p	V (m/s)	P/L	R_c/R_p
544	0.018	1.357	1,514	0.636	2.500
715	0.050	1.929	1,815	0.779	2.643
848	0.100	2.036	2,382	0.982	3.071
1,033	0.243	1.821	2,985	1.071	3.500
1,195	0.400	2.107	3,093	1.125	3.750
1,260	0.436	1.821	3,733	1.286	4.643

TABLE 8.7. Penetration depth and crater radius of $L/D = 10$ tungsten alloy (D17) projectiles striking HzB20 (German armor steel) targets; $L = 2.8$ cm or $L = 6.0$ cm (marked with *) [98].

V (m/s)	P/L	R_c/R_p	L	V (m/s)	P/L	R_c/R_p	L
407	0.020	1.500	*	1,497	0.768	1.857	
646	0.072		*	1,835	1.133	2.167	*
829	0.148	1.917	*	2,117	1.264	2.393	
835	0.157	1.900	*	2,549	1.400	2.643	
885	0.178	1.900	*	2,814	1.457	2.893	
975	0.258	1.833	*	3,397	1.536	3.250	
1,273	0.582	1.750		3,760	1.607	3.607	
1,348	0.653	1.867	*				

It becomes immediately apparent on looking at the data that in all cases save one (steel into hard steel) the penetrations exceed the hydrodynamic limit at high impact speeds, that is – $P/L > \sqrt{\rho_p/\rho_t}$ even when we expect $R_t > Y_p$. Table 8.5 shows the hydrodynamic limits.

Figures 8.5–8.7 show a comparison of the Tate model to the data in Tables 8.6–8.13 as well as a modified version of the Tate model to be discussed below. For the given values of the parameters, in all cases the Tate model penetration is less than that seen in the data at high velocities. Save for aluminum striking lead, in all cases $R_t > Y_p$, and so the Tate model predicts penetration below the hydrodynamic limit and it is not possible for the calculated penetration to reach the penetrations observed in the experiments.

TABLE 8.8. Penetration depth and crater radius of $L/D = 10$ tungsten alloy (D17) projectiles striking steel 52 targets; $L = 2.8$ cm or $L = 6.0$ cm (marked with *) [98].

V (m/s)	P/L	R_c/R_p	L	V (m/s)	P/L	R_c/R_p	L
434	0.045	1.933	*	1,696	1.180	2.333	*
523	0.055	2.133	*	1,756	1.262	2.417	*
722	0.153	1.933	*	1,910	1.368	2.500	
723	0.155	1.933	*	1,912	1.397	2.750	*
993	0.450	1.933	*	1,952	1.367	2.667	*
1,137	0.621	1.821		2,004	1.457	2.679	
1,183	0.640	2.067	*	2,100	1.464	2.929	
1,201	0.680	1.967	*	2,232	1.483	3.033	*
1,346	0.833	2.083	*	2,484	1.596	3.214	
1,400	0.939	2.071		2,820	1.650	3.571	
1,464	0.987	2.250	*	3,107	1.643	3.929	
1,478	0.962	2.133	*	3,263	1.679	3.964	
1,591	1.100	2.250	*	3,578	1.714	4.286	

TABLE 8.9. Penetration depth and crater radius of $L/D = 10$ tungsten alloy (D17) projectiles striking aluminum alloy targets (with Al BHN); $L = 2.8$ cm or $L = 6.0$ cm (marked with *) [98].

V (m/s)	P/L	R_c/R_p	BHN	L	V (m/s)	P/L	R_c/R_p	BHN	L
240	0.325		77		1,615	2.250		77	
369	0.682		77		1,652	2.268	2.300	75	*
460	0.918	1.071	83		1,660	2.286	2.286	83	
527	0.900		77		1,772	2.468	2.393	83	
627	1.200		77		1,775	2.429	2.214	83	
630	1.033	1.250	75	*	1,838	2.646	2.250	83	
849	1.202	1.467	75	*	1,848	2.637	2.450	75	*
886	1.114	1.429	83		1,865	2.786		77	
912	1.150		77		2,000	3.121		77	
1,117	1.504		77		2,036	3.114		77	
1,170	1.679	1.714	83		2,148	3.207	2.750	83	
1,221	1.650		77		2,162	3.198	2.800	75	*
1,244	2.175	1.714	83		2,494	3.539	2.929	75	
1,258	1.714		77		2,645	3.571	3.464	83	
1,427	2.182		77		2,868	3.554	3.607	83	
1,468	2.158	2.033	75	*	3,392	3.607	4.536	83	
1,525	2.079		77		3,495	3.571	4.714	83	

8.6. Tate's Later Modifications

Tate in subsequent papers addressed some of the shortcomings in predictive ability with the Tate model [181, 182]. Tate felt that the main difficulties in the model were in reproducing the initial transient impact and the final transient phases of penetration. Thus, he added two terms to the penetration depth to address these

TABLE 8.10. Penetration depth and crater radius of $L/D = 10$ steel (C110W1) projectiles striking HzB20 (German armor steel) targets; $L = 2.5$ cm, $L = 4.3$ cm (marked with **), or $L = 5.4$ cm (marked with *) [98].

V (m/s)	P/L	R_c/R_p	L	V (m/s)	P/L	R_c/R_p	L
560	0.007	1.209	**	1,820	0.484	2.280	
594	0.011	1.185	*	2,154	0.649	2.395	**
600	0.008	1.080		2,252	0.664	2.560	
800	0.026	1.419	**	2,264	0.677	2.488	**
888	0.030	1.721	**	2,513	0.740	2.560	
897	0.033	1.814	**	2,552	0.872	3.148	*
1,160	0.106	2.222	*	2,614	0.778	2.833	*
1,161	0.092	2.040		2,820	0.849	2.953	**
1,225	0.141	2.074	*	3,087	0.872	2.960	
1,263	0.149	1.953	**	3,108	0.874	2.977	**
1,535	0.330	2.116	**	3,593	0.916	3.360	
1,793	0.467	2.204	*	3,722	0.920	3.520	

TABLE 8.11. Penetration depth and crater radius of $L/D = 10$ steel (C110W1) projectiles striking St 52 steel targets; $L = 2.5$ cm, $L = 4.3$ cm (marked with **), or $L = 5.4$ cm (marked with *) [98].

V (m/s)	P/L	R_c/R_p	L	V (m/s)	P/L	R_c/R_p	L
595	0.030	1.628	**	1,957	0.658	2.558	**
747	0.041	1.944	*	2,348	0.846	2.685	*
783	0.050	1.981	*	2,552	0.872	3.148	*
965	0.072	2.040		2,596	0.928	2.920	
1,115	0.167	2.047	**	2,629	0.919	3.023	**
1,171	0.196	2.240		2,673	0.895	3.140	**
1,419	0.378	2.204	*	3,172	0.992	3.560	
1,801	0.558	2.442	**	3,459	1.016	3.680	

TABLE 8.12. Penetration depth and crater radius of $L/D = 10$ Sseel (C110W1) projectiles striking St 37 steel targets; $L = 2.5$ cm, $L = 4.3$ cm (marked with **), or $L = 5.4$ cm (marked with *) [98].

V (m/s)	P/L	R_c/R_p	L	V (m/s)	P/L	R_c/R_p	L
663	0.050	1.778	*	2,231	0.821	2.767	**
985	0.160	1.977	**	2,463	0.926	3.296	*
1,180	0.244	2.160		2,762	0.963	3.326	**
1,369	0.381	2.047	**	3,236	1.012	3.800	
1,589	0.539	2.222	*	3,528	1.052	4.280	
1,697	0.628	2.256	**	3,548	1.020	4.040	

TABLE 8.13. Penetration depth and crater radius of $L/D = 10$ aluminum alloy projectiles striking lead targets; $L = 6.35$ cm [180].

V (m/s)	P/L	R_c/R_p	V (m/s)	P/L	R_c/R_p
500	0.38	2.6	1,598	0.77	5.1
860	0.55	2.7	1,677	0.75	5.2
994	0.60	3.2	1,689	0.69	5.2
1,018	0.65	3.6	1,762	0.70	5.7
1,036	0.66	3.2	1,844	0.71	5.5
1,201	0.69	3.6	1,902	0.70	6.0
1,360	0.75	4.1	1,957	0.71	5.9
1,393	0.70	4.5	2,012	0.65	6.3
1,402	0.67	4.4	2,168	0.65	7.1
1,463	0.74	5.0	2,240	0.66	6.8
1,476	0.75	4.6	2,347	0.65	7.2

phases,

$$P = P_{tr} + P_{\text{Tate model}} + P_{af}, \tag{8.48}$$

where "tr" refers to initial transient at impact and "af" refers to "after flow," which is how Tate referred to the final transient phase when the projectile rapidly decelerates and comes to rest. For modeling the transient state of penetration, the acoustical impedances for two impacting materials are used to determine a penetration velocity, as found in Eq. (4.162),

$$u_{tr} = \min\left(\frac{\rho_p c_{0p}}{\rho_p c_{0p} + \rho_t c_{0t}} V, v_c\right). \tag{8.49}$$

Tate limits the initial transient penetration velocity based on the eroding penetration transition velocity. The transient velocity holds at the penetration front until rarefaction waves arrive from the corner of the projectile to release the stress state. The speed of the bulk release is related to what he refers to as a plastic shear wave velocity c_{ps}, and the time of interest is

$$t_{tr} = \frac{R_p}{c_{ps}}, \tag{8.50}$$

where R_p is the radius of the projectile. The initial transient phase penetration term is

$$P_{tr} = u_{tr} t_{tr} = \frac{R_p}{c_{ps}} \min\left(\frac{\rho_p c_{0p}}{\rho_p c_{0p} + \rho_t c_{0t}} V, v_c\right). \tag{8.51}$$

Tate in his papers discusses values of c_{ps} of 170 and 230 m/s, but in the plots here 1,000 m/s has been used as it better fits the data. For the late phase of penetration, the idea is that at the very end of penetration the residual projectile approaches a length equal to the initial projectile diameter. It is known that P/L increases with decreasing L/D so an $L/D = 1$ type penetration is added onto the end. Tate relates this additional penetration to the crater radius, for which he provides the following equation to be discussed in Chapter 9,

$$\left(\frac{R_c}{R_p}\right)^2 = 1 + \frac{2\rho_p(V - u)^2}{R_t}. \tag{8.52}$$

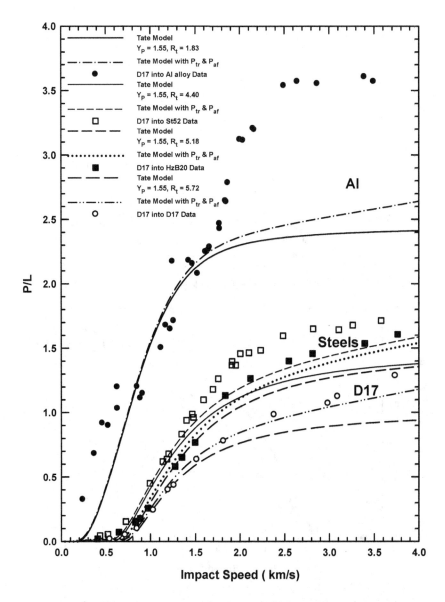

FIGURE 8.5. Depth of penetration of $L/D = 10$ tungsten alloy projectiles impacting an aluminum alloy, two steel alloys, and the same tungsten alloy. Tate model, the modified Tate model, and data are compared.

He then says that the additional penetration at the end of the penetration process is proportional to the crater radius using the formula

$$P_{\text{af}} = \frac{R_c}{2} \left(\left(1 + \frac{3\rho_t u^2}{2R_t} \right)^{1/3} - 1 \right) \qquad (8.53)$$

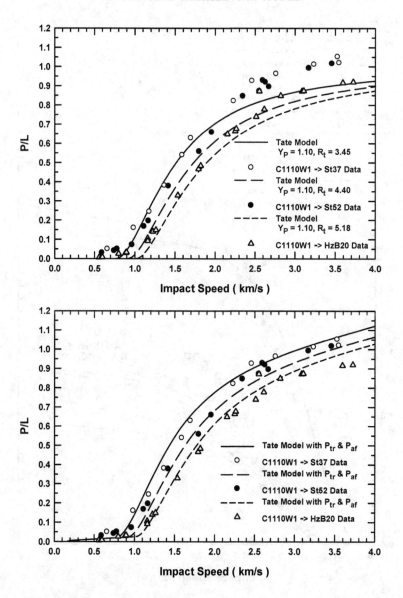

FIGURE 8.6. Depth of penetration of $L/D = 10$ steel alloy projectiles impacting three steel alloys. The upper figure compares the Tate model and data; the lower figure compares the modified Tate model and data.

(though he uses the R_t computed from Eq. 8.63 below rather than the fit R_t; we use the fit R_t in the plots here). In this expression, u is computed using the initial impact velocity V. These terms then give the total penetration, and results of this modified Tate model are shown in the plots.

Figure 8.5 shows the comparisons with the data where a tungsten alloy is the projectile material. There are four targets: an aluminum alloy, two steel alloys, and the same tungsten alloy. The reasonable agreement can be seen, with the only

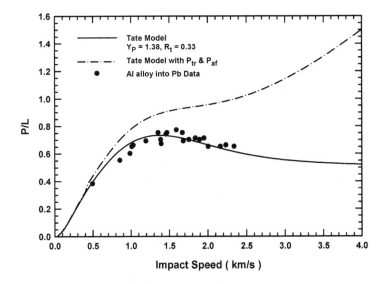

FIGURE 8.7. Depth of penetration of $L/D = 10$ aluminum alloy projectiles impacting lead. Tate model, the modified Tate model, and data are compared.

extreme deviance being the aluminum data. It is also apparent that the modifications help, as the penetration is typically too low. The large penetration into the aluminum alloy by the tungsten alloy projectile is due to secondary penetration, which will be discussed in Chapter 10. Figure 8.6 shows penetration into three steel alloys, and they have been divided into the Tate model (upper figure) and the modified Tate model (lower figure) to avoid clutter in the graph. Since the projectile and target materials are the same, the hydrodynamic limit is 1 and the Tate model cannot exceed this penetration. Again, the result of the modifications is apparent in increasing the penetration depth. Finally, Fig. 8.7 are data from Tate to demonstrate an effect that was predicted by the model: namely, that if $Y_p > R_t$, there can be a decrease in penetration for increasing velocity. Here a strong aluminum alloy impacts lead, which is very weak. The values of R_t and Y_p were chosen to give a good fit to the data [180]. Those are the values used here, though it can be seen that the modifications to the model are not desirable in this case. The value of Y_p that fits the data seems very large. In the remainder of the chapter (and book) when we refer to the Tate model we mean the original Tate–Alekseevskii model, which does not included the final-penetration-depth modifications.

8.7. The Link between R_t and the Cavity Expansion

So far we have treated R_t as a fit parameter. However, there are expressions to compute R_t. It is clear from Table 8.4 that the target resistance R_t is greater than the material strength of the target. We expect there to be a relationship between R_t and the target material flow stress, but R_t also takes into account penetration geometry effects.

In the end, all theoretical attempts to determine R_t utilize a cavity expansion argument, whether static or dynamic. The argument goes like this. Suppose a hemisphere is opening a penetration channel, with $\theta = 0$ the central axis. The hemisphere represents either the nose of a hemispherical projectile that is penetrating without deforming or the cavity that is opened up by an eroding projectile as it penetrates with a mushroom nose. It can be viewed that the projectile is opening up a suite of annular wedges, θ to $\theta + d\theta$, each of which is modeled as a spherical cavity. For penetration speed u, in spherical coordinates the velocity is $\vec{u} = u\hat{e}_z = u_r\hat{e}_r + u_\theta\hat{e}_\theta$. The unit normal in the z direction is

$$\hat{z} = \cos(\theta)\hat{e}_r - \sin(\theta)\hat{e}_\theta, \tag{8.54}$$

and the instantaneous velocity of the hemisphere centered at the origin as it penetrates in the z direction is

$$u_r = u\cos(\theta), \quad u_\theta = -u\sin(\theta). \tag{8.55}$$

To compute the axial force acting on the hemisphere, we use the stress from the spherical cavity expansion (Eq. 6.118),

$$\tilde{\sigma}_{rr}(V) = B_3 + \rho_t V^2. \tag{8.56}$$

The force is found by integrating the traction over the hemisphere. The traction is given by

$$\vec{t} = \tilde{\sigma}_{rr}(u\cos(\theta))\,\hat{e}_r. \tag{8.57}$$

The total force is

$$\begin{aligned}
F_z &= \int_{\text{hemisphere}} \vec{t}\cdot\hat{z}\,dA \\
&= 2\pi R_c^2 \int_0^{\pi/2} \tilde{\sigma}_{rr}(u\cos(\theta))\,\hat{e}_r\cdot\hat{z}\sin(\theta)\,d\theta \\
&= 2\pi R_c^2 \int_0^{\pi/2} (B_3 + \rho_t u^2\cos^2(\theta))\cos(\theta)\sin(\theta)\,d\theta \\
&= \pi R_c^2\left(B_3 + \frac{1}{2}\rho_t u^2\right).
\end{aligned} \tag{8.58}$$

R_c is the cavity or crater radius. It should be noted that the derivation of the force in the target was not based on the force required to open a spherical cavity from zero radius to the current radius. Rather, the cavity expansion is simply supplying the force as a hemisphere of a constant diameter penetrates. Since the cavity wall stress required to open the cavity is independent of cavity radius, we expect the stress computed from the cavity expansion to be fairly robust. Dividing by the frontal area of the cavity, we get

$$\frac{F_z}{A} = B_3 + \frac{1}{2}\rho_t u^2. \tag{8.59}$$

Viewing this expression as the resistance to penetration, by comparing with Tate's $\frac{1}{2}\rho_t u^2 + R_t$ (Eq. 8.19), we conclude

$$R_t = B_3 = \frac{2Y_t}{3}\left(1 + \ln\left\{\frac{2/Y_t + 4/(3K_t)}{1/\mu_t + 4/(3K_t)}\right\}\right), \tag{8.60}$$

where Y_t, μ_t, and K_t are the target's flow stress, shear modulus, and bulk modulus, respectively. Similar arguments using cylindrical cavity expansions to open stacks of cylindrically center-opened pancakes approximately yield

$$R_t = B_2 = \frac{Y_t}{2}\left(1 + \ln\left\{\frac{2/Y_t + 1/K_t}{1/\mu_t + 1/K_t}\right\}\right). \tag{8.61}$$

Thus, R_t arises from the zero velocity cavity expansion driving stress term B_2 or B_3. In Chapter 10 we will present a totally different approach to computing R_t that does not use the cavity expansion driving stress; rather, the cavity expansion is only used to determine the extent of the elastic-plastic zone α.

This paragraph contains examples of use of the spherical cavity expansion to obtain R_t. Tate suggests the following expressions for Y_p and R_t based both on theoretical arguments and experimental data [182],

$$Y_p = 1.7\sigma_p, \tag{8.62}$$

$$R_t = Y_t\left[\frac{2}{3} + \ln\left(\frac{0.57E_t}{Y_t}\right)\right], \tag{8.63}$$

where Y_t and σ_p are the dynamic compressive yield strengths of the target and the projectile, respectively, and E_t is Young's modulus of the target. Often, a larger-strain flow stress is used in lieu of an initial yield value. This expression is a variant of the equation to open a spherical cavity in an infinite elastic-plastic space of material due to Bishop, Hill and Mott, Eq. (6.168),

$$p_\infty = \frac{2Y_t}{3}\left\{1 + \ln\left(\frac{2E_t}{3Y_t}\right)\right\} + \frac{2\pi^2 H_t}{27}, \tag{8.64}$$

which holds for an elastically incompressible ($\nu = \frac{1}{2}$) material where there is a linear work hardening with slope H_t (i.e., $d\sigma_{11}/d\varepsilon_{11} = H_t$ after yield in a uniaxial stress test). Tate's suggested formula leads to a larger target resistance than that of Bishop et al., in the absence of work hardening (note that the 2/3 is in a different location with respect to the parentheses in these equations). In the analytic models developed here we do not separately include hardening; rather, we suggest using an average flow stress for Y_t to capture the behavior of the material after yield. Forrestal and Longcope [68] calculated a quasi-static pressure required to open a spherical cavity from zero initial radius at low velocity; for the case of an elastic-perfectly plastic material, their expression is

$$R_t = 2Y_t\left[\frac{1}{3} + \ln\left(\frac{R}{a}\right)\right], \quad \left(\frac{a}{R}\right)^3 = \beta - \frac{2(1-2\nu_t)}{3(1-\nu_t)}\beta^2\left[1 - \ln(\beta)\right], \quad \beta = 3(1-\nu_t)\frac{Y_t}{E_t}, \tag{8.65}$$

where ν_t is Poisson's ratio for the target. If the β^2 term is neglected (it is negligible when Y_t/E_t is small), this formulation reduces to the low-velocity cavity expansion result B_3 (assuming Y is much smaller than K, Eq. 6.120),

$$R_t = \frac{2}{3}Y_t\left[1 + \ln\left(\frac{E_t}{3(1-\nu_t)Y_t}\right)\right] \approx B_3, \tag{8.66}$$

as expected. If elastic incompressibility is assumed ($\nu_t = \frac{1}{2}$), $R_t \approx B_3$ further reduces to Eq. (8.64) without the work hardening term.

As an example of using the cylindrical cavity expansion to obtain R_t, we have for $Y \ll K$,

$$B_2 \approx \frac{Y}{2}\left(1 + \ln\left\{\frac{2/Y}{1/\mu + 1/K}\right\}\right) = \frac{Y}{2}\left(1 + \ln\left\{\frac{2E}{(5-4\nu)Y}\right\}\right). \tag{8.67}$$

TABLE 8.14. Various expressions for R_t.

R_t	Equation	Steel (GPa)	Aluminum (GPa)	Tungsten (GPa)
Y_t		1.2	0.34	1.2
$Y_t\left[\frac{2}{3}+\ln\left(\frac{0.57E_t}{Y_t}\right)\right]$	(8.63)	6.26	1.85	6.94
$\frac{2}{3}Y_t\left(1+\ln\left(\frac{2E_t}{3Y_t}\right)\right)$	(8.64)	4.57	1.34	5.02
$2Y_t\left[\frac{1}{3}+\ln\left(R/a\right)\right]$	(8.65)	4.31	1.28	4.75
B_3	(8.60)	4.29	1.28	4.74
$\frac{2}{3}Y_t\left[1+\ln\left(\frac{E_t}{3(1-\nu_t)Y_t}\right)\right]$	(8.66)	4.28	1.28	4.74
$\frac{Y_t}{\sqrt{3}}\left\{1+\ln\left[\frac{\sqrt{3}E_t}{(5-4\nu_t)Y_t}\right]\right\}$	(8.68)	3.69	1.09	4.03
B_2	(8.61)	3.28	0.97	3.62
$\frac{7}{3}Y_t\ln(\alpha(u))$	(8.69)	5.19	1.43	5.46
Experimental (Table 8.3)		5.18	1.83	5.72

In our cavity expansion derivation we assumed a Tresca yield surface; if a von Mises yield condition weren used instead, then, as shown in Chapter 3, the $Y/2$ factor would be replaced by $Y/\sqrt{3}$, yielding the result of Rosenberg, Marmor, and Mayseless [166],

$$R_t = \frac{Y_t}{\sqrt{3}}\left\{1+\ln\left[\frac{\sqrt{3}E_t}{(5-4\nu_t)Y_t}\right]\right\}. \tag{8.68}$$

In Chapter 10, using a different approach, we will derive a formula for penetration resistance for the Walker–Anderson model,

$$R_t = \frac{7}{3}Y_t\ln(\alpha(u)),, \tag{8.69}$$

where $\alpha(u)$ is the ratio of the radial extent of the plastic flow field in the target divided by the crater radius; $\alpha(u)$ comes from the cylindrical cavity expansion and depends on the penetration velocity and the elastic and yield properties of the target material (Eq. 6.49). Equation (8.69) will be derived in Chapter 10 (Eq. 10.59). Tables 8.14 and 8.15 show a numerical comparison of these different formulae for computing R_t. The values of the various constants used are those found in Tables 8.4 and 8.16. Also, for Eq. (8.69), which includes dynamic information, material properties from Tables 8.4 and 8.16 were included, as was a penetration velocity that was determined from the Tate model using the values in Table 8.4. The experimental values are those that Tate obtained from his fits to the various penetration scenarios and are thus determined from ballistic experiments and not laboratory materials tests. The R_t/Y_t for the experimental values are Tate's values divided by our assumed flow stress. The same flow stress was chosen for the steel and the tungsten so that it was possible to see the difference due to elastic properties. It is seen that R_t/Y_t ranges from 3 to 6 for these specific examples. In an earlier paper, Tate gives a rule of thumb that R_t is roughly 4.5 times the dynamic yield strength of the target material [179].

TABLE 8.15. Various expressions for R_t/Y_t.

R_t/Y_t	Steel	Aluminum	Tungsten
$\frac{2}{3} + \ln\left(\frac{0.57E_t}{Y_t}\right)$	5.22	5.43	5.78
$\frac{2}{3}\left(1 + \ln\left(\frac{2E_t}{3Y_t}\right)\right)$	3.81	3.95	4.18
$2\left[\frac{1}{3} + \ln\left(R/a\right)\right]$	3.59	3.77	3.96
B_3/Y_t	3.57	3.75	3.95
$\frac{2}{3}\left[1 + \ln\left(\frac{E_t}{3(1-\nu_t)Y_t}\right)\right]$	3.57	3.75	3.95
$\frac{1}{\sqrt{3}}\left\{1 + \ln\left[\frac{\sqrt{3}E_t}{(5-4\nu_t)Y_t}\right]\right\}$	3.07	3.22	3.39
B_2/Y_t	2.73	2.86	3.01
$\frac{7}{3}\ln(\alpha(u))$	4.32	4.21	4.55
Experimental (Table 8.3)	4.32	5.38	4.77

TABLE 8.16. Values used in Tables 8.14 and 8.15.

Material	E_t (GPa)	ν_t	Y_t (GPa)	u (m/s)	α
Steel	200	0.29	1.2	683	6.39
Aluminum	70	0.33	0.34	833	6.08
Tungsten	350	0.29	1.2	586	7.02

8.8. Target Resistance and the R_t Dilemma

So far, we have been discussing the Tate model and all its positive aspects. Now we need to point out its primary weakness, which we have had clues about so far but can present in a more direct way. To do that, normalized depths of penetration P/L of tungsten alloy projectiles into hard steel targets as a function of impact velocity are shown in Fig. 8.8. Some of the data have already been listed in this chapter; additional data points are from Hohler and Stilp [96] and Sorensen et al. [172]. The solid line is for $L/D = 10$ (including the data above) while the dashed line is for $L/D \geq 15$ projectiles. The curves represent least-squares curve fits through the experimental data. The steady-state hydrodynamic limit is given by $(P/L)_{\text{hyd}} = \sqrt{\rho_p/\rho_t}$. Note that the projectile diameter appears nowhere in the Tate model and the L/D effect cannot be exhibited by the Tate model.

When R_t is computed using the standard Tate model by matching the experimental final depth of penetration, the target resistances from the P/L data of the upper part of Fig. 8.8 are shown as the lower part of the figure as a function of impact velocity. A constant flow stress of 2.0 GPa was used for Y_p. R_t is strongly velocity dependent. Note that as P/L approaches the hydrodynamic limit, R_t approaches Y_p; also, note that R_t must be less than Y_p for the penetration performance to exceed the hydrodynamic limit; hence, it is seen that R_t drops below 2 GPa in the vicinity of 2.8 km/s. We conclude,

GENERAL PRINCIPLE 8.3. R_t is not a material constant.

However, we know from the phases of penetration that one reason R_t is decreasing is that at high impact speeds the original Tate model does not include the final transient and so all the penetration needs to come from the steady state

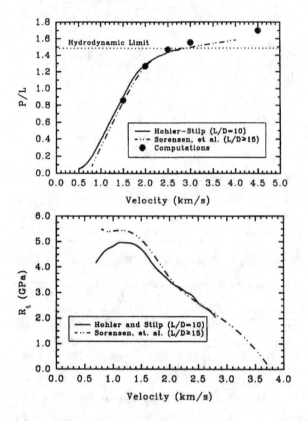

FIGURE 8.8. Normalized depth of penetration vs. impact velocity (upper) and R_t for the Tate model based on final penetration for hardened steel using the data of the upper figure (lower). The computations (upper) were performed with CTH. (From Anderson, Littlefield, and Walker [8].)

phase. That R_t depends on projectile L/D (using the original Tate model) shows that even if R_t were viewed as velocity dependent it cannot be viewed as a material parameter as it depends on the geometry of the impact event.

Let us go back and ask the question: if we just focus on the steady-state portion of penetration, does R_t still depend on other things? Based on Fig. 8.4, we expect that there will be an R_t dependence on penetration depth or penetration velocity (implying larger R_t for lower penetration velocity), but that dependence may be due to transient terms, since the 1.5 km/s impact speed is in the quasi-steady-state regime where there is a continual decrease in penetration speed. Thus, let us look for impact scenarios where there are no transients in the steady-state phase of penetration. Such a regime is impacts above 2.0 km/s. It will be necessary to take penetration vs. time data, and not rely on the final depth of penetration.

Experiments to study the penetration velocity at high velocities were performed in reverse ballistic mode where the projectile remained stationary and the target was accelerated to the impact velocity by a two-stage light gas gun, as reported by Anderson et al. [11]. Projectiles were right-circular cylinders; in some cases, the projectiles were composite assemblies where a long leading portion ($L/D \approx 16$)

TABLE 8.17. Projectile and target materials with hydrodynamic limit.

Projectile material	ρ_p (g/cm^3)	Target material	ρ_t (g/cm^3)	$(P/L)_{\text{hyd}}$ $= \sqrt{\rho_p/\rho_t}$	Relative strength
Cu	8.96	Al	2.71	1.82	Low → Low
Steel	7.86	Al	2.71	1.70	High → Low
Au	19.30	Al	2.71	2.67	Low → Low
W	17.45	Steel	7.85	1.49	High → High

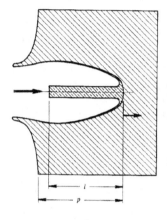
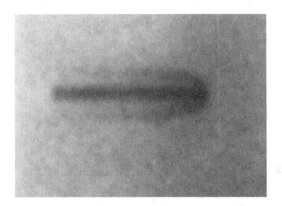

FIGURE 8.9. Schematic of penetration process (left) and an X-ray shadowgraph of a gold projectile penetrating an aluminum target at 6.7 μs after impact at 3.15 km/s (right). (From Anderson et al. [11].)

was followed by differently shaped elements. Only the penetration data relating to the leading portion are considered here. In the remaining cases the projectiles consisted of right-circular cylinders of $L/D \approx 21$. Projectile–target combinations used in these tests were gold into aluminum, copper into aluminum, steel into aluminum, and tungsten into steel (Table 8.17). Impact velocities ranged from 1.5 to 4 km/s. In all the experiments the projectile yaw angles were less than 1.5°.

The penetration process was observed by four independent flash X-ray stations. The radiographs yield the penetration depth and projectile residual length at distinct times after impact. Figure 8.9 shows a gold projectile penetrating an aluminum target at 6.7 μs after an impact at 3.15 km/s. The impact velocity was independently measured by X-ray just prior to impact. Projectile diameters varied, but were typically about 0.9 mm. The target diameters approximate the maximum that would still provide sufficient transparency to X-rays for each particular target material; they ranged from roughly 1.3 cm for steel targets to roughly 2.4 cm for aluminum targets. Experimental errors for individual data points are estimated to be about ±20 m/s for impact velocity, ±0.2 mm for penetration depth and consumed projectile length, ±0.2 μs for times after impact, and ±0.5° for angle of attack at impact. Table 8.18 shows some of the data.

TABLE 8.18. Examples of experimental data with least-squares penetration velocity (all tests impact at $t = 0$).

Proj. mat.	Target mat.	D (mm)	V (km/s)	Time (μs)	Depth (mm)	ΔL (mm)	u (km/s)	$v - u$ (km/s)
Au	Al	0.76	3.15	4.6	9.9	4.2	2.14	0.94
				6.7	14.0	6.2		
				9.7	21.0	9.2		
Au	Al	0.76	3.16	4.0	9.4	3.9	2.23	0.96
				6.9	15.1	6.6		
Steel	Al	0.97	3.89	4.8	11.7	6.5	2.42	1.44
				10.4	25.1	15.2		
Steel	Al	0.97	3.93	1.9	4.6	3.1	2.46	1.34
				4.8	12.1	6.2		
				7.6	18.7	10.2		
W	Steel	0.91	2.10	8.8	10.0	8.6	1.11	0.98
				13.0	14.2	12.7		
W	Steel	0.91	3.10	5.0	8.9	7.0	1.76	1.34
				11.7	20.5	15.5		
W	Steel	0.91	3.13	4.3	7.2	5.8	1.73	1.30
				8.0	14.2	10.3		
				12.2	21.0	15.9		

To determine the penetration velocity, a linear least-squares fit was computed to the penetration-time data shown in Table 8.18 that assumed the line went through the $(0,0)$ point. A least-squares fit was also done to the consumed length of the projectile. The time derivative of the linear least-squares fit to the penetration-time data gives a penetration speed u and the time derivative of the least squares fit to the consumed length gives $v - u$. These values are displayed in Table 8.19. The next column shows $R_t - Y_p$ from the Tate model based on the u and $v - u$ values. The next column shows the penetration velocity compared to the hydrodynamic penetration velocity (where reported impact speed V is used to compute $u_{\text{hyd}} = u/(V(1 + \sqrt{\rho_t/\rho_p}))$). The next column shows the penetration efficiency $dP/dL_e = u/(v - u)$, normalized by the hydrodynamic penetration, where u and $v - u$ were used from the least-squares fits. If the penetration were hydrodynamic these last two columns would equal 1. That it is consistently less than 1 shows that the hydrodynamic penetration velocity is not reached at velocities of 2 and 3 km/s, and hence that target strength is still playing an important role at these impact speeds. The last two columns are displayed in Figs. 8.10 and 8.11 from the original paper, though there are slight differences to the values in Table 8.19, since the original data were reanalyzed for this discussion. The plot of u/u_{hyd} shows as 1 the hydrodynamic limit, and demonstrates that the steady state penetration speed is less, showing the effect of target shear strength. The plot of $dP/dL_e/(P/L)_{\text{hyd}} = (u/(v-u))/\sqrt{\rho_p/\rho_t}$ is again less than 1 (save for an outlier at 4 km/s for steel into aluminum). Marked on this plot is the normalized depth of penetration for $L/D = 20$ tungsten alloy into armor steel, showing the difference between the amount of penetration that occurs due to the steady state data and the final depth of penetration data. Some data for speeds less than 2 km/s are shown for comparison.

TABLE 8.19. Experimental data with least-squares penetration velocity and comparison with hydrodynamic result.

Proj. mat.	Target mat.	D (mm)	V (km/s)	u (km/s)	$v - u$ (km/s)	$R_t - Y_p$ (GPa)	u /u_{hyd}	dP/dL_e /$(P/L)_{\mathrm{hyd}}$
Cu	Al	0.95	2.49	1.49	1.00	1.48	0.93	0.82
Cu	Al	0.94	2.50	1.51	0.99	1.23	0.94	0.85
Cu	Al	0.94	3.01	1.87	1.13	0.94	0.96	0.91
Cu	Al	0.94	3.04	1.87	1.15	1.21	0.95	0.89
Au	Al	0.76	3.15	2.14	0.94	2.26	0.93	0.86
Au	Al	0.76	3.16	2.23	0.96	2.18	0.97	0.87
Steel	Al	0.97	3.89	2.42	1.44	0.26	0.99	0.98
Steel	Al	0.97	3.93	2.48	1.34	−1.24	1.00	1.08
W	Steel	0.91	2.06	1.09	0.95	3.20	0.88	0.77
W	Steel	0.91	2.07	1.09	0.96	3.41	0.88	0.76
W	Steel	0.91	2.08	1.06	0.99	4.22	0.85	0.71
W	Steel	0.91	2.10	1.08	1.08	5.61	0.85	0.67
W	Steel	0.91	2.10	1.11	0.98	3.53	0.88	0.76
W	Steel	0.91	3.07	1.73	1.31	3.25	0.94	0.88
W	Steel	0.91	3.09	1.73	1.36	4.41	0.93	0.85
W	Steel	0.91	3.10	1.76	1.33	3.47	0.95	0.88
W	Steel	0.91	3.13	1.73	1.30	3.02	0.92	0.89

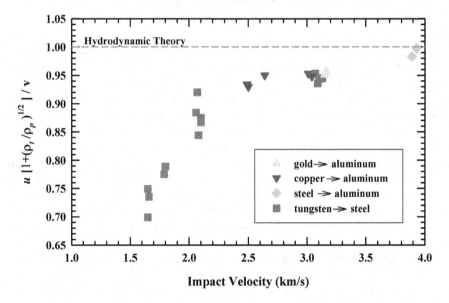

FIGURE 8.10. Normalized penetration velocity, obtained from the experimental data, vs. impact velocity. (From Anderson et al. [11].)

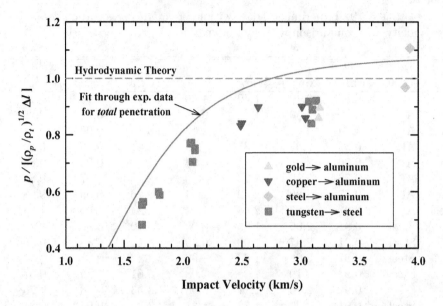

FIGURE 8.11. Normalized penetration efficiency, obtained from the experimental data, vs. impact velocity. The solid curve represents the *total* penetration of tungsten alloy $L/D = 20$ projectiles into armor steel from data. (From Anderson et al. [11].)

Of particular interest to us is the behavior of $R_t - Y_p$ as a function of impact speed. For copper impacting aluminum, at an average impact speed of 2.50 km/s the average $R_t - Y_p = 1.36$ GPa and at an average velocity of 3.03 km/s the average $R_t - Y_p = 1.08$ GPa. Thus, we see a decline in R_t as impact speed increases. For gold impacting aluminum, the average impact speed is 3.16 km/s and the average $R_t - Y_p = 2.22$ GPa. For a similar impact speed of copper impacting aluminum and gold impacting aluminum, there is a difference in $R_t - Y_p$ of 1.14 GPa, which is presumably due to the strength of the copper projectile vs. the strength of the gold projectile, which is low; there may also be a small effect of scale, since the copper projectile is 24% larger in diameter than the gold projectile; hence the target may appear relatively stronger in the gold–aluminum impact because of the smaller scale than the copper–aluminum impact. For the tungsten impacting steel, though there is scatter, for an average impact speed of 2.08 km/s the average $R_t - Y_p = 3.99$ GPa and for an average impact speed of 3.10 km/s the average $R_t - Y_p = 3.54$ GPa, again showing a decrease in R_t as a impact speed increases. Thus, we see a decrease in R_t vs. penetration speed (which is a function of impact speed), and in this case there are no transient terms, just steady-state penetration. These data show that the target resistance due to strength decreases as a function of increasing penetration speed. This feature of target resistance will be a result of the approach taken to computed target resistance due to strength in the Walker–Anderson model developed in Chapter 10.

We are left with a dilemma. R_t is often used in the armor community to rank the ballistic performance of materials, with R_t being computed from final depth of penetration data. The question arises whether this notion of penetration resistance

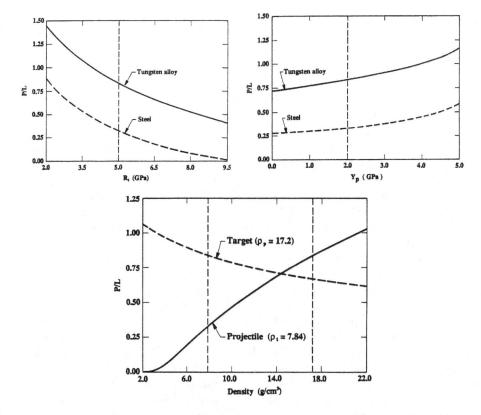

FIGURE 8.12. Tate model's influence of target resistance (upper left), projectile strength (upper right), and density (lower) on normalized penetration. Unless otherwise noted, $\rho_p = 17.2\ \text{g/cm}^3$, $\rho_t = 7.84\ \text{g/cm}^3$, $Y_p = 2\ \text{GPa}$, and $R_t = 5\ \text{GPa}$. (From Anderson, Walker, and Hauver [16].)

is a bad idea (it is not) or whether transient and geometric effects not included in the Tate model are important and can mask the R_t (they can). It turns out that the velocity dependence of R_t is important and the transient terms are also important. Thus, in the next two chapters we will develop a model that includes a fundamental and direct computation of target resistance and the transient terms.

8.9. Sensitivity of the Tate Model to Various Parameters

To explore the effect of the various inputs into the Tate model, we examine impacts at 1.5 km/s. In Fig. 8.12, Y_p is held constant at 2 GPa and P/L is plotted as a function of the target resistance R_t for two different density projectiles: one representing a tungsten alloy, the other a steel alloy. It is evident that penetration performance is strongly affected by target resistance. For example, a 20% increase in target resistance (5 to 6 GPa) results in a 16% decrease in penetration depth.

The penetration as a function of projectile strength Y_p is also plotted, assuming a constant target resistance R_t of 5 GPa. Penetration is weakly dependent upon penetrator strength, though near $Y_p = 5\ \text{GPa} = R_t$ the slope increases rather dramatically, as $Y_p > R_t$ can have rigid penetration. $Y_p = 0$ represents a "fluid" impactor. It takes an increase in penetrator strength of approximately 100% (2 to 4 GPa) to counteract an increase in target resistance of only 17%.

Penetration efficiency is strongly affected by the density of the penetrator. In Fig. 8.12, the effects of projectile density and target density are plotted explicitly, holding the target density constant at $7.84\ \text{g/cm}^3$ and projectile density constant at $17.2\ \text{g/cm}^3$, respectively. For example, changing the density of the projectile from steel to tungsten (assuming that $Y_p = 2$ GPa and $R_t = 5$ GPa remain constant) results in a 150% increase in penetration performance. The sensitivity of P/L to changes in target density is less dramatic than to changes in projectile density; these results show, for this case, that a relative change in projectile density has three times as much influence as a relative change in target density on the penetration.

Though not shown in the parameter study, the Tate model is most sensitive to the impact velocity. In general, unless R_t and Y_p are comparable or the velocity is high, the penetration is most sensitive to the impact velocity, the sensitivities to the projectile density and target strength are comparable, and the penetration is least sensitive to target density.

8.10. The Minimum Speed for Penetration

When $R_t > Y_p$, in the Tate model we have a critical velocity v_c, such that if $v < v_c$ there is no penetration ($v < v_c \Rightarrow \frac{1}{2}v^2 + Y_p < R_t$). Our development of the Tate model and even the upcoming Walker–Anderson model does not deal with the transition from nonpenetrating impacts to penetrating impacts. The models compute target resistance with approaches that assume penetration is occurring inside a thick target with confinement on the sides. When the impact first begins, the confinement is less, as the projectile is at the flat surface of the target. We would like to determine the corresponding target resistance at the beginning of the penetration process to determine at what impact speed penetration first occurs.

To examine the minimum stress for penetration, we will look at the stress within an elastic solid, assuming a blunt impact strikes the top surface. Though this is a dynamic problem, we will look at the static solution and view this as a lowest target resistance that needs to be overcome. The dynamic load for the blunt impactor will be taken from our result for the Taylor anvil, $\tilde{\sigma}_{zz} = \frac{3}{4}\rho_p v^2 + Y_p$. Our analysis of the stress state in the target starting point is Boussinesq's point loading problem in three dimensions, where for an elastic half space a load P is applied at the origin in the positive z direction (see, e.g., [111] or [169]). The stress distribution in the elastic half space ($z \geq 0$) is

$$\sigma_{xx}(x,y,z) = -\frac{P}{2\pi R^2}\left[\frac{3x^2 z}{R^3} - (1 - 2\nu)\left(\frac{z}{R} - \frac{R}{R+z} + \frac{x^2(2R+z)}{R(R+z)^2}\right)\right],$$

$$\sigma_{zz}(x,y,z) = -\frac{3Pz^3}{2\pi R^5}, \qquad \sigma_{yy}(x,y,z) = \sigma_{xx}(y,x,z),$$

$$\sigma_{xy}(x,y,z) = -\frac{P}{2\pi R^2}\left[\frac{3xyz}{R^3} - (1 - 2\nu)\frac{xy(2R+z)}{R(R+z)^2}\right], \tag{8.70}$$

$$\sigma_{yz}(x,y,z) = -\frac{3Pyz^2}{2\pi R^5}, \qquad \sigma_{xz}(x,y,z) = \sigma_{yz}(y,x,z),$$

where

$$R^2 = x^2 + y^2 + z^2 = r^2 + z^2, \quad r^2 = x^2 + y^2. \tag{8.71}$$

(Briefly, in this section, P and R have different definitions than the usual ones in the text, to conform with standard usage for this specific elasticity problem.) One can readily confirm that this solution satisfies the static momentum balance equations $(\partial \sigma_{ji}/\partial x_j = 0)$ everywhere save at the origin, where it is singular. If we integrate the traction (given by $t_i = \sigma_{ji}n_j$, where $\vec{n}$ is the normal to the surface) over a cylindrical pillbox that collapses to the origin, with flat top and bottom parallel to the x–y plane at $z = \pm z_{pb}$ and a cylindrical surface $x^2 + y^2 = r_{pb}^2$, we confirm that the load is that of a point load $\int_{\text{pillbox}} t_z dA = \int_{\text{pillbox}} \sigma_{jz}n_j dA = -P$ at the origin.

Now we use the solution to the Boussinesq problem to determine the stress distribution in the half space based on a distributed load. Our load distribution will be that of uniform pressure of $\tilde{\sigma}_{zz}$ over the disk $z = 0$, $x^2 + y^2 \leq R_p^2$, giving the total load $P = \pi R_p^2 \tilde{\sigma}_{zz}$. We will only determine the stress along the centerline of the target, which allows a considerable simplification of the convolution integrals. In particular the xx component of the stress along the z-axis is (σ_{xx} in the right-hand-side convolution integral is the Boussinesq solution)

$$
\sigma_{xx}(0,0,z) = \frac{1}{\pi R_p^2} \int_0^{2\pi} \int_0^{R_p} \sigma_{xx}(x - r\cos(\theta), y - r\sin(\theta), z)\, r\, dr\, d\theta
$$

$$
= \frac{1}{\pi R_p^2} \int_0^{2\pi} \int_0^{R_p} -\frac{P}{2\pi R^2}\left[\frac{3(r\cos(\theta))^2 z}{R^3} \right.
$$

$$
\left. -(1-2\nu)\left(\frac{z}{R} - \frac{R}{R+z} + \frac{(r\cos(\theta))^2(2R+z)}{R(R+z)^2} \right) \right] r\, dr\, d\theta
$$

$$
= -\frac{1}{\pi R_p^2} \int_0^{R_p} \frac{P}{R^2}\left[\frac{3r^2 z}{2R^3} - (1-2\nu)\left(\frac{z}{R} - \frac{R}{R+z} + \frac{r^2(2R+z)}{2R(R+z)^2} \right) \right] r\, dr
$$

$$
= -\frac{P}{\pi R_p^2} \int_0^{R_p} \frac{z}{2R^3}\left[-\frac{3z^2}{R^2} + 2(1+\nu) \right] r\, dr
$$

$$
= -\tilde{\sigma}_{zz}(0)\left\{ \frac{z^3}{2(z^2 + R_p^2)^{3/2}} - \frac{(1+\nu)z}{(z^2 + R_p^2)^{1/2}} + \frac{1}{2} + \nu \right\}.
$$

$$(8.72)$$

The zz component of the stress along the z-axis is

$$
\sigma_{zz}(0,0,z) = \frac{1}{\pi R_p^2} \int_0^{2\pi} \int_0^{R_p} \sigma_{zz}(x - r\cos(\theta), y - r\sin(\theta), z)\, r\, dr\, d\theta
$$

$$
= \frac{1}{\pi R_p^2} \int_0^{2\pi} \int_0^{R_p} -\frac{3Pz^3}{2\pi R^5} r\, dr\, d\theta = -\frac{P}{\pi R_p^2} \int_0^{R_p} \frac{3z^3 r}{R^5}\, dr \qquad (8.73)
$$

$$
= -\tilde{\sigma}_{zz}(0)\left(1 - \frac{z^3}{(z^2 + R_p^2)^{3/2}} \right).
$$

The shear terms integrate to zero along the z-axis. Thus, the equivalent stress along the z-axis stress is

$$
\sqrt{3J_2} = |\sigma_{xx}(0,0,z) - \sigma_{zz}(0,0,z)| = \tilde{\sigma}_{zz}(0)\left| \frac{3z^3}{2(z^2 + R_p^2)^{3/2}} - \frac{(1+\nu)z}{(z^2 + R_p^2)^{1/2}} + \nu - \frac{1}{2} \right|.
$$

$$(8.74)$$

The equivalent stress begins at the projectile–target interface $z = 0$ at $\tilde{\sigma}_{zz}(0)(\frac{1}{2}-\nu)$, then grows to a maximum, then decreases to zero as z gets large. For $\nu = 0.29$, the maximum occurs at $z = 0.63R_p$, where the absolute value expression on the right-hand side of the equation is 0.67. Hence, for the entire target area to be elastic, the largest $\tilde{\sigma}_{zz}(0)$ can be is $\sqrt{3J_2}/0.67 = 1.49Y$. Thus, we expect the target to have elastic response and no penetration when $\tilde{\sigma}_{zz}(0) = \frac{3}{4}\rho_p v^2 + Y_p \leq 1.49Y_t$. We have determined a low speed cutoff for penetration.

As the load increases it produces a yielding region in the interior of the target, but outside of this yielding region the target response is elastic. Hence, the target can support a larger load than the above with no penetration. Though the elastic solution no longer exactly applies, if we use it as a guide and accept an extent of yielding matching the projectile radius, then we have yielding, for $\nu = 0.29$, the range of $0.25R_p < z < 1.25R_p$ and $\tilde{\sigma}_{zz}(0) < 1.99Y_t$. If we allow the extent of the plastic region to go all the way to the target interface, then $\tilde{\sigma}_{zz}(0) < Y_t/(\frac{1}{2} - \nu) = 4.76Y_t$. These values all correspond to R_t is some sense. A likely cutoff speed for penetration is at least $\tilde{\sigma}_{zz}(0) = \frac{3}{4}\rho_p v^2 + Y_p \leq 2Y_t$. It is interesting (and perhaps coincidental) that when the equivalent stress reaches yield at the target surface, the value of R_t is close to that computed by other means.

The target stress distribution also reveals that there is a minimum thickness of target material required to support the projectile load, on the order of a projectile diameter. Thus, extremely hard but thin coatings on a target surface do not improve the ballistic resistance, because the target interior has the highest stresses and needs to support the load delivered by the projectile, not the surface.

8.11. Sources

In addition to the Tate and Alekseevskii papers [1, 179, 180, 181, 182] and data from Hohler and Stilp [94, 95, 96, 97, 98], material is this chapter also drew on computations and data in Anderson, Walker, and Hauver [16] for Section 8.4, discussions of R_t in Anderson and Walker [12] and Anderson, Littlefield, and Walker [8], and data in Anderson, Orphal, Franzen, and Walker [11] for Section 8.8.

Exercises

(8.1) Program the Tate model. Demonstrate that it works by making some penetration vs. impact speed curves for tungsten impacting steel and steel impacting aluminum.

(8.2) Derive Eqs. (8.36) and (8.37) for hydrodynamic penetration.

(8.3) Hohler and Stilp [94] state that they arrived at Y_p values of 1.1 ± 0.15 GPa for steel and 1.55 ± 0.15 GPa for D17 by examining the residual lengths of the impactors at low impact velocities, where the penetrations are very small. In that regime, the impact experiment is nearly like a Taylor anvil test. Looking at the photographs in their paper, the residual lengths are given in Table 8.20. Given these values, using the Taylor anvil theory of Chapter 5, determine whether the values of Y_p for D17 and C110W2 are reasonable.

(8.4) Show that the projectile erosion rate $v - u$ is a nondecreasing function of impact velocity in the Tate model.

(8.5) For penetration, when the projectile and target material are the same, the penetration speed is no longer half the impact velocity, as it was in one-dimensional impact. In the Tate model, how much does the penetration

TABLE 8.20. Residual lengths.

Projectile material	Target material	V (m/s)	L_0 (cm)	L_f (cm)
D17	HzB20	363	2.8	1.6
D17	St52	434	6.0	3.0
D17	HzB20	506	2.8	0.95
D17	HzB20	662	2.8	0.45
C110W2	St37	320	2.5	1.6
C110W2	HzB20	735	5.4	1.3

speed deviate from $v/2$? Work examples of tungsten into steel and steel into steel at 1.5 km/s.

(8.6) For the Tate equation, show that

$$\rho_p(v - u) + \rho_t u = \sqrt{\rho_p \rho_t v^2 + 2(\rho_p - \rho_t)(R_t - Y_p)}. \qquad (8.75)$$

Next, if the velocity v is sufficiently above the critical velocity, then show the following approximation:

$$\sqrt{\rho_p \rho_t v^2 + 2(\rho_p - \rho_t)(R_t - Y_p)} \approx \sqrt{\rho_p \rho_t} v + \frac{(\rho_p - \rho_t)(R_t - Y_p)}{\sqrt{\rho_p \rho_t} v}. \qquad (8.76)$$

Finally show that it is possible to write the instantaneous penetration efficiency as

$$\frac{dP}{dL_e} \approx \frac{\sqrt{\rho_p} - \left(1/\sqrt{\rho_p} + 1/\sqrt{\rho_t}\right)(R_t - Y_p)/v}{\sqrt{\rho_t} + \left(1/\sqrt{\rho_p} + 1/\sqrt{\rho_t}\right)(R_t - Y_p)/v}. \qquad (8.77)$$

This expression shows how the instantaneous penetration efficiency approaches the hydrodynamic limit.

(8.7) The penetration efficiency of eroding projectiles is $dP/dL_e = u/(v - u)$. We want to define a penetration efficiency that applies to projectiles, whether or not they erode, and is similar in value to dP/dL_e for eroding penetration. In general, penetration efficiency is a measure of how much projectile momentum is lost in exchange for a certain amount of penetration. For a cylindrical projectile, the momentum m is given by $m = \rho_p A L v$, where A is the cross-sectional area, L is the length, and v is the velocity. The change in momentum is $dm/dt = \rho_p A v dL/dt + \rho_p A L dv/dt$. If this expression is normalized by the momentum, and the change in penetration is normalized by the length of the projectile, then show that a penetration efficiency that meets our criteria is

$$\text{Efficiency} \equiv \frac{m}{L}\left|\frac{dP}{dm}\right| = -\frac{u}{dL/dt + (L/v)dv/dt} \qquad (8.78)$$

(the absolute value is so that the expression will be positive) and in particular for the Tate model with eroding penetration the predicted efficiency is

$$\text{Efficiency} = \frac{u}{v - u + Y_p/(\rho_p v)}. \qquad (8.79)$$

Plot this efficiency and dP/dL_e for tungsten into steel from $u = 0$ to some reasonable speed (like 2 km/s) and show that the penetration efficiency and

dP/dL_e are close. Next show that for nondeforming projectile penetration the penetration efficiency is

$$\text{Efficiency} = \frac{\rho_p u^2}{(1/2)\rho_t u^2 + R_t}. \tag{8.80}$$

Show that the penetration efficiency is bounded by $2\rho_p/\rho_t$. Since rigid penetration is unbounded as the velocity increases, it is not penetration efficiency that limits eroding projectile penetration to the hydrodynamic limit at hypervelocity impacts, but the fact that the projectile erodes away and the penetration process ends. Show that at the eroding–rigid transition v_r the eroding efficiency and the rigid efficiency match, and so the penetration efficiency curve for the Tate model is continuous. Plot the penetration efficiency of tungsten striking aluminum (with $R_t < Y_p$) over the same velocity range as in the previous plot and compare.

(8.8) For C110W2 steel impacting HzB20 steel at 1.5 km/s, compute the penetration velocity and the instantaneous penetration efficiency. Are the values much different than for tungsten into steel?

(8.9) Tate provided explicit solutions for P/L in his paper for the special case of the projectile and target having the same density and R_t/Y_p equal to 1, 3, and 5 [179]. The simplest is for $R_t/Y_p = 1$, where

$$z_i(v) = L_0 \left(1 - \exp\left\{\left[\left(\frac{v}{V}\right)^2 - 1\right]\frac{\rho V^2}{4Y_p}\right\}\right). \tag{8.81}$$

This equation gives the current location of the projectile–target interface z_i based on the projectile back-end velocity. Derive it. What is the final P/L? What is the effect of the projectile deceleration? (That is, how does this result compare to the hydrodynamic result?) This case would be difficult to realize in that it would require, for example, a really strong steel alloy impacting a very weak steel alloy. The other cases are more likely, given material strengths, but the mathematics is more tedious.

(8.10) What is the change in penetration efficiency with respect to target density divided by the change in penetration efficiency with respect to projectile density? What does it say?

(8.11) Two tests were conducted examining the penetration of tungsten alloy projectiles into titanium alloy Ti-6Al-4V targets (density of 4.41 g/cm^3). The tungsten alloy projectile of diameter 4.93 mm was made of the same material as used in the experiments described in this chapter with S-7 steel in Fig. 8.1. In a first test $L/D = 16$ and in a second test $L/D = 14$. The titanium alloy target had a 76-mm square cross section. Impact velocities were 1,428 and 1,431 m/s, respectively. The depth of penetration and the projectile length were recorded at two times for each test using flash X-rays. The final depth of penetration was determined from posttest measurements and was adjusted for L/D. The penetration data are plotted and listed in Fig. 8.13.

Compute the instantaneous R_t, as was done for the S-7 impact example in this chapter. Also compute the average R_t. In armor we often hear of people searching for a material that is as strong as steel but half the weight. Is titanium that material?

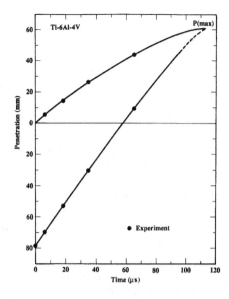

t	z_i	z_p
(μs)	(mm)	(mm)
0.0	0.0	−78.8
6.2	5.6	−70.0
18.3	14.4	−52.9
34.8	26.5	−30.1
65.0	44.3	10.0
115[a]	61.5	$(= P)$

[a]Estimated

FIGURE 8.13. Experimental results for $L/D = 16$ tungsten alloy projectiles striking Ti-6Al-4V targets at 1.43 km/s. (From Anderson, Walker, and Hauver [16].)

(8.12) Show that if $u \le u_{\text{hyd}}$ then

$$\frac{dP/dL_e}{(P/L)_{\text{hyd}}} \le \frac{u}{u_{\text{hyd}}}.$$

(8.13) Define k as

$$k = 1 - \left(\frac{dP/dL_e}{(P/L)_{\text{hyd}}}\right)^2.$$

This k measures deviation from hydrodynamic penetration, where $k = 0$ means hydrodynamic penetration. Show that for the Tate equation,

$$R_t - Y_p = k\frac{1}{2}\rho_p(v - u)^2 = \frac{k}{1 - k}\frac{1}{2}\rho_t u^2.$$

(8.14) Assume that the projectile–target interface and the projectile back end decelerate linearly with respect to time – that is, that $\partial u/\partial t < 0$ and $\partial v/\partial t < 0$ are constant. For there to be a maximum in $R_t - Y_p$ during penetration, as in the lower plot of Fig. 8.8, show that

$$\frac{1}{u}\frac{\partial u}{\partial t} < \frac{1}{v}\frac{\partial v}{\partial t} \tag{8.82}$$

and that equality replaces the less than if $R_t - Y_p$ is constant. Hence, conclude that the local maximum requires that the deceleration of u must be greater than that predicted by the Tate model. Compare with the results in Chapter 7 on the L/D effect.

(8.15) The Tate model does not show an L/D effect because D does not enter into the model. However, show that there is also another difference. In the 1 to 2 km/s regime, the hydrocode computations (Figs. 7.21 and 7.22) show that $v - u$ increases as a function of penetration depth during the "steady state" phase of penetration. Show that for the Tate model $v - u$ strictly decreases as a function of penetration depth. Thus, the Tate model is not, for example, matching the $L/D = 10$ event and then missing the details for smaller and larger L/Ds.

(8.16) In Section 8.8 we showed a strong dependence of R_t on velocity when R_t was found by matching the final depth of penetration with the Tate model. With the recognition that primary penetration is comprised of three phases and that the Tate model essentially is a model of the steady-state phase, we may wonder if our firm statement that R_t is not a material property was premature, and that to be fair we should look only at the behavior of R_t during the steady-state phase of penetration, though this value is hard to determine experimentally. The departure from idealized steady-state no-strength behavior is given by the difference $R_t - Y_p = (1/2)\rho_p(v - u)^2 - (1/2)\rho_t u^2$. Show the final transient phase of penetration has an R_t that is a (potentially large) negative number and so it may be that the final transient is causing R_t to decrease at higher velocities. Given the data in Table 8.19, do we see a velocity dependence in R_t for the steady-state regime of penetration? (Unfortunately, these data also include the initial transient.) If you have access to a hydrocode, you can perform penetration computations such as are described in this chapter for 1.5, 3.0, and 4.5 km/s of tungsten into steel and then compute the depth-averaged R_t as described in Eq. (8.47) (assume a constant Y_p). Do you find a velocity dependence? As a comment, a difficulty with this computation, particularly from the data, is that the two terms on the right-hand side are similar in size and we are looking at their difference.

(8.17) Often Bernoulli's law is derived with a slightly weaker assumption. Namely, the flow is not assumed to be incompressible, but the pressure is assumed to be only a function of density. Typically, the pressure depends upon two thermodynamics parameters such as the density and the temperature, but if the loading path is known (it could be isothermal or isentropic, for example), then the pressure can be written as a function of the density. (If the material is such that the pressure depends only on the density, $p = p(\rho)$, the material is called barotropic.) In particular, this implies that there exists a function

$$P(p) = \int_{p_0}^{p} \frac{dp}{\rho(p)} \tag{8.83}$$

and, differentiating this integral (using the Leibniz rule applied to the upper limit of integration), we obtain

$$\frac{\partial P(p)}{\partial x_i} = \frac{1}{\rho}\frac{\partial p}{\partial x_i} \quad \Rightarrow \quad \mathrm{grad}(P(p)) = \frac{1}{\rho}\mathrm{grad}(p). \tag{8.84}$$

Use this to show the more general form of Bernoulli's law,

$$\frac{1}{2}|\vec{v}|^2 + P(p) + \Psi = \text{constant}. \tag{8.85}$$

Now ask the question: how much difference can this assumption make? For example, if we assume the material is steel and assume that $p = K(1 - \rho_0/\rho)$, then what strain and what pressure is required for P to differ from p/ρ by 5%? Argue that, given a pressure p, this error estimate is conservative.

The Crater and Ejecta

We now use results of the cavity expansion to derive an equation for radius of the crater formed during an impact. We are developing a crater radius equation because the Walker–Anderson model, derived in the next chapter, needs the crater radius. We then explore the liberation of target material due to penetration, referred to as crater ejecta, and its size scale effect.

9.1. Axial Change in Momentum for the Target

We assume that an eroding projectile of initial length L has a velocity v and penetrates into the target at a velocity u. For eroding penetration, an increment of penetration has a corresponding time increment of

$$\Delta t = \frac{\Delta P}{u} = \frac{\Delta L}{v - u}. \tag{9.1}$$

We will assume the projectile begins as a right-circular cylinder. To begin the discussion, assume a hemisphere is penetrating at a velocity u along the z-axis. The hemisphere represents the cavity that is opened up by an eroding projectile as it penetrates. In the previous chapter we computed the force required for this hemisphere to penetrate, Eq. (8.58),

$$F_z = \pi R_c^2 \left(B_3 + \frac{1}{2} \rho_t u^2 \right). \tag{9.2}$$

Given the force, which is a constant, the total axial momentum change in the target for the penetration event is given by the time integral of the force,

$$\Delta \text{momentum}_{\text{target}} = F_z \Delta t = \pi R_c^2 \left(B_3 + \frac{1}{2} \rho_t u^2 \right) \frac{\Delta L}{v - u}. \tag{9.3}$$

9.2. Axial Change in Momentum for the Projectile

It might be thought that the change in momentum of the projectile is simple – it is given by the initial momentum of the projectile. However, it is not so straightforward. It is clear we need to be careful in that, as we showed in Chapter 4, when an elastic projectile bounces off a rigid wall it delivers twice its initial momentum to the wall. Thus, we need to determine the residual velocity of the projectile debris as the projectile erodes.

The projectile is traveling at a velocity v and the penetration occurs at a velocity u. Moving into the frame of reference where the interface between the target and projectile is at rest, the projectile material flows into the interface stagnation point at speed $v - u$. If we assume that the material is simply turned about by the hemispherical crater, then it will maintain the speed $v - u$ as it turns $180°$ and heads back the other direction. Back in the laboratory frame of impact, the eroded projectile debris thus have a net velocity of $-(v - u) + u = 2u - v$. The change in velocity of the projectile material is $v - (2u - v) = 2(v - u)$ and, when multiplied by the projectile mass, gives the momentum change of the projectile material.

There is another subtlety with the projectile. We assumed that the projectile does not decelerate during the penetration event. This is not true. To maintain this assumption, we need to assume that the projectile is being pushed with its flow stress Y_p in order to counteract the deceleration that would otherwise occur. This effect corresponds to a momentum change equal to the cross-sectional area of the projectile multiplied by the flow stress multiplied by the length of the penetration event.

These two calculations combine to give us the axial projectile change in momentum (where m_p is the initial mass of the projectile),

$$\Delta\text{momentum}_{\text{projectile}} = m_p \frac{\Delta L}{L} 2(v - u) + \pi R_p^2 Y_p \Delta t$$
$$= \pi R_p^2 \Delta L \left(2\rho_p(v - u) + \frac{Y_p}{v - u} \right). \tag{9.4}$$

9.3. Radial Momentum and the Crater Radius

We now equate the axial momentum change for the target and the axial momentum change for the projectile,

$$\pi R_c^2 \left(B_3 + \frac{1}{2}\rho_t u^2 \right) \frac{\Delta L}{v - u} = \pi R_p^2 \Delta L \left(2\rho_p(v - u) + \frac{Y_p}{v - u} \right), \tag{9.5}$$

which gives

$$\left(\frac{(R_c)_0}{R_p} \right)^2 = \frac{Y_p + 2\rho_p(v - u)^2}{B_3 + (1/2)\rho_t u^2}. \tag{9.6}$$

The crater radius in this axial momentum balance has been identified with $(R_c)_0$ to emphasize that this axial momentum balance is based on the instant of penetration. However, the final crater radius is a radial expansion, not an axial one. As eroded material mushrooms out it continues to push radially. Though at the instant of penetration the speed for the cavity expansion is u, as the crater radius expands laterally the pressure applied by the projectile erosion products drops until the material resists further expansion, but this stoppage of expansion occurs when there is no longer any radial velocity and the stress is at or below the zero-velocity cavity expansion stress B_3. Thus, if we assume the applied pressure expands the crater radius to this point, we replace $B_3 + (1/2)\rho_t u^2$ with B_3 in the denominator to obtain

$$\left(\frac{R_c}{R_p} \right)^2 = \frac{Y_p + 2\rho_p(v - u)^2}{B_3}, \tag{9.7}$$

with the recognition that $R_c \geq R_p$. This equation requires a penetration velocity u to compute the crater radius. We use the Tate equation with the $R_t = B_3$,

$$\frac{1}{2}\rho_p(v - u)^2 + Y_p = \frac{1}{2}\rho_t u^2 + B_3. \tag{9.8}$$

This equation agrees quite well with data shown in Figs. 9.1 and 9.2 for the data sets given in Chapter 8. The values of the material constants are presented in Table 10.4 and are summarized in Table 9.1. This expression uses reasonable values for the material properties and good agreement with experiment is achieved.

TABLE 9.1. Values used in crater radius calculations (from Tables 8.4 and 10.4).

		Tate		Walker–Anderson	
Material	ρ (g/cm^3)	Y_p (GPa)	R_t (GPa)	Y_p (GPa)	Y_t (GPa)
Steel C110W2	7.85	1.1		0.9	
Steel 37	7.85		3.45		0.85
Steel 52	7.85		4.40		0.95
Steel HzB20	7.85		5.18		1.30
W alloy D17	17.0	1.55	5.72	1.3	1.30
Al alloy 1	2.85		1.83[a]		0.45
Al alloy 2	2.7	1.38		0.40	
Pb	11.2		0.33		0.045

[a]See footnote to Table 8.3.

It is conceivable that for some impacts the crater radius expansion occurs well after the instant of penetration, where the crater radius continues to grow because of inertia in the target material near the crater wall. In this situation, the crater radius used in a penetration model will differ from the late time expansion radius R_c and be more aligned with the early time crater radius $(R_c)_0$. This behavior is indeed seen in the next chapter for aluminum penetrating lead (Fig. 10.12).

The expression for crater radius is similar to that stated by Tate [182],

$$\left(\frac{R_c}{R_p}\right)^2 = 1 + \frac{2\rho_p(v-u)^2}{R_t}. \tag{9.9}$$

The similarity is especially clear when $R_t = B_3$. The only major difference is forcing the expression to go through 1 at 0 km/s by adjusting the constant term. To compute the penetration velocity u, Tate used his equation $\frac{1}{2}\rho_p(v-u)^2 + Y_p = \frac{1}{2}\rho_t u^2 + R_t$. The results of this equation are also shown in Figs. 9.1 and 9.2. The values of the material properties are shown in Table 9.1. Again, there is good agreement with experiment.

As can be seen, the equations give a reasonable account of the crater radius. The kink in the curve at lower velocities occurs when $u = v_c$ or $u = v_r$. The crater radius is a nondecreasing function of the impact velocity v since $v - u$ is a nondecreasing function of the impact velocity, because

$$\frac{\partial(v-u)}{\partial v} = 1 - \frac{\partial u}{\partial v} = \frac{\rho_t u}{\rho_p(v-u) + \rho_t u} \geq 0 \tag{9.10}$$

for the Tate equation. Strictly speaking, the derivation for crater radius above does not hold for velocities below v_c or v_r. In the case where $Y_p < B_3$, the above derivation sets $0 = 0$, since $u = 0$ for $v < v_c$, and hence there is no penetration. However, choosing to use the same equation after the division by u gives a continuous crater radius plot and seemingly reasonable values. In the case where $Y_p > B_3$, the projectile rigidly penetrates for $v < v_r$. Thus, one approach is to assume that $R_c = R_p$ in this regime. Only one set of data falls under this criterion, and that is the aluminum alloy impacting lead data set.

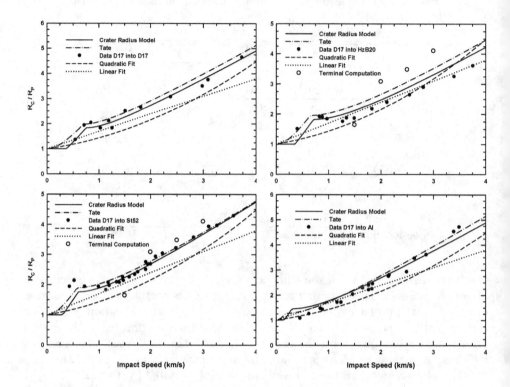

FIGURE 9.1. Crater radius data and model results for $L/D = 10$ tungsten alloy (D17) projectiles striking tungsten alloy (D17), HzB20 armor steel, St 52 steel, and aluminum alloy targets. The model results are from Eq. (9.7) developed here, Eq. (9.9) from Tate, and the empirical relations Eqs. (9.14) and (9.15). The open circles are $2g$ from Tables 7.11 and 9.2.

Experimentally, rigid penetration typically opens a channel that is the diameter of the projectile, as might be expected since there are no erosion products pushing out against the cavity. However, it is interesting to ask what the radius might be if the same assumption about later time expansion were used in the derivation of the crater radius for the case of the rigid projectile . Following the same reasoning as for the eroding computation above, the force over area is given by $F_z/A = B_3 + \frac{1}{2}\rho_t v^2$, so $(R_c)_0 = R_p$ would be assumed but the later expansion would say

$$\left(\frac{R_c}{(R_c)_0}\right)^2 = \left(\frac{R_c}{R_p}\right)^2 = \frac{B_3 + (1/2)\rho_t v^2}{B_3}. \tag{9.11}$$

Since (Eq. 8.29)

$$v_r^2 = \frac{2(Y_p - B_3)}{\rho_t}, \tag{9.12}$$

at v_r the crater radius is

$$\left(\frac{R_c}{R_p}\right)^2_{v=v_r} = \frac{Y_p}{B_3}, \tag{9.13}$$

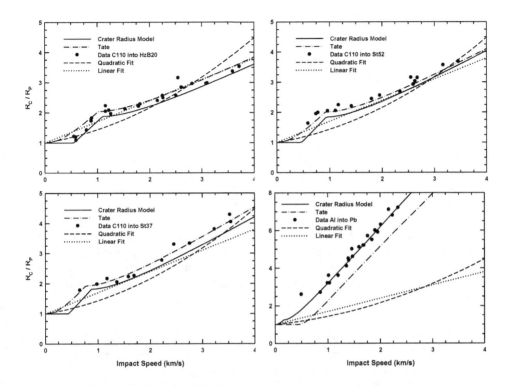

FIGURE 9.2. Crater radius data and model results for $L/D = 10$ steel (C110W1) projectiles striking HzB20 armor, St 52 steel, and St 37 steel targets. The lower-right figure is for $L/D = 10$ aluminum alloy projectiles striking lead targets. The model results are from Eq. (9.7) developed here, Eq. (9.9) from Tate, and the empirical relations Eqs. (9.14) and (9.15).

which equals the same value given by Eq. (9.7) when $v = v_r = u$. Using this expression for the crater radius leads to a continuous crater radius vs. impact velocity curve.

It is interesting to note that the crater radius data appear to have a kink similar to the equations at both v_c and v_r.

Finally we compare the crater radius to the terminal phase of penetration. Though it is difficult to examine the crater expansion directly, similar expansion behavior should be seen at the bottom of the crater during the terminal phase when the projectile completely erodes. In that case, the crater in the direction of penetration is still in motion. From numerical simulation results contained in Table 7.11 of a tungsten alloy penetrating steel, of interest are the values labeled g, which give the transient depth of penetration in terms of projectile diameter for the terminal phase. Thus, we might expect a correlation between $P_{\text{terminal}} \approx gD_p = 2gR_p$ and the crater radius (see Eq. 7.21). The values are shown rewritten in Table 9.2. The table also shows the value of the crater radius given by Eqs. (9.7) and (9.14) for the two most similar cases shown in Fig. 9.1 (upper right and lower left).

TABLE 9.2. Impact of tungsten alloy into steel target (numerical results of g from Table 7.11, crater radius model results, and quadratic curve fit).

Impact velocity (km/s)	Terminal penetration (in terms of R_p) $2g$	R_c/R_p D17 into HzB20	R_c/R_p D17 into St 52	R_c/R_p Quadratic curve fit
1.5	1.64	2.12	2.24	1.76
2.0	3.08	2.48	2.68	2.16
2.5	3.48	2.88	3.17	2.63
3.0	4.10	3.31	3.68	3.17
4.5	6.04	4.69	5.29	5.24

The values are also plotted in those graphs for comparison. Based on the assumed constitutive model for the steel target, it is expected that the correspondence is most similar for the experimental case of the impact into armor steel (HzB20). The trends are similar, showing there is some similarity in these two expansion events.

9.4. Two Empirical Relations

In addition to these expressions for crater radius, it is also possible to produce curve fits to the data. As an example, consider the left plot in Fig. 9.3, where the normalized crater radius is plotted vs. impact velocity for $L/D = 22.9$, $L = 12.18$ cm or 15.58 cm, tungsten rods into steel targets; the crater data are given in Table 9.3 [170]. The crater radius is well approximated by a quadratic fit in terms of the impact velocity that is required to pass through the value 1 at 0 km/s,

$$\frac{R_c}{R_p} = 1 + 0.2869V + 0.1457V^2 \quad (V \text{ in km/s}). \tag{9.14}$$

Hohler and Stilp's experimental data for tungsten alloy and steel projectiles impacting armor steel targets for different target hardnesses (BHN 255 and BHN 295) and projectiles with two different aspect ratios ($L/D = 1$ and 10) are presented in Chapters 8 and 10 (see Tables 10.2 and 10.3). In addition to data already presented, they have data for tungsten alloy (WA) projectiles with a density of 17.0 and 17.6 g/cm³. Based on their data, a second curve fit was obtained, relating normalized crater radius with the impact velocity. The various symbols in the right plot of Fig. 9.3 denote the different materials and different projectile aspect ratios. The Hohler–Stilp data indicate that the normalized crater radius is linear with impact velocity, in contrast to Silsby's data (left plot). Within the data scatter, little distinction can be made between the steel and tungsten alloy data, or between the two target hardnesses. The $L/D = 1$ crater data are slightly below the $L/D = 10$ data, but, for engineering purposes, we will assume that the relationship between impact velocity and crater radius is independent of aspect ratio. A linear least-squares curve fit through the $L/D = 10$ data, assuming R_c/R_p is 1 at $V = 0$, is

$$\frac{R_c}{R_p} = 1 + 0.70V \quad (V \text{ in km/s}). \tag{9.15}$$

Both these data fits were shown in all the plots in Figs. 9.1 and 9.2.

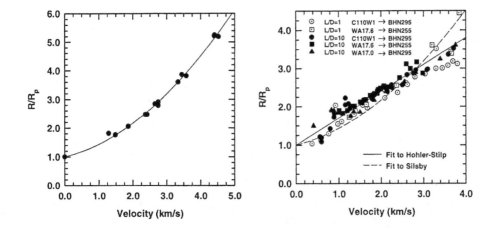

FIGURE 9.3. Normalized crater radius vs. impact velocity. Left, Silsby data [**170**]; right, Hohler and Stilp data [**10**]. (From Walker and Anderson [**199**].)

TABLE 9.3. Data of $L/D = 23$ Kennametal W10 tungsten alloy (90W-7Ni-3Fe) projectiles, swaged 18% (tensile yield 0.2% offset 1.17 GPa and tensile ultimate 1.19 GPa, compression yield at 0.2% offset 1.20 GPa and 1.32 GPa at 1% offset, hardness Rockwell C 40.6, density 17.3 g/cm³) impacting RHA with Brinell hardness ranging from 270.4 for 15.24 cm plate to 231.6 for 20.32 cm inch plate. Projectiles are at two scales: $L = 15.8$ cm, $D = 0.68$ cm and $L = 12.2$ cm, $D = 0.53$ cm (marked with *). [**170**]

Impact velocity (m/s)	Yaw (°)	P/L	R_c/R_p	Target thickness (cm)	
1,291	0	0.513	1.82	15.24	
1,494	0	0.719	1.76	15.24	
1,865	0.7	1.119	2.05	20.32	
2,365	0.6	1.356	2.46	20.32	*
2,409	0.5	1.415	2.47	20.32	*
2,653	0.7	1.466	2.84	20.32	
2,742	0.5	1.463	2.79	15.24	
2,746	2.4	1.448	2.91	20.32	*
3,335	2.6	1.524	3.60	15.24	
3,449	2.6	1.550	3.85	20.32	
3,580	0.2	1.549	3.81	15.24	*
4,398	4.7	1.586	5.22	15.24	
4,415	2.5	1.592	5.25	20.32	
4,525	2.2	1.591	5.19	20.32	*

Equation (9.14) has been used extensively to estimate the crater radius for many projectile and target materials for the Walker–Anderson penetration model. The crater radius is calculated from the initial impact velocity and is assumed constant throughout the calculation.

9.5. Where Does the Material Go?

For metal targets, because of their ductility, most of the crater volume is due to moving material aside, not detaching it from the target as ejecta. For large aspect ratio projectiles, little of the target material that occupied the crater is ejected as debris during the penetration process, usually less than 20%.

To learn more about the cratering and ejecta process, we consider data of Denardo and Nysmith at NASA Ames of impacts of aluminum spheres into aluminum targets at speeds ranging up to 7 km/s [54], augmented by work of Walker, Chocron, and Grosch at SwRI [204]. In the experiments, a range of sizes of aluminum spheres struck 2024-T4 and 2024-T351 aluminum (a relatively strong aluminum) and 1100-O aluminum (essentially pure, soft aluminum) targets. The diameters of the Al 2017-T4 impacting spheres were 0.1588, 0.3175, 0.635, and 1.27 cm in [54] and the diameter of the Al 6061-T6 impacting spheres was 3.0 cm [204]. These sizes correspond to a change in length scale of a factor of 18.9. Images of a 3.0-cm-diameter sphere impact into Al 2024-T351 are shown in Fig. 9.4.

In Section 3.15 we discussed linear size scaling, where P/L is independent of L. The same is true of D_c/D_p. Similarly, ejecta mass divided by projectile mass is independent of size scale, as is the momentum enhancement (to be defined) divided by impactor momentum, in linear size scaling. Data from these tests are presented in Fig. 9.5: the top row shows the penetration normalized by the projectile diameter P/L and the crater diameter normalized by projectile diameter D_c/D_p. The normalized crater depths overlay each other, save for the lower speed 1.27 cm spheres; hence they show linear size scaling. There is a slight size-scale dependence in the crater diameter, meaning that as the projectile size increases, D_c/D_p spreads, showing a dependence on initial projectile diameter at higher impact speeds. Interestingly, the crater diameter does not show a target strength dependence, while the penetration depth data do. The middle left image of Fig. 9.5 shows the aspect ratio of the craters, namely P/R_c. It is seen that for the 2024 aluminum the aspect ratio tends to 0.8, while for the 1100-O aluminum it tends towards 1. Thus, the craters for 2024 have a shallow aspect to them.

The targets were weighed before and after each test. These data provide the amount of material that was liberated from the target in the experiments. The middle right of Fig. 9.5 shows that the majority of the crater volume is not due to liberated material. The volume of target material liberated from the target is 20 to 40% of the volume of the crater. Impacts over a similar velocity range into 1100-O aluminum produced less than 10% of the crater volume being ejecta. The majority of the target material is simply moved aside or forward, away from the penetration channel, or into the crater lips for the Al 1100-O. Most of the deformation is plastic; there is minimal permanent elastic compression in the target, as can be seen by looking at the compression figures after the projectile has passed in Chapter 7 (Fig. 7.18). The fraction of target ejecta for larger L/D projectiles will be even less. At lower velocities, less than 2 km/s, the percentage of liberated material decreases.

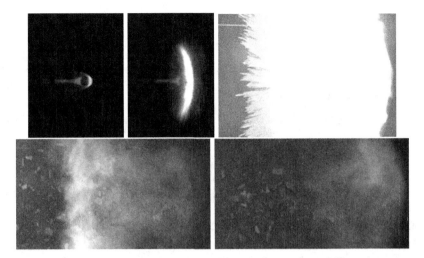

FIGURE 9.4. Aluminum sphere in flight at 5.77 km/s, shot from the two-stage light gas gun in Fig. 1.4, self illuminated from roughly 50 Torr atmosphere, impacting an Al 2024-T351 target, with subsequent impact flash and crater ejecta. (From Walker, Chocron, and Grosch [205].)

GENERAL PRINCIPLE 9.1. *For metal targets, the majority of the target material that originally occupied the crater region is simply moved aside, enlarging the outer dimensions of the target. At hypervelocities, 20 to 40% is ejected from the crater as debris, with less at lower impact speeds.*

The bottom left image shows the ejecta mass normalized by the projectile mass, where the plot is with respect to impact speed squared. The normalized ejecta mass has a strong dependence on the projectile diameter. For the smaller diameter spheres, each impacting sphere diameter has a different slope in the ejecta mass plot, showing that linear geometric size scaling does not apply. The larger sphere values do fall on a straight line; hence for larger sizes there is no size dependence in the amount of material liberated. The mass liberation is seen to be linear in impact speed squared for each diameter for Al 2024, implying a penetration stress or impactor kinetic energy dependence. To further emphasize the change in diameter size-dependent behavior, the bottom left image of Fig. 9.5 shows the normalized ejecta mass vs. $V^2\sqrt{D_p}$. The smaller diameter data collapses onto a straight line for the Al 2024-T4. This result is intriguing since, if the horizontal axis variable is multiplied by density, the dimensions of a stress times the square root of a length, or fracture toughness, result. However, the plot clarifies that if the same size scaling for ejecta mass occurred for the 3.0-cm-diameter projectile as occurred for the smaller projectiles, then the ejecta mass would be *twice* what it is. Thus, a large change in behavior has occurred. This change will be explored in the next section in the context of a damage model. For Al 1100-O, however, it appears that better agreement between the two sizes is the left plot with no size dependence.

Careful examination of the craters from the 3.0-cm-diameter impactors into Al 2024-T351 reveals numerous shear surface initiation sites, which are lines on the crater surface, including at the bottom of the crater. The shear bands extend into

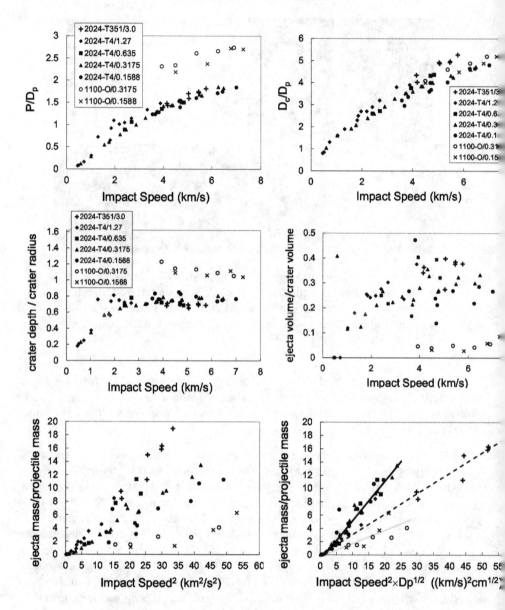

FIGURE 9.5. Crater depth and diameter (upper left and right), crater aspect ratio (middle left) and fraction of crater volume due to ejecta material (middle right), and ejecta mass (lower left and right) of impacting aluminum spheres (diameters in cm) into stated aluminum alloy (data from [**54, 204**]). (From Walker, Chocron, and Grosch [**205**].)

the target material as (not necessarily flat) planes. Towards the sides of the crater these shear bands move into the target at the classical 45° to the surface plane, but closer to the bottom of the crater the angle is more shallow. The intersection of these shear planes leads to the material liberation as fragments – that is, it is the intersection of shear planes that leads to fragment formation and material separation as ejecta from the target. Around the exterior of the crater is a shelf of liberated material approximately 0.9 to 1.1 cm deep, but no lip formation (Fig. 9.6). This shelf is primarily formed by a tensile separation, and not the formation and intersection of shearing planes, as it appears to have ductile dimples implying a void nucleation and growth in tension, as opposed to the interior crater failure surface and recovered fragments surfaces which appear to be relatively smooth shearing planes. Based on the impact high-speed video footage and recovered fragments, most fragments from the crater are relatively flat, that is, two directions are larger than the third. The largest fragments have a lateral extent on the order of the size of the impactor, indicating the lateral spacing of the shear bands initiation sites (assuming they initiate on the crater surface) seen in the crater bottom are separated at least this far (Fig. 9.6). The flatness of the fragments also implies that the propagation of the shear bands or planes into the target material is shallower than 45° (with the angle measured with respect to the tangent plane to the surface), as it appears the fragments are formed as flakes when these shear bands intersect beneath the crater surface.

There is a direct correlation between the size-scale behavior of the crater volume and the ejecta mass size-scale behavior. We assume the crater volume is a half ellipsoid based on the measured crater diameter and crater depth, $V_c = \pi D_c^2 P/6$. We further assume that the crater volume partitions into a part owing to material plastic deformation, denoted V_Y, and a part due to ejecta mass, V_e. With the projectile volume given by V_p,

$$\frac{V_c}{V_p} = \frac{V_Y + V_e}{V_p} = \frac{V_Y}{V_e} + \frac{\rho V_e}{\rho V_p} = \frac{V_p}{V_e} + \frac{m_e}{m_p}, \tag{9.16}$$

where m_e and m_p are the masses of the ejecta and projectile, respectively. Then,

$$\frac{V_Y}{V_p} = \frac{V_c}{V_p} - \frac{m_e}{m_p} = \left(\frac{D_c}{D_p}\right)^2 \frac{P}{D_p} - \frac{m_e}{m_p}. \tag{9.17}$$

V_Y/V_p is shown in the top left of Fig. 9.7. Since the points overlay, the data says that V_Y/V_p is roughly a function of the impact speed V only – that is, $V_Y/V_p = f(V)$. Since it is a function of v only, the nonlinear scaling behavior of crater volume for smaller radii impactors is due solely to the nonlinear scaling behavior of the ejecta mass. The pushed-aside volume controlled by plastic flow does not have a nonlinear size-scaling behavior, as it only depends on velocity, and thus the nonlinear size-scaling behavior of the crater volume is due to mass loss. The conclusion regarding V_Y agrees with our thinking that flow stress does not show a size-scale dependence, so the resulting plastic deformation should not, either. This example ties back the original discussion about scaling in impact and the notion that the lack of direct geometric scaling is likely due to damage processes occurring in the target materials.

Given the square root of the projectile diameter used in Fig. 9.5 in looking at ejecta mass scaling, we are encouraged to look to fracture mechanics, which scale as the square root of the crack length. It should be pointed out that fracture mechanics is not a local damage model; rather, it is a structural failure model. A representative length of the plastic zone around the crack tip from fracture mechanics is

$$r_p = \frac{1}{\pi} \left(\frac{K_I}{Y}\right)^2. \tag{9.18}$$

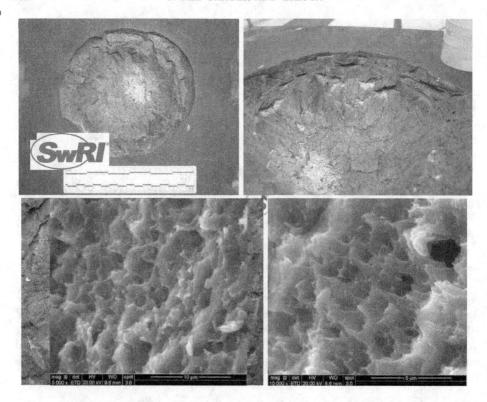

FIGURE 9.6. Impact crater for the 5.77 km/s impact showing
shearing planes intersecting the crater surface, removed material
from intersection of shearing planes, and upper shelf of material removed (left). Also shown is the shearing surface on a piece of material that was separated from target but was geometrically trapped,
showing dimples indicative of ductile fracture (right). The length
bar is 5 μm, so the scanning electron microscope image is 12.75 μm
across. (From Walker, Chocron, and Grosch [**205**].)

If we use the critical stress intensity factor K_{Ic} in this expression, then we arrive at
a characteristic length of plastic deformation associated with the fracture failure of
the material; hence a material property with units of length. For Al 2024-T4, using
a typical flow stress of 500 MPa and a fracture toughness of 25 MPa$\sqrt{m}$, that length
is approximately $r_p = 0.8$ mm. This value of r_p could be describing a length scale
which is interesting and relevant for our impact problem, given the transition in
behavior observed in the ejecta mass as a function of impactor diameter, though
we cannot draw a conclusion about the transition in behavior from the fracture
mechanics arguments we have presented here. In the bottom right of Fig. 9.5, two
solid lines are drawn in the ejecta mass vs. $V^2\sqrt{D_p}$, one for 2024 and one for 1100
aluminum. The slopes of the lines differ by a factor of three, which is similar to the
difference in the fracture toughness of these materials, as Al 1100-O has a fracture
toughness of about $K_{Ic} = 75$ MPa$\sqrt{m}$, suggesting that crater ejecta mass correlates
with the inverse fracture toughness of the target material for the smaller impactor

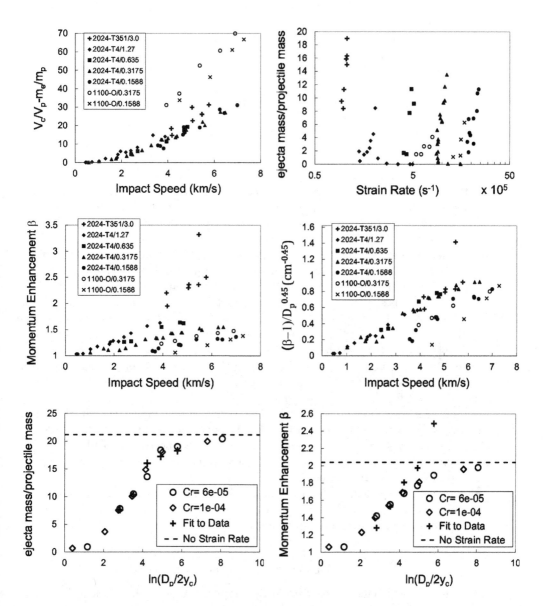

FIGURE 9.7. V_Y/V_p vs. impact speed (from Eq. (9.17)) (top left) and normalized ejecta mass vs. strain rate $\dot{\varepsilon}_p \approx V/P$ (top right), momentum enhancement (middle left and right), and computational comparison with the shear-band-based model for ejecta mass (lower left) and momentum enhancement (lower right), all for impacting spheres (diameters in cm) into stated alloys. Note that on the bottom row there are five points on each plot (data from [**54, 204, 205**]). (From Walker, Chocron, and Grosch [**205**].)

sizes; more brittle materials have more crater ejecta, while more ductile materials have less.

Also measured in these tests is the amount of momentum transferred to the target fixture. The target was attached to a pendulum and the maximum swing was used to determine the momentum transferred to the target. This momentum is greater than the projectile momentum because of the ejecta. It is quantified by the *momentum enhancement* β,

$$\beta = \frac{\text{target momentum}}{m_p V} \geq 1. \tag{9.19}$$

The momentum enhancement is shown in the middle row of Fig. 9.7. The plots, both in terms of impact speed, show that momentum enhancement is linear in impact speed for a given projectile diameter. This fact alone is interesting, given that ejecta mass behaves as impact speed squared. Thus, there is not a linear relation between momentum enhancement and ejecta mass. The left plot shows that there is a strong size-scaling effect in momentum enhancement, just as there was in ejecta mass. For all but the smallest (0.15875 cm diameter) size projectiles, empirically the additional momentum $\beta - 1$ scales on impactor diameter to the 0.45 power, as is shown in the center-right plot (this is the power used in the plot; it is between 0.4 and 0.5). The fact that there is no change in behavior in the momentum enhancement size scaling in the region of 1 cm is quite intriguing since there was a quantitative change in the ejecta mass scaling. Since less material is being liberated than we expected because of the transition in ejecta mass loss from nonlinear size scaling to linear size scaling, we know that the average speed of ejecta has increased for the larger scale size.

There is one anomalous point of high momentum enhancement. In this test a chain holding the aluminum block to the heavy steel pendulum broke. This test demonstrates that two bodies in contact but weakly connected can lead to large momentum enhancements, perhaps up to double, as is seen with elastic body impacts of significantly different sizes. The value is not representative of the aluminum material impacts (though dividing the result by two gives a lower bound on the momentum enhancement), but it provides insight on possible effects of impacting bodies comprised of consolidated pieces of material that are in contact but weakly bound.

9.6. A Shear Band Motivated Damage Model

The size-scale effects observed in this impact response imply a length scale or a time scale in the failure process. Locally, this size scale in dynamic problems can be reflected in a strain-rate dependence of the material behavior. Given the lack of a size-scale effect in the crater depth results, we conclude that the strain rate dependence of the flow stress is not greatly influencing the results, since increasing strength at smaller projectile diameters (hence higher strain rates) would decrease the depth of penetration, something that is not observed. Thus, we look at a strain rate effect for the damage. Looking at the craters after the tests shows that material is liberated as ejecta fragments when shear planes intersect. Given this behavior, we explore a failure model based on shear bands or planes. The overarching idea is that shear bands form once a certain strain is reached and then the band slips until it has displaced a certain distance at which point the material fails (separates). Thus, there is a length scale that is not observable by examining the virgin material under a microscope, but is a length that a shear band slides before the material separates (fails). Since hydrocodes do not directly address discontinuities in the material displacements, the shear band is modeled assuming a constant strain rate for a region and we determine when the critical displacement of the shear band is reached. This step will convert the nonlocal shear band slippage to a local failure model that can be evaluated using just the strain rate. The specific development follows the fragmentation studies of Kipp, Grady, Olsen, and Mott [**80, 83, 84, 120**]. Fragmentation will be discussed in Chapter 11.

We begin by assuming the material is undergoing a shearing strain at a constant strain rate $\dot{\varepsilon}_{12}$. A velocity in the y direction that depends on the x location is assumed, so that the velocity field, before the shear band begins to slide, is

$$v_y(x, y) = 2\dot{\varepsilon}_{12} \cdot x. \tag{9.20}$$

The rate of deformation tensor is $D_{12} = \dot{\varepsilon}_{12}$. Plastic shearing occurs until a critical strain is reached at which the shear band begins to slip. That time will be taken as time zero. At this time, the slip along $x = 0$ begins. The slip will be denoted by $y_s(t)$ with $y_s(0) = 0$. We assume that the material to the right of the slip moves as a rigid body to a point denoted $x_r(t)$, and that $x_r(t)$ marks how far the rigid region has propagated into the region undergoing shearing deformation. The velocity field is thus given by

$$v_y(x, y) = \begin{cases} 2\dot{\varepsilon}_{12}x_r(t), & 0 \le x \le x_r(t), \\ 2\dot{\varepsilon}_{12}x, & x > x_r(t). \end{cases} \tag{9.21}$$

In the plastically shearing region, where $x > x_r(t)$, the plastic shearing stress is assumed from the von Mises flow, $\sigma_{12} = Y/\sqrt{3}$. The shear stress on the shear band will be assumed to be a linear decreasing function of displacement until the critical length y_c is reached,

$$\sigma_{12}(0, y) = \frac{Y}{\sqrt{3}}\left(1 - \frac{y_s(t)}{y_c}\right). \tag{9.22}$$

(If one views the material to the left of the y-axis to be shearing also, then the critical length is twice the stated value. The slip displacement is then twice the stated value, so Eq. (9.22) still holds. We will continue our reasoning just considering the half space, though in the end we will report in terms of $2y_c$.) The momentum balance in the y direction is integrated over the rigidly moving length from 0 to $x_r(t)$ to give

$$\frac{\partial \sigma_{12}}{\partial x} = \rho \frac{\partial v_y}{\partial t} \quad \Rightarrow \quad \sigma_{12}(x_r(t), y) - \sigma_{12}(0, y) = \rho x_r(t)\frac{\partial v_y(0, y)}{\partial t}. \tag{9.23}$$

This says

$$\frac{Y}{\sqrt{3}} - \frac{Y}{\sqrt{3}}\left(1 - \frac{y_s(t)}{y_c}\right) = \rho x_r(t)\frac{\partial}{\partial t}(2\dot{\varepsilon}_{12}x_r(t)), \tag{9.24}$$

which yields

$$\frac{Y}{\sqrt{3}}\frac{y_s(t)}{y_c} = 2\rho\dot{\varepsilon}_{12}x_r(t)\frac{dx_r}{dt}. \tag{9.25}$$

The slip motion at the shear band is given by

$$\frac{dy_s}{dt} = v_y(0, y_s) = 2\dot{\varepsilon}_{12}x_r(t). \tag{9.26}$$

Inserting this into Eq. (9.25) yields

$$\frac{Y}{\sqrt{3}}\frac{y_s(t)}{y_c} = \frac{\rho}{2\dot{\varepsilon}_{12}}\frac{dy_s}{dt}\frac{d^2y_s}{dt^2}. \tag{9.27}$$

A solution of the form $y_s(t) = at^b$ is assumed for this nonlinear ordinary differential equation, and the result is

$$y_s(t) = \frac{1}{9\sqrt{3}}\frac{Y}{\rho}\frac{\dot{\varepsilon}_{12}}{y_c}t^3 = \frac{1}{18}\frac{Y}{\rho}\frac{\dot{\varepsilon}_p}{y_c}t^3. \tag{9.28}$$

The second equality writes the shear strain rate in terms of the equivalent plastic strain rate, $\dot{\varepsilon}_p = 2\dot{\varepsilon}_{12}/\sqrt{3}$, for convenience. The shear band slips enough to lead to material failure when $y_s(t_f) = y_c$, so the failure time is

$$t_f = \left(18\frac{\rho}{Y}\frac{y_c^2}{\dot{\varepsilon}_p}\right)^{1/3}. \tag{9.29}$$

This allows a calculation of a strain to failure of the form[1]

$$\varepsilon_p^f = \varepsilon_{p0}^f + \dot{\varepsilon}_p t_f = \varepsilon_{p0}^f + \dot{\varepsilon}_p^{2/3}y_c^{2/3}\left(18\frac{\rho}{Y}\right)^{1/3} = \varepsilon_{p0}^f + C_r\dot{\varepsilon}_p^{2/3}. \tag{9.30}$$

We will take C_r as a constant and assume a representative value of flow stress Y. The relationship between C_r and y_c is

$$C_r = y_c^{2/3}\left(18\frac{\rho}{Y}\right)^{1/3}, \qquad y_c = \frac{C_r^{3/2}}{3\sqrt{2}}\sqrt{\frac{Y}{\rho}}. \tag{9.31}$$

The expression for the failure strain shows the relationship between the failure length scale, given by how far the shear band slips, and the strain rate. Damage is accumulated in the model, as was done in Chapter 3, $D = \int(1/\varepsilon_p^f)d\varepsilon_p$, and when the damage value of $D = 1$ is reached the strength Y is set equal to zero.

This shear band model is a mechanical model of nonlocal behavior, namely the slip of the shear band over a certain finite distance. This model gives rise to a local evaluation of strain rate by the assumption that over a certain region the strain rate is constant. The strain rate term arises, as can be seen by examining the expression for C_r, due to inertia, since if the density of the material were zero $C_r = 0$ and there would be no strain rate effect (the shear band slip would move at infinite speed, leading to immediate failure). The strain rate expression is not directly from a local material constitutive model and is thus a failure model with nonlocal origin, owing to slipping occurring over the finite critical length y_c.

We are interested in the physical scale of events that underlie this nonlocal development. Using an analytic flow field for penetration presented in the next chapter, the strain rate in the crater beneath the projectile is $\dot{\varepsilon}_p = 2u/R$, where u is the penetration speed and R is the crater radius. For aluminum into aluminum impacts, the penetration speed is half the impact speed and we will take the crater depth to be indicative of the radius; hence a characteristic strain rate is $\dot{\varepsilon}_p = V/P$. From this expression, characteristic strain rates are shown in Fig. 9.7 (upper right).

---

[1]Following the reasoning begun in Kipp and Grady [120] and continued in Grady and Olsen [84], Grady [80] arrives essentially at Eq. (9.30) as an estimate of the fracture strain for tensile expanding ring fragmentation based on a linear reduction in tensile stress as a function of crack opening displacement and a total energy to tensile fracture. His equation is in terms of work to failure Γ, not y_c. For us the work to failure on the slipping shear band is $\Gamma = \int_0^{y_c} \vec{t}\cdot d\vec{x} = \int_0^{y_c} \sigma_{12}dy = Yy_c/2\sqrt{3}$ and his Eq. (8.11) differs from ours by a factor of 3 within the cube root.

Extensive computations, described below, determined the value $C_r = 6.0 \times 10^{-5}$ s$^{2/3}$ for Al 2024-T351, a value that leads to computations that match extremely well the ejecta mass experimental results. Based on the Johnson–Cook constants for Al 2024-T351, for a strain of 0.2 a representative flow stress is 500 MPa, and with a density of 2.785 g/cm^3 the critical length is $2y_c = 0.0093$ cm$= 93$ μm. The lateral motion of x_r that occurs during the shear band slip – that is, the rigidly moving material at the time of failure – is (from above)

$$\frac{x_r(t_f)}{y_c} = \left(\frac{Y/\sqrt{3}}{2\rho}\right)^{1/3} \frac{1}{\dot{\varepsilon}_p^{2/3} y_c^{2/3}} = \frac{\sqrt{3}}{C_r \dot{\varepsilon}_p^{2/3}}. \tag{9.32}$$

With the lower end strain rate of interest to us being around 1×10^5 s^{-1} (corresponding to the 3-cm-diameter impactor) and the value $C_r = 6.0 \times 10^{-5}$ s$^{2/3}$, we find the maximum lateral extent is $x_r(t_f)/y_c \approx 13.4$. On the upper end of strain rates for our smallest impactor 0.1588 cm, around 2×10^6 s^{-1}, the extent is $x_r(t_f)/y_c \approx 1.8$. Thus, the physical scale that this model corresponds to is around 0.1 mm to 1 mm. We now have three length scales: $r_p = 0.8$ mm, $2y_c = 0.093$ mm, and the dimple length in the scanning electron microscope (SEM), 0.005 mm, but none of them seems directly comparable, all differing by a factor of 10.

For the static failure model, we use the model described in Eq. (3.201) in Chapter 3, so that $\varepsilon_{p0}^f = \varepsilon_p^f(\eta, \zeta)$. The specific constants presented with that model development were for Al 2024-T351. The requirement that damage only occurs and accumulates when there is a tensile stress is an important feature, which prevents damage (but not plastic flow) directly below the projectile in the crater. Damage does accumulate to the sides of the crater, leading to the observed flatter craters.

The above strain-rate dependent damage model was implemented in CTH. The Johnson–Cook flow stress model using constants in Appendix C was used to model the strength of Al 2024-T351 (with the exception that thermal softening was suppressed). Exploratory computations showed that model results matched data for the value of $C_r = 6.0 \times 10^{-5}$ s$^{2/3}$. Comparisons of the model[2] with data are shown in the bottom row of Fig. 9.7, where the comparison is in terms of $D_p/2y_c$, all at the impact speed of 5.77 km/s. To do the comparison at the same impact speed for each impactor diameter, linear least square fits were performed for the ejecta mass in terms of impact speed squared for each projectile diameter and linear least-squares fits were performed for the momentum enhancement in terms of impact speed, again for each projectile diameter. These fits were then evaluated at 5.77 km/s to provide ejecta mass and β for the impactor diameters tested. The value of $C_r = 6.0 \times 10^{-5}$ s$^{2/3}$ was used to plot the data in terms of $D_p/2y_c$, which is the same value that produces the excellent results shown in the plot. A series of computations was also performed using $C_r = 1 \times 10^{-4}$ s$^{2/3}$, and shows how the computational model scales in terms of C_r, which is the only size scale dependent variable in the damage model (and is the only new constant introduced by the shear band model). It is seen that this model transitions from a

---

[2]Important details regarding the computations can be found in [**205**], including convergence questions and the fact that there is a lot of variability in the computations as is typical when fracture occurs, since cracks initiate at different locations and run along different paths when the initial conditions are slightly perturbed (which was intentionally done to provide an ensemble of results for analysis). Computations were axisymmetric (hence two dimensional) with uniform zoning throughout the mesh and had at least 60 computational cells across the projectile radius.

small amount of ejecta mass to a maximum amount of ejecta mass over a relatively narrow region, essentially from $\ln(D_p/2y_c)$ from 1 to 5, or D_p from $2.72 \cdot 2y_c$ to $148 \cdot 2y_c$, or in particular over the range 0.025 cm to 1.38 cm. It agrees very well with the experimental data. For momentum enhancement, however, the computations do not reflect the continued increase in terms of scale. The computations are well below the experimental $\beta = 2.50$ for the 3 cm impactor; close examination of the results also shows that the slopes differ for small diameters. Thus, there is still work to do in correctly computing momentum enhancement. With extensive fragmentation we have exited the realm of continuum mechanics, and it could be that the post-failed constitutive model is not adequate or there are issues in the numerical algorithms addressing relative motion of separate pieces of material.

9.7. Damage Saturation

One explanation for the transition in behavior from scaling that depends on projectile diameter to scaling that no longer depends on projectile diameter is that there is a saturation in the damage process within the material, meaning that for larger scale size there is sufficient time for the damage process to complete. This implies that crater ejecta is more easily liberated for the larger diameter impactor shots. A corresponding implication is that the material, which has traveled along the crater wall, is launched out of the crater at a higher speed. A larger average speed of ejecta material agrees with the results of these experiments, which is implied by a continued increase in momentum enhancement (as a function of projectile diameter) even though ejecta mass has asymptoted.

One way to think of the saturation is to look at how the shear band slip failure scales. If the material is undergoing deformation at a constant strain rate, then once the strain ε_{p0}^f is reached the subsequent required time to reach failure is t_f as given by Eq. (9.29). Beginning at a base scale, we go to another scale, denoted with a *, with the scale factor λ. Hence, $L^* = \lambda L$ (where L is a length) and $t^* = \lambda t$. The failure time scales as

$$t_f^* = \left(18 \frac{\rho}{Y} \frac{y_c^2}{\dot{\varepsilon}_p^*} \right)^{1/3} = \left(18 \frac{\rho}{Y} \frac{y_c^2}{(\dot{\varepsilon}_p/\lambda)} \right)^{1/3} = \lambda^{1/3} t_f. \tag{9.33}$$

The impact event has a characteristic time t_c, primarily controlled by inertia and material strength, where $t_c^* = \lambda t_c$. The fraction of event time required for failure scales as

$$\left(\frac{t_f}{t_c} \right)^* = \frac{1}{\lambda^{2/3}} \frac{t_f}{t_c}. \tag{9.34}$$

Thus, as the scale size gets larger the fraction of time required for fracture decreases, leading to more damage. This explains why the larger target appears weaker and more damage occurs for larger scales. The damage saturation is due to this time dependent part of the damage to occur (relatively) more quickly at large scale and so the ejecta mass is controlled by the static failure strain of the material.

With damage saturation there is sufficient time for the damage process to complete. Even though complete, the earlier the relative time of failure the more easily the crater ejecta is liberated for the 3-cm-diameter impactor shots. This could explain why the momentum enhancement is still on its same nonlinear size-scaling behavior at 3.0-cm-diameter impactors, even though the ejecta mass has ceased its nonlinear size scaling.

The additive form of the shear-band-based model determines a diameter range over which damage goes from minimal damage to saturated maximal damage. Given that the Al 2024-T351 damage model has ε_{p0}^f ranging between 0.2 and 0.5, it seems reasonable that damage will have saturated when the additive term only increases the failure strain by 0.1 and there will be minimal failure when the total failure strain is 2. Since the strain rate scales as $1/R_p$, we conclude that the size range in question is $(1.5/0.1)^{1/(2/3)} = 58$, where the 2/3 power is the form of the strain rate in the additive damage model. Thus, the size scale for the transition is fairly narrow and can be estimated without performing hydrocode computations.

9.8. Sources

Much of this chapter is new material, though some is from Tate [182], Walker and Anderson [197, 199], and Walker and Chocron and Grosch [202, 204, 205].

Exercises

(9.1) Given Eq. (9.9) for the initial crater radius and assuming that the penetration velocity is given by Tate's equation with $R_t = B_3$, show that for $v > v_r$

$$1 \leq \left(\frac{(R_c)_0}{R_p} \right)^2 = \frac{Y_p + 2\rho_p(v-u)^2}{B_3 + (1/2)\rho_t u^2} \leq 4, \tag{9.35}$$

and hence that the maximum initial crater radius $(R_c)_0$ is $2R_p$.

(9.2) For $Y_p < B_3$, find the value of the crater radius expression at $v = v_c$ and show that $R_c < 2$ there.

(9.3) In an earlier paper, Tate showed with data he had, in line with original observations of Christman and Gehring, that the volume of the crater appears to be directly proportional to the kinetic energy of the impactor [180]. Thus,

$$\frac{1}{2}\rho_p \pi R_p^2 L V^2 = k\pi R_c^2 P. \tag{9.36}$$

Desiring the constant k to be related to the strength, comparison with his data implied $k \approx R_t/2$. This assumption gives the expression for crater radius as

$$\left(\frac{R_c}{R_p} \right)^2 = \frac{\rho_p V^2}{R_t(P/L)}. \tag{9.37}$$

Show that this expression gives good results for the aluminum alloy into lead data, but not particularly good results for the tungsten into steel data. (Just pick three velocities for each, and perform the calculation using P/L from the experimental data.)

(9.4) On July 4, 2005, the Deep Impact mission impactor struck comet Temple 1 at 10.2 km/s. The Deep Impact impactor massed 350 kg and was roughly a sphere of copper. Assume the copper strength was 250 MPa, and that the comet nucleus surface was made of fully dense ice with density 0.91 g/cm^3 with strength 2 MPa. What is the predicted late-time crater radius based on the approach in this chapter? (It is likely that the comet material is quite porous, which leads to more complicated mechanics than assumed. The ejecta from the impact obscured the crater so it was not measured by the Deep Impact flyby spacecraft. Six years later, the Stardust spacecraft flew by Temple 1 and the impact site was identified. Though there is no clearly

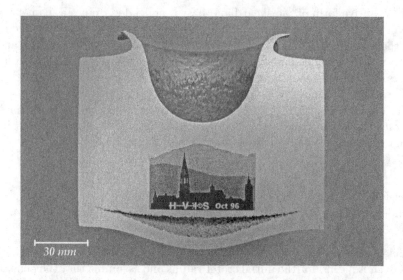

FIGURE 9.8. Crater and spall plane produced by a 1-cm-diameter aluminum sphere striking an aluminum target block at 7 km/s distributed by Ernst-Mach-Institut to attendees of the 1996 Hypervelocity Impact Symposium in Freiburg, Germany.

demarcated crater rim, it appears the outer radius is 75 m. It also appears the crater has a central mound (uplift).)

(9.5) At the 1996 Hypervelocity Impact Symposium held in Freiburg, Germany, the hosting institution, the Ernst-Mach-Instit gave the aluminum target shown in Fig. 9.8 to each attendee as a memento. It was struck by a 1-cm-diameter aluminum sphere at 7.0 km/s. Measure the crater depth and radius and compare to the data in this chapter. Is the target a strong or soft aluminum? Notice the smooth crater and lack of apparent shear bands. The spall plane at the back is clearly a ductile void-growth-linking failure owing to its dimpled surface. Assuming that energy decays geometrically as $1/r^2$, argue that stress and strain decay as $1/r$ (hint: see Exercise 6.18). From this behavior estimate a rough spall stress based on the impact stress and the penetration stress. Which one is more reasonable?

(9.6) Suppose that instead of Eq. (9.22) the following expression is used to describe the separation stress for the shear band model,

$$\sigma_{12}(0, y) = \frac{Y}{\sqrt{3}} \left\{ 1 - \left(\frac{y_s(t)}{y_c} \right)^n \right\}, \tag{9.38}$$

where $n > 0$ is arbitrary. Solve the shear band model assuming this expression, and show the surprising result that the same form of Eq. (9.29) and Eq. (9.30) still ensue, with only differences in the numerical constants. Thus, this type of model gives rise to failure strain depending on the 2/3 power of strain rate, independent of the specifics of the separation.

(9.7) Compute the critical length r_p for Al 1100-O and compare to Al 2024-T4.

The Walker–Anderson Model

We are now in a position to present the central penetration model of this book, specifically for the case of metals into metals. The Walker–Anderson model is a first principles physics-based model where a series of assumption are made, many of which were motivated by analysis of results of numerical simulations. The model uses material properties for the projectile and target and computes the penetration depth vs. time during the penetration event. The model, based on the momentum balance from Newton's second law, includes transient effects and in particular reproduces the early and late-time behavior seen in the impact event. A desire was to retain the overall simplicity of the Tate–Alekseevski model, and the model developed has the satisfying behavior of reducing to the original Tate model upon removing its more sophisticated geometry.

10.1. The Centerline Momentum Balance

The projectile and target, considered axisymmetric, lie along the z-axis which is the axis of symmetry. The location of the interface between the projectile and the target is $z_i(t)$, with $z_i(0) = 0$. The rear of the projectile is $z_p(t)$, and $z_p(0) = -L_0$, where L_0 is the initial length of the projectile. The velocity along the centerline in the projectile and target is $v_z(z)$. With these definitions, the interface velocity u and the velocity of the back end of the projectile v are given by

$$u = \frac{dz_i}{dt} = v_z(z_i),$$

$$v = \frac{dz_p}{dt} = v_z(z_p).$$

(10.1)

The target is assumed to be semi-infinite. Figure 10.1 provides a schematic of the projectile and target at a time $t > 0$.

A central theme of the model is the use of the momentum balance along the z-axis, which will be referred to as the *centerline momentum balance*, the basic derivation of which is as follows. The z component of the general momentum conservation equation (Eq. 2.24) is

$$\rho \frac{Dv_z}{Dt} = \rho \left(\frac{\partial v_z}{\partial t} + v_x \frac{\partial v_z}{\partial x} + v_y \frac{\partial v_z}{\partial y} + v_z \frac{\partial v_z}{\partial z} \right) = \frac{\partial \sigma_{xz}}{\partial x} + \frac{\partial \sigma_{yz}}{\partial y} + \frac{\partial \sigma_{zz}}{\partial z}. \quad (10.2)$$

Assuming axial symmetry along the z-axis, on the axis itself $x = y = 0$ and $v_x = v_y = 0$ by symmetry. Along the axis, since the x and y directions are equivalent, we may write

$$\frac{\partial \sigma_{xz}}{\partial x} + \frac{\partial \sigma_{yz}}{\partial y} = 2 \frac{\partial \sigma_{xz}}{\partial x}. \quad (10.3)$$

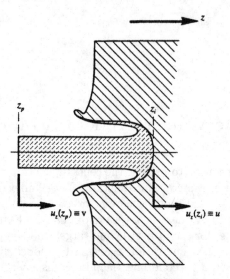

FIGURE 10.1. Schematic of projectile and target with coordinate notation. (From Walker and Anderson [199].)

The momentum balance along the centerline thus simplifies to

$$\rho\frac{\partial v_z}{\partial t} + \frac{1}{2}\rho\frac{\partial(v_z)^2}{\partial z} - \frac{\partial\sigma_{zz}}{\partial z} - 2\frac{\partial\sigma_{xz}}{\partial x} = 0. \tag{10.4}$$

We now integrate the momentum balance equation along the centerline through the target and projectile, from the back of the projectile $z_p(t)$ to the end of the target at $z = +\infty$. Assuming that the change in density in both the target and projectile is negligible (it is not being assumed that the materials are incompressible, merely that the density changes are sufficiently small that the density terms can be pulled out of the integrals), then integration of Eq. (10.4) gives

$$\rho_p\int_{z_p}^{z_i}\frac{\partial v_z}{\partial t}dz + \rho_t\int_{z_i}^{+\infty}\frac{\partial v_z}{\partial t}dz + \frac{1}{2}\rho_p\left.v_z^2\right|_{z_p}^{z_i} + \frac{1}{2}\rho_t\left.v_z^2\right|_{z_i}^{+\infty}$$

$$- \left.\sigma_{zz}\right|_{z_p}^{+\infty} - 2\int_{z_p}^{+\infty}\frac{\partial\sigma_{xz}}{\partial x}dz = 0. \tag{10.5}$$

Since the target material for large z is not participating in the penetration event, $v_z(+\infty) = 0$ and $\sigma_{zz}(+\infty) = 0$ (the stress decays as $1/z^2$ for large z, per Eq. 8.73). The rear surface of the projectile is a free surface and hence stress free, so $\sigma_{zz}(z_p) = 0$. Combining these facts with the definitions in Eq. (10.1) yields

$$\rho_p\int_{z_p}^{z_i}\frac{\partial v_z}{\partial t}dz + \rho_t\int_{z_i}^{+\infty}\frac{\partial v_z}{\partial t}dz + \frac{1}{2}\rho_p\left(u^2 - v^2\right) - \frac{1}{2}\rho_t u^2 - 2\int_{z_p}^{+\infty}\frac{\partial\sigma_{xz}}{\partial x}dz = 0. \tag{10.6}$$

To integrate the equation further, assumptions must be made concerning the velocity profile $v_z(z)$ and the shear stress behavior.

Also of interest to us are the integrals of the centerline momentum balance over just the projectile and target. Computing these integrals will make it straightforward in later chapters to change the constitutive model considered in the target and to adjust to targets of finite thickness. For example, if the integral along the centerline is taken only through the projectile, then an explicit expression for the stress at the projectile–target interface can be written as

$$\sigma_{zz}(z_i) = \rho_p \int_{z_p}^{z_i} \frac{\partial v_z}{\partial t} dz + \frac{1}{2}\rho_p \left(u^2 - v^2\right) - 2 \int_{z_p}^{z_i} \frac{\partial \sigma_{xz}}{\partial x} dz. \tag{10.7}$$

Similarly, for the integration of the momentum balance only through the target material one obtains

$$\sigma_{zz}(z_i) = -\rho_t \int_{z_i}^{+\infty} \frac{\partial v_z}{\partial t} dz + \frac{1}{2}\rho_t u^2 + 2 \int_{z_i}^{+\infty} \frac{\partial \sigma_{xz}}{\partial x} dz. \tag{10.8}$$

The last term on the right-hand side represents the resistance of the target due to shear stress. It is of central importance in modeling penetration. We refer to it as S_t,

$$S_t = -2 \int_{z_i}^{+\infty} \frac{\partial \sigma_{xz}}{\partial x} \bigg|_{x=0} dz. \tag{10.9}$$

We place a minus sign in the definition so that S_t is positive. Similarly, there is a shear stress term in the projectile,

$$S_p = 2 \int_{z_p}^{z_i} \frac{\partial \sigma_{xz}}{\partial x} \bigg|_{x=0} dz. \tag{10.10}$$

In the model development, we will not address or include this term, essentially assuming $S_p = 0$, though there will be a comment on it later.

As an interesting application, if the velocity profile in the target is given by $v_z(z,t) = u(t)f(z - z_i(t))$, where $f(0) = 1$ and $f(+\infty) = 0$, then Eq. (10.8) gives

$$\sigma_{zz}(z_i) = -\left(\frac{1}{2}\rho_t u^2 + S_t\right) - \dot{u} \int_0^\infty f(z)dz. \tag{10.11}$$

Thus the stress at the nose tip of the projectile has a Poncelet form of penetration resistance, where S_t depends only on shear stress, plus some transient terms. Since the projectile is decelerating, the transient terms decrease the magnitude of the stress. The integral multiplying $\dot{u}$ roughly equals the crater radius, as shown below.

10.2. The Model

There are six physical constructions that are made in the Walker–Anderson model. In terms of their importance, there are three major and three minor ones. This model explicitly incorporates ideas in elastic-plastic behavior of both the target and projectile. The following model developments were made based on the examination of numerical (hydrocode) simulations of large L/D projectile impacts that were performed for numerous impact simulations at different impact velocities and with various materials, such as those described in Chapter 7. The three major model constructions or assumptions are

(1) A velocity profile along the centerline in both the projectile and the target is specified.
(2) The back end of the projectile is decelerated by elastic waves, with a magnitude proportional to the yield strength of the projectile. These waves reflect off the rear of the projectile, and this free-surface reflection decelerates the rear of the projectile. At the front of the projectile, they

reflect off the elastic-plastic interface within the projectile rather than the projectile–target interface.

(3) A shear behavior in the target material is specified. We will link the shear behavior to the velocity profile in the target in assumption 1. The accuracy of the model depends on correctly predicting this shear stress term. A velocity field in the target combined with rigid plasticity will determine its functional form and a cylindrical cavity expansion will determine the extent of plastic flow. This assumption will allow a computation of S_t.

With suitable expressions resulting from assumptions 1 and 3, the axial momentum equation can then be integrated to obtain an equation of motion for the location of the interface between the target and the projectile. Assumption 2 provides an equation for the deceleration of the rear of the projectile.

In the case of eroding penetration, these assumptions, with the momentum equation, require additional pieces of the information. The three minor assumptions are

(4) An expression for the elastic-plastic interface location within the eroding projectile, presently based on an assumed no-slip condition at the crater wall.

(5) An expression that relates the crater radius to the impact velocity (developed in Chapter 9).

(6) An initial interface velocity, typically obtained from one-dimensional impact considerations (as derived in Chapter 5).

Each of these constructions or assumptions will now be explored in detail, and the resulting equations will be derived.

10.3. A Velocity Profile in the Projectile

The solid line in Fig. 10.2 is the velocity profile along the centerline of an $L/D = 10$ tungsten alloy projectile impacting a semi-infinite target from a numerical simulation. The velocity in the projectile is constant over most of the projectile length, except for a small region near the projectile–target interface. This velocity profile will be approximated by a bilinear expression (two straight-line segments). With s denoting the extent of the plastic zone along the axis over which the velocity changes (with constant slope), the velocity in the projectile is

$$v_z(z) = \begin{cases} u - \dfrac{v - u}{s}(z - z_i) & z_i - s \le z \le z_i, \\ v & z_p \le z < z_i - s. \end{cases} \tag{10.12}$$

As v, u, s, and $z_i(t)$ only depend on time, the partial derivative with respect to time of this velocity profile is

$$\frac{\partial v_z}{\partial t}(z) = \begin{cases} \dot{u} - \left(\dfrac{\overset{\cdot}{v - u}}{s}\right)(z - z_i) + \dfrac{v - u}{s}u & z_i - s \le z \le z_i, \\ \dot{v} & z_p \le z < z_i - s. \end{cases} \tag{10.13}$$

The dot represents differentiation with respect to time; the dot over $(v - u)/s$ (the first expression in the second term on the right-hand side of Eq. 10.13) implies that the whole term is differentiated with respect to time. Equation (10.6) requires the

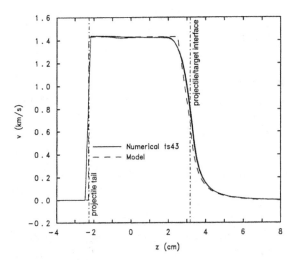

FIGURE 10.2. Axial velocity profile along projectile–target center-line. The dashed lines are from Eqs. (10.12) and (10.29) and the solid line is from a CTH simulation with the same geometry at the same time. (From Walker and Anderson [199].)

integration of Eq. (10.13) over the range

$$\int_{z_p}^{z_i} \frac{\partial v_z}{\partial t} dz = \dot{v}(L - s) + \dot{u}s + (v - u)u + \left(\frac{\dot{v} - \dot{u}}{s} \right) \frac{s^2}{2}. \tag{10.14}$$

Figure 10.5 depicts the projectile at two times, t and $t + \Delta t$. Some of the relevant notation is indicated in the figure.

10.4. A Velocity Profile in the Target

The behavior of the velocity in the target is more complicated because ultimately we are interested in the velocity everywhere within the target, not simply along the centerline. Velocity fields in numerical simulations have a hemispherical behavior (Fig. 10.3), and because of this a flow field is written in spherical geometry with coordinates (r, ϕ, θ). Incompressible flows can be obtained from the curl of a vector potential, since the divergence of the curl of any vector field is zero. The cylindrical symmetry of the problem implies no ϕ dependence and $v_\phi = 0$. There is no radial component to the velocity at the top of the flow field (i.e., $v_r = 0$ when $\theta = \pi/2$) and the flow is purely radial directly underneath the nose of the projectile (i.e., $v_\theta = 0$ when $\theta = 0$). A vector potential yielding such a velocity field is

$$\vec{A} = f(r) \sin(\theta) \, \hat{e}_\phi, \tag{10.15}$$

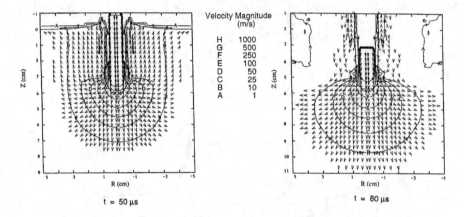

FIGURE 10.3. Velocity field during penetration from a CTH simulation of tungsten impacting steel at 1.5 km/s (projectile is moving down). Arrows indicate the direction of the velocity and the contours the speed, showing the hemispherical nature of the flow. (From Walker and Anderson [199].)

as can be seen by taking the curl in spherical coordinates (Appendix A),

$$\vec{v}(r,\theta) = \mathrm{curl}(\vec{A}) = \left[\frac{1}{r\sin(\theta)} \frac{\partial}{\partial\theta} (A_\phi \sin(\theta)) - \frac{1}{r\sin(\theta)} \frac{\partial A_\theta}{\partial\phi} \right] \hat{e}_r$$

$$+ \left[\frac{1}{r\sin(\theta)} \frac{\partial A_r}{\partial\phi} - \frac{1}{r} \frac{\partial}{\partial r} (rA_\phi) \right] \hat{e}_\theta + \left[\frac{1}{r} \frac{\partial}{\partial r} (rA_\theta) - \frac{1}{r} \frac{\partial A_r}{\partial\theta} \right] \hat{e}_\phi \quad (10.16)$$

$$= \frac{2f}{r} \cos(\theta)\hat{e}_r - \frac{1}{r} \frac{d(rf)}{dr} \sin(\theta)\hat{e}_\theta$$

or

$$v_r(r,\theta) = \frac{2f}{r}\cos(\theta), \qquad v_\theta(r,\theta) = -\frac{1}{r}\frac{d(rf)}{dr}\sin(\theta). \quad (10.17)$$

This formulation has the advantage of there being only one unknown function f.

Along the axis of symmetry, the velocity is $v_r(r,0) = 2f/r$. Thus, the velocity along the centerline determines $f(r)$ and so completely determines $\vec{v}$. The function f can be anything, but let us consider some powers in r,

$$f(r) = ar + b + \frac{c}{r} + \frac{d}{r^2}, \quad (10.18)$$

where a, b, c, and d are constants. This gives rise to terms of the form a, b/r, c/r^2, and d/r^3 in the velocity,

$$v_r(r,\theta) = \left(2a + \frac{2b}{r} + \frac{2c}{r^2} + \frac{2d}{r^3} \right)\cos(\theta), \quad v_\theta(r,\theta) = \left(-2a - \frac{b}{r} + \frac{d}{r^3} \right)\sin(\theta). \quad (10.19)$$

The c/r term drops out of the v_θ, which can greatly simplify algebraic expressions. The simplest form occurs when b and d are set to zero,

$$v_r(r,\theta) = \left(2a + \frac{2c}{r^2} \right)\cos(\theta), \quad v_\theta(r,\theta) = -2a\sin(\theta). \quad (10.20)$$

Geometrically this flow field is comprised of a $1/r^2$ radial flow – the volume conserving radial expansion – scaled by the coefficient $c\cos(\theta)$, and a pure translation along the z-axis, with speed $2a$.

The boundary condition we wish to apply is to have the surface of the crater bottom moving in the $+z$ direction with velocity penetration velocity u. Mathematically, this can be written

$$\vec{v} \cdot \vec{n} = u\cos(\theta) \quad \text{on} \quad r = R, \tag{10.21}$$

where $\vec{n}$ is the normal to the surface of the hemisphere representing the crater. Secondly, on a larger hemisphere in the target, we want there to be no flow through the surface – that is, a certain size hemisphere within the target contains all the plastic flow. Mathematically this condition can be written

$$\vec{v} \cdot \vec{n} = 0 \quad \text{on} \quad r = \alpha R. \tag{10.22}$$

Radius αR is the extent of plastic flow in the target. For the hemisphere the normal is in the radial direction, $\vec{n} = \hat{e}_r$, and so only the radial component of the velocity is involved in the boundary condition. Owing to the form of the solution it is sufficient to satisfy the boundary conditions along the centerline; then they are satisfied everywhere on the hemispherical surface. Thus, on the centerline the target's velocity at the point of contact with the projectile nose must equal the projectile's velocity, and on the centerline the radial velocity component must be zero at the outer edge of the flow field,

$$v_r(R, 0) = u, \qquad v_r(\alpha R, 0) = 0, \tag{10.23}$$

or

$$f(R) = uR/2, \qquad f(\alpha R) = 0. \tag{10.24}$$

The only material flow out of the region occurs through the $\theta = \pi/2$ surface, in the opposite direction of penetration. In the specific case of Eq. (10.19) the boundary condition becomes

$$2a + \frac{2b}{R} + \frac{2c}{R^2} + \frac{2d}{R^3} = u, \qquad 2a + \frac{2b}{\alpha R} + \frac{2c}{(\alpha R)^2} + \frac{2d}{(\alpha R)^3} = 0. \tag{10.25}$$

With $b = d = 0$, these boundary conditions yield the $f(r)$ term in the potential,

$$f(r) = \frac{u}{2(\alpha^2 - 1)} \left\{ \frac{(\alpha R)^2}{r} - r \right\}, \tag{10.26}$$

and the velocity field

$$v_r = \frac{u}{(\alpha^2 - 1)} \left(\frac{\alpha^2 R^2}{r^2} - 1 \right) \cos(\theta), \qquad v_\theta = \frac{u}{\alpha^2 - 1} \sin(\theta). \tag{10.27}$$

Figure 10.4 displays the velocity field, showing a strong similarity to the velocity field seen in Fig. 10.3. This flow field is the simplest velocity field for hemispherical-nosed penetration. It is comprised of a $(\cos(\theta)/r^2)\hat{e}_r$ source flow plus a constant velocity in the z direction with magnitude $-u/(\alpha^2 - 1)$. The addition of these two velocities results in a velocity field that has a shearing surface on the hemisphere boundary, where $v_r(\alpha R, \theta) = 0$ but $v_\theta(\alpha R, \theta) \neq 0$. It is possible to remove this shearing surface by using additional terms in the potential to force $v_\theta(\alpha R, \theta) = 0$. However, doing so subsequently gives rise to much more tedious calculations. In the following, the shearing surface will not be considered.

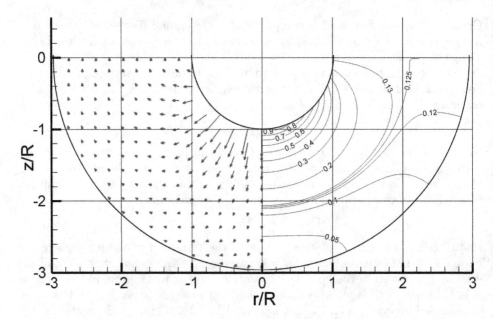

FIGURE 10.4. Velocity field from Eq. (10.27) with $\alpha = 3$; the left side is velocity vectors and the right side is velocity magnitude contours. The penetration velocity is $-1 \cdot \hat{z}$.

We mention that this flow field has rotation. Computing the vorticity of the flow yields $\vec{\mathrm{curl}} = \{2f/r - \partial^2(rf)/\partial r^2\}(\sin(\theta)/r)\hat{e}_\varphi = \{V\alpha^2 R^2 \sin(\theta)/((\alpha^2 - 1)r^3)\}\hat{e}_\varphi$. The $\vec{v} = \cos(\theta)\hat{e}_r/r^2$ (from $f = 1/r$) yields the rotation; irrotational (or gradient potential) flow only occurs for linear combinations of $f = r$ and $f = 1/r^2$, corresponding to spatially constant velocity and a $1/r^3$ dependent velocity.

Given the hemispherical flow field above for target response, to obtain the explicit expression for the flow field along the centerline we let

$$r(z) = z - z_i(t) + R, \tag{10.28}$$

where R is the *crater* radius, and then the velocity in the target along the centerline is

$$v_z(z) = \begin{cases} \dfrac{u}{\alpha^2 - 1}\left[\left(\dfrac{\alpha R}{r(z)}\right)^2 - 1\right] & R \le r(z) < \alpha R, \\ 0 & r(z) \ge \alpha R. \end{cases} \tag{10.29}$$

This profile is plotted in Fig. 10.2 as the dashed line along with the projectile velocity profile, Eq. (10.12), nearly matching the centerline velocity profile of the numerical simulation. Although the target is assumed to be semi-infinite, the velocity profile is only over a finite domain. The furthest extent in the target is $z = z_i + (\alpha - 1)R$. The term αR represents the extent of the plastic zone in the target, and so, for example, its time derivative equals the speed at which the plastic zone expands. Obtaining the extent as well as the rate of increase of the plastic zone in the target is nontrivial. We will use the location of the elastic-plastic interface that we derived in the cavity expansion.

Just as in the projectile case, the integral of the partial of velocity with respect to time is needed over the range $(z_i, +\infty)$,

$$
\frac{\partial v_z}{\partial t}(z) = \frac{\dot{u}}{\alpha^2 - 1}\left\{\left(\frac{\alpha R}{r(z)}\right)^2 - 1\right\} + \dot{z}_i\frac{2u}{(\alpha^2 - 1)}\frac{(\alpha R)^2}{(r(z))^3}
$$

$$
+ \dot{\alpha}\frac{2u\alpha}{(\alpha^2 - 1)^2}\left\{1 - \left(\frac{R}{r(z)}\right)^2\right\}, \qquad R \leq r \leq \alpha R,
$$

(10.30)

$$
\int_{z_i}^{+\infty}\frac{\partial v_z}{\partial t}dt = \int_R^{\alpha R}\frac{\partial v_z}{\partial t}dr = \dot{u}R\frac{\alpha - 1}{\alpha + 1} + \dot{\alpha}\frac{2Ru}{(\alpha + 1)^2} + u^2.
$$

(10.31)

It has been assumed that the crater radius R is constant. It is relatively straightforward to derive the equations in the case that the radius is not constant (Exercise 10.1); however, experience shows that the results of the model are not very sensitive to the crater radius.

Another way to visualize the above flow field is to look at how material moves as the projectile passes by – that is, streamlines. To explicitly compute these, we need to be in a stationary frame where we assume that α is constant. We take the velocity $\vec{v}$ above and subtract off $\vec{u} = u\cos(\theta)\hat{e}_r - u\sin(\theta)\hat{e}_\theta$ to get into a stationary frame. There,

$$
\hat{v} = \vec{v} - \vec{u} = \frac{d}{dt}(r\hat{e}_r) = \frac{dr}{dt}\hat{e}_r - r\frac{d\theta}{dt}\hat{e}_\theta = \hat{v}_r\hat{e}_r + \hat{v}_\theta\hat{e}_\theta,
$$

(10.32)

and so, where the prime means differentiation with respect to r,

$$
\frac{d\theta}{dr} = \frac{\hat{v}_\theta/r}{\hat{v}_r} = -\frac{1}{r}\left(\frac{rf' + f - ur}{2f - ur}\right)\frac{\sin(\theta)}{\cos(\theta)} = -\frac{1}{2}\left((\ln(f - ur/2))' + \frac{1}{r}\right)\frac{\sin(\theta)}{\cos(\theta)}.
$$

(10.33)

This equation can be explicitly integrated to give

$$
\frac{\sin^2(\theta_0)}{\sin^2(\theta)} = \frac{f(r) - ur/2}{f(r_0) - ur_0/2}\cdot\frac{r}{r_0},
$$

(10.34)

where (r_0, θ_0) is the initial point on the streamline. For our specific flow field, this becomes

$$
\frac{\sin^2(\theta_0)}{\sin^2(\theta)} = \frac{R^2 - r^2}{R^2 - r_0^2} \quad\Rightarrow\quad r^2 = R^2 + (r_0^2 - R^2)\frac{\sin^2(\theta_0)}{\sin^2(\theta)}.
$$

(10.35)

There is a one to one relation between θ and r, and r decreases as θ increases. The flow field is initially engaged when $r_0 = \alpha R$.

10.5. The Deceleration of the Rear of the Projectile

The projectile is decelerated by elastic waves. If we were being completely consistent, these elastic waves would be represented as small velocity fluctuations in the assumed projectile velocity profile, traveling up and down the length of the projectile. Finely zoned numerical simulations clearly show the "step" deceleration of the rear of the projectile, as we discussed in Chapter 7. However, to keep the model simple, only the average momentum behavior of the projectile is described by the velocity profile, Eq. (10.12), and a second equation is written to describe the deceleration of the rear of the projectile.

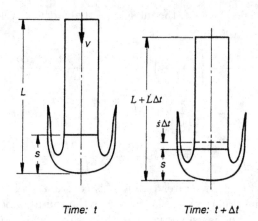

FIGURE 10.5. Projectile at times t and $t + \Delta t$ depicting notation.
(From Walker and Anderson [199].)

This problem was first presented in Chapter 5 for the Taylor anvil problem. Repeating the arguments, the compressive wave travels from the elastic-plastic interface, reflects as a tensile wave off the rear free surface, and returns to the elastic-plastic interface, where it reflects as a compressive wave, and the cycle continues. In numerical simulations, only the first several elastic step decelerations are discernible in the tail velocity. Complicated wave interactions between the elastic-plastic interface and the projectile–target interface smear the elastic deceleration wave, and so after the first few reflections the deceleration waves are more continuous, resulting in a smoother deceleration history for the projectile tail.

For the first step, consider the compressive elastic wave as it travels from the elastic-plastic interface to the back of the projectile. The material is initially uncompressed and the Lagrangian uniaxial stress bar wave velocity is c_E. The transit time from elastic-plastic interface to the rear of the projectile is

$$\Delta t_c = \frac{L - s}{c_E}. \tag{10.36}$$

The subscript c refers to the compressive nature of the wave. For the return trip, the projectile is shorter and compressed. For the rarefaction wave the Lagrangian wave velocity is $c_E - \Delta v$. However, assuming the change in velocity Δv is small compared to c_E, u and v, the amount of time for the return trip from the back of the projectile to the elastic-plastic interface is

$$\Delta t_r = \frac{L - s - (v - u)(\Delta t_c + \Delta t_r) - \dot{s}(\Delta t_c + \Delta t_r)}{c_E}. \tag{10.37}$$

Solving for Δt_r,

$$\Delta t_r = \frac{L - s - (v - u)\Delta t_c - \dot{s}\Delta t_c}{c_E + (v - u) + \dot{s}}. \tag{10.38}$$

The total time for the round trip is

$$\Delta t_c + \Delta t_r = \Delta t_c + \frac{L - s - (v - u)\Delta t_c - \dot{s}\Delta t_c}{c_E + (v - u) + \dot{s}}$$
$$= \frac{c_E \Delta t_c + L - s}{c_E + (v - u) + \dot{s}} = 2\frac{L - s}{c_E + (v - u) + \dot{s}}. \tag{10.39}$$

As was shown in Chapter 5, the change in particle velocity at the free rear surface of the projectile is twice the particle velocity of the wave:

$$\Delta v = -2c_E \frac{\sigma_p}{E_p}. \tag{10.40}$$

Thus, the rear surface deceleration can be approximated as

$$\frac{dv}{dt} \approx \frac{\Delta v}{\Delta t_c + \Delta t_r} = -c_E \frac{\sigma_p}{E_p} \frac{c_E + (v - u) + \dot{s}}{L - s} = -\frac{\sigma_p}{\rho_p(L - s)} \left(1 + \frac{v - u}{c_E} + \frac{\dot{s}}{c_E} \right). \tag{10.41}$$

This equation is essentially Eq. (5.28), modified for the fact that the projectile–target interface is moving at a velocity u.

An integral approach to achieve the same deceleration equation is as follows. The travel time of the wave is the distance divided by the wave speed. Due to erosion and change in the size of the plastic zone, the distance of the return trip is less. If Δt is the time of the round trip of the wave, then

$$c_E \Delta t = L - s + L - \int_0^{\Delta t} (v - u)dt - \left(s + \int_0^{\Delta t} \dot{s} dt \right) \tag{10.42}$$

or

$$c_E \Delta t \left\{ 1 + \frac{1}{c_E \Delta t} \int_0^{\Delta t} (v - u)dt + \frac{1}{c_E \Delta t} \int_0^{\Delta t} \dot{s} dt \right\} = 2(L - s). \tag{10.43}$$

To obtain a continuous approximation for the deceleration of the rear of the projectile, Eqs. (10.40) and (10.43) give

$$\frac{dv}{dt} \approx \lim_{\Delta t \to 0} \frac{\Delta v}{\Delta t} = \lim_{\Delta t \to 0} \left[-\frac{c_E^2 \sigma_p}{E_p(L - s)} \left(1 + \frac{1}{c_E \Delta t} \int_0^{\Delta t} (v - u)dt + \frac{1}{c_E \Delta t} \int_0^{\Delta t} \dot{s} dt \right) \right]$$

$$= -\frac{\sigma_p}{\rho_p(L - s)} \left(1 + \frac{u - v}{c_E} + \frac{\dot{s}}{c_E} \right). \tag{10.44}$$

Thus, both these arguments arrive at the same deceleration equation for the rear of the projectile. The back-end deceleration depends on a reduced length of the projectile owing to the wave reflecting off the elastic-plastic boundary within the projectile, with adjustments for the erosion rate and the change in extent of the plastic region in the projectile. Using the elastic-plastic boundary for the wave reflection, instead of the projectile–target interface, produces better agreement with results from numerical simulations and was essential to obtaining reasonable final-phase penetration results.

10.6. The Stress in the Target and the Penetration Resistance

The target resistance is the central and most difficult part of all penetration models. The idea behind what is done here is as follows. We first suppose that the three-dimensional flow field in the target can be determined. Then, assuming the target is behaving in an elastic-plastic manner, the stresses can be calculated from the von Mises flow rule. These stresses are then used to calculate S_t.

For plastic flow associated with a von Mises yield surface, the stress deviators and the plastic rate of deformation are related by (Eq. 3.99):

$$s_{ij} = (1/\Lambda)D_{ij}^p, \tag{10.45}$$

where Λ is a constant. Unfortunately, for a full elastic-plastic problem, it is necessary to solve for the rate of deformation tensor and then separate it out into elastic and plastic parts. The separation depends on the deformation history and is very tedious to perform (that is why today we use large-scale numerical simulations for problems of this type). However, it is possible to make a major simplification by assuming *rigid plasticity*. As described in Chapter 3, the assumption in rigid plasticity is that all deformation is plastic, and thus that D_{ij}^p can be computed from the current velocity field. *We make the rigid plastic assumption only to calculate the shear stress term.* It is important to keep in mind that compressibility will still be included in our solution through the use of a cavity expansion – the assumption of rigid plasticity is only used in this particular part of the model. In Chapter 3, for rigid plasticity we derived (Eq. 3.102)

$$s_{ij} = \Lambda D_{ij} = \frac{\sqrt{\frac{2}{3}} Y_t D_{ij}}{\sqrt{D_{ij}D_{ij}}}, \quad \text{where} \quad D_{ij} = D_{ij}^p = \frac{1}{2}\left(\frac{\partial v_i}{\partial x_j} + \frac{\partial v_j}{\partial x_i}\right). \tag{10.46}$$

By assuming rigid plasticity we no longer need to know any information about the deformation history; the stress tensor depends only on the current rate of deformation. We now proceed with finding S_t, which requires the derivative of the shear stress $\sigma_{xz} = s_{xz}$ along the centerline $x = y = 0$ (Eq. 10.9). As the assumed velocity field is incompressible, the following relations hold directly along the line of symmetry,

$$\frac{\partial v_x}{\partial x} = \frac{\partial v_y}{\partial y} = -\frac{1}{2}\frac{\partial v_z}{\partial z}, \tag{10.47}$$

giving for the inner product of the deformation rate along the centerline,

$$D_{ij}D_{ij} = \frac{3}{2}\left(\frac{\partial v_z}{\partial z}\right)^2 + 2\left(\frac{\partial v_y}{\partial x}\right)^2 + \left(\frac{\partial v_x}{\partial z} + \frac{\partial v_z}{\partial x}\right)^2, \tag{10.48}$$

where the symmetry between x and y has been used.

The specific shear stress of interest is

$$s_{xz} = \frac{Y_t}{\sqrt{6 D_{ij}D_{ij}}}\left(\frac{\partial v_x}{\partial z} + \frac{\partial v_z}{\partial x}\right). \tag{10.49}$$

When the derivative is calculated, since it is along the centerline, a number of simplifications arise. First, since $D_{ij}D_{ij}$ is even with respect to x and y, the derivative of D_{ij} with respect to x along $x = 0$ is zero. This gives

$$\left.\frac{\partial s_{xz}}{\partial x}\right|_{x=0} = \frac{Y_t}{\sqrt{6 D_{ij}D_{ij}}}\left(\frac{\partial^2 v_x}{\partial x \partial z} + \frac{\partial^2 v_z}{\partial x^2}\right)\bigg|_{x=0}. \tag{10.50}$$

Second, $v_x \equiv 0$ along the centerline, so $\partial v_x/\partial z = 0$ there. Third, v_z is even in x, and so $\partial v_z/\partial x = 0$ along the centerline. Fourth, v_y is even in x and so $\partial v_y/\partial x = 0$. These last three observations allow calculation of D_{ij} along the centerline. Finally, if we exchange the order of the partials for the mixed term in Eq. (10.50), Eq. (10.47) may be used. All these observations lead to

$$\left.\frac{\partial s_{xz}}{\partial x}\right|_{x=0} = \frac{\frac{1}{3}Y_t}{|\partial v_z/\partial z|}\left(\frac{\partial^2 v_z}{\partial x^2} - \frac{1}{2}\frac{\partial^2 v_z}{\partial z^2}\right)\bigg|_{x=0}. \tag{10.51}$$

It is expected that $\partial v_z/\partial x < 0$ – that is, the velocity of the flow field in the target decreases monotonically as one moves into the target away from the projectile nose.

Thus, the absolute value in the denominator can be replaced with $-\partial v_z/\partial x$, yielding

$$S_t = -2 \int_R^{\alpha R} \frac{\partial s_{xz}}{\partial x}\bigg|_{x=0} dr$$

$$= \int_R^{\alpha R} \frac{Y_t}{3}\left[-\frac{\partial}{\partial z}\left\{ \ln\left(-\frac{\partial v_z}{\partial z} \right) \right\} + 2\left(\frac{\partial^2 v_z}{\partial x^2} \bigg/ \frac{\partial v_z}{\partial z} \right) \right]_{x=0} dr,$$

(10.52)

where $r(z)$ was defined in Eq. (10.28).

To evaluate the integral in Eq. (10.52) requires information about the flow field. Initially, the only assumption we will make is that we will use our potential flow

$$v_r(r,\theta) = \frac{2f}{r}\cos(\theta), \qquad v_\theta(r,\theta) = -\frac{1}{r}\frac{d(rf)}{dr}\sin(\theta).$$

(10.53)

With

$$\cos(\theta) = z/r, \quad \sin(\theta) = x/r, \quad r = \sqrt{z^2 + x^2},$$

(10.54)

the z component of velocity is given by

$$v_z = v_r\cos(\theta) - v_\theta\sin(\theta) = \frac{2f}{r}\left(\frac{z}{r}\right)^2 + \left(f' + \frac{f}{r}\right)\left(\frac{x}{r}\right)^2.$$

(10.55)

The prime denotes differentiation with respect to r. The denominator term of the far right-hand side of Eq. (10.52) is

$$\frac{\partial v_z}{\partial z}\bigg|_{x=0} = \frac{\partial}{\partial z}\left(\frac{2fz^2}{z^3} \right) = 2\frac{d}{dz}\left(\frac{f}{z} \right),$$

(10.56)

since $r = z$ when $x = 0$. The numerator term is given by

$$\frac{\partial^2 v_z}{\partial x^2}\bigg|_{x=0} = \frac{\partial^2}{\partial x^2}\left(\frac{2fz^2}{r^3} + \left(f' + \frac{f}{r}\right)\left(\frac{x}{r}\right)^2 \right)\bigg|_{x=0}$$

$$= 2\frac{d}{dx}\left(\frac{f'xz^2}{r^4} - \frac{3fxz^2}{r^5} \right)\bigg|_{x=0} + 2\left(f' + \frac{f}{r} \right)\frac{1}{r^2}\bigg|_{x=0}$$

$$= 2\left(\frac{f'z^2}{r^4} - \frac{3fz^2}{r^5} \right)\bigg|_{x=0} + 2\left(f' + \frac{f}{r} \right)\frac{1}{r^2}\bigg|_{x=0}$$

(10.57)

$$= 2\left(\frac{f'}{z^2} - \frac{3f}{z^3} \right) + 2\left(f' + \frac{f}{z} \right)\frac{1}{z^2}$$

$$= \frac{4}{z}\frac{\partial}{\partial z}\left(\frac{f}{z} \right),$$

where the fact that the evaluation is along the line $x = 0$ has been used several times. Thus, we have the remarkable result that

$$\left(\frac{\partial^2 v_z}{dx^2} \bigg/ \frac{\partial v_z}{\partial z} \right)\bigg|_{x=0} = \frac{4}{z}\frac{\partial}{\partial z}\left(\frac{f}{z} \right) \bigg/ 2\frac{\partial}{\partial z}\left(\frac{f}{z} \right) = \frac{2}{z},$$

(10.58)

and, for a constant Y_t, inserted into the integrand of the right-hand side integral of Eq. (10.52) yields

$$S_t = -\frac{Y_t}{3}\ln\left|\frac{\partial v_z}{\partial z}\right|_{z=R}^{z=\alpha R} + \frac{2Y_t}{3}\int_R^{\alpha R}\frac{2}{z}dz = -\frac{Y_t}{3}\ln\left|\frac{\partial v_z}{\partial z}\right|_{z=R}^{z=\alpha R} + \frac{4Y_t}{3}\ln(\alpha).$$

(10.59)

What is truly remarkable about this result is that few assumptions on the flow field have been made: 1) the flow field is monotonically decreasing along the axis, and 2) that it has a certain reasonable hemispherical behavior. From these assumptions follow the result that the shear term in the momentum balance only depends on the extent of the flow field (α) and the slopes of the velocity at the front end and back end of the flow field. We emphasize that this result is for a hemispherical flow field and rigid plastic assumptions, and though it may seem that an arbitrary velocity profile along the centerline could be used, the flow field properties should reflect the appropriate behavior. For example, the derivation implies, because of the first term on the right-hand side of Eq. (10.59), that the axial plastic flow velocity cannot tangentially touch the zero axis, as the existence of the spatial derivative of the centerline velocity within the logarithm shows that it is not permissible for the derivative of the flow field to become zero anywhere.[1]

To finish the evaluation of S_t, we use our specific assumed velocity field with the flow field (Eq. 10.27). The slope of the velocity profile in the target is

$$\frac{\partial v_z}{\partial z} = -2\frac{u}{\alpha^2 - 1}\frac{(\alpha R)^2}{z^3} \tag{10.60}$$

and evaluating this at $z = R$ and $z = \alpha R$ gives

$$S_t = -2\int_R^{\alpha R}\left.\frac{\partial s_{xz}}{\partial x}\right|_{x=0} dz = -\frac{Y_t}{3}\ln\left(\frac{1}{\alpha^3}\right) + \frac{4Y_t}{3}\ln(\alpha) = \frac{7}{3}\ln(\alpha)Y_t. \tag{10.61}$$

This completes the assumed stress behavior. The final result is simple and elegant, and a minimum of assumptions were made along the way. In addition, if Y_t is not constant, then the specific flow field yields $-\partial\{\ln(-\partial v_z/\partial z)\}/\partial z = 3/z$ and so

$$S_t = -2\int_R^{\alpha R}\left.\frac{\partial s_{xz}}{\partial x}\right|_{x=0} dz = \frac{7}{3}\int_R^{\alpha R}\frac{Y_t}{r}dr. \tag{10.62}$$

This shear stress integral S_t is related to the resistance of the target to penetration and is thus a central result in penetration mechanics, identifying the resistance to penetration due to strength.

We make some comments about S_t. It is possible to explicitly compute this integral in hydrocode computations of penetration. One of the things that is discovered is that the value of S_t is not entirely independent of Y_p, as the projectile strength appears to affect the exact shape of the crater and the tractions at the surface of the crater, both of which affect the shear stress in the target near the projectile–target interface. Also, in some cases the shear stress term S_p for the projectile is not negligible. In some hydrocode computations S_t does not decrease as rapidly as a function of velocity as S_t in our analytic model when α is computed from the cavity expansion, but the overall decrease of $S_t - S_p$ is similar for both hydrocode and analytic model (where $S_p = 0$ for the analytic model). However, as will be seen below, the large decrease in α in terms of penetration speed is required to match the penetration data of nondeforming hemispherical-nosed projectiles.

Later we will need the strain rates and rigid plasticity stress deviators in spherical coordinates. Recalling

$$v_r(r, \theta) = \frac{2f}{r}\cos(\theta), \qquad v_\theta(r, \theta) = -\frac{1}{r}\frac{d(rf)}{dr}\sin(\theta), \qquad v_\phi = 0, \tag{10.63}$$

---

[1] We revisit this restriction after explicitly computing the strain near the end of the chapter.

and that there is no ϕ dependence due to our assumed cylindrical symmetry, the rate of deformation tensor components are given by (Appendix A)

$$D_{rr} = \frac{\partial v_r}{\partial r} = 2\left(\frac{f}{r}\right)' \cos(\theta) = 2\left(\frac{f'}{r} - \frac{f}{r^2}\right)\cos(\theta),$$

$$D_{\theta\theta} = \frac{1}{r}\frac{\partial v_\theta}{\partial \theta} + \frac{v_r}{r} = \left(\frac{f}{r^2} - \frac{f'}{r}\right)\cos(\theta) = -\frac{1}{2}D_{rr},$$

$$D_{\phi\phi} = \frac{1}{r\sin(\theta)}\frac{\partial v_\phi}{\partial \phi} + \frac{v_\theta}{r}\cot(\theta) + \frac{v_r}{r} = \left(\frac{f}{r^2} - \frac{f'}{r}\right)\cos(\theta) = -\frac{1}{2}D_{rr},$$

$$D_{r\theta} = \frac{1}{2}\left[\frac{1}{r}\frac{\partial v_r}{\partial \theta} + \frac{\partial v_\theta}{\partial r} - \frac{v_\theta}{r}\right] = -\frac{f''}{2}\sin(\theta),$$

$$D_{r\phi} = \frac{1}{2}\left[\frac{1}{r\sin(\theta)}\frac{\partial v_r}{\partial \phi} + \frac{\partial v_\phi}{\partial r} - \frac{v_\phi}{r}\right] = 0,$$

$$D_{\theta\phi} = \frac{1}{2}\left[\frac{1}{r\sin(\theta)}\frac{\partial v_\theta}{\partial \phi} + \frac{1}{r}\frac{\partial v_\phi}{\partial \theta} - \frac{v_\phi}{r}\cot(\theta)\right] = 0.$$

(10.64)

Given these, we have

$$\dot{\varepsilon}_p^2 = \frac{2}{3}D_{ij}D_{ij} = D_{rr}^2 + \frac{4}{3}D_{r\theta}^2. \tag{10.65}$$

It will be helpful to separate out the r dependence and the θ dependence in the rate of deformation terms, and to that end we define

$$\bar{D}_{rr}(r) = 2\left(\frac{f}{r}\right)' = -\frac{2u}{r^3}\frac{(\alpha R)^2}{(\alpha^2 - 1)}, \quad \bar{D}_{r\theta}(r) = -\frac{1}{2}f'' = -\frac{u}{2r^3}\frac{(\alpha R)^2}{(\alpha^2 - 1)}, \tag{10.66}$$

so that

$$D_{rr}(r, \theta) = \bar{D}_{rr}(r)\cos(\theta), \quad D_{r\theta}(r, \theta) = \bar{D}_{r\theta}(r)\sin(\theta). \tag{10.67}$$

It will be assumed that $\bar{D}_{rr} < 0$ (strict). Thus,

$$\frac{2}{3}D_{ij}D_{ij} = \bar{D}_{rr}^2\cos^2(\theta) + \frac{4}{3}\bar{D}_{r\theta}^2\sin^2(\theta) \tag{10.68}$$

and the stress deviators, based on the rigid plasticity assumption, are

$$s_{rr} = \frac{2Y_t}{3}\frac{\bar{D}_{rr}(r)\cos(\theta)}{\sqrt{\bar{D}_{rr}^2\cos^2(\theta) + (4/3)\bar{D}_{r\theta}^2\sin^2(\theta)}}, \quad s_{\theta\theta} = s_{\phi\phi} = -\frac{1}{2}s_{rr},$$

$$s_{r\theta} = \frac{2Y_t}{3}\frac{\bar{D}_{r\theta}(r)\sin(\theta)}{\sqrt{\bar{D}_{rr}^2\cos^2(\theta) + (4/3)\bar{D}_{r\theta}^2\sin^2(\theta)}}, \quad s_{r\phi} = s_{\theta\phi} = 0.$$

(10.69)

These expressions will be used in later sections.

10.7. The Momentum Balance Equation

Inserting the terms obtained in the previous paragraphs into the centerline momentum expressions, we obtain for the interface stress as evaluated based on the integration along the centerline through the projectile (Eq. 10.7)

$$\sigma_{zz}(z_i) = \rho_p\dot{v}\left(L - s\right) + \rho_p\dot{u}s - \frac{1}{2}\rho_p(v - u)^2 + \rho_p\left(\frac{v - u}{s}\right)\frac{s^2}{2}. \tag{10.70}$$

Similarly, the expression for the interface stress based on the integration of the centerline momentum balance through the target is (Eq. 10.8)

$$\sigma_{zz}(z_i) = -\rho_t \dot{u} R \frac{\alpha - 1}{\alpha + 1} - \rho_t \dot{\alpha} \frac{2Ru}{(\alpha + 1)^2} - \left\{ \frac{1}{2} \rho_t u^2 + \frac{7}{3} \ln(\alpha) Y_t \right\}. \qquad (10.71)$$

The stress at the interface depends on two transient expressions and a target resistance that includes an inertial and strength term. Equating the two interface stress terms yields

$$\rho_p \dot{v} (L - s) + \dot{u} \left\{ \rho_p s + \rho_t R \frac{\alpha - 1}{\alpha + 1} \right\} + \rho_p \left(\frac{v - u}{s} \right) \frac{s^2}{2} + \rho_t \dot{\alpha} \frac{2Ru}{(\alpha + 1)^2}$$
$$= \frac{1}{2} \rho_p (v - u)^2 - \left\{ \frac{1}{2} \rho_t u^2 + \frac{7}{3} \ln(\alpha) Y_t \right\}. \qquad (10.72)$$

The deceleration of the tail of the projectile is given by Eq. (10.41), and the time rate of change of the length of the projectile is the difference between the penetration speed and the speed of the tail,

$$\dot{v} = -\frac{\sigma_p}{\rho_p (L - s)} \left\{ 1 + \frac{v - u}{c_E} + \frac{\dot{s}}{c_E} \right\}, \qquad (10.73)$$

$$\dot{L} = -(v - u). \qquad (10.74)$$

Equations (10.72) to (10.74) are the three central equations of the Walker–Anderson model. In subsequent sections it will be described how to compute α, s, and R, which are the extents of the plastic flow field within the target and projectile and the crater radius (α and R have already had a chapter each devoted to them).

The use of Eq. (10.73) allows Eq. (10.72) to be written in a form that highlights more clearly the transient terms,

$$\dot{u} \left\{ \rho_p s + \rho_t R \frac{\alpha - 1}{\alpha + 1} \right\} + \rho_p \left(\frac{v - u}{s} \right) \frac{s^2}{2} + \rho_t \dot{\alpha} \frac{2Ru}{(\alpha + 1)^2} - \sigma_p \frac{\dot{s}}{c_E}$$
$$= \frac{1}{2} \rho_p (v - u)^2 + \sigma_p \left\{ 1 + \frac{v - u}{c_E} \right\} - \left\{ \frac{1}{2} \rho_t u^2 + \frac{7}{3} \ln(\alpha) Y_t \right\}. \qquad (10.75)$$

Notice that the right-hand side of this equation is very much like the Tate equation.

It is informative to take a certain limit of the above equations. If the two measures of spatial extent, R and s are allowed to go to zero – that is, $R \to 0$ and $s \to 0$, and the Young's modulus for the projectile is allowed to become very large, so $c_E \to \infty$ – then Eqs. (10.72)–(10.74) become

$$-\rho_p \dot{v} L + \frac{1}{2} \rho_p (v - u)^2 = \frac{1}{2} \rho_t u^2 + \frac{7}{3} \ln(\alpha) Y_t,$$

$$\dot{v} = -\frac{\sigma_p}{\rho_p L}, \qquad (10.76)$$

$$\dot{L} = -(v - u).$$

These equations are the original Tate model with the target resistance given by

$$R_t = S_t = \frac{7}{3} \ln(\alpha) Y_t. \qquad (10.77)$$

Thus, one interpretation of R_t is that it is the shear stress penetration resistance S_t. R_t, as such, does not include transient terms. Since S_t is velocity dependent we conclude that R_t is velocity dependent, in agreement with our earlier observations. Also note that Y_p in the Tate model is identified with the flow stress σ_p of the projectile. In summary,

THEOREM 10.1. *(Walker–Anderson reduces to Tate) In the limit where the three-dimensional aspects of the model are removed (the projectile and crater radius go to zero), the extent of the plastic zone in the projectile is set to zero, and the sound speed in the projectile goes to infinity, the Walker–Anderson model reduces to the Tate model with $R_t = S_t$.*

The model predicts an R_t that varies with the extent of the plastic flow zone and the flow stress Y_t of the target. (An alternate view is that the second, third, and fourth terms in Eq. (10.71), along with the term $(7/3)\ln(\alpha)Y_t$, define a time-varying R_t, as suggested by Anderson, Walker, and Hauver [12, 16].)

Returning to Eq. (10.71), the value of the stress at the projectile–target interface along the centerline, it is important enough to state as a theorem:

THEOREM 10.2. *(Stress at projectile–target interface centerline) For a material undergoing a negligible density change, the steady-state stress at the interface between the target and projectile along the axis of symmetry is given by*[2]

$$\tilde{\sigma}_{zz}(z_i) = -\sigma_{zz}(z_i) = \frac{1}{2}\rho_t u^2 + S_t = \frac{1}{2}\rho_t u^2 + \frac{7}{3}\ln(\alpha)Y_t. \tag{10.79}$$

When hydrocode computations are performed, the stress at the nose of the projectile is always given by these values.

10.8. The Extent of the Plastically Flowing Region in the Target

We will compute the plastic extent in the target using the cylindrical cavity expansion derived in Chapter 6. There, the expansion velocity c of the elastic-plastic interface was computed assuming a constant cavity opening velocity u. Thus, the extent of the plastically deforming region in the target is given by

$$\alpha = \frac{ct}{ut} = \frac{c}{u}. \tag{10.80}$$

This was computed in Eq. (6.52). The explicit expression for α as a function of the penetration velocity u is

$$\left(1 + \frac{\rho_t u^2}{Y_t}\right)\sqrt{K_t - \rho_t \alpha^2 u^2} = \left(1 + \frac{\rho_t \alpha^2 u^2}{2G_t}\right)\sqrt{K_t - \rho_t u^2}. \tag{10.81}$$

However, as discussed below, the bulk modulus is stiffened to allow the model to address hypervelocity impacts, as described in Eq. (6.137).

---

[2]Our derivation of the centerline momentum balance does not apply if the density changes a great deal. The physics of blunt body (such as a hemispherical nose) supersonic penetration into a gas are that a shock wave forms in front of the penetrating body and travels at the same speed as the body, and the gas behind the shock is nearly going the same speed. Instead of using the centerline momentum balance, which assumes negligible density change and a smooth velocity change, on the target side of the equation the Hugoniot jump conditions should be applied. Thus, for a highly compressible material, we have $U \approx u$ and the Hugoniot jump conditions give

$$\tilde{\sigma}_{zz}(z_i) = -\sigma_{zz}(z_i) = p(z_i) \approx \rho_t u^2, \tag{10.78}$$

where ρ_t is the initial target density. Depending on the material, a target resistance term due to shear may also be added. Note the difference between the consolidated material and the highly compressible material is no factor of $\frac{1}{2}$ on the inertial $\rho_t u^2$ term. Examples with gases and fluids are discussed in the exercises.

10.9. The Extent of the Plastically Flowing Region in the Projectile

The extent of the plastic flow in the projectile is defined by the length s at the front of the projectile (measured from the projectile–target interface). For simplicity, the shear term will be ignored. The method to determine the extent of plastic flow in the projectile is based on the following observation from many numerical simulations with many materials: the slope of the velocity profile along the centerline is smooth at the material interface (e.g., Fig. 10.2). In other words, the velocity profile is not only continuous, the first derivative along the axis is also continuous. Why might this behavior hold? We now show that a no slip boundary condition for an interface between incompressible materials implies continuity of the slope.

THEOREM 10.3. *When two incompressible materials meet at a no slip boundary, the spatial slopes of the velocity in the direction normal to the boundary are equal.*

To show this, pick a point on the interface between the two materials. We assume that a tangent plane to the interface at that point exists. Pick two orthogonal vectors of unit length which span the tangent plane, $\vec{t}_1$ and $\vec{t}_2$. The normal to the tangent plane can be found by the cross product, $\vec{n} = \vec{t}_1 \times \vec{t}_2$. These three vectors form an orthonormal set which spans 3-space, and we change coordinate systems into the coordinate system in which these are the coordinate directions, with variables t_1, t_2, and n.

At the point in question, one material is on the $+$ side ($n > 0$) and one material is on the $-$ side ($n < 0$). The velocity for the material above the plane is

$$\vec{V}^+ = V_1^+ \vec{t}_1 + V_2^+ \vec{t}_2 + V_n^+ \vec{n}, \tag{10.82}$$

with a similar expression for the velocity below the plane. Since the interface is no slip, the velocities along the interface match everywhere. For the point in question, this requires that both the velocities match and the partials of the velocity in the tangent plane match,

$$V_1^+ = V_1^-, \qquad V_2^+ = V_2^-, \tag{10.83}$$

and

$$\frac{\partial V_1^+}{\partial t_1} = \frac{\partial V_1^-}{\partial t_1}, \qquad \frac{\partial V_2^+}{\partial t_2} = \frac{\partial V_2^-}{\partial t_2}. \tag{10.84}$$

Incompressibility of the flow field says that the divergence of the velocity field is zero, or

$$\frac{\partial V_1^+}{\partial t_1} + \frac{\partial V_2^+}{\partial t_2} + \frac{\partial V_n^+}{\partial n} = 0. \tag{10.85}$$

Thus, the partials in the tangent plane of an incompressible velocity field determine the normal partial and give

$$\frac{\partial V_n^+}{\partial n} = -\frac{\partial V_1^+}{\partial t_1} - \frac{\partial V_2^+}{\partial t_2} = -\frac{\partial V_1^-}{\partial t_1} - \frac{\partial V_2^-}{\partial t_2} = \frac{\partial V_n^-}{\partial n}. \tag{10.86}$$

Therefore, the normal component of the spatial derivatives of the velocity fields must match – or, in other words, the spatial slope of the velocity in the direction normal to the interface is the same for both materials.

This theorem recognizes that for thick targets the target material must be pushed to the side for penetration to occur. For thin targets (or near the back of thick ones) target material can be pushed backwards and matching the slopes becomes a poor assumption.

It is now possible to derive an expression for s as well as for its time derivative. The slope of the assumed velocity profile in the target was calculated in Eq. (10.60). When evaluated at the interface, and set equal to the change in velocity in the projectile divided by the length s, then

$$\frac{u - v}{s} = -2\frac{u}{\alpha^2 - 1}\frac{\alpha^2}{R}. \qquad (10.87)$$

The centerline momentum balance requires both s and $d\left[(u - v)/s\right]/dt$. Equation (10.87) gives

$$s = \frac{R}{2}\left(\frac{v}{u} - 1\right)\left(1 - \frac{1}{\alpha^2}\right) \qquad (10.88)$$

and

$$\left(\frac{v - u}{s}\right) = \frac{4}{(\alpha^2 - 1) R}\left\{\frac{\alpha^2 \dot{u}}{2} - \frac{u\alpha\dot{\alpha}}{\alpha^2 - 1}\right\}. \qquad (10.89)$$

These results are placed into the momentum balance equation.

For low L/Ds and large impact velocities, such as hypervelocities, the crater radius R can be large enough that the resulting plastic zone s from Eq. (10.88) is larger than the projectile length L. This situation is not allowable in the model, and so we achieve $s < L$ always, for example, by initializing $s(0) = 0$ and letting s grow according to a physical wave speed but requiring that s never exceeds $0.95L$, hence in this scenario Eqs. (10.88) and (10.89) are not used to compute s and $\dot{s}$.

10.10. Initial Impact Conditions

The quantities that change need to have their initial conditions specified. In particular, the variables of interest are v, u, L, s, and α. Two of these values are relatively straightforward, v and L. In particular, $v(0)$ equals the initial velocity of the projectile (though see some discussion below). $L(0)$ is the initial length of the projectile; this length includes the nose of the projectile.

One approach to determining u is to use the Hugoniot jump conditions. For a linear U_s–u_p expression of the form $U = c_0 + ku$, one can use the explicit expression in Eq. (5.109),

$$u(0) = \frac{2C}{B + \sqrt{B^2 - 4AC}} \quad \text{or} \quad u(0) = \frac{1}{2A}\left(B - \sqrt{B^2 - 4AC}\right),$$

$$A = k_p - k_t\frac{\rho_t}{\rho_p}, \quad B = 2k_p v(0) + c_{0p} + c_{0t}\frac{\rho_t}{\rho_p}, \quad C = c_{0p} v(0) + k_p v(0)^2. \qquad (10.90)$$

Both expressions yield the same answer, with the advantage of the first expression being that it works when the projectile and target materials are the same. Another approach is to compute the initial velocity from the acoustical impedance match as given in Eq. (4.162), which is the same as assuming that $k = 0$ for both the target and the projectile in the previous expression,

$$u(0) = \frac{\rho_p c_{0p}}{\rho_p c_{0p} + \rho_t c_{0t}} v(0). \qquad (10.91)$$

This approach was used in the example computations in this chapter. (One could alternatively use the longitudinal wave speed c_L in this expression.) For large L/D projectile penetration, the final results are not very sensitive to the initial value of u – that is, the model results achieve the same quasi-steady state velocity independent of the initial penetration velocity. However, to be specific, for typical material properties for a $L/D = 10$ tungsten alloy projectile impacting an armor steel at 1.5 km/s, for $u(0)$ from 500 m/s to $u(0) = 1,100$ m/s in 100 m/s increments, the model P/L are 0.808, 0.838, 0.860, 0.877, 0.890, 0.900, and 0.909. The acoustical impedance match $u(0)$ (Eq. 10.91) is roughly 1,000 m/s. (See also Exercise 5.2.)

Some important points arise when considering the initial values of α and s. One approach is to use the cavity expansion to obtain $\alpha(0)$ and use Eq. (10.88) to compute $s(0)$. This approach was used in the examples in this chapter because it is simple, but it has a potential drawback, especially for low L/D projectiles. In particular, if $\alpha(0) \neq 1$ and $s(0) \neq 0$, the initial momentum of the system may not equal the initial momentum of the projectile. To be specific, if $u(0) > 0$ and $\alpha(0) > 1$, then there is initial momentum in the target. Correspondingly, if $s(0) > 0$, the initial momentum in the projectile is less than that of the striking projectile. Though these offset somewhat, they likely do not exactly match, and it is the altered initial system momentum that is conserved throughout the impact calculation, not the initial momentum of the projectile. The momentum along the centerline in the projectile (left) and target (right) are

$$\int_{z_p}^{z_i} \rho_p v_z(z)dz = \rho_p v \left(L - \frac{s}{2}\right) + \rho_p u \frac{s}{2}, \quad \int_{z_i}^{+\infty} \rho_t v_z(z)dz = \rho_t R u \frac{\alpha - 1}{\alpha + 1}. \quad (10.92)$$

At the time of impact, the difference of the momentum from the intended momentum $\rho_p L v(0)$ is (all the terms are at $t = 0$)

$$\int_{z_p}^{\infty} \rho v_z dz - \rho_p L v = \rho_t R u \frac{\alpha - 1}{\alpha + 1} + \rho_p (u - v)\frac{s}{2} = \rho_t R u \frac{\alpha - 1}{\alpha + 1} - \rho_p R \frac{(v - u)^2}{4u}\left(1 - \frac{1}{\alpha^2}\right). \quad (10.93)$$

Using the initial value of u from the Hugoniot jump conditions for $L/D = 10$ projectiles typically leads to an initial momentum error of 1 to 2%. One reason we accept this small error in the examples is that we also do not address other embedment details such as the initial growth of the crater. An approach to making this initial system momentum match the initial projectile momentum is to set Eq. (10.93) equal to zero and solve for $u(0)$, recalling $\alpha = \alpha(u)$; another approach is to compute $u(0)$ as above and then adjust $v(0)$ to obtain the correct system momentum (Exercise 10.14). However, we can show that setting Eq. (10.93) equal to zero implies that $u(0) < u_{\text{hyd}}$, which does not agree with numerical simulations (see Fig. 10.9). Initial momentum error problems are apparent when one varies target density, since increasing target density increases system momentum which for low L/D projectiles can lead to increasing penetration with increasing target density, which is unrealistic. In studies like this, it is necessary to carefully set the initial conditions.

In addition to the fact the initial conditions combine to affect the initial system momentum, if $\alpha(0) > 1$, then the initial $u(0)$ cannot be arbitrarily set if the velocity derivative is matched at the interface, since if $u(0)$ is too small it is possible that the computed $s(0) > L$, which is not allowed in the model.

Another approach to avoid error in the initial system momentum is to set $\alpha(0) = 1$ and $s(0) = 0$ and to allow these to grow at some rate, such as $\dot{\alpha} = c_K/R$ and $\dot{s} = c_E$, until they reach the values given by the cavity expansion and matching slopes, respectively.[3] We use this approach for low L/D projectiles in hypervelocity

---

[3]A more sophisticated derivation is to introduce $\tilde{\alpha}$, with $\tilde{\alpha}(0) = 1$, and compute $d\tilde{\alpha}/dt = \alpha(u)u/R$, and then use $\alpha = \min(\tilde{\alpha}(t), \alpha(u))$ for α in the model equations. The argument is that

TABLE 10.1. Elastic and shock material properties used in this chapter.

Material	ρ (g/cm^3)	K (GPa)	E (GPa)	G (GPa)	ν	c_o (km/s)	k
Aluminum (6061-T6)	2.71	72.8	72.8	27.3	0.33	5.35	1.34
Armor steel	7.85	166.7	206.8	76.9		4.50	1.49
Tungsten alloy	17.3	302.1	327.5	124.1		3.85	1.44
Tungsten alloy D17	17.0		410.0		0.30	4.03	1.24
Armor steel HzB20	7.85				0.29	4.57	1.49
St 52 Steel	7.85				0.29	4.57	1.49
St 37 Steel	7.85				0.29	4.57	1.49
C110W1 Steel	7.85		205.0			4.57	
Aluminum alloy 1	2.85				0.33	5.35	1.34
Aluminum alloy 2	2.70		72.0			5.35	
Lead	11.2				0.44	2.05	1.46

impacts. Additionally, using this approach, $u(0)$ can be arbitrarily set to any value. Note, however, that the α in the strength term should still come from the cavity expansion, as we do not expect the target to have zero initial strength.

10.11. Examples

The behavior of the model will now be demonstrated with a series of examples. So far, we have not completely described how we compute S_t, in that we have not yet described how we compute α. We obtain the extent of the plastically deforming region through the cavity expansion. The plastically flowing region is demarcated by a boundary between material that is only elastically deformed and material that has undergone both elastic and plastic deformation. In a cavity expansion at constant expansion velocity u (the penetration velocity), we have that the interface location moves outward at speed c. Assuming that the cavity has expanded to the crater radius (i.e., the time t is such that $ut = R$), the location of the boundary between the elastic and elastic-plastic region is given by ct. In terms of the crater radius, that location is given at $ct/R = c/u = \alpha$, where α was defined and calculated in Chapter 6. As was shown there, α decreases as a function of velocity, and hence so does S_t, as we concluded in Chapter 7. In this section we will demonstrate that such behavior is required to match final depth of penetration data. Table 10.1 lists the material constants used in the example problems.

Although the original intent of the model was for eroding large L/D projectile penetration problems, examining nondeforming (rigid) projectile penetration where the projectile has a hemispherical nose allows us to focus on target response. For rigid-body penetration, $v = u$, and Eq. (10.72) simplifies to

$$\dot{u}\left\{\rho_p L + \rho_t R\frac{\alpha - 1}{\alpha + 1}\right\} + \rho_t\dot{\alpha}\frac{2Ru}{(\alpha + 1)^2} = -\left\{\frac{1}{2}\rho_t u^2 + \frac{7}{3}\ln(\alpha)Y_t\right\}. \quad (10.94)$$

As a first step, we examine what happens if a constant α is chosen. Figure 10.6 shows steel projectiles striking aluminum (data from [71], listed in Table 12.6, and from [65]), comparing the data to three constant values of α, two chosen based

---

the initial speed of the elastic-plastic interface from the cavity expansion is $c = \alpha(u)u$, and so set $d\tilde{\alpha}/dt = c/R$ until $\alpha = \alpha(u)$, the value from the cavity expansion.

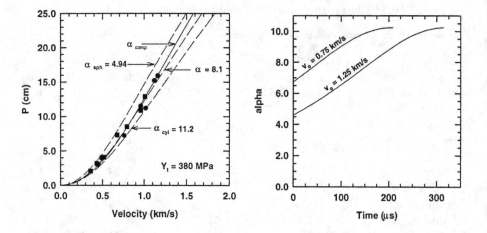

FIGURE 10.6. Comparison of the Walker–Anderson model to experimental data for rigid penetration (dashed lines are for constant α, solid line uses the cavity expansion α) (left) and history of α for two impact velocities when the penetration model uses the cavity expansion α (right). (From Walker and Anderson [199].)

on low velocity cavity expansion values (a spherical $\alpha = (2E_t/3Y_t)^{1/3} = 4.94$, Eq. 6.104 with $1/K$ negligible compared to $1/Y$ and $\nu = \frac{1}{2}$; a cylindrical $\alpha = (\sqrt{3}E_t/2(1+\nu_t)Y_t)^{1/2} = 11.2$, Eq. 6.54 with $1/K = 0$ for incompressible plastic region and 2 replaced by $\sqrt{3}$ for von Mises rather than Tresca) and one based on a value chosen to best match the data ($\alpha = 8.1$). In the experiments, hemispherical-nosed, maraging-steel projectiles, 74.7 mm long ($D = 7.10$ mm), were fired into 6061-T651 aluminum targets. The projectile remained rigid during penetration. For these penetrations, the crater diameter equals the projectile diameter. A reasonable value for the flow stress of the aluminum target is $Y_t = 380$ MPa.

It is seen that the constant α model values do not have a fast enough increase in the depth of penetration, as compared to the data. To match both the low and high velocity depth of penetration, a velocity-dependent α is needed. Using α from the cylindrical cavity expansion (Eq. 6.52),

$$\left(1 + \frac{\rho_t u^2}{Y_t}\right)\sqrt{K_t - \rho_t \alpha^2 u^2} = \left(1 + \frac{\rho_t \alpha^2 u^2}{2G_t}\right)\sqrt{K_t - \rho_t u^2}, \qquad (10.95)$$

where K_t and G_t are the bulk and shear modulus of the target, respectively, results in good agreement with the depth of penetration curve over the whole experimental range. The time evolution of α is shown in Fig. 10.6 for two different impact velocities. The extent of the plastic zone grows with time since the penetration velocity decreases as penetration proceeds.

Next we will show the same behavior for eroding penetration. Here we also need to specify the crater diameter, and in this example we use the quadratic fit given by Eq. (9.14),

$$R = R_p(1 + 0.2869v + 0.1457v^2), \quad v \text{ in km/s}. \qquad (10.96)$$

R is calculated from the initial impact velocity and is held constant throughout the calculation. Though this specific crater radius as a function of the impact velocity

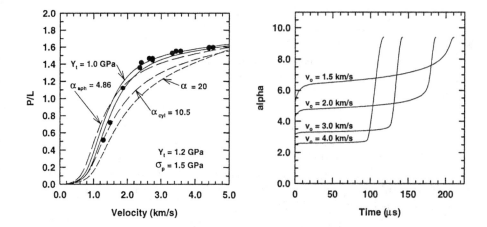

FIGURE 10.7. Comparison of the Walker–Anderson model to experimental data for eroding penetration of Silsby [**170**] (dashed
lines use constant α, solid lines use the cavity expansion α) (left)
and history of α for four impact velocities when the model uses the
cavity expansion α (right). (From Walker and Anderson [**199**].)

is empirical, it was shown in the previous chapter to agree with data as well as the
derived expression for crater radius. In the last set of examples in this chapter,
results from this curve fit will be compared to the expression derived in the last
chapter and it will be seen that, except in one case, the penetration results are
nearly identical. In the case where the results differ, it will be seen that the curve
fit gives better results (and the reasons for this will be discussed). Based on our
experience, the quadratic fit applies to a number of materials and is recommended
for eroding penetration in the absence of additional information.

The penetration data from Silsby presented in the previous chapter (Table 9.3)
for $L/D = 22.9$, $L = 12.18$ or 15.58 cm tungsten alloy projectiles striking steel
are plotted in Fig. 10.7. The dashed curves are for constant α. The values of
α calculated from the spherical and cylindrical cavity expansion expressions two
paragraphs above are $\alpha = 4.86$ and 10.5, respectively. A value of $Y_t = 1.2$ GPa
represents an estimate of the flow stress of armor steel at fairly large strain after
some work hardening has occurred. The smaller the value of α, the more rapid
the increase in penetration in the 0.5 to 2 km/s region; but, for $\alpha = 4.86$, the
increase occurs at too low a velocity as compared to the experimental data (for
smaller values of α, the rapid increase in penetration depth occurs even sooner
with velocity). As the value of α becomes larger, the initial rise in the curve shifts
to the right, but becomes too shallow. This again shows the need to use an α that
depends on the penetration velocity, and in fact, it was the serious discrepancy of
tungsten penetration into steel that led to the development of Eq. (10.95).

At some point the rate of cavity expansion is greater than the low-pressure
elastic wave speed and α will become less than 1. As discussion in Section 6.7,
Eq. (10.95) requires an additional modification to increase the stiffness of the bulk
modulus K_t with increasing velocity. Physically, the bulk modulus increases with

material compression; and material compression increases with penetration velocity. We use a heuristic expression to estimate how K_t increases with increasing penetration velocity, as follows. Under uniaxial strain conditions, the shock velocity is related to the particle velocity by the equation $U = c_0 + ku$. The ambient bulk modulus $K_0 = \rho_0 c_0^2$, and we can relate the dynamic bulk modulus of the target with the shock velocity (Eq. 6.139),

$$K_t \approx \rho_0 u_s^2 = K_0 \left(1 + k\frac{u}{c_0}\right)^2. \tag{10.97}$$

Equation (10.97) is for one-dimensional impact. The concern here is with cylindrical cavity expansion, and the geometric divergence would suggest that Eq. (10.97) is probably too stiff with penetration velocity, so the square root of the term within the parenthesis is taken (Eq. 6.137),

$$K_t \sim K_0 \left(1 + k\frac{u}{c_0}\right). \tag{10.98}$$

The solid line in Fig. 10.7 (left) represents the use of Eq. (10.95) with Eq. (10.98). Agreement is fairly good over the entire velocity range. Again, an α that decreases with increasing penetration velocity is required to match the inflection of the S curve of penetration. The figure also shows α as a function of time for four different impact velocities. In contrast to rigid penetration, where α increases continuously with time, α is nearly constant for most of the penetration history: α does change rapidly during the final deceleration phase of penetration as v approaches u and v and u both go to zero as the projectile comes to rest. These results with α say that the extent of the normalized plastic zone decreases with increasing penetration velocity, as was shown in Chapter 7 through numerical simulations. The numerical simulations show that the physical extent of the plastic zone, as represented by αR, actually increases with increasing penetration velocity, but the crater radius grows at a faster rate than the plastic zone, thereby resulting in a decrease in α.

The assumption of nearly constant density that was assumed in the model integration is still valid. However, there are still compressibility effects, and these come in through the cavity expansion through α. Compressibility cannot be ignored in the resistance of the target material to penetration, especially as the impact velocity increases.

As a third set of examples, from Hohler–Stilp data, tungsten alloy and steel projectiles impacting armor steel targets [10] for different target hardnesses (BHN 255 and BHN 295) and projectiles with two different aspect ratios ($L/D = 1$ and 10) are considered. The tungsten alloy (WA) projectiles had a density of 17.0 and 17.6 g/cm^3. The steel alloy (C110W1) projectile was reported to have a hardness of BHN 230.

The left side of Fig. 10.8 shows normalized penetration data for $L/D = 10$ projectiles into two hardnesses of armor steel. For the tungsten alloy projectiles, $Y_p = 2.0$ GPa; for the steel projectiles, $Y_p = 1.2$ GPa. The triangles are for an armor steel with a hardness of BHN 255 ($Y_t = 1.2$ GPa), and the data represented by the square data are for an armor steel with BHN 295 ($Y_t = 1.4$ GPa). The values for Y_p and Y_t represent reasonable dynamic values for the projectile and target materials. The right side of Fig. 10.8 depicts similar data for $L/D = 1$ tungsten alloy and steel projectiles into armor steel. Material constants were the same as for the $L/D = 10$ cases. Data are in Tables 10.2 and 10.3.

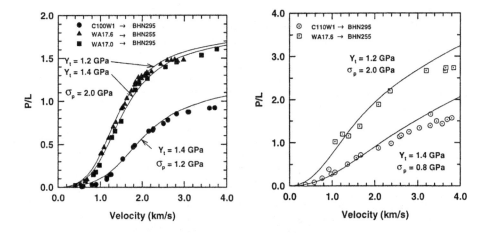

FIGURE 10.8. Normalized depth of penetration vs. impact velocity for $L/D = 10$ (left) and $L/D = 1$ (right) tungsten alloy and steel projectiles; the Walker–Anderson model is solid lines, data from Hohler and Stilp [98]. Note the L/D effect. (From Walker and Anderson [199].)

TABLE 10.2. Impact data for $L/D = 1$ C110W1 steel projectiles striking an armor steel target (BHN = 295) [98].

Velocity (m/s)	P/L	L (mm)	P (mm)	Vol (cm^3)	R_c/P
374	0.033	9.2	0.3	0.011	13.95
579	0.083	12	1.0	0.07	5.78
762	0.182	5.5	1.0	0.02	3.09
922	0.387	15	5.8	2.8	2.62
983	0.272	9.2	2.5	0.22	2.59
1,072	0.317	12	3.8	0.63	2.34
1,361	0.500	9.2	4.6	0.8	1.98
1,566	0.650	12	7.8	2.7	1.65
1,670	0.709	5.5	3.9	0.21	1.30
2,095	0.875	12	10.5	4.6	1.38
2,390	0.982	5.5	5.4	0.59	1.34
2,593	1.141	9.2	10.5	3.2	1.15
2,627	1.163	9.2	10.7	3.3	1.13
2,800	1.250	9.2	11.5	3.7	1.08
3,046	1.337	9.2	12.3	4.8	1.11
3,302	1.391	9.2	12.8	5.3	1.10
3,313	1.655	5.5	9.1	1.8	1.07
3,451	1.491	5.5	8.2	1.3	1.06
3,580	1.436	5.5	7.9	1.3	1.12
3,700	1.527	5.5	8.4	1.4	1.06
3,804	1.564	5.5	8.6	1.4	1.03

TABLE 10.3. Impact data for $L/D = 1$ tungsten alloy $\rho = 17.6 \text{ g/cm}^3$) projectiles with $L = 9$ mm striking HzB,A steel target (BHN = 255) [98].

Velocity (m/s)	P/L	P (mm)	Vol (cm^3)	R_c/P
1,067	1.022	9.2	0.9	0.74
1,244	1.200	10.8	1.4	0.73
1,385	1.156	10.4	2.1	0.94
1,648	1.378	12.4	2.9	0.85
2,093	1.889	17.0	4.8	0.68
2,368	2.200	19.8	7.1	0.66
3,203	2.667	24.0	12.3	0.65
3,650	2.667	24.0	14.6	0.71
3,663	2.733	24.6	15.3	0.70
3,845	2.733	24.6	17.5	0.75

Overall, agreement is quite good between the model and the experimental data. For the $L/D = 10$ data, the model overpredicts the depth of penetration at very low velocities ($v < 1.0 \text{ km/s}$). The blunt nose shape of the projectiles used in the experiments is the likely source of this discrepancy. Walker and Anderson [198], using numerical simulations, show that it takes approximately two projectile diameters of eroding penetration before the effects of nose shape disappear. Although nose shape effects were only investigated at an impact velocity of 1.5 km/s in the study, it was demonstrated that a hemispherical nose penetrates more easily (deeper penetration at equivalent times) than a blunt nose projectile at early times.

The analytical model does quite well at the low impact velocities for the $L/D = 1$ projectiles. Craters for small L/D impacts tend to be hemispherical in shape, so the hemispherical flow field of the analytical model appears to represent a good approximation of the flow patterns for small L/D impacts. The model tends to overpredict penetration for the $L/D = 1$ steel projectiles at the higher impact velocities. A curve fit for crater radius based on $L/D = 1$ impactor data instead of Eq. (10.96) (based on $L/D = 10$ data) had little influence on the normalized depth of penetration, and does not change this behavior of the model (i.e., it is not a crater radius effect).

10.12. Comparison of Velocities and Projectile Residual Length

The original motivation for the model was to match the velocity profiles seen in the target and projectile. If the model is working well, good agreement with the time-dependent interface and rear projectile velocities is expected. These velocities, vs. depth of penetration and vs. time, are shown as the dotted lines in Fig. 10.9. The solid curves are from the numerical simulation. The value of α is calculated from the cavity expansion, $Y_t = 1.15$ GPa, and $Y_p = 1.8$ GPa (chosen to match the 7.17-cm depth of penetration of the simulation). The Walker–Anderson model provides a reasonable approximation of the initial transient phase of penetration. The final deceleration of the tail occurs a little too late at the end of penetration, but the deceleration of the tail of the projectile does have approximately the correct slope (this can be contrasted with the deceleration of the tail in the Tate model, which

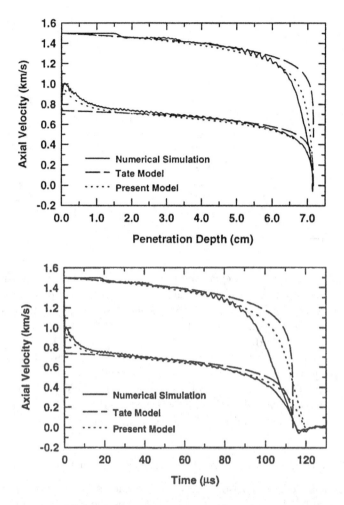

FIGURE 10.9. Comparison of penetration (nose) and tail velocities vs. depth of penetration (above) and time (below) from numerical simulation, Tate model, and Walker–Anderson model ("present model"): $L/D = 10$ tungsten alloy projectile into an armor steel, $v_0 = 1.5$ km/s. (From Walker and Anderson [199].)

is far too steep). As already discussed, the deceleration at the end of penetration approximately parallels that of the numerical simulations.

It is noted that the penetration velocity of the model lies slightly below that of the numerical simulation in Fig. 10.9. This is a consequence of requiring that the total depths of penetration be the same for comparison purposes. On the other hand, using $Y_t = 1.20$ GPa, and $\sigma_p = 1.8$ GPa, the quasi-steady state penetration velocities of the model agrees with those of the numerical simulation, but the projectile penetrates to 7.4 cm (instead of 7.17 cm, a difference of 3%). It might be thought that slightly increasing the initial penetration velocity of the model would shift the penetration velocity slightly higher and therefore provide better agreement with

TABLE 10.4. Strength values used in the model computations. Y_0 estimates are based on the experimental BHN [97], save for Al alloy 2, which is an ultimate tensile strength from [180].

Material	ρ (g/cm^3)	BHN (kp/mm^2)	Y_0 estimate (GPa)	Y_p (GPa)	Y_t (GPa)	Y/Y_0
Steel C110W2	7.85	230±30	0.77	0.9		1.17
Steel 37	7.85	135±20	0.45		0.85	1.89
Steel 52	7.85	180±20	0.60		0.95	1.58
Steel HzB20	7.85	295±35	0.98		1.30	1.33
W alloy D17	17.0	294±10	0.76	1.3	1.30	1.71*
Al alloy 1	2.85	75 − 85	0.31		0.45	1.45
Al alloy 2	2.7		0.165**	0.40		
Pb	11.2				0.045	

the numerical simulation. However, the penetration velocity emerges from the initial transient with approximately the same quasi-steady-state penetration velocity. This is expected, since the physics of the model (e.g., the densities and strengths of the projectile and the target, and the initial impact velocity) essentially dictate the quasi-steady flow conditions.

The model predicts a finite residual projectile at the end of penetration, although the projectile has eroded more than observed in experiments at the same impact velocity. The model predicts approximately 0.4 cm of projectile remaining at the bottom of the crater. This is approximately half of that observed in experiments. The Tate model predicts nearly complete erosion of the projectile.

10.13. Comparison to the Hohler–Stilp Data from Chapter 8

As a final set of comparison, we compare the model results to the Hohler–Stilp data presented in Chapter 8. Table 10.1 gives the elastic material properties that were used. Table 10.4 gives the strength values used. As can be seen, the strength values used compare reasonably to the statically measured values. Also, the ability of the model to match penetration data over a wide range of velocities is again evident. The Walker–Anderson model results are shown in Figs. 10.10–10.12, where the crater radius curve fit is used (Eq. 10.96), as is the equation for the crater radius derived in Chapter 9 (Eq. 9.7), repeated here, based on the striking velocity,

$$\left(\frac{R_c}{R_p}\right)^2 = \frac{Y_p + 2\rho_p(v-u)^2}{B_3} \quad \text{with} \quad \frac{1}{2}\rho_p(v-u)^2 + Y_p = \frac{1}{2}\rho_t u^2 + B_3. \quad (10.99)$$

It is seen that the results are extremely close, save for the aluminum into lead case in Fig. 10.12. The fact that the final crater radius leads to an overprediction of results argues that the correct radius to use in the model is the radius that is near the eroding penetrating front of the projectile and the late-time crater radius is not the radius to use. The fact that the curve fit radius gives better penetration depth results for aluminum striking lead than the actual radius (see Fig. 9.2 to see how well Eq 9.7 matches the crater radius) is another reason why we often use the empirical expression for crater radius in calculations.

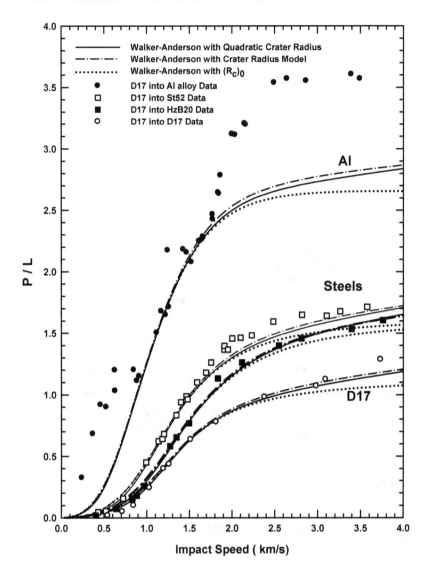

FIGURE 10.10. $L/D = 10$ tungsten alloy projectiles impacting an aluminum alloy, two steel alloys, and the same tungsten alloy compared to the Walker–Anderson model with the crater radius fit (solid) and the crater radius model (dashed).

10.14. Comparison to Hypervelocity Penetration vs. Time Data

Target strength and, to a lesser degree, projectile strength influence penetration velocity even at impact velocities over 2 and 3 km/s. Figure 10.13 compares the results of the Walker–Anderson model with the experimental data of penetration depth and projectile tail location vs. time for both the 2.085 and 3.10 km/s tungsten-into-steel tests reported in Section 8.8. Values for the flow stresses need to account for strain hardening and rate effects since the analytical model uses a single value for the flow stress. For these calculations, the flow stress of the tungsten

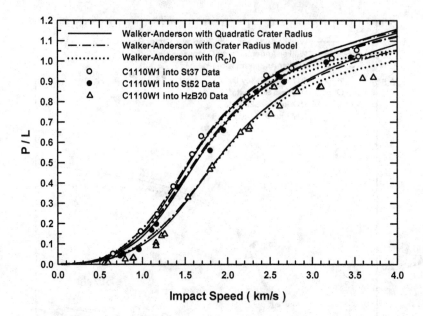

FIGURE 10.11. $L/D = 10$ steel alloy projectiles impacting three steel alloys compared to the Walker–Anderson model with the crater radius fit (solid) and the crater radius model (dashed).

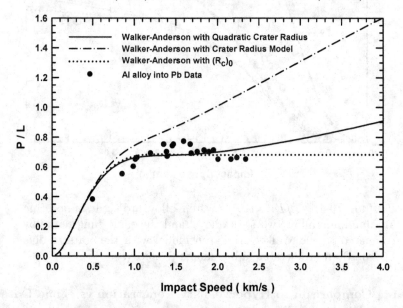

FIGURE 10.12. $L/D = 10$ aluminum alloy projectiles impacting lead compared to the Walker–Anderson model with the crater radius fit (solid) and the crater radius model (dashed).

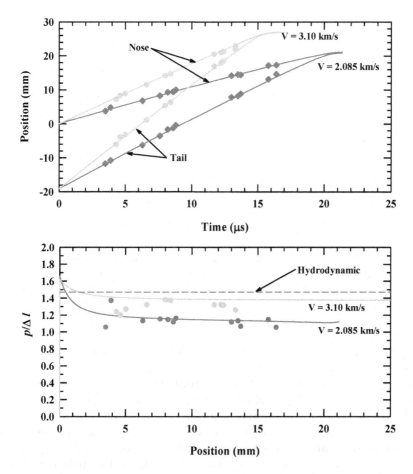

FIGURE 10.13. Comparison of Walker–Anderson analytical penetration model (lines) with experimental data (circles) for a tungsten alloy projectile striking a steel target at two impact velocities, 2.085 km/s and 3.10 km/s. The upper plot shows position vs. time and the lower plot shows dP/dL_e vs. penetration depth. (From Anderson et al. [11].)

was taken as 1.7 GPa and the flow stress for the steel target was taken as 1.5 GPa. The figure shows the nose and tail positions of the projectile vs. time. The lines are the results calculated from the model and the symbols depict the corresponding experimental data. The figure also shows the penetration efficiency $\Delta P/\Delta L_e$ vs. penetration depth. The model matches the 2.085 km/s data very well. The agreement between the model and the 3.10 km/s data is not quite as good, perhaps because the model uses a single value for target flow stress for both impact velocities, which would not reflect differences in strain hardening, strain-rate, and thermal softening effects between the two impact velocities. However, the agreement is good enough that the model can be used with confidence to perform a parameter study to analyze the influence of strength on the steady-state penetration.

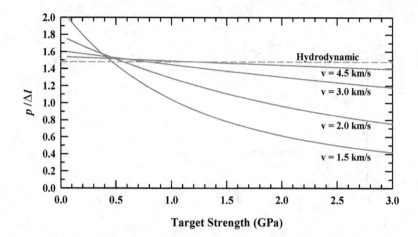

FIGURE 10.14. Steady-state-phase penetration efficiency vs. target flow stress at different impact velocities for tungsten ($Y_p = 1.7$ GPa) into steel of various strengths. (From Anderson et al. [11].)

Figure 10.14 shows dP/dL_e vs. target flow stress, for impact velocities between 1.5 and 4.5 km/s. In this figure, dP and dL_e are averaged over the quasi-steady portions of the analytical results. The analytical model shows, for an impact velocity of 1.5 km/s, that dP/dL_e decreases very rapidly with increasing target strength, decreasing by almost 33% as the target strength is increased from 0.5 to 1.0 GPa. In contrast, for an impact velocity of 4.5 km/s, dP/dL_e does not deviate substantially from hydrodynamic values over the full range of target strengths from 0 to 3.0 GPa. For a target strength of about 1.2 GPa (typical of RHA), it takes an impact velocity of about 4.5 km/s to achieve hydrodynamic penetration. For a stronger target, like a high-hard steel ($Y_t \approx 2$ GPa), penetration is less than hydrodynamic at 4.5 km/s. Hard targets such as ceramics require even higher velocities to achieve hydrodynamic-like penetration.

Penetration efficiency dP/dL_e for the steady-state phase of primary penetration is plotted as a function of impact velocity for a tungsten alloy projectile into an armor steel target ($Y_t = 1.5$ GPa) for two tungsten alloy flow stresses: zero-strength ($Y_p = 0$) and 1.7 GPa (Fig. 10.15). The curve for the zero-strength (hydrodynamic) projectile is below the curve of the projectile that has strength. The target strength affects the steady-state penetration phase for impact velocities up to 7 km/s. These two curves reinforce the experimental observations that target and projectile strengths have a measurable effect on steady-state penetration at considerably higher impact velocities than one might assume from an examination of the final depth of penetration (Fig. 7.20).

10.15. Tungsten into Aluminum: Rigid and Secondary Penetration

The above comparisons show good agreement with all the data but the tungsten into aluminum. Both at low velocities and high velocities, the experimental values exceed the model predicted values.

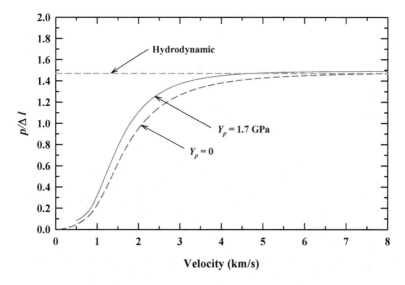

FIGURE 10.15. Primary phase penetration efficiency vs. impact velocity for fixed steel target strength ($Y_t = 1.5\,\text{GPa}$) and two values of tungsten alloy penetrator strength ($Y_p = 1.7\,\text{GPa}$ and $Y_p = 0$). (From Anderson et al. [11].)

At low velocities, the explanation lies in the fact that the tungsten projectile is strong enough that it can penetrate as a rigid body. As will be demonstrated in the next chapter, the rigid-eroding transition for a hemispherical nosed projectile corresponds roughly to the impact velocity when $(1/2)\rho_t v^2 \approx Y_p$. For the tungsten alloy into aluminum, this velocity is roughly 980 m/s for the values $Y_p = 1.3\,\text{GPa}$ and $\rho_t = 2.85\,\text{g/cm}^3$ and 730 m/s for $Y_p = 760\,\text{MPa}$ with the same aluminum density. Thus, we would expect the projectile to penetrate in a rigid fashion at these lower velocities, giving rise to the deeper penetrations. Figure 10.16 shows the rigid penetration curve (assuming a hemispherical nose) and, as can be seen, there is reasonable agreement at the velocities below 750 m/s. Thus, at these lower velocities, the strength of the tungsten is high enough that rigid penetration occurs.

At high velocities there is anomalously deep penetration, also seen in comparing to the Tate model. In 1961, Allen and Rogers [2] suggested a model for high velocity penetration. They proposed that the flow of projectile material into and out of the projectile–target interface could be modeled as a reversal of a streamline in steady-state penetration. In streamline reversal, the direction of the flow changes, but the velocity magnitude remains constant. We already used this assumption about $v - u$ for the eroded debris remaining constant after the turn in the crater to calculate the eroded projectile material velocity when we did the energy rate balance during the crater radius derivation. However, at that time we did not pay attention to the sign. Allen and Rogers noted that for the case of hydrodynamic penetration, if the projectile has a higher density than the target, the eroded projectile material has a residual velocity into the target. In particular, if v_d is used to denote the velocity of the eroded projectile debris material, then the velocity in the laboratory frame

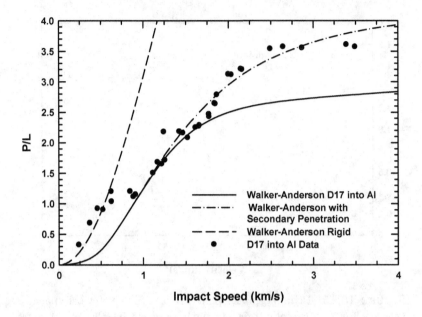

FIGURE 10.16. $L/D = 10$ D17 alloy projectiles impacting aluminum alloy targets compared to the Walker–Anderson model, assuming rigid penetration, and including secondary penetration.

is the debris speed exiting the stationary crater plus the penetration speed[4]

$$v_d = -(v - u) + u = 2u - v. \tag{10.100}$$

For example, in the hydrodynamic approximation this gives

$$v_d = \frac{1 - \sqrt{\rho_t/\rho_p}}{1 + \sqrt{\rho_t/\rho_p}} v. \tag{10.101}$$

For like-into-like materials at high speed, we expect eroded projectile material to be left sitting on the crater wall with zero residual velocity. If the penetration velocity is low, then we expect eroded projectile material to be thrown backward with respect to the stationary Lagrangian frame. However, if the projectile is much stronger and denser than the target, then the eroded projectile material will have a velocity forward in the Lagrangian frame and thus it may have additional penetration capability into the target. Allen and Rogers named the additional target penetration due to the eroded projectile material "secondary" or "residual" penetration. They reported posttest experimental data for depth of penetration that they believed showed this secondary penetration for high velocity (> 2 km/s) impacts of gold projectiles into Al 7075-T6 targets.

Let us explore some consequences of this argument. First, note that the debris cannot penetrate faster than the original projectile. This result is evident since $u < v$ and thus $u - v < 0$ and addition of u gives $u_d < v_d = 2u - v < u$. A conclusion is that the debris will not be competing with the original projectile penetration event, and if there is secondary penetration it will occur after the initial

---

[4]The fact that there are energy losses in the turning process is discussed in Exercise 7.8.

projectile penetration is over. Next, the length of the eroded projectile debris is given by

$$L_d(t) = \int_0^t (u - (2u - v))dt = \int_0^t (v - u)dt = L_e(t) = L_0 - L(t). \qquad (10.102)$$

Thus, the length of the eroded projectile debris material is equal to the amount of eroded projectile material:

$$L_d(t) + L(t) = L_0. \qquad (10.103)$$

This result is surprising enough to justify

THEOREM 10.4. *The current length of the projectile plus the length of spatially distributed eroded debris is constant, equal to the original length of the projectile.*

Thus, for a projectile that fully erodes, at the end of the penetration event the length of the eroded debris is equal to the original length of the projectile:

$$L_d = L_0. \qquad (10.104)$$

Therefore, we can expect a projectile of zero strength (because it is eroded debris) with the same length as the original length of the impactor to strike the target at the reduced velocity of $2u - v$, where u is an average penetration velocity of the original penetration event. This debris can then give rise to secondary penetration.

As we have shown, the penetration velocity is not constant, but decreases over the penetration event. A measure of average penetration velocity is to assume

$$\frac{u_{\text{ave}}}{v - u_{\text{ave}}} = \frac{P}{L} \Rightarrow u_{\text{ave}} = \frac{P/L}{P/L + 1}v. \qquad (10.105)$$

This value can then be used to determine the velocity of the debris with respect to the target:

$$v_d = \frac{P/L - 1}{P/L + 1}v. \qquad (10.106)$$

Thus, for initial penetrations below $P/L = 1$ the eroded debris is moving away from the target and there is no secondary penetration, while for initial penetrations above $P/L = 1$ the eroded debris is moving into the target and there may be secondary penetration by the projectile debris.

Now that we know the length of the eroded debris and its velocity, we need to address the question of the state of the projectile. If the projectile were solid and cylindrical, we could perform a second computation at the new velocity and get a depth of penetration. However, that is not the state of the projectile. In the first instance, the eroded projectile debris is not cohesive, so the strength is zero: $Y_p = 0$. Next, it is in some sort of tubular shape and is having some interaction with the walls of the crater. Referencing the work of Franzen and Schneidewind on tubular penetrators [73], they show for manufactured tubular penetrators of intact material, that at high velocities the penetration velocity can roughly be obtained by considering a penetrator with the outside diameter of the tube and a reduced density of the material, assuming that the material is evenly distributed over the frontal area of the tube. To use this result, it is necessary for us to determine the radius of the tube of debris. If we assume that the erosion products are, very roughly, spread somewhere between the radius of the projectile and the radius of the crater, then a radius of the debris is

$$R_d = \frac{R_p + R_c}{2}. \qquad (10.107)$$

The adjusted density of the projectile would be

$$\rho_{\text{secondary}} = \rho_p \left(\frac{R_p}{R_d} \right)^2 = \frac{4\rho_p}{(1 + R_c/R_p)^2}. \tag{10.108}$$

This density and projectile radius, with projectile length L and impact speed v_d, are now inserted into the Walker–Anderson model to compute secondary penetration. When this is done the results are similar to what is seen in the tungsten into aluminum data in Fig. 10.16. Thus, the anomalous deeper penetration, which has been seen in other materials at hypervelocities, is due to the added penetration that occurs by the projectile debris.

THEOREM 10.5. *(secondary penetration) If $P/L > 1$, then eroded projectile debris can produce secondary penetration into the target. For metals, if the density of the projectile material is much larger than the target, this secondary penetration can be significant at hypervelocities (> 2 km/s).*

10.16. Plastic Strain in the Target

From the flow field and the rigid plastic assumption, it is possible to determine the plastic deformation in the target. In general this requires tracking a point in the target and locally performing the integration over time. To obtain an analytic form, we simplify the problem and assume a steady penetration, where α is constant and the penetration speed u is constant, and only consider the centerline. The velocity field is our standard expression,

$$v(r) = \frac{u}{\alpha^2 - 1} \left\{ \left(\frac{\alpha R}{r} \right)^2 - 1 \right\}. \tag{10.109}$$

We use the fact that along the axis of symmetry the equivalent plastic strain is the absolute value of the logarithmic strain (see Exercise 3.15). To compute the logarithmic strain, we first consider the material displacement X. A point of material on the z-axis begins to move when the velocity field first touches it. The Lagrangian location of the point of interest is $z = z_0 + X(z_0, t)$. The velocity of the point is given by the flow field, which is advancing at a constant rate,

$$\frac{dX}{dt} = v(r) = v(\alpha R - ut + X). \tag{10.110}$$

The velocity field first touches the point when $r = \alpha R$, and as time progresses r decreases, even though z for the point is increasing. The projectile, and hence the flow field, moves at a constant velocity u to the right. As we have noted,

$$r = \alpha R - ut + X \qquad \text{and hence} \qquad \frac{dX}{dt} = u + \frac{dr}{dt}. \tag{10.111}$$

Equation (10.110) becomes

$$\frac{dr}{dt} = -u + v(r) \quad \Rightarrow \quad \frac{\alpha^2 - 1}{\alpha^2} \cdot \frac{dr}{1 - (R/r)^2} = -udt. \tag{10.112}$$

This result can be integrated to give

$$\frac{\alpha^2 - 1}{\alpha^2} \left\{ \frac{1}{2} \ln \left(\frac{r - R}{r + R} \right) - \frac{1}{2} \ln \left(\frac{\alpha - 1}{\alpha + 1} \right) + \frac{r}{R} - \alpha \right\} = -\frac{ut}{R}. \tag{10.113}$$

From this the displacement $X = r + ut - \alpha R$ at a given time can be determined. We could use this to compute the strain, but it is somewhat simpler not to use the final result, but rather Eq. (10.110). To compute the Lagrangian engineering strain, we look for the change in length of an initial gage length Δz_0 – that is, we examine the displacement for two nearby points separated by Δz_0. By the assumption of steady flow, the displacement $X(z_0, t)$ is self-similar in the sense that $X(z_0, t) = X(0, t - z_0/u)$. The strain is

$$
\begin{aligned}
\varepsilon_{\text{eng}} &= \frac{\partial X(z_0, t)}{\partial z_0} = \lim_{\Delta z_0 \to 0} \frac{X(z_0 + \Delta z_0, t) - X(z_0, t)}{\Delta z_0} \\
&= \lim_{\Delta z_0 \to 0} \frac{X(z_0, t - \Delta z_0/u) - X(z_0, t)}{\Delta z_0} = \frac{\partial X(z_0, t)}{\partial t} \left(\frac{-1}{u} \right) = -\frac{v(r(z_0, t))}{u}
\end{aligned}
\tag{10.114}
$$

Thus (recall that $u = v(R)$),

$$
\varepsilon_{\text{eng}} = -\frac{v(r)}{u} = -\frac{1}{\alpha^2 - 1} \left\{ \left(\frac{\alpha R}{r} \right)^2 - 1 \right\}.
\tag{10.115}
$$

This result is surprisingly simple. For this compression problem, the initial engineering strain is 0 at $r = \alpha R$ and the (minimum) engineering strain is -1 at $r = R$, corresponding to "full" compression. The equivalent plastic strain is

$$
\varepsilon_p = |\ln(1 + \varepsilon_{\text{eng}})| = \left| \ln \frac{\alpha^2}{\alpha^2 - 1} \left\{ 1 - \left(\frac{R}{r} \right)^2 \right\} \right| = \ln(1 - 1/\alpha^2) - \ln(1 - (R/r)^2).
\tag{10.116}
$$

As we move along the centerline flow field from the outer edge of the field at $r = \alpha R$ to the point at the bottom of the crater (projectile nose location) at $r = R$, the equivalent plastic strain increases. At the bottom of the crater itself, the strain is infinite. Examining Eq. (10.113) reveals that it reaches infinite value when time reaches infinity. To obtain an estimate of the strain at finite time, note that ut is just the penetration depth P and Eq. (10.113) says

$$
\ln \left(\frac{r/R - 1}{r/R + 1} \right) - \ln \left(\frac{\alpha - 1}{\alpha + 1} \right) = 2 \left(\alpha - \frac{r}{R} \right) - \frac{2\alpha^2}{\alpha^2 - 1} \frac{P}{R}.
\tag{10.117}
$$

If we assume that the time is late enough that r/R is close to 1, we obtain

$$
\frac{r}{R} \approx 1 + 2 \left(\frac{\alpha - 1}{\alpha + 1} \right) \exp \left(2(\alpha - 1) - \frac{2\alpha^2}{\alpha^2 - 1} \frac{P}{R} \right)
\tag{10.118}
$$

and the equivalent plastic strain is

$$
\varepsilon_p \approx 2 \ln \left(\frac{\alpha + 1}{2\alpha} \right) - 2(\alpha - 1) + \frac{2\alpha^2}{\alpha^2 - 1} \frac{P}{R}.
\tag{10.119}
$$

For a specific case, assume $\alpha = 5$ and $R = 2R_p = D$, then for $P/D = 10$ the numbers are $r/D \approx 1 + 3.56 \times 10^{-6}$, the engineering strain $\varepsilon_{\text{eng}} \approx -1$, and $\varepsilon_p = 11.8$ for the material that has been moved while the projectile advanced a distance P. Thus, the equivalent plastic strain can be very large in front of the projectile. For an application of the strain, such as including work hardening in the target resistance, it may be simpler to work with the engineering strain, as is done in Exercises 10.20 to 10.22.

The local equivalent plastic strain rate along the centerline is easily computed from Eq. (10.65),

$$\dot{\varepsilon}_p = \sqrt{\frac{2}{3}D_{ij}D_{ij}} = |D_{rr}| = \left|\frac{\partial v}{\partial r}\right| = \frac{2u(\alpha R)^2}{(\alpha^2 - 1)r^3}. \tag{10.120}$$

The values $u = 750$ m/s, $R = 1$ cm, and $\alpha = 5$ give an equivalent plastic strain rate of 1.5×10^5 s^{-1}, in good agreement with Chapter 7's numerical simulations. The effect of strain rate on target resistance is explored in Exercise 10.23.

We would finally like to compute S_t for a general constitutive model $Y_t = Y(\varepsilon_p, \dot{\varepsilon}_p, T)$. Introducing $\tilde{\varepsilon} = -\varepsilon_{eng}$ and solving $\tilde{\varepsilon} = v(r)/u$ for r gives

$$r = \alpha R / \sqrt{1 + (\alpha^2 - 1)\tilde{\varepsilon}}. \tag{10.121}$$

The equivalent plastic strain and equivalent plastic strain rate (using Eq. 10.120) are

$$\varepsilon_p(\tilde{\varepsilon}) = -\ln(1 - \tilde{\varepsilon}) = \ln\left(\frac{1}{1 - \tilde{\varepsilon}}\right), \quad \dot{\varepsilon}_p(\tilde{\varepsilon}) = \frac{2u}{\alpha R(\alpha^2 - 1)}(1 + (\alpha^2 - 1)\tilde{\varepsilon})^{3/2}. \tag{10.122}$$

Changing variables gives

$$S_t(u, \alpha) = \frac{7}{3}\int_R^{\alpha R}\frac{Y_t}{r}dr = \frac{7}{6}\int_0^1 Y(\varepsilon_p(\tilde{\varepsilon}), \dot{\varepsilon}_p(\tilde{\varepsilon}), T(\tilde{\varepsilon}))\frac{\alpha^2 - 1}{1 + (\alpha^2 - 1)\tilde{\varepsilon}}d\tilde{\varepsilon}, \tag{10.123}$$

$$T(\tilde{\varepsilon}) = \frac{1}{\rho C_v}\int_0^{\tilde{\varepsilon}} Y(\varepsilon_p(\tilde{\varepsilon}), \dot{\varepsilon}_p(\tilde{\varepsilon}), T(\tilde{\varepsilon}))d\tilde{\varepsilon},$$

though through the cavity expansion $\alpha = \alpha(u)$ there is still an R dependence in S_t due to the strain rate term. If both integrals are numerically computed simultaneously starting from $\tilde{\varepsilon} = 0$, then Y need only be stored for the current integration step. At the $\tilde{\varepsilon} = 1$ endpoint the integrand is singular, since ε_p is singular; however, for typical work hardening expressions the integral is finite (Exercise 10.24).

Using the strain determination we can address another question that may have arisen for some, namely that since the approach for computing target resistance should work for any reasonable velocity field, how do we handle a centerline velocity that tangentially becomes zero – that is, it has $\partial v/\partial r = 0$ when $v(\alpha r) = 0$ – hence causing a singularity in the evaluation of Eq. (10.59),

$$S_t = -\frac{Y_t}{3}\ln\left|\frac{\partial v}{\partial r}\right|_{r=R}^{r=\alpha R} + \frac{4Y_t}{3}\ln(\alpha)? \tag{10.124}$$

Let us consider a specific case, namely

$$v(r) = u\frac{(\alpha R - r)^2}{(\alpha^2 - 1)R^2}. \tag{10.125}$$

This centerline velocity produces an infinite result in the integration of Eq. (10.124) as written at $r = \alpha R$ since

$$\frac{\partial v}{\partial r} = -2u\frac{(\alpha R - r)}{(\alpha^2 - 1)R^2}. \tag{10.126}$$

However, owing to the very small value of v when r is near αR, we ask the question as to what value of r leads to an initial yield as the velocity field progresses into the target. Setting ε_p to the yield value $\varepsilon_p = Y_t/E = v(r)/u$ gives

$$\alpha R - r = \sqrt{\alpha^2 - 1}R\sqrt{Y_t/E}, \tag{10.127}$$

and when this value of r is placed in Eq. (10.124) for the outer flow field extent the expression inside the natural logarithm term is

$$\left.\frac{\partial v}{\partial r}\right|_{r=R}^{r=r} = \sqrt{\frac{Y_t}{E}}\sqrt{\frac{\alpha+1}{\alpha-1}}. \tag{10.128}$$

$\sqrt{E/Y_t}$ is close to the cylindrical cavity expansion α and so the whole expression is roughly $1/\alpha$. Thus, the integral for S_t again is a multiplier of $\ln(\alpha)$, and the target shear resistance term is of the form that we expect. In Chapter 12 we again discuss robustness of the S_t calculation for different flow fields, including singular ones.

10.17. Finding α using the Dynamic Plasticity Approach

In this section another approach to determining the plastic extent in the target α is outlined. The dynamic plasticity approach was described at the end of Section 3.16 and was used to determine the plastic extent in the Taylor anvil problem in Chapter 5. The approach is as follows. From a velocity field that depends on geometric parameters (α in this case), the accelerations are determined. Using the momentum balance, stresses that formally produce that acceleration field are determined. When these stresses satisfy the yield condition and boundary conditions, they are referred to as admissible. Finally, the dissipation is maximized to determine the rate of change of the geometric variables ($\dot{\alpha}$), given the constraint that the stress lies within or on the yield surface (i.e., remains admissible). Though it might be thought that the symmetries of the problem would produce a nice solution, it will be seen to be quite complex.

To use notation similar to Chapter 5, we introduce

$$\Gamma = \frac{u}{\alpha^2-1} \quad \Rightarrow \quad \alpha^2\Gamma = \Gamma+u \quad \Rightarrow \quad 2\alpha\dot{\alpha} = \frac{\dot{u}\Gamma - u\dot{\Gamma}}{\Gamma^2}. \tag{10.129}$$

The derivative $\dot{\Gamma}$ will be the adjustable parameter in the optimization, rather than $\dot{\alpha}$. The assumed velocity field in the target is

$$\vec{v} = \Gamma\left\{\left(\frac{\alpha R}{r}\right)^2 - 1\right\}\cos(\theta)\hat{e}_r + \Gamma\sin(\theta)\hat{e}_\theta. \tag{10.130}$$

The acceleration is complicated by the fact that the velocity field is written in a moving spherical coordinate frame. The motion of the projectile is most easily seen by writing spherical coordinates r and θ in cylindrical coordinates z and x, where x is used as the cylindrical radial coordinate to avoid confusion with r. With z_i the location of the nose of the projectile (i means interface),

$$r = \sqrt{x^2 + (z - z_i(t) + R)^2}, \quad \cos(\theta) = \frac{z - z_i(t) + R}{r}, \quad \sin(\theta) = \frac{x}{r}. \tag{10.131}$$

Since z_i is a function of time, its derivative with respect to time is not zero (in particular, it is the penetration velocity u). That means that when the acceleration is computed using the Eulerian expression in Appendix A (Eq. A.191) there are terms in $\partial\vec{v}/\partial t$ that are due to this time dependence. We compute these terms explicitly first, since there are subtleties due to the derivatives of the basis vectors. Beginning with

$$\frac{\partial\vec{v}}{\partial z_i}\frac{dz_i}{dt} = u\left(\frac{\partial\vec{v}}{\partial r}\frac{\partial r}{\partial z_i} + \frac{\partial\vec{v}}{\partial\theta}\frac{\partial\theta}{\partial z_i}\right) = u\left(-\cos(\theta)\frac{\partial\vec{v}}{\partial r} + \frac{\sin(\theta)}{r}\frac{\partial\vec{v}}{\partial\theta}\right), \tag{10.132}$$

from Eq. (A.182)

$$\frac{\partial \hat{e}_r}{\partial r} = 0, \quad \frac{\partial \hat{e}_r}{\partial \theta} = \hat{e}_\theta, \quad \frac{\partial \hat{e}_\theta}{\partial r} = 0, \quad \frac{\partial \hat{e}_\theta}{\partial \theta} = -\hat{e}_r, \tag{10.133}$$

a straightforward computation using our flow field Eq. (10.130) yields

$$\frac{\partial \vec{v}}{\partial z_i} \frac{dz_i}{dt} = u\Gamma \frac{(\alpha R)^2}{2r^3}\{(1 + 3\cos(2\theta))\hat{e}_r + \sin(2\theta)\hat{e}_\theta\}, \tag{10.134}$$

where trigonometric half angle $(2\cos^2(\theta) = 1 + \cos(2\theta))$ and double angle formulas have been used. This expression takes care of the time derivatives of the r, θ, $\hat{e}_r$, and $\hat{e}_\theta$ dependencies in the partial derivative with respect to time in the spherical acceleration term in Eq. (A.191), and including these terms when performing the rest of the acceleration computation with our velocity field yields

$$a_r = -\alpha^4 \Gamma^2 \frac{R^2}{r^3}\left\{\left(\frac{R}{r}\right)^2 - \frac{1}{2}\right\} + \left[\dot{\Gamma}\left\{\left(\frac{R}{r}\right)^2 - 1\right\} + \dot{u}\left(\frac{R}{r}\right)^2\right]\cos(\theta)$$

$$\quad - \alpha^4 \Gamma^2 \frac{R^2}{r^3}\left\{\left(\frac{R}{r}\right)^2 - \frac{3}{2}\right\}\cos(2\theta), \tag{10.135}$$

$$a_\theta = \dot{\Gamma}\sin(\theta) + \alpha^4 \Gamma^2 \frac{R^2}{2r^3}\sin(2\theta).$$

Again, trigonometric identities have been used. Notice that only the $\cos(\theta)$ and $\sin(\theta)$ terms depend on $\dot{\Gamma}$ or $\dot{u}$; in other words, they are the only part of the acceleration that can be adjusted by the maximization process, since Γ and α are given at the given time.

To provide guidance for making approximations, we recall that the admissible stress that maximizes dissipation will be close to the rigid plastic stresses. Computing the rigid plastic stresses for our assumed flow field yields

$$s_{rr} - s_{\theta\theta} = -Y_t \cdot \frac{2\sqrt{3}\cos(\theta)}{\sqrt{12\cos^2(\theta) + \sin^2(\theta)}} \approx -Y_t \cdot \left\{\frac{47}{36}\cos(\theta) - \frac{11}{36}\cos(2\theta)\right\}. \tag{10.136}$$

The approximation is chosen to have the same basis as seen in the accelerations. Though making this choice limits some of the freedom in the maximization, it facilitates explicit expression of the stress.

Given the basis for these functions, we assume the admissible stress can be written as

$$\sigma^a_{rr} = S_1(r) + S_2(r)\cos(\theta) + S_3(r)\cos(2\theta), \quad \sigma^a_{r\theta} = S_4(r)\sin(\theta) + S_5(r)\sin(2\theta). \tag{10.137}$$

Similarly define by comparing with Eq. (10.135)

$$a_r = A_1(r) + A_2(r)\cos(\theta) + A_3(r)\cos(2\theta), \quad a_\theta = A_4(r)\sin(\theta) + A_5(r)\sin(2\theta). \tag{10.138}$$

We assume that $\sigma^a_{\theta\theta} = \sigma^a_{\varphi\varphi}$ and, based on the rigid plastic result, $\sigma^a_{rr} - \sigma^a_{\theta\theta}$ equals the approximate right-hand side of Eq. (10.136). The admissible stress formally satisfies the momentum balance, given the prescribed accelerations. With our assumptions the momentum balance is

$$\frac{\partial \sigma_{rr}^a}{\partial r} + \frac{1}{r}\frac{\partial \sigma_{r\theta}^a}{\partial \theta} + \frac{1}{r}[2(\sigma_{rr}^a - \sigma_{\theta\theta}^a) + \sigma_{r\theta}^a \cot(\theta)] = \rho_t a_r,$$

$$\frac{\partial \sigma_{r\theta}^a}{\partial r} + \frac{1}{r}\frac{\partial \sigma_{\theta\theta}^a}{\partial \theta} + \frac{3\sigma_{r\theta}^a}{r} = \rho_t a_\theta. \tag{10.139}$$

Plugging these into the momentum balance yields five equations, one for each of the linearly independent basis terms. These equations are

$$(1; i) \quad S_1' + \frac{S_5}{r} = \rho_t A_1,$$

$$(\cos(\theta); ii) \quad S_2' + \frac{2S_4}{r} - \frac{47}{18}\frac{Y_t}{r} = \rho_t A_2,$$

$$(\cos(2\theta); iii) \quad S_3' + \frac{3S_5}{r} + \frac{11}{18}\frac{Y_t}{r} = \rho_t A_3, \tag{10.140}$$

$$(\sin(\theta); iv) \quad S_4' - \frac{S_2}{r} + \frac{3S_4}{r} - \frac{47}{36}\frac{Y_t}{r} = \rho_t A_4,$$

$$(\sin(2\theta); v) \quad S_5' - \frac{2S_3}{r} + \frac{3S_5}{r} + \frac{11}{18}\frac{Y_t}{r} = \rho_t A_5.$$

There are two simpler subsystems in this system of equations, (iii) and (v), and (ii) and (iv). The latter two equations combine and the former two combine to give, respectively,

$$r^2 S_4'' + 4r S_4' + 2S_4 = \frac{47}{18}Y_t + \rho_t A_2 r + \rho_t (A_4 r)' r,$$

$$r^2 S_5'' + 4r S_5' + 6S_5 = -\frac{11}{9}Y_t + 2\rho_t A_3 r + \rho_t (A_5 r)' r. \tag{10.141}$$

These equations allow solving for S_4 and S_5; then S_2 and S_3 can be explicitly written down; then S_1 is determined by an integration. Defining

$$C(r) = \cos\left(\frac{\sqrt{15}}{2}\ln(r/R)\right), \quad S(r) = \sin\left(\frac{\sqrt{15}}{2}\ln(r/R)\right), \tag{10.142}$$

the solution to these equations is

$$S_1 = \frac{\sqrt{15}c_4 + 3c_3}{12r^{3/2}}C(r) + \frac{3c_4 - \sqrt{15}c_3}{12r^{3/2}}S(r) + \frac{1}{5}\rho_t \Gamma^2\left(\frac{\alpha R}{r}\right)^4 + \frac{11}{54}\ln(r/R)Y_t + c_5,$$

$$S_2 = \frac{2c_1}{r} + \frac{c_2}{r^2} + \rho_t(\dot{\Gamma} + \dot{u})R^2\frac{1 + 2\ln(r/R)}{r} - \rho_t \dot{\Gamma}r + \frac{47}{18}Y_t,$$

$$S_3 = \frac{\sqrt{15}c_4 + 3c_3}{4r^{3/2}}C(r) + \frac{3c_4 - \sqrt{15}c_3}{4r^{3/2}}S(r) + \frac{1}{10}\rho_t \Gamma^2\left(\frac{\alpha R}{r}\right)^4,$$

$$S_4 = \frac{c_1}{r} + \frac{c_2}{r^2} + \rho_t(\dot{\Gamma} + \dot{u})R^2\frac{\ln(r/R)}{r} + \frac{47}{36}Y_t,$$

$$S_5 = \frac{c_3}{r^{3/2}}C(r) + \frac{c_4}{r^{3/2}}S(r) - \rho_t \alpha^4 \Gamma^2\left(\frac{R}{r}\right)^2\left\{\frac{1}{5}\left(\frac{R}{r}\right)^2 - \frac{1}{2}\right\} - \frac{11}{54}Y_t, \tag{10.143}$$

where c_1 through c_5 are five constants. With $\dot{\Gamma}$ and $\dot{u}$, there are seven unknowns. However, we assume that $\dot{u}$ comes from the Walker–Anderson centerline momentum balance equation, so there are six unknowns to be addressed by the maximization.

Two of these unknowns are determined by the radial stress boundary conditions along the centerline, namely

$$S_1(R)+S_2(R)+S_3(R) = \sigma_{zz}(z_i), \quad S_1(\alpha R)+S_2(\alpha R)+S_3(\alpha R) = \sigma_{rr}(\alpha R) \quad (10.144)$$

(as shown in Exercise 6.6, $-|\sigma_{\text{HEL}}| < \sigma_{rr}(\alpha R) \leq -2Y_t/3$, but we will choose the upper limit for specificity and simplicity). Technically, the other four constants are determined by maximizing the dissipation rate $\sigma_{ij}^a D_{ij}$ while insisting that the yield condition is never exceeded by the stress $J_2(\sigma_{ij}^a) \leq Y_t^2/3$ (hence the stress is admissible). Since $\sigma_{rr}^a - \sigma_{\theta\theta}^a$ is set by assumption (Eq. 10.136), the maximization reduces to maximizing

$$\sigma_{r\theta}^a D_{r\theta} = -(S_4(r) + 2\cos(\theta)S_5(r))\sin(\theta) \cdot u\frac{\alpha^2}{2(\alpha^2-1)}\frac{R^2}{r^3}\sin(\theta). \quad (10.145)$$

This is done by making $\sigma_{r\theta}^a$ as negative as possible, subject to admissibility.

When the J_2 constraint is considered, with $\sigma_{rr}^a - \sigma_{\theta\theta}^a$ given by Eq. (10.136),

$$|\sigma_{r\theta}^a| = |S_4(r) + 2\cos(\theta)S_5(r)|\sin(\theta) \leq \frac{Y_t}{\sqrt{3}}\left\{1 - \left(\frac{47}{36}\cos(\theta) - \frac{11}{36}\cos(2\theta)\right)^2\right\}^{1/2}.$$
$$(10.146)$$

This is the constraint that must be satisfied for admissibility. Looking at it along the centerline, by expanding the right-hand side near $\theta = 0$ and dividing by θ, gives

$$|S_4(r) + 2S_5(r)| \leq \frac{Y_t}{6}. \quad (10.147)$$

Near the top of the flow field, $\theta = \pi/2$, and the constraint is

$$|S_4(r)| \leq \frac{Y_t}{\sqrt{3}}\frac{5\sqrt{47}}{36} \approx \frac{Y_t}{\sqrt{3}}. \quad (10.148)$$

Looking at $S_4(r)$ ad $S_5(r)$, and recalling there are four constants to be determined, shows that there is not a simple analytic solution to this problem, and we will not pursue it.

However, we would like to recover something from this effort. Let's assume $\alpha(0) = 1$. Looking at all the $S_i(r)$ we see that they will be dominated by Γ^2, which is infinite at $t = 0$. The only expression that can offset this value is $\dot{\Gamma}$. Thus, we assume that at early time

$$\dot{\Gamma} + k_1\Gamma^2 \approx 0, \quad (10.149)$$

where k_1 is a constant that we will need to determine, hopefully through a dynamic plasticity argument. This equation can be explicitly integrated to give $\Gamma = 1/(k_1t)$, and solving for α,

$$\alpha(t) = \sqrt{1 + k_1 u(t)t} = \sqrt{1 + \frac{2\,\dot{\alpha}(0)\,u(t)\,t}{u(0)}}, \quad (10.150)$$

where k_1 was found by differentiation of the expressions for α. Careful work with the S_i and the constraints shows that they do not determine $\dot{\alpha}(0)$. We can use the centerline momentum balance (Eq. 10.72) to arrive at $\dot{\alpha}(0)$, though it will be discovered it is necessary to know $\dot{s}(0)$. Equation (10.72) at $t = 0$ with $s(0) = 0$

and $\alpha(0) = 1$ yields

$$\rho_p \dot{v}(0)L - \rho_p(v(0) - u(0))\frac{\dot{s}(0)}{2} + \rho_t \dot{\alpha}(0)\frac{Ru(0)}{2} = \frac{1}{2}\rho_p(v(0) - u(0))^2 - \frac{1}{2}\rho_t u^2(0).$$
$$(10.151)$$

The deceleration of the tail is given by (Eq. 10.73)

$$\dot{v}(0) = -\frac{\sigma_p}{\rho_p L}\left\{1 + \frac{v(0) - u(0)}{c_E} + \frac{\dot{s}(0)}{c_E}\right\}. \qquad (10.152)$$

Using the dynamic plasticity solution for the Taylor anvil problem, $s = \frac{1}{2}(v - u)t$, and combining these expressions yields

$$\frac{\dot{\alpha}(0)R}{u(0)} = -1 + \frac{2}{\rho_t u^2}\left(\frac{3}{4}\rho_p(v - u)^2 + \sigma_p\left\{1 + \frac{3}{2}\frac{(v - u)}{c_E}\right\}\right)\Bigg|_{t=0} = \frac{1}{2}Rk_1. \quad (10.153)$$

This provides a reasonable $\dot{\alpha}(0)$ and it gives a reasonable overall solution to the penetration when extrapolated to large α. In this case the target resistance depends on the history of the penetration and not just the current penetration velocity, as occurs when we use the cavity expansion to determine α.

It is interesting to note that if we set the shear stress terms $S_4(r)$ and $S_5(r)$ equal to their approximate rigid plastic value at R and αR, as implied by Eqs. (10.147) and (10.148) (the actual values don't matter, as will be seen), then this provides four equations and we can now find all six unknowns by the solution of a set of linear equations. The result is not easy to write because of the large number of terms. Specifically, though, we are interested when α is close to 1 (i.e., early times); using REDUCE to compute the solution to the linear system and then to compute a Taylor series for the $\dot{\Gamma}$ term, we obtain

$$\rho_t R\dot{\Gamma} = \left(\sigma_{zz}(z_i) - \sigma_{rr}(\alpha R) + \frac{1}{2}\rho_t u^2\right)(\alpha - 1)^{-2} + O((\alpha - 1)^{-1}). \qquad (10.154)$$

Notice that Y_t does not appear in this leading term from the dynamic plasticity solution for α near 1. Since

$$\dot{\Gamma} = -\frac{1}{2}u\dot{\alpha}(\alpha - 1)^{-2} + O((\alpha - 1)^{-1}), \qquad (10.155)$$

we have

$$-\frac{\dot{\alpha}}{2} = \frac{1}{\rho_t R}\left(\sigma_{zz}(z_i) - \sigma_{rr}(\alpha R) - \frac{1}{2}\rho_t u^2\right). \qquad (10.156)$$

Again using the dynamic plasticity Taylor anvil solution $s = \frac{1}{2}(v - u)t$ and $s(0) = 0$, then the interface stress (Eq. 10.70) yields

$$\sigma_{zz}(z_i) = \rho_p \dot{v}L - \frac{1}{2}\rho_p(v - u)^2 - \rho_p(v - u)\frac{\dot{s}}{2} = -\sigma_p\left\{1 + \frac{3}{2}\frac{(v - u)}{c_E}\right\} - \frac{3}{4}\rho_p(v - u)^2,$$
$$(10.157)$$

where everything is evaluated at $t = 0$. With $\alpha(0) = 1$, we obtain

$$\frac{\dot{\alpha}(0)R}{u(0)} = -1 + \frac{2}{\rho_t u^2}\left(\frac{3}{4}\rho_p(v - u)^2 + \sigma_p\left\{1 + \frac{3}{2}\frac{(v - u)}{c_E}\right\} - \frac{2Y_t}{3}\right)\Bigg|_{t=0}, \quad (10.158)$$

nearly the same as Eq. (10.153), with the only difference due to the $\sigma_{zz}(\alpha R)$ term.

10.18. Sources

The original sources for the Walker–Anderson model are Walker and Anderson [**197, 199**]. New material here further elucidates the model, compares to additional data, and addresses secondary penetration. The hypervelocity test data with computations examining strength and the hydrodynamic limit are from Anderson, Orphal, Franzen, and Walker [**11**].

Exercises

(10.1) Program up the Walker–Anderson model. Reproduce the tungsten D17 impact into HzB20 steel curve (Fig. 10.10) from 0 to 4 km/s. Next, produce the same curve for titanium alloy Ti 6Al-4V. What does the comparison of these two curves imply about the relative role of strength and density in penetration? Some people say they are looking for a material with "the strength of steel but half the density." Is Ti 6Al-4V it? (The people usually want the same or half the cost. Also, real armors have finite thickness and titanium's brittleness is an issue.)

(10.2) Show that the solution to Boussinesq's problem predicts that the stresses decay in the elastic half space as $1/r^2$ and thus conclude that $\sigma_{zz}(+\infty) = 0$ in the target.

(10.3) It turns out for a certain value of r that the hemispherical flow field has a velocity magnitude (speed) contour that is independent of θ. What is r? What is the magnitude of the velocity? Does this value agree with contours plotted in Fig. 10.4?

(10.4) Assuming that the crater radius can change adds only two transient terms to the Walker–Anderson model, one through the crater inertia and the other through the extent of the plastic zone. Assuming that $R(t)$ is a function of time, calculate the two terms

$$\int_{z_i}^{+\infty} \frac{\partial v_z}{\partial t} dt \quad \text{and} \quad \left(\frac{v - u}{s} \right) \tag{10.159}$$

and write down the centerline momentum balance for the case where the crater radius is not a constant.

(10.5) Beginning with the results of the previous exercise, assume that $\alpha(u)$ is a function of the penetration velocity and the crater radius $R(v - u)$ is a function of the erosion rate (for example, as derived in the last chapter). Write the centerline momentum balance of the Walker Anderson model. This form is rarely employed, since the crater radius is usually assumed constant and the expression for $d\alpha/du$ is not trivial.

(10.6) Another form of potential flow that gives rise to incompressible flow is gradient flow. When the Laplacian of a function is zero, $\nabla^2 f = 0$, the flow field given by $\vec{v} = \text{grad}(f)$ is incompressible, since the divergence of the gradient is zero,

$$\text{div}(\vec{v}) = \text{div}(\text{grad}(f)) = \nabla^2 f = 0. \tag{10.160}$$

Show that for the four terms in the potential of Eq. (10.18), the velocities resulting from two of the terms can be arrived at by a gradient flow, while two of the terms cannot be the result of a gradient flow. For the two parts that can be derived from the gradient of a function, what is $f(r, \theta, \phi)$?

(10.7) The hemispherical flow field is comprised of two parts, a constant velocity in the z direction with magnitude $-v/(\alpha^2 - 1)$ and the radial (times a cosine) source term $(\cos(\theta)/r^2)\hat{e}_r$ with magnitude $v(\alpha R)^2/(\alpha^2 - 1)$. Consider just this latter term. Plot it. What is its behavior like? There is flow out from the top hemisphere ($z > 0$) at infinity and flow in at infinity for the bottom hemisphere. What is the rate of flow?

(10.8) The assumed flow field does have an effect on the constant 7/3 multiplier on our computed target resistance. This value arises from the assumed $1/r^2$ behavior of the flow field. This exercise will look at a different flow field, arising from the flow of an ideal (i.e., nonviscous, clearly unlike our plastically deforming target material) fluid around a sphere, which has $1/r^3$ behavior. To begin, assume that

$$f(r) = ar + \frac{b}{r^2}, \tag{10.161}$$

where a and b are constants. Derive the velocity from the curl of the vector potential $A_\varphi = f(r)\sin(\theta)$. Determine the values of a and b needed to satisfy the boundary conditions $\vec{v} \cdot \vec{n} = V\cos(\theta)$ on $r = R$ and $\vec{v} \cdot \vec{n} = 0$ on $r = \alpha R$, where $\vec{n}$ is the normal to the surface of the sphere. Compute the rate of deformation tensor D_{ij} and compute the subsequent value for the target resistance integral. Compare this value to that obtained with the potential in the text (i.e., compare to 7/3).

(10.9) The flow field of Exercise 10.8 is one of those that also arises as the gradient of a scalar potential. Find the potential.

(10.10) Show that for the assumed velocity field of Eq. (10.19), if the boundary conditions Eq. (10.25) are satisfied with $u > 0$, then

$$\int_R^{\alpha R} v_\theta(r, \pi/2)\, 2\pi r\, dr > 0; \tag{10.162}$$

that is, that material exits at the top of the hemispherical region.

(10.11) Compute the integral of the shear stress term in a slightly different fashion by transforming the rigid-plastic stress in spherical coordinates to Cartesian coordinates, take the derivative, and integrate. Show that it gives the same results.

(10.12) Using the momentum balance in cylindrical coordinates in the z direction (Appendix A) and assuming axial symmetry, show

$$S_t = -2\int_{z_i}^{+\infty} \frac{\partial \sigma_{rz}}{\partial r}\bigg|_{r=0} dz. \tag{10.163}$$

Similarly, in spherical coordinates, use the radial component momentum balance along the $\theta = 0$ to show

$$S_t = -\int_{r_i}^{+\infty} \frac{1}{r}\left(2\frac{\partial \sigma_{\theta r}}{\partial \theta} + 2\sigma_{rr} - \sigma_{\theta\theta} - \sigma_{\phi\phi}\right)\bigg|_{\theta=0} dr. \tag{10.164}$$

This term only depends on stress deviators. (Hint: both require taking limits – first $r \to 0$ and second $\theta \to 0$.)

(10.13) Suppose that the flow field yields "good behavior" of the shear stress, namely that it is zero along the centerline and, in spherical coordinates, can be expanded as a Taylor series in the form

$$\sigma_{r\theta} = f_1(r)\theta + f_2(r)\theta^2 + \cdots. \tag{10.165}$$

Show that the target resistance shear integral equals

$$S_t = 2Y\ln(\alpha) - 2\int_R^{\alpha R} \frac{1}{r}\frac{\partial \sigma_{\theta r}}{\partial \theta}\bigg|_{\theta=0} dr. \tag{10.166}$$

Use this form and the rigid plastic stress deviators for the hemispherical flow field to show $S_t = \frac{7}{3}\ln(\alpha)Y_t$.

(10.14) Show the $u(0)$ that gives rise to the correct initial system momentum in the case that $\alpha(0) > 1$ and s is computed by matching slopes is

$$u(0) = \frac{\sqrt{\rho_p}(\alpha(0)+1)v(0)}{\sqrt{\rho_p}(\alpha(0)+1) + 2\sqrt{\rho_t}\alpha(0)}. \tag{10.167}$$

What initial penetration velocity results if the $\alpha \to 1$ limit is taken?

(10.15) The target resistance is $\frac{1}{2}\rho_t u^2 + \frac{7}{3}Y_t\ln(\alpha)$. Plot the target resistance and show it does not greatly change for steel and aluminum targets over the penetration range of 0 to 1 km/s, because of the fact that the S_t term is decreasing as the inertial term increases. How much does it change for a tungsten target? What about an beryllium alloy?

(10.16) Theorem 10.5 on secondary penetration did not say "If $P/L > 1$ for primary penetration ..." Why is there no need to include the primary penetration qualifier?

(10.17) Can the secondary penetration impact velocity v_d exceed the primary penetration velocity u?

(10.18) Should we expect to see secondary penetration for tungsten striking steel? Why or why not? Given the data in this book, do we see any evidence of it?

(10.19) Figure 10.17 shows results from an experiment of a tungsten projectile with density $\rho_p = 19.3$ g/cm^3 striking a Zelux target, which is a plastic material with density $\rho_t = 1.2$ g/cm^3 from Orphal and Anderson [153]. The impact velocity was 3.61 km/s. The experiment was performed in the reverse ballistic mode, meaning the projectile is at rest and the target is in flight. The tungsten projectile had a diameter $D = 0.508$ mm, length $L = 50.8$ mm, and thus $L/D = 100$. The diameter of the target was 25.4 mm; the target length was 114.3 mm. Primary instrumentation for the experiment was comprised flash X-rays. The first X-ray shows the projectile before impact and the remaining three X-rays show the tungsten penetrating the Zelux, at at 12.2, 24.8, and 37.2 μs after the first frame (the impact time was not determined). The eroded tungsten is obvious in the X-rays. The outer diameter of the eroded tungsten debris is seen to be quite constant and about equal to twice the diameter of the projectile.

Assuming the figure reproduction in the book is fairly accurate, measure the length of the residual projectile and the remaining projectile. What is $L_d(t) + L(t)$? Is it constant? Does this experiment confirm the reasoning leading up to secondary penetration?

In the reference frame where the Zelux target is at rest, what is the penetration velocity u? (Hint: do a least-squares fit to the front location data.) Similarly, what is the velocity of the back end of the projectile debris (i.e., v_d)? Is the debris in motion relative to the target? Compare with

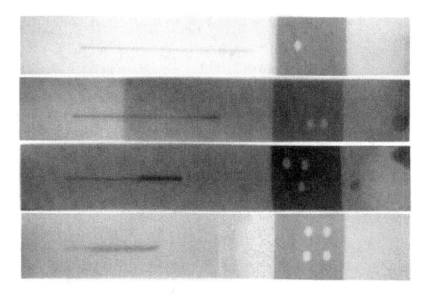

FIGURE 10.17. X-ray images of the experiment. (From Orphal and Anderson [153].)

TABLE 10.5. Summary of experimental results: $v_0 = 3.61$ km/s, $L_0 = 50.8$ mm [153].

X-ray	Time (μs)	z_i (mm)	z_d (mm)	L (mm)	L_{d1} (mm)	$L_{d1} + L$ (mm)	L_{d2} (mm)	$L_{d2} + L$ (mm)
2	12.2	28.2	22.9	44.1	5.3		5.3	
3	24.8	65.7	52.5	35.0	13.2		13.6	
4	37.2	96.1	75.4	27.5	20.7		19.5	

Eq. (10.100). Will it lead to further penetration of the Zelux? How do these velocities compare to that predicted by the hydrodynamic theory?

Compare your measurements from the X-rays to those in Table 10.5. The table lists the length of the remaining projectile L and the length of the eroded projectile debris L_d, measured two ways: first as the difference between the penetration depth and the depth of the rear of the debris $L_{d1} = z_i - z_d$; and second as a direct measurement L_{d2}. Determine the sums to compare with L_0.

(10.20) For linear work hardening with $Y = A + B|\varepsilon_{\text{eng}}|$, show that the target resistance is given by

$$R_t = \frac{7}{3}A\ln(\alpha) + \frac{7}{3}B\left(\frac{1}{2} - \frac{\ln(\alpha)}{\alpha^2 - 1}\right). \tag{10.168}$$

For $\alpha = 5$ show that the parenthetical term on the right is 0.43, which corresponds to using $Y = A + 0.27B$ in $R_t = \frac{7}{3}Y\ln(\alpha)$ (for that α). Hence, the "flow stress" corresponds to Y at 27% engineering strain, again for the specific α.

(10.21) For $n = \frac{1}{2}$ work hardening given by $Y = A + B|\varepsilon_{eng}|^{1/2}$, show that the target resistance is given by

$$R_t = \frac{7}{3}A\ln(\alpha) + \frac{7}{3}B\left(1 - \frac{\cos^{-1}(1/\alpha)}{\sqrt{\alpha^2 - 1}}\right). \tag{10.169}$$

(Hint: try the change of variable $r = \alpha R \cos(\theta)$.) For $\alpha = 5$, show that the parenthetical term on the right is 0.72, which corresponds to using $Y = A + 0.45B$ in $R_t = \frac{7}{3}Y\ln(\alpha)$ (for that α). Hence, the "flow stress" corresponds to Y at 20% engineering strain for $\alpha = 5$.

(10.22) The hardening exponent $n = \frac{1}{m}$, where m is an integer, is a special case that is explicitly integrable. Use a symbolic computation package to determine the target resistance for $n = \frac{1}{3}$ and $n = \frac{1}{4}$ when work hardening is given by $Y = A + B|\varepsilon_{eng}|^n$. According to Macsyma 2.7.0,

$$n = \frac{1}{3}, \quad \beta = (\alpha^2 - 1)^{1/3}, \quad R_t = \frac{7}{3}A\ln(\alpha) + \frac{7}{3}B\left(\frac{3}{2} - \frac{1}{4\beta}\left\{\ln\left(\frac{(\beta+1)^2}{\beta^2 - \beta + 1}\right)\right.\right.$$
$$\left.\left. +2\sqrt{3}\tan^{-1}\left(\frac{2\beta - 1}{\sqrt{3}}\right) + \frac{\pi}{\sqrt{3}}\right\}\right),$$

$$n = \frac{1}{4}, \quad \beta = (\alpha^2 - 1)^{1/4}, \quad R_t = \frac{7}{3}A\ln(\alpha) + \frac{7}{3}B\left(2 - \frac{\sqrt{2}}{4\beta}\left\{\ln\left(\frac{\beta^2 + \sqrt{2}\beta + 1}{\beta^2 - \sqrt{2}\beta + 1}\right)\right.\right.$$
$$\left.\left. +2[\tan^{-1}(\sqrt{2}\beta + 1) + \tan^{-1}(\sqrt{2}\beta - 1)]\right\}\right). \tag{10.170}$$

For $\alpha = 5$, show that the parenthetical term on the right is 0.90 and 1.02, for $n = \frac{1}{3}$ and $n = \frac{1}{4}$, respectively, which corresponds to using $Y = A + 0.56B$ and $Y = A + 0.64B$ in $R_t = \frac{7}{3}Y\ln(\alpha)$ (for that α). Hence, the "flow stress" corresponds to Y at 18% and 16% engineering strain for $\alpha = 5$, again respectively. (Notice the progression in this and the previous two problems: since $|\varepsilon_{eng}| \leq 1$ and $0 < n \leq 1$, as n decreases the weighting on B increases.)

(10.23) A common form of strengthening due to strain rate is $Y = A(1 + C\ln(\dot{\varepsilon}_p/\dot{\varepsilon}_0))$, where $\dot{\varepsilon}_0 = 1\,s^{-1}$. Using the equivalent plastic strain rate, compute the increase in target resistance due to such a strain rate term,

$$R_t = \frac{7}{3}A\ln(\alpha)\left\{1 + C\ln\left(\frac{2u\sqrt{\alpha}}{(\alpha^2 - 1)R\dot{\varepsilon}_0}\right)\right\}. \tag{10.171}$$

For $u = 750$ m/s, $R = 1$ cm, and $\alpha = 5$, show the term multiplying C is 9.5; hence $Y = A(1 + 9.5C)$ in $R_t = \frac{7}{3}Y\ln(\alpha)$. Show this value corresponds to the strain rate at $r/R = \sqrt{\alpha}$.

(10.24) In this exercise you show for work hardening of the form $|\varepsilon_p|^n$, $n > 0$, that the integral Eq. (10.123) is well behaved. Note that $\dot{\varepsilon}_p$ is bounded, so only focus on the work hardening. Show up to a multiplicative bounding constant that the S_t integral is finite if

$$\int_0^1 |\varepsilon_p|^n d\varepsilon = \int_0^1 |-\ln(1-\varepsilon)|d\varepsilon = \int_0^1 |\ln(x)|^n dx \tag{10.172}$$

is finite. For this equation, show that

$$\int_0^1 |\varepsilon_p|^n d\varepsilon = \int_0^{1/e} |\ln(x)|^n dx + \int_{1/e}^1 |\ln(x)|^n dx < I_m + I_0, \tag{10.173}$$

where $I_k = \int_0^1 |\ln(x)|^k dx$, k is an integer, and m is an integer larger than n. Now show that $I_0 = 1$, $I_k = kI_{k-1}$, and hence that $I_k = k!$ (k factorial).

(10.25) Show that the target resistance term for linear thermal softening with $Y = A(1 - (T^*)^m)$ is given by

$$R_t = \frac{7A}{6} \int_0^1 \exp\left(-\frac{A\varepsilon_p}{\rho C_v \Delta T}\right) \frac{\alpha^2 - 1}{1 + (\alpha^2 - 1)\varepsilon_p} d\varepsilon_p$$

$$\approx \frac{7}{3} A \ln(\alpha) - \frac{7}{3} \frac{A^2}{\rho C_v \Delta T} \left(\frac{1}{2} - \frac{\ln(\alpha)}{\alpha^2 - 1}\right),$$

(10.174)

where $\Delta T = T_{\text{melt}} - T_{\text{room}}$.

(10.26) (Stoke's formula) The next four problems ask what we might learn from fluid mechanics about resistance to penetration. In this problem we compute the drag on a sphere moving at very low velocity through a viscous fluid. The result is referred to as Stoke's formula. We assume the sphere is at rest and that a steady state fluid flow exists around the sphere. The assumed stress for a viscous fluid is

$$\sigma_{ij} = -p\delta_{ij} + \lambda D_{kk}\delta_{ij} + 2\mu D_{ij},$$

(10.175)

where λ and μ are the viscosity coefficients (notice that the stress is directly proportional to the rate of deformation tensor, as opposed to our plasticity assumption, where the term is divided by the magnitude of the rate of deformation tensor so that the stress magnitude is constant – hence "rate independent"). To compute the flow field, assume the flow can be given by the curl of $\vec{A} = f(r)\cos(\theta)\hat{e}_\varphi$. In the steady-state low velocity assumption, the acceleration term $Dv_i/Dt = \partial v_i/\partial t + v_j \partial v_i/\partial x_j$ drops out, since it is steady state ($\partial/\partial t = 0$) and the product of v_j and $\partial v_i/\partial x_j$ terms are small. With this assumption, show for the momentum balance $\vec{\nabla} \cdot \sigma = 0$ to be satisfied it is necessary that

$$\frac{1}{2} r f'''' + 2f''' - \frac{2f''}{r} = 0,$$

(10.176)

where f is a function of r and prime means differentiation with respect to r. (Hint: for the momentum balance to be satisfied, p must be integrable, which can only occur if the mixed partial of p, $\partial^2 p/\partial r\partial\theta$, obtained by appropriate differentiation of $(\vec{\nabla} \cdot \sigma)_r = 0$ and $(\vec{\nabla} \cdot \sigma)_\theta = 0$, is the same.) Thus show that for solutions $f = r^n$ that $n = 0, 1$, and -2. Thus, all the terms in Eq. (10.18) work except for c. Now solve the problem for a flow velocity at $z = \pm\infty$ of $-u\hat{z}$ and zero velocity at the surface of the sphere (no-slip boundary conditions are the standard assumption in viscous flow), showing that $a = -u/2$, $b = 3Ru/4$, and $d = -R^3u/4$, where R is the radius of the sphere; hence

$$v_r = -u\left(1 - \frac{3R}{2r} + \frac{R^3}{2r^3}\right)\cos\theta, \quad v_\theta = u\left(1 - \frac{3R}{4r} - \frac{R^3}{4r^3}\right)\sin\theta.$$

(10.177)

The pressure is integrated to give

$$p = \frac{2b\mu}{r^2}\cos\theta + p_\infty = \frac{3R\mu u}{2r^2}\cos\theta + p_\infty,$$

(10.178)

where p_∞ is the pressure at infinity. Given the flow field, now compute the stress deviators at the surface of the sphere. Show that they are all zero except

$$s_{r\theta} = \frac{3\mu u}{2R} \sin\theta. \tag{10.179}$$

Finally, compute the force exerted on the sphere by integrating the z component of the traction over the surface. Unlike our penetration problems, here there are forces on both the front of the sphere and the back of the sphere, since a cavity is not opened up. The traction is given by

$$t_i = n_j \sigma_{ji} \quad \Rightarrow \quad \vec{t} = \sigma_{rr}\hat{e}_r + \sigma_{r\theta}\hat{e}_\theta. \tag{10.180}$$

The z component is given by

$$\begin{aligned}
t_z = \vec{t}\cdot\hat{z} = \vec{t}\cdot(\cos\theta\,\hat{e}_r - \sin\theta\,\hat{e}_\theta) &= \sigma_{rr}\cos\theta - \sigma_{r\theta}\sin\theta \\
&= -p\cos\theta - s_{r\theta}\sin\theta = -\frac{3\mu u}{2R} - p_\infty\cos\theta.
\end{aligned} \tag{10.181}$$

Thus, the total force (the drag) on the sphere is the surface integral over the whole sphere of this equation, giving a value of

$$F_z = -6\pi R\mu u. \tag{10.182}$$

This result is Stoke's formula, which states that the resistance to the flow at very low velocities is linear with the velocity u.

(10.27) Show that if the resisting force for penetration behaves as $F = -kv^q$, where k is a constant and q is a real number, then a projectile will come to rest in finite time if $q < 1$ and it will never come to rest if $q \geq 1$. Use this result to show that a sphere with an initial speed in a fluid does not come to rest, nor will a projectile in the Walker–Anderson model where the target has zero strength come to rest. Hence, going to a resistance that is linear in u rather than u^2 does not stop projectiles.

(10.28) (Newtonian impact theory) There is a simplified theory for gas loading of surfaces at hypersonic speeds loosely based on Newton's attempt to model gases, referred to as Newtonian impact theory. The reasoning goes as follows. In the kinetic theory of gases it is possible to show that the root mean square speed of a molecule is $\sqrt{3kT/m}$, where k is Boltzmann's constant and m is the mass of a molecule. Show that this formula gives the RMS speed of a molecule of "air" as 507 m/s or Mach 1.5. Thus, for large Mach number flow (hypersonic) the interaction between particles can be ignored and the momentum transfer to the surface can be computed by considering inelastic collisions of individual molecules with the surface.

Assume that the gas molecule's impact is inelastic, that the normal component of the impact is absorbed without rebound by the surface, and that the tangential velocity is unaffected by the impact. Show for a relative velocity of u that the momentum transferred to a surface with obliquity θ to the impinging flow ($\theta = 0$ means the surface is perpendicular to the flow direction), area A, and in a time Δt is given by

$$\Delta\text{momentum} = u\cos(\theta)\cdot\text{mass} = u\cos(\theta)\rho_0 Au\Delta t\cos(\theta). \tag{10.183}$$

If we normalize this result by the surface area and divide by Δt we obtain the pressure,

$$p = \rho_0 u^2 \cos^2(\theta). \tag{10.184}$$

Notice this result says that with this model of gaseous behavior the stress along the centerline is twice what we would have expected from our negligible-density-change approximation in integrating the centerline momentum balance. Given the pressure on the surface, go back to the ideal gas law (Exercise 5.16) to obtain an approximate density of the gas assuming there is a large energy change and that the inelastic collision only heats the gas. Show that roughly

$$E = \frac{1}{2}V^2, \qquad \frac{\rho}{\rho_0} = \frac{2}{\gamma - 1}. \tag{10.185}$$

In fluid mechanics, the drag force past an object is written as

$$F = C_D A \frac{1}{2}\rho u^2, \tag{10.186}$$

where C_D is the drag coefficient, A is the projected area of the body perpendicular to the flow, ρ is the density of the fluid, and u is the flow speed. Show that at hypervelocities the Newtonian impact theory of gases predicts $C_D = 1$. (Hint: see Exercise 10.26 to make sure you are only integrating the traction component in the z direction and in this case only integrate on the front hemisphere; the back hemisphere is in a vacuum since at hypervelocities a cavity is opened up in the gas.) As an aside, it has been experimentally observed on the surface of a sphere at these high velocities that $p(\theta) = p(0)\cos^2(\theta)$. This theory roughly agrees with the experimental results for the drag on a sphere in hypersonic (defined as above Mach 5) flight, data taken with spheres launched by two-stage light gas guns, where C_D asymptotes to around 0.9 for large Mach and large Reynolds number Re,

$$Re = \frac{\rho D v}{\mu} = \frac{D v}{\nu}, \tag{10.187}$$

where $\mu = \eta$ is the dynamic viscosity and $\nu = \mu/\rho$ is the kinematic viscosity. The kinematic viscosity of air is 0.15 cm^2/s and water 0.010 cm^2/s.

(10.29) Assume that a hemispherical nosed projectile moving at hypersonic speeds through a compressible gas forms a shock wave in front of the sphere. Using the Hugoniot results of Exercise 5.17 for an ideal gas, assume a steady-state behavior and no density change between the back of the shock and the nose of the projectile to show using the centerline momentum balance that for high speeds the centerline pressure at the nose of the projectile is

$$p = \rho_0 v^2 \frac{\gamma + 3}{2(\gamma + 1)}. \tag{10.188}$$

For a monatomic gas, the value is $p = (7/8)\rho_0 v^2$ and for a diatomic gas, the value is $p = (11/12)\rho_0 v^2$. Compare with the previous two exercises.

As a comment, it might be thought that we have collected enough information to model meteorite penetration of the atmosphere. Unfortunately, we have not. Meteorites have strength, and a solid rock with no other processes would penetrate the air with no erosion. However, when hypersonic bodies with strength enter the Earth's atmosphere, the surface heats and ablates and subsurface vaporization and gas infiltration into fractures lead to fragmentation. Bright light from heating the air and meteorite during atmospheric entry and the subsequent explosion is referred to as a *bolide*.

(10.30) (Exercise from Chapter 12) Though we formulated the force on the nose in terms of angles, it is also possible to formulate the force by integrating along the length of the nose. For the cylindrical cavity expansion approach, where the projectile moves in the $+z$ direction and the nose is at the origin, show that the total force found by integrating the full length of the nose is

$$F_z = \int_{-L_n}^{0} 2\pi r (B + A\rho_t v_r^p)\frac{dr}{dz}dz = 2\pi \int_{-L_n}^{0} \left\{ Brdr + A\rho_t r \left(-v\frac{dr}{dz}\right)^p \frac{dr}{dz}dz \right\}$$

$$= -\pi \left(BR^2 + 2A\rho_t v^p \int_{-L_n}^{0} r \left|\frac{dr}{dz}\right|^{p+1} dz \right).$$

(10.189)

The force as a function of the nose shape begs interest in what the cylindrical cavity expansion predicts as the best nose shape for penetration. To perform such an analysis, it is best to think of the axial extent as a function of radius along the nose, $z = z(r)$, since the radial extent of the nose is fixed at R_p, while the axial extent of the nose would be a variable length L_n. Written this way, show that the integral for the force becomes

$$F_z = -\pi \left(BR^2 + 2A\rho_t v^p \int_{-L_n}^{0} r \left|\frac{dz}{dr}\right|^{-p} dr \right).$$
(10.190)

From the calculus of variations, when making an integral stationary (either a local maximum or minimum) implies Euler's equation,

$$\delta I = \delta \int_{r_1}^{r_2} f(z, z', r)dr = 0 \quad \Rightarrow \quad \frac{\partial f}{\partial z} - \frac{d}{dr}\left(\frac{\partial f}{\partial z'}\right) = 0,$$
(10.191)

where $z' = dz/dr$. Applied to our specific problem, the variable part of the integral in Eq. (10.190) is $f(z, z', r) = r|z'|^{-p}$. Show Euler's equation gives

$$-\frac{d}{dr}\left(\frac{\partial f}{\partial z'}\right) = p\frac{d}{dr}\left(r|z'|^{-p-1}\right) = 0 \quad \Rightarrow \quad z' = \text{const} \cdot r^{1/(p+1)}.$$
(10.192)

Integrating this expression and solving for the nose tip and nose/shank matching conditions gives

$$z(r) = -L_n \left(\frac{r}{R}\right)^{(p+2)/(p+1)} \quad \Rightarrow \quad r(z) = R\left(-\frac{z}{L_n}\right)^{(p+1)/(p+2)}.$$
(10.193)

Show that this nose has a greater angle of intersection at the nose tip than the conical nose, and then the profile flattens out as the shank of the projectile is approached, so it is somewhat like an ogive. Placing this nose shape into the force integral yields

$$F_z = -\pi \left(BR^2 + 2\left(\frac{p+1}{p+2}\right)^{p+1} A\rho_t v^p \frac{R^{p+2}}{L_n^p} \right)$$

(10.194)

$$= -\pi R^2 \left(B + 2\left(\frac{p+1}{p+2}\right)^{p+1} A\rho_t (v\tan(\beta))^p \right),$$

where we have written $\tan(\beta) = R/L_n$. Show that this force has less magnitude than the conical nose for the same α, assuming $p \geq 1$ (there's something to show here because of the 2).

Finite Targets

Armors have finite extent, both in thickness and in lateral directions. This chapter discusses the back surface bulging, breakout, behind armor debris, and edge effects for targets. We explore the role of target and projectile strength in the context of finite targets. It is assumed that the projectile erodes in this chapter, or that it has a hemispherical nose; rigid perforation is discussed in the next chapter.

Projectiles perforating finite targets lead to complicated failure modes in the target. Understanding material and target failure is one of the most difficult research topics in penetration mechanics. Some general failure modes are shown in Fig. 11.1:

(1) Plugging is when the projectile punches out a clean hole in the target. The target material fails in a simple shear slip motion on the failure surface. Typically a well defined cylinder of target material – a "plug" – can be recovered after the impact event.

(2) Petalling is when a thin target tears and the pie-shaped pieces deform to allow the projectile to pass through. Petalling is a three-dimensional failure mode, since it involves radial tears.

(3) Fracture and fragmentation are a general failure of the target and can be combined with plugging or petalling. Failure of the target material often happens in a broad region near the back of the target (often producing "spall rings") and target fragments of various sizes are launched and travel downrange with the perforating projectile. The residual projectile and liberated target fragments are referred to as *behind armor debris.*

Modeling and predicting these events are an ongoing research topic. We develop models that give reasonable results.

We now present definitions related to target failure and perforation. We begin with

DEFINITION 11.1. (ballistic limit velocity) The *ballistic limit velocity* V_{bl} of a target is the maximum striking velocity at which a target is not completely penetrated by a given projectile and below which only partial penetration occurs.

For completeness, this definition requires another: complete penetration. Though the concept may seem intuitive, various definitions have been used over the years by various organizations. The complication is that the projectile may be stopped in the target but debris or spall be thrown from the back of the target. Thus, the definition we typically use in our testing is

DEFINITION 11.2. (complete penetration) A target is completely penetrated (or perforated) when, after the experiment, light is visible through an 0.0508 cm (0.020 in) thick aluminum witness sheet located 15 cm (6 in) behind the target (i.e., there is a hole all the way through the witness plate).

  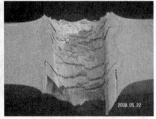

FIGURE 11.1. Various aluminum target failures: left is a target back surface showing deformation with a crack on the verge of plugging; center shows petalling of a target back surface, demonstrating the asymmetry of failure; and right shows a sectioned target with a penetration channel and then breakout including spall rings at the bottom of the target.

This definition addresses the problem of debris being thrown off the back of the target, since debris thrown with sufficient speed will perforate the thin aluminum witness sheet whether the projectile makes it through the target or not. In our modeling we will compute the velocity at which penetration just occurs and we will not include a back witness sheet in our computations. The opposite of complete penetration is partial penetration:

DEFINITION 11.3. (partial penetration) A target undergoes *partial penetration* when it is deformed but not completely penetrated.

When the goal is to stop a round, a partial penetration is a pass and a complete penetration is a fail. *No penetration* is different from partial penetration or complete penetration: it means that the target is not permanently deformed or damaged by the impact. Hence, no penetration occurs in elastic impacts.

Though the ballistic limit velocity definition is easily applicable to penetration models that are deterministic, impact testing often shows scatter in the results. In particular, there can be what is referred to as a *zone of mixed results*, where a partial penetration occurs at a higher velocity than a complete penetration. Because of this experience, other terminology has been arisen:

DEFINITION 11.4. (V_{50}) The V_{50} is the velocity at which a projectile has a 50% chance of completely penetrating the target.

Similarly, the V_{05} is the velocity at which a projectile has a 5% chance of completely penetrating the target. By inference, the ballistic limit velocity is V_0. Having said that, rarely are there enough data for these percentages to be meaningful. As an example, the U. S. Army Research Laboratory (formerly the Ballistic Research Laboratory) typically uses the following approach to determine the V_{50} and a corresponding standard deviation. After a number of experiments are performed in the vicinity of the V_{50}, the list is examined for a grouping of 2 partial penetrations and 2 complete penetrations occurring over an impact velocity range of 18.3 m/s (60 ft/s) or 3 partial penetrations and 3 complete penetrations over a range of 27.4 m/s (90 ft/s). It is required that there are no mixed results outside the velocity range used. The V_{50} is then defined to be the average (mean) of the 4 or 6 test velocities and the standard deviation is defined to be the standard deviation of these numbers about the V_{50} (the mean). Another approach to computing the V_{50} is described in Exercise 11.15.

We will take it as given that *if all initial conditions (material, geometry, and impact conditions) are the same* in an experiment or model *then no zone of mixed results will occur and the ballistic limit velocity V_{bl} will equal the V_{50}.* Because of this assumption, we will sometimes use the terms interchangeably. This statement means we believe the impact event is deterministic and the scatter observed in testing is due to variations in materials, geometry, and/or impact conditions, not intrinsic stochastic processes.

11.1. A Velocity Field for Back Surface Bulging

This section presents a velocity field for a target of finite thickness. We extend the flow field approach to include back surface bulging and breakout. The expressions for the velocity field are obtained by multiplicatively blending the potentials of a velocity field for deep penetration and a velocity field for simple shear deformation. This new velocity field is then placed into the centerline momentum balance, thus including back surface effects. Back surface bulging from this modified analytical model demonstrates close agreement with large-scale numerical simulations.

Others researchers have constructed piecewise flow fields and used volume-conserving geometric constructions to study penetration and flow. Original work in this approach, which divided the deforming target into three flowing regions, applied plastic dissipation arguments and examined back surface bulging was performed by Ravid and Bodner [158], with failure modes by Awerbuch and Bodner [22] and Ravid and Bodner [158]. That approach was combined with the hemispherical flow field and the Walker–Anderson model by Ravid et al. [159] with further description in Chocron, Anderson, and Walker [39]. Though we will not present it here, examples can be found in the original papers as well as in Walker, Bigger, and Chocron [200].

11.1.1. A Deep Penetration Velocity Potential. We begin with velocity fields that are incompressible. In particular, beginning with the development in Chapter 10, a vector potential $\vec{A}$ in spherical coordinates always points in the ϕ direction. The curl is then

$$\vec{v} = curl(A(r,\theta)\hat{e}_\phi) = \frac{1}{r\sin(\theta)}\frac{\partial}{\partial\theta}(A\sin(\theta))\hat{e}_r - \frac{1}{r}\frac{\partial}{\partial r}(rA)\hat{e}_\theta. \tag{11.1}$$

The hemispherical velocity field for penetration is (Eq. 10.27)

$$A(r,\theta) = \frac{ur}{2(\alpha^2-1)}\left\{\left(\frac{\alpha R}{r}\right)^2 - 1\right\}\sin(\theta),$$

$$v_r = \frac{u}{(\alpha^2-1)}\left\{\left(\frac{\alpha R}{r}\right)^2 - 1\right\}\cos(\theta), \qquad v_\theta = \frac{u}{(\alpha^2-1)}\sin(\theta). \tag{11.2}$$

This flow field gave good results for semi-infinite targets and the resulting Walker–Anderson penetration model gives excellent penetration predictions. Once the plastically flowing region reaches the back surface, however, a somewhat simpler deep penetration flow field will be used to blend with a soon to be developed shear field.

This velocity field does not have zero radial velocity at the outer edge of the plastically flowing region $r = \alpha R$ (R is the crater radius), but is similar in that it also has a $1/r^2$ behavior:

$$A_{deep}(r, \theta) = \frac{uR^2}{2r} \sin(\theta), \qquad v_r = u\left(\frac{R}{r}\right)^2 \cos(\theta), \qquad v_\theta = 0. \qquad (11.3)$$

11.1.2. A Shearing Velocity Potential. Back surface bulging occurs as the velocity field transitions from deep penetration to simple shear. A potential of the form

$$\vec{A}(r, \theta, \phi) = \frac{uR}{2} f(r \sin(\theta)/R) \, \hat{e}_\phi = \frac{uR}{2} f(x/R) \, \hat{e}_\phi \qquad (11.4)$$

will be considered, where f is an arbitrary function that has the properties needed for simple shear. $\theta = 0$ is the z-axis and the distance from the z-axis is $x(r, \theta) = r\sin(\theta)$. x will often be used to simplify the appearance of expressions and it is used instead of the letter r because r is the spherical coordinate r. When the curl is taken, the velocity field is

$$v_r = \frac{u}{2}\left\{\left(\frac{R}{x}\right) f + f'\right\} \cos(\theta), \qquad v_\theta = -\frac{u}{2}\left\{\left(\frac{R}{x}\right) f + f'\right\} \sin(\theta). \qquad (11.5)$$

The z and x velocity components are

$$v_z = v_x \cos(\theta) - v_\theta \sin(\theta) = \frac{u}{2}\left\{\left(\frac{R}{x}\right) f + f'\right\}, \qquad v_x = v_r \sin(\theta) + v_\theta \cos(\theta) = 0. \qquad (11.6)$$

This form of the potential has a velocity in the z direction only. The deformation is simple shear.

Directly under the crater, the material only moves in the z direction. Thus we want,

$$f(x/R) = x/R \quad \Rightarrow \quad v_z = u, \qquad x \leq R \qquad (11.7)$$

(see Exercise 11.1). For example, extending the potential to $x > R$ in the following form leads to discontinuous shear flow,

$$f(x/R) = R/x \quad \Rightarrow \quad v_z = 0, \qquad x > R. \qquad (11.8)$$

This potential is continuous, but not differentiable. The velocity field describes a cylindrical plug rigidly protruding from the rest of the target material. Though such plugs do form, more typical is a smooth back surface bulge formation. We want to extend the potential function to $x > R$ so that it is relatively smooth at $x = R$ and, at large x/R, has no z velocity. Mathematically, the requirements are

$$f(1) = 1, \qquad f'(1) = 1, \qquad f''(1) = 0, \qquad f'''(1) = 0, \ldots$$
$$f(x/R) \approx cR/x \quad \text{for large } x/R, \quad c \text{ an arbitrary constant.} \qquad (11.9)$$

Rational functions that satisfy these requirements that are third order and fourth order in R/x are

$$f(x/R) = 6(R/x) - 8(R/x)^2 + 3(R/x)^3, \qquad x > R,$$
$$f(x/R) = 10(R/x) - 20(R/x)^2 + 15(R/x)^3 - 4(R/x)^4, \qquad x > R. \qquad (11.10)$$

The corresponding respective velocities for $x > R$ are

$$v_z = u\left(\frac{R}{x}\right)^3 \left\{4 - 3\frac{R}{x}\right\}, \qquad v_z = u\left(\frac{R}{x}\right)^3 \left\{10 - 15\frac{R}{x} + 6\left(\frac{R}{x}\right)^2\right\}. \qquad (11.11)$$

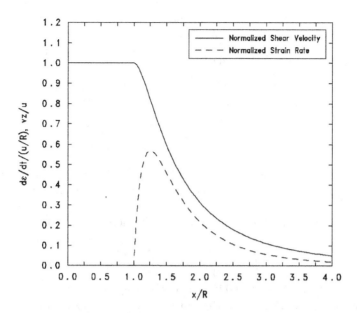

FIGURE 11.2. Shear velocity and strain rate for the third order potential.

Figure 11.2 shows these shearing velocities for the third order flow field, where the center of the flow is rigid. Higher order expressions spread out the flow – the shearing is less localized (Exercise 11.3). The fourth order expression was used in the computational examples in this chapter.

11.1.3. Target Resistance through Plastic Work Rate. Penetration models require a resistance by the target to plastic flow. In the Walker–Anderson model, the resistance was calculated on the centerline using the assumed flow and rigid plastic assumptions. The shear resistance calculation in Chapter 10 relies on derivative information at the centerline to provide information about the flow in the rest of the flowing region. For the shear velocity field described above, that approach will not work. The flow along the centerline contains no information about shearing because the velocity is defined in a piecewise fashion. A different approach to calculating the resistance must therefore be adopted. First, the plastic dissipation in the deforming volume will be calculated. This dissipation will be viewed as a work rate, and a force will be calculated by dividing the work rate by the rate at which target material is displaced. The resistance along the centerline will then be obtained by dividing the force by the presented area of the crater.

The energy dissipation due to plastic flow for a volume V undergoing plastic flow is (Eq. 3.109)

$$\dot{W}_p = \int_V Y_t \dot{\varepsilon}_p dV, \tag{11.12}$$

where Y_t is the target material flow stress and $\dot{\varepsilon}_p$ is the equivalent plastic strain rate. For the deep penetration potential, Eq. (11.3), the equivalent plastic strain rate is (Eq. 10.65)

$$\dot{\varepsilon}_p = \sqrt{\frac{2}{3} D_{ij} D_{ij}} = \frac{2uR^2}{r^3} \sqrt{1 - \frac{11}{12} \sin^2(\theta)}. \tag{11.13}$$

The region of plastic flow is the hemisphere from inner radius R to outer radius $\tilde{\alpha}R$,

$$\dot{W}_p = 2\pi u R^2 Y_t \ln(\tilde{\alpha}) \left\{ 1 + \frac{1}{\sqrt{132}} \ln\left(\sqrt{11} + \sqrt{12}\right) \right\}. \tag{11.14}$$

The target resistance from the plastic dissipation is the work rate divided by the rate of the target volume displaced,

$$S_t^W = \frac{\dot{W}_p}{\pi R^2 u} = 2Y_t \ln(\tilde{\alpha}) \left\{ 1 + \frac{1}{\sqrt{132}} \ln\left(\sqrt{11} + \sqrt{12}\right) \right\} \approx 2.3332 \ln(\tilde{\alpha}) Y_t. \tag{11.15}$$

This resistance is nearly identical to the $(7/3)\ln(\tilde{\alpha})Y_t$ expression derived in the original Walker–Anderson model based on integrating the normal derivative of the shear stress.[1]

We now perform the target resistance calculation for the shear velocity fields. First,

$$\dot{\varepsilon}_p = \frac{2}{\sqrt{3}} |D_{xz}| = \frac{1}{\sqrt{3}} \left| \frac{\partial v_z}{\partial x} \right|. \tag{11.18}$$

With the thickness of the material in front of the projectile written $z_b - z_i = (\tilde{\alpha} - 1)R$, where z_i is the position of the position of the projectile nose and z_b is the position of the back of the target,

$$\dot{W}_p = \frac{Y_t}{\sqrt{3}} 2\pi (\tilde{\alpha} - 1) R^2 \left(u + \frac{1}{R} \int_R^\infty v_z dx \right). \tag{11.19}$$

The corresponding target resistance is

$$S_t^W = \frac{\dot{W}_p}{\pi R^2 u} = \frac{Y_t}{\sqrt{3}} (\tilde{\alpha} - 1) \left(2 + \frac{2}{Ru} \int_R^\infty v_z dx \right) = \frac{Y_t}{\sqrt{3}} (\tilde{\alpha} - 1) \left(1 + \int_1^\infty \frac{f(a)}{a} da \right). \tag{11.20}$$

For the third order potential, the resistances is

$$S_t^W = \frac{4}{\sqrt{3}} Y_t (\tilde{\alpha} - 1) \approx 2.309 Y_t (\tilde{\alpha} - 1). \tag{11.21}$$

The resistance value is similar to that for the thick target resistance for small $\tilde{\alpha}$, since $\ln(\tilde{\alpha}) \approx \tilde{\alpha} - 1$ and $7/3 = 2.333$. It is seen that the more spread out the shear flow is, the more the work required to deform the material. In the least amount of shear flow, which is a cylindrical shearing surface, the resistance is $2Y_t(\tilde{\alpha} - 1)/\sqrt{3}$, or half that of the third order potential flow (set $v_z = 0$ in Eq. 11.19 or, equivalently, $f = 1/a$ in Eq. 11.20).

---

[1]In this derivation we used the "deep penetration" potential of Eq. (11.3) which is slightly different from the hemispherical flow field of Chapter 10 restated in Eq. (11.2). When the Chapter 10 flow field is used, which differs from the "deep potential" flow used in this chapter to blend with the shear in that it has the outer surface at $r = \alpha R$ which the flow does not penetrate, the resulting equivalent plastic strain rate is

$$\dot{\varepsilon}_p = \frac{2uR^2}{r^3} \frac{\alpha^2}{\alpha^2 - 1} \sqrt{1 - \frac{11}{12} \sin^2(\theta)}, \tag{11.16}$$

which leads to a target resistance of

$$S_t^W = 2Y_t \frac{\alpha^2}{\alpha^2 - 1} \ln(\tilde{\alpha}) \left\{ 1 + \frac{1}{\sqrt{132}} \ln\left(\sqrt{11} + \sqrt{12}\right) \right\} \approx 2.333 \frac{\alpha^2}{\alpha^2 - 1} \ln(\tilde{\alpha}) Y_t, \tag{11.17}$$

which as a target resistance due to strength has the nice property that $\lim_{\alpha \to 1} S_t^W = 1.167 Y_t$, rather than zero; at $\alpha = 5$ there is a 4% difference and at $\alpha = 10$ there is a 1% difference.

The equivalent plastic strain rate for the third order potential is

$$\dot{\varepsilon}_p = 4\sqrt{3}\frac{u}{R}\left(\frac{R}{x}\right)^4\left(1 - \frac{R}{x}\right). \tag{11.22}$$

The normalized equivalent plastic strain rate as a function of x/R is plotted in Fig. 11.2. The maximum equivalent plastic strain rate is

$$(\dot{\varepsilon}_p)_{\max}(x/R = 5/4) = \sqrt{3}\left(\frac{4}{5}\right)^5\frac{u}{R} = 0.568\frac{u}{R}. \tag{11.23}$$

Thus, for the bulging flow field to be derived in the next section, a maximum strain would be expected near the shear strain's maximum location, with a strain less than, for example, the product of 0.568 and the depth of penetration/R, for the third order potential.

11.1.4. Blended Potential. The velocity field for the finite target penetration is found by multiplicatively blending the potentials for deep penetration and shear flow,

$$\vec{A}(r, \theta) = A_{deep}^{\lambda}\, A_{shear}^{1-\lambda}\, \hat{e}_\phi. \tag{11.24}$$

The curl of the blended potential gives the velocity

$$v_r = \frac{1}{r}\left\{(1-\lambda)\left(\frac{A_{deep}}{A_{shear}}\right)^{\lambda}\frac{\partial A_{shear}}{\partial\theta} + \lambda\left(\frac{A_{shear}}{A_{deep}}\right)^{1-\lambda}\frac{\partial A_{deep}}{\partial\theta} + A_{deep}^{\lambda}\, A_{shear}^{1-\lambda}\cot(\theta)\right\},$$

$$v_\theta = -\left\{(1-\lambda)\left(\frac{A_{deep}}{A_{shear}}\right)^{\lambda}\frac{\partial A_{shear}}{\partial r} + \lambda\left(\frac{A_{shear}}{A_{deep}}\right)^{1-\lambda}\frac{\partial A_{deep}}{\partial r} + \frac{A_{deep}^{\lambda}\, A_{shear}^{1-\lambda}}{r}\right\}. \tag{11.25}$$

The velocity using Eqs. (11.24) and (11.25) is

$$v_r = \frac{u}{2}\left(\frac{R}{r}\right)^{2\lambda}\left(\frac{f}{x/R}\right)^{1-\lambda}\left\{(1-\lambda)\left(\frac{x}{R}\right)\frac{f'}{f} + 1 + \lambda\right\}\cos(\theta),$$

$$v_\theta = -\frac{u}{2}\left(\frac{R}{r}\right)^{2\lambda}\left(\frac{f}{x/R}\right)^{1-\lambda}\left\{(1-\lambda)\left(\frac{x}{R}\right)\frac{f'}{f} + 1 - \lambda\right\}\sin(\theta). \tag{11.26}$$

The velocity components have been written in a fashion that emphasizes their similarities: they only differ by the sign in front and the sign of the second λ term in the bracket, and by the sine and cosine multipliers. The x and z components of the velocity are

$$v_z = \frac{u}{2}\left(\frac{R}{r}\right)^{2\lambda}\left(\frac{f}{x/R}\right)^{1-\lambda}\left[\left\{(1-\lambda)\left(\frac{x}{R}\right)\frac{f'}{f} + 1\right\} + \lambda\left(\cos^2(\theta) - \sin^2(\theta)\right)\right],$$

$$v_x = \lambda u\left(\frac{R}{r}\right)^{2\lambda}\left(\frac{f}{x/R}\right)^{1-\lambda}\cos(\theta)\sin(\theta). \tag{11.27}$$

Placement of the polynomials from the previous section into these expressions gives the velocity fields. In the target directly under the crater characterized by $r\sin(\theta) = $

$x \le R$, where f is given by Eq. (11.7),

$$\vec{v} = u \left(\frac{R}{r}\right)^{2\lambda} \{\cos(\theta)\,\hat{e}_r - (1-\lambda)\sin(\theta)\hat{e}_\theta\}, \quad x \le R,$$

$$v_z = u \left(\frac{R}{r}\right)^{2\lambda} \left(1 - \lambda\sin^2(\theta)\right), \quad v_x = \lambda u \left(\frac{R}{r}\right)^{2\lambda} \cos(\theta)\sin(\theta), \quad x \le R.$$

(11.28)

Outside the plugging region, $r\sin(\theta) > R$, for the third order potential the particular expressions needed to determine the velocities are

$$\frac{f}{x/R} = 6\left(\frac{R}{x}\right)^2 - 8\left(\frac{R}{x}\right)^3 + 3\left(\frac{R}{x}\right)^4, \quad \frac{x}{R}\left(\frac{f'}{f}\right) = \frac{-6 + 16\,(R/x) - 9\,(R/x)^2}{6 - 8\,(R/x) + 3\,(R/x)^2}.$$

(11.29)

Placement of Eqs. (11.29) in Eqs. (11.26) and (11.27) gives the velocity of the new flow field.

11.1.5. The Blending Parameter λ. The value of λ determines how much of the potential is "deep" as opposed to "shear." The volume of the material in the truncated hemisphere with radius $\tilde{R}$ and height T is $V = \pi T(\tilde{R}^2 - T^2/3)$. The weighting is obtained by dividing this volume by the volume of the original hemisphere,

$$\lambda = \frac{3}{2}\frac{T}{\tilde{R}} - \frac{1}{2}\left(\frac{T}{\tilde{R}}\right)^3$$

(11.30)

When the distance from the nose to the back surface is greater than or equal to the size of a defined hemisphere $\tilde{R}$, then $\lambda = 1$, yielding deep penetration. As the distance to the back surface decreases, then λ decreases until it reaches zero at zero distance, where the velocity field is simple shear. In practice, the formula is used with $\tilde{R} = (\alpha - 1)R$, where α is the extent of the plastically flowing region in the target obtained from the cavity expansion solution, and T is the distance from the nose of the projectile to the target back surface.

11.1.6. Modifications to the Penetration Model. In an impact scenario, the Walker–Anderson model for a semi-infinite target is run until the predicted flow field reaches the back surface – that is, while $z_i + (\alpha - 1)R \le z_b$. Once the plastic flow field reaches the back surface, the model then transitions to the back surface bulge flow field. Owing to the similarities of Eqs. (11.15) and (11.21) and the behavior of the velocity in the target along the centerline, only minor modifications need to be made to the Walker–Anderson model to adjust for a finite thickness target. (One can carefully redo the target velocity integrations, but it does not affect the results greatly.) It is necessary to track the location of the back surface of the target. This is done by determining the velocity at the back surface from the plugging region of the flow field,

$$v_z(z_b) = u_{back} = u \left(\frac{R}{z_b - z_i + R}\right)^{2\lambda}, \quad z_b - z_i \le (\alpha - 1)R.$$

(11.31)

The location of the back surface, z_b, is then updated by integrating this expression with respect to time. The modifications to the model for the back surface can be effected by replacing

$$-\left\{\frac{1}{2}\rho_t u^2 + \frac{7}{3}Y_t \ln(\alpha)\right\} \quad \text{with} \quad -\left\{\frac{1}{2}\rho_t (u - u_{back})^2 + \frac{7}{3}Y_t \ln(\tilde{\alpha})\right\},$$

(11.32)

where
$$\tilde{\alpha} = \frac{z_b - z_i + R}{R} = 1 + \frac{T}{R}. \tag{11.33}$$

The change in the target resistance term reflects the fact that there is less material being deformed in a plastic fashion. No other change occurs because $(7/3)\ln(\alpha)Y_t$ is similar to the expressions obtained in Eq. (11.21). With these changes, Eq. (10.72) becomes

$$\rho_p \dot{v}(L-s) + \dot{u}\left\{\rho_p s + \rho_t R\frac{\alpha-1}{\alpha+1}\right\} + \rho_p \frac{d}{dt}\left(\frac{v-u}{s}\right)\frac{s^2}{2} + \rho_t \dot{\alpha}\frac{2Ru}{(\alpha+1)^2}$$
$$= \frac{1}{2}\rho_p(v-u)^2 - \left\{\frac{1}{2}\rho_t(u-u_{back})^2 + \frac{7}{3}\ln(\tilde{\alpha})Y_t\right\} \tag{11.34}$$

for the finite thickness target. All other equations of the model are the same.

11.1.7. Back Surface Bulging. The velocities in Eqs. (11.27), (11.28), and (11.29) are used to determine the motion of the back surface bulge. In particular, a line of tracer points is placed along the back surface and these points' location are updated each time step based on the velocity at each point, supplied by the velocity field. This level of detail is computed to determine the shape of the back surface and also, as will be seen below, to determine the back surface strain for a failure criterion for breakout of the projectile from the target (i.e., complete penetration or perforation).

To examine the model, an $L/D = 10$ tungsten alloy projectile impacting a steel target at 1.5 km/s was compared to a CTH calculation. The values used in the analytic penetration model were $\rho_t = 7.85$ g/cm^3, $Y_t = 1.25$ GPa, $\rho_p = 17.6$ g/cm^3, $Y_p = 2.0$ GPa. The elastic and shock wave properties required in the model were as stated and given in Appendix C. In the CTH calculations, the Johnson–Cook model was used for the constitutive response, with the published values for 4340 steel and tungsten alloy (also in Appendix C). The Mie–Grüneisen equation of state used the densities above and Appendix C values for shock parameters. The target was 8-cm thick and the initial projectile length was 10 cm.

Figure 11.3 shows the comparison of the CTH calculation with the model. Displayed is the complete prediction of the Walker–Anderson model, including projectile–target interface location, crater diameter, projectile length, and target back surface shape. The lines displaying the model results are overlaid on a CTH calculation, where the steel target and tungsten alloy projectile has been shaded (tungsten debris from the CTH calculation can be seen along the crater wall). The comparison times are 90, 100, 110, and 120 μs. The agreement is excellent. The shape of the bulge from the analytic model matches those of the hydrocode calculation, as do the crater shape and the projectile shape. The excellent agreement between the analytic model bulge shape and that obtained from the full numerical simulation by CTH allows us to infer that the flow field assumed for the material motion in the target accurately reflects the motion of the material.

11.1.8. Strain-Based Failure Criteria. The final frames of the comparison with CTH show failure of the target and breakout. To continue the model analysis it is necessary to invoke a failure model to allow the target to fail and the projectile to perforate the target. Robust and accurate failure criteria are difficult, as discussed in Chapters 3 and 9. A target may fail through a variety of failure modes, such as tensile failure or shear bands, which are manifested in different regions of the

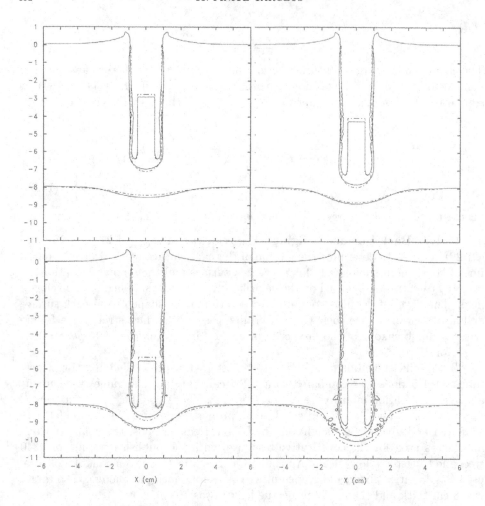

FIGURE 11.3. Comparison of analytic model bulging (dotted line) with CTH calculation (solid line region) at 90, 100, 110, and 120 μs. (From Walker [**191**].)

target. A material may have several competing damage modes; the damage mode that leads to failure (material separation) first depends upon the strain rate and strain history. It is extremely difficult for simple analytical models to handle, in a consistent way, failure as different as petaling, plugging, ductile failure, brittle shear, adiabatic shear localization, fragmentation, etc.

A simple failure model is saying failure occurs when the equivalent plastic strain reaches a critical value ε_f. A nice feature of the analytic model is that the velocity flow fields are defined throughout the target. Since the flow field is known, it is possible to evaluate the equivalent plastic strain rate at any point of the target,

$$\dot{\varepsilon}_p = \sqrt{\frac{2}{3} D_{ij} D_{ij}}. \tag{11.35}$$

Unfortunately, the evaluation of Eq. (11.35) is complicated for a general location within the target. Then a specific point (say $\vec{X}_q$) in the material is followed the velocity field – that is, $d\vec{x}_q/dt = \vec{v}(\vec{x}_q(t), t), \vec{x}_q(0) = \vec{X}_q$. It is straightforward, however, to evaluate the equivalent plastic strain rate along the projectile–target centerline, where the expression for the flow field is algebraically simple (Exercise 11.5),

$$\dot{\varepsilon}_p = \frac{2\lambda u}{r} \left(\frac{R}{r} \right)^{2\lambda}, \tag{11.36}$$

where R, r, and λ are the radius of the crater, the position of the point where the plastic strain is calculated, and the blending parameter, defined in Eq. (11.30), respectively. Integration of Eq. (11.36) at a specific target point along the centerline over time provides the equivalent plastic strain there.

To determine the strain in other regions of the target, it is simplest to use the velocity field to follow the motion of specified locations and then compute the strains from these motions. In particular, at the beginning of the computation a row of points is placed on the back surface with spacing Δx between the points. A second row of points is placed offset into the target a distance Δx from the first row. During the computation of the analytic model, the velocity at these points is computed from the flow field and their location vs. time is then determined. For each square surrounded by the four points the strain is calculated based on finite differences and from that the equivalent strain is computed. (Another approach is to produce the analytic expression for $\dot{\varepsilon}_p$ using symbolic manipulators that write software for evaluation.) Since rigid plasticity is assumed, the equivalent strain is the equivalent plastic strain.

However, an even simpler approach is as follows. The failure parameter that arises most naturally from the flow field model is the back surface strain in the target – that is, the Lagrangian strain along the back surface. The failure criterion is the back surface strain exceeding a failure strain ε_f. Thus, given a specific point $\vec{x}(t)$, which lies on the back surface at $t = 0$, the motion of this point can be computed through $d\vec{x}/dt = \vec{v}(\vec{x}(t), t)$. Following multiple points, it is then possible to approximately compute the Lagrangian strain along the back surface – for example,

$$\varepsilon(t) = \frac{|\vec{x}_1(t) - \vec{x}_2(t)|}{|\vec{x}_1(0) - \vec{x}_2(0)|} - 1, \tag{11.37}$$

where $|\ |$ is the length of the vector, where $\vec{x}_1$ and $\vec{x}_2$ are initially nearby points on the back surface. For axisymmetry, the principal strain in the r–z plane is found this way, while the hoop strain is $\varepsilon_{\theta\theta}(t) = r(t)/r(0) - 1$. It is typically necessary to use an experimental data point, such as the ballistic limit velocity, to determine a value for ε_f. Once calibrated for a material, the ε_f is viewed as a material "constant" for other targets of the same material where the failure criterion is used. Typically, at early time the maximum strain is on the centerline, but at some point in the penetration event the back surface strain near the maximum of the shear flow field (Eq. 11.23) becomes the maximum strain for the rest of the back surface bulging until failure occurs. Once the failure criterion is reached, the target resistance is assumed to be zero and perforation of the target occurs.

11.1.9. Residual Velocity. When a failure model identifies failure of the target, it is possible that the projectile will still have a residual velocity. This is computed by finding the momentum of all the material along the centerline (the residual projectile and the residual target plug) and then finding a velocity for all that material. Thus, for the bilinear velocity profile in the projectile, upon failure the centerline momentum in the projectile is

$$p_p = \rho_p \{vL + \frac{1}{2}(u - v)s\} \tag{11.38}$$

and the centerline momentum in the remaining target material is

$$p_t = \frac{\rho_t u T}{\alpha^2 - 1} \left(\alpha^2 \frac{R}{R+T} - 1 \right). \tag{11.39}$$

The final residual projectile of the velocity is then

$$v_r = \frac{p_p + p_t}{\rho_p L + \rho_t T}. \tag{11.40}$$

11.2. Model Bulge and Breakout and Experimental Comparisons

The Ernst-Mach Institut performed experiments taking flash X-ray photographs of a tungsten alloy projectile penetrating a steel target to determine the positions of the projectile–target interface and the projectile tail as well as the instantaneous length of the projectile at specific instances of time during the penetration process. Being able to determine the front of the projectile through the target by X-ray limits the target thickness. The brick-shaped steel targets were 4.0 cm in the direction the X-ray imaging was performed and 7.0 cm in the other direction, with thicknesses of 2.90 cm and 4.95 cm (Fig. 11.4). The high-hard armor steel (HzB-W) targets had a density 7.85 g/cm^3, Vickers hardness 440, and ultimate tensile strength 1.45 GPa. The tungsten alloy projectiles were length 5 cm and diameter 4 mm ($L/D = 12.5$) and impacted the steel targets at velocities 1.25 and 1.70 km/s. The projectiles were a tungsten sintered alloy (92.5% W, 4.85% Ni, 2.4% Fe, 0.2% Co) with a density of 17.6 g/cm^3, an ultimate tensile strength of 1.20 GPa, and an elongation to failure of 10%. The ballistic limit was determined to be 1.25 km/s for the 2.90 cm thick target ($T/D = 7.25$) and 1.70 km/s for the 4.95 cm thick target ($T/D = 12.4$). One X-ray image per test was taken at different delay times to record the projectiles' penetration. Figure 7.3 shows one of these images from the 600 kV flash X-ray. The positions of the projectile nose and tail, z_i and z_p, were taken from the X-ray images, as was the length of the projectile L. Because each of these measurements is independent, there is a consistency check in that $L = z_i - z_p$ (where $\Delta z_p = z_p - z_p(0)$ and $z_p(0) = -L_0$). Tables 11.1 and 11.2 give the results for a series of experiments, where the various position measurements were multiplied by ($V_{\text{desired}}/V_{\text{exp}}$) to adjust for slight variations in impact velocity.

The analytical model is compared with the above data for both 1.25 and 1.70 km/s in Fig. 11.5. On the plot, long-dashed lines show the initial front and back locations of the plates. For these computations, the material properties for the projectile and target are summarized in Table 11.3. The values of the strength for both the projectile and target are 100 MPa above the stated ultimate strengths. (When the ultimate strengths are used, CTH predicts perforation for both cases and the analytic model predicts maximum back surface strains of 81% and 27% for the 2.90 and 4.95 cm thick target, respectively, implying that the thinner target would be perforated as discussed below. When both the target and projectile strengths are increased by 100 MPa to the values shown in Table 11.3, CTH predicts stops and the analytical model predicts maximum back surface strains of 36%

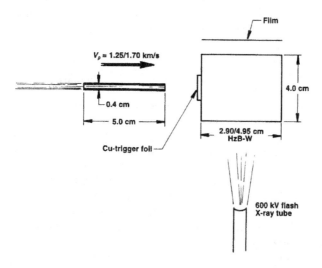

FIGURE 11.4. Experimental geometry. (From Anderson et al. [**6**].)

TABLE 11.1. Penetration of 4-mm diameter tungsten alloy $L/D =$ 12.5 projectiles into high-hard steel of thickness 2.90 cm at 1.25 km/s ($T/D = 7.25$)

Time (μs)	Penetration depth z_i (cm)	Projectile back end displacement Δz_p (cm)	Length L (cm)
10.0	0.58	1.19	4.39
19.9	1.08	2.34	3.74
24.9	1.31	2.99	3.32
30.0	1.57	3.49	3.08
34.9	1.66	4.13	2.53
35.0	1.68	4.16	2.52
39.9	2.07	4.61	2.46
44.9	2.13	5.28	1.85
50.0	2.19	5.54	1.65
55.0	2.32	6.33	0.99

and 19%, respectively, also implying stops. The failure model in CTH is a simple maximum tensile stress of 2 GPa.) Also on the plot are CTH comparisons with exactly the same constitutive models – the response is elastic-perfectly plastic with no hardening, rate effects, or thermal softening. It is seen that there is reasonable agreement, thus showing that the analytic model and the two-dimensional axisymmetric CTH calculation are similar, which is partial verification of both codes. There is additional information from the analytic model on each plot, in the form of a short-dashed line. Initially this line indicates the location of the front of the plastic zone in the target; once the line reaches the back surface, the projectile is aware of the back surface, and the bulge model is invoked. As time continues, the short-dashed line shows the location of the back surface. The difference between the short-dashed line and the solid line is the remaining ligament predicted by the

TABLE 11.2. Penetration of 4-mm diameter tungsten alloy $L/D =$ 12.5 projectiles into high-hard steel of thickness 4.95 cm at 1.70 km/s ($T/D = 12.4$).

Time (μs)	Penetration depth z_i (cm)	Projectile back end displacement Δz_p (cm)	Length L (cm)
10.4	0.92	1.68	4.24
15.3	1.34	2.59	3.75
20.5	1.75	3.46	3.29
20.7	1.72	3.36	3.36
25.3	2.16	4.18	2.98
28.3	2.38	4.69	2.69
30.5	2.62	5.20	2.42
40.4	3.24	6.54–6.74	1.5–1.7
40.6	3.33		
60.3	4.56		

TABLE 11.3. Input parameters used for the analytical model and CTH (L/D is not a model input, but follows from projectile length and diameter).

Projectile		Target	
Density	17.6 g/cm^3	Density	7.85 g/cm^3
Yield strength	1.3 GPa	Yield strength	1.55 GPa
Bar wave speed	3.85 km/s	Shear modulus	76.9 GPa
Initial length	5.0 cm	Bulk sound speed	4.57 km/s
Diameter	0.4 cm	Slope U_s–u_p	1.49
L/D	12.5	Strain to failure	0.75

analytic model. It is seen that CTH predicts slightly deeper penetration. However, these results are near the target V_{50}, which is very sensitive to any small differences in material properties, impact velocities, and specifics of the models.

Hohler and Stilp performed experiments where $L/D = 10$ tungsten alloy projectiles struck hard steel targets of two hardnesses and two thicknesses. The experimental results are shown in Table 11.5. The Walker–Anderson penetration model, combined with the back surface velocity field and the back surface strain failure model Eq. (11.37), was used to compare with the measured ballistic limit velocity and residual velocities. This velocity field, combined with back surface strains, gives reasonable ballistic limit velocities for the two thicknesses of BHN 415 steel targets, namely 1,005 m/s for the 20.9 mm ($T/D = 3.6$) thick target and 1,455 m/s for the 40.0 mm ($T/D = 6.9$) thick target. For these V_{bl} values the strength of the steel Y_t in the model is 1.875 GPa and the critical failure strain is 65%. BHN 415 is a very hard steel, hence the large Y_t value. The result is shown in Fig. 11.6.

Of interest is the sensitivity of the ballistic limit velocity to both the strength of the projectile and the strength of the target. Figure 11.7 shows that the ballistic limit velocity is not very sensitive to the strength of the projectile, with a range of $Y_p = 1$ to 2 GPa in strength, resulting in a range of ballistic limit velocity

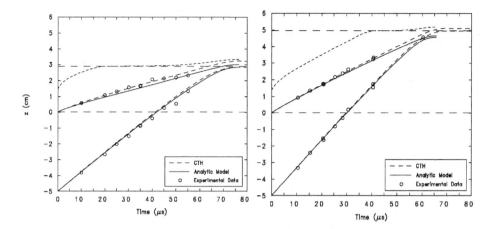

FIGURE 11.5. Comparison of the model, CTH computations, and experiments of the nose and tail positions; 1.25 km/s (left) and 1.70 km/s (right). Long-dashed lines indicate initial location of the front and back of the target. Short-dashed lines are from the analytic model; initially it shows the centerline location of the elastic-plastic interface in the target $z_i + (\alpha - 1)R$, and then the centerline location of the back surface of the target z_b.

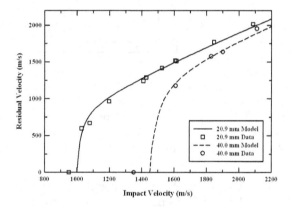

FIGURE 11.6. Ballistic limit data for $L/D = 10$ tungsten alloy projectiles striking BHN 415 steel targets of thickness 20.9 mm and 40.0 mm ($T/D = 3.6$ and 6.9) and results of the velocity field and back surface strain failure criterion. (From Walker, Bigger, and Chocron [200].)

of 1,022 m/s to 964 m/s. The ballistic limit velocity is more sensitive to target strength, with the range of $Y_t = 1.5$ to 2.5 GPa, leading to a range in ballistic limit velocities from 901 to 1,149 m/s. The 20.6 mm BHN 460 target had similar residual velocities to the 415 target and so we would expect a not much greater ballistic limit velocity; for example, the model predicts $V_{bl} = 1,084$ m/s for $Y_t = 2.2$ GPa. As shown in Fig. 11.6, the model predicts good residual velocity for the 40.0 mm

TABLE 11.4. Input parameters used for the analytical model.

Projectile		Target	
Density	17.0 g/cm^3	Density	7.85 g/cm^3
Yield strength	1.3 GPa	Yield strength	1.875 GPa
Bar wave speed	4.03 km/s	Bulk, Shear modulus	166.7 GPa, 76.9 GPa
Initial length	5.8 cm	Bulk sound speed	4.50 km/s
Diameter	0.58 cm	Slope U_s–u_p	1.49
L/D	10	Strain to failure	0.65

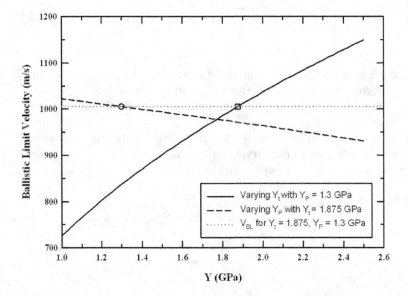

FIGURE 11.7. The effect of target and projectile strength on the ballistic limit according to the velocity field combined with maximum back surface strain model for the thinner target.

BHN 415 target. For the 40.0 mm BHN 460 targets, experimentally no perforations were reported up through 2,084 m/s. The model predicts $V_{bl} < 1,600$ m/s. However, the yaws are large and all but one exceeded what is called critical yaw, as discussed in Exercise 11.12, implying that the back of the projectile could be striking the side of the crater, resulting in different projectile–target interaction and decreased penetration. (Yaw is the angle the projectile makes with its flight direction.) We mention this fact because in other respects the model has good predictions for this and other data.

The use of a failure strain of 65% may seem large, and so we examine the sensitivity of the ballistic limit velocity to the failure strain (Fig. 11.8). For this specific geometry and $Y_t = 1.875$ GPa, the back surface is immediately engaged, so that if the failure strain is extremely small, the target immediately fails. However, the model essentially saturates at a failure strain of 22%, where $V_{bl} = 980$ m/s; after this value, the V_{bl} increases slowly with increasing failure strain, and does not exceed 1,020 m/s. For $\varepsilon_f = 65\%$, $V_{bl} = 1,005$ m/s.

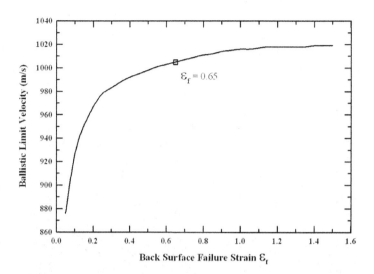

FIGURE 11.8. The effect of critical target back surface strain on the ballistic limit according to the velocity field combined with maximum back surface strain model for the 20.9 mm target and $Y_t = 1.875$ GPa.

11.3. Multiple Plates

If there are separated target plates, the projectile behavior after perforation of intermediate plates must be examined. It is necessary to determine the residual projectile length and velocity when it reaches the next target plate. When a target plate fails, the projectile perforates and no longer encounters extensive resistance from the target. Projectile erosion continues for a short time because of the large velocity gradient in the projectile's plastically deforming region near the nose. After the front of a projectile perforates a plate, the nose velocity increases, the tail velocity decreases, and the projectile tends towards a uniform velocity. The time to reach a uniform velocity is primarily a function of the size of the plastic zone at the front of the projectile and not the overall length of the projectile. Elastic waves travel back and forth across the plastic zone: the distance s_f is traversed twice for waves to reach the free surface at the projectile nose, and each time the wave reaches the free surface the particle velocity change is $2\Delta u$, so the time to equilibrate to a uniform velocity is

$$\Delta \bar{t} = \frac{v_f - u_f}{2\Delta u}\left(\frac{2s_f}{c_K}\right) = \frac{\rho_p(v_f - u_f)}{\sigma_p}s_f, \tag{11.41}$$

where the back end projectile velocity is v_f, the nose of the projectile (penetration) velocity is u_f, and the extent of the plastic zone in the projectile is s_f. The elastic wave speed is being approximated by the bulk wave speed. For a tungsten alloy projectile, $\Delta u = \sigma_p/(\rho_p c_K) \approx 25$ m/s, and so if $v_f - u_f = 750$ m/s (tungsten striking steel at 1.5 km/s), then $\Delta \bar{t} = 30 s_f/c_K$. For modeling purposes an approximation is $s_f \approx R_p$, where R_p is the radius of the projectile.

TABLE 11.5. Impact of $L/D = 10$ D17 Tungsten alloy projectile striking armor steel plate (data from Hohler and Stilp [10]).

Target thickness (cm)	Target BHN	V (m/s)	v_r (m/s)	L_f/L	Impact yaw (°)
2.09	415	950	0	0.11	0.0
2.09	415	1,028	598	0.25	2.0
2.09	415	1,077	670	0.25	1.3
2.09	415	1,197	964	0.42	3.2
2.09	415	1,408	1,245	0.45	6.4
2.09	415	1,427	1,292	0.53	2.7
2.09	415	1,613	1,513	0.60	3.3
2.09	415	1,845	1,771	0.61	1.7
2.09	415	2,086	2,017	0.63	2.6
2.06	460	945	0	0.12	0
2.06	460	1,064	587	0.26	1.3
2.06	460	1,188	941	0.39	2.7
2.06	460	1,640	1,510	0.59	0.2
2.06	460	1,824	1,737	0.64	3.7
2.06	460	2,076	2,009	0.61	1.9
4.00	415	1,350	0		0
4.00	415	1,609	1,179	0.21	1.6
4.00	415	1,828	1,579	0.27	2.3
4.00	415	1,901	1,637	0.26	0.4
4.00	415	2,110	1,952	0.32	1.3
4.00	460	1,360	0		0
4.00	460	1,598	0		3.1
4.00	460	1,816	0		2.3
4.00	460	1,893	0		4.5
4.00	460	1,934	0		2.9
4.00	460	2,084	0		3.9

Between failure of a plate and reaching a uniform velocity in the projectile, no force acts on the projectile and so the total momentum is constant, allowing the uniform velocity to be calculated. Assuming that the length of the projectile at target failure is L_f, then, with the bilinear velocity distribution in the projectile, the uniform velocity of the projectile after equilibration is

$$\bar{v} = \frac{L_f - s_f}{L_f} v_f + \frac{s_f}{L_f} \left(\frac{v_f + u_f}{2} \right). \tag{11.42}$$

In the continuum approximation, the nose velocity u of the projectile during free flight increases linearly with time and the tail velocity v decreases linearly with time,

$$u(t) = u_f + \left(\frac{\bar{v} - u_f}{\Delta \bar{t}}\right)(t - t_f), \quad v(t) = v_f + \left(\frac{\bar{v} - v_f}{\Delta \bar{t}}\right)(t - t_f), \quad t_f < t < t_f + \Delta \bar{t},$$

$$(11.43)$$

where t_f is a time of the target failure. The difference between the front and back end velocity gives the erosion rate, and integration of the erosion rate over time provides the decrease in projectile length. If the space between the plates is large enough that the projectile equilibrates, then the length of projectile that is lost after the perforation is

$$\Delta L = -\int_{t_f}^{t_f + \Delta \bar{t}} (v - u)dt = -\frac{1}{2}\frac{\rho_p(v_f - u_f)^2}{\sigma_p}s_f.$$

$$(11.44)$$

Using values as above for tungsten impacting steel gives that the eroded length is 2.8 projectile radii (assuming $s_f = R_p$) or 1.4 projectile diameters. When computations for spaced plates are performed, it is important to reduce the projectile length between plates in this fashion. For detailed descriptions of two- and six-plate target calculations, see Chocron et al. [40].

11.4. Ductility vs. Strength Influences on V_{bl} and V_r

We continue our examination of projectiles by focusing on the effect of projectile and target strength. Intuitively we expect a projectile with a larger flow stress to be a more effective penetrator than a weaker projectile. We expect a target with a larger flow stress to be a more effective armor than a weaker target. However, for many metals, increasing strength (for example, through work hardening or heat treating) comes at the cost of decreasing ductility. In finite targets, decreasing ductility equates to a decrease in the failure strain. In this chapter we see that increasing the target strength at the expense of ductility does not necessarily increase its performance; in fact, it can get worse. Also we see that increasing the strength of the projectile does not have our expected behavior unless we also decrease the ductility, since otherwise the projectile can open a larger crater which can lead to more projectile deceleration.

Experiments performed at the Ernst-Mach-Institut investigated the effect of projectile and target hardness on the ballistic limit velocity and residual projectile properties (velocity and length). Projectiles were fabricated from five steels and one tungsten alloy. All projectiles were cylindrical with a flat nose, with a length L of 5.80 cm, and a diameter D of 0.58 cm ($L/D = 10$). For each steel, heat treatments to different hardnesses were performed. Including the tungsten alloy projectile, 17 different material and/or hardness projectile types were used for the test series. The density and Vickers hardness were determined for each projectile type. The tensile strength, ultimate tensile strength (UTS), and the strain at failure were determined for some of the steels and the tungsten alloy. The properties are listed in Table 11.6.

The targets consisted of two different armor steels, designated RHA1 and RHA2, with different hardnesses. The plate thickness T was 2.06 cm; thus, T/D for the experiments was 3.55. Material properties of the two target steels are also given in Table 11.6.

Experiments were conducted at $0°$ obliquity and impact yaw was less than critical yaw for all tests. The impact velocity V and the residual velocity V_r were

TABLE 11.6. Material properties and V_{bl}.

Mat.	Density (g/cm^3)	Vickers hardness (kg/mm^2)	Tensile strength (MPa)	UTS (MPa)	Elong. (%)	V_{bl} RHA1 (m/s)	V_{bl} RHA2 (m/s)
Steel 1	7.85	212±10	495±10	565±10	16.2±0.7	1,415±20	1,395±
Steel 1	7.85	425±15	1,290±10	1,440±10	28.3	1,395±15	1,405±
Steel 2	7.85	465±15	1,530±10	1,575±10	13.0±0.5	1,405±15	1,400±
Steel 2	7.85	750±15	2,465±10	2,555±20	7	1,280±20	1,280±
Steel 3	7.85	527±10				1,380±10	
Steel 3	7.85	562±10				1,338±5	
Steel 3	7.85	600±10				1,327±14	
Steel 3	7.85	640±10				1,320±5	
Steel 4	8.20	260±10	635±10	855±10	19.5±1.5	1,410±10	
Steel 4	8.20	280±10				1,390±5	
Steel 4	8.20	340±10				1,385±5	
Steel 4	8.20	440±10				1,385±5	
Steel 4	8.20	525±10				1,380±5	
Steel 4	8.20	550±10	1,440±10	1,800±10	11±1	1,357±10	
Steel 5	8.83	227±10				1,382±10	
Steel 5	8.83	286±10				1,352±10	
D17	17.0	300±10	580±10	807±8	11	940±15	940±
RHA1	7.85	486±10	1,350±20	1,430±20	8±1		
RHA2	7.85	550±10	1,510±25	1,620±25	6±1		

determined by flash X-rays. V_r/V vs. V are plotted for 5 of the 17 projectile types in Fig. 11.9. Closed symbols represent data against the RHA1 target steel, and open symbols against the RHA2 steel. The curve fits, denoted by the lines, will be discussed below. The ballistic limit velocity V_{bl} was estimated for each of the 17 penetrator types into the RHA1 target material. V_{bl} was also determined for the first two steel types and the tungsten alloy into the RHA2 material. These values, along with their uncertainties, are given in Table 11.6.

The ballistic limit velocities (Table 11.6) are nearly identical for the two target materials, within the estimate of the uncertainty. At first, this appears to be contradictory, since penetration efficiency is a strong function of target hardness. However, the weaker target material (RHA1) is more ductile than the stronger material. Crater profiles for the two target materials at approximately the same impact velocities are shown in Fig. 11.10.

In Fig. 11.10 (top row), where the impact velocity is below V_{bl}, the penetration depth is greater and there is more bulging of the back of the target for the RHA1 material than for the RHA2 material, consistent with expectations. At just above the ballistic limit velocity (middle row), which is 1,280 m/s for both materials with this projectile type, a spall ring of target material has failed and separated from the back of the RHA2 target and a bulge persists in the RHA1 target. At higher impact velocities (bottom row), the RHA1 target also loses the bulged region owing to failure, but the failed region at the rear of the target is not as extensive as it is for the RHA2 material. This damaged region at the rear of the target is even more noticeable for the D17 target (not shown). We conclude that the harder RHA2

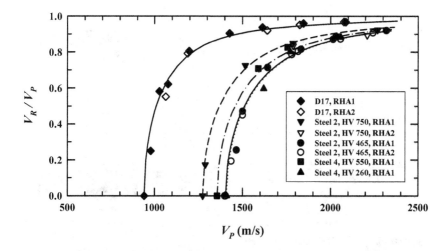

FIGURE 11.9. Residual velocity, normalized by the impact velocity, vs. impact velocity. (From Anderson et al. [7].)

steel target impedes penetration more than RHA1 target material at the beginning
of penetration, but the RHA2 target material then loses its ballistic performance
advantage over the RHA1 target material because of differences in failure at the rear
target surface during breakout, which in our modeling we would represent with a
smaller failure strain. Two target materials have essentially the same ballistic limit
velocity for the T/D ratio of the experiments; this experimental result is geometry
specific; we would not expect the two target materials to have equal ballistic limit
velocities for other T/D ratios.

The residual lengths of the projectiles after exiting the targets were also measured from the flash radiographs. L_r/L vs. V are plotted in Fig. 11.11 for 6 of
the 17 projectile types; the results for the two target steels are again shown as the
closed and opened symbols respectively. A least-squares cubic polynomial was used
to highlight the trends. The experimental results show that the residual length is a
strong function of projectile hardness, with the harder projectile having the longer
residual length. However, there is little effect of target hardness on residual length,
though again this result probably only applies to this T/D ratio.

Lambert and Jonas [219] suggested an equation for the residual velocity in
terms of the impact velocity and ballistic limit velocity,

$$V_r = \begin{cases} a\left(V^p - V_{bl}^p\right)^{1/p}, & V > V_{bl}, \\ 0, & V \le V_{bl}, \end{cases} \tag{11.45}$$

where V_{bl} is the ballistic limit velocity and a, p, and V_{bl} are curve fit parameters.
Equation (11.45) can be rearranged into

$$\frac{V_r}{V_{bl}} = a\left[\left(\frac{V}{V_{bl}}\right)^p - 1\right]^{1/p}, \qquad V > V_{bl}. \tag{11.46}$$

The form of Eq. (11.46) suggests that we plot V_r/V_{bl} vs. V/V_{bl}, and this is done in
Fig. 11.12 for all 17 projectile types; a different symbol is used to denote each of
the 6 distinct materials (5 steels and one tungsten alloy). All the experimental data

RHA1 RHA2

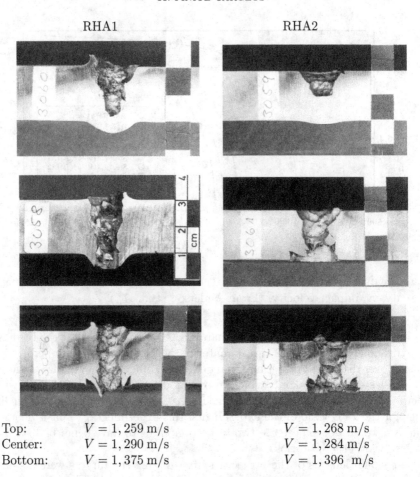

Top:	$V = 1,259$ m/s	$V = 1,268$ m/s
Center:	$V = 1,290$ m/s	$V = 1,284$ m/s
Bottom:	$V = 1,375$ m/s	$V = 1,396$ m/s

FIGURE 11.10. Sectioned craters for the Steel 2, HV750 projectile into the RHA1 (left) and RHA2 (right) targets – note the spall rings on three of the targets. (From Anderson et al. [7].)

collapse to a single curve. A somewhat better curve fit, rather than the Lambert–Jonas equation, was suggested by Grabarek [79] in the form

$$\frac{V_r}{V_{bl}} = \frac{V_{bl}}{V}(a_0\sqrt{x} + a_1 x + a_2 x^2) = \frac{a_0\sqrt{x} + a_1 x + a_2 x^2}{1 + x}, \quad x = \frac{V}{V_{bl}} - 1, \quad x < 2.5.$$
(11.47)

Values are $a_0 = 1.6$, $a_1 = 1.3$, and $a_2 = 0.90$. These coefficients are slightly different than those found by Grabarek using a completely different data set including different T/D ratios, suggesting that Eq. (11.47) may be applicable to other large L/D projectile data. The solid line in Fig. 11.12 is the regression fit. The equation can be written slightly differently by multiplying both sides by V_{bl}/V to give

$$\frac{V_r}{V} = \frac{a_0\sqrt{x} + a_1 x + a_2 x^2}{(1 + x)^2}, \quad x = \frac{V}{V_{bl}} - 1, \quad x \leq 2.5. \tag{11.48}$$

The lines in Fig. 11.9 were generated using Eq. (11.48) and V_{bl} for each of the specific materials, all using the same a_i values as given above.

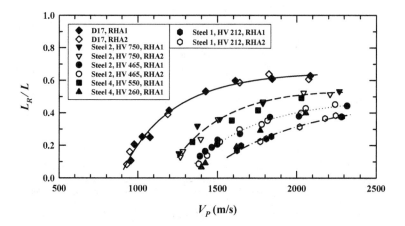

FIGURE 11.11. Normalized residual projectile length vs. impact velocity. (From Anderson et al. [7].)

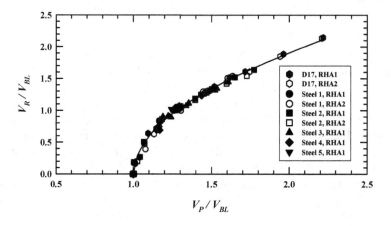

FIGURE 11.12. Residual velocity, normalized by the ballistic limit velocity, vs. a reduced impact velocity. (From Anderson et al. [7].)

The ballistic limit velocities with their uncertainties for the 16 steel projectile variants and the two target types are plotted vs. projectile hardness in Fig. 11.13. The dotted vertical lines represent the hardnesses of the two target steels, although it has already been noted that because of failure of the plate at breakout, the harder steel target behaves, in an average sense, essentially similar to the less hard target material. It is observed that the ballistic limit velocity drops when the projectile hardness is greater than the target hardness, though this is likely a coincidence related to the specific thickness of the target, since Y_p compares to R_t or S_t in our models, not directly to Y_t.

CTH was used to conduct a parametric study to investigate the effect of projectile hardness on the residual lengths and velocities of the projectile. The simulations were for the same geometry as the experiments – that is, a $L/D = 10$ projectile impacting a target with $T/D = 3.55$. An impact velocity of 1.5 km/s was selected for the study, which is above the ballistic limit velocity for all the steel materials by

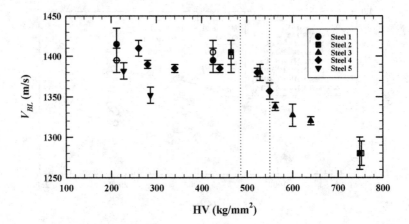

FIGURE 11.13. Ballistic limit velocity vs. projectile hardness (Vickers); solid markers RHA1, open markers RHA2. (From Anderson et al. [7].)

roughly 100 to 200 m/s. The calculated projectile residual velocities and lengths were then compared to the experimentally measured values.

Simulations were used to investigate the trends observed experimentally. The flow stress of the materials was calculated from the Johnson–Cook constitutive model, where the strain hardening and strain-rate hardening coefficients were set to zero in the Johnson–Cook model, but the thermal softening term was retained. An estimate of the flow stress was obtained from the hardness using the relationship $Y = \text{HV} \cdot g/3$, where Y is the flow stress, HV is the Vickers hardness, and g is the acceleration due to gravity. If the tensile and ultimate strengths are averaged for the materials in Table 11.6, then the calculated Vickers hardness is generally within 10% of the measured value. The steel target plate was modeled with $Y_t = 1.65$ GPa, while the steel projectile flow stress Y_p was varied parametrically from 0.75 to 2.25 GPa in increments of 0.25 GPa. This nominally corresponds to a range of HV from 280 to 700 kg/mm^2, which covers the interesting range of hardnesses displayed in Fig. 11.13.

Plots of six of the computational results are shown at 75 μs after impact in Fig. 11.14; the dark-shaded material is projectile material and the lighter-shaded material is target material. The images show that the length of the residual projectile increases with increasing projectile strength, in agreement with the experimental results. Since the frames are all at the same time, the front of the projectile location also shows that the residual velocity of the projectile is decreasing.

The normalized residual length and normalized residual velocity are plotted vs. projectile strength in the upper plot of Fig. 11.15. Experimental data for different projectiles were grouped by approximately the same impact velocity. Open symbols denote V_r/V and closed symbols denote L_r/L. The octagons connected by the lines represent the computational results. As already observed, for constant impact velocity, the residual length of the projectile increases with increased projectile strength for both the experiments and the computations. The experimental data also show that the residual velocity increases with an increase in projectile

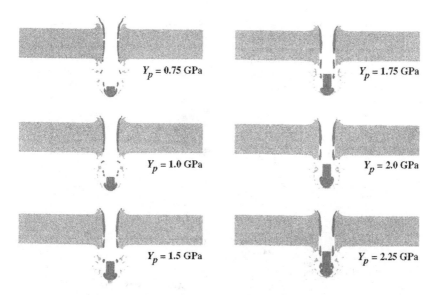

FIGURE 11.14. Images from numerical simulations of the experiment at $75\,\mu s$ after impact, each with increasing constant projectile strength; the same projectile failure strain (large) was used for each. (From Anderson et al. [**7**].)

strength. This observation is consistent with a decrease in the ballistic limit velocity with an increase in projectile strength. In contrast, the numerical simulations show a decrease in residual velocity with an increase in projectile strength. The Steel 3 material may be an exception to these observations; within the data scatter, both V_r/V and L_r/L appear to be relatively constant over the range of projectile strengths investigated.

The numerical simulations can be used to understand the underlying cause for the computational trend that shows V_r/V decreases with increasing projectile strength. The penetration (nose) and tail velocities for two of the simulations ($Y_p = 1.0$ and 2.25 GPa) are plotted as a function of nose position in the upper plot of Fig. 11.16. The stronger projectile initially penetrates the target at a higher velocity. The tail of the projectile decelerates owing to elastic waves; the step decreases are easily discernible in the figure. The step Δv is given by $\Delta v = 2Y_p/\rho c_E$, where ρ for the steel projectiles was 7.84 g/cm^3 and c_E is approximately $5{,}200$ m/s. For $Y_p = 2.25$ GPa, $\Delta v = 110$ m/s and for $Y_p = 1.0$ GPa, $\Delta v = 50$ m/s. These are the magnitudes of the step decreases in the tail velocities in Fig. 11.16. Therefore, as strength increases, the projectile undergoes a larger deceleration because of its strength and, according to the simulations, the residual velocity decreases as projectile strength increases.

Why is there a discrepancy between the simulations and the experimental results for V_r/V as a function of Y_p? It is due to the fact that both the target and projectile are considered to be "infinitely" ductile in the simulations. (Even when a failure model or failure strain is not specified, material can separate in computations because of interface tracking effects in Eulerian computations and the deletion of highly distorted elements in Lagrangian computations, so there is some implicit

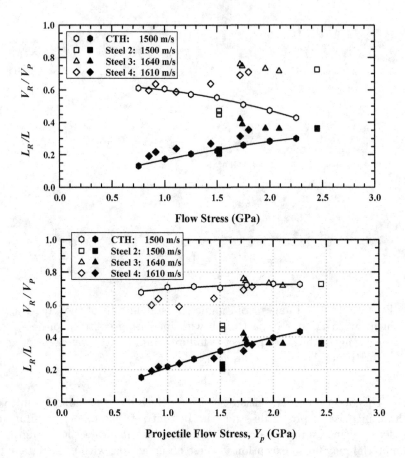

FIGURE 11.15. Normalized residual velocities and normalized residual lengths vs. projectile flow stress Y_p. The computations in the upper plot used a uniformly large failure strain for the projectile material and in the lower plot the computations used a failure strain that decreased for increasing projectile strength, holding $Y_p\,\varepsilon_f = 1.5$ GPa constant. CTH numerical simulation results are connected by lines. (From Anderson et al. [7].)

large failure strain.) In general, harder materials are more brittle and thus, as the projectile material is made stronger, it cannot support as large a mushroom head. This leads to a narrower crater and more efficient penetration, resulting in a lower ballistic limit velocity. Simulations where the projectile strain-to-failure was parametrically varied predict deeper penetration into a semi-infinite target as the strain-to-failure is decreased [164].[2] Numerical simulations using CTH were conducted to investigate the effects of projectile material failure as a function of accumulated equivalent plastic strain. When the equivalent plastic strain reaches a user-specified value ε_f, the shear strength is set to zero. (In CTH, the Johnson–Cook damage model was used with all the damage coefficients set to zero except

---

[2]Similarly, differences in material failure are typically invoked to explain the deeper penetration by depleted uranium projectiles vs. tungsten alloy projectiles at ordnance velocities.

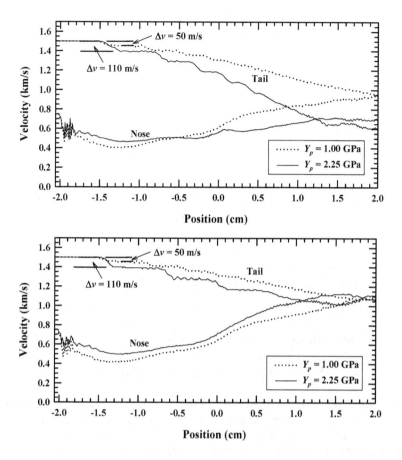

FIGURE 11.16. Penetration and tail velocities vs. nose position for two projectiles with differing strengths. For the upper figure the computations use the same large failure strain for the projectile material, while for the lower plot the failure strains were 150% and 67%, for target strengths $Y_p = 1$ and 2.25 GPa, respectively, thus holding $Y_p \varepsilon_f = 1.5$ GPa constant. Position -2.06 cm is the initial front face location and position 0 is the initial back surface location of the target. (From Anderson et al. [7].)

for D_1, which was varied parametrically.) The first parametric study consisted of holding the projectile flow stress constant at 1.5 GPa while varying the strain-to-failure for the projectile (no strain-to-failure was specified for the target material). The normalized residual projectile lengths and velocities are plotted in Fig. 11.17; the open symbols denote normalized residual velocities, and the closed symbols denote the normalized residual projectile lengths. By ε_f of 4.0, the values of V_r/V and L_r/L are the same for when the failure strain is set to infinity. It is observed, except for small values of ε_f, that both L_r/L and V_r/V decrease as failure strain increases.

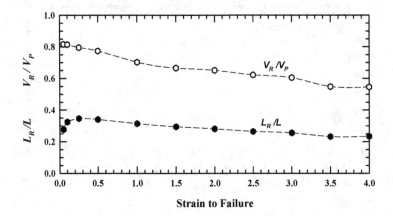

FIGURE 11.17. Normalized residual velocities and normalized residual lengths vs. projectile strain-to-failure for a constant projectile strength ($Y_p = 1.5$ GPa). (From Anderson et al. [7].)

To carry this analysis further, it is assumed that the specific plastic work (plastic work per unit volume) available is approximately a constant,

$$W_p = \int_0^{\varepsilon_p} Y_p d\varepsilon_p = Y_p \varepsilon_f = \text{constant} = 1.5\,\text{GPa} = 1.5\,\frac{\text{kJ}}{\text{cm}^3}. \qquad (11.49)$$

In words, the area under the stress-strain curve up to failure is assumed to be constant. Thus, if we have a higher projectile strength Y_p then there will be a corresponding decrease in projectile failure strain ε_f. This assumption is commensurate with the observation that as the yield strength of a material is increased, materials generally become less ductile. CTH results holding the specific plastic work to failure constant at $W_p = 1.5$ GPa give computational values for V_r/V and L_r/L that agree with the experimental data, as shown in the lower plot of Fig. 11.15, plotted as a function of flow stress Y_p (hence $\varepsilon_f = W_p/Y_p$).

The nose and tail velocities for the same two examples shown in the upper plot of Fig. 11.16 are shown again in the lower plot, now assuming the specific plastic work to failure is constant at $W_p = 1.5$ GPa. The elastic decrements in the tail velocities are the same in both upper and lower plots. Initially, for the first 0.5 cm or so of penetration, the penetration velocities in both plots are the same. However, as the projectile nose mushrooms, the effect of the failure strain is manifested; the projectile with a finite erosion strain penetrates faster, presumably because it is not necessary to open as large a crater and/or to continue loading such a large mushroom projectile head. Since the projectile penetrates faster, there is less erosion and therefore the length of the projectile is greater at the same point of penetration. Faster penetration also means that it takes less time for the projectile to perforate the target and thus less time is available for elastic waves to reduce the tail velocity (fewer elastic waves arrive at the back surface, reducing the speed). These effects contribute to a higher residual velocity. The residual velocity for the 1 GPa projectile increased from 0.95 km/s (infinite failure strain) to 1.07 km/s ($\varepsilon_f = 1.5$). For the 2.25 GPa projectile, the residual velocity increased from 0.65 km/s (infinite failure strain) to 1.08 km/s ($\varepsilon_f = \frac{2}{3}$).

To summarize, the residual length of the projectile increases measurably with increasing projectile hardness both experimentally and in numerical simulations. However, matching the experimental observation that residual velocity increases with hardness requires introducing a finite failure strain for the projectile that decreases as projectile strength increases. These results suggest that the improved ballistic performance of harder projectiles (e.g., harder projectiles producing a lower V_{bl}) is a consequence of changes is the projectile failure and not just the increase in projectile strength.

11.5. Fragmentation and Behind Armor Debris

When projectiles perforate a target, target material is often liberated. In some cases, the target material has fragmented and debris is thrown. This debris, along with projectile debris, is called *behind armor debris*. Armored vehicles have fabric or composite liners attached to the interior of a vehicle, referred to as *spall liners*, to protect the crew by stopping this debris.

In Chapter 3 tensile tests were described to determine the strength of materials. When these tests are performed, the specimens fracture at one location, even when the dog bone specimens do not have a reduced center section – that is, even though failure could technically occur at any point in the gauge length. Why is that? A little thought, given our experience with wave propagation, tells us that elastic waves inform the entire specimen of where the first necking occurs, and so the stress is concentrated there and further deformation localizes in the region of the initial necking.

When high velocity penetration occurs, elastic wave information may not be fast enough to prevent necking and subsequent failure at multiple locations. As a specific example, Fig. 11.18 shows the results of the impact of a roughly 1.27 cm $L/D = 1$ heavy metal alloy impactor striking a nominally $T/D = 2$ steel target plate at 2.68 km/s. Figure 11.19 shows the same impactor hitting a nominally $T/D = 4$ target plate at 3.12 km/s. These tests were performed at SwRI using the two-stage light gas gun in Fig. 1.4 for Lawrence Livermore National Laboratory. Initially the impact flash is seen, then the bulge formation, and then fragmentation of the remaining ligament of the back surface: the projectile penetrates into the steel first, pushing material laterally out of the way, until there is less material between the projectile nose and the back surface and the bulge begins to rapidly grow. The characteristic fragment size differs in the two events. In particular, by observation a nominal fragment size for the thinner target is 1 cm across and for the larger target is 2.5 cm across.

One approach to fragmentation is the work of Grady on spallation and fragmentation [82]. Grady was interested in understanding the characteristic size s of fragments during the tensile expansion of a material when it fragments into pieces. We present his approach in the next paragraphs.

We assume there are multiple fractures spaced a distance s apart. Information about fractures is communicated by the mechanical sound speed of the material. When an edge fractures, the stress relief is communicated by the rarefaction wave to the rest of the fragment, thus preventing further fractures. The time it takes for information from the fracture to get from an edge to the center point, hence

FIGURE 11.18. Perforation of a steel $T/D = 2$ target at 2.68 km/s by an $L/D = 1$ heavy metal alloy impactor; $4.34\,\mu s$ between images.

preventing fracture there, is

$$t = \frac{s/2}{c_K}, \tag{11.50}$$

where we using the elastic bulk sound speed as a representative sound speed. Thus, for this size fragment, the fracture event for expanding material cannot occur any earlier than this, otherwise the required information to define the size s of the fragment could not be communicated. Grady refers to this minimal time as a "communication horizon" condition.

In Chapter 7, Table 7.7, we compared energy stored elastically and energy dissipated plastically. Grady's idea was to look at the energy stored elastically and determine the elastic strain required for that elastic strain energy density to match

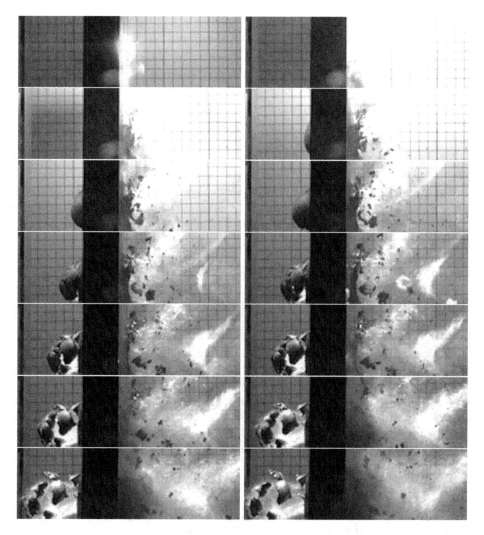

FIGURE 11.19. Perforation of a steel $T/D = 4$ target at 3.12 km/s by an $L/D = 1$ heavy metal alloy impactor; 8.68 μs between images.

the energy density dissipated through plastic flow through the failure strain,

$$\frac{1}{2}K\varepsilon^2 = Y\varepsilon_f \qquad \Rightarrow \qquad \varepsilon = \sqrt{\frac{2Y\varepsilon_f}{K}}. \tag{11.51}$$

The elastic strain is a volumetric strain, $\varepsilon = \varepsilon_v$, while the failure strain is an equivalent plastic strain. We assume the material is uniformly expanding, so $\dot{\varepsilon}$ is constant and $\varepsilon = \dot{\varepsilon}t$. Using the time in Eq. (11.50), which is the minimum time for communication for this fragment size (from the outer edge to the center), we obtain for a fragment size

$$s = 2c_K t = 2c_K\frac{\varepsilon}{\dot{\varepsilon}} = \sqrt{\frac{8Y\varepsilon_f}{\rho}}\,\frac{1}{\dot{\varepsilon}}. \tag{11.52}$$

Thus, we have an expression for a representative fragment size that depends inversely on strain rate. Looking at our figures, for Fig. 11.18 we measure a rough speed of the back surface at the time of fragmentation (roughly 17 μs) of 2.11 km/s and a bulge height of 2.54 cm; for Fig. 11.19 we measure a rough speed of the back surface at fragmentation time (roughly 26 μs) at 977 m/s and a bulge height of 3.175 cm, so the respective estimated strain rates are

$$\dot{\varepsilon} \approx \frac{u}{R} \approx \frac{2{,}110 \text{ m/s}}{0.0254 \text{ m}} = 8.3 \times 10^4 \text{ s}^{-1}, \qquad \dot{\varepsilon} \approx \frac{u}{R} \approx \frac{977 \text{ m/s}}{0.03175 \text{ m}} = 3.1 \times 10^4 \text{ s}^{-1}.$$
(11.53)

Computations with the Walker–Anderson model with bulge and breakout, with $Y = 1$ GPa and $\varepsilon_f = 0.7$, gave the perforation speed as 2,374 m/s and 1,317 m/s, respectively. With $Y = 1$ GPa and $\varepsilon_f = 0.7$, for this steel, $\sqrt{8Y\varepsilon_f/\rho} = 845$ m/s and the respective representative fragment sizes are

$$s = 1 \text{ cm} \quad \text{and} \quad s = 2.7 \text{ cm},$$
(11.54)

which agree with the estimates above for these two impact scenarios.

The above is for ductile void growth and fracture. In his development, Grady uses a critical strain for ductile spall, instead of our use of ε_f, which he estimates to be $\varepsilon_c = 0.15$. Owing to the square root in Eq. (11.52), the corresponding predicted size with this fracture strain differs by a factor of $\sqrt{0.7/0.15} = 2.16$.

He then extends these ideas to fragmentation based on brittle fracture and on the breakup of fluids. The difference for these two cases is that now, instead of comparing energy densities, it is necessary to consider the elastic energy inside the body of volume V and the energy on the outer surface with area A. For brittle fracture, the critical stress intensity factor or fracture toughness of the material K_c provides an energy to fracture the surface. In classical fracture mechanics the required surface energy density G_c equals the fracture toughness squared divided by the modulus, roughly

$$G_c = \frac{K_c^2}{K}.$$
(11.55)

This energy is spread over both sides of the crack surface; hence the energy required is $\frac{1}{2}G_c A$ and to get the required elastic energy density,

$$\frac{1}{2}K\varepsilon^2 V = \frac{1}{2}G_c A \qquad \Rightarrow \qquad \varepsilon = \frac{K_c}{K}\sqrt{\frac{A}{V}}.$$
(11.56)

This expression is a little more complicated than the ductile fracture version in that A/V has a length term in it, so it is necessary to pick a shape. If we idealize to a sphere or cube, $A/V = 4\pi R^2/(4/3)\pi R^3 = 3/R = 6/s$ or $A/V = 6s^2/s^3 = 6/s$. We obtain for the characteristic fragment size (compare Eq. 11.52)

$$s = 2c_K t = 2c_K \frac{\varepsilon}{\dot{\varepsilon}} = 2c_K \frac{K_c}{K}\sqrt{\frac{6}{s}}\frac{1}{\dot{\varepsilon}} \qquad \Rightarrow \qquad s = \left(\frac{\sqrt{24}K_c}{\rho c_K \dot{\varepsilon}}\right)^{2/3}.$$
(11.57)

If we use a generic $K_c = 50$ MPa$\sqrt{\text{m}}$ for a fracture toughness, the values of s we get for the above geometry are 1.9 mm and 3.6 mm, respectively, which do not agree with the observed sizes (which agrees with the fact that the fragmentation appears ductile, not brittle). The same argument can be applied to a liquid, only now the surface tension γ provides the energy, which is the energy for one side of the surface, so

$$\frac{1}{2}K\varepsilon^2 V = \gamma A \qquad \Rightarrow \qquad \varepsilon = \sqrt{\frac{2\gamma A}{KV}}. \qquad (11.58)$$

The same reasoning for a sphere or cube yields

$$s = 2c_K t = 2c_K \frac{\varepsilon}{\dot{\varepsilon}} = 2c_K \sqrt{\frac{12\gamma}{Ks}} \frac{1}{\dot{\varepsilon}} \qquad \Rightarrow \qquad s = \left(\frac{48\gamma}{\rho\dot{\varepsilon}^2}\right)^{1/3}. \qquad (11.59)$$

The three expressions for s derived above are referred to as the Grady–Kipp model of fragmentation.

The above arguments are end-state arguments and do not provide a notion of how the fragmentation develops. To address that, Kipp and Grady present a different approach building on work originally done by Mott [80, 120]. The math is similar to that presented for the shear-band model in Section 9.6, though this is in tension, while that was in shear. It is easiest to visualize in the case of a stretching rod. Suppose that the material stretches in the x direction at a uniform strain rate $\dot{\varepsilon}$. With the assumption of uniform strain rate, we have for the velocity field $v(x,t)$,

$$\frac{\partial v(x,t)}{\partial x} = \dot{\varepsilon} \qquad \Rightarrow \qquad v(x,t) = \dot{\varepsilon}x, \qquad (11.60)$$

where we have assumed that $v(0,t) = 0$. If no fracture occurred, then we would have $v(x,t) = \dot{\varepsilon}x$ for all time. However, suppose we begin the material separation process at $x = 0$, $t = 0$, and that the strength of the material linearly decreases from a value Y to 0 at a crack opening displacement of x_c. We assume that the material separates from the wall at $x = 0$, that there is no material from $x = 0$ to $x = x_\ell$, that there is a region that is no longer stretching (hence rigid) from x_ℓ to x_r, and that then it is stretching for $x > x_r$ with no knowledge of the fracture surface. The mass in the rigid region is equal to the density times length the stretching material would have occupied if there was no fracture – that is, rigid mass $= \rho x_r$. (The rod cross section reduces during the stretching, but the area appears on both sides of the momentum balance equation and cancels out, so we will not include it.) The stretching material experiences the tensile flow stress Y; hence that is the load on the right end of the rigid material. The momentum balance for the rigid material is

$$\rho x_r \frac{d}{dt}v(x_r,t) = \rho x_r \frac{d}{dt}(\dot{\varepsilon}x_r) = Y - Y(1 - x_\ell/x_c) \quad \Rightarrow \quad \rho x_r \frac{dx_r}{dt} = \frac{Y}{\dot{\varepsilon}}\frac{x_\ell}{x_c}. \quad (11.61)$$

The fracture end of the rigid rod material moves at the speed $\dot{\varepsilon}x_r$; hence

$$\frac{dx_\ell}{dt} = \dot{\varepsilon}\,x_r. \qquad (11.62)$$

Initially $x_\ell(0) = 0$ and $x_r(0) = 0$, so these equations combine to give

$$\frac{\rho}{\dot{\varepsilon}}\frac{dx_\ell}{dt}\frac{d^2x_\ell}{dt^2} = Y\frac{x_\ell}{x_c} \qquad \Rightarrow \qquad x_\ell(t) = \frac{Y\dot{\varepsilon}t^3}{18\rho x_c}. \qquad (11.63)$$

From this we get the location of the moving front between the plastically stretching and rigid material,

$$x_r(t) = \frac{1}{\dot{\varepsilon}}\frac{dx_\ell}{dt} = \frac{Yt^2}{6\rho x_c}. \qquad (11.64)$$

The time of fracture is when $x_\ell(t_f) = x_c$ and, if we look at the size of the rigid region at the time of fracture as indicative of the size of our fragments as before, we obtain

$$s = 2x_r(t_f) = \left(\frac{12Yx_c}{\rho\dot\varepsilon^2}\right)^{1/3} = \left(\frac{24\Gamma}{\rho\dot\varepsilon^2}\right)^{1/3}, \tag{11.65}$$

where $\Gamma = \frac{1}{2}Yx_c$ is the work to fracture. The primary purpose of this derivation is to show that a more causal mechanics argument leads to the same functional form as the end-state results above and that the strain need not be a volumetric strain but can be an equivalent plastic strain.

Once fracture occurs, there is a qualitative difference in the wave propagation behavior. We solve for this for the case $x_c = 0$, corresponding to instantaneous fracture; hence the momentum balance of Eq. (11.61) is

$$\rho x_r \frac{d}{dt}(\dot\varepsilon x_r) = Y \quad \Rightarrow \quad x_r(t) = \sqrt{\frac{2Yt}{\rho\dot\varepsilon}}. \tag{11.66}$$

This wave-speed behavior, where the wave speed is proportional to $1/\sqrt{t}$, is similar to solutions to the diffusion equation and is referred to as a "Mott" wave, due to Mott's work during World War II.

In behind armor debris and other fragmentation events the fragment sizes are not all the same; there is extensive research literature on the statistical distribution of fragment size and mass. Here we have just estimated the nominal size. In the back surface bulge and breakout model developed above or in a hydrocode, we can compute the equivalent plastic strain rate in the material at failure and use it to estimate a nominal fragment size. The strain rate can be computed at different locations to determine a representative fragment size based on the geometry of the velocity field. It is then possible to bin fragments into sizes and determine the number that comprise the volume of the plug. Given the specific location in the failed plug region with its specific velocity, it is then possible to compute a post-failure trajectory for each fragment. More discussion of fragmentation can be found in [53, 80].

11.6. Sources

The velocity field for back surface bulging was developed by Walker [191]. Penetration data and examples in Section 11.2 are from Anderson, Hohler, Walker, and Stilp [6] and Walker, Bigger, and Chocron, Anderson, Walker and Ravid [200]. The behavior of the projectile between plates began with Chocron et al. [40], but the specific results are new. The discussion of projectile and target strength is due to Anderson, Hohler, Walker, and Stilp [7].

Exercises

(11.1) Show that the potential given by Eqs. (11.7) and (11.9) gives rise to simple shear deformation with the stated properties.

(11.2) Show for the residual velocity formula Eq. (11.39) that $0 \le p_t \le \rho_t uT$, as we would expect. (Hint: for the lower limit notice the back surface bulge model is not invoked unless $\alpha > 1 + T/R_c$.)

(11.3) Show the third and fourth order in R/x rational functions satisfy the requirements for the shear flow as shown in Eq. (11.9). Plot and compare the corresponding velocities.

(11.4) For the fourth order rational potential for shear in Eq. (11.10), compute the corresponding results to Eqs. (11.21), (11.22), (11.23), and (11.29).

(11.5) For the analytic flow field, show that the equivalent plastic strain rate in the region directly in front of the crater (i.e., $x \le R$) is

$$\dot{\varepsilon}_p \equiv \sqrt{\frac{2}{3} D_{ij} D_{ij}} = \frac{2\lambda u}{r} \left(\frac{R}{r}\right)^{2\lambda} \sqrt{\cos^2(\theta) + \frac{(1-2\lambda)^2}{12} \sin^2(\theta)}. \qquad (11.67)$$

Show that it reduces Eq. (11.36) for the centerline equivalent plastic strain rate.

(11.6) With the Walker–Anderson analytic model with bulge and breakout or a hydrocode, examine work presented by Mescall and Papirno, where a hardened Rc 52 blunt cylindrical projectile of diameter 0.559 cm and $L/D = 2$ was impacted into a 4340 steel $T/D = 1$ target plate (10×10 cm) with two different hardnesses: Rc 52 and Rc 15 [**144, 145**]. The harder Rc 52 targets were perforated at 747 m/s, when a conical shear surface formed a plug, but at impact speeds below that there was minimal deformation of the target plate. The softer Rc 15 targets showed considerable crater formation and deformation from impact at lower impact speeds, but a plug leading to perforation was not formed till the impact speed reached 823 m/s. Thus, the same material (4340) showed a decreased ballistic performance with an increased strength. What failure strains produce these results? Postulate a relation between yield stress and failure strain and look to find the optimum performances, good and bad, for the armor plate.

(11.7) In his development of the fragmentation model, Grady initially included kinetic energy as well as elastic potential energy [**82**]. However, he showed it was a small fraction of the energy, and so will you in this exercise. Assume that the material of interest is expanding uniformly in all directions at a constant strain rate $\dot{\varepsilon} = \dot{\varepsilon}_v = -\dot{\rho}/\rho > 0$. We focus on a sphere of material and move to a coordinate system where its center is at rest. Locally the velocity in the radial direction can be written $v_r(r) = Vr/R$, where V is a constant and R is a representative radius (later to be chosen to be the radius of the sphere when fragmentation occurs – it is, however, a constant). Show that for a spherically symmetric expansion,

$$\dot{\varepsilon} = D_{ii} = \frac{\partial v_r}{\partial r} + \frac{2v_r}{r} = 3\frac{V}{R}. \qquad (11.68)$$

We want to determine the amount of available energy if the material were to fragment at a given time t after expansion began into spheres of radius R. Show that for the sphere, the kinetic energy density (kinetic energy per unit volume) is

$$T = \rho E_{KE} = \frac{1}{vol} \int_{sphere} \frac{1}{2} \rho v^2 \, dvol = \frac{3}{10} \rho V^2 = \frac{1}{30} \rho \dot{\varepsilon}^2 R^2 = \frac{1}{120} \rho \dot{\varepsilon}^2 s^2. \qquad (11.69)$$

The last step recognizes that for spall fragments, the characteristic size is the diameter of the sphere, so $s = 2R$. Next, determine the amount of internal energy stored in elastic expansion. The stress in the material as a function of

TABLE 11.7. Experimental spall data from Grady.

Material	Spall type	Fracture toughness (MPa $\sqrt{m}$)	Strength (GPa)	Experimental spall stress (GPa)
Al 6061-T6	Brittle	25–30		0.8–1.5
Titanium	Brittle	40–70		2.1–3.9
4340 Steel	Brittle	40-80		2.5–5.4
Soft aluminum	Ductile		0.015–0.3	0.5–1.1
Copper	Ductile		0.025	1.0–2.5
Titanium-6Al-4V	Ductile		0.8–0.9	4.1–5.0

time is
$$\sigma = K\varepsilon = K\dot{\varepsilon}t. \tag{11.70}$$
Thus, show that the energy density that is stored elastically is
$$U = \rho E_e = \frac{1}{2}\frac{\sigma^2}{K} = \frac{1}{2}K\dot{\varepsilon}^2 t^2. \tag{11.71}$$
For Grady's communication horizon condition, where fracture occurs at its earliest time $t = R/c_K = s/(2c_K)$, show that internal elastic energy at that time is
$$U = \frac{1}{8}\rho\dot{\varepsilon}^2 s^2. \tag{11.72}$$
Thus, the available internal energy dominates the available kinetic energy by a factor of 15. (Kinetic energy was ignored in the derivation of the main text and this exercise shows that it is reasonable to ignore it.)

(11.8) In this exercise the stress at failure for the Grady fragmentation model is determined [82]. Equation (11.70) gives the stress during the expansion and its value at the time of fragmentation will be identified as the spall stress. Show that for brittle and ductile cases the spall stresses are
$$\sigma = (3\rho c_K K_c^2 \dot{\varepsilon})^{1/3} \quad \text{and} \quad \sigma = \sqrt{2KY\varepsilon_c} = \sqrt{2\rho c_K^2 Y\varepsilon_c} \tag{11.73}$$
respectively. To obtain a feel for the size of fragments and the stress involved, assume a strain rate of $\dot{\varepsilon} = 10^5$ s^{-1}, compute the spall size s and spall stress σ for Al 6016-T6, titanium, and 4340 steel. Do the same for a soft aluminum, copper, and the Ti-6Al-4V alloy. Compare with Table 11.7.

(11.9) In the Grady fragmentation model, show that the strain rate, stress, and fragment size at which the transition from brittle to ductile occurs is
$$\dot{\varepsilon} = \sqrt{\frac{8K^2(Y\varepsilon_c)^3}{9\rho K_c^4}}, \quad s = \frac{3K_c^2}{KY\varepsilon_c}, \quad \sigma = \sqrt{2KY\varepsilon_c}. \tag{11.74}$$
Are strain rates lower than the transition strain rate the realm of ductile or brittle fragmentation?

(11.10) Given the Grady fracture model, would we expect tensile SHPB tests to produce multiple necks?

(11.11) The analytic model with the back surface bulge and breakout does not include any information about the failure strain of the projectile, it only includes projectile strength. What would you expect the predictions to be the problem in Section 11.4 of $L/D = 10$ tungsten rods perforating $T/D = 3.55$ thick steel plates? If you have programmed up the model, run some test cases and see what happens.

(11.12) Critical yaw is the yaw such that the side of the projectile at its back end just strikes the target at the side of the crater opened by the front of the projectile at the strike face of the target. Work out a formulate for critical yaw γ_c in terms of the projectile diameter, projectile length, and crater diameter. Do any of the yaws in Table 11.5 exceed the critical yaw? (Hint: you need to compute the crater diameter.)

(11.13) (Compare with 7.19) Assume that one has projectiles that are right circular cylinders of tungsten of constant mass m_p with various L/D. Assume that the empirical penetration equation for tungsten into steel (Eq. 7.23) also predicts the target thickness that would be penetrated - for example, add $2D$ to the depth P predicted by that equation to taken into account the back surface effects, so that the V_{50} thickness is given by $T = P + 2D$, where P is given by Eq. (7.23). Show for a fixed target thickness T, when L/D is varied but the mass of the projectile is constant, that the V_{50} has a minimum. Derive an equation for the L/D at the minimum and for V_{50} at the minimum.

(11.14) (Targets with finite lateral extent.) Another way targets can have finite extent is for them to have finite extent in the lateral direction. Experiments with square targets were conducted at the Army Research Laboratory by L. Pridgeon in 1995. The projectiles were 65 g, $L/D = 20$ tungsten alloy rods with a hemispherical nose, with nominal diameter 0.610 cm and length 12.37 cm. The projectile X21 alloy (92.98% W, 4.91% Ni, 2.11% Fe) has a density of 18.1 g/cm^3, a yield strength of 1.09 GPa (at 0.2% offset), and an ultimate tensile strength of 1.13 GPa at 10.6% elongation. The hardness was measured as Rc 41.5±1.0. The impact velocity for the tests was 1.50 ± 0.01 km/s. The projectiles were launched at RHA target blocks 15.2 cm on a side. Five tests were conducted; the impact point was moved from the center of the target block towards an edge in 1.27-cm increments; the impact points relative to the target block geometry are shown in the upper half plane in Fig. 11.20 (left). Total impact yaw was less than 0.7° for all the tests. The depths of penetration normalized by the original length of the projectile are plotted in Fig. 11.20 (right). The dotted line represents the average $P/L = 0.786$ for the four tests that were at least 3.81 cm from the boundary. The data in Fig. 11.20 (right) are plotted in terms of the distance from the target edge d_e normalized by the projectile diameter D. Also shown are computations performed with CTH [17], but with $L/D = 10$ projectiles, with impact locations in Fig. 11.20 (left) and P/L results in Fig. 11.20 (right). The computations were three-dimensional.

Though the tests and computations were for tungsten alloy impacting steel, our cavity expansion results of Chapter 8 show similar plastic flow extents for aluminum and steel given the same expansion penetration velocity.

Using the above information, argue the following result:

GENERAL PRINCIPLE 11.5. *(edge effects) For metal targets, the final depth of penetration is affected by edge confinement when projectile impact is less than seven projectile diameters from the target boundary.*

Essentially, at this distance the plastic deformation field begins to interact with the target boundary. Also, computations at higher velocities reveal

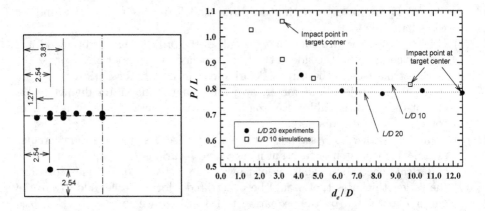

FIGURE 11.20. Relative impact points for experimental (upper half plane) and computational (lower half plane) study are shown to the left, and P/L vs. normalized distance from target edge is shown to the right (the two dotted lines are from the center impacts). (From Anderson, Walker, and Sharron [17].)

that the influence of the boundary decreases as impact velocity increases. For ordnance-like impact velocities, target strength combined with lateral confinement dominates target resistance and the penetration response. As the velocity increases, strength has less of a role and the inertial terms dominate the stress and hence the reduction in confinement has less effect; strength does have an effect, however. See Littlefield, Anderson, Partom, and Bless [134] for more discussion of finite targets (in this case, cylinders struck in the center) and Anderson and Chocron [4] for more discussion of velocity effects.

    Another observation from the simulations is that as the impact point gets closer to the target boundary, there is a slow drift of the penetrating projectile towards the free surface. The projectile, however, does not turn and exit the side of the target, even for the case where the impact point was only 1.27 cm from the target edge. This is in contrast to the case of rigid body penetration, where the entire projectile reacts to any imbalance of forces by rotation about the projectile center of mass and can lead to large drift and even J-hook-shaped trajectories.

(11.15) Another approach to determining the V_{50} is to take the experimentally determined residual velocities and do a least-squares fit to the Lambert equation (Eq. 11.45) for variables a, p, and V_{bl}. The V_{50} is then the fit ballistic limit velocity V_{bl}. This approach can be useful, especially if there are no data points near the ballistic limit velocity. Though simple in appearance, this equation can be very hard to fit because of its nonlinearity, part of which is due to p and part of which is due to the fact that the residual velocity is zero for all experimental points where no perforation occurs. Using a least squares method and assuming $p = 2$ (to simplify the least-squares calculation so that a and V_{bl} are the only two unknowns and it can be linearized by taking the logarithm of each side before performing the least squares), find the ballistic limit of the data in Table 11.5 the 20.9 mm thick BHN 460 target.

Nondeforming (Rigid) Impactors

This chapter explores penetration by projectiles that maintain their shape, so in particular the penetration speed equals the tail speed of the projectile ($u = v$). It might be thought that this topic would be simpler than eroding penetration, since it is only necessary to address target response. However, eroding penetration leads to a hemispherical penetration front, which simplifies the response of the target.

We first present blunt penetration of thin targets. Then we look at flow fields for pointed projectiles for use in the Walker–Anderson model. After that we look at an alternate approach of directly using the cavity expansion. We finish with a discussion of the rigid-eroding transition.

In this chapter the crater radius always equals the projectile radius, hence $R = R_p$.

12.1. Thin Plate Perforation by Blunt Rigid Projectiles

If the projectile is a blunt right-circular cylinder, a possible response is that the target plate simply shears, forming a plug. However, simple shear is not really what happens; Jeng and Goldsmith drilled holes, aligned with the projectile flight direction, in a plate in a radial line moving away from the shot line. Placing a pin in the plate after the impact shows that the rotation near the perforation is around $20°$ [103] (this experiment essentially measures $\partial u_r/\partial z$, where u_r is the radial deflection). Hence there is also bending of the plate. However, to simplify the discussion, we assume simple shear.

From the back surface bulge and breakout model of the previous chapter, we use the third order shear flow of Eq. (11.11). With the thin target material assumed a rigid plug in front of the blunt projectile, the centerline momentum balance is

$$(\rho_p L + \rho_t T)\dot{v} = -\frac{7}{3}Y_t \frac{T}{R}, \tag{12.1}$$

with the approximate resistance from Eq. (11.21). Since the right-hand side is constant, using $\dot{v} = v\,dv/dx$ this is readily integrated to give for the ballistic limit velocity, where the final projectile speed is zero as the target fails,

$$(\rho_p L + \rho_t T)\frac{1}{2}\left(\frac{\rho_p L}{\rho_p L + \rho_t T}V_{bl}\right)^2 = \frac{7}{3}Y_t\frac{T}{R}P_f. \tag{12.2}$$

The initial speed was reduced to conserve momentum as we assume the plate material in front of the projectile is put into motion with the same speed as the projectile. For the ballistic limit velocity the deflection of the target is equal to its deflection (or penetration) at failure P_f. Solving,

$$V_{bl}^2 = \frac{14}{3}\frac{\rho_p L + \rho_t T}{(\rho_p L)^2}Y_t T\frac{x_f}{R} = \frac{14}{3}(1+\chi)\chi\frac{Y_t}{\rho_t}\frac{P_f}{R}, \tag{12.3}$$

where

$$\chi = \frac{\rho_t T}{\rho_p L} \tag{12.4}$$

is the ratio of the areal density of the plate material compared to the areal density of the blunt projectile.

Using the third order flow field we can compute the deflection at failure by either using the strain on the back surface, as was used in the examples in the previous chapter, or given the simplicity of the simple shear flow we can explicitly find the equivalent plastic strain. The strain along the back surface is

$$\varepsilon = \lim_{\Delta x \to 0} \frac{\sqrt{(x + \Delta x - x)^2 + (u_z(x + \Delta x) - u_z(x))^2} - \Delta x}{\Delta x} = \sqrt{1 + \left(\frac{\partial u_z}{\partial x}\right)^2} - 1. \tag{12.5}$$

For the third order flow field, the maximum strain occurs at $x = 5R/4$ and thus integrating the maximum equivalent plastic strain rate for the shearing flow with respect to time gives

$$\varepsilon_p = \frac{1}{\sqrt{3}} \left| \frac{\partial u_z}{\partial x} \right|_{\max} = \sqrt{3} \left(\frac{4}{5}\right)^5 \frac{P(t)}{R} = 0.568 \frac{P(t)}{R}. \tag{12.6}$$

When the strain reaches a maximum allowed value, failure occurs and we assume there is no longer any penetration resistance. If we compare the strain to failure using equivalent plastic strain ε_p^f and using the strain along the back surface ε_f we have that the penetration at which failure occurs is

$$\frac{P_f}{R} = \frac{1}{3} \sqrt{(1 + \varepsilon_f)^2 - 1} \left(\frac{5}{4}\right)^5, \qquad \frac{P_p^f}{R} = \frac{\varepsilon_p^f}{\sqrt{3}} \left(\frac{5}{4}\right)^5. \tag{12.7}$$

For the typical failure strains that we use, such as 0.65, the penetrations at failure are similar, namely $P_f/R = 1.34$ and $P_p^f/R = 1.15$. They are independent of impact speed.

Before presenting results, we make the following modification. In our shear flow, the maximum deformation – hence the location of failure – occurs at $\frac{5}{4}R$. However, experimentally for blunt projectile perforating thin plates, the shearing plug is essentially the radius of the projectile. Thus, we adjust the projectile so that it has $\frac{4}{5}$ times the radius so that the maximum shearing and failure location is at the projectile radius. (However, the $(1 + \chi)$ term from the initial condition momentum balance is not adjusted.) Maintaining the mass of the rigid projectile adjusts χ as

$$\chi' = \frac{\rho_t T}{\rho_p L'} = \frac{\rho_t T}{\rho_p L} \frac{L}{L'} = \chi \frac{L}{L'} = \chi \frac{R'^2}{R^2} = \chi \left(\frac{4}{5}\right)^2. \tag{12.8}$$

Combining these terms together and using the back surface strain yields the ballistic limit

$$V_{bl} = \sqrt{\chi(1 + \chi)} \left(\frac{14}{9}\right)^{1/2} \left(\frac{5}{4}\right)^{3/2} \sqrt{\frac{Y_t}{\rho_t}} \{(1 + \varepsilon_f)^2 - 1\}^{1/4}$$

$$= 1.74 \sqrt{\chi(1 + \chi)} \sqrt{\frac{Y_t}{\rho_t}} \{(1 + \varepsilon_f)^2 - 1\}^{1/4}. \tag{12.9}$$

TABLE 12.1. Blunt body penetration of thin plates (rigid steel projectiles, $\rho_t = 7.85 \text{ g/cm}^3$; $\varepsilon_f = \varepsilon_p^f = 0.65$ for all targets). For experimental V_{bl}, the lower value is a nonperforating speed, while the upper value is a speed where either a plug was launched or a perforation occurred.

| | Target | | Projectile | | | | | |
ρ_t (g/cm³)	Y_t (MPa)	T (mm)	L (mm)	D (mm)	L/D	T/D	χ	Ref.
7.85	800	10.5	282	30.0	9.40	0.35	0.0372	[66]
7.85	800	5.3	268	30.8	8.70	0.17	0.0198	[66]
7.85	800	10.5	89.1	30.8	2.89	0.34	0.1178	[66]
2.85	270	3.3	25.4	12.6	2.01	0.26	0.0472	[207]
2.70	200	3.18	38.1	12.7	3.00	0.25	0.0287	[103]
2.70	200	4.76	38.1	12.7	3.00	0.37	0.0430	[103]
2.70	200	6.35	38.1	12.7	3.00	0.50	0.0573	[103]

| | | max V | min V | | | v_{bl} | v_{bl} | v_{bl} |
$\sqrt{Y_t/\rho_t}$ (m/s)	χ	no perf (m/s)	perf (m/s)	P_f (mm)	P_f/R_p	(12.9) (m/s)	(12.10) (m/s)	(12.13) (m/s)
319	0.0372	114	127			125	116	122
319	0.0198	84	99			90	84	99
319	0.1178	169	178			231	214	179
308	0.0472	119	122	8.3	1.32	136	126	127
272	0.0287	93	97	12.0	1.89	93	86	95
272	0.0430	99	114	10.8	1.70	115	106	109
272	0.0573	123	126	8.1	1.28	134	124	120

If we use the equivalent plastic strain strain failure criterion, then the expression is

$$V_{bl} = \sqrt{\chi(1+\chi)} \left(\frac{14}{9}\right)^{1/2} \left(\frac{5}{4}\right)^{3/2} \sqrt{\frac{Y_t}{\rho_t}} (\sqrt{3}\varepsilon_p^f)^{1/2} = 2.29\sqrt{\chi(1+\chi)} \sqrt{\frac{Y_t \varepsilon_p^f}{\rho_t}}.$$
(12.10)

In this case, the plastic work to failure is given by $Y_t \varepsilon_p^f$, so we see the direct relationship between strength and ductility. It is interesting to note that the dependence on density of the target is quite weak and essentially nonexistent for small χ, since using the definition of χ,

$$V_{bl} = 1.74\sqrt{1+\chi} \sqrt{\frac{TY_t}{\rho_p L}} \left\{(1+\varepsilon_f)^2 - 1\right\}^{1/4}, \quad V_{bl} = 2.29\sqrt{1+\chi} \sqrt{\frac{TY_t \varepsilon_p^f}{\rho_p L}}.$$
(12.11)

The ballistic limit is linearly dependent on $\sqrt{Y_t T}$.

We compare to results of the impact of rigid steel projectiles into both steel and aluminum plates, as given in Table 12.1. Data are from Forrestal and Hanchak [66], Wei and Zhang [207], and Jeng and Goldsmith [103]. The table shows details about the nominally rigid, right-circular cylinder impactor and the thin plate. The two ballistic limit expressions we developed do reasonably well at matching the ballistic limit. The deflection is always greater than that predicted by our shear flow, indicative of the bending that we are not including.

Though the agreement seems reasonable, there is an outlier, which suggests more examination is warranted. A power-law curve fit in terms of χ through the steel points produces a power of 0.32 and through the three same-material aluminum points a power of 0.38. These argue that for small χ, hence thin targets, a slightly better result could be obtained using a power of $1/3$. For small χ, fits to the data for the steel and the aluminum projectiles yield

$$\text{Steel} \quad V_{bl} \approx (365 \text{ m/s})\chi^{0.33}; \qquad \text{Al} \quad V_{bl} \approx (320 \text{ m/s})\chi^{0.33}. \tag{12.12}$$

Using the data we have, for small χ we obtain an empirical expression of the form

$$V_{bl} = \sqrt{\frac{Y_t}{\rho_t}}\{(1+\varepsilon_f)^2 - 1\}^{1/4}\chi^{1/3}, \quad V_{bl} = 1.42\sqrt{\frac{Y_t\varepsilon_p^f}{\rho_t}}\chi^{1/3}, \tag{12.13}$$

where the numerical constant is to agree with the data. The results of this expression are also shown in Table 12.1. Notice, however, that our material property data are really not much different between the aluminum and the steel; thus it is hard to know if this empirical form is very accurate for other alloys or metals.

An interesting fact falls out of these expressions: with a nondeforming blunt projectile, the projectile radius does not appear. When a deforming projectile hits a rigid plate, the projectile radius does not appear (Chapter 5, Taylor anvil). But when both projectile and target deform, the projectile radius does appear in the results, which we referred to as the L/D effect.

12.2. Flow Fields for Pointed Projectiles

So far we have discussed flow fields for a hemispherical-nosed projectile, for simple shear, and multiplicative blends of them to do back surface bulge. As a first step in pursuing flow fields for other axisymmetric nose shapes, we recall the potential for the hemispherical flow in spherical coordinates, Eq. (10.15), $\vec{A} = A_\varphi \hat{e}_\varphi$ and

$$A_\varphi = f(r)\sin(\theta) = \frac{V\sin(\theta)}{2(\alpha^2-1)}\left\{\frac{(\alpha R)^2}{r} - r\right\} = \frac{Vr\sin(\theta)}{2}\frac{\alpha^2}{(\alpha^2-1)}\left\{\left(\frac{R}{r}\right)^2 - \frac{1}{\alpha^2}\right\}. \tag{12.14}$$

In this section the penetration speed ($u = v$) will be denoted by V. The spherical $\hat{e}_\varphi$ points in the same direction as the cylindrical $\hat{e}_\theta$, so the identical vector potential written in cylindrical coordinates is $\vec{A} = A_\theta\hat{e}_\theta$, where for the rest of this chapter we use cylindrical coordinates and $r = \sqrt{x^2 + y^2}$ is the cylindrical r, which equals the spherical radius times the sine of the spherical θ,

$$A_\theta = \frac{Vr}{2}\frac{\alpha^2}{(\alpha^2-1)}\left\{\frac{1}{\xi^2} - \frac{1}{\alpha^2}\right\}, \qquad \xi(r,z) = \frac{\sqrt{r^2+z^2}}{R}. \tag{12.15}$$

Values for ξ range from $\xi = 1$ on the nose of the hemispherical-nosed projectile to $\xi = \alpha$ on the outer edge of the flow field. Our flow field is given by $\vec{v} = \text{curl}(\vec{A})$ and in cylindrical coordinates is

$$\begin{aligned}
v_r &= -\frac{\partial A_\theta}{\partial z} = V\frac{\alpha^2}{(\alpha^2-1)}\frac{r}{\xi^3}\frac{\partial \xi}{\partial z}, \\
v_z &= \frac{1}{r}\frac{\partial(rA_\theta)}{\partial r} = V\frac{\alpha^2}{(\alpha^2-1)}\left(\frac{1}{\xi^2} - \frac{1}{\alpha^2} - \frac{r}{\xi^3}\frac{\partial \xi}{\partial r}\right).
\end{aligned} \tag{12.16}$$

We now use A_θ in Eq. (12.15), but view $\xi(r, z)$ as a function to be determined with the property that $\xi = 1$ on the projectile nose and $\xi = \alpha$ on the flow-field outer boundary. Since $\xi = 1$ on the nose, the gradient of ξ is normal to the nose. The boundary condition that target material does not intrude into the projectile is that the velocity component normal to the nose matches the normal-to-the-nose velocity component of the penetration velocity $V\hat{e}_z$ – that is,

$$\vec{\mathrm{grad}}(\xi) \cdot \vec{v}|_{\xi=1} = \vec{\mathrm{grad}}(\xi) \cdot V\hat{e}_z = V\frac{\partial \xi}{\partial z}. \tag{12.17}$$

Computing,

$$\vec{\mathrm{grad}}(\xi) \cdot \vec{v}|_{\xi=1} = \begin{pmatrix} \partial\xi/\partial r \\ \partial\xi/\partial z \end{pmatrix} \cdot \frac{V\alpha^2}{(\alpha^2 - 1)} \left. \begin{pmatrix} \frac{r}{\xi^3}\frac{\partial\xi}{\partial z} \\ \frac{1}{\xi^2} - \frac{1}{\alpha^2} - \frac{r}{\xi^3}\frac{\partial\xi}{\partial r} \end{pmatrix} \right|_{\xi=1} = V\frac{\partial\xi}{\partial z}, \tag{12.18}$$

and hence the boundary condition at the nose is satisfied for a general $\xi(r, z)$. The boundary condition at the outer edge of the flow field, where $\xi = \alpha$, is that there is no material exiting the flow field, and

$$\vec{\mathrm{grad}}(\xi) \cdot \vec{v}|_{\xi=\alpha} = \begin{pmatrix} \partial\xi/\partial r \\ \partial\xi/\partial z \end{pmatrix} \cdot \frac{V\alpha^2}{(\alpha^2 - 1)} \left. \begin{pmatrix} \frac{r}{\xi^3}\frac{\partial\xi}{\partial z} \\ \frac{1}{\xi^2} - \frac{1}{\alpha^2} - \frac{r}{\xi^3}\frac{\partial\xi}{\partial r} \end{pmatrix} \right|_{\xi=\alpha} = 0, \tag{12.19}$$

as required.

As a first examination of a different ξ, consider $\xi(r, z) = \frac{1}{R}(r^b + z^b)^{1/b}$. For $b = 1$ we have a 45° half angle cone, for $b = 2$ we have the hemispherical flow, and for $b \to \infty$ we have a blunt nose. Assuming rigid plastic stresses for the flow field, we can compute the penetration resistance. For $b = 1$,

$$\left.\frac{\partial\sigma_{rz}}{\partial r}\right|_{r=0} = \left.\frac{\partial}{\partial r}\left(\frac{\sqrt{\frac{2}{3}}Y_t D_{rz}}{\sqrt{D_{ij}D_{ij}}}\right)\right|_{r=0} = -\frac{6}{7\sqrt{7}}\frac{Y_t}{z}, \tag{12.20}$$

and for $b > 1$, with the help of a symbolic manipulator, we have for r,

$$\frac{\partial\sigma_{rz}}{\partial r} = -\frac{Y_t}{6}\left(\frac{3}{z} + \frac{b^2 + b - 2}{z^{b-1}r^{2-b}}\right) + O(r^{\min(b, 2b-2)}). \tag{12.21}$$

This allows computation of S_t as

$$S_t = -2\int_R^{\alpha R}\left.\frac{\partial\sigma_{rz}}{\partial r}\right|_{r=0} dz = \begin{cases} \frac{12}{7\sqrt{7}}Y_t\ln(\alpha), & b = 1, \\ +\infty, & 1 < b < 2, \\ \frac{7}{3}Y_t\ln(\alpha), & b = 2, \\ Y_t\ln(\alpha), & b > 2. \end{cases} \tag{12.22}$$

This shows that the stress calculation for the hemispherical flow field ($b = 2$) is a special case, and that rigid plastic stresses for other flow fields may not give good shear stress expressions in the vicinity of the centerline. Following our experience with the shear flow, a more robust approach for determining target resistance is to compute S_t^W,

$$S_t^W = \frac{1}{\pi R^2 V}\int Y_t \dot{\epsilon}_p dv = \text{coefficient} \times Y_t\frac{\alpha^2}{\alpha^2 - 1}\ln(\alpha). \tag{12.23}$$

Numerical results for the coefficients for various values for b are given in Table 12.2. Also shown are the coefficients of $Y_t\ln(\alpha)$ from Eq. (12.22) for comparison. The issue with the $b = 1$ evaluation for S_t^W and the limit as $b \to 1^+$ from the right will

TABLE 12.2. S_t Coefficients for flow fields based on $\xi = \frac{1}{R}(r^b + z^b)^{1/b}$.

b	1	1^+	1.5	1.9	2	2.1	3	150	∞
S_t^W	3.382	2.227	2.346	2.327	2.333	2.342	2.466	4.377	4.647
S_t	0.648	∞	∞	∞	$2.33\overline{3}$	1	1	1	1

be discussed later. For the $b = \infty$ case, $\xi = \frac{1}{R}\max(r, z)$ and there is a velocity discontinuity along the conical $r = z$ surface; the exact value of the coefficient is $2(7^{3/2} - 8)/9$ plus $4/\sqrt{3}$ for the interior shearing surface along $r = z$ (Exercise 12.3). Computing these required the rate of deformation tensor D,

$$D_{rr} = \frac{\partial v_r}{\partial r} = \frac{C}{\xi^3}\left(\frac{\partial \xi}{\partial z} - 3\frac{r}{\xi}\frac{\partial \xi}{\partial z}\frac{\partial \xi}{\partial r} + r\frac{\partial^2 \xi}{\partial r \partial z}\right), \qquad D_{\theta\theta} = \frac{v_r}{r} = \frac{C}{\xi^3}\frac{\partial \xi}{\partial z},$$

$$D_{zz} = \frac{\partial v_z}{\partial z} = \frac{C}{\xi^3}\left(-2\frac{\partial \xi}{\partial z} + 3\frac{r}{\xi}\frac{\partial \xi}{\partial r}\frac{\partial \xi}{\partial z} - r\frac{\partial^2 \xi}{\partial z \partial r}\right), \qquad C = V\frac{\alpha^2}{\alpha^2 - 1},$$

$$D_{rz} = \frac{1}{2}\left(\frac{\partial v_r}{\partial z} + \frac{\partial v_z}{\partial r}\right) = \frac{C}{2\,\xi^3}\left\{r\frac{\partial^2 \xi}{\partial z^2} - 3\frac{r}{\xi}\left(\frac{\partial \xi}{\partial z}\right)^2 - 3\frac{\partial \xi}{\partial r} + 3\frac{r}{\xi}\left(\frac{\partial \xi}{\partial r}\right)^2 - r\frac{\partial^2 \xi}{\partial r^2}\right\}.$$

$$\tag{12.24}$$

It is interesting that as we tend towards the pointed nose with nose length $L_n/R = 1$, the coefficient for target resistance given by S_t^W does not change very much from the 7/3 value of the hemispherical flow.

    Pointed projectiles introduce a new complication. In the impact event, for the target flow of the hemisphere above and all other values of b, the material in front of the projectile is brought up to the speed of the projectile, since

$$v_z(0, R) = V\frac{\alpha^2}{(\alpha^2 - 1)}\left(\frac{1}{\xi^2} - \frac{1}{\alpha^2} - \frac{r}{\xi^3}\frac{\partial \xi}{\partial r}\right)\bigg|_{r=0, \xi=1} = V. \tag{12.25}$$

The flow smoothly goes around the projectile and the material does not separate, even for deep penetration. However, such no-separation behavior is not what we expect for a projectile with a pointed nose. The point causes the target material to separate (i.e., fail). The target material is not fully brought up to the projectile speed; as it fails it is moved to the side. To reflect this behavior and use the same A_θ as above, it is necessary to have a singularity at the point of the nose in our potential. This singularity at the nose can be achieved by

$$\xi(r, z) = \frac{r}{f(z)}, \tag{12.26}$$

where the shape of the nose is $r = f(z)$, $-L_n \leq z \leq 0$ (L_n is the length of the nose), and the point of the nose is at $(0, 0)$. With this ξ the velocity is

$$\vec{v} = -V\frac{\alpha^2}{(\alpha^2 - 1)}\frac{1}{\xi}\frac{df}{dz}\hat{e}_r - \frac{V}{(\alpha^2 - 1)}\hat{e}_z. \tag{12.27}$$

The singularity at the nose tip is manifested in $v_r(0, 0)$ being nonzero and not well defined – its value depends on the direction of approach to the nose tip, given by ξ. The value of v_z at the tip is $-V/(\alpha^2 - 1)$; it is not V.

    The flow field, as assumed, has three boundaries: the projectile nose, where $\xi = 1$ (hence $r = f(z)$); an outer boundary, where $\xi = \alpha$ (hence $r = \alpha f(z)$), that no material crosses; and the exit region where material leaves the region along the

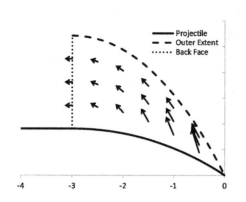 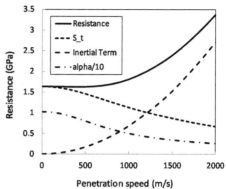

FIGURE 12.1. Left: flow field for the ogive nose, with CRH = 3, $L_n = 3$, $R = 0.91$, and $\alpha = 3$. Velocity vectors at $z = -\frac{n}{2}$. Right: target resistance, with components S_t^W and $\frac{1}{4}\rho_t v^2$, and $\alpha/10$, for CRH = 3 ogive penetrating Al 6061-T6. (From Walker [196].)

plane $z = -L_n$ between $r = R$ and $r = \alpha R$ (Fig. 12.1 left). To match the behavior of the hemispherical flow field, the velocity on the exit surface should be the same as in the hemispherical case, namely $\vec{v}\,|_{z=-L_n} = -\frac{V}{\alpha^2-1}\hat{e}_z$. That only occurs, however, if $f'(-L_n) = 0$, since otherwise $v_r \neq 0$ on the exit plane.

To determine the target resistance, we need the rate of deformation tensor,

$$D_{rr} = \frac{\partial v_r}{\partial r} = V\frac{\alpha^2}{(\alpha^2-1)}\frac{1}{r\,\xi}\frac{df}{dz} = -\frac{v_r}{r} = -D_{\theta\theta}, \quad D_{zz} = \frac{\partial v_z}{\partial z} = 0,$$

$$D_{rz} = \frac{1}{2}\left(\frac{\partial v_r}{\partial z} + \frac{\partial v_z}{\partial r}\right) = \frac{V}{2}\frac{\alpha^2}{(\alpha^2-1)}\left\{\frac{1}{r}\left(\frac{df}{dz}\right)^2 + \frac{1}{\xi}\frac{d^2f}{dz^2}\right\}$$

(12.28)

From this we can compute the equivalent plastic strain rate, after some algebra,

$$\dot{\varepsilon}_p = \sqrt{\frac{2}{3}D_{ij}D_{ij}} = \frac{V}{r}\frac{\alpha^2}{\alpha^2-1}\frac{1}{\sqrt{3}}\left\{\frac{4}{\xi^2}\left(\frac{df}{dz}\right)^2 + \left[\left(\frac{df}{dz}\right)^2 + f\frac{d^2f}{dz^2}\right]^2\right\}^{1/2}. \quad (12.29)$$

Since in cylindrical coordinates the volume differential is $2\pi r\,dr\,dz$, it is seen that the work rate integral is well behaved.

Most pointed projectiles have ogive noses. The measure of the ogive is given by the caliber radius head, CRH. In the r–z plane the nose shape is given by the arc of a circle with radius $R_h = 2\cdot\text{CRH}\cdot R$, so the radius is equal to CRH times the diameter or caliber of the projectile; hence the origin of the terminology. The circle is centered at $(r, z) = (R - R_h, -L_n)$, where L_n is the length of the nose. This arc goes from the base of the ogive nose at $(R, -L_n)$, where it is tangent to the cylindrical body of the projectile shaft, through the nose tip at $(0, 0)$. The equation of the circle is

$$(r + (2C - 1)R)^2 + (z + L_n)^2 = 4C^2R^2, \quad (12.30)$$

TABLE 12.3. Target resistance coefficient for ogive-nosed projectiles.

			Coefficient for $\alpha =$		
CRH	β	L_n/R	5	7	9
2	41.4°	2.65	1.755	1.912	2.083
2.5	36.9°	3	1.671	1.804	1.951
3	33.6°	3.32	1.613	1.728	1.858
3.5	31.0°	3.61	1.569	1.671	1.788
4	29.0°	3.87	1.535	1.627	1.733

where $C = $ CRH, yielding a nose length for the ogive of $L_n = R\sqrt{4C - 1}$. This equation allows a determination of $f(z)$ and its derivatives for the ogive,

$$f(z) = -(2C - 1)R + \sqrt{4C^2R^2 - (z + L_n)^2}, \qquad \beta = \mathrm{acos}(1 - 1/2C),$$

$$\frac{df}{dz} = -\frac{z + L_n}{\sqrt{4C^2R^2 - (z + L_n)^2}}, \qquad \frac{d^2f}{dz^2} = -\frac{4C^2R^2}{(4C^2R^2 - (z + L_n)^2)^{3/2}}. \tag{12.31}$$

The value of β represents the half angle the pointed nose makes with the axis of symmetry. We determine the target resistance through

$$S_t^W = \frac{1}{\pi R^2 V} \int Y_t \dot{\varepsilon}_p dv = \text{coefficient} \times Y_t \ln(\alpha), \tag{12.32}$$

which needs to be done numerically. Values for various CRH are given in Table 12.3. Though the coefficient is not a constant times $\ln(\alpha)$, we will use a coefficient of 1.85 for CRH $= 3$ ogive-nosed projectiles.

The conical-nosed projectile brings with it certain challenges. The problem is that the cone has two singular points, the nose tip and the back end of the projectile where the flow needs to transition to flow moving along the projectile shaft with no radial component. To see what the issue is, we introduce a projectile nose that approximates the cone but does not have the discontinuous-slope edge at the back. The front is the cone of slope $a = R/L_n$, where $\beta = \mathrm{atan}(R/L_n)$ is the cone's half angle, and the entire nose is

$$f(z) = \begin{cases} -az, & -L_n + \Delta z \le z \le 0, \\ -az - \frac{a}{4\Delta z}(z + L_n - \Delta z)^2, & -L_n \le z - \Delta z < -L_n + \Delta z. \end{cases} \tag{12.33}$$

The edge now has a continuous derivative at the back end of the projectile, with a parabolic shape of length $2\Delta z$ centered at the back of the cone. Performing the integral for S_t^W in Eq. (12.32), over the length $z = -\Delta z - L_n$ to the nose tip $z = 0$, provides the results given in Table 12.4. It is seen that the singularity at the back of the cone leads to increasing resistance. We can also determine the extra amount that is coming from the back surface by taking a limit as $\Delta z \to 0$. Both f and f'

TABLE 12.4. Target resistance coefficients for conical-nosed projectile (Eq. 12.33), with $L_n/R = 3$ and $\beta = 18.4°$.

		Coefficient for $\Delta z =$						Partition for $\Delta z = 0$		
α		0.5	0.4	0.3	0.2	0.1	0	=	Volume	+	Back Surface
S_t^W	5	1.84	1.91	1.99	2.08	2.19	2.31	=	1.32	+	1.00
S_t^W	7	2.02	2.10	2.20	2.31	2.43	2.57	=	1.36	+	1.21
S_t^W	9	2.21	2.31	2.42	2.54	2.68	2.83	=	1.41	+	1.42

are bounded, so the f'' term dominates Eq. (12.29),

$$
\begin{aligned}
S_t^W\big|_{back} &= \lim_{\Delta z \to 0} \frac{1}{\pi R^2 V} \int_R^{\alpha R} \int_{-L_n-\Delta z}^{-L_n+\Delta z} Y_t \dot\varepsilon_p dvol \\
&= \lim_{\Delta z \to 0} \frac{1}{\pi R^2 V} \int_R^{\alpha R} \int_{-L_n-\Delta z}^{-L_n+\Delta z} Y_t \frac{V}{r} \frac{\alpha^2}{\alpha^2-1} \frac{1}{\sqrt{3}} f \left|\frac{d^2 f}{dz^2}\right| 2\pi r dz dr \\
&= \lim_{\Delta z \to 0} \frac{1}{\pi R^2 V} \int_R^{\alpha R} \int_{-L_n-\Delta z}^{-L_n+\Delta z} \frac{Y_t}{\sqrt{3}} \frac{V}{r} \frac{\alpha^2}{\alpha^2-1} R \frac{a}{2\Delta z} 2\pi r dz dr \\
&= \frac{1}{\pi R^2 V} \int_R^{\alpha R} \frac{Y_t}{\sqrt{3}} \frac{V}{r} \frac{\alpha^2}{\alpha^2-1} Ra 2\pi r dr = \frac{1}{\pi R^2 V} \int_R^{\alpha R} \frac{Y_t}{\sqrt{3}} |v_r|_{z=-L_n} 2\pi r dr \\
&= \frac{2Y_t}{\sqrt{3}} \frac{\alpha^2}{\alpha^2-1} a(\alpha - 1),
\end{aligned}
$$

(12.34)

where we have noted that this case is the same as that of a general von Mises shearing surface that dissipates energy as $Y_t|\Delta \vec{v}|/\sqrt{3}$, where $\Delta \vec{v}$ is the change in velocity along the surface (the velocity normals to the surface from each side need to match, as they do here, with $v_z = -V/(\alpha^2 - 1)$ on both sides).

The conical flow is unlike the hemispherical or the ogival flow in that $v_r \neq 0$ on the back surface, and this behavior is not what we expect as experimentally the cavity opened by a nondeforming projectile in strong metals at speeds less than 1 km/s does not exceed the diameter of the projectile. The amount this surface contributes to the target resistance is shown in the last column of Table 12.4. It is seen that it is quite large and it grows more quickly than the volume dissipation because of the linear dependence on α in Eq. (12.34). The total resistance is larger than the resistance of the hemispherical nose, which implies this flow field is not representative of the flow around a conical-nosed projectile. It is interesting in the flow of Table 12.2 that the back edge does not influence the results for $b > 1$, as indicated in the table by the limit from the right $b = 1^+$. The value for $b = 1$ in the table includes the nonzero shearing surface at the back of the cone.

We can explicitly compute S_t^W, where the nose is conical with slope $a = R/L_n$, where $\beta = \text{atan}(R/L_n)$ is the half angle. Thus, $r = f(z) = -az$. The equivalent plastic strain rate is

$$
\dot\varepsilon_p = V \frac{\alpha^2}{\alpha^2-1} \frac{a^2}{\sqrt{3}} \frac{\sqrt{4z^2+r^2}}{r^2}.
$$

(12.35)

We integrate over the region of the flow field to obtain the penetration resistance due to strength,

$$S_t^W \big|_{vol} = \frac{\dot{W}}{V\pi R_p^2} = \frac{1}{V\pi R_p^2} \int_{-R_p/a}^{0} \int_{az}^{\alpha az} Y_t \, \dot{\varepsilon}_p 2\pi r \, dr \, dz$$

$$= \frac{\alpha^2}{\alpha^2 - 1} \frac{Y_t}{\sqrt{3}} \left\{ \sqrt{\alpha^2 a^2 + 4} - \sqrt{a^2 + 4} + 2\ln\left[\frac{(\sqrt{\alpha^2 a^2 + 4} - 2)(\sqrt{a^2 + 4} + 2)}{\alpha a^2} \right] \right\}.$$

$$(12.36)$$

It is possible to develop flow fields with more general forms of A_θ, which do a better job of matching the exit boundary conditions, but they are fairly tedious and it is not possible to exactly match the hemispherical outflow conditions if $f'(-L_n) \neq 0$ – that is, if there is a discontinuous slope at the back of the projectile. However, we will not pursue the topic further. Based on the values in the table, we move forward assuming a 7/3 multiplier on $Y_t \ln(\alpha)$ as providing the target resistance for conical noses.

12.3. Examples with the Walker–Anderson Model

Now that we have determined the target strength coefficient for target resistance due to target strength, we need to discuss the inertial term. In the derivation of the Walker–Anderson model, the target resistance along the centerline is (Eq. 10.5)

$$\sigma_{zz}(z_i) = -\rho_t \int_{z_i}^{+\infty} \frac{\partial v_z}{\partial t} dz + + \frac{1}{2}\rho_t \, v_z^2 \big|_{z_i}^{+\infty} + 2 \int_{z_i}^{+\infty} \frac{\partial \sigma_{xz}}{\partial x} dz. \qquad (12.37)$$

As discussed above, we no longer expect that the target material just off the centerline to be accelerated up to the penetration speed, since the material will fail and separate at the nose tip. If the velocity profile in the target is given by $v_z(z,t) = \bar{u}(t)f(z - z_i(t))$, where $f(0) = 1$ and $f(+\infty) = 0$, then with $\dot{z}_i = u = v$ being the penetration speed,

$$\sigma_{zz}(z_i) = -\left(\rho_t \bar{u}(u - \frac{1}{2}\bar{u}) + S_t \right) - \dot{\bar{u}} \int_0^\infty f(z)dz, \qquad (12.38)$$

where we expect $\bar{u} < u$ (for the hemispherical flow $\bar{u} = u$). The pointed-projectile flow fields of the previous section are not very helpful here, as they are not defined for $z > 0$. In that sense, they are not realistic, since we expect that information is communicated in front of the projectile and that there is motion in the target there (the penetration is not supersonic with respect to the metal wave speed). But if we were to assume that the magnitude of the speed at the nose tip from the flow field should represent the axial velocity of the target there, then we would have $\bar{u} \approx u\tan(\theta_n)$. However, this expression does not have the right asymptote for a hemispherical-nosed projectile, where $\bar{u} = u$. Perhaps a better estimate is a speed $\bar{u} = u\sin(\varphi)$. Using this, for the $L_n/R = 3$ cone the coefficient on $\rho_t u^2$ is 0.27, while for the CRH = 3 ogive the coefficient on $\rho_t u^2$ is 0.40. Clearly, though, the actual failure and separation of the material affects the material speed at the nose tip. Given the lack of a precise model, we use a coefficient of $\frac{1}{4}$ in modeling pointed projectiles, to contrast with the $\frac{1}{2}$ for hemispherical-nosed projectiles.

As examples, Fig. 12.2 shows results from the Walker–Anderson model for $L/D = 10$ rigid projectiles for thick penetration, where the target resistance is

TABLE 12.5. Al 6061-T6 target material properties.

ρ_0 (g/cm^3)	Y_t (MPa)	G (GPa)	c_0 (km/s)	k
2.71	380	27.3	5.35	1.34

modified as discussed above. Table 12.6 lists the results of the experiments shown in Fig. 12.2 performed by Forrestal, Okajima, and Luk [71] and Table 12.5 shows the material properties. Specifically, the resistances are

$$\text{Hemispherical}: \quad \frac{1}{2}\rho_t u^2 + \frac{7}{3}\ln(\alpha)Y_t \quad \text{(this curve is the same as Fig. 10.6)};$$

$$\text{Conical}, \ \alpha = 18.35°: \quad \frac{1}{4}\rho_t u^2 + \frac{7}{3}\ln(\alpha)Y_t;$$

$$\text{Ogive, CRH} = 3: \quad \frac{1}{4}\rho_t u^2 + 1.85\ln(\alpha)Y_t.$$

$$(12.39)$$

The curves in the figure show that these target resistances used in the Walker–Anderson model give good agreement with the data. There were two versions of the hemispherical-nosed projectiles in experiments, one with a reduced radius after the hemispherical nose but longer shaft (to make the same mass) to demonstrate that friction along the side of the projectile has little effect on penetration.

In the target resistance, $\alpha(u)$ is a decreasing function of penetration velocity while $\frac{1}{4}\rho_t u^2$ is an increasing function of penetration velocity. For the CRH $= 3$ ogive nose and for Al 6061-T6 properties in Table 12.5, the target resistance $\frac{1}{4}\rho_t u^2 + 1.85\ln(\alpha)Y_t$ lies between 1.62 and 1.66 GPa for speeds from 0 to 700 m/s (Fig. 12.1 right). For increasing speeds, the inertial term contributes more and the total resistance is 1.81, 2.39, and 3.38 GPa for 1, 1.5, and 2 km/s, respectively. Since the target resistance is almost constant at lower speeds, we use

$$\rho_p L \dot{u} = -(\frac{1}{4}\rho_t u^2 + 1.85 Y_t \ln(\alpha(u))) \approx 1.85 Y_t \ln(\alpha_0). \quad (12.40)$$

Recalling that $\dot{u} = du/dt = (du/dx)(dx/dt) = u\,du/dx$, integration yields

$$\frac{P}{L} = \frac{\rho_p V^2}{2 \cdot 1.85 \cdot Y_t \ln(\alpha_0)}, \quad (12.41)$$

where V is the striking velocity, and

$$V_{bl} = \sqrt{2 \cdot 1.85 \cdot Y_t \ln(\alpha_0) T/(\rho_p L)}, \quad \alpha_0 = \alpha(0) = \sqrt{\frac{2/Y_t + 1/K}{1/\mu + 1/K}}, \quad (12.42)$$

where T is the thickness of the target, L is the total initial length of the projectile, P is the penetration depth of the projectile, and the zero velocity limit of the cavity expansion α is from Eq. (6.54) with K and μ the target bulk and shear modulus. Here, the ballistic limit is a function of the square root of the strength times the thickness of the target. Writing the result in terms of the relative density,

$$V_{bl} = \sqrt{2 \cdot 1.85 \cdot (Y/\rho_t) \ln(\alpha_0)\chi}, \quad \chi = \frac{\rho_t T}{\rho_p L}. \quad (12.43)$$

Written this way, the ballistic limit scales with the square root of the relative areal density and target specific strength. When Eq. (12.41) is evaluated over a thickness

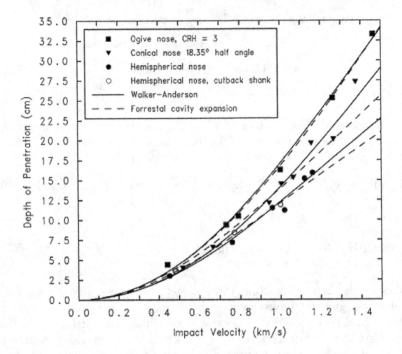

FIGURE 12.2. Data for projectiles with three nose shapes [71]; Walker–Anderson model results (solid line) where the coefficients on the target resistance terms are as shown in Eq. (12.39); results directly using the cavity expansion (next section, dashed line), including a friction term of $\mu = 0.15$ for the hemispherical-nosed projectiles, $\mu = 0.08$ for the conical-nosed projectiles, and $\mu = 0.03$ for the ogive-nosed projectile.

that does not bring the projectile to rest, some arithmetic yields the residual velocity as

$$V_r = \sqrt{V^2 - V_{bl}^2} \tag{12.44}$$

for the perforation problem. There is no failure strain in these equations because we assume the material separates and fails at the projectile nose; hence a failure strain may be reflected in the $\frac{1}{4}$ inertial coefficient.

To compare with data, we consider data compiled by Forrestal, Børvik, Warren, and Chen of fifteen ballistic limits of APM2 projectiles impacting finite-thickness plates made of five aluminum alloys [64]. They provide a uniaxial stress-strain curve for each alloy in the form

$$Y(\varepsilon) = \begin{cases} E\varepsilon & 0 \leq \varepsilon < Y_0/E \\ Y_0(E\varepsilon/Y_0)^n, & \varepsilon \geq Y_0/E. \end{cases} \tag{12.45}$$

To obtain a flow stress for our formula, we evaluate each at $\varepsilon = 0.20$ – that is, 20% strain. The mechanical properties are shown in Table 12.7.

The 7.62-mm APM2 (.30 cal armor piercing M2) is a 10.7 g round of length 3.53 cm and diameter 7.62 to 7.84 mm. It has a 5.25 g, sharp CRH = 3 ogive-nosed armor piercing Rc 63 steel core of length 2.74 cm and diameter from 5.88

TABLE 12.6. Hard steel projectiles impacting thick Al 6061-T651 [71].

Impact speed (m/s)	Penetration (cm)	Impact speed (m/s)	Penetration (cm)
Hemispherical-nosed, m_p=23.3 g, L=7.47 cm, $2R$=0.71 cm		Conical nose, m_p=23.7–0.1+0.4 g, $2R$=0.71 cm, L=8.18 cm, cone half angle β=18.35°	
450	3.0	510	4.0
760	7.2	660	6.6
960	11.5	940	12.1
1,020	11.2	1,000	14.5
1,120	15.2	1,060	15.3
1,160	15.9	1,150	19.6
		1,260	20.1
		1,370	27.3
Hemispherical-nosed with cutback shank, m_p=23.7+0.3 g, $2R$=0.71/0.64 cm, L=9.50 cm		Ogive nosed, m_p=24.8±0.1 g, $2R$=0.71 cm, L=8.29 cm, CRH=3	
480	3.6	440	4.4
770	8.4	730	9.4
1,000	11.9	790	10.5
		1,000	16.3
		1,260	25.3
		1,460	33.3

to 6.17 mm. The density of the steel core is 7.85 g/cm^3. There is variation in the round, but typically there is 0.78 g of lead in the nose, up to 0.50 g of lead in the base, and 4.2 g of gilding metal for the full metal jacket (i.e., the metal jacket encloses the round). Flow stress curves for the hard core and gilding metal jacket materials were presented in Chapter 3, though they are not used in this analysis. It will also be noted that we never use the mass of the core, since with our centerline momentum balance the length and density of the core enter in to the analysis, not the mass.

Table 12.8 and Fig. 12.3 shows the V_{bl} from the model for the APM2 core into the alloy, the experimental value, and then the core reduced by 1.03 (3%, discussed below) to adjust for the APM2 full bullet. This last value is then compared with the experimental result. As the table indicates, there is excellent agreement between the model and the experiments. (Difference = (Model V_{bl} − Exp.V_{bl})/Exp.V_{bl}.)

Additional data from Forrestal et al. describe ballistic limit velocity of the APM2 and the APM2 core against a specific aluminum at four obliquities [64]. The core-only experiments were performed with the APM2 core launched by itself in a plastic sabot. These results, for the 20 mm thick target, show that the ballistic limit is linearly dependent on the line-of-sight thickness (Table 12.9). Since the ballistic limit for normal impact tracks with the square root of thickness, we see that for oblique impact the performance is not tracking with line-of-sight thickness

TABLE 12.7. Properties of aluminum alloys used in APM2 ballistic limit experiments [64, 67].

Al alloy	ρ (g/cm^3)	E (GPa)	ν	K (GPa)	G (GPa)	Y_0 (MPa)	n	$Y(0.20)$ (MPa)
2139-T8	2.790	73	0.33	71.6	27.4	414	0.087	564
5083-H116	2.660	71	0.33	69.6	26.7	240	0.108	373
5083-H131	2.660	70	0.33	68.6	26.3	276	0.084	384
6061-T651	2.703	69	0.33	67.6	25.9	262	0.085	367
6082-T651	2.710	69	0.33	67.6	25.9	265	0.060	336
7075-T651	2.804	71	0.33	69.6	26.7	520	0.060	634

TABLE 12.8. Ballistic limit velocity of 7.62-mm caliber APM2 data [64] compared with APM2 core V_{bl} from Eq. (12.43).

Al alloy	T (mm)	Y (MPa)	α_0	Eq. (12.43) core V_{bl} (m/s)	Exp. full V_{bl} (m/s)	Eq. (12.43) core $V_{bl}/1.03$ (m/s)	Diff. (%)
5083-H116	20.0	373	10.19	546	492	530	7.7
5083-H116	40.0	373	10.19	772	722	749	3.8
5083-H116	60.0	373	10.19	945	912	918	0.6
5083-H131	26.0	384	9.97	628	588	610	3.7
5083-H131	37.8	384	9.97	758	712	736	3.3
5083-H131	50.9	384	9.97	879	876	854	−2.6
5083-H131	54.7	384	9.97	911	890	885	−0.6
5083-H131	57.2	384	9.97	932	927	905	−2.4
6061-T651	25.7	367	10.12	613	596	595	−0.2
6061-T651	26.0	367	10.12	616	583	598	2.6
6061-T651	38.8	367	10.12	753	754	731	−3.0
6061-T651	51.2	367	10.12	865	883	840	−4.9
6082-T651	20.0	336	10.58	522	501	507	1.2
7075-T651	20.0	634	7.82	670	628	650	3.6
7075-T651	40.0	634	7.82	947	909	920	1.2

and that it is more difficult to penetrate the armor at obliquity for a given thickness. Thus, the behavior is

$$V_{bl}(T,\gamma) = V_{bl}(T,0)/\cos(\gamma) = V_{bl}(T/\cos(\gamma),0)/\sqrt{\cos(\gamma)}. \qquad (12.46)$$

However, this behavior is likely just for this specific thickness divided by projectile diameter, as the entry and exit each have some effect in addition to the line-of-sight thickness effect. Also from the table it is seen that the core alone performance is nearly as much as the full APM2 round. These data are the origin of the 3% improvement for the full round as compared to the core alone that was used in the comparisons in this section.

Finally we show some deep penetration data. During perforation and computation of V_{bl}, we are not concerned about the entry or exit from the target because the irregularities of nose embedment during entry and exit cancel each other out. Also, we did not worry about the entry for the $L/D = 10$ cases, because of the

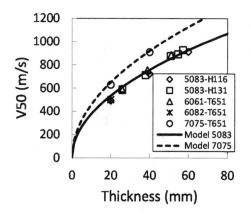
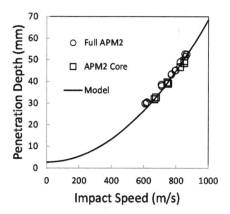

FIGURE 12.3. Left: V_{50} data compared to V_{bl} for the representative Al 5083-H131 and Al 7075-T651. Right: APM2 penetration into Al 6061-T6 compared model. (From Walker [196].)

TABLE 12.9. Ballistic limit velocity of 7.62-mm caliber APM2 and APM2 core striking 2.0-cm-thick Al 6082-T621 [64].

Obliquity γ		$0°$	$15°$	$30°$	$45°$
V_{bl} full APM2 bullet	(m/s)	501	516	580	718
V_{bl} APM2 core	(m/s)	514	535	597	723
(Core V_{bl}/full V_{bl}) − 1	(%)	2.59	3.68	2.93	0.70
Core V_{bl}/core $V_{bl}(0°)$		1	1.041	1.161	1.407
$1/\cos(\gamma)$		1	1.035	1.155	1.414

length of the projectile. For the APM2 and APM2 core, however, we need to address entry details because during the initial embedment only part of the nose tip is involved so the resistance is less. This effect was handled by increasing the depth of penetration by $0.27L_n$ for the APM2 core for full embedment, which is an added depth of $0.27 \cdot 3.32 \cdot (6.17/2)$ mm = 2.79 mm. Experiments were done with both the full APM2 as well as a round where the front end of the APM2 was removed so the tip of the core was exposed, while the rest of the jacket was left on to engage the rifling during launch. The experimental results in Table 12.10 from Chocron, Grosch, and Anderson [42] and in Fig. 12.3 (left) compare to

$$P = \frac{\rho_p V^2}{2 \cdot 1.85 \cdot Y \ln(\alpha_0)} L + 0.27L_n, \tag{12.47}$$

with comparisons for core and comparison with the 3% added depth adjustment.[1] Until the speed is high enough that the full nose of the core embeds at $P = L_n = 1.02$ cm, which is 340 m/s in this case, the curve in Fig. 12.3 (left) overpredicts the depth of penetration.

The left side of Fig. 12.4 shows Walker–Anderson results for the APM2 core and also shows the results where the entire APM2 projectile was modeled as two

---

[1]Perhaps the depth adjustment for the full APM2 should be 6% since the velocity is squared, and when done adds 3% to the difference error, which is still very good agreement.

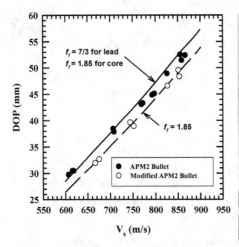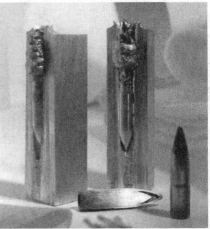

FIGURE 12.4. Comparison of analytical to experimental results for the modified and standard 7.62-mm APM2 bullets (left) and two sectioned aluminum targets struck by APM2s, showing the jacket material stripped and left behind during penetration of thick targets; also in the photograph are an APM2 and a sectioned APM2 (right).

TABLE 12.10. Depth of penetration results for 7.62-mm APM2 striking 6.3-cm thick Al 6061-T6 targets (data from [42]).

Full round						Modified round		
V_0 (m/s)	P (cm)	Diff. (%)	V_0 (m/s)	P (cm)	Diff. (%)	V_0 (m/s)	P (cm)	Diff. (%)
609	2.98	−6.1	794	4.50[a]	1.3	666	3.19[b]	0.2
618	3.00	−4.2	799	4.52	2.0	675	3.27[b]	0.2
620	3.05	−5.2	826	4.90	0.2	745	3.96[b]	−0.8
708	3.86[a]	−4.6	856	5.26	−0.2	751	3.90[b]	2.3
710	3.79	−2.3	859	5.15	2.6	826	4.67[a]	2.1
770	4.32	−0.4	866	5.25	2.2	851	4.85[a]	3.9
772	4.34[a]	−0.4				851	4.95	1.8

[a]Core was intact but fractured (observed in posttest examination)
[b]Core fracture status not recorded

materials, a lead nose with the 7/3 coefficient and, once the lead was eroded away, the rigid ogive-nosed steel core penetrating with a 1.85 coefficient [42]. The figure also shows two impacted targets after sectioning; it is clear the penetration process is complex, as the metal jacket has been stripped and left behind in these impacts. Also shown in the figure are a full AMP2 with its jacket and a sectioned APM2, showing the various parts (the lead at the nose melted out during the grinding process and is no longer present).

This reasoning can be extended to other materials. For steel target material and penetration speeds from 0 to 700 m/s, $\frac{1}{4}\rho_t u^2 + 1.85 \ln(\alpha) Y_t$ lies between 3.54 and

3.67 GPa for $Y_t = 800$ MPa; it lies between 4.22 and 4.31 GPa for $Y_t = 1$ GPa; and it lies between 4.87 and 4.97 GPa for $Y_t = 1.2$ GPa. Thus the constant deceleration approximation of Eq. (12.40) is valid, and the subsequent results for penetration and ballistic limit can be used.

We can extend the usefulness of the constant deceleration approximation to both the conical- and hemispherical-nosed projectiles by multiplying the constant target resistance $\frac{1}{4}\rho_t u^2 + 1.85 \ln(\alpha)Y_t$ by $(7/3)/1.85$ to give $0.315\rho_t u^2 + \frac{7}{3}\ln(\alpha)Y_t$, which is again roughly constant. For the conical-nosed projectile, this exceeds the target resistance by $0.065\rho_t u^2$ while for the hemispherical projectile it is less by $0.185\rho_t u^2$. Thus, we can solve for these cases with

$$\rho_p L \dot{u} = -(c_1 \rho_t u^2 + \frac{7}{3} Y_t \ln(\alpha_0)), \tag{12.48}$$

which can be integrated to give

$$\frac{P}{L} = \frac{\rho_p}{2c_1} \ln\left\{1 + \frac{c_1 \rho_t V^2}{(7/3)Y_t \ln(\alpha_0)}\right\} + c_2 \frac{L_n}{L},$$

$$V_{bl} = \sqrt{\frac{7}{3} Y_t \ln(\alpha_0)} \sqrt{\frac{1}{c_1 \rho_t}\left\{\exp\left(\frac{2c_1 \rho_t T}{\rho_p L}\right) - 1\right\}} = \sqrt{\frac{7}{3}\frac{Y_t}{\rho_t}\ln(\alpha_0)\frac{\exp(2c_1\chi) - 1}{c_1}},$$

$$V_r = \exp\left(-c_1 \frac{\rho_t T}{\rho_p L}\right)\sqrt{V^2 - V_{bl}^2} = \exp(-c_1\chi)\sqrt{V^2 - V_{bl}^2},$$

$$\tag{12.49}$$

where $c_1 = -0.065$ for a conical-nosed projectile and $c_1 = 0.185$ for a hemispherical-nosed projectile, and $c_2 < 1$ is the nose embedment adjustment, reflecting the lack of entry/exit symmetry on the penetration event, which was 0.27 for the CRH $= 3$ ogive. These explicit formulas should be used for less than 1 km/s. Going higher, one should use the $\alpha(u)$ form of the target resistance in the Walker–Anderson model.

12.4. Direct Use of the Cavity Expansion

Another approach to determining the target resistance is to directly use the cavity expansion, as we did for the hemispherical-nosed projectile in Eq. (8.58). This approach has been widely employed by Forrestal and colleagues and as an example we cite Forrestal, Okajima, and Luk [71].

In our cavity expansion solution in Chapter 6, we developed the following form for the normal stress on the cavity wall,

$$\tilde{\sigma}_n = B_i + \rho_t c_{0t}^{2-p} u^p, \tag{12.50}$$

where B_i and p are constants, c_0 is the bulk sound speed, and u is the cavity expansion velocity. There are two approaches to directly applying the cavity expansion result to the nondeforming projectile penetration problem. One is to view the target as made of thin slices of material that expand radially from the axis of symmetry, and thus can be modeled using the cylindrical cavity expansion. As the nose of the projectile pushes into the target, it pushes material away normal to the central axis of symmetry and cylindrical cavity expansion stress is integrated along the nose of the projectile. We will write the traction on the surface from the cylindrical cavity expansion as $t_r(v_r)$, where the r is the cylindrical radial direction. The second approach is to assume that the projectile pushes material in a direction normal to the surface of the nose and to integrate the spherical cavity expansion stress over the nose of the projectile. We will write the traction from the spherical cavity expansion as $t_n(v_n)$, where n means normal to the surface. In both these techniques, the force is not exactly right in that there is slippage between the layers that is not included, nor is the deformation in the target exactly that assumed by the two forms of cavity expansion.

The nose shape is given by $r = f(z)$. In this section we have a different assumption about the location of the nose, namely that it has been translated $+L_n$ in the z direction, running from the nose tip at $z = L_n$ back to the base at $z = 0$, where it connects to the shaft of the projectile. Hence $f(0) = R$, $f(L_n) = 0$, $f' = df/dz < 0$, the surface area increment is

$$dS = 2\pi r d\ell = 2\pi f \sqrt{dr^2 + dz^2} = 2\pi f \sqrt{1 + f'^2}\, dz, \qquad (12.51)$$

and the surface normal in the r–z plane is

$$\vec{n} = \frac{1}{f\sqrt{1 + f'^2}}(f\hat{e}_r - f f' \hat{e}_z). \qquad (12.52)$$

We define θ to be the angle between the axial axis $\hat{e}_z$ and $\vec{n}$, so $\cos(\theta) = \vec{n} \cdot \hat{e}_z = -f'/\sqrt{1 + f'^2}$. We see that $\sin(\theta) = \vec{n} \cdot \hat{e}_r = 1/\sqrt{1 + f'^2}$ and $\tan(\theta) = -1/f'$.

The projectile moves along the z-axis at a velocity $v_z = v$. The component of the velocity in the surface normal direction is

$$v_n = v\hat{e}_z \cdot \vec{n} = v\cos(\theta). \qquad (12.53)$$

This value is used for the spherical cavity expansion. The velocity component in the radial r direction for cylindrical cavity expansion can be found by letting $\alpha \to \infty$ in our potential flow of Eq. (12.27), yielding

$$\vec{v} = -v\frac{1}{\xi}\frac{df}{dz}\hat{e}_r = -v\frac{f}{r}\frac{df}{dz}\hat{e}_r. \qquad (12.54)$$

Thus, the radial speed along the nose ($\xi = 1$) is

$$v_r(z) = v\cot(\theta). \qquad (12.55)$$

To determine the force exerted in the z direction, we integrate over the whole surface. We need the z component of the traction for the z force. For the spherical cavity expansion, we are using wedges of our hemispherical flow field; the traction is given by $t_z = \cos(\theta)t_n$ and the force is given by

$$F_z = \int_S t_z(v_n)dS = \int_S \cos(\theta)t_n(v_n)dS = \int_S \cos(\theta)t_n(v\cos(\theta))dS, \qquad (12.56)$$

where the traction is a function of the appropriate velocity. For the cylindrical expansion, we determine the force through a work rate argument. Initially we assume a conical nose. Consider a penetration depth ΔP. If the conical nose is completely embedded in the target then the work to move thin sheets with a radial cavity expansion is (the radial speed v_r is constant for the cone)

$$W_{\Delta P} = \left(\int_0^R t_r 2\pi r dr\right)\Delta P = \pi R^2 t_r \Delta P. \qquad (12.57)$$

The work can also be computed in the axial direction by integrating over the conical surface, where $R/L_n = \cot(\theta)$ and $1 + \tan^2(\theta) = 1/\cos^2(\theta)$,

$$W_{\Delta P} = F_z \Delta P = \left(\int_S t_z dS \right) \Delta P = t_z \pi R \sqrt{R^2 + L_n^2} = t_z \pi R^2 \frac{1}{\cos(\theta)} \Delta P. \quad (12.58)$$

Setting these two expressions equal to each other gives $t_z = \cos(\theta)t_r$. The force for the cylindrical cavity expansion is then

$$F_z = \int_S t_z(v_r) dS = \int_S \cos(\theta)t_r(v_r) dS = \int_S \cos(\theta)t_r(v \cot(\theta)) dS. \quad (12.59)$$

Thus, the two cavity expansions produce essentially the same form for the force, the only difference being the specific stress of the cavity expansion and the expansion velocity.

Both the cylindrical and spherical cavity expansion stress expressions begin with a constant B (i.e., independent of expansion speed), and as the expression for a constant is the same in Eqs. (12.56) and (12.59), we integrate that first,

$$\int_S B \cos(\theta) dS = 2\pi B \int_0^{L_n} (-ff') \, dz = \pi f^2(0) B = \pi R^2 B. \quad (12.60)$$

This is a surprising result: the integrated constant part of the traction term in Eqs. (12.56) and (12.59) is independent of nose shape. (In fact, the nose shape did not enter into the computation in Eq. 12.57.) The result is a concern because we expect $B_2 \leq B \leq B_3$ and we know that $B_2 < B_3 < S_t$ at low speeds, in particular, at low speeds,

$$\frac{B_2}{S_t} \approx \frac{3}{4}\frac{B_3}{S_t} < \frac{B_3}{S_t} = \frac{\frac{2}{3}Y(1+3\ln(\alpha_0))}{\frac{7}{3}Y\ln(\alpha)} = \frac{2}{7}\frac{1+3\ln(\alpha_{S0})}{\ln(\alpha_{C0})} \approx \frac{2}{7}\frac{1+3\ln(5)}{\ln(10)} \approx \frac{3}{4}, \quad (12.61)$$

where S refers to spherical cavity expansion (for B_3) and C refers to the cylindrical cavity expansion (for S_t). The Walker–Anderson model gave good predictions for the penetration of a hemispherical-nosed projectile, so we suspect the target resistance from directly using the cavity expansion will be too low, which is the case. This problem is addressed in [**71**] by including friction between the projectile and the target, proportional to the normal traction with a friction coefficient μ. In particular, the tangential traction due to friction is given by $t_t = \mu t_n$, which will have a component in the z direction of $t_z = \sin(\theta)t_t$. The frictional force in the z direction is

$$F_z^\mu = \mu \int_S \sin(\theta)t_n(v_n) dS = \mu \int_S \sin(\theta)t_n(v \cos(\theta)) dS,$$

$$F_z^\mu = \mu \int_S \sin(\theta)t_n(v_r) dS = \mu \int_S \sin(\theta)t_r(v_r) dS = \mu \int_S \sin(\theta)t_r(v \cot(\theta)) dS, \quad (12.62)$$

for the spherical cavity expansion and cylindrical cavity expansion expressions, respectively. The friction force affects both the constant B term in the force as well as the velocity term.

We are now in a position to integrate velocity term of Eq. (12.56) over the nose. Initially we use the spherical cavity expansion, where $p = 2$, and some tedious computations yield[2]

Hemi: $F_z/\pi R^2 = \left(1 + \dfrac{\mu\pi}{2}\right) B_3 + \dfrac{1}{2}\left(1 + \dfrac{\mu\pi}{4}\right)\rho_t v^2;$

Blunt: $F_z/\pi R^2 = B_3 + \rho_t v^2$ (no friction included);

Ogive: $F_z/\pi R^2 = \left(1 + \mu\{4\beta C^2 - (2C-1)\sqrt{4C-1}\}\right) B_3$

$$+ \left(\frac{8C-1}{24C^2} + \mu\left\{\beta C^2 - \frac{(2C-1)(6C^2 + 4C - 1)\sqrt{4C-1}}{24C^2}\right\}\right)\rho_t v^2;$$

Cone: $F_z/\pi R^2 = (1 + \mu\cot(\beta)) B_3 + (1 + \mu\cot(\beta))\sin^2(\beta)\rho_t v^2,$

$$(12.63)$$

where the ogive CRH $= C = R_h/2R$ and nose-tip half angle $\beta = \cos^{-1}(1 - 1/(2C))$, and for the cone β is the half angle.

For the same cone with the conical cavity expansion, where $p = 1.65$, we obtain[3]

Cone: $F_z/\pi R^2 = (1 + \mu\cot(\beta))B_2 + (1 + \mu\cot(\beta))\tan^p(\beta)\rho_t c_{0t}^{2-p} v^p.$　　(12.64)

It is clear that the spherical and conical terms give different answers for the penetration force. Better agreement for the thick target data in Table 12.6 is achieved with the spherical cavity expansion, and that is what is shown below.

Given the functional expression for the force, it is now possible to integrate to determine the penetration or the residual velocity. For thin targets, shallow penetration, or low L/D impactors, the details of the initial penetration or breakout matter (e.g., there is a penetration regime before the nose of the projectile is fully embedded in the target, where the resisting force is not the total force). Newton's equation of motion is

$$m_p \frac{dv}{dt} = -F_z(v). \qquad (12.65)$$

---

[2]For the blunt nose we simply integrate over the flat front of the projectile with $\theta = 0$. For the ogive nose, in the r–z plane the nose shape is given by the arc of a circle with radius R_h centered at $(r, z) = (R - R_h, 0)$. This arc goes from the base of the ogive nose at $(R, 0)$, where it is tangent to the cylindrical body of the projectile, through the nose tip at $(0, L_n)$. Thinking of the larger circle centered at $(r, z) = (R - R_h, 0)$, the angle θ between the line $r = R - R_h$ (which points in the $\hat{e}_z$ direction) and a given point on the nose surface equals the angle θ that defines the angle between the surface normal and the z-axis. (This θ is not the spherical coordinate system θ unless $R_h = R$.) This allows for the relatively simple $dA = 2\pi r dz/\sin(\theta)$ with $r = R - R_h + R_h \sin(\theta)$ and $z = R_h \cos(\theta)$, hence $dz = -R_h \sin(\theta)d\theta$ and $dA = -2\pi r R_h d\theta$. With CRH $= C = R_h/2R$, the integration goes from the nose base $\theta = \pi/2$ to the nose tip $\theta = \theta_n$, where $\sin(\theta_n) = 1 - R/R_h = 1 - 1/2C$ (i.e., where $r = 0$; notice $r = R_h(\sin\theta - \sin\theta_n)$). The nose-tip half angle $\beta = \pi/2 - \theta_n$. The integral is performed using these relations and completed using trigonometric identities, or by noting that $\cos(\theta_n) = (1/2C)\sqrt{4C-1}$, yielding a nose length for the ogive of $L_n = R_h \cos(\theta_n) = R\sqrt{4C-1}$.

[3]We explicitly mention here, as was stated in Chapter 6, that for a cylindrical cavity expansion Forrestal uses a fit to $p = 2$ rather than our best fit of $p = 1.65$. When $p = 2$ it is possible to integrate the ogive nose in closed form, and he obtains $F_z/\pi R^2 = B_2 + B_C\left\{8C^2\ln\left(\frac{C}{C-1/2}\right) - 4C - 1\right\}\rho_t v^2$, where $B_C = 1.837$. In a similar fashion, Forrestal multiplies the spherical velocity term by $B_S = 1.041$ based on his curve fit to the cavity expansion for Al 6061-T651 with $Y = 400$ MPa, $K = 68.95$ GPa, and $\rho = 2.71$ g/cm^3. It turns out that for penetration examples the two formulations for the cylindrical cavity expansion yield similar results.

With $\dot{v} = (dv/dz)(dz/dt) = v\,dv/dz$, integration yields

$$P = m_p \int_V^0 \frac{v}{F_z(v)}\,dv \qquad (12.66)$$

for a final depth of penetration or

$$m_p \int_V^{V_r} \frac{v}{F_z(v)}\,dv = T \qquad (12.67)$$

for a perforated target. For $F_z/\pi R^2 = \tilde{B} + \tilde{A}\rho_t v^2$, the integral can be explicitly integrated to give

$$\frac{P}{\tilde{L}} = \rho_p \int_0^V \frac{v}{\tilde{B} + \tilde{A}\rho_t v^2}\,dv = \frac{\rho_p}{2\rho_t \tilde{A}} \ln\left(1 + \frac{\tilde{A}}{\tilde{B}}\rho_t V^2\right), \qquad \text{where} \quad \tilde{L} = \frac{m_p}{\pi R^2}. \qquad (12.68)$$

$\tilde{L}$ is the approximate length of the projectile. Typically $\tilde{L}$ is lightly less than L to account for the nose mass being less than the corresponding right-circular cylinder. (Note the distinction with the Walker–Anderson model, which is based on the centerline momentum balance; hence the centerline length L is used. In the direct cavity expansion approach, a force over the nose of the projectile is computed and that decelerates the mass of the projectile.). The perforation result is

$$\frac{T}{\tilde{L}} = \rho_p \int_{V_r}^V \frac{v}{\tilde{B} + \tilde{A}\rho_t v^2}\,dv = \frac{\rho_p}{2\rho_t \tilde{A}} \ln\left(\frac{\tilde{B} + \tilde{A}\rho_t V^2}{\tilde{B} + \tilde{A}\rho_t V_r^2}\right). \qquad (12.69)$$

When the residual speed is set equal to zero the ballistic limit results, yielding

$$V_{bl} = \left[\frac{\tilde{B}}{\rho_t \tilde{A}}\left\{\exp\left(\frac{2\rho_t \tilde{A}}{\rho_p}\frac{T}{\tilde{L}}\right) - 1\right\}\right]^{1/2} \qquad (12.70)$$

and

$$V_r = \sqrt{\frac{\tilde{B}}{\tilde{B} + \tilde{A}\rho_t V_{bl}^2}}\sqrt{V^2 - V_{bl}^2}. \qquad (12.71)$$

Thus, the residual velocity in terms of the striking velocity is in the Lambert form with a power of 2.

Figure 12.2 shows comparisons to data for thick penetration. Table 12.6 provides the data [71]. Two versions of the hemispherical-nosed projectiles were employed in experiments, one with a reduced radius but longer shaft (to make the same mass), which demonstrated that friction along the side of the projectile does not affect penetration. The friction coefficients were adjusted to give the best agreement with the data; they are $\mu = 0.15$ for the hemispherical-nosed projectiles, $\mu = 0.08$ for the conical-nosed projectiles, and $\mu = 0.03$ for the ogive-nosed projectiles. The specific forces for the results in the figure are

Hemispherical nose, $\mu = 0.15$: $F_z/\pi R^2 = 1.24B_3 + 0.59\rho_t v^2$;

Conical nose, half angle $\beta = 18.35°$, $\mu = 0.08$: $F_z/\pi R^2 = 1.24B_3 + 0.123\rho_t v^2$;

Ogive nose, CRH $= 3$, $\beta = 33.6°$, $\mu = 0.03$: $F_z/\pi R^2 = 1.135B_3 + 0.115\rho_t v^2$.

$$(12.72)$$

The values for the Al 6061-T6 target are given in Table 12.5. It is clear that the approach is able to yield good agreement with the data, the only anomaly being that

TABLE 12.11. Transition velocity for impacts of projectiles into Al 6061 targets (last row is impact of 7.62 mm APM2 into RHA).

Nose shape	Constant multiplying $\rho_t v^2$ term	Y_p (GPa)	Computed transition velocity (m/s)	Experimental transition velocity (m/s)	Reference
Blunt	1	1.3	690	650	Fig. 10.10
Hemispherical	0.5	1.5	1,050	1,200	Fig. 12.2
Ogive	0.25	1.5	1,490	1,500	Fig. 7.1
Ogive	0.25	2.25	1,820	1,750	Fig. 7.1
Ogive	0.25	2.5	1,920	> 860	Fig. 12.4
Ogive (into steel)	0.25	2.5	1,130	> 850	(SwRI)

all the methods fail to match the lowest data point for the ogive-nosed projectile, where specifics of embedment matter.

12.5. Projectile Eroding-Noneroding Transition Velocity

Given the perspective of forces being applied all around the surface of the nose, it is possible to present a heuristic argument for when erosion might begin – that is, the velocity at which the projectile transitions from nondeforming penetration to eroding penetration. The argument goes as follows. In the context of the cavity expansion model, we have the stress at the surface of the cavity as $\sigma_n = B_i + \rho_t u^2$. During penetration, this is applied as a traction normal to the surface of the nose. The velocity term varies depending on the direction of the surface normal compared to the penetration direction. The B_i term does not vary over the surface of the nose and hence it gives rise to a uniform spherical stress (pressure) in the projectile nose. Since yielding is independent of pressure, only the u^2 term needs to be considered. Thus we conjecture that the eroding-noneroding transition speed is when

$$\text{constant} \cdot \rho_t v^2 \approx Y_p \quad \Rightarrow \quad V_{\text{transition}} = \sqrt{\frac{Y_p}{\text{constant} \cdot \rho_t}}, \qquad (12.73)$$

where the constant is the appropriate one for the nose shape. Based on our discussion for pointed flow fields, we used a constant of $\frac{1}{4}$ for pointed projectiles, $\frac{1}{2}$ for hemispherical-nosed projectiles, and 1 for blunt projectiles. As examples, Table 12.11 shows the results for impacts into Al 6061 described in other places in the book, where the projectile material flow stresses are estimates based upon yield values. The target must be softer than the projectile for this argument to work – that is, at low speeds the projectile is assumed to penetrate the target in a nondeforming fashion. It is surprising that the strength of the target material has no effect (based on this argument).

12.6. Sources

The direct use of the cavity expansion in Section 12.4 is due to Forrestal, Okajima, and Luk [71]; the rest of the chapter is new, with some pointed projectile material appearing in Walker [196].

Exercises

(12.1) The simplest model for rigid penetration of blunt impactors is to assume that a plug of the same diameter as the impactor forms and shears out with the von Mises shearing stress $Y/\sqrt{3}$. For a plate of thickness T with penetration $P(t)$ and speed u show that the target resistance, assuming the plug has the same speed as the projectile, is

$$S_t^W = \frac{2(T - P(t))Y_t}{\sqrt{3}R}, \qquad 0 \le P(t) \le T. \tag{12.74}$$

Assuming the initial momentum in the projectile and plug equals the initial momentum in the projectile before striking, find the ballistic limit velocity and the residual velocity. Why does failure strain not enter into the expressions? Compare the result to the data in Table 12.1.

(12.2) We use the centerline momentum balance in the Walker–Anderson model and hence always use the total projectile length in the above comparisons, except that we did not do so for the cut back shaft data. This contrasts with the direct cavity expansion models, where we worked with the projectile mass. To understand where the "constant mass" projectile arises in the centerline momentum balance, consider the static problem of two long cylinders with a common axis whose ends are touching and are loaded at the two opposing far ends. Show that the "correction" term arises from the shear derivative term. Based on the solution to Boussinesq's problem (Section 8.10), over what length scale would you expect the shear stress terms to be important? (Hint: if a force F_z is applied at a distance to the two far ends of the cylinders, the stress at each end will be $\sigma_{zz} = F_z/A = F_z/\pi R^2$. The static centerline momentum balance gives

$$\sigma_{zz}(z_+) - \sigma_{zz}(z_-) + \int_{z_-}^{z_+} 2\frac{\partial \sigma_{xz}}{\partial x}dz = \frac{F_z}{\pi}\left(\frac{1}{R_+^2} - \frac{1}{R_-^2}\right) + \int_{z_-}^{z_+} 2\frac{\partial \sigma_{xz}}{\partial x}dz = 0, \tag{12.75}$$

so this shear term and $F_z = m_p dv/dt$ can be inserted in the centerline momentum balance.)

(12.3) Write down the velocity field for the $b = \infty$ case for the ξ in Table 12.2. Show that the velocity vector component perpendicular to the $r = z$ surface matches for the flow field above and below the $r = z$ line. Determine the difference of the velocity component parallel to the surface and from it determine the shear dissipation across the surface. Determine the dissipation in the regions. Sum these terms and show that you agree with the S_t^W entry in the table and the exact value given in the corresponding text (which list the coefficients).

(12.4) Show that the streamlines for the pointed projectile are given by $r^2 = f^2(z) + (1 - 1/\alpha^2)r_0^2$ for $-L_n < z < f^{-1}(r_0/\alpha)$, where $r = r_0$ is the value before interacting with the flow field. (Hint: see how streamlines were computed for the hemispherical nose in Chapter 10.)

(12.5) Compare the difference between the CRH $= 3$ ogive and the hemispherical-nosed penetration resistances for the Walker–Anderson model and the direct use of the cavity expansion for speeds of 0 to 1.5 km/s into an aluminum. Similarly, compare the resistances for an $L_n/R = 3$ conical-nosed penetrator over the same speed range.

TABLE 12.12. Ballistic limit velocity of 12.7-mm caliber APM2 data [67] and difference with APM2 core V_{bl} from Eq. (12.43)/1.03.

Al alloy	T (mm)	Exp. full V_{bl} (m/s)	Diff. (%)	Al alloy	T (mm)	Exp. full V_{bl} (m/s)	Diff. (%)
5083-H131	51.4	656	−0.7	2139-T8	39.0	657	0.8
5083-H131	57.2	703	−2.2	2139-T8	39.9	667	0.4
5083-H131	63.7	735	−1.3	2139-T8	40.9	677	0.1
5083-H131	70.5	786	−2.9	2139-T8	52.1	785	−2.5
5083-H131	76.4	831	−4.4	2139-T8	52.1	790	−3.2
6061-T651	50.9	667	−4.7	2139-T8	53.8	796	−2.3
6061-T651	57.7	716	−5.4	2139-T8	57.2	819	−2.1
6061-T651	64.3	749	−4.6	2139-T8	64.1	873	−2.8
6061-T651	77.0	846	−7.5				

(12.6) Plot the conical penetration forces from Eq. (12.63) for β, ranging from 0 to 90° for 500 m/s for an aluminum target. Find the nose cone angle at which it matches the hemispherical-nosed penetrator resistance.

(12.7) Forrestal, Lim, and Chen [67] provide ballistic limit data for the 12.7 mm APM2 (i.e., the .50 cal APM2). Table 12.12 contains the experimental data. Table 12.7 has the material property data. The projectile data they provide are full projectile length, diameter, and mass values of 58.7 mm, 12.98 mm, and 45.9 g, respectively, and for the steel core are 47.5 mm, 10.9 mm, and 25.9 g, respectively. Complete the same table as in Table 12.8 for these data (i.e., produce all the same columns). (Ans: for Al 2139-T8, $\alpha_0 = 8.40$; the differences between the experimental data and the core V_{bl} adjusted by 3% are given, as a check.)

(12.8) The data in Table 12.13 reported in Jones et al. [113] are for two calibers of APM2 ammunition and three calibers of fragment simulating projectiles (FSPs) striking RHA steel, an aluminum alloy, and a magnesium alloy with two heat treats (AZ31B). Given these data: 1) Does this data scale? That is, is the performance increase at larger scales what we would expect based on the smaller scales? 2) Can one argue that all strong metals perform the same? How do specific strengths compare, for example (and perhaps consider titanium)? 3) There is really only one outlier in this data set; identify and discuss it.

   The reported densities are 7.83, 2.66, and 1.77 g/cm^3 for the RHA, Al 5083 and AZ31B, respectively. AZ31B-O (hot rolled and fully annealed) is reported to have a yield of 152 ± 1 MPa and an ultimate of 256 ± 2 MPa, with an elongation of 21.5 or 11.5% for a 7.6 mm or 31.5 mm plate, and the AZ31B-H24 (cold rolled and partially annealed) is reported to have a yield of 174 ± 5 MPa and an ultimate of 263 ± 2 MPa, with an elongation of 19.0 or 9.5% for a 7.8 mm or 76.5 mm plate, respectively.

(12.9) Using the Walker–Anderson ogive-nosed projectile penetration model, compare to the APM2 data in Table 12.13.

TABLE 12.13. Ballistic limits for RHA, Al 5083-H131 and a Magnesium Alloy (AZ31B) with two tempers struck by APM2 and FSP projectiles [113].

Target material	Thickness (mm)	Areal den. (kg/m^2)	Proj. diam. (mm)	V_{50} (m/s)
Projectile type			APM2	
RHA	7.1			524
Al 5083-H131	21.0	55.7	7.62	506
AZ31B-O	31.5			511
RHA	17.2			649
Al 5083-H131	50.9	135	12.7	626
AZ31B-O	76.5			649
RHA	17.2			914
Al 5083-H131	50.9	135	7.62	853
AZ31B-H24	76.5			856
Projectile type			FSP	
RHA	1.8			366
Al 5083-H131	5.2	13.7	5.59	396
AZ31B-O	7.6			417
AZ31B-H24	7.8			421
RHA	7.1			718
Al 5083-H131	21.0	55.7	12.7	663
AZ31B-O	31.5			639
RHA	17.2			878
Al 5083-H131	50.9	135	20.0	1,125
AZ31B-H24	76.5			897

FIGURE 12.5. Fragment simulating projectile (FSP).

(12.10) Use the explicit formula for the hemispherical nose to compare with the fragment simulating projectile data in Table 12.13. Now, if you have programmed up the Walker–Anderson model, see what results you get with that model. For the latter a failure strain is required. How do they compare? (FSPs have $L/D = 1.17$ with a beveled asymmetric nose, which is a flat centered strip of width $0.46D$ from one side to the other and drops off with a 35° slope flat bevel to either side. Made of steel, the 5.59-mm-diameter

FSP has mass 1.1 g and Rc 24–30, the 12.7-mm FSP has mass 13.4 g and Rc 29–31, and the 20-mm FSP has mass 53.8 g and Rc 29–31 [**113**]. See Fig. 12.5.)

(12.11) Based on the model described in this chapter for ogive-nosed projectiles (Eq. 12.43), suppose you had the ballistic limit for a given material armor thickness for a given length and density projectile. Write down an expression for V_{bl} for different target thicknesses and different (assumed geometrically similar) length projectiles. Pick a few examples from Tables 12.8 and 12.12 to show your result agrees with those data.

(12.12) Show that the volume of an ogive nose with CRH $= C$ is

$$vol = \pi R^3 \left\{ \frac{1}{3}\sqrt{4C-1}(12C^2 - 4C + 1) - 4\beta C^2(2C - 1) \right\}. \tag{12.76}$$

For extra credit, confirm the footnote expression for the penetration resistance of a tangent ogive for a cylindrical cavity expansion.

(12.13) Show that if the spherical cavity expansion radius were used in $S_t = \frac{7}{3}\ln(\alpha)Y_t$ instead of the cylindrical cavity expansion radius, then

$$1.07 > \frac{B_3}{S_t(\alpha_S)} \approx \frac{2}{7}\frac{1 + 3\ln(\alpha_S)}{\ln(\alpha_S)} > 0.96 \quad \text{for} \quad 4 \le \alpha_S \le 15. \tag{12.77}$$

Thus, the choice to use the cylindrical cavity expansion in the Walker–Anderson model is important to the model's success; otherwise the target resistance would not be large enough, as is the case for the direct use of the cavity expansion.

(12.14) Show that if $P(t)$ is the current depth of penetration, then the derivative of the specific kinetic energy of a nondeforming projectile with respect to penetration depth is the projectile acceleration – that is, $d(\frac{1}{2}v^2)/dP(t) = d^2P(t)/dt^2$.

(12.15) An exercise for this chapter has been placed at the end of Chapter 10: Exercise 10.30 (for space reasons).

(12.16) (Exercise from Appendix A) Show that the γ_{ij}^k can be determined from the Γ_{ij}^k as follows:

$$\frac{\partial \hat{e}_{y_j}}{\partial y_i} = \gamma_{ij}^k \hat{e}_{y_k} = \frac{\partial}{\partial y^i}\left(\frac{1}{h_j}\frac{\partial}{\partial y^j}\right) = -\frac{1}{h_j^2}\frac{\partial h_j}{\partial y_i}\frac{\partial}{\partial y^j} + \frac{1}{h_j}\frac{\partial}{\partial y^i}\left(\frac{\partial}{\partial y^j}\right) \quad \text{(no sum on } j)$$

$$= -\frac{1}{h_j^2}\frac{\partial h_j}{\partial y_i}\frac{\partial}{\partial y^j} + \frac{1}{h_j}\Gamma_{ij}^k\frac{\partial}{\partial y^k} = \sum_k\left(-\frac{1}{h_j}\frac{\partial h_j}{\partial y_i}\delta_{jk} + \frac{h_k}{h_j}\Gamma_{ij}^k\right)\hat{e}_{y_k} \quad \text{(no sum on } j)$$

$$\Rightarrow \qquad \gamma_{ij}^k = -\frac{1}{h_j}\frac{\partial h_j}{\partial y_i}\delta_{jk} + \frac{h_k}{h_j}\Gamma_{ij}^k \quad \text{(no sum on } j, k). \tag{12.78}$$

Use the specific results for orthonormal bases in Eq. (A.245) to confirm this yields Eq. (A.54).

(12.17) (Exercise from Appendix A, for those knowing about differential forms) From Cartan's moving frame approach, with the 1-form $\omega_j^k = \gamma_{ij}^k dy^i$ (see footnote page 617) show that ω is skew symmetric and show the Euclidean structural equation $d\omega_j^k + \omega_i^k \wedge \omega_j^i = \Omega_j^k = 0$ gives the six equations in Eq. (A.300), up to multiplicative h_i terms. The 2-form $\Omega_j^k = -\frac{1}{h_j h_k}\sum_{i<m} R_{jkim}\, dy^i \wedge dy^m$ (no sum on j, k) is the curvature, the obstruction to being flat (Euclidean). (The dual forms: $\theta^i = h_i dy^i$ (no sum on i), so $\theta^i(\hat{e}_{y_j}) = \delta_j^i$ and $d\theta^i = -\omega_k^i \wedge \theta^k$.)

Yarns, Fabrics, and Fiber-Based Composites

Fabrics are an extremely important element of body armors and other armors. Understanding fabrics requires understanding how yarns deform. In this chapter, we explicitly solve the problem of a single yarn being struck transversely and then develop a model for the ballistic response of fabric sheets. The model addresses the effect of the anisotropy of a plain-woven fabric sheet. We then examine an extension of the ideas to fiber-based composite panels. Fabrics and fiber-based composites perform well ballistically due to large deflections, which allow the yarns to decelerate the projectile in tension, the mode of deformation where a fiber has its best mechanical performance.

Fibers are long, thin molecules bound in bundles. Fibers of interest for ballistic applications, such as Kevlars® and ultra-high-molecular-weight polyethylene (Spectra®/Dyneema®), typically behave in a linear elastic fashion until they break at a failure stress σ_f and a failure strain ε_f. Fibers are bundled together into yarns. Yarns are woven to make fabric sheets. Yarns are described in terms of denier, a unit of mass per length of the yarn. Specifically, denier is the number of grams of yarn per 9,000 m. Fibers of interest for ballistics typically are in yarns that have deniers on the order of 1,000. There are three material properties typically reported for yarns or fibers: Young's modulus, breaking strength, and breaking strain. These various properties are shown in Table 13.1 for a variety of fibers. We will use our standard units for a yarn's Young's modulus and strength, such as pascals or dyne/cm^2. However, in the textile industry the modulus and strength are sometimes reported differently; modulus is referred to as initial modulus and strength as tenacity. They are often reported in units of grams per denier (gpd). For example, a test records the hanging mass required to break the yarn at the surface of the Earth, and this value is divided by the yarn's denier to produce the tenacity. The stress in the fibers in the yarn is σ, the cross-sectional area of the fibers of the yarn is A, the hanging test mass is m_t, and the gravitational acceleration is g, so that $\sigma A = m_t g$. The fiber density is $\rho_f = \text{denier}/(AL_d)$, where $L_d = 9,000$ m is the defining length of denier. Thus, the stress is

$$\sigma = \frac{m_t g}{A} = \frac{g \cdot \text{denier}}{A} \frac{m_t}{\text{denier}} = g L_d \rho_f \frac{m_t}{\text{denier}}. \tag{13.1}$$

The term m_t/denier is the gpd value, and to obtain stress in typical stress units we multiply the gpd by $g L_d \rho_f$. In particular, to get GPa we multiply gpd by $0.08826\rho_f$, where ρ_f is the fiber density in g/cm^3. We also mention that sometimes the mass of the yarn per length is measured in tex, which is the mass of the yarn in grams per 1,000 m, so 1 tex = 9 denier, and with this unit the force to break is measured in newtons, so N/tex = 0.08826 gpd.

TABLE 13.1. Some fibers/yarns and their properties ([**52**] with minor corrections). These values should be viewed as approximate, as there is significant variation.

Fiber	Denier = g per 9,000 m (denier)	Density (g/cm^3)	E (GPa)	c_E (km/s)	Strength (GPa)	Failure strain (%)
PBO		1.56	169	10.4	5.20	3.10
Spectra	1,000	0.97	120	11.1	2.57	3.50
Kevlar KM2	600	1.44	82.6	7.57	3.40	3.55
Kevlar KM2	850	1.44	73.7	7.15	3.34	3.80
Kevlar 129	840	1.44	99.1	8.30	3.24	3.25
Kevlar 29	1,500	1.44	74.4	7.19	2.90	3.38
Kevlar 29	200	1.44	91.1	7.95	2.97	2.95
Kevlar 29	1,000	1.44	78.8	7.40	2.87	3.25
Kevlar 49	1,140	1.44	120	9.13	3.04	2.40
Nylon		1.135	9.57	2.90	0.91	
E-glass		2.55	74	5.39	3.5	4.7
Carbon fiber		1.80	227	11.2	3.80	1.76

We begin with a mathematical digression to describe the first Piola–Kirchhoff stress tensor. Though this may seem tedious at first, it greatly simplifies the work in developing our analytic model. As an added bonus, you end up learning about the Piola–Kirchhoff stress. In the next chapter, we use this added knowledge to explore wave propagation in finite elastic materials such as rubber.

13.1. Deformation, Strain, and the First Piola–Kirchhoff Stress

The stress we have used in the book up to this point is the Cauchy stress. The Cauchy stress is defined with respect to the current spatial coordinate – that is, the Eulerian coordinates. The stress is simple to work with and is symmetric. However, there are other reference coordinates in which the stress tensor can be written. Fabrics in particular work as an armor material because of large deflections, but the strains are small. Because of their large deflections and their essentially two-dimensional natures, it is advantageous to work with fabrics in terms of coordinates describing their original undeformed configuration, which is typically a flat plane. However, choosing to work in the original undeformed configuration requires a change in the stress tensor.

We work in Cartesian coordinates and begin with some definitions. First, the original coordinates of the material will be labeled with capital letters, $X_1 = X$, $X_2 = Y$, $X_3 = Z$. Thus, we have a coordinate system (X, Y, Z) for the body based on its configuration before it deforms: the coordinate system (X, Y, Z) does not change with time – it is fixed. A fixed coordinate system based on initial material configuration is referred to as the Lagrangian frame. An alternate labeling might have been x_{0i} for X_i; however, the double subscript seems tedious and it is important to remember what coordinates we are discussing. The coordinates (x, y, z) are the current ones – that is, the location of the material after deformation is $x = x(X, Y, Z)$, etc. We now introduce the important deformation gradient.

DEFINITION 13.1. (deformation gradient) The *deformation gradient* F takes an incremental vector $d\vec{X}$ in the original configuration and coordinates to the incremental vector $d\vec{x}$ in the current configuration and coordinates, $d\vec{x} = Fd\vec{X}$. In Cartesian coordinates, the deformation gradient F is[1]

$$F_{ij} = \frac{\partial x_i}{\partial X_j} = \delta_{ij} + \frac{\partial u_i}{\partial X_j}, \tag{13.2}$$

or in matrix notation,

$$F = \begin{pmatrix} F_{11} & F_{12} & F_{13} \\ F_{21} & F_{22} & F_{23} \\ F_{31} & F_{32} & F_{33} \end{pmatrix} = \begin{pmatrix} \frac{\partial x}{\partial X} & \frac{\partial x}{\partial Y} & \frac{\partial x}{\partial Z} \\ \frac{\partial y}{\partial X} & \frac{\partial y}{\partial Y} & \frac{\partial y}{\partial Z} \\ \frac{\partial z}{\partial X} & \frac{\partial z}{\partial Y} & \frac{\partial z}{\partial Z} \end{pmatrix} = \begin{pmatrix} 1 + \frac{\partial u_x}{\partial X} & \frac{\partial u_x}{\partial Y} & \frac{\partial u_x}{\partial Z} \\ \frac{\partial u_y}{\partial X} & 1 + \frac{\partial u_y}{\partial Y} & \frac{\partial u_y}{\partial Z} \\ \frac{\partial u_z}{\partial X} & \frac{\partial u_z}{\partial Y} & 1 + \frac{\partial u_z}{\partial Z} \end{pmatrix}, \tag{13.3}$$

where u_i is the displacement $u_i = x_i - X_i$.

As a first example of the use of the deformation gradient, consider the density in our new coordinate system. If we are interested in a specific piece of material of volume v in the deformed coordinates, its mass is

$$m = \int_v \rho \, dx_1 dx_2 dx_3. \tag{13.4}$$

To perform the integral in the original configuration coordinates we have

$$m = \int_v \rho \left| \frac{\partial x}{\partial X} \right| dX_1 dX_2 dX_3 = \int_V \rho \det(F) dX_1 dX_2 dX_3, \tag{13.5}$$

where $\det(F)$ is the Jacobian of the transformation and V is the volume of material in the original coordinates. The mass was originally given by

$$m = \int_V \rho_0 dX_1 dX_2 dX_3. \tag{13.6}$$

Since our region of integration is arbitrary, we conclude that the two integrands must equal (see also Eq. A.138)

$$\rho_0 = \rho \det(F). \tag{13.7}$$

Thus, we have explicitly computed the density in the deformed configuration. Mass is intrinsically conserved in the Lagrangian (original) frame and this is the conservation of mass equation.

As a second example of the use of the deformation gradient, we write down two strain tensors. The deformation gradient provides a differential length in the deformed configuration based on a differential length in the original configuration,

$$dx_i = \frac{\partial x_i}{\partial X_j} dX_j = F_{ij} dX_j \quad \text{or} \quad d\vec{x} = Fd\vec{X}. \tag{13.8}$$

For the change in length caused by the deformation, where dS is the original length of a vector and ds is the deformed length,

$$(ds)^2 - (dS)^2 = d\vec{x}^T d\vec{x} - d\vec{X}^T d\vec{X} = d\vec{X}^T \{F^T F - I\} d\vec{X} = d\vec{x}^T \{I - (F^{-1})^T F^{-1}\} d\vec{x}. \tag{13.9}$$

---

[1] Important facts and subtleties of the deformation gradient, including behavior in other coordinate systems, are discussed in the first section of the next chapter and Appendix A.

The expressions in brackets {} define the strain tensors: the Lagrangian strain E, or Green strain, which corresponds (for small strain) to the engineering strain $(L - L_0)/L_0$, where L_0 is the original length and L the deformed length,

$$E = \frac{1}{2}(F^T F - I); \qquad (13.10)$$

and the Eulerian strain E^*, which corresponds (for small strain) to the strain $(L - L_0)/L$, where the gage length is the deformed length,

$$E^* = \frac{1}{2}(I - (F^{-1})^T F^{-1}) = \frac{1}{2}(I - B^{-1}), \quad \text{where} \quad B = FF^T, \qquad (13.11)$$

and where $I = (\delta_{ij})$ is the identity matrix. The factors of $\frac{1}{2}$ appear in these definitions to match these strain tensors to the small strain tensor when the strain is small – that is, when ds and dS are close –

$$\text{strain} = \frac{ds - dS}{dS} = \frac{ds - dS}{dS}\frac{ds + dS}{ds + dS} = \frac{(ds)^2 - (dS)^2}{dS(ds + dS)} \approx \frac{1}{2}\frac{(ds)^2 - (dS)^2}{(dS)^2}. \qquad (13.12)$$

To be precise, the far right-hand side of the above equation is the definition of the Lagrangian strain: there is a unique symmetric quadratic form E that gives rise to the far right-hand side. It should also be pointed out that this measure of strain is quadratic: it is based on the square of the distance, since getting the linear distance requires taking square roots, which are difficult to work with. For small strain, the strain is approximately $(L - L_0)/L_0$, but the Lagrangian strain is $(L - L_0)/L_0 + (1/2)((L - L_0)/L_0)^2$ and for large strain the quadratic term becomes very large. As written, the Lagrangian strain E is in the original configuration and the Eulerian strain E^* is in the deformed configuration. For fabrics we work in the original configuration, and so we write out the strain components for later use,

$$E_{ij} = \frac{1}{2}(F_{ik}^T F_{kj} - \delta_{ij}) = \frac{1}{2}(F_{ki}F_{kj} - \delta_{ij}) = \frac{1}{2}\left\{\left(\delta_{ki} + \frac{\partial u_k}{\partial X_i}\right)\left(\delta_{kj} + \frac{\partial u_k}{\partial X_j}\right) - \delta_{ij}\right\}$$

$$= \frac{1}{2}\left(\frac{\partial u_i}{\partial X_j} + \frac{\partial u_j}{\partial X_i} + \frac{\partial u_k}{\partial X_i}\frac{\partial u_k}{\partial X_j}\right), \qquad (13.13)$$

where the first two terms are the small strain tensor and the third is the nonlinear term.

A major goal is to write everything in the original Lagrangian configuration and coordinate system, including the stress and boundary conditions. Regardless of the coordinate system used for reference, the force, which lives in the current configuration and coordinate system (laboratory frame), has the same direction and magnitude. Thus, for a given piece of surface we have

$$\vec{f}^T = \int_{\partial v} \vec{n}^T \sigma\, da = \int_{\partial V} \vec{N}^T S\, dA \quad \text{or} \quad f_i = \int_{\partial v} n_j \sigma_{ji}\, da = \int_{\partial V} N_j S_{ji}\, dA, \qquad (13.14)$$

where $\vec{n}$ is the normal to the deformed surface with a corresponding surface area da and $\vec{N}$ is the normal to the original surface in the Lagrangian (original) coordinates of the material and its corresponding area dA. σ is the Cauchy stress that we have used so far in the book: it takes a surface normal in the current configuration and coordinate system and produces a traction in the current configuration and coordinate system. S is the stress we are looking for: it takes a surface normal in the original configuration and coordinate system and produces a traction in the current configuration and coordinate system. Thus, we need to understand the relationship between the surface normals and area elements in the original undeformed configuration and in the deformed configuration and corresponding coordinate systems (though, in the Cartesian case, the coordinate systems are the same).

We will first go through the argument in two dimensions as it is simpler than the three-dimensional manipulations. In two dimensions the linear surface segment will be denoted as $\vec{L}$. The surface normal with the correct length will be denoted as $\vec{A}$, where $|\vec{A}| = |\vec{L}|$. The surface normal vector of unit length is $\vec{N}$ so $\vec{A} = |\vec{L}|\vec{N}$. In the plane, the normal is a clockwise rotation by $90°$, which is accomplished by the rotation matrix $R = \left(\begin{smallmatrix} 0 & 1 \\ -1 & 0 \end{smallmatrix}\right)$ and hence $\vec{A} = R\vec{L}$. Under the deformation gradient mapping we have

$$\vec{a} = F(\vec{A}) = R\vec{\ell} = RF\vec{L} = RFR^T\vec{A} = \begin{pmatrix} F_{22} & -F_{21} \\ -F_{12} & F_{11} \end{pmatrix} \vec{A} = \det(F)F^{-T}\vec{A}. \quad (13.15)$$

Thinking of this in terms of an increment of length, the equation becomes

$$d\vec{a} = \vec{n}d\ell = F(d\vec{A}) = F(\vec{N}dL) = \det(F)F^{-T}\vec{N}dL. \quad (13.16)$$

In three dimensions, the unit surface normal affiliated with the area da can be represented as a parallelogram formed by two differential vectors:

$$\vec{n}\,da = d\vec{x}^{\,1} \times d\vec{x}^{\,2} \quad \text{or} \quad n_i\,da = \varepsilon_{ijk}dx_j^1 dx_k^2. \quad (13.17)$$

A superscript is used to distinguish the two vectors to avoid confusion with the subscripts, which identify components. We want to identify this unit vector in the deformed configuration with its da with a unit vector in the undeformed configuration with its corresponding area dA. There we have

$$\vec{N}\,dA = d\vec{X}^1 \times d\vec{X}^2 \quad \text{or} \quad N_i dA = \varepsilon_{ijk}dX_j^1 dX_k^2. \quad (13.18)$$

To go from one differential vector to another we use the deformation gradient,

$$n_i da = \varepsilon_{ijk}dx_j^1 dx_k^2 = \varepsilon_{ijk}F_{jl}F_{km}dX_l^1 dX_m^2. \quad (13.19)$$

Multiplying both sides by F_{ip} gives

$$n_i F_{ip} da = \varepsilon_{ijk}F_{ip}F_{jl}F_{km}dX_l^1 dX_m^2. \quad (13.20)$$

Now, the determinant of F is

$$\det(F) = \varepsilon_{ijk}F_{1i}F_{2j}F_{3k} = \varepsilon_{ijk}F_{i1}F_{j2}F_{k3}. \quad (13.21)$$

In this notation, we are expanding about the rows or columns 1, 2, and 3. However, we can write the rows or columns in various orders. Specifically, if the ordering is an even permutation of the columns (e.g., 2, 3, 1 or 3, 1, 2), then the determinant results. If the ordering is an odd permutation of the columns (e.g., 1, 3, 2 or 2, 1, 3), then the negative of the determinant results. If a column is repeated, then the result is zero. Thus,[2]

$$\varepsilon_{ijk}F_{ip}F_{jl}F_{km} = \varepsilon_{plm}\det(F). \quad (13.23)$$

---

[2]This is exactly the same argument as is used to prove $\det(AB) = \det(A)\det(B)$, namely

$$\det(AB) = \varepsilon_{ijk}A_{1\ell}B_{\ell i}A_{2m}B_{mj}A_{3n}B_{nk} = A_{1\ell}A_{2m}A_{3n}\varepsilon_{ijk}B_{\ell i}B_{mj}B_{nk}$$
$$= A_{1\ell}A_{2m}A_{3n}\varepsilon_{\ell mn}\det(B^T) = \varepsilon_{\ell mn}A_{1\ell}A_{2m}A_{3n}\det(B) = \det(A)\det(B). \quad (13.22)$$

Since $\det(F) = \rho_0/\rho$, we obtain

$$n_i F_{ip} da = \varepsilon_{plm} \frac{\rho_0}{\rho} dX_l^1 dX_m^2 = \frac{\rho_0}{\rho} N_p dA. \qquad (13.24)$$

In vector notation this is

$$\vec{n}^T F da = \frac{\rho_0}{\rho} \vec{N}^T dA \quad \text{or} \quad \vec{n}^T da = \frac{\rho_0}{\rho} \vec{N}^T F^{-1} dA. \qquad (13.25)$$

By taking the transpose, we see that this is exactly the same equation as the two-dimensional result in Eq. (13.16),

$$d\vec{a} = \vec{n}\, da = d\vec{x}^{\,1} \times d\vec{x}^{\,2} = F(d\vec{A}) = F(\vec{N} dA)$$
$$= F(d\vec{X}^1 \times d\vec{X}^2) = F d\vec{X}^1 \times F d\vec{X}^2 = \det(F) F^{-T} \vec{N} dA, \qquad (13.26)$$

where the intermediate expressions are reminders of the notation. Thus the unit normal in the original undergoes a transformation through left multiplication by $(\rho_0/\rho)F^{-T}$. Notice $\det((\rho_0/\rho)F^{-T}) = (\rho_0/\rho)^3 \det(F^{-T}) = (\rho_0/\rho)^2$, which need not equal 1. This change need not be a rotation even if the determinant equals 1: da and dA need not be equal;[3] explicitly, $da/dA = \det(F)|F^{-T}\vec{N}|$, the value of which depends on the direction $\vec{N}$. We place Eq. (13.25) in Eq. (13.14) for force equality to obtain what we are looking for in terms of the stress in the original coordinates,

$$\vec{f}^T = \int_{\partial v} \vec{n}^T \sigma \, da = \int_{\partial V} \frac{\rho_0}{\rho} \vec{N}^T F^{-1} \sigma \, dA = \int_{\partial V} \vec{N}^T S \, dA. \qquad (13.27)$$

Since this should hold true for all surface elements and all unit vectors, we obtain the new stress tensor S, called the *first Piola–Kirchhoff stress*,[4]

$$S = \frac{\rho_0}{\rho} F^{-1} \sigma \quad \text{or equivalently} \quad \sigma = \frac{\rho}{\rho_0} F S = \frac{1}{\det(F)} F S. \qquad (13.28)$$

Since the Cauchy stress tensor is symmetric, and the deformation gradient F is in general not symmetric, it follows that in general S is not symmetric. The first Piola–Kirchhoff stress tensor represents the force per unit undeformed area while the Cauchy stress represents the force per unit deformed area. The first Piola–Kirchhoff stress S has similar behaviors to the deformation gradient F: it takes a vector (in this case a surface normal) in the undeformed configuration and coordinate system and produces a vector (in this case the traction) in the current configuration and coordinate system.

Given the first Piola–Kirchhoff stress, we now want to determine the equation of motion for the stress in the original, undeformed coordinate system. Using the

---

[3]To be explicit, $\vec{t}^T = \vec{n}^T \sigma$ is the force in the current configuration per unit area in the current configuration while $\vec{t}'^T = \vec{N}^T S$ is the force in the current configuration per unit area in the original configuration. Note $\vec{t} \neq \vec{t}'$ and $\vec{n} \neq (\rho_0/\rho)F^{-T}\vec{N}$. For example, a pure dilation $F = \alpha I$ has $(\rho_0/\rho)F^{-T}\vec{N} = \alpha^2 \vec{N} \neq \vec{n}$ (it is not unit length) and, as will be shown in the next two equations in the main text, $\sigma = (\rho/\rho_0)FS = S/\alpha^2$, so $\vec{t}'^T = \vec{N}^T S = \vec{n}^T \alpha^2 \sigma \neq \vec{n}^T \sigma = \vec{t}^T$ (because F has no rotation, $\vec{n}$ and $\vec{N}$ are both unit vectors pointing in the same direction).

[4]Care is needed while reading the literature because some authors prefer to work with the transpose of S rather than the first Piola–Kirchhoff stress S. Fortunately, they usually use a different name, though the difference may be slight. The reason they do that is to avoid our proliferation of transposes: the tractions are computed by right multiplication with the normal vector and the conjugate strain rate is $\dot{F}$, not $\dot{F}^T$.

divergence theorem acting on the momentum balance in the deformed coordinates, we have

$$f_i = \int_V \frac{\partial S_{ji}}{\partial X_j} dV = \int_v \frac{\partial \sigma_{ji}}{\partial x_j} dv$$

$$= \int_v \rho \left(\frac{Dv_i}{Dt} - b_i \right) dv = \int_V \rho \left| \frac{dx}{dX} \right| \left(\frac{Dv_i}{Dt} - b_i \right) dV = \int_V \rho_0 \left(\frac{Dv_i}{Dt} - b_i \right) dV.$$

(13.29)

Since the volumes are arbitrary, $\partial \sigma_{ji}/\partial x_j + \rho b_i = \rho Dv_i/Dt$ implies

$$\frac{\partial S_{ji}}{\partial X_j} + \rho_0 b_i = \rho_0 \frac{Dv_i}{Dt}.$$

(13.30)

In the Lagrangian frame, the acceleration term is

$$\frac{Dv_i}{Dt} = \frac{\partial v_i}{\partial t} + \frac{\partial v_i}{\partial X}\frac{dX}{dt} + \frac{\partial v_i}{\partial Y}\frac{dY}{dt} + \frac{\partial v_i}{\partial Z}\frac{dZ}{dt} = \frac{\partial v_i}{\partial t},$$

(13.31)

since the original coordinates do not change and thus have no time dependence. The initial density ρ_0 does not depend on time, finally yielding

$$\frac{\partial S_{ji}}{\partial X_j} + \rho_0 b_i = \rho_0 \frac{\partial v_i}{\partial t} = \frac{\partial}{\partial t}(\rho_0 v_i) \quad \text{or} \quad \vec{\nabla} \cdot S + \rho_0 \vec{b} = \rho_0 \frac{\partial \vec{v}}{\partial t},$$

(13.32)

where $\vec{\nabla}$ is in terms of original Lagrangian coordinates $\vec{X}$. Thus we have derived the momentum balance for the Lagrangian frame using the first Piola–Kirchhoff stress. Dropping the body force, this equation is in conservation form and thus is directly applicable to the jump condition form with $A = \rho_0 v_i$ and $B = -S_{ji}$, so

$$[-S_{ji}] = U[\rho_0 v_i],$$

(13.33)

where the brackets [ ] denote the discontinuous change in value across the wave (shock) front, as described in Chapter 2.

As a final task we determine the energy rate balance in the original configuration, following the approach in Appendix A. The energy within a given amount of material is comprised of internal energy (here E is the specific energy, or energy per unit mass) and kinetic energy,

$$\tilde{e}_V = \int_V \left(\rho_0 E + \frac{1}{2}\rho_0 v^2 \right) dV \quad \Rightarrow \quad \frac{d\tilde{e}_V}{dt} = \int_V \left\{ \rho_0 \frac{\partial E}{\partial t} + \frac{1}{2}\rho_0 \frac{\partial(v_i v_i)}{\partial t} \right\} dV.$$

(13.34)

There are four sources of energy that can be supplied to our arbitrary volume of material. Two are on the boundary. First, there is a power input to the arbitrary volume of material found by the continuum analog of force times velocity. Second, there can be heat conduction, which allows energy to enter or leave without any motion of material. We will denote the heat flux by $\vec{q}$ (positive when flowing out of the volume). The third source of power input is work done by body forces, which operate over the volume of material. The fourth source of energy is internal energy released within the volume, such as through chemical reactions, which is an energy source at a given rate r. The energy per unit time (power) input terms to the arbitrary volume are

$$\frac{d\tilde{e}_V}{dt} = \int_{\partial V} \vec{t} \cdot \vec{v} dA - \int_{\partial V} \vec{q} \cdot \vec{n} dA + \int_V \rho_0 \vec{b} \cdot \vec{v} dV + \int_V \rho_0 r dV.$$

(13.35)

With $\vec{t} \cdot \vec{v} = t_i v_i = S_{ji} v_i N_j$, use of the divergence theorem and the momentum balance gives

$$
\frac{d\tilde{e}_V}{dt} = \int_V \left\{ \frac{\partial(S_{ji} v_i)}{\partial X_j} - \frac{\partial q_i}{\partial X_i} + \rho_0 b_i v_i + \rho_0 r \right\} dV
$$

$$
= \int_V \left\{ S_{ji} \frac{\partial v_i}{\partial X_j} + v_i \left(\frac{\partial S_{ji}}{\partial X_j} + \rho_0 b_i \right) - \frac{\partial q_i}{\partial X_i} + \rho_0 r \right\} dV \qquad (13.36)
$$

$$
= \int_V \left\{ S_{ji} \frac{\partial F_{ij}}{\partial t} + v_i \rho_0 \frac{\partial v_i}{\partial t} - \frac{\partial q_i}{\partial X_i} + \rho_0 r \right\} dV.
$$

Setting the two expressions for $d\tilde{e}_V/dt$ equal and recognizing that the volume V is arbitrary, we obtain the energy balance equation,

$$
\rho_0 \frac{\partial E}{\partial t} = S_{ji} \frac{\partial F_{ij}}{\partial t} - \frac{\partial q_i}{\partial X_i} + \rho_0 r. \qquad (13.37)
$$

The leading term on the right-hand side is

$$
S_{ij} \frac{\partial F_{ji}}{\partial t} = S_{ij} \frac{\partial F_{ij}^T}{\partial t} = S_{ij} \dot{F}_{ij}^T, \qquad (13.38)
$$

and S and $\dot{F}^T$ are referred to as conjugate variables. (Similarly, σ and D are conjugate based on the energy balance in the current configuration.) We finish with the conservation form,

$$
\frac{\partial}{\partial t}(\rho_0(E + \frac{1}{2} v_i v_i)) - \frac{\partial(S_{ji} v_i)}{\partial X_j} = 0 \quad \text{or} \quad \frac{\partial}{\partial t}(\rho_0(E + \frac{1}{2}|\vec{v}|)) - \text{div}(S\vec{v}) = 0, \quad (13.39)
$$

where the divergence is with respect to Lagrangian coordinates $\vec{X}$, where body, source, and heat terms have been dropped.

We finish this section by writing down the Hugoniot jump conditions $[B] = U[A]$. For mass conservation, $\partial(\rho \det(F))/\partial t = 0)$; hence $A = \rho \det(F)$ and $B = 0$. Supposing the material is initially in motion with velocity v_r and then a wave traveling at speed U changes the particle velocity to v_ℓ, the displacement is

$$
x(X, t) = \begin{cases} X + v_r t, & X > Ut, \\ X + v_r(X/U) + v_\ell(t - X/U), & X \le Ut. \end{cases} \qquad (13.40)
$$

From this the deformation gradient is computed,

$$
F_{11} = \frac{\partial x}{\partial X} = \begin{cases} 1, & X > Ut, \\ 1 + (v_r - v_\ell)/U, & X \le Ut. \end{cases} \qquad (13.41)
$$

We have $[\rho \det(F)] = 0$; hence

$$
\rho_r \cdot 1 = \rho_\ell \det(F) = \rho_\ell F_{11} = \rho_\ell(1 + (v_r - v_\ell)/U) \quad \Rightarrow \quad \frac{\rho_r}{\rho_\ell} = 1 - \frac{v_\ell - v_r}{U}. \quad (13.42)
$$

This is the jump condition for the mass. For momentum, repeating from Eq. (13.33), $A = \rho_r v$ and $B = -S$, so the jump condition is

$$
-S_\ell + S_r = \rho_r U(v_\ell - v_r). \qquad (13.43)
$$

Finally, for energy $A = \rho_r(E + \frac{1}{2}v^2)$ and $B = -Sv$, and the jump condition is

$$
-S_\ell v_\ell + S_r v_r = \rho_r U(E_\ell + \frac{1}{2}v_\ell^2 - E_r - \frac{1}{2}v_r^2). \qquad (13.44)
$$

With similar manipulation to Chapter 2, though in Lagrangian coordinates it is simpler, this transforms to

$$E_\ell - E_r = \frac{1}{2}(-S_\ell - S_r)\left(\frac{1}{\rho_r} - \frac{1}{\rho_\ell}\right). \tag{13.45}$$

Summarizing these jump conditions,

THEOREM 13.2. *(Rankine–Hugoniot jump conditions, Lagrangian frame) The relationship between density, stress, energy, and particle velocity v in front of and behind a discontinuous shock moving at velocity U is given by*

$$\text{Mass:} \qquad \frac{\rho_r}{\rho_\ell} = 1 - \frac{v_\ell - v_r}{U}, \tag{13.46}$$

$$\text{Momentum:} \quad -S_\ell + S_r = \rho_r U(v_\ell - v_r), \tag{13.47}$$

$$\text{Energy:} \qquad E_\ell - E_r = \frac{1}{2}(-S_\ell - S_r)\left(\frac{1}{\rho_r} - \frac{1}{\rho_\ell}\right), \tag{13.48}$$

$$\text{Entropy:} \qquad s_\ell \geq s_r. \tag{13.49}$$

Comparing with Theorem 2.5, we see the only difference is that in the Lagrangian frame the shock speed relative to the material is U and in the Eulerian frame it is $U - v_r$. Since only one-dimensional deformation is occurring, the normal area is not changing and $\tilde{\sigma} = -S$.

13.2. The Behavior of a Single Yarn under Impact

In this section we study the response of a single yarn. A yarn is comprised of many fibers, typically spun together. In uniaxial stress, the stress σ as a function ε of strain in the yarn with a linear elastic stress-strain relation can be written

$$\sigma(\varepsilon) = (E\varepsilon)_+, \qquad \text{where} \qquad (a)_+ = \begin{cases} a, & a \geq 0, \\ 0, & a < 0. \end{cases} \tag{13.50}$$

E is Young's modulus. This equation represents the fact that the yarn supports stress in tension but not in compression. We are interested in two additional situations, namely where the yarn is initially in tension and where the yarn is initially slack. The first case can be represented by an initial strain $\varepsilon_0 > 0$ in the yarn,

$$\sigma(\varepsilon + \varepsilon_0) = (E(\varepsilon + \varepsilon_0))_+, \qquad T_0 = \text{initial tension} = E\varepsilon_0 > 0. \tag{13.51}$$

The second case is where the yarn is initially slack, and is given by

$$\sigma(\varepsilon + \varepsilon_0) = (E(\varepsilon + \varepsilon_0))_+, \qquad T_0 = 0, \qquad \varepsilon_0 < 0. \tag{13.52}$$

The strain energy stored in the yarn is

$$\rho E_e = e(\varepsilon) = \int_0^\varepsilon \sigma d\varepsilon = (\frac{1}{2}E\varepsilon^2)_+. \tag{13.53}$$

In the following, the Lagrangian strain is always with two subscripts, E_{ij}, to avoid confusion with E (the Young's modulus), E_e (the specific energy), and E_0, an initial strain.

We will work directly from the strain energy in what follows, and we will use as our strain the Lagrangian strain E_{11}, with the possibility of an initial strain E_0. Thus, the energy in the yarn will be given by $e(E_{11} + E_0)$. For the majority of the following derivation we are not assuming that the stress-strain response is linear. The stress-strain response is given by stress $= e'(E_{11} + E_0)$, where prime denotes the derivative, where $e'(\varepsilon) = 0$ for $\varepsilon \leq 0$, where $e'(\varepsilon)$ is a strictly increasing function for $\varepsilon \geq 0$, and an initial tension in the yarn is denoted by $T_0 = e'(E_0)$. For the specific case of linear stress-strain relation, $e'(E_{11} + E_0) = E \cdot (E_{11} + E_0)$.

We make a few comments about the Lagrangian strain. The axial strain E_{11} is the axial strain in the yarn, since this is the axial strain in terms of the initial configuration. Written out,

$$E_{11} = \frac{1}{2}(F_{1j}^T F_{j1}-1) = \frac{1}{2}(F_{j1}F_{j1}-1) = \frac{\partial u_x}{\partial X} + \frac{1}{2}\left(\frac{\partial u_x}{\partial X}\right)^2 + \frac{1}{2}\left(\frac{\partial u_y}{\partial X}\right)^2 + \frac{1}{2}\left(\frac{\partial u_z}{\partial X}\right)^2.$$
(13.54)

If we think of the yarn as a collection of springs all lined up along the x axis, and we think of the Lagrange strain as given by $(L - L_0)/L_0$, where L is the deformed length of a spring and $L_0 = \Delta x_0$ is the initial length of the spring, then the strain in the spring connecting the points i and $i+1$ is

$$\varepsilon_{11} = \frac{L - L_0}{L_0} = \frac{\sqrt{(L_0 + \Delta x)^2 + (\Delta y)^2 + (\Delta z)^2} - L_0}{L_0},$$
(13.55)

where Δ refers to the change in each coordinate. In the limit as the length of the springs goes to zero, with the displacement written as u_i,

$$\lim_{\Delta x_0 \to 0} \varepsilon_{11} = \sqrt{\left(1 + \frac{\partial u_x}{\partial X}\right)^2 + \left(\frac{\partial u_y}{\partial X}\right)^2 + \left(\frac{\partial u_z}{\partial X}\right)^2} - 1 = \sqrt{1 + 2E_{11}} - 1 \approx E_{11},$$
(13.56)

where the approximation is for small strains but possibly large displacements. For small strains the $\partial u_x/\partial X$ term dominates its square, so we typically drop the $(\partial u_x/\partial X)^2$ term.

In small-strain elasticity, the Cauchy stress can be derived by differentiating the stored elastic energy density by the strain tensor (Exercise 3.27). We would like to do something similar here, and the question is what to differentiate with. With e as the energy, assuming that we differentiate by a yet to be determined h_{ij},

$$\frac{\partial e}{\partial t} = \frac{\partial e}{\partial h_{ij}} \frac{\partial h_{ij}}{\partial t} = \text{stress} \times \text{strain rate} = S_{ij}\frac{d}{dt}F_{ij}^T.$$
(13.57)

The last expression comes from the energy rate balance (Eqs. 13.36 and 13.38) and we see that $h_{ij} = F_{ij}^T$. Thus, the first Piola–Kirchhoff stress can be derived by differentiation of the elastic strain energy density with respect to the transpose of the deformation gradient,

$$S_{ij} = \frac{\partial e}{\partial F_{ij}^T} = \frac{\partial e}{\partial F_{ji}} = \frac{\partial e}{\partial E_{kl}} \frac{\partial E_{kl}}{\partial F_{ji}}.$$
(13.58)

To explicitly compute this derivative we need

$$\frac{\partial E_{ij}}{\partial F_{kl}} = \frac{1}{2}\frac{\partial (F_{mi}F_{mj})}{\partial F_{kl}} = \frac{1}{2}(\delta_{mk}\delta_{il}F_{mj} + F_{mi}\delta_{mk}\delta_{jl}) = \frac{1}{2}(\delta_{il}F_{kj} + \delta_{jl}F_{ki}). \quad (13.59)$$

We then have, using the symmetry of E_{ij},

$$S_{ij} = \frac{\partial e}{\partial E_{kl}}\frac{1}{2}(\delta_{ki}F_{jl} + \delta_{li}F_{jk}) = \frac{\partial e}{\partial E_{il}}\frac{1}{2}F_{jl} + \frac{\partial e}{\partial E_{ki}}\frac{1}{2}F_{jk} = \frac{\partial e}{\partial E_{il}}F_{jl} = \frac{\partial e}{\partial E_{il}}F_{lj}^T$$
(13.60)

or, using matrix notation, where $(\partial e/\partial E)_{ij} = \partial e/\partial E_{ij}$,

$$S = \frac{\partial e}{\partial E} F^T. \tag{13.61}$$

For our single yarn, $e = e(E_{11} + E_0)$, so the stress tensor for motion in the x-z plane can be written[5]

$$
\begin{aligned}
S &= \begin{pmatrix} e'(E_{11}+E_0) & 0 & 0 \\ 0 & 0 & 0 \\ 0 & 0 & 0 \end{pmatrix} F^T = \begin{pmatrix} e'(E_{11}+E_0) & 0 & 0 \\ 0 & 0 & 0 \\ 0 & 0 & 0 \end{pmatrix} \begin{pmatrix} F_{11} & 0 & F_{31} \\ 0 & 1 & 0 \\ 0 & 0 & 1 \end{pmatrix} \\
&= \begin{pmatrix} e'(E_{11}+E_0)F_{11} & 0 & e'(E_{11}+E_0)F_{31} \\ 0 & 0 & 0 \\ 0 & 0 & 0 \end{pmatrix},
\end{aligned}
\tag{13.62}
$$

where $F_{11} = 1 + \partial u_x/\partial X$ and $F_{31} = \partial u_z/\partial X$. Writing the components,

$$S_{11} = e'(E_{11}+E_0)F_{11}, \quad S_{13} = e'(E_{11}+E_0)F_{31}, \quad \text{all other } S_{ij} = 0. \tag{13.63}$$

We now have the stress terms required to evaluate the behavior of a single yarn. Explicitly, originally the yarn lays along the x-axis with original Lagrangian coordinate X. Our boundary condition, at $X = 0$, is an applied velocity V in the z direction, which could be viewed as being applied by the normal impact of a projectile at the center of a yarn that extends to infinity in either direction. Just as in our plate impact and cavity expansion work, this impact results in a two-wave structure, with a longitudinal wave with all motion in the x direction traveling along the yarn at a speed c_1 and a transverse wave which imparts a z velocity to the yarn traveling at a speed c_2. The jump condition $[-S_{ji}] = c[\rho_0 v_i]$ for the initial longitudinal wave is given by

$$[-S_{11}] = c_1 \rho_0 [v_x] \quad \Leftrightarrow \quad -S_{11L} + T_0 = c_1 \rho_0 v_{xL}, \tag{13.64}$$

or specifically

$$-e'(E_{11L}+E_0)F_{11L} + T_0 = c_1 \rho_0 v_{xL}. \tag{13.65}$$

Here, c_1 is the speed of the first wave, and it is to be determined by the analysis. For the subsequent transverse wave, we have two jump conditions that must be simultaneously satisfied, namely

$$
\begin{aligned}
[-S_{11}] = c_2 \rho_0 [v_x] \quad &\Leftrightarrow \quad -S_{11T} + S_{11L} = c_2 \rho_0 (v_{xT} - v_{xL}), \\
[-S_{13}] = c_2 \rho_0 [v_z] \quad &\Leftrightarrow \quad -S_{13T} = c_2 \rho_0 v_z,
\end{aligned}
\tag{13.66}
$$

or specifically

$$
\begin{aligned}
-e'(E_{11T}+E_0)F_{11T} + e'(E_{11L}+E_0)F_{11L} &= c_2 \rho_0 (v_{xT} - v_{xL}), \\
-e'(E_{11T}+E_0)F_{31T} &= c_2 \rho_0 v_z.
\end{aligned}
\tag{13.67}
$$

Here, c_2 is the speed of the transverse wave in the original Lagrangian coordinate system, which in this case will be different from the speed of the wave in the laboratory coordinate system (denoted U_T), which will be computed later. To connect the particle velocity and the strain, we explicitly write down the velocity of the yarn and then integrate it to get displacement. The following velocity is written down, realizing that changes in wave speed only occur at the wave fronts where the jumps occur; everywhere else the material velocity is constant. The initial tensile

---

[5]See Section 14.1 for a discussion of why we choose to work with the Piola–Kirchhoff stress and not just use the Green strain with the Cauchy stress.

wave produces material inflow towards the origin, $v_{xL} < 0$, but since material does not pile up at the origin it is clear that upon passage of the second (transverse) wave the x velocity must be zero: $v_{xT} = 0$. The velocity in the yarn is

$$v_x(X,t) = \begin{cases} 0, & X > c_1t, \\ v_{xL}, & c_2t < X \le c_1t, \\ 0, & X \le c_2t; \end{cases} \qquad v_z(X,t) = \begin{cases} 0, & X > c_2t, \\ V, & X \le c_2t, \end{cases} \qquad (13.68)$$

where c_1 and c_2 are wave speeds that are to be determined. From the velocity, the displacements can be found,

$$u_x(X,t) = \begin{cases} 0, & X > c_1t, \\ v_{xL}(t - X/c_1), & c_2t < X \le c_1t, \\ v_{xL}(X/c_2)(1 - c_2/c_1), & X \le c_2t; \end{cases} \qquad (13.69)$$

$$u_z(X,t) = \begin{cases} 0, & X > c_2t, \\ V(t - X/c_2), & X \le c_2t. \end{cases}$$

From the displacements, the spatial derivatives can be found,

$$\frac{\partial u_x}{\partial X}(X,t) = \begin{cases} 0 & X > c_1t, \\ -v_{xL}/c_1 & c_2t < X \le c_1t, \\ (v_{xL}/c_2)(1 - c_2/c_1) & X \le c_2t; \end{cases} \qquad (13.70)$$

$$\frac{\partial u_z}{\partial X}(X,t) = \begin{cases} 0 & X > c_2t, \\ -V/c_2 & X \le c_2t. \end{cases}$$

Notice that $\partial u_x/\partial X$ changes sign as we move along the axis, since the original motion of longitudinal wave is tensile but then the yarn "bunches up" along the x-axis (though it is spread out in the z direction), so the derivative changes sign. Given this deformation, the deformation gradient is

$$F_L = \begin{pmatrix} 1 - v_{xL}/c_1 & 0 & 0 \\ 0 & 1 & 0 \\ 0 & 0 & 1 \end{pmatrix}, \qquad F_T = \begin{pmatrix} 1 + v_{xL}/c_2 - v_{xL}/c_1 & 0 & 0 \\ 0 & 1 & 0 \\ -V/c_2 & 0 & 1 \end{pmatrix}. \qquad (13.71)$$

Combined with the velocity profile, the third jump condition (the second part of Eq. 13.67) can be written

$$e'(E_{11T} + E_0)\frac{1}{c_2} = c_2\rho_0. \qquad (13.72)$$

When the second part of Eq. (13.67) is inserted into the second jump condition (the first part of Eq. 13.67) along with the velocity profile, we obtain

$$-e'(E_{11T} + E_0)F_{11T} + e'(E_{11L} + E_0)F_{11L} = -c_2\rho_0 v_{xL} = -e'(E_{11T} + E_0)\frac{v_{xL}}{c_2}. \qquad (13.73)$$

This equation gives

$$-e'(E_{11T}+E_0)\left(1 + \frac{v_{xL}}{c_2} - \frac{v_{xL}}{c_1}\right) + e'(E_{11L}+E_0)\left(1 - \frac{v_{xL}}{c_1}\right) = -e'(E_{11T}+E_0)\frac{v_{xL}}{c_2}, \qquad (13.74)$$

which leads to the result that

$$e'(E_{11T} + E_0) = e'(E_{11L} + E_0) \quad \Rightarrow \quad E_{11T} = E_{11L}. \qquad (13.75)$$

FIGURE 13.1. Multiple images from Jameson et al. [**102**] of a single nylon yarn being impacted, each image slightly offset in the vertical direction. The yarn material moves directly to the left after passage of the longitudinal wave (the yarn has been marked to aid in observing material motion). Motion to the left ends upon arrival of the transverse wave, whereupon all motion is in the vertical direction. (Original photograph courtesy of Phil Cunniff.)

This quite amazing result says that the strain in the yarn does not change when the transverse wave passes (Fig. 13.1). We note at this point that the motion in the initial longitudinal wave is related to the strain as

$$E_{11} = -\frac{v_{xL}}{c_1} + \frac{1}{2}\left(\frac{v_{xL}}{c_1}\right)^2 \quad \Rightarrow \quad \frac{v_{xL}}{c_1} = -(\sqrt{1 + 2E_{11}} - 1) = -\varepsilon_{11}. \quad (13.76)$$

Given our analysis so far, the three jump conditions in terms of the three unknowns are (where the second equation is the equality of the strain)

$$-e'\left(-\frac{v_{xL}}{c_1} + \frac{1}{2}\left(\frac{v_{xL}}{c_1}\right)^2 + E_0\right)\left(1 - \frac{v_{xL}}{c_1}\right) + T_0 = c_1\rho_0 v_{xL},$$

$$2\frac{v_{xL}}{c_2} - \left(\frac{v_{xL}}{c_1}\right)^2 + \left(\frac{v_{xL}}{c_2} - \frac{v_{xL}}{c_1}\right)^2 + \left(\frac{V}{c_2}\right)^2 = 0, \quad (13.77)$$

$$e'\left(-\frac{v_{xL}}{c_1} + \frac{1}{2}\left(\frac{v_{xL}}{c_1}\right)^2 + E_0\right) = \rho_0 c_2^2.$$

These equations can be solved to find the inflow velocity of the yarn towards the impact point (v_{xL}) and the longitudinal (c_1) and transverse (c_2) wave speeds. Our results so far can be summarized:

THEOREM 13.3. *For the impact on a single yarn, a tensile wave is formed pulling the yarn directly towards the initial impact point. The tensile wave is followed by a transverse wave that does not change the strain in the yarn but ends the yarn's motion towards the initial impact point and begins a translation in the direction of impact.*

There is an immediate corollary:

GENERAL PRINCIPLE 13.4. *For a single yarn, the yarn will either break immediately upon impact or it will not break.*

This result follows, since the strain in the yarn does not change throughout the event.

Let us write down a solution to Eqs. (13.77). Though we would prefer to parameterize with V, the natural parameter for this system is E_{11}. Thus, one solves the system for various E_{11} until the desired V is obtained. It is always required that $E_{11} > 0$ and $E_{11} + E_O > 0$. Reordering, rearranging, and some arithmetic yields the following sequence of equations for determining c_2, c_1, v_{xL}, and V, in that order, for a given E_{11},

$$\rho_0 c_2^2 = e'\left(E_{11} + E_0\right),$$
$$\left(\sqrt{1 + 2E_{11}} - 1\right)c_1^2 = c_2^2\sqrt{1 + 2E_{11}} - T_0/\rho_0,$$
$$v_{xL}/c_1 = 1 - \sqrt{1 + 2E_{11}}, \tag{13.78}$$
$$\left(\frac{V}{c_1}\right)^2 = -\frac{v_{xL}}{c_1}\left(\frac{c_2}{c_1}\right)\left\{2 + \left(\frac{c_1}{c_2} - 2\right)\frac{v_{xL}}{c_1}\right\},$$
$$U_T = c_2 + v_{xL}(1 - c_2/c_1)$$

U_T is the transverse wave speed in the fixed laboratory reference frame and we include the expression for completeness; it will be derived below. Following these steps allows determination of the various speeds for a given strain for any elastic constitutive model for the yarn.

Let us consider these equations with the simplification that the stress-strain response is linear – that is, $e'(E_{11}) = E \cdot E_{11}$. With this constitutive response there exists a bar wave speed $c_E = \sqrt{E/\rho_0}$ that will be used as a reference speed. Some arithmetic yields

$$\left(\frac{c_2}{c_E}\right)^2 = E_{11} + E_0,$$
$$\left(\frac{c_1}{c_E}\right)^2 = \frac{E_{11} + E_0 - T_0/E}{2E_{11}}(1 + \sqrt{1 + 2E_{11}}) + E_{11} + E_0,$$
$$\frac{v_{xL}}{c_1} = 1 - \sqrt{1 + 2E_{11}}, \tag{13.79}$$
$$\left(\frac{V}{c_E}\right)^2 = -\frac{v_{xL}}{c_1}\left(\frac{c_1}{c_E}\right)\left(\frac{c_2}{c_E}\right)\left\{2 + \left(\frac{c_1}{c_2} - 2\right)\frac{v_{xL}}{c_1}\right\},$$
$$\frac{U_T}{c_E} = \frac{c_2}{c_E} + \frac{v_{xL}}{c_1}\left(\frac{c_1}{c_E} - \frac{c_2}{c_E}\right),$$

where the last two expressions are computed by inserting the various terms. To obtain a better feel for the response, let us make the small strain approximation, where we assume E_{11} is small compared to 1. Then

$$\left(\frac{c_2}{c_E}\right)^2 = E_{11} + E_0, \qquad \left(\frac{c_1}{c_E}\right)^2 \approx \frac{E_{11} + E_0 - T_0/E}{E_{11}}, \qquad \frac{v_x L}{c_1} \approx -E_{11},$$

$$\left(\frac{V}{c_E}\right)^2 \approx \sqrt{E_{11}}\sqrt{E_{11} + E_0 - T_0/E}\left(2\sqrt{E_{11} + E_0} - \sqrt{E_{11}}\sqrt{E_{11} + E_0 - T_0/E}\right).$$

$$(13.80)$$

If the yarn is initially in tension, then $T_0 = E \cdot E_0 > 0$ and these expressions are

$$\left(\frac{c_2}{c_E}\right)^2 = E_{11} + E_0, \qquad \left(\frac{c_1}{c_E}\right)^2 \approx 1, \qquad \frac{v_x L}{c_1} \approx -E_{11},$$

$$(13.81)$$

$$\left(\frac{V}{c_E}\right)^2 \approx E_{11}\left(2\sqrt{E_{11} + E_0} - E_{11}\right), \qquad \frac{U_T}{c_E} \approx \sqrt{E_{11} + E_0} - E_{11}.$$

If the yarn is originally slack, $T_0 = 0$ and $E_0 < 0$, then these expressions become

$$\left(\frac{c_2}{c_E}\right)^2 = E_{11} + E_0, \qquad \left(\frac{c_1}{c_E}\right)^2 \approx \frac{E_{11} + E_0}{E_{11}}, \qquad \frac{v_x L}{c_1} \approx -E_{11},$$

$$\left(\frac{V}{c_E}\right)^2 \approx \sqrt{E_{11}}\left(E_{11} + E_0\right)\left(2 - \sqrt{E_{11}}\right), \qquad \frac{U_T}{c_E} \approx \sqrt{E_{11} + E_0}\left(1 - \sqrt{E_{11}}\right).$$

$$(13.82)$$

The most common case of interest for the linear stress-strain response is when the yarn is neither strained nor loose to begin with – that is, $E_0 = 0$ and $T_0 = 0$. Again the solution is in terms of the parameter of the strain E_{11}. Some arithmetic yields (Exercise 13.5)

$$\left(\frac{c_1}{c_E}\right)^2 = \frac{1}{2} + \frac{1}{2}\sqrt{1 + 2E_{11}} + E_{11},$$

$$\left(\frac{c_2}{c_E}\right)^2 = E_{11},$$

$$\frac{v_x L}{c_1} = 1 - \sqrt{1 + 2E_{11}},$$

$$\left(\frac{V}{c_E}\right)^2 = -\frac{v_x L}{c_1}\left(\frac{c_1}{c_E}\right)\left(\frac{c_2}{c_E}\right)\left\{2 + \left(\frac{c_1}{c_2} - 2\right)\frac{v_x L}{c_1}\right\}$$

$$= \sqrt{E_{11}}\left(\sqrt{1 + 2E_{11}} - 1\right)\sqrt{\frac{1}{2} + \frac{1}{2}\sqrt{1 + 2E_{11}} + E_{11}}$$

$$\times \left\{2\sqrt{1 + 2E_{11}} + \left(1 - \sqrt{1 + 2E_{11}}\right)\sqrt{\frac{1}{2E_{11}} + \frac{\sqrt{1 + 2E_{11}}}{2E_{11}} + 1}\right\}.$$

$$(13.83)$$

Table 13.2 shows values from the full, exact solution. Most fibers of interest in ballistics have a failure strain less than 4%, so the table lists greater detail for small strains. For the large strains in the table, there are no known fibers that operate in this regime, but they are included to show the behavior of the solution as well as their relevance to initially slack yarns as described below. The second column,

TABLE 13.2. Values from the single yarn solution.

E_{11}	$\lambda - 1$	V/c_E	c_1/c_E	v_{xL}/c_E	c_2/c_E	U_T/c_E	$\theta(°)$	"ε_G"$/E_{11}$
0.001	0.001	0.0079	1.0007	−0.0010	0.0316	0.0307	14.4	32.6
0.005	0.005	0.0262	1.0037	−0.0050	0.0707	0.0661	21.6	15.2
0.010	0.010	0.0439	1.0075	−0.0100	0.1000	0.0910	25.7	11.0
0.015	0.015	0.0593	1.0112	−0.0151	0.1225	0.1092	28.5	9.19
0.020	0.020	0.0734	1.0148	−0.0201	0.1414	0.1241	30.6	8.10
0.025	0.025	0.0867	1.0185	−0.0252	0.1581	0.1369	32.4	7.35
0.030	0.030	0.0993	1.0221	−0.0302	0.1732	0.1481	33.8	6.80
0.035	0.034	0.1114	1.0258	−0.0353	0.1871	0.1582	35.2	6.37
0.040	0.039	0.1231	1.0294	−0.0404	0.2000	0.1675	36.3	6.03
0.045	0.044	0.1345	1.0330	−0.0455	0.2121	0.1760	37.4	5.74
0.050	0.049	0.1455	1.0365	−0.0506	0.2236	0.1839	38.3	5.50
0.075	0.072	0.1975	1.0541	−0.0763	0.2739	0.2174	42.3	4.68
0.100	0.095	0.2457	1.0713	−0.1023	0.3162	0.2442	45.2	4.19
0.150	0.140	0.3358	1.1046	−0.1548	0.3873	0.2868	49.5	3.60
0.200	0.183	0.4207	1.1365	−0.2082	0.4472	0.3209	52.7	3.24
0.300	0.265	0.5821	1.1969	−0.3171	0.5477	0.3758	57.2	2.81
0.400	0.342	0.7371	1.2533	−0.4282	0.6325	0.4203	60.3	2.55
0.500	0.414	0.8885	1.3066	−0.5412	0.7071	0.4588	62.7	2.36
0.600	0.483	1.0377	1.3571	−0.6558	0.7746	0.4931	64.6	2.22
0.700	0.549	1.1853	1.4052	−0.7717	0.8367	0.5244	66.1	2.10
0.800	0.612	1.3318	1.4513	−0.8888	0.8944	0.5534	67.4	2.01
0.900	0.673	1.4775	1.4955	−1.0070	0.9487	0.5805	68.6	1.93
1.000	0.732	1.6226	1.5382	−1.1260	1.0000	0.6060	69.5	1.86

labeled $\lambda - 1$, provides an indication of the difference between the large strain E_{11}, which has nonlinear terms, and the linear strain as shown in Eq. (13.56); λ is the stretch (language from the next chapter) and happens, in this case, to equal $-v_{xL}/c_1$ (Eq. 13.76). Recall that our constitutive model has stress linear in E_{11}.

From the full, exact solution, it is possible to obtain a small strain approximation, assuming E_{11} is negligible compared to 1:

$$\frac{c_1}{c_E} = 1, \quad \frac{c_2}{c_E} = \sqrt{E_{11}}, \quad \frac{v_{xL}}{c_1} = -E_{11}, \quad \frac{V}{c_E} = (E_{11})^{3/4}\sqrt{2 - \sqrt{E_{11}}}. \quad (13.84)$$

This result agrees with the small strain approximation of Smith, McCrackin, and Schiefer [171] and formulations attributed to Smith [50]. A nice feature of this result is that now it is possible to nearly write the transverse wave speed explicitly in terms of the applied velocity. In particular,

$$c_2 = \frac{c_E^{1/3} V^{2/3}}{(2 - c_2/c_E)^{1/3}} \approx \left(\frac{c_E}{2}\right)^{1/3} V^{2/3}. \quad (13.85)$$

A value that is straightforward to measure from photographs is the angle that the yarn, after its transverse deflection, makes with the original yarn direction. To compute this value, it is necessary to determine the location of the transverse wave. The location of the front in the original Lagrangian coordinate space is $c_2 t$. Before arriving at that point, the material has moved a distance of $v_{xL}(t - t^*)$, where t^* is the arrival time of the longitudinal wave at that point. Thus, $t^* = c_2 t/c_1$ and the distance the yarn has translated towards the centerline is $v_{xL}(1 - c_2/c_1)t$. This distance allows us to compute the transverse wave speed in the laboratory coordinates as

$$U_T = c_2 + v_{xL}(1 - c_2/c_1). \tag{13.86}$$

Computing this for the solution gives

$$\frac{U_T}{c_E} = \sqrt{E_{11}}\sqrt{1 + 2E_{11}} + (1 - \sqrt{1 + 2E_{11}})\sqrt{\frac{1}{2} + \frac{1}{2}\sqrt{1 + 2E_{11}} + E_{11}} \tag{13.87}$$

$$\approx \sqrt{E_{11}} - E_{11}.$$

The angle is then

$$\tan(\theta) = \frac{V}{U_T} = \frac{V}{c_2 + v_{xL}(1 - c_2/c_1)} \approx \frac{(E_{11})^{1/4}\sqrt{2 - \sqrt{E_{11}}}}{1 - \sqrt{E_{11}}}, \tag{13.88}$$

where the approximation at the right is for small strains. As an observation, notice that for a general stress-strain response,

$$\sigma_{11T} = \frac{\rho}{\rho_0} F_{11T} S_{11T} = \frac{\rho}{\rho_0} F_{11T}\, e'(E_{11T} + E_0) F_{11T}, \tag{13.89}$$

and so

$$U_T = c_2 F_{11T} = \sqrt{\frac{e'(E_{11T} + E_0)}{\rho_0}} F_{11T} = \sqrt{\frac{\sigma_{11T}}{\rho}}, \tag{13.90}$$

which is the classical physics result that the transverse wave speed is given by the square root of the tension in the yarn divided by the density.

An interesting fact for a yarn that is not initially in tension (i.e., $T_0 = 0$) is that the angle in the laboratory frame determines E_{11}, regardless of the yarn constitutive model $e(E_{11})$ or the value of $E_0 \leq 0$. This can be seen because the ratio c_2/c_1 does not depend on E_0 or the constitutive model in Eq. (13.78). Since $E_{11} + E_0 > 0$, more initial slackness increases E_{11} and hence increases the angle, as we would expect. If the angle is known experimentally, Table 13.2 can be used to determine the strain, because the related values of E_{11} and θ are correct, even if there is slack in the yarn or the stress-strain relationship is not linear. With E_{11}, the value of V then determines c_1 and then c_2, which then determines the axial stress in the yarn. At that point, given a constitutive model, it is possible to determine E_0.

The last column of the table is the ratio of an external geometric measure of strain, not including the inflow, and the actual strain in the yarn ("ε_G"/E_{11}). The external geometric measure of strain is what would be measured from a photograph of the yarn during the impact were no inflow included, "ε_G"$= (\sqrt{h^2 + R^2} - R)/R = \sqrt{(V/U_T)^2 + 1} - 1$. In this expression h is the height of the deflection and the extent of the deflection is R. It is telling that this ratio is not very close to 1; the inflow of the yarn greatly reduces the strain and stress in the yarn, and thus it is important to include the inflow to obtain the correct values. Other geometric approximations to the strain such as $(1/2)(h/R)^2 = (1/2)(V/U_T)^2$ and $(1/2)(V/c_2)^2$ produce similar ratios and are not good strain approximations because they do not include the inflow.

We now comment on the main role of the nonlinearity in the large strain E_{ij}. Given the single yarn solution and Table 13.2, we can write down the deformation gradient F_T. For a yarn strain of 2%, F_T and the Lagrangian strain are

$$F_T = \begin{pmatrix} 0.878 & 0 & 0 \\ 0 & 1 & 0 \\ -0.519 & 0 & 1 \end{pmatrix}, \quad (E_{ij}) = \frac{1}{2}(F_T^T F_T - I) = \begin{pmatrix} 0.020 & 0 & -0.260 \\ 0 & 0 & 0 \\ -0.260 & 0 & 0 \end{pmatrix}.$$

$$(13.91)$$

We compare the large strain, which contains nonlinear terms, to the linear small strain approximation, which drops the nonlinear terms,

$$\varepsilon = \left(\frac{1}{2} \left\{ \frac{\partial u_i}{\partial X_j} + \frac{\partial u_j}{\partial X_i} \right\} \right) = \frac{1}{2}(F_T^T + F_T - 2I) = \begin{pmatrix} -0.122 & 0 & -0.260 \\ 0 & 0 & 0 \\ -0.260 & 0 & 0 \end{pmatrix}.$$

$$(13.92)$$

We see how much error is introduced into the strain due to rotation by not including the nonlinear terms, as $E_{11} = 0.020$, as expected, while the linear small strain has $\varepsilon_{11} = -0.122$, a wildly different value. If the small strain is computed for a pure rotation, the result is $\varepsilon_{11} = \cos(\theta) - 1$. For our specific yarn example of 2% strain, the rotation is $\tan^{-1}(V/c_2) = \tan^{-1}(0.519) = 27.4°$ and so the rotation error $\cos(27.4°) - 1 = -0.112$ is the primary error that occurs when not including the nonlinear terms in the strain, and it dwarfs the actual strain by a factor of 6.

Finally, the force required to maintain the transverse speed V as constant is

$$F_z = 2S_{13}A = 2\rho_0 c_2 V A, \tag{13.93}$$

where A is the cross-sectional area of the fibers in the yarn and the factor of 2 appears since it is assumed the yarn stretches in both directions from the centerline. The force does not depend on deflection. This independence of deflection is true for all the wave problems we have examined so far, but for the yarn it seems a surprise. If the yarn was held at the ends and a static deflection was occurring, the force would depend on the deflection. However, the problem of uniform wave motion does not depend on deflection.

This last observation completes our discussion of the single yarn problem. We have developed an explicit solution which has interesting properties, in particular that the strain remains the same after the initial longitudinal strain.

We point out that Eq. (13.83) substantially overpredicts reported projectile impact speeds that break yarns if the failure strain of the yarn is inserted for E_{11}. It appears the result is due to the fact that yarns are typically struck by fragment simulating projectiles or right-circular cylinders, both of which have flat impact surfaces, so the boundary condition is not the same as the prescribed velocity at a point. When the yarn is struck, it has been hypothesized that the yarn bounces. If the yarn bounces off the surface at the impact velocity (essentially an elastic impact), then by symmetry one can envision the wave solution we have developed traveling inward and outward along the yarn from the projectile edge. The inward traveling waves meet over the center of the projectile, where the stress and hence strain in the yarn double. Thus, the appropriate strain to use is half the failure strain, and for a blunt projectile striking a yarn the impact speed at which the yarn breaks is given by

$$V_{\text{break}} = c_E(\varepsilon_f/2)^{3/4}\sqrt{2 - \sqrt{\varepsilon_f/2}}. \tag{13.94}$$

For a yarn with 4% failure strain, there is a 40% reduction in impact speed that breaks the yarn when using Eq. (13.94) vs. using Eq. (13.84). The lower speeds agree with experiment. As an additional side note, the break is no longer immediate, but it quickly occurs, as it is now given by the length of time for the longitudinal wave in the yarn to travel half the width of the blunt projectile face. A discussion of the single yarn breaking speed can be found in Walker and Chocron [201].

We might use these results to study the impact of a projectile on a fabric. Looking at the deflection of a fabric sheet it, might be thought that one need only consider the cross of fibers that interact with the projectile nose and ignore the other fibers. However, a few experimental results on impact of fabrics say we will need to look more closely at fabric deflection:

(1) The single yarn solution predicts failure immediately. Experimental work shows that the fabric deflects a certain distance before failure occurs.

(2) It is not immediately obvious how the weight of the fabric would be involved in failure. Thicker fabrics do a better job of stopping projectiles, but the single yarn solution says that the yarn breaks based on impact velocity and fiber properties, not yarn properties (i.e., the cross section of the yarn does not enter in to the yarn solution; but see Exercise 13.3).

(3) If we decide to use Eq. (13.93) (the yarn force) to decelerate the projectile, assuming it always holds so that the deceleration depends on velocity in a linear fashion, then the yarn has infinite deflection (it never comes to rest).

Given these observations, we move to examining the deflection of a fabric sheet.

13.3. Static Deflection of a Fabric Sheet Composed of 0/90 Yarns

When a projectile strikes a fabric, the anisotropic nature of the fabric sheet is evident in the deformation pattern. Figure 13.2 shows the deformation of a plate made of Spectra material with the yarns in alternating layers of $0°$ and $90°$ to the horizontal. The fabric in the plate is not woven; rather, the yarns lie next to each other and are held in place by the matrix material. A first step for us in modeling the ballistic impact event will be to model the static deflection of a fabric and recover the pyramidal deflection. This problem differs from the dynamic problem in that for the static problem we will assume the boundary is held fixed.

For a 0/90 fabric, yarns run in both directions. Given a mass of fabric, half the yarns run in the x direction and half the yarns run in the y direction. The potential energy density e will be composed of half the sum of the potential energy of the yarns in each direction. We will assume the yarns are linearly elastic and they do not slip over each other – that is, our fabric sheet is essentially a field of springs, as in Fig. 13.3. Thus, assuming there is no initial tension or strain in the yarns and that the yarns are linearly elastic, the potential energy density for the fabric is

$$\tilde{e} = \frac{1}{2}\left(e(E_{11}) + e(E_{22})\right) = \frac{E}{4}((E_{11}^2)_+ + (E_{22}^2)_+), \qquad (13.95)$$

where e is the energy in the yarn, as developed in the previous section. This strain energy applies to an ideal fabric with a plain weave and no slippage between fibers. Using Eq. (13.61), the first Piola–Kirchhoff stress is

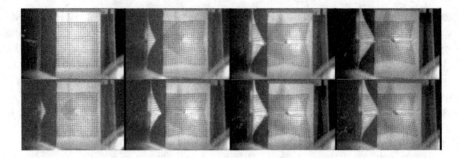

FIGURE 13.2. (0/90) Spectra panel struck by an APM2 projectile fired backwards (to provide a blunt impactor) at 357 m/s. The camera sits off to the side of the panel and its direct view is to the left in each frame. A mirror oriented at 45° sits behind the panel to allow a view of the back of the panel, where a grid has been drawn to assist in observing deformation. Notice the pyramid shape of the deformed material as well as the inflow along the fiber directions, typical of fabric and low resin fiber-based composites.

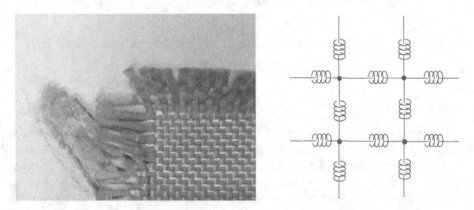

FIGURE 13.3. Image of Kevlar fabric made of yarns in a plain weave, with frayed edges showing how yarns are made of fibers (left) and idealized fabric model: an extended collection of connected springs (right). (Right from Walker [192].)

$$
S = \frac{\partial \tilde{e}}{\partial E} F^T = \frac{1}{2} \begin{pmatrix} e'(E_{11}) & 0 & 0 \\ 0 & e'(E_{22}) & 0 \\ 0 & 0 & 0 \end{pmatrix} F^T = \frac{E}{2} \begin{pmatrix} E_{11} & 0 & 0 \\ 0 & E_{22} & 0 \\ 0 & 0 & 0 \end{pmatrix} F^T
$$

$$
= \frac{E}{2} \begin{pmatrix} E_{11}F_{11} & E_{11}F_{21} & E_{11}F_{31} \\ E_{22}F_{12} & E_{22}F_{22} & E_{22}F_{32} \\ 0 & 0 & 0 \end{pmatrix}.
$$

(13.96)

Again, the first Piola–Kirchhoff stress tensor is not symmetric.

Once the stresses have been explicitly found, it is possible to determine the equations of motion for the fabric with respect to the initial coordinates. The local density of the fabric $\hat{\rho}$ is given by the ratio of its areal density $\tilde{\rho}$ and its thickness $\tilde{T}$,

$$\hat{\rho} = \frac{\tilde{\rho}}{\tilde{T}}. \tag{13.97}$$

In fact, in the model usage the density $\hat{\rho}$ is set equal to the fiber density ρ_{fiber} and the $\tilde{T}$ is computed from that assignment. The equation of motion is

$$\hat{\rho} \frac{d^2 u_i}{dt^2} = \frac{\partial S_{ji}}{\partial X_j}. \tag{13.98}$$

The physical thickness obtained by measurement in the laboratory may be larger owing to empty regions in the weave and the way the strands and layers cross and lay on each other.

13.3.1. Approximate Static Deflection Solution. The approximate analytical solution for static deflection has the features seen in tests with fabrics – a roughly pyramidal shape with curvature along the principal axes (see Fig. 13.2). A first step in finding the approximate analytical solution is solving the problem where all the motion is out of plane; that is, where $u_x = u_y = 0$. This motion is sometimes referred to as anti-plane shear. The solution will be for static deformation of a single fabric sheet under loading conditions similar to those experienced in a ballistic event and similar to those seen in static tests. The solution to the strictly out of plane problem will then be used to find an approximate solution to the complete problem.[6]

Assuming that the motion is strictly out of plane says,

$$F = \begin{pmatrix} 1 & 0 & 0 \\ 0 & 1 & 0 \\ F_{31} & F_{32} & 1 \end{pmatrix} = \begin{pmatrix} 1 & 0 & 0 \\ 0 & 1 & 0 \\ \frac{\partial u_z}{\partial X} & \frac{\partial u_z}{\partial Y} & 1 \end{pmatrix}, \tag{13.99}$$

and the nonzero strains are

$$E_{11} = \frac{F_{31}^2}{2}, \qquad E_{22} = \frac{F_{32}^2}{2}. \tag{13.100}$$

This leads to the stress

$$S = \frac{E}{4} \begin{pmatrix} F_{31}^2 & 0 & 0 \\ 0 & F_{32}^2 & 0 \\ 0 & 0 & 0 \end{pmatrix} F^T = \frac{E}{4} \begin{pmatrix} F_{31}^2 & 0 & F_{31}^3 \\ 0 & F_{32}^2 & F_{32}^3 \\ 0 & 0 & 0 \end{pmatrix}$$

$$= \frac{E}{4} \begin{pmatrix} (\partial u_z/\partial X)^2 & 0 & (\partial u_z/\partial X)^3 \\ 0 & (\partial u_z/\partial Y)^2 & (\partial u_z/\partial Y)^3 \\ 0 & 0 & 0 \end{pmatrix}. \tag{13.101}$$

The equation of motion in the out of plane z direction is

$$\hat{\rho} \frac{d^2 u_z}{dt^2} = \frac{\partial S_{13}}{\partial X_1} + \frac{\partial S_{23}}{\partial X_2} = \frac{E}{4} \left\{ \frac{\partial}{\partial X} \left(\frac{\partial u_z}{\partial X} \right)^3 + \frac{\partial}{\partial Y} \left(\frac{\partial u_z}{\partial Y} \right)^3 \right\}. \tag{13.102}$$

For the static case, with the boundary held fixed and the fabric indented at the center, Eq. (13.102) can be solved explicitly. On the boundary it is assumed that $u_z(R, 0) = 0$ and $u_z(0, R) = 0$. Part of the problem is finding the boundary curve

---

[6]Unfortunately, the nice explicit solution we find does not extend to fabric with slack, which prevents including yarn inflow. Since inflow occurs in an impact, we will compensate by increasing the transverse wave speed (Eq. 13.136). Another approach to modeling fabric impacts by Phoenix and Porwal [155] assumes isotropy in the plane and includes inflow from all directions.

that connects $(R,0)$ and $(0,R)$ on which $u_z(X,Y) = 0$. The deformed height at the center is denoted $h = u_z(0,0)$. Considering only the first quadrant introduces boundary conditions along the x- and y-axis,

$$\frac{\partial u_z}{\partial X} = 0 \quad \text{for} \quad X = 0;$$

$$\frac{\partial u_z}{\partial Y} = 0 \quad \text{for} \quad Y = 0; \tag{13.103}$$

$$u_z = 0 \quad \text{along a plane curve to be determined.}$$

Since some type of symmetry is expected in the problem, an assumed form of solution is introduced. The form of r below is quite general (notice that r does not have the dimensions of length unless $\beta = 1$),

$$u_z(X,Y) = u(r), \qquad r = X^\beta + Y^\beta. \tag{13.104}$$

As a general equation of the form below, the static problem, Eq. (13.102), reduces to

$$\frac{\partial}{\partial X}\left(\frac{\partial u_z}{\partial X}\right)^\alpha + \frac{\partial}{\partial Y}\left(\frac{\partial u_z}{\partial Y}\right)^\alpha = \alpha\beta^\alpha (u')^{\alpha-1}\{X^{\alpha\beta-\alpha-1}((\beta-1)u' + \beta X^\beta u'')$$

$$+ Y^{\alpha\beta-\alpha-1}((\beta-1)u' + \beta Y^\beta u'')\} = 0, \tag{13.105}$$

where the prime denotes differentiation with respect to r. This equation becomes an equation in r if the power $\alpha\beta - \alpha - 1 = 0$, or

$$\beta = 1 + 1/\alpha. \tag{13.106}$$

For the particular case of the linear stress-strain relation we are using, $\alpha = 3$ and $\beta = 4/3$. This leads to the term in brackets equaling

$$\frac{2}{\alpha}u' + \frac{\alpha+1}{\alpha}ru'' = 0. \tag{13.107}$$

This ordinary differential equation has the solution

$$u(r) = C_1 r^{1-2/(\alpha+1)} + C_2, \tag{13.108}$$

where C_1 and C_2 are arbitrary constants, and are found based on the boundary conditions. Equation (13.108) solves the out of plane fabric deformation problem. The boundary curve on which the deflection is zero is $r = R^{1+1/\alpha}$. The solution in general is

$$u_z(X,Y) = h\left\{1 - \left((X/R)^{1+1/\alpha} + (Y/R)^{1+1/\alpha}\right)^{1-2/(\alpha+1)}\right\}$$

$$= h\left\{1 - \sqrt{(X/R)^{4/3} + (Y/R)^{4/3}}\right\}. \tag{13.109}$$

The solution satisfies the boundary conditions on the axes, Eq. (13.103). For the taut fabric, there is a slight curvature to the pyramidal shape, as seen in Fig. 13.2. Given the solution, it is possible to compute the derivatives of interest,

$$\frac{\partial u_z}{\partial X} = -\frac{\alpha-1}{\alpha}\frac{h}{R}\frac{(X/R)^{1/\alpha}}{\{(X/R)^{1+1/\alpha} + (Y/R)^{1+1/\alpha}\}^{2/(1+\alpha)}}$$

$$= -\frac{2}{3}\frac{h}{R}\frac{(X/R)^{1/3}}{\sqrt{(X/R)^{4/3} + (Y/R)^{4/3}}}. \tag{13.110}$$

The partial with respect to Y is similar, with X and Y exchanged in the expression.

In addition to the out-of-plane deflection of the fabric, there is also in-plane motion. To obtain an in-plane displacement, it is assumed that the out-of-plane deflection is given and we then solve for the in-plane deflection (essentially a perturbation solution). In particular, the deformation gradient is

$$F = \begin{pmatrix} F_{11} & F_{12} & 0 \\ F_{21} & F_{22} & 0 \\ F_{31} & F_{32} & 1 \end{pmatrix} = \begin{pmatrix} 1 + \frac{\partial u_x}{\partial X} & \frac{\partial u_x}{\partial Y} & 0 \\ \frac{\partial u_y}{\partial X} & 1 + \frac{\partial u_y}{\partial Y} & 0 \\ \frac{\partial u_z}{\partial X} & \frac{\partial u_z}{\partial Y} & 1 \end{pmatrix} \tag{13.111}$$

and so the stress is

$$S = \frac{E}{2} \begin{pmatrix} E_{11}\left(1 + \frac{\partial u_x}{\partial X}\right) & E_{11}\frac{\partial u_y}{\partial X} & E_{11}\frac{\partial u_z}{\partial X} \\ E_{22}\frac{\partial u_x}{\partial Y} & E_{22}\left(1 + \frac{\partial u_y}{\partial Y}\right) & E_{22}\frac{\partial u_z}{\partial Y} \\ 0 & 0 & 0 \end{pmatrix}. \tag{13.112}$$

The equation of motion for the in-plane motion in the x direction is then

$$\hat{\rho}\frac{d^2 u_x}{dt^2} = \frac{\partial S_{11}}{\partial X_1} + \frac{\partial S_{21}}{\partial X_2} = \frac{E}{2}\left\{\frac{\partial}{\partial X}\left(\left(1 + \frac{\partial u_x}{\partial X}\right)E_{11}\right) + \frac{\partial}{\partial Y}\left(\frac{\partial u_x}{\partial Y}E_{22}\right)\right\}. \tag{13.113}$$

E_{11} was given above in Eq. (13.54), and E_{22} is the same, only with the derivatives with respect to Y instead of X. We will assume that a number of these terms are small and can be ignored. In particular, it will be assumed that $\partial u_x/\partial X$ is small compared to 1, and that $\partial u_x/\partial Y$ and $\partial u_y/\partial X$ are small compared to $\partial u_x/\partial X$, reducing the equation to

$$\hat{\rho}\frac{d^2 u_x}{dt^2} \approx \frac{E}{2}\frac{\partial}{\partial X}(E_{11}). \tag{13.114}$$

Thus, the static in-plane displacement is given by $\partial E_{11}/\partial X = 0$. This equation can be integrated to give

$$E_{11} = \frac{\partial u_x}{\partial X} + \frac{1}{2}\left(\frac{\partial u_z}{\partial X}\right)^2 = f(Y), \tag{13.115}$$

where f is an arbitrary function of Y and where our representation of E_{11} includes our assumptions about small terms. As written, this equation cannot be integrated in closed form. To do so, we expand the $\partial u_x/\partial X$ about its value on the outer boundary curve. First, the value of X on the outer boundary curve of the region is

$$X_R(Y) = R\left\{1 - (Y/R)^\beta\right\}^{1/\beta}. \tag{13.116}$$

Then define $b_R(Y)$ to equal the X partial derivative of u_z along the outer boundary curve

$$b_R(Y) = \frac{\partial u_z}{\partial X}(X_R(Y), Y). \tag{13.117}$$

Specifically,

$$b_R(Y) = -\frac{\alpha - 1}{\alpha}\frac{h}{R}\left\{1 - (Y/R)^{1+1/\alpha}\right\}^{1/(\alpha+1)} = -\frac{2}{3}\frac{h}{R}\left\{1 - (Y/R)^{4/3}\right\}^{1/4}. \tag{13.118}$$

Then,

$$
\begin{aligned}
E_{11} &= \frac{\partial u_x}{\partial X} + \frac{1}{2}\left(b_R(Y) + \frac{\partial u_z}{\partial X} - b_R(Y)\right)^2 \\
&= \frac{\partial u_x}{\partial X} + b_R(Y)\frac{\partial u_z}{\partial X} - \frac{1}{2}b_R^2(Y) + \frac{1}{2}\left(\frac{\partial u_z}{\partial X} - b_R(Y)\right)^2.
\end{aligned}
\tag{13.119}
$$

It will be assumed that the last term is small, which is true for some of the strained region, in particular for larger X and Y, to be made more precise later. This assumption then reduces the equation for the in plane strain, where we have lumped the $(1/2)b_R^2(Y)$ term in the arbitrary function of Y denoted $\tilde{f}$, to

$$
\frac{\partial u_x}{\partial X} + b_R(Y)\frac{\partial u_z}{\partial X} = \tilde{f}(Y).
\tag{13.120}
$$

This can now be integrated with respect to X,

$$
u_X(X,Y) + b_R(Y)\,u_z(X,Y) = \tilde{f}(Y)X + g(Y),
\tag{13.121}
$$

where g is an arbitrary function of Y. On the inner boundary $X = 0$ by symmetry and $u_x = 0$, and so

$$
g(Y) = b_R(Y)\,u_z(0,Y).
\tag{13.122}
$$

Evaluating on the outer boundary where displacements u_x and u_z are zero gives

$$
\tilde{f}(Y) = -\frac{g(Y)}{x_R(Y)}.
\tag{13.123}
$$

Combining these results gives for the in plane x displacement

$$
u_x(X,Y) = -b_R(Y)\,u_z(X,Y) + b_R(Y)u_z(0,Y)\left(1 - \frac{X}{X_R(Y)}\right).
\tag{13.124}
$$

Specifically,

$$
\begin{aligned}
u_x(X,Y) = -\frac{2}{3}\frac{h^2}{R}\left\{1 - (Y/R)^{4/3}\right\}^{1/4}&\left\{\left(1 - (Y/R)^{2/3}\right)\left[1 - \frac{X/R}{\left\{1-(Y/R)^{4/3}\right\}^{3/4}}\right]\right. \\
&\left. +\sqrt{(X/R)^{4/3} + (Y/R)^{4/3} - 1}\right\}.
\end{aligned}
\tag{13.125}
$$

By symmetry, the u_y solution is $u_y(X,Y) = u_x(Y,X)$.

We are interested in computing strains. To that end,

$$
\frac{\partial u_x}{\partial X} = -b_R(Y)\frac{\partial u_z}{\partial X} - \frac{b_R(Y)}{X_R(Y)}u_z(0,Y).
\tag{13.126}
$$

The original definition of strain gives

$$
E_{11} = \frac{\partial u_x}{\partial X} + \frac{1}{2}\left(\frac{\partial u_z}{\partial X}\right)^2 = -b_R(Y)\left(\frac{\partial u_z}{\partial X} + \frac{u_z(0,Y)}{X_R(Y)}\right) + \frac{1}{2}\left(\frac{\partial u_z}{\partial X}\right)^2,
\tag{13.127}
$$

yielding

$$
E_{11} = \frac{2}{3}\left(\frac{h}{R}\right)^2\left\{\frac{(X/R)^{2/3}}{3r/R^{4/3}} - \frac{2(1 - (Y/R)^{4/3})^{1/4}(X/R)^{1/3}}{3\sqrt{r}/R^{2/3}} + \frac{1 - (Y/R)^{2/3}}{\sqrt{1-(Y/R)^{4/3}}}\right\},
\tag{13.128}
$$

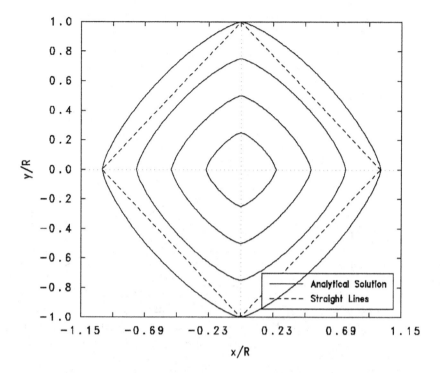

FIGURE 13.4. Curves of constant z deflection. Dashed lines are straight lines to highlight the solution curvature.   (From Walker [192].)

where again $r = X^{4/3} + Y^{4/3}$. Along the x-axis $(Y = 0)$ this reduces to

$$E_{11} = \frac{2}{9} \left(\frac{h}{R} \right)^2 \left\{ \left(\frac{R}{X} \right)^{2/3} - 2 \left(\frac{R}{X} \right)^{1/3} + 3 \right\}. \tag{13.129}$$

Examples of the solution are given in Figs. 13.4 and 13.5. Figure 13.4 shows the view from above the fabric sheet, showing the contours on which the z displacement is constant. These curves are given by

$$X^{4/3} + Y^{4/3} = \text{constant}. \tag{13.130}$$

The behavior shows the pyramidal shape seen in Fig. 13.2. The outer curve is the boundary curve, one of the original unknowns, on which the $u_z(X, Y) = 0$. The dashed lines are straight lines to give a better feel for the curvature in the solution.

Figure 13.5 shows the solution along the x-axis compared to a full transient numerical solution (similar to [167]). Along the axis the x displacement is (from Eq. 13.125)

$$u_x(X, 0) = \frac{2 h^2}{3 R} \left\{ \frac{X}{R} - \left(\frac{X}{R} \right)^{2/3} \right\}. \tag{13.131}$$

The curvature is similar to that seen in Fig. 13.2. Agreement between the two solutions is excellent. Again, the straight lines (dashed) are for comparison. The general pyramidal shape seen in tests is evident in these solutions.

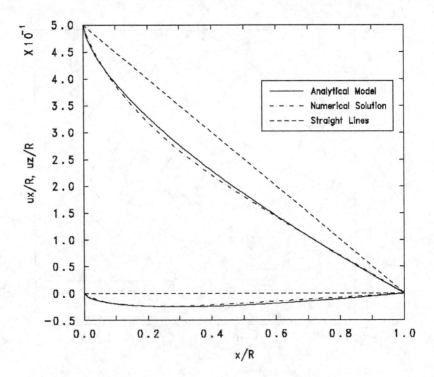

FIGURE 13.5. Analytic solution along x-axis compared to full numerical solution. (From Walker [192].)

Figure 13.6 shows the strain E_{11} along the x-axis for $h/R = 1/4$. The solid curve is with the complete solution, Eq. (13.129). The dot-dash curve uses just the out-of-plane solution, which is given by the first term in brackets of Eq. (13.129). The dashed curve is the strain of the straight line approximation given by $(1/2)(h/R)^2$, a little over 3% for the specified deflection. It is seen that the strains for the out-of-plane solution differ from the full solution, so any calculation requiring those strains, such as the force deflection calculation below or an energy calculation, should use the complete approximate solution. The solution is singular at the origin. The approximate strains are less than the straight line approximation near the outer edge and they increase as the indentation point is approached.

Equation (13.129) explains why fabrics break at a time later than at impact. Impact tests show that h/R is often constant during an impact event. As the impact event proceeds, the radius R expands. However, x is fixed, since it is the radius of the projectile. Thus, as the impact progresses, X/R decreases, and the strains slowly increase until the fiber failure strain is reached.

13.3.2. Force vs. Deflection. A force vs. deflection curve identifies how a clamped piece of fabric resists a center indentation. The force required to deflect the fabric to a height h along the axis of symmetry is given by

$$F_z = \int_{\partial A} S_{j3} n_j \tilde{T} dl, \tag{13.132}$$

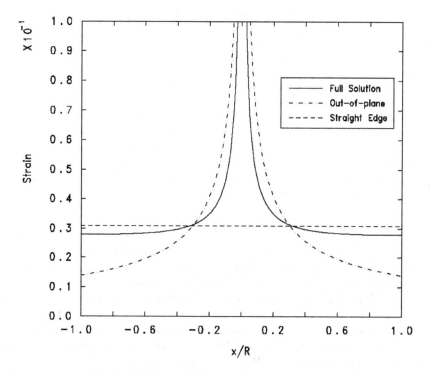

FIGURE 13.6. Strains along the x-axis for full analytic solution (solid line), out of plane solution (dot-dash line), and straight line solution (dashed line). (From Walker [192].)

where ∂A is the boundary of the region and dl is the arc length on the boundary. If we choose to use a circle of radius $\tilde{R}$ as our boundary, and we use the symmetry of the problem, we have

$$
\begin{aligned}
F_z &= \int_{\partial A} (S_{13}\cos(\theta) + S_{23}\sin(\theta)\}\tilde{T}\tilde{R}d\theta \\
&= 8\tilde{T}\int_0^{\pi/2} S_{13}\cos(\theta)\tilde{R}d\theta = 8\tilde{T}\int_0^{\pi/2} S_{13}X d\theta.
\end{aligned}
\tag{13.133}
$$

(Alternately, the loading surface could be taken to be that described by the curve $X^{4/3} + Y^{4/3} = \tilde{R}^{4/3}$.) With

$$
S_{13} = \frac{E}{2}E_{11}\frac{\partial u_z}{\partial X},
\tag{13.134}
$$

inserting the results for the two terms on the right-hand side (Eq. 13.109 and 13.129) into the integral and evaluating the integral shows an $\tilde{R}$ dependence, but an approximate numerical evaluation near the outer boundary gives

$$
F_z \approx -\frac{8}{9}\frac{h^3}{R^2}\frac{E}{2}\tilde{T}.
\tag{13.135}
$$

The force depends on the cube of the deflection, where the fabric is static with its edges held in place. Such cubic behavior is also seen in shell and membrane theories.

13.4. The Ballistic Limit of a Fabric

So far, our analysis of yarns and fabrics has been exact. Now, we will begin making assumptions and approximations to yield a closed-form ballistic limit expression.

13.4.1. Transverse Wave Speed in the Fabric. To determine how much of the fabric is involved in the ballistic event, it is necessary to compute a transverse wave speed. For our purposes, we will approximate the transverse speed in the fabric with the transverse speed of a single yarn, computed earlier (c_2). However, we will make a further approximation. As will be seen in the next section, it will be possible to write an explicit formula for the ballistic limit if c_2 depends linearly on $\sqrt{V}$. Because of this, we will

$$\text{replace} \quad c_2 = \frac{1}{2^{1/3}} c_E^{1/3} V^{2/3} \quad \text{with} \quad c_2 \approx c_E^{1/2} V^{1/2} = \sqrt{c_E V}. \tag{13.136}$$

This substitution is driven by what we will be able to explicitly solve, but a qualitative argument in its favor is as follows: this substitution essentially increases the size of the fabric involved in the impact, reducing the strain. However, in our fabric model we have not included the longitudinal inflow of yarns towards the impact point, which also reduces the strain. Thus, the larger radius will compensate for our not including the inflow in our force vs. deflection calculation: if the angle of the pyramid is desired, use Eq. (13.85), but for the ballistic limit calculation to follow use the right-hand side of Eq. (13.136).

Another approximation we will make is that the angle of the fabric pyramid remains the same throughout the impact event – that is, h/R remains constant and the shape of pyramid is self similar. This behavior is observed in photographs of the impact of fabrics. Thus, we have

$$\frac{h}{R} = \frac{Vt}{c_2 t} = \sqrt{\frac{V}{c_E}}, \tag{13.137}$$

where V is the impact velocity and c_E is the uniaxial stress wave speed in the fiber.

13.4.2. Fabric Ballistic Limit. The final step in determining the ballistic limit is to combine the force vs. deflection curve and the region of affected fabric with a momentum balance. Our expression for force vs. deflection for the fabric is a function of deflection, but we expect based on our experience with the single yarn that the force, for a while at least, is independent of deflection. Therefore, we will assume that the force is independent of deflection for the entire deceleration event, and we will write the force as

$$F_z = -\frac{8}{9} \left(\frac{h}{R} \right)^2 \frac{E}{2} \tilde{T} h_{\text{final}}. \tag{13.138}$$

Thus, we will use our force vs. deflection expression and assume that the shape of the deformation pyramid remains the same (i.e., h/R remains constant at its initial value) and that the force is given by the maximum deflection value. Initially, we do not know that value, but we will be able to solve for it.

The equation of motion for a projectile impacting a fabric is given by

$$(m_p + m_f) \frac{dv}{dt} = F_z = -\frac{8}{9} \left(\frac{h}{R} \right)^2 \frac{E}{2} \tilde{T} h_{\text{final}}, \tag{13.139}$$

where m_p is the mass of the projectile, m_f is the mass of the fabric directly involved in the region of the projectile, and v is the projectile velocity. If the shape of the deformed fabric remains self-similar as deflection increases, then h/R is constant, as described above. Use of $dv/dt = v\,dv/dh$ leads to

$$v^2 - V^2 = -\frac{8}{9}\left(\frac{h}{R}\right)^4 c_E^2 \frac{\tilde{\rho}}{m_p + m_f} R_{\text{final}} R. \tag{13.140}$$

Since the fabric motion just comes to rest at the ballistic limit, Eq. (13.140) becomes

$$V_{bl}^2 = \frac{8}{9}\left(\frac{h}{R}\right)^4 c_E^2 \frac{\tilde{\rho}}{m_p + m_f} R_{bl}^2, \tag{13.141}$$

where R_{bl} is the radial extent of the fabric involved in the impact as it comes to rest at the ballistic limit velocity. Equations (13.137) and (13.140) allow a calculation of R_{bl},

$$\frac{R_{bl}}{R_p} = \sqrt{\frac{9(m_p + m_f)}{8\tilde{\rho}R_p^2}} = \sqrt{\frac{9\pi}{8}\left(\frac{1}{\chi} + \beta\right)}, , \tag{13.142}$$

where both sides have been divided by R_p, the radius of the projectile, assumed to be a right-circular cylinder impacting on end. χ is the areal density of the fabric times the presented area of the projectile A_p divided by the mass of the projectile,

$$\chi = \frac{\tilde{\rho}A_p}{m_p}, \tag{13.143}$$

that is, the ratio of the areal densities of the fabric and projectile. Specifically for a projectile which is a right-circular cylinder,

$$\chi = \frac{\tilde{\rho}\pi R_p^2}{m_p} = \frac{\rho_{\text{fiber}}\tilde{T}}{\rho_p L}. \tag{13.144}$$

With this definition, $m_f = \beta\chi m_p$. χ has been found by Cunniff to be an important variable in scaling ballistic response of fabrics [52]. β is the area multiplier of the fabric carried along by the projectile, and it will be taken to be $\beta = (1.6)^2 = 2.56$, i.e., an additional 60% projectile radius of fabric beyond the presented area of the projectile is carried along directly by the projectile ($m_f = \beta\pi R_p^2\tilde{\rho}$). The initial momentum transfer bringing this amount of fabric up to speed is included in the ballistic limit expression,

$$V(0) = \frac{m_p}{m_p + m_f} V_{\text{impact}}. \tag{13.145}$$

Assuming the fabric breaks at a strain ε_f, the ballistic limit equation for the fabric is obtained through the strain equation from the explicit static solution, Eq. (13.129), and from the h/R expression, Eq. (13.137),

$$V_{bl} = \frac{9}{2}(1 + \beta\chi)\,c_E\varepsilon_f\left\{\left(\frac{R_{bl}}{R_p}\right)^{2/3} - 2\left(\frac{R_{bl}}{R_p}\right)^{1/3} + 3\right\}^{-1}. \tag{13.146}$$

Since R_{bl}/R is a function of χ and β only, the material properties of the fabric enter into this expression by the multiplicative term $c^* = c_E\varepsilon_f$. Our result predicts that the ballistic limit is directly proportional to both the fiber (yarn) failure strain and to the fiber (yarn) sound speed. c^* has units of velocity and will later also be shown to be related to the elastic strain energy to failure. The slope of V_{50} vs. χ for large χ is proportional to β. For very small χ the behavior is $V_{bl} \approx 3c_E\varepsilon_f\chi^{1/3}$.

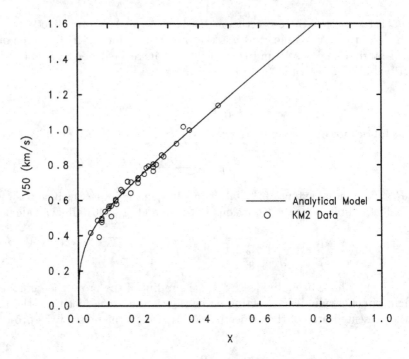

FIGURE 13.7. Analytical ballistic limit prediction compared with V_{50} data for Kevlar KM2. (Figure from Walker [192].)

Figure 13.7 compares the analytic V_{bl} curve with V_{50} data [52] for a woven Kevlar KM2 (850 denier yarns), with a fiber wave speed $c_E = 7,154\,\mathrm{m/s}$ and a breaking strain $\varepsilon_f = 3.8\%$. These data are for a wide range of projectile sizes, from 2 to 64 grain (0.25 to 4.15 g), and for 12 to 44 sheets of fabric (for the particular fabric used, 44 sheets have $\tilde{\rho} = 1.0\,\mathrm{g/cm^2}$). Though this specific comparison case is with Kevlar KM2, we will show below that this model provides good agreement for many fabrics.

Equation (13.142) deserves comment. It predicts that the radius R_{bl} of fabric material involved in the deformation at the ballistic limit – that is, when the projectile comes to rest just as the strain in the fabric at the point of maximal strain (the fabric in contact with the outer radius of the projectile) reaches the fiber failure strain – does not depend on the fabric elastic properties. In particular, it depends only on the areal density of the fabric, the amount of fabric inertially involved, and the mass and face area of the projectile; it does not depend on failure strain or wave speed of the fiber. The deflection of the fabric h and the ballistic limit velocity do depend on the material properties of the fabric. Given two different fabrics of the same areal density, Eq. (13.142) predicts the same R_{bl}, though the ballistic limit velocities differ if $c_E \varepsilon_f$ differ (Eq. 13.146).

An approach using the single yarn behavior, rather than the anisotropic sheet as was done in the development of the above ballistic limit result, and hence taking into account the yarn inflow, yields something of the form (Exercise 13.3)

$$V_{bl} = (1 + \beta\chi)V_{\text{break}}, \tag{13.147}$$

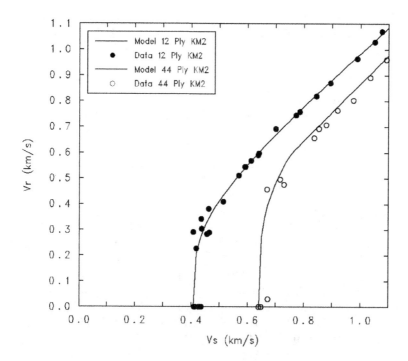

FIGURE 13.8. Residual velocity vs. impact velocity for a 4.15 g $L/D = 1$ right-circular cylinder striking 12 and 44 plies of Kevlar KM2 (data from Cunniff [51]).

where V_{break} is given in Eq. (13.94). This result lacks the small χ nonlinear behavior seen in the data but roughly bounds the ballistic limit from above. In fact, Eq. (13.146) for large χ asymptotes to a straight line and for $\beta = 2.56$ the $\chi = 0$ intercept is roughly $2c_E \varepsilon_f$, which is typically slightly larger than V_{break}.

13.4.3. Residual Velocity. When a fabric fails, it tears and there is less momentum transferred to the fabric than might be expected otherwise. Examining the experimental residual velocity, a limiting form at high striking velocity of the residual velocity is $v_r = V/(1 + \chi\beta(V_{bl}/V)^2)$. Essentially, the velocity is being reduced by a smaller amount of interaction with the fabric, and the amount of fabric involved, as reflected by β, decreases as the impact velocity increases. To link this high velocity form with the ballistic limit velocity, we use a Lambert form $(v_r = a(V^p - V_{bl}^p)^{1/p})$ with the exponential between 3 and 4; for example,

$$v_r = \min\left\{(V^{3.5} - V_{bl}^{3.5})^{1/3.5}, \frac{V}{1 + \chi\beta(V_{bl}/V)^2}\right\}. \qquad (13.148)$$

This expression is compared to data for a 64 grain (4.15 g) $L/D = 1$ right-circular cylinder of steel striking 12 and 44 plies of KM2 in Fig. 13.8. The agreement is very good. For these, $\chi = 0.040$ and 0.146. The data are given in Table 13.3. This type of data is what provides the V_{50} data for the fabrics; it can be seen that there is a zone of mixed results. In particular, the lower χ value corresponds to the lowest V_{50} in Fig. 13.7.

TABLE 13.3. Residual velocity vs. impact velocity for a 4.15 g
$L/D = 1$ right-circular cylinder striking 12 and 44 plies of Kevlar
KM2 [51].

No. of plies	V (m/s)	V_r (m/s)	No. of plies	V (m/s)	V_r (m/s)
12	405.7	290.2	12	784.3	758.0
12	408.7	0.0	12	843.1	818.4
12	410.9	0.0	12	892.8	869.3
12	416.4	226.8	12	987.6	963.2
12	423.7	0.0	12	1,049.1	1,029.0
12	433.4	0.0	12	1,074.7	1,069.9
12	434.0	341.1			
12	435.9	303.0	44	638.6	0.0
12	454.2	282.6	44	646.8	0.0
12	460.3	380.1	44	668.4	456.6
12	462.1	288.0	44	671.2	30.5
12	512.1	408.4	44	714.8	495.6
12	567.5	510.5	44	728.5	475.5
12	589.2	544.4	44	835.8	655.9
12	591.6	544.4	44	852.2	691.9
12	611.4	567.8	44	877.8	706.8
12	637.0	589.2	44	918.7	763.5
12	639.5	596.5	44	974.5	802.5
12	699.2	691.0	44	1,033.6	890.6
12	771.2	744.6	44	1,091.2	959.8

13.4.4. Deflection vs. Time. Though the expression for ballistic limit gives
excellent agreement with V_{50} data, as we mentioned, it predicts less deflection
than seen in tests. If the deflection of the fabric sheet vs. time is of interest, an
approach that works well is to use momentum conservation for the projectile and
the amount of fabric moving out of plane [5]. It is necessary to assume an angle for
the pyramid, but good experimental results are found for around $\theta = 25°$, which
is what we obtained as an angle for yarns for low strains (Table 13.2). We assume
that there is inflow of fabric material, so we will compute the area of the faces of
the pyramid and multiply that by the initial areal density of the fabric sheet, and
we will assume the material in the fabric pyramid is moving at the same velocity
as the projectile.

To compute the area of the pyramid, the height of the pyramid is h and the
length of the edge of the pyramid is $h/\sin(\theta)$. The vertices of the pyramid in (x, y, z)
are the apex $(0, 0, h)$ and the corners $(h \cot(\theta), 0, 0)$, $(0, h \cot(\theta), 0)$, $(0, -h \cot(\theta), 0)$,
and $(0, -h \cot(\theta), 0)$. The first three vertices define one of the four faces, and we
can use them to define the vectors connecting the corners to the apex to compute
the area through a cross product. The normal to one face of the pyramid is

$$(-h \cot(\theta), 0, h) \times (0, -h \cot(\theta), h) = (h^2 \cot(\theta), h^2 \cot(\theta), h^2 \cot^2(\theta)) \quad (13.149)$$

and the area of the face is given by 1/2 the length of this vector. Since there are four faces to the fabric pyramid, the total area of the fabric pyramid is

$$A_{pyr} = 2h^2 \cot(\theta)\sqrt{2 + \cot^2(\theta)} = 2h^2 \cot(\theta)\sqrt{1 + \csc^2(\theta)}, \tag{13.150}$$

where the definition of the cosecant is $1/\text{sine}$. Thus, conservation of momentum says

$$(m_p + \tilde{\rho}A_{pyr})v = \text{constant} = m_p V. \tag{13.151}$$

Explicitly, we have

$$v(h) = \frac{m_p V}{m_p + 2h^2 \tilde{\rho} \cot(\theta)\sqrt{1 + \csc^2(\theta)}}. \tag{13.152}$$

To obtain the displacement as a function of time, we note that $v = dh/dt$ and we integrate this expression to obtain

$$h + \frac{2}{3}h^3 \frac{\tilde{\rho}}{m_p} \cot(\theta)\sqrt{1 + \csc^2(\theta)} = V(t - t_0). \tag{13.153}$$

The t_0 is explicitly mentioned because it takes some initial time for the pyramid to form. This expression provides good results for the height. It will be noted that it predicts the projectile never comes to rest; to model the whole deceleration typically involves including the fabric interaction with the test frame or other restraints. If a projectile perforates the fabric, these expressions will be good for early time, but as fabric sheets fail they fall out of the momentum balance and the projectile velocity does not decrease as much as predicted.

13.5. Fiber-Based Composites

Composite panels made from fabrics and resin are also important components in modern light armors (Fig. 13.9). Experimental work with panels made from fabrics and resin has indicated that for a few sheets of fabric, "dry is better," meaning that fabric sheets with no resin (dry) outperform the equivalent areal density of a fabric/resin composite. Simply put, for low relative areal densities of fabric, the loss in fabric material by adding resin in order to maintain the same areal density leads to a loss in performance of the armor system. However, as the relative areal density increases, the fabric with resin composite panel begins to exhibit bending stiffness, and its performance increases. Experimentally it has been observed that the crossover in ballistic limit performance is in the region of $\chi = 1/\beta$ (see comments below). However, as the number of fabric sheets increases, additional mechanisms are involved. There are five mechanisms that come into play as resin is added to fabric:

(1) As noted above, removal of fibers when resin is added (to maintain the same areal density) tends to reduce the ballistic limit, since (especially for thin sheets) the ballistic resistance is in the tensile strength of the fibers.

(2) The increase in resin leads to a resistance to bending in the fabric/resin composite (as opposed to the dry fabric that only has tensile membrane stresses), thus increasing the panel's resistance to deformation and thus increasing the ballistic limit;

(3) The newly acquired bending strength leads to an increase in the transverse wave speed, thus reducing the strain and increasing the ballistic limit;

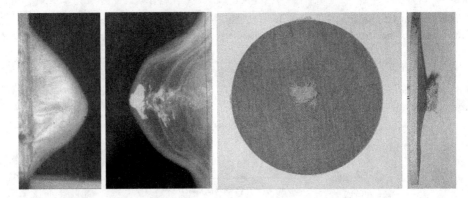

FIGURE 13.9. Kevlar composite with an alumina face struck by 7.62-mm LPS round (mild steel core). First and second orthogonal images are at similar time for a nonperforating strike; the left image is optical and the right image is an X-ray. There is delamination and large deflection of the composite and large deformation of the projectile. The two right images are posttest photographs of a perforating strike, showing final deformation.

(4) The now harder panel deforms the projectile, leading to an increase in the presented area of the projectile and thus leading to an increase in the ballistic limit;

(5) It is possible that for larger amounts of resin the harder panel will rigidly hold the fabric, allowing the fibers to be sheared by the projectile rather than failing in tension (their optimal failure mode), thus leading to a reduction in ballistic limit.

These mechanisms all have a role, but here the focus will be on the first two mechanisms, where it is possible to quantitatively estimate the magnitude of the effect. The reduction in fabric (to maintain the same areal density when resin is added) reduces the modulus of the fabric as $E = E_f(1 - r)$, where r is the mass fraction of the resin, which leads to a reduction in the ballistic limit as $\sqrt{1 - r}$, since the Young's modulus enters into the ballistic limit equation through a square root term (in c_E in Eq. 13.146). Resistance to bending in a beam or plate is proportional to the moment of inertia of the beam or plate, which for a constant cross section is proportional to the cube of the thickness. For the fabric with resin, the thickness of the plate will be assumed proportional to the areal density and so it is possible to write heuristically the increase in stiffness of the plate in the form

$$E = E_f(1 - r)(1 + \gamma(r)\chi^3), \qquad (13.154)$$

where γ is a function of r that is to be determined and the $1 - r$ term comes from the reduction in fabric to maintain the same areal density when resin is added. Experimentally, the ballistic limit of the fabric with resin and the dry fabric often occurs in the vicinity of $\chi = 0.3$ to 0.5, and so here it will be approximated by $\chi = 1/\beta = 0.39$. (This assumption is intriguing, since it says that the crossover in ballistic limit behavior occurs when $\beta A_p \tilde{\rho} = m_p$ – that is, it occurs at the point where the mass of the fabric or fabric/resin panel inertially involved in the impact event equals the mass of the projectile.) These assumptions give $\gamma(r) = r\beta^3/(1-r)$,

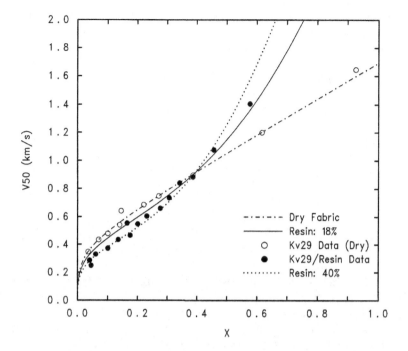

FIGURE 13.10. Ballistic limit curves for dry Kevlar 29 fabric and Kevlar 29/resin composite panel; points are data from [52]; curves are from Eqs. (13.146) and (13.155). (From Walker [193].)

which gives as an expression for ballistic limit velocity (again, the adjustment comes through c_E in Eq. 13.146)

$$V_{bl}(\chi, r) = \sqrt{1 - r + r(\beta\chi)^3} \; V_{bl}(\chi, 0). \tag{13.155}$$

Equation (13.155) applied to Eq. (13.146) is shown in Fig. 13.10 for 1,000 denier Kevlar 29, $c_E = 7,400$ m/s and $\varepsilon_f = 3.25\%$, and the composite panel is 18% resin ($r = 0.18$). The curves agree well with the data taken for a PVB/phenolic composite using Kevlar 29 [52]. The experimental degradation in ballistic limit velocity is greater than predicted for small χ. Also shown is a curve with $r = 0.4$, to give an idea of the mass fraction required to match the small χ ballistic limit data. Overall, the agreement is good.

13.6. Behavior of Other Fibers

Phil Cunniff of the US Army Natick Soldier Research, Development and Engineering Center presented extensive data at the 18th International Symposium on Ballistics on fabrics and composites [51]. Our model agrees extremely well with the Kevlar 29 fabric data he presented, but as we saw for our composite model agreement was not perfect, so we present his curve fit for that case:

$$V_{bl} = (179.17 \text{ m/s})(1.277)^{\sec(\theta)-1} \exp(2.8198\chi^{0.5692}). \tag{13.156}$$

Numerous tests were also performed at obliquity, and this curve fit includes the obliquity θ. For 45° impact angle, the curve fit predicts the ballistic limit increases by a factor of 1.11. For Kevlar 29 fabric, Cunniff had an obliquity factor of 1.0707

TABLE 13.4. Performance relative to Kevlar 29 of fabrics and fiber-based composites in terms of V_{50} ($c^* = c_f \varepsilon_f$ save for the parenthetical value for Spectra where Eq. 13.161 was used).

Fiber	U^* (m/s)	Fabric	Composite	H_{U^*}	c^* (m/s)	H_{c^*}
PBO	813	X		1.30	323	1.34
Spectra	801(672)	X	X+	1.08	389 (335)	1.62 (1.4)
Kevlar KM2	682	X	X+	1.09	272	1.13
Kevlar 129	672	X	X	1.08	270	1.12
Kevlar 29	623	X	X	1	240	1
Kevlar 49	612	X		0.98	219	0.91
Nylon	482	X	X	0.77		
E-glass	559(482)		X	0.77	253	1.05
Carbon fiber	593(375)		X	0.60	197	0.82

instead of 1.277, indicating an increase in the ballistic limit for 45° impact angle by a factor of 1.03. Thus, obliquity does not have a large effect on fabric V_{50}, a general result of experiments.

For the fabric materials, a number of Kevlars (Table 13.2 lists deniers), a Spectra 1,000 fabric and a nylon 6,6 were tested. Composite materials included Kevlar 29/poly(vinyl-butyral)/phenolic, Spectra 900 fabric/vinylester, Spectra shield/ Kraton (this is a 0/90 unidirectional layup that is then pressed to form solid plate), E-glass/polyester, carbon fiber/epoxy, and nylon/poly(viny-butyral)/phenolic. For each of the composites the resin content was approximately 15 to 18% by weight, with the exception that the E-glass and carbon-fiber composites were approximately 30% by weight. Plots of the test results are in Cunniff [51, 52].

Through examining V_{50} data for different fibers, Cunniff determined that the denier of the specific fiber does not greatly affect the ballistic performance. He also found a quantity, which he called U^*, that appears to correctly predict the relative V_{50} performance of the fabric or composite material,[7]

$$U^* = \left(\frac{\sigma_f \varepsilon_f}{2\rho} c_E \right)^{1/3}. \tag{13.157}$$

This expression has the units of velocity: the first term is the specific strain energy to failure and the second term is the elastic wave speed. Cunniff's extensive data set shows good correlation: for example, using a performance ratio H,

$$V_{bl} = H_{U^*} (V_{bl})_{\text{Kevlar 29}}, \quad \text{where} \quad H_{U^*} = \frac{U^*}{(U^*)_{\text{Kevlar 29}}}, \tag{13.158}$$

for both fabrics and composites (where the fabric V_{bl} is used for fabrics, and the composite V_{bl} is used for composites, respectively). Specific values are in Table 13.4 (in particular for 850 denier KM2 and 1,000 denier Kevlar 29). Cunniff obtained the nylon U^* value based on other data and reasoning. An X in the "Fabric" or "Composite" column indicates that fabric and/or composite data was compared to Eqs. (13.157) and (13.158). In the table, when scaling by U^* did not provide good

---

[7]Actually, Cunniff defines the left-hand side of Eq. (13.157) as $(U^*)^{1/3}$; we change the power in the definition so that U^* has dimensions of velocity.

agreement, Cunniff provided experimental fits (here in parentheses) for the fabric data. A + in the "Composite" column indicates two cases where the composite performed better than the performance ratio would indicate (the performance ratio indicates fabric performance in these cases).

In our ballistic limit equation, the deciding factor between fibers is $c^* = c_E \varepsilon_f$. This expression is not the same as U^*: for example, assuming a linear stress-strain response to failure yields $U^* = c_E \varepsilon_f^{2/3}/2^{1/3} = c_E^{1/3}(c^*)^{2/3}/2^{1/3}$, and the failure strain enters as a power 2/3 rather than 1. The $c_E \varepsilon_f$ expression was based on an assumed linear strain to failure for the yarn. However, it should be kept in mind that the difference in powers could be due to the approximation in Eq. (13.136) we made so that we could have an explicit solution. For a new fiber being considered, if the failure strain, strength, and Young's modulus do not imply linear to failure behavior, the expression in the ballistic limit should be replaced with one based on the strain energy to failure. If there is some nonlinearity in the yarn, so that the failure strain is not the strength divided by the Young's modulus, one approach to finding the strain energy to failure is to assume a quadratic form of the stress vs. strain relation:

$$\sigma = E\varepsilon + F\varepsilon^2. \tag{13.159}$$

F (here just a constant) can be determined by the values of stress and strain at failure: $F = (\sigma_f/\varepsilon_f - E)/\varepsilon_f$. Integrating to obtain the strain energy to fail the yarn yields

$$e_{ef} = \int_0^{\varepsilon_f} \sigma d\varepsilon = \frac{1}{6}E\varepsilon_f^2 + \frac{1}{3}\sigma_f \varepsilon_f. \tag{13.160}$$

Thus, in the ballistic limit expression for nonlinear stress-strain response,

$$\text{replace } c^* = c_E \varepsilon_f \quad \text{with} \quad c^* = \sqrt{\frac{2e_{ef}}{\rho}} = \left(\frac{1}{3\rho}(E\varepsilon_f^2 + 2\sigma_f \varepsilon_f)\right)^{1/2}. \tag{13.161}$$

The coefficient becomes the square root of two times the specific strain energy to failure. Values of c^* are shown in Table 13.4, as well as H_{c^*} values by comparing to Kevlar 29, to compare to how well this measure does in predicting performance. For the most part, the performance ratios from the two methods are similar, and notice in particular that, though the performance values differ for KM2, the model shows excellent agreement with the data as seen above.

We make three final comments about the data. First, PBO (Zylon) fresh from the factory is an excellent ballistic fabric (Exercise 13.15). Second, it will be noticed that though Spectra performs well, it does not perform as well as might be expected based on its fiber properties (Exercise 13.14). Third, E-glass (fiberglass) and carbon fiber composites do not perform well ballistically and are worse than expected. Overall, however, the models do a good job in describing the ballistic performance of the fabrics and composites.

13.7. One, Two, Many

It is an amazing fact that a homogeneous distribution of yarns around the 360°, where they are connected together when they cross, yields only three different possibilities as to response: 0° corresponding to one yarn, 0°/90° corresponding to two yarns, and a homogeneous membrane corresponding to three or more yarns. Here we derive the corresponding moduli.

If we are interested in the strain in a yarn that is originally oriented at an angle θ to the X-axis, then that strain is given by pre- and post-multiplication with the direction vector $\vec{N}^T = (\cos(\theta), \sin(\theta), 0)$. Hence, with E_{ij} the Lagrangian strain tensor,

$$E_\theta = \vec{N}^T(E_{ij})\vec{N} = E_{11}\cos^2(\theta) + 2E_{12}\cos(\theta)\sin(\theta) + E_{22}\sin^2(\theta) \quad (13.162)$$

is the strain in the yarn originally oriented at the angle θ. The energy in this yarn is given by $e_\theta = \frac{1}{2}E_f E_\theta^2$. For p yarns uniformly oriented around the circle with angles $\theta_n = \pi n/p$, where $p \geq 3$ and $n = 0, 1, ..., p-1$, the energy is given by[8]

$$e = \frac{1}{p}\sum_{n=0}^{p-1} e_{\theta_n} = \frac{1}{p}\sum_{n=0}^{p-1} \frac{1}{2}E_f E_{\theta_n}^2 = \frac{E_f}{8}(3E_{11}^2 + 2E_{11}E_{22} + 3E_{22}^2 + 4E_{12}^2)$$

$$= E\left\{\frac{3}{8}(E_{11} + E_{22})^2 - \frac{1}{2}(E_{11}E_{22} - E_{12}^2)\right\} = \frac{3E_f}{8}I_1^2(E_{ij}) + \frac{E_f}{2}I_2(E_{ij}),$$
$$(13.164)$$

where I_1 and I_2 are the first and second invariants of the Lagrangian strain tensor (E_{ij}). For small strain elasticity, the strain energy is

$$e = \frac{\lambda}{2}(\varepsilon_{ii})^2 + \mu\varepsilon_{ij}\varepsilon_{ij} = (\frac{\lambda}{2} + \mu)I_1^2(\varepsilon) + 2\mu I_2(\varepsilon). \quad (13.165)$$

Comparing the forms, we see that

$$\frac{\lambda}{2} + \mu = \frac{3E_f}{8}, \qquad \mu = \frac{E_f}{4}, \quad (13.166)$$

and, using the various expressions between elastic moduli,

$$\lambda = \mu = \frac{E_f}{4}, \qquad E = \frac{5E_f}{8}, \qquad \nu = \frac{1}{4} \quad (13.167)$$

are the elastic moduli for the homogeneous membrane.

13.8. General Anisotropy

At this point we discuss anisotropic materials a little more generally. The linear elastic relationship between the Cauchy stress and small strain in general can be written

$$\sigma_{ij} = c_{ijkl}\varepsilon_{kl}, \quad (13.168)$$

where c_{ijkl} is referred to as the stiffness matrix. c_{ijkl} represents 81 constants. However, since the Cauchy stress tensor and the small strain tensor are symmetric,

---

[8]We used the following sums that can be obtained, for example, with Mathematica,

$$\sum_{n=0}^{p-1}\cos^4(\theta_n) = \sum_{n=0}^{p-1}\sin^4(\theta_n) = \frac{3p}{8}, \qquad \sum_{n=0}^{p-1}\cos^2(\theta_n)\sin^2(\theta_n) = \frac{p}{8},$$

$$\sum_{n=0}^{p-1}\cos^3(\theta_n)\sin(\theta_n) = \sum_{n=0}^{p-1}\cos(\theta_n)\sin^3(\theta_n) = 0, \qquad \theta_n = \frac{\pi n}{p}, \quad p \geq 3. \quad (13.163)$$

it is possible to represent the relationship in terms the stiffness matrix C:

$$
\begin{pmatrix} \sigma_{11} \\ \sigma_{22} \\ \sigma_{33} \\ \sigma_{23} \\ \sigma_{31} \\ \sigma_{12} \end{pmatrix} = \begin{pmatrix} C_{11} & C_{12} & C_{13} & C_{14} & C_{15} & C_{16} \\ C_{21} & C_{22} & C_{23} & C_{24} & C_{25} & C_{26} \\ C_{31} & C_{32} & C_{33} & C_{34} & C_{35} & C_{36} \\ C_{41} & C_{42} & C_{43} & C_{44} & C_{45} & C_{46} \\ C_{51} & C_{52} & C_{53} & C_{54} & C_{55} & C_{56} \\ C_{61} & C_{62} & C_{63} & C_{64} & C_{65} & C_{66} \end{pmatrix} \begin{pmatrix} \varepsilon_{11} \\ \varepsilon_{22} \\ \varepsilon_{33} \\ 2\varepsilon_{23} \\ 2\varepsilon_{31} \\ 2\varepsilon_{12} \end{pmatrix}, \tag{13.169}
$$

where

$$
C_{11} = c_{1111}, \quad C_{12} = c_{1122}, \quad C_{16} = c_{1112} = c_{1121}, \quad \text{etc.} \tag{13.170}
$$

The symmetric tensors reduce the stiffness matrix to 36 constants. It is typically assumed that there exits an elastic strain energy that is quadratic in strain from which the stresses can be derived – that is,

$$
\sigma_{ij} = \frac{\partial(\rho_0 E_e)}{\partial \varepsilon_{ij}}, \quad \text{where} \quad \rho_0 E_e = \frac{1}{2} c_{ijkl} \varepsilon_{ij} \varepsilon_{kl}. \tag{13.171}
$$

Materials where the stresses can be obtained in this fashion are referred to as *hyperelastic*. This assumption implies that the stiffness matrix C is symmetric since, for example,

$$
\sigma_{12} = \frac{1}{2} c_{12ij} \varepsilon_{ij} + \frac{1}{2} c_{ij12} \varepsilon_{ij} \quad \text{and} \quad \sigma_{33} = \frac{1}{2} c_{33ij} \varepsilon_{ij} + \frac{1}{2} c_{ij33} \varepsilon_{ij}, \tag{13.172}
$$

the dependence of σ_{12} on ε_{33} from the first equation is $\frac{1}{2}(c_{1233} + c_{3312})$ and the dependence on σ_{33} on ε_{12} from the second equation is $\frac{1}{2}(c_{3312} + c_{1233})$, thus demonstrating the symmetry of C (there is a needed extra factor of 2 from $\varepsilon_{12} = \varepsilon_{21}$). Thus, for general anisotropic materials, c_{ijkl} can have 21 independent constants. Rather than discussing anisotropic materials in general, we focus on orthotropic materials, which have three planes of elastic symmetry, where a shear strain only gives rise to the corresponding shear stress. Hence, the symmetric stiffness matrix becomes

$$
\begin{pmatrix} \sigma_{11} \\ \sigma_{22} \\ \sigma_{33} \\ \sigma_{23} \\ \sigma_{31} \\ \sigma_{12} \end{pmatrix} = \begin{pmatrix} C_{11} & C_{12} & C_{13} & 0 & 0 & 0 \\ C_{12} & C_{22} & C_{23} & 0 & 0 & 0 \\ C_{13} & C_{23} & C_{33} & 0 & 0 & 0 \\ 0 & 0 & 0 & C_{44} & 0 & 0 \\ 0 & 0 & 0 & 0 & C_{55} & 0 \\ 0 & 0 & 0 & 0 & 0 & C_{66} \end{pmatrix} \begin{pmatrix} \varepsilon_{11} \\ \varepsilon_{22} \\ \varepsilon_{33} \\ 2\varepsilon_{23} \\ 2\varepsilon_{31} \\ 2\varepsilon_{12} \end{pmatrix}, \tag{13.173}
$$

and there are only nine independent constants. As an example, a fiber-based composite where fibers run in the 0 and 90° directions is orthotropic, and there is no coupling between normal and shearing directions. To understand the relationship between stress and strain, it is much simpler to describe the compliance matrix – the inverse of C – which relates the strain to the stress,

$$
\begin{pmatrix} \varepsilon_{11} \\ \varepsilon_{22} \\ \varepsilon_{33} \\ 2\varepsilon_{23} \\ 2\varepsilon_{31} \\ 2\varepsilon_{12} \end{pmatrix} = \begin{pmatrix} 1/E_1 & -\nu_{21}/E_2 & -\nu_{31}/E_3 & 0 & 0 & 0 \\ -\nu_{12}/E_1 & 1/E_2 & -\nu_{32}/E_3 & 0 & 0 & 0 \\ -\nu_{13}/E_1 & -\nu_{23}/E_2 & 1/E_3 & 0 & 0 & 0 \\ 0 & 0 & 0 & 1/G_{23} & 0 & 0 \\ 0 & 0 & 0 & 0 & 1/G_{31} & 0 \\ 0 & 0 & 0 & 0 & 0 & 1/G_{12} \end{pmatrix} \begin{pmatrix} \sigma_{11} \\ \sigma_{22} \\ \sigma_{33} \\ \sigma_{23} \\ \sigma_{31} \\ \sigma_{12} \end{pmatrix}. \tag{13.174}
$$

When the material is pulled in uniaxial stress in the 1 direction, for example, then $\varepsilon_{11} = \sigma_{11}/E_1$ and E_1 is the Young's modulus in the 1 direction. In the orthogonal directions the strains are given by $\varepsilon_{22} = -\nu_{12}\varepsilon_{11}$ and $\varepsilon_{33} = -\nu_{13}\varepsilon_{11}$. In general, the Poisson's ratios are defined as $\nu_{ji} = -\varepsilon_{ii}/\varepsilon_{jj}$ (no sum on i, j, $i \neq j$) during uniaxial stress loading in the j direction. Since the compliance matrix is symmetric, $\nu_{ij}/E_i = \nu_{ji}/E_j$ (no sum on i, j, $i \neq j$). The stiffness matrix C is the inverse of the compliance matrix (as in Eq. 3.17 for isotropic materials),

$$C = \frac{1}{D}\begin{pmatrix} E_1(1-\nu_{23}\nu_{32}) & E_1(\nu_{21}+\nu_{23}\nu_{31}) & E_1(\nu_{31}+\nu_{32}\nu_{21}) & 0 & 0 & 0 \\ E_2(\nu_{12}+\nu_{13}\nu_{32}) & E_2(1-\nu_{13}\nu_{31}) & E_2(\nu_{32}+\nu_{31}\nu_{12}) & 0 & 0 & 0 \\ E_3(\nu_{13}+\nu_{12}\nu_{23}) & E_3(\nu_{23}+\nu_{21}\nu_{13}) & E_3(1-\nu_{12}\nu_{21}) & 0 & 0 & 0 \\ 0 & 0 & 0 & G_{23} & 0 & 0 \\ 0 & 0 & 0 & 0 & G_{31} & 0 \\ 0 & 0 & 0 & 0 & 0 & G_{12} \end{pmatrix},$$

$$D = 1 - \nu_{12}\nu_{21} - \nu_{13}\nu_{31} - \nu_{23}\nu_{32} - \nu_{12}\nu_{23}\nu_{31} - \nu_{13}\nu_{32}\nu_{21}.$$

$$(13.175)$$

For the fabric model as developed above, examination of these equations reveals that we have assumed $\nu_{ij} = 0$, all ij, $E_1 = E_2 = E/2$, $E_3 = 0$, and $G_{ij} = 0$. If we had been interested in an isotropic fabric sheet, again assuming that through-thickness behavior was not of interest, then the elastic constants would have included a possibly nonzero $\nu_{12} = \nu_{21}$ and a consistent nonzero shear modulus $G_{12} = E/(4(1+\nu_{12}))$. If, on the other hand, the interest had been in including a matrix material, such could be included by using appropriate moduli and Poisson's ratio to reflect the homogeneous matrix material and the anisotropic fibers. If we have an isotropic resin or matrix material, its volume fraction is $\phi = (\rho_f/\rho_{\text{matrix}})r$, where r is the mass fraction as above. The strain energy of an isotropic matrix material is

$$e_{\text{matrix}} = \frac{1}{2}\sigma_{ij}\varepsilon_{ij} = \frac{1}{2}\lambda_{\text{matrix}}(\varepsilon_{ii})^2 + \mu_{\text{matrix}}\varepsilon_{ij}\varepsilon_{ij}. \qquad (13.176)$$

In our large deflection formulation the strain energy would be

$$e = (1-\phi)e_{\text{fiber}} + \phi e_{\text{matrix}}$$

$$= (1-\phi)\frac{E_f}{4}(E_{11}^2 + E_{22}^2) + \phi\frac{E_{\text{matrix}}}{2(1+\nu_{\text{matrix}})}\left\{\frac{\nu_{\text{matrix}}}{(1-2\nu_{\text{matrix}})}(E_{ii})^2 + E_{ij}E_{ij}\right\},$$

$$(13.177)$$

where for large deflection E_{ij} is used instead of ε_{ij}. This rule of mixtures approach is not entirely consistent, but it allows the computation of the various terms including a matrix.

With the introduction of the 6×6 elastic coefficient matrix C, the topic comes up of how this transforms under a rotation. If we write the vector notation as

$$\vec{\sigma} = (\vec{\sigma_{ij}}) = \begin{pmatrix} \sigma_{11} \\ \sigma_{22} \\ \sigma_{33} \\ \sigma_{23} \\ \sigma_{13} \\ \sigma_{12} \end{pmatrix}, \qquad \vec{\varepsilon} = (\vec{\varepsilon_{kl}}) = \begin{pmatrix} \varepsilon_{11} \\ \varepsilon_{22} \\ \varepsilon_{33} \\ \varepsilon_{23} \\ \varepsilon_{13} \\ \varepsilon_{12} \end{pmatrix}, \qquad (13.178)$$

where we are using $\sigma_{31} = \sigma_{13}$ (but the actual index order matters in what follows) and note that the factor of 2 is not in $\bar{\varepsilon}$. We will write the rotation from original coordinates to a primed coordinates as

$$\sigma' = R\sigma R^T \rightarrow \vec{\sigma}' = \bar{R}\vec{\sigma}, \qquad \varepsilon' = R\varepsilon R^T \rightarrow \vec{\varepsilon}' = \bar{R}\vec{\varepsilon}, \qquad (13.179)$$

where R is a rotation matrix (i.e., $R^T R = I$). The 6×6 transformation matrix $\bar{R}$ is not a rotation matrix; it is given by

$$\bar{R} = \begin{pmatrix} R_{11}^2 & R_{12}^2 & R_{13}^2 & 2R_{12}R_{13} & 2R_{11}R_{13} & 2R_{11}R_{12} \\ R_{21}^2 & R_{22}^2 & R_{23}^2 & 2R_{22}R_{23} & 2R_{21}R_{23} & 2R_{21}R_{22} \\ R_{31}^2 & R_{32}^2 & R_{33}^2 & 2R_{32}R_{33} & 2R_{31}R_{33} & 2R_{31}R_{32} \\ R_{21}R_{31} & R_{22}R_{32} & R_{23}R_{33} & R_{22}R_{33}+R_{23}R_{32} & R_{21}R_{33}+R_{23}R_{31} & R_{21}R_{32}+R_{22}R_{31} \\ R_{11}R_{31} & R_{12}R_{32} & R_{13}R_{33} & R_{12}R_{33}+R_{13}R_{32} & R_{11}R_{33}+R_{13}R_{31} & R_{11}R_{32}+R_{12}R_{31} \\ R_{11}R_{21} & R_{12}R_{22} & R_{13}R_{23} & R_{12}R_{23}+R_{13}R_{22} & R_{11}R_{23}+R_{13}R_{21} & R_{11}R_{22}+R_{12}R_{21} \end{pmatrix}.$$
$$(13.180)$$

Examining the specific values shows that $\bar{R}$ decomposes into four 3×3 blocks and

$$\bar{R} = \begin{pmatrix} R_{ij}^2 & 2R_{ik}R_{jl} \\ R_{ik}R_{jl} & R_{ik}R_{jl}+R_{il}R_{jk} \end{pmatrix}, \qquad \text{where} \quad (\vec{\sigma}_{ij}) = \bar{R}\,(\vec{\sigma}_{kl}). \qquad (13.181)$$

The same equation holds for the strain vector.

Hooke's law is

$$\vec{\sigma} = C \begin{pmatrix} I & 0 \\ 0 & 2I \end{pmatrix} \vec{\varepsilon} \quad \text{and} \quad \vec{\sigma}' = C' \begin{pmatrix} I & 0 \\ 0 & 2I \end{pmatrix} \vec{\varepsilon}', \qquad (13.182)$$

where I represents a 3×3 identity matrix. With the rotation in Eq. (13.179),

$$C' = \bar{R}C \begin{pmatrix} I & 0 \\ 0 & 2I \end{pmatrix} \bar{R}^{-1} \begin{pmatrix} I & 0 \\ 0 & \frac{1}{2}I \end{pmatrix}. \qquad (13.183)$$

At first it might be thought that computing $\bar{R}^{-1}$ would be quite difficult, since $\bar{R}$ is not a rotation matrix; however, it is based on a rotation matrix, so the inverse is found by viewing $\bar{R}$ as a function of R, $\bar{R} = \bar{R}(R)$. Then $\bar{R}^{-1} = \bar{R}(R^T)$ – that is, the indices of R in $\bar{R}$ are swapped –

$$\bar{R}^{-1} = \begin{pmatrix} R_{ji}^2 & 2R_{ki}R_{lj} \\ R_{ki}R_{lj} & R_{ki}R_{lj}+R_{li}R_{kj} \end{pmatrix}, \qquad \text{where} \quad \bar{R}^{-1}(\vec{\sigma}_{ij})' = (\vec{\sigma}_{kl}). \qquad (13.184)$$

Now,

$$\begin{pmatrix} I & 0 \\ 0 & 2I \end{pmatrix} \bar{R}^{-1} \begin{pmatrix} I & 0 \\ 0 & \frac{1}{2}I \end{pmatrix} = \begin{pmatrix} I & 0 \\ 0 & 2I \end{pmatrix} \begin{pmatrix} R_{ji}^2 & 2R_{ki}R_{lj} \\ R_{ki}R_{lj} & R_{ki}R_{lj}+R_{li}R_{kj} \end{pmatrix} \begin{pmatrix} I & 0 \\ 0 & \frac{1}{2}I \end{pmatrix}$$
$$= \begin{pmatrix} R_{ji}^2 & R_{ki}R_{lj} \\ 2R_{ki}R_{lj} & R_{ki}R_{lj}+R_{li}R_{kj} \end{pmatrix}$$
$$= \bar{R}^T.$$
$$(13.185)$$

This surprising result yields

$$C' = \bar{R}C\bar{R}^T, \qquad (13.186)$$

even though $\bar{R}$ is not an orthonormal rotation matrix [189].

13.9. Sources

The discussion of the Piola–Kirchhoff stress follows Malvern [138]. The single yarn impact discussion is new, though it follows in concept original work by Smith [171]. The static solution for the fabric sheet and the fabric V_{50} are due to Walker [192]. The composite model is from Walker [193]. Further details can be found in Walker [195]. The data presented in this chapter are due to Cunniff; one is directed to the proceedings of the 18th International Symposium on Ballistics, which include seven papers by him on fabrics, including the ones we explicitly reference [51, 52] (as well as the original model paper by Walker [192]).

Exercises

(13.1) We did not define a U_L corresponding to c_1 like we defined a U_T corresponding to c_2. Why?

(13.2) From Fig. 13.1, showing the deflection of the nylon yarn, measure the relative wave speed V/c_1, the strain in the yarn, and the angle θ. From the single yarn solution, show that the various values are consistent (the strain in the yarn, the inflow velocity, etc.). Based on the properties listed for nylon in Table 13.1, estimate the impact speed of the projectile and the absolute amounts of the velocities c_1, v_{xL}, c_2, and U_T.

(13.3) Develop a fabric ballistic limit expression based on the yarn solution (see Eq. 13.94). The necessary observation is that the way the fabric areal density enters is through the reduction in speed based on momentum conservation during the initial impact (as was done in Eq. 13.145). To get something reasonable, it is necessary to introduce some empiricism to account for the fact that the yarns do not immediately break and thus there is momentum transferred to an extent of fabric larger than the projected projectile area, as with β in Eq. (13.145) and its preceding sentences. Compare this model to the fabric model results for KM2 and Kevlar 29. How do the sound speed and failure strain enter into the model? Compare to U^* and c^*.

(13.4) Table 13.5 shows V_{50} data for Kevlar 129 plain weave fabric [51]. Compare the data and the fabric V_{bl} model (with a plot, for example).

TABLE 13.5. V_{50} data for Kevlar 129 plain weave fabric.

χ	V_{50} (m/s)	χ	V_{50} (m/s)	χ	V_{50} (m/s)
0.0418	415.3	0.1053	530.0	0.2018	697.9
0.0664	459.8	0.1273	615.1	0.2546	772.0
0.0802	501.1	0.1328	570.8	0.2632	789.0
0.1046	589.9	0.1660	650.7	0.3321	877.5

(13.5) Derive Eq. (13.83) from Eq. (13.77).

(13.6) Show that for the small strain approximation for a single yarn,

$$\frac{U_T}{c_E} = \frac{c_2}{c_E} - \left(\frac{c_2}{c_E}\right)^2 + \left(\frac{c_2}{c_E}\right)^3. \tag{13.187}$$

(13.7) The small strain formulation for the single yarn problem attributed to Smith is developed as follows. It is assumed that $c_1 = c_E$ (thus the stress-strain law is slightly different from what we use). It follows from Theorem 4.1 that $v_{xL} = -c_E\varepsilon$, where ε is the strain in the yarn. With the deformed yarn having density $\rho = \rho_0/(1 + \varepsilon)$, if the transverse wave speed in the laboratory reference frame is assumed to have speed $\sqrt{\sigma/\rho}$, then adjusting for material motion gives $U_T = c_E(\sqrt{\varepsilon(1 + \varepsilon)} - \varepsilon)$. Using Eq. (13.86), we compute $c_2 = c_E\sqrt{\varepsilon/(1 + \varepsilon)}$. Assuming the tranverse wave does not change the strain in the yarn (as we showed in our derivation), at a given time t we can now use a geometric argument that the length of yarn that has been deflected by the transverse wave, taking strain into account, is $c_2(1 + \varepsilon)t$, and this length equals the length in laboratory coordinates of $\sqrt{V^2 + U_T^2}t$. Setting these lengths equal yields

$$V = c_E\sqrt{\varepsilon(2\sqrt{\varepsilon(1 + \varepsilon)} - \varepsilon)}. \tag{13.188}$$

Work through the steps of this derivation and show that, for small strains, these results and the yarn solution in the text (Eqs. 13.83, 13.84, 13.87) are nearly identical.

(13.8) Assuming $c_1 \approx c_E$ and $v_{xL} \approx -c_E E_{11}$, show Eq. (13.84) using the following geometric construction: the initial length of involved yarn is $L_0 = c_1 t$ and the final length is $L = \sqrt{(Vt)^2 + ((c_2 + v_{xL})t)^2} + (c_1 - (c_2 + v_{xL}))t$, the exact result that $c_2/c_E = \sqrt{E_{11}}$ still holds, and $E_{11} = L/L_0 - 1$. Next, use the geometric construction to compute V/c_E for an initially strained yarn with stress $T_0 = EE_0 > 0$. Show that if $E_{11} = E_{11T} + E_0$ is the total strain in the yarn, then

$$\left(\frac{V}{c_E}\right)^2 = (E_{11} - E_0)\left\{2\sqrt{E_{11}} - (E_{11} - E_0)\right\}. \tag{13.189}$$

(13.9) In Fig. 13.6, do all the strains cross at the same place? If so, what is x/R, and if not, what are the x/R for the various intersections?

(13.10) Given the projectile deceleration in Eq. (13.152), compute a resulting force vs. displacement acting on the projectile.

(13.11) Given a strain energy density

$$e = \sum_{ij} \frac{1}{2}k_{ij}E_{ij}^2 = \frac{1}{2}(k_{11}E_{11}^2 + k_{12}E_{12}^2 + \ldots + k_{33}E_{33}^2), \tag{13.190}$$

show that the resulting first Piola–Kirchhoff stress is

$$S = \begin{pmatrix} k_{11}E_{11} & \frac{1}{2}(k_{12} + k_{21})E_{12} & \frac{1}{2}(k_{13} + k_{31})E_{13} \\ \frac{1}{2}(k_{21} + k_{12})E_{21} & k_{22}E_{22} & \frac{1}{2}(k_{23} + k_{32})E_{23} \\ \frac{1}{2}(k_{31} + k_{13})E_{31} & \frac{1}{2}(k_{32} + k_{23})E_{32} & k_{33}E_{33} \end{pmatrix} F^T. \tag{13.191}$$

The effect of (k_{ij}) has been made symmetric, even if it initially was not. Why?

(13.12) Compute U^* for all the fiber data in Table 13.2. Compare the results for the various deniers of the same fiber type. Are they similar?

(13.13) Cunniff came up with his U^* parameter by performing a Buckingham π theorem analysis and then comparing to lots of data to determine the form of U^*. If you know how it is done, perform the same π theorem analysis. Assume the properties of the yarn are given by density, elastic modulus, failure stress, and failure strain, and assume other relevant geometric parameters for the fabric and for the projectile. Show that your nondimensional variables basis allows a representation of U^*/c_E and c^*/c_E (both forms – one will involve adding two nondimensional expressions together and taking a root). Also show that χ can be written in terms of your basis.

(13.14) Some speculation on the performance of Spectra fiber being less than expected is that it may be undergoing thermal softening. Under high-rate loading, heat generated locally would not be dissipated and would result in a rise in temperature of the fiber. This exercise is for you to determine if there is enough thermal energy involved. Recall that we assume elastic deformation does not raise temperature, just inelastic deformation. Compute, based on Table 13.1, the expected amount of inelastic deformation and its associated energy change and temperature rise in straining to failure assuming elastic-perfectly plastic behavior based on the properties in Table 13.1. Do you think it is enough to significantly affect the strength? (Heat capacity is roughly 2 J/g K, melting temperature is roughly 150°C.)

(13.15) Find out why PBO fabric is not currently in use in body armors (for example, find relevant news articles describing recalls).

(13.16) If you could increase a fiber's modulus, strength, or failure strain by 20%, which one would lead to the greatest increase in ballistic limit performance? How much increase in V_{bl} occurs for each? To be specific, answer this question by producing a table of the change in (both forms of) c^* and U^* when changing each of the above material properties by the given amount.

(13.17) In the context of the models for ballistic limit and residual velocity, is the ballistic limit increased or decreased if the fabric sheet is split into two separated sheets of half the areal density? Does the result depend on χ?

(13.18) Two exotic materials that are sometimes discussed as potential armor materials are spider silk and graphene. According to Koski et al. [122], the density of spider silk is 1.3 g/cm^3 and for a variety of spiders the longitudinal sound speed ranges from 6.5 to 7.0 km/s and Young's modulus in the direction of the fiber ranges from 55 to 62 GPa. The reported strengh at higher strain rates of spider silks is typically in the range of 1 to 2 GPa with breaking strains around 20% (e.g., Gosline et al. [78]). According to Lee et al. [130], the Young's modulus of graphene is 1 TPa (terra = 10^{12}) with a second order term $D = -2$ TPa (i.e., $\sigma = E\varepsilon + D\varepsilon^2$), its strength is 130 GPa and its strain at failure is 25%, all based on atomic force microscope deflections of single layer graphene flakes. Given these properties (assume the density of graphene is the same as graphite), discuss the predicted yarn breaking speeds of both these materials as well as their expected performance in fabric armors. The amount of fabric in current body armors is determined by yarn failure. Does either of these materials transition from failure limited to deflection limited (i.e., the amount of material is not controlled by the strength/strain at failure but rather by the allowable deflection behind the armor)?

(13.19) Suppose we formulated a loose fabric by introducing a nonlinear stress-strain relation for the yarn. For example, assuming a failure stress σ_f and a failure strain ε_f, the following expression has the same failure stress and strain energy to failure as a linear stress-strain relation:

$$\sigma_{11} = \sigma_f \left(\frac{E_{11}}{\bar{\varepsilon}_f}\right)^\gamma, \quad \bar{\varepsilon}_f = \frac{(\gamma+1)\varepsilon_f}{2}. \tag{13.192}$$

This expression matches the failure stress at a modified failure strain $\bar{\varepsilon}_f$ (found by equating the failure strain energy of this expression to that of the linear case: show this). This constitutive model has a yarn breaking at the same energy and the same stress as does the linear yarn, but it takes a greater strain to fail, thus introducing a looseness in the fabric through the power γ. If $\gamma = 1$ then the original linear stress-strain relation results. The amount of "looseness" depends on the failure strain. For example, if the failure strain is that of Kevlar KM2, $\varepsilon_f = 0.038$ (3.8%), then a power $\gamma = 5$ corresponds to a "looseness" of $((\gamma+1)/2 - 1)\epsilon_f = 0.076$ (7.6%) in the fabric – thus, large powers of γ do not lead to really loose fabrics.

Show that the potential energy density is written

$$e = \frac{\sigma_f}{2(\gamma+1)\bar{\varepsilon}_f^\gamma}(E_{11}^{\gamma+1} + E_{22}^{\gamma+1}). \tag{13.193}$$

Show that the complete first Piola–Kirchhoff stress tensor is

$$S = \frac{\sigma_f}{2\bar{\varepsilon}_f^\gamma}\begin{pmatrix} E_{11}^\gamma F_{11} & E_{11}^\gamma F_{21} & E_{11}^\gamma F_{31} \\ E_{22}^\gamma F_{12} & E_{22}^\gamma F_{22} & E_{22}^\gamma F_{32} \\ 0 & 0 & 0 \end{pmatrix} = \frac{\sigma_f}{2\bar{\varepsilon}_f^\gamma}\begin{pmatrix} E_{11}^\gamma & 0 & 0 \\ 0 & E_{22}^\gamma & 0 \\ 0 & 0 & 0 \end{pmatrix}F^T. \tag{13.194}$$

For the anti-plane shear (out of plane) strain problem, show that

$$S = \frac{\sigma_f}{2^{\gamma+1}\bar{\varepsilon}_f^\gamma}\begin{pmatrix} F_{31}^{2\gamma} & 0 & 0 \\ 0 & F_{32}^{2\gamma} & 0 \\ 0 & 0 & 0 \end{pmatrix}F^T = \frac{\sigma_f}{2^{\gamma+1}\bar{\varepsilon}_f^\gamma}\begin{pmatrix} F_{31}^{2\gamma} & 0 & F_{31}^{2\gamma+1} \\ 0 & F_{32}^{2\gamma} & F_{32}^{2\gamma+1} \\ 0 & 0 & 0 \end{pmatrix} \tag{13.195}$$

and

$$S = \frac{\sigma_f}{2^{\gamma+1}\bar{\varepsilon}_f^\gamma}\begin{pmatrix} \left(\frac{\partial u_z}{\partial x}\right)^{2\gamma} & 0 & \left(\frac{\partial u_z}{\partial x}\right)^{2\gamma+1} \\ 0 & \left(\frac{\partial u_z}{\partial y}\right)^{2\gamma} & \left(\frac{\partial u_z}{\partial y}\right)^{2\gamma+1} \\ 0 & 0 & 0 \end{pmatrix}. \tag{13.196}$$

Defining $\alpha = 2\gamma + 1$, show that the equation of motion in the out of plane z direction is

$$\hat{\rho}\frac{d^2 u_z}{dt^2} = \frac{\sigma_f}{2^{\gamma+1}\bar{\varepsilon}_f^\gamma}\left\{\frac{\partial}{\partial x}\left(\frac{\partial u_z}{\partial x}\right)^\alpha + \frac{\partial}{\partial y}\left(\frac{\partial u_z}{\partial y}\right)^\alpha\right\}. \tag{13.197}$$

Show that the general solution to the static problem is

$$u_z(x, y) = h\left\{1 - \left((x/R)^{1+1/\alpha} + (y/R)^{1+1/\alpha}\right)^{1-2/(\alpha+1)}\right\}. \tag{13.198}$$

Now show that the looseness of the fabric leads to straighter edges in the deformation: a linear fabric edge has curvature during deformation, but as the fabric becomes looser show that the edges of the pyramid become straighter. Show that in the loose fabric limit, the faces of the pyramid become flat:

$$\gamma \to +\infty \Rightarrow u_z(x, y) \to h\{1 - |x|/R - |y|/R\}. \tag{13.199}$$

However, for finite γ, this straight-sided pyramid solution does not satisfy the boundary conditions along the $x = 0, y = 0$ edge (Eq. 13.103) and is therefore not a solution to the center-indented problem.

(13.20) The in-plane part of the nonlinear analysis of the previous exercise is in Appendix D: Exercise D.1 (moved there for space reasons).

(13.21) Notice instead of Eq. (13.119) to define strain, we could have rather used our own approximation to the strain used in the derivation, thus yielding a self-consistent result,

$$E_{11} = \frac{\partial u_x}{\partial x} + b_R(y)\frac{\partial u_z}{\partial x} - \frac{1}{2}b_R^2(y) = -b_R(y)\left(\frac{b_R(y)}{2} + \frac{u_z(0, y)}{x_R(y)}\right). \qquad (13.200)$$

What does this expression yield? Comment.

(13.22) By doing the integral, show that a "continuous" distribution of yarns, where

$$e = \frac{1}{\pi}\int_0^\pi \frac{1}{2}E_f E_\theta^2 d\theta, \qquad (13.201)$$

gives rise to the same homogeneous in the plane elastic constants as does an equally distributed around the circle finite number of yarns with at least three yarns (Eq. 13.167).

(13.23) Show that for an isotropic material, in Eq. (13.175) $D = (1 + \nu)^2(1 - 2\nu)$ and that

$$\varepsilon_{ij} = \frac{1 + \nu}{E}\sigma_{ij} - \frac{\nu}{E}\sigma_{kk}\delta_{ij}. \qquad (13.202)$$

(13.24) For an anisotropic material, a failure or a yield surface is often denoted by something of the form (variations are due to Hill and Tsai and Wu), $f(\sigma_{ij}) = \sum F_{ij}\sigma_{ij}^2 + \sum G_{ij}\sigma_{ij} - \tilde{Y}$, where F_{ij} and G_{ij} are constants (not the deformation gradient). Suppose that we know the uniaxial stress failure or yield stress in each direction in both tension and compression, and call it Y_{1c} and Y_{1t}, etc. Show that in order for the failure or yield surface to be closed in the π plane, it is necessary that

$$Y_{1t} > \frac{Y_{2c}Y_{3c}}{Y_{2c} + Y_{3c}}, \quad Y_{1c} > \frac{Y_{2t}Y_{3t}}{Y_{2t} + Y_{3t}}, \qquad (13.203)$$

and similarly for Y_{2c} etc., cycling through the indices. Thus, it is not possible to have a closed surface where there are two strong directions and one much weaker direction; specifically, if the strengths are the same in compression and tension and the 2 and 3 directions are the same, $Y_1 > \frac{1}{2}Y_2$. (Hint: use the last part of Exercise 3.21 and work in the π plane. Since the failure or yield surface is a quadratic, to be closed it must be an ellipse, and so the failure points must form a convex hull. Draw the line connecting Y_{2c} to Y_{3c} and see what it says about Y_{1t}.)

(13.25) In Appendix A, Section A.2 it is argued that conservation of angular momentum implies that the Cauchy stress tensor is symmetric. Why does this argument not apply to the first Piola–Kirchhoff stress tensor? Or, phrased another way, what step or part of this argument breaks down when the original Lagrangian frame is the frame of reference? (There is a simple, one sentence explanation of the difference.)

Rotation, Stretch, and Finite Elasticity

In the previous chapter, the Lagrangian strain and the Piola–Kirchhoff stress were introduced as ways to address large rotations. In this chapter we further explore the large strain perspective and work to understand deformation better. We look at constitutive models in isotropic finite elasticity (finite means the strains are large) and look at wave propagation in these models. We finish by considering corotational stress rates, which particularly arise in the context of modern numerical solution methods.

14.1. The Deformation Gradient

Our most basic measure of deformation is the deformation gradient F, where in Cartesian coordinates

$$F_{ij} = \frac{\partial x_i}{\partial X_j} \quad \text{and} \quad d\vec{x} = F d\vec{X}. \tag{14.1}$$

$\vec{X}$ is the original configuration and $\vec{x}$ is the current deformed configuration, which is in the laboratory reference frame. The deformation gradient has some interesting properties, but we emphasize here the fact that it takes a vector in the original configuration to a vector in the current configuration. In some sense, these spaces are not the same. Sometimes it is helpful to emphasize this fact by writing $F = {}_xF_X$, so $d\vec{x} = {}_xF_X d\vec{X}$, which is intended to imply the left side of F is in the current x configuration and the right side of F is in the original X configuration. This helps remind us that $F^T F = ({}_xF_X)^T {}_xF_X = {}_XF_x^T {}_xF_X$ makes sense, as $F^T F$ lives in the original configuration and $d\vec{X}$ can multiply both sides, and $FF^T = {}_xF_X({}_xF_X)^T = {}_xF_X{}_XF_x^T$ makes sense, as FF^T lives in the current configuration and $d\vec{x}$ can multiply both sides, but $F^2 = {}_xF_X{}_xF_X$ does not make sense because ${}_xF_X d\vec{x}$ does not make sense. Notice the variations $F = {}_xF_X$, $F^T = {}_XF_x^T$, $F^{-1} = {}_XF_x^{-1}$, and $F^{-T} = {}_xF_X^{-T}$.

F is not the only object to straddle two configurations. The first Piola–Kirchhoff stress does also. Recall that for a surface normal in the original configuration $\vec{N}$ the first Piola–Kirchhoff stress takes that normal to a traction vector $\vec{t}'$ in the current configuration that is a force in the current configuration per unit area in the original configuration, $\vec{t}'^T = \vec{N}^T S$. Hence, $S = {}_XS_x$, using the same notation (it may be clearer to take the transpose to see $S^T = {}_xS_X^T$). The Cauchy stress lives in the current configuration (the Eulerian or laboratory frame), and we have

$$\sigma = \frac{\rho}{\rho_0} F S = \frac{\rho}{\rho_0} {}_xF_X{}_XS_x = {}_x\sigma_x, \tag{14.2}$$

as expected.

Because of this behavior straddling two configurations, some authors state that F is not a tensor. It is helpful in this context to see how the argument goes to show that the Cauchy stress is a tensor. If $\vec{n}$ is a normal vector in the current configuration and $\vec{n}^* = Q\vec{n}$ represents a change in the current configuration coordinates by Q at the specific point, where Q is a rotation so $Q^{-1} = Q^T$, then, recalling $\vec{t}^T = \vec{n}^T \sigma$,

$$\vec{t}^{*T} = \vec{n}^{*T}\sigma^* \quad \Rightarrow \quad (Q\vec{t})^T = (Q\vec{n})^T\sigma^* \quad \Rightarrow \quad \vec{t}^T Q^T = \vec{n}^T Q^T \sigma^*$$
$$\Rightarrow \quad \vec{t}^T = \vec{n}^T Q^T \sigma^* Q \quad \Rightarrow \quad \sigma = Q^T \sigma^* Q \quad \Rightarrow \quad \sigma^* = Q\sigma Q^T. \tag{14.3}$$

Thus, the Cauchy stress transforms as a tensor. But this shows the complication with F and S: each side of the object lives in a different configuration. One cannot assume that the coordinate change in both configurations is the same and thus that the corresponding "tensor transformation" of Eq. (14.3) follows. Examining the coordinate change derivation for Cauchy stress, F's transformation is

$$F^* = {}_xQ_x \; {}_xF_X \; {}_XQ_X^T, \tag{14.4}$$

where ${}_xQ_x$ is the coordinate change for the current configuration and ${}_XQ_X$ is the coordinate change for the original configuration. These coordinate changes can be independent. In fact, it is usually assumed there are no changes to the original configuration (the original configuration is "invariant", ${}_XQ_X = I$), so that $F^* = QF$, where $Q = {}_xQ_x$. This does not look like a tensor transformation as it is a one-sided action on F.

In the previous chapter, why did we choose to work with the Piola–Kirchhoff stress and not just use the large Green (Lagrangian) strain with the Cauchy stress when we worked with yarns and fabrics? For example, a direct approach to using large strains and the Cauchy stress would be to say something like $\sigma = \lambda\mathrm{trace}(E)I + 2\mu E$, where $E = \frac{1}{2}(F^T F - I)$ is the large strain tensor. However, the left-hand side of the equation, $\sigma = {}_x\sigma_x$, and the right-hand side, $E = {}_XE_X$ since $F^T F = {}_XF_x^T \; {}_xF_X$, are in different configurations. Tensors need to be in the same configuration to be set equal in an equation.

To specifically understand what the issue is, suppose we stretch a material with stretch α in the X direction and then rotate it $90°$ counterclockwise. Then

$$F = R\begin{pmatrix} \alpha & 0 \\ 0 & 1 \end{pmatrix} = \begin{pmatrix} 0 & -1 \\ 1 & 0 \end{pmatrix}\begin{pmatrix} \alpha & 0 \\ 0 & 1 \end{pmatrix} = \begin{pmatrix} 0 & -1 \\ \alpha & 0 \end{pmatrix}. \tag{14.5}$$

The Lagrangian (Green) strain E is

$$E = {}_XE_X = \frac{1}{2}(F^T F - I) = \frac{1}{2}\begin{pmatrix} \alpha^2 - 1 & 0 \\ 0 & 0 \end{pmatrix}. \tag{14.6}$$

Recall E originates from $d\vec{x} - d\vec{X} = (F - I)d\vec{X}$ and we see that the change in length in E aligns in the X-axis direction.

However, the Cauchy stress σ should be aligned with the y-axis, which is the current configuration of the material after the rotation. We also have available to us the Eulerian strain E^*. It is based on $d\vec{x} - d\vec{X} = (I - F^{-1})d\vec{x}$. We have

$$E^* = \frac{1}{2}(1 - B^{-1}) = \frac{1}{2}(1 - (FF^T)^{-1}) = \frac{1}{2}(1 - F^{-T}F^{-1}) = \frac{1}{2}(1 - {}_xF_X^{-T} \; {}_XF_x^{-1}) = {}_xE_x^*. \tag{14.7}$$

The Eulerian strain is in the current configuration space by construction. In our example,

$$E^* = {}_xE^*_x = \frac{1}{2}(I - (FF^T)^{-1}) = \frac{1}{2}\begin{pmatrix} 0 & 0 \\ 0 & 1 - 1/\alpha^2 \end{pmatrix}. \tag{14.8}$$

In E^*, the deformation is aligned with the y-axis, as expected.

From the first Piola–Kirchhoff stress S we can produce the second Piola–Kirchhoff stress $S^{(2)} = SF^{-T}$, which by construction is symmetric,

$$S^{(2)} = SF^{-T} = {}_xS_x \, {}_xF_X^{-T} = {}_xS_X^{(2)}, \tag{14.9}$$

and we see that the second Piola–Kirchhoff stress is in the original configuration. Acceptable (and reasonable) forms for linear elasticity are

$$\sigma = \lambda\text{trace}(E^*)I + 2\mu E^*, \qquad S^{(2)} = \lambda\text{trace}(E)I + 2\mu E. \tag{14.10}$$

(The relations are linear, but the large strains are not linear, as they have nonlinear geometric terms.) Both sides of each equation are in the correct configuration, either current (left) or original (right). The two equations are not the same, however. One might think that $E^* = RER^T$, but this is not the case even in our specific example. A computation shows

$$E = F^T E^* F \quad \text{and} \quad E^* = F^{-T} E F^{-1}. \tag{14.11}$$

Since E and E^* are not related by $E^* = QEQ^{-1}$ for some Q, they are not the same tensor; indeed, E corresponds to the definition of strain given by $(L-L_0)/L_0$ and E^* corresponds to $(L - L_0)/L$, where L is length. Similarly, since $\sigma = (\rho/\rho_0)FS^{(2)}F^T$, σ and $S^{(2)}$ are not the same tensor.

The reason we used the Piola–Kirchhoff stress is that it was the natural one to work with when using E as the strain measure. In the next two sections we will find the average rotation in F, which relates the current and original configurations. There are additional subtleties in computing F in the cylindrical and spherical coordinate systems that are described in Appendix A (beginning on page 614).

14.2. Deformation: Rotation and Stretch in Two Dimensions

We first work in two dimensions because very nice explicit expressions result. For a given deformation, characterized by the deformation gradient F, when a small circle is drawn around the point of interest, the deformation takes that circle to an ellipse. This fact is due to our assumption that the deformation is a differentiable function of the original configuration. Let us determine how F deforms a circle of radius 1 drawn at the point $\vec{X}$. Let $\Delta\vec{X}^T = (\cos(\varphi) \;\; \sin(\varphi))$. As the parameter φ ranges from 0 to 2π, a circle is drawn around the point $\vec{X}$ in the original configuration. In the deformed geometry, this circle goes to $F\Delta\vec{X}$. As an example, consider a simple stretch of the material in the X direction so that $x = 2X$, $y = Y$. Then

$$\begin{pmatrix} x \\ y \end{pmatrix} = F\Delta\vec{X} = F\begin{pmatrix} \cos(\varphi) \\ \sin(\varphi) \end{pmatrix} = \begin{pmatrix} 2 & 0 \\ 0 & 1 \end{pmatrix}\begin{pmatrix} \cos(\varphi) \\ \sin(\varphi) \end{pmatrix} \quad \Rightarrow \quad \frac{1}{4}x^2 + y^2 = 1, \tag{14.12}$$

where the last equation was arrived at by multiplying through by F^{-1} and taking the dot product of each side of the equation with itself, or in general,

$$(x \;\; y) \, F^{-T}F^{-1}\begin{pmatrix} x \\ y \end{pmatrix} = (x \;\; y) \, (FF^T)^{-1}\begin{pmatrix} x \\ y \end{pmatrix} = (x \;\; y) \, B^{-1}\begin{pmatrix} x \\ y \end{pmatrix} = 1, \tag{14.13}$$

where we have defined $B = FF^T$. In the deformed geometry we have an ellipse. For the specific example, the semi-major axis of the ellipse aligns with the x-axis.

Consider another example: a shearing deformation, where $x = X + Y$ and $y = Y$. Then

$$F = \begin{pmatrix} 1 & 1 \\ 0 & 1 \end{pmatrix} \quad \Rightarrow \quad B = FF^T = \begin{pmatrix} 2 & 1 \\ 1 & 1 \end{pmatrix} \tag{14.14}$$

and hence

$$B^{-1} = \left(FF^T \right)^{-1} = \begin{pmatrix} 1 & -1 \\ -1 & 2 \end{pmatrix} \quad \Rightarrow \quad x^2 - 2xy + 2y^2 = 1. \tag{14.15}$$

For this ellipse, the semi-major axis does not align with a coordinate axis.

The next topic to consider is rotation. It intuitively seems that deformation can be decomposed into a displacement, a rotation, and some sort of deformation. We will compute an average rotation by comparing the original $\Delta \vec{X}$ with $\Delta \vec{x} = F \Delta \vec{X}$. In particular, this can be done by taking a cross product. Again, with $\Delta \vec{X}^T = (\cos(\varphi) \;\; \sin(\varphi))$ and for a specific φ,

$$\theta(\varphi) = \sin^{-1} \left(\Delta \vec{X} \times \frac{\Delta \vec{x}}{|\Delta \vec{x}|} \cdot \hat{z} \right). \tag{14.16}$$

A calculation provides

$$\theta(\varphi) = \sin^{-1} \left(\frac{F_{21} \cos^2(\varphi) - F_{12} \sin^2(\varphi) + (F_{22} - F_{11}) \sin(\varphi) \cos(\varphi)}{\sqrt{(F_{11} \cos(\varphi) + F_{12} \sin(\varphi))^2 + (F_{21} \cos(\varphi) + F_{22} \sin(\varphi))^2}} \right). \tag{14.17}$$

The average rotation produced by F is

$$\theta_a = \frac{1}{2\pi} \int_{-\pi}^{\pi} \theta(\varphi) d\varphi. \tag{14.18}$$

The rest of this paragraph is taken up in evaluating this integral. Since $\sin(\theta) = y/\sqrt{x^2 + y^2}$ and $\tan(\theta) = y/x$, a computation determining x produces

$$\theta(\varphi) = \tan^{-1} \left(\frac{F_{21} \cos^2(\varphi) - F_{12} \sin^2(\varphi) + (F_{22} - F_{11}) \sin(\varphi) \cos(\varphi)}{F_{11} \cos^2(\varphi) + (F_{12} + F_{21}) \sin(\varphi) \cos(\varphi) + F_{22} \sin^2(\varphi)} \right). \tag{14.19}$$

Half angle formulas produce

$$\theta(\varphi) = \tan^{-1} \left(\frac{F_{21} - F_{12} + (F_{21} + F_{12}) \cos(2\varphi) + (F_{22} - F_{11}) \sin(2\varphi)}{F_{11} + F_{22} + (F_{11} - F_{22}) \cos(2\varphi) + (F_{12} + F_{21}) \sin(2\varphi)} \right). \tag{14.20}$$

Letting

$$r = \sqrt{(F_{11} - F_{22})^2 + (F_{12} + F_{21})^2}, \quad \cos(\zeta) = \frac{F_{12} + F_{21}}{r}, \quad \sin(\zeta) = \frac{F_{11} - F_{22}}{r} \tag{14.21}$$

yields

$$\theta(\varphi) = \tan^{-1} \left(\frac{F_{21} - F_{12} + r \cos(2\varphi + \zeta)}{F_{11} + F_{22} + r \sin(2\varphi + \zeta)} \right). \tag{14.22}$$

Letting $\psi = 2\varphi$ leads to the average rotation integral Eq. (14.18) becoming

$$\theta_a = \frac{1}{4\pi} \int_{-2\pi}^{2\pi} \tan^{-1} \left(\frac{F_{21} - F_{12} + r \cos(\psi + \zeta)}{F_{11} + F_{22} + r \sin(\psi + \zeta)} \right) d\psi. \tag{14.23}$$

Recalling that the tangent of the angle is given by y/x, the arctangent is being taken of

$$x = F_{11} + F_{22} + r\sin(\psi + \zeta), \quad y = F_{21} - F_{12} + r\cos(\psi + \zeta). \qquad (14.24)$$

This is a circle centered at $(x, y) = (F_{11} + F_{22}, F_{21} - F_{12})$. The integral is the average of a double wrap around this circle of the angle of the point on the circle with the origin. The average around the circle is simply the angle to the center. Hence,

$$\theta_a = \tan^{-1}\left(\frac{F_{21} - F_{12}}{F_{11} + F_{22}}\right). \qquad (14.25)$$

This θ_a is the average rotation in the deformation gradient F.

As an example, suppose F is a pure rotation through an angle $\bar{\theta}$, with

$$F = R(\bar{\theta}) = \begin{pmatrix} \cos\bar{\theta} & -\sin\bar{\theta} \\ \sin\bar{\theta} & \cos\bar{\theta} \end{pmatrix}. \qquad (14.26)$$

Computing with Eq. (14.25) gives

$$\theta_a = \tan^{-1}(\tan(\bar{\theta})) = \bar{\theta} \qquad (14.27)$$

and hence the average rotation is the applied rotation as expected.

Next, if F is symmetric, then Eq. (14.25) yields $\theta_a = 0$. Thus, when the deformation gradient is symmetric there is no average rotation. With the form of the rotation matrix (Eq. 14.26) and the fact that the tangent Eq. (14.25) is the ratio of the sine to the cosine, we can explicitly write down the average rotation matrix, which is

$$R(\theta_a) = \frac{1}{A}\begin{pmatrix} F_{11} + F_{22} & -(F_{21} - F_{12}) \\ F_{21} - F_{12} & F_{11} + F_{22} \end{pmatrix}, \quad A = \sqrt{(F_{11} + F_{22})^2 + (F_{21} - F_{12})^2}, \qquad (14.28)$$

and by inspection

$$R(\theta_a) = \frac{1}{A}(\text{trace}(F)I + F - F^T) \qquad \text{(two dimensions)}. \qquad (14.29)$$

We expect that the deformation decomposes into a rotation and a deformation that is average rotation free, which we write as

$$F = RU, \qquad (14.30)$$

where U is called the right stretch since it is on the right in this matrix multiplication and $R = R(\theta_a)$ is the rotation of the average rotation of F. We now explicitly compute

$$U = R^T F = \frac{1}{A}\begin{pmatrix} F_{11}^2 + F_{21}^2 + F_{11}F_{22} - F_{12}F_{21} & F_{11}F_{12} + F_{21}F_{22} \\ F_{11}F_{12} + F_{21}F_{22} & F_{12}^2 + F_{22}^2 + F_{11}F_{22} - F_{12}F_{21} \end{pmatrix}, \qquad (14.31)$$

or, by careful examination,

$$U = \frac{1}{A}(F^T F + \det(F)I) \qquad \text{(two dimensions)}. \qquad (14.32)$$

U is symmetric and hence its average rotation is zero.[1] Thus, we have found explicit expressions for the decomposition of the deformation gradient into a stretch U followed by a rotation R.

It is also possible to do the decomposition of F with the rotation performed first followed by the stretch. This order is written

$$F = VR, \tag{14.33}$$

where V is the left stretch. Again $R = R(\theta_a)$, since the average rotation comes from F. We explicitly compute V,

$$V = FR^T = \frac{1}{A}\begin{pmatrix} F_{11}^2 + F_{12}^2 + F_{11}F_{22} - F_{12}F_{21} & F_{11}F_{21} + F_{12}F_{22} \\ F_{11}F_{21} + F_{12}F_{22} & F_{21}^2 + F_{22}^2 + F_{11}F_{22} - F_{12}F_{21} \end{pmatrix}, \tag{14.34}$$

which is symmetric and by examination is

$$V = \frac{1}{A}(FF^T + \det(F)I) \qquad \text{(two dimensions)}. \tag{14.35}$$

The stretch U and V are similar, in the linear algebra sense, as $RUR^T = V$. The stretch tensor U is in the original configuration (the Lagrangian frame $\vec{X}$), while the stretch tensor V is in the current configuration (the Eulerian frame $\vec{x}$). The decomposition of F into RU and VR is referred to as the *polar decomposition*.

Now that we have a good understanding of the average rotation R, let's examine the stretch U in a little more detail. A computation, using the symmetry of U, yields

$$U^2 = U^T U = (R^T F)^T (R^T F) = F^T F \quad \Rightarrow \quad U = \sqrt{F^T F} = \sqrt{C}. \tag{14.36}$$

Hence, U is the matrix square root of $F^T F$. That is really quite extraordinary. It turns out that, just as there are two square roots of a real number (e.g., $\sqrt{4} = \pm 2$), there are four square roots of a 2×2 symmetric matrix. The one that we have in U is the positive definite one (its eigenvalues are positive), as can be confirmed by $\det(F) > 0$, $\det(R) = 1$, and $\text{trace}(U) = A > 0$. Similarly,

$$V^2 = VV^T = (FR^T)(FR^T)^T = FF^T \quad \Rightarrow \quad V = \sqrt{FF^T} = \sqrt{B}. \tag{14.37}$$

Let's return to the example of the shear deformation gradient $F = \begin{pmatrix} 1 & 1 \\ 0 & 1 \end{pmatrix}$. This deformation gradient F is not symmetric, nor can it be diagonalized by a rotation matrix (in fact, it cannot be diagonalized at all). From the above explicit expressions we can write down R, U, and V,

$$R(\theta_a) = \frac{1}{\sqrt{5}}\begin{pmatrix} 2 & 1 \\ -1 & 2 \end{pmatrix}, \quad U = \frac{1}{\sqrt{5}}\begin{pmatrix} 2 & 1 \\ 1 & 3 \end{pmatrix}, \quad V = \frac{1}{\sqrt{5}}\begin{pmatrix} 3 & 1 \\ 1 & 2 \end{pmatrix}. \tag{14.38}$$

We have decomposed the deformation into a stretch U followed by a rotation R. The initial stretch takes a small circle into an ellipse. The ellipse is determined by

$$\begin{pmatrix} x & y \end{pmatrix} U^{-2} \begin{pmatrix} x \\ y \end{pmatrix}, \quad U^{-2} = \begin{pmatrix} 2 & -1 \\ -1 & 1 \end{pmatrix} \quad \Rightarrow \quad 2x^2 - 2xy + y^2 = 1. \tag{14.39}$$

---

[1] Note that the stretch U is typically not diagonal as that does not contain enough information; in two dimensions F has four components and the rotation is determined by one angle; hence there are three degrees of freedom in the symmetric stretch U. In three dimensions there are nine components in F and the rotation is determined by three angles, and the remaining six degrees of freedom are represented in the symmetric stretch U.

The semi-major axis of the ellipse is in the direction of the largest eigenvector of U. The eigenvalues and eigenvectors of U are

$$\frac{\sqrt{5}+1}{2}, \quad \begin{pmatrix} 2 \\ 1+\sqrt{5} \end{pmatrix}; \qquad \frac{\sqrt{5}-1}{2}, \quad \begin{pmatrix} 1+\sqrt{5} \\ -2 \end{pmatrix}. \tag{14.40}$$

Hence the angle of the semi-major axis of the U ellipse with respect to the X-axis is given by $\theta_U = \tan^{-1}((1+\sqrt{5})/2) = 58.3°$ (positive is counterclockwise). The average rotation of F is

$$\theta_a = \tan^{-1}\left(-\frac{1}{2}\right) = -26.6°. \tag{14.41}$$

The final angle of the semi-major axis of the ellipse of F with the X-axis is hence $\theta_U + \theta_a = 58.3° - 26.6° = 31.7°$. As a check, recall from Eq. (14.13) that the ellipse of F is determined by B, but $B = V^2$, and hence the final ellipse is defined by V, whose eigenvalues and eigenvectors are

$$\frac{\sqrt{5}+1}{2}, \quad \begin{pmatrix} 1+\sqrt{5} \\ 2 \end{pmatrix}; \qquad \frac{\sqrt{5}-1}{2}, \quad \begin{pmatrix} -2 \\ 1+\sqrt{5} \end{pmatrix}. \tag{14.42}$$

The angle with the x-axis (which is the same as the X-axis) is $\theta_V = \tan^{-1}(2/(1+\sqrt{5})) = 31.7°$, as expected. The ellipse of V is the same as the ellipse of F because for $F = VR$ the rotation is done first, which just rotates the original circle $\Delta \vec{X}$; then the circle is deformed into an ellipse by the stretch V. (The angles for arbitrary F are worked out in Exercise 14.35.)

Given R represents the average rotation between the original configuration and the current deformed state, what is the physical meaning of U? Recall from our definition of Lagrangian strain in Section 13.1 that $F^T F$ provides, when pre- and post-multiplied by $d\vec{X}$, the exact square of the length ds^2. Does taking the square root $U = \sqrt{F^T F}$ give us a measure of length ds? Defining the unit vector in the same direction $\vec{N} = d\vec{X}/|d\vec{X}|$, the question we ask is whether $\vec{N}^T U \vec{N}$ equals ds/dS. The answer: not exactly, but it is close. As an example, consider the shear example of the previous paragraph. The exact deformed length ratio going around the circle is $ds/dS = \sqrt{\vec{N}^T F^T F \vec{N}} = \sqrt{1 + \sin(2\varphi) + \sin^2(\varphi)}$. The length ratio given by $\vec{N}^T U \vec{N} = (2 + \sin(2\varphi) + \sin^2(\varphi))/\sqrt{5}$. These expressions are not the same, but they are close. They are equal at their extremes, where $\tan(2\varphi) = -2$, and otherwise the latter expression from U is less than the actual length ratio; it can be less by as much as 11% as we go around the circle and on average is less by 5.6%. Thus, U is approximately ds/dS, so $U - I$ is approximately the linear strain $(ds - dS)/dS$. This compares and contrasts with the Lagrangian strain $E = \frac{1}{2}(F^T F - I) = (ds^2 - dS^2)/dS^2$. Notice that $E = \frac{1}{2}(U^2 - I) = U - I + \frac{1}{2}(U - I)^2$ and hence $U - I$ and E are nearly identical for U close to I, but for large strains E is larger (it behaves quadratically as ds^2). In Exercise 14.40, a number of strain measures are introduced, one of which is $E^{(1)} = U - I$. There, $E^{(2)} = E = E^{(1)} + \frac{1}{2}(E^{(1)})^2$.

As a final topic, the notion of angle allows us to define shear. Shear is when there is a relative angle between $d\vec{X}$ and $d\vec{x}$ after the average rotation has been subtracted out. Thus, a measure of shear, which is an invariant of the deformation gradient F, is

$$\theta_{\text{shear}} = \frac{1}{2\pi} \int_{-\pi}^{\pi} |\theta(\varphi) - \theta_a| d\varphi. \tag{14.43}$$

Considering the case of an F that in the limit tends towards $\left(\begin{smallmatrix} 1 & 0 \\ 0 & 0 \end{smallmatrix}\right)$ so that $\Delta\vec{x} = \left(\begin{smallmatrix} 1 \\ 0 \end{smallmatrix}\right)$ yields, by calculation, that the maximum shear is $\theta_{\text{shear}} = \pi/4$.

There are invariants of F that reflect only one-sided rotation – that is, $F^* = QF$. (To be explicit, the trace of F is not an invariant of the one-sided rotation action.) Since the rotation of a vector does not change the vector's length, the sum of square of each column of F is an invariant of this action. This invariant for each column can be summed to provide another invariant. Hence, $F_{11}^2 + F_{12}^2 + F_{21}^2 + F_{22}^2$ is an invariant of the action of one-sided rotation. The determinant $\det(F^*) = \det(QF) = \det(Q)\det(F) = \det(F)$ is also an invariant. An algebraic combination of these two yields a one-sided rotation invariant of F that measures shear,

$$I_{\text{shear 2D}} = \frac{F_{11}^2 + F_{22}^2 + F_{12}^2 + F_{21}^2}{2\det(F)} - 1 = \frac{(F_{11} - F_{22})^2 + (F_{12} + F_{21})^2}{2\det(F)} \geq 0.$$

$$(14.44)$$

Computations show that there is a one to one relationship between this invariant and the shear angle – that is, $\theta_{shear} = f(I_{\text{shear 2D}})$. If $I_{\text{shear 2D}} > 0$, then there is some shear in the deformation characterized by F. By inspection of Eq. (14.44) it is seen that the only deformation that has no shear is a dilation (uniform expansion or compression) with a rotation – that is, $F = cR$ for some constant scalar c. This means that when there is no shear, F takes a circle in the original configuration to a circle in the deformed configuration.

14.3. Deformation: Stretch and Rotation in Three Dimensions

The discussion so far has been in two dimensions. In three dimensions the polar decomposition is

$$F = RU, \quad F = VR,$$

$$(14.45)$$

where F is the deformation gradient, R is a rotation matrix, and U and V are the right and left stretch, respectively. Direct computation yields

$$F^T F = (RU)^T RU = U^T R^T RU^T = U^T U = U^2,$$
$$FF^T = VR(VR)^T = VRR^T V^T = VV^T = V^2.$$

$$(14.46)$$

It is possible to take the square root of a matrix by diagonalizing it and taking the square roots of the eigenvalues. Since $F^T F$ and FF^T are symmetric and positive definite, their eigenvalues are real and positive, and we always choose the positive square roots to yield the unique square root that is positive definite (see Exercise 14.28),

$$U = \sqrt{F^T F}, \quad V = \sqrt{FF^T}.$$

$$(14.47)$$

In these expressions, R turns out to be the same rotation as can be shown using the uniqueness of the square root (Exercise 14.28), and R is

$$R = FU^{-1} = V^{-1}F.$$

$$(14.48)$$

U and V are related by

$$U = R^T V R, \quad V = RUR^T,$$

$$(14.49)$$

and hence U and V are the same deformation (in fact, the same tensor); they only differ by the fact that U is the deformation as represented in the original configuration and V is the deformation as represented in the current (deformed) configuration. U is called the right stretch and V the left stretch, where right and left refer to the side on which they multiply R. Hence, we either perform a right stretch and then rotate, or we rotate and perform a left stretch. The decomposition into a symmetric stretch and a rotation is, again, referred to as the *polar decomposition* of F.

In three dimensions there is not a nice formula like Eqs. (14.17) and (14.18) for the rotation. The reason is that in three dimensions the cross product between $\Delta \vec{X}$ and $\Delta \vec{x}$ does not define a unique rotation connecting $\Delta \vec{X}$ and $\Delta \vec{x}$. However, the rotation R is still the average rotation, and in three dimensions we define the average rotation as the rotation matrix that minimizes H_1,

$$\min_R H_1 = \min_R \int_S \left| \frac{\Delta \vec{x}}{|\Delta \vec{x}|} - R\Delta \vec{X} \right|^2 dS, \tag{14.50}$$

where the integral is performed over the sphere defined by $|\Delta \vec{X}| = 1$. (This approach also works in two dimensions and the same angle results.) Expanding this expression,

$$H_1 = \int_S \left| \frac{F\Delta \vec{X}}{|F\Delta \vec{X}|} - R\Delta \vec{X} \right|^2 dS = \int_S \left\{ 2 - \frac{1}{|F\Delta \vec{X}|} \Delta \vec{X}^T (F^T R + R^T F) \Delta \vec{X} \right\} dS. \tag{14.51}$$

Both terms on the right produce the same number and so our minimization problem to find R is equivalent to maximizing $H_2 = 4\pi - \frac{1}{2} H_1$,

$$\max_R H_2 = \max_R \int_S \frac{\Delta \vec{X}^T F^T R \Delta \vec{X}}{|F\Delta \vec{X}|} dS. \tag{14.52}$$

A geometric picture of maximizing $H_2/4\pi$ is that we are maximizing the average cosine of the angle between $\Delta \vec{X}$ and $\Delta \vec{x}/|\Delta \vec{x}|$, hence trying to make the average angle over all directions as small as possible.

We will use what we know about the polar decomposition at this point, writing $F = \bar{R}U$, where $\bar{R}$ is the rotation matrix resulting from the polar decomposition. U is symmetric and hence diagonalizable by a rotation matrix Q as $U = QDQ^T$. Let $\Delta \vec{Y} = Q\Delta \vec{X}$. These statements yield

$$H_2 = \int_S \frac{\Delta \vec{X}^T U^T \bar{R}^T R \Delta \vec{X}}{|RU\Delta \vec{X}|} dS = \int_S \frac{\Delta \vec{Y}^T DQ\bar{R}^T RQ^T \Delta \vec{Y}}{|D\Delta \vec{Y}|} dS, \tag{14.53}$$

where the integral is performed over the sphere $|\Delta \vec{Y}| = 1$. We define $\hat{R} = Q\bar{R}^T RQ^T$, yielding

$$H_2 = \int_S \frac{\Delta \vec{Y}^T D\hat{R} \Delta \vec{Y}}{|D\Delta \vec{Y}|} dS, \tag{14.54}$$

and note that if the polar decomposition angle $\bar{R}$ is the angle that minimizes H_1 in Eq. (14.50) (hence maximizes H_2), then $\hat{R} = I$. Thus our aim now is to show that $\bar{R} = I$ produces a local maximum of H_2. This is done by first writing down an explicit expression for the rotation matrix of angle θ_a about the unit vector pointing along the axis of the rotation $\vec{n}$ (sometimes written $\vec{\theta}_a = \theta_a \vec{n}$),

$$R(\theta_a, \vec{n}) = \exp(\theta_a N) = I + \sin(\theta_a)N + (1 - \cos(\theta_a))N^2, \quad N = \begin{pmatrix} 0 & -n_3 & n_2 \\ n_3 & 0 & -n_1 \\ -n_2 & n_1 & 0 \end{pmatrix}. \tag{14.55}$$

Notice that $N\vec{x} = \vec{n} \times \vec{x}$ for an arbitrary vector $\vec{x}$ (Exercise 14.2 elucidates R). To show that the integral is a local maximum, it is first necessary to show that for any rotation the derivative of the integral evaluated at $\hat{R} = I$ is zero. The rotation axis can be any direction and now we assume the rotation axis is fixed. The derivative of the rotation matrix evaluated at $\theta_a = 0$ is

$$\frac{d\hat{R}}{d\theta_a}\bigg|_{\theta_a=0} = N. \tag{14.56}$$

Thus, we want to show that

$$\frac{dH_2}{d\theta_a}\bigg|_{\hat{R}=I} = \int_S \frac{\Delta\vec{Y}^T DN\Delta\vec{Y}}{|D\Delta\vec{Y}|} dS = 0. \tag{14.57}$$

That will be accomplished by decomposing N into three terms,

$$N = n_1 N_1 + n_2 N_2 + n_3 N_3, \quad \text{where, for example,} \quad N_3 = \begin{pmatrix} 0 & -1 & 0 \\ 1 & 0 & 0 \\ 0 & 0 & 0 \end{pmatrix}, \tag{14.58}$$

and showing that each integral separately is zero. We will show it for one case; the others are the same and only require a component permutation in $\Delta\vec{Y}$ to achieve. With the parameterization of the unit sphere in Appendix A,

$$\Delta\vec{Y} = \begin{pmatrix} \cos(\varphi)\sin(\theta) \\ \sin(\varphi)\sin(\theta) \\ \cos(\theta) \end{pmatrix}, \quad dS = \sin(\theta)d\varphi d\theta, \quad 0 \le \varphi < 2\pi, \quad 0 \le \theta \le \pi, \tag{14.59}$$

notice that

$$\frac{\partial(\Delta\vec{Y})}{\partial\varphi} = \begin{pmatrix} -\sin(\varphi)\sin(\theta) \\ \cos(\varphi)\sin(\theta) \\ 0 \end{pmatrix} = N_3 \Delta\vec{Y}. \tag{14.60}$$

We have

$$2\Delta\vec{Y}^T DN_3 \Delta\vec{Y} = \Delta\vec{Y}^T DN_3 \Delta\vec{Y} + (\Delta\vec{Y}^T DN_3 \Delta\vec{Y})^T$$
$$= \Delta\vec{Y}^T DN_3 \Delta\vec{Y} + \Delta\vec{Y}^T N_3^T D^T \Delta\vec{Y} \tag{14.61}$$
$$= \Delta\vec{Y}^T (DN_3 - N_3 D) \Delta\vec{Y}.$$

An explicit computation shows that

$$D^2 N_3 - N_3 D^2 = (D_{11} + D_{22})(DN_3 - N_3 D). \tag{14.62}$$

Hence we have an exact differential in φ as part of our integrand,

$$2\int_S \frac{\Delta\vec{Y}^T DN_3 \Delta\vec{Y}}{|D\Delta\vec{Y}|} dS = \frac{1}{D_{11} + D_{22}} \int_S \frac{\Delta\vec{Y}^T (D^2 N_3 - N_3 D^2) \Delta\vec{Y}}{(\Delta\vec{Y}^T D^2 \Delta\vec{Y})^{1/2}} \sin(\theta)d\theta d\varphi$$
$$= \frac{-2}{D_{11} + D_{22}} \int_0^\pi (\Delta\vec{Y}^T D^2 \Delta\vec{Y})^{1/2}|_0^{2\pi} \sin(\theta)d\theta = 0 \tag{14.63}$$

Thus, each of the three rotational axes terms is zero and we conclude $dH_2/d\theta_a|_{\hat{R}=I} = 0$, which is what we were trying to show. The next necessary step in showing H_2 is a local maximum is showing $d^2 H_2/d\theta_a^2|_{\hat{R}=I} \le 0$. To this end,

$$\frac{d^2\hat{R}}{d\theta_a^2}\bigg|_{\theta_a=0} = N^2, \tag{14.64}$$

and hence the second derivative of the functional is

$$\left.\frac{d^2 H_2}{d\theta_a^2}\right|_{\hat{R}=I} = -\int_S \frac{(ND\Delta\vec{Y})^T N\Delta\vec{Y}}{|D\Delta\vec{Y}|}dS, \qquad (14.65)$$

where $N^T = -N$ was used. D is diagonal with positive entries; thus there are essentially three "extreme" values of D, with the understanding that these values can be multiplied by a positive constant. For each of these values we compute this second derivative,

$$D = I \quad \Rightarrow \quad \left.\frac{d^2 H_2}{d\theta_a^2}\right|_{\hat{R}=I} = -\int_S |N\Delta\vec{Y}|^2 dS = -\frac{8\pi}{3},$$

$$D = \begin{pmatrix} 1 & 0 & 0 \\ 0 & 1 & 0 \\ 0 & 0 & 0 \end{pmatrix} \quad \Rightarrow \quad \left.\frac{d^2 H_2}{d\theta_a^2}\right|_{\hat{R}=I} = -\int_S \frac{(ND\Delta\vec{Y})^T N\Delta\vec{Y}}{\sin(\theta)}dS = -\frac{\pi^2(1+n_3^2)}{2},$$

$$D = \begin{pmatrix} 1 & 0 & 0 \\ 0 & 0 & 0 \\ 0 & 0 & 0 \end{pmatrix} \quad \Rightarrow \quad \left.\frac{d^2 H_2}{d\theta_a^2}\right|_{\hat{R}=I} = -\int_S \frac{(ND\Delta\vec{Y})^T N\Delta\vec{Y}}{|\cos(\varphi)|\sin(\theta)}dS = -2\pi(1-n_1^2).$$

$$(14.66)$$

We conclude that $d^2 H_2/d\theta_a^2|_{\hat{R}=I} \leq 0$; hence $\hat{R} = I$ is a local maximum of H_2, thus showing that $R = \bar{R}$ given by the polar decomposition $\bar{R}U$ minimizes H_1 (or, equivalently, maximizes H_2). Thus the rotation R resulting from the polar decomposition RU is an average rotation in the sense of minimizing H_1.

What if $F = VR$ had been used in this argument rather than $F = RU$? It turns out that exactly the same minimization occurs; hence it demonstrates that R is the same for both decompositions. In outline, one first shows that the eigenvalues of U and V are the same by showing that the invariants of $F^T F$ and FF^T are the same: I_1 because trace$(F^T F) =$ trace(FF^T), I_2 because of this and because trace$((F^T F)^2) =$ trace$(F^T FF^T F) =$ trace$(FF^T FF^T) =$trace$((FF^T)^2)$, and I_3 because det$(F^T F) =$ det(FF^T). Because of this, $V = \bar{Q}D\bar{Q}^T$ for the same diagonal matrix D as occurs for U, though the Q differs, hence $\bar{Q}$. If we define $F = V\bar{R}$, $\Delta\vec{Y} = \bar{Q}^T \bar{R}\Delta\vec{X}$, and $\hat{R} = \bar{Q}^T R\bar{R}^T \bar{Q}$, then exactly the same H_2 formula results, and the maximization of H_2 occurs at $\hat{R} = I$, showing that $R = \bar{R}$ for $F = V\bar{R}$ minimizes the integral H_1. There is only one R from minimizing the integral, so we conclude that R is the same for $F = RU$ and $F = VR$.

Looking at $_X F^T F_X$ and $_x FF_x^T$, we see that $U = _X U_X$ is in the original configuration and $V = _x V_x$ is in the current configuration. These give $R = _x F_{XX} U_X^{-1} = _x R_X$ and we concluded that R, which is the average rotation of F, connects the original and current configurations. Since $V = RUR^T$, U and V are the same tensor representing the deformation, but they are in different configurations.

When the strains and rotations are small, we can explicitly compute approximations for R and U. Writing $F = I + \hat{F}$, then with u_i the displacement, we have in Cartesian coordinates

$$\hat{F}_{ij} = \frac{\partial u_i}{\partial X_j} \approx \frac{\partial u_i}{\partial x_j} = \frac{1}{2}\left(\frac{\partial u_i}{\partial x_j} + \frac{\partial u_j}{\partial x_i}\right) + \frac{1}{2}\left(\frac{\partial u_i}{\partial x_j} - \frac{\partial u_j}{\partial x_i}\right) = \varepsilon_{ij} + \omega_{ij}. \qquad (14.67)$$

The symmetric part of $\hat{F}$ is ε, the small strain, and skew symmetric part of $\hat{F}$ is ω, the small rotation, where $\hat{\theta} = \frac{1}{2}\mathrm{curl}(\vec{u}) \approx \theta_a$ and $\omega_{ij} = -\varepsilon_{ijk}\theta_k$. Then

$$F \approx I + \varepsilon + \omega \qquad (14.68)$$

and, since the small strain and small rotation are small compared to I, the nonlinear terms are dropped in

$$F^T F = I + \hat{F} + \hat{F}^T + \hat{F}^T \hat{F} \approx I + \hat{F} + \hat{F}^T = I + 2\varepsilon. \qquad (14.69)$$

Expanding the square root in a Taylor series,

$$U = \sqrt{F^T F} \approx I + \varepsilon, \qquad (14.70)$$

confirming for small strain that $U - I$ is the strain. Similarly,

$$R = FU^{-1} \approx (I + \hat{F})(I - \frac{1}{2}(\hat{F} + \hat{F}^T)) \approx I + \frac{1}{2}(\hat{F} - \hat{F}^T) \approx I + \omega. \qquad (14.71)$$

In the next section we will see a velocity version of the rotation matrix ω called the spin. Due to ω being skew symmetric, pre- and post-multiplying by $d\vec{X}$ yields zero, indicating no change in length by the rotation.

As we did in two dimensions, here in three dimensions we make some comments on shear deformation. Again, one-sided rotation invariants of F are $F_{ij}F_{ij}$ and $\det F$. From these we build

$$I_{\text{shear 3D}} = \frac{F_{ij}F_{ij}}{3(\det(F))^{2/3}} - 1 \geq 0. \qquad (14.72)$$

When $I_{\text{shear 3D}} = 0$, then there is no shear and $F = cR$ for some rotation matrix and the image of F is a sphere. If F takes a sphere in the original configuration to an ellipsoid that is not a sphere in the deformed configuration, then some of the deformation is shear.

14.4. Stress and Strain in Original and Current Configurations

In this section we collect and summarize information about the two configurations, the original configuration and the current configuration (Eulerian or laboratory frame) and how they are linked. The link is through the deformation gradient F. This is very particular mapping, and when objects are ascribed physical meaning they are transformed differently by this mapping, as will be seen in the following discussion.

We can write a commutative diagram describing the polar decomposition of F. A commutative diagram is one where all paths connecting the same vertices, through composition of the functions identified on the edge labels, produce the same result (i.e., the paths commute). Matrices describe linear operators through matrix multiplication – for example, $d\vec{x} = F(d\vec{X}) = Fd\vec{X} = RUd\vec{X} = VRd\vec{X}$. Specifically there are two vectors we have operated on with F: the length segment $d\vec{X}$ and the area normal $d\vec{A}$. The latter behaves differently when acted upon by F, as $d\vec{a} = F(d\vec{A}) = \det(F)F^{-T}d\vec{A}$ (Eq. 13.26); here $d\vec{a} = \vec{n}da$ and $d\vec{A} = \vec{N}dA$. The commutative diagrams of F acting on these two vectors are

$$
\begin{array}{ccc}
d\vec{X} & \xrightarrow{\;U\;} & Ud\vec{X} \\
{\scriptstyle R}\downarrow & & \downarrow{\scriptstyle R} \\
Rd\vec{X} & \xrightarrow{\;V\;} & d\vec{x}
\end{array}
\qquad\qquad
\begin{array}{ccc}
d\vec{A} & \xrightarrow{\;\det(U)U^{-1}\;} & R^T d\vec{a} \\
{\scriptstyle R}\downarrow & & \downarrow{\scriptstyle R} \\
Rd\vec{A} & \xrightarrow{\;\det(V)V^{-1}\;} & d\vec{a}
\end{array}
$$

The upper row of the diagram is in the original configuration (though possibly deformed), while the lower row in the diagram has material that has been rotated to the current configuration. The left column is undeformed, the right is deformed.[2]

The first Piola–Kirchhoff stress, which was used extensively in the previous chapter, took a surface normal in the original undeformed configuration to a traction in the current deformed configuration. We now ask what a stress tensor would look like that takes a unit normal in the original undeformed configuration to a traction $\vec{T}$ in the original undeformed configuration. Thinking of a point mass in the original configuration being accelerated by a force – the pseudo-force $\vec{F}$ in the original configuration accelerates over the original configuration distance – which the actual force in the laboratory frame accelerates over the actual distance $d\vec{x} = Fd\vec{X}$, we have $\vec{f} = F\vec{F}$, where the pseudo-force in the original configuration is denoted $\vec{F}$, not to be confused with the deformation gradient F. Taking the force from Eq. (13.14) for a very small volume,

$$\vec{F}^T = (F^{-1}\vec{f})^T = \vec{f}^T F^{-T} = (\int_{\partial V} \vec{N}^T S\, dA) F^{-T} \approx \int_{\partial V} \vec{N}^T S F^{-T} dA, \quad (14.73)$$

where F is evaluated at a point in that small volume until the last term, where the smoothness of F was used to pull it into the integral of very small volume. The stress we are looking for is referred to as the *second Piola–Kirchhoff stress* and is

$$S^{(2)} = SF^{-T} = \frac{\rho_0}{\rho}F^{-1}\sigma F^{-T}. \quad (14.74)$$

$S^{(2)}$ is symmetric since σ is symmetric, as taking the transpose shows. This tensor produces a traction in the original configuration – the pseudo-force in the original configuration in terms of the unit surface area in the original undeformed configuration,

$$\vec{T}^T = \vec{N}^T S^{(2)}. \quad (14.75)$$

With this tensor and the Cauchy and first Piola–Kirchhoff stress tensors, we now have three stress tensors and three tractions, and a summary of the relations between them is

$$\sigma = \frac{\rho}{\rho_0}FS, \quad S = S^{(2)}F^T, \quad \sigma = \frac{\rho}{\rho_0}FS^{(2)}F^T, \quad \frac{\rho_0}{\rho} = \det(F),$$

$$\vec{t}^T = \vec{n}^T\sigma, \quad \vec{t}'^T = \vec{N}^T S, \quad \vec{T}^T = \vec{N}^T S^{(2)}, \quad \vec{n} = \frac{F^{-T}\vec{N}}{|F^{-T}\vec{N}|}, \quad (14.76)$$

$$\vec{t} = \frac{dA}{da}\vec{t}', \quad \vec{t}' = F\vec{T}, \quad \vec{t} = \frac{dA}{da}F\vec{T}, \quad \frac{da}{dA} = \det(F)|F^{-T}\vec{N}|.$$

The reason we worked with the first Piola–Kirchhoff stress in the previous chapter, rather than the second, is that the second has a more complicated momentum balance, $\partial(S^{(2)}_{jk}F_{ik})/\partial X_j + \rho_0 b_i = \rho_0 \partial v_i/\partial t$.

---

[2]In general coordinate systems as discussed in Appendix A.7, vectors that transform like differentials are called *covariant* while vectors that transform like gradients are called *contravariant*. Thinking of differential geometry terminology to help clarify this setting, F connects $d\vec{X}$ and $d\vec{x}$ and so is like a transformation connecting covariant vectors, while $(\mathrm{grad}_x(\phi))_i = \partial\phi/\partial x_i = (\partial\phi/\partial X_j)(\partial X_j/\partial x_i) = F^{-T}_{ij}(\mathrm{grad}_X(\phi))_j$ and hence F^{-T} is like a transformation between contravariant vectors. Thus, $d\vec{X}$ and $d\vec{x}$ are covariant in their behavior, while $d\vec{A}$ and $d\vec{a}$ are close to being contravariant in their behavior (up to the multiplicative scalar $\det(F)$).

These relations can be summarized in commutative diagrams. Thinking of the stress tensor as an operator, where we multiply on the right with a vector and a new vector is produced (actually we would multiply on the left by a transposed vector, but for this diagram the second Piola–Kirchhoff stress and the Cauchy stress are both symmetric), the diagrams are

$$
\begin{array}{ccc}
d\vec{A} & \xrightarrow{S^{(2)}} & d\vec{F} \\
\scriptstyle\det(F)F^{-T}\downarrow & & \downarrow\scriptstyle F \\
d\vec{a} & \xrightarrow{\sigma} & d\vec{f}
\end{array}
\qquad
\begin{array}{ccc}
\vec{N} & \xrightarrow{S^{(2)}} & \vec{T} \\
\scriptstyle\vec{n}=\frac{F^{-T}\vec{N}}{|F^{-T}\vec{N}|}\downarrow & & \downarrow\scriptstyle\vec{t}=\frac{F\vec{T}}{\det(F)|F^{-T}\vec{N}|} \\
\vec{n} & \xrightarrow{\sigma} & \vec{t}
\end{array}.
$$

The left diagram is written in a differential form, which is to be interpreted in the context of a surface integral that touches the same material: on the upper row the integral is in the original configuration and on the lower row it is in the current configuration. The right diagram has nonlinear maps from the upper row (original configuration) to the lower row (current configuration) so they are explicitly defined, rather than just showing the matrix that multiplies. In these diagrams, we can always go up and down, since F is invertible; we can always go to the right; however, we may not go to the left as the stress (and later strain) need not be invertible.

The commutative diagrams for the first Piola–Kirchhoff stress are

$$
\begin{array}{ccc}
d\vec{A}^{T} & \xrightarrow{S} & d\vec{f}^{T} \\
\scriptstyle\det(F)F^{-1}\downarrow & & \|\scriptstyle I \\
d\vec{a}^{T} & \xrightarrow{\sigma} & d\vec{f}^{T}
\end{array}
\qquad
\begin{array}{ccc}
\vec{N}^{T} & \xrightarrow{S} & \vec{t'}^{T} \\
\scriptstyle\vec{n}^{T}=\frac{\vec{N}^{T}F^{-1}}{|\vec{N}^{T}F^{-1}|}\downarrow & & \downarrow\scriptstyle\frac{1}{\det(F)|\vec{N}^{T}F^{-1}|} \\
\vec{n}^{T} & \xrightarrow{\sigma} & \vec{t}^{T}
\end{array}
$$

Here the multiplication is on the left by the transpose, so we show the vectors as a transpose as a reminder. The stress S is not symmetric. Again the left diagram is written in a differential form and is interpreted in the context of a surface integral that touches the same material: on the upper row it is in the original configuration and on the lower row it is in the current configuration.

We also want to think of the strain as a linear operator, but that leads to the question of what it takes a vector to – that is, what is $Ed\vec{X}=d\vec{Y}$ or $E^{*}d\vec{x}=d\vec{y}$? Essentially, these produce a measuring stick that allows computation of $ds^{2}-dS^{2}$ using a dot product. The measuring stick is referred to as the *dual*: $d\vec{y}$ and $d\vec{Y}$ are the duals of E^{*} and E, respectively,

$$
\tfrac{1}{2}(ds^{2}-dS^{2})=d\vec{X}^{T}Ed\vec{X}=d\vec{Y}^{T}d\vec{X}=d\vec{x}^{T}E^{*}d\vec{x}=d\vec{y}^{T}d\vec{x}. \tag{14.77}
$$

What is the relation between these measuring sticks (formally inverting E)?

$$
d\vec{y}=E^{*}d\vec{x}=E^{*}Fd\vec{X}=E^{*}FE^{-1}d\vec{Y}=(F-F^{-T})(F^{T}(F-F^{-T}))^{-1}d\vec{Y}=F^{-T}d\vec{Y}. \tag{14.78}
$$

A commutative diagram for the strain is

$$
\begin{array}{ccc}
d\vec{X} & \xrightarrow{E} & d\vec{Y} \\
\scriptstyle F\downarrow & & \downarrow\scriptstyle F^{-T} \\
d\vec{x} & \xrightarrow{E^{*}} & d\vec{y}
\end{array}
$$

Following the two paths in the diagram from lower left $d\vec{x}$ to lower right $d\vec{y}$ immediately gives $E^* = F^{-T}EF^{-1}$. (For this diagram we are multiplying on the right.) The vertical maps in the strain diagram are not the same as the vertical maps in the second Piola–Kirchhoff stress diagram (roughly speaking, the left and right maps are swapped); in the stress diagram, going from lower left $d\vec{a}$ to lower right $d\vec{f}$ produces $\sigma = (1/\det(F))FS^{(2)}F^T$. Even though these diagrams connect original configuration stress or strain with current configuration stress or strain, the stress tensor and the strain tensor are different objects and under the mapping F from original to current configurations they behave differently. Strain takes a length segment to the dual length segment. Stress takes an area normal to a length segment (which is how force behaves). Roughly speaking, vectors go from the original configuration to the current configuration by multiplication with F or F^{-T}. The fact that both the stress and strain tensors take a vector that transforms in one way under F to a vector that transforms the other way under F is the "deep" reason that the objects are bracketed by a transform and its *transpose* rather than a transform and its *inverse* – for example, F and F^T bracket $S^{(2)}$ to produce σ (up to a constant multiplier). That stress and strain transform differently raises interesting questions when constitutive models set stress equal to a function of strain, as in Eq. (14.10). Those two expressions cannot represent exactly the same constitutive model, since the same transformation cannot be applied to both sides of one equation to produce the other.

14.5. Rate of Deformation and Rotation Rate

The motion of a rigid body is completely determined by the translation of a reference point and the rotation of the body (hence the "6 degrees of freedom" terminology). To be explicit, the current location of a point $\vec{x}$ that started at location $\vec{X}$ is given by

$$\vec{x}(\vec{X}, t) = \vec{x}_r(t) + R_r(t)(\vec{X} - \vec{X}_r), \tag{14.79}$$

where R is the rotation matrix and the subscript r means reference point. We assume that $\vec{X}_r$ does not change. The velocity of a given point is

$$\vec{v}(\vec{X}, t) = \frac{d\vec{x}}{dt} = \frac{d\vec{x}_r}{dt} + \frac{dR_r}{dt}(\vec{X} - \vec{X}_r). \tag{14.80}$$

This expression is written in terms of the original configuration, but using Eq. (14.79) the velocity can be written in the laboratory frame as

$$\vec{v}(\vec{x}, t) = \frac{d\vec{x}_r}{dt} + \frac{dR_r}{dt}R_r^T(\vec{x} - \vec{x}_r). \tag{14.81}$$

We refer to $\dot{R}R^T$ as the *rotation rate*, as it describes the rotation in the laboratory (Eulerian) frame,

$$\Omega_r = \frac{dR_r}{dt}R_r^T = \dot{R}_r R_r^T, \tag{14.82}$$

where the dot above the rotation is the differentiation with respect to time. A feature of the rotation rate is that it is skew symmetric,

$$\Omega^T = -\Omega, \tag{14.83}$$

since

$$0 = \frac{dI}{dt} = \frac{d}{dt}(RR^T) = \dot{R}R^T + R\dot{R}^T = \Omega + \Omega^T. \qquad (14.84)$$

Given the reference to the original configuration, we now explore the rate of deformation tensor that we have been using and also examine rotation rates. In the Eulerian (laboratory) frame, we begin with the *velocity gradient*

$$L_{ij} = \frac{\partial v_i}{\partial x_j}. \qquad (14.85)$$

The velocity gradient can be computed without reference to an original configuration, since it depends on the current velocity in the Eulerian frame. However, consider the velocity gradient in terms of the deformation gradient:

$$L_{ij} = \frac{\partial v_i}{\partial x_j} = \frac{\partial v_i}{\partial X_k}\frac{\partial X_k}{\partial x_j} = \frac{\partial}{\partial X_k}\left(\frac{dx_i}{dt}\right)\frac{\partial X_k}{\partial x_j} = \frac{d}{dt}\left(\frac{\partial x_i}{\partial X_k}\right)\frac{\partial X_k}{\partial x_j} = \left(\frac{dF}{dt}\right)_{ik}(F^{-1})_{kj} \qquad (14.86)$$

or

$$L = \dot{F}F^{-1} \quad \text{or} \quad \dot{F} = LF. \qquad (14.87)$$

This is a very important relationship, as it relates deformation rates in the Eulerian frame (through the velocity gradient L) to the original frame.[3] Using the polar decomposition $F = RU$ yields

$$L = (\dot{R}U + R\dot{U})U^{-1}R^T = \dot{R}R^T + R\dot{U}U^{-1}R^T. \qquad (14.88)$$

With the rotation rate

$$\Omega = \dot{R}R^T \qquad (14.89)$$

the velocity gradient is

$$L = \Omega + R\dot{U}U^{-1}R^T. \qquad (14.90)$$

Computing Ω requires knowledge of the initial configuration, since R is the rotation from the original configuration. The *rate of deformation* tensor D is

$$D = \frac{1}{2}(L + L^T) = \frac{1}{2}R(\dot{U}U^{-1} + U^{-1}\dot{U})R^T \qquad (14.91)$$

and the *spin* tensor W is

$$W = \frac{1}{2}(L - L^T) = \Omega + \frac{1}{2}R(\dot{U}U^{-1} - U^{-1}\dot{U})R^T. \qquad (14.92)$$

If a deformed material undergoes a rigid rotation, $\dot{U} = 0$, and hence

$$D = 0, \qquad W = \Omega. \qquad (14.93)$$

Thus we explicitly see the role of deformation in the spin and that the rate of deformation tensor has the rotation rate removed.

There are two polar decompositions; what happens if $F = VR$ is used instead of $F = RU$, as in the previous paragraph? Calculations yield

$$L = \dot{V}V^{-1} + V\Omega V^{-1},$$

$$D = \frac{1}{2}(\dot{V}V^{-1} + V^{-1}\dot{V} + V\Omega V^{-1} - V^{-1}\Omega V), \qquad (14.94)$$

$$W = \frac{1}{2}(\dot{V}V^{-1} - V^{-1}\dot{V} + V\Omega V^{-1} + V^{-1}\Omega V).$$

---

[3]Though we use the same word, *gradient*, to describe F and L, realize that they are very different objects, as F straddles two configurations, while L lives entirely in the current configuration.

These equations appear more complicated because, in this instance of the polar decomposition, the stretch is performed after the rotation. Thus, even when there is a "rigid body" rotation, where no deformation occurs, V changes. It changes because V is in the Eulerian frame and as the body rotates in that frame the specific values of the components of the stretch tensor change. Specifically, since $V = RUR^T$,

$$\dot{V} = \dot{R}UR^T + R\dot{U}R^T + RU\dot{R}^T$$
$$= \dot{R}R^T RUR^T + R\dot{U}R^T + RUR^T R\dot{R}^T \qquad (14.95)$$
$$= \Omega V + R\dot{U}R^T - V\Omega.$$

When the terms that lead to change in V due to rotation are moved to the left-hand side of the expression to be with $\dot{V}$, the Green–Naghdi corotational rate $\overset{\circ}{V}$ ensues. We denote corotational rates with an open circle above the variable V,

$$\overset{\circ}{V}_{GN} = \dot{V} + V\Omega - \Omega V = R\dot{U}R^T. \qquad (14.96)$$

Corotational rates will be discussed in detail in the last section of this chapter. For a rigid body rotation, $\dot{U} = 0$ and so $\overset{\circ}{V} = 0$, which can be used to compute $V(t)$ when $\Omega(t)$ is known.

14.6. Finite Strain Elasticity

We now turn to a more abstract discussion. In the following it is important to remember the following assumption:

DEFINITION 14.1. (original configuration is invariant) In discussions of changes to reference frames, it is assumed the original configuration and the original coordinate system never change – that is, changes of frame do not affect the original configuration. Objects that live entirely in the original configuration, such as $S^{(2)}$ and E, are called *invariant*.

The most basic measure of smooth deformation is the deformation gradient F. Thus, for a smooth elastic deformation the Cauchy stress can be written $\sigma = f(F)$. The notation means $\sigma_{ij} = f_{ij}(F_{11}, F_{12}, ..., F_{33})$, where f_{ij} is an arbitrary function of the components of the deformation gradient for each ij. However, for the expression to have the proper frame invariance, there are relationships between the f_{ij}. If Q is an arbitrary rotation that is applied after the deformation, then the deformation gradient is affected as $d\vec{x}^* = Qd\vec{x} = QFd\vec{X}$ and hence $F^* = QF$. Since the Cauchy stress is a tensor, this same arbitrary rotation will produce $\sigma^* = Q\sigma Q^T$. For the functional form for the stress to be frame indifferent, we need it to work in the presence of this arbitrary rotation – that is, we need to have $\sigma^* = f(F^*)$, or that $Q\sigma Q^T = f(QF)$, which yields $\sigma = Q^T f(QF)Q$. Notice in particular that this reasoning says that the function f must satisfy $Qf(F)Q^T = f(QF)$ for any rotation Q. With the right-stretch polar decomposition $F = RU$, specifically choosing the special case $Q = R^T$ yields

$$\sigma = Rf(U)R^T. \qquad (14.97)$$

This form must hold for any constitutive model where the stress is a function of the deformation gradient. As seen in the previous paragraph, this result shows that a function of the stretch in the original configuration must be rotated to produce the Cauchy stress.

In our development for yarns and fabrics, we essentially continue this argument by noting that

$$S = \frac{\rho_0}{\rho}F^{-1}\sigma = \frac{\rho_0}{\rho}U^{-1}R^T Rf(U)R^T = g(U)R^T = g(U)U^{-T}F^T = h(E)F^T,$$

(14.98)

where h is an arbitrary function of E, since $E = \frac{1}{2}(F^T F - I) = \frac{1}{2}(U^2 - I)$. This is seen in Eq. (13.96). Since F is invertible, we conclude that $h(E)$ is symmetric and $S^{(2)} = h(E)$.

Continuing from the paragraph before the last, the left-stretch polar decomposition $F = VR$ yields $U = R^T VR$ and hence $R^T \sigma R = f(R^T VR)$. We now assume that the material is isotropic. This means an arbitrary rotation Q applied to the deformation leads to the same functional form applied to the stress. Hence, $Q(R^T \sigma R)Q^T = f(Q(R^T VR)Q^T)$. With the special choice $Q = R$, we conclude that

$$\sigma = f(V) \quad \text{for an isotropic material.} \tag{14.99}$$

We will now begin an argument as to the functional form of f by first assuming that the left-stretch tensor V is diagonal. We pick a special Q, one that rotates about the x-axis by 180°,

$$Q = \begin{pmatrix} 1 & 0 & 0 \\ 0 & \cos(\pi) & \sin(\pi) \\ 0 & -\sin(\pi) & \cos(\pi) \end{pmatrix} = \begin{pmatrix} 1 & 0 & 0 \\ 0 & -1 & 0 \\ 0 & 0 & -1 \end{pmatrix}. \tag{14.100}$$

Since V is diagonal, $QVQ^T = V$, and applying this rotation to our isotropy condition $Q\sigma Q^T = f(QVQ^T)$ yields $Q\sigma Q^T = f(QVQ^T) = f(V) = \sigma$, or

$$Q\sigma Q^T = \begin{pmatrix} \sigma_{11} & -\sigma_{12} & -\sigma_{13} \\ -\sigma_{21} & \sigma_{22} & \sigma_{23} \\ -\sigma_{31} & \sigma_{32} & \sigma_{33} \end{pmatrix} = \sigma = \begin{pmatrix} \sigma_{11} & \sigma_{12} & \sigma_{13} \\ \sigma_{21} & \sigma_{22} & \sigma_{23} \\ \sigma_{31} & \sigma_{32} & \sigma_{33} \end{pmatrix}. \tag{14.101}$$

Hence, $\sigma_{12} = 0$ and $\sigma_{13} = 0$. Rotating about one of the other axes also leads to $\sigma_{23} = 0$ and thus the conclusion that when the left-stretch tensor is diagonal, so is the Cauchy stress tensor. Thus, for a diagonal stretch we have

$$\sigma = \begin{pmatrix} f_1(V_1, V_2, V_3) & 0 & 0 \\ 0 & f_2(V_1, V_2, V_3) & 0 \\ 0 & 0 & f_2(V_1, V_2, V_3) \end{pmatrix}. \tag{14.102}$$

We now choose to write these functions as sums of a certain form. In particular, we write them as

$$\begin{pmatrix} f_1 \\ f_2 \\ f_3 \end{pmatrix} = \begin{pmatrix} 1 & V_1^2 & 1/V_1^2 \\ 1 & V_2^2 & 1/V_2^2 \\ 1 & V_3^2 & 1/V_3^2 \end{pmatrix} \begin{pmatrix} g_1(V_1, V_2, V_3) \\ g_2(V_1, V_2, V_3) \\ g_3(V_1, V_2, V_3) \end{pmatrix}. \tag{14.103}$$

Since the V_i are arbitrary positive numbers (stretches are always positive), the center matrix can be inverted and the g_i are well defined. The stress can then be written

$$\sigma = \begin{pmatrix} 1 & 0 & 0 \\ 0 & 1 & 0 \\ 0 & 0 & 1 \end{pmatrix} g_1 + \begin{pmatrix} V_1^2 & 0 & 0 \\ 0 & V_2^2 & 0 \\ 0 & 0 & V_3^2 \end{pmatrix} g_2 + \begin{pmatrix} 1/V_1^2 & 0 & 0 \\ 0 & 1/V_2^2 & 0 \\ 0 & 0 & 1/V_3^2 \end{pmatrix} g_3. \tag{14.104}$$

From the polar decomposition, $V = \sqrt{FF^T} = \sqrt{B}$. Notice that the second and third matrices in Eq. (14.104) are B and B^{-1} when V and hence B are diagonal. Since the Cauchy stress is symmetric, every stress state can be achieved by a diagonal stress rotated in the proper fashion. Thus, $R\sigma R^T = g_1 I + g_2 RBR^T + g_3 RB^{-1}R^T$ for diagonal B yields $\sigma = g_1 I + g_2 B + g_3 B^{-1}$ for any B. We consider the functions g_i as functions of the invariants $I_i(B)$ of B and label them β_i or α_i to obtain

$$\sigma = \beta_0 I + \beta_+ B + \beta_- B^{-1} \quad \text{or} \quad \sigma = \alpha_0 I + \alpha_1 B + \alpha_2 B^2, \tag{14.105}$$

where the second form is by the Cayley–Kirchhoff theorem, which states that a matrix satisfies its own characteristic equation,

$$B^3 - I_1(B)B^2 - I_2(B)B - I_3(B)I = 0 \quad \Rightarrow \quad B^2 = I_1(B)B + I_2(B)I + I_3(B)B^{-1}. \tag{14.106}$$

Thus, we have developed a very simple functional representation of the stress for an isotropic elastic material.

In the previous chapter in Eq. (13.58) we computed the first Piola–Kirchhoff stress by differentiating a strain energy density e (or potential) by the deformation gradient F^T. At the time, since we were interested in working in the original configuration, we viewed e as a function of $E = \frac{1}{2}(U^2 - I)$. Now, we want to do that same but view e as a function of $B = V^2$ instead of as a function of U,

$$S_{ij} = \frac{\partial e}{\partial F_{ij}^T} = \frac{\partial e}{\partial F_{ji}} = \frac{\partial e}{\partial B_{kl}} \frac{\partial B_{kl}}{\partial F_{ji}}. \tag{14.107}$$

To explicitly compute this derivative we need

$$\frac{\partial B_{kl}}{\partial F_{ji}} = \frac{\partial (F_{km}F_{lm})}{\partial F_{ji}} = \delta_{kj}\delta_{mi}F_{lm} + F_{km}\delta_{lj}\delta_{mi} = \delta_{kj}F_{li} + \delta_{lj}F_{ki}. \tag{14.108}$$

We then have, using the symmetry of B,

$$S_{ij} = \frac{\partial e}{\partial B_{kl}}(\delta_{kj}F_{li} + \delta_{lj}F_{ki}) = \frac{\partial e}{\partial B_{jl}}F_{li} + \frac{\partial e}{\partial B_{kj}}F_{ki} = 2F_{ik}^T \frac{\partial e}{\partial B_{kj}}, \tag{14.109}$$

or, using matrix notation, where $(\partial e / \partial B)_{ij} = \partial e / \partial B_{ij}$,

$$S = 2F^T \frac{\partial e}{\partial B}. \tag{14.110}$$

This gives

$$\sigma = \frac{\rho}{\rho_0} FS = \frac{2}{\det(F)} FF^T \frac{\partial e}{\partial B} = \frac{2}{\sqrt{\det(B)}} B \frac{\partial e}{\partial B}. \tag{14.111}$$

For an isotropic material, e is a function of the three invariants,

$$e(B) = e(I_1(B), I_2(B), I_3(B)), \tag{14.112}$$

and the various terms can be computed as in Exercise 3.50 to yield

$$\sigma = \frac{2}{\sqrt{\det(B)}} \left\{ \det(B) \frac{\partial e}{\partial I_3} I + \left(\frac{\partial e}{\partial I_1} - \text{trace}(B) \frac{\partial e}{\partial I_2} \right) B + \frac{\partial e}{\partial I_2} B^2 \right\}. \tag{14.113}$$

Using the Cayley–Hamilton theorem, this can also be written

$$\sigma = \frac{2}{\sqrt{\det(B)}} \left\{ \left(I_2(B) \frac{\partial e}{\partial I_2} + \det(B) \frac{\partial e}{\partial I_3} \right) I + \frac{\partial e}{\partial I_1} B + \det(B) \frac{\partial e}{\partial I_2} B^{-1} \right\}. \tag{14.114}$$

Comparing these with Eq. (14.105) shows agreement in the form, though for the stress originating from e the functions α_i and β_i are no longer completely arbitrary. Since the invariants of $F^T F$ are the same as those of $B = FF^T$ (see page 563), e can be a function of either. Potential derivable elastic materials are hyperelastic.

14.7. Blatz–Ko and Mooney–Rivlin Constitutive Models

We now present three finite elasticity constitutive models that fall into the above form. The simplest is a Blatz–Ko model for a foamed, polyurethane elastomer [27],

$$\sigma = \mu_0 \left(I - \frac{1}{\sqrt{\det(B)}} B^{-1} \right). \tag{14.115}$$

If the constitutive model is written in terms of the Eulerian strain $E^* = \frac{1}{2}(I - B^{-1})$,

$$\sigma = \mu_0 \left(1 - \frac{1}{\sqrt{\det(B)}} \right) I + \frac{2\mu_0}{\sqrt{\det(B)}} E^*, \tag{14.116}$$

and we see the similarities to our previous $\sigma = \lambda \text{trace}(E^*)I + 2\mu E^*$. However, in the following we will work with Eq. (14.115), as it is easier.

We are interested in the wave behavior of this material. Uniaxial compression can be written in terms of stretches as

$$F = \begin{pmatrix} \lambda & 0 & 0 \\ 0 & 1 & 0 \\ 0 & 0 & 1 \end{pmatrix}, \qquad B = FF^T = \begin{pmatrix} \lambda^2 & 0 & 0 \\ 0 & 1 & 0 \\ 0 & 0 & 1 \end{pmatrix}, \tag{14.117}$$

where conservation of mass says $\rho \det(F) = \rho_0$ hence $\lambda = \rho_0/\rho$. This constitutive model is strictly mechanical, so the shock wave is determined by the mass and momentum balances, which say

$$\frac{\rho_0}{\rho} = 1 - \frac{u}{U} \quad \Rightarrow \quad u = U(1 - \lambda),$$

$$-\sigma_{11} = \rho_0 U u \quad \Rightarrow \quad -\mu_0 \left(1 - \frac{1}{\lambda^3}\right) = \rho_0 U^2 (1 - \lambda) \quad \Rightarrow \quad U = \sqrt{\frac{\mu_0}{\rho_0}} \sqrt{\frac{1 + \lambda + \lambda^2}{\lambda^3}}. \tag{14.118}$$

For a small stretch we recover the low pressure longitudinal sound speed as $c_L = \sqrt{3\mu_0/\rho_0}$ and therefore infer that the initial Poisson's ratio is $\nu_0 = \frac{1}{4}$ (see Exercise 14.7) and the initial Young's modulus is $E_0 = \frac{5}{2}\mu_0$. As to whether the material will form a shock, the first equation in Theorem 4.12 in terms of stretch is

$$\frac{d}{d\lambda} \left(\frac{\tilde{\sigma}}{1 - \lambda} \right) = \frac{d}{d\lambda} \left(\frac{\sigma_{11}}{\lambda - 1} \right) \leq 0. \tag{14.119}$$

For the Blatz–Ko model in uniaxial strain, we have

$$\frac{d}{d\lambda} \left(\frac{\sigma_{11}}{\lambda - 1} \right) = \frac{d}{d\lambda} \left(\frac{\mu_0(1 - 1/\lambda^3)}{\lambda - 1} \right) = -\mu_0 \left(\frac{1}{\lambda^2} + \frac{2}{\lambda^3} + \frac{3}{\lambda^4} \right) < 0. \tag{14.120}$$

Therefore this uniaxial strain response is strictly superlinear and it forms shocks in compression.

We next consider a shear wave into an initially undeformed material. As in Exercise 3.39, consider the deformation $x = X$, $y = Y + u_2(t - X/U)$ for $t > X/U$, and $z = Z$, where u_2 is the transverse particle velocity, x_i refers to current coordinates, and X_i refers to original coordinates. Since the momentum balance represents three independent equations, there are in fact three independent jump conditions to be satisfied. For longitudinal waves, we typically only discuss one of them because uniaxial strain compression does not give rise to shear stress, though in the previous chapter we did simultaneously solve two of the jump conditions for the transverse wave in the yarn. Mass conservation for a shear wave is automatic because the volume does not change. Following Exercise 2.11, for the shear wave traveling in the x direction with material motion in the y direction the three jump conditions are

$$[-\sigma_{11}] = 0, \qquad [-\sigma_{12}] = \rho_0 U[u_2], \qquad [-\sigma_{13}] = 0. \qquad (14.121)$$

For the deformation described,

$$F = \begin{pmatrix} 1 & 0 & 0 \\ -u_2/U & 1 & 0 \\ 0 & 0 & 1 \end{pmatrix}, \qquad B = FF^T = \begin{pmatrix} 1 & -u_2/U & 0 \\ -u_2/U & 1 + (u_2/U)^2 & 0 \\ 0 & 0 & 1 \end{pmatrix}. \qquad (14.122)$$

The center equation of the three momentum jump conditions yields

$$-\sigma_{12} = \mu_0 u_2/U = \rho_0 U u_2 \quad \Rightarrow \quad U = \sqrt{\frac{\mu_0}{\rho_0}}. \qquad (14.123)$$

This equation must be solved simultaneously with the other two momentum jump conditions. The first jump condition is

$$-\sigma_{11} = \mu_0 (u_2/U)^2 = 0. \qquad (14.124)$$

It is not possible to simultaneously solve these two equations with a nontrivial solution, and thus there is not a simple shear shock wave. The situation is analogous to the uniaxial stress wave considered in Chapter 4: though for low stress there is a sound speed for shear waves, for finite strains additional material motion will occur and a simple wave does not result. It is due to the Poynting effect (Exercise 14.5).

We next consider rubber and rubber-like materials. The rubber Silastic RTV 521 has an initial density of 1.372 g/cm^3 and a U_s–u_p behavior of $U = 1.84 + 1.44u$ km/s (Appendix C). Thus, a bulk modulus is $K = \rho_0 c_0^2 = 4.65$ GPa. A shear modulus for rubber is 205 kPa. The ratio is $\mu/K = 4.4 \times 10^{-5}$. For a metal, $\mu/K = 3(1 - 2\nu)/2(1 + \nu) \approx 0.46$. For rubber the bulk modulus is so much larger than the shear modulus that, at low pressure, a reasonable approximation is to treat it as incompressible. Of course we know rubber is not incompressible, but like water in fluid mechanics it is a reasonable approximation for studying certain situations where the pressure is low and the volumetric deformation is small compared to the total deformation. Phrased this way, we also recognize that when steel undergoes large deformation at low pressure the majority of the deformation is plastic, which is assumed volume conserving, hence incompressible.

Our next example is perhaps the most famous finite elastic model. The Mooney–Rivlin model for rubber assumes that the material is incompressible – that is, $\det(F) = 1 = \det(B)$. It then writes the stress

$$\sigma = -\hat{p}I + \beta_+ B + \beta_- B^{-1}, \qquad (14.125)$$

where β_+ and β_- are constants and the shear modulus is given by $\mu_0 = \beta_+ - \beta_-$ (Exercise 14.3, hence $\beta_+ > \beta_-$). The specific case when $\beta_- = 0$ is referred to as a neo-Hookean material. To be specific, in an example used later, coefficients for one rubber material are $\beta_+ = 185.8$ kPa and $\beta_- = -19.35$ kPa [26] (in this paper the authors state that typically $-0.125 \leq \beta_-/\beta_+ \leq -0.1$). The incompressibility assumption has some far reaching consequences. If one compresses a cube of material with equal pressure on all the sides, it will not deform, yet the pressure in the material clearly increases. Hence, the pressure is not determined by material deformation but rather is determined by the loading conditions; the pressure $\hat{p}$ is indeterminate (from the deformation) in the model and is determined by solving the specific boundary value problem of interest. Imagine now an incompressible material undergoing a flyer plate impact test. Since the material cannot compress, it must immediately transfer information about the fact that its front face has been impacted to its back face, and thus the longitudinal sound wave and shock wave speeds are infinite. Interestingly, though, the indeterminate nature of $\hat{p}$ allows a finite strain shear wave, in spite of the Poynting effect. F and B are the same as in Eq. (14.122), and with $\beta_+ - \beta_- = \mu_0$, the $[-\sigma_{12}] = \rho_0 U[u_2]$ jump condition yields $U = \sqrt{\mu_0/\rho_0}$. However, in this case the jump condition $[-\sigma_{11}] = 0$ yields

$$\sigma_{11} = -\hat{p} + \beta_+ + \beta_-(1 + (u_2/U)^2) = 0, \qquad (14.126)$$

which can be solved by $\hat{p} = \beta_+ + \beta_-(1 + (u_2/U)^2)$, yielding $\sigma_{11} = 0$, $\sigma_{22} = \mu_0(u_2/U)^2$, $\sigma_{33} = -\beta_-(u_2/U)^2$, in addition to $\sigma_{12} = -\mu_0 u_2/U$ behind the wave front. The pressure behind the front is $p = -\frac{1}{3}\sigma_{ii} = -\frac{1}{3}(\beta_+ - 2\beta_-)(u_2/U)^2 \neq \hat{p}$.

The fact that the Mooney–Rivlin does well in low pressure response of the material implies that it should be able to address wave propagation behavior in scenarios where there is little confinement, such as uniaxial stress. As in Chapter 4, we use Eq. (14.118) to define the relationship between stretch and particle speed, $u = U(1 - \lambda)$, since we cannot use mass conservation. In uniaxial stress, incompressibility implies $\lambda_2 = \lambda_3 = 1/\sqrt{\lambda}$; hence

$$F = \begin{pmatrix} \lambda & 0 & 0 \\ 0 & 1/\sqrt{\lambda} & 0 \\ 0 & 0 & 1/\sqrt{\lambda} \end{pmatrix}, \qquad B = FF^T = \begin{pmatrix} \lambda^2 & 0 & 0 \\ 0 & 1/\lambda & 0 \\ 0 & 0 & 1/\lambda \end{pmatrix}, \qquad (14.127)$$

where λ is the stretch along the loading axis. Placing these values into $\sigma_{22} = 0$ allows a determination of $\hat{p}$ that in turn leads to the stress along the loading axis,

$$\sigma_{11} = \beta_+ \left(\lambda^2 - \frac{1}{\lambda} \right) + \beta_- \left(\frac{1}{\lambda^2} - \lambda \right) = \left(\beta_+ - \frac{\beta_-}{\lambda} \right) \left(\lambda^2 - \frac{1}{\lambda} \right)$$

$$= \left(\beta_+ - \frac{\beta_-}{\lambda} \right) (\lambda - 1) \frac{1 + \lambda + \lambda^2}{\lambda}. \qquad (14.128)$$

The response is linear in $\lambda - 1$ for stretches near 1. The momentum jump condition yields

$$U = \sqrt{\frac{\beta_+ - \beta_-/\lambda}{\rho_0}} \sqrt{\frac{1 + \lambda + \lambda^2}{\lambda}}. \qquad (14.129)$$

For the low pressure uniaxial stress (bar) wave speed we have $c_E = \sqrt{3\mu_0/\rho_0}$; hence the initial Young's modulus is $E_0 = 3\mu_0$ and Poisson's ratio is $\nu = \frac{1}{2}$. A similar argument as in the Blatz–Ko case says this stress-strain relation is superlinear in compression when $\beta_- \leq 0$. Perhaps shocks could form, but realize there is lateral motion in the material behind the wave front, which is not in equilibrium after the passage of the shock.

We now compute the strain energy of the Mooney–Rivlin constitutive model.[4] The conjugate strain rate to the Cauchy stress is the rate of deformation D; hence for the energy term we compute $\sigma_{ij} D_{ij}$. We recall the velocity gradient $L = \dot{F} F^{-1}$ (Eq. 14.87) and $D = \frac{1}{2}(L + L^T)$ to write D in terms of F. The strain energy

---

[4]This is the reverse of what is usually done, where the material is assumed hyperelastic: a potential (Eq. 14.135) is introduced and is differentiated (Eq. 14.114) to obtain the Mooney–Rivlin constitutive material.

is computed term by term. Since the Mooney–Rivlin material is incompressible, $\text{trace}(D) = 0$, hence $-\hat{p}\delta_{ij}D_{ij} = -\hat{p}D_{ii} = 0$. For the second term, we use (and in the following) a number of times the fact that, for any symmetric matrix B and any matrix D decomposed as $D = \frac{1}{2}(L + L^T)$, $B_{ij}D_{ij} = B_{ij}L_{ij}$, so

$$B_{ij}D_{ij} = B_{ij}L_{ij} = F_{ik}F_{kj}^T\dot{F}_{il}F_{lj}^{-1} = F_{ik}F_{lj}^{-1}F_{jk}\dot{F}_{il} = F_{ik}\delta_{lk}\dot{F}_{il}$$

$$= F_{il}\dot{F}_{il} = \frac{d}{dt}\left(\frac{1}{2}F_{ij}F_{ij}\right) = \frac{d}{dt}\left(\frac{1}{2}F_{ij}F_{ji}^T\right) = \frac{d}{dt}\left(\frac{1}{2}\text{trace}(B)\right). \tag{14.130}$$

The third term, $B_{ij}^{-1}D_{ij}$, is more complicated. As a first step we use the Cayley–Hamilton theorem to write

$$I_3(B)B^{-1} = B^2 - I_1(B)B - I_2(B)I, \tag{14.131}$$

where the $I_i(B)$ are the invariants of B. For our specific case, $I_3(B) = \det(B) = 1$ and we have already computed the two right-hand terms of the right-hand side, so all that is left to compute is $B_{ij}^2D_{ij}$, again using the above mentioned symmetry of B and the decomposition of D and $\dot{B}$,

$$B_{ij}^2D_{ij} = B_{ij}^2L_{ij} = B_{ik}F_{kl}F_{lj}^T\dot{F}_{im}F_{mj}^{-1} = B_{ik}F_{kl}F_{mj}^{-1}F_{jl}\dot{F}_{im} = B_{ik}F_{kl}\delta_{ml}\dot{F}_{im}$$

$$= B_{ik}F_{km}\dot{F}_{mi}^T = B_{ik}\frac{d}{dt}\left(\frac{1}{2}B_{ki}\right) = \frac{d}{dt}\left(\frac{1}{4}B_{ik}B_{ki}\right) = \frac{d}{dt}\left(\frac{1}{4}\text{trace}(B^2)\right). \tag{14.132}$$

Beginning with Eq. (14.131), we have

$$B_{ij}^{-1}D_{ij} = B_{ij}^2D_{ij} - I_1(B)B_{ij}D_{ij} - I_2(B)\delta_{ij}D_{ij}$$

$$= \frac{d}{dt}\left(\frac{1}{4}\text{trace}(B^2)\right) - I_1(B)\frac{d}{dt}\left(\frac{1}{2}\text{trace}(B)\right) \tag{14.133}$$

$$= \frac{d}{dt}\left(\frac{1}{4}\text{trace}(B^2) - \frac{1}{4}(\text{trace}(B))^2\right) = \frac{d}{dt}\left(\frac{1}{2}I_2(B)\right).$$

A straightforward computation for a 3×3 matrix using cofactor expansion shows that in general $I_1(B^{-1}) = \text{trace}(B^{-1}) = -I_2(B)/I_3(B)$ and, since for our case $\det(B) = 1$, we conclude that

$$B_{ij}^{-1}D_{ij} = -\frac{d}{dt}\left(\frac{1}{2}\text{trace}(B^{-1})\right). \tag{14.134}$$

Integrating these expressions over time from an initial stress-free zero energy configuration, these computations allow us to write the strain energy in the Mooney–Rivlin solid as

$$e = \frac{\beta_+}{2}(I_1(B) - 3) + \frac{\beta_-}{2}(I_2(B) + 3) = \frac{\beta_+}{2}(\text{trace}(B) - 3) - \frac{\beta_-}{2}(\text{trace}(B^{-1}) - 3). \tag{14.135}$$

Writing the strain energy brings up an interesting point about rubber-like materials. Experimentally, when we stretch a metal in tension, the temperature of the metal drops. This was discussed in Chapter 5: Eq. (5.152) gives the temperature along the adiabat of a metal as $T_s = T_0 \exp(-\Gamma_0\varepsilon)$ and so $dT_s/d\varepsilon|_{\varepsilon=0} = -\Gamma_0 T_0 < 0$. However, if you stretch a rubber band it warms up (this behavior is easily confirmed for large stretches by touching a stretched rubber band to your face – you can feel that it is warmer). Thus metals and rubber materials differ in some essential way. Rubber materials are often referred to as entropic materials because the internal energy is not stored in the deformation of a lattice; rather, they are similar to an ideal gas in that all the energy storage is thermal. Examining Eq. (5.153), when there is no "cold compression curve" the stress is directly multiplied by the temperature.[5] Thus, the thermal behavior of the Mooney–Rivlin model, assuming there is no elastically stored energy, is

$$\sigma = -\hat{p}I + \frac{T}{T_0}\beta_+ B + \frac{T}{T_0}\beta_- B^{-1}, \qquad (14.136)$$

where T is absolute temperature and T_0 is the temperature at which the constants were measured. To see what entropic elasticity implies, for the response in uniaxial stress, the stress along the loading axis is

$$\sigma_{11} = \frac{T}{T_0}\beta_+\left(\lambda^2 - \frac{1}{\lambda}\right) + \frac{T}{T_0}\beta_-\left(\frac{1}{\lambda^2} - \lambda\right). \qquad (14.137)$$

Suppose that the rubber material is stretched in tension and a constant stress load is applied. To maintain the stress, upon heating the material will contract, since the modulus is increasing. This behavior is referred to as a Gough–Joule thermoelastic effect [89]. It is exactly opposite of what occurs in metals, where the heating of the metal expands the lattice through thermal expansion without changing the stress.

Returning to adiabatic stretching, the temperature of the rubber increases because internally the rubber is composed of polymer chains. As load is applied, these chains straighten out; hence order increases and the entropy decreases. For the expansion to be adiabatic it is necessary for the thermal motion, hence temperature and entropy, to increase to counteract the decrease in entropy due to increased order. Another way to view this is that since no elastic energy is stored in the deformation, all the work performed deforming the rubber material must show up as thermal energy, which is recoverable upon unloading. That the temperature must increase is another Gough–Joule thermoelastic effect and is again opposite to what happens in elastic tensile deformation of metals, where the temperature drops as the lattice expands (Eq. 5.152).

There is one final topic that arises in finite elasticity, namely the nonuniqueness of solutions. The example we present has been experimentally demonstrated for rubber. Suppose a thin rubber sheet is loaded biaxially. Incompressibility gives $\lambda_3 = 1/\lambda_1\lambda_2$ and B as

$$F = \begin{pmatrix} \lambda_1 & 0 & 0 \\ 0 & \lambda_2 & 0 \\ 0 & 0 & 1/\lambda_1\lambda_2 \end{pmatrix}, \qquad B = FF^T = \begin{pmatrix} \lambda_1^2 & 0 & 0 \\ 0 & \lambda_2^2 & 0 \\ 0 & 0 & 1/\lambda_1^2\lambda_2^2 \end{pmatrix}. \qquad (14.138)$$

Being traction free through the thin direction gives $\sigma_{33} = 0$, which allows a determination of $\hat{p}$, which then allows a determination of the stresses in the in-plane

---

[5]Another approach to this result is to identify our strain energy e as the Helmholtz free energy f. Thus, $e = f = u - Ts$, where u is an internal energy, T is absolute temperature, and s is the entropy. With λ as a representative stretch, $df = (\partial u/\partial\lambda)d\lambda - T(\partial s/\partial\lambda)d\lambda - s\,dT$. By assumption, there is no internal strain energy storage, so the first term is zero and for an isothermal loading the third term is zero, and we conclude that the isothermal modulus of the material is $-T(\partial s/\partial\lambda)$, which shows the direct T dependence. Since the stretch straightens out the chains, the entropy decreases and the modulus is positive.

directions as

$$\sigma_{11} = \beta_+ \left(\lambda_1^2 - \frac{1}{\lambda_1^2 \lambda_2^2} \right) + \beta_- \left(\frac{1}{\lambda_1^2} - \lambda_1^2 \lambda_2^2 \right),$$

$$\sigma_{22} = \beta_+ \left(\lambda_2^2 - \frac{1}{\lambda_1^2 \lambda_2^2} \right) + \beta_- \left(\frac{1}{\lambda_2^2} - \lambda_1^2 \lambda_2^2 \right).$$

(14.139)

The force applied to each edge of the square sheet is the area of the face multiplying the stress, so $F_1 = \lambda_2 \lambda_3 A_0 \sigma_{11}$ and $F_2 = \lambda_1 \lambda_3 A_0 \sigma_{22}$, where A_0 is the initial area of face. Setting these forces equal to each other gives $\lambda_2 \sigma_{11} = \lambda_1 \sigma_{22}$; algebraic manipulation yields

$$(\lambda_1 - \lambda_2) \left\{ \beta_+ \left(\lambda_1 \lambda_2 + \frac{1}{\lambda_1^2 \lambda_2^2} \right) + \beta_- \left(\lambda_1^2 \lambda_1^2 - \frac{\lambda_1^2 + \lambda_1 \lambda_2 + \lambda_2^2}{\lambda_1^2 \lambda_2^2} \right) \right\} = 0. \quad (14.140)$$

One loading curve is $\lambda_1 = \lambda_2$. However, there is the possibility of another loading curve, when $\{\} = 0$. For negative values of β_-, solutions exist for the $\{\} = 0$ equation that are like $\lambda_2 \approx |\beta_+/\beta_-|/\lambda_1$ hyperbolas in the λ_1–λ_2 plane. The $\lambda_1 = \lambda_2$ and $\{\} = 0$ solutions intersect at $\beta_+ \left(\lambda^6 + 1 \right) + \beta_- \left(\lambda^8 - 3\lambda^2 \right) = 0$. For example, for the constants stated above, $\beta_+ = 185.8$ kPa and $\beta_- = -19.35$ kPa, the intersection point is at $\lambda_1 = \lambda_2 = 3.106$. This intersection point is a bifurcation point because the solution of the equal loading problem bifurcates into multiple paths at it. If one loads the rubber sheet from the initial condition of $\lambda_1 = \lambda_2 = 1$ up to this point, then the solution can continue loading on the $\lambda_1 = \lambda_2$ curve, or it can turn to the right and load on the λ_1-increasing, λ_2-decreasing path, or turn to the left and load on the λ_1-decreasing, λ_2-increasing path (these two latter paths are mirror images of each other and, since the $\{\} = 0$ equation is differentiable, we have that the $\lambda_1 = \lambda_2$ line and the $\{\} = 0$ curve intersect at right angles). Though mechanically it can take any path, we ask ourselves which path is most "stable." For a ball that can sit on top of a table or on the floor, we view the floor as the most stable position: on the floor, the ball has the lowest potential energy or available stored energy owing to its position. If we apply the same reasoning to the stretched rubber sheet, we compute the stored energy in the rubber sheet from Eq. (14.135) and obtain

$$e = \frac{\beta_+}{2} \left(\lambda_1^2 + \lambda_2^2 + \frac{1}{\lambda_1^2 \lambda_2^2} - 3 \right) - \frac{\beta_-}{2} \left(\frac{1}{\lambda_1^2} + \frac{1}{\lambda_2^2} + \lambda_1^2 \lambda_2^2 - 3 \right). \quad (14.141)$$

as the energy per unit volume. Taking the gradient of the energy in the λ_1-λ_2 plane and evaluating along the $\lambda_1 = \lambda_2$ loading line gives

$$\text{grad}(e)|_{\lambda_1 = \lambda_2} = \lambda(\beta_+ - \lambda^2 \beta_-) \left(1 - \frac{1}{\lambda^6} \right) (\hat{\lambda}_1 + \hat{\lambda}_2). \quad (14.142)$$

This gradient points along the $\lambda_1 = \lambda_2$ loading path. Each component is positive and thus the $\lambda_1 = \lambda_2$ loading path has the maximum rate of potential energy increase. The two paths bifurcating off to the left and right have lower potential energies, and thus we expect them to be more stable.

Batra, Müller, and Strehlow, in experiments with a 0.4 mm thick rubber sheet of area 5 cm×5 cm (cut from a balloon) with the Mooney–Rivlin constants above show that this bifurcation behavior occurs [26]. They did this by attaching weights to the outside of the rubber sheet with wires, which were hung over a frame (Fig. 14.1). The expected bifurcating load is $F_1 = A_0 \lambda(\beta_+ - \lambda^2 \beta_+)(1 - 1/\lambda^6) = 23.11$ N. Up to a load of 23.12 N on each face, the deformation is square. With a load of 23.3 N on each face, the deformation became rectangular with an interior

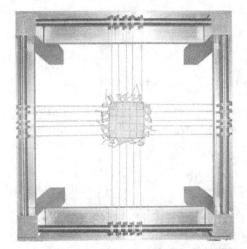

FIGURE 14.1. Experiments with an 0.4 mm thick rubber sheet. Left shows initial configuration where the stretches match, the middle square 1 cm on a side; right shows loading after the bifurcation point, with the middle square 3.05 cm×4.06 cm. This experiment also demonstrates Saint-Venant's principle for boundary conditions. (From Batra, Müller and Strehlow [26].)

square of 1 cm×1 cm deformed to roughly 3.05 cm×4.06 cm (hence these numbers are the stretches). Placing the two dissimilar stretches into the $\{\} = 0$ expression gives $\beta_-/\beta_+ = -0.081$, which is close to the value $\beta_-/\beta_+ = -0.104$ obtained through other means.

A natural question arises as to whether the nonuniqueness of the solution is due to geometric effects or to the constitutive model. We set $\{\} = 0$ in Eq. (14.140) along the $\lambda_1 = \lambda_2$ loading and obtain

$$\frac{\beta_-}{\beta_+} = -\frac{\lambda^6 + 1}{\lambda^2(\lambda^6 - 3)}. \tag{14.143}$$

There is a bifurcation when this equation holds. For $\lambda \geq 1$ (tension) the right-hand side begins at 1 for $\lambda = 1$ and grows to $+\infty$ on the vertical asymptote at $\lambda = \sqrt[6]{3}$. Then it comes from $-\infty$ on the other side of $\lambda = \sqrt[6]{3}$ and approaches 0 as $\lambda \to +\infty$. We conclude that $-\infty < \beta_-/\beta_+ < 0$ and $1 \leq \beta_-/\beta_+ < +\infty$. Since $\beta_+ - \beta_- = \mu_0 > 0$, for $\beta_+ > 0$ we have $\beta_-/\beta_+ < 1$. The overlap for the inequalities allows us to conclude that for $\beta_+ > 0$, then for any value of $\beta_- < 0$, there is a bifurcation; the smaller $|\beta_-|$, the larger the stretch $\lambda > \sqrt[6]{3}$ at which the bifurcation occurs. If $\beta_+ < 0$, then $\beta_-/\beta_+ > 1$, then the overlap of the inequalities allows us to conclude that there is bifurcation for any $\beta_- < \beta_+ < 0$. In this case the λ for the bifurcation lies between 1 and $\sqrt[6]{3}$. Putting the pieces together, there is a unique solution with no bifurcation if $0 \leq \beta_- < \beta_+$ and there is a bifurcation if $\beta_- < 0$. Thus we can say that the nonuniqueness is due to the constitutive model, since the choice of the constitutive model determines whether there is nonuniqueness. In particular, a neo-Hookean material ($\beta_- = 0$) has a unique solution.

However, a little more scrutiny is warranted, in particular as to what constitutes a "reasonable" constitutive model. Let us write down our Mooney–Rivlin model for both uniaxial and biaxial stress loading,

$$\text{uniaxial} \quad \sigma_{11} = \beta_+ \left(\lambda^2 - \frac{1}{\lambda} \right) + \beta_- \left(\frac{1}{\lambda^2} - \lambda \right),$$

$$\text{biaxial} \quad \sigma_{11} = \beta_+ \left(\lambda^2 - \frac{1}{\lambda^4} \right) + \beta_- \left(\frac{1}{\lambda^2} - \lambda^4 \right).$$

(14.144)

If $\beta_+ > 0$ and $\beta_- = 0$, then the implication is that, for large stretches, the stress is roughly the same. This result does not seem reasonable; we expect the stress in the biaxial loading to be considerably larger than the uniaxial loading for large strain. Also, notice that if $\beta_- > 0$, then at some (perhaps large) strain, for the biaxial stress loading the stress reaches a maximum and then decreases to below zero. (In addition, if $\beta_- = 0$ we do not observe the "hard, easy, hard" pattern for inflating balloons that is observed; see Exercise 14.14.) If we expect the constitutive model to be applicable for these large strains, then these behaviors are not reasonable. We conclude that we expect $\beta_- < 0$. Thus, a "reasonable" constitutive model leads to bifurcation and nonuniqueness. Similar bifurcation behavior exists for the Blatz–Ko constitutive model (Exercise 14.11) and the linear stress-strain relation (Exercise 14.12).

14.8. Incremental Constitutive Models and Corotational Stress Rates

We now come full circle to our initial constitutive modeling from Chapter 3. There we studied small deformation elasticity and incremental plasticity, with metals being the primary application. It will be recalled that none of our examples included rotation. In fact, we introduced the Lagrangian strain (E_{ij}) in Chapter 13 to allow us to model the large deflection of yarns and fabrics, which included rotation. The finite elasticity discussed in the previous two chapters has been such that there was an explicit expression for the stress tensor in terms of a strain tensor. Modern computations for inelastic deformation are typically carried out in an incremental fashion; hence a rate form of the constitutive model is used. In incremental theories an explicit relation between stress and deformation does not exist. Thus, we need to discuss how to perform incremental updates to the stress in the context of finite rotations.

One approach is to always work in the original configuration. That, however, is not an approach that interests us. Our yield surfaces were written in terms of the Cauchy stress and our preference is to utilize our work with the Cauchy stress and to work in the Eulerian frame. However, in our development we will reference back to the original configuration because it is necessary to do so to actually know the rotation. In the following a dot above an object and d/dt are used concurrently to represent differentiation with respect to time.

As a first path to a rate dependent formulation working in the current configuration, for isotropic elastic materials in small strain (but large rotations) we have

$$\sigma = f(V) = g(E^*) = \lambda \text{trace}(E^*)I + 2\mu E^*, \quad (14.145)$$

where $E^* = \frac{1}{2}(I - B^{-1})$. Our plan is to differentiate this expression with respect to time. As a first step (recalling $\dot{F} = LF$ from Eq. 14.87),

$$0 = \frac{d}{dt}(F^{-1}F) = \frac{d}{dt}(F^{-1})F + F^{-1}\dot{F} = \frac{d}{dt}(F^{-1})F + F^{-1}LF. \quad (14.146)$$

Hence

$$\frac{d}{dt}(F^{-1}) = -F^{-1}L. \tag{14.147}$$

This result gives

$$\dot{E}^* = \frac{dE^*}{dt} = \frac{d}{dt}\frac{1}{2}(I - B^{-1}) = -\frac{1}{2}\left(\frac{d}{dt}F^{-1}\right)^T F^{-1} - \frac{1}{2}F^{-T}\frac{d}{dt}(F^{-1})$$

$$= \frac{1}{2}(F^{-1}L)^T F^{-1} + \frac{1}{2}F^{-T}F^{-1}L = \frac{1}{2}(L^T B^{-1} + B^{-1}L) \tag{14.148}$$

$$= \frac{1}{2}(L + L^T) - L^T E^* - E^*L = D - L^T E^* - E^*L.$$

The time derivative of E^* is not a simple expression. It involves strain terms pre- and post-multiplied by the velocity gradient. Moving all the Eulerian strain terms to one side of the equation yields

$$\overset{\circ}{E}^*_{CR} = \frac{dE^*}{dt} + L^T E^* + E^*L = D. \tag{14.149}$$

The circle above E^* at the left is the notation for the corotational rate, which is defined in the center equality. This particular corotational rate is called the *Cotter–Rivlin rate*, indicated in the subscript, and we have shown the remarkable result that the Cotter–Rivlin corotational rate of the Eulerian strain is equal to the rate of deformation tensor, $\overset{\circ}{E}^*_{CR} = D$.

What does the corotational rate accomplish (a formal definition is provided later)? A clue is provided by considering the relation between the Lagrangian strain E and the Eulerian strain E^*, namely $E^* = F^{-T}EF^{-1}$. The time derivative of E is

$$\frac{dE}{dt} = \frac{d}{dt}\frac{1}{2}(F^T F - I) = \frac{1}{2}(\dot{F}^T F + F^T \dot{F}) = \frac{1}{2}(F^T L^T F + F^T LF) = F^T DF. \tag{14.150}$$

Thus,

$$\overset{\circ}{E}^*_{CR} = (F^{-T}\dot{E}F^{-1})_{CR} = D = F^{-T}(F^T DF)F^{-1} = F^{-T}\dot{E}F^{-1}. \tag{14.151}$$

Thus, the Cotter–Rivlin corotational rate is a way of computing, in the Eulerian frame ($\overset{\circ}{E}^*_{CR} = \dot{E}^* + L^T E^* + E^*L$), the rate of change that is achieved by computing the rate change in the original frame and moving the result to the current configuration in the same manner. In symbols we have

$$E \rightarrow F^{-T}EF^{-1} = E^* \quad \text{and} \quad \dot{E} \rightarrow F^{-T}\dot{E}F^{-1} = \overset{\circ}{E}^*_{CR}. \tag{14.152}$$

We can now determine the derivative of the stress in Eq. (14.145), which was our original goal. Taking the Cotter–Rivlin corotational rate of both sides of the equation,

$$\overset{\circ}{\sigma}_{CR} = \frac{d\sigma}{dt} + L^T\sigma + \sigma L = \lambda\text{trace}(D)I + 2\mu D + 2\lambda\{\text{trace}(E^*)D - \text{trace}(E^*D)I\}, \tag{14.153}$$

where the trace term was written using $\overset{\circ}{E}^* = D$. If the strain is small compared to 1, the term in curly brackets drops out,

$$\overset{\circ}{\sigma}_{CR} = \frac{d\sigma}{dt} + L^T\sigma + \sigma L \approx \lambda\text{trace}(D)I + 2\mu D. \tag{14.154}$$

This provides an incremental formula for updating the stress given the rate of deformation tensor. Explicitly, we can now compute $\dot{\sigma}$ when σ, L and $D = \frac{1}{2}(L + L^T)$, all objects in the Eulerian current configuration, are known. As will be discussed below, however, this incremental equation is not typically used.

In a more abstract sense, we have a relationship between the Cauchy stress in the current configuration and a stress Σ^{CR} in the original configuration, where

$$\sigma = F^{-T}\Sigma^{CR}F^{-1}. \tag{14.155}$$

The notation Σ^{CR} is to remind us that this is a stress; it is entirely in the original configuration – that is, $\Sigma^{CR} =_X \Sigma^{CR}_X$ – but it is not the second Piola–Kirchhoff stress ($\sigma = (\rho/\rho_0)FS^{(2)}F^T$, so the pre- and post-multiplying terms differ and the ratio of current to initial densities is missing). The Cotter–Rivlin corotational stress rate performs the computation

$$\overset{\circ}{\sigma}_{CR} = \dot{\sigma} + L^T\sigma + \sigma L = (F^{-T}\overset{\circ}{\Sigma}{}^{CR}F^{-1})_{CR} = F^{-T}\dot{\Sigma}^{CR}F^{-1}. \tag{14.156}$$

We now consider three more examples of writing the Cauchy stress in terms of the original configuration based on our previous finite elasticity work,

$$\sigma = Rf(U)R^T, \quad \sigma = Fg(E)F^T, \quad \sigma = \frac{\rho}{\rho_0}Fh(E)F^T. \tag{14.157}$$

The middle expression is new here, but it immediately follows, since $E = \frac{1}{2}(U^2 - I)$ and $\rho/\rho_0 = 1/\det(U)$. Isotropy is not assumed. In these expressions, the R, R^T and F, F^T pre- and post-multiplication rotate the stress from the original configuration to the current one. The only one of these expressions that uses (or needs) the polar decomposition is the first, where R and U come from the polar decomposition.

Thinking about what each of these stress expressions represents, the first expression produces a Cauchy stress, which has been rotated into the same orientation as the original configuration; hence $f(U) = \sigma^{GN}$. The second is a stress in the original configuration, so we write $g(E) = \Sigma^O$ (it is not the second Piola–Kirchhoff stress, as the ratio of densities is missing). The last is the second Piola–Kirchhoff stress, $h(E) = S^{(2)}$. Thus, another way to write these expressions is

$$\sigma = R\sigma^{GN}R^T, \quad \sigma = F\Sigma^O F^T, \quad \sigma = \frac{\rho}{\rho_0}FS^{(2)}F^T. \tag{14.158}$$

These have a generic form of

$$\sigma = M_1\Sigma M_2, \tag{14.159}$$

where M_1 and M_2 are matrices and typically $M_2 = M_1^T$ or $M_2 = M_1^{-1}$. To determine a generic corotational rate,

$$\frac{d\sigma}{dt} = \dot{M}_1\Sigma M_2 + M_1\dot{\Sigma}M_2 + M_1\Sigma\dot{M}_2 = \dot{M}_1M_1^{-1}\sigma + M_1\dot{\Sigma}M_2 + \sigma M_2^{-1}\dot{M}_2. \tag{14.160}$$

This yields

$$\overset{\circ}{\sigma}_M = \frac{d\sigma}{dt} - \dot{M}_1M_1^{-1}\sigma - \sigma M_2^{-1}\dot{M}_2 = M_1\dot{\Sigma}M_2. \tag{14.161}$$

A commutative diagram for this expression is (compare with Eq. 14.152)

$$
\begin{array}{ccc}
\Sigma & \xrightarrow{\ \ d/dt\ \ } & \dot{\Sigma} \\
{\scriptstyle M_1\Sigma M_2}\Big\downarrow & & \Big\downarrow{\scriptstyle M_1\dot{\Sigma}M_2} \\
\sigma & \xrightarrow[\text{corotational rate}]{} & \overset{\circ}{\sigma}_M
\end{array}
$$

The upper row is related to the original configuration and the lower row is the current configuration. In some mechanics literature, terminology from differential geometry is invoked to describe this process. One begins in the current configuration Eulerian frame and "pulls back" σ to Σ in the original configuration using the inverse of the map $\sigma = M_1 \Sigma M_2$ (i.e., $\Sigma = M_1^{-1} \sigma M_2^{-1}$). Assuming objects in the original configuration are invariant, one then computes the time derivative $\dot{\Sigma}$ in the original configuration. Then one uses the map to "push forward" the time derivative from the original configuration to the current configuration, giving $M_1 \dot{\Sigma} M_2$.

In our examples, there are only three M_1: R, F, and F^{-T}. Each takes the original orientation or configuration to the current configuration. Using the polar decomposition, they are RI, RU, and RU^{-1} (or IR, VR, and $V^{-1}R$), respectively, so each performs the same rotation but applies different deformation. Computing the terms required for the second term in the corotational rate gives,

$$
\begin{aligned}
M_1 = R: && \dot{M}_1 M_1^{-1} &= \dot{R} R^T = \Omega, \\
M_1 = F: && \dot{M}_1 M_1^{-1} &= \dot{F} F^{-1} = L, \\
M_1 = F^{-T}: && \dot{M}_1 M_1^{-1} &= (d(F^{-1})/dt)^T F^T = -L^T.
\end{aligned}
\tag{14.162}
$$

For most of our examples, $M_2 = M_1^T$ and so $M_2^{-1} \dot{M}_2 = M_1^{-T} \dot{M}_1^T = (\dot{M}_1 M_1^{-1})^T$. This allows us to quickly compute the corresponding corotational rates to our stress expressions. For the first expression, $\sigma = R \sigma^{GN} R^T$, so $M_1 = R$ and we obtain the *Green–Naghdi* corotational rate,

$$
\mathring{\sigma}_{GN} = \frac{d\sigma}{dt} - \Omega \sigma - \sigma \Omega^T = \frac{d\sigma}{dt} - \Omega \sigma + \sigma \Omega = R \dot{\sigma}^{GN} R^T = R \dot{f}(U) R^T, \tag{14.163}
$$

where we note that

$$
\Omega = \dot{R} R^T = -R \dot{R}^T = -\Omega^T, \tag{14.164}
$$

that is, Ω is skew symmetric, since $d(RR^T)/dt = 0$. The rate of change of the Cauchy stress is computed by $d\sigma/dt = \Omega \sigma - \sigma \Omega + R \dot{f}(U) R^T$, and this rate includes the adjustment due to rotation (notice that if $\dot{R} \neq 0$, then $d\sigma/dt \neq 0$, even if $\dot{f} = 0$). For the second stress form, $\sigma = F \Sigma^O F^T$, $M_1 = F$ and the resulting *Oldroyd* corotational stress rate is

$$
\mathring{\sigma}_O = \frac{d\sigma}{dt} - L\sigma - \sigma L^T = F \dot{\Sigma}^O F^T = F \dot{g}(E) F^T. \tag{14.165}
$$

Again, one can solve for $d\sigma/dt$, given σ, L, F, and $\dot{g}$. We also see that the Cotter–Rivlin rate fits this generic form, as $\sigma = F^{-T} \Sigma^{CR} F^{-1}$; hence $M_1 = F^{-T}$, and so

$$
\mathring{\sigma}_{CR} = \frac{d\sigma}{dt} + L^T \sigma + \sigma L = F^{-T} \dot{\Sigma}^{CR} F^{-1}. \tag{14.166}
$$

For the third stress form, $\sigma = (\rho/\rho_0) F h(E) F^T$, we choose $M_1 = F$ and $M_2 = (\rho/\rho_0) F^T$. Using the conservation of mass, $d\rho/dt = -\rho \operatorname{div}(\vec{v}) = -\rho D_{ii}$, we need

$$
M_2^{-1} \dot{M}_2 = \frac{\rho_0}{\rho} F^{-T} \left(\frac{\dot{\rho}}{\rho_0} F^T + \frac{\rho}{\rho_0} \dot{F}^T \right) = \frac{\dot{\rho}}{\rho} I + L^T = -D_{ii} I + L^T \tag{14.167}
$$

to yield the *Truesdell* corotational rate,

$$
\mathring{\sigma}_T = \frac{d\sigma}{dt} - L\sigma - \sigma L^T + \operatorname{trace}(D)\, \sigma = \frac{\rho}{\rho_0} F S^{(2)} F^T = \frac{\rho}{\rho_0} F \dot{h}(E) F^T. \tag{14.168}
$$

As a final rate that does not fit the form, the arithmetic average of the Cotter–Rivlin and Oldroyd rates is the *Jaumann* corotational rate,

$$\overset{\circ}{\sigma}_J = \frac{d\sigma}{dt} - W\sigma + \sigma W = \frac{1}{2}(F^{-T}\dot{\Sigma}^J F^{-1} + F\dot{\Sigma}^J F^T). \tag{14.169}$$

The mapping between the original configuration and the current configuration is more complicated to write down than the other maps, $\sigma = \frac{1}{2}(F^{-T}\Sigma^J F^{-1} + F\Sigma^J F^T)$, but it is the same in concept (see Exercise 14.38).

We have now developed five corotational rates. To compare them, we write them out in detail,

$$\overset{\circ}{\sigma}_{CR} = \dot{\sigma} + L^T\sigma + \sigma L = \dot{\sigma} - W\sigma + \sigma W + D\sigma + \sigma D = F^{-T}\dot{\Sigma}^{CR}F^{-1},$$

$$\overset{\circ}{\sigma}_{GN} = \dot{\sigma} - \Omega\sigma - \sigma\Omega^T = \dot{\sigma} - \Omega\sigma + \sigma\Omega = R\dot{\sigma}^{GN}R^T,$$

$$\overset{\circ}{\sigma}_O = \dot{\sigma} - L\sigma - \sigma L^T = \dot{\sigma} - W\sigma + \sigma W - D\sigma - \sigma D = F\dot{\Sigma}^O F^T,$$

$$\overset{\circ}{\sigma}_T = \dot{\sigma} - L\sigma - \sigma L^T + \text{trace}(D)\,\sigma = \dot{\sigma} - W\sigma + \sigma W - D\sigma - \sigma D + D_{ii}\sigma$$

$$= \frac{\rho}{\rho_0}F\dot{S}^{(2)}F^T,$$

$$\overset{\circ}{\sigma}_J = \dot{\sigma} - W\sigma + \sigma W = \frac{1}{2}(F^{-T}\dot{\Sigma}^J F^{-1} + F\dot{\Sigma}^J F^T). \tag{14.170}$$

Each of these corotational rates allows computation of the stress rate mainly in the current configuration. By examining the far right-hand side of each expression we can see the mapping between the original and current configurations. If $D = 0$ all of these reduce to the Jaumann rate. As a reminder the velocity gradient L can be decomposed into $L = D + W$, where $D = \frac{1}{2}(L + L^T)$ is the (symmetric) rate of deformation tensor and $W = \frac{1}{2}(L - L^T)$ is the (skew symmetric) spin tensor, which in Cartesian coordinates is $W_{ij} = \frac{1}{2}(\frac{\partial v_i}{\partial x_j} - \frac{\partial v_j}{\partial x_i})$. Recalling that $L = \dot{F}F^{-1}$, using the polar decomposition $F = RU$, a relationship between Ω and W is (Eq. 14.90)

$$L = \dot{F}F^{-1} = (\dot{R}U + R\dot{U})U^{-1}R^T = \Omega + R\dot{U}U^{-1}R^T = D + W. \tag{14.171}$$

When any one of D, $\dot{E}$, or $\dot{U}$ is zero, the other two are also: $D = 0$ represents rigid rotation, and in that case $L = W = \Omega$.

Using E as our strain in the original configuration, the corresponding strains and strain rates are, recalling $F = VR$, $E = F^T E^* F$, and $\dot{E} = F^T DF$ (Eq. 14.150),

$$(\overset{\circ}{E^*})_{CR} = (F^{-T}\dot{E}F^{-1})_{CR} = F^{-T}\dot{E}F^{-1} = F^{-T}F^T DFF^{-1} = D,$$

$$(V\overset{\circ}{E^*}V)_{GN} = (R\dot{E}R^T)_{GN} = R\dot{E}R^T = RF^T DFR^T = VDV,$$

$$(V^2\overset{\circ}{E^*}V^2)_O = (F\dot{E}F^T)_O = F\dot{E}F^T = FF^T DFF^T = V^2DV^2,$$

$$(\frac{1}{\det(V)}V^2 E^*V^2)_T = (\frac{\rho}{\rho_0}F\dot{E}F^T)_T = \frac{\rho}{\rho_0}F\dot{E}F^T = \frac{\rho}{\rho_0}FF^T DFF^T = \frac{1}{\det(V)}V^2DV^2,$$

$$(\frac{1}{2}\{F^{-T}EF^{-1} + FEF^T\})_J = \frac{1}{2}\{(F^{-T}\overset{\circ}{E}F^{-1})_{CR} + (F\overset{\circ}{E}F^T)_O\} = \frac{1}{2}(D + V^2DV^2);$$

also $\quad (\overset{\circ}{E^*})_{GN} = V^{-1}DV^{-1} \quad$ and $\quad (\overset{\circ}{E^*})_J = \frac{1}{2}(V^{-2}D + DV^{-2}). \tag{14.172}$

The first three strains are $\frac{1}{2}(V^2 - I)V^n$, for $n = -2$, 0, 2. For small strain, hence small stretch, V is close to I and all these strains and strain rates are approximately equal to the Eulerian strain E^* and to the rate of deformation D, respectively.

Since $S^{(2)}$ lives entirely in the original configuration, it is invariant. However, the first Piola–Kirchhoff stress S needs a corotational rate. Beginning with $S = h(E)F^T = S^{(2)}F^T$,

$$\frac{dS}{dt} = \dot{h}(E)F^T + h(E)\dot{F}^T = \dot{h}(E)F^T + SF^{-T}\dot{F}^T = \dot{h}(E)F^T + SL^T, \quad (14.173)$$

and so a corotational rate for the first Piola–Kirchhoff is

$$\overset{\circ}{S} = \frac{dS}{dt} - SL^T = \dot{S}^{(2)}F^T = \dot{h}(E)F^T. \quad (14.174)$$

The rate fits our general form with $M_1 = I$ and $M_2 = F^T$.

There are two pages to go in the main text of this book and we finally define *corotational rate*:

DEFINITION 14.2. A *corotational* object is one that transforms correctly under an arbitrarily moving reference frame, and a *corotational rate* is a rate that transforms correctly under an arbitrarily moving reference frame, or more generally adheres to the principle of material frame indifference. Other terms used to describe this behavior are *objective* and *frame indifferent*.

We already used material frame indifference in developing our finite elasticity formulations, since we said that the constitutive model had to have certain behaviors under rotations. If we assume the Eulerian coordinate frame is arbitrarily moved according to a time-dependent translation and rotation, such as $\vec{x}^* = \vec{c}(t) + Q(t)\vec{x}$, where $\vec{c}$ is a displacement and Q is a rotation, then we expect the mechanics that occurs not to depend on the coordinate frame. (Note that we are not arbitrarily moving materials; we are arbitrarily moving the reference frame in which the materials and their corresponding stress and strain live.) The stress tensor behaves as $\sigma^* = Q\sigma Q^T$. However, the time derivative of the stress is not indifferent to the change in frame, since it does not transform as a tensor under the change in frame,

$$\frac{d\sigma^*}{dt} = \frac{d}{dt}(Q\sigma Q^T) = \dot{Q}\sigma Q^T + Q\dot{\sigma}Q^T + Q\sigma\dot{Q}^T \neq Q\frac{d\sigma}{dt}Q^T. \quad (14.175)$$

To compute the frame change behavior of the corotational rates, we need, using $F^* = QF$,

$$L^* = \dot{F}^*(F^*)^{-1} = (\dot{Q}F)F^{-1}Q^T + Q\dot{F}F^{-1}Q^T = \dot{Q}Q^T + QLQ^T. \quad (14.176)$$

Now the velocity gradient L is a tensor, but it does not transform as a tensor (i.e., as QLQ^T) under the prescribed time-dependent change in frame and hence is not frame indifferent. Choosing the Oldroyd corotational stress rate as an example, we now show that this rate does transform correctly,

$$\begin{aligned}
\overset{\circ}{\sigma}_O^* &= \frac{d\sigma^*}{dt} - L^*\sigma^* - \sigma^*(L^*)^T \\
&= \dot{Q}\sigma Q^T + Q\dot{\sigma}Q^T + Q\sigma\dot{Q}^T - (\dot{Q}Q^T + QLQ^T)Q\sigma Q^T - Q\sigma Q^T(Q\dot{Q}^T + QL^TQ^T) \\
&= Q\dot{\sigma}Q^T - QL\sigma Q^T - Q\sigma L^TQ^T \\
&= Q(\dot{\sigma} - L\sigma - \sigma L^T)Q^T \\
&= Q\overset{\circ}{\sigma}_O Q^T.
\end{aligned}$$

$$(14.177)$$

Another approach to this same calculation is available to us based on what we know about the Oldroyd rate in terms of its mapping back to the original configuration,

$$\overset{\circ}{\sigma}_O^* = F^* \overset{\circ}{\Sigma}{}^O (F^*)^T = QF \overset{\circ}{\Sigma}{}^O F^T Q^T = Q \overset{\circ}{\sigma}_O Q^T. \tag{14.178}$$

Thus, $\overset{\circ}{\sigma}_O$ behaves as a tensor under the prescribed frame change. Hence, it is a corotational rate.

As another example, the rate of deformation tensor is corotational or frame indifferent, since

$$D^* = \frac{1}{2}(L^* + (L^*)^T) = \frac{1}{2}(\dot{Q}Q^T + QLQ^T + Q\dot{Q}^T + QL^T Q^T)$$
$$= Q\frac{1}{2}(L + L^T)Q^T = QDQ^T, \tag{14.179}$$

where $d(QQ^T)/dt = 0$ was used. Thus, if a corotational stress rate is set equal to a linear function of the rate of deformation tensor, then the resulting rate form constitutive model behaves correctly under arbitrary Eulerian frame coordinate changes.

This observation leads to the incremental rates that are actually used in modern computational mechanics for metal plasticity. Of our five example rates, two of them have the desirable feature that the corotational rate of the stress deviator is deviatoric – that is, trace($\overset{\circ}{s}$) = 0. The two rates are the Jaumann and the Green–Naghdi rates. In impact computations the incremental stress update is typically only used for the stress deviators, since the pressure is computed directly from the density and internal energy through an equation of state and is not directly updated through a rate equation. Hence, working with only the stress deviators is preferred. What is done in computations is these stress deviator rates are set equal to the deviator form of linear elasticity. The two approaches are

$$(\overset{\circ}{s})_J = \frac{ds}{dt} - Ws + sW = 2\mu d = 2\mu(D - \frac{1}{3}\text{trace}(D)I),$$
$$(\overset{\circ}{s})_{GN} = \frac{ds}{dt} - \Omega s + s\Omega = 2\mu d = 2\mu(D - \frac{1}{3}\text{trace}(D)I). \tag{14.180}$$

Thus, to compute the stress increment for isotropic elasticity, we first compute the rate of deformation tensor, then determine its deviator d_{ij}, and then compute the new stress deviator s_{ij} by solving, for the Jaumann rate,

$$\frac{ds}{dt} = Ws - sW + 2\mu d. \tag{14.181}$$

For the Green–Naghdi rate Ω replaces W. The word "solve" is used here since there are explicit ways to do this update or implicit ways – the terms $Ws - sW$ produce a matrix, whose components are various W_{ij} terms, that needs to be inverted if the method is implicit. This equation takes care of the material rotation. Elasticity models where the stress rate is determined through a modulus times the strain rate are called *hypoelastic*. One then takes the stress deviator result and checks the yield surface and adjusts the stress if plastic flow has occurred, using the radial return or a variant method. The Jaumann rate form is simplest since W is easy to compute in the Eulerian frame and no history needs to be recorded (other than the history that is stored in the current stress state). The Green–Naghdi rate requires computing F, which is either done by maintaining knowledge of the original configuration or by updating F through the expression $\dot{F} = LF$. Both the Jaumann and Green–Naghdi approaches are extensively used. Discussion and criticism of one vs. the other can be found in the literature. The technical details of the discussions come down to the actual plasticity or inelasticity model in use (as an example, see [**62**]).

Like L, the spin tensor W and the rotation rate Ω are not frame indifferent and behave as

$$W^* = \frac{1}{2}(L^* - (L^*)^T) = \frac{1}{2}(\dot{Q}Q^T + QLQ^T - Q\dot{Q}^T - QL^TQ^T) = \dot{Q}Q^T + QWQ^T,$$

$$\Omega^* = (d(QR)/dt)(QR)^T = (\dot{Q}R + Q\dot{R})R^TQ^T = \dot{Q}Q^T + Q\Omega Q^T.$$

$$\text{(14.182)}$$

Using these expressions, the other corotational stress rates can be shown to be objective/material frame indifferent in a similar fashion. There are lots of corotational stress rates; linear combinations of the five presented here are again corotational, and even a sixth corotational stress rate is presented in the exercises. Recall, however, that we began our study of corotational rates by the direct differentiation of the Eulerian strain E^* with respect to time, without any theoretical encumbrances.

And we are done.

14.9. Sources

The approach to rotation is new. The static finite elasticity material follows discussions in Beatty [**27**], Gurtin, Fried, and Anand [**89**], and Müller and Strehlow [**149**].

Exercises

(14.1) If $S_{ij}^{(2)} = \partial e/\partial E_{ij}$ is symmetric, where $e = e(E_{ij})$ is the strain energy, show that $S = S^{(2)}F^T$ follows from $S_{ij} = \partial e/\partial F_{ij}^T$.

(14.2) The explicit rotation matrices used in this chapter can be developed through matrix exponentials. The exponential of a matrix A is defined by a power series expansion,

$$\exp(A) = I + A + \frac{1}{2!}A^2 + \frac{1}{3!}A^3 + \cdots. \quad \text{(14.183)}$$

As a warmup, show from this definition that in two dimensions,

$$R(\theta) = \exp(\theta N) = \cos(\theta)I + \sin(\theta)N = \begin{pmatrix} \cos(\theta) & -\sin(\theta) \\ \sin(\theta) & \cos(\theta) \end{pmatrix}, \quad N = \begin{pmatrix} 0 & -1 \\ 1 & 0 \end{pmatrix}. \quad \text{(14.184)}$$

Now show the expression in Eq. (14.55) for three dimensions. (Hint: show $N^3 = -N$.) Show that $R(\theta_a, \vec{n})$ is a rotation matrix. (Hint: since N commutes with itself, it is possible to do a matrix multiply of exponentials as an exponential of the sum of the matrices; recall $N^T = -N$.) Now show that $N\vec{x} = \vec{n} \times \vec{x}$. Write $R(\theta_a, \vec{n})$ as cross products with $\vec{n}$ rather than matrix multiplies with N. Show that the rotation is about the $\vec{n}$-axis by showing $R\vec{n} = \vec{n}$. Write a three vector orthogonal frame with two vectors $\vec{n}_1$ and $\vec{n}_2$ such that $\vec{n}_1 \times \vec{n}_2 = \vec{n}$, and show for an arbitrary vector $\vec{x} = a\vec{n}_1 + b\vec{n}_2 + c\vec{n}$ that $R\vec{x}$ is such that the rotation axis component does not change but we have a rotation in the plane defined by $\vec{n}_1$ and $\vec{n}_2$. (Hint: it is possible to determine all the cross products $\vec{n} \times \vec{n}_1$, etc., given what we know about the orthogonal frame of $\vec{n}$s.)

(14.3) Show that in the Blatz–Ko constitutive model that μ_0 is the shear modulus and in the Mooney–Rivlin constitutive model that $\mu_0 = \beta_+ - \beta_-$ is the shear modulus. (Hint: start by writing down B for a small shear deformation.)

(14.4) Using either of the expressions in Eq. (14.105), and assuming that zero strain implies zero stress, show that for small strain and small rotation, if the relation between the Cauchy stress and the small strains is linear then this expression implies the form of isotropic linear elasticity as shown in Eq. (3.17) with only two independent constants. (Hint: at some point you need to compute the linearized invariants of B; if you get to the end of your argument and you have failed to do so, you missed something important.)

(14.5) (Poynting effect) Assume for an isotropic elastic material the stress from the first expression in Eq. (14.105). For simple shear deformation, where $x = X + \gamma Y$, $y = Y$ and $z = Z$, show that

$$\sigma_{11} - \sigma_{22} = \gamma \sigma_{12}. \tag{14.185}$$

This result is independent of the constitutive model and depends only on isotropy and elasticity. Conclude that a simple shear deformation state cannot produce a pure shear stress state. This fact is referred to as the *Poynting effect*. (When finite deformations are considered, simple shear deformation leads to nonzero diagonal entries in E and E^*, unlike in small strain elasticity theory, and this is the origin of the effect.)

(14.6) Determine the uniaxial stress-strain curve for the Blatz–Ko constitutive model. Compare to the Mooney–Rivlin constitutive model uniaxial stress-strain curve when both have the same initial Young's modulus E_0 and $\beta_-/\beta_+ = -0.1$. When stress-strain data are returned from the laboratory they are usually in two forms. The first form is "engineering stress vs. engineering strain" (or sometimes just "stress vs. strain") and is the load divided by the sample's initial cross-sectional area vs. the change in length divided by initial length $\Delta L / L_0$. Show that this form is the first Piola–Kirchhoff stress vs. stretch $\lambda - 1$. The second form is called "true stress vs. true strain" and is the load divided by the sample's current cross-sectional area vs. the logarithmic strain $\int dL/L$. (The area is determined by direct measurement or by assumptions such as incompressibility and uniform deformation.) Show that this form corresponds to the Cauchy stress vs. the natural log of the stretch $\ln(\lambda)$. Present your results in both forms.

(14.7) For small strains, Poisson's ratio is defined as $\nu_{ji} = -\varepsilon_{ii}/\varepsilon_{jj}$, where there is no sum on i, j, and $i \neq j$. The change in volume is given by $\Delta vol / vol = \varepsilon_{11} + \varepsilon_{22} + \varepsilon_{33}$. For a large strain Poisson's ratio, consider a box with sides of length x, y, z, volume $vol = xyz$, related to stretches $x = \lambda_1 X$, $y = \lambda_2 Y$, $z = \lambda_3 Z$. Then

$$\frac{dvol}{vol} = \frac{dx}{x} + \frac{dy}{y} + \frac{dz}{z} = \frac{d\lambda_1}{\lambda_1} + \frac{d\lambda_2}{\lambda_2} + \frac{d\lambda_3}{\lambda_3} = d\ln(\lambda_1) + d\ln(\lambda_2) + d\ln(\lambda_3). \tag{14.186}$$

Thus, we are led to using the differential logarithmic strain in defining Poisson's ratio,

$$\nu_{ji} = -\frac{d\ln(\lambda_i)}{d\ln(\lambda_j)}, \quad \text{where } i \neq j \tag{14.187}$$

and the loading is uniaxial stress along the j-axis. For an isotropic material, all the Poisson's ratios are the same. Using this definition, show that the

Poisson's ratio for the Blatz–Ko constitutive model is $\nu(\lambda) = \frac{1}{4}$ and for the Mooney–Rivlin model is $\nu(\lambda) = \frac{1}{2}$. In both cases the Poisson's ratio is a constant, even in large deformation.

(14.8) Derive the general wave formulas for the longitudinal and shear waves for an isotropic elastic solid. For the longitudinal wave, show that the invariants for a stretch λ along the wave propagation direction are $I_1(B) = 2 + \lambda^2$, $I_2(B) = -1 - 2\lambda^2$, and $I_3(B) = \lambda^2$. Writing each β_i as $\beta_i(\lambda) = \beta_i(I_1(B), I_2(B), I_3(B))$, and assuming that $\sigma_{11}(1) = 0$, show that

$$U = \sqrt{\frac{\sigma_{11}(\lambda)}{\rho_0(\lambda - 1)}} = \sqrt{\frac{\beta_0(\lambda) + \beta_+(\lambda)\lambda^2 + \beta_-(\lambda)/\lambda^2}{\rho_0(\lambda - 1)}},$$

$$c_L = \sqrt{\frac{1}{\rho_0}\left(\frac{\partial\sigma_{11}}{\partial\lambda}\right)_{\lambda=1}} \qquad (14.188)$$

$$= \left\{\frac{2}{\rho_0}\left[\mu_0 + \left(\frac{\partial}{\partial I_1} - 2\frac{\partial}{\partial I_2} + \frac{\partial}{\partial I_3}\right)(\beta_0 + \beta_+ + \beta_-)\right]_{B=I}\right\}^{1/2}.$$

In a similar fashion, for a shear wave propagating in the 1 direction with material velocity u_2 in the 2 direction, with $\gamma = u_2/U$, show that $I_1(B) = 3 + \gamma^2$, $I_2(B) = -3 - \gamma^2$, $I_3(B) = 1$, $\sigma_{12}(\gamma) = -(\beta_+(\gamma) - \beta_-(\gamma))\gamma \equiv -\mu(\gamma)\gamma$, and assuming $\sigma_{ij}(0) = 0$ for all ij, show the shear shock speed is given by

$$U = \sqrt{\frac{\beta_+(\gamma) - \beta_-(\gamma)}{\rho_0}} = \sqrt{\frac{\mu(\gamma)}{\rho_0}}, \qquad (14.189)$$

subject to the constraint that

$$\sigma_{11}(\gamma) = \beta_0(\gamma) + \beta_+(\gamma) + (1 + \gamma^2)\beta_-(\gamma) = 0, \qquad (14.190)$$

as otherwise a simple shock wave does not form. The shear sound speed is given by evaluation at $\gamma = 0$,

$$c_s = \sqrt{\frac{\beta_+(0) - \beta_-(0)}{\rho_0}} = \sqrt{\frac{\mu_0}{\rho_0}}. \qquad (14.191)$$

(14.9) Plot the solution set of Eq. (14.140) for the constants in the text. Explicitly identify the bifurcation point (what values of stretch?).

(14.10) (From Müller and Strehlow [149]) Assume in the rubber sheet loading problem that the forces are not equal, $F_1 = \xi F_2$. Plot the solution curves for different loads for $\xi = 0.999$, $\xi = 0.99$, and $\xi = 0.9$. Müller and Strehlow refer to the unequal load as "symmetry breaking." The bifurcation point goes away and a (starting point dependent) unique solution results.

(14.11) For biaxial stress with equal force loading on the sides (same boundary conditions as the Mooney–Rivlin bifurcation example, except that here the material is not incompressible), work the bifurcation problem for the Blatz–Ko constitutive model Eq. (14.115) and derive the equation corresponding to Eq. (14.140) for this model. Show bifurcation occurs in tension at stretch $3^{3/10} \approx 1.39$.

(14.12) Bifurcation and multiple solutions also occur for isotropic linear elasticity constitutive model when compressing a cube to large strains. If the energy is written $e = \frac{\lambda}{2}(E_{kk})^2 + \mu E_{ij}E_{ij}$, show that

$$S = (\lambda \text{trace}(E)I + 2\mu E)F^T, \tag{14.192}$$

where E is the large strain. (See Eqs. 13.61, 14.9 and 14.10.) For loading a cube,

$$F = \begin{pmatrix} a_1 & 0 & 0 \\ 0 & a_2 & 0 \\ 0 & 0 & a_3 \end{pmatrix}, \qquad a_i > 0 \quad \text{all } i, \tag{14.193}$$

show that the first Piola–Kirchhoff stress is

$$S = \left\{ \frac{\lambda}{2}(a_1^2 + a_2^2 + a_3^2 - 3) - \mu \right\} \begin{pmatrix} a_1 & 0 & 0 \\ 0 & a_2 & 0 \\ 0 & 0 & a_3 \end{pmatrix} + \mu \begin{pmatrix} a_1^3 & 0 & 0 \\ 0 & a_2^3 & 0 \\ 0 & 0 & a_3^3 \end{pmatrix}. \tag{14.194}$$

Let $A = \frac{\lambda}{2}(a_1^2 + a_2^2 + a_3^2 - 3) - \mu$ and show that when the stress is equal on the cube faces (which means the forces on the faces are the same, since it is always a cube in initial configuration space) we have

$$(a_i - a_j)\left\{ A + \mu(a_i^2 + a_i a_j + a_j^2) \right\} = 0. \tag{14.195}$$

Thus, either $a_i = a_j$ or $A + \mu(a_i^2 + a_i a_j + a_j^2) = 0$. We expect that, with three equations and three unknowns, there will only be a finite number of solutions if all the a_i differ. Hence, for loading paths we expect that at least two a_i are the same. Assuming this, show that there are four solution paths, where the four solutions are given by the conditions $a_1 = a_2 = a_3$, $a_1 = a_2 \neq a_3$, $a_1 = a_3 \neq a_2$, and $a_2 = a_3 \neq a_1$. These four solution curves meet at a bifurcation point. Find it. What is the strain at the bifurcation point for a Poisson's ratio of 0.3? All these states can thus be reached by uniformly loading the cube from its initial ambient state to the bifurcation point and then following the loading curve of interest. (Hint: for $\nu = 0.3$, bifurcation occurs at smaller-in-magnitude strain than seen in the previous exercise for Blatz–Ko.)

(14.13) Unique Piola–Kirchhoff stresses do not imply unique Cauchy stresses. Assume we load a cube, as in the previous problem. Assume the Piola–Kirchhoff stress S is diagonal, with terms $f(a)$ for stretch a, and the stress is zero for the ambient state (meaning $f(1) = 0$, as it was above). Show that for one Cauchy stress to be due to only one strain, $df(a)/da > 2f(a)/a$, and hence for $f(a) = \frac{1}{n}K(a^n - 1)$ show that a given Cauchy stress can only be achieved by one strain when $n \geq 1$ (hence the solution is unique for this loading) and at least one Cauchy stress exists that can be achieved by two or more strains when $n < 1$. (This result does not say there will not be a bifurcation and hence a nonunique solution if the deformed shape is no longer a cube for uniform force loadings, as was seen in the previous exercise.)

(14.14) A classic problem in finite elasticity is inflating a spherically symmetric balloon made of the Mooney–Rivlin material. We go through the steps to perform the inflation in this exercise.

(a) As a first step, compute B, using R as the original Lagrangian coordinates and r as the current Eulerian coordinates assuming spherical symmetry,

$$
B = \begin{pmatrix} (\partial r/\partial R)^2 & 0 & 0 \\ 0 & (r/R)^2 & 0 \\ 0 & 0 & (r/R)^2 \end{pmatrix} = \begin{pmatrix} (R/r)^4 & 0 & 0 \\ 0 & (r/R)^2 & 0 \\ 0 & 0 & (r/R)^2 \end{pmatrix}
$$

$$
\approx \begin{pmatrix} 1/\lambda^4 & 0 & 0 \\ 0 & \lambda^2 & 0 \\ 0 & 0 & \lambda^2 \end{pmatrix},
$$

(14.196)

where λ is the stretch and incompressibility has been used, and the balloon is assumed thin so that the stretch is uniform through the thickness.

(b) Write down the stress and determine $\hat{p}$ through the boundary condition $\sigma_{rr}(b) = 0$.

(c) Show that the Cauchy stress static momentum balance in spherical coordinates, assuming spherical symmetry, is

$$
\frac{\partial(r^2 \sigma_{rr})}{\partial r} = 2r\sigma_{\theta\theta}.
$$

(14.197)

Show that for a thin-walled balloon this is approximately $\partial \sigma_{rr}/\partial r = (2/r)\sigma_{\theta\theta}$. Assuming the stress inside the balloon is $\sigma_{rr}(a) = -p_a$ and the stress outside the balloon $\sigma_{rr}(b) = 0$, with balloon thickness t, show that $\sigma_{\theta\theta} = (r/2t)p_a$.

(d) Now solve for p_a and put it in terms of the original thickness t_0 and the original radius A,

$$
p_a = \frac{2t_0}{\lambda A}\left(1 - \frac{1}{\lambda^6}\right)(\beta_+ - \lambda^2\beta_-).
$$

(14.198)

Plot the pressure vs. stretch and show that it has the expected behavior of initially difficult, then easier, then again more difficult (and, if you've never inflated a balloon by mouth, buy one and do it to confirm).

(14.15) For the Mooney–Rivlin model for rubber, compute the in-plane longitudinal wave speed for a thin sheet. (Hint: begin by showing that F is given by the diagonal matrix with entries λ, 1, and $1/\lambda$, then use $\sigma_{33} = 0$ to determine $\hat{p}$, etc.)

(14.16) Look up the heat capacity of rubber and determine the rise in temperature for uniaxial stress stretches 2, 5, and 10.

(14.17) Show that E^* is corotational (frame indifferent) but $\dot{E}^*$ is not.

(14.18) Show that the Jaumann rate is corotational.

(14.19) Show that the linear combination of two corotational stress rates is a corotational stress rate, or more specifically, that if $\overset{\circ}{\sigma}_1$ and $\overset{\circ}{\sigma}_2$ are two corotational stress rates, then $\overset{\circ}{\sigma} = (1 - a)\overset{\circ}{\sigma}_1 + a\overset{\circ}{\sigma}_2$, where a is also constant, is a corotational stress rate. Thus, there are uncountably many corotational stress rates and determining the appropriate one to use comes back to the constitutive model for the material that is being modeled.

(14.20) Suppose that the deformation is simply a rotation plus stretch,

$$
\vec{x} = RU\vec{x}_0 = \begin{pmatrix} \cos(\omega t) & -\sin(\omega t) & 0 \\ \sin(\omega t) & \cos(\omega t) & 0 \\ 0 & 0 & 1 \end{pmatrix} \begin{pmatrix} (1+at) & 0 & 0 \\ 0 & 1 & 0 \\ 0 & 0 & 1 \end{pmatrix} \vec{x}_0,
$$

(14.199)

giving the polar decomposition $F = RU$, where R is the rotation and U is the right stretch, and R and U are functions of t and a (assumed small) and

ω are constants. Small-strain elasticity is assumed. The following questions will demonstrate that a corotational stress rate is required to obtain the correct stress state.

(a) First compute the correct answer to this problem by stretching in the x direction with $F_{11} = 1 + at$, and then rotating by the angle ωt. (Hint: here and below you will need to compute L to compute D, and the resulting strain is logarithmic.)

(b) Now, compute the wrong answer by setting $d\sigma/dt$ " $=$ " $\lambda \text{trace}(D)I + 2\mu D$, where the rate of deformation D is for the rotation and stretch combined. Show that the answer is not correct.

(c) Compute the spin W and the rotation rate Ω. (In this case W and Ω are the same because the deformation is simple.)

(d) Using your answer to part (a), show for the combined rotation and stretch that using the corotational stress rate produces the correct result – that is, verify that the following equation holds,

$$R\frac{d\sigma_0}{dt}R^T = \lambda \text{trace}(D)I + 2\mu D = \overset{\circ}{\sigma} = \frac{d\sigma}{dt} - W\sigma + \sigma W. \tag{14.200}$$

For the more complicated case of simple shear, where $W \neq \Omega$, see [62].

(14.21) Using $L = \dot{F}F^{-1}$, compute this corotational stress rate directly from the definition of the first Piola–Kirchhoff stress,

$$\overset{\circ}{\sigma} = \frac{d\sigma}{dt} - L\sigma - \frac{d\rho/dt}{\rho}\sigma = \frac{\rho}{\rho_0}F\frac{dS}{dt}. \tag{14.201}$$

(14.22) Using $L = \dot{F}F^{-1}$, given that $\sigma_{ij}D_{ij}$ in the energy rate balance identifies D as the conjugate variable to the Cauchy stress σ, show that the conjugate rate variable to the first Piola–Kirchhoff stress S is $\frac{d}{dt}F^T$ and the conjugate rate variable to the second Piola–Kirchhoff stress $S^{(2)}$ is $\frac{d}{dt}E$.

(14.23) In this exercise you will establish the relationship between the rate of deformation tensor D and the small-strain strain rate $\dot{\varepsilon}$, where $\varepsilon_{ij} = \frac{1}{2}(\partial u_i/\partial x_j + \partial u_j/\partial x_i)$, where u_i is the displacement and x_j is the current coordinate. Define a displacement gradient in terms of the current coordinate as $\ell_{u\,ij} = \partial u_i/\partial x_j$; hence $\varepsilon = \frac{1}{2}(\ell_u + \ell_u^T)$. Differentiating ℓ_u with respect to time is complicated because of the current coordinates in the derivative. Using $d(F^{-1})/dt = -F^{-1}L$, show that

$$\frac{d}{dt}\ell_u = L - \ell_u L \quad \text{and hence} \quad D = \dot{\varepsilon} + \frac{1}{2}(\ell_u L + (\ell_u L)^T). \tag{14.202}$$

Thus, they are close for small strain.

(14.24) It would be nice if the rate of deformation tensor D was the time derivative of a strain tensor, just as the conjugate strain rate for the Piola–Kirchhoff stress is the time derivative of F^T. In this exercise we consider two possibilities,

$$\varepsilon_C = E^{(0)} = \frac{1}{2}\ln(C) = \frac{1}{2}\ln(F^T F), \quad \varepsilon_B = \frac{1}{2}\ln(B) = \frac{1}{2}\ln(FF^T). \tag{14.203}$$

Show that these strains are plausible candidates: if F is a stretch with the only nonzero terms on the diagonal, show that $D = d\varepsilon_C/dt = d\varepsilon_B/dt$; hence the rate of deformation tensor is in some sense a derivative of a logarithmic strain (as stated in Chapter 3). For a general deformation, the logarithm of C or B is determined by diagonalizing the matrix and then taking the logarithm of the eigenvalues, in a similar fashion to computing the square root. Now consider the simple shear deformation $x = X + \gamma Y$ and $y = Y$. Compute F and D. Compute the eigenvalues of C and B (they are the same) and the corresponding matrix Q that diagonalizes the respective matrix – for example, $C = F^T F = Q \Lambda Q^T$. An expression that greatly consolidates the algebra is $\Gamma = \frac{1}{2}(\gamma + \sqrt{\gamma^2 + 4})$; show that $\lambda_+ = \Gamma^2$ and $\lambda_- = 1/\Gamma^2$. Show that the expressions are

$$\text{for } C, \quad Q = \frac{1}{\sqrt{1+\Gamma^2}} \begin{pmatrix} 1 & -\Gamma \\ \Gamma & 1 \end{pmatrix}, \quad \Lambda = \begin{pmatrix} \Gamma^2 & 0 \\ 0 & 1/\Gamma^2 \end{pmatrix},$$

$$\text{for } B, \quad Q = \frac{1}{\sqrt{1+\Gamma^2}} \begin{pmatrix} 1 & \Gamma \\ -\Gamma & 1 \end{pmatrix}, \quad \Lambda = \begin{pmatrix} 1/\Gamma^2 & 0 \\ 0 & \Gamma^2 \end{pmatrix}. \tag{14.204}$$

Show that

$$\varepsilon_C = \frac{1}{2}\ln(C) = \frac{1}{2}Q\ln(\Lambda)Q^T = \frac{\ln(\Gamma)}{1+\Gamma^2}\begin{pmatrix} -(\Gamma^2-1) & 2\Gamma \\ 2\Gamma & \Gamma^2-1 \end{pmatrix},$$

$$\varepsilon_B = \frac{1}{2}\ln(B) = \frac{1}{2}Q\ln(\Lambda)Q^T = \frac{\ln(\Gamma)}{1+\Gamma^2}\begin{pmatrix} \Gamma^2-1 & 2\Gamma \\ 2\Gamma & -(\Gamma^2-1) \end{pmatrix}. \tag{14.205}$$

Now compute $d\varepsilon_C/dt$ and $d\varepsilon_B/dt$ and show that $D \neq d\varepsilon_C/dt$ and $D \neq d\varepsilon_B/dt$, so our hope is not realized. As a comment, the underlying (deep) reason they are not equal is that for deformations in which the eigenvectors change direction (i.e., Q is not constant), B and $\dot{B}$ do not commute – that is, $B\dot{B} \neq \dot{B}B$. Using power series it is straightforward to show that with $A = f(B)$, $\dot{A} = f'(B)\dot{B}$ when B and $\dot{B}$ commute; otherwise they are not equal. This is why generally $d\ln(B)/dt \neq B^{-1}\dot{B}$ for matrices, which defeats our hopes. The same is true for C. As a final piece, though, show that if B commutes with $\dot{B}$ and B commutes with L, then $d\varepsilon_B/dt = D$. (Hint: show $\dot{B}B^{-1} = L + BL^T B^{-1} = 2D$.)

(14.25) (For those who have had more linear algebra) Following the previous exercise, consider the strain

$$\varepsilon_F = \frac{1}{2}(\ln(F) + \ln(F)^T). \tag{14.206}$$

Show that this strain has $d\varepsilon_F/dt = D$ for F diagonal, for a rigid rotation, *and* for the simple shear strain considered in the previous exercise (for this you will need to compute the matrix function of a Jordan block: specifically, look for the matrix A such that $\exp(A) = I + A + \frac{1}{2!}A^2 + \dots = F$, where a hint is to look for a nilpotent A – that is, where $A^n = 0$ for some n). This result, though, is due to $F = \begin{pmatrix} 1 & \gamma \\ 0 & 1 \end{pmatrix}$ having constant eigenvectors (hence $F\dot{F} = \dot{F}F$), even though $F^T F$ does not. Show that equality fails for $F = \begin{pmatrix} 1 & \gamma \\ 0 & \gamma \end{pmatrix}$, where the eigenvectors are functions of γ and $F\dot{F} \neq \dot{F}F$.

(14.26) Show that, in two dimensions, requiring FR^T or $R^T F$ to be symmetric, assuming R is a rotation matrix, yields Eq. (14.25) (i.e., require $V_{12} = V_{21}$, for example)

(14.27) Show, using the definition of F and the chain rule, that for two subsequent deformations $F_{\text{total}} = F_{\text{2nd}}F_{\text{1st}}$. Next, using the simple shears $F_1 = \begin{pmatrix} 1 & 1 \\ 0 & 1 \end{pmatrix}$ and $F_2 = \begin{pmatrix} 1 & 2 \\ 0 & 1 \end{pmatrix}$, with $F_1 = R_1 U_1$ and $F_2 = R_2 U_2$, show that while $F_2 = F_1^2$, we have $R_2 \neq R_1^2$, $U_2 \neq U_1^2$, and $F_2 = F_1^2 = R_2 U_2 = R_1 U_1 R_1 U_1 \neq R_1^2 U_1^2$.

(14.28) (Polar decomposition) The polar decomposition is central to much theory of finite strains. This exercise walks through the details and finishes with an important result. Assume $\det(F) > 0$. Show that $F^T F$ and FF^T are symmetric and positive definite (positive definite means that for a matrix A, $\vec{x}^T A \vec{x} > 0$ for all nonzero vectors $\vec{x}$). (Hint: the length of any nonzero vector is positive.) Show that all eigenvalues of $F^T F$ and FF^T are strictly positive. Next, using the spectral decomposition, define the square roots of these matrices by diagonalizing and taking the square roots of the eigenvalues and then transforming back. Show that the square root that is positive definite (i.e., taking the positive square root of each eigenvalue) is symmetric and unique (this is the subtlest step in this exercise, but it is important; there is a very subtle aspect to the argument highlighted in the next exercise). This positive definite square root is what we mean when we say square root from now on. Starting with $U = \sqrt{F^T F}$ and $V = \sqrt{FF^T}$, show for $F = RU$ and $F = VR$ that R is a rotation matrix – that is, $R^T R = RR^T = I$ – and hence that U and V are correctly defined (here work with each decomposition separately). Now show that the polar decomposition is unique – for example, that if $F = \tilde{V}\tilde{R}$ and $F = VR$ then $\tilde{V} = V$ and $\tilde{R} = R$ (this step is simple since you have shown the square root is unique and V is invertible). Finally, since the decomposition is unique, we have $F = RU = RUR^T R = \tilde{V}R = VR$, since $\tilde{V}$ is symmetric and positive definite, and thus you have shown result, which is not obvious, that the rotation matrix R is the same for both the left and right polar decompositions – that is, that the same amount of rotation occurs whether you do the rotation before or after the stretch, though the stretches are not the same. Note, however, that though the stretches U and V are not the same, they are similar in the linear algebra sense in that V is just U rotated by R; hence U and V have the same invariants and the same eigenvalues.

(14.29) For the 2×2 identity matrix I, it seems obvious that $\sqrt{I} = I$. However, show for symmetric matrix $A = \begin{pmatrix} \cos(\phi) & \sin(\phi) \\ \sin(\phi) & -\cos(\phi) \end{pmatrix}$ that $A^2 = I$. Thus, there are lots of square roots of I and not just the four you might expect by the outline of the argument in the previous exercise. Explain why A is not a square root of interest to us and why there are so many square roots of I. Hint: calculate the eigenvalues of A; the heart of the large number of square roots is that I has a repeated eigenvalue and hence the normalized eigenvectors are not unique; give the right words to the following computations,

$$Q \begin{pmatrix} 1 & 0 \\ 0 & -1 \end{pmatrix} Q^T = \begin{pmatrix} \cos(2\theta) & -\sin(2\theta) \\ -\sin(2\theta) & -\cos(2\theta) \end{pmatrix} \quad \text{and} \quad QIQ^T = I$$

$$(14.207)$$

for the matrix of eigenvector columns $Q = \begin{pmatrix} \cos(\theta) & \sin(\theta) \\ -\sin(\theta) & \cos(\theta) \end{pmatrix}$.

(14.30) Determine the positive definite square root of a symmetric 2×2 matrix that is positive definite (i.e., eigenvalues are positive). Hints: method 1 – by brute force,

$$\begin{pmatrix} a & b \\ b & c \end{pmatrix} \begin{pmatrix} a & b \\ b & c \end{pmatrix} = \begin{pmatrix} A & B \\ B & C \end{pmatrix}.$$

$$(14.208)$$

Show there are four equations that result from this (including one from the determinants),

$$a^2 + b^2 = A, \quad ab + bc = B, \quad b^2 + c^2 = C, \quad (ac - b^2)^2 = AC - B^2. \quad (14.209)$$

Squaring the second and combining with the square root of the forth, then combining with the first and third, yields b, which in turn yields

$$\begin{pmatrix} a & b \\ b & c \end{pmatrix} = \frac{1}{\sqrt{A + C + 2\sqrt{AC - B^2}}} \begin{pmatrix} A + \sqrt{AC - B^2} & B \\ B & C + \sqrt{AC - B^2} \end{pmatrix}. \quad (14.210)$$

Method 2 – use the Cayley–Hamilton theorem. For 2×2 matrices, the invariants are $I_1(A) = \text{trace}(A) = \lambda_1 + \lambda_2$ and $I_2(A) = -\det(A) = -\lambda_1\lambda_2$, where the λ_i are the eigenvalues. Using the notation of a for square root of A, so $a^2 = A$, show that $I_2(a) = -\sqrt{-I_2(A)}$ and $I_1^2(a) = I_1(A) - 2I_2(a)$. By the Cayley–Hamilton theorem for a,

$$a = \frac{1}{I_1(a)}(a^2 - I_2(a)I) = \frac{1}{\sqrt{\text{trace}(A) + 2\sqrt{\det(A)}}}(A + \sqrt{\det(A)}I). \quad (14.211)$$

For whichever method you use, show that both eigenvalues are positive; hence, this is the square root we want (of the four square roots). Show your result agrees with Eqs. (14.32) and (14.35).

(14.31) Determine R, U, and V for a simple shear deformation $x = X + \gamma Y$ and $y = Y$. Also determine E and E^*.

(14.32) If only things were so simple for 3×3 matrices! Unfortunately it is not simple to write the $I_i(a)$ in terms of the $I_i(A)$, so it is necessary to compute the eigenvalues of A. Once that is done it is easy to compute the invariants of a, and the Cayley–Hamilton theorem yields

$$a = (A - I_2(a)I)^{-1}(I_1(a)A + I_3(a)I). \quad (14.212)$$

Write $a = rA^2 + sA + tI$. Show with two uses of the Cayley–Hamilton theorem that $a^2 = A$ when

$$r^2\{I_1^2(A) + I_2(A)\} + 2rsI_1(A) + 2rt + s^2 = 0,$$
$$r^2\{I_1(A)I_2(A) + I_3(A)\} + 2rtI_2(A) + 2st = 1, \quad (14.213)$$
$$r^2 I_1(A)I_3(A) + 2rsI_3(A) + t^2 = 0.$$

If one types solve($\{f, g, h\}, \{r, s, t\}$) in REDUCE, where f, g, h are these quadratics, REDUCE returns an answer where r is the root of an 8th degree polynomial and s and t are given as polynomials of that root. This is what we expect, since there are three square roots of eigenvalues to take; hence $2^3 = 8$ square roots of A (only one square root is positive definite). When specific values are used for the $I_i(A)$, it returns eight sets r, s, t, which requires determining which one leads to a positive definite a. Hence, it ends up being necessary to compute the eigenvalues of A. Once that is done, suppose that Q diagonalizes A and a (by Eq. 14.28 the same rotation diagonalizes both). Computing $Q^T(a = rA^2 + sA + tI)Q$ yields

$$\begin{pmatrix} \sqrt{\lambda_1} & 0 & 0 \\ 0 & \sqrt{\lambda_2} & 0 \\ 0 & 0 & \sqrt{\lambda_3} \end{pmatrix} = r\begin{pmatrix} \lambda_1^2 & 0 & 0 \\ 0 & \lambda_2^2 & 0 \\ 0 & 0 & \lambda_3^2 \end{pmatrix} + s\begin{pmatrix} \lambda_1 & 0 & 0 \\ 0 & \lambda_2 & 0 \\ 0 & 0 & \lambda_3 \end{pmatrix} + tI. \quad (14.214)$$

We see that r, s, t are the coefficients of a parabola $q(x) = rx^2 + sx + t$, where $q(\lambda_i) = \sqrt{\lambda_i}$. Thus, we have explicitly solved for

$$\begin{pmatrix} r \\ s \\ t \end{pmatrix} = \begin{pmatrix} \lambda_1^2 & \lambda_1 & 1 \\ \lambda_2^2 & \lambda_2 & 1 \\ \lambda_3^2 & \lambda_3 & 1 \end{pmatrix}^{-1} \begin{pmatrix} \sqrt{\lambda_1} \\ \sqrt{\lambda_2} \\ \sqrt{\lambda_3} \end{pmatrix}. \tag{14.215}$$

(14.33) Given the elegance of the expressions for R and U in two dimensions, it might be thought that something similar must hold in three dimensions, especially considering the form of the rotation as given in Eq. (14.55) – that is $R = I + a(F - F^T) + b(F - F^T)^2$ must hold for scalar a and b. This exercise is to show that though $F - F^T$ defines a rotation axis by virtue of being skew symmetric and by comparison to N in Eq. (14.55), it is in general not the axis of rotation of R in $F = RU$. We resort to a numerical case. Let

$$F = \begin{pmatrix} 1 & 1 & 0 \\ 0 & 1 & 1 \\ 0 & 0 & 1 \end{pmatrix},$$

$$R = \begin{pmatrix} 0.871 & 0.483 & -0.086 \\ -0.388 & 0.785 & 0.483 \\ 0.301 & -0.388 & 0.871 \end{pmatrix}, \quad U = \begin{pmatrix} 0.871 & 0.483 & -0.086 \\ 0.483 & 1.268 & 0.397 \\ -0.086 & 0.397 & 1.355 \end{pmatrix}. \tag{14.216}$$

Show that $F = RU$, $U^2 = F^T F$, and $RR^T = I$. Now show that the rotation axis defined by $F - F^T$, comparing with Eq. (14.55), is $(1, 0, 1)/\sqrt{2}$. Show, again using Eq. (14.55), that the rotation axis of R is $\vec{n}^T = (0.675, 0.300, 0.675)$. (As a check, confirm $\theta_a = 0.702$ radians.) Show that $R\vec{n} = \vec{n}$, as expected (why?). Another way to show this result is to note that if $\vec{n}$ is the rotation axis and it coincides with a coordinate axis, so that it points in the same direction as the inferred one from $F - F^T$ (like $\hat{z}$ and $F = \begin{pmatrix} 1 & 1 & 0 \\ 0 & 1 & 0 \\ 0 & 0 & 1 \end{pmatrix}$), if we do a rotation Q then $Q\vec{n}$ in general points in a different direction than the inferred direction from $QF - F^T Q^T$.

(14.34) To finish off this line of thought about the unlikely existence of a simple 3D expression for U, consider the following. If we use the Cayley–Hamilton theorem to derive Eq. (14.32), show that the key step in the 2D expression is that we can compute the trace of U using the fact that in two dimensions $2 \det(U) = \text{trace}^2(U) - \text{trace}(U^2)$ (show this). The key is that $\det(U) = \sqrt{\det(U^2)}$, so since we know U^2 we can explicitly compute $\text{trace}(U)$. What happens in three dimensions? Consider Eq. (3.39). The problem is that we don't know $\text{trace}(U^3)$. However, an examination of Eq. (3.39) suggests that $\text{trace}(\sigma^4)$ is dependent on the traces of the lower powers of σ, where σ is an arbitrary 3×3 matrix that need not be symmetric. Show that

$$6I_1(\sigma^4) - I_1^4(\sigma) + 6I_1^2(\sigma)I_1(\sigma^2) - 8I_1(\sigma)I_1(\sigma^3) - 3I_1^2(\sigma^2) = 0, \tag{14.217}$$

where $I_1(\sigma) = \text{trace}(\sigma)$. (Hint: the last line of Eq. 3.39 can be obtained through the Cayley–Hamilton theorem by taking the trace of the characteristic equation in terms of the tensor. Now multiply by σ before taking the trace, take the trace, and use the last line of Eq. 3.39 to remove the determinant.) Combine this expression with the last line in Eq. (3.39) to produce a quartic equation for $\text{trace}(U)$ in terms of $\text{trace}(U^2)$, $\text{trace}(U^4)$, and $\det(U)$, all of which we know. When this quartic is solved, we can write down U, which requires the inversion of a matrix, from Eq. (14.212).

(14.35) In two dimensions, with θ_U the angle the semimajor axis of the ellipse U makes with the X-axis and θ_V the angle the semimajor axis of the ellipse V (hence F) makes with the x-axis (x and X coincide), show that

$$\tan(\theta_U) = \frac{G + F_{22}'^2 - F_{11}^2 - (F_{21}^2 - F_{12}^2)}{2(F_{11}F_{12} + F_{21}F_{22})}, \quad \tan(\theta_V) = \frac{G + F_{22}^2 - F_{11}^2 + F_{21}^2 - F_{12}^2}{2(F_{11}F_{21} + F_{12}F_{22})},$$

where $G = \sqrt{(F_{11} + F_{22})^2 + (F_{12} - F_{21})^2}\sqrt{(F_{11} - F_{22})^2 + (F_{12} + F_{21})^2}$,

(14.218)

and then explicitly show using the tangent angle addition formula that $\tan(\theta_U + \theta_a) = \tan(\theta_V)$, where θ_a is the average rotation of F. (Hint: the eigenvalues of U and V are the same; use a symbolic manipulator.)

(14.36) Show that the rotation R obtained in the polar decomposition is not the same rotation that diagonalizes the large strain tensor E, specifically by considering a simple shear of 1. For what types of deformations does R diagonalize E?

(14.37) From the previous exercise we know that the rotation R and the matrix q that diagonalizes $B = FF^T$ are not the same. Let q diagonalize B so that $q^T Bq = \Lambda$, where Λ is the diagonal matrix of eigenvalues. Define $\sigma = q\Sigma^q q^T$ and $\Omega_B = \dot{q}q^T$. Show that

$$\mathring{\sigma}_q = \frac{d\sigma}{dt} - \Omega_B\sigma + \sigma\Omega_B = q\dot{\Sigma}^q q^T \qquad (14.219)$$

is a corotational stress rate. (Hint: first show that $B^* = QBQ^T$, where Q is the arbitrary frame rotation applied to the current configuration. Since the eigenvalues of B are not changed by the frame change, $q^T Bq = \Lambda = (q^*)^T B^* q^*$ implies $(q^*)^T Q = q^T$ gives $q^* = Qq$. Show that $\Omega_B^* = \dot{Q}Q^T + Q\Omega_B Q^T$. Finish by showing $\mathring{\sigma}_q^* = Q\mathring{\sigma}_q Q^T$.) Show that it preserves deviators – that is, trace$(\mathring{s}_q) = 0$. This corotational rate is unlike the others described in this chapter in that it does not connect to the original configuration. A corresponding strain rate is determined by choosing the Eulerian strain E^* (different use of $*$) as the strain of interest in the current configuration, so then $E^q = q^T E^* q$. Show $E^q = \frac{1}{2}(I - \Lambda^{-1})$, $\dot{E}^q = \frac{1}{2}\Lambda^{-2}\dot{\Lambda}$ (since Λ is diagonal), and $\dot{E}_q^* = \frac{1}{2}q\Lambda^{-2}\dot{\Lambda}q^T$. Thus, E^q is a principal strain and $\dot{E}^q$ is a principal strain rate. Computationally this approach is expensive, since it appears necessary to compute B, q, and Λ, as well as store the previous time step Λ to compute $\dot{\Lambda}$, so it seems this corotational rate provides little benefit.

(14.38) On page 583 it is stated that the arithmetic average of the Cotter–Rivlin and Oldroyd rates, giving the Jaumann corotational rate, is the same in concept as the others. In two dimensions, show that the Jaumann rate cannot be written as $\mathring{\sigma}_J = A\dot{\Sigma}^J A^T$ for some 2×2 A by showing that it cannot be done for $F = \begin{pmatrix} a & 0 \\ 0 & b \end{pmatrix}$, where $a \neq b$. However, the real issue is not what the form is but rather whether the mapping $\sigma = \frac{1}{2}(F^{-T}\Sigma^J F^{-1} + F\Sigma^J F^T)$ is invertible. Show $2F^T\sigma F^{-T} = \Sigma^J C^{-1} + C\Sigma^J$, where $C = F^T F$. The left-hand side is invertible since F is invertible. For the right-hand side, show that in two dimensions you can set up a 3×3 matrix A, where $A(\Sigma_{11}^J\ \Sigma_{22}^J\ \Sigma_{12}^J)^T =$ corresponding $2F^T\sigma F^{-T}$ vector transpose, where

$$A = \begin{pmatrix} C_{11} + C_{11}^{-1} & 0 & C_{12} + C_{12}^{-1} \\ 0 & C_{22} + C_{22}^{-1} & C_{12} + C_{12}^{-1} \\ \frac{1}{2}(C_{12} + C_{12}^{-1}) & \frac{1}{2}(C_{12} + C_{12}^{-1}) & \frac{1}{2}(C_{11} + C_{11}^{-1} + C_{22} + C_{22}^{-1}) \end{pmatrix}. \qquad (14.220)$$

Show

$$\det(A) = \frac{1}{2}\left\{\det(C) + \frac{1 + C_{ij}C_{ij}}{\det(C)}\right\}\text{trace}(C + C^{-1}). \qquad (14.221)$$

Show this is not zero; hence the mapping is invertible.

(14.39) Show that $(U^{-2})^{\cdot} = -2F^{-1}DF^{-T}$ and use it to show $(\dot{E}^*)_{GN} = V^{-1}DV^{-1}$. Using the definition of the Jaumann rate and $\dot{E}^*$ show that $(\overset{\circ}{E}{}^*)_J = D - DE^* - E^*D = \frac{1}{2}(V^{-2}D + DV^{-2})$ (results shown in Eq. 14.172). Draw the corresponding commutative diagram for $(\dot{E}^*)_{GN}$, identifying all four tensors and four mappings, the lower left tensor being E^*.

(14.40) In the literature you will find other large measures of strain defined. For example, with U the right stretch,

$$E^{(m)} = \begin{cases} \frac{1}{m}(U^m - I), & m \neq 0, \\ \ln(U), & m = 0, \end{cases} \qquad (14.222)$$

where the logarithm of a matrix is found by diagonalizing it, taking the logarithm of the eigenvalues along the diagonal term by term, then rotating back into the original configuration. $E^{(2)}$ in this notation is our Lagrangian or Green strain E. Is the Eulerian strain E^* equal to $E^{(-2)}$? (Ans: no; $E^{*(m)} = RE^{(m)}R^T = \frac{1}{m}(V^m - I)V^{-2}$.) Show that all these strains $E^{(m)}$ can be written as a function of $E = E^{(2)}$ by explicitly finding the relation – that is, finding g, where $E^{(m)} = g(E)$. Thus, strain is really the same in every case; we are just choosing to represent it differently. In particular, if $\sigma = Rf(E^{(m)})R^T$, then it can also be written $\sigma = Rh(E)R^T$, where E is the Lagrangian strain and $h(E) = f(g(E))$. Using a Taylor series expansion about $U = I$, show that all these strain measures are similar when the principal stretches are close to 1 (i.e., U close to I). These other expressions for strain would be chosen for convenience in working with them or because they provided a simpler functional relationship between stress-strain data, though notice that we can always then write the stress as a function of E. From a measurable quantities perspective, what is measured in the laboratory is initial gage length L_0 and final gage length L, and their ratio is the stretch. Show that for m an integer, for a uniaxial stretch the nonzero component of $E^{(m)}$ is $(\lambda - 1)f(\lambda)/m$, where $f(1)/m = 1$ (explicitly determine f). One might argue that the directly measured strain is most similar to $E^{(1)}$, since for uniaxial loading the principal loading direction component of $E^{(1)} = \lambda - 1 = (L - L_0)/L_0$. A challenge in experiments is to obtain homogeneous (the same everywhere) deformation over the gage length, though modern optical measurement methods try to overcome this difficulty by determining the surface displacements for a region of interest, rather than just the ends, allowing computation of local surface strains, from which interior strains are inferred. The other laboratory measured quantity is force, and for large deformations it is especially important to compute the "true stress" – that is, to adjust the force by the current specimen cross-sectional area to determine the stress, and again optical methods are convenient in determining the cross-sectional area. As to whether there is a "better" strain, the underlying issue is that our measure of length, based on Pythagoras' theorem, is always the same: $ds^2 = d\vec{x}^T d\vec{x} \Rightarrow U = \sqrt{F^T F}$ is the sum of the squares of the lengths in each direction, and our notion of strain is to compare deformed length ds to original length dS. Is there a different way to measure distance, so that you might actually end up with a "different" strain? (You might look up Finsler geometry.) Is there a reason to do so?

(14.41) The preceding exercise sets the stage for a discussion of strain in generalized continua, alluded to in the footnote on page 14 and in Section 2.12. Suppose we choose $E^{(1)} = U - I$ as our measure of strain. Using the polar decomposition $F = RU$, show that the strain can be written $E^{(1)} = R^T F - I = F^T R - I$, where the two expressions are transposes of each other. Now, the Cosserat brothers suggested that the rotation be unrelated to the rotation derived from F, in other words, replace $R(\vec{\theta}_a)$ with $R(\vec{\phi})$, where $\vec{\phi}$ is a rotation obtained in some other fashion, often referred to as the *microrotation* $\vec{\phi}$. Define strains $\epsilon^A = R(\vec{\phi})^T F - I$ and its transpose $\epsilon^B = F^T R(\vec{\phi}) - I$. These strains may no longer be symmetric. Show that for small deformations these two expressions lead to the two linearized forms

$$\epsilon_{ij}^A = \frac{\partial u_i}{\partial x_j} + \varepsilon_{ijk}\phi_k \quad \text{and} \quad \epsilon_{ij}^B = \frac{\partial u_j}{\partial x_i} - \varepsilon_{ijk}\phi_k, \qquad (14.223)$$

corresponding to the approximations $\epsilon^A = \hat{F} - \omega$ and $\epsilon^B = \hat{F}^T + \omega$, where now $\vec{\omega} = \frac{1}{2}\text{curl}(\vec{u})$ is replaced by $\vec{\phi}$ and therefore $\omega = -\varepsilon_{ijk}\omega_k$ by $-\varepsilon_{ijk}\phi_k$ (see Eqs. 14.55, 14.67). The B form is typically used in Cosserat elasticity. A constitutive model for the Cauchy stress is

$$\sigma_{ij} = \lambda \epsilon_{kk}^B \delta_{ij} + (\mu + \kappa)\epsilon_{ij}^B + \mu \epsilon_{ji}^B, \qquad (14.224)$$

with λ, μ, and κ constants. Show that the equation can be written in terms of the small strain ε, the small rotation $\vec{\omega}$, and the microrotation $\vec{\phi}$ as

$$\sigma_{ij} = \lambda \varepsilon_{kk}\delta_{ij} + (2\mu + \kappa)\varepsilon_{ij} + \kappa \varepsilon_{ijk}(\omega_k - \phi_k), \qquad (14.225)$$

showing that asymmetry of the Cauchy stress occurs when the microrotation $\vec{\phi}$ does not agree with the small rotation $\vec{\omega}$ from F (in this context called the macrorotation). It also shows that μ may not be the classical shear modulus. The relation between the rotation and the couple stress is obtained through the definition of the *wryness tensor*, which is the spatial gradient of $\vec{\phi}$, measuring how the rotation changes in spatial directions,

$$\kappa_{ij} = \partial \phi_i / \partial x_j \quad \text{and} \quad \kappa = \hat{F}(\vec{x}, \vec{\phi}) = (\phi_{i;j}) \quad \text{(Appendix A notation)}. \qquad (14.226)$$

For Cosserat elasticity a constitutive equation for the couple stress is

$$m_{ij} = \alpha \kappa_{kk}\delta_{ij} + \beta \kappa_{ij} + \gamma \kappa_{ji}, \qquad (14.227)$$

where α, β, and γ are constants which have a length scale squared in them (compared to the other elastic moduli). The equation for the asymmetric Cauchy stress as well as the couple stress, with associated boundary conditions, are required to solve a problem.

Conservation Laws and Curvilinear Coordinates

This appendix reviews indicial notation, the conservation laws, and the various differential expressions in cylindrical and spherical coordinates. Central to understanding curvilinear coordinate systems is determining the derivatives of the basis vectors.

A.1. Indicial Notation

Throughout the book we use indicial notation, which means that for repeated indices there is a summation, from 1 to 3. For example, the inner product (also called the dot product) between two vectors $\vec{a}$ and $\vec{b}$ is

$$\vec{a} \cdot \vec{b} = a_1 b_1 + a_2 b_2 + a_3 b_3 = a_i b_i. \tag{A.1}$$

The far right side of the equation has a summation from $i = 1$ to 3 since the index is repeated. Similarly, the multiplication of a matrix A and a vector $\vec{b}$ can be written

$$\vec{a} = A\vec{b} \quad \Rightarrow \quad a_i = A_{i1}b_1 + A_{i2}b_2 + A_{i3}b_3 = A_{ij}b_j, \tag{A.2}$$

where there is a summation over the index j for index $i = 1$, 2 or 3. The index i has an implied relation to the basis of the coordinate system. The vector nature can be explicitly called out by writing

$$\begin{aligned} \vec{a} = A\vec{b} \quad \Rightarrow \quad a_i \hat{e}_{x_i} &= a_1 \hat{e}_{x_1} + a_2 \hat{e}_{x_2} + a_3 \hat{e}_{x_3} \\ &= A_{ij}b_j \hat{e}_{x_i} = A_{11}b_1 \hat{e}_{x_1} + A_{12}b_2 \hat{e}_{x_1} + \cdots + A_{33}b_3 \hat{e}_{x_3}, \end{aligned} \tag{A.3}$$

where there is an implied summation over both i and j from 1 to 3, so the far right-hand side has 9 terms. The $\hat{e}_{x_i}$ are the unit (length 1) coordinate basis vectors.

We will use notation where the vector components in an assumed basis for a vector are listed as a column – that is,

$$\vec{v} = v_i \hat{e}_i = \begin{pmatrix} \hat{e}_{x_1} & \hat{e}_{x_2} & \hat{e}_{x_3} \end{pmatrix} \begin{pmatrix} v_1 \\ v_2 \\ v_3 \end{pmatrix}. \tag{A.4}$$

The row to column rule for left to right multiplication is assumed to hold. Here we have extended the notation in that the row vector objects are unit vectors and not numbers. Sometimes we have two different coordinate systems and we want to transform vectors (such as velocity or acceleration) or tensors (such as stress or strain) between them. We assume that we have a unit coordinate basis for each coordinate system, one represented by x_i and one represented by y_i. The transformation matrix of direction cosines is given by

$$(_xQ_y)_{ij} = \hat{e}_{x_i} \cdot \hat{e}_{y_j}, \tag{A.5}$$

which in matrix notation looks like

$$_xQ_y = \begin{pmatrix} \hat{e}_{x_1} \cdot \hat{e}_{y_1} & \hat{e}_{x_1} \cdot \hat{e}_{y_2} & \hat{e}_{x_1} \cdot \hat{e}_{y_3} \\ \hat{e}_{x_2} \cdot \hat{e}_{y_1} & \hat{e}_{x_2} \cdot \hat{e}_{y_2} & \hat{e}_{x_2} \cdot \hat{e}_{y_3} \\ \hat{e}_{x_3} \cdot \hat{e}_{y_1} & \hat{e}_{x_3} \cdot \hat{e}_{y_2} & \hat{e}_{x_3} \cdot \hat{e}_{y_3} \end{pmatrix} \quad \Rightarrow \quad \begin{pmatrix} \hat{e}_{x_1} \\ \hat{e}_{x_2} \\ \hat{e}_{x_3} \end{pmatrix} = {}_xQ_y \begin{pmatrix} \hat{e}_{y_1} \\ \hat{e}_{y_2} \\ \hat{e}_{y_3} \end{pmatrix}. \quad (A.6)$$

The leading subscript x and the trailing subscript y on Q are because one of the hardest things about rotation matrices is remembering which way the transformation works. Since the two bases use orthonormal basis vectors, Q is orthonormal so that $_xQ_y^{-1} = {}_xQ_y^T = {}_yQ_x$. As written,

$$(v_x)_i \hat{e}_{x_i} = (v_x)_i ({}_xQ_y)_{ij} \hat{e}_{y_j} = (v_y)_j \hat{e}_{y_j}. \quad (A.7)$$

When the coordinates are written as columns in the respective basis, this equation is

$$\vec{v}_x^T \, _xQ_y = \vec{v}_y^T \quad \Rightarrow \quad \vec{v}_y = {}_xQ_y^T \vec{v}_x = {}_yQ_x \vec{v}_x. \quad (A.8)$$

This equation defines the transformation between coordinates (components) in this basis, with the reverse transformation

$$\vec{v}_x = {}_xQ_y \vec{v}_y. \quad (A.9)$$

In a similar fashion, two-dimensional tensors such as stress transform as

$$_y\sigma_y = {}_x Q_y^T \, _x\sigma_x \, _xQ_y \quad \text{and} \quad _x\sigma_x = {}_x Q_y \, _y\sigma_y \, _xQ_y^T. \quad (A.10)$$

A.2. The Three Conservation Laws

In this section we derive the forms used in the text of the three conservation laws – conservation of mass, conservation of momentum, and conservation of energy. The derivations will be in the Cartesian coordinate system. In this section, we are interested in a volume V of material and the surface that surrounds the volume, which we will denote with S. We use two mathematical results. One is the divergence theorem, which in Cartesian coordinates can be written for a scalar function f and for one variable x_i as

$$\int_V \frac{\partial f}{\partial x_i} dvol = \int_S f n_i da, \quad (A.11)$$

where again the surface S is the boundary of the volume V, n_i is the i-th component of the unit normal to the surface $\vec{n}$, and da is the surface area differential. Notice that this result is simply the generalization of the fundamental theorem of the calculus

$$\int_a^b \frac{\partial f}{\partial x} dx = f(b) - f(a) \quad (A.12)$$

to a volume instead of a line segment. By inserting f_i rather than a scalar function f and summing over i, the more common form of the divergence theorem is produced, namely

$$\int_V \text{div}(\vec{f}) dvol = \int_V \frac{\partial f_i}{\partial x_i} dvol = \int_S f_i n_i da = \int_S \vec{f} \cdot \vec{n} da = \int_S \vec{f} \cdot d\vec{a}. \quad (A.13)$$

The second result we use is due to Leibniz, and has to do with differentiating an integral. In one dimension this result is

$$\frac{d}{dt} \int_{a(t)}^{b(t)} f(t, s) ds = \int_{a(t)}^{b(t)} \frac{\partial f(t, s)}{\partial t} ds + f(t, b(t)) \frac{db}{dt} - f(t, a(t)) \frac{da}{dt}. \quad (A.14)$$

We generalize this result as follows. If a volume $V(t)$ of interest changes with time and is contained within the surface $S(t)$ described by $\vec{x}_s(t)$, then for a function f the integral

$$\int_{V(t)} f(t, \vec{x}) dvol = \int_{V(t)} f(t, x_1, x_2, x_3) dx_1 dx_2 dx_3 \qquad (A.15)$$

is strictly a function of time. We differentiate the integral to obtain

$$\begin{aligned}
\frac{d}{dt} \int_{V(t)} f(t, \vec{x}) dvol &= \int_{V(t)} \frac{\partial f(t, \vec{x})}{\partial t} dvol + \int_{S(t)} f(t, \vec{x}) \frac{d\vec{x}_s}{dt} \cdot d\vec{a} \\
&= \int_{V(t)} \frac{\partial f(t, \vec{x})}{\partial t} dvol + \int_{S(t)} f(t, \vec{x}) \vec{v} \cdot d\vec{a},
\end{aligned} \qquad (A.16)$$

where we have used the assumption that the surface bounds the volume and is determined by the volume's motion, and thus the velocity is given by $\vec{v} = d\vec{x}_s/dt$.

Conservation of mass is the assumption that mass is neither created nor destroyed. We begin by assuming we have an arbitrary amount of mass of density $\rho(\vec{x}, t)$ that is enclosed by a time-dependent volume $V(t)$. Thus,

$$\int_{V(t)} \rho \, dvol = \text{constant} \qquad (A.17)$$

and differentiating using Eq. (A.16) gives

$$\frac{d}{dt} \int_{V(t)} \rho \, dvol = \int_{V(t)} \frac{\partial \rho}{\partial t} dvol + \int_{S(t)} \rho \vec{v} \cdot d\vec{a} = 0. \qquad (A.18)$$

An application of the divergence theorem Eq. (A.13) gives

$$\int_{V(t)} \frac{\partial \rho}{\partial t} dvol + \int_{V(t)} \frac{\partial(\rho v_i)}{\partial x_i} dvol = 0. \qquad (A.19)$$

Since the volume is arbitrary, this equation implies

$$\frac{\partial \rho}{\partial t} + \frac{\partial(\rho v_i)}{\partial x_i} = 0. \qquad (A.20)$$

The second term on the left can be expanded, and the definition of the total derivative or material derivative (Chapter 2) can be used to write

$$\frac{\partial \rho}{\partial t} + \frac{\partial(\rho v_i)}{\partial x_i} = \frac{\partial \rho}{\partial t} + v_i \frac{\partial \rho}{\partial x_i} + \rho \frac{\partial v_i}{\partial x_i} = \frac{D\rho}{Dt} + \rho \frac{\partial v_i}{\partial x_i} = 0. \qquad (A.21)$$

These equations are the two forms of the conservation of mass,

$$\text{Conservation of mass:} \quad \frac{\partial \rho}{\partial t} + \text{div}(\rho \vec{v}) = 0, \quad \frac{D\rho}{Dt} + \rho \, \text{div}(\vec{v}) = 0, \qquad (A.22)$$

written in general coordinates.

The conservation of mass provides for a very interesting result, which simplifies the momentum and energy conservation derivations. The result is referred to as the Reynolds transport theorem [138], and it says that for a function f and a volume $V(t)$,

$$\frac{d}{dt} \int_{V(t)} \rho f \, dvol = \int_{V(t)} \rho \frac{Df}{Dt} dvol, \qquad (A.23)$$

where the volume encloses a region of material and the velocity at the surface of the region is given by the material velocity. To see that this theorem holds, perform the differentiation using Eq. (A.16), apply the divergence theorem, expand, use the mass balance, and use the definition of the material derivative:

$$\frac{d}{dt}\int_{V(t)} \rho f \, dvol = \int_{V(t)} \frac{\partial(\rho f)}{\partial t} dvol + \int_{S(t)} \rho f \, \vec{v} \cdot d\vec{a}$$

$$= \int_{V(t)} \frac{\partial(\rho f)}{\partial t} dvol + \int_{V(t)} \frac{\partial(\rho f v_j)}{\partial x_j} dvol$$

$$= \int_{V(t)} \left\{ \rho\frac{\partial f}{\partial t} + f\left(\frac{\partial \rho}{\partial t} + \frac{\partial(\rho v_j)}{\partial x_j}\right) + \rho v_j\frac{\partial f}{\partial x_j} \right\} dvol \quad (A.24)$$

$$= \int_{V(t)} \left\{ \rho\frac{\partial f}{\partial t} + \rho v_j\frac{\partial f}{\partial x_j} \right\} dvol$$

$$= \int_{V(t)} \rho\frac{Df}{Dt} dvol.$$

We will use this result in the next two derivations. (It should be pointed out that the velocity relationship between the volume and the surface is required for this argument. This theorem does not hold for arbitrary volume changes with time such as are employed by some arbitrary Lagrangian-Eulerian (ALE) schemes.)

For the conservation of momentum, we begin with Newton's law of $\vec{F} = m\vec{a}$ and consider an arbitrary volume of material enclosed by a surface. The force acting on the volume is given by the integral of the traction over the surface and the contribution of the body force,

$$\vec{F} = \int_{S(t)} \vec{t} \, da + \int_{V(t)} \rho\vec{b} dvol. \quad (A.25)$$

With the relation of the traction to the stress tensor as $t_i = n_j\sigma_{ji}$, an application of the divergence theorem yields

$$\vec{F} = \int_{V(t)} \frac{\partial \sigma_{ji}}{\partial x_j}\hat{e}_{x_i} da + \int_{V(t)} \rho\vec{b} dvol. \quad (A.26)$$

For the change in momentum of the material side of the equation, the momentum $\vec{p}$ is given by

$$\vec{p} = \int_{V(t)} \rho\vec{v} dvol \quad (A.27)$$

and use of the Reynolds transport theorem on each component of the velocity vector yields

$$\frac{d\vec{p}}{dt} = \frac{d}{dt}\int_{V(t)} \rho\vec{v} dvol = \int_{V(t)} \rho\frac{D\vec{v}}{Dt} dvol. \quad (A.28)$$

Setting the forces $\vec{F}$ equal to the change in momentum $d\vec{p}/dt$ and recognizing the integral is over an arbitrary volume yields the conservation of momentum

$$\frac{\partial \sigma_{ji}}{\partial x_j} + \rho b_i = \rho\frac{Dv_i}{Dt}. \quad (A.29)$$

The left term of the left-hand side is often viewed as the gradient of a tensor and sometimes denoted $\vec{\nabla} \cdot \sigma$, which allows a general coordinate statement of the conservation of momentum,

$$\text{Conservation of momentum:} \quad \vec{\nabla} \cdot \sigma + \rho\vec{b} = \rho\frac{D\vec{v}}{Dt}. \quad (A.30)$$

In Cartesian coordinates, notice that

$$\vec{\nabla} \cdot \sigma = \mathrm{div} \begin{pmatrix} \sigma_{xx} \\ \sigma_{yx} \\ \sigma_{zx} \end{pmatrix} \hat{e}_x + \mathrm{div} \begin{pmatrix} \sigma_{xy} \\ \sigma_{yy} \\ \sigma_{zy} \end{pmatrix} \hat{e}_y + \mathrm{div} \begin{pmatrix} \sigma_{xz} \\ \sigma_{yz} \\ \sigma_{zz} \end{pmatrix} \hat{e}_z, \tag{A.31}$$

or that the vector component of each direction is given by the divergence of the traction components that point in the same direction (not tractions on the same face, though for a symmetric stress tensor they are the same). In other coordinate systems there are additional terms in $\vec{\nabla} \cdot \sigma$, since there are derivatives of the unit coordinate vectors, and they do not always point in the same direction.

Finally we come to the energy balance. The energy within a given amount of material is comprised of internal energy (here E is the specific energy, or energy per unit mass) and kinetic energy,

$$\tilde{e}_V = \int_{V(t)} \tilde{e} dvol = \int_{V(t)} \left(\rho E + \frac{1}{2} \rho v^2 \right) dvol. \tag{A.32}$$

The use of $\tilde{e}_V$ is to be consistent with the use in Chapter 2 of $\tilde{e}$ to represent the total energy per unit volume. The change in time of the energy is, using the Reynolds transport theorem, Eq. (A.23),

$$\frac{d\tilde{e}_V}{dt} = \frac{d}{dt} \int_{V(t)} \left(\rho E + \frac{1}{2} \rho v^2 \right) dvol = \int_{V(t)} \left\{ \rho \frac{DE}{Dt} + \frac{1}{2} \rho \frac{D(v_i v_i)}{Dt} \right\} dvol. \tag{A.33}$$

There are four sources of energy that can be supplied to our arbitrary volume of material. Two are on the boundary. First, there is a power input to the arbitrary volume of material found by the continuum analog of force times velocity. Second, there can be heat conduction, which allows energy to enter or leave without any motion of material. We will denote the heat flux by $\vec{q}$ (positive when flowing out of the volume). The third source of power input is work done by body forces, which operate over the volume of material. The fourth source of energy is internal energy released within the volume, such as through chemical reactions, which is an energy source at a given rate r. The energy per unit time (power) input terms to the arbitrary volume are

$$\frac{d\tilde{e}_V}{dt} = \int_{S(t)} \vec{t} \cdot \vec{v} da - \int_{S(t)} \vec{q} \cdot \vec{n} da + \int_{V(t)} \rho \vec{b} \cdot \vec{v} dvol + \int_{V(t)} \rho r dvol. \tag{A.34}$$

Use of the divergence theorem and the momentum balance gives

$$\begin{aligned}
\frac{d\tilde{e}_V}{dt} &= \int_{V(t)} \left\{ \frac{\partial(\sigma_{ji} v_i)}{\partial x_j} - \frac{\partial q_i}{\partial x_i} + \rho b_i v_i + \rho r \right\} dvol \\
&= \int_{V(t)} \left\{ \sigma_{ji} \frac{\partial v_i}{\partial x_j} + v_i \left(\frac{\partial \sigma_{ji}}{\partial x_j} + \rho b_i \right) - \frac{\partial q_i}{\partial x_i} + \rho r \right\} dvol \\
&= \int_{V(t)} \left\{ \sigma_{ji} \frac{\partial v_i}{\partial x_j} + v_i \rho \frac{Dv_i}{Dt} - \frac{\partial q_i}{\partial x_i} + \rho r \right\} dvol \\
&= \int_{V(t)} \left\{ \sigma_{ji} \frac{\partial v_i}{\partial x_j} + \frac{1}{2} \rho \frac{D(v_i v_i)}{Dt} - \frac{\partial q_i}{\partial x_i} + \rho r \right\} dvol.
\end{aligned} \tag{A.35}$$

This expression provides for the change in energy of the system.

When the time rate of change of energy (Eq. A.33) is set equal to the supplied energy rate (power) (Eq. A.35), recognizing the volumes are arbitrary, the energy balance equation is obtained

$$\rho\frac{DE}{Dt} = \sigma_{ji}\frac{\partial v_i}{\partial x_j} - \frac{\partial q_i}{\partial x_i} + \rho r, \tag{A.36}$$

which can be written,

$$\text{Conservation of energy:} \qquad \rho\frac{DE}{Dt} = \sigma_{ij}D_{ij} - \text{div}(\vec{q}) + \rho r \tag{A.37}$$

in general coordinates if the stress tensor is symmetric (Exercise 2.3).

The angular momentum is defined as

$$\text{angular momentum} = \int_V \rho\vec{r}\times\vec{v}\,dvol, \tag{A.38}$$

where $\vec{r} = \vec{x}-\vec{x}_0$ for a fixed location $\vec{x}_0$. In classical mechanics of particles it is shown in the event the particles interact through forces acting in the direction of the line connecting them that the change in angular momentum equals the torque applied to the system (the torque is $\vec{r}\times\vec{F}$, where $\vec{F}$ is the applied force). (Conservation of angular momentum is not a separate conservation law but follows from linear momentum conservation with the assumption on the forces between particles.) The continuum mechanics analogue is

$$\int_S \vec{r}\times\vec{t}\,dS + \int_V \rho\vec{r}\times\vec{b}\,dvol = \frac{d}{dt}\int_V \rho\vec{r}\times\vec{v}\,dvol. \tag{A.39}$$

Reynolds transport theorem applied to the right-hand side yields

$$\frac{d}{dt}\int_V \rho\vec{r}\times\vec{v}\,dvol = \int_V \rho\frac{d}{dt}(\vec{r}\times\vec{v})\,dvol$$

$$= \int_V \rho\left(\vec{v}\times\vec{v} + \vec{r}\times\frac{D\vec{v}}{Dt}\right)dvol = \int_V \rho\vec{r}\times\frac{D\vec{v}}{Dt}\,dvol. \tag{A.40}$$

Applying the divergence theorem to the surface integral term in Eq. (A.39) yields

$$\int_S \vec{r}\times\vec{t}\,dS = \int_S \varepsilon_{ijk}r_j t_k\hat{e}_i\,dS = \int_S \varepsilon_{ijk}r_j n_\ell\sigma_{\ell k}\hat{e}_i\,dS$$

$$= \int_V \varepsilon_{ijk}\frac{\partial(r_j\sigma_{\ell k})}{\partial x_\ell}\hat{e}_i\,dvol = \int_V \varepsilon_{ijk}\left(\delta_{j\ell}\sigma_{\ell k} + r_j\frac{\partial\sigma_{\ell k}}{\partial x_\ell}\right)\hat{e}_i\,dvol \tag{A.41}$$

$$= \int_V \left(\varepsilon_{ijk}\sigma_{jk}\hat{e}_i + \vec{r}\times(\vec{\nabla}\cdot\sigma)\right)dvol,$$

where ε_{ijk} is the permutation symbol ($\varepsilon_{ijk} = 1$ if ijk is an even permutation, $\varepsilon_{ijk} = -1$ if ijk is an odd permutation, $\varepsilon_{ijk} = 0$ if ijk has repeated indices). If we place these two results back into Eq. (A.39) and use the momentum conservation equation Eq. (A.30), nearly all the terms fall out, leaving

$$\int_V \varepsilon_{ijk}\sigma_{jk}\hat{e}_i\,dvol = 0. \tag{A.42}$$

With an arbitrary volume V, this equation yields three equations,

$$\sigma_{12} = \sigma_{21}, \quad \sigma_{13} = \sigma_{31}, \quad \sigma_{23} = \sigma_{32}, \tag{A.43}$$

or that the Cauchy stress tensor is symmetric, $\sigma_{ij} = \sigma_{ji}$. Thus, angular momentum conservation implies symmetry of the Cauchy stress tensor.

A.3. Curvilinear Geometry and Differential Operators

We will always be working in Euclidean three-dimensional space, but we have different ways of representing the geometry through different coordinate systems (i.e., different parameterizations of space). Given a geometry, it is possible to compute the metric tensor that describes how distance is computed for the specific parameterization. We write the location in space as $\vec{x} = \vec{x}(y_1, y_2, y_3)$, where (y_1, y_2, y_3) are the specific parameters. Given $\vec{x}$, the metric tensor is

$$g_{ij}(\vec{x}) = \frac{\partial \vec{x}}{\partial y_i} \cdot \frac{\partial \vec{x}}{\partial y_j}. \tag{A.44}$$

The inner (dot) product is taken between the two vectors and the vector notion is intentional, since if the unit coordinate vectors do not always point in the same direction (as occurs in cylindrical and spherical coordinates) it is necessary to include the coordinate unit vectors in the differentiation. With the metric tensor, the length of a line segment can be computed in terms of the parameters as

$$(ds)^2 = g_{ij} dy_i dy_j. \tag{A.45}$$

We will only work with parameterizations that describe orthogonal coordinate systems, so all the off-diagonal terms in the metric tensor are zero and

$$(ds)^2 = g_{11}(dy_1)^2 + g_{22}(dy_2)^2 + g_{33}(dy_3)^2. \tag{A.46}$$

For convenience we define

$$h_i = \sqrt{g_{ii}} \qquad \text{(no sum on } i) \tag{A.47}$$

and the length of a line segment can be written as

$$(ds)^2 = (h_1 dy_1)^2 + (h_2 dy_2)^2 + (h_3 dy_3)^2. \tag{A.48}$$

This equation implies that there is a natural choice of unit vectors that are orthogonal to each other, in particular

$$d\vec{x} = h_1 \hat{e}_{y_1} dy_1 + h_2 \hat{e}_{y_2} dy_2 + h_3 \hat{e}_{y_3} dy_3, \tag{A.49}$$

and $(ds)^2 = d\vec{x} \cdot d\vec{x}$. We observe that in each direction, the change in length is given by $h_i dy_i$ (no sum on i). The volume differential is given by

$$dvol = \sqrt{\det(g)} dy_1 dy_2 dy_3 = h_1 h_2 h_3 dy_1 dy_2 dy_3. \tag{A.50}$$

The volume element is a deformed cube with side lengths h_1, h_2, and h_3. If $\vec{x}$ is known, then we can explicitly determine the unit basis vectors, since

$$d\vec{x} = \frac{\partial \vec{x}}{\partial y_i} dy_i, \tag{A.51}$$

and, by comparing terms in Eqs. (A.49) and (A.51), the coordinate unit vectors are defined as

$$\hat{e}_{y_i} = \frac{1}{h_i} \frac{\partial \vec{x}}{\partial y_i} = \frac{\partial \vec{x}/\partial y_i}{|\partial \vec{x}/\partial y_i|} = \frac{\partial \vec{x}/\partial y_i}{\sqrt{(\partial \vec{x}/\partial y_i) \cdot (\partial \vec{x}/\partial y_i)}} = \frac{\partial \vec{x}/\partial y_i}{\sqrt{g_{ii}}} \quad \text{(no sum on } i). \tag{A.52}$$

By construction, these unit vectors are orthonormal (meaning they are orthogonal and of length 1), a fact that will be used often in the subsequent development and allows the inner product to always be given by Eq. (A.1).

Since the unit basis vectors do not always point in the same direction, the various partial derivatives of the basis vectors are nonzero. Since the derivative of a vector is a vector in our three-dimensional Euclidean space, it can be written as a linear combination of the three basis vectors. Hence, we may write

$$\frac{\partial}{\partial y_i}\hat{e}_{y_j} = \gamma_{ij}^k \hat{e}_{y_k}. \tag{A.53}$$

The coefficients γ_{ij}^k are called Christoffel symbols.[1] Near the end of this section we will compute the γ_{ij}^k in terms of the h_i. The result (Eq. A.133) is

$$\gamma_{ii}^j = -\frac{1}{h_j}\frac{\partial h_i}{\partial y_j} \quad \text{(no sum on } i, j), \quad i \neq j,$$

$$\text{all other } \gamma_{ij}^k = 0 \tag{A.54}$$

$$\gamma_{ji}^j = \frac{1}{h_i}\frac{\partial h_j}{\partial y_i} \quad \text{(no sum on } i, j), \quad i \neq j,$$

$$\text{or} \quad \gamma_{ij}^k = \frac{1}{h_j}\frac{\partial h_i}{\partial y_j}\delta_{ik} - \frac{1}{h_k}\frac{\partial h_i}{\partial y_k}\delta_{ij} \quad \text{(no sum on } i, j, k).$$

The set of the 27 γ_{ij}^k are called the *connection* since they describe how the nearby basis vectors are connected together.

As a first use of our definitions, we compute acceleration. Equation (A.52) says that the velocity of a particle is given by

$$\vec{v} = \frac{d\vec{x}(\vec{y}(t))}{dt} = h_1\frac{dy_1}{dt}\hat{e}_{y_1} + h_2\frac{dy_2}{dt}\hat{e}_{y_2} + h_3\frac{dy_3}{dt}\hat{e}_{y_3} = v_{y_1}\hat{e}_{y_1} + v_{y_2}\hat{e}_{y_2} + v_{y_3}\hat{e}_{y_3}, \tag{A.55}$$

so $v_{y_i} = \frac{1}{h_i}\frac{dy_i}{dt}$. (See the specifics in the cylindrical coordinate section for more discussion of the velocity.) There are two perspectives, the Lagrangian and the Eulerian. In the Lagrangian frame, a particle location is identified by its initial location and time, $\vec{x}(y_1(0), y_2(0), y_3(0), t)$. Because a specific set of y_i describe the location $\vec{x}$, for a Lagrangian particle, the y_i are functions of time. Since the basis vectors $\hat{e}_{y_i}$ are functions of y_i, they are also functions of time for a Lagrangian particle. Hence the Lagrangian acceleration is given by

$$\vec{a}_L = \frac{\partial \vec{v}}{\partial t} = \frac{\partial v_{y_i}}{\partial t}\hat{e}_{y_i} + v_{y_i}\frac{\partial \hat{e}_{y_i}}{\partial t} = \frac{\partial v_{y_i}}{\partial t}\hat{e}_{y_i} + v_{y_i}\frac{\partial \hat{e}_{y_i}}{\partial y_j}\frac{\partial y_j}{\partial t}$$

$$= \frac{\partial v_{y_i}}{\partial t}\hat{e}_{y_i} + v_{y_i}\sum_j \gamma_{ji}^k \hat{e}_{y_k}\frac{v_{y_j}}{h_j} \tag{A.56}$$

(the sum with respect to j is explicitly stated because there are more than two j indices in the rightmost product terms and hence the Einstein summation convention does not apply; there are still implied sums with respect to i and k in that product). Inserting the values of the Christoffel symbols gives

$$a_{y_1} = \frac{\partial v_{y_1}}{\partial t} + \frac{v_{y_1}v_{y_2}}{h_1 h_2}\frac{\partial h_1}{\partial y_2} + \frac{v_{y_1}v_{y_3}}{h_1 h_3}\frac{\partial h_1}{\partial y_3} - \frac{v_{y_2}^2}{h_2 h_1}\frac{\partial h_2}{\partial y_1} - \frac{v_{y_3}^2}{h_3 h_1}\frac{\partial h_3}{\partial y_1},$$

$$a_{y_2} = \frac{\partial v_{y_2}}{\partial t} + \frac{v_{y_2}v_{y_1}}{h_2 h_1}\frac{\partial h_2}{\partial y_1} + \frac{v_{y_2}v_{y_3}}{h_2 h_3}\frac{\partial h_2}{\partial y_3} - \frac{v_{y_1}^2}{h_1 h_2}\frac{\partial h_1}{\partial y_2} - \frac{v_{y_3}^2}{h_3 h_2}\frac{\partial h_3}{\partial y_2}, \tag{A.57}$$

$$a_{y_3} = \frac{\partial v_{y_3}}{\partial t} + \frac{v_{y_3}v_{y_1}}{h_3 h_1}\frac{\partial h_3}{\partial y_1} + \frac{v_{y_3}v_{y_2}}{h_3 h_2}\frac{\partial h_3}{\partial y_2} - \frac{v_{y_1}^2}{h_1 h_3}\frac{\partial h_1}{\partial y_3} - \frac{v_{y_2}^2}{h_2 h_3}\frac{\partial h_2}{\partial y_3}.$$

---

[1] However, it should be pointed out that they depend on the specifics of the tangent space basis vectors, and in this section we assume the basis vectors are orthonormal, which is not typically assumed in modern differential geometry. In our section on general coordinates we will have a slightly different definition of Christoffel symbols and will denote those by the typical symbol Γ_{ij}^k.

From the Eulerian perspective, we need to compute the material derivative $D\vec{v}/Dt$, where the velocity is a function of the current location $\vec{x}$ and time, which translates into a dependence on y_i and t, and where the convection term appears as

$$\vec{a}_E = \frac{D\vec{v}}{Dt} = \frac{\partial\vec{v}}{\partial t} + \frac{\partial\vec{v}}{\partial y_i}\frac{\partial y_i}{\partial t} = \frac{\partial\vec{v}}{\partial t} + \sum_i \frac{\partial\vec{v}}{\partial y_i}\frac{v_{y_i}}{h_i}, \tag{A.58}$$

but notice that for the partial with respect to t term the unit basis vectors in $\vec{v}$ do not depend on t. Thus,

$$\vec{a}_E = \frac{\partial v_{y_i}}{\partial t}\hat{e}_{y_i} + \sum_i \left(\frac{\partial v_{y_j}}{\partial y_i}\hat{e}_{y_j} + v_{y_j}\frac{\partial\hat{e}_{y_j}}{\partial y_i}\right)\frac{v_{y_i}}{h_i} = \vec{a}_L + \left(\sum_i \frac{v_{y_i}}{h_i}\frac{\partial v_{y_j}}{\partial y_i}\right)\hat{e}_{y_j}, \tag{A.59}$$

where the last equality is by comparing terms with the Lagrangian acceleration.

We now will present the various differential operators in these coordinate systems. It is simplest to go back to the original definitions to determine the expressions. The gradient of a function is defined as the vector that, when the inner-product with a vector of length 1 is taken, produces the change of the function in that direction. We know, using the chain rule, that the change in f in the given direction is

$$\frac{df}{ds} = \frac{\partial f}{\partial y_i}\frac{dy_i}{ds}. \tag{A.60}$$

Thus, we want

$$\frac{df}{ds} = \text{grad}(f) \cdot \frac{d\vec{x}(\vec{y}(s))}{ds} = \{\text{grad}(f)_{y_1}\hat{e}_{y_1} + \text{grad}(f)_{y_2}\hat{e}_{y_2} + \text{grad}(f)_{y_3}\hat{e}_{y_3}\} \cdot \frac{d\vec{x}}{ds}, \tag{A.61}$$

where s is arc length (i.e., $|d\vec{x}/ds| = 1$). The unit vector in terms of the three coordinate unit vectors is

$$\frac{d\vec{x}(\vec{y}(s))}{ds} = h_1\frac{dy_1}{ds}\hat{e}_{y_1} + h_2\frac{dy_2}{ds}\hat{e}_{y_2} + h_3\frac{dy_3}{ds}\hat{e}_{y_3}. \tag{A.62}$$

Thus, comparing Eqs. (A.60, A.61, A.62), the gradient of the function f is seen to be

$$\text{grad}(f) = \vec{\nabla}f = \frac{1}{h_1}\frac{\partial f}{\partial y_1}\hat{e}_{y_1} + \frac{1}{h_2}\frac{\partial f}{\partial y_2}\hat{e}_{y_2} + \frac{1}{h_3}\frac{\partial f}{\partial y_3}\hat{e}_{y_i} = f_{;i}\hat{e}_{y_i}. \tag{A.63}$$

The ∇ symbol, said "del" or "nabla," is typically used to denote the gradient. Another notation, to be described below, is to indicate the differentiation with a semicolon (;), hence the gradient is $\vec{\nabla}f = f_{;i}\hat{e}_{y_i}$.

The divergence of a vector function $\vec{f}$ (also referred to as a vector field $\vec{f}$) is defined to be

$$\text{div}(\vec{f}) = \vec{\nabla}\cdot\vec{f} = \lim_{vol\to 0}\frac{\int_{\text{surface}}\vec{f}\cdot d\vec{A}}{vol}. \tag{A.64}$$

Considering the deformed cube formed with edges h_1dy_1, h_2dy_2, and h_3dy_3, and looking at the six faces, we have

$$\int_{\text{cube faces}} \vec{f}\cdot d\vec{A} \approx (h_2h_3dy_2dy_3f_1)_+ - (h_2h_3dy_2dy_3f_1)_- + (h_1h_3dy_1dy_3f_2)_+$$

$$- (h_1h_3dy_1dy_3f_2)_- + (h_1h_2dy_1dy_2f_3)_+ - (h_1h_2dy_1dy_2f_3)_-, \tag{A.65}$$

where $+$ indicates the $\vec{x} + h_idy_i\hat{e}_{y_i}$ (no sum on i) face of the cube (though notice the evaluation is at $y_i + dy_i$) and $-$ indicates the $\vec{x}$ side of the cube (where the evaluation is at y_i). If we consider one of these opposite faces, notice that with the volume of the deformed cube given by $vol = h_1h_2h_3dy_1dy_2dy_3$,

$$\lim_{dy_1 \to 0} \frac{(h_2h_3dy_2dy_3f_1)_+ - (h_2h_3dy_2dy_3f_1)_-}{h_1h_2h_3dy_1dy_2dy_3}$$

$$= \lim_{dy_1 \to 0} \frac{(h_2h_3f_1)_+ - (h_2h_3f_1)_-}{h_1h_2h_3(dy_1)} = \frac{1}{h_1h_2h_3} \frac{\partial(h_2h_3f_1)}{\partial y_1}. \quad (A.66)$$

Performing the same limit for the other two opposite faces with the corresponding dy_i going to zero, the divergence is found to be

$$\text{div}(\vec{f}) = \vec{\nabla} \cdot \vec{f} = \frac{1}{h_1h_2h_3} \left\{ \frac{\partial(h_2h_3f_1)}{\partial y_1} + \frac{\partial(h_1h_3f_2)}{\partial y_2} + \frac{\partial(h_1h_2f_3)}{\partial y_3} \right\}. \quad (A.67)$$

The expressions for the gradient and divergence are not as similar as we might have thought based on the Cartesian expressions. To understand why, let us take the gradient of a vector function. The result will be a matrix, and to track the bases we introduce the notation for a matrix A that $A = A_{ij}\hat{e}_{y_i} \otimes \hat{e}_{y_j}$. Thus,

$$\vec{\nabla}\vec{f} = \sum_i \frac{1}{h_i} \frac{\partial}{\partial y_i}(f_j\hat{e}_{y_j}) \otimes \hat{e}_{y_i} = \sum_i \frac{1}{h_i} \left\{ \frac{\partial f_j}{\partial y_i}\hat{e}_{y_j} + f_j\frac{\partial \hat{e}_{y_j}}{\partial y_i} \right\} \otimes \hat{e}_{y_i}$$

$$= \sum_i \frac{1}{h_i} \left\{ \frac{\partial f_j}{\partial y_i}\hat{e}_{y_j} + f_j\gamma_{ij}^k\hat{e}_{y_k} \right\} \otimes \hat{e}_{y_i} = \sum_i \frac{1}{h_i} \left\{ \frac{\partial f_j}{\partial y_i} + f_k\gamma_{ik}^j \right\} \hat{e}_{y_j} \otimes \hat{e}_{y_i}, \quad (A.68)$$

where indices repeated twice are summed, the i summation is called out explicitly since there are three i indices, and the last step swapped the j and k arbitrary summation indices. We refer to this derivative as the *absolute derivative* since it is independent of the coordinate system employed and we will use a semicolon (;) to indicate it in indicial notation,

$$(\vec{\nabla}\vec{f})_{ij} = f_{i;j} = \frac{1}{h_j} \left\{ \frac{\partial f_i}{\partial y_j} + f_k\gamma_{jk}^i \right\} \quad \text{(no sum on } j). \quad (A.69)$$

To see all the terms, we write out the gradient of the vector $\vec{u}$. The matrix form is

$$\vec{\nabla}\vec{u} = (u_{i;j}) = \hat{F}(\vec{x}, \vec{u}) = \begin{pmatrix} \frac{1}{h_1}\frac{\partial u_1}{\partial y_1} & \frac{1}{h_2}\frac{\partial u_1}{\partial y_2} & \frac{1}{h_3}\frac{\partial u_1}{\partial y_3} \\ \frac{1}{h_1}\frac{\partial u_2}{\partial y_1} & \frac{1}{h_2}\frac{\partial u_2}{\partial y_2} & \frac{1}{h_3}\frac{\partial u_2}{\partial y_3} \\ \frac{1}{h_1}\frac{\partial u_3}{\partial y_1} & \frac{1}{h_2}\frac{\partial u_3}{\partial y_2} & \frac{1}{h_3}\frac{\partial u_3}{\partial y_3} \end{pmatrix}$$

$$+ \begin{pmatrix} \frac{u_2}{h_1h_2}\frac{\partial h_1}{\partial y_2} + \frac{u_3}{h_1h_3}\frac{\partial h_1}{\partial y_3} & -\frac{u_2}{h_2h_1}\frac{\partial h_2}{\partial y_1} & -\frac{u_3}{h_3h_1}\frac{\partial h_3}{\partial y_1} \\ -\frac{u_1}{h_1h_2}\frac{\partial h_1}{\partial y_2} & \frac{u_1}{h_2h_1}\frac{\partial h_2}{\partial y_1} + \frac{u_3}{h_2h_3}\frac{\partial h_2}{\partial y_3} & -\frac{u_3}{h_3h_2}\frac{\partial h_3}{\partial y_2} \\ -\frac{u_1}{h_1h_3}\frac{\partial h_1}{\partial y_3} & -\frac{u_2}{h_2h_3}\frac{\partial h_2}{\partial y_3} & \frac{u_1}{h_3h_1}\frac{\partial h_3}{\partial y_1} + \frac{u_2}{h_3h_2}\frac{\partial h_3}{\partial y_2} \end{pmatrix}, \quad (A.70)$$

where we chose $\vec{u}$ instead of $\vec{f}$ to make a less cluttered formula (later we discuss $\hat{F}(\vec{x}, \vec{u}) = F - I = \vec{\nabla}\vec{u}$). The divergence is the trace of this vector derivative, namely

$$\text{div}(\vec{f}) = \vec{\nabla} \cdot \vec{f} = \text{trace}(\vec{\nabla}\vec{f}) = f_{i;i} = \sum_i \frac{1}{h_i} \left\{ \frac{\partial f_i}{\partial y_i} + f_j\gamma_{ij}^i \right\}. \quad (A.71)$$

Placing the γ_{ij}^k terms from Eq. (A.54) into Eq. (A.71) (or comparing to Eq. A.70) confirms this is equivalent to Eq. (A.67). Thus, the added complexity in Eq. (A.67) is because it includes information about the change in basis vectors.

We combine the gradient and the divergence to compute the Laplacian of a scalar function f,

$$\Delta f = \nabla^2(f) = \text{div}(\text{grad}(f))$$

$$= \frac{1}{h_1 h_2 h_3} \left\{ \frac{\partial}{\partial y_1} \left(\frac{h_2 h_3}{h_1} \frac{\partial f}{\partial y_1} \right) + \frac{\partial}{\partial y_2} \left(\frac{h_1 h_3}{h_2} \frac{\partial f}{\partial y_2} \right) + \frac{\partial}{\partial y_3} \left(\frac{h_1 h_2}{h_3} \frac{\partial f}{\partial y_3} \right) \right\}. \tag{A.72}$$

Alternatively,

$$\Delta f = \text{trace}(\vec{\nabla}\vec{\nabla}f) = f_{;ii} = \sum_i \left(\frac{1}{h_i} \frac{\partial f}{\partial y_i} \right)_{;i} = \sum_{i,j} \frac{1}{h_i} \left\{ \frac{\partial}{\partial y_i} \left(\frac{1}{h_i} \frac{\partial f}{\partial y_i} \right) + \frac{1}{h_j} \frac{\partial f}{\partial y_j} \gamma^i_{ji} \right\}. \tag{A.73}$$

The curl of a vector function or vector field $\vec{f}$ is the only operator where it is necessary to know the orientation of the coordinate system. It will be assumed that the three coordinate unit vectors as ordered form a right-handed coordinate system, where $\hat{e}_{y_1} \times \hat{e}_{y_2} = \hat{e}_{y_3}$, or more generally $\hat{e}_{y_i} = \varepsilon_{ijk}\hat{e}_{y_j} \times \hat{e}_{y_k}$, where $\times$ is the cross product. Given an arbitrary unit vector $\vec{n}$, the curl is defined as

$$\text{curl}(\vec{f}) \cdot \vec{n} = \lim_{area \to 0} \frac{\oint_{\text{curve}} \vec{f} \cdot d\vec{s}}{area}, \tag{A.74}$$

where the curve bounds the area (around which the integral is taken) and $\vec{n}$ is the limit normal to the area considered. To compute the curl, the value for each coordinate direction will be calculated. For y_1, consider a deformed square curve in the y_2–y_3 plane, which is traversed counterclockwise when looking down on it from above (looking at the plane in the opposite direction to $\hat{e}_{y_1}$, where the direction is due to the right-handed nature of the coordinate unit vectors). For this curve,

$$\oint_{\text{square}} \vec{f} \cdot d\vec{s} \approx (h_3 dy_3 f_3)_{y_2+} - (h_2 dy_2 f_2)_{y_3+} - (h_3 dy_3 f_3)_{y_2-} + (h_2 dy_2 f_2)_{y_3-}, \tag{A.75}$$

where the subscript outside the parentheses indicates the side of the square on which the evaluation occurs. When this line integral is divided by the area of the deformed square $h_2 h_3 dy_2 dy_3$, the result is

$$\lim_{dy_2, dy_3 \to 0} \frac{1}{h_2 h_3} \left\{ \frac{(h_3 f_3)_{y_2+} - (h_3 f_3)_{y_2-}}{dy_2} - \frac{(h_2 f_2)_{y_3+} - (h_2 f_2)_{y_3-}}{dy_3} \right\}$$

$$= \frac{1}{h_2 h_3} \left\{ \frac{\partial(h_3 f_3)}{\partial y_2} - \frac{\partial(h_2 f_2)}{\partial y_3} \right\}. \tag{A.76}$$

The other three directions are similar. The only complication is that, looking from above, when $\vec{n} = \hat{e}_{y_2}$ the rotation of the curve in the plane is in the clockwise direction, and so the sign must change to work. Adding the three results together for each of the three coordinate unit vectors, the curl is

$$\text{curl}(\vec{f}) = \vec{\nabla} \times \vec{f} = \frac{1}{h_2 h_3} \left\{ \frac{\partial(h_3 f_3)}{\partial y_2} - \frac{\partial(h_2 f_2)}{\partial y_3} \right\} \hat{e}_{y_1} - \frac{1}{h_1 h_3} \left\{ \frac{\partial(h_3 f_3)}{\partial y_1} - \frac{\partial(h_1 f_1)}{\partial y_3} \right\} \hat{e}_{y_2}$$

$$+ \frac{1}{h_1 h_2} \left\{ \frac{\partial(h_2 f_2)}{\partial y_1} - \frac{\partial(h_1 f_1)}{\partial y_2} \right\} \hat{e}_{y_3} = \sum_{ijk} \varepsilon_{ijk} \frac{1}{h_j h_k} \frac{\partial(h_k f_k)}{\partial y_j} \hat{e}_{y_i}. \tag{A.77}$$

The similarity to the Cartesian expression as a determinant is evident (Exercise A.5). See Exercise A.6 for the absolute derivative version.

Looking back at the Eulerian acceleration in Eq. (A.59), where the basis vectors do not depend on time, we see that with matrix multiplication,

$$\vec{a}_E = \frac{\partial \vec{v}}{\partial t} + (\vec{\nabla}\vec{v})\vec{v} \quad \text{or} \quad a_i = \frac{\partial v_{y_i}}{\partial t} + v_{i;j}v_j. \tag{A.78}$$

For the momentum balance, we examine the surface integral of the tractions over the deformed cube, and then we normalize by the volume of the deformed cube to obtain the divergence of the stress tensor. We have

$$
\begin{aligned}
\int_{\text{cube faces}} \vec{t}da = & \ (h_2h_3dy_2dy_3\{\sigma_{11}\hat{e}_{y_1} + \sigma_{12}\hat{e}_{y_2} + \sigma_{13}\hat{e}_{y_3}\}) + \\
& - (h_2h_3dy_2dy_3\{\sigma_{11}\hat{e}_{y_1} + \sigma_{12}\hat{e}_{y_2} + \sigma_{13}\hat{e}_{y_3}\}) - \\
& + (h_1h_3dy_1dy_3\{\sigma_{21}\hat{e}_{y_1} + \sigma_{22}\hat{e}_{y_2} + \sigma_{23}\hat{e}_{y_3}\}) + \\
& - (h_1h_3dy_1dy_3\{\sigma_{21}\hat{e}_{y_1} + \sigma_{22}\hat{e}_{y_2} + \sigma_{23}\hat{e}_{y_3}\}) - \\
& + (h_1h_2dy_1dy_2\{\sigma_{31}\hat{e}_{y_1} + \sigma_{32}\hat{e}_{y_2} + \sigma_{33}\hat{e}_{y_3}\}) + \\
& - (h_1h_2dy_1dy_2\{\sigma_{31}\hat{e}_{y_1} + \sigma_{32}\hat{e}_{y_2} + \sigma_{33}\hat{e}_{y_3}\}) -.
\end{aligned}
\tag{A.79}
$$

When this is divided by the volume of the deformed cube $vol = h_1h_2h_3dy_1dy_2dy_3$, and taking the limit as the $dy_i \to 0$, the following result is obtained (using the result on the divergence in Eq. A.67),

$$
\vec{\nabla}\cdot\sigma = \lim_{dy_i\to 0} \frac{1}{vol} \int_{\text{cube faces}} \vec{t}da = \text{div}\begin{pmatrix}\sigma_{11}\\\sigma_{21}\\\sigma_{31}\end{pmatrix}\hat{e}_{y_1} + \text{div}\begin{pmatrix}\sigma_{12}\\\sigma_{22}\\\sigma_{32}\end{pmatrix}\hat{e}_{y_2} + \text{div}\begin{pmatrix}\sigma_{13}\\\sigma_{23}\\\sigma_{33}\end{pmatrix}\hat{e}_{y_3}
$$

$$
+ \frac{\sigma_{11}}{h_1}\frac{\partial\hat{e}_{y_1}}{\partial y_1} + \frac{\sigma_{12}}{h_1}\frac{\partial\hat{e}_{y_2}}{\partial y_1} + \frac{\sigma_{13}}{h_1}\frac{\partial\hat{e}_{y_3}}{\partial y_1} + \frac{\sigma_{21}}{h_2}\frac{\partial\hat{e}_{y_1}}{\partial y_2} + \frac{\sigma_{22}}{h_2}\frac{\partial\hat{e}_{y_2}}{\partial y_2} + \frac{\sigma_{23}}{h_2}\frac{\partial\hat{e}_{y_3}}{\partial y_2}
$$

$$
+ \frac{\sigma_{31}}{h_3}\frac{\partial\hat{e}_{y_1}}{\partial y_3} + \frac{\sigma_{32}}{h_3}\frac{\partial\hat{e}_{y_2}}{\partial y_3} + \frac{\sigma_{33}}{h_3}\frac{\partial\hat{e}_{y_3}}{\partial y_3}. \tag{A.80}
$$

This expression is used in the momentum balance. It has not been assumed that the stress tensor is symmetric in this derivation. To be explicit, here the column notation means, for example,

$$
\begin{aligned}
\text{div}\begin{pmatrix}\sigma_{11}\\\sigma_{21}\\\sigma_{31}\end{pmatrix} &= \text{div}(\sigma_{11}\hat{e}_{y_1} + \sigma_{21}\hat{e}_{y_2} + \sigma_{31}\hat{e}_{y_3}) \\
&= \frac{1}{h_1h_2h_3}\left\{\frac{\partial(h_2h_3\sigma_{11})}{\partial y_1} + \frac{\partial(h_1h_3\sigma_{21})}{\partial y_2} + \frac{\partial(h_1h_2\sigma_{31})}{\partial y_3}\right\}.
\end{aligned}
\tag{A.81}
$$

Using the expression for the derivatives of the basis vectors (Eq. A.133) that is derived later in the section, the expression for the momentum balance can be expanded out (sadly, nothing cancels),

$$
\vec{\nabla}\cdot\sigma = \left\{ \frac{1}{h_1}\frac{\partial\sigma_{11}}{\partial y_1} + \frac{\sigma_{11}-\sigma_{22}}{h_1h_2}\frac{\partial h_2}{\partial y_1} + \frac{\sigma_{11}-\sigma_{33}}{h_1h_3}\frac{\partial h_3}{\partial y_1} \right.
$$

$$
+ \frac{1}{h_2}\frac{\partial\sigma_{21}}{\partial y_2} + \frac{\sigma_{21}+\sigma_{12}}{h_1h_2}\frac{\partial h_1}{\partial y_2} + \frac{\sigma_{21}}{h_2h_3}\frac{\partial h_3}{\partial y_2}
$$

$$
\left. + \frac{1}{h_3}\frac{\partial\sigma_{31}}{\partial y_3} + \frac{\sigma_{31}+\sigma_{13}}{h_1h_3}\frac{\partial h_1}{\partial y_3} + \frac{\sigma_{31}}{h_2h_3}\frac{\partial h_2}{\partial y_3} \right\}\hat{e}_{y_1}
$$

$$+\left\{\frac{1}{h_1}\frac{\partial\sigma_{12}}{\partial y_1}+\frac{\sigma_{12}+\sigma_{21}}{h_1 h_2}\frac{\partial h_2}{\partial y_1}+\frac{\sigma_{12}}{h_1 h_3}\frac{\partial h_3}{\partial y_1}\right.$$

$$+\frac{1}{h_2}\frac{\partial\sigma_{22}}{\partial y_2}+\frac{\sigma_{22}-\sigma_{11}}{h_1 h_2}\frac{\partial h_1}{\partial y_2}+\frac{\sigma_{22}-\sigma_{33}}{h_2 h_3}\frac{\partial h_3}{\partial y_2}$$

$$\left.+\frac{1}{h_3}\frac{\partial\sigma_{32}}{\partial y_3}+\frac{\sigma_{23}+\sigma_{32}}{h_2 h_3}\frac{\partial h_2}{\partial y_3}+\frac{\sigma_{32}}{h_1 h_3}\frac{\partial h_1}{\partial y_3}\right\}\hat{e}_{y_2} \tag{A.82}$$

$$+\left\{\frac{1}{h_1}\frac{\partial\sigma_{13}}{\partial y_1}+\frac{\sigma_{13}+\sigma_{31}}{h_1 h_3}\frac{\partial h_3}{\partial y_1}+\frac{\sigma_{13}}{h_1 h_2}\frac{\partial h_2}{\partial y_1}\right.$$

$$+\frac{1}{h_2}\frac{\partial\sigma_{23}}{\partial y_2}+\frac{\sigma_{23}+\sigma_{32}}{h_2 h_3}\frac{\partial h_3}{\partial y_2}+\frac{\sigma_{23}}{h_1 h_2}\frac{\partial h_1}{\partial y_2}$$

$$\left.+\frac{1}{h_3}\frac{\partial\sigma_{33}}{\partial y_3}+\frac{\sigma_{33}-\sigma_{11}}{h_1 h_3}\frac{\partial h_1}{\partial y_3}+\frac{\sigma_{33}-\sigma_{22}}{h_2 h_3}\frac{\partial h_2}{\partial y_3}\right\}\hat{e}_{y_3}.$$

The other approach to this is to take the gradient of the stress tensor,

$$\vec{\nabla}\sigma=\sum_i\frac{1}{h_i}\frac{\partial}{\partial y_i}(\sigma_{jk}\hat{e}_{y_j}\otimes\hat{e}_{y_k})\otimes\hat{e}_{y_i}$$

$$=\sum_i\frac{1}{h_i}\left\{\frac{\partial\sigma_{jk}}{\partial y_i}\hat{e}_{y_j}\otimes\hat{e}_{y_k}+\sigma_{jk}\frac{\partial\hat{e}_{y_j}}{\partial y_i}\otimes\hat{e}_{y_k}+\sigma_{jk}\hat{e}_{y_j}\otimes\frac{\partial\hat{e}_{y_k}}{\partial y_i}\right\}\otimes\hat{e}_{y_i}$$

$$=\sum_i\frac{1}{h_i}\left\{\frac{\partial\sigma_{jk}}{\partial y_i}\hat{e}_{y_j}\otimes\hat{e}_{y_k}+\sigma_{jk}\gamma_{ij}^\ell\hat{e}_{y_\ell}\otimes\hat{e}_{y_k}+\sigma_{jk}\hat{e}_{y_j}\otimes\gamma_{ik}^\ell\hat{e}_{y_\ell}\right\}\otimes\hat{e}_{y_i} \tag{A.83}$$

$$=\sum_i\frac{1}{h_i}\left\{\frac{\partial\sigma_{jk}}{\partial y_i}+\sigma_{\ell k}\gamma_{i\ell}^j+\sigma_{j\ell}\gamma_{i\ell}^k\right\}\hat{e}_{y_j}\otimes\hat{e}_{y_k}\otimes\hat{e}_{y_i}.$$

The definition in terms of indices, using the (;) to indicate differentiation, is

$$(\vec{\nabla}\sigma)_{jki}=\sigma_{jk;i}=\frac{1}{h_i}\left\{\frac{\partial\sigma_{jk}}{\partial y_i}+\sigma_{\ell k}\gamma_{i\ell}^j+\sigma_{j\ell}\gamma_{i\ell}^k\right\}\quad\text{(no sum on }i\text{)}. \tag{A.84}$$

We take the trace, or contraction, of this to obtain

$$(\vec{\nabla}\cdot\sigma)_i=\text{trace}(\vec{\nabla}\sigma)_i=(\vec{\nabla}\sigma)_{jij}=\sigma_{ji;j}=\sum_j\frac{1}{h_j}\left\{\frac{\partial\sigma_{ji}}{\partial y_j}+\sigma_{\ell i}\gamma_{j\ell}^j+\sigma_{j\ell}\gamma_{j\ell}^i\right\}. \tag{A.85}$$

The next topic we discuss is the deformation gradient, which will be denoted F. The deformation gradient is a measure of how much the object in space has deformed. The deformation gradient shows how a small initial $d\vec{x}_0$ ($=d\vec{X}$ in Chapters 13 and 14) differential is deformed to $d\vec{x}_1$ through

$$d\vec{x}_1=Fd\vec{x}_0, \tag{A.86}$$

where F is a 3×3 matrix. Suppose that the initial configuration in space is parameterized by y_i. We can write the initial $d\vec{x}_0$ as

$$d\vec{x}_0=\begin{pmatrix}dx_1\\dx_2\\dx_3\end{pmatrix}_0=\begin{pmatrix}F_{11}&F_{12}&F_{13}\\F_{21}&F_{22}&F_{23}\\F_{31}&F_{32}&F_{33}\end{pmatrix}_0\begin{pmatrix}dy_1\\dy_2\\dy_3\end{pmatrix}=F_0 d\vec{y}, \tag{A.87}$$

where F_0 is a 3×3 matrix. In general F_0 could be a full matrix, but for our specific case it is diagonal because of our orthogonal coordinate system (Eq. A.49).

Similarly, the final configuration could be parameterized by z_i and we have

$$d\vec{x}_1 = F_1 d\vec{z}. \tag{A.88}$$

To be explicit, F_0 and F_1 are diagonal and are given by

$$F_0 = \begin{pmatrix} _y h_1 & 0 & 0 \\ 0 & _y h_2 & 0 \\ 0 & 0 & _y h_3 \end{pmatrix}, \quad F_1 = \begin{pmatrix} _z h_1 & 0 & 0 \\ 0 & _z h_2 & 0 \\ 0 & 0 & _z h_3 \end{pmatrix}, \tag{A.89}$$

or

$$(F_k)_{ij} = {}_k h_i \delta_{ij} \quad \text{(no sum on } i), \tag{A.90}$$

where the prefix on the h_i is to remind us which coordinate system (i.e., parameterization) is in use. Now if there is a relationship between the parameters at the initial configuration and at the final configuration, such that $z_i = z_i(y_1, y_2, y_3)$, then writing the Jacobian of this relationship as

$$F_J = \begin{pmatrix} \frac{\partial z_1}{\partial y_1} & \frac{\partial z_1}{\partial y_2} & \frac{\partial z_1}{\partial y_3} \\ \frac{\partial z_2}{\partial y_1} & \frac{\partial z_2}{\partial y_2} & \frac{\partial z_2}{\partial y_3} \\ \frac{\partial z_3}{\partial y_1} & \frac{\partial z_3}{\partial y_2} & \frac{\partial z_3}{\partial y_3} \end{pmatrix} \tag{A.91}$$

yields

$$d\vec{z} = F_J d\vec{y}. \tag{A.92}$$

The last four equations yield

$$d\vec{x}_1 = F_1 F_J F_0^{-1} d\vec{x}_0 \quad \Rightarrow \quad F = F_1 F_J F_0^{-1}, \tag{A.93}$$

thus explicitly yielding the deformation gradient for this situation. Since F_1 and F_0 are diagonal, we can easily compute F,

$$F = \begin{pmatrix} \frac{h_1 \partial z_1}{H_1 \partial y_1} & \frac{h_1 \partial z_1}{H_2 \partial y_2} & \frac{h_1 \partial z_1}{H_3 \partial y_3} \\ \frac{h_2 \partial z_2}{H_1 \partial y_1} & \frac{h_2 \partial z_2}{H_2 \partial y_2} & \frac{h_2 \partial z_2}{H_3 \partial y_3} \\ \frac{h_3 \partial z_3}{H_1 \partial y_1} & \frac{h_3 \partial z_3}{H_2 \partial y_2} & \frac{h_3 \partial z_3}{H_3 \partial y_3} \end{pmatrix} \quad \text{or} \quad F_{ij} = \frac{_z h_i}{_y h_j} \frac{\partial z_i}{\partial y_j} \quad \text{(no sum on } i, j), \tag{A.94}$$

where again the prefix on the h_i is to identify whether it corresponds to the initial parameterization of $d\vec{x}_0$ by y_i ($H_i = {}_y h_i$) or the final parameterization of $d\vec{x}_1$ by z_i ($h_i = {}_z h_i$). F is a very interesting object in that it takes the differential $d\vec{x}_0$ written in terms of $\hat{e}_{y_i}$ and produces a differential $d\vec{x}_1$ written in terms of $\hat{e}_{z_i}$. Thus, the right side of F is written in terms of the y coordinate system and the left side is written in terms of the z coordinate system, and the coordinate systems can be different. More importantly, though, the right-hand side of F is in the material's original configuration, while the left-hand side is in the material's current configuration.[2] This means that $F^T F$ is a tensor that is written in terms of the material's original configuration (it will be in the y coordinate system), which is the correct configuration for computing the Lagrangian strain E, and it means that $B = FF^T$ is a tensor written in terms of the material's current configuration (it will be in the z coordinate system), which is the correct configuration for computing the Eulerian strain E^*. One can compute the deformation gradient from

$$F = \text{grad}_{\vec{x}_0}(\vec{x}_1), \tag{A.95}$$

---

[2]To emphasize this point, Malvern writes the indices F_{kM} rather than F_{km} (capital for initial, lower case for current) and we will too when we explicitly write this form of F for the three common coordinate systems. However, this is just a friendly reminder of which configuration is being dealt with when we take the products $F^T F$ and FF^T; whether the two sides are actually different coordinate systems is a different question and is denoted by the y and z basis vectors.

recognizing that the object acted on is a vector, rather than a function, and care is taken in differentiating and tracking the basis vectors.

However, often in mechanics we know the displacement from the original position, so that $\vec{x}_1 = \vec{x}_0 + \vec{u}$, where the original location is $\vec{x}_0$ and the displacement is $\vec{u}$. We will assume both are written in terms of the parameterization y_i. We now have

$$d\vec{x}_1 = d(\vec{x}_0 + \vec{u}) = d\vec{x}_0 + d\vec{u}. \tag{A.96}$$

We have already discussed $d\vec{x}_0$, so now we need to discuss $d\vec{u}$. In particular,

$$d\vec{u} = d(u_i \hat{e}_{y_i}) = (du_i)\hat{e}_{y_i} + u_i d\hat{e}_{y_i} = \frac{\partial u_i}{\partial y_j}\hat{e}_{y_i} dy_j + u_i \frac{\partial \hat{e}_{y_i}}{\partial y_j} dy_j, \tag{A.97}$$

so it is necessary to compute the derivatives of $\hat{e}_{y_i}$. This is an important computation, and for now we will only state the result, since it will break the flow of the discussion (it is derived Eqs. A.123 to A.133),

$$\frac{\partial \hat{e}_{y_i}}{\partial y_j} = \frac{1}{h_i}\frac{\partial h_j}{\partial y_i}\hat{e}_{y_j} - \left(\sum_{k=1}^{3}\frac{1}{h_k}\frac{\partial h_i}{\partial y_k}\hat{e}_{y_k}\right)\delta_{ij} \quad \text{(no sum on } i, j\text{)}. \tag{A.98}$$

Using this result, if we write

$$d\vec{u} = \tilde{F}(\vec{x}_0, \vec{u})d\vec{y}, \quad \text{so} \quad \tilde{F}_{ij}(\vec{x}_0, \vec{u}) = \hat{e}_{y_i} \cdot \frac{\partial \vec{u}}{\partial y_j}, \tag{A.99}$$

then

$$
\begin{aligned}
\tilde{F}_{ij}(\vec{x}_0, \vec{u}) &= \hat{e}_{y_i} \cdot \frac{\partial \vec{u}}{\partial y_j} = \hat{e}_{y_i} \cdot \sum_{m=1}^{3}\left(\frac{\partial u_m}{\partial y_j}\hat{e}_{y_m} + u_m \frac{\partial \hat{e}_{y_m}}{\partial y_j}\right) \\
&= \frac{\partial u_i}{\partial y_j} + \hat{e}_{y_i} \cdot \sum_{m=1}^{3} u_m \left(\frac{1}{h_m}\frac{\partial h_j}{\partial y_m}\hat{e}_{y_j} - \left(\sum_{k=1}^{3}\frac{1}{h_k}\frac{\partial h_m}{\partial y_k}\hat{e}_{y_k}\right)\delta_{mj}\right) \\
&= \frac{\partial u_i}{\partial y_j} + \sum_{m=1}^{3} u_m \left(\frac{1}{h_m}\frac{\partial h_j}{\partial y_m}\delta_{ij} - \left(\sum_{k=1}^{3}\frac{1}{h_k}\frac{\partial h_m}{\partial y_k}\delta_{ik}\right)\delta_{mj}\right) \\
&= \frac{\partial u_i}{\partial y_j} + \sum_{m=1}^{3} u_m \left(\frac{1}{h_m}\frac{\partial h_i}{\partial y_m}\delta_{ij} - \frac{1}{h_i}\frac{\partial h_m}{\partial y_i}\delta_{mj}\right) \\
&= \frac{\partial u_i}{\partial y_j} + \left(\sum_{k=1}^{3}\frac{u_k}{h_k}\frac{\partial h_i}{\partial y_k}\right)\delta_{ij} - \frac{u_j}{h_i}\frac{\partial h_j}{\partial y_i} \quad \text{(no sum on } i, j\text{)},
\end{aligned}
\tag{A.100}
$$

or in matrix form

$$\tilde{F}(\vec{x}_0, \vec{u}) = \begin{pmatrix} \frac{\partial u_1}{\partial y_1} & \frac{\partial u_1}{\partial y_2} & \frac{\partial u_1}{\partial y_3} \\ \frac{\partial u_2}{\partial y_1} & \frac{\partial u_2}{\partial y_2} & \frac{\partial u_2}{\partial y_3} \\ \frac{\partial u_3}{\partial y_1} & \frac{\partial u_3}{\partial y_2} & \frac{\partial u_3}{\partial y_3} \end{pmatrix}$$

$$+ \begin{pmatrix} \frac{u_2}{h_2}\frac{\partial h_1}{\partial y_2} + \frac{u_3}{h_3}\frac{\partial h_1}{\partial y_3} & -\frac{u_2}{h_1}\frac{\partial h_2}{\partial y_1} & -\frac{u_3}{h_1}\frac{\partial h_3}{\partial y_1} \\ -\frac{u_1}{h_2}\frac{\partial h_1}{\partial y_2} & \frac{u_1}{h_1}\frac{\partial h_2}{\partial y_1} + \frac{u_3}{h_3}\frac{\partial h_2}{\partial y_3} & -\frac{u_3}{h_2}\frac{\partial h_3}{\partial y_2} \\ -\frac{u_1}{h_3}\frac{\partial h_1}{\partial y_3} & -\frac{u_2}{h_3}\frac{\partial h_2}{\partial y_3} & \frac{u_1}{h_1}\frac{\partial h_3}{\partial y_1} + \frac{u_2}{h_2}\frac{\partial h_3}{\partial y_2} \end{pmatrix}. \tag{A.101}$$

Given that everything is written in terms of the y_i coordinate system, the deformation gradient is given by ($I = (\delta_{ij})$ is the identity matrix)

$$d\vec{x}_1 = (F_0 + \tilde{F}(\vec{x}_0, \vec{u}_0))F_0^{-1}d\vec{x}_0 \quad \Rightarrow \quad F(\vec{x}, \vec{u}) = I + \tilde{F}(\vec{x}_0, \vec{u}_0)F_0^{-1}, \tag{A.102}$$

and, since F_0 is diagonal, it is easy to perform the computation and we obtain

$$F(\vec{x}, \vec{u})_{ij} = \delta_{ij} + \frac{1}{h_j} \left\{ \frac{\partial u_i}{\partial y_j} + \left(\sum_{k=1}^{3} \frac{u_k}{h_k} \frac{\partial h_i}{\partial y_k} \right) \delta_{ij} - \frac{u_j}{h_i} \frac{\partial h_j}{\partial y_i} \right\} \quad \text{(no sum on } i, j\text{)}$$

(A.103)

or, defining the displacement gradient $\hat{F} = F - I$, the matrix form is

$$\hat{F}(\vec{x}, \vec{u}) = \begin{pmatrix} \frac{1}{h_1} \frac{\partial u_1}{\partial y_1} & \frac{1}{h_2} \frac{\partial u_1}{\partial y_2} & \frac{1}{h_3} \frac{\partial u_1}{\partial y_3} \\ \frac{1}{h_1} \frac{\partial u_2}{\partial y_1} & \frac{1}{h_2} \frac{\partial u_2}{\partial y_2} & \frac{1}{h_3} \frac{\partial u_2}{\partial y_3} \\ \frac{1}{h_1} \frac{\partial u_3}{\partial y_1} & \frac{1}{h_2} \frac{\partial u_3}{\partial y_2} & \frac{1}{h_3} \frac{\partial u_3}{\partial y_3} \end{pmatrix}$$

$$+ \begin{pmatrix} \frac{u_2}{h_1 h_2} \frac{\partial h_1}{\partial y_2} + \frac{u_3}{h_1 h_3} \frac{\partial h_1}{\partial y_3} & -\frac{u_2}{h_2 h_1} \frac{\partial h_2}{\partial y_1} & -\frac{u_3}{h_3 h_1} \frac{\partial h_3}{\partial y_1} \\ -\frac{u_1}{h_1 h_2} \frac{\partial h_1}{\partial y_2} & \frac{u_1}{h_2 h_1} \frac{\partial h_2}{\partial y_1} + \frac{u_3}{h_2 h_3} \frac{\partial h_2}{\partial y_3} & -\frac{u_3}{h_3 h_2} \frac{\partial h_3}{\partial y_2} \\ -\frac{u_1}{h_1 h_3} \frac{\partial h_1}{\partial y_3} & -\frac{u_2}{h_2 h_3} \frac{\partial h_2}{\partial y_3} & \frac{u_1}{h_3 h_1} \frac{\partial h_3}{\partial y_1} + \frac{u_2}{h_3 h_2} \frac{\partial h_3}{\partial y_2} \end{pmatrix}.$$

(A.104)

We recognize this as the gradient of a vector computed in Eq. (A.70), $\hat{F}(\vec{x}, \vec{u}) = \vec{\nabla}\vec{u} = (u_{i;j})$. In indicial notation the displacement gradient is

$$\hat{F}(\vec{x}, \vec{u})_{ij} = u_{i:j} = \frac{1}{h_j} \hat{e}_{y_i} \cdot \frac{\partial \vec{u}}{\partial y_j}$$

(A.105)

$$= \frac{1}{h_j} \left\{ \frac{\partial u_i}{\partial y_j} + \left(\sum_{k=1}^{3} \frac{u_k}{h_k} \frac{\partial h_i}{\partial y_k} \right) \delta_{ij} - \frac{u_j}{h_i} \frac{\partial h_j}{\partial y_i} \right\} \quad \text{(no sum on } i, j\text{)},$$

as well as Eq. (A.69). Since $\vec{u}$ is written in the original y coordinate system, this $F = I + \hat{F}$ is entirely in the original y coordinate system (both sides). Since both the $\hat{e}_{y_i}$ and $\hat{e}_{z_i}$ are orthonormal sets of basis vectors, they are related by a rotation. Thus, F as described in Eq. (A.94) and F as described in Eq. (A.103) are related by multiplication by a rotation matrix $_zQ_y$, in that $F_{A.94} = {}_zQ_y F_{A.103}$. Of course, if the two coordinate systems are the same – for example, if they are both the same Cartesian coordinate system – then the F are componentwise the same (no rotation is required and $Q = I$). Please notice, though, that the right-hand side of F is still in the original configuration and the left-hand side of F is in the current configuration.

A subtle but important fact now comes up that is easiest to describe by example. In both the cylindrical and spherical coordinate systems, a basis vector points in the direction of the material point, since $\vec{x} = r\hat{e}_r + z\hat{e}_z$ and $\vec{x} = r\hat{e}_r$, respectively. When a material undergoes a motion that involves rotation in these coordinate systems, the $\hat{e}_r$ basis vector *no longer points in the same direction*, and thus going from the original configuration to the current configuration implies a change in coordinate system (i.e., basis). To see the effect, consider motion in the plane and suppose that the entire motion is a counterclockwise rotation by angle β. Using Eq. (A.94), it is easy to compute F by noting the $r = R$ and $\theta = \Theta + \beta$, and $F = I$ (for coordinates we are using capitals for original configuration, and lower case for current configuration). However, let us suppose we wanted to use Eq. (A.103). As a first step, we compute $\vec{x}_1 - \vec{x}_0 = \vec{u} = u_R \hat{e}_R + u_\Theta \hat{e}_\Theta = R(\cos(\beta) - 1)\hat{e}_R + R\sin(\beta)\hat{e}_\Theta$, then (Eq. A.173)

$$F = I + \hat{F} = \begin{pmatrix} \cos(\beta) & -\sin(\beta) \\ \sin(\beta) & \cos(\beta) \end{pmatrix}.$$

(A.106)

These two F are the same object but their specific component values are not the same because of the fact that this latter F is written entirely in terms of the original coordinate system. With $F = F_{ij}\hat{e}_{z_i} \otimes \hat{e}_{y_j}$ denoting the basis of F, then F from Eq. (A.94) is

$$F = 1 \cdot \hat{e}_r \otimes \hat{e}_R + 0 \cdot \hat{e}_r \otimes \hat{e}_\Theta + 0 \cdot \hat{e}_\theta \otimes \hat{e}_R + 1 \cdot \hat{e}_\theta \otimes \hat{e}_\Theta \qquad (A.107)$$

(the · is just multiplication), and F from Eq. (A.103) is

$$F = \cos(\beta) \cdot \hat{e}_R \otimes \hat{e}_R - \sin(\beta) \cdot \hat{e}_R \otimes \hat{e}_\Theta + \sin(\beta) \cdot \hat{e}_\Theta \otimes \hat{e}_R + \cos(\beta) \cdot \hat{e}_\Theta \otimes \hat{e}_\Theta. \quad (A.108)$$

The current and original basis vectors are related by a rotation, $\hat{e}_r = \cos(\beta)\hat{e}_R + \sin(\beta)\hat{e}_\Theta$, and $\hat{e}_\theta = -\sin(\beta)\hat{e}_R + \cos(\beta)\hat{e}_\Theta$, or in the notation of Section A.3,

$$\begin{pmatrix} \hat{e}_r \\ \hat{e}_\theta \end{pmatrix} = \begin{pmatrix} \cos(\beta) & \sin(\beta) \\ -\sin(\beta) & \cos(\beta) \end{pmatrix} \begin{pmatrix} \hat{e}_R \\ \hat{e}_\Theta \end{pmatrix} = {}_rQ_R \begin{pmatrix} \hat{e}_R \\ \hat{e}_\Theta \end{pmatrix}. \qquad (A.109)$$

To change to the new basis on the left,

$$F = \begin{pmatrix} \hat{e}_R \\ \hat{e}_\Theta \end{pmatrix}^T {}_RF_R \begin{pmatrix} \hat{e}_R \\ \hat{e}_\Theta \end{pmatrix} = ({}_rQ_R^T \begin{pmatrix} \hat{e}_r \\ \hat{e}_\theta \end{pmatrix})^T {}_RF_R \begin{pmatrix} \hat{e}_R \\ \hat{e}_\Theta \end{pmatrix} = \begin{pmatrix} \hat{e}_r \\ \hat{e}_\theta \end{pmatrix}^T {}_rQ_R\, {}_RF_R \begin{pmatrix} \hat{e}_R \\ \hat{e}_\Theta \end{pmatrix}$$
$$(A.110)$$

(where the subscripts on F denote the coordinate system) and we see that ${}_rQ_R F_{A.108} = {}_rQ_R\, {}_RF_R = {}_rF_R = F_{A.107} = I$. In this example, no harm is done, since computing E and E^* leads to zero strain for this rigid rotation, but if there were deformation, though FF^T is in the current configuration, it is in the original coordinate system rather than the current coordinate system, which can lead to errors in application.

Let's do a more complicated example for further clarification. Consider a stretch of α in the X direction in polar coordinates. The stretch takes the point $(X, Y) = (R\cos(\Theta), R\sin(\Theta))$ to the point $(x, y) = (\alpha R\cos(\Theta), R\sin(\Theta)) = (r\cos(\theta), r\sin(\theta))$, so that $r = (1 + (\alpha^2 - 1)\cos^2(\Theta))^{1/2}R$ and $\tan(\theta) = \tan(\Theta)/\alpha$. A direct computation using Eq. (A.94) yields

$$_rF_R = \begin{pmatrix} r/R & -(R/r)(\alpha^2 - 1)\cos(\Theta)\sin(\Theta) \\ 0 & \alpha R/r \end{pmatrix}. \qquad (A.111)$$

This result should make us pause. There is no rotation of the material's configuration in this deformation, yet F is not symmetric, which implies that the polar decomposition includes a rotation. However, in this case, there is an apparent rotation strictly due to the coordinate system change and not any rotation of the body. To see this fact, let us compute F from Eq. (A.103). The displacement in terms of the original coordinate systems is $\vec{x}_1 - \vec{x}_0 = \vec{u} = (\alpha - 1)R\cos^2(\Theta)\hat{e}_R - (\alpha - 1)R\cos(\Theta)\sin(\Theta)\hat{e}_\Theta$, and the computation yields

$$_RF_R = I + \hat{F} = I + (\alpha - 1)\begin{pmatrix} \cos^2(\Theta) & -\cos(\Theta)\sin(\Theta) \\ -\cos(\Theta)\sin(\Theta) & \sin^2(\Theta) \end{pmatrix}. \qquad (A.112)$$

This F is symmetric, saying that there is no rotation in the deformation. These two versions of F are related by a rotation, namely the counterclockwise rotation $\beta = \theta - \Theta$ that occurs in changing the coordinate system from (R, Θ) to (r, θ). This can be seen by noting that $\cos(\theta) = \alpha\cos(\Theta)R/r$, $\sin(\theta) = \sin(\Theta)R/r$, computing the rotation matrix ${}_rQ_R$ as above and then showing by multiplying it out that ${}_rQ_R^T\, {}_rF_R = {}_RF_R$. The point being made here is that in this coordinate system a deformation that has no rotation leads to a change of the coordinate system that involved rotation. To see that the $_rF_R$ version's polar decomposition also does not include a configuration rotation; notice that $_rF_R^T {}_rF_R = (_rQ_R {}_RF_R)^T {}_rQ_R {}_RF_R = {}_RF_{RR}^T F_R = {}_RF_R^2$, and hence $U = {}_RF_R$, and the corresponding rotation matrix for $_rF_R$ is strictly due to the change in coordinates (basis vectors) arising from the deformation. In general, the rotation in the polar decomposition of F can be comprised of both deformation and change in coordinates, and the rotation due to change in coordinates can be different for every point. Finally, notice that when F is written entirely in the current coordinates, $(_rF_r)^T = (_rQ_R {}_RF_R {}_rQ_R^T)^T = {}_rQ_R {}_RF_R {}_rQ_R^T = {}_rF_r$ and it is symmetric, hence exhibiting no material rotation. To compute E in the original coordinate system, where the Piola–Kirchhoff stress lives, we can use $F^TF = {}_RF_R^T {}_RF_R = {}_rF_R^T {}_rF_R$, and to compute E^* in the current coordinate system, where the Cauchy stress lives, we can use $FF^T = {}_rF_R {}_rF_R^T = {}_rF_r {}_rF_r^T$.

As seen above, when Eq. (A.103) is used to compute F, then F is entirely in the y coordinate system. Written there, it is correct for computing the Lagrangian strain E in the original coordinates (which requires F^TF), but it is not correct (if there are rotations) for computing the Eulerian strain E^* or $B = FF^T$ in the current coordinates for finite strain elasticity. We can use our above result to obtain an expression in the z coordinate system by noting that, if the displacements are known, then F in the current coordinate system is given by writing $\vec{x}_0 = \vec{x}_1 - \vec{u}$, where now $\vec{u}$ is written in terms of the current coordinate system parameterized by z_i. Then,

$$d\vec{x}_0 = d\vec{x}_1 - d\vec{u} = (F_1 + \tilde{F}(\vec{x}_1, -\vec{u}))d\vec{z}, \qquad (A.113)$$

and with $d\vec{x}_1 = F_1 d\vec{z}$ we obtain

$$F = F_1(F_1 + \tilde{F}(\vec{x}_1, -\vec{u}))^{-1} = (I + \hat{F}(\vec{x}_1, -\vec{u}))^{-1}. \qquad (A.114)$$

With $\vec{u}$ written in the z coordinate system, this F will be in the current coordinate system[3] (both sides) and can be used to compute $B = FF^T$ in the current coordinates (note that it differs from F in Eq. A.94 by multiplication by a rotation matrix on the right), but is not appropriate for computing F^TF in the original coordinates.

The deformation gradient shows us how the original length element $d\vec{x}_0$ deforms into a new length element $d\vec{x}_1$. If we call the original length dS and the new length ds then

$$(ds)^2 - (dS)^2 = d\vec{x}_1^T d\vec{x}_1 - d\vec{x}_0^T d\vec{x}_0 = d\vec{x}_0^T F^TF d\vec{x}_0 - d\vec{x}_0^T d\vec{x}_0$$
$$= d\vec{x}_0^T(F^TF - I)d\vec{x}_0. \qquad (A.115)$$

The definition of the Lagrangian (or Green) strain tensor E is

$$(ds)^2 - (dS)^2 = 2d\vec{x}_0^T E\, d\vec{x}_0 \quad \Rightarrow \quad E = \frac{1}{2}(F^TF - I) = \frac{1}{2}(C - I), \quad \text{where } C = F^TF, \qquad (A.116)$$

and where the factor of 2 arises to match E for small strain with the small strain tensor, $((ds)^2 - (dS)^2)/(dS)^2 = (ds - dS)(ds + dS)/(dS)^2 \approx 2(ds - dS)/dS = 2\varepsilon$. The strain E is dimensionless, is in the original configuration, and is a measure of how the deformation has changed the length. In terms of the displacement gradient,

$$E = \frac{1}{2}(F^TF - I) = \frac{1}{2}(I + \hat{F})^T(I + \hat{F}) - \frac{1}{2}I = \frac{1}{2}(\hat{F}^T + \hat{F} + \hat{F}^T\hat{F}). \qquad (A.117)$$

---

[3]Reiterating, $F = F_{ij}\hat{e}_{z_i} \otimes \hat{e}_{y_j}$ in Eq. (A.94) while $F = F_{ij}\hat{e}_{y_i} \otimes \hat{e}_{y_j}$ in Eq. (A.103) and $F = F_{ij}\hat{e}_{z_i} \otimes \hat{e}_{z_j}$ in Eq. (A.114), where the F_{ij} are the appropriate values for the various bases.

This strain tensor is symmetric and can be computed once $\hat{F}$ or F is computed. For a small displacement gradient, the linearized small strain tensor is obtained by discarding the nonlinear term $\hat{F}^T \hat{F}$,

$$\varepsilon = (\varepsilon_{ij}) = \frac{1}{2}(\hat{F}^T + \hat{F}). \tag{A.118}$$

The Lagrangian strain is a measure of strain that corresponds to the engineering strain $\varepsilon = (L - L_0)/L_0$. The large Eulerian strain E^* that corresponds to $\varepsilon = (L - L_0)/L$ and is in the current configuration is given by

$$(ds)^2 - (dS)^2 = d\vec{x}_1^T d\vec{x}_1 - d\vec{x}_0^T d\vec{x}_0 = d\vec{x}_1^T (I - (F^{-1})^T F^{-1}) d\vec{x}_1 = 2 d\vec{x}_1^T E^* d\vec{x}_1$$

$$\Rightarrow \quad E^* = \frac{1}{2}(I - (F^{-1})^T F^{-1}) = \frac{1}{2}(I - B^{-1}), \quad \text{where } B = FF^T. \tag{A.119}$$

The velocity gradient is given by

$$L = \hat{F}(\vec{x}, \vec{v}) = \vec{\nabla} \vec{v} = (v_{i;j}), \tag{A.120}$$

where the velocity $\vec{v}$ replaces the displacement $\vec{u}$ in the displacement gradient expression (though it may be better to write $L = \hat{F}(\vec{x}_1, \vec{v})$, since it is understood that the L and $\vec{v}$ are in the current coordinates). The rate of deformation tensor is

$$D = \frac{1}{2}(L + L^T) \tag{A.121}$$

and is often used as the (logarithmic) strain rate in solid mechanics. In indicial coordinates it is

$$D_{ij} = \frac{1}{2} \left\{ \frac{1}{h_j} \hat{e}_{y_i} \cdot \frac{\partial \vec{v}}{\partial y_j} + \frac{1}{h_i} \hat{e}_{y_j} \cdot \frac{\partial \vec{v}}{\partial y_i} \right\} \quad \text{(no sum on } i, j), \tag{A.122}$$

so it is the same expression as ε_{ij}, where the velocity $\vec{v}$ replaces the displacement $\vec{u}$.

We now come back to computing the derivatives of the basis vectors, $(\partial/\partial y_i)\hat{e}_{y_j} = \gamma_{ij}^k \hat{e}_{y_k}$.[4] Though it is simplest to compute the unit coordinate vector components in Cartesian coordinates and then see how they behave, it is possible to explicitly compute these expressions in terms of the h_i. The computation is done componentwise. First, since $\hat{e}_{y_i} \cdot \hat{e}_{y_i} = 1$ (no sum on i), we have

$$\frac{\partial \hat{e}_{y_i}}{\partial y_j} \cdot \hat{e}_{y_i} = 0 \quad \text{(no sum on } i), \text{ all } i, j. \tag{A.123}$$

We now use explicit indicies to make the next computation clear. From the definition of the unit coordinate vectors we obtain the following two derivatives:

$$\frac{\partial \hat{e}_{y_1}}{\partial y_2} = \frac{\partial}{\partial y_2} \left(\frac{1}{h_1} \frac{\partial \vec{x}}{\partial y_1} \right) = \frac{\partial}{\partial y_2} \left(\frac{1}{h_1} \right) \frac{\partial \vec{x}}{\partial y_1} + \frac{1}{h_1} \frac{\partial^2 \vec{x}}{\partial y_2 \partial y_1}$$

$$= -\frac{1}{h_1} \frac{\partial h_1}{\partial y_2} \hat{e}_{y_1} + \frac{1}{h_1} \frac{\partial^2 \vec{x}}{\partial y_2 \partial y_1}. \tag{A.124}$$

Similarly,

$$\frac{\partial \hat{e}_{y_2}}{\partial y_1} = -\frac{1}{h_2} \frac{\partial h_2}{\partial y_1} \hat{e}_{y_2} + \frac{1}{h_2} \frac{\partial^2 \vec{x}}{\partial y_1 \partial y_2}. \tag{A.125}$$

---

[4]Another approach to computing γ_{ij}^k, using Γ_{ij}^k and the machinery developed in Section A.7, is outlined in Exercise 12.16. In Cartan's moving frame approach, $d\hat{e}_{y_j} = \omega_j^k \hat{e}_{y_k}$ and $\omega_j^k = \gamma_{ij}^k dy^i$.

Multiplying the first equation by h_1 and the second equation by h_2 and subtracting, assuming the order of the partial derivatives can be reversed, yields

$$h_1 \frac{\partial \hat{e}_{y_1}}{\partial y_2} - h_2 \frac{\partial \hat{e}_{y_2}}{\partial y_1} = -\frac{\partial h_1}{\partial y_2} \hat{e}_{y_1} + \frac{\partial h_2}{\partial y_1} \hat{e}_{y_2}. \tag{A.126}$$

Taking the inner product with $\hat{e}_{y_1}$, using orthonormality and Eq. (A.123), this equation yields

$$\frac{\partial \hat{e}_{y_2}}{\partial y_1} \cdot \hat{e}_{y_1} = \frac{1}{h_2} \frac{\partial h_1}{\partial y_2} \quad \Rightarrow \quad \frac{\partial \hat{e}_{y_j}}{\partial y_i} \cdot \hat{e}_{y_i} = \frac{1}{h_j} \frac{\partial h_i}{\partial y_j} \quad \text{(no sum on } i, j), \quad i \neq j, \tag{A.127}$$

where we have inferred the general case from the specific. Next, since $\hat{e}_{y_1} \cdot \hat{e}_{y_2} = 0$, taking a derivative produces

$$\frac{\partial \hat{e}_{y_1}}{\partial y_1} \cdot \hat{e}_{y_2} = -\frac{\partial \hat{e}_{y_2}}{\partial y_1} \cdot \hat{e}_{y_1} \tag{A.128}$$

(this result allows us to switch indices on the $\hat{e}_{y_i}$ by multiplying by -1.) Combining Eq. (A.128) with (A.127) yields

$$\frac{\partial \hat{e}_{y_1}}{\partial y_1} \cdot \hat{e}_{y_2} = -\frac{1}{h_2} \frac{\partial h_1}{\partial y_2} \quad \Rightarrow \quad \frac{\partial \hat{e}_{y_i}}{\partial y_i} \cdot \hat{e}_{y_j} = -\frac{1}{h_j} \frac{\partial h_i}{\partial y_j} \quad \text{(no sum on } i, j), \quad i \neq j. \tag{A.129}$$

The difference between Eqs. (A.127) and (A.129) is in the location of the indices. So far, the components have been computed if at least two of the indices are the same, so now the case of all indices being different is addressed. Taking the inner product of Eq. (A.126) with $\hat{e}_{y_3}$ yields

$$\frac{1}{h_2} \frac{\partial \hat{e}_{y_1}}{\partial y_2} \cdot \hat{e}_{y_3} = \frac{1}{h_1} \frac{\partial \hat{e}_{y_2}}{\partial y_1} \cdot \hat{e}_{y_3} \tag{A.130}$$

(this result allows us to switch indices between $\hat{e}_{y_i}$ and the derivative y_j and its corresponding h_j.) This result, with the appropriate generalization to different configurations, along with the appropriate use of Eq. (A.128), again for different unit coordinate vectors and derivatives, will now be chained together in an alternating fashion:

$$\frac{1}{h_2} \frac{\partial \hat{e}_{y_1}}{\partial y_2} \cdot \hat{e}_{y_3} = \frac{1}{h_1} \frac{\partial \hat{e}_{y_2}}{\partial y_1} \cdot \hat{e}_{y_3} = -\frac{1}{h_1} \frac{\partial \hat{e}_{y_3}}{\partial y_1} \cdot \hat{e}_{y_2} = -\frac{1}{h_3} \frac{\partial \hat{e}_{y_1}}{\partial y_3} \cdot \hat{e}_{y_2}$$

$$= \frac{1}{h_3} \frac{\partial \hat{e}_{y_2}}{\partial y_3} \cdot \hat{e}_{y_1} = \frac{1}{h_2} \frac{\partial \hat{e}_{y_3}}{\partial y_2} \cdot \hat{e}_{y_1} = -\frac{1}{h_2} \frac{\partial \hat{e}_{y_1}}{\partial y_2} \cdot \hat{e}_{y_3}. \tag{A.131}$$

The first expression is equal to its negative (since space has three dimensions, and thus $(-1)^3 = -1$) and so it is zero and all the terms in the chain are zero. We conclude that

$$\frac{\partial \hat{e}_{y_i}}{\partial y_j} \cdot \hat{e}_{y_k} = 0 \quad i, j, k \text{ all different.} \tag{A.132}$$

Now we have all the components of the $\partial \hat{e}_{y_i}/\partial y_j$ and it is possible to write down the result. A little thought shows there are two cases, and the result is

$$\frac{\partial \hat{e}_{y_i}}{\partial y_i} = -\frac{1}{h_j} \frac{\partial h_i}{\partial y_j} \hat{e}_{y_j} - \frac{1}{h_k} \frac{\partial h_i}{\partial y_k} \hat{e}_{y_k} \quad \text{(no sum on } i, j, k), \quad i, j, k \text{ all different,}$$

$$\frac{\partial \hat{e}_{y_i}}{\partial y_j} = \frac{1}{h_i} \frac{\partial h_j}{\partial y_i} \hat{e}_{y_j} \quad \text{(no sum on } i, j), \quad i \neq j, \tag{A.133}$$

which implies Eqs. (A.54) and (A.98) as used above. To get a better sense of these derivatives of the basis functions, let $\vec{E}_y$ be the column vector of unit basis functions; then the Christoffel symbols γ_{ij}^k are

$$\frac{\partial}{\partial y_i}\vec{E}_y = \frac{\partial}{\partial y_i}\begin{pmatrix}\hat{e}_{y1}\\\hat{e}_{y2}\\\hat{e}_{y3}\end{pmatrix} = (\gamma_{ij}^k)\vec{E}_y = \begin{pmatrix}\gamma_{i1}^1 & \gamma_{i1}^2 & \gamma_{i1}^3\\\gamma_{i2}^1 & \gamma_{i2}^2 & \gamma_{i2}^3\\\gamma_{i3}^1 & \gamma_{i3}^2 & \gamma_{i3}^3\end{pmatrix}\begin{pmatrix}\hat{e}_{y1}\\\hat{e}_{y2}\\\hat{e}_{y3}\end{pmatrix}. \tag{A.134}$$

Thus, for γ_{ij}^k, where j is the row index and k is the column index,

$$(\gamma_{1j}^k) = \begin{pmatrix}0 & -\frac{1}{h_2}\frac{\partial h_1}{\partial y_2} & -\frac{1}{h_3}\frac{\partial h_1}{\partial y_3}\\\frac{1}{h_2}\frac{\partial h_1}{\partial y_2} & 0 & 0\\\frac{1}{h_3}\frac{\partial h_1}{\partial y_3} & 0 & 0\end{pmatrix}, \quad (\gamma_{2j}^k) = \begin{pmatrix}0 & \frac{1}{h_1}\frac{\partial h_2}{\partial y_1} & 0\\-\frac{1}{h_1}\frac{\partial h_2}{\partial y_1} & 0 & -\frac{1}{h_3}\frac{\partial h_2}{\partial y_3}\\0 & \frac{1}{h_3}\frac{\partial h_2}{\partial y_3} & 0\end{pmatrix},$$

$$(\gamma_{3j}^k) = \begin{pmatrix}0 & 0 & \frac{1}{h_1}\frac{\partial h_3}{\partial y_1}\\0 & 0 & \frac{1}{h_2}\frac{\partial h_3}{\partial y_2}\\-\frac{1}{h_1}\frac{\partial h_3}{\partial y_1} & -\frac{1}{h_2}\frac{\partial h_3}{\partial y_2} & 0\end{pmatrix}. \tag{A.135}$$

There are up to 12 nonzero entries. Each matrix is skew symmetric: $\gamma_{ij}^k = -\gamma_{ik}^j$, so there are really only six expressions that need to be computed. Note that $\gamma_{ij}^j = 0 \neq \gamma_{ji}^j$ ($i \neq j$, no sum on j); hence γ_{ij}^k are not symmetric in ij, in contrast to Γ_{ij}^k.

We finally make some remarks about integration. Since for matrices $\det(A^T) = \det(A)$ and $\det(AB) = \det(A)\det(B)$, the fact that $g = F_0^T F_0$ says that the volume differential for the configuration $\vec{x}_0$ is

$$dvol(\vec{x}_0) = \sqrt{\det(g)}\,dy_1 dy_2 dy_3 = \det(F_0)dy_1 dy_2 dy_3. \tag{A.136}$$

For an arbitrary volume V that deforms to contain the same mass, the mass in the volume is

$$\int_{V(\vec{x}_0)}\rho_0 dvol(\vec{x}_0) = \int_{V(\vec{x}_1)}\rho\, dvol(\vec{x}_1) = \int_{V(\vec{z})}\rho\det(F_1)dz_1 dz_2 dz_3$$

$$= \int_{V(\vec{y})}\rho\det(F_1)\det(F_J)dy_1 dy_2 dy_3 \tag{A.137}$$

$$= \int_{V(\vec{x}_0)}\rho\det(F_1)\det(F_J)\det(F_0)^{-1}\,dvol(\vec{x}_0).$$

As the volume was arbitrary and since $\det(F_1)\det(F_J)\det(F_0)^{-1} = \det(F)$ (from Eq. A.93),

$$\rho_0 = \rho\det(F) = \rho\det(F(\vec{x},\vec{u})) = \rho\det(I + \hat{F}(\vec{x},\vec{u})) \tag{A.138}$$

relates the initial density to the deformed configuration density.

These results complete our discussion of the various differential terms and operators in arbitrary orthogonal coordinates. The three specific cases used in the text will now be examined.

A.4. Cartesian Coordinates

It is assumed that Cartesian coordinates are understood, and the only reason we list it is to compare with the cylindrical and spherical cases. In Cartesian coordinates, a given point is identified by

$$\vec{x} = x\hat{e}_x + y\hat{e}_y + z\hat{e}_z. \tag{A.139}$$

In our notation from the previous section, $y_1 = x = x_1$, $y_2 = y = x_2$, and $y_3 = z = x_3$. The unit coordinate vectors are fixed, and in Cartesian coordinates are

$$\hat{e}_x = \begin{pmatrix} 1 \\ 0 \\ 0 \end{pmatrix}_{\text{Cart.}}, \quad \hat{e}_y = \begin{pmatrix} 0 \\ 1 \\ 0 \end{pmatrix}_{\text{Cart.}}, \quad \hat{e}_z = \begin{pmatrix} 0 \\ 0 \\ 1 \end{pmatrix}_{\text{Cart.}} . \tag{A.140}$$

Of course, this is true by definition, since it says $\hat{e}_x = 1 \cdot \hat{e}_x + 0 \cdot \hat{e}_y + 0 \cdot \hat{e}_z$, for example (it is what the column notation means, Eq. A.4). Written in the order (x, y, z), the unit coordinate vectors produce a right-handed system as $\hat{e}_x \times \hat{e}_y = \hat{e}_z$. The unit direction vectors always point in the same direction,

$$\frac{\partial \hat{e}_{x_i}}{\partial x_j} = 0, \quad \text{all } i, j \quad \text{or} \quad \gamma_{ij}^k = 0, \quad \text{all } i, j, k. \tag{A.141}$$

The metric g for the Cartesian coordinates is

$$g_{xx} = 1, \quad g_{yy} = 1, \quad g_{zz} = 1, \text{ and all off-diagonal } g_{ij} = 0,$$
$$h_x = \sqrt{g_{xx}} = 1, \quad h_y = \sqrt{g_{yy}} = 1, \quad h_z = \sqrt{g_{zz}} = 1,$$
$$(ds)^2 = g_{ij} dy_i dy_j = (dx)^2 + (dy)^2 + (dz)^2,$$
$$dvol = \sqrt{\det(g)} dy_1 dy_2 dy_3 = dx dy dz. \tag{A.142}$$

Working with the results of the previous section, the h_i, and the behavior of the unit coordinate direction vectors, it is possible to write the various differential operators from the previous section. We have

$$\text{grad}(f) = \vec{\nabla} f = (f_{;i}) = \frac{\partial f}{\partial x} \hat{e}_x + \frac{\partial f}{\partial y} \hat{e}_y + \frac{\partial f}{\partial z} \hat{e}_z,$$

$$\text{div}(\vec{f}) = \vec{\nabla} \cdot \vec{f} = f_{i;i} = \frac{\partial f_x}{\partial x} + \frac{\partial f_y}{\partial y} + \frac{\partial f_z}{\partial z},$$

$$\Delta f = \nabla^2(f) = \text{div}(\text{grad}(f)) = f_{;ii} = \frac{\partial^2 f}{\partial x^2} + \frac{\partial^2 f}{\partial y^2} + \frac{\partial^2 f}{\partial z^2},$$

$$\text{curl}(\vec{f}) = \vec{\nabla} \times \vec{f} = \left\{ \frac{\partial f_z}{\partial y} - \frac{\partial f_y}{\partial z} \right\} \hat{e}_x + \left\{ \frac{\partial f_x}{\partial z} - \frac{\partial f_z}{\partial x} \right\} \hat{e}_y + \left\{ \frac{\partial f_y}{\partial x} - \frac{\partial f_x}{\partial y} \right\} \hat{e}_z,$$
$$\tag{A.143}$$

and for the momentum balance we have (the left-hand side excluding the body forces is denoted $\vec{\nabla} \cdot \sigma = (\sigma_{ji;j})$)

$$\frac{\partial \sigma_{xx}}{\partial x} + \frac{\partial \sigma_{yx}}{\partial y} + \frac{\partial \sigma_{zx}}{\partial z} + \rho b_x = \rho a_x,$$

$$\frac{\partial \sigma_{xy}}{\partial x} + \frac{\partial \sigma_{yy}}{\partial y} + \frac{\partial \sigma_{zy}}{\partial z} + \rho b_y = \rho a_y, \tag{A.144}$$

$$\frac{\partial \sigma_{xz}}{\partial x} + \frac{\partial \sigma_{yz}}{\partial y} + \frac{\partial \sigma_{zz}}{\partial z} + \rho b_z = \rho a_z,$$

where $\vec{b}$ is the body force, $\vec{a} = D\vec{v}/Dt$ is the acceleration, and it has not been assumed that the stress tensor is symmetric.

The details of the velocity computation are described in the cylindrical geometry section. The velocity in the Lagrangian frame is

$$\vec{v} = \frac{D\vec{x}(\vec{y}_0, t)}{Dt} = \frac{\partial}{\partial t}(x\hat{e}_x + y\hat{e}_y + z\hat{e}_z) = \frac{\partial x}{\partial t}\hat{e}_r + \frac{\partial y}{\partial t}\hat{e}_y + \frac{\partial z}{\partial t}\hat{e}_z = v_x\hat{e}_x + v_y\hat{e}_y + v_z\hat{e}_z. \tag{A.145}$$

In the Eulerian computation, there is an advection term $(\partial/\partial y_i)(\partial y_i/\partial t)$, and the velocity vector is

$$
\vec{v} = \frac{D\vec{x}(\vec{y})}{Dt} = \left(\frac{\partial x}{\partial t}\frac{\partial}{\partial x} + \frac{\partial y}{\partial t}\frac{\partial}{\partial y} + \frac{\partial z}{\partial t}\frac{\partial}{\partial z} \right)(x\hat{e}_x + y\hat{e}_y + z\hat{e}_z)
$$

$$
= \frac{\partial x}{\partial t}\hat{e}_x + \frac{\partial y}{\partial t}\hat{e}_y + \frac{\partial z}{\partial t}\hat{e}_z = v_x\hat{e}_x + v_y\hat{e}_y + v_z\hat{e}_z. \tag{A.146}
$$

The velocity in each reference frame is the same.

The acceleration is more complicated, but it is the same overall approach. For the Lagrangian frame there is no advection term,

$$
\vec{a} = \frac{D\vec{v}}{Dt} = \frac{\partial\vec{v}}{\partial t}, \tag{A.147}
$$

or

$$
a_x = \frac{\partial v_x}{\partial t}, \qquad a_y = \frac{\partial v_y}{\partial t}, \qquad a_z = \frac{\partial v_z}{\partial t}. \tag{A.148}
$$

For the Eulerian frame, there is an advection term, and

$$
\vec{a} = \frac{D\vec{v}}{Dt} = \frac{\partial\vec{v}}{\partial t} + \left(\frac{\partial x}{\partial t}\frac{\partial}{\partial x} + \frac{\partial y}{\partial t}\frac{\partial}{\partial y} + \frac{\partial z}{\partial t}\frac{\partial}{\partial z} \right)\vec{v}
$$

$$
= \frac{\partial\vec{v}}{\partial t} + \left(v_x\frac{\partial}{\partial x} + v_y\frac{\partial}{\partial y} + v_z\frac{\partial}{\partial z} \right)\vec{v} = a_x\hat{e}_x + a_y\hat{e}_y + a_z\hat{e}_z. \tag{A.149}
$$

Here the unit vectors do not depend upon time or position, yielding

$$
a_x = \frac{\partial v_x}{\partial t} + v_x\frac{\partial v_x}{\partial x} + v_y\frac{\partial v_x}{\partial y} + v_z\frac{\partial v_x}{\partial z},
$$

$$
a_y = \frac{\partial v_y}{\partial t} + v_x\frac{\partial v_y}{\partial x} + v_y\frac{\partial v_y}{\partial y} + v_z\frac{\partial v_y}{\partial z}, \tag{A.150}
$$

$$
a_z = \frac{\partial v_z}{\partial t} + v_x\frac{\partial v_z}{\partial x} + v_y\frac{\partial v_z}{\partial y} + v_z\frac{\partial v_z}{\partial z}.
$$

Finally we write down the displacement gradient

$$
\hat{F}(\vec{x}, \vec{u}) = \vec{\nabla}\vec{u} = (u_{i;j}) = \begin{pmatrix} \hat{F}_{xx} & \hat{F}_{xy} & \hat{F}_{xz} \\ \hat{F}_{yx} & \hat{F}_{yy} & \hat{F}_{yz} \\ \hat{F}_{zx} & \hat{F}_{zy} & \hat{F}_{zz} \end{pmatrix} = \begin{pmatrix} \frac{\partial u_x}{\partial x} & \frac{\partial u_x}{\partial y} & \frac{\partial u_x}{\partial z} \\ \frac{\partial u_y}{\partial x} & \frac{\partial u_y}{\partial y} & \frac{\partial u_y}{\partial z} \\ \frac{\partial u_z}{\partial x} & \frac{\partial u_z}{\partial y} & \frac{\partial u_z}{\partial z} \end{pmatrix} \tag{A.151}
$$

and the small strain tensor $\frac{1}{2}(\hat{F} + F)$,

$$
\varepsilon_{xx} = \frac{\partial u_x}{\partial x}, \quad \varepsilon_{yy} = \frac{\partial u_y}{\partial y}, \quad \varepsilon_{zz} = \frac{\partial u_z}{\partial z},
$$

$$
\varepsilon_{xy} = \frac{1}{2}\left(\frac{\partial u_x}{\partial y} + \frac{\partial u_y}{\partial x} \right), \quad \varepsilon_{yz} = \frac{1}{2}\left(\frac{\partial u_y}{\partial z} + \frac{\partial u_z}{\partial y} \right), \quad \varepsilon_{xz} = \frac{1}{2}\left(\frac{\partial u_x}{\partial z} + \frac{\partial u_z}{\partial x} \right). \tag{A.152}
$$

The velocity gradient L is given by $L = \hat{F}(\vec{x}, \vec{v}) = \vec{\nabla}\vec{v} = (v_{i;j})$ and the rate of deformation tensor $D = \frac{1}{2}(L+L^T)$ has D_{ij} the same as ε_{ij}, where the displacement $\vec{u}$ is replaced by the velocity $\vec{v}$. If the new, deformed locations (x, y, z) are written in terms of the original locations (X, Y, Z) – for example where $x = x(X, Y, Z)$ – then the deformation gradient is given by

$$
F = \begin{pmatrix} F_{xX} & F_{xY} & F_{xZ} \\ F_{yX} & F_{yY} & F_{yZ} \\ F_{zX} & F_{zY} & F_{zZ} \end{pmatrix} = \begin{pmatrix} \frac{\partial x}{\partial X} & \frac{\partial x}{\partial Y} & \frac{\partial x}{\partial Z} \\ \frac{\partial y}{\partial X} & \frac{\partial y}{\partial Y} & \frac{\partial y}{\partial Z} \\ \frac{\partial z}{\partial X} & \frac{\partial z}{\partial Y} & \frac{\partial z}{\partial Z} \end{pmatrix}, \tag{A.153}
$$

where the capitalization on the subscripts indicates original coordinates.

A.5. Cylindrical Coordinates

In cylindrical coordinates, the x–y plane is represented by polar coordinates and the out-of-plane direction is represented by z. The relationship between the cylindrical and Cartesian coordinates is

$$r = \sqrt{x^2 + y^2}, \qquad x = r\cos(\theta), \tag{A.154}$$

$$\theta = \tan^{-1}(y/x), \qquad y = r\sin(\theta), \tag{A.155}$$

$$z = z, \qquad z = z, \tag{A.156}$$

where $r \geq 0$ and $0 \leq \theta < 2\pi$. In cylindrical coordinates, a given point is identified by

$$\vec{x} = r\hat{e}_r + z\hat{e}_z. \tag{A.157}$$

In our notation from Section A.3, $y_1 = r$, $y_2 = \theta$, and $y_3 = z$. It is easiest to understand the behavior of the three coordinate direction unit vectors by determining what they are in Cartesian coordinates; however, once the behavior of the coordinate vectors is determined there will be no further reference to Cartesian coordinates. From Eq. (A.52), the direction unit vectors are

$$\hat{e}_r = \begin{pmatrix} \cos(\theta) \\ \sin(\theta) \\ 0 \end{pmatrix}_{\text{Cart.}}, \quad \hat{e}_\theta = \frac{\partial \hat{e}_r}{\partial \theta} = \begin{pmatrix} -\sin(\theta) \\ \cos(\theta) \\ 0 \end{pmatrix}_{\text{Cart.}}, \quad \hat{e}_z = \begin{pmatrix} 0 \\ 0 \\ 1 \end{pmatrix}_{\text{Cart.}}. \tag{A.158}$$

Written in the order (r, θ, z), the unit coordinate vectors produce a right-handed system as $\hat{e}_r \times \hat{e}_\theta = \hat{e}_z$. The relations between the Cartesian unit vectors and the cylindrical unit vectors can be represented as

$$\begin{pmatrix} \hat{e}_r \\ \hat{e}_\theta \\ \hat{e}_z \end{pmatrix} = {}_{\text{Cyl.}}Q_{\text{Cart}} \begin{pmatrix} \hat{e}_x \\ \hat{e}_y \\ \hat{e}_z \end{pmatrix} = \begin{pmatrix} \cos(\theta) & \sin(\theta) & 0 \\ -\sin(\theta) & \cos(\theta) & 0 \\ 0 & 0 & 1 \end{pmatrix} \begin{pmatrix} \hat{e}_x \\ \hat{e}_y \\ \hat{e}_z \end{pmatrix},$$

$$\begin{pmatrix} \hat{e}_x \\ \hat{e}_y \\ \hat{e}_z \end{pmatrix} = {}_{\text{Cart.}}Q_{\text{Cyl.}} \begin{pmatrix} \hat{e}_r \\ \hat{e}_\theta \\ \hat{e}_z \end{pmatrix} = \begin{pmatrix} \cos(\theta) & -\sin(\theta) & 0 \\ \sin(\theta) & \cos(\theta) & 0 \\ 0 & 0 & 1 \end{pmatrix} \begin{pmatrix} \hat{e}_r \\ \hat{e}_\theta \\ \hat{e}_z \end{pmatrix}. \tag{A.159}$$

The complexity of the cylindrical and spherical coordinate systems is that the unit direction vectors are not constant. We also find

$$\frac{\partial \hat{e}_r}{\partial \theta} = \hat{e}_\theta, \quad \frac{\partial \hat{e}_\theta}{\partial \theta} = -\hat{e}_r, \quad \frac{\partial \hat{e}_r}{\partial r} = \frac{\partial \hat{e}_\theta}{\partial r} = 0, \quad \frac{\partial \hat{e}_z}{\partial y_i} = 0, \quad \frac{\partial \hat{e}_{y_i}}{\partial z} = 0, \quad \text{all } i,$$

or $\quad \gamma^\theta_{\theta r} = -\gamma^r_{\theta\theta} = 1, \quad$ all other $\gamma^k_{ij} = 0$ including $\gamma^\theta_{r\theta} = 0$. $\tag{A.160}$

When coordinates are written in terms of Cartesian basis vectors, as in Eq. (A.154), it is simplest to compute the metric g from Eq. (A.44), because the derivatives of the Cartesian basis vectors are zero. For example,

$$g_{\theta\theta} = \frac{\partial \vec{x}}{\partial \theta} \cdot \frac{\partial \vec{x}}{\partial \theta} = \frac{\partial x_i}{\partial \theta} \frac{\partial x_i}{\partial \theta} = r^2 \sin^2(\theta) + r^2 \cos^2(\theta) + 0 = r^2. \tag{A.161}$$

The same computation performed within cylindrical coordinates still uses Eq. (A.44), though when using non-Cartesian bases the derivatives of the basis vectors are not zero and must be included. For example,

$$g_{\theta\theta} = \frac{\partial \vec{x}}{\partial \theta} \cdot \frac{\partial \vec{x}}{\partial \theta} = \frac{\partial(r\hat{e}_r + z\hat{e}_z)}{\partial \theta} \cdot \frac{\partial(r\hat{e}_r + z\hat{e}_z)}{\partial \theta} = r\hat{e}_\theta \cdot r\hat{e}_\theta = r^2. \tag{A.162}$$

The metric g for the cylindrical coordinates is

$$g_{rr} = 1, \quad g_{\theta\theta} = r^2, \quad g_{zz} = 1, \text{ and all off-diagonal } g_{ij} = 0,$$
$$h_r = \sqrt{g_{rr}} = 1, \quad h_\theta = \sqrt{g_{\theta\theta}} = r, \quad h_z = \sqrt{g_{zz}} = 1,$$
$$(ds)^2 = g_{ij}dy_i dy_j = (dr)^2 + r^2(d\theta)^2 + (dz)^2, \quad \text{(A.163)}$$
$$dvol = \sqrt{\det(g)}dy_1 dy_2 dy_3 = r\,dr\,d\theta\,dz.$$

Working with the results of Section A.3, the h_i, and the behavior of the unit coordinate direction vectors, it is possible to write the various differential operators from the previous section. We have

$$\text{grad}(f) = \vec{\nabla}f = (f_{;i}) = \frac{\partial f}{\partial r}\hat{e}_r + \frac{1}{r}\frac{\partial f}{\partial \theta}\hat{e}_\theta + \frac{\partial f}{\partial z}\hat{e}_z,$$

$$\text{div}(\vec{f}) = \vec{\nabla} \cdot \vec{f} = f_{i;i} = \frac{1}{r}\frac{\partial(rf_r)}{\partial r} + \frac{1}{r}\frac{\partial f_\theta}{\partial \theta} + \frac{\partial f_z}{\partial z},$$

$$\Delta f = \nabla^2(f) = \text{div}(\text{grad}(f)) = f_{;ii} = \frac{1}{r}\frac{\partial}{\partial r}\left(r\frac{\partial f}{\partial r}\right) + \frac{1}{r^2}\frac{\partial^2 f}{\partial \theta^2} + \frac{\partial^2 f}{\partial z^2}, \quad \text{(A.164)}$$

$$\text{curl}(\vec{f}) = \vec{\nabla} \times \vec{f} = \left\{\frac{1}{r}\frac{\partial f_z}{\partial \theta} - \frac{\partial f_\theta}{\partial z}\right\}\hat{e}_r + \left\{\frac{\partial f_r}{\partial z} - \frac{\partial f_z}{\partial r}\right\}\hat{e}_\theta$$
$$+ \left\{\frac{1}{r}\frac{\partial(rf_\theta)}{\partial r} - \frac{1}{r}\frac{\partial f_r}{\partial \theta}\right\}\hat{e}_z,$$

and for the momentum balance we have (the left-hand side excluding the body forces is denoted $\vec{\nabla} \cdot \sigma = (\sigma_{ji;j})$)

$$\frac{\partial \sigma_{rr}}{\partial r} + \frac{1}{r}\frac{\partial \sigma_{\theta r}}{\partial \theta} + \frac{\partial \sigma_{zr}}{\partial z} + \frac{\sigma_{rr} - \sigma_{\theta\theta}}{r} + \rho b_r = \rho a_r,$$
$$\frac{\partial \sigma_{r\theta}}{\partial r} + \frac{1}{r}\frac{\partial \sigma_{\theta\theta}}{\partial \theta} + \frac{\partial \sigma_{z\theta}}{\partial z} + \frac{\sigma_{r\theta} + \sigma_{\theta r}}{r} + \rho b_\theta = \rho a_\theta, \quad \text{(A.165)}$$
$$\frac{\partial \sigma_{rz}}{\partial r} + \frac{1}{r}\frac{\partial \sigma_{\theta z}}{\partial \theta} + \frac{\partial \sigma_{zz}}{\partial z} + \frac{\sigma_{rz}}{r} + \rho b_z = \rho a_z,$$

where $\vec{b}$ is the body force, $\vec{a} = D\vec{v}/Dt$ is the acceleration, and it has not been assumed that the stress tensor is symmetric.

To calculate the velocity, it is necessary to choose whether we are using the Lagrangian or Eulerian frame. The velocity in both cases is obtained using the material derivative $D\vec{x}/Dt$,

$$\frac{D}{Dt} = \frac{\partial}{\partial t} + \frac{\partial z_i}{\partial t}\frac{\partial}{\partial z_i} \quad \text{(A.166)}$$

(see Chapter 2), where z_i represents whatever variables are being used to spatially describe the material, current coordinates y_i, or original coordinates y_{0i}. The computations will yield the same result for the velocity (but not the acceleration), but we will explicitly go through the computations to highlight the differences. For Lagrangian coordinates, we have $\vec{x} = \vec{x}(\vec{y}_0, t)$ (where there is no current position dependence, just original position dependence). The unit vectors have a dependence $\hat{e}_{y_i} = \hat{e}_{y_i}(\vec{x}(\vec{y}_0, t))$ and thus have a time dependence. There will be no advection term because $(\partial/\partial y_{0i})(\partial y_{0i}/\partial t) = 0$ as $\partial y_{0i}/\partial t = 0$, since the original Lagrangian material coordinates are fixed in space and do not change with time.

Using $\partial\hat{e}_r/\partial t = (\partial\hat{e}_r/\partial\theta)(\partial\theta/\partial t) = (\partial\theta/\partial t)\hat{e}_\theta$, the velocity is

$$
\begin{aligned}
\vec{v} = \frac{D\vec{x}(\vec{y}_0, t)}{Dt} &= \frac{\partial}{\partial t}(r\hat{e}_r + z\hat{e}_z) = \frac{\partial r}{\partial t}\hat{e}_r + r\frac{\partial\theta}{\partial t}\frac{\partial\hat{e}_r}{\partial\theta} + \frac{\partial z}{\partial t}\hat{e}_z \\
&= \frac{\partial r}{\partial t}\hat{e}_r + r\frac{\partial\theta}{\partial t}\hat{e}_\theta + \frac{\partial z}{\partial t}\hat{e}_z = v_r\hat{e}_r + v_\theta\hat{e}_\theta + v_z\hat{e}_z.
\end{aligned}
\tag{A.167}
$$

For Eulerian coordinates we have $\vec{x} = \vec{x}(\vec{y}) = \vec{x}$ (where there is no time dependence) and the unit vectors also have no time dependence as $\hat{e}_{y_i} = \hat{e}_{y_i}(\vec{x}(\vec{y}))$. In the Eulerian computation, there is an advection term $(\partial/\partial y_i)(\partial y_i/\partial t)$. Using $(\partial\hat{e}_r/\partial\theta) = \hat{e}_\theta$ and noting that $r\partial\theta/\partial t = v_\theta$, the Eulerian velocity vector is given by

$$
\begin{aligned}
\vec{v} = \frac{D\vec{x}(\vec{y})}{Dt} &= \left(\frac{\partial r}{\partial t}\frac{\partial}{\partial r} + \frac{\partial\theta}{\partial t}\frac{\partial}{\partial\theta} + \frac{\partial z}{\partial t}\frac{\partial}{\partial z}\right)(r\hat{e}_r + z\hat{e}_z) \\
&= \frac{\partial r}{\partial t}\hat{e}_r + r\frac{\partial\theta}{\partial t}\hat{e}_\theta + \frac{\partial z}{\partial t}\hat{e}_z = v_r\hat{e}_r + v_\theta\hat{e}_\theta + v_z\hat{e}_z.
\end{aligned}
\tag{A.168}
$$

The velocity in each reference frame is the same.

The acceleration is more complicated, but it is the same overall approach. For the Lagrangian frame there is no advection term,

$$
\vec{a} = \frac{D\vec{v}}{Dt} = \frac{\partial\vec{v}}{\partial t},
\tag{A.169}
$$

and, recalling the time dependence of the unit vectors,

$$
\begin{aligned}
a_r &= \frac{\partial v_r}{\partial t} - \frac{v_\theta^2}{r}, \\
a_\theta &= \frac{\partial v_\theta}{\partial t} + \frac{v_\theta v_r}{r}, \\
a_z &= \frac{\partial v_z}{\partial t}.
\end{aligned}
\tag{A.170}
$$

For the Eulerian frame, there is an advection term

$$
\begin{aligned}
\vec{a} = \frac{D\vec{v}}{Dt} = \frac{\partial\vec{v}}{\partial t} &+ \left(\frac{\partial r}{\partial t}\frac{\partial}{\partial r} + \frac{\partial\theta}{\partial t}\frac{\partial}{\partial\theta} + \frac{\partial z}{\partial t}\frac{\partial}{\partial z}\right)\vec{v} \\
&= \frac{\partial\vec{v}}{\partial t} + \left(v_r\frac{\partial}{\partial r} + \frac{v_\theta}{r}\frac{\partial}{\partial\theta} + v_z\frac{\partial}{\partial z}\right)\vec{v} = a_r\hat{e}_r + a_\theta\hat{e}_\theta + a_z\hat{e}_z.
\end{aligned}
\tag{A.171}
$$

Here the unit vectors do not depend upon time, but they do depend on the current position, yielding

$$
\begin{aligned}
a_r &= \frac{\partial v_r}{\partial t} + v_r\frac{\partial v_r}{\partial r} + \frac{v_\theta}{r}\frac{\partial v_r}{\partial\theta} - \frac{v_\theta^2}{r} + v_z\frac{\partial v_r}{\partial z}, \\
a_\theta &= \frac{\partial v_\theta}{\partial t} + v_r\frac{\partial v_\theta}{\partial r} + \frac{v_\theta}{r}\frac{\partial v_\theta}{\partial\theta} + \frac{v_\theta v_r}{r} + v_z\frac{\partial v_\theta}{\partial z}, \\
a_z &= \frac{\partial v_z}{\partial t} + v_r\frac{\partial v_z}{\partial r} + \frac{v_\theta}{r}\frac{\partial v_z}{\partial\theta} + v_z\frac{\partial v_z}{\partial z}.
\end{aligned}
\tag{A.172}
$$

Finally, we write down the displacement gradient

$$
\hat{F}(\vec{x}, \vec{u}) = \vec{\nabla}\vec{u} = (u_{i;j}) = \begin{pmatrix} \hat{F}_{rr} & \hat{F}_{r\theta} & \hat{F}_{rz} \\ \hat{F}_{\theta r} & \hat{F}_{\theta\theta} & \hat{F}_{\theta z} \\ \hat{F}_{zr} & \hat{F}_{z\theta} & \hat{F}_{zz} \end{pmatrix} = \begin{pmatrix} \frac{\partial u_r}{\partial r} & \frac{1}{r}\frac{\partial u_r}{\partial\theta} - \frac{u_\theta}{r} & \frac{\partial u_r}{\partial z} \\ \frac{\partial u_\theta}{\partial r} & \frac{1}{r}\frac{\partial u_\theta}{\partial\theta} + \frac{u_r}{r} & \frac{\partial u_\theta}{\partial z} \\ \frac{\partial u_z}{\partial r} & \frac{1}{r}\frac{\partial u_z}{\partial\theta} & \frac{\partial u_z}{\partial z} \end{pmatrix}
\tag{A.173}
$$

and the small strain tensor $\frac{1}{2}(\hat{F} + F)$,

$$\varepsilon_{rr} = \frac{\partial u_r}{\partial r}, \quad \varepsilon_{\theta\theta} = \frac{1}{r}\frac{\partial u_\theta}{\partial \theta} + \frac{u_r}{r}, \quad \varepsilon_{zz} = \frac{\partial u_z}{\partial z},$$

$$\varepsilon_{r\theta} = \frac{1}{2}\left(\frac{1}{r}\frac{\partial u_r}{\partial \theta} + \frac{\partial u_\theta}{\partial r} - \frac{u_\theta}{r}\right), \quad (A.174)$$

$$\varepsilon_{\theta z} = \frac{1}{2}\left(\frac{\partial u_\theta}{\partial z} + \frac{1}{r}\frac{\partial u_z}{\partial \theta}\right), \quad \varepsilon_{rz} = \frac{1}{2}\left(\frac{\partial u_r}{\partial z} + \frac{\partial u_z}{\partial r}\right).$$

The velocity gradient L is given by $L = \hat{F}(\vec{x}, \vec{v}) = \vec{\nabla}\vec{v} = (v_{i;j})$ and the rate of deformation tensor $D = \frac{1}{2}(L+L^T)$ has D_{ij} the same as ε_{ij}, where the displacement $\vec{u}$ is replaced by the velocity $\vec{v}$. If the deformation is explicitly written in the new coordinates (r, θ, z) in terms of the original coordinates (R, Θ, Z) – for example, $r = r(R, \Theta, Z)$ – then the deformation gradient is

$$F = \begin{pmatrix} F_{rR} & F_{r\Theta} & F_{rZ} \\ F_{\theta R} & F_{\theta\Theta} & F_{\theta Z} \\ F_{zR} & F_{z\Theta} & F_{zZ} \end{pmatrix} = \begin{pmatrix} \frac{\partial r}{\partial R} & \frac{1}{R}\frac{\partial r}{\partial \Theta} & \frac{\partial r}{\partial Z} \\ r\frac{\partial\theta}{\partial R} & \frac{r}{R}\frac{\partial\theta}{\partial\Theta} & r\frac{\partial\theta}{\partial Z} \\ \frac{\partial z}{\partial R} & \frac{1}{R}\frac{\partial z}{\partial\Theta} & \frac{\partial z}{\partial Z} \end{pmatrix}. \quad (A.175)$$

In this form, $F^T F$ is in the original coordinate system, appropriate for Lagrangian strain E, and $B = FF^T$ is in the current coordinate system, appropriate for Eulerian strain E^*.

A.6. Spherical Coordinates

In the standard representation of spherical coordinates, the polar angle (corresponding to latitude) is represented by θ, where $\theta = 0$ represents the north pole and $\theta = \pi$ represents the south pole (these correspond to the z-axis). The azimuthal angle or longitude is represented by φ. Notice the use of θ differs from its use in the cylindrical coordinate system (be aware that some authors use a different notation). In the presentation here, we use $y_1 = r$, $y_2 = \theta$, and $y_3 = \varphi$ because the variable ordering (r, θ, φ) leads to a right-handed coordinate system as $\hat{e}_r \times \hat{e}_\theta = \hat{e}_\varphi$. However, an ordering of (r, φ, θ) seems to most naturally correspond to (x, y, z), and so some of our equation and term orderings correspond to a (r, φ, θ) ordering. However, the computations are performed using (r, θ, φ). We will be explicit on subscripts to avoid confusion.

The relationship between the spherical and Cartesian coordinates is

$$r = \sqrt{x^2 + y^2 + z^2}, \qquad x = r\cos(\varphi)\sin(\theta), \quad (A.176)$$

$$\varphi = \tan^{-1}(y/x), \qquad y = r\sin(\varphi)\sin(\theta), \quad (A.177)$$

$$\theta = \tan^{-1}(\sqrt{x^2 + y^2}/z), \qquad z = r\cos(\theta), \quad (A.178)$$

where $r \geq 0$, $0 \leq \varphi < 2\pi$, and $0 \leq \theta \leq \pi$. In spherical coordinates, a given point is identified by

$$\vec{x} = r\hat{e}_r. \quad (A.179)$$

The three direction unit vectors can be written in Cartesian coordinates as

$$\hat{e}_r = \begin{pmatrix} \cos(\varphi)\sin(\theta) \\ \sin(\varphi)\sin(\theta) \\ \cos(\theta) \end{pmatrix}_{\text{Cart.}}, \qquad \hat{e}_\theta = \frac{\partial \hat{e}_r}{\partial \theta} = \begin{pmatrix} \cos(\varphi)\cos(\theta) \\ \sin(\varphi)\cos(\theta) \\ -\sin(\theta) \end{pmatrix}_{\text{Cart.}},$$

$$\hat{e}_\varphi = \frac{\partial \hat{e}_r / \partial \varphi}{|\partial \hat{e}_r / \partial \varphi|} = \begin{pmatrix} -\sin(\varphi) \\ \cos(\varphi) \\ 0 \end{pmatrix}_{\text{Cart.}},$$

$$(A.180)$$

so

$$\begin{pmatrix} \hat{e}_r \\ \hat{e}_\theta \\ \hat{e}_\varphi \end{pmatrix} = {}_{\text{Sph.}}Q_{\text{Cart.}} \begin{pmatrix} \hat{e}_x \\ \hat{e}_y \\ \hat{e}_z \end{pmatrix} = \begin{pmatrix} \cos(\varphi)\sin(\theta) & \sin(\varphi)\sin(\theta) & \cos(\theta) \\ \cos(\varphi)\cos(\theta) & \sin(\varphi)\cos(\theta) & -\sin(\theta) \\ -\sin(\varphi) & \cos(\varphi) & 0 \end{pmatrix} \begin{pmatrix} \hat{e}_x \\ \hat{e}_y \\ \hat{e}_z \end{pmatrix},$$

$$\begin{pmatrix} \hat{e}_x \\ \hat{e}_y \\ \hat{e}_z \end{pmatrix} = {}_{\text{Cart.}}Q_{\text{Sph.}} \begin{pmatrix} \hat{e}_r \\ \hat{e}_\theta \\ \hat{e}_\varphi \end{pmatrix} = \begin{pmatrix} \cos(\varphi)\sin(\theta) & \cos(\varphi)\cos(\theta) & -\sin(\varphi) \\ \sin(\varphi)\sin(\theta) & \sin(\varphi)\cos(\theta) & \cos(\varphi) \\ \cos(\theta) & -\sin(\theta) & 0 \end{pmatrix} \begin{pmatrix} \hat{e}_r \\ \hat{e}_\theta \\ \hat{e}_\varphi \end{pmatrix}.$$

$$(A.181)$$

Again, the complexity of the spherical coordinate system as contrasted to the Cartesian coordinate system is that the unit direction vectors are not constant. By differentiating the above we find

$$\frac{\partial \hat{e}_r}{\partial r} = 0, \qquad \frac{\partial \hat{e}_r}{\partial \theta} = \hat{e}_\theta, \qquad \frac{\partial \hat{e}_r}{\partial \varphi} = \sin(\theta)\hat{e}_\varphi,$$

$$\frac{\partial \hat{e}_\varphi}{\partial r} = 0, \qquad \frac{\partial \hat{e}_\varphi}{\partial \theta} = 0, \qquad \frac{\partial \hat{e}_\varphi}{\partial \varphi} = -\sin(\theta)\hat{e}_r - \cos(\theta)\hat{e}_\theta, \qquad (A.182)$$

$$\frac{\partial \hat{e}_\theta}{\partial r} = 0, \qquad \frac{\partial \hat{e}_\theta}{\partial \theta} = -\hat{e}_r, \qquad \frac{\partial \hat{e}_\theta}{\partial \varphi} = \cos(\theta)\hat{e}_\varphi, \quad \text{or}$$

$$\gamma^\theta_{\theta r} = -\gamma^r_{\theta\theta} = 1, \quad \gamma^\varphi_{\varphi r} = -\gamma^r_{\varphi\varphi} = \sin(\theta), \quad \gamma^\varphi_{\varphi\theta} = -\gamma^\theta_{\varphi\varphi} = \cos(\theta), \text{ all other } \gamma^k_{ij} = 0.$$

The metric g for spherical coordinates is

$$g_{rr} = 1, \quad g_{\varphi\varphi} = r^2\sin^2(\theta), \quad g_{\theta\theta} = r^2, \text{ and all off-diagonal } g_{ij} = 0,$$

$$h_r = \sqrt{g_{rr}} = 1, \quad h_\varphi = \sqrt{g_{\varphi\varphi}} = r\sin(\theta), \quad h_\theta = \sqrt{g_{\theta\theta}} = r,$$

$$(ds)^2 = g_{ij}dx_i dx_j = (dr)^2 + r^2\sin^2(\theta)(d\varphi)^2 + r^2(d\theta)^2, \qquad (A.183)$$

$$dvol = \sqrt{\det(g)}\,dy_1 dy_2 dy_3 = r^2\sin(\theta)drd\theta d\varphi.$$

Working with the results of Section A.3, the h_i, and the behavior of the unit coordinate direction vectors, it is possible to write down the various differential operators from the previous section. We have

$$\text{grad}(f) = \vec{\nabla}f = (f_{;i}) = \frac{\partial f}{\partial r}\hat{e}_r + \frac{1}{r\sin(\theta)}\frac{\partial f}{\partial \varphi}\hat{e}_\varphi + \frac{1}{r}\frac{\partial f}{\partial \theta}\hat{e}_\theta,$$

$$\text{div}(\vec{f}) = \vec{\nabla}\cdot\vec{f} = f_{i;i} = \frac{1}{r^2}\frac{\partial(r^2 f_r)}{\partial r} + \frac{1}{r\sin(\theta)}\frac{\partial f_\varphi}{\partial \varphi} + \frac{1}{r\sin(\theta)}\frac{\partial(\sin(\theta)f_\theta)}{\partial \theta},$$

$$\Delta f = \nabla^2(f) = \text{div}(\text{grad}(f)) = f_{;ii} = \frac{1}{r^2}\frac{\partial}{\partial r}\left(r^2\frac{\partial f}{\partial r}\right) + \frac{1}{r^2\sin^2(\theta)}\frac{\partial^2 f}{\partial \varphi^2}$$

$$+ \frac{1}{r^2\sin(\theta)}\frac{\partial}{\partial \theta}\left(\sin(\theta)\frac{\partial f}{\partial \theta}\right),$$

$$(A.184)$$

$$\text{curl}(\vec{f}) = \vec{\nabla} \times \vec{f} = \left\{ \frac{1}{r\sin(\theta)} \frac{\partial(\sin(\theta)f_\varphi)}{\partial\theta} - \frac{1}{r\sin(\theta)} \frac{\partial f_\theta}{\partial\varphi} \right\} \hat{e}_r$$
$$+ \left\{ \frac{1}{r} \frac{\partial(rf_\theta)}{\partial r} - \frac{1}{r} \frac{\partial f_r}{\partial\theta} \right\} \hat{e}_\varphi + \left\{ \frac{1}{r\sin(\theta)} \frac{\partial f_r}{\partial\varphi} - \frac{1}{r} \frac{\partial(rf_\varphi)}{\partial r} \right\} \hat{e}_\theta,$$

and for the momentum balance we have (the left-hand side excluding the body forces is denoted $\vec{\nabla} \cdot \sigma = (\sigma_{ji;j})$)

$$\frac{\partial\sigma_{rr}}{\partial r} + \frac{1}{r\sin(\theta)} \frac{\partial\sigma_{\varphi r}}{\partial\varphi} + \frac{1}{r} \frac{\partial\sigma_{\theta r}}{\partial\theta} + \frac{1}{r}\{2\sigma_{rr} - \sigma_{\varphi\varphi} - \sigma_{\theta\theta} + \sigma_{\theta r}\cot(\theta)\} + \rho b_r = \rho a_r,$$

$$\frac{\partial\sigma_{r\varphi}}{\partial r} + \frac{1}{r\sin(\theta)} \frac{\partial\sigma_{\varphi\varphi}}{\partial\varphi} + \frac{1}{r} \frac{\partial\sigma_{\theta\varphi}}{\partial\theta} + \frac{1}{r}\{2\sigma_{r\varphi} + \sigma_{\varphi r} + (\sigma_{\theta\varphi} + \sigma_{\varphi\theta})\cot(\theta)\} + \rho b_\varphi = \rho a_\varphi,$$

$$\frac{\partial\sigma_{r\theta}}{\partial r} + \frac{1}{r\sin(\theta)} \frac{\partial\sigma_{\varphi\theta}}{\partial\varphi} + \frac{1}{r} \frac{\partial\sigma_{\theta\theta}}{\partial\theta} + \frac{1}{r}\{2\sigma_{r\theta} + \sigma_{\theta r} + (\sigma_{\theta\theta} - \sigma_{\varphi\varphi})\cot(\theta)\} + \rho b_\theta = \rho a_\theta,$$

$$(A.185)$$

where $\vec{b}$ is the body force, $\vec{a} = D\vec{v}/Dt$ is the acceleration, and it has not been assumed that the stress tensor is symmetric.

The details of the following computations were described in the cylindrical co-ordinates section, and here we state results. The velocity in Lagrangian coordinates is

$$\vec{v} = \frac{D\vec{x}(\vec{y}_0, t)}{Dt} = \frac{\partial}{\partial t}(r\hat{e}_r) = \frac{\partial r}{\partial t}\hat{e}_r + r\frac{\partial\varphi}{\partial t}\frac{\partial\hat{e}_r}{\partial\varphi} + r\frac{\partial\theta}{\partial t}\frac{\partial\hat{e}_r}{\partial\theta}$$
$$= \frac{\partial r}{\partial t}\hat{e}_r + r\sin(\theta)\frac{\partial\varphi}{\partial t}\hat{e}_\varphi + r\frac{\partial\theta}{\partial t}\hat{e}_\theta = v_r\hat{e}_r + v_\varphi\hat{e}_\varphi + v_\theta\hat{e}_\theta.$$
$$(A.186)$$

The velocity in Eulerian coordinates is

$$\vec{v} = \frac{D\vec{x}(\vec{y})}{Dt} = \left(\frac{\partial r}{\partial t}\frac{\partial}{\partial r} + \frac{\partial\varphi}{\partial t}\frac{\partial}{\partial\varphi} + \frac{\partial\theta}{\partial t}\frac{\partial}{\partial\theta} \right)(r\hat{e}_r)$$
$$= v_r\hat{e}_r + r\sin(\theta)\frac{\partial\varphi}{\partial t}\hat{e}_\varphi + r\frac{\partial\theta}{\partial t}\hat{e}_\theta = v_r\hat{e}_r + v_\varphi\hat{e}_\varphi + v_\theta\hat{e}_\theta.$$
$$(A.187)$$

Again, the velocities in both reference frames are the same.

The acceleration is more complicated. For the Lagrangian frame there is no advection term,

$$\vec{a} = \frac{D\vec{v}}{Dt} = \frac{\partial\vec{v}}{\partial t},$$
$$(A.188)$$

and, recalling the time dependence of the unit vectors,

$$a_r = \frac{\partial v_r}{\partial t} - \frac{v_\varphi^2}{r} - \frac{v_\theta^2}{r},$$

$$a_\varphi = \frac{\partial v_\varphi}{\partial t} + \frac{v_r v_\varphi}{r} + \frac{v_\theta v_\varphi}{r}\cot(\theta),$$
$$(A.189)$$

$$a_\theta = \frac{\partial v_\theta}{\partial t} + \frac{v_\theta v_r}{r} - \frac{v_\varphi^2}{r}\cot(\theta).$$

For the Eulerian frame, there is an advection term

$$
\vec{a} = \frac{D\vec{v}}{Dt} = \frac{\partial \vec{v}}{\partial t} + \left(\frac{\partial r}{\partial t} \frac{\partial}{\partial r} + \frac{\partial \varphi}{\partial t} \frac{\partial}{\partial \varphi} + \frac{\partial \theta}{\partial t} \frac{\partial}{\partial \theta} \right) \vec{v}
$$

$$
= \frac{\partial \vec{v}}{\partial t} + \left(v_r \frac{\partial}{\partial r} + \frac{v_\varphi}{r \sin(\theta)} \frac{\partial}{\partial \varphi} + \frac{v_\theta}{r} \frac{\partial}{\partial \theta} \right) \vec{v} = a_r \hat{e}_r + a_\varphi \hat{e}_\varphi + a_\theta \hat{e}_\theta.
$$

(A.190)

Here the unit vectors do not depend upon time, but they do depend on the current position, yielding

$$
a_r = \frac{\partial v_r}{\partial t} + v_r \frac{\partial v_r}{\partial r} + \frac{v_\varphi}{r \sin(\theta)} \frac{\partial v_r}{\partial \varphi} - \frac{v_\varphi^2}{r} + \frac{v_\theta}{r} \frac{\partial v_r}{\partial \theta} - \frac{v_\theta^2}{r},
$$

$$
a_\varphi = \frac{\partial v_\varphi}{\partial t} + v_r \frac{\partial v_\varphi}{\partial r} + \frac{v_\varphi}{r \sin(\theta)} \frac{\partial v_\varphi}{\partial \varphi} + \frac{v_\varphi v_\theta}{r} \cot(\theta) + \frac{v_\theta}{r} \frac{\partial v_\varphi}{\partial \theta} + \frac{v_\varphi v_r}{r},
$$

(A.191)

$$
a_\theta = \frac{\partial v_\theta}{\partial t} + v_r \frac{\partial v_\theta}{\partial r} + \frac{v_\varphi}{r \sin(\theta)} \frac{\partial v_\theta}{\partial \varphi} - \frac{v_\varphi^2}{r} \cot(\theta) + \frac{v_\theta}{r} \frac{\partial v_\theta}{\partial \theta} + \frac{v_\theta v_r}{r}.
$$

Finally we write down the displacement gradient $\hat{F}(\vec{x}, \vec{u}) = \vec{\nabla}\vec{u} = (u_{i;j})$,

$$
\hat{F}(\vec{x}, \vec{u}) = \begin{pmatrix} \hat{F}_{rr} & \hat{F}_{r\theta} & \hat{F}_{r\varphi} \\ \hat{F}_{\theta r} & \hat{F}_{\theta\theta} & \hat{F}_{\theta\varphi} \\ \hat{F}_{\varphi r} & \hat{F}_{\varphi\theta} & \hat{F}_{\varphi\varphi} \end{pmatrix} = \begin{pmatrix} \frac{\partial u_r}{\partial r} & \frac{1}{r}\frac{\partial u_r}{\partial \theta} - \frac{u_\theta}{r} & \frac{1}{r\sin(\theta)}\frac{\partial u_r}{\partial \varphi} - \frac{u_\varphi}{r} \\ \frac{\partial u_\theta}{\partial r} & \frac{1}{r}\frac{\partial u_\theta}{\partial \theta} + \frac{u_r}{r} & \frac{1}{r\sin(\theta)}\frac{\partial u_\theta}{\partial \varphi} - \frac{u_\varphi}{r}\cot(\theta) \\ \frac{\partial u_\varphi}{\partial r} & \frac{1}{r}\frac{\partial u_\varphi}{\partial \theta} & \frac{1}{r\sin(\theta)}\frac{\partial u_\varphi}{\partial \varphi} + \frac{u_r}{r} + \frac{u_\theta}{r}\cot(\theta) \end{pmatrix}
$$

(A.192)

and the small strain tensor $\frac{1}{2}(\hat{F} + F)$,

$$
\varepsilon_{rr} = \frac{\partial u_r}{\partial r}, \qquad \varepsilon_{\theta\theta} = \frac{1}{r}\frac{\partial u_\theta}{\partial \theta} + \frac{u_r}{r},
$$

$$
\varepsilon_{\varphi\varphi} = \frac{1}{r\sin(\theta)}\frac{\partial u_\varphi}{\partial \varphi} + \frac{u_r}{r} + \frac{u_\theta}{r}\cot(\theta),
$$

$$
\varepsilon_{r\theta} = \frac{1}{2}\left(\frac{1}{r}\frac{\partial u_r}{\partial \theta} + \frac{\partial u_\theta}{\partial r} - \frac{u_\theta}{r} \right),
$$

(A.193)

$$
\varepsilon_{r\varphi} = \frac{1}{2}\left(\frac{1}{r\sin(\theta)}\frac{\partial u_r}{\partial \varphi} + \frac{\partial u_\varphi}{\partial r} - \frac{u_\varphi}{r} \right),
$$

$$
\varepsilon_{\varphi\theta} = \frac{1}{2}\left(\frac{1}{r}\frac{\partial u_\varphi}{\partial \theta} + \frac{1}{r\sin(\theta)}\frac{\partial u_\theta}{\partial \varphi} - \frac{u_\varphi}{r}\cot(\theta) \right).
$$

The velocity gradient L is given by $L = \hat{F}(\vec{x}, \vec{v}) = \vec{\nabla}\vec{v} = (v_{i;j})$ and the rate of deformation tensor $D = \frac{1}{2}(L + L^T)$ has D_{ij} the same as ε_{ij}, where the displacement $\vec{u}$ is replaced by the velocity $\vec{v}$. If the deformation is explicitly written in the new coordinates (r, θ, φ) in terms of the original coordinates (R, Θ, Φ) – for example, $r = r(R, \Theta, \Phi)$ – then the deformation gradient is

$$
F = \begin{pmatrix} F_{rR} & F_{r\Theta} & F_{r\Phi} \\ F_{\theta R} & F_{\theta\Theta} & F_{\theta\Phi} \\ F_{\varphi R} & F_{\varphi\Theta} & F_{\varphi\Phi} \end{pmatrix} = \begin{pmatrix} \frac{\partial r}{\partial R} & \frac{1}{R}\frac{\partial r}{\partial \Theta} & \frac{1}{R\sin(\Theta)}\frac{\partial r}{\partial \Phi} \\ r\frac{\partial \theta}{\partial R} & \frac{r}{R}\frac{\partial \theta}{\partial \Theta} & \frac{r}{R\sin(\Theta)}\frac{\partial \theta}{\partial \Phi} \\ r\sin(\theta)\frac{\partial \varphi}{\partial R} & \frac{r\sin(\theta)}{R}\frac{\partial \varphi}{\partial \Theta} & \frac{r\sin(\theta)}{R\sin(\Theta)}\frac{\partial \varphi}{\partial \Phi} \end{pmatrix}.
$$

(A.194)

In this form, $F^T F$ is in the original coordinate system, appropriate for Lagrangian strain E, and $B = FF^T$ is in the current coordinate system, appropriate for Eulerian strain E^*.

A.7. General Coordinates

The mathematically inclined may have been hoping to see words like covariant and contravariant. Such concepts are not necessary, though, when the space is Euclidean and the coordinate systems are orthonormal, since transformations between coordinate systems are handled by Eqs. (A.8) and (A.10). The derivations presented in previous sections of this appendix for orthonormal bases are intuitive and easy to follow, but they provide the foundation for this last section, where more general coordinate systems are discussed. We choose to use the notation found in differential and Riemannian geometry texts. In modern differential geometry, by convention, the coordinates are indexed with a superscript: (y^1, y^2, y^3). A central idea is that the mathematicians do not know what $\vec{x}(y^1, y^2, y^3)$ is. We have always explicitly stated what $\vec{x}$ is – for example, in Cartesian coordinates $\vec{x} = x\hat{e}_x + y\hat{e}_y + z\hat{e}_z$, in cylindrical coordinates $\vec{x} = r\hat{e}_r + z\hat{e}_z$, and in spherical coordinates $\vec{x} = r\hat{e}_r$. Mathematicians assume they know the parameters y^i and the metric g, but not how the object (manifold) appears, and that motivates some of the notation. The basis vectors for the tangent space are written as $\frac{\partial}{\partial y^i}$, indicating that they point in the direction of changing y^i. Thus,

$$\text{tangent space basis vector} = \frac{\partial}{\partial y^i} = \frac{\partial \vec{x}}{\partial y_i} = h_i \hat{e}_{y_i} \quad \text{(no sum on } i\text{)}. \tag{A.195}$$

The notation $\frac{\partial}{\partial y^i}$ represents the basis vector, even if $\vec{x}$ is not known. The tangent space basis vectors span the tangent space, written $T_p M$ for tangent to the manifold M at point p, which for us is a three-dimensional Euclidean vector space. Thus, in particular, for us the tangent space is spanned by $\{\frac{\partial}{\partial y^1}, \frac{\partial}{\partial y^2}, \frac{\partial}{\partial y^3}\}$. There is no assumption that the coordinate directions are orthogonal, and hence there is no effort to normalize the basis vectors. Mathematicians dislike putting arrows above vectors to identify them, so we refer to vector fields with a capital letter, as they do. This will help us recall whether the vector field is in simple orthonormal coordinates or in our generalized coordinates. As an example, consider the velocity vector field $\vec{v} = V$. Then

$$\vec{v} = v_i \hat{e}_{y_i} = V = b^i \frac{\partial}{\partial y^i}. \tag{A.196}$$

We immediately see that the coefficients v_i are unlikely to numerically equal b^i as $b^i = v_i/h_i$ (no sum on i). The inner product between two vectors $X = a^i \frac{\partial}{\partial y^i}$ and $Y = b^j \frac{\partial}{\partial y^j}$ is given by the metric

$$g(X, Y) = g_{ij} a^i b^j, \tag{A.197}$$

where the components of the metric tensor are

$$g_{ij} = g\left(\frac{\partial}{\partial y^i}, \frac{\partial}{\partial y^j}\right). \tag{A.198}$$

When we know $\vec{x}$, g_{ij} is computed with Eq. (A.44). g is called a metric because the length of a vector X is given by $\sqrt{g(X, X)}$. As an inner product, g allows us to measure the angle between two vectors, $\cos(\theta) = g(X, Y)/\sqrt{g(X, X) \cdot g(Y, Y)}$. g is understood to be a function of where you are in space – hence of the y^k – but that dependence is not explicitly shown; we do not write $g(y^1, y^2, y^3, X, Y)$. The metric is essentially a measure of distance based on Pythagoras' theorem, since it is a quadratic form. $g(X, Y)$ is bilinear in X and Y, meaning that it is linear in each X and Y, so it can be expanded as

$$g(X,Y) = g\left(a^i \frac{\partial}{\partial y^i}, b^j \frac{\partial}{\partial y^j}\right) = a^i g\left(\frac{\partial}{\partial y^i}, b^j \frac{\partial}{\partial y^j}\right) = a^i b^j g\left(\frac{\partial}{\partial y^i}, \frac{\partial}{\partial y^j}\right) = a^i b^j g_{ij}.$$
(A.199)

This inner product produces the same numerical answer in the situation where the metric tensor is diagonal, as what we had before since $\vec{u} \cdot \vec{v} = u_i v_i = g_{ij} a^i b^j = \Sigma_i h_i^2 (u_i/h_i)(v_i/h_i) = u_i v_i$.

One of the things we did not discuss was the notion of change of variables, because all our differential operators were computed from their original definitions and not by changing variables from a reference frame where they were known. Here, however, the change of variables is central to certain terminology. Given two parameterizations of space, y^i and $(y')^i$, to change tangent space basis vectors from one to the other requires

$$\frac{\partial}{\partial (y')^i} = \frac{\partial y^j}{\partial (y')^i} \frac{\partial}{\partial y^j}.$$
(A.200)

Given this change, it can be seen that for a vector X,

$$X = (a')^i \frac{\partial}{\partial (y')^i} = (a')^i \frac{\partial y^j}{\partial (y')^i} \frac{\partial}{\partial y^j} = a^j \frac{\partial}{\partial y^j},$$
(A.201)

or that the coefficients transform as (switching the prime and original, as they are arbitrary identifications)

$$\text{Contravariant transformation:} \quad (a')^j = a^i \frac{\partial (y')^j}{\partial y^i}.$$
(A.202)

This type of transformation is referred to as *contravariant*. Vectors that are written in terms of the tangent basis vectors $\frac{\partial}{\partial y^i}$ are called contravariant vectors, or first-order $(1, 0)$ tensors. The components are identified with a superscript. Tensors are objects that transform appropriately when changing coordinates.

To find the gradient of a function $f(y^i)$, we go to its definition. The gradient vector can be written $\text{grad}(f) = \nabla f = a^i \frac{\partial}{\partial y^i}$. Our object is to determine the a^i. In this next equation, the first equality is simply the chain rule, while the second equality is the definition of the gradient (compare to Eq. A.61),

$$\frac{df}{ds} = \frac{\partial f}{\partial y^i} \frac{dy^i}{ds} = g\left(\nabla f, \frac{dy}{ds}\right) = g_{ij} a^i \frac{dy^j}{ds}.$$
(A.203)

Noting that the i and j indices can be swapped on the far right side, this equation says

$$\frac{\partial f}{\partial y^i} = g_{ji} a^j.$$
(A.204)

Denoting the components of g^{-1} by g^{ij}, where g^{-1} is the inverse of g, hence $g^{ik} g_{kj} = \delta^i_j$, yields

$$g^{ik} \frac{\partial f}{\partial y^i} = g^{ik} g_{ji} a^j = g^{ki} g_{ij} a^j = \delta^k_j a^j = a^k,$$
(A.205)

where the symmtery of g and g^{-1} has been used. Hence,

$$a^j = g^{ij} \frac{\partial f}{\partial y^i}.$$
(A.206)

Thus, the gradient of a scalar function in the tangent space is

$$\nabla f = g^{ij} \frac{\partial f}{\partial y^i} \frac{\partial}{\partial y^j}. \tag{A.207}$$

In particular, this equation is the contravariant form of the gradient.

Another way to consider the gradient is to consider the differential df,

$$df = \frac{\partial f}{\partial y^i} dy^i = f_{;i} dy^i, \tag{A.208}$$

where the second equality is semicolon (;) notation for differentiation (note the $\frac{1}{h_i}$ multiplicative difference with Eq. A.63 due to the normalization there). One way to think of this expression is to consider the dy^i as small changes in each coordinate, and thus we obtain the corresponding small change in the scalar function f from numerically evaluating Eq. (A.208). However, another way to view this equation is to say that we do not know what the dy^i are, and so the dy^i are symbolic placeholders. In this sense, the dy^i can be viewed as basis vectors. As such, they span what is called the cotangent space $T_p^* M$ at the point p, which for us is again a three-dimensional Euclidean space. Thus, the three-dimensional cotangent space is spanned by $\{dy^1, dy^2, dy^3\}$. Going further, saying that both expression Eq. (A.207) and Eq. (A.208) represent the same object, the gradient, then

$$\nabla f = df \quad \Leftrightarrow \quad g^{ij} \frac{\partial f}{\partial y^i} \frac{\partial}{\partial y^j} = \frac{\partial f}{\partial y^i} dy^i. \tag{A.209}$$

Since f can be arbitrary,

$$\text{cotangent space basis vector} = dy^i = g^{ij} \frac{\partial}{\partial y^j} \quad (\Rightarrow \frac{\partial}{\partial y^i} = g_{ij} dy^j). \tag{A.210}$$

For an orthonormal basis, $dy^i = \hat{e}_i / h_i$ (no sum on i).

Though we presented a relationship, through the metric, between the tangent and cotangent space basis vectors, it is not necessary to do so. The terminology "cotangent" space or "dual" space arises because it can be viewed as the linear functionals operating on the tangent space – that is., they are the linear functions that, given a vector, produce a scalar. The basis of the cotangent space can be defined by viewing the $dy^i()$ as functions that operate on vectors, where in particular $dy^i(\frac{\partial}{\partial y^j}) = \delta_j^i$. Thus, if the linear functional were $f(X) = \beta_i dy^i(X)$, where $X = a^i \frac{\partial}{\partial y^i}$, then

$$f(X) = \beta_i dy^i(X) = \beta_i dy^i(a^j \frac{\partial}{\partial y^j}) = \beta_i a^j dy^i(\frac{\partial}{\partial y^j}) = \beta_i a^j \delta_j^i = \beta_i a^i. \tag{A.211}$$

This algebraic definition of the dual space is consistent with our relationship between the tangent and cotangent basis vectors in that

$$g\left(dy^i, \frac{\partial}{\partial y^j}\right) = g\left(g^{ik} \frac{\partial}{\partial y^k}, \frac{\partial}{\partial y^j}\right) = g^{ik} g\left(\frac{\partial}{\partial y^k}, \frac{\partial}{\partial y^j}\right) = g^{ik} g_{kj} = \delta_j^i. \tag{A.212}$$

Vectors can be written in terms of the cotangent basis dy^i. Vectors written with cotangent basis vectors are called covariant vectors, or first-order $(0,1)$ tensors, and the components are identified with a subscript. Thus, if we have the same vector $X = a^i \frac{\partial}{\partial y^i} = \alpha_i dy^i$, then the relationship between the components of the contravariant form and the covariant form is

$$\alpha_i = g_{ij} a^j, \qquad a^i = g^{ij} \alpha_j. \tag{A.213}$$

These actions are called lowering the index and raising the index, respectively, and the action is performed by the metric tensor or its inverse. A convention is to write a covariant vector, which can be viewed as a 1-form (a specific type of differential form), with a Greek letter: $\omega = \alpha_i dy^i$. To be consistent with our treatment of 2-tensors, we could use a_i instead of α_i to represent the lowered covariant component of the vector; however, for clarity, we use a Greek letter. A convention to assist in keeping all these indices straight is to only sum over a repeated index pair when one index is a superscript and one is a subscript.

Under a change of variables, the cotangent basis changes as follows,

$$(dy')^i = \frac{\partial(y')^i}{\partial y^j} dy^j. \tag{A.214}$$

Hence, for a covariant vector $\omega = \alpha_i dy^i$, the transformed components are

$$\omega = (\alpha')_i (dy')^i = (\alpha')_i \frac{\partial(y')^i}{\partial y_j} dy^j = \alpha_j dy^j \tag{A.215}$$

and the components transform as

$$\text{Covariant transformation:} \quad (\alpha')_i = \alpha_j \frac{\partial y^j}{\partial(y')_i}. \tag{A.216}$$

As an aside, the coordinate components y^i themselves are neither covariant nor contravariant, as a transformation for them is a direct substitution – that is, $(y')^i = (y'(y^j))^i$. However, they are typically written with indices as superscripts.

Tensors are objects that transform in certain ways based on how they are written in terms of the basis functions. Tensors such as the stress tensor are written in terms of basis functions. Thus, in our orthonormal (ON) coordinate systems of the previous sections, $\sigma = \sigma_{ij,ON} \hat{e}_{y^i} \otimes \hat{e}_{y^j}$, where $\otimes$ helps remind us of the location in the matrix (order matters). In our generalized coordinate system, we have $\sigma = \sigma^{ij} \frac{\partial}{\partial y^i} \otimes \frac{\partial}{\partial y^j}$ and so $\sigma^{ij} = \sigma_{ij,ON}/h_i h_j$ (no sum on i, j). This example is a contravariant tensor, and the components are not numerically equal to their orthonormal counterpart though they represent the same stress. 2-tensors come in three types: contravariant $(2,0)$ tensors, mixed $(1,1)$ tensors, and covariant $(0,2)$ tensors. The tensor components are indicated by superscripts, a superscript and a subscript, and subscripts, respectively. They transform as we would expect, linking transformations together:

$$\text{Contravariant transformation:} \quad (\sigma')^{ij} = \sigma^{kl} \frac{\partial(y')^i}{\partial y^k} \frac{\partial(y')^j}{\partial y^l},$$

$$\text{Mixed transformation:} \quad (\sigma')^j_i = \sigma^l_k \frac{\partial y^k}{\partial(y')^i} \frac{\partial(y')^j}{\partial y^l}, \tag{A.217}$$

$$\text{Covariant transformation:} \quad (\sigma')_{ij} = \sigma_{kl} \frac{\partial y^k}{\partial(y')^i} \frac{\partial y^l}{\partial(y')^j}.$$

An operation performed on 2-tensors is taking the trace of the tensor, also called a contraction. Two indices are set equal, indicating they are summed over. Since only a raised and lowered index pair can be summed over, if both the indices are raised or both are lowered, it is necessary to raise or lower one of them before performing the sum. Thus,

$$\text{trace}(\sigma) = g_{ij}\sigma^{ji} = \sigma^i_i = g^{ij}\sigma_{ji}. \tag{A.218}$$

In mathematical geometry theory, an absolute derivative is defined, meaning that it is independent of a coordinate system. With $X = a^i \frac{\partial}{\partial y^i}$ a vector (it need only be defined at the point of interest for a specific computation) and $Y = b^j \frac{\partial}{\partial y^j}$ a vector field, essentially the directional derivative (i.e., the change in Y in the X direction) is identified by the symbol $\nabla_X Y$,

$$
\begin{aligned}
\nabla_X Y &= \lim_{t \to 0} \frac{Y(y^1 + ta^1, y^2 + ta^2, y^3 + ta^3) - Y(y^1, y^2, y^3)}{t} \\
&= \frac{\partial}{\partial y^i}\left(b^j \frac{\partial}{\partial y^j}\right) a^i = a^i \frac{\partial b^j}{\partial y^i} \frac{\partial}{\partial y^j} + a^i b^j \frac{\partial}{\partial y^i}\left(\frac{\partial}{\partial y^j}\right) \\
&= a^i \frac{\partial b^j}{\partial y^i} \frac{\partial}{\partial y^j} + a^i b^j \Gamma_{ij}^k \frac{\partial}{\partial y^k} = a^i \left\{ \frac{\partial b^j}{\partial y^i} + b^k \Gamma_{ik}^j \right\} \frac{\partial}{\partial y^j}.
\end{aligned}
\tag{A.219}
$$

In going from the second to the third line, the derivative of the bases gave rise to the Christoffel symbols of the second kind, Γ_{ij}^k, which are defined as

$$
\frac{\partial}{\partial y^i}\left(\frac{\partial}{\partial y^j}\right) = \Gamma_{ij}^k \frac{\partial}{\partial y^k}.
\tag{A.220}
$$

In this equation, $\frac{\partial}{\partial y^i}$ means to take a partial derivative, and $\frac{\partial}{\partial y^j}$ and the $\frac{\partial}{\partial y^k}$ are tangent space basis vectors. Thus, the change in basis function $\frac{\partial}{\partial y^j}$ when moving in the y^i direction is written in terms of all the basis functions that span the tangent space. The expression in brackets in Eq. (A.219) is referred to as the *absolute derivative* or the *covariant derivative*. In components, it is indicated by a subscript semicolon (;) preceding the index created by the differentiation,

$$
b_{;i}^j = \frac{\partial b^j}{\partial y^i} + b^k \Gamma_{ik}^j.
\tag{A.221}
$$

(Note the $\frac{1}{h_i}$ multiplicative difference with Eq. A.69, due to the normalization). This object transforms like a $(1, 1)$ tensor and so can be written

$$
\nabla Y = b_{;i}^j dx^i \otimes \frac{\partial}{\partial y^j} = \left\{ \frac{\partial b^j}{\partial y^i} + b^k \Gamma_{ik}^j \right\} dx^i \otimes \frac{\partial}{\partial y^j},
\tag{A.222}
$$

explicitly showing the tensor in terms of the basis. Hence the term "covariant derivative:" it has taken an object (a contravariant vector, in this case) and added a covariant index through differentiation. "Absolute" means that this process is independent of a coordinate system. Thus, we have defined a differentiation procedure that takes into account the fact that the basis vectors do not always point in the same direction. Notice how the notation works for $\nabla_X Y$ in terms of ∇Y, $dx^i(X) = dx^i(a^j \frac{\partial}{\partial y^j}) = a^j dx^i(\frac{\partial}{\partial y^j}) = a^j \delta_j^i = a^i$.

The directional derivatives of the tangent space bases are referred to as the *connection*,

$$
\nabla_{\frac{\partial}{\partial y^i}} \frac{\partial}{\partial y^j} = \frac{\partial}{\partial y^i}\left(\frac{\partial}{\partial y^j}\right) = \Gamma_{ij}^k \frac{\partial}{\partial y^k},
\tag{A.223}
$$

where again the Γ_{ij}^k are the Christoffel symbols of the second kind. (We immediately see that for the Cartesian coordinate system on Euclidean space, all the Christoffel symbols equal zero as the basis vectors are unit length and always point in the same direction.)

The velocity of material can be computed in a couple of ways, either with parameters or components. Since we assume space is not moving, each point of space is determined by the parameters $\{y^i\}$ that give the appropriate $\vec{x}$. The Eulerian approach uses the advection. The velocity $V = b^i \frac{\partial}{\partial y^i}$ is then

$$V = \left(\frac{\partial y^i}{\partial t} \frac{\partial}{\partial y^i} \right)(\vec{x}) = \frac{\partial y^i}{\partial t} \frac{\partial \vec{x}}{\partial y^i} = \frac{dy^i}{dt} \frac{\partial}{\partial y^i} = b^i \frac{\partial}{\partial y^i} \quad \Rightarrow \quad b^i = \frac{dy^i}{dt}. \qquad (A.224)$$

This result seems almost trivial, utilizing the definition of the basis vectors. It identifies the components of the velocity vector as $b^i = dy^i/dt$.

The Lagrangian approach is more challenging. It assumes we have explicitly written the particle motion in terms of the local tangent basis vectors, where the location and motion are given by the components and bases,

$$\vec{x} = x^i(y^1(t), y^2(t), y^3(t)) \left. \frac{\partial}{\partial y^i} \right|_{(y^1(t), y^2(t), y^3(t))}. \qquad (A.225)$$

Knowing the $x^i(y^1, y^2, y^3)$ should not be trivialized; it says that the location (hence parameters) for an origin has been chosen and that, given the current location in space, the direction to the origin has been written in the local tangent space and the appropriate distance to the origin in terms of the local metric has been determined. Differentiation with respect to time, recalling the $\frac{\partial}{\partial y^i}$ depend on time, yields

$$\begin{aligned}
V = \frac{d\vec{x}}{dt} &= \frac{d}{dt} \left\{ x^i \frac{\partial}{\partial y^i} \right\} = \frac{\partial x^i}{\partial y^j} \frac{dy^j}{dt} \frac{\partial}{\partial y^i} + x^i \frac{d}{dt} \left(\frac{\partial}{\partial y^i} \right) \\
&= \frac{\partial x^i}{\partial y^j} \frac{dy^j}{dt} \frac{\partial}{\partial y^i} + x^i \frac{\partial}{\partial y^j} \left(\frac{\partial}{\partial y^i} \right) \frac{dy^j}{dt} \\
&= \frac{\partial x^i}{\partial y^j} \frac{dy^j}{dt} \frac{\partial}{\partial y^i} + x^i \frac{dy^j}{dt} \Gamma^k_{ji} \frac{\partial}{\partial y^k} = \frac{dy^j}{dt} \left\{ \frac{\partial x^i}{\partial y^j} + x^k \Gamma^i_{jk} \right\} \frac{\partial}{\partial y^i} = \frac{dy^i}{dt} \frac{\partial}{\partial y^i} = b_i \frac{\partial}{\partial y^i}.
\end{aligned}$$
$$(A.226)$$

In going from the first to second equality on the last line, the arbitrary i and k summation indices were swapped. As shown at the end of the third line of Eq. (A.226), we already knew the answer, namely that $b^i = dy^i/dt$, since the Eulerian and Lagrangian approach produce the same velocity. We recognize the term in brackets on the last line as the absolute derivative, and since there are no time derivatives, only spatial derivatives related to x^i in the absolute derivative, we conclude that

$$x^i_{;j} = \frac{\partial x^i}{\partial y^j} + x^k \Gamma^i_{jk} = \delta^i_j \quad \text{or} \quad \nabla \vec{x} = I \quad \Rightarrow \quad x^i - x^i_0 + \int_{y_0}^{y} x^k \Gamma^i_{jk} dy^j = y^i - y^i_0,$$
$$(A.227)$$

as $dx^i = \frac{\partial x^i}{\partial y^j} dy^j$. These are nine coupled linear partial differential equations that can be solved to determine x^i as a function of y^j in terms of the Christoffel symbols (i.e., the connection) or, as shown, three integral equations involving line integrals.

Now consider the case of acceleration. The tangent space basis vectors are functions of y^i. For a two-dimensional situation, each tangent space can be viewed as a sheet of paper with its own coordinate system. Each point has its own sheet of paper. Since our coordinate system (chart) is smooth, these sheets of paper smoothly transition to each other. Thus, as a particle follows a path in the plane, its numerical velocity components change as it moves from point to point both owing to acceleration and owing to the change in basis vectors. In fact, a particle not undergoing any acceleration could still have its numerical velocity components changing from point to point owing to changing basis vectors of the tangent space. Thus, we need to compute the acceleration, taking these changes into account. Since the parameterization (the coordinates) is smooth, it is possible to understand how these basis functions change as one moves in any direction.

For the acceleration, there are two forms, Lagrangian and Eulerian. In the Lagrangian situation, we have velocity $V = b^i \frac{\partial}{\partial y^i}$, where $b^i = b^i(y^1(t), y^2(t), y^3(t))$, where we identify the material point by its original location at $t = 0$. However, the tangent basis is in the current location of the material point, not the original location, so the tangent basis is a function of the current location in space, hence for the basis functions $y^i = y^i(t)$ and

$$
\begin{aligned}
A_L &= \frac{\partial V}{\partial t} = \frac{\partial}{\partial t} \left\{ b_i \frac{\partial}{\partial y^i} \right\} = \frac{\partial b^i}{\partial t} \frac{\partial}{\partial y^i} + b^i \frac{\partial}{\partial t} \left(\frac{\partial}{\partial y^i} \right) \\
&= \frac{\partial b^i}{\partial t} \frac{\partial}{\partial y^i} + b^i \frac{\partial}{\partial y^j} \left(\frac{\partial}{\partial y^i} \right) \frac{dy^j}{dt} \\
&= \frac{\partial b^i}{\partial t} \frac{\partial}{\partial y^i} + b^i b^j \Gamma^k_{ji} \frac{\partial}{\partial y^k} = \left\{ \frac{\partial b^i}{\partial t} + b^k b^j \Gamma^i_{jk} \right\} \frac{\partial}{\partial y^i}.
\end{aligned}
\tag{A.228}
$$

In the last step the i and k indices were swapped as they are arbitrary summation indices. We used $b^i = dy^i/dt$ from Eq. (A.224). Typically we know $b^i(t)$ in Lagrangian scenarios, hence the way it is written. However, we could have assumed we knew $y^j(t)$, and then written $\frac{\partial b^i}{\partial t} = \frac{\partial b^i}{\partial y^j} \frac{dy^j}{dt}$, leading to

$$
\begin{aligned}
A_L &= \frac{\partial V}{\partial t} = \frac{\partial b^i}{\partial y^j} \frac{dy^j}{dt} \frac{\partial}{\partial y^i} + b^i b^j \Gamma^k_{ji} \frac{\partial}{\partial y^k} = b^j \left\{ \frac{\partial b^i}{\partial y^j} + b^k \Gamma^i_{jk} \right\} \frac{\partial}{\partial y^i} \\
&= b^j b^i_{;j} \frac{\partial}{\partial y^i} = \nabla_V V.
\end{aligned}
\tag{A.229}
$$

Thus, the acceleration is the directional derivative of the velocity in the velocity direction. In one dimension this looks like $\frac{\partial v}{\partial t} = \frac{\partial v}{\partial x} \frac{\partial x}{\partial t} = v \frac{\partial v}{\partial x}$. Our Lagrangian acceleration vector A_L components have changing values because of both the change in velocity and the change in the tangent basis vectors,

$$
a^i = \frac{\partial b^i}{\partial t} + b^k b^j \Gamma^i_{jk} = b^j \left\{ \frac{\partial b^i}{\partial y^j} + b^k \Gamma^i_{jk} \right\} = b^j b^i_{;j}.
\tag{A.230}
$$

For the Eulerian acceleration, the frame is fixed and material flows through it. The velocity is a function of the current location in space, instead of initial location, and time. The velocity is $b^i(y^1, y^2, y^3, t)$, the basis vectors do not depend upon time, and the acceleration has the change in V and the advection terms:

$$
\begin{aligned}
A_E &= \frac{DV}{Dt} = \frac{\partial V}{\partial t} + \left(\frac{\partial y^j}{\partial t} \frac{\partial}{\partial y^j} \right)(V) = \frac{\partial V}{\partial t} + \left(b^j \frac{\partial}{\partial y^j} \right) \left(b^i \frac{\partial}{\partial y^i} \right) \\
&= \frac{\partial b^i}{\partial t} \frac{\partial}{\partial y^i} + b^j \frac{\partial b^i}{\partial y^j} \frac{\partial}{\partial y^i} + b^j b^i \Gamma^k_{ji} \frac{\partial}{\partial y^k} = \left\{ \frac{\partial b^i}{\partial t} + b^j \left(\frac{\partial b^i}{\partial y^j} + b^k \Gamma^i_{jk} \right) \right\} \frac{\partial}{\partial y^i} \\
&= \left\{ \frac{\partial b^i}{\partial t} + b^j b^i_{;j} \right\} \frac{\partial}{\partial y^i} = \frac{\partial b^i}{\partial t} \frac{\partial}{\partial y^i} + \nabla_V V = \frac{\partial V}{\partial t} + \nabla_V V.
\end{aligned}
\tag{A.231}
$$

The $b^j (\partial / \partial y^j)$ is the advection term, where we used $b^i = dy^i / dt$ from Eq. (A.224). In components, the Eulerian acceleration A_E is

$$a^i = \frac{\partial b^i}{\partial t} + b^j \frac{\partial b^i}{\partial y^j} + b^k b^j \Gamma^{\,i}_{jk} = \frac{\partial b^i}{\partial t} + b^j b^i_{;j}. \tag{A.232}$$

The next, rather lengthy, reasoning is to allow us to compute the Christoffel symbols simply by knowing the metric tensor (analogous to finding γ^k_{ij} in terms of h_i, pages 617–619). Suppose we have a particle moving in the Eulerian frame and that the acceleration is zero. Then we would expect the velocity vector to always point in the same direction. If there is no time dependence of the velocity field (i.e., $\partial V / \partial t = 0$), then the acceleration $\nabla_V V = 0$. No acceleration implies that as a particle path is traversed, the velocity V always points in the same direction. This notion has been generalized to the notion of parallel translation of a vector along a curve, as follows. If a curve is defined by $c(t)$ in the coordinate space and a vector field W is given, then the derivative of the vector field W along the curve is given by

$$\frac{DW}{dt} = \nabla_{\frac{dc}{dt}} W = \frac{dc^i}{dt} \left\{ \frac{\partial w^j}{\partial y^i} + w^k \Gamma^{\,j}_{ik} \right\} \frac{\partial}{\partial y^j}. \tag{A.233}$$

With the derivative $(dc^i / dt)(\partial w^j / \partial y^i) = dw^j / dt$, this expression becomes

$$\frac{DW}{dt} = \left\{ \frac{dw^j}{dt} + w^k \Gamma^{\,j}_{ik} \frac{dc^i}{dt} \right\} \frac{\partial}{\partial y^j}. \tag{A.234}$$

This expression now includes only the components of the vector along the curve; thus for a vector along a curve (rather than a vector field)

$$W(t) = w^j(t) \left. \frac{\partial}{\partial y^j} \right|_{y=c(t)} \tag{A.235}$$

we can compute the derivative of the vector along the curve using Eq. (A.234). In Cartesian coordinates in Euclidean space, a constant length vector undergoing parallel motion has constant coefficients, the Christoffel symbols are zero, and hence $DW/dt = 0$. In general, then, parallel translation of a vector is defined as when the vector components satisfy $DW/dt = 0$, or

$$\text{Parallel translation:} \quad \frac{dw^j}{dt} + w^k \Gamma^{\,j}_{ik} \frac{dc^i}{dt} = 0 \quad \text{along} \quad y = c(t). \tag{A.236}$$

The notion is that the components of the vector are adjusted to compensate for the changes in the basis. Given the initial three components of the vector W, Eq. (A.236) is a first order ordinary differential equation system to compute the components along the curve $c(t)$ and thus provides a unique parallel translation for a given initial vector for a given path. The notion of parallel translation allows us to link the Christoffel symbols with the metric. To make the link, there are two assumptions (these assumptions are fundamental in Riemannian geometry). Assumption #1: the Christoffel symbols are symmetric in the indices ij : $\Gamma^k_{ij} = \Gamma^k_{ji}$ (referred to as *torsion free*). This assumption is obvious from our tangent space basis for Euclidean space based on $\vec{x}$, since it is just a statement that partial derivatives commute. Assumption #2: the connection is *compatible* with the metric, which means that vectors translated along a curve maintain their lengths and inner products, in particular that the length of a vector undergoing parallel translation does not change – that is, $g(W, W) = $ constant. That the connection is compatible with the metric leads to the following result. Pick a contravariant orthonormal basis for the vector at the point $c(0)$ along a curve $c(t)$ and write it as B_i. Orthonormal means $g(B_i, B_j) = \delta_{ij}$. These basis vectors can be parallel translated, and the assumption about the metric being compatible means that the translates $B_i(t)$ of these vectors remain orthonormal. Thus, two vectors, written in terms of these basis vectors, $X = a^i(t)B_i(t)$ and $Y = b^j(t)B_j(t)$, have the following properties,

$$g(X,Y) = g(a^i B_i, b^j B_j) = a^i b^j g(B_i, B_j) = a^i b^j \delta_{ij} = a^i b_j. \tag{A.237}$$

The a^i and b^j are the coefficients that multiply the B_i, and thus $b^j = b^i \delta_{ij} = b_j$ and, to be explicit, $a^i b_j = a^1 b^1 + a^2 b^2 + a^3 b^3$. (The superscript is lowered through the delta function, rather than the metric, because we are not using the standard tangent space basis but rather an orthonormal set of vectors.) We have, for example,

$$\frac{DX}{dt} = \frac{D}{dt}(a^i B_i) = \frac{da^i}{dt} B_i + a^i \frac{DB_i}{dt} = \frac{da^i}{dt} B_i, \tag{A.238}$$

and so

$$g\left(\frac{DX}{dt}, Y\right) + g\left(X, \frac{DY}{dt}\right) = g\left(\frac{da^i}{dt} B_i, b^j B_j\right) + g\left(a^i B_i, \frac{db^j}{dt} B_j\right)$$

$$= \frac{da^i}{dt} b^j g(B_i, B_j) + a^i \frac{db^j}{dt} g(B_i, B_j) = \frac{da^i}{dt} b^j \delta_{ij} + a^i \frac{db^j}{dt} \delta_{ij} \tag{A.239}$$

$$= \frac{da^i}{dt} b_i + a^i \frac{db^i}{dt} = \frac{d}{dt}(a^i b_j) = \frac{d}{dt} g(X,Y).$$

This result is rather profound. It leaves us with the following equation, which results from compatibility, namely if we choose a curve so that $dc/dt|_{t=0} = \frac{\partial}{\partial y^k}$ (a basis vector) then $\frac{d}{dt} = \frac{\partial}{\partial y^k} \frac{dy^k}{dt} = \frac{\partial}{\partial y^k}$ (a derivative, no sum on k) and letting X and Y be the i and j basis vectors yields

$$\frac{\partial}{\partial y^k} g\left(\frac{\partial}{\partial y^i}, \frac{\partial}{\partial y^j}\right) = g\left(\frac{D}{dt}\left(\frac{\partial}{\partial y^i}\right), \frac{\partial}{\partial y^j}\right) + g\left(\frac{\partial}{\partial y^i}, \frac{D}{dt}\left(\frac{\partial}{\partial y^j}\right)\right)$$

$$= g\left(\nabla_{\frac{\partial}{\partial y^k}} \frac{\partial}{\partial y^i}, \frac{\partial}{\partial y^j}\right) + g\left(\frac{\partial}{\partial y^i}, \nabla_{\frac{\partial}{\partial y^k}} \frac{\partial}{\partial y^j}\right) = g\left(\Gamma_{ki}^m \frac{\partial}{\partial y^m}, \frac{\partial}{\partial y^j}\right) \tag{A.240}$$

$$+ g\left(\frac{\partial}{\partial y^i}, \Gamma_{kj}^m \frac{\partial}{\partial y^m}\right) = \Gamma_{ki}^m g\left(\frac{\partial}{\partial y^m}, \frac{\partial}{\partial y^j}\right) + \Gamma_{kj}^m g\left(\frac{\partial}{\partial y^i}, \frac{\partial}{\partial y^m}\right),$$

which can be written

$$\frac{\partial g_{ij}}{\partial y^k} = \Gamma_{ki}^m g_{mj} + \Gamma_{kj}^m g_{im} = \Gamma_{ki,j} + \Gamma_{kj,i}. \tag{A.241}$$

Thus, the connection completely determines all the derivatives of all orders of the metric when the metric and connection are compatible. Knowing the metric g at one location and knowing the connection everywhere determines the metric everywhere given adequate differentiability. By cycling through the indices and using the symmetry of the Christoffel symbols ($\Gamma_{ij}^k = \Gamma_{ji}^k$) and the metric, we obtain

$$\Gamma_{ij,k} = [ij, k] = \frac{1}{2}\left(\frac{\partial g_{ik}}{\partial y^j} + \frac{\partial g_{jk}}{\partial y^i} - \frac{\partial g_{ij}}{\partial y^k}\right) = g_{km}\Gamma_{ij}^m,$$

$$\Gamma_{ij}^k = \left\{ \begin{matrix} k \\ ij \end{matrix} \right\} = g^{km} \frac{1}{2}\left(\frac{\partial g_{im}}{\partial y^j} + \frac{\partial g_{jm}}{\partial y^i} - \frac{\partial g_{ij}}{\partial y^m}\right) = g^{km}\Gamma_{ij,m}. \tag{A.242}$$

The $\Gamma_{ij,k}$ are the Christoffel symbols of the first kind and the Γ_{ij}^k are the Christoffel symbols of the second kind. Older notation for the Christoffel symbols is indicated in the first equality. The analog of this equation for orthonormal coordinates is Eq. (A.54). The Christoffel symbols are not tensors, and their transformation law is given by

$$(\Gamma')_{ij}^k = \Gamma_{lm}^n \frac{\partial y^l}{\partial (y')^i} \frac{\partial y^m}{\partial (y')^j} \frac{\partial (y')^k}{\partial y^n} + \frac{\partial^2 y^l}{\partial (y')^i \partial (y')^j} \frac{\partial (y')^k}{\partial y^l}. \tag{A.243}$$

The second term on the right-hand side is the reason it does not transform as a tensor. (The torsion tensor defined as $T_{ij}^k = \Gamma_{ij}^k - \Gamma_{ji}^k$ transforms correctly as a $(1,2)$ tensor, since the incorrectly transforming term subtracts out. For us the torsion tensor is always zero, but it can be nonzero in non-Riemannian geometries.) A more general statement of the connection being compatible with the metric, or that it is metric, is, with $Z = c^i \frac{\partial}{\partial y^i}$ and three instances of Eq. (A.241) added together,

$$D_Z g(X,Y) = c^i \frac{\partial}{\partial y^i} g(X,Y) = g(\nabla_Z X, Y) + g(X, \nabla_Z Y). \tag{A.244}$$

It is a fact from Riemannian geometry that there is only one torsion-free connection compatible with the metric [154, 173].

As examples, we compute some Christoffel symbols. For our diagonal metric $g = h_1^2 dy^1 \otimes dy^1 + h_2^2 dy^2 \otimes dy^2 + h_3^2 dy^3 \otimes dy^3$, some manipulation yields

$$\Gamma_{ji}^j = \Gamma_{ij}^j = \frac{1}{h_j} \frac{\partial h_j}{\partial y^i} \quad \text{and} \quad \Gamma_{ii}^k = -\frac{h_i}{h_k^2} \frac{\partial h_i}{\partial y^k} \quad \text{(no sum on } i, j, k), \quad i \neq k,$$
$$\Gamma_{ij}^k = 0 \quad \text{(all } i, j, k \text{ different),} \tag{A.245}$$

or in detail,

$$\Gamma_{11}^1 = \frac{1}{h_1} \frac{\partial h_1}{\partial y^1}, \quad \Gamma_{11}^2 = -\frac{h_1}{h_2^2} \frac{\partial h_1}{\partial y^2}, \quad \Gamma_{11}^3 = -\frac{h_1}{h_3^2} \frac{\partial h_1}{\partial y^3}, \quad \Gamma_{12}^1 = \frac{1}{h_1} \frac{\partial h_1}{\partial y^2},$$

$$\Gamma_{12}^2 = \frac{1}{h_2} \frac{\partial h_2}{\partial y^1}, \quad \Gamma_{12}^3 = 0, \quad \Gamma_{13}^1 = \frac{1}{h_1} \frac{\partial h_1}{\partial y^3}, \quad \Gamma_{13}^2 = 0, \quad \Gamma_{13}^3 = \frac{1}{h_3} \frac{\partial h_3}{\partial y^1},$$

$$\Gamma_{22}^1 = -\frac{h_2}{h_1^2} \frac{\partial h_2}{\partial y^1}, \quad \Gamma_{22}^2 = \frac{1}{h_2} \frac{\partial h_2}{\partial y^2}, \quad \Gamma_{22}^3 = -\frac{h_2}{h_3^2} \frac{\partial h_2}{\partial y^3}, \quad \Gamma_{23}^1 = 0, \quad \Gamma_{23}^2 = \frac{1}{h_2} \frac{\partial h_2}{\partial y^3},$$

$$\Gamma_{23}^3 = \frac{1}{h_3} \frac{\partial h_3}{\partial y^2}, \quad \Gamma_{33}^1 = -\frac{h_3}{h_1^2} \frac{\partial h_3}{\partial y^1}, \quad \Gamma_{33}^2 = -\frac{h_3}{h_2^2} \frac{\partial h_3}{\partial y^2}, \quad \Gamma_{33}^3 = \frac{1}{h_3} \frac{\partial h_3}{\partial y^3}. \tag{A.246}$$

Specifically, for the cylindrical coordinate system ($y^r = r$, $y^\theta = \theta$, $y^z = z$, $h_r = 1$, $h_\theta = r$, $h_z = 1$, and (not used, but for completeness) $x^r = r$, $x^\theta = 0$, $x^z = z$),

$$\Gamma_{r\theta}^\theta = \Gamma_{\theta r}^\theta = \frac{1}{r}, \quad \Gamma_{\theta\theta}^r = -r, \quad \text{all other } \Gamma_{ij}^k = 0 \tag{A.247}$$

(compare these to the orthonormal frame Christoffel symbols Eq. A.160, which are

$$\gamma_{\theta r}^\theta = -\gamma_{\theta\theta}^r = 1, \quad \text{all other } \gamma_{ij}^k = 0 \text{ including } \gamma_{r\theta}^\theta = 0), \tag{A.248}$$

and for the spherical coordinate system ($y^r = r$, $y^\theta = \theta$, $y^\varphi = \varphi$, $h_r = 1$, $h_\theta = r$, $h_\varphi = r\sin(\theta)$ and (not used, but for completeness) $x^r = r$, $x^\theta = 0$, $x^\varphi = 0$),

$$\Gamma_{r\theta}^\theta = \Gamma_{\theta r}^\theta = \frac{1}{r}, \quad \Gamma_{r\varphi}^\varphi = \Gamma_{\varphi r}^\varphi = \frac{1}{r}, \quad \Gamma_{\theta\theta}^r = -r, \quad \Gamma_{\theta\varphi}^\varphi = \Gamma_{\varphi\theta}^\varphi = \cot(\theta),$$

$$\Gamma_{\varphi\varphi}^r = -r\sin^2(\theta), \quad \Gamma_{\varphi\varphi}^\theta = -\sin(\theta)\cos(\theta), \quad \text{all other } \Gamma_{ij}^k = 0 \tag{A.249}$$

(again compare with the orthonormal frame Christoffel symbols Eq. A.182

$$\gamma^\theta_{\theta r} = -\gamma^r_{\theta\theta} = 1, \quad \gamma^\varphi_{\varphi r} = -\gamma^r_{\varphi\varphi} = \sin(\theta), \quad \gamma^\varphi_{\varphi\theta} = -\gamma^\theta_{\varphi\varphi} = \cos(\theta), \text{ all other } \gamma^k_{ij} = 0).$$
(A.250)

Through a direct computation we can now compute the covariant derivative of a covariant vector field. We need an intermediate result. Observing that

$$\frac{\partial(g^{mj}g_{jl})}{\partial y^i} = \frac{\partial\delta^m_l}{\partial y^i} = 0 = \frac{\partial g^{mj}}{\partial y^i}g_{jl} + g^{mj}\frac{\partial g_{jl}}{\partial y^i},$$
(A.251)

and thus,

$$\frac{\partial g^{mj}}{\partial y^i}g_{jl} = -g^{mj}\frac{\partial g_{jl}}{\partial y^i} = -g^{mj}(\Gamma^n_{ij}g_{nl} + \Gamma^n_{il}g_{jn}) = -g^{mj}g_{nl}\Gamma^n_{ij} - \Gamma^m_{il},$$
(A.252)

multiplying both sides by g^{lk} gives

$$\frac{\partial g^{mj}}{\partial y^i}g_{jl}g^{lk} = -g^{mj}g_{nl}g^{lk}\Gamma^n_{ij} - g^{lk}\Gamma^m_{il}$$
(A.253)

or

$$\frac{\partial g^{mk}}{\partial y^i} = -g^{mj}\Gamma^k_{ij} - g^{lk}\Gamma^m_{il}.$$
(A.254)

With a covariant vector field

$$\omega = \alpha_i dy^i = \alpha_i g^{ij}\frac{\partial}{\partial y^j},$$
(A.255)

we obtain

$$\begin{aligned}
\nabla_X\omega &= a^i\left\{\frac{\partial(\alpha_m g^{mj})}{\partial y^i} + \alpha_m g^{mk}\Gamma^j_{ik}\right\}\frac{\partial}{\partial y^j} \\
&= a^i\left\{\frac{\partial\alpha_m}{\partial y^i}g^{mj} + \alpha_m\frac{\partial g^{mj}}{\partial y^i} + \alpha_m g^{mk}\Gamma^j_{ik}\right\}\frac{\partial}{\partial y^j} \\
&= a^i\left\{\frac{\partial\alpha_m}{\partial y^i}g^{mj} + \alpha_m(-g^{mn}\Gamma^j_{in} - g^{lj}\Gamma^m_{il}) + \alpha_m g^{mk}\Gamma^j_{ik}\right\}\frac{\partial}{\partial y^j} \\
&= a^i\left\{\frac{\partial\alpha_m}{\partial y^i}g^{mj} - \alpha_m g^{lj}\Gamma^m_{il}\right\}\frac{\partial}{\partial y^j} = a^i\left\{\frac{\partial\alpha_m}{\partial y^i}g^{mj} - \alpha_n g^{mj}\Gamma^n_{im}\right\}\frac{\partial}{\partial y^j} \\
&= a^i\left\{\frac{\partial\alpha_m}{\partial y^i} - \alpha_n\Gamma^n_{im}\right\}g^{mj}\frac{\partial}{\partial y^j} \\
&= a^i\left\{\frac{\partial\alpha_m}{\partial y^i} - \alpha_n\Gamma^n_{im}\right\}dy^m.
\end{aligned}$$
(A.256)

Thus, the directional derivative of a covariant vector is a covariant tensor with coefficients similar to Eq. (A.221); only the sign on the Christoffel symbol is reversed. Implicitly we have defined a $(0,2)$ covariant tensor of the form

$$\nabla\omega = \alpha_{j;i}dx^i \otimes dy^j = \left\{\frac{\partial\alpha_j}{\partial y^i} - \alpha_k\Gamma^k_{ij}\right\}dx^i \otimes dy^j,$$
(A.257)

where

$$\alpha_{j;i} = \frac{\partial\alpha_j}{\partial y^i} - \alpha_k\Gamma^k_{ij}.$$
(A.258)

We have thus defined covariant differentiation for both a contravariant vector and a covariant vector.

It is now possible to differentiate second order tensors of all types. As a first, example, for a contravariant $(2,0)$ tensor we have

$$
\begin{aligned}
\nabla \sigma &= \frac{\partial \sigma}{\partial y^k} dy^k = \frac{\partial}{\partial y^k} \left\{ \sigma^{ij} \frac{\partial}{\partial y^i} \otimes \frac{\partial}{\partial y^j} \right\} \otimes dy^k \\
&= \left[\frac{\partial \sigma^{ij}}{\partial y^k} \frac{\partial}{\partial y^i} \otimes \frac{\partial}{\partial y^j} + \sigma^{ij} \frac{\partial}{\partial y^k} \left\{ \frac{\partial}{\partial y^i} \right\} \otimes \frac{\partial}{\partial y^j} + \sigma^{ij} \frac{\partial}{\partial y^i} \otimes \frac{\partial}{\partial y^k} \left\{ \frac{\partial}{\partial y^j} \right\} \right] \otimes dy^k \\
&= \left\{ \frac{\partial \sigma^{ij}}{\partial y^k} \frac{\partial}{\partial y^i} \otimes \frac{\partial}{\partial y^j} + \sigma^{ij} \Gamma_{ki}^{\,m} \frac{\partial}{\partial y^m} \otimes \frac{\partial}{\partial y^j} + \sigma^{ij} \frac{\partial}{\partial y^i} \otimes \Gamma_{kj}^{\,m} \frac{\partial}{\partial y^m} \right\} \otimes dy^k \\
&= \left\{ \frac{\partial \sigma^{ij}}{\partial y^k} + \sigma^{mj} \Gamma_{km}^{\,i} + \sigma^{im} \Gamma_{km}^{\,j} \right\} \frac{\partial}{\partial y^i} \otimes \frac{\partial}{\partial y^j} \otimes dy^k \\
&= \sigma^{ij}_{\;;k} \frac{\partial}{\partial y^i} \otimes \frac{\partial}{\partial y^j} \otimes dy^k .
\end{aligned}
$$

$$(A.259)$$

Essentially, differentiation occurs on each component individually, as follows:

$$
\begin{aligned}
\sigma^{ij}_{\;;k} &= \frac{\partial \sigma^{ij}}{\partial y^k} + \sigma^{mj} \Gamma_{km}^{\,i} + \sigma^{im} \Gamma_{km}^{\,j}, \\
\sigma^{j}_{i;k} &= \frac{\partial \sigma^j_i}{\partial y^k} - \sigma^j_m \Gamma_{ki}^{\,m} + \sigma^m_i \Gamma_{km}^{\,j}, \\
\sigma_{ij;k} &= \frac{\partial \sigma_{ij}}{\partial y^k} - \sigma_{mj} \Gamma_{ki}^{\,m} - \sigma_{im} \Gamma_{kj}^{\,m}.
\end{aligned}
$$

$$(A.260)$$

(Note the $\frac{1}{h_i}$ multiplicative difference with Eq. A.84 due to the normalization there.) This derivative is covariant differentiation $\nabla \sigma$, and the differentiation takes a $(2,0)$ tensor to a $(2,1)$ tensor (a 3-tensor), a $(1,1)$ tensor to a $(1,2)$ tensor, and a $(0,2)$ tensor to a $(0,3)$ tensor. The first two are mixed tensors and the last is a covariant tensor. There is a special case of interest, and that is the covariant derivative of the metric tensor. In mixed mode, the metric tensor is simply $g^j_i = \delta^j_i$, a constant, and so we expect its covariant derivative to be zero, and $\delta^j_{i;k} = 0 - \delta^j_m \Gamma_{ik}^{\,m} + \delta^m_i \Gamma_{mk}^{\,j} = -\Gamma_{ik}^{\,j} + \Gamma_{ik}^{\,j} = 0$, as expected. A tensor that is zero in one reference frame is zero in all of them. As further verification, when the covariant form and the contravariant form are employed and covariantly differentiated, the definition of the Christoffel symbols shows that they are zero also, as expected. Thus, $\nabla g = 0$.

For completeness, we determine the covariant derivative of a scalar function. With $X = a^i \frac{\partial}{\partial y^i}$ a vector (it need only be defined at the point of interest for a specific computation),

$$
\nabla_X f = \lim_{t \to 0} \frac{f(y^1 + ta^1, y^2 + ta^2, y^3 + ta^3) - f(y^1, y^2, y^3)}{t} = a^i \frac{\partial f}{\partial y^i}. \qquad (A.261)
$$

Thus, the covariant derivative of a scalar function is simply the gradient. Componentwise, the covariant derivative of a scalar function produces a covariant vector,

$$
f_{;i} = \frac{\partial f}{\partial y^i}. \qquad (A.262)
$$

In texts the covariant derivative in the X direction (the directional derivative) of a scalar function f is written in equivalent ways,

$$
\nabla_X f = df(X) = D_X f = X(f) = Xf, \qquad (A.263)
$$

with the various notations being (hopefully) intuitive. All of these expressions produce a scalar. Also be aware that the covariant derivative is not the Lie derivative, which is also discussed in geometry texts. For a scalar they are the same – that is, $L_X f = \nabla_X f$ – but for a vector field the Lie derivative equals the Lie bracket, $L_X Y = XY - YX = [X, Y] \neq \nabla_X Y$. The Lie bracket $[X, Y]$ denotes the unique vector Z such that for any scalar function f,

$$D_Z f = D_X(D_Y f) - D_Y(D_X f)$$

$$= a^i \frac{\partial}{\partial y^i} \left(b^j \frac{\partial f}{\partial y^j} \right) - b^i \frac{\partial}{\partial y^i} \left(a^j \frac{\partial f}{\partial y^j} \right) = \left(a^i \frac{\partial b^j}{\partial y^i} - b^i \frac{\partial a^j}{\partial y^i} \right) \frac{\partial f}{\partial y^j}, \tag{A.264}$$

hence the parenthetical term must be zero for each j, and the vector Z is

$$[X, Y] = XY - YX = \left(a^i \frac{\partial b^j}{\partial y^i} - b^i \frac{\partial a^j}{\partial y^i} \right) \frac{\partial}{\partial y^j}. \tag{A.265}$$

They are clearly not equal, evident by the lack of any geometric information as contained in the Christoffel symbols and the fact that X needs to be a vector field and not just known at the point–the Lie derivative includes a derivative of X.

We now do some linear algebra to obtain an expression for the contraction of the Christoffel symbols. A determinant can be written by expanding in cofactors. For example, if we chose to expand about the ith row of the metric matrix,

$$\det(g) = g_{i1}C_{i1} + g_{i2}C_{i2} + \cdots + g_{in}C_{in}. \tag{A.266}$$

In the various cofactors, the term g_{ij} for the specified row i does not appear, because the cofactors are computed by taking the determinant of the minor, where the specific row and column are excluded. Thus,

$$\frac{\partial \det(g)}{\partial g_{ij}} = C_{ij}. \tag{A.267}$$

For an invertible matrix, the inverse terms can be found by cofactor expansion and are

$$(g^{-1})_{ij} = \frac{C_{ji}}{\det(g)}. \tag{A.268}$$

Since the metric g is symmetric, these results combine to say that

$$g^{ij} = \frac{1}{\det(g)} \frac{\partial \det(g)}{\partial g_{ij}}. \tag{A.269}$$

Now,

$$\frac{\partial \det(g)}{\partial y^k} = \frac{\partial \det(g)}{\partial g_{ij}} \frac{\partial g_{ij}}{\partial y^k} = \det(g)g^{ij} \frac{\partial g_{ij}}{\partial y^k} = \det(g)g^{ij} \left(\Gamma_{ik,j} + \Gamma_{jk,i} \right) = 2\det(g)\Gamma^i_{ik},$$
$$\tag{A.270}$$

and this result can be rewritten a number of ways, including the second equality,

$$\Gamma^i_{ik} = \frac{1}{2\det(g)} \frac{\partial \det(g)}{\partial y^k} = \frac{1}{\sqrt{\det(g)}} \frac{\partial \sqrt{\det(g)}}{\partial y^k}, \tag{A.271}$$

and will be used in the next paragraph.

As might be guessed, the divergence of a vector field $X = a^i \frac{\partial}{\partial y^i}$ is the trace of ∇X. In terms of coordinates this is, with the last equality using Eq. (A.271),

$$\operatorname{div}(X) = \operatorname{trace}(\nabla X) = a^i_{;i} = \frac{\partial a^i}{\partial y^i} + a^j \Gamma^i_{ij} = \frac{1}{\sqrt{\det(g)}} \frac{\partial(\sqrt{\det(g)}\, a^i)}{\partial y^i}. \tag{A.272}$$

This result allows us to compute the Laplacian of a scalar function f, where we use the covariant form of the gradient, as the result is immediate,

$$\Delta f = \nabla^2 f = \text{div}(\text{grad}(f)) = \text{trace}(\nabla df) = \text{trace}(\nabla\nabla f) = f_{;i}^i = (g^{ij}f_{;j})_{;i} = g^{ij}f_{;ji},$$
$$(A.273)$$

where after computing the covariant gradient of f, we can raise the index, take the covariant derivative, then take the trace or we can take the covariant derivative, raise the index, then take the trace, as indicated in the last two equalities. Explicitly,

$$\Delta f = f_{;i}^i = g^{ij}f_{;ji} = g^{ij}\left(\frac{\partial f}{\partial y^j}\right)_{;i} = g^{ij}\left\{\frac{\partial^2 f}{\partial y^i \partial y^j} - \Gamma_{ij}^k \frac{\partial f}{\partial y^k}\right\}.$$

$$\Delta f = f_{;i}^i = (g^{ij}f_{;j})_{;i} = g_{;i}^{ij}f_{;j} + g^{ij}f_{;ji} = g^{ij}f_{;ji} \quad \text{since } \nabla g = g_{;k}^{ij} = 0 \text{ (page 640).}$$
$$(A.274)$$

The first term is what is seen in Cartesian coordinates, and the second term is the adjustment due to the connection. From Eq. (A.272), we can also write

$$\Delta f = \frac{1}{\sqrt{\det(g)}} \frac{\partial}{\partial y^i}\left(\sqrt{\det(g)}\, g^{ij}\frac{\partial f}{\partial y^j}\right).$$
$$(A.275)$$

To compute the curl, we enter the world of differential forms with minimal explanation. With the 1-form (which is a covariant vector) $\omega = \alpha_i dy^i$, the curl is

$$\text{curl}(\omega) = d\omega = \left(\frac{\partial \alpha_i}{\partial y^j}dy^j\right) \wedge dy^i = \left(\frac{\partial \alpha_2}{\partial y^1} - \frac{\partial \alpha_1}{\partial y^2}\right) dy^1 \wedge dy^2$$
$$+ \left(\frac{\partial \alpha_3}{\partial y^1} - \frac{\partial \alpha_1}{\partial y^3}\right) dy^1 \wedge dy^3 + \left(\frac{\partial \alpha_3}{\partial y^2} - \frac{\partial \alpha_2}{\partial y^3}\right) dy^2 \wedge dy^3, \quad (A.276)$$

where we have used the properties of forms that $dy^i \wedge dy^j = -dy^j \wedge dy^i$ (hence $dx^1 \wedge dx^1 = 0$) and $d^2(\text{anything}) = 0$. In three dimensions, there are three basis vectors for 2-forms (as indicated in Eq. A.276), and so the 2-form space is a three-dimensional Euclidean vector space. Thus, there exists a relationship between the 2-form basis and the contravariant vector basis, just as there was between 1-form basis (covariant vector basis) and contravariant vector basis. There, we knew a class of vectors (gradients) in both spaces and we were able to determine the relationship between the bases. Here, we infer the relationship using the fact that $\text{div}(\text{curl}(f)) = 0$. Notice that we trivially have $\text{curl}(df) = 0$. Looking at Eqs. (A.272) and (A.276), and requiring $\text{div}(\text{curl}(w)) = 0$, we infer (up to a multiplicative constant)

$$\frac{\partial}{\partial y^1} = \sqrt{\det(g)}\, dy^2 \wedge dy^3, \quad \frac{\partial}{\partial y^2} = -\sqrt{\det(g)}\, dy^1 \wedge dy^3, \quad \frac{\partial}{\partial y^3} = \sqrt{\det(g)}\, dy^1 \wedge dy^2.$$
$$(A.277)$$

This is the Hodge star ($*$) operator which maps the $dy^i \wedge dy^j$ to dy^i,

$$* dy^i \wedge dy^j = \varepsilon_{ijk}\frac{1}{\sqrt{\det(g)}}\frac{\partial}{\partial y^k} = \varepsilon_{ijk}\frac{1}{\sqrt{\det(g)}}g_{\ell k}dy^\ell.$$
$$(A.278)$$

This operator yields the cross product: a straightforward calculation (Exercise A.25) shows that if we begin with two covariant vectors ω_1 and ω_2, then $g(\omega_i, *\, \omega_1 \wedge \omega_2) = 0$; thus $*\, \omega_1 \wedge \omega_2$ is perpendicular to both ω_1 and ω_2. Further, some computation yields $g(\frac{\partial}{\partial y^1}, *\frac{\partial}{\partial y^2} \wedge \frac{\partial}{\partial y^3}) = \sqrt{\det(g)}$, showing $dvol = \sqrt{\det(g)}\, dy^1 \wedge dy^2 \wedge dy^3$, thus showing the aforementioned multiplicative constant is 1. Given these relations for the basis, the curl of a contravariant vector $X = a^i \frac{\partial}{\partial y^i}$ can be written as a contravariant vector as

$$\text{curl}(X) = \frac{1}{\sqrt{\det(g)}} \left\{ \left(\frac{\partial(g_{3i}a^i)}{\partial y^2} - \frac{\partial(g_{2i}a^i)}{\partial y^3} \right) \frac{\partial}{\partial y^1} \right.$$
$$\left. - \left(\frac{\partial(g_{3i}a^i)}{\partial y^1} - \frac{\partial(g_{1i}a^i)}{\partial y^3} \right) \frac{\partial}{\partial y^2} + \left(\frac{\partial(g_{2i}a^i)}{\partial y^1} - \frac{\partial(g_{1i}a^i)}{\partial y^2} \right) \frac{\partial}{\partial y^3} \right\}. \quad (A.279)$$

Working with one of the components to expand and collect terms yields (with algebra between each line),

$$\frac{\partial(g_{3i}a^i)}{\partial y^2} - \frac{\partial(g_{2i}a^i)}{\partial y^3} = g_{3i}\frac{\partial a^i}{\partial y^2} + a^i g_{3m}\Gamma^m_{2i} - g_{2i}\frac{\partial a^i}{\partial y^3} - a^i g_{2m}\Gamma^m_{3i}$$
$$= g_{3i}\left(\frac{\partial a^i}{\partial y^2} + a^m \Gamma^i_{2m} \right) - g_{2i}\left(\frac{\partial a^i}{\partial y^3} + a^m \Gamma^i_{3m} \right) \quad (A.280)$$
$$= g_{3i}a^i_{;2} - g_{2i}a^i_{;3}.$$

Each term only depends on the metric g and on the Christoffel symbols, but no derivatives of Christoffel symbols, which has been true of all our operators. Thus

$$\text{curl}(X) = \frac{1}{\sqrt{\det(g)}} \left\{ (g_{3i}a^i_{;2} - g_{2i}a^i_{;3})\frac{\partial}{\partial y^1} - (g_{3i}a^i_{;1} - g_{1i}a^i_{;3})\frac{\partial}{\partial y^2} \right.$$
$$\left. + (g_{2i}a^i_{;1} - g_{1i}a^i_{;2})\frac{\partial}{\partial y^3} \right\} = \sum_{ijkm} \frac{1}{\sqrt{\det(g)}}\varepsilon_{ijk}a^m_{;i}g_{mj}\frac{\partial}{\partial y^k}, \quad (A.281)$$

where ε_{ijk} is the permutation symbol ($= +1, -1$, or 0 for even, odd, or repeated permutations, respectively).

We are now in a position to write down the conservation equations. The mass balance in the Eulerian frame is, where $V = b^i \frac{\partial}{\partial y^i}$,

$$\frac{\partial \rho}{\partial t} + \text{div}(\rho V) = \frac{\partial \rho}{\partial t} + (\rho b^i)_{;i} = \frac{\partial \rho}{\partial t} + \frac{\partial(\rho b^i)}{\partial y^i} + \rho b^j \Gamma^i_{ij} = 0,$$
$$\frac{D\rho}{Dt} + \rho\,\text{div}(V) = \frac{d\rho}{dt} + \rho(b^i)_{;i} = \frac{\partial \rho}{\partial t} + \frac{\partial \rho}{\partial y^i}\frac{dy^i}{dt} + \rho\left(\frac{\partial b^i}{\partial y^i} + b^j \Gamma^i_{ij} \right) = 0. \quad (A.282)$$

Since $b^i = dy^i/dt$, we see that these are the same.

We now look at the derivative of the stress tensor used in the momentum balance. We essentially want the divergence of a tensor, as before. Here we find it by taking the trace (a contraction) of the absolute derivative of the stress,

$$(\text{trace}(\nabla\sigma))^j = (\text{div}(\sigma))^j = \sigma^{ij}_{;i} = \frac{\partial \sigma^{ij}}{\partial y^i} + \sigma^{mj}\Gamma^i_{mi} + \sigma^{im}\Gamma^j_{mi}$$
$$= \frac{1}{\sqrt{\det(g)}}\frac{\partial(\sqrt{\deg(g)}\,\sigma^{ij})}{\partial y^i} + \sigma^{im}\Gamma^j_{mi}. \quad (A.283)$$

In words, the contraction of the covariant derivative of the contravariant stress tensor is a contravariant vector. The momentum balance can now be written as

$$\rho A = \text{div}(\sigma) + \rho B = \text{trace}(\nabla\sigma) + \rho B, \quad \text{or in components as} \quad \rho a^i = \sigma^{ji}_{;j} + \rho b^i, \quad (A.284)$$

where $A = a^i \frac{\partial}{\partial y^i}$ is the acceleration, $B = b^i \frac{\partial}{\partial y^i}$ is the body force, and $\sigma = \sigma^{ij}$ is the stress tensor.

For the energy balance, we need to understand the power input $\vec{t} \cdot \vec{v}$ on the surface, which is given by $g(T, V)$ in our more general setting. To determine it requires clarification of the stress tensor. The stress tensor is a linear function that returns the traction vector when given a normal vector. In our notation, with

$$\sigma = \sigma_i^j dx^i \otimes \frac{\partial}{\partial y^j}, \tag{A.285}$$

for unit normal $N = a^i \frac{\partial}{\partial y^i}$, where $|N| = \sqrt{g(N,N)} = \sqrt{g_{ij} a^i a^j} = 1$, the contravariant traction T is given by

$$T = \sigma(N) = a^i \sigma_i^j \frac{\partial}{\partial y^j}. \tag{A.286}$$

The term corresponding to $\vec{t} \cdot \vec{v}$ is thus $g(T, V) = g(\sigma(N), V)$, and some manipulation gives

$$g(\sigma(N), V) = g(a^i \sigma_i^j \frac{\partial}{\partial y^j}, b^k \frac{\partial}{\partial y^k}) = a^i \sigma_i^j b^k g_{jk} = a^i \sigma_{ik} b^k = a^i g_{\ell i} \sigma_k^\ell b^k = g(\sigma(V), N). \tag{A.287}$$

The divergence theorem converts the surface integral of $g(\sigma(V), N)$ to a volume integral of $\operatorname{div}(\sigma(V))$, which is

$$\operatorname{div}(\sigma(V)) = (\sigma_j^i b^j)_{;i} = \frac{\partial(\sigma_j^i b^j)}{\partial y^i} + \sigma_j^i b^j \Gamma_{ik}^k = \sigma_j^i \frac{\partial b^j}{\partial y^i} + b^j \left(\frac{\partial \sigma_j^i}{\partial y^i} + \sigma_j^i \Gamma_{ik}^k \right)$$

$$= \sigma_j^i \left(b_{;i}^j - b^k \Gamma_{ik}^j \right) + b^j \left(\frac{\partial \sigma_j^i}{\partial y^i} + \sigma_j^i \Gamma_{ik}^k \right) \tag{A.288}$$

$$= \sigma_j^i b_{;i}^j + b^j \left(\frac{\partial \sigma_j^i}{\partial y^i} + \sigma_j^i \Gamma_{ik}^k - \sigma_k^i \Gamma_{ij}^k \right) = \sigma_j^i b_{;i}^j + b^j \sigma_{j;i}^i$$

$$= \sigma_j^i (\nabla V)_i^j + b^j (\operatorname{trace}(\nabla \sigma))_j = \sigma_j^i (\nabla V)_i^j + b^j (\operatorname{div}(\sigma))_j.$$

The covariant derivative of vectors and a tensor was used in this derivation. Through the momentum balance the last term (including a body force, if present) will be replaced with $(\operatorname{div}(\sigma))_j = \rho a_j$. Next, the material derivative of the specific kinetic energy $\frac{1}{2}|V|^2$ is

$$\frac{1}{2} \frac{D}{Dt} |V|^2 = \frac{1}{2} \left(\frac{\partial}{\partial t} + b^i \frac{\partial}{\partial y^i} \right) g(V, V) = \frac{1}{2} \frac{\partial}{\partial t} g(V, V) + \frac{1}{2} b^i \frac{\partial}{\partial y^i} g(V, V)$$

$$= g(\frac{\partial b^i}{\partial t} \frac{\partial}{\partial y^i}, V) + g(\frac{DV}{dt}, V) = g(A, V) = g_{ij} a^i b^j = a_j b^j, \tag{A.289}$$

where the $\partial/\partial t$ only operates on the time part of the coefficient $b^i(y^1, y^2, y^3, t)$, the absolute derivative result (moving the derivative inside g) was shown in Eq. (A.239), since $DV/dt = b^i \partial V/\partial y^i$, and the symmetry of $g(A, B) = g(B, A)$ was used. Once again, just as in the derivation of the energy balance on page 603, the two $\rho b^j a_j$ terms cancel in the final expression. We are left with, for the conservation of energy,

$$\rho \frac{DE}{Dt} = \sigma_j^i (\nabla V)_i^j - \operatorname{div}(Q) + \rho r \quad \text{or} \quad \rho \left(\frac{\partial E}{\partial t} + b^i \frac{\partial E}{\partial y^i} \right) = \sigma_j^i b_{;i}^j - q_{;i}^i + \rho r, \tag{A.290}$$

where E is the specific internal energy, $\sigma = \sigma_i^j$ is the stress tensor, $V = b^i \frac{\partial}{\partial y^i}$ is the velocity, $Q = q^i \frac{\partial}{\partial y^i}$ is the heat flux, and r is the rate of internal energy production.

To compute the deformation and displacement gradients, we recall that in our notation we are looking for $d(x + U) = F dx$, which gives $dU = (F - I)dx = \hat{F}dx$, where $U = u^i \frac{\partial}{\partial y^i}$ is the displacement and x is the original location. The left-hand side of this expression is

$$dU = \frac{\partial}{\partial y^j}\left(u^i \frac{\partial}{\partial y^i}\right) dy^j = \frac{\partial u^i}{\partial y^j}\frac{\partial}{\partial y^i}dy^j + u^i \Gamma^k_{ji}\frac{\partial}{\partial y^k}dy^j. \tag{A.291}$$

The right-hand side is

$$\hat{F}dx = \hat{F}^i_j \frac{\partial}{\partial y^i}dy^j. \tag{A.292}$$

Viewing the dy^j as arbitrary, we can set the two sides equal to each other as

$$\left(\frac{\partial u^i}{\partial y^j} + u^k \Gamma^i_{jk}\right)\frac{\partial}{\partial y^i} \otimes dy^j = u^i_{;j}\frac{\partial}{\partial y^i} \otimes dy^j = \hat{F}^i_j\frac{\partial}{\partial y^i} \otimes dy^j. \tag{A.293}$$

Thus, we have the simple result that

$$\hat{F} = \nabla U, \quad \text{or} \quad F = I + \nabla U, \quad \text{or in indicial notation} \quad F^i_j = \delta^i_j + u^i_{;j}. \tag{A.294}$$

Recall that the deformation gradient F is used to measure the change in length of a vector $d\vec{S}$ to $d\vec{s} = F d\vec{S}$ due to the deformation. Now, with $dS = a^i \frac{\partial}{\partial y^i}$, we obtain $ds = F(dS) = (\delta^i_j + \hat{F}^i_j)\frac{\partial}{\partial y^i} \otimes dy^j (dS) = (\delta^i_j + \hat{F}^i_j)\frac{\partial}{\partial y^i}a^j$. Computing the large strain tensor is related to the length difference between these vectors, now given by $g(ds, ds) - g(dS, dS)$. The strain tensor as a covariant $(0, 2)$ tensor is[5]

$$E = \frac{1}{2}(g(F, F) - g(I, I)) = \frac{1}{2}(g(I, \nabla U) + g(\nabla U, I) + g(\nabla U, \nabla U)). \tag{A.295}$$

To clarify the notation, in indicial notation we have

$$\begin{aligned}
E &= \frac{1}{2}\left\{g\left(\delta^k_i \frac{\partial}{\partial y^k} \otimes dy^i, u^l_{;j}\frac{\partial}{\partial y^l} \otimes dy^j\right)\right. \\
&\quad \left. + g\left(u^k_{;i}\frac{\partial}{\partial y^k} \otimes dy^i, \delta^l_j \frac{\partial}{\partial y^l} \otimes dy^j\right) + g\left(u^k_{;i}\frac{\partial}{\partial y^k} \otimes dy^i, u^l_{;j}\frac{\partial}{\partial y^l} \otimes dy^j\right)\right\} \\
&= \frac{1}{2}\left\{g\left(\frac{\partial}{\partial y^i}, u^l_{;j}\frac{\partial}{\partial y^l}\right) + g\left(u^k_{;i}\frac{\partial}{\partial y^k}, \frac{\partial}{\partial y^j}\right) + g\left(u^k_{;i}\frac{\partial}{\partial y^k}, u^l_{;j}\frac{\partial}{\partial y^l}\right)\right\} dy^i \otimes dy^j \\
&= \frac{1}{2}\left\{g_{il}u^l_{;j} + g_{kj}u^k_{;i} + g_{kl}u^k_{;i}u^l_{;j}\right\} dy^i \otimes dy^j \\
&= \frac{1}{2}\left\{u_{i;j} + u_{j;i} + g^{kl}u_{l;i}u_{k;j}\right\} dy^i \otimes dy^j = E_{ij}dy^i \otimes dy^j \\
&= \frac{1}{2}\left\{g^{jl}u^k_{;j} + g^{ik}u^l_{;i} + g_{mn}g^{ik}g^{jl}u^m_{;i}u^n_{;j}\right\}\frac{\partial}{\partial y^k} \otimes \frac{\partial}{\partial y^l} = E^{kl}\frac{\partial}{\partial y^k} \otimes \frac{\partial}{\partial y^l}.
\end{aligned} \tag{A.296}$$

The last two lines are examples of raising and lowering indices (see Exercise A.23). We have now computed all the differential operators.

We are now in a position to ask a geometry question: what h_i when used in an orthonormal coordinate system give rise to flat space? In other words, how can we

---

[5]In Section A.3 the vectors used to measure the length change, which are in the tangent space, are written in terms of the orthonormal basis. Thus, the length $(ds)^2$ is simply given by the sum of the squares of each of the components. This fact allows for $(ds)^2 = d\vec{s}^T d\vec{s} = d\vec{S}^T F^T F d\vec{S}$ so that $E = (1/2)(F^T F - I)$, and the metric does not explicitly appear (it implicitly appears through the development of the orthonormal basis).

tell when we are given a set of (h_1, h_2, h_3) as functions of coordinates (y^1, y^2, y^3) that the resulting geometry is Euclidean? This question is what Riemann asked in his famous 1854 lecture, though in his case it was for a general metric g. His answer was to state that specifically the question is: when does there exist a change of variables $x^i(y^j)$ so that

$$g'_{ij} = g_{mn} \frac{\partial y^m}{\partial x^i} \frac{\partial y^n}{\partial x^j} = \delta_{ij}, \tag{A.297}$$

where g_{mn} is the metric in the y^j coordinates? (That is, we need the metric tensor to transform correctly into Cartesian coordinates.) Riemann showed that this change of variables could be found when certain integrability conditions were satisfied, namely that all components of what is now called the Riemann curvature tensor are zero. To explicitly write the requirements, geometrically there are two tensors of interest, the torsion and the curvature,

$$\begin{aligned} T(X, Y) &= \nabla_X Y - \nabla_Y X - [X, Y], \\ R(X, Y)Z &= \nabla_X(\nabla_Y Z) - \nabla_Y(\nabla_X Z) - \nabla_{[X,Y]}Z. \end{aligned} \tag{A.298}$$

The brackets denote the Lie derivative, and to compute T and R in local coordinates we notice that $[\frac{\partial}{\partial y^i}, \frac{\partial}{\partial y^j}] = 0$. For a change in variables from the Euclidean space both these tensors are zero. Torsion being zero gives $\Gamma_{ij}^k = \Gamma_{ji}^k$. Curvature being zero gives

$$R_{ijkl} = \frac{\partial \Gamma_{lj,i}}{\partial y^k} - \frac{\partial \Gamma_{kj,i}}{\partial y^l} + g^{mn}(\Gamma_{kj,n}\Gamma_{li,m} - \Gamma_{lj,n}\Gamma_{ki,m}) = 0. \tag{A.299}$$

(Be aware that different authors use different sign conventions and index conventions in the curvature; ours follows [123] and [173].) A fair amount of computation allows us to determine when the h_i are acceptable to represent a change in coordinates (i.e., they represent space being flat or Euclidean):

$$\begin{aligned} R_{1212} &= -h_1 h_2 \left\{ \frac{\partial}{\partial y^2}\left(\frac{1}{h_2}\frac{\partial h_1}{\partial y^2}\right) + \frac{\partial}{\partial y^1}\left(\frac{1}{h_1}\frac{\partial h_2}{\partial y^1}\right) + \frac{1}{h_3^2}\frac{\partial h_1}{\partial y^3}\frac{\partial h_2}{\partial y^3} \right\}, \\ R_{1313} &= -h_1 h_3 \left\{ \frac{\partial}{\partial y^3}\left(\frac{1}{h_3}\frac{\partial h_1}{\partial y^3}\right) + \frac{\partial}{\partial y^1}\left(\frac{1}{h_1}\frac{\partial h_3}{\partial y^1}\right) + \frac{1}{h_2^2}\frac{\partial h_1}{\partial y^2}\frac{\partial h_3}{\partial y^2} \right\}, \\ R_{2323} &= -h_2 h_3 \left\{ \frac{\partial}{\partial y^3}\left(\frac{1}{h_3}\frac{\partial h_2}{\partial y^3}\right) + \frac{\partial}{\partial y^2}\left(\frac{1}{h_2}\frac{\partial h_3}{\partial y^2}\right) + \frac{1}{h_1^2}\frac{\partial h_2}{\partial y^1}\frac{\partial h_3}{\partial y^1} \right\}, \\ R_{2131} &= \frac{h_1}{h_3}\frac{\partial h_1}{\partial y^3}\frac{\partial h_3}{\partial y^2} - h_2 h_1 \frac{\partial}{\partial y^3}\left(\frac{1}{h_2}\frac{\partial h_1}{\partial y^2}\right) = R_{1213}, \\ R_{1232} &= \frac{h_2}{h_3}\frac{\partial h_2}{\partial y^3}\frac{\partial h_3}{\partial y^1} - h_1 h_2 \frac{\partial}{\partial y^3}\left(\frac{1}{h_1}\frac{\partial h_2}{\partial y^1}\right) = -R_{1223}, \\ R_{1323} &= \frac{h_3}{h_2}\frac{\partial h_3}{\partial y^2}\frac{\partial h_2}{\partial y^1} - h_1 h_3 \frac{\partial}{\partial y^2}\left(\frac{1}{h_1}\frac{\partial h_3}{\partial y^1}\right), \end{aligned} \tag{A.300}$$

where the specific indices displayed are to help identify the symmetries in the expressions ($R_{ijkl} = -R_{jikl}$, $R_{ijkl} = -R_{ijlk}$, $R_{ijkl} = R_{klij}$, and so if $i = j$ or $k = l$ then $R_{ijkl} = 0$). Thus, in order for the h_i to be valid and produce a flat Euclidean space, the h_i must be such that these curvature components are zero. We can easily verify that for the Cartesian, cylindrical, and spherical coordinate systems the curvature tensor components are zero.

A final comment about curvature. One approach to solving problems in mechanics is to find a stress tensor that satisfies the momentum balance and the boundary tractions and then use the material's constitutive model to determine the strain tensor that gives rise to that stress tensor. In two-dimensional elasticity, this is the approach of the Airy stress function, which is the foundation of the complex variables method. For the computed stress to be a valid solution, the strain tensor needs to have originated from a displacement field. For the small strain tensor, relations between the components of the strain tensor that imply the strain tensor field originated from a displacement vector field are called *compatibility conditions*. These are based on the fact that for smooth displacements, partial derivatives commute. Explicitly, if ε_{ij} are the small strains then the strain compatibility conditions in Cartesian coordinates are $\varepsilon_{ij,kl} + \varepsilon_{kl,ij} - \varepsilon_{ik,jl} - \varepsilon_{jl,ik} = 0$, where the terms behind the comma mean partial derivatives with respect to that variable. When these equations (there are 81 equations but only 6 are linearly independent) are satisfied, then the strain came from a displacement field u_i; otherwise it did not and the strains are not compatible. In a similar fashion, if we are given the large strain E_{ij}, then we can construct the metric $g = 2E + I$, compute the Christoffel symbols for the metric g, and then compute the Riemann curvature tensor. If the curvature tensor is zero, then the large strain came from an allowable displacement U. The way to think about this compatibility condition is as follows. If we have material in the plane and deform it as much as we like, but it still stays in the plane, then mathematically there is no curvature, just a change of variables in the plane, because all the material is still in the plane and the plane is flat. Similarly, in three dimensions, if we begin with material and deform it as much as we like, the material still lies in three-dimensional Euclidean space. Hence, the resulting strain produces a metric that is flat, since the deformation is simply a change of variables in three-dimensional Euclidean space.

A.8. Sources

The material in this appendix is typically covered in continuum mechanics books, often in appendices. The specific one consulted here was Malvern [138]. The geometry discussion in Section A.7 follows presentations in Kreyszig [123], Spivak [173], and Petersen [154].

Exercises

(A.1) Starting with the fact that the traction $\vec{t}$ and the surface normal $\vec{n}$ are vectors and defining the Cauchy stress by $\vec{t}^T = \vec{n}^T \sigma$, show that the Cauchy stress is a tensor. (Hint: $\vec{t}^* = Q\vec{t}$ for an arbitrary rotation Q and likewise for $\vec{n}$. Given the definition of the Cauchy stress, show that $\sigma^* = Q\sigma Q^T$.)

(A.2) With the small strain tensor defined as $\varepsilon_{ij} = (1/2)(\partial u_i/\partial x_j + \partial u_j/\partial x_i)$, where $\vec{u}$ (the displacement) and $\vec{x}$ (the current laboratory frame coordinates) are vectors, show that ε is a tensor. (Hint: for arbitrary rotation Q, $u_i^* = Q_{ij}u_j$ and $x_i = Q_{ij}^T x_j^*$. Show through direct computation using $\partial/\partial x_i^* = (\partial x_j/\partial x_i^*)\partial/\partial x_j$ that $\varepsilon^* = Q\varepsilon Q^T$.)

(A.3) What happens if we have a general transformation A between coordinates and not a rotation Q? Two challenges arise to our understanding. First, show by example that $\vec{n}^* = A\vec{n}$ may not have unit length. Next, assuming

$\vec{n}^* = A\vec{n}$ and $\vec{t}^T = \vec{n}^T\sigma$, show that $\sigma^{*T} = A\sigma^T A^{-1}$. The second challenge: show by example that if σ is symmetric, σ^* need not be symmetric.

(A.4) Show that in general (by plugging in the appropriate expressions in Section A.3),

$$\text{div}(\text{curl}(\vec{f})) = 0,$$
$$\text{curl}(\text{grad}(f)) = 0. \tag{A.301}$$

(A.5) Write the curl in Eq. (A.77) in general coordinates as a determinant.

(A.6) Show that the curl in an orthogonal coordinate system can be written

$$\text{curl}(\vec{f}) = \sum_j \hat{e}_{y_j} \frac{1}{h_j} \frac{\partial}{\partial y_j}(\times f_k \hat{e}_{y_k}) = f_{k;j}\, \hat{e}_{y_j} \times \hat{e}_{y_k} = \varepsilon_{ijk} f_{k;j}\, \hat{e}_{y_i}. \tag{A.302}$$

(A.7) Work through the details of the differential operators (grad, div, Laplacian, curl) terms in cylindrical coordinates.

(A.8) Work through the details of the acceleration in cylindrical coordinates.

(A.9) Work through the details of the differential operators (grad, div, Laplacian, curl) terms in spherical coordinates.

(A.10) Work through the details of the acceleration in spherical coordinates.

(A.11) Write the mass conservation equation in cylindrical coordinates.

(A.12) Write the mass conservation equation in spherical coordinates.

(A.13) Discuss the sign of

$$\int_{S(t)} \vec{t} \cdot \vec{v}\, da \tag{A.303}$$

and why, as written, it corresponds to power going *into* the volume bounded by the surface. For example, you might consider a mechanical spring undergoing deformation, in either compression or tension (notice there are essentially four regimes to consider, based on the sign of the traction component and the sign of the velocity component).

(A.14) The only other coordinate systems the author has used are those based on ellipses, for elliptical cracks or ellipsoidal inclusions. For three-dimensional ellipsoidal geometry, the foci of the ellipsoid are assumed to lie on the z-axis at $z = \pm c$. The angle φ is the angle about the z-axis (the azimuthal angle). With

$$x = c\sqrt{(\xi^2 - 1)(1 - \eta^2)}\cos(\varphi),$$
$$y = c\sqrt{(\xi^2 - 1)(1 - \eta^2)}\sin(\varphi), \tag{A.304}$$
$$z = c\xi\eta,$$

$1 \le \xi < \infty$, $-1 \le \eta \le 1$, $0 \le \varphi < 2\pi$. Ellipsoids from rotation of an ellipse about the z-axis are given by $\xi = \text{const}$ and semimajor axes of $a = c\sqrt{\xi^2 - 1}$ and $b = c\xi$, and hyperboloids are given by $\eta = \text{const}$ with semimajor axis of $a = c\sqrt{1 - \eta^2}$ and $b = c\eta$. Compute the h_i and the Laplacian for this coordinate system.

(A.15) This exercise works further with the ellipsoidal coordinate system, as defined in the previous exercise in Eq. (A.304). First, determine the three basis vectors $\hat{e}_\xi$, $\hat{e}_\eta$, and $\hat{e}_\varphi$ in terms of the Cartesian basis vectors (essentially this answer is given in Q below). Once that is done, determine the derivatives of the basis vectors. You can use Eq. (A.135). Another approach is as follows, using a computer algebra system for assistance, as the manipulations can be quite tedious. Write the column vector of unit basis vectors as $\vec{E}_{\text{ell.}} = (\hat{e}_\xi \ \hat{e}_\eta \ \hat{e}_\varphi)^T$ and the corresponding column of Cartesian unit vectors as $\vec{E}_{\text{Cart.}} = (\hat{e}_1 \ \hat{e}_2 \ \hat{e}_3)^T$. The relation between them is $\vec{E}_{\text{ell.}} = {}_{\text{ellipsoidal}}Q_{\text{Cart.}}\vec{E}_{\text{Cart.}}$. To simplify the next expression, let $Q = {}_{\text{ellipsoidal}}Q_{\text{Cart.}}$. Then the derivative of the basis vectors with respect to ξ is

$$\frac{\partial \vec{E}_{\text{ell.}}}{\partial \xi} = \frac{\partial Q}{\partial \xi} \vec{E}_{\text{Cart.}} = \frac{\partial Q}{\partial \xi} Q^{-1} \vec{E}_{\text{ell.}} = (\gamma^j_{\xi i}) \vec{E}_{\text{ell.}}. \tag{A.305}$$

Thus, the coefficients of the derivatives of the basis vectors (the orthonormal Christoffel symbols) in the ellipsoidal basis are given by the entries of $(\partial Q/\partial \xi)Q^{-1}$. To be specific, in REDUCE, this is done by defining MATRIX Q(3,3);, then defining all nine entries one by one with Q(1,1)=xi*sqrt(1-eta**2)/sqrt(xi**2-eta**2), for example. From determining the basis functions you know

$$Q^T = \begin{pmatrix} \frac{\xi\sqrt{1-\eta^2}}{\sqrt{\xi^2-\eta^2}}\cos(\varphi) & -\frac{\eta\sqrt{\xi^2-1}}{\sqrt{\xi^2-\eta^2}}\cos(\varphi) & -\sin(\varphi) \\ \frac{\xi\sqrt{1-\eta^2}}{\sqrt{\xi^2-\eta^2}}\sin(\varphi) & -\frac{\eta\sqrt{\xi^2-1}}{\sqrt{\xi^2-\eta^2}}\sin(\varphi) & \cos(\varphi) \\ \frac{\eta\sqrt{\xi^2-1}}{\sqrt{\xi^2-\eta^2}} & \frac{\xi\sqrt{1-\eta^2}}{\sqrt{\xi^2-\eta^2}} & 0 \end{pmatrix}. \tag{A.306}$$

Then df(Q,xi)*minv(Q);, etc., yields all the basis derivatives; alternatively you could use df(Q,xi)*tp(Q) since in this case $Q^{-1} = Q^T$.

Now show that (by taking the dot product of the ellipsoidal basis vectors with Eq. (A.304), for example)

$$\vec{x} = \frac{c\xi\sqrt{\xi^2-1}}{\sqrt{\xi^2-\eta^2}}\hat{e}_\xi + \frac{c\eta\sqrt{1-\eta^2}}{\sqrt{\xi^2-\eta^2}}\hat{e}_\eta. \tag{A.307}$$

Determine the velocity vector (i.e., in terms of $\partial\xi/\partial t$, etc.), the Lagrangian acceleration, and the Eulerian acceleration.

We mention that symbolic manipulators can deal with basis vector derivatives in a symbolic way; for example, in REDUCE you can say let df(exi,xi) = x*eeta + y*etheta;, where x and y are the appropriate Christoffel symbols and where you have stated that depend exi, xi, eta, theta. Then in any expression where you have a vector and differentiate it, REDUCE will insert the appropriate basis vector derivative results.

(A.16) We never discussed how, when given h_1, h_2, h_3, to find $\vec{x} = x_1\hat{e}_{y_1} + x_2\hat{e}_{y_2} + x_3\hat{e}_{y_3}$, where $x_i = x_i(y_1, y_2, y_3)$. Show that the equation is

$$\hat{F}(\vec{x}, \vec{x}) = I, \tag{A.308}$$

($\hat{F}$ is defined in Eq. A.104) which is nine equations for three unknowns and is thus consistent with the six equations the h_i need to satisfy to describe flat space (all $R_{ijkl} = 0$ in Eq. A.300). (This corresponds to $\nabla\vec{x} = I$ in Eq. A.227 in general coordinates for finding x^i.) Show that if $h_1 = 1$, then $x_1 = y_1 + f(y_2, y_3)$ and x_2 and x_3 have no y_1 dependence. Show that Eq. (A.308) is satisfied when $h_1 = 1$, $h_2 = r$, $h_3 = 1$ with the $\vec{x}$ for cylindrical coordinates.

(A.17) Consider a torsion (twist) about the z-axis in cylindrical coordinates, with a twist rate α. In particular this means that material point (R, Θ, Z) goes to

$(r, \theta, z) = (R, \Theta + \alpha Z, Z)$. Show that in terms of the original coordinates the displacement is $\vec{u} = R(\cos(\alpha z) - 1)\hat{e}_R + R\sin(\alpha z)\hat{e}_\Theta$. Compute the deformation gradient F in the two forms $_r F_R$ (Eq. A.94) and $_R F_R$ (Eq. A.103). Show they differ by a rotation of angle αz (hence this is the rotation of the coordinate system). Compute the actual rotation of the material configuration by computing the polar decomposition of $_R F_R$ (either right or left – see Exercises 14.30 and 14.31) and show that this rotation angle is not αz. (Comparing to the simple shear described in Exercise 14.31, the shear $\gamma = \alpha r$ reflects the fact that the gage length in the z direction does not change but the displacement around the circular arc is αr, which increases with radius.)

(A.18) Compute a general expression in orthonormal coordinates for each of the Lagrangian and Eulerian accelerations based on the general Eqs. (A.230) and (A.231).

(A.19) Show that the forms of grad, div, curl, and the Laplacian for general metrics in Section A.7 agree with the derivations in Section A.3 for curvilinear coordinates.

(A.20) For one on-diagonal and one off-diagonal element, show that the general expression for the displacement gradient $\hat{F}$, Eq. (A.293), agrees with the specific one in Eq. (A.104) for an orthonormal coordinate system. (Hint: this exercise uses Christoffel symbols, the relations between $\frac{\partial}{\partial y^j}$, dy^j, and $\hat{e}_{y_j}$, and that $u_i = u^i/h_i$ (no sum on i).)

(A.21) For the cylindrical coordinate system, ($h_r = 1, h_\theta = r, h_z = 1$) and for the spherical coordinate system ($h_r = 1, h_\theta = r, h_\varphi = r\sin(\theta)$). Thus, consider coordinate systems of the form $h_r = 1$, $h_\theta = r$, $h_3 = h(r, \theta, y^3)$. Find the general solution for h that leads to flat space and show the cylindrical and spherical coordinate systems fall under the general solution. What coordinate systems might the more general expressions describe?

(A.22) Show that the x^i for the Cartesian, cylindrical, and spherical coordinate systems, with the corresponding Christoffel symbols, satisfy Eq. (A.227).

(A.23) Show that $g_{ik}a^k_{;j} = a_{i;j} = \alpha_{i;j}$, which in words says that covariant differentiation followed by lowering the index produces the same result as lowering the index followed by covariant differentiation. Hence ∇a is well defined. (Hint: to avoid recomputing relations between the metric tensor and the Christoffel symbols, at some point use $\nabla g = 0$.)

(A.24) (Compatibility equations) For small strains, when only linear terms are kept – so, for example, $E_{ij} \approx \varepsilon_{ij}$ – show that the curvature being zero (Eq. A.299) implies the compatibility condition in the last paragraph of the text,

$$\varepsilon_{ij,kl} + \varepsilon_{kl,ij} - \varepsilon_{ik,jl} - \varepsilon_{jl,ik} = 0, \tag{A.309}$$

where the terms behind the comma mean partial derivatives with respect to that variable – that is, $\varepsilon_{ij,kl} = \frac{\partial^2 \varepsilon_{ij}}{\partial y_k \partial y_l}$. (Hint: recall $g = 2E + I$, compute $\Gamma_{ij,k}$, remember to discard all product of ε terms, and swap indicies.)

(A.25) Show (page 642), with two covariant vectors $\omega_1 = \alpha_i dy^i$ and $\omega_2 = \beta_j dy^j$, that $g(\omega_i, *\omega_1 \wedge \omega_2) = 0$; thus $*\omega_1 \wedge \omega_2$ is perpendicular to both ω_1 and ω_2. Show for an orthogonal (not orthonormal) system, where the volume of the cube is given by $\hat{e}_{y_1} \cdot \hat{e}_{y_2} \times \hat{e}_{y_3} = 1$, that $g(\hat{e}_{y_1}, *\hat{e}_{y_2} \wedge \hat{e}_{y_3}) = 1$. Show, for a general coordinate system, that $g(dy^1, *dy^2 \wedge dy^3) = 1/\sqrt{\det(g)}$.

Units, Conversions, and Constants

B.1. Consistent Units

There are four commonly used systems of units in the impact physics community. These four sets are all internally consistent:

Quantity	cgs	MKS	shock units	English
Time	second (s)	second (s)	microsecond (μs$=10^{-6}$ s)	second (s)
Length	centimeter (cm)	meter (m)	centimeter (cm)	inch (in)
Mass	gram (g)	kilogram (kg)	gram (g)	pound mass$/g$ (lbf s^2/in)
Pressure	dyne/cm^2 ($=$g/cm s^2)	pascal (Pa $=$ kg/m s^2)	megabar (Mbar $=$ g/cm μs^2 $=$ 100 GPa)	pound per square inch (psi $=$ lbf/in^2)
Energy	erg (g cm^2/s^2)	joule (J $=$ kg m^2/s^2)	mega-mega-erg ($= 10^{12}$ erg $=$ 100 kJ)	inch pound (in-lbf)

Though the following is not a consistent set of units, it is useful to note that if velocity is written in units of km/s and density is written in g/cm^3, then $\tilde{\sigma} = \rho U u$ has units of GPa.

In the United States, English units are widely used in engineering applications. The awkwardness of English units is with mass vs. weight. When a person says they "weigh 150 pounds," what they mean is the person of 150 lbm exerts a force on the scale of 150 lbf – that is, they include the acceleration of gravity in their statement as $F = mg$. Explicitly, 1 lbf/1 lbm $= g$, where g is the acceleration of gravity at the surface of the Earth. Thus, since the units of stress σ_{ij} are given in lbf in $\partial \sigma_{ji}/\partial x_j = \rho D v_i/Dt$, to get the correct units for the density it is necessary to multiply the density in lbm/in^3 by lbf/(lbm 386.4 in/s^2). (If instead of inches the unit of length were feet, then lbm would be multiplied by lbf/(lbm 32.2 ft/s^2); this unit of mass is called a slug and 1 slug$=$1 lbf s^2/ft$=$14.594 kg.) Density becomes 1 g/cm$^3 = 9.36 \times 10^{-5}$ lbf s^2/in$^4 = 1.94$ lbf s^2/ft$^4 = 1.94$ slug/ft^3. In this book all the computations were performed in cgs or MKS units.

B.2. Useful Conversions

Mass

1 lbm = 1 pound = 453.5924 g

1 kg = 2.2046 lbm

1 lbm = 1 lbf/g, where g = 32.2 ft/s^2 = 386.4 in/s^2

1 g = 15.432 grains

1 lbm = 7,000 grains

Distance

1 in = 2.54 cm

1 ft = 12 in

Density

1 g/cm^3 = 62.428 lbm/ft^3 = 1.94 slug/ft^3 = 1.94 lbf s^2/ft^4 = 9.36×10^{-5} lbf s^2/in^4

Pressure and stress

1 Pa = 1 kg/m s^2 = 10 dynes/cm^2 = 10 g/cm s^2

1 bar = 10^5 Pa = 14.5 psi = 14.5 lbf/in^2

1 kbar = 10^8 Pa = 100 MPa = 14.5 ksi

1 Mbar = 10^{11} Pa = 100 GPa = 10^{12} dynes/cm^2 = 14.5 Msi

1 psi = 1 lbf/in^2 = 6894.757 Pa = 6.895 kPa

1 ksi = 1,000 psi = 6.895 MPa

1 Msi = 10^6 psi = 6.895 GPa

Energy

1 J = 1 kg m^2/s^2 = 10^7 erg = 10^7 g cm^2/s^2

1 cal = 4.186 J

1 Btu = 1,054.8 J

1 "ton of TNT equivalence" $\equiv$ 4.184×10^9 J

Areal Density

1 g/cm^2 = 10 kg/m^2 = 2.05 psf = 2.05 lbm/ft^2

1 psf = 1 lbm/ft^2 = 0.488 g/cm^2 = 4.88 kg/m^2

Temperature

K = °C (Centigrade or Celsius) + 273.15 K

°F = 1.8 × °C + 32 °F

°R = °F + 459.69 °R

1 eV = $e \cdot \frac{1J}{1C}$ = 1.6021 × 10^{-19} J = 11604.45 K (= 1 eV/k_B)

B.3. Constants of Interest

Acceleration of gravity at Earth's surface $g \approx$ 980 cm/s^2 = 32.2 ft/s^2 = 386 in/s^2

Absolute zero: −273.16 °C

1 atmosphere pressure = 1.01325 × 10^5 Pa = 14.7 psi

Universal Gas Constant R = 8.3143 × 10^7 erg/(K mole) = 8.3143 J/(K mole)

Avogadro's number N_A = 6.022 × 10^{23} atoms/mole

Boltzmann's constant (which is R per molecule – that is, $k_B = R/N_A$)

$$k_B = 1.38054 \times 10^{-16} \text{ erg/K}$$

Electron charge: e = 4.803 × 10^{-10} esu = 1.6021 × 10^{-19} C (=coulomb)

Ratio of the specific heats for air $\gamma = C_p/C_v$ = 1.4

Molecular weight of air = 28.97

Elastic, Shock, and Strength Properties of Materials

This appendix lists elastic, Hugoniot, and strength properties of materials. These data are from compilations of compilations, so the entries can be many steps from the original source. As such, for critical applications you should always go to the literature for the most up to date information, or at least to confirm these values.

First, a small number of material strength values are listed, from the original Johnson–Cook paper [105], from [104], and from [99]; the RHA values are from Meyer and Kleponis [146].

The second table is of low pressure elastic properties. The density ρ (computed from molar volume and atomic weight), bulk modulus K and melt temperature data are (save for diamond) from Young [214]. The shear modulus G, Poisson's ratio ν, linear expansion coefficient α and constant pressure heat capacity C_v are from the great survey article by Gschneidner [88] (watch out for atypical units for the moduli). Parentheses are used when Gschneidner used them to indicate estimates. In some cases Gschneidner reported the values for various configurations of the same material; we have marked the one we chose in the parenthetical comment on the name of the element. Please recall the excellent approximation for the heat capacity of metals by $C_p = 3R$, where $R = 8.31434\,\mathrm{J/mole\,K}$ is the ideal gas constant. To obtain the heat capacity in terms of $\mathrm{J/g\,K}$, divide $3R$ by the atomic weight in grams per mole of the metal.

The third table contains properties from ultrasound measurements, flyer plate impact experiments, and explosive shock experiments. The primary reference is the Los Alamos data collection edited by Marsh [139] and the Los Alamos GMX-6 group published compilation [125]. The longitudinal wave speed c_L and the shear wave speed c_s are from ultrasound measurement and are all from Marsh. The density, c_0, s, and q values are from Marsh and GMX-6, with preference given to Marsh in the case of discrepancies. The Grüneisen parameter Γ_0 is from the GMX-6 table. C_p is from the marvelous bookmark distributed at the 1989 American Physical Society Shock Compression of Condensed Matter Topical Conference held in Albuquerque, New Mexico. The materials marked with an asterisk (*) in this table are the reference Hugoniots that were used in impedance matching experiments to obtain other Hugoniots (there are nine of them). A listing in this table with sound speeds but no Hugoniot data implies there are Hugoniot data in Marsh for the material they are not very linear in the (u, U) plane. These values are the American laboratory values. Corresponding Russian laboratory values can be found in [72]. Ice comes from [177].

TABLE C.1. Large strain and high strain rate material strength properties (Johnson–Cook parameters).

Material	ρ_0 (g/cm³)	C_p (J/g K)	T_{Melt} (K)	Hardness Rockwell	A (MPa)	B (MPa)	n	C	m
OFHC Copper	8.96	0.383	1,356	F-30	90	292	0.31	0.025	1.09
Cartridge Brass	8.52	0.385	1,189	F-67	112	505	0.42	0.009	1.68
Nickel 200	8.90	0.446	1,726	F-79	163	648	0.33	0.006	1.44
1100 Aluminum	2.70	0.905	923		116	69	0.58	0.016	1.13
2024-T351 Aluminum	2.77	0.875	775	B-75	265	426	0.34	0.015	1.00
6061-T6 Aluminum	2.70	0.896	925	B-58	324	114	0.42	0.002	1.34
7039 Aluminum	2.77	0.875	877	B-76	337	343	0.41	0.010	1.00
Armco Iron	7.89	0.452	1,811	F-72	175	380	0.32	0.060	0.55
Carpenter Electrical Iron	7.89	0.452	1,811	F-83	290	339	0.40	0.055	0.55
1006 Steel	7.89	0.452	1,811	F-94	350	275	0.36	0.022	1.00
4340 Steel	7.83	0.477	1,793	C-30	792	510	0.26	0.014	1.03
2-inch-thick RHA[a]			1,783		780	780	0.106	0.004	1.00
S-7 Tool Steel	7.75	0.477	1,763	C-50	1539	477	0.18	0.012	1.00
Tungsten Alloy (7%Ni, 3%Fe)[b]	17.0	0.134	1,723	C-47	1506	177	0.12	0.016	1.00
Depleted Uranium (0.75%Ti)	18.6	0.117	1,473	C-45	1079	1120	0.25	0.007	1.00
Polycarbonate	1.20	1.340	533		76	69	1.00	0.000	1.85
Polyethersulfone	1.37	1.120	617		97	69	1.00	0.009	1.85

[a] [146] determines A from an RHA yield in terms of thickness t of $Y = 0.8772 - 0.1428 \ln(t/2.54\,\text{cm})$ GPa based on data $Y = 1.14, 0.94, 0.82,$ and 0.69 GPa for thicknesses 3/16, 1/2, 1.5 and 4 inch, respectively.
[b] Swaged 25%; melt temperature based on matrix.

The tables, from difference sources, in some ways overlap, allowing low temperature Grüneisen parameter calculations from the first and bulk modulus, shear modulus, and Poisson ratio calculations from the ultrasound data of the second. Such comparisons can be used to better understand the variation in properties for a given material. Density gives an indication of the material tested in the shock experiments. The materials are listed in atomic number order to facilitate observation of trends based on atomic structure.

Table C.2. Element properties.

Element		Name	ρ (g/cm³)	K (GPa)	T_{Melt} (K)	G (GPa)	ν	α (×10⁻⁶) (1/K)	C_v (J/gK)
1	H	Hydrogen	0.044 (0K)	0.186 (0K)	13.8				3.291
1	D	Deuterium	0.101 (0K)	0.335 (0K)	18.69				1.630
2	He-3	Helium-3	0.082 (0K)	0.0027 (0K)	liquid				1.010
2	He-4	Helium-4	0.145 (0K)	0.0078 (0K)	liquid				0.502
3	Li	Lithium	0.547 (4K)	12.6 (0K)	453.7	4.22	0.362	45	3.291
4	Be	Beryllium	1.858	110	1,562	143±4	0.039±0.011	11.5	1.630
5	B	Boron	2.463	196	2,365	203	0.089	8.3	1.010
6	C	Carbon (graphite)	2.271	35.8	4,530	3.25±0.06	0.27±0.06	3.8±3.1	0.718
6	C	Carbon (diamond)	3.536	556		450±100	0.18±0.08	1.19±0.01	0.502
7	N	Nitrogen	0.516 (8K)	2.16 (8K)	63.15				
8	O	Oxygen	0.769 (7K)	3.60 (7K)	54.36				
9	F	Fluorine	0.985 (23K)		53.48				
10	Ne	Neon	1.507 (0K)	1.10 (0K)	24.56				
11	Na	Sodium	1.015 (4K)	7.34 (4K)	371	3.4	0.315	70.6±0.6	1.134
12	Mg	Magnesium	1.736	34.1	923	17.3	0.28	25.7±0.7	0.983
13	Al	Aluminum	2.698	72.8	934	26.6±0.1	0.34±0.01	23.1±0.5	0.864
14	Si	Silicon	2.329	97.7	1,687	39.7	0.44	3.07±0.07	0.690
15	P	Phosphorous (red)	2.708	32.5	900	(7.2)	(0.335)	(66.5)	(0.575)
16	S	Sulfur (orthorhombic)	2.065	7.61	388	7.22	0.343	64.1±0.1	0.599
17	Cl	Chlorine	1.069 (22K)	14.2	172.2				
18	Ar	Argon	1.771 (0K)	2.86 (0K)	83.8				
19	K	Potassium	0.905 (0K)	3.70 (4K)	337	1.3	0.35	83.0	0.688
20	Ca	Calcium	1.545 (0K)	18.4 (0K)	1,115	7.4	0.31	22.4±0.1	0.643
21	Sc	Scandium	2.997	54.6	1,814	(31.3)	(0.269)	10.0	(0.561)
22	Ti	Titanium	4.500	106	1,943	39.3±0.5	0.345±0.005	8.35±0.15	0.517
23	V	Vanadium	6.123	155	2,183	46.5±0.1	0.36	8.3	0.480

C. MATERIAL CONSTANTS

Element	Name	ρ (g/cm^3)	K (GPa)	T_{Melt} (K)	G (GPa)	ν	α ($\times 10^{-6}$) (1/K)	C_v (J/gK)
24	Cr	Chromium	7.192	160	2,136	117	0.209	8.4	0.443
25	Mn	Manganese	7.475	90.4	1,519	76	0.24	22.6±0.3	0.468
26	Fe	Iron	7.877	163	1,811	81.4±0.6	0.279±0.013	11.7	0.441
27	Co	Cobalt	8.836	186	1,768	76.3	0.334	12.4	0.413
28	Ni	Nickel	8.906	179	1,728	75.0	0.30±0.01	12.7±0.2	0.435
29	Cu	Copper	8.938	133	1,358	41.1±1.5	0.345±0.005	16.7±0.3	0.375
30	Zn	Zinc	7.139	64.8	693	37.1	0.29	29.7	0.369
31	Ga	Gallium	5.985 (4K)	56.9	303	37.4	0.235	18.1±0.2	0.362
32	Ge	Germanium	5.327	74.9	1,212	39	0.27	5.75	0.314
33	As	Arsenic	5.785	63.1	1,090	(14.6)	(0.335)	4.28±0.42	0.329
34	Se	Selenium	4.809	9.1	494	(21.7)	(0.338)	36.9±0.1	0.315
35	Br	Bromine	2.081 (5K)	12.2	266				
36	Kr	Krypton	3.092 (0K)	3.34 (0K)	115.8				
37	Rb	Rubidium	1.625 (4K)	2.92 (4K)	313	1.00	(0.356)	88.1±1.9	0.317
38	Sr	Strontium	2.647 (0K)	12.4 (0K)	1,042	(5.22)	(0.304)	20	0.296
39	Y	Yttrium	4.472	41	1,795	25.8±0.4	0.258±0.008	12.0	0.295
40	Zr	Zirconium	6.507	94.9	2,128	34.1±0.8	0.34	5.78±0.07	0.280
41	Nb	Niobium	8.579	169	2,742	37.4±0.1	0.35±0.03	7.07±0.05	0.266
42	Mo	Molybdenum	10.228	261	2,896	116	0.30	4.98±0.15	0.247
43	Tc	Technectium	11.345		2,428	(142)	(0.293)	(8.06)	(0.243)
44	Ru	Ruthenium	12.371	303	2,607	160	0.286	9.36±0.27	0.234
45	Rh	Rhodium	12.428	282	2,236	147±3	0.27	8.40±0.10	0.240
46	Pd	Palladium	12.432	189	1,828	51.1	0.375±0.015	11.5±0.4	0.239
47	Ag	Silver	10.503	98.8	1,235	28.6±0.7	0.37	19.2±0.4	0.227
48	Cd	Cadmium	8.647	49.8	594	24.1	0.30	30.6±1.3	0.218
49	In	Indium	7.286	39.3	430	3.7	0.46	31.4±1.4	0.218
50	Sn	Tin (white)	7.287	55	505	20.4	0.33	21.2	0.213

Element		Name	ρ (g/cm^3)	K (GPa)	T_{Melt} (K)	G (GPa)	ν	α ($\times 10^{-6}$) (1/K)	C_v (J/gK)
51	Sb	Antimony	6.693	41.1	904	20.0	0.31 ± 0.06	10.9	0.206
52	Te	Tellurium	6.237	23.3	723	15.4	0.33	16.77 ± 0.03	0.199
53	I	Iodine	2.576(110K)	13.6	387				
54	Xe	Xenon	3.779(0K)	3.65(0K)	161.4				
55	Cs	Cesium	2.001(4K)	2.10(0K)	302	(0.65)	(0.356)	97	0.214
56	Ba	Barium	3.634(0K)	9.30(0K)	1,002	4.9	0.28	18.8 ± 0.8	0.189
57	La	Lanthanum	6.146	26.6	1,191	14.9	0.288	10.4	0.187
58	Ce	Cerium (γ)	6.769	20.8	1,071	12.0	0.248	8.5	0.193
59	Pr	Praseodymium	6.774	28.7	1,204	13.5	0.305	6.79	0.191
60	Nd	Neodymium	7.009	32.7	1,294	14.5	0.306	9.98	0.189
61	Pm	Promethium	7.160	38	1,315	(17)	(0.278)	(9.0)	(0.187)
62	Sm	Samarium	7.518	36	1,347	12.6	0.352	10.4	0.192
63	Eu	Europium	5.244	16.2	1,095	(5.9)	(0.286)	33.1	0.170
64	Gd	Gadolinium	7.902	37.6	1,586	22.2	0.259	8.28	0.231
65	Tb	Terbium	8.230	38.4	1,629	22.8	0.261	10.3	0.181
66	Dy	Dysprosium	8.553	40.5	1,685	25.4	0.243	10.0	0.172
67	Ho	Holmium	8.796	40.4	1,747	26.7	0.255	10.7	0.163
68	Er	Erbium	9.066	45	1,802	36.5	0.238	12.3	0.166
69	Tm	Thulium	9.323	45.6	1,818	(30)	(0.235)	13.3	0.158
70	Yb	Ytterbium	6.966	14.6	1,092	7.0	0.284	24.96 ± 0.04	0.146
71	Lu	Lutetium	9.841	47.4	1,936	33.8	(0.233)	8.12	0.154
72	Hf	Hafnium	13.281	108	2,504	52.9	0.30	6.01 ± 0.16	0.142
73	Ta	Tantalum	16.677	191	3,293	68.6	0.35	6.55 ± 0.05	0.139
74	W	Tungsten	19.414	308	3,695	153 ± 4	0.284 ± 0.004	4.59 ± 0.03	0.132
75	Re	Rhenium	21.017	360	3,459	178	0.293	6.63 ± 0.06	0.136

Element		Name	ρ (g/cm^3)	K (GPa)	T_{Melt} (K)	G (GPa)	ν	α ($\times 10^{-6}$) (1/K)	C_v (J/gK)
76	Os	Osmium	22.589		3,306	(210)	(0.285)	4.7±0.1	(0.130)
77	Ir	Iridium	22.561	358	2,720	210	0.26	6.63±0.12	0.131
78	Pt	Platinum	21.461	277	2,042	61.0	0.38±0.01	8.95±0.05	0.130
79	Au	Gold	19.291	166	1,338	27.5±0.3	0.425±0.010	14.1±0.1	0.124
80	Hg	Mercury	14.771(77K)	28.2	234	100	0.364	61	0.124
81	Tl	Thallium	11.869	33.7	577	2.7	0.46	29.4±1.0	0.122
82	Pb	Lead	11.347	41.7	601	5.4	0.44	29.0±0.3	0.121
83	Bi	Bismuth	9.807	33.2	545	12.8	0.33	13.41±0.09	0.123
84	Po	Polonium	9.098		527	(9.5)	(0.338)	23.0±1.5	(0.122)
85	At	Astatine							
86	Rn	Radon	4.396		202				
87	Fr	Francium				(0.62)	(0.356)	(102)	(0.125)
88	Ra	Radium	5.501		973	(6.0)	(0.304)	(20.2)	(0.118)
89	Ac	Actinium	10.068		1,324	(13.5)	(0.269)	(14.9)	(0.118)
90	Th	Thorium	11.719	56.9	2,028	27.8±0.2	0.285±0.015	11.2±0.4	0.116
91	Pa	Protactinium	15.423	157	1,845	(39.0)	(0.282)	(7.3)	(0.122)
92	U	Uranium	19.058	111	1,408	74±4	0.245±0.005	12.6±0.4	0.113
93	Np	Neptunium	20.453	118	912	(39.8)	(0.255)	27.5	(0.116)
94	Pu	Plutonium	20.271	47.6	913	43.7	(0.15)	55	0.124
95	Am	Americium	13.787	45	1,449				
96	Cm	Curium	13.688	25	1,618				
97	Bk	Berkelium	14.672	30	1,323				
98	Cf	Californium	15.217	50	1,173				
99	Es	Einsteinium	8.839		1,133				
100	Fm	Fermium							

TABLE C.3. Hugoniot data.

Element		Name	ρ_0 (g/cm³)	c_L (km/s)	c_s (km/s)	c_0 (km/s)	k	q (s/km)	Γ_0	C_p (J/g K)	
3	Li	Lithium	0.53			4.645	1.133		0.81	3.41	
4	Be	Beryllium	1.85	13.15	8.97	7.998	1.13		1.16	0.18	
5	B	Boron	2.338	13.9	9						
6	C	Carbon (g)	1.0			0.79	1.30				
6	C	Carbon (d)	3.191			7.81	1.43				
7	N	Nitrogen	0.82	0.88	0						
8	O	Oxygen	1.202			1.88	1.34				
11	Na	Sodium	0.968			2.629	1.223		1.17	1.23	
12	Mg	Magnesium	1.74	5.74	3.15	4.492	1.263		1.42	1.02	
16	S	Sulfur	2.02			3.223	0.959				
19	K	Potassium	0.86			1.95	1.19		1.23	0.76	
20	Ca	Calcium	1.547	4.39	2.49	3.63	0.94		1.2	0.66	
21	Sc	Scandium	3.195	5.57	3.07						
22	Ti	Titanium	4.528	6.16	3.19	5.22	0.767		1.09		< 17.5 GPa, $U = 5.74$ km/s
	Ti	Titanium	4.528			4.877	1.049		1.09		above transition
23	V	Vanadium	6.099	6.15	2.69	5.077	1.19		1.29		
24	Cr	Chromium	7.117			5.173	1.473		1.19		
26	Fe	Iron*	7.85	5.94	3.26	3.574	1.920	−0.068	1.69	0.45	above $U = 5.0$ km/s
27	Co	Cobalt	8.82	5.73	3.04	4.752	1.28		1.97		
28	Ni	Nickel	8.875	5.79	3.13	4.59	1.44		1.93	0.44	
29	Cu	Copper*	8.93	4.76	2.33	3.940	1.489		1.96	0.4	
30	Zn	Zinc	7.139			3.03	1.55		1.96	0.39	
32	Ge	Germanium	5.328			1.75	1.75		0.56		> 30 GPa, $U = 4.20$ km/s
37	Rb	Rubidium	1.53			1.134	1.272		1.06	0.36	
38	Sr	Strontium	2.628	2.7	1.45	1.7	1.23		0.41		> 15 GPa, $U = 3.63$ km/s
39	Y	Yttrium	4.579	4.38	2.52						

Element	Name	ρ_0 (g/cm^3)	c_L (km/s)	c_s (km/s)	c_0 (km/s)	k	q (s/km)	Γ_0	C_p (J/gK)	
40 Zr	Zirconium	6.506	4.77	2.39	3.757	1.018		1.09		< 26 GPa, $U = 4.63$ km/s above transition
40 Zr	Zirconium	6.505			3.296	1.271		1.09		
41 Nb	Niobium	8.586			4.46	1.2		1.47	0.27	
42 Mo	Molybdenum	10.208	6.44	3.48	5.14	1.22		1.52	0.25	
45 Rh	Rhodium	12.429	6	3.64	4.79	1.38		1.88		
46 Pd	Palladium	11.991	4.68	2.33	4.01	1.55		2.26	0.24	
47 Ag	Silver	10.490	3.71	1.66	3.27	1.55		2.38	0.24	
48 Cd	Cadmium	8.639	3.2	1.65	2.48	1.64		2.27		
49 In	Indium	7.279			2.419	1.536		1.8		
50 Sn	Tin	7.287	3.43	1.77	2.59	1.49		2.11	0.22	
51 Sb	Antimony	6.7			1.983	1.652		0.6		
55 Cs	Cesium	1.826			1.048	1.043	0.051	1.62	0.24	
56 Ba	Barium	3.705	2.16	1.28	0.7	1.6		0.55		> 11.5 GPa, $U = 2.54$ km/s
57 La	Lanthanum	6.138	2.69	1.51						
58 Ce	Cerium	6.743	2.33	1.34						
59 Pr	Praseodymium	6.756	2.74	1.51						
60 Nd	Neodymium	6.98	2.84	1.6						
62 Sm	Samarium	7.461	2.89	1.64						
64 Gd	Gadolinium	7.861	2.95	1.69						
65 Tb	Terbium	8.209	2.94	1.69						
66 Dy	Dysprosium	8.41	3.07	1.78						
67 Ho	Holmium	8.734	3.21	1.86						
68 Er	Erbium	9.015	3.13	1.84						
69 Tm	Thulium	9.291	3.02	1.77						
70 Yb	Ytterbium	7.019	1.94	1.12						
72 Hf	Hafnium	12.89	3.86	3.12	2.954	1.121		0.98		< 40 GPa, $U = 3.86$ km/s above transition
72 Hf	Hafnium	12.89			2.453	1.353		0.98		

Element		Name	ρ_0 (g/cm^3)	c_L (km/s)	c_s (km/s)	c_0 (km/s)	k	q (s/km)	Γ_0	C_P (J/g K)
73	Ta	Tantalum	16.656	4.16	2.09	3.43	1.19		1.6	0.14
74	W	Tungsten	19.235	5.22	2.89	4.04	1.23		1.54	0.13
75	Re	Rhenium	21.021	5.3	2.89	4.21	1.34		2.44	
77	Ir	Iridium	22.484			3.916	1.457		1.97	
78	Pt	Platinum*	21.44	4.08	1.76	3.633	1.472		2.40	0.13
79	Au	Gold*	19.24	3.25	1.19	3.056	1.572		2.97	0.13
80	Hg	Mercury	13.54			1.75	1.72		1.96	0.14
81	Tl	Thallium	11.84			1.862	1.523		2.25	
82	Pb	Lead	11.346	2.25	0.89	2.051	1.46		2.77	0.13
83	Bi	Bismuth	9.836	2.49	1.43	1.826	1.473		1.1	0.12
90	Th	Thorium	11.68	2.95	1.57	2.18	1.24		1.26	
92	U	Uranium	18.93	3.45	2.12	2.51	1.51		1.56	0.12
Alloys										
		Brass	8.45	4.41	2.13	3.74	1.43		2.04	0.38
		1100 Aluminum*	2.714	6.38	3.16	5.392	1.341		2.25	
		2024 Aluminum*	2.785	6.36	3.16	5.328	1.338		2.00	0.89
		6061 Aluminum	2.703	6.40	3.15	5.35	1.34		2.00	0.89
		7075 Aluminum	2.804			5.20	1.36			
		921T Aluminum*	2.833	6.29	3.11	5.041	1.420		2.10	
		Magnesium–Li(14%)–Al(1%)	1.391	6.35	4.17	4.46	1.22		1.45	
		Magnesium AZ31B*	1.775	5.7	3.05	4.516	1.256		1.43	
		Stainless Steel 304	7.89	5.77	3.12	4.569	1.49		2.17	0.44
		U-Mo(3%)*	18.45	3.31	1.85	2.565	1.531		2.03	
Woods										
		Birch	0.693			0.65	1.44			
		White Fir	0.355			0.37	1.30			

Name	ρ_0 (g/cm³)	c_L (km/s)	c_s (km/s)	c_0 (km/s)	k	q (s/km)	Γ_0	C_p (J/gK)	
Synthetics									
Adiprene	1.094			2.332	1.536		1.48		< 24 GPa, $U = 7.0$ km/s
Epoxy	1.186	2.63	1.16	2.73	1.493		1.13		above transition
Epoxy	1.186			3.234	1.255		1.13		
Lucite	1.181			2.26	1.816		0.75		
Neoprene	1.439			2.785	1.419		1.39		
Paraffin	0.918	2.18	0.83	2.908	1.56		1.18		
Phenoxy	1.178			2.266	1.698		0.55		
Polyamide (nylon)	1.14	2.53	1.08	2.57	1.849	−0.081	1.07		
Polycarbonate (lexan)	1.193	2.18	0.88	2.57	1.31				
PMMA (plexiglass)[a]	1.186	2.72	1.36	2.598	1.516		0.97	1.2	
Polyethylene	0.916	2.04	0.66	2.901	1.481		1.64	2.3	
Polyrubber	1.01			0.852	1.865		1.5		
Polystyrene	1.046	2.31	1.14	2.746	1.319		1.18	1.2	
Polyurethane	1.264	2.39	1.03	2.486	1.577		1.55		< 22 GPa, $U = 6.5$ km/s
Silastic (RTV-521)[b]	1.372			1.84	1.44		1.4		
Teflon	2.153	1.23	0.41	1.841	1.707		0.59	1.02	
Compounds									
LiF	2.638			5.15	1.35		2	1.5	
Periclase (MgO)	3.584	9.71	6.02	6.597	1.369		1.32		> 20 GPa, $U = 7.45$ km/s
Quartz	2.204	5.96	3.77	0.794	1.695		0.9		Stishovite > 40 GPa
NaCl	2.165			3.528	1.343		1.6	0.87	transition ignored
Water	1.0	1.48	0	1.647	1.921	−0.096		4.19	< 1.15 GPa, $T = 263$ K
Ice	0.918	3.61		3.61	0.92				$u > 1.59$ km/s, shocks to liquid
Ice	0.918	3.61		1.7	1.44				at STP ($P = 1$ atm, $T = 0°$C)
Air	1.29×10⁻³	0.332	0					1.0	

[a]PMMA = polymethylmethylacrylate
[b]RTV = room temperature vulcanizing

Figure Acknowledgments

The following figures are courtesy of Southwest Research Institute, with photographs and figures by Charlie Anderson, Darrell Barnette, Sidney Chocron, Kathryn Dannemann, Tim Holmquist, Bill Livermore, Ian McKinney, Tom Moore, Art Nicholls, Larry Walther, and the author: 1.1, 1.2, 1.3, 1.4, 1.5, 1.6, 3.2, 3.7, 3.8, 3.10, 3.13, 4.13, 5.2 left, 5.3, 5.5, 5.6, 7.6, 11.1, 13.2, and 13.9.

The following figures and photographs were produced specifically for this text by the author with the assistance of Charlie Anderson, Rory Bigger, Sidney Chocron, Trent Kirchdoerfer, Dick Sharron, Mike Shearn, and Jim Spencer: 2.1, 2.2, 2.3, 3.1, 3.3 left, 3.4, 3.5, 3.6, 3.12, 3.14, 4.1, 4.2, 4.3, 4.4, 4.5, 4.6, 4.7, 4.8, 4.9, 4.10, 4.11, 4.12, 5.1, 5.2 right, 5.4, 5.7, 5.8, 5.9, 5.10, 5.13, 5.14, 5.15, 5.16, 5.19, 6.1, 6.3, 6.4, 6.5, 6.6, 6.7, 6.8, 6.9, 6.10, 6.11, 6.12, 6.13, 6.14, 6.15, 6.16, 6.17 right, 7.1, 8.5, 8.6, 8.7, 9.1, 9.2, 9.8, 10.4, 10.10, 10.11, 10.12, 10.16, 11.2, 11.5, 11.7, 11.8, 11.18, 11.19, 12.1 left, 12.2, 12.3 left, 12.4, 12.5, 13.3 left, and 13.8.

The following figures, with their journal reference following and copyrighted the year of their publication, are reprinted with permission of Elsevier, appearing in the *International Journal of Impact Engineering, Nuclear Engineering and Design,* and *Procedia Engineering*: 3.9 [**38**], 5.18 [**206**], 6.2 [**199**], 6.17 (left) [**199**], 7.2 [**156**] (original figures courtesy of Andy Piekutowski), 7.3 [**6**], 7.5 [**126**], 7.7 [**126**], 7.8 [**12**], 7.9 [**12**], 7.10 [**12**], 7.11 [**12**], 7.12 [**12**], 7.13 [**8**], 7.14 [**8**], 7.15 [**8**], 7.16 [**12**], 7.17 [**8**], 7.18 [**8**], 7.19 [**194**], 7.20 [**15**], 7.21 [**15**], 7.22 [**14**], 7.23 [**14, 15**], 7.24 [**15**], 7.25 [**14**], 8.1 [**16**], 8.2 [**16**], 8.3 [**16**], 8.4 [**16**], 8.8 [**8**], 8.9 [**11**], 8.10 [**11**], 8.11 [**11**], 8.12 [**16**], 8.13 [**16**], 9.3 [**199**], 9.4 [**204**], 9.5 [**205**], 9.6 [**205**], 9.7 [**205**], 10.1 [**199**], 10.2 [**199**], 10.3 [**199**], 10.5 [**199**], 10.6 [**199**], 10.7 [**199**], 10.8 [**199**], 10.9 [**199**], 10.13 [**11**], 10.14 [**11**], 10.15 [**11**], 10.17 [**153**], 11.4 [**6**], 11.9 [**7**], 11.10 [**7**], 11.11 [**7**], 11.12 [**7**], 11.13 [**7**], 11.14 [**7**], 11.15 [**7**], 11.16 [**7**], and 11.17 [**7**].

The following figures are reprinted with permission of the International Ballistics Society and appeared in various *Proceedings of the International Symposium on Ballistics*: 11.3 [**191**], 11.20 [**17**], 12.1 right [**196**], 12.3 right [**196**], 13.3 right [**192**], 13.4 [**192**], 13.5 [**192**], 13.6 [**192**], 13.7 [**192**], and 13.10 [**193**].

The following figures are reprinted with permission from AIP Publishing from proceedings of the Shock Compression of Condensed Matter conferences (first two) and from the *Journal of Applied Physics* (last one), and are copyrighted on the respective proceedings and journal publication dates: 7.4 [**13**], 11.6 [**200**], and 5.12 [**178**].

The following two figures are reprinted with permission from Springer Verlag from the book in the reference, copyrighted that year: 5.11 [**114**] and 5.17 left [**75**].

The following figures are reprinted with permission from SAGE Publications, who holds the copyright for *Textile Research Journal* and *Mathematics and Mechanics of Solids*: 13.1 [**102**] (original photograph courtesy of Phil Cunniff) and 14.1 [**26**].

The following figure is reprinted with permission from JVE International Ltd., who holds the copyright for *Journal of Vibroengineering*: 3.3 right [**45**].

The following figures are reprinted with permission and courtesy of Gerry Kerley: 2.4 [**117**], 5.17 right [**116**], and 5.20 [**115**].

The following figures are reprinted from U. S. government reports in the public domain: 3.11 [**212**] and 4.11 [**128**].

Exercises

(D.1) Continue the nonlinear analysis of Exercise 13.19 and show that for the in-plane displacement, the deformation gradient is

$$F = \begin{pmatrix} F_{11} & F_{12} & 0 \\ F_{21} & F_{22} & 0 \\ F_{31} & F_{32} & 1 \end{pmatrix} = \begin{pmatrix} 1 + \frac{\partial u_x}{\partial x} & \frac{\partial u_x}{\partial y} & 0 \\ \frac{\partial u_y}{\partial x} & 1 + \frac{\partial u_y}{\partial y} & 0 \\ \frac{\partial u_z}{\partial x} & \frac{\partial u_z}{\partial y} & 1 \end{pmatrix} \tag{D.1}$$

and so the stress is

$$S = \frac{\sigma_f}{2\bar{\varepsilon}_f^\gamma} \begin{pmatrix} E_{11}^\gamma \left(1 + \frac{\partial u_x}{\partial x}\right) & E_{11}^\gamma \frac{\partial u_y}{\partial x} & E_{11}^\gamma \frac{\partial u_z}{\partial x} \\ E_{22}^\gamma \frac{\partial u_x}{\partial y} & E_{22}^\gamma \left(1 + \frac{\partial u_y}{\partial y}\right) & E_{22}^\gamma \frac{\partial u_z}{\partial y} \\ 0 & 0 & 0 \end{pmatrix}. \tag{D.2}$$

The equation of motion for the in-plane motion in the x direction is then

$$\hat{\rho} \frac{d^2 u_x}{dt^2} = \frac{\sigma_f}{2\bar{\varepsilon}_f^\gamma} \left\{ \frac{\partial}{\partial x} \left(\left(1 + \frac{\partial u_x}{\partial x}\right) E_{11}^\gamma \right) + \frac{\partial}{\partial y} \left(\frac{\partial u_x}{\partial y} E_{22}^\gamma \right) \right\}. \tag{D.3}$$

Again assume that $\partial u_x/\partial x$ is small compared to 1, that $\partial u_x/\partial y$ and $\partial u_y/\partial x$ are small compared to $\partial u_x/\partial x$, so the equation reduces to

$$\hat{\rho} \frac{d^2 u_x}{dt^2} \approx \frac{\sigma_f}{2\bar{\varepsilon}_f^\gamma} \frac{\partial}{\partial x} (E_{11}^\gamma) \qquad \Rightarrow \qquad \frac{\partial}{\partial x} (E_{11}^\gamma) = 0. \tag{D.4}$$

Show that this is exactly the same as Eq. (13.114), and so

$$E_{11} = \frac{\partial u_x}{\partial x} + \frac{1}{2} \left(\frac{\partial u_z}{\partial x} \right)^2 = f(y), \tag{D.5}$$

where f is an arbitrary function of y with the power $1/\gamma$ in it and where our representation of E_{11} includes our assumptions about small terms is a solution.

Bibliography

1. V. P. Alekseevskii, *Penetration of a rod into a target at high velocity*, Combustion, Explosion and Shock Waves **2** (1966), 63–66, translated from the Russian.
2. W. A. Allen and J. W. Rogers, *Penetration of a rod into a semi-infinite target*, Journal of the Franklin Institute **272** (1961), 275–284.
3. C. E. Anderson, Jr., I. S. Chocron, and A. E. Nicholls, *Damage modeling for Taylor impact simulations*, Journal de Physique IV **134** (2006), 331–337.
4. C. E. Anderson, Jr. and S. Chocron, *Effect of lateral confinement on penetration efficiency as a function of impact velocity*, Computational Ballistics II (Billerica, MA) (V. Sanchez-Galvez, C. A. Brebbia, A. A. Motta, and C. E. Anderson, Jr., eds.), WIT Press, 2005, pp. 149–158.
5. _____, *Experimental results and a simple theory for the early deflection-time history of a ballistic fabric*, Shock Compression of Condensed Matter: 2009 (Melville, NY) (M. L. Elert, W. T. Buttler, M. D. Furnish, W. W. Anderson, and W. G. Proud, eds.), AIP Press, 2009, pp. 1457–1460.
6. C. E. Anderson, Jr., V. Hohler, J. D. Walker, and A. J. Stilp, *Time-resolved penetration of long rods into steel targets*, International Journal of Impact Engineering **16** (1995), no. 1, 1–18.
7. _____, *The influence of projectile hardness on ballistic performance*, International Journal of Impact Engineering **22** (1999), no. 6, 619–632.
8. C. E. Anderson, Jr., D. L. Littlefield, and J. D. Walker, *Long-rod penetration, target resistance, and hypervelocity impact*, International Journal of Impact Engineering **14** (1993), no. 1–4, 1–12.
9. C. E. Anderson, Jr. and B. L. Morris, *The ballistic performance of confined Al_2O_3 ceramic tiles*, International Journal of Impact Engineering **12** (1992), no. 2, 167–187.
10. C. E. Anderson, Jr., B. L. Morris, and D. L. Littlefield, *A penetration mechanics database*, *SwRI Report 3593/001*, Southwest Research Institute, San Antonio, TX, 1992, Prepared for DARPA.
11. C. E. Anderson, Jr., D. L. Orphal, R. R. Franzen, and J. D. Walker, *On the hydrodynamic approximation for long-rod penetration*, International Journal of Impact Engineering **22** (1999), no. 1, 23–42.
12. C. E. Anderson, Jr. and J. D. Walker, *An examination of long-rod penetration*, International Journal of Impact Engineering **11** (1991), no. 4, 481–501.
13. _____, *An analytical expression for P/L for WA rods into armor steel*, Shock Compression of Condensed Matter: 1995 (Woodbury, NY) (S. C. Schmidt and W. C. Tao, eds.), AIP Press, 1996, pp. 1135–1138.
14. C. E. Anderson, Jr., J. D. Walker, S. J. Bless, and Y. Partom, *On the L/D effect for long-rod penetrators*, International Journal of Impact Engineering **18** (1996), no. 3, 247–264.
15. C. E. Anderson, Jr., J. D. Walker, S. J. Bless, and T. R. Sharron, *On the velocity dependence of the L/D effect for long-rod penetrators*, International Journal of Impact Engineering **17** (1995), 13 – 24.
16. C. E. Anderson, Jr., J. D. Walker, and G. E. Hauver, *Target resistance for long-rod penetration into semi-infinite targets*, Nuclear Engineering and Design **138** (1992), 93–104.
17. C. E. Anderson, Jr., J. D. Walker, and T. R. Sharron, *The influence of edge effects on penetration*, 17th International Symposium on Ballistics (Midrand, South Africa) (C. van Niekerk, ed.), vol. 3, 1998, pp. 33–40.

18. C. E. Anderson, Jr., J. S. Wilbeck, P. S. Westine, U. S. Lindholm, and A. B. Wenzel, *A short course in penetration mechanics course notes*, Southwest Research Institute, San Antonio, TX, 1985.

19. T. Antoun, D. R. Curran, S. V. Razorenov, L. Seaman, G. I. Kanel, and A. V. Utkin, *Spall fracture*, Springer, New York, 2003.

20. J. R. Asay, L. C. Chhabildas, R. J. Lawrence, and M. A. Sweeney, *Impactful times: memories of 60 years of shock wave research at Sandia National Laboratories*, Springer, Berlin, 2017.

21. J. R. Asay and J. Shahinpoor (eds.), *High-pressure shock compression of solids*, Springer-Verlag, New York, 1993.

22. J. Awerbuch and S. R. Bodner, *Analysis of the mechanics of perforation of projectiles in metallic plates*, International Journal of Solids and Structures **10** (1974), 671–684.

23. Y. Bao and T. Wierzbicki, *A comparative study on various ductile crack formation criteria*, Journal of Engineering Materials and Technology **126** (2004), 314–324.

24. _____, *On fracture locus in the equivalent strain and stress triaxiality space*, International Journal of Mechanical Sciences **46** (2004), 81–98.

25. _____, *On the cut-off value of negative triaxility for fracture*, Engineering Fracture Mechanics **72** (2005), 1049–1069.

26. R. C. Batra, I. Mueller, and P. Strehlow, *Treloar's biaxial tests and Kearsley's bifurcation in rubber sheets*, Mathematics and Mechanics of Solids **10** (2005), 705–713.

27. M. F. Beatty, *Topics in finite elasticity: hyperelasticity of rubber, elastomers, and biological tissues – with examples*, Applied Mechanics Reviews **40** (1987), no. 12, 1699–1734.

28. G. Birkhoff, D. P. MacDougall, E. M. Pugh, and G. Taylor, *Explosives with lined cavities*, Journal of Applied Physics **19** (1948), 563–582.

29. D. R. Bland, *On shock structure in a solid*, Journal of the Institute of Mathematics and its Applications **1** (1965), 56–75.

30. P. W. Bridgman, *The physics of high pressure*, G. Bell and Sons, London, 1949.

31. _____, *Studies in large plastic flow and fracture (with special emphasis on the effects of hydrostatic pressure)*, McGraw-Hill, New York, 1952.

32. R. Brun (ed.), *High temperature phenomena in shock waves*, Springer, Berlin, 2012.

33. B. M. Butcher and C. H. Karnes, *Dynamic compaction of porous iron*, Journal of Applied Physics **40** (1969), no. 7, 2967–2976.

34. H. B. Callen, *Thermodynamics*, John Wiley & Sons, Inc., New York, 1960.

35. G. F. Carrier and C. E. Pearson, *Partial differential equations: Theory and technique*, 2nd ed., Academic Press, San Diego, CA, 1988.

36. M. M. Carroll and A. C. Holt, *Static and dynamic pore-collapse relations for ductile porous materials*, Journal of Applied Physics **43** (1972), no. 4, 1,626–1,636.

37. L. C. Chhabildas, L. Davison, and Y. Horie (eds.), *High-pressure shock compression of solids VIII: the science and technology of high-velocity impact*, Springer-Verlag, Berlin, 2005.

38. S. Chocron, C. E. Anderson, Jr., D. J. Grosch, and C. H. Popelar, *Impact of the 7.62-mm APM2 projectile against the edge of a metallic target*, International Journal of Impact Engineering **25** (2001), 423–437.

39. S. Chocron, C. E. Anderson, Jr., and J. D. Walker, *A consistent plastic flow approach to model penetration and failure of finite-thickness metallic targets*, 18th International Symposium on Ballistics (San Antonio, TX) (W. G. Reinecke, ed.), vol. 2, Technomic Publishing Co., 1999, pp. 761–768.

40. S. Chocron, C. E. Anderson, Jr., J. D. Walker, and M. Ravid, *A unified model for long-rod penetration in multiple metallic target plates*, International Journal of Impact Engineering **28** (2003), no. 4, 391–411.

41. S. Chocron, B. Erice, and C. E. Anderson, *A new plasticity and failure model for ballistic application*, International Journal of Impact Engineering **38** (2011), 755–764.

42. S. Chocron, D. J. Grosch, and C. E. Anderson, Jr., *DOP and V_{50} predictions for the 0.30-cal APM2 projectile*, 18th International Symposium on Ballistics (San Antonio, TX) (W. G. Reinecke, ed.), Technomic Publishing Co., 1999, pp. 769–776.

43. R. M. Christensen, *The theory of materials failure*, Oxford University Press, Oxford, UK, 2013.

44. D. R. Christman and J. W. Gehring, *Analysis of high-velocity projectile penetration mechanics*, Journal of Applied Physics **37** (1966), no. 4, 1,579–1,587.

45. C. C. Chung, K. L. Lee, and W. F. Pan, *Finite element analysis on the response of 6061-T6 aluminum alloy tubes with a local sharp cut under cyclic bending*, Journal of Vibroengineering **18** (2016), no. 7, 4,276–4,284.

46. M. G. Cockcroft and D. J. Latham, *Ductility and the workability of metals*, Journal of the Institute of Metals **96** (1968), 33–39.

47. R. Courant and K. O. Friedrichs, *Supersonic flow and shock waves*, Interscience Publishers (reprint by Springer-Verlag 1985), New York, 1948.

48. F. S. Crawford, Jr., *Waves*, McGraw-Hill, New York, 1968.

49. R. J. M. Crozier and S. C. Hunter, *Similarity solution for the rapid uniform expansion of a cylindrical cavity in a compressible elastic-plastic solid.*, Quarterly Journal of Mechanics and Applied Mathematics **23** (1970), no. 3, 349–363.

50. P. M. Cunniff, *An analysis of the system effects of woven fabrics under ballistic impact*, Textile Research Journal **62** (1992), no. 9, 495–509.

51. _____, *A design tool for the development of fragmentation protective body armor*, 18th International Symposium on Ballistics (San Antonio, TX) (W. G. Reinecke, ed.), vol. 2, Technomic Publishing Co., 1999, pp. 1295–1302.

52. _____, *Dimensionless parameters for optimization of textile-based body armor systems*, 18th International Symposium on Ballistics (San Antonio, TX) (W. G. Reinecke, ed.), vol. 2, Technomic Publishing Co., 1999, pp. 1303–1310.

53. L. Davison, D. E. Grady, and M. Shahinpoor (eds.), *High-pressure shock compression of solids II: dynamic fracture and fragmentation*, Springer, New York, 1996.

54. B. P. Denardo and C. R. Nysmith, *Momentum transfer and cratering phenomena associated with the impact of aluminum spheres into thick aluminum targets at velocities to 24000 ft/s*, Proceedings of the AGARD-NATO Specialists, Vol. 1: The Fluid Dynamic Aspect of Space Flight (New York) (S. C. Schmidt, J. W. Shaner, G. A. Samara, and R. Ross, eds.), Gordon and Breach Science Publishers, 1964, pp. 389–402.

55. S. Dey, T. Børvik, O. S. Hopperstad, and M. Langseth, *On the influence of fracture criterion in projectile impact of steel plates*, Computational Materials Science **38** (2006), 176–191.

56. G. E. Duvall, *Some properties and applications of shock waves*, Response of Metals to High Velocity Deformation (New York) (P. G. Shewnon and V. F. Zackay, eds.), Interscience Publishers, 1961, Chapter 4, pp. 165–203.

57. _____, *Shock waves and equations of state*, Dynamic Response of Materials to Intense Impulsive Loading (Wright Patterson Air Force Base, OH) (P. C. Chou and A. K. Hopkins, eds.), Air Force Materials Laboratory, 1973, Chapter 4.

58. R. J. Eichelberger, *Experimental test of the theory of penetration by metallic jets*, Journal of Applied Physics **27** (1956), no. 1, 63–68.

59. R. J. Eichelberger and J. W. Gehring, *Effects of meteoroid impacts on space vehicles*, ARS Journal **32** (1962), 1,583–1,591.

60. A. C. Eringen, *Microcontinuum field theories I: foundations and solids*, Springer, New York, 1999.

61. W. Fickett and W. C. Davis, *Detonation*, University of California Press, Berkeley, CA, 1979.

62. D. P. Flanagan and L. M. Taylor, *An accurate numerical algorithm for stress integration with finite rotations*, Computer Methods in Applied Mechanics and Engineering **62** (1987), 305–320.

63. P. S. Follansbee, *Fundamentals of strength*, John Wiley & Sons, Hoboken, NJ, 2014.

64. M. J. Forrestal, T. Børvik, T. L. Warren, and W. Chen, *Perforation of 6082-T651 aluminum plates with 7.62 mm APM2 bullets at normal and oblique impacts*, Experimental Mechanics **54** (2014), 471–481.

65. M. J. Forrestal, N. S. Brar, and V. K. Luk, *Penetration of strain-hardening targets with rigid spherical-nose rods*, Journal of Applied Mechanics **58** (1991), 7–10.

66. M. J. Forrestal and S. J. Hanchak, *Perforation experiments on HY-100 steel plates with 4340 R_c 38 and maraging T-250 steel rod projectiles*, International Journal of Impact Engineering **22** (1999), 923–933.

67. M. J. Forrestal, B. Lim, and W. Chen, *A scaling law for APM2 bullets and aluminum armor plates*, Experimental Mechanics **59** (2019), no. 1, 121–123.

68. M. J. Forrestal and D. B. Longcope, *Target strength of ceramic materials for high-velocity penetration*, Journal of Applied Physics **67** (1990), no. 8, 3669–3672.

69. M. J. Forrestal and V. K. Luk, *Dynamic spherical cavity-expansion in a compressible elastic-plastic solid*, Journal of Applied Mechanics **55** (1988), 275–279.

70. _____, *Dynamic spherical cavity-expansion of strain-hardening materials*, Journal of Applied Mechanics **58** (1991), 1–6.

71. M. J. Forrestal, K. Okajima, and V. K. Luk, *Penetration of 6061-T651 aluminum targets with rigid long rods*, Journal of Applied Mechanics **55** (1988), 755–760.

72. V. E. Fortov, L. V. Al'tshuler, R. F. Trunin, and A. I. Funtikov (eds.), *High-pressure shock compression of solids VII: shock waves and extreme states of matter*, Springer-Verlag, New York, 2004.

73. R. R. Franzen and P. N. Schneidewind, *Observations concerning the penetration mechanics of tubular hypervelocity penetrators*, International Journal of Impact Engineering **11** (1991), 289–303.

74. Y. C. Fung, *Foundations of solid mechanics*, Prentice Hall International, London, 1965.

75. A. I. Funtikov, *Phase diagram of iron*, High-Pressure Shock Compression of Solids VII: shock Waves and Extreme States of Matter (New York) (V. E. Fortov, L. V. Al'tshuler, R. F. Trunin, and A. I. Funtikov, eds.), Springer-Verlag, 2004, pp. 225–246.

76. U. Gerlach, *Microstructural analysis of residual projectiles – a new method to explain penetration mechanisms*, Metallurgical Transactions A **17** (1986), no. 3, 435–442.

77. W. Goldsmith, *Impact*, Edward Arnold, London, 1960.

78. J. M. Gosline, P. A. Guerette, C. S. Ortlepp, and K. N. Savage, *The mechanical design of spider silks: from fibroin sequence to mechanical function*, Journal of Experimental Biology **202** (1999), 3295–3303.

79. C. L. Grabarek, *Penetration of armor by steel and high density penetrators (BRL Memorandum Report No. 2134)*, U.S. Army Ballistics Research Laboratory, Aberdeen Proving Ground, MD, 1971.

80. D. Grady, *Fragmentation of rings and shells: the legacy of N. F. Mott*, Springer, Berlin, 2006.

81. _____, *Physics of shock and impact volume 1: Fundamentals and dynamic failure*, IOP Publishing, Bristol, 2017.

82. D. E. Grady, *The spall strength of condensed materials*, Journal of the Mechanics and Physics of Solids **36** (1988), no. 3, 353–384.

83. D. E. Grady and M. E. Kipp, *The growth of unstable thermoplastic shear with applications to steady-wave shock compression in solids*, Journal of the Mechanics and Physics of Solids **35** (1987), no. 1, 95–118.

84. D. E. Grady and M. L. Olsen, *A statistics and energy based theory of dynamic fragmentation*, International Journal of Impact Engineering **29** (2003), 293–306.

85. R. A. Graham, *Solids under high-pressure shock compression*, Springer-Verlag, New York, 1993.

86. G. T. Gray, III, S. R. Chen, W. Wright, and M. F. Lopez, *Constitutive equations for annealed metals under compression at high strain rates and high temperatures (LA-12669-MS)*, Los Alamos National Laboratory, Los Alamos, NM, 1994.

87. G. T. (R.) Gray, III, *Classic split-Hopkinson pressure bar testing*, ASM, 2000.

88. K. A. Gschneidner, Jr., *Physical properties and interrelationships of metallic and semimetallic elements*, Solid State Physics (New York) (F. Seitz and D. Turnbull, eds.), vol. 16, Academic Press, 1964, pp. 275–426.

89. M. E. Gurtin, E. Fried, and L. Anand, *The mechanics and thermodynamics of continua*, Cambridge University Press, New York, 2010.

90. J. W. Hancock and A. C. MacKenzie, *On the mechanisms of ductile failure in high-strength steels subjected to multi-axial stress states*, Journal of the Mechanics and Physics of Solids **24** (1976), 147–169.

91. D. B. Hayes, *Introduction to stress wave phenomena (SLA-73-0801)*, Sandia National Laboraties, Albuquerque, NM, 1973.

92. W. Herrmann, *Constitutive equation for the dynamic compaction of ductile porous materials*, Journal of Applied Physics **40** (1969), no. 6, 2490–2499.

93. R. Hill, *The mathematical theory of plasticity*, Oxford University Press, Oxford, UK, 1950.

94. V. Hohler and A. J. Stilp, *Penetration of steel and high density rods in semi-infinite steel targets*, 3rd Int. Symp. on Ballistics (Karlsruhe, Germany), 1977, pp. H3/1–12.

95. _____, *Influence of length-to-diameter ratio in the range from 1 to 32 on the penetration performance of rod projectiles*, 8th International Symposium on Ballistics (Orlando, Florida) (W. G. Reinecke, ed.), Avco Systems Division, 1984, pp. IB/13–19.

96. _____, *Hypervelocity impact of rod projectiles with L/D from 1 to 32*, International Journal of Impact Engineering **5** (1987), no. 38721, 323–331.

97. _____, *Long-rod penetration mechanics*, High Velocity Impact Dynamics (New York) (J. A. Zukas, ed.), John Wiley & Sons, 1990, pp. 394–395.

98. _____, *Penetration data*, A Penetration Mechanics Database (SwRI Report 3593/001) (San Antonio, TX) (C. E. Anderson, Jr., B. L. Morris, and D. L. Littlefield, eds.), 1992.

99. T. J. Holmquist, D. W. Templeton, and K. D. Bishnoi, *Constitutive modeling of aluminum nitride for large strain, high-strain rate, and high-pressure applications*, International Journal of Impact Engineering **25** (2000), 211–231.

100. H. G. Hopkins, *Dynamic expansion of spherical cavities in metals*, vol. 1, North-Holland Publishing Co., Amsterdam, 1960, Chapter 3.

101. W. M. Isbell, *Shock waves: measuring the dynamic response of materials*, Imperial College Press, London, 2005.

102. J. W. Jameson, G. M. Stewart, D. R. Petterson, and F. A. Odell, *Dynamic distribution of strain in textile materials under high-speed impact: part iii: strain-time-position history in yarns*, Textile Research Journal **32** (1962), no. 10, 858–860.

103. S. T. Jeng and W. Goldsmith, *Shear and bending phenomena in normal projectile impacts on thin targets*, Computational Techniques for Contact, Impact, Penetration and Perforation of Solids (New York) (L. E. Schwer, N. J. Salamon, and W. K. Liu, eds.), vol. AMD 103, ASME, 1989, pp. 223–233.

104. G. R. Johnson, *Material characterization for warhead computations*, Tactical Missile Warheads (Washington, D.C.) (J. Carleone, ed.), American Institute of Aeronautics and Astronautics, 1993, pp. 165–197.

105. G. R. Johnson and W. H. Cook, *A constitutive model and data for metals subjected to large strains, high strain rates, temperatures and pressures*, 7th International Symposium on Ballistics (The Hague, Netherlands), 1983, pp. 541–547.

106. _____, *Fracture characteristics of three metals subjected to various strains, strain rates, temperatures and pressures*, Engineering Fracture Mechanics **21** (1985), no. 1, 31–48.

107. G. R. Johnson, T. J. Holmquist, C. E. Anderson, Jr., and A. E. Nicholls, *Strain-rate effects for high-strain-rate compuations*, Journal de Physique IV **134** (2006), 391–396.

108. J. N. Johnson, *Single-particle model of a solid: the Mie–Grüneisen equation*, American Journal of Physics **36** (1968), no. 10, 917–919.

109. J. N. Johnson and F. L. Addessio, *Tensile plasticity and ductile fracture*, Journal of Applied Physics **64** (1988), no. 12, 6699–6712.

110. J. N. Johnson and R. Chéret (eds.), *Classic papers in shock compression science*, Springer-Verlag, New York, 1998.

111. K. L. Johnson, *Contact mechanics*, Cambridge University Press, Cambridge, UK, 1985.

112. W. Johnson, *Henri Tresca as the originator of adiabatic heat lines*, International Journal of Mechanical Sciences **29** (1987), no. 5, 301–310.

113. T. L. Jones, R. D. DeLorme, M. S. Burkins, and W. A. Gooch, *Ballistic evaluation of magnesium alloy AZ31B (ARL-TR-4077)*, Army Research Laboratory, Aberdeen, MD, 2007.

114. N. N. Kalitkin and L. V. Kuzmina, *Wide-range characteristic thermodynamic curves*, High-Pressure Shock Compression of Solids VII: shock Waves and Extreme States of Matter (New York) (V. E. Fortov, L. V. Al'tshuler, R. F. Trunin, and A. I. Funtikov, eds.), Springer-Verlag, 2004, pp. 109–176.

115. G. I. Kerley, *Multiphase equation of state for iron (SAND93-0027)*, Sandia National Laboratories, Albuquerque, NM, 1993.

116. _____, *Equation of state for copper and lead (KTS02-1)*, Kerley Technical Services, Appomattox, VA, 2002.

117. _____, *Equations of state for hydrogen and deuterium (SAND2003-3613)*, Sandia National Laboratories, Albuquerque, NM, 2003.

118. _____, *The linear U_s–u_p in shock-wave physics (KTS06-1)*, Kerley Technical Services, Appomattox, VA, 2006.

119. R. Kinslow (ed.), *High-velocity impact phenomena*, Academic Press, New York, 1970.

120. M. E. Kipp and D. E. Grady, *Dynamic fracture growth and interaction in one dimension*, Journal of the Mechanics and Physics of Solids **33** (1985), no. 4, 399–415.

121. H. Kolsky, *Stress waves in solids*, Dover Publications, New York, 1963.

122. K. J. Koski, P. Akhenblit, K. McKiernan, and J. L. Yarger, *Non-invasive determination of the complete elastic moduli of spider silks*, Nature Materials (2013), 1–6, DOI:10.1038/NMAT3549.

123. E. Kreyszig, *Introduction to differential geometry and Riemannian geometry*, University of Toronto Press, Toronto, Canada, 1968.

124. R. D. Krieg and D. B. Krieg, *Accuracies of numerical solution methods for the elastic-perfectly plastic model*, ASME Journal of Pressure Vessel Technology **99** (1977), 510–515.

125. Los Alamos National Laboratory, *GMX-6 Hugoniot data*, High-Pressure Shock Compression of Solids (New York) (J. R. Asay and J. Shahinpoor, eds.), Springer-Verlag, 1993, Appendix C.

126. J. Lankford, C. E. Anderson, Jr., S. A. Royal, and J. P. Riegel, III, *Penetration erosion phenomenology*, International Journal of Impact Engineering **18** (1996), no. 5, 565–578.

127. J. Lankford, H. Couque, A. Bose, and C. E. Anderson, Jr., *Microstructural dependence of high strain rate deformation and damage development in tungsten heavy alloys*, Shock Waves and High-Strain Rate Phenomena in Materials (New York) (M. A. Meyers, L. E. Murr, and K. P. Staudhammer, eds.), Marcel Dekker, 1991.

128. G. V. Latham, M. Ewing, F. Press, G. Sutton, J. Dorman, Y. Nakamura, N. Toksoz, F. Duennebier, and D. Lammlein, *Passive seismic experiment*, Apollo 14 Preliminary Science Report (Washington, D.C.) (NASA Manned Spacecraft Center, ed.), US Government Printing Office, 1971, pp. 133–161.

129. P. D. Lax, *Hyperbolic systems of conservation laws and the mathematical theory of shock waves*, SIAM, Philadelphia, PA, 1973.

130. C. Lee, X. Wei, J. W. Kysar, and J. Hone, *Measurement of the elastic properties and intrinsic strength of monolayer graphene*, Science **321** (2008), 385–388, DOI: 10.1126/science.1157996.

131. R. J. LeVeque, *Numerical methods for conservation laws*, 2nd ed., Birkhauser, Basel, Switzerland, 1992.

132. C. C. Lin. and L. A. Segel, *Mathematics applied to deterministic problems in the natural sciences*, Macmillan Publishing Co., New York, 1974.

133. U. S. Lindholm, *Deformation maps in the region of high dislocation velocity*, High Velocity Deformation of Solids (New York) (K. Kawata and J. Shiori, eds.), Springer Verlag, 1978, pp. 26–35.

134. D. L. Littlefield, C. E. Anderson, Jr., Y. Partom, and S. J. Bless, *The penetration of steel targets finite in radial extent*, International Journal of Impact Engineering **19** (1997), no. 1, 49–62.

135. A. E. H. Love, *A treatise on the mathematical theory of elasticity*, Dover, New York, 1944.

136. J. Lubliner, *Plasticity theory*, Macmillan Publishing Co., New York, 1990.

137. V. K. Luk, M. J. Forrestal, and D. E. Amos, *Dynamic spherical cavity expansion of strain-hardening materials*, Journal of Applied Mechanics **58** (1991), 1–6.

138. L. E. Malvern, *Introduction to the mechanics of a continuous medium*, Prentice-Hall, Englewood Cliffs, NJ, 1969.

139. S. P. Marsh, *LASL shock Hugoniot data*, University of California Press, Berkeley, CA, 1980.

140. G. A. Maugin and A. V. Metrikine (eds.), *Mechanics of generalized continua: one hundred years after the Cosserats*, Springer, New York, 2010.

141. P. McConnell, J. W. Sheckherd, J. S. Perrin, and R. A. Wullaert, *Experience in subsized specimen testing*, The Use of Small-Scale Specimens for Testing Irradiated Material (Philadelphia, PA) (W. R. Corwin and G. E. Lucas, eds.), ASTM, 1986, pp. 353–368.

142. J. M McGlaun, S. L. Thompson, and M. G. Elrick, *CTH: a three-dimensional shock wave physics code*, International Journal of Impact Engineering **10** (1990), 351–360.

143. R. G. McQueen, S. P. Marsh, J. W. Taylor, J. N. Fritz, and W. J. Carter, *The equation of state of solids from shock wave studies*, High-Velocity Impact Phenomena (New York) (R. Kinslow, ed.), Academic Press, 1970.

144. J. Mescall, *Materials issues in computer simulation of penetration mechanics*, Computational Aspects of Penetration Mechanics (New York) (J. Chandra and J. E. Flaherty, eds.), Springer-Verlag, 1983, pp. 47–62.

145. J. Mescall and R. Papirno, *Spallation in cylinder-plate impact*, Experimental Mechanics **14** (1974), 257–266.

146. H. W. Meyer, Jr. and D. S. Kleponis, *An analysis of parameters for the Johnson–Cook strength model for 2-in-thick rolled homogeneous armor (ARL-TR-2528)*, Army Research Laboratory, Aberdeen Proving Ground, MD, 2001.

147. M. A. Meyers, *Dynamic behavior of materials*, Wiley-Interscience, New York, 1994.

148. C. E. Morris (ed.), *Los Alamos shock wave profile data*, University of California Press, Berkeley, CA, 1982.

149. I. Mueller and P. Strehlow, *Rubber and rubber balloons, paradigms of thermodynamics*, Springer-Verlag, Berlin, 2004.

150. W. J. Nellis, *Dynamic compression of materials: metallization of fluid hydrogen at high pressures*, Reports on Progress in Physics **69** (2006), 1479–1580.

151. _____, *Ultracondensed matter by dynamic compression*, Cambridge University Press, Cambridge, UK, 2017.

152. J. Nuckolls, J. Emmett, and L. Wood, *Laser-induced thermonuclear fusion*, Physics Today **26** (1973), no. 8, 46–53.

153. D. L. Orphal and C. E. Anderson, Jr., *Streamline reversal in hypevelocity penetration*, International Journal of Impact Engineering **23** (1999), 699–710.

154. P. Petersen, *Riemannian geometry*, 2nd ed., Springer, New York, 2006.

155. S. L. Phoenix and P. K. Porwal, *A new membrane model for the ballistic impact response and V_{50} performance of multi-ply fibrous systems*, International Journal of Solids and Structures **40** (2003), 6723–6765.

156. A. J. Piekutowski, M. J. Forrestal, K. L. Poormon, and T. L. Warren, *Pentration of 6061-T6511 aluminum targets by ogive-nose steel projectiles with striking velocities between 0.5 and 3.0 km/s*, International Journal of Impact Engineering **23** (1999), no. 1, 723–733.

157. E. J. Rapacki, K. Frank, Jr., R. B. Leavy, M. J. Keele, and J. J. Prifti, *Armor steel hardness influence on kinetic energy penetration*, 15th International Symposium on Ballistics (Jerusalem, Israel) (M. Mayseless and S. R. Bodner, eds.), vol. 1, 1995, pp. 323–330.

158. M. Ravid and S. R. Bodner, *Dynamic perforation of viscoplastic plates by rigid projectiles*, International Journal of Engineering Science **21** (1983), no. 6, 577–591.

159. M. Ravid, S. R. Bodner, J. D. Walker, S. Chocron, C. E. Anderson, Jr., and J. P. Riegel, III, *Modifications of the Walker–Anderson penetration model to include exit failure modes and fragmentation*, 17th International Symposium on Ballistics (Midrand, South Africa) (C. van Niekerk, ed.), vol. 3, 1998, pp. 267–274.

160. J. E. Reaugh, A. C. Holt, M. L. Wilkins, B. J. Cunningham, B. L. Hord, and A. S. Kusubov, *Impact studies of five ceramic materials and Pyrex*, International Journal of Impact Engineering **23** (1999), 771–782.

161. F. Reif, *Fundamentals of statistical and thermal physics*, McGraw-Hill, New York, 1965.

162. J. S. Rinehart, *Stress transients in solids*, Hyperdynamics, Santa Fe, NM, 1975.

163. D. Rittel, *Dynamic shear failure of materials*, Dynamic Failure of Materials and Structures (New York) (A. Shukla, G. Ravichandran, and Y. D. S. Rajapakse, eds.), Springer-Verlag, 2010, pp. 29–61.

164. Z. Rosenberg and E. Dekel, *On the relation between deformation modes and the penetration capability of long-rods*, International Workshop on New Models and New Codes on Shock Wave Processes in Condensed Media (St. Catherine's College, Cambridge, UK), 1997.

165. _____, *Terminal ballistics*, Springer-Verlag, Berlin, 2012.

166. Z. Rosenberg, E. Marmor, and M. Mayseless, *On the hydrodynamic theory of long-rod penetration*, International Journal of Impact Engineering **10** (1990), 483–486.

167. D. Roylance and S. S. Wang, *Penetration mechanics of textile structures*, Ballistics Materials and Penetration Mechanics (Amsterdam, Netherlands) (R. C. Laible, ed.), Elsevier, 1980, pp. 273–292.

168. W. Rudin, *Real and complex analysis*, McGraw-Hill, New York, 1974.

169. M. H. Sadd, *Elasticity: theory, applications, and numerics*, Elsevier Butterworth-Heinemann, Burlington, MA, 2005.

170. G. F. Silsby, *Penetration of semi-infinite steel targets by tungsten long rods at 1.3 to 4.5 km/s*, 8th International Symposium on Ballistics (Orlando, FL), vol. 2, 1984, pp. TB31–36.

171. J. C. Smith, F. L. McCrackin, and H. F. Schiefer, *Stress-strain relationships in yarns subjected to rapid impact loading, part V: wave propagation in long textile yarns impacted transversely*, Textile Research Journal **28** (1958), no. 4, 288–302.

172. B. R. Sorensen, K. D. Kimsey, G. F. Silsby, D. R. Scheffler, T. M. Sherrick, and W. S. deRosset, *High velocity penetration of steel targets*, International Journal of Impact Engineering **11** (1991), no. 1, 107–119.

173. M. Spivak, *A comprehensive introduction to differential geometry*, 3rd ed., vol. 2, Publish or Perish, Houston, TX, 1999.

174. D. J. Steinberg, *Equation of state and strength properties of selected materials (UCRL-MA-106439)*, Lawrence Livermore National Laboratory, Livermore, CA, 1991.

175. D. J. Steinberg and C. M. Lund, *A constitutive model for strain rates from 10^{-4} to 10^6 s^{-1}*, Journal of Applied Physics **64** (1989), no. 4, 1528–1533.

176. E. J. Sternglass and D. A. Stuart, *An experimental study of the propagation of transient longitudinal deformations in elastoplastic media*, Journal of Applied Mechanics **20** (1953), 427–434.

177. S. T. Stewart and Ahrens, *A new H_2O ice Hugoniot: implications for planetary impact events*, Shock Compression of Condensed Matter: 2003 (New York) (M D. Furnish, Y. M. Gupta, and J. W. Forbes, eds.), Amer. Inst. of Phys., 2004, pp. 1478–1483.

178. J. W. Swegle and D. E. Grady, *Shock viscosity and the prediction of shock wave rise times*, Journal of Applied Physics **58** (1985), no. 2, 692–701.

179. A. Tate, *A theory for the deceleration of long rods after impact*, Journal of the Mechanics and Physics of Solids **15** (1967), 387–399.

180. _____, *Further results in the theory of long rod penetration*, Journal of the Mechanics and Physics of Solids **17** (1969), 141–150.

181. _____, *Long rod penetration models–part I. A flow field model for high speed long rod penetration*, International Journal of Engineering Science **28** (1986), no. 8, 535–548.

182. _____, *Long rod penetration models–part II. Extensions to the hydrodynamic theory of penetration*, International Journal of Engineering Science **28** (1986), no. 9, 599–612.

183. G. I. Taylor, *The use of flat-ended projectiles in determining dynamic yield stress*, Proceedings of the Royal Society of London, Series A **194** (1948), 298–299.

184. _____, *The scientific papers of G. I. Taylor, volume 1: the mechanics of solids*, Cambridge University Press, Cambridge, UK, 1958.

185. G. I. Taylor and H. Quinney, *Plastic distortion of metals*, Philosophical Transactions of the Royal Society, Series A **230** (1931), 323–362.

186. X. Teng and T. Wierzbicki, *Evaluation of six fracture models in high velocity perforation*, Engineering Fracture Mechanics **73** (2006), 1653–1678.

187. P. F. Thomason, *Ductile fracture of metals*, Pergamon Press, Oxford, UK, 1990.

188. F. R. Tuler and B. M. Butcher, *A criterion for the time dependence of dynamic fracture*, International Journal of Fracture Mechanics **4** (1968), no. 4, 431–437.

189. J. D. Walker, *Waves and cracks in composites (Ph.D. Thesis)*, University of Utah, Salt Lake City, UT, 1988.

190. _____, *On maximum dissipation for dynamic plastic flow (SwRI Report 07-9753)*, Southwest Research Institute, San Antonio, TX, 1995.

191. _____, *An analytic velocity field for back surface bulging*, 18th International Symposium on Ballistics (San Antonio, TX) (W. G. Reinecke, ed.), vol. 2, Technomic Publishing Co., 1999, pp. 1,239–1,246.

192. _____, *Constitutive model for fabrics with explicit static solution and ballistic limit*, 18th International Symposium on Ballistics (San Antonio, TX) (W. G. Reinecke, ed.), vol. 2, Technomic Publishing Co., 1999, pp. 1,231–1,238.

193. _____, *Ballistic limit of fabrics with resin*, 19th International Symposium on Ballistics (Thun, Switzerland) (I. R. Crewther, ed.), vol. 3, Vetter Druck AG, 2001, pp. 1,409–1,414.

194. _____, *Hypervelocity penetration modeling: momentum vs. energy and energy transfer mechanisms*, International Journal of Impact Engineering **26** (2001), 809–822.

195. _____, *Analytic expression for the deformation and ballistic limit of fabrics*, International Conference on Composite Materials–2003 (San Diego, CA), Society of Manufacturing Engineers, 2003, SME Technical Paper EM03-336.

196. _____, *Target flow fields and penetration resistance for pointed projectiles*, 31st International Symposium on Ballistics (Hyderabad, India), 2019.

197. J. D. Walker and C. E. Anderson, Jr., *A nonsteady-state model for penetration*, 13th International Symposium on Ballistics (Stockholm, Sweden) (A. Persson, K. Andersson, and E. B. Bjorck, eds.), vol. 3, National Defence Research Establishment, 1992, pp. 9–16.

198. _____, *The influence of initial nose shape in eroding penetration*, International Journal of Impact Engineering **15** (1994), no. 2, 139–148.

199. _____, *A time-dependent model for long-rod penetration*, International Journal of Impact Engineering **16** (1995), no. 1, 19–48.

200. J. D. Walker, R. P. Bigger, and S. Chocron, *Comparison of breakout modes in analytic penetration modeling*, Shock Compression of Condensed Matter: 2009 (Melville, NY) (M. L. Elert, W. T. Buttler, M. D. Furnish, W. W. Anderson, and W. G. Proud, eds.), AIP Press, 2009, pp. 1,439–1,442.

201. J. D. Walker and S. Chocron, *Why impacted yarns break at lower speed than classical theory predicts*, Journal of Applied Mechanicss **78** (2011), 051012-1–7.

202. J. D. Walker and S. Chocron, *Damage modeling, scaling, and momentum enhancement for asteroid and comet nucleus deflection*, Procedia Engineering **103** (2015), 636–641.

203. J. D. Walker, S. Chocron, and W. Gray, *Analytical models for foam, ice and ablator impacts into space shuttle thermal tiles*, 22nd International Symposium on Ballistics (Vancouver, British Columbia, November 14-18, 2005) (W. Flis and B. Scott, eds.), vol. 2, DEStech Publications, 2005, pp. 1,196–1,203.

204. J. D. Walker, S. Chocron, and D. J. Grosch, *Size scaling of crater size, ejecta mass, and momentum enhancement due to hypervelocity impacts into 2024-T4 and 2024-T351 aluminum*, 2019 Hypervelocity Impact Symposium (D. Littlefield, ed.), no. HVIS2019-049, V001T04A004, ASME, 2020.

205. _____, *Size scaling of hypervelocity-impact ejecta mass and momentum enhancement: experiments and a nonlocal-shear-band-motivated strain-rate-dependent failure model*, International Journal of Impact Engineering **135** (2020), 103388-1–14.

206. J. D. Walker, S. Chocron, J. H. Waite, and T. Brockwell, *The vaporization threshold: hypervelocity impacts of ice grains into a titanium Cassini spacecraft instrument chamber*, Procedia Engineering **103** (2015), 628–635.

207. G. Wei and W. Zhang, *Penetration of thin aluminum alloy plates by blunt projectiles: an experimental and numerical investigation*, Shock Compression of Condensed Matter: 2013 (Melville, NY) (M. L. Elert, W. T. Buttler, M. D. Furnish, W. W. Anderson, and W. G. Proud, eds.), vol. 500, AIP Press, 2014, pp. 1–4.

208. T. Wierzbicki, Y. Bao, Y.-W. Lee, and Y. Bai, *Calibration and evaluation of seven fracture models*, International Journal of Mechanical Sciences **47** (2005), 719–743.

209. M. L. Wilkins, *Third progress report of light armor program (UCRL-50460)*, Lawrence Radiation Laboratory, Livermore, CA, 1968.

210. _____, *Computer simulation of dynamic phenomena*, Springer, Berlin, 1999.

211. M. L. Wilkins and M. W. Guinan, *Impact of cylinders on a rigid boundary*, Journal of Applied Physics **44** (1973), no. 3, 1200–1206.

212. M. L. Wilkins, R. D. Streit, and J. E. Reaugh, *Cumulative-strain-damage model of ductile fracture: simulation and prediction of engineering fracture tests (UCRL-53058)*, Lawrence Livermore National Laboratory, Livermore, CA, 1980.

213. T. W. Wright, *The physics and mathematics of adiabatic shear bands*, Cambridge University Press, Cambridge, UK, 2002.

214. David A. Young, *Phase diagrams of the elements*, University of California Press, Berkeley, CA, 1991.

215. Y. B. Zel'dovich and Y. P. Raizer, *Physics of shock waves and high-temperature hydrodynamic phenomena*, Academic Press, New York, 1967.

216. M. W. Zemansky and R. H. Dittman, *Heat and thermodynamics*, 6th ed., McGraw-Hill, New York, 1981.

217. F. J. Zerilli and R. W. Armstrong, *Dislocation-mechanics-based constitutive relations for material dynamics calculations*, Journal of Applied Physics **61** (1987), no. 5, 1816–1825.

218. J. A. Zukas (ed.), *High velocity impact dynamics*, John Wiley & Sons, 1990.

219. J. A. Zukas, T. Nicholas, H. F. Swift, L. B. Greszczuk, and D. R. Curran, *Impact dynamics*, John Wiley & Sons, New York, 1982.

Index

Printed in the United States
by Baker & Taylor Publisher Services